ALGEBRAIC FORMULAS

Factoring cubes:

$a^3 + b^3 = (a + b)(a^2 - ab + b^2)$

$a^3 - b^3 = (a - b)(a^2 + ab + b)^2)$

Quadratic formula: Solution to $ax^2 + bx + c = 0$ is

$$x = \frac{-b \pm \sqrt{b^2 - 4ac}}{2a}$$

Equation of a line:

$$y = mx + b$$

where m is the slope and b is the y intercept.

GEOMETRIC FORMULAS

Circle of Radius R

$$\text{circumference} = c = 2\pi R$$
$$\text{area} = A = \pi R^2$$

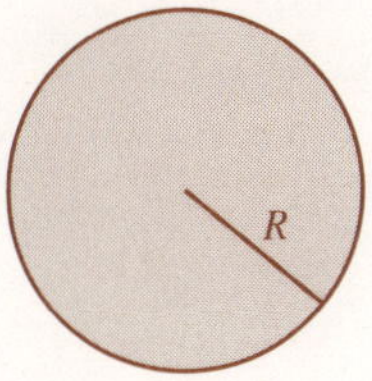

Sphere of Radius R

$$\text{surface area} = A = 4\pi R^2$$
$$\text{volume} = V = \tfrac{4}{3}\pi R^3$$

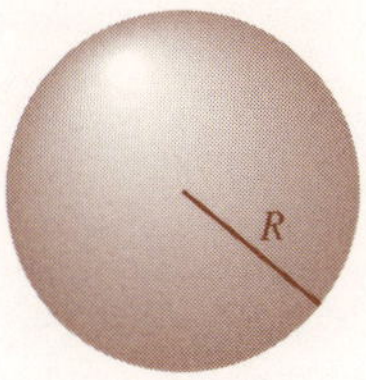

Right Circular Cylinder of Height h and Radius R

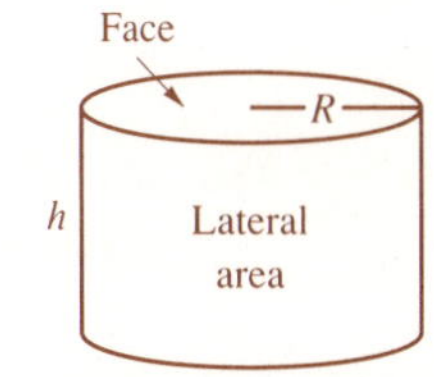

$$\text{lateral area} = 2\pi Rh$$
$$\text{volume} = \pi R^2 h$$

Area of Triangle

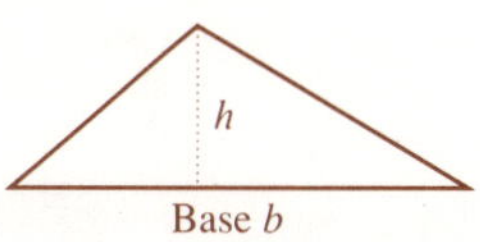

$$A = \tfrac{1}{2}\text{ base} \times \text{altitude} = \tfrac{1}{2}bh$$

Pythagorean Theorem

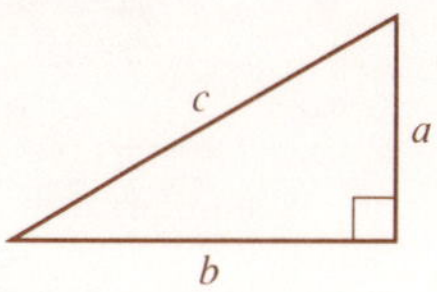

for any right triangle

$$a^2 + b^2 = c^2$$

APPLIED TRIGONOMETRY

APPLIED TRIGONOMETRY

Michael N. Payne
College of Alameda

Wayne. R. Anderson
Sacramento City College

HBJ

HARCOURT BRACE JOVANOVICH, PUBLISHERS
and its subsidiary, ACADEMIC PRESS
San Diego • New York • Chicago • Austin • Washington, D.C.
London • Sydney • Tokyo • Toronto

This book is dedicated by Michael Payne to
Ronald, Joan, Kenneth, and Linda

and by Wayne Anderson to
My mother and father, Mildred and Raymond

Cover Photo: This pin-point-sized star was created on January 13, 1986, when researchers investigating nuclear fusion at the Lawrence Livermore National Laboratory in California fired the world's most powerful laser, Nova, and hit a tiny fuel capsule. In the fifty-trillionths of a second that it lasted, this tiny fusion burst gave off ten trillion neutrons. Exercises involving lasers occur in Section 2.7 of this text. (*Photograph by Gary Stone. Courtesy of Lawrence Livermore National Lab News Bureau Office.*)

Fig. 1.4-11 courtesy of the U.S. Department of Agriculture; Figs. 1.4-12, 1.4-13, 1.4-14, 2.3-2, 2.5-39, 2.7-12, 9.2-5a courtesy of NASA; Figs. 1.5-8, 1.7-6 courtesy of Yerkes Observatory; Fig. 1.6-7 courtesy of Lick Observatory, University of California, Santa Cruz; Fig. 1.6-8 courtesy of Kitt Peak National Observatory; Fig. 1.7-7 courtesy of Palomar Observatory, California Institute of Technology; Fig. 1.7-8 courtesy of National Optical Astronomy Observatories; Fig. 2.1 UPI/Bettman Newsphotos; Fig. 2.7-5 courtesy of Graham Field Company; Fig. 2.7-9 courtesy of U.S. Geological Survey; Fig. 2.7-13 courtesy of High Altitude Observatory; Fig. 5.1-4 from Ohanian, PHYSICS, Fig. 14.14, p. 347, W. W. Norton & Company, used with permission; Fig. 5.1-9 A. W. Ambler/Photo Researchers Inc.; Fig. 5.3-10 UPI/Bettman Newsphotos; Fig. 9.2-5b Alvin Staffan/Photo Researchers Inc.; Fig. 9.2-5c The Bettman Archive; Fig. 9.2-13a Jim & Julie Bruton/Photo Researchers Inc.; 9.2-13b Carleton Ray/Photo Researchers Inc.; 9.2-13b Pat Ross/Jeroboam; 9.2-14 Leonard Lessin/Photo Researchers Inc.

ISBN: 0-15-502911-8

Library of Congress Catalog Card Number: 88-81042

Printed in the United States of America

Preface

TO THE STUDENT

Trigonometry is one of the oldest and most practical branches of mathematics. Early mathematicians used trigonometry to solve concrete problems in such fields as architecture, astronomy, surveying, and time-keeping. Over the centuries, trigonometry has become even more important, and its use now extends to physics, music, archaeology, chemistry, biomedicine, geology, social science, engineering, and the rest of mathematics. This practical aspect of trigonometry guides the development of the subject, the examples, and the exercises in this book. Throughout, we emphasize the importance of solving *real-world* problems.

SPIRAL DEVELOPMENT

The approach of this book is to address key ideas in stages rather than covering them exhaustively at a single sitting. This spiral approach allows you to master fairly extensive subjects, such as the inverse functions, in easily comprehensible units. It also lets you *make use of new topics as you learn them*. (The validity of this approach should be clear to anyone who has tried to master new computer software: the sooner you can use the knowledge, the easier it is to assimilate it!) The spiral approach requires that the first three chapters be rather long because we introduce most of the core principles of trigonometry in them. The succeeding chapters extend this core material and should, therefore, be much easier to cover.

THE EXAMPLES

In addition to the basic examples, we explore the applications of trigonometry to photography, architecture, biology, archaeology, physics, music, and others. These applications are all *self-contained* and no prior knowledge of the subject is necessary. The large number of applications are there to provide choice; it is not necessary to do all of them.

Most examples and some proofs conclude with **Now It's Your Turn**, in which you are asked to work a problem similar to the one we demonstrated. This participation on your part is *extremely important*. If you cannot work the problem posed, you do not really understand the example, and you are not ready to go on. Working out these extra examples will slow down your reading, but remember that in mathematics, as in music, sports, science, and art, learning requires *participation*. It does not occur simply by watching someone else perform.

THE EXERCISES

Do not be intimidated by the number of problems in this book. No one expects you to do them all! They are there to give you and your professor a wide choice. Many of the exercises cover additional topics not discussed in the text, such as vector components and Euler's formulas. These are self-discovery problems, to be used at the professor's discretion.

We have grouped the exercises as follows:

- **Drill.** The first exercises in each section are usually the basic problems necessary to develop the mechanics of calculation. If you can't do these drill problems, you won't be able to work out the applications. We have included a large number of drill problems to provide lots of practice. Do as many of these as you need; once you understand the basics, go on to the more challenging (and interesting) problems.

- **Starred Exercises.** Exercises with a *single star* (*) require more thought or more extensive calculations than the drill problems or make use of several different principles. Exercises with a *double star* (**) are very challenging. Plan on spending considerable thought and effort.

- **Looking Ahead.** These are self-discovery exercises dealing with topics that will be covered later in the book. We believe that you will understand something much better if you figure it out for yourself, and we encourage you to use these exercises as often as possible.

- **3-Dimensional Problems.** Most people have difficulty visualizing things in three dimensions, largely because they have not learned to reduce a three-dimensional case to a two-dimensional case. We illustrate this in several examples and then include many such exercises. Because work is very important in many fields, we encourage you to do as many of these problems as possible.

THE PATH OF π

Section 1.8, *The Path of* π, is for your enjoyment. We hope that you will find it stimulating, and that you will want to read some of the references cited in the footnotes. It is a section for enrichment or for extra projects, and it need not be covered in class.

CALCULATORS VERSUS TABLES

Every student of mathematics should be able to use a pocket calculator. Using calculators rather than tables to evaluate trigonometric functions eliminates drudgery and makes it possible to solve interesting and realistic problems. You should probably get the cheapest one that will do what you need. In addition to the trigonometric functions (sin, cos, tan), be sure it has a "1/X" key. (Any calculator having the trigonometric functions probably has everything you will need for your lower-division mathematics courses.)

ACKNOWLEDGMENTS

We would like to thank the staff at Harcourt Brace Jovanovich for their support and assistance, especially our editor Richard Wallis, production manager Mary Douglas, and style editor Ruth S. Cottrell. Special thanks go to Forrest Newman of Cosumnes River College and Michael Richardson of Sacramento City College for patiently checking (and correcting where necessary) the exercises. In addition, we wish to thank the following people who reviewed the manuscript and whose many helpful suggestions have been incorporated in the book: Lee R. Clancy, Cypress College; William Drezdon, Oakton Community College; Karen Egedy, Louisiana State University; Stanley J. Guberud, Milwaukee School of Engineering; Linda Holden, Indiana University; Elice Price, Tarrant Junior College; Judith Rollstin, University of New Mexico; Thomas Spradley, American River College; Monty Strauss, Texas Tech University; Arlene Watkins, University of Arizona; Kay Weiss, Kansas City Community College; Timmothy Wilson, Honolulu Community College; Karen E. Zak, US Naval Academy. Finally, we appreciate the suggestions and help of the following people: Bruce Pierini, Cecil Draper, Tamera Fields, Joe Hartman, Christopher Johnson, James Lucky, Adrian Martinez, Judy Pogan, and Frank Shermoen.

Michael Payne
Wayne Anderson

Contents

◆ *Enrichment sections. These sections can be read by the student but are not expected to be covered by the instructor. They are for enrichment and motivation of the student.*

▶ *Optional sections.*

APPENDICES

1

Introduction to Trigonometry

1.1 THE DAWN OF TRIGONOMETRY

Thousands of years ago, how did people determine the seasons of the year? Why did they divide the day into 24 hours? How could they measure the distance to the moon? Did they know the sizes of the earth and the moon? People solved these and other such problems using what we now call *trigonometry*.

For example, it was customary to tell the seasons of the year and the hours of the day using shadows cast by a pole or tree. Any such erect object whose shadow was used for this purpose was called a *gnomon*, from the Greek word for "inspector" (see Figure 1.1-1). When the noontime shadow was longest, the sun was lowest in the sky and it was the *winter solstice*; when the shadow was shortest, it was the *summer solstice*.

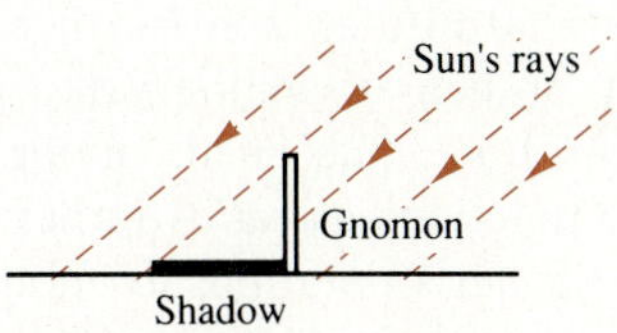

Figure 1.1-1

A gnomon (any erect object) casts a shadow. The length of the shadow depends on the angle of the sun's rays.

The word *trigonometry* comes from the ancient Greek *tri* (three) + *gonia* (angle) + *metron* (measure). Although the derivation of this word seems to limit trigonometry to the study of triangles, the subject is really much wider. It includes the measurement of angles in general and plays an important role in many theoretical and practical applications of mathematics.

Historically, trigonometry was first used in astronomy, where distances that could not be measured directly were calculated by means of angles and lines that could be so measured. The beginnings of these techniques can be traced back to Babylon and ancient Egypt. The following list chronicles significant occurrences in the development of trigonometry.

1. Ahmoses (Ahmes) (approximately 1650 B.C.) Many people believe that trigonometry's beginnings go back to a papyrus called the Rhind Papyrus after its former owner, A. Henry Rhind. It was written by an Egyptian named Ahmes, and some of the information in his writings goes back as far as 1800 B.C.

2. Hipparchus (approximately 180–125 B.C.) The Greek astronomer Hipparchus is often called the founder of trigonometry. He was born in Bithynia, Asia Minor, studied in Alexandria, Egypt, and did most of his work on the island of Rhodes, where he developed trigonometry as an aid in measuring angles and lines in connection with astronomy. Although none of his papers have survived, most of his work was transmitted through Theon of Alexandria, a fourth-century A.D. writer. (This early work was actually *spherical* trigonometry, the study of triangles on the surface of a sphere. Because of the study of astronomy, spherical trigonometry preceded plane trigonometry.)

3. Claudius Ptolemy (approximately A.D. 85–150) An improvement in trigonometry is credited to Claudius Ptolemy, the great Greek astronomer. In his major work, *Almagest* (*The Thirteen Books of Mathematical Compilation*), Ptolemy continued the work of Hipparchus in spherical trigonometry and introduced plane trigonometry and numerical calculations. Ptolemy developed trigonometry by the use of chords of arcs that were the predecessors of our sine, cosine, and tangent functions.

4. Johann Müller (Regiomontanus) (A.D. 1436–1476) The German mathematician, Johann Müller, who was called Regiomontanus, translated and published Ptolemy's *Almagest*. As a result of this work, he wrote *De Triangulis Omnimodis* (*On Triangles of All Kinds*). In *De Triangulis Omnimodis*, Regiomontanus systematically summarized much of the previous work and also established many connections between spherical and plane triangles. The only trigonometric functions he introduced were the sine and the cosine, but he did compute a table of tangents. Most importantly, he established trigonometry as a separate branch of mathematics, not just a tool of astronomy.

5. Leonhard Euler (1707–1783) It was the work of Leonhard Euler, a Swiss, that firmly established trigonometry as an independent branch of mathematics. While his achievements are too numerous to mention here, one of his major accomplishments was to represent the trigonometric functions as *ratios* rather than chords as Ptolemy did. In his work *Introductio*, Euler's notation brought great order to mathematics. His symbols sin, cos, e (the base for natural logarithms), $i = \sqrt{-1}$, and π (for the ratio of the circumference of a circle to its diameter) are still used today.

Other mathematicians have contributed to the development of trigonometry. During the eighteenth century, **John Bernoulli (1667–1748)**, along with Euler, developed trigonometry as a means of analysis. **Abraham de Moivre (1667–1754)** applied trigonometry to the study of complex numbers. **Johann H. Lambert (1728–1777)** extended trigonometry to include hyperbolic functions. **Joseph Fourier (1768–1830)** showed that many physical phenomena could be expressed as infinite sums of sines and cosines, and these are now called the Fourier series. In the nineteenth century, **Niels Abel (1802–1829)** generalized Fourier's trigonometric functions and introduced elliptical functions.

Although trigonometry had its beginnings in astronomy, it was used in everyday life, in the telling of time and seasons, surveying, and measuring. The expansion of mathematical symbols and the application of trigonometry to the physical sciences led to outstanding results, both in mathematics and in our understanding of the universe.

1.2 ANGLES

1.2-1 Trigonometric Angles

Definition of a Trigonometric Angle

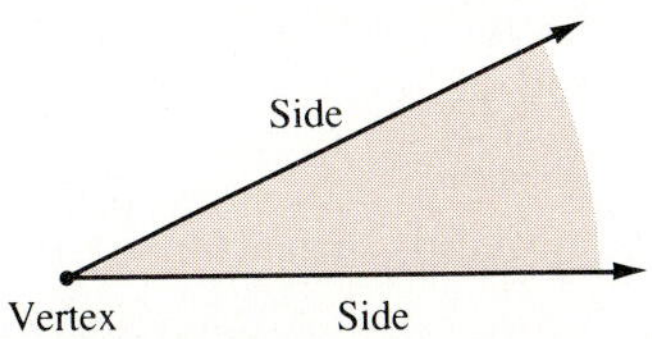

Figure 1.2-1
Geometric angle

Since much of trigonometry deals with triangles, the concept of an angle is central to its study. In geometry you probably defined an angle as the "opening" between two rays (half-lines) originating from a common point called the *vertex*. The two rays form the sides of the angle (see Figure 1.2-1).

This geometric definition of an angle is not quantitative; it does not tell us how large the opening is. To introduce *size* to an angle, we can define it in terms of a *rotation*. We call an angle defined in terms of rotation a **trigonometric angle**.

> **DEFINITION OF TRIGONOMETRIC ANGLE**
>
> **When two rays have a common initial point (vertex) *O*, a *trigonometric angle* is the amount of rotation required to turn one ray (the initial side) so that it coincides with the other ray (the terminal side).**

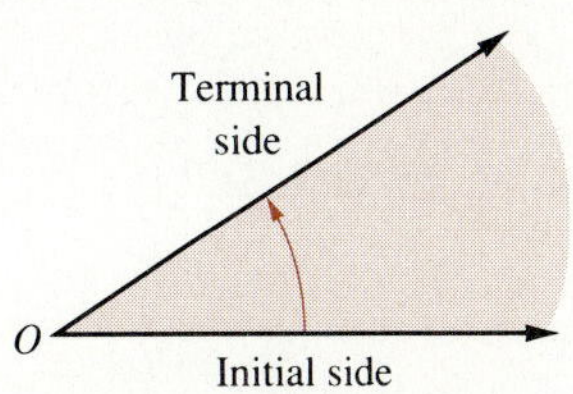

(a) Positive angle (counter-clockwise rotation)

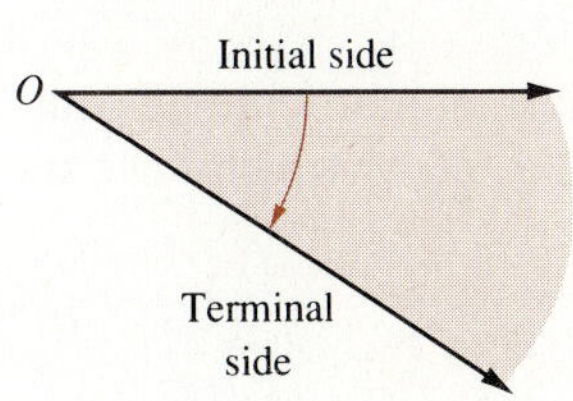

(b) Negative angle (clockwise rotation)

Figure 1.2-2
Positive and negative angles

A rotation can be in either of two directions—clockwise or counterclockwise—and we must choose one of these directions as positive and the other as negative. Choosing a direction as positive or negative is called a **sign convention**.

> **SIGN CONVENTION FOR ANGLES**
>
> **An angle is called *positive* if the direction of rotation is *counterclockwise* and *negative* if the direction of rotation is *clockwise*.**

The angle in Figure 1.2-2a is a positive angle, and the one in Figure 1.2-2b is a negative angle.

Measure of an Angle

The measure of an angle tells us how large the angle is. A common unit of measure for a trigonometric angle is the revolution. One full turn of a ray about its endpoint is called *one revolution*.

Example 1

Figure 1.2-3 shows examples of positive and negative angles and their sizes.

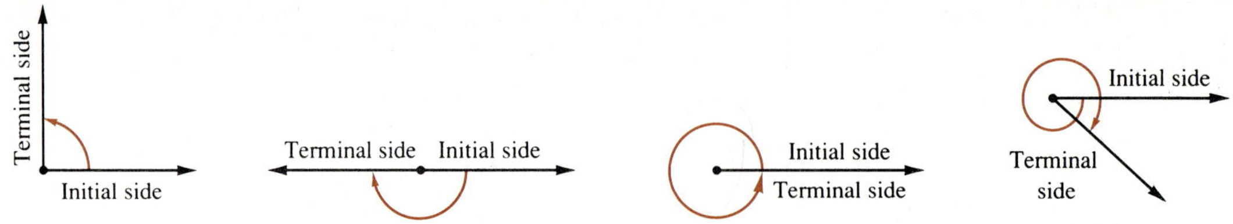

Figure 1.2-3

(a) Positive angle, $\frac{1}{4}$ revolution ***(b) Negative angle, $\frac{1}{2}$ revolution*** ***(c) Positive angle, 1 revolution*** ***(d) Negative angle, $1\frac{1}{8}$ revolutions***

NOW IT'S YOUR TURN

Using the same initial side as in Figure 1.2-3, sketch the following angles: (a) $+\frac{3}{4}$ revolution and (b) $-\frac{3}{4}$ revolution. (Answer: See the end of Section 1.2.)

Example 2

Through what angle does the hour hand of a clock (nondigital) turn from 4 P.M. to 6 P.M.?

Solution

The hour hand makes one revolution every 12 hours. Therefore, in the 2-hour interval from 4 P.M. to 6 P.M., it turns through $\frac{2}{12} = \frac{1}{6}$ of a revolution. ■

NOW IT'S YOUR TURN

Through what angles does the hour hand turn from 8 A.M. to 11 A.M.? (Answer: $\frac{1}{4}$ revolution)

The definition of a trigonometric angle is an extension of the definition of a geometric angle. But, as the following example illustrates, one geometric angle can represent *more than one* trigonometric angle.

Example 3

Consider the geometric angle AOB in Figure 1.2-4. This geometric angle could represent any of the trigonometric angles in Figure 1.2-5, all of which have the same initial and terminal sides.

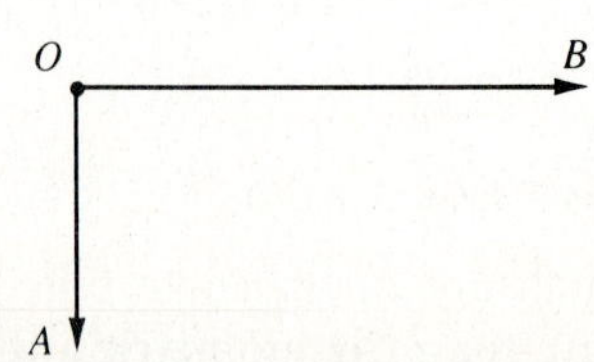

Figure 1.2-4

Geometric angle

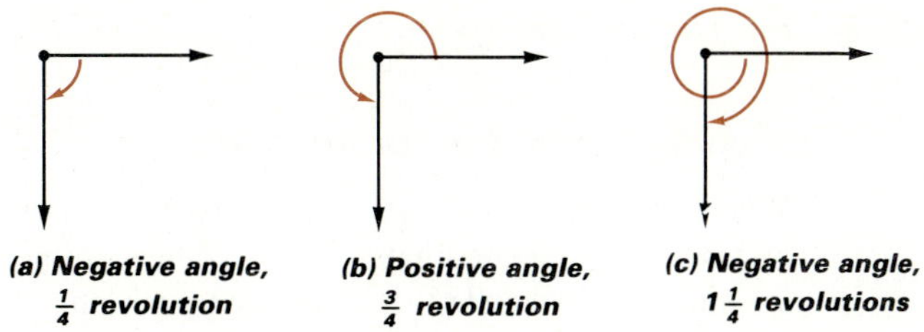

Figure 1.2-5

All three trigonometric angles represent the same geometric angle in Figure 1.2-4.

NOW IT'S YOUR TURN

Infinitely many trigonometric representations are actually possible for the geometric angle in Figure 1.2-4. Find some more using OB as the initial side. (Answer: See the end of Section 1.2.)

Some Special Angles

The following angles occur very frequently; you should get to know them because you will see them often.

DEFINITION OF COTERMINAL ANGLES

Trigonometric angles that have the same initial side and the same terminal side are called *coterminal angles*.

Right Angle. An angle having a rotation of $\frac{1}{4}$ of a revolution is a *right angle*; that is, a right angle is formed when the initial and terminal sides are *perpendicular* (see Example 1a).

Straight Angle. An angle having a rotation of $\frac{1}{2}$ of a revolution is a *straight angle* (see Example 1b).

Coterminal Angles. In Figure 1.2-6, the angle α (alpha) is positive. But, as we have just seen in Example 3, we can construct infinitely many other angles, both positive and negative, having the *same initial and terminal sides*. We call such angles **coterminal angles**.

Example 4

The angles α (alpha) and β (beta) in Figure 1.2-6 are coterminal.

Figure 1.2-6

Coterminal angles The angles α and β are coterminal because they have the same initial side and the same terminal side.

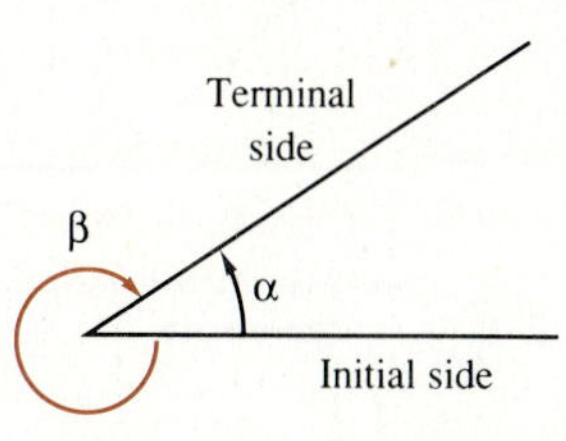

NOW IT'S YOUR TURN

Find other angles that are coterminal with α.
(Answer: See the end of Section 1.2.)

1.2-2 Angles on Graphs

Standard Position for an Angle

When working with angles, we often need to place them on a graph on a cartesian (x, y) coordinate system. When doing this, it is useful to draw them in a standard way, called the **standard position**.

DEFINITION OF STANDARD POSITION

A trigonometric angle is in *standard position* on a cartesian (x, y) coordinate system when it is drawn with its vertex at the origin and its initial side along the positive x-axis.

Example 5

The angles in Figure 1.2-7 are in standard position.

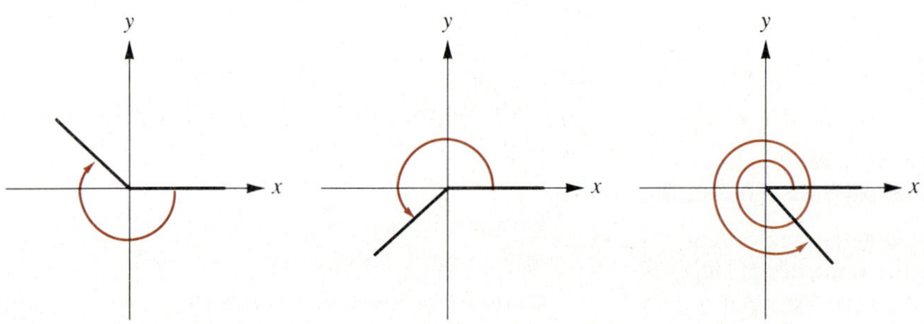

Figure 1.2-7

Example 6

The angles in Figure 1.2-8 are *not* in standard position.

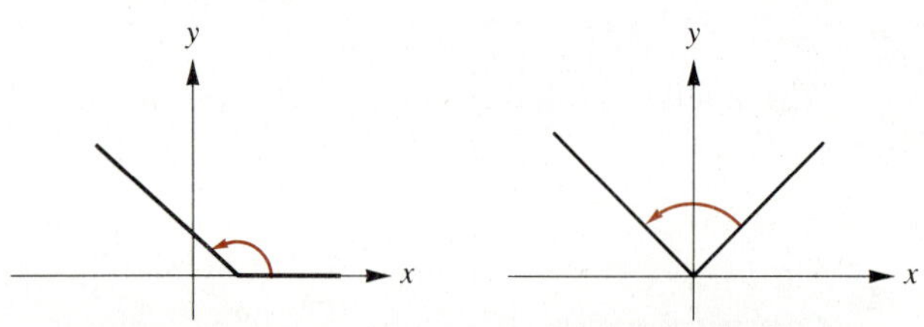

Figure 1.2-8 ***(a) The vertex is not at the origin.*** ***(b) The initial side is not along the positive x-axis.***

NOW IT'S YOUR TURN

Which of the four angles in Figure 1.2-9 are in standard position? For those that are *not*, explain why not. (Answer: See the end of Section 1.2.)

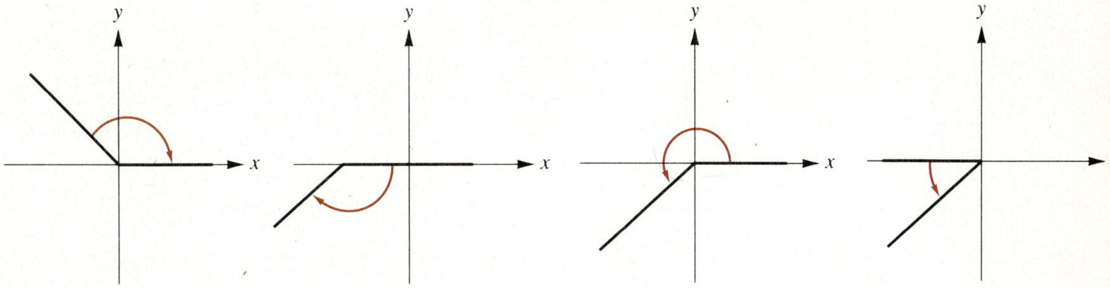

Figure 1.2-9 *(a)* *(b)* *(c)* *(d)*

Quadrant of an Angle

> **QUADRANT OF AN ANGLE**
>
> If the terminal side of an angle in standard position lies in a certain quadrant, then we say that the *angle lies in that quadrant*.

The x and y axes divide the plane into four regions called **quadrants**, as shown in Figure 1.2-10. Region I is called the first quadrant, region II the second quadrant, and so on. When an angle is in standard position, the location of its *terminal side* determines what quadrant it is in.

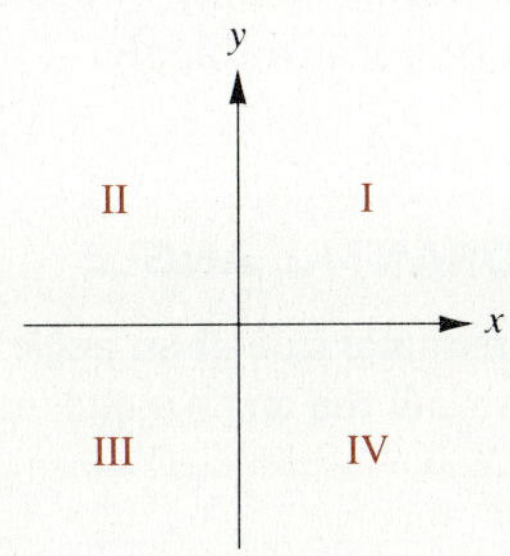

Figure 1.2-10
The four quadrants

Example 7

Figure 1.2-11 shows angles in various quadrants.

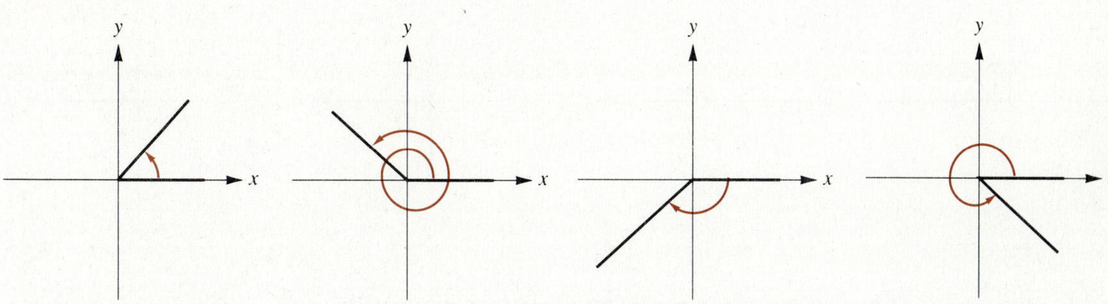

(a) First-quadrant angle ***(b) Second-quadrant angle*** ***(c) Third-quadrant angle*** ***(d) Fourth-quadrant angle***

Figure 1.2-11

NOW IT'S YOUR TURN

In Figure 1.2-12, in which quadrant does each angle lie? (Answer: See the end of Section 1.2.)

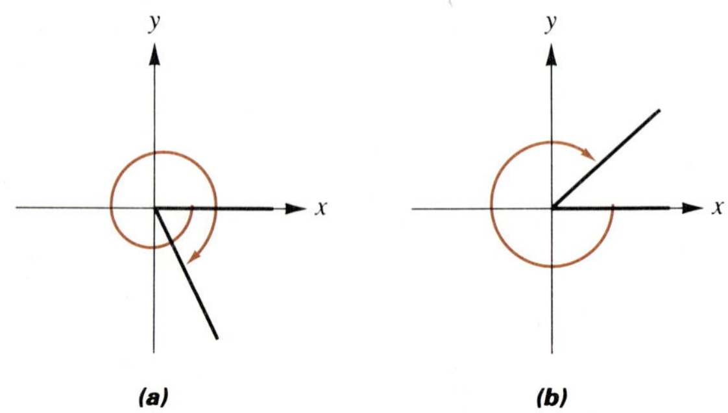

Figure 1.2-12

Quadrantal Angles

The terminal side of an angle in standard position does not always end *in* one of the quadrants; it may end *on* one of the axes. When this happens, we call the angle a **quadrantal angle**.

> **QUADRANTAL ANGLE**
>
> **If the terminal side of an angle in standard position coincides with the x axis or the y axis, we call the angle a *quadrantal angle*.**

Example 8

The angles in Figure 1.2-13 are all quadrantal angles.

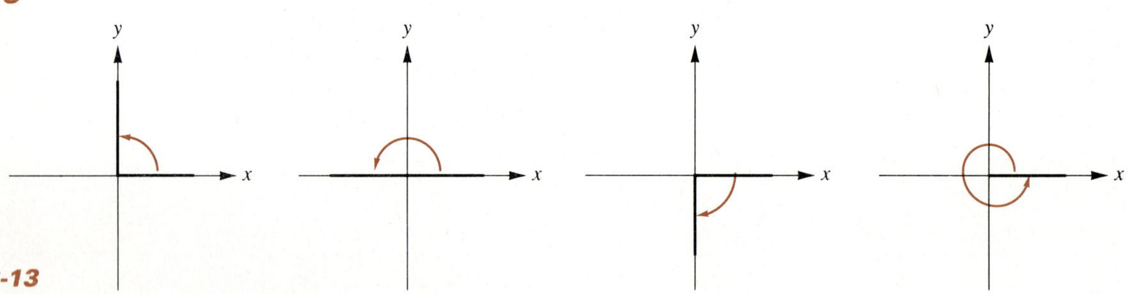

Figure 1.2-13

ANSWERS TO "NOW IT'S YOUR TURN"

ANSWER TO EXAMPLE 1

Figure 1.2-14 shows two possible answers.

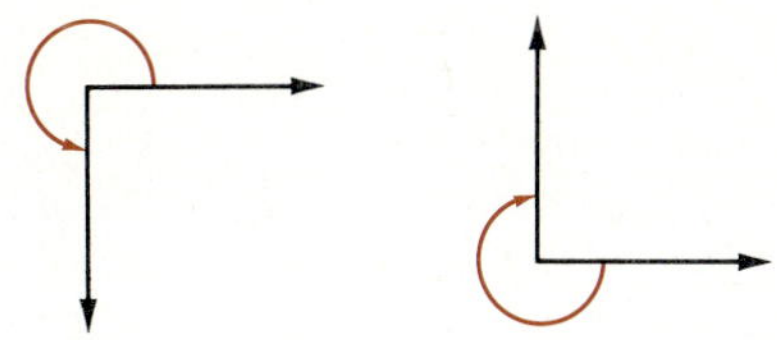

Figure 1.2-14 *(a)* $+\frac{3}{4}$ *revolution* *(b)* $-\frac{3}{4}$ *revolution*

ANSWER TO EXAMPLE 3

Two answers are $+1\frac{3}{4}$ revolutions and $-2\frac{1}{4}$ revolutions.

ANSWER TO EXAMPLE 4

Figure 1.2-15 shows one possible answer. There are infinitely many more.

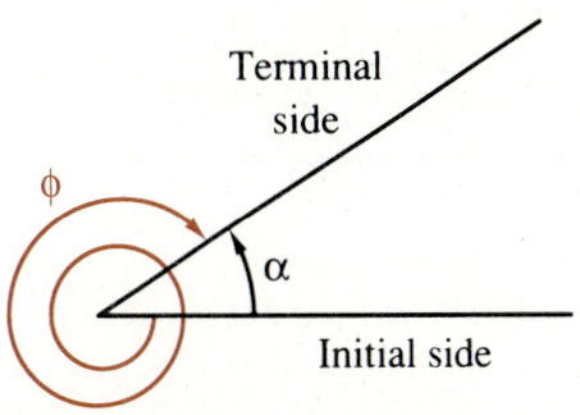

Figure 1.2-15
The angle ϕ is coterminal with α.

ANSWER TO EXAMPLE 6

(a) No. The initial side is not along the $+x$ axis. (b) No. The vertex is not at the origin. (c) Yes. (d) No. The initial side is along the *negative* rather than the positive x axis.

ANSWER TO EXAMPLE 7

(a) Quadrant IV (b) Quadrant I

EXERCISES 1.2

In Exercises 1–10, sketch the following angles in standard position and indicate rotation with a curved arrow.

1. $\frac{1}{2}$ revolution clockwise
2. $\frac{1}{4}$ revolution clockwise
3. $\frac{3}{4}$ revolution counterclockwise
4. $\frac{1}{2}$ revolution counterclockwise
5. 1 revolution clockwise
6. $1\frac{1}{4}$ revolutions clockwise
7. $1\frac{3}{4}$ revolutions counterclockwise
8. 2 revolutions counterclockwise
9. $2\frac{3}{4}$ revolutions clockwise
10. $1\frac{1}{2}$ revolutions clockwise

In Exercises 11–20, decide whether the angle is in standard position. If not, give the reason. Also indicate whether the angle is positive or negative.

11.

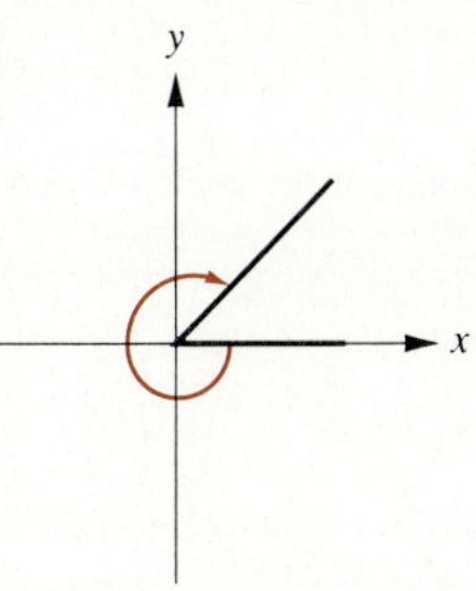

12.

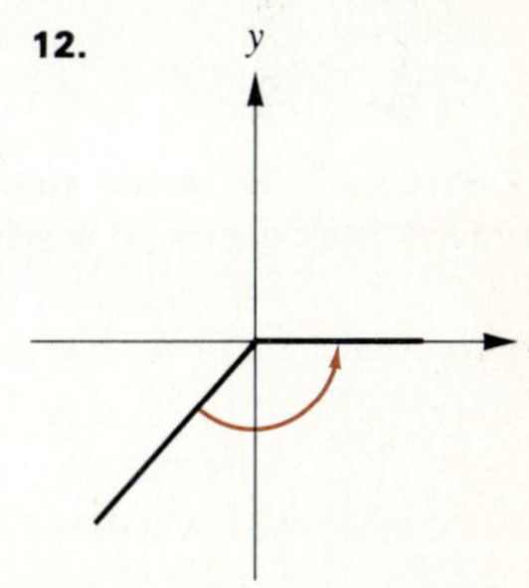

13.

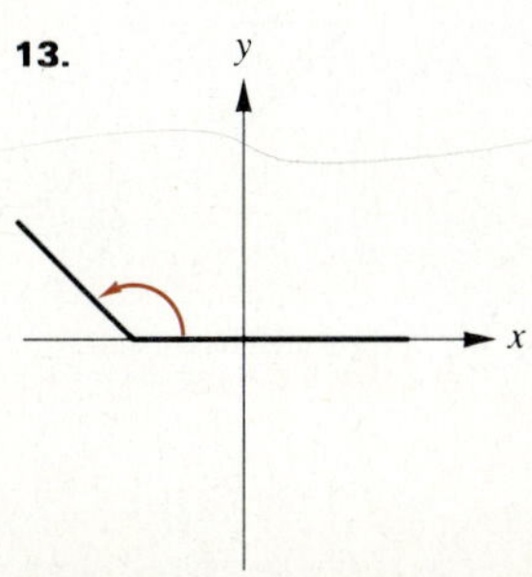

14.

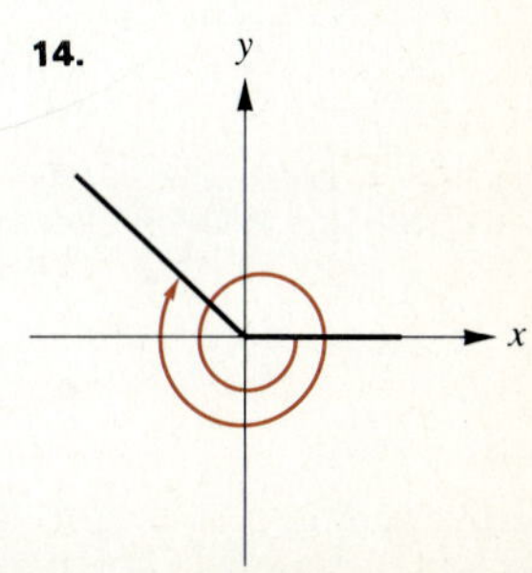

15.

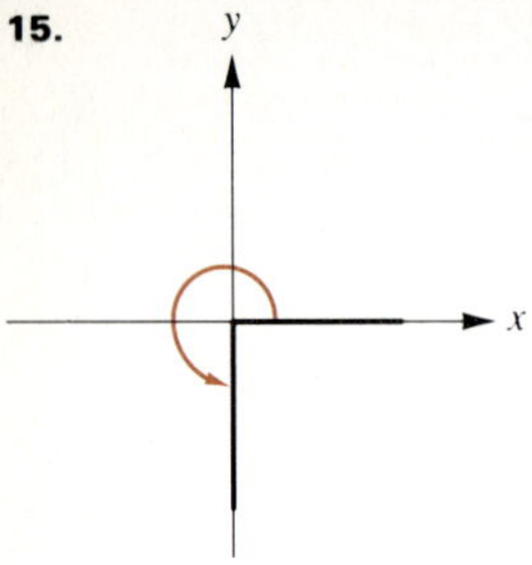

16.

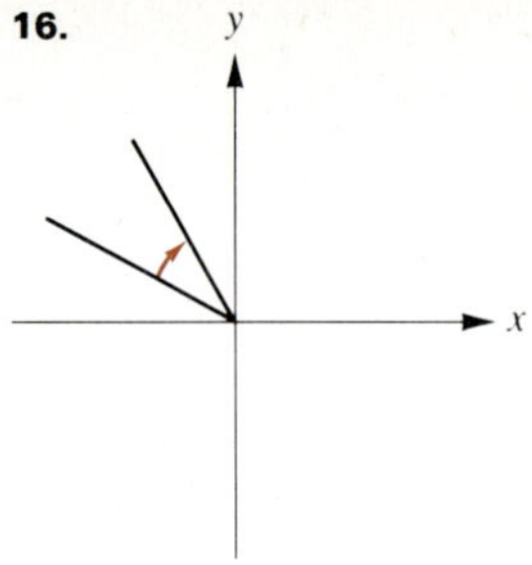

17.

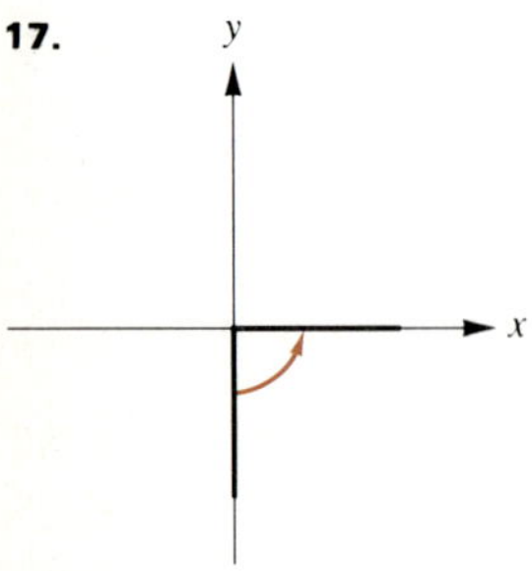

18.

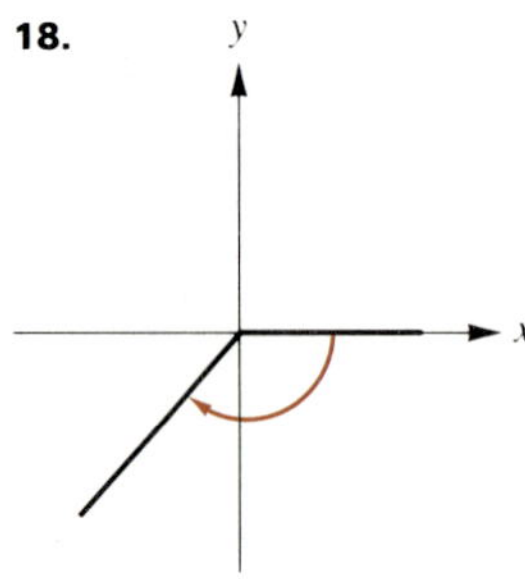

19.

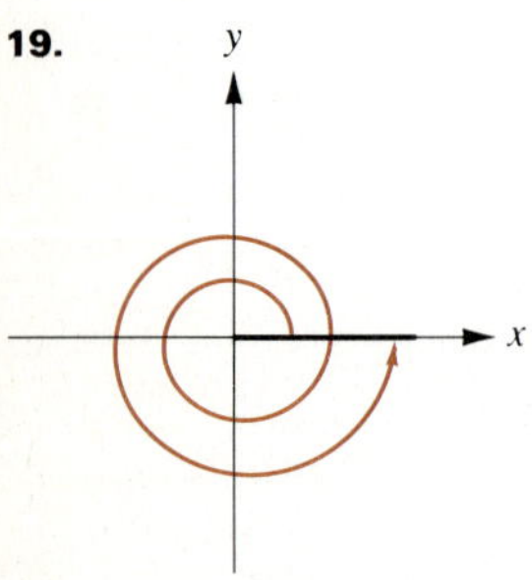

20.

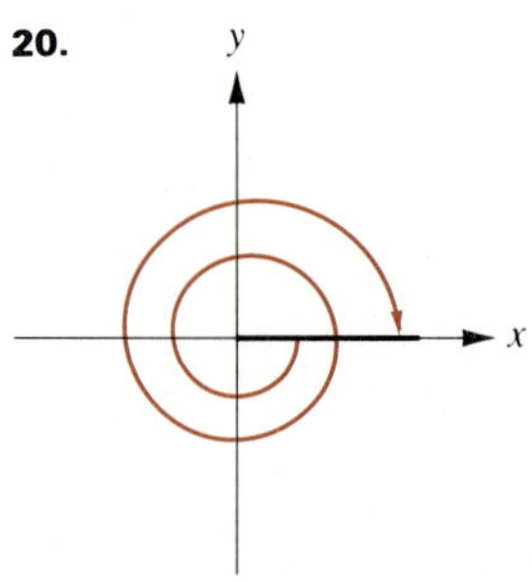

In Exercises 21–26, decide which quadrant the angle is in or whether it is a quadrantal angle.

21.

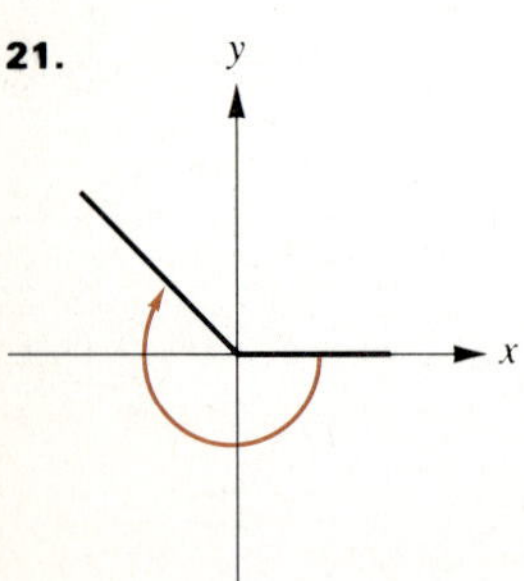

22.

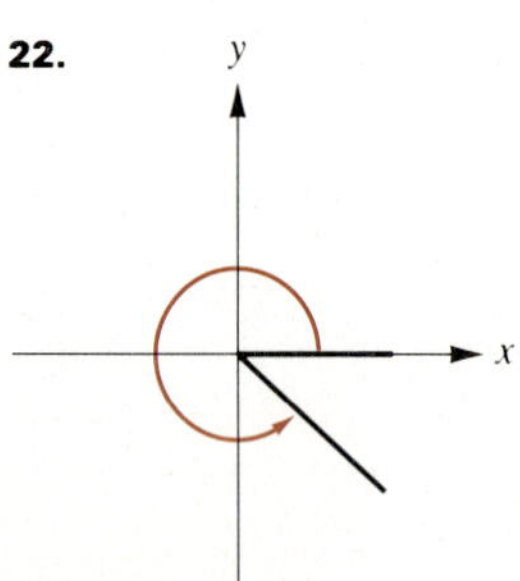

23.

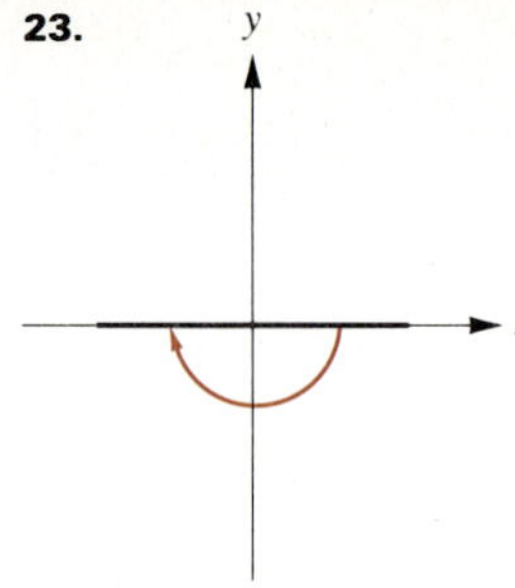

24.

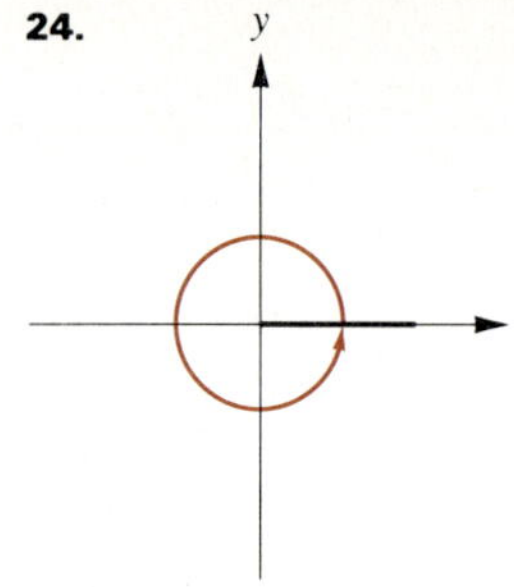

25.

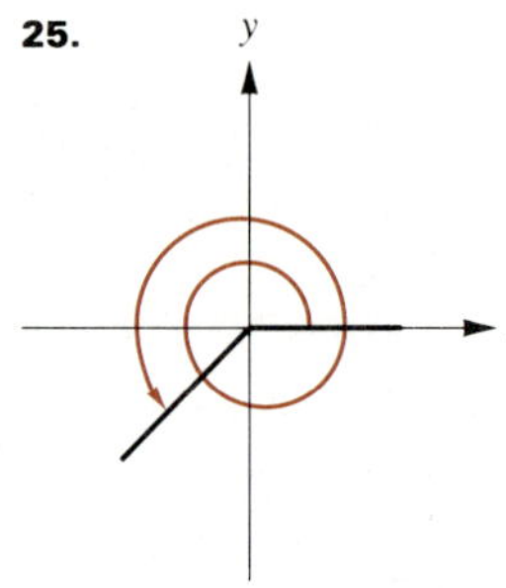

26.

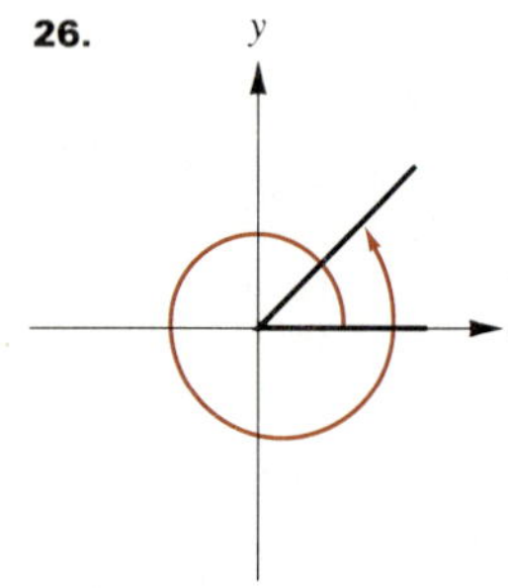

In Exercises 27–30, find a negative angle that is coterminal with the given positive angle.

27. $\frac{1}{2}$ revolution

28. $\frac{3}{4}$ revolution

29. $\frac{1}{4}$ revolution

30. 1 revolution

In Exercises 31–40, find 3 angles (positive and/or negative) that are coterminal with the given angle.

31. $\frac{1}{3}$ revolution

32. $-\frac{1}{2}$ revolution

33. $1\frac{1}{2}$ revolutions

34. $\frac{3}{4}$ revolution

35. -1 revolution

36. 0 revolutions

37. $1\frac{1}{4}$ revolutions

38. $5\frac{1}{2}$ revolutions

39. $-\frac{1}{4}$ revolution

40. 198 revolutions

In Exercises 41–45, through what angle do the hour, minute, and second hands of a clock move in each of the time intervals given?

41. From noon to 3 P.M.

42. From noon to 3 A.M.

43. From 6 P.M. to 6:30 P.M.

***44.** From 1:45 A.M. to 3:30 A.M.

***45.** From 2:15 P.M. to 6:45 A.M.

In Exercises 46–50, through what angle does the earth turn in the time intervals given? *Hint:* How long does it take for one revolution of the earth? Can you apply the method of Example 2?

46. From midnight to noon

47. From midnight to 6 A.M.

48. From midnight to 6 P.M.

49. From 3 P.M. to 6 A.M.

50. From 2 A.M. to 5:30 P.M.

In Exercises 51–55, what time is it when the hour hand of a clock has turned through the given angles, starting at midnight?

51. $\frac{1}{2}$ revolution

52. $\frac{1}{4}$ revolution

53. $1\frac{1}{4}$ revolutions

54. $3\frac{1}{2}$ revolutions

***55.** $1\frac{1}{12}$ revolutions

In Exercises 56–58, starting at midnight on the morning of New Year's Day, 1989, through what angle has the earth moved around the sun at the beginning of each of the dates given? Assume exactly 365 days per year.

***56.** January 2 ***57.** February 15 ***58.** November 10

***59.** When a 45-rpm (revolutions per minute) phonograph record is playing through what angle does it turn in $\frac{1}{5}$ minute? In $\frac{1}{90}$ minute?

60. How many revolutions (or fractions of revolutions) does the earth make around the sun in: **(a)** one day **(b)** one 30-day month **(c)** one week? *Hint:* How long does it take the earth to make one complete revolution around the sun?

***61.** Starting at midnight on the morning of New Year's Day, 1989, what date is it when the earth has gone through $\frac{1}{5}$ of a revolution around the sun?

In Exercises 62–63, in which quadrants can $-A$, $2A$, and $A/2$ lie?

62. **(a)** A is a positive angle in quadrant I.
(b) A is a positive angle in quadrant II.

63. **(a)** A is a positive angle in quadrant III.
(b) A is a positive angle in quadrant IV.

1.3 MEASURE OF ANGLES

In Section 1.2 we described angles in terms of revolutions and fractions of revolutions. However, the units of measure most often used are the *degree* and the *radian*. In this section, we discuss these two units of angular measure and the relationship between them.

1.3-1 Degrees

Definition of the Degree

The right angle defined in the previous section is one of the oldest and simplest units of angular measure. However, navigation by compass required a smaller unit, so the right angle was divided into eight "points" that were used by sailing ships. To record the locations of the stars accurately, Babylonian astronomers needed an even smaller unit to measure an angle.

The Babylonians used a 360-day year and added 5 festival days. The reason for this rather unusual system is that the Babylonians counted in base 60 (a *sexagesimal* system)* rather than in our base 10. By using a 360-day year, they avoided awkward fractions (360 is exactly 6×60). (We do something similar with our leap year. We pretend that the year has 365 days when it really has about $365\frac{1}{4}$ days, and then we add a day—February 29—every fourth year to

* We still do not know why the Babylonians counted in base 60. For an interesting sampling of some of the suggested reasons, see Georges Ifrah, *From One to Zero*, Penguin Books, 1987, pp. 49–51.

avoid dealing with an extra $\frac{1}{4}$ day each year.) The Babylonians noticed that certain bright stars revolve around the North Star once each year. Therefore, in their 360-day year, these stars made $\frac{1}{360}$ of a complete revolution each day. Consequently they divided one revolution into 360 equal units, each unit being called a *degree* and written °.

One complete revolution generates an angle with measure 360°. Therefore, 1° is $\frac{1}{360}$ of a complete revolution. Although 1° is a small angle, you can easily see things that subtend (intercept) such an angle. For example, the full moon subtends an angle of approximately $\frac{1}{2}°$ (see Figure 1.3-1), whereas a penny held about $3\frac{1}{2}$ feet away subtends a 1° angle (see Figure 1.3-2).

Figure 1.3-1

Angle subtended by the full moon

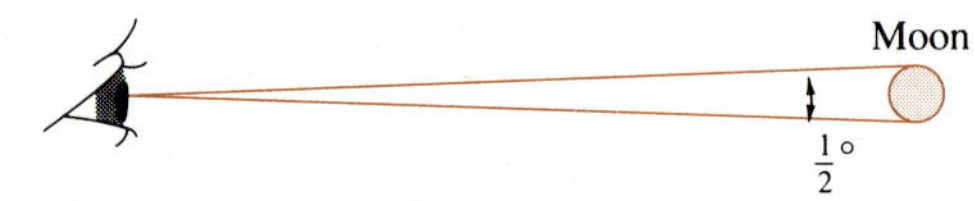

Figure 1.3-2

Angle subtended by a penny at $3\frac{1}{2}$ feet from the eye

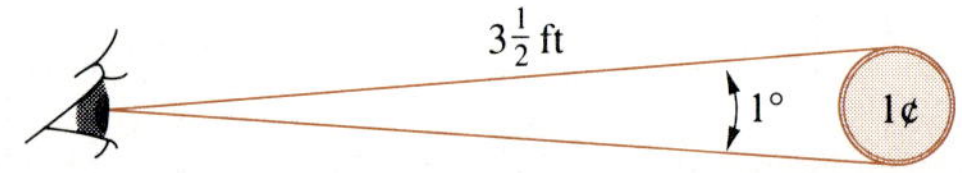

Quadrantal Angles in Degrees

As Example 1 shows, the quadrantal angles introduced in Section 1.2 can be represented in degrees.

Example 9

Convert the following quadrantal angles from revolution measure to degree measure. (a) $\frac{1}{2}$ rev, (b) $\frac{3}{4}$ rev, (c) $-\frac{1}{4}$ rev.

Solution

Since 1 rev = 360°, we have

(a) $\frac{1}{2}$ rev $= \frac{1}{2}$ of $360° = 180°$.

(b) $\frac{3}{4}$ rev $= \frac{3}{4}$ of $360° = \frac{3}{4}(360°) = 270°$.

(c) $-\frac{1}{4}$ rev $= -\frac{1}{4}(360°) = -90°$. ■

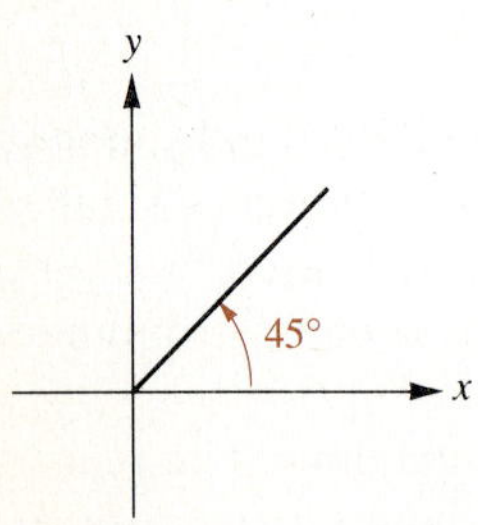

Figure 1.3-3

$\frac{1}{8}$ rev = 45°

NOW IT'S YOUR TURN

Convert $+\frac{1}{8}$ rev to degree measure and sketch the angle in standard form.
(Answer: 45°, as shown in Figure 1.3-3.)

Figure 1.3-4 illustrates several common quadrantal angles in both revolutions and degrees.

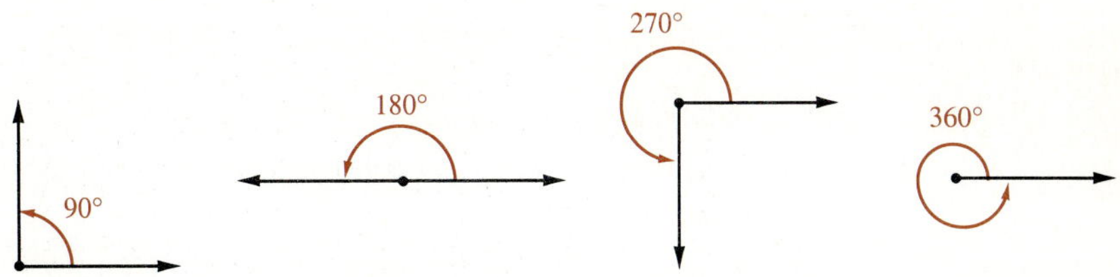

Figure 1.3-4 $\frac{1}{4}$ **revolution** $\frac{1}{2}$ **revolution** $\frac{3}{4}$ **revolution** **1 revolution**

> **SOME IMPORTANT DEFINITIONS**
>
> **An acute angle** is an angle measuring less than 90° but more than 0°.
>
> **An obtuse angle** is an angle measuring more than 90° but less than 180°.
>
> **Complementary angles** are two positive angles whose sum is 90°.
>
> **Supplementary angles** are two positive angles whose sum is 180°.

We illustrate these four angles in Figure 1.3-5, where α (alpha) and β (beta) represent angles.

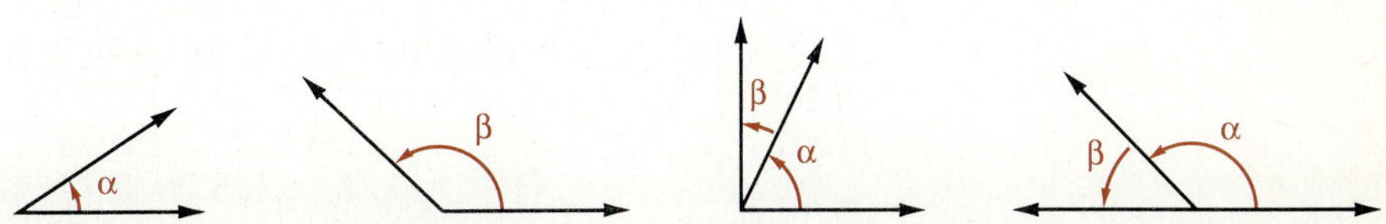

Figure 1.3-5 Acute angle $0° < \alpha < 90°$ Obtuse angle $90° < \beta < 180°$ Complementary angles $\alpha + \beta = 90°$ Supplementary angles $\alpha + \beta = 180°$

Minutes and Seconds

For even smaller angles, we divide the degree into 60 equal divisions, called *minutes*, and we further divide the minute into 60 equal divisions, called *seconds*. (This division into 60 units is due, as you may have guessed, to the Babylonians and their base-60 counting system.) We use the symbol ′ for minutes and ″ for seconds. For example, 15 minutes = 15′, and 35 seconds = 35″.

> **DEFINITION OF MINUTES AND SECONDS**
>
> $60' = 1° \quad \longleftrightarrow \quad 1' = \frac{1}{60}$ of a degree $= \frac{1}{60}°$
>
> $60'' = 1' \quad \longleftrightarrow \quad 1'' = \frac{1}{60}$ of a minute $= \frac{1}{60}'$

To give you some idea of the size of these units, a penny viewed face-on would have to be 215 ft away (roughly two-thirds the length of a football field) to subtend an angle of 1′, but 2.4 mi away to subtend an angle of 1″.

Example 10

Convert $\frac{13}{54}$ rev to degree-minute measure.

Solution

Since 1 rev = 360°, it follows that $\frac{13}{54}$ rev is $\frac{13}{54}$ of 360°. Therefore,

$$\frac{13}{54}\text{ rev} = \left(\frac{13}{54}\right)(360^\circ) = 86\tfrac{2}{3}^\circ = 86^\circ + \tfrac{2}{3}^\circ$$

But there are 60′ in 1°, so $\frac{2}{3}^\circ$ must be $\frac{2}{3}$ of 60′.

$$\tfrac{2}{3}^\circ = \tfrac{2}{3}(60') = 40'$$

Therefore, $\frac{13}{54}\text{ rev} = 86\frac{2}{3}^\circ = 86^\circ 40'$ ■

NOW IT'S YOUR TURN

Convert $\frac{1}{7}$ rev to degree-minute measure. (Answer: 51°26′)

Example 11

Find the angle that is supplementary to 113°20′45″.

Solution

Call the angle we want to find α. Because α and the given angle are supplementary, their sum must be 180°, so

$$\alpha + 113^\circ 20' 45'' = 180^\circ$$

$$\alpha = 180^\circ - 113^\circ 20' 45''$$

To see how to subtract one angle from another, we can write the angles the way you wrote numbers when you first learned to subtract.

$$\alpha = 180^\circ - 113^\circ 20' 45'' = \begin{array}{r} 180^\circ 00' 00'' \\ -\underline{113^\circ 20' 45''} \end{array}$$

We cannot subtract the minutes and seconds or we would get negative numbers, so we must make these numerators larger than the denominators. We know that

$$180^\circ = 179^\circ + 1^\circ = 179^\circ + 60'$$

and 60′ can be written as

$$60' = 59' + 1' = 59' + 60''$$

Therefore, combining these two gives us

$$180^\circ = 179^\circ + 1^\circ = 179^\circ + 60' = 179^\circ + 59' + 60'' = 179^\circ 59' 60''$$

Now we can subtract as follows:

$$\begin{array}{r} 180°00'00'' \\ -113°20'45'' \\ \hline \end{array} \quad \text{is equivalent to} \quad \begin{array}{r} 179°59'60'' \\ -113°20'45'' \\ \hline 66°39'15'' \end{array}$$

Therefore, the angle supplementary to 113°20′45″ is the angle 66°39′15″. ■

NOW IT'S YOUR TURN

Find the angle complementary to 61°43′18″. (Answer: 28°16′42″)

Using Calculators

Most pocket calculators do not use minutes and seconds. Instead they treat angles less than 1° as *decimal parts of a degree.* For example, 15′, which is $\frac{1}{4}°$, would be 0.25°. Therefore, it is usually necessary to convert an angle from minute-second form to decimal form before entering it in a calculator. As Example 12 shows, we can use the definitions of minute and second to make these conversions.

Example 12

Convert 58°18′ to decimal form.

Solution

58°18′ means exactly the same thing as 58° + 18′. There are 60′ in 1°, so $1' = \frac{1}{60}°$. Therefore, $18' = 18 \times \frac{1}{60}° = \frac{18}{60}° = 0.3°$. Thus

$$(58°18' = 58° + 18' = 58° + \frac{18}{60}° = 58° + 0.3° = 58.3° \ ■$$

After you get an angle in decimal form from your calculator, you may need to convert it to degree-minute form as shown in Example 13.

Example 13

Convert 15.2° to degrees and minutes.

Solution

15.2° is exactly the same as 15° + 0.2°. Because 1° = 60′, 0.2 of a degree must be 0.2 of 60′.

$$0.2° = 0.2(60') = 12'$$

Therefore,

$$15.2° = 15° + 0.2° = 15° + 12' = 15°\ 12' \ ■$$

NOW IT'S YOUR TURN

(a) Convert 48°11′ to decimal form with three-figure accuracy. (b) Convert 19.6° to degree-minute form. (Answers: 48.2°, 19°36′)

SIGNIFICANT FIGURES

When a number represents a measured quantity, it has limited accuracy due to limitations in measurement. The number of *significant figures* (or *digits*) tells us how accurately the number is known. To determine the number of significant figures, count the number of nonzero integers. For example, 1.456, 8397, 92.18, 732.9, and 5.731×10^8 all have four significant digits. If zero occurs as a digit, we treat it as follows:

1. If there is *no decimal point*, then 0 is significant *only* if it is *between* two nonzero integers. For example, 701 has three significant digits, 710 has only two, and 700 only one.

2. If there *is a decimal point*, 0 is significant only if it comes *after other significant digits*. If 0 comes before any significant digits (such as 0.002), we call it a *place-holder* and it is not significant. For example, 501, 50.1, 5.01, 0.501, 510., 5.10, 0.510, and 0.0510 all have three significant digits, but 0.051 and 0.00051 have only two.

Example

The number of significant digits (or figures) for each number is shown in brackets.

62.01 [4], 0.00003 [1], 1.00003 [6]
14050 [4], 14050. [5], 14050.00 [7]
93,000,000 [2], 186,000 [3]

DECIMAL ACCURACY

We often ask for numbers accurate to a certain number of decimals, which means that we want that number of *decimal places* showing. For example, 89.0015 has four decimals places, 788.1 has one decimal place, and 0.00091 has five decimal places.

ROUNDING OFF

When a number is written with too many digits, you must round it off to the proper number of *significant digits*. First, determine the number of significant digits. Then look at the digit right *after* the last significant digit. If that digit is ≥ 5, add 1 to the last significant digit and eliminate all the rest; if it is <5, leave the last significant digit as it is and eliminate all the rest. (If this digit *equals* 5, some people add 1 only if the previous digit is odd and add nothing if it is even; this always makes the final digit even. We shall *not* use that convention in this book.) For example, we round each number to three significant digits.

$46.88 \rightarrow 46.9$ $\quad$ $1.024 \rightarrow 1.02$
$3705 \rightarrow 3710$ $\quad$ $586.49 \rightarrow 586$
$0.03095 \rightarrow 0.0310$ (*not* 0.031)

SIGNIFICANT FIGURES IN ARITHMETIC

When doing arithmetic, the number of significant figures in the final answer can be *no greater than the number of significant digits in the least significant number used.* For example, $36.18 \times 1.2 = 43.42$. This product must be rounded to two significant digits because 1.2 has only two such digits. Therefore, we *must* express the answer as $36.18 \times 1.2 = 43$. Likewise, $7.0957/4.1 = 1.7307 = 1.7$. Some other examples are

$$34{,}837.1 \times 0.0028 = 97.544 = 98 \quad \blacksquare \quad \text{(because 0.0028 has two significant digits)}$$

$$9801.0/3.10 = 3161.7 = 3160 \quad \blacksquare \quad \text{(because 3.10 has three significant digits)}$$

$$(205)(13) = 2665 = 2700 \quad \blacksquare \quad \text{(because 13 has two significant digits)}$$

This rounding off process is sometimes painful, especially when calculators express answers in up to ten digits. Nevertheless, no matter what your calculator says, you must round your answers to the correct number of significant digits. *It is wrong to express an answer with more digits than are significant because doing so implies an accuracy that is not there.* For example, suppose an engineer measured the minimum strength of a cable to be 450 lb to two-figure accuracy, but he carelessly wrote 450.0 lb on a report. According to his *measurements*, he knows only that the cable's minimum strength is somewhere between 454 lb and 445 lb; it *could* be just 445 lb. Yet what he *wrote* implies that he knows its minimum strength is between 450.04 lb and 449.95 lb, so it could be *no less* than 449.95 lb. If someone believed what he wrote and put 449.0 lb on the cable, he could face a lawsuit.

EXACT NUMBERS

Measurements are *never* exact, but mathematicians often deal with ideal abstract numbers such as 2, $\frac{3}{17}$, $\sqrt{2}$, π, e, and so on. When these occur, they represent perfectly *exact* numbers, which means that they have infinitely many significant digits. Therefore, $\sqrt{2}$ is exact, but 1.414 is only an approximation for $\sqrt{2}$. We often write this approximation as $\sqrt{2} \approx 1.414$. Likewise, π is exact, but $\pi \approx 3.14$; e is exact, but $e \approx 2.7183$; $\frac{3}{7}$ is exact, but $\frac{3}{7} \approx 0.428571$, and so on.

CALCULATORS AND SIGNIFICANT FIGURES

Calculators usually display nine or ten digits. You can set your calculator to display only as many significant digits as you want (see your instruction manual). However, even though you set your calculator to display just three digits, for example, it is *operating using many more.* If you set your calculator to display three digits and enter $\sqrt{2}$, it shows 1.41 on the screen, but when performing calculations it is *using* 1.414213562. This fact can lead to discrepancies in answers.

(*continued*)

Suppose we multiply $\sqrt{2}\,\sqrt{3}$ to three significant digits. If we enter these numbers in the calculator as $\sqrt{2}$ and $\sqrt{3}$ and then multiply, we get 2.45. But if we enter each square root as a three-digit decimal—1.41 (for $\sqrt{2}$) and 1.73 (for $\sqrt{3}$)—we get 2.44. Notice the discrepancy in the third digit. This discrepancy occurs because in the first calculation, the calculator was *using* $\sqrt{2} = 1.414213562$ and $\sqrt{3} = 1.732050808$, even though it showed you only 1.41 and 1.73 on the screen. This gives $\sqrt{2}\,\sqrt{3} = (1.414213562)(1.732050808) = 2.449489743 = 2.45$. In the second case it was using only 1.41 and 1.73, giving $\sqrt{2}\,\sqrt{3} = (1.41)(1.73) = 2.439300000 = 2.44$.

Such a discrepancy can be especially bothersome with π. If you set your calculator to three digits and enter π, it will show 3.14, but it will *use* 3.141592654. Therefore, if you enter 3.14 *manually* for π, you will get a slightly different answer than you will if you let the calculator enter π for you. Be aware of possible discrepancies when you compare your answers with ours, especially when converting between radians and degrees.

EXERCISES

1. To how many digits is each number significant?

(a) 750 **(b)** 75.0
(c) −0.00128 **(d)** 4800010
(e) 4 000 000 000 **(f)** 103.000
(g) 103,000 **(h)** 10.0
(i) 0.408 **(j)** 0.00000003

2. $\pi = 3.141592654$

(a) To how many significant figures is π expressed here? To how many decimal places?
(b) Round π to the following number of significant figures. (i) 2, (ii) 4, (iii) 7.

3. $\sqrt{5} = 2.236067977$

(a) Round this number to the following number of significant figures. (i) 5, (ii) 8, (iii) 2.
(b) To how many decimal places have we expressed it?

4. Round to three significant figures: 821,490.

5. Carry out the indicated arithmetic and express the final answer to the correct number of significant figures.

(a) (72.3)(11.048) **(b)** $(18.5)(1.146)\sqrt{2}$
(c) 70578/3.417 **(d)** $\frac{1}{2}(89)(0.03455)$
(e) (987.082)(3) **(f)** 3.14159/2.0

6. In measuring a rectangular field, the surveyor records the following figures: length = 243.6 ft, width = 140 ft. What area should this surveyor properly record for this field?

7. The area of a triangle is $\frac{1}{2}bh$ ($\frac{1}{2}$ base × height). If you measure the base as 124.0 cm and the height as 87.1 cm, what is the area?

8. The volume of a sphere is $\frac{4}{3}\pi r^3$. If you measure the radius as 1.2 m and use $\pi = 3.14159$, what is the volume?

***9.** Make a plausible argument to show that the following sum is accurate to three significant figures: 75.8 + 0.3 = 76.1
Hint: How does the 0.3 affect the accuracy of 75.8?

10. If you measure that $x = 5.64$, you really know that x lies *between* two numbers. What are those numbers?

The Mil and the Grad

Two units of angular measure that are not often used in science and engineering are the *mil* and the *grad*. The mil, frequently used in the military, is defined to be $\frac{1}{64{,}000}$ of a revolution. The grad, used in some calculators, is defined to be $\frac{1}{100}$ of a right angle, which means that 100 grad = 90°.

1.3-2 The Radian

DEFINITION OF THE RADIAN

A *radian* is an angle which, if its vertex is at the center of a circle, intercepts an arc on that circle equal in length to the radius of the circle.

Definition of the Radian

The degree is the unit of angular measure used in such fields as navigation and surveying. In scientific work and advanced mathematics, it is much more convenient to use a circular system of angular measurement, and then the standard unit is called the *radian*.* The idea of the radian is based on the fact that if an angle has its vertex at the center of a circle, it will intercept an arc length on that circle. The larger the angle, the greater the arc length it will intercept. If an angle intercepts an *arc length just equal to the radius* of the circle, then we define that angle to be **one radian** (1 rad).

Figure 1.3-6a illustrates +1 radian (1 rad). From the definition of a radian, we see in Figure 1.3-6b that if a positive angle intercepts an arc length equal to twice the radius, the angle has measure +2 rad. Similarly, if the angle is in a clockwise direction, then it has measure −2 rad.

There are no units used to denote a radian, and quite often the word radian is not even used. So when we say, for example, that an angle has measure 2 (or −2), it is *understood* to mean 2 (or −2) rad.

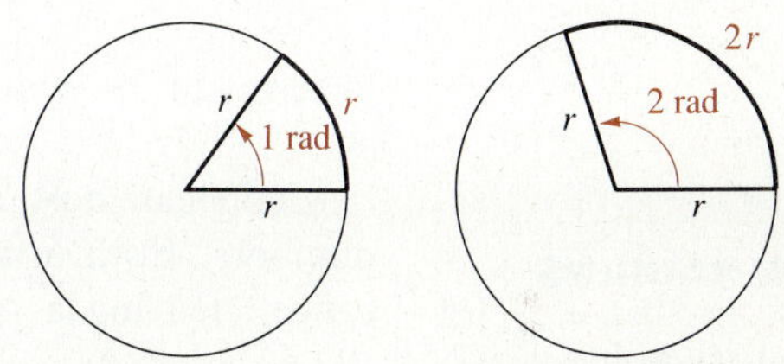

Figure 1.3-6

Radian measure of an angle ***(a) 1 radian*** ***(b) 2 radians***

Radian Measure and Arc Length

A radian is the same size no matter what the size of the circle, as you can see by looking at Figure 1.3-7. In this figure, angle POQ is 1 rad, which means that for each of the three concentric circles, POQ intercepts an arc length equal to the radius of that circle. Therefore,

$$\text{arc } AB = OA$$

$$\text{arc } A_1B_1 = OA_1$$

$$\text{arc } A_2B_2 = OA_2$$

We can rearrange these three equations slightly to obtain the ratio of the arc length to the radius of each circle.

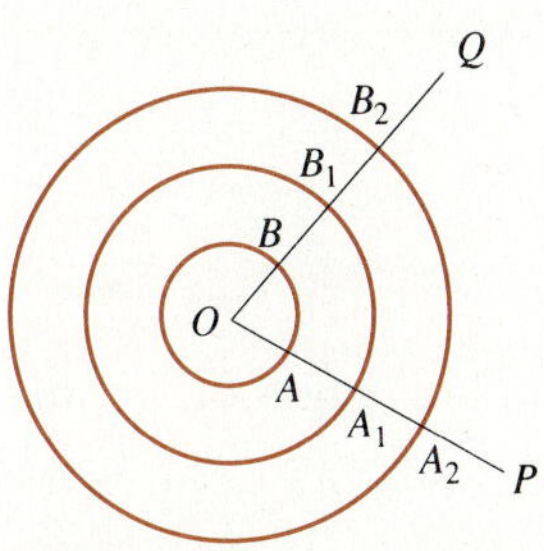

Figure 1.3-7

Angle POQ is 1 rad. Notice that the size of this angle is the same no matter what the size of the circle.

* The name *radian* first appeared in print in 1873 at Queens College, Belfast, Ireland, in examination questions set by James Thomson, a brother of the famous scientist Lord Kelvin (William Thomson).

$$\frac{\text{arc } AB}{OA} = 1 \qquad \frac{\text{arc } A_1B_1}{OA_1} = 1 \qquad \frac{\text{arc } A_2B_2}{OA_2} = 1$$

Now notice that *each of these three ratios is equal to 1 because the angle is 1 rad.* If the angle were 2 rad, then the arc length intercepted by each would be twice the radius and the ratio of arc length to radius would be 2 for each circle. If the angle were 3 rad, then the arc length intercepted would be 3 times the radius and the ratio of arc length to radius would be 3 for each circle, and so on. In other words, *the arc length divided by the radius is equal to the number of radians in the angle.* As you can see, for *any angle* having its vertex at the center of a circle, *the ratio of the arc length subtended on that circle to the radius of the circle is equal to the measure (in radians) of that angle.*

We can generalize this very important result as follows. In Figure 1.3-8, angle AOB is formed by two radii of the circle. We call such an angle a *central angle* because its vertex is at the center of the circle. The symbol θ (Greek theta) represents the measure of angle AOB, in this case in radians. Because θ is a central angle, the results we have just seen tell us that

$$\theta \text{ (rad)} = \frac{\text{arc length subtended by } \theta}{\text{radius of circle}}$$

In terms of the symbols of Figure 1.3-8, we can write this equation more briefly as $\theta = S/r$.

You can now see clearly why angles in radians are pure, dimensionless numbers. Both S and r have units of length; in the ratio S/r the length units cancel, leaving a pure number.

ANGLE IN RADIANS

If θ is a central angle intercepting an arc length S on a circle of radius r, then θ in radians is

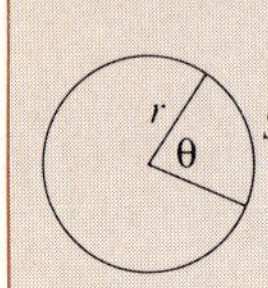

$$\theta = S/r$$

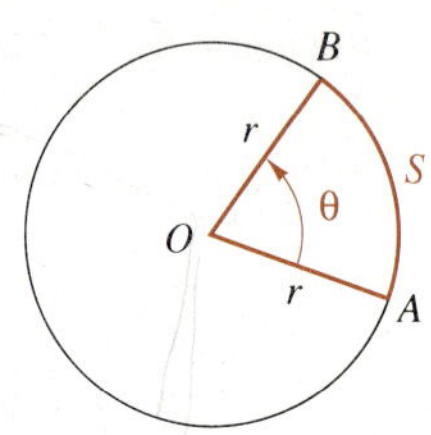

Figure 1.3-8
Central angle

Example 14

If a central angle of a 6.0-in. diameter circle subtends an arc length of 5.0 in., find this angle in radians.

Solution

First make a sketch illustrating the problem (Figure 1.3-9). We just showed that if θ is a central angle, then

$$\theta\text{(rad)} = \frac{\text{arc length subtended by } \theta}{\text{radius of circle}} = \frac{S}{r}$$

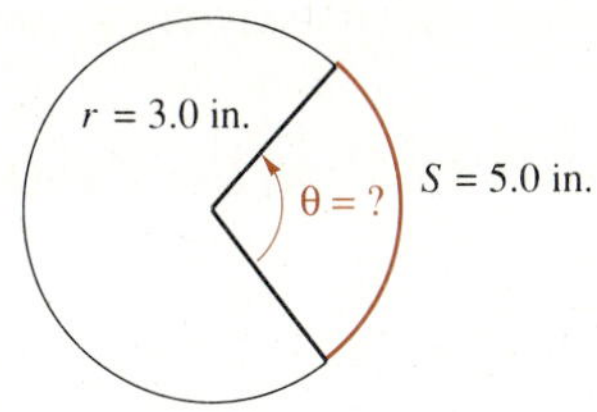

Figure 1.3-9

As you can see in Figure 1.3-9, $S = 5.0$ in. and $r = 3.0$ in. (one-half the diameter), so

$$\theta(\text{rad}) = \frac{S}{r} = \frac{5.0\text{ in.}}{3.0\text{ in.}} = 1.7 \;\blacksquare$$

We can leave the answer this way or write it as 1.7 rad because radians have no units. As you can see the inches cancel, leaving a pure, dimensionless number for θ. Notice that we rounded the answer to two significant digits because the other numbers have this degree of accuracy (see the section on Significant Figures).

NOW IT'S YOUR TURN

What arc length will a 1.40-rad angle subtend on a circle of radius 13.4 cm? (Answer: 18.7 cm)

Example 15

In laying out a new highway, a civil engineer must design a circular curve 76 ft long in which cars change direction by 1.3 rad. What must be the radius of this circular section of highway?

Solution

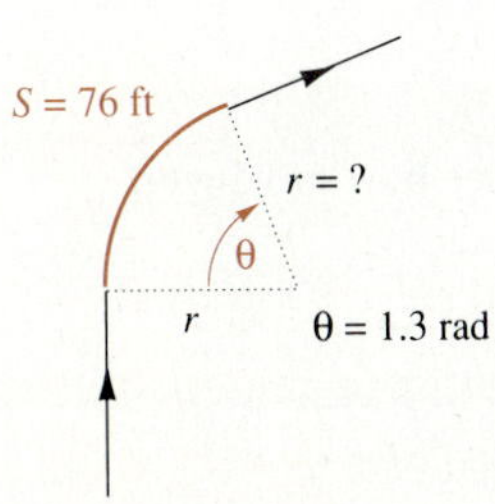

Figure 1.3-10
A car moves through an arc length of 76 ft while going through an angle of 1.3 rad.

First sketch the highway curve (Figure 1.3-10). The cars change direction by the angle $\theta = 1.3$ rad as they drive around the 76-ft section of arc length. Because θ is in radians,

$$\theta = \frac{S}{r} \quad \rightarrow \quad r = \frac{S}{\theta} \qquad (\text{solve for } r)$$

$$r = \frac{S}{\theta} = \frac{76\text{ ft}}{1.3\text{ rad}} = 58\text{ ft} \;\blacksquare$$

Notice that the answer is 58 ft, not 58 ft/rad, because radians have no units.

NOW IT'S YOUR TURN

If this curved section of highway is later extended to 186 ft but has the same radius, by what angle will cars change direction as they drive around it? (Answer: 3.2 rad)

1.3-3 Degree–Radian Conversions

The Basic Conversion Factor

We have seen that angles are usually measured in either degrees or radians. Therefore, it is often necessary to make conversions from one angular measure to the other.

To see how to do this, let's consider a complete revolution, or 360°, as shown in Figure 1.3-11. To find the number of radians in one complete revolution, we can apply the formula $\theta = S/r$ to this figure. In this case, the angle is one complete revolution so S is the circumference of the circle. In geometry you learned that the circumference c of a circle of radius r is $c = 2\pi r$, where $\pi = 3.14159\ldots$. (See Section 1.8 for a discussion of this fascinating number π.) Thus, for a complete revolution, $S = c = 2\pi r$, which gives us

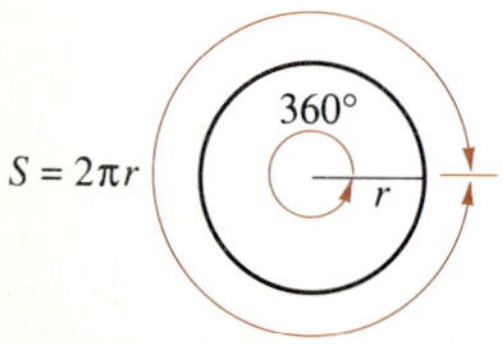

Figure 1.3-11

Arc length subtended by one complete revolution

$$\theta(\text{rad}) = \frac{S}{r} = \frac{2\pi r}{r} = 2\pi$$

which means that 1 rev is 2π rad.

$$1 \text{ revolution} = 2\pi \text{ radians}$$

We also know that there are 360° in a complete revolution, so it must be true that 2π rad = 360°. Dividing by 2 gives

$$\pi \text{ radians} = 180°$$

▸ **This preceding result is probably the easiest way to remember how to convert between radians and degrees.**

Example 16

How many degrees are in 1.00 rad?

Solution

We know that π rad = 180°, so (180°/π rad) = 1. Multiplication by 1 never changes the magnitude of a quantity, so multiplication by (180°/π rad) will not change the magnitude of 1 rad, but multiplication by 1 *will change its units* by cancelling the radians and leaving degrees (see the section on Conversion of Units).

$$1 \text{ rad} \times \frac{180°}{\pi \text{ rad}} = 1 \text{ rad} \times \frac{180°}{3.14 \text{ rad}} = 57.3° \ \blacksquare$$

Example 17

How many radians are there in 1.00°?

Solution

As in Example 16, we can use π rad = 180°, but this time we want to cancel the units of degrees in the numerator of 1.00° and leave radians. Consequently we should multiply 1.00° by (π rad/180°), which is equal to 1 and yet has the degrees in the denominator to cancel the degree in the numerator (see Conversion of Units). Thus

$$1.00° \times \frac{\pi \text{ rad}}{180°} = 0.0174 \text{ rad} \quad ■$$

Example 18

(a) Convert the following angles to radian measure: 30°, −45°, 330°.

(b) Convert the following angles to degree measure: $\frac{2\pi}{3}$, $-\frac{5\pi}{7}$ rad, 4.00.

Solution

(a) In each case, we want to cancel the degrees in the numerator and replace them with radians, so we should multiply by (π rad/180°).

$$30° \times \frac{\pi \text{ rad}}{180°} = \frac{\pi}{6} \text{ or } \frac{\pi}{6} \text{ rad}$$

$$-45° \times \frac{\pi \text{ rad}}{180°} = -\frac{\pi}{4} \text{ or } -\frac{\pi}{4} \text{ rad}$$

$$330° \times \frac{\pi \text{ rad}}{180°} = \frac{11\pi}{6} \text{ or } \frac{11\pi}{6} \text{ rad}$$

Notice that we left π in the answers to express the angles *in terms of* π. This is done very frequently. You will often see 30° expressed as $\frac{\pi}{6}$ rather than 0.5236.

(b) First, we must realize that all of these angles are in radians, even though the radians are not always written. To make the conversions, we want to cancel the radians in the numerator and replace them with degrees, so we must multiply by (180°/π rad).

$$\frac{2\pi}{3} \times \frac{180°}{\pi \text{ rad}} = 120°$$

$$\frac{-5\pi}{7} \text{ rad} \times \frac{180°}{\pi \text{ rad}} = -\frac{900°}{7} = -129°$$

$$4.00 \times \frac{180°}{\pi \text{ rad}} = \frac{720°}{\pi} = \frac{720°}{3.14} = 229° \quad ■$$

NOW IT'S YOUR TURN

(a) How many radians are in a right angle?

(b) How many degrees are in 3.72 rad? (Answer: $\frac{\pi}{2}$, 213°)

CONVERSION OF UNITS

The conversion of units is difficult for many students, but it is very easy if you remember one simple fact: Multiplication by 1 does not change the value of a quantity. The trick is to multiply by factors (called *conversion factors*) that are equal to 1 and yet cancel the given units and replace them with the desired units. This process is sometimes called the factor-dimensional method or the factor-units method. We illustrate it with several examples.

Example 1

Convert 6.00 yr to units of days.

Solution

We need a conversion factor between days and years. Because 365 days = 1 yr,

$$\frac{365 \text{ days}}{1 \text{ yr}} = 1 \qquad \text{and} \qquad \frac{1 \text{ yr}}{365 \text{ days}} = 1$$

Therefore, multiplying by either of these conversion factors does not change the value of the quantity 6 yr, but it will change the *units*. We want to cancel "yr" and replace it with "days," so we multiply by the conversion factor with "yr" in the denominator.

$$6.00 \text{ yr} \times \left(\frac{365 \text{ days}}{1 \text{ yr}}\right) = (6)(365) \text{ days} = 2190 \text{ days}$$

Example 2

Convert 3.00 mi to units of centimeters.

Solution

We need conversion factors for miles to kilometers, kilometers to meters, and meters to centimeters. Because 1 mi = 1.61 km,

$$\frac{1.61 \text{ km}}{1 \text{ mi}} = 1$$

Because 1000 m = 1 km,

$$\frac{1000 \text{ m}}{1 \text{ km}} = 1$$

and because 100 cm = 1 m,

$$\frac{100 \text{ cm}}{1 \text{ m}} = 1$$

Using these conversion factors to cancel miles and end up with centimeters, we have

$$3.00 \text{ mi} \times \underbrace{\left(\frac{1.61 \text{ km}}{1 \text{ mi}}\right)}_{\text{mi} \to \text{km}} \times \underbrace{\left(\frac{1000 \text{ m}}{1 \text{ km}}\right)}_{\text{km} \to \text{m}} \times \underbrace{\left(\frac{100 \text{ cm}}{1 \text{ m}}\right)}_{\text{m} \to \text{cm}} = (3)(1.61)(1000)(100) \text{ cm}$$

$$= 4.83 \times 10^5 \text{ cm}$$

Example 3

Convert 60.0 mph to units of feet per second.

Solution

We need conversion factors for miles to feet, hours to minutes, and minutes to seconds. Because 5280 ft = 1 mi,

$$\frac{5280 \text{ ft}}{1 \text{ mi}} = 1$$

Because 1 hr = 60 min,

$$\frac{1 \text{ hr}}{60 \text{ min}} = 1$$

Because 60 s = 1 min,

$$\frac{1 \text{ min}}{60 \text{ s}} = 1$$

Using these conversion factors to replace the miles by feet and hours by seconds gives us

$$60.0 \text{ mph} = 60.0 \frac{\text{mi}}{\text{hr}} \times \underbrace{\left(\frac{5280 \text{ ft}}{1 \text{ mi}}\right)}_{\text{mi} \to \text{ft}} \times \underbrace{\left(\frac{1 \text{ hr}}{60 \text{ min}}\right)}_{\text{hr} \to \text{min}} \times \underbrace{\left(\frac{1 \text{ min}}{60 \text{ s}}\right)}_{\text{min} \to \text{s}}$$

$$= \frac{(60)(5280) \text{ ft}}{(60)(60) \text{ s}} = 88 \frac{\text{ft}}{\text{s}} = 88 \text{ ft/s}$$

When using this method, you *must* write your units in *every step*. If the units come out right, your answer has to be correct unless you have used an incorrect conversion factor. If you do not write in the units as you go, you are defeating the whole purpose of this method because you cannot see if the units are coming out right!

(continued)

EXERCISES

Use the factor-dimensional method to make the following conversions. For the conversion factor between the metric and English systems, use *only* the fact that 1 in. = 2.54 cm.

1. 2.0 m to millimeters
2. 7.5 km to millimeters
3. 17 yd to inches
4. 1.0 hr to seconds
5. 1.0 ft to centimeters
6. 2.0 m to inches
7. 1.0 mi to kilometers
8. 1.00 yr to seconds
9. 55 mph to feet per minute
10. 186,000 mi/s to meters per second (Do you recognize this number?)
11. 88 km/hr to inches per second
12. 32 ft/s^2 to meters per second^2
13. 100.0 m^2 to centimeters^2
14. 1 ft^2 to inches^2
15. 1.00 cubic yd = ? cubic in.?
16. 1 square yd = ? sq in.?
17. 1.00 yd^3 = ? centimeters^3
18. A solid box has as its rectangular dimensions 4.0 ft × 2.0 yd × 18 in. Therefore, its volume is $V = (4.0 \text{ ft})(2.0 \text{ yd})(18 \text{ in.})$. Convert these rather unconventional units to (a) cubic feet and (b) cubic centimeters.
19. The area of a triangle is $\frac{1}{2}$ the base times the height, or $\frac{1}{2}bh$. If you measure the base to be 16.0 cm and the height to be 285 mm, calculate the area using those units and then convert to meters^2.
20. Convert 9.8 m/s^2 to units of kilometers/hours^2. (This is the acceleration of a freely falling body.)
21. Convert 2.4° to units of seconds.
22. Convert one-quarter revolution to units of minutes.
23. Convert 7500′ to units of revolutions.

Example 19

Through what angle in radians and degrees does the earth spin on its axis in 1.0 hr?

Solution

The earth makes 1 rev (2π rad) each day (24 hr). Therefore, in 1.0 hr it makes $\frac{1}{24}$ of a complete revolution or

$$\frac{1}{24} \times 2\pi \text{ rad} = \frac{\pi}{12} \text{ rad}$$

Converting this to degrees,

$$\frac{\pi}{12} \text{ rad} \times \frac{180°}{\pi \text{ rad}} = 15° \quad ■$$

NOW IT'S YOUR TURN

Through what angle in radians and degrees does the earth move around the sun in 1 day? (Answer: 0.017 rad, 0.99°)

Example 20

How many radians are in 46°10′?

Solution

First we must convert this angle to decimal form, and then we can use the same type of conversion used in Example 17. Converting to decimal form,

$$46°10' = 46° + 10' = 46° + \left(\frac{10}{60}\right)^{\circ} = 46° + 0.167° = 46.167°$$

Now we convert to radians.

$$46.167° \times \frac{\pi \text{ rad}}{180°} = 0.806 \text{ rad} \quad ■$$

where we have used $\pi = 3.14$.

NOW IT'S YOUR TURN

Convert 21°49′ to radians. (Answer: 0.381 rad)

Table 1.3-1 gives the degree measure and the corresponding radian measure of some commonly used angles. Do not attempt to memorize this table! Instead, make each conversion yourself to be sure that you understand where it comes from, and use the table to check your work.

Table 1.3-1

Some commonly used angles in radian and degree measure

Degrees	0°	15°	30°	45°	60°	90°	120°	135°	150°	180°	270°	360°
Radians	0	$\frac{\pi}{12}$	$\frac{\pi}{6}$	$\frac{\pi}{4}$	$\frac{\pi}{3}$	$\frac{\pi}{2}$	$\frac{2\pi}{3}$	$\frac{3\pi}{4}$	$\frac{5\pi}{6}$	π	$\frac{3\pi}{2}$	2π

Conversions Using Calculators

Modern pocket calculators will calculate with angles in either degrees or radians, but you must tell your calculator which units you are using! To do this, look for a key or switch that says RAD or DEG or something similar and set it to the type of units you want to use. Some calculators display a small RAD or DEG on the screen.

▸ **Radian Mode.** With a calculator set to the radian mode, angles are entered and displayed as pure numbers. For example, 1.42 rad is entered (and displayed) simply as 1.42. Be careful to convert angles expressed in terms of π to decimal numbers before entering them! For example, $\frac{\pi}{6} = \frac{3.1416}{6} = 0.5236$, so we would enter 0.5236.

▸ **Degree Mode.** With a calculator set to the degree mode, angles are entered and displayed as pure numbers that represent degrees (the ° symbol is not entered or displayed). For example, 75° is entered and displayed simply as 75.

The biggest problem in using the degree mode is handling fractions of a degree, such as 128°09′21″. Most calculators use fractions of a degree expressed in decimal form rather than in minutes and seconds. Therefore, *you must convert fractions of a degree to decimals* and enter them that way. For example, 20°15′ must be entered as 20.25, meaning 20.25°.

Many calculators have a π key. When you press this key, the calculator automatically enters π to as many decimal places as the calculator can handle. If your calculator has such a key, you can use it to enter π rather than entering 3.14159 . . . . **Caution**: If you use such a key, your answers may differ slightly from those in the examples, where we have used $\pi \approx 3.14$.

Some calculators convert directly from degrees to radians and radians to degrees. Consult your instruction manual to find out if your calculator does this, and, if so, how to do it. Even if your calculator will do this for you, it is still very important to know how to make these conversions without a special calculator. **Caution**: If your calculator makes such conversions for you automatically, it uses π to as many decimal places as it can handle, just like the π key we discussed. Therefore your answers may once again differ slightly from ours.

Quadrants Revisited

In Section 1.2 we learned to determine in what quadrant an angle is located using revolutions. We also need to be able to figure the quadrant using degrees and radians. Using Table 1.3-1, we can express the four quadrantal angles in terms of degrees and radians. For example, $\frac{1}{4}$ of a revolution is $\frac{1}{4}$ of 360°, or 90°. From the table, we see that 90° is $\frac{\pi}{2}$ in radian measure. In a similar way, we can determine other quadrantal angles and summarize the results, as shown in Figure 1.3-12.

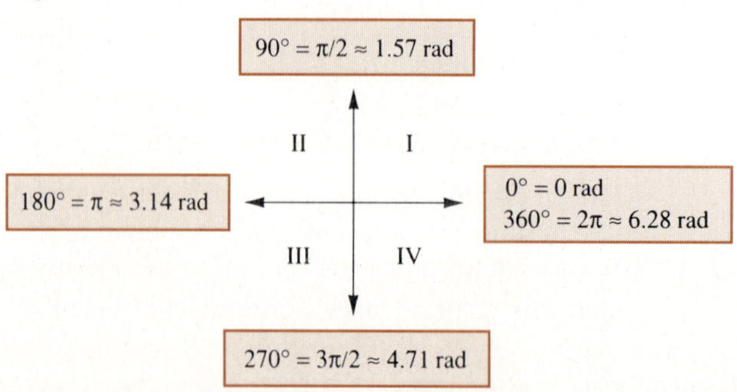

Figure 1.3-12

Quadrantal angles in degrees and radians

Example 21

In which quadrant is each of the following angles located? (a) $\frac{7\pi}{4}$, (b) 2.36, (c) 0.93, (d) -6.24.

Solution

First express the angle as a quadrantal angle plus an angle smaller than 90° ($\frac{\pi}{2}$).

(a) $\frac{7\pi}{4} = \frac{6\pi}{4} + \frac{\pi}{4} = \frac{3\pi}{2} + \frac{\pi}{4}$. From Figure 1.3-12, we can see that $\frac{3\pi}{2}$ is the quadrantal angle between quadrants III and IV. The additional $\frac{\pi}{4}$ is not great enough to put the angle past quadrant IV, so the angle lies in quadrant IV.

(b) 2.36 is between 1.57 and 3.14, so this angle lies in quadrant II.

(c) 0.93 is between 0 and 1.57, so this angle lies in quadrant I.

(d) -6.24 is slightly less than a complete revolution (6.28 rad) yet more than $\frac{3}{4}$ of a revolution (4.71 rad), but it is in a clockwise direction (see Figure 1.3-13). Thus this angle lies in quadrant I. ■

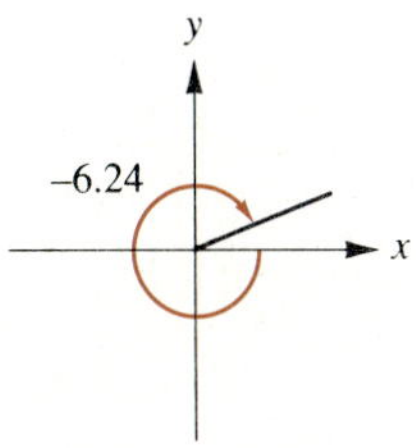

Figure 1.3-13

NOW IT'S YOUR TURN

In what quadrant does each angle lie? (a) 3.57, (b) $\frac{11\pi}{12}$, (c) -4.10.
(Answers: III, II, II)

Example 22

In what quadrant does each angle lie? (a) 171°, (b) −40°, (c) −315°.

Solution

(a) 171° is greater than 90° but less than 180°, so it lies in quadrant II.

(b) −40° is less than $\frac{1}{4}$ of a revolution (90°), but it is in a *clockwise* direction so it lies in quadrant IV.

(c) −315° more than $\frac{3}{4}$ of a revolution (270°) and less than a full revolution (360°), but it is in the *clockwise* direction so the angle lies in quadrant I. ■

NOW IT'S YOUR TURN

In what quadrant does each angle lie? (a) 210°, (b) −200°. (Answers: III, II)

EXERCISES 1.3-1

In Exercises 1–8, find the degree measure of each angle.

1. $\frac{2}{3}$ rev counterclockwise

2. $\frac{1}{6}$ rev clockwise

3. $\frac{5}{8}$ rev clockwise

4. $\frac{3}{10}$ rev counterclockwise

5. $\frac{5}{12}$ rev counterclockwise

6. $1\frac{3}{4}$ rev clockwise

7. $\frac{1}{3}$ rev clockwise

8. $\frac{7}{5}$ rev counterclockwise

In Exercises 9–19, find the angle that is complementary to the given angle.

9. 30°
10. 60°
11. 45°
12. 0°
13. 90°
14. 15°42′
15. 74°35′
16. 19°30′25″
17. 84°17′10″
18. 46.25°
19. 81.59°

In Exercises 20–29, find the angle that is supplementary to the given angle.

20. 27°
21. 121°8′
22. 14″
23. 49°32′31″
24. 107°15′54″
25. 125°10″
26. 59°33″
27. 77.9°
28. 63.75°
29. 17′30″

In Exercises 30–38, convert the given angle to decimal notation. Carry each answer to two decimal places.

30. 26°15′
31. 179°45′
32. −14°30′
33. −110°10′
34. 13°21′5″
35. 345°50′15″
36. 10°20″
37. 180°12″
38. −45°6″

In Exercises 39–44, convert the given angles to degree-minute-second form, rounding to the nearest whole second.

***39.** 7.23°
***40.** 196.18°
***41.** −51.36°
42. −126.5°
***43.** 37.934°
***44.** −74.505°

In Exercises 45–50, find two angles, one positive and one negative, that are coterminal with the given angle.

45. 125°
46. −170°
47. −180°
48. 315°
49. 210.8°
50. 295°18′

51. Draw a circle using a compass or template. Next carefully draw an angle with its vertex at the center of this circle so that the angle cuts off (intercepts) an arc length on the circle equal in length to the circle's radius. Do this again for three or four circles of different sizes and then compare the sizes of the angles in each case. Do the sizes depend on the size of the circle?

In Exercises 52–55, starting at midnight on the morning of January 1, 1989, through what angle in degrees has the earth moved around the sun by midnight on the morning of each date given?

***52.** January 10
***53.** March 25
***54.** October 30
***55.** December 13

***56.** Starting at midnight on the morning of January 1, 1989, what date is it when the earth has moved around the sun through a 20° angle?

57. How many grad are in 1.00°?

58. One grad is how many degrees?

59. Express 15° in grad.

60. If an angle is 250 grad, how many degrees is it?

61. If an angle is 4000. mil, how many degrees is it?

62. If an angle is 120 grad, how many mil is it?

63. How many grad are in a 5000.-mil angle?

64. How many mil are in a right angle?

65. How many grad are in a circle? How many mil?

In Exercises 66–68, in which quadrants can $-A$, $2A$, and $A/2$ lie?

66. A is a positive angle:
(a) in quadrant I.
(b) in quadrant II, greater than 135°.

67. A is a positive angle:
(a) in quadrant IV.
(b) in quadrant III, less than 225°.

***68.** A is a negative angle:
(a) in quadrant II.
(b) in quadrant III.
(c) in quadrant IV.

EXERCISES 1.3-2

In Exercises 1–4, let S be the arc length and θ the central angle of a circle of radius r or diameter d.

1. Find θ in degrees if $r = 40.0$ cm and $S = 30.0$ cm.

2. Find θ if $d = 565$ m and $S = 782$ m.

3. Find r if $\theta = 1.8$ rad and $S = 2.6$ yd.

4. Find S if $d = 200.0$ ft and $\theta = \frac{1}{10}$.

In Exercises 5–7, use the relationship $\theta = S/r$ for Figure 1.3-8 to draw freehand but *carefully* and *to scale* each of the angles. *Hint:* You will have to estimate carefully or measure when the arc length equals r, $2r$, etc.

5. (a) $\theta = 1$ rad, **(b)** $\theta = 2$ rad

6. (a) $\theta = 3$ rad, **(b)** $\theta = 4$ rad

7. (a) $\theta = 6$ rad, **(b)** $\theta = 7$ rad

8. If a central angle of a circle of radius 8.0 cm intercepts an arc of 1.0 cm, what is this angle in radians?

9. How large in radians would a central angle of a circle have to be to intercept arc lengths equal to **(a)** the radius, **(b)** 85% of the radius, **(c)** 1.50 times the radius?

10. If a ship sailing along the earth's equator travels 2000.0 mi, through what angle in radians has this ship moved relative to the earth's center?

11. While walking along a circular path, a person walks a distance of 68 ft. A person standing at the center of this circle has to move her head through 1.75 rad to follow the walking person. What is the radius of this path?

12. If a little bug riding on the rim of a 12-in. diameter LP record turns through an angle of 2.2 rad, through what distance has the bug moved?

13. You want to cut a wedge from an 18-in. diameter pizza so that the crust of this wedge is 10.0 in. long at the rounded edge. At what central angle should you cut this piece of pizza?

14. A circular spool 8.0 in. in diameter has 650 ft of thin wire wound around its outer rim. Through what angle will this spool turn as all the wire is pulled off?

15. For the wire and spool in Exercise 14, what length of wire is pulled off when the spool turns through 4.6 rad?

16. You are decorating the curved underside of an arched doorway using a single line of 1.0-in. Mexican tiles (see Figure 1.3-14). The arched part of this doorway is the curve of a circular arc that woud subtend 0.75 rad on a circle 5.00 ft in diameter. How many tiles will you need for this job?

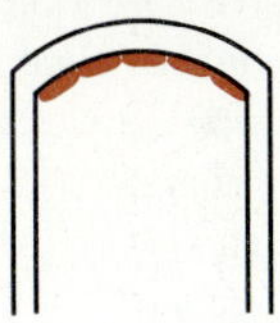

Figure 1.3-14

17. Radian-Degree Conversions

(a) How many radians are in 1.00 rev? ***Hint:*** Use the definition of the radian. What is the arc length subtended in this case?
(b) How many radians are in 180°? ***Hint:*** Use the results of (a).
(c) How many radians are in a right angle? ***Hint:*** See (a) and (b).

EXERCISES 1.3-3

In Exercises 1–6, convert each degree measure to radian measure. Give the answers in terms of π.

1. 50.0° **2.** −120°
3. 225° **4.** −135°
5. −45° **6.** 315°

In Exercises 7–14, convert each degree measure to radian measure. Express each answer to two decimal places.

7. 11° **8.** −67°
9. 200.0° **10.** 390°
11. −17.8° **12.** 25.3°
13. 18°36′ **14.** −74°40′

In Exercises 15–29, convert each radian measure to decimal degrees.

15. $\frac{\pi}{15}$ **16.** $\frac{5\pi}{12}$
17. $\frac{11\pi}{18}$ **18.** $\frac{\pi}{30}$
19. 5π **20.** $\frac{-3\pi}{5}$
21. $\frac{-7\pi}{9}$ **22.** $\frac{-7\pi}{3}$
23. 2.00 **24.** −4.00
25. 1.110π ***26.** $\pi - 3.00$
***27.** $\frac{2}{\pi}$ ***28.** $\pi - \frac{1}{\pi}$
***29.** π^2

In Exercises 30–42, determine in which quadrant each angle is located.

30. $\frac{4\pi}{3}$ **31.** $\frac{-5\pi}{3}$
32. $\frac{7\pi}{4}$ **33.** 3.5
34. −3.5 **35.** 235°
36. −214° **37.** 372°
38. −400° **39.** 753°
40. −942° **41.** 9.1π
42. −1.72

In Exercises 43–56, convert the angles for use in a calculator.

43. 31°41′ **44.** 125.87°
45. 92°5′18″ **46.** 4°45′35″
47. 55′14″ **48.** 10°36″
49. π **50.** $\frac{\pi}{6}$
51. 3.57 **52.** $\pi + \pi^2$
53. $\frac{\pi^3}{3}$ **54.** $(\frac{\pi}{3})^3$
55. $\pi + 2$ **56.** $\pi + \frac{4}{\pi}$

In Exercises 57–60, through what angle in degrees and in radians does the hour hand on a clock move in the time intervals given?

57. Between noon and 3 P.M.
58. Between 1 P.M. and 2 P.M.
59. Between 1 A.M. and 2 P.M.
***60.** From to 9:15 P.M.

In Exercises 61–64, through what angle in degrees and in radians does the earth turn on its axis in the time intervals given?

61. In 1.00 hr

62. In 8.0 hr

63. Between 1 P.M. and 6 P.M.

64. Between 1 P.M. and 5 A.M.

In Exercises 65–68, starting at 12 noon, what time is it after the minute hand of a clock has turned through the angle given?

65. 720°

66. $\frac{\pi}{4}$

67. 3π

68. 135°

69. In which quadrants can $-\theta$, 2θ, and $\theta/2$ lie?

(a) θ is a positive angle less than $\frac{\pi}{4}$.
(b) θ is a positive angle less than $\frac{3\pi}{4}$ in quadrant II.
***(c)** θ is a negative angle in quadrant I.
***(d)** θ is a negative angle greater than $-\pi$.

70. Which is *greater*?

(a) 1.0 or 1.0°
(b) 100° or $\frac{\pi}{2}$
(c) 2.0 or 90°
(d) 3.0 or 90°
(e) 3.0 or 180°
(f) 6.0 or 360°
(g) 7.0 or 360°
(h) 1.0 or 60°

71. If a central angle of a 10.0-ft diameter circle intercepts an arc length of 11 ft, what is this angle in degrees?

72. An ant riding at the edge of an 8.0-in. diameter record turns through a 20.0° angle. How far has this ant moved in this time?

***73.** You are ordering multicolored 8-in.-long bricks to use as the border of your 12-ft-diameter circular garden. If the wedge subtended by the red bricks has a 30° central angle and that subtended by the orange bricks has a 1.2 rad central angle, how many bricks of each color should you order?

74. A thin 25-cm steel rod is bent into a circular arc subtending an angle of 15° from the center of the circle. What is the radius of this circle?

***75.** A thin 140-cm-long brass wire is bent into the outline of a wedge of a circle. If this wedge subtends a 30.0° central angle, find its radius and the length of the curved part of the wedge.

76. A bricklayer is using a 20.0-in-diameter winch to lift pallets of bricks to the roof of a building 45 ft above the ground, as shown in Figure 1.3-15. The rope that lifts each pallet is wrapped around the outer rim of the winch.

(a) Through how many degrees must the bricklayer turn the winch to lift the pallet by 1.0 ft?
(b) Through how many revolutions must the winch turn to lift a pallet from the ground to the roof?

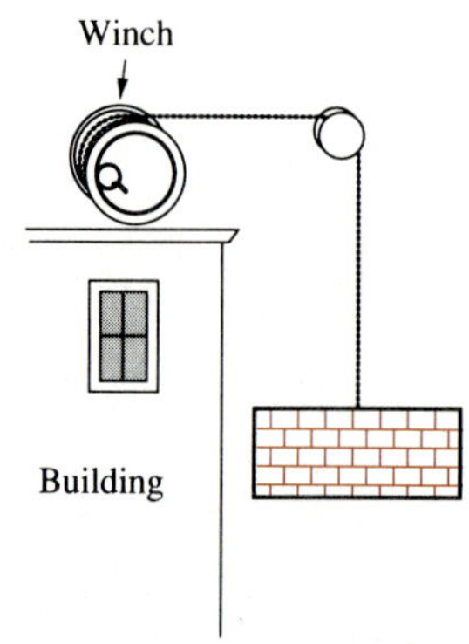

Figure 1.3-15

77. When laying out a railway track curve, the engineer must change the train's direction by 105° over a distance of 1750 ft. What must be the radius of this curve? ***Hint:*** See Example 15.

78. **(a)** How many grad are in 1.00 rad?
(b) One grad is how many radians?

79. **(a)** How many mil are in 1.00 rad?
(b) One mil is how many radians?

80. If an angle is 1.4 rad, how many grad is it?

81. If an angle is 0.750 rad, how many mil is it?

1.4 CENTRAL ANGLE AND ARC LENGTH

In many uses of trigonometry, it is essential to express angles in radian measure. In this section, we shall discuss several of these applications, including arc length and the area of the wedge of a circle.

Consider a circle of radius r, as shown in Figure 1.4-1. The central angle θ subtends (intercepts) an arc length S. In Section 1.3, we showed that if θ is in radians, then $\theta = S/r$. Solving this for the arc length S, we have $S = r\theta$.

The central angle θ subtends an arc length S on the circle.

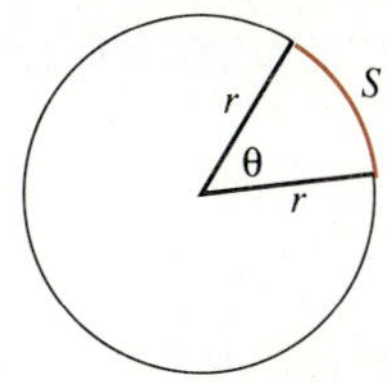

Figure 1.4-1
Central angle and arc length

$$S = r\theta$$

- **It is important to remember that radians are pure, dimensionless numbers having no units. Therefore, S and r have the same units because θ has no units.**
- **It is also important to realize that the formula $S = r\theta$ can be used only if *θ is in radians*. If the angle θ is given in degrees, you must convert it to radians before using this formula.**

Example 23

If a $\frac{\pi}{5}$ central angle intercepts an arc on a circle of radius 8.0 cm, how long will that arc be?

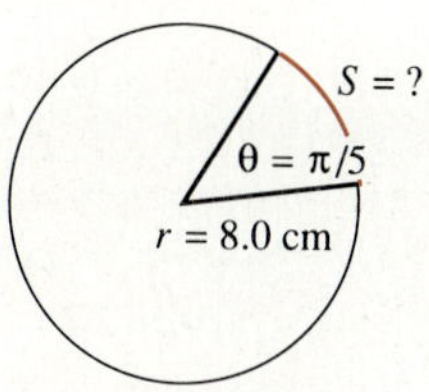

Figure 1.4-2

Solution

First make a clear sketch showing what you know and what you are looking for (Figure 1.4-2). In this case, we know that $r = 8.0$ cm and $\theta = \frac{\pi}{5}$ and we want to find S. Solving for S gives

$$S = r\theta = (8.0 \text{ cm})\left(\tfrac{\pi}{5}\right) = \frac{8\pi \text{ cm}}{5} = \frac{8(3.14) \text{ cm}}{5} = 5.0 \text{ cm} \quad \blacksquare$$

Example 24

How long an arc will a 140° central angle subtend on a circle 3.6 ft in radius?

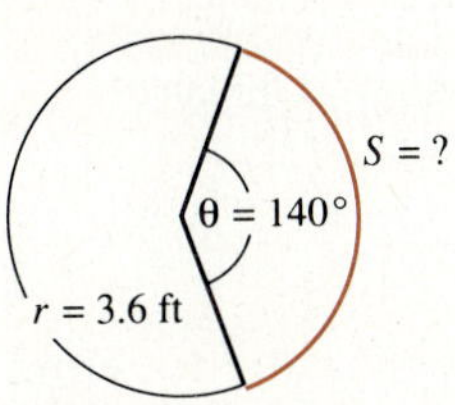

Figure 1.4-3

Solution

First make a sketch of the problem (Figure 1.4-3).

As in Example 23, we can use the formula $S = r\theta$, but in this case we must first convert θ to radians. **Why?**

$$140° \times \frac{\pi \text{ rad}}{180°} = \frac{7\pi}{9}$$

Now we can find S.

$$S = r\theta = (3.6 \text{ ft})(7\pi/9) = 2.8\pi \text{ ft} = (2.8)(3.14) \text{ ft} = 8.8 \text{ ft} \quad \blacksquare$$

SCIENTIFIC NOTATION

Very large and very small numbers are cumbersome to work with because of all the zeros. For example, the speed of light is 300,000,000 m/s, and a year contains approximately 31,500,000 s. Therefore, the distance that light travels in a year (called a *light-year*) is

$$1 \text{ light-year} = (300{,}000{,}000 \text{ m/s})(31{,}500{,}000 \text{ s}) = 9{,}450{,}000{,}000{,}000{,}000 \text{ m}$$

There is a better way to write this! We start by expressing numbers in terms of powers of 10. Notice that

$$\begin{aligned} 1. &= 10^0 \\ 10. &= 10^1 \\ 100. &= 10^2 \\ 1000. &= 10^3 \\ 10{,}000. &= 10^4 \\ &\;\;\ldots \end{aligned}$$

We treat numbers other than 1 in basically the same way. To see how to do this, consider the following numbers:

$$\begin{aligned} 200 &= 2. \times 100 = 2. \times 10^2 \\ 82{,}000 &= 82 \times 1000 = (8.2 \times 10^1)(10^3) = 8.2 \times 10^4 \\ 186{,}000 &= (186)(1000) = (1.86 \times 10^2)(10^3) = 1.86 \times 10^5 \end{aligned}$$

In each case, we have written these numbers in the same form: a decimal with only one number in front of the decimal point multiplied by 10 to a power. We call this way of writing numbers the *standard form* (or *standard position*) of the number. *Notice that the number of places we would have to move the decimal point to put the number in standard form is the exponent of 10*. For example, 93,000,000 in standard form is

$$93{,}000{,}000 = 93 \times 10^6 = (9.3 \times 10^1)(10^6) = 9.3 \times 10^7$$

The exponent is 7 because we would have to move the decimal point of 93,000,000 seven places to the left to put this number in standard form:

$$9\underbrace{3{,}000{,}000}_{7 \text{ places}} = 9.3 \times 10^7$$

We would write 4700 as

$$4\;\underbrace{7\;0\;0}_{3 \text{ places}} = 4.7 \times 10^3$$

We can express small numbers in a similar way.

$$1. = 10^0$$

$$0.1 = \frac{1}{10} = 10^{-1}$$

$$0.01 = \frac{1}{100} = 10^{-2}$$

$$0.001 = \frac{1}{1000} = 10^{-3}$$

$$0.0001 = \frac{1}{10{,}000} = 10^{-4}$$

$$\dots$$

We can use this notation to write small numbers in the *same standard form* we used for large ones, as the following examples show.

$$0.004 = 4. \times 0.001 = 4. \times 10^{-3}$$

$$0.00006 = 6. \times 0.00001 = 6. \times 10^{-5}$$

Notice again that the number of decimal places we would have to move the decimal point to put the number in *standard position* is the exponent of 10. If we have to move the decimal point to the left, the exponent is positive, whereas if we have to move the decimal point to the right, the exponent is negative. Applying this notation to small numbers gives

$$0.00302 = 0.\underbrace{0\ \ 0\ \ 3}_{3\text{ places}}\ \ 0\ \ 2 = 3.02 \times 10^{-3}$$

and

$$0.00000000123 = 0.\underbrace{0\ \ 0\ \ 0\ \ 0\ \ 0\ \ 0\ \ 0\ \ 0\ \ 1}_{9\text{ places}}\ \ 2\ \ 3 = 1.23 \times 10^{-9}$$

We call this method of writing numbers *scientific notation*.

USE OF SCIENTIFIC NOTATION

Scientific notation is useful because it allows us to write large and small numbers compactly. But it is even more useful for *multiplication* and *division* of such numbers.

To multiply or divide two numbers, first put them in scientific notation. Then multiply or divide by using the laws of exponents (add exponents when you multiply and subtract them when you divide). The following examples illustrate this.

(*continued*)

1. Let us calculate the length of the light-year with which we opened this section.

$$\begin{aligned}1 \text{ light-year} &= (300{,}000{,}000 \text{ m/s})(31{,}500{,}000 \text{ s}) \\ &= (3.00 \times 10^8 \text{ m/s})(3.15 \times 10^7 \text{ s}) \\ &= (3.00)(3.15) \times 10^8 \times 10^7 \text{ m} = 9.45 \times 10^{8+7} \text{ m} \\ &= 9.45 \times 10^{15} \text{ m}\end{aligned}$$

2. $$\begin{aligned}0.000310 \times 0.0000000470 &= 3.10 \times 10^{-4} \times 4.70 \times 10^{-8} \\ &= (3.10)(4.70) \times 10^{-4} \times 10^{-8} = 14.6 \times 10^{-4-8} \\ &= 14.6 \times 10^{-12}\end{aligned}$$

But 14.6 is not yet in standard form. To put it that way, write it as $14.6 = 1.46 \times 10^1$. Then the answer becomes

$$14.6 \times 10^{-12} = 1.46 \times 10^1 \times 10^{-12} = 1.46 \times 10^{1-12} = 1.46 \times 10^{-11}$$

3. $$\begin{aligned}\frac{17{,}000{,}000}{240{,}000} &= \frac{1.7 \times 10^7}{2.4 \times 10^5} = \left(\frac{1.7}{2.4}\right) \times \left(\frac{10^7}{10^5}\right) \\ &= 0.71 \times 10^{7-5} = \underbrace{0.71}_{7.1 \times 10^{-1}} \times 10^2 \\ &= 7.1 \times 10^{-1} \times 10^2 = 7.1 \times 10^1\end{aligned}$$

Because this last number is not very large, we would usually write it as 71 and not use scientific notation for the final answer, although the use of scientific notation was certainly very helpful in calculating that answer.

4. $$\begin{aligned}\frac{46{,}000}{0.0000314} &= \frac{4.6 \times 10^4}{3.14 \times 10^{-5}} = \left(\frac{4.6}{3.14}\right) \times \left(\frac{10^4}{10^{-5}}\right) \\ &= 1.5 \times 10^{4-(-5)} = 1.5 \times 10^9\end{aligned}$$

EXERCISES

1. Express each number in the standard form for scientific notation.

(a) 750,000 **(b)** 25,100
(c) 4,700,000,000 yr **(d)** 810
(e) 0.000 543 **(f)** 0.98
(g) 0.000 000 000 25 **(h)** 6.07
(i) 470×10^6 **(j)** 0.82×10^3
(k) 609×10^6 **(l)** 0.012×10^{-6}
(m) 3800×10^{-9} **(n)** 0.00069×10^{12}

2. Write each of the following numbers without the powers of 10.

(a) 9.2×10^3 **(b)** 7.2×10^{-2}
(c) 6.02×10^{23} **(d)** 16×10^4
(e) 0.0142×10^{-5} **(f)** 10^{-10}
(g) 3.15×10^7 **(h)** 0.0532×10^8
(i) 4900×10^{-6}

3. Carry out the indicated operations using scientific notation. Be sure to express your final answer in scientific notation to the correct number of significant figures!

- **(a)** $(1.34 \times 10^9)(6.63 \times 10^4)$
- **(b)** $(5.90 \times 10^{-22})(6.8 \times 10^{13})$
- **(c)** $(2.31 \times 10^7)/(1.44 \times 10^3)$
- **(d)** $(7.888 \times 10^{-10})/(9.11 \times 10^3)$
- **(e)** $(1.2 \times 10^4)/(3.34 \times 10^{-8})$
- **(f)** 38,000,000/49,300
- **(g)** 45,000/876,000
- **(h)** (0.000 51)/9800
- **(i)** 23,100,000/0.004571
- **(j)** (0.000 000 047)/(0.000 002 91)
- **(k)** $(1.6 \times 10^{-19})/(9.11 \times 10^{-31})$
- **(l)** (0.000 0231)(0.00443)/(87,000)
- **(m)** (98,000)(104,000,000)/(25,000)
- **(n)** $[(86,400)^2\ (19,000,000)]/[(0.00542)(12,000)]$

Example 25

What distance does the earth move each day as it goes around the sun in a (nearly) circular orbit?

Solution

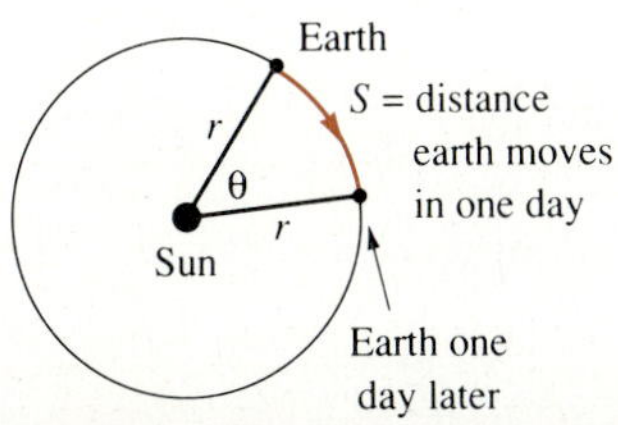

Figure 1.4-4

The earth moving around the sun

As usual, we start with a clear figure showing what we know and what we are trying to find (Figure 1.4-4). In this figure, r is the distance from earth to sun, which is 93.0 million mi $= 9.30 \times 10^7$ mi, and θ is the angle the earth moves through in 1 day. Because it takes the earth 365 days to make 1 rev around the sun, each day must be $\frac{1}{365}$ of a complete revolution, so $\theta = \frac{1}{365}$ rev $= (\frac{1}{365})(2\pi \text{ rad}) = 0.0172$ rad. We want to find S, the *distance* the earth moves each day. Since $\theta = S/r$, solving for S gives $S = r\theta$, where we *must* use θ in radians.

$$S = r\theta = (9.30 \times 10^7 \text{ mi})(0.0172 \text{ rad}) = 1.60 \times 10^6 \text{ mi} \ ■$$

(See the section on Scientific Notation.)

NOW IT'S YOUR TURN

(a) What arc length will a $\frac{\pi}{9}$ central angle subtend on a circle 6.2 m in radius? (Answer: 2.2 m)

(b) What arc length will a 52.0° central angle subtend on a circle 30.0 mm in diameter? (Answer: 13.6 mm)

(c) Through what angle in radians around the sun will the earth have moved when it has travelled a distance of 158 million mi? (Answer: 1.70 rad)

Example 26

If a 30.0-in. diameter bicycle wheel makes 4.00 rev, how far has the bicycle travelled?

Solution

First make a sketch showing the wheel in its starting place and then 4 rev later (Figure 1.4-5). In this sketch, notice that the center of the wheel is always directly

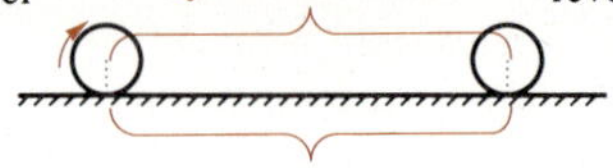

Figure 1.4-5

above the point of contact with the ground. Therefore, *the distance travelled by the center of the wheel must be exactly the same as the arc length rolled by the wheel's rim*. Therefore, the distance the bicycle travels is also equal to the arc length rolled by the wheel's rim. Now, the formula $S = r\theta$ tells us the arc length rolled, and therefore it also tells us the distance the bicycle moves. Before we can use this formula, we must convert θ from 4.00 rev to radians.

$$\theta = (4.00 \text{ rev})\left(\frac{2\pi \text{ rad}}{1 \text{ rev}}\right) = 8\pi \text{ rad}$$

Now we can find the distance moved

$$S = r\theta = (15.0 \text{ in.})\,(8\pi \text{ rad}) = 377 \text{ in.}$$

This number is awkwardly large, so it is more convenient to convert it to feet. This gives

$$S = (377 \text{ in.})\left(\frac{1 \text{ ft}}{12 \text{ in.}}\right) = 31.4 \text{ ft} \ \blacksquare$$

Example 26 illustrated how the formula $S = r\theta$ can be used to find the relationship between the revolutions of a wheel and the distance it has travelled. The following example extends this idea to explain how the mileage gauge of your car works.

Example 27
Measuring Automobile Mileage

The odometer (mileage gauge) of your car tells how many miles you have driven, but it doesn't count the number of miles directly. Instead, it counts the number of revolutions of your car's wheels and then converts the revolutions to mileage, presuming a standard size tire. It works this way: as we showed in the last example, the distance travelled by the center of the tire is equal to the arc length rolled by the tire's rim, providing that the tire does not slip on the pavement. But the arc length rolled is $r\theta$, where r is the tire's radius and θ is the angle the tire turns through. And, of course, the distance the center of the tire has travelled is also the distance d the car has moved. Therefore

$$\underbrace{\text{distance tire travels}}_{d} = \underbrace{\text{arc length rolled by tire}}_{r\theta}$$

$$d = r\theta \ \blacksquare$$

With Standard Tires

A typical car tire is 24 in. in diameter (measure yours to check this). How many revolutions of such a tire are necessary for the odometer to read 1.0 mi?

Solution

Solving the formula $d = r\theta$ for θ, we get $\theta = d/r$. We know that $r = 12$ in. = 1.0 ft; $d = 1.0$ mi = 5280 ft. So the angle θ the wheels will turn is

$$\theta = \frac{d}{r} = \frac{5280 \text{ ft}}{1.0 \text{ ft}} = 5280 \text{ rad}$$

Converting this result to revolutions gives us

$$\theta = (5280 \text{ rad})\left(\frac{1 \text{ rev}}{2\pi \text{ rad}}\right) = 840 \text{ rev} \quad ■$$

When your odometer has counted 840 rev, your mileage gauge reads 1.0 mi.

With Oversize Tires

Suppose you now put on oversize tires, of diameter 28 in. When your odometer reads 1.0 mi, how many miles have you *really* driven?

Solution

We can use the formula $d = r\theta$ to calculate the distance travelled d, where r is now 14 in. Converting this to feet gives us

$$r = (14 \text{ in.})\left(\frac{1 \text{ ft}}{12 \text{ in.}}\right) = 1.17 \text{ ft}$$

But what is θ? Your odometer reads 1.0 mi, so it must have made 840 rev (5280 rad). Therefore, the distance travelled by the car with the larger tires is

$$d = r\theta = (1.17 \text{ ft})(5280 \text{ rad}) = 6180 \text{ ft}$$

Converting this to miles gives

$$d = (6180 \text{ ft})\left(\frac{1 \text{ mi}}{5280 \text{ ft}}\right) = 1.2 \text{ mi} \quad ■$$

When your odometer reads 1.0 mi, you have really driven 1.2 mi with the oversize tires.

NOW IT'S YOUR TURN

If a car has standard 30.0-in. diameter tires, how many revolutions will the odometer count when you have driven 1.0 mi? If you put on oversize 32-in. diameter tires, how far will you really have driven when your odometer reads 1.0 mi? (Answers: 670 rev, 1.1 mi)

Example 28

Latitude and Longitude

In geography, we locate a point on the earth's surface by giving two angles for that point (see Figure 1.4-6).

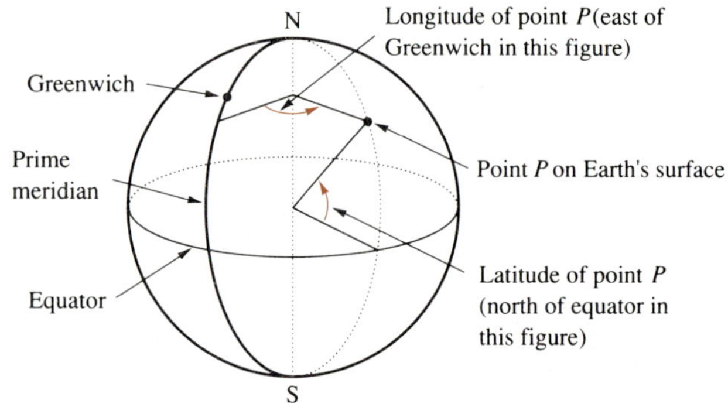

Figure 1.4-6

Latitude and longitude

1. The first angle is the angle of the point north (or south) of the equator. This angle is called the *latitude* of the point. Points north of the equator are called north latitude, whereas those south are called south latitude.

2. The second angle is the angle of the point east (or west) of a standard north-south reference line called the prime meridian. This angle is called the *longitude* of the point. The prime meridian is arbitrarily chosen to be the line (actually a semicircle) from the North Pole to the South Pole, passing through Greenwich, England.

In summary, the latitude of a point tells how many degrees north or south it is from the equator, whereas the longitude tells how many degrees east or west it is from the prime meridian.

Dallas, Texas, is at 97° west longitude and 33° north latitude, whereas Mexico City is at nearly the same longitude but 19°30′ north latitude. How far north of Mexico City is Dallas?

Solution

Because both cities are on nearly the same longitude, we can draw their locations as shown in Figure 1.4-7. We are looking for the distance S in this figure.

Figure 1.4-7 may be hard to visualize in three-dimensions. When dealing with three-dimensional problems, it is usually helpful to try to express them as two-dimensional figures if possible. In this case, we can focus in on the section of Figure 1.4-7 showing the cities and angles, as shown in Figure 1.4-8.

If θ is in radians, then $S = r\theta$, where r is the radius of the earth, which is 3960 mi. From Figure 1.4-8 we can see that

$$\theta = 33° - 19°30' = 32°60' - 19°30' = 13°30' = 13.5°$$

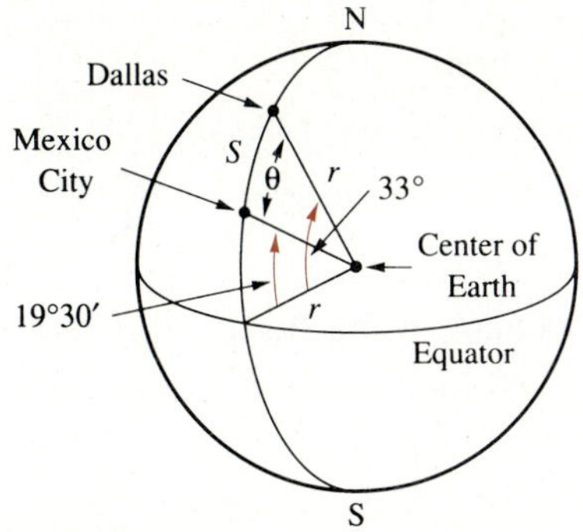

Figure 1.4-7

Dallas and Mexico City

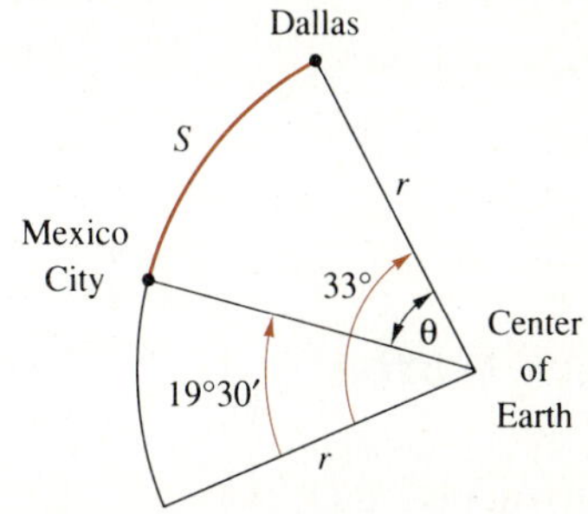

Figure 1.4-8

Two-dimensional reduction of Figure 1.4-7

Converting this to radians (**why?**), we have

$$\theta = (13.5°)\left(\frac{\pi \text{ rad}}{180°}\right) = 0.24 \text{ rad}$$

Therefore, the distance between these cities is

$$S = r\theta = (3960 \text{ mi})\ (0.24 \text{ rad}) = 950 \text{ mi} \quad ■$$

NOW IT'S YOUR TURN

How far north of the equator is Mexico City? (Answer: 1350 mi)

Example 29
Area of a Circular Sector

Suppose you cut a wedge, or *sector*, of central angle θ from a circle, as in Figure 1.4-9. Find the area of this sector.

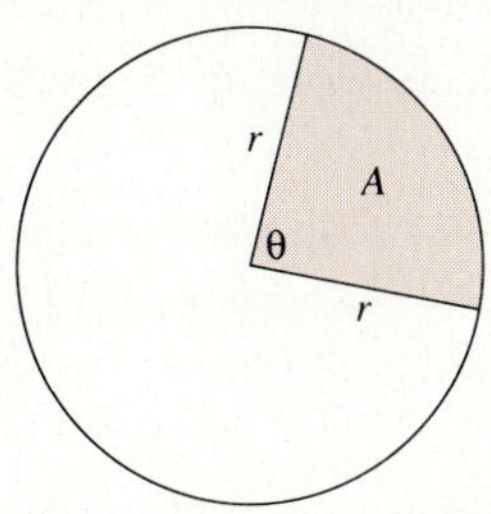

Figure 1.4-9

Circular sector

Solution

To see how to begin, first let's consider some specific cases in Figure 1.4-9. If, for example, θ is $\frac{1}{3}$ of a revolution, then the area A will be $\frac{1}{3}$ of the area of the entire circle; if θ is $\frac{3}{4}$ of a revolution, then the area A will be $\frac{3}{4}$ of the entire area; if θ is $\frac{2}{7}$

of a revolution, then A will be $\frac{2}{7}$ of the total area and so forth. We can write this as a simple proportion.

$$\frac{\theta}{1 \text{ rev}} = \frac{A_{\text{wedge}}}{A_{\text{circle}}}$$

Expressing θ in radians, using $A_{\text{wedge}} = A$ and $A_{\text{circle}} = \pi r^2$, this last equation gives us

$$\frac{\theta}{2\pi} = \frac{A}{\pi r^2}$$

Solving for the area A, we get the important formula

$$A = \tfrac{1}{2} r^2 \theta \quad ■$$

AREA OF A SECTOR

The area A of a circular sector with radius r and central angle $\boldsymbol{\theta}$ measured in radians is given by

$$A = \tfrac{1}{2} r^2 \theta$$

Example 30

A landscape architect is designing a circular floral display for a public park. The entire circle will be 10.0 ft in diameter and will be divided into wedges with different types of flowers. For the wedge devoted to hyacinths, the architect has enough plants to fill 28 ft^2. In her design, what angle should she specify for the hyacinth wedge?

Solution

First sketch the floral display (Figure 1.4-10). The area of the hyacinth wedge is 28 ft^2 and the radius is 5.0 ft (one-half the diameter). We can solve the formula $A = \frac{1}{2} r^2 \theta$ to find θ. ■

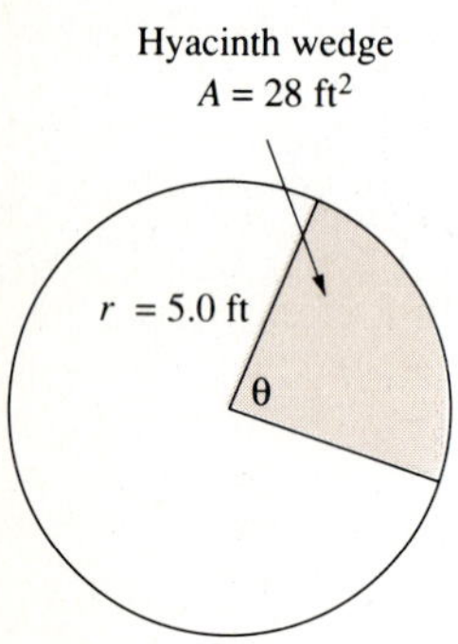

Figure 1.4-10
Circular floral display

$$\theta = \frac{2A}{r^2} = \frac{2(28 \text{ ft}^2)}{(5.0 \text{ ft})^2} = 2.2 \text{ rad}$$

Most architects work in degrees rather than radians, so we should convert to degrees.

$$\theta = (2.2 \text{ rad}) \left(\frac{180°}{\pi \text{ rad}} \right) = 130° \text{ for hyacinths} \quad ■$$

NOW IT'S YOUR TURN

If the caladium wedge is to have an area of 11 ft^2, what should be its angle? (Answer: 50°)

EXERCISES 1.4

In Exercises 1–6, let S be the arc length and θ the central angle of a circle of radius r or diameter d.

1. Find d if $S = 15\pi$ m and $\theta = 300.0°$.
2. Find θ in degrees if $r = 3.0$ ft and $S = 48$ in.
3. Find r if $S = 96.0$ cm and $\theta = \frac{\pi}{4}$.
4. Find S if $r = 10.0$ ft and $\theta = 48°$.
5. Find d if $\theta = 32.6°$ and $S = 21.40$ cm.
6. Find θ in degrees if $d = 15$ in. and $S = 60.0$ in.
7. If the end of a 3.5-ft pendulum swings through a 16° angle, through how long an arc has the bottom end of the pendulum moved?
8. An archaeoastronomer is studying the Big Horn Medicine Wheel in Wyoming's Big Horn Mountains. This ancient wheel, which was made by arranging hundreds of stones and may have astronomical significance, has 28 nearly equally spaced spokes (Figure 1.4-11). The arc lengths between spokes average approximately 3.4 ft. What is the diameter of this wheel and what is the average angle between adjacent spokes?

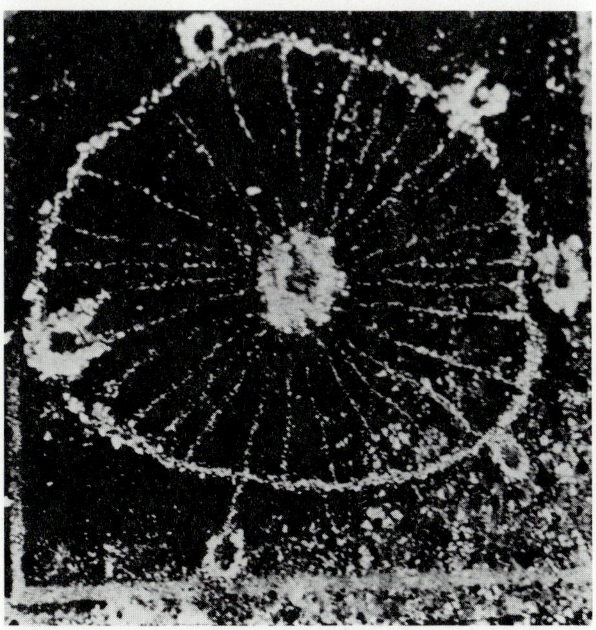

Figure 1.4-11 ***Big Horn Medicine Wheel***

9. If it takes 800.0 equal steps to walk all the way around a garden path that completely surrounds a pie-shaped *wedge* with a 50.0° central angle, how many steps will it take to walk just the *curved* part?

*10. A brick-lined flower bed in a park is in the form of a pie-shaped wedge of central angle 30.0°, with the curved part being 20.0 ft long. A new gardener decides to redesign this bed into a circle using the old bricks as the border. What is the largest radius this new bed can have?

*11. A curved wire subtends an angle of 120° on a circle 18.0 in. in diameter. If this wire is now bent into a complete circle, what will be the radius of this circle?

*12. If a circle has an area of 100.0 m^2, what arc length does a 10.0° central angle subtend?

*13. If the semicircumference (half the circumference) of a certain circle is equal to its diameter plus an extra arc length, find the central angle subtended by that arc.

14. Through what distance does Mt. Kilimanjaro, on the equator, move in 1.0 hr due to our planet's spin on its axis? In 1.0 day? ***Hint:*** What angle does the earth turn through in 1.0 hr?

15. How far does the earth move around the sun in 1.0 week?

*16. The planet Neptune is 30.1 times as far from the sun as the earth, and it takes 165 earth-years for Neptune to make a complete orbit around the sun. Through what angle in degrees does it move around the sun in 1.00 earth-year? Through how many earth-miles does Neptune travel in 1.00 earth-year?

17. You are designing a winch to lift heavy blocks of marble to the top of a building. What must be the diameter of this winch if it is to lift the marble 0.50 in. for every 1.0° you turn it?

18. A roller skate with 2.0-in.-diameter wheels coasts 7 ft 5 in. before stopping. How many turns do its wheels make?

19. If a wheel makes 3.40 rev while moving 2.40 m along the ground, what is its radius?

20. A child's tricycle has a large wheel 16 in. in diameter and two small wheels each 7.0 in. in diameter. How many revolutions does each wheel make while the tricycle travels 25 ft?

21. A typical paint roller is a circular cylinder 3.00 in. in diameter and 9 in. long. How long a section of paint do you roll when you turn this roller through 785°?

22. If a pickup truck has 30.0-in.-diameter tires, how many revolutions do its wheels make while it drives 1.0 mi?

23. A car has 28-in.-diameter tires, and its wheels make 2500 rev. What does the odometer read?

24. As an automotive engineer, you are designing the wheel-odometer system for a new model car. How large should the standard tires be so that 1.0 mi of driving will result from 920 rev of the car's wheels?

25. If a car has 26-in.-diameter tires, how many turns of the wheels are equivalent to an odometer reading of $\frac{1}{10}$ mi? When the wheels have turned 100 times, what will the odometer read?

26. Suppose you put oversize 30.0-in. diameter tires on the car in Exercise 25. What will the odometer read after the wheels have turned 100 times?

****27.** If you put 30.0-in. tires on a car with 24-in. standard tires, how far will you have driven when the odometer reads 100.0 mi?

****28.** On a car with standard 24-in. tires, what size tires do you need so that an odometer reading of 500.0 mi means you have driven 650 mi?

****29.** You buy a car with standard 24-in. tires but immediately change to 30-in. tires. You must take your car in for warranty servicing after the first 500.0 mi of actual driving. At what *odometer* reading should you take your car in for this service?

30. Chicago is at north latitude 41°59′. How far north of the equator is it?

31. Los Angeles is 2350 mi north of the equator. What is its latitude?

32. Buenos Aires, Argentina, is at 35° south latitude. How far is it from the equator?

33. Aswan is 500 mi south of Alexandria in Egypt. Find the *difference* in the latitude of these two cities.

34. Kisumu, Kenya, is at 0.0° latitude and 35° east longitude, whereas Singapore is at almost the same latitude but at 104° east longitude. How far is Kisumu from the equator? If you flew directly from Kisumu to Singapore, how many miles would you cover?

***35.** If you start at Kisumu (see Exercise 34) and travel 500.0 mi directly westward and then 700.0 mi directly southward, find the latitude and longitude of the point you would reach.

****36.** Suppose you make the same journey as in Exercise 35, except you do it in reverse order, first travelling 700.0 mi southward and then 500 mi westward. Would you wind up at the same place as in Exercise 35? Do not attempt to calculate a numerical answer, but justify your answer on *geometric* grounds. A clear three-dimensional sketch should help.

37. Utica, New York is at about 43°6′ north latitude. Philadelphia, Pennsylvania, is at the same longitude but at about 40.0° north latitude. The distance between these cities has been surveyed to be 214 mi. Use this information to calculate the radius of the earth.

38. Seattle, Washington, is at 47°30′ north latitude. Rio de Janeiro, Brazil, is at 23° south latitude but at about the same longitude. How far is it from Seattle to Rio de Janeiro?

***39.** **(a)** How many miles are there in each 1.0° of latitude? Does this answer depend on longitude?
(b) Can you give a single numerical answer for the length of each degree of longitude? Does this answer depend on latitude?

40. Nautical Mile. The nautical mile is defined to be the distance between two points on the equator (or any great circle) that are separated by 1.00′ of longitude. (The more familiar mile, 5280 ft, is the statute mile.)

Figure 1.4-12 ***Copernicus***

(a) How many feet are in a nautical mile?
(b) How many nautical miles are in a statute mile?
(c) If two cities are 100.0 statute miles apart, how many nautical miles apart are they?

***41.** How many feet would there be in a *Martian* nautical mile (a strange name since Mars has no oceans!) if we used the same nautical mile definition as on earth? (See Exercise 40.)

42. Apollo 11, the first manned moon landing, set down on Mare Tranquilitatus on July 16, 1969, at approximately 2.0° north latitude. How far from the lunar equator did Neil Armstrong take his "one giant step for mankind"?

43. The 1976 Viking mission to Mars landed in two places, Chryse on July 20 at 22.4° north latitude and Utopia on September 3 at 47.9° north latitude. How far is each site from the Martian equator?

44. The great impact crater Copernicus on our moon lies at a north latitude of 10.0° (Figure 1.4-12). How far is Copernicus from the moon's equator?

45. Olympus Mons, the largest volcano on Mars (and possibly in the solar system), extends over nearly 10° of Martian latitude (Figure 1.4-13). How many kilometers wide is this great volcano?

46. The great rift valley Valles Marineris (Valley of the Mariner) is nearly at the equator of Mars and extends over approximately

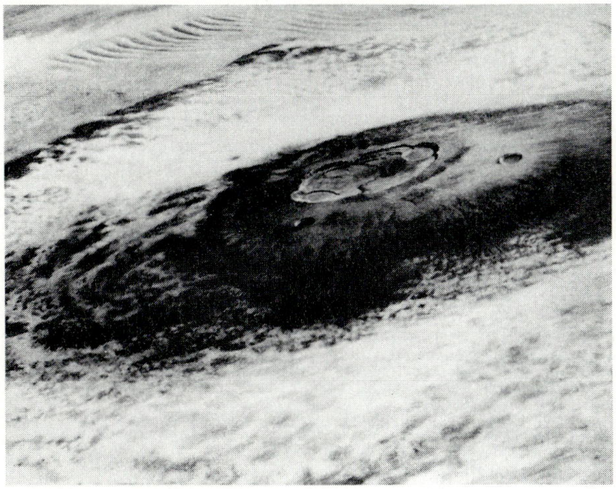

Figure 1.4-13 ***Olympus Mons Volcano***

Figure 1.4-14 ***Valles Marineris***

70° of longitude (see Figure 1.4-14). How long is this valley? How does its length compare to that of the United States?

***47.** Using the definition of latitude in Example 28, calculate how many miles there are in 1.0° of latitude on our moon and on Mars?

48. You want to design a globe to represent the earth such that 10.0° of latitude will cover a distance of 3.0 in. on its surface. What should be the radius of this globe?

49. Gears.** Gears are used to transfer power from one part of a machine to another through a series of rotating disks with interlocking teeth (Figure 1.4-15). Because of the differences in size of the two gears, the spin rate of one part of a machine can be slowed down or increased. In Figure 1.4-15, the smaller gear has a radius of 12 cm and the larger a radius of 30.0 cm. If the smaller turns through an angle of 80.0°, through what angle does the larger gear turn. ***Hint: Since the teeth of the gears mesh, what must be true about the arc lengths rolled by each gear?

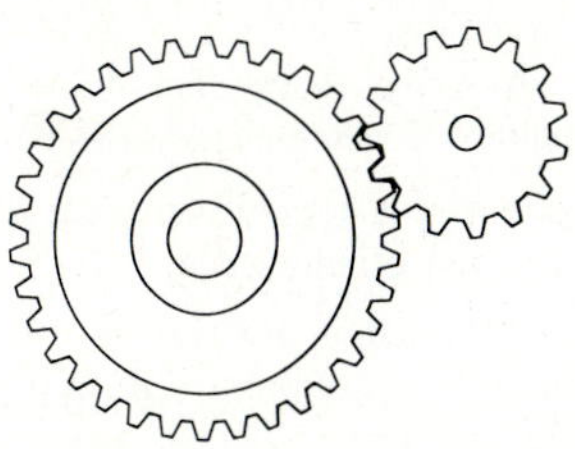

Figure 1.4-15

50. Two intermeshing gears have radii 20.0 cm and 32 cm. How many turns does the smaller gear make for every turn of the larger gear? (See Exercise 49.)

51. Three gears, A, B, and C, mesh as in Figure 1.4-16. Gear A has a radius of 14.0 cm, B has a radius of 3.0 cm, and C has a radius of 11.0 cm. When A has turned 10 times, how many times have B and C turned? Would the answer for gear C have been any different if it had been connected *directly* to gear A without B in between? If C must turn counterclockwise, in what direction must A turn? (see Exercise 49.)

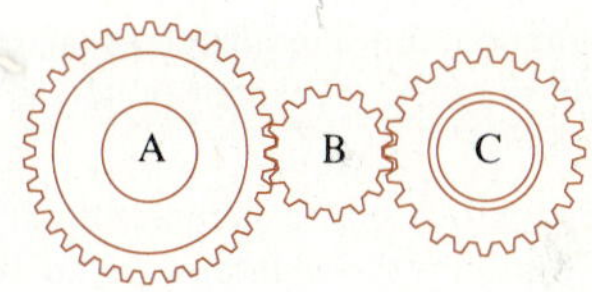

Figure 1.4-16

***52.** An arrangement of two gears, A and B, is used to lift an elevator as in Figure 1.4-17. If gear A is 10.0 ft in diameter and B is 3.0 ft in diameter, through what angle must A turn to lift the elevator by 1.50 ft? If B were a different size, would that change the answer? In which direction, clockwise or counterclockwise, must A turn to lift the elevator? (See Exercise 49.)

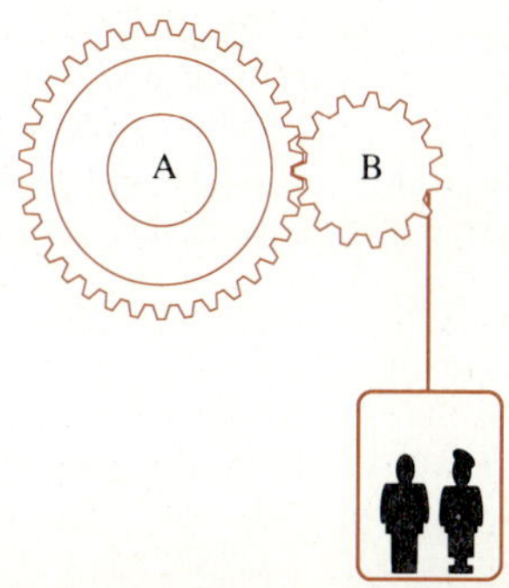

Figure 1.4-17

*53. **Belt Drives.** In many machines, it is common for one wheel to turn another through "friction drive" by a belt fitted around the outer rim of both wheels, as in Figure 1.4-18. (The fanbelt on your car and many phonograph turntables operate this way.) If wheel A has a diameter of 25 cm and B has a diameter of 18 cm, through what angle does B turn when A makes 1.0 rev? *Hint:* As A turns through an angle, what must be true about the distance the belt moves and the arc length turned by A and B?

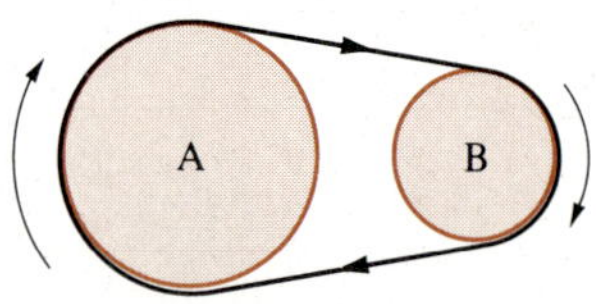

Figure 1.4-18

54. For the belt-drive in Exercise 53, what distance does the belt move when the larger wheel has turned through a 30.0° angle?

*55. Design a belt-drive (see Exercise 53) with one wheel 8.0 in. in diameter. This wheel must make five turns for every two turns of the other wheel.

56. **(a) If the radius of a circle is increased by 20%, by what percent does the arc length subtended by a 60° central angle change?
(b) If the area of a circle is increased by 20%, by what percent does the arc length subtended by a 60° central angle change?
(c) If the circumference of a circle is increased by 20%, by what percent does the arc length subtended by a 60° central angle change?

57. **(a) If the radius of a sphere increases by 20%, what happens to the arc length on the equator subtended by a 60° central angle?
(b) If the surface area of a sphere increases by 20%, what happens to the arc length on the equator subtended by a 60° central angle?
(c) If the volume of a sphere increases by 20%, what happens to the arc length on the equator subtended by a 60° central angle?

*58. **(a)** If a circle has a circumference of 500.0 cm, how far apart are two points on the rim separated by a central angle of 155°?
(b) If a circle has an area of 20.0 ft^2, how far apart are two points on the rim separated by a central angle of 28°?

*59. **(a)** If a sphere has a circumference of 500.0 cm, how far apart are two points on the equator separated by a central angle of 155°?
(b) If a sphere has a surface area of 75.0 ft^2, how far apart are two points on the equator separated by a central angle of 81.0°?
(c) If a sphere has a volume of 100.0 m^3, how far apart are two points on the equator separated by a central angle of 52.0°?

For Exercises 60–83, see Examples 29 and 30.

60. Find the area of a circular sector having a radius of 3.00 cm and central angle 28°.

61. A circle 68.5 in. in diameter has a wedge of central angle 126° cut out. What is the area of the remaining portion of the circle?

62. If the area of a circular sector is 187 ft^2 and the central angle is $\frac{3\pi}{4}$, find the radius of the circle.

63. Make the necessary *measurements* on Figure 1.4-19 to calculate the area of the shaded wedge.

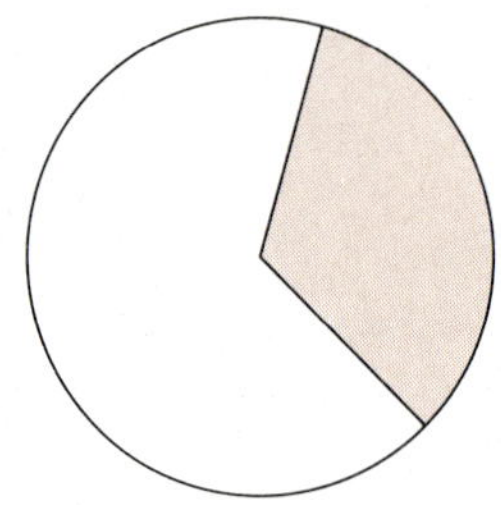

Figure 1.4-19

*64. A circular sector of radius 6.5 ft subtends an arc length of 11.3 ft. What is its area in square feet?

*65. If a circular wedge of angle 2.0 rad has a perimeter of 200.0 ft, what is its area?

*66. Many engineers use the formula $A = mr^2\theta$ to calculate the area of a circular sector of radius r, in which the central angle θ is measured in degrees.

(a) Find the value of m to three significant digits.
(b) Test your answer for m by using it to calculate the area of a sector of radius 2.00 m and central angle 30.0°. Then calculate this area by converting the angle to radians. Do you get the same answer both ways?

*67. A 10.0-acre farm field is in the shape of a wedge of radius 1000.0 ft. Find total length of fencing needed to surround this field.

*68. A 15.0-hectare wedge-shaped field has central angle 40.0°. Find the total length of fencing needed to surround it.

*69. A 3.00-m long wire is bent into a wedge of radius $\frac{3}{4}$ m. Find the angle and the area of this wedge.

**70. A wire circle 400.0 cm in diameter is reshaped into a wedge of angle 150°. Find the area of this wedge. Is it the same as that of the original circle?

*71. A uniform circular metal sheet weighs 100.0 oz. If a wedge of central angle 37.0° is cut from this circle, what will the wedge weigh? *Hint:* Since the sheet is uniform, the weight is proportional to the area.

72. A circular wedge has an angle of 35° and radius 15 cm. If the angle is doubled, what happens to the area of the wedge?

73. A wedge has an angle of 66° and radius 10.0 ft. If the radius is doubled, what happens to the area of the wedge?

***74.** A rubber wedge has an angle of 1.9 rad and radius 2500 mm. If you stretch this without changing the angle so that the area is doubled, what happens to the radius?

***75.** In Exercise 74, if you stretch the original wedge so that the area is doubled but the *radius stays constant*, what happens to the angle?

76. A 150° wedge has a radius of 60.0 cm.

(a) If the angle increases by 20%, by what percent does the area change?

(b) If the radius is increased by 20%, by what percent does the area change?

***77.** A wedge has a radius of 18 ft and angle 45°. If the area increases by 20% while the radius stays the same, by what percent does **(a)** the angle change? **(b)** the perimeter change?

***78.** A landscape architect wants to increase the area of a wedge-shaped flower bed by 30% to make room for more plants without changing the 80.0° angle. If the original radius was 10.0 ft, by what percent should he increase the radius?

79. Find the area of the shaded section of the wedge in Figure 1.4-20. ***Hint:*** The shaded area is the difference between two areas.

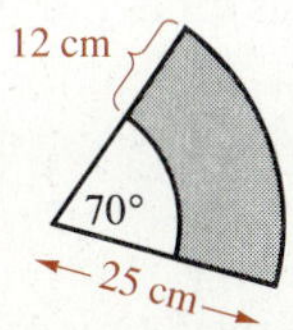

Figure 1.4-20

***80.** The angle of the composite wedge in Figure 1.4-21 is 0.90 rad. Find *a* and *b* so that both sections will have equal areas. What is the area of each section?

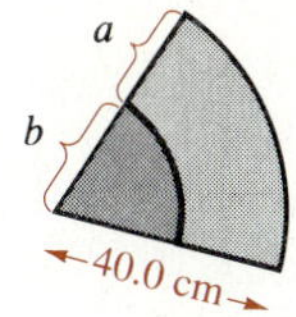

Figure 1.4-21

81. In Exercise 80, find the area of each section if $a = b$.

***82.** A tiled patio in a public park is to have a total area of 4000.0 m^2 in the shape of a 120° wedge, as shown in Figure 1.4-22. This patio will be divided into three sections using different types of tile, but with *equal areas for each section.*

(a) Find *a*, *b*, and *c*, the width of each section.

(b) What is the total outside perimeter of this patio?

(c) Sketch this patio, showing the three sections *to scale.*

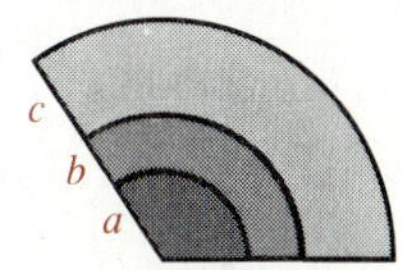

Figure 1.4-22

***83.** A right-circular aluminum cylinder is 60.0 cm long and 20.0 cm in diameter. A 30.0° wedge is cut out, as shown in Figure 1.4-23. Find the volume that is left after this wedge is removed.

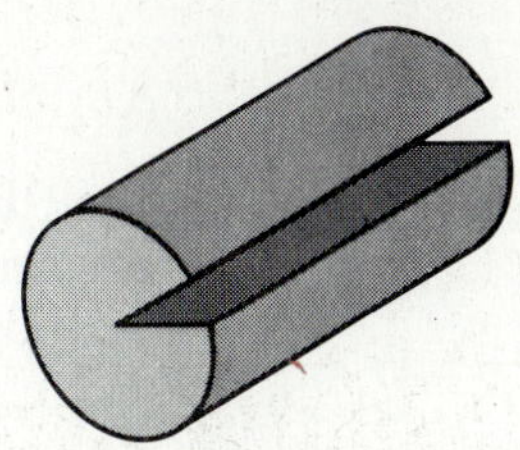

Figure 1.4-23

1.5 ANGULAR SPEED

1.5-1 Linear and Angular Motion

Linear Speed

You are no doubt familiar with speed, which is a measure of how *fast* something is moving. To find speed, just divide the distance travelled by the time taken to travel that distance. For example, if you drive 40 mi in 1 hr, your speed is (40 mi)/(1 hr) = 40 mi/hr (40 mph). If you drive 40 mi in 2 hr, your speed is (40 mi)/(2 hr) = 20 mi/hr.

$$\text{speed} = \frac{\text{distance travelled}}{\text{time to travel this distance}}$$

Using symbols for speed (v), distance (d), and time (t), this formula can be written more briefly as

$$v = \frac{d}{t}$$

Note that this formula is strictly valid only when the speed is uniform (not changing). For example, if you uniformly jog 10 mi in 2.5 hr, your speed will be

$$v = \frac{d}{t} = \frac{10 \text{ mi}}{2.5 \text{ hr}} = 4.0 \text{ mi/hr} \quad \blacksquare$$

This speed is often called the *linear* speed because the distance you move is along a line (not necessarily a *straight* line—it could be a curve). *The units of linear speed will always be a distance divided by a time*, such as miles per hour (mi/hr or mph), feet per second (ft/s), and so on.

Angular Speed

Something that is going in a circle has a linear speed because it is moving through a distance. But it is also moving through an angle so it has an angular speed as well. By analogy with linear speed, we define the **angular speed** to be the angle an object moves through divided by the time it takes to move through this angle.

$$\text{angular speed} = \frac{\text{angle moved through}}{\text{time to move through this angle}}$$

We usually use the Greek letter ω (omega) to represent angular speed, θ to represent the angle, and t for time (see Figure 1.5-1). In terms of these symbols, we can write this equation more briefly as

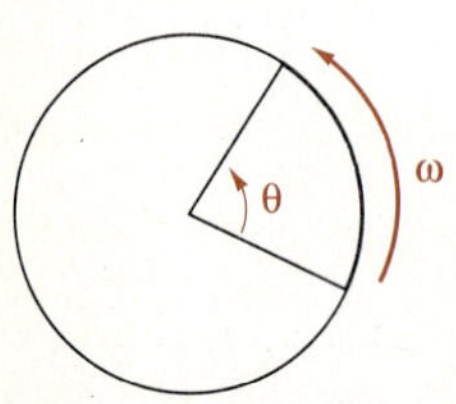

Figure 1.5-1

Angular speed

$$\omega = \frac{\theta}{t}$$

As with linear speed, this formula is strictly valid only when the angular speed is uniform.

As you can see by looking at the formula, *the units of angular speed are always an angle divided by a time*, such as radians per second (rad/s), degrees per second (°/s), revolutions per minute (rpm), and so on.

Example 31

Find the angular speed of the earth as it spins on its axis in radians per hour (rad/hr).

Solution

The earth spins once per day, which is 2π rad in 24 hr. Therefore, applying the formula for angular speed

$$\omega = \frac{\theta}{t} = \frac{2\pi \text{ rad}}{24 \text{ hr}} = 0.26 \text{ rad/hr} \ ■$$

NOW IT'S YOUR TURN

Find the angular speed of the earth as it goes around the sun in radians per day. (Answer: 0.017 rad/day)

Example 32

How many minutes will it take for a 45-rpm record to turn through a 75° angle?

Solution

ω = 45 rpm = 45 rev/min

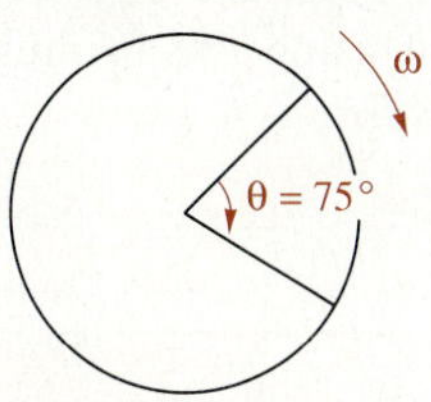

Figure 1.5-2
45-rpm record

First sketch the record (Figure 1.5-2). We want the time, so we must solve the equation $\omega = \theta/t$ for t. This gives $t = \theta/\omega$. The angular speed is $\omega = 45$ rpm, and the angle $\theta = 75°$. Both ω and θ must be in the same angular measure, so we must convert one of them. Converting ω we get

$$\omega = \left(\frac{45 \text{ rev}}{\text{min}}\right)\left(\frac{360°}{1 \text{ rev}}\right) = 1.6 \times 10^4 \text{ °/min}$$

Now we can find the time t.

$$t = \frac{\theta}{\omega} = \frac{75°}{1.6 \times 10^4 \text{ °/min}} = 4.7 \times 10^{-3} \text{ min} \ ■$$

NOW IT'S YOUR TURN

How many minutes will it take a $33\frac{1}{3}$-rpm LP record to turn through 110°? (Answer: 9.2×10^{-3} min)

1.5-2 Relating Angular Speed and Linear Speed

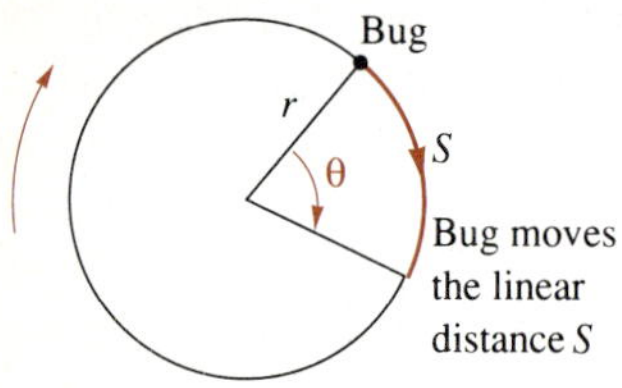

Figure 1.5-3

Bug riding on an LP record

Suppose that a bug is riding on the rim of an LP record. The record is turning, giving the bug an *angular* speed which is the same as that of the record, $33\frac{1}{3}$ rpm. But the bug also moves a *linear* distance S with the rim as the record turns (Figure 1.5-3). The bug's *angular speed* is

$$\omega = \frac{\theta}{t}$$

The bug's *linear speed* is

$$v = \frac{S}{t}$$

S is the arc length subtended by the central angle θ. In Section 1.4, we saw that $S = r\theta$. Therefore, we can express the linear speed as

$$v = \frac{S}{t} = \frac{r\theta}{t} = r\left(\frac{\theta}{t}\right)$$

We recognize that θ/t is just the angular speed ω, so

$$v = r\omega$$

This speed is tangent to the circle, so it is often called the *tangential speed* (see Figure 1.5-4).

▸ **Important Note.** When we derived this formula, we used the equation $S = r\theta$, which is true only if θ is in radians. Therefore, in using the formula $v = r\omega$, the angular speed ω *must always be in radian measure.*

Summarizing these results:

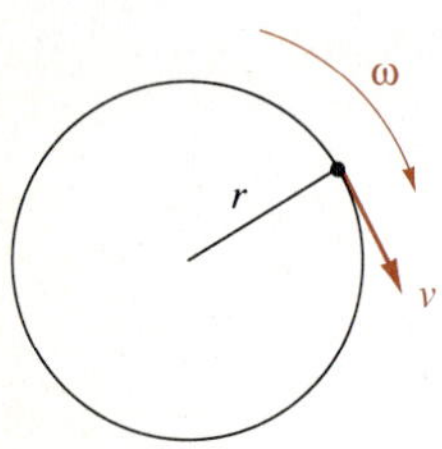

Figure 1.5-4

Angular speed and linear speed. As the object rotates with an angular speed ω, it also has a linear speed v tangent to the circle. The linear speed is $v = r\omega$.

RELATING ANGULAR AND LINEAR SPEED

The linear (tangential) speed v of a moving point on a circle of radius r is given by

$$v = r\omega$$

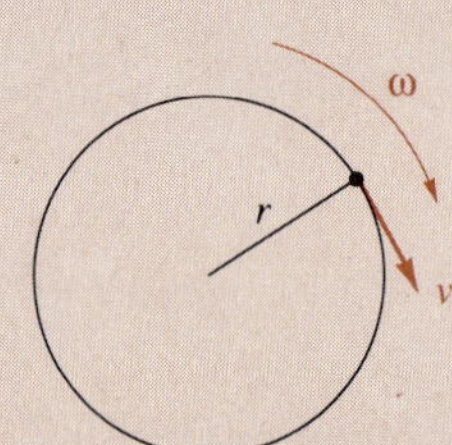

where ω is the angular speed expressed in radians per unit of time.

Example 33

How fast is a bug moving if it sits at the rim of an LP record while the record is playing?

Solution

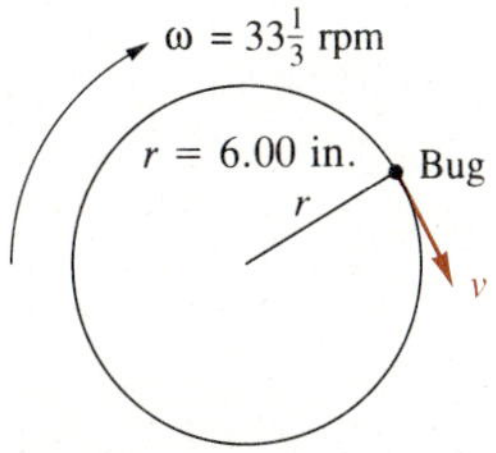

Figure 1.5-5

LP record with bug

First sketch the LP record (Figure 1.5-5). A typical LP record is 12.0 in. in diameter and revolves at $33\frac{1}{3}$ rpm. We want to know how fast the bug is *moving*, not turning, so we want the *linear* speed, not the angular speed. The linear speed is given by $v = r\omega$, where $\omega = 33\frac{1}{3}$ rpm $= 33.33$ rev/min and $r = 6.00$ in. But we must first convert ω to radian measure. **Why?** So

$$\omega = \left(\frac{33.33 \text{ rev}}{\text{min}}\right)\left(\frac{2\pi \text{ rad}}{1 \text{ rev}}\right) = 209 \text{ rad/min}$$

Now we can find the linear speed v:

$$v = r\omega = (6.00 \text{ in.})\left(\frac{209 \text{ rad}}{\text{min}}\right) = 1250 \text{ in./min}$$

Notice that the radians do not appear in the final answer because radians are pure, dimensionless numbers having no units.

NOW IT'S YOUR TURN

Where should the bug snooze so that he is moving at only 450 in./min? (Answer: 2.15 in. in from the center of the LP record)

Example 34

You have been hired to design a conveyor belt to move at a uniform speed of 60.0 ft/min for an assembly line. It will be driven by a wheel of diameter 30.0 in., as in Figure 1.5-6. How fast must this wheel rotate?

Solution

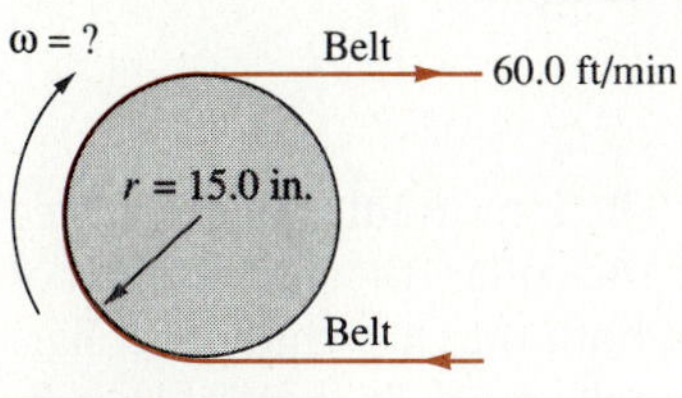

Figure 1.5-6

Conveyor belt

We want to know how fast the wheel must *rotate*, so we are looking for the *angular* speed ω. Since the belt moves around the rim of the wheel, the tangential speed of the rim must be the same as the belt's speed, which is 60.0 ft/min. Solving the equation $v = r\omega$ for ω gives

$$\omega = \frac{v}{r}$$

The radius r is $\frac{1}{2}$ the diameter, or 15.0 in. Because the linear speed is in feet per minute, we must convert the 15.0 in. to feet.

$$r = (15.0 \text{ in.})\left(\frac{1 \text{ ft}}{12.0 \text{ in.}}\right) = 1.25 \text{ ft}$$

Now we can calculate ω:

$$\omega = \frac{v}{r} = \frac{60.0 \text{ ft/min}}{1.25 \text{ ft}} = 48.0 \text{ rad/min}$$

In industry, angular speeds are usually expressed in rpm (rev/min), so we shall convert our answer to those units.

$$\omega = \left(\frac{48.0 \text{ rad}}{\text{min}}\right)\left(\frac{1 \text{ rev}}{2\pi \text{ rad}}\right) = 7.64 \text{ rev/min} = 7.64 \text{ rpm} \blacksquare$$

The wheel must rotate at 7.64 rpm to drive the belt at 60 ft/min.

Example 35
Phonograph Turntable

Many phonograph turntables are belt-driven (Figure 1.5-7). Power is provided to turn a shaft. A rubber belt connects this shaft to a collar beneath the turntable, causing the turntable to spin. A typical shaft and collar might have diameters of 1.5 in. and 10 in., respectively, making $a = 0.75$ in. and $b = 5.0$ in. in Figure 1.5-7b. At what angular speed should the motor turn the shaft so the turntable can play LP records?

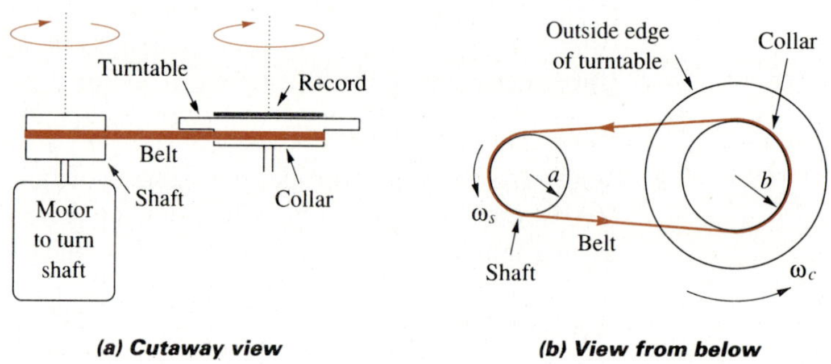

Figure 1.5-7
Phonograph turntable

Solution

We want to find ω_s, the angular speed of the shaft. The important point to notice is this: because the belt does not slip on the shaft or the collar, the linear speed of the belt must be the same as the tangential speed of both the shaft and the collar. This means that *both the shaft and the collar have the same tangential speeds.*

$$v_{\text{shaft}} = v_{\text{collar}}$$

or, more briefly,

$$v_s = v_c$$

Recalling that $v = r\omega$, we can express the linear speeds in terms of the angular speeds:

$$v_s = a\omega_s \qquad \text{and} \qquad v_c = b\omega_c$$

Because $v_s = v_c$, we get

$$a\omega_s = b\omega_c$$

Solving for the angular speed of the shaft, ω_s gives us

$$\omega_s = \left(\frac{b}{a}\right)(\omega_c)$$

where ω_c is the angular speed of the turntable, which must be $33\frac{1}{3}$ rpm for an LP record. Substituting the appropriate numbers for a, b, and ω_c, we have

$$\omega_s = \left(\frac{b}{a}\right)(\omega_c) = \left(\frac{5.0 \text{ in.}}{0.75 \text{ in.}}\right)(33.33 \text{ rpm}) = 220 \text{ rpm} \blacksquare$$

If the shaft spins at 220 rpm, the turntable will spin at $33\frac{1}{3}$ rpm.

Multiple-speed turntables operate in basically the same way, except they have a different size collar for each speed.

NOW IT'S YOUR TURN

If the shaft of a turntable is designed to spin at 150 rpm, what must be the radius of the collar in order to play LP records properly? (Answer: 3.4 in.)

Example 36

Speedometer

As we saw in Example 27, your car's odometer counts the revolutions of the wheels and converts them to miles, presuming standard size tires. The speedometer works in much the same way, except it counts the number of revolutions per second of the wheels and converts that to miles per hour, also presuming standard size tires. During any interval of time t, the *forward* speed of the center of the tire is

$$v_{\text{tire}} = \frac{\text{distance center of tire moves forward in time } t}{\text{time } t}$$

and the *tangential* speed of the tire's rim is

$$v_{\text{rim}} = \frac{\text{arc length rolled by tire in time } t}{\text{time } t}$$

In the odometer example, we showed that the distance that the center of the tire moves forward is the *same* as the arc length that the tire rolls (if the tire does not slip). Therefore, v_{tire} and v_{rim} must be equal because these two distances are the same.

$$v_{\text{tire}} = v_{\text{rim}}$$

Because the car and the center of the tire must be moving together, it follows that $v_{\text{car}} = v_{\text{tire}}$. But we just showed that $v_{\text{tire}} = v_{\text{rim}}$, so v_{car} must also equal v_{rim}.

$$v_{\text{car}} = v_{\text{rim}}$$

We know that the tangential speed of the rim is $v_{\text{rim}} = r\omega$. Calling v_c the speed of the car, we have $v_c = v_{\text{rim}} = r\omega$.

$$v_c = r\omega$$

where r is the radius of the tire and ω is the tire's angular speed in *radian* measure.

With Standard Tires

If your car has standard 24-in.-diameter tires, how many revolutions per second are the wheels making when you are driving at a freeway speed of 55 mph?

Solution

We just showed that the speed of the car is the same as the tangential speed of the rim of the tires, which means that $v_c = r\omega$. Solving for ω gives us

$$\omega = \frac{v_c}{r}$$

Both r and v_c must be expressed in the same length units, so we must make some conversions.

$$r = 12 \text{ in.} = 1.0 \text{ ft} \qquad \text{and} \qquad v_c = 55 \text{ mph}$$

$$v_c = (55 \text{ mph})\left(\frac{88 \text{ ft/s}}{60 \text{ mph}}\right) \quad \textbf{(using 88 ft/s = 60 mph to convert miles per hour to ft/s)}$$

$$= 81 \text{ ft/s}$$

Now we can find ω.

$$\omega = \frac{v_c}{r} = \frac{81 \text{ ft/s}}{1.0 \text{ ft}} = 81 \text{ s}^{-1} = 81 \text{ rad/s}$$

$$= \left(81 \frac{\text{rad}}{\text{s}}\right)\left(\frac{1 \text{ rev}}{2\pi \text{ rad}}\right) \quad \textbf{(convert rad/s to rev/s)}$$

$$\omega = 13 \text{ rev/s} \;\blacksquare$$

Therefore, when your wheels are spinning at 13 rev/s, your speedometer is calibrated to read 55 mph (see "Conversion of Units" on pp. 24–26 to learn how to show that 88 ft/s = 60 mph).

With Oversize Tires

If you replace your standard tires with 30-in.-diameter oversize tires, how fast are you really going when your speedometer reads 55 mph?

Solution

If your speedometer reads 55 mph, the previous example tells us that your wheels must be turning at 13 rev/s (81 rad/s) because the speedometer presumes standard size tires. The speed of the car is

$$v_c = r\omega$$

where r is now the 15-in. radius of the larger oversize tires.

$$v_c = r\omega = (15 \text{ in.})(81 \text{ rad/s}) = 1220 \text{ in./s}$$

$$= \left(1220 \frac{\text{in.}}{\text{s}}\right)\left(\frac{1 \text{ ft}}{12 \text{ in.}}\right) \quad \textbf{(convert in./s to ft/s)}$$

$$= (102 \text{ ft/s})\left(\frac{60 \text{ mph}}{88 \text{ ft/s}}\right) \quad \textbf{(convert ft/s to mph)}$$

$$v_c = 70 \text{ mph} \;\blacksquare$$

When your speedometer reads 55 mph, you are really travelling 70 mph! Beware the flashing red light behind you!

NOW IT'S YOUR TURN

If your car has standard 28-in.-diameter tires: (a) How many revolutions per second are your wheels making when you are travelling at 45 mph? (b) If you put on 30.0-in. oversize tires, how fast are you really travelling when your speedometer reads 45 mph? (Answers: (a) 9.0 rev/s, (b) 48 mph)

EXERCISES 1.5

1. At top speed, many home lathes spin at 5000 rpm. Express this angular speed in radians per second and degrees per second.

2. If a wheel turns through 25° in 3.8 s, what is its angular speed in degrees per second and radians per second?

3. Standard shop motors spin at 1725 rpm and 3450 rpm.

(a) Through what angle, in revolutions and in degrees, will the faster motor turn in 3.00 min?

(b) How many minutes will it take the slower motor to turn through 125,000 rev?

4. Find the angular speeds in radians per second of the second, minute, and hour hands on a clock.

5. The planet Mercury takes 88 days to circle the sun. What is its angular speed in radians per day?

6. A typical LP record takes about 25 min to play each side. How many turns does it make in this time?

7. How long does it take an LP record to turn through 1.0°? Through 1.0 rad?

8. The rock singer on an LP record holds a note for 38 s. Through how many degrees does the LP turn while he holds this note?

9. Trivia to Amaze Your Friends. How long does it take the earth to spin on its axis through an angle of 1.0°? 1.0′? 1.0″? 1.0 rad?

10. Time-Exposure Photography. You are taking time-exposure photos of the night sky. Through what angle do the stars move during a 5.0 min exposure?

11. Through what angle does the earth move around the sun during summer vacation, June 15 to September 1?

12. Man in the Moon. Our moon spins on its axis in the same direction and at the same rate as it goes around the earth. (This synchronous orbit is not an accident but is due to the gravitational pull of the earth on irregularities in the moon's structure). Using careful drawings, prove that this synchronous orbit causes the moon always to keep the same face toward the earth. ***Hint:*** Draw the moon every $\frac{1}{4}$ of its orbit, showing its motion around the earth and the motion of the side facing earth.

13. The photos in Figure 1.5-8 show the Kruger–60 double-star system in which the smaller star's motion is clearly visible. Make the necessary measurements *on the photo* to calculate the angular speed of the smaller star.

14. Precession of the Equinoxes. The earth's axis of spin now points toward the North Star, Polaris, but slowly *precesses* like a spinning top, taking 25,780 years for a complete cycle (see Figure 1.5-9). Through what angle does the axis precess in a century? How long does it take it to precess through 1.00°?

****15.** Suppose that today Mars is 60.0° ahead of the earth in its 1.90-yr orbit around the sun in the same direction as earth. When will the earth and Mars first be aligned along a straight line from the sun?

****16.** Find all the times of day at which the minute and hour hands of a clock *overlap* each other.

****17.** Find all the times of day at which the minute and hour hands of a clock are perpendicular.

18. How fast is the rim of a 25 cm diameter wheel moving if the wheel is turning at 2.57 rad/s? If it is turning at 49°/s? If it is turning at 0.13 rev/s?

19. A gnat is having a picnic midway between the rim and center of a 10.0-in.-diameter 78-rpm record. How long does it take this gnat to move 5.0 in.?

20. Bruce Pierini has carelessly revved up his airplane so much that the tip of the 9.0-ft long propellor has reached the speed of sound (1100 ft/s). How many revolutions per minute does this propellor make?

21. A typical chemistry lab centrifuge spins at 3600 rpm with the bottoms of the test tubes 4.0 in. from the center of rotation. How fast in feet per second are these bottoms moving?

Figure 1.5-8 ***Kruger–60***

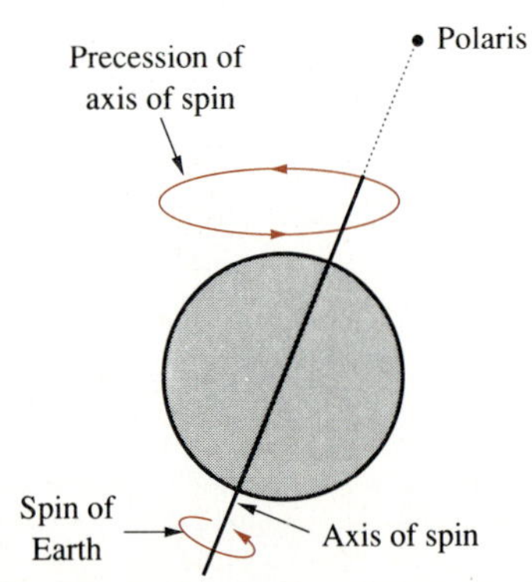

Figure 1.5-9

22. Find the linear speed, in miles per hour, of Singapore, which is practically on the earth's equator, due to our planet's axial spin.

23. Find the linear speed, in miles per hour, of the earth around the sun.

24. Trivia to Amaze Your Friends! As the earth goes around the sun, through how many miles does it move each second of time?

25. Skylab. Skylab, our first manned space station, was launched in 1973. It moved in a circular orbit 273 mi above the earth's surface and went around our planet once every 1.55 hr. Find Skylab's angular and linear speeds in radians per second and miles per hour, respectively.

26. Communications Satellites. We use communications satellites, such as Telstar, by bouncing radiowaves off them. To be useful, they must always be above the same point on the earth's equator and orbit at 22,500 mi above the earth's surface. (This type of motion is called *geosynchronous*.) Find the angular and linear speeds of all communications satellites in revolutions per day and miles per hour, respectively.

27. Pulsars. The pulsar (a spinning neutron star) in the Crab Nebula spins 30 rev/s and is about 10 mi in diameter. How fast is matter at the equator of this pulsar moving, in miles per second?

***28.** Two mice, Cheezy and Breezy, start at the same point on a 60.0-in.-diameter circular track and run around the track in opposite directions. Cheezy runs at 5.00 in./s while Breezy runs at 8.00 in./s. Through what *angle* (in degrees) will each have run when they first meet?

***29.** Two spiders, Hairy and Scarey, are riding on the edges of different phonograph records. Hairy is on a 12.0-in. LP and Scarey is on a 9.00-in. 45.0-rpm record. When Hairy has travelled a distance of 100.0 in., how far has Scarey gone, and through what angle, in radians, has each turned?

30. An ant, Fleetfeet, is at the edge of a 12-in. LP record while it is playing, and the ant is running at a speed of 500.0 in./min along the LP's rim. Find Fleetfeet's angular speed in radians per second if she is running:

(a) In the same direction the LP is turning.
(b) In the opposite direction the LP is turning.

***31. Stylus Tracking Speed.** Measure the inner and outer radii of the *playing area* of an LP record and use these to calculate the tracking speed in inches per second of the stylus on the grooves at the beginning and at the end of the record.

32. Distortion on LP Records. You have probably noticed that LP's have more tracking distortion at the end of a record than at the beginning. To see why, calculate the distance the stylus moves along the outer and the inner grooves while playing 1 s of music. In which case is the music more compressed in the grooves? Where would you expect more distortion? Measure the necessary quantities on an actual record.

***33. Tape Cassette.** When a tape cassette is playing, the tape guides spin and drive the tape at a constant speed of $1\frac{7}{8}$ ips (in./s) over the playing heads, forcing the tape off and on the take-up reels, as shown in Figure 1.5-10. For a standard tape deck, the tape guides are 2.0 mm in diameter, while the inner and outer diameters of the take-up reels are $\frac{7}{8}$ in. and 2 in., respectively.

(a) What must be the angular speed of the tape guides?
(b) Find the angular speeds of the take-up reels when they are (i) full of tape and (ii) just at the end of the tape.
(c) How many feet are there on a standard 60.0-min (30.0 min in each direction) cassette?

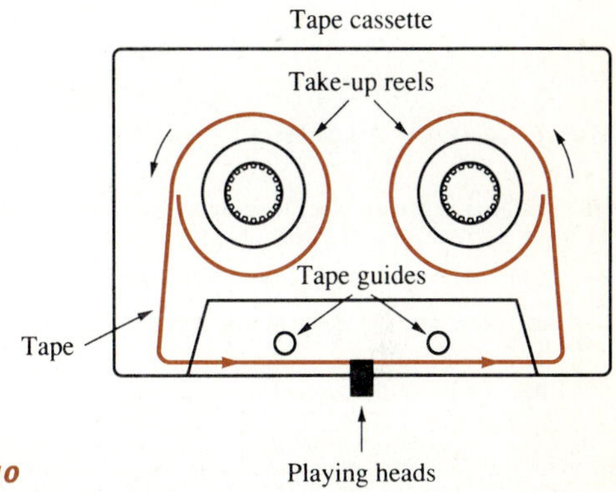

Figure 1.5-10

34. Compact Discs. Compact discs (CD's) operate differently from LP records. With LP's, the record turns at a constant *angular* speed of $33\frac{1}{3}$ rpm, which means that the *tangential* speed of the stylus on the record changes from the outside to the inside of the record. But when a CD is playing, the turntable adjusts the angular speed of the disc so that the laser beam maintains a constant tangential speed of about 1.25 m/s as the laser reads from the *inside to the outside* of the disc. Measurements on a CD show that the inner and outer diameters of the playing area are 5.0 cm and 11.4 cm, respectively. At what angular speeds must the turntable spin when the laser is tracking at the outside and at the inside of the playing area?

35. An 18.0-in.-diameter winch is used to lift a heavy bucket of concrete to a rooftop, as shown in Figure 1.5-11. When the winch is turning at 5.00 rpm, how fast is the bucket moving?

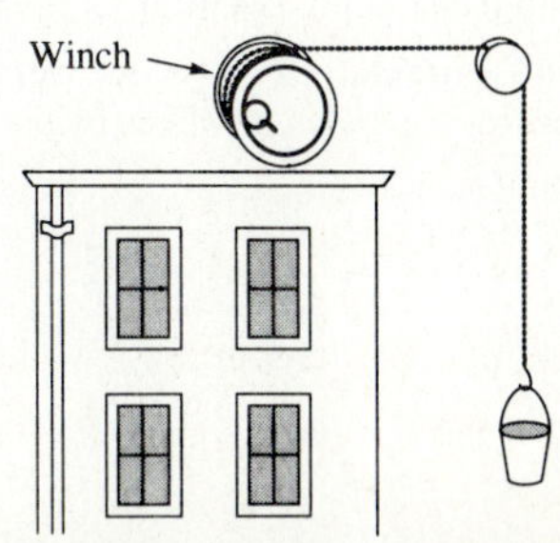

Figure 1.5-11

36. Gear A, of radius 40.0 cm and spinning at 150 rpm, meshes with gear B, of radius 15 cm. How fast will B spin?

***37.** Gears A and B mesh, as shown in Figure 1.5-12, to lift a stone. Gear A is driven at 15.0 rpm in the direction shown. If A is 12.0 cm in diameter and B is 38.0 cm in diameter, how fast is the stone moving? Is it rising or falling?

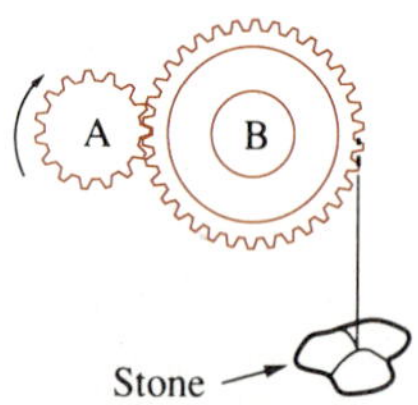

Figure 1.5-12

***38.** Gear B is driven clockwise by a motor at 50.0 rpm and turns gears A and C, as shown in Figure 1.5-13. If A has radius 45 in., B has radius 20.0 in., and C has radius 30.0 in., find:

(a) The speeds of the stone and the bucket. Are they rising or falling?
(b) The angular speed of each gear.

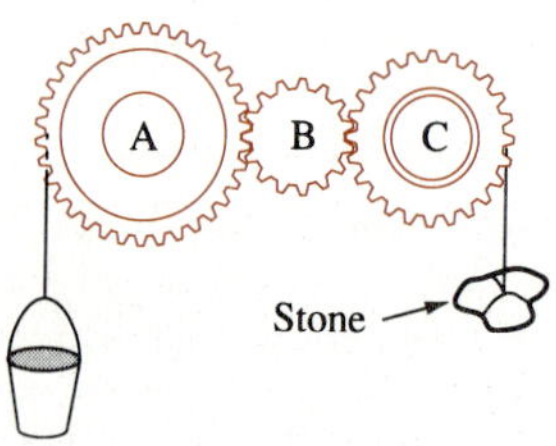

Figure 1.5-13

Exercises 39–43 refer to the turntable in Example 35.

39. How fast is the belt moving?

40. If the collar were 6.00 in. in diameter instead of 5.00 in., how fast would the shaft have to spin?

41. If the same shaft were to spin at 260 rpm, what would the radius of the collar have to be?

42. What size should the collar be to play 45-rpm records rather than LP's?

43. What should be the angular speed of the shaft if the collar has a radius of 10.0 cm and the shaft has a radius of 2.00 cm?

***44. Two-Speed Turntable.** A two-speed turntable, for 45-rpm and $33\frac{1}{3}$-rpm records, operates just like the one in Example 35, except it has two collars of different sizes. The same shaft with the same angular speed is used for both speeds. In a certain turntable, the shaft is 1.00 in. in diameter and the larger collar is 8.00 in. in diameter.

(a) *Without doing any calculations*, figure out which collar is for the 45-rpm speed and which is for the $33\frac{1}{3}$-rpm speed.
(b) At what angular speed must the shaft rotate to spin the turntable at $33\frac{1}{3}$ rpm? At 45 rpm? What must be the diameter of the collar for each speed?

***45. Three-Speed Turntable.** Suppose you want to convert the turntable in Exercise 44 to a three-speed unit capable of playing your grandfather's treasured 78-rpm Caruso records. What should be the diameter of the new collar?

For Exercises 46–52, see Example 36.

46. If your car has standard 26-in.-diameter tires, how many revolutions per minute will the wheels make when you are going 60.0 mph?

47. What should be the diameter of a tractor's tires so that they will turn at 120 rpm when the tractor is going 30 mph?

48. If your car's standard 28-in.-diameter tires are turning at 275 rpm, how fast is the car going?

***49.** You change your car's standard 24-in. tires to 22-in. tires. How fast are you going when your speedometer reads 50.0 mph?

***50.** You have replaced your car's standard 26-in. tires with 30.0 in. tires and want to drive within the 55-mph speed limit. What speed should your *speedometer* show for you to be driving at 55 mph?

***51.** Repeat Exercise 50 for 20.0-in. (instead of 30-in.) tires.

****52.** Your car normally has 24-in. tires. What size tires should you put on your car if you want your car to be going:

(a) 70.0 mph when the speedometer reads 60.0 mph?
(b) 50.0 mph when the speedometer reads 60.0 mph?

53. If a bicycle has 30.0-in.-diameter tires, how fast in revolutions per second are the wheels turning when you are riding at 10.0 mph?

***54.** A child's tricycle has one 17-in.-diameter wheel and two 7.0-in.-diameter wheels. How fast is each wheel turning in rpm when the tricycle is going 2.0 ft/s?

***55.** A tractor has two different size wheels. If the larger 5.0-ft diameter wheel is turning at 20.0 rpm and the smaller wheels are turning at 45 rpm, find the diameter of the smaller wheels and the speed of the tractor.

1.6 SMALL-ANGLE APPROXIMATION

S = arc length subtended by the central angle θ

d = length of chord of θ

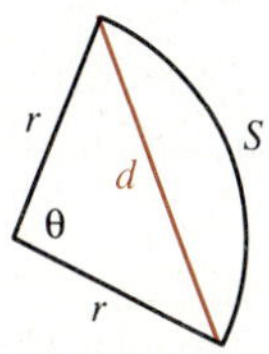

Figure 1.6-1

Arc length and chord of angle θ S = arc length subtended by the central angle θ. d = length of chord of θ.

Most people think that scientists and engineers solve their elaborate formulas exactly. In practice, this rarely happens! The formulas in real-world applications are often frightfully complicated and can sometimes be solved only for special situations using simplifying approximations. One very useful such approximation is the *small-angle approximation*.

To calculate a central angle in radians, we divide the circular arc length S which that angle subtends by the circle's radius r: $\theta = S/r$. Notice in Figure 1.6-1 that the arc length S is *not* the same as the length of the chord d. However, look at Figure 1.6-2 and notice what happens as θ gets smaller and smaller.

As you can see, for small angles the arc length S and the chord d are nearly the same. And the smaller θ is, the closer these two are to being the same. So, for very small angles—that is, *angles much smaller than 1 rad*—we can approximate the arc length by the chord. Now, since θ is *exactly* equal to S/r, and S is approximately equal to d, it follows that θ is approximately equal to d/r.

$$\theta = \frac{S}{r} \quad \text{and} \quad S \approx d \quad \rightarrow \quad \theta \approx \frac{d}{r}$$

We call this approximation the *small-angle approximation* because it is true only when θ is much less than 1 rad.

As θ gets smaller and smaller, d and S approach the same value.

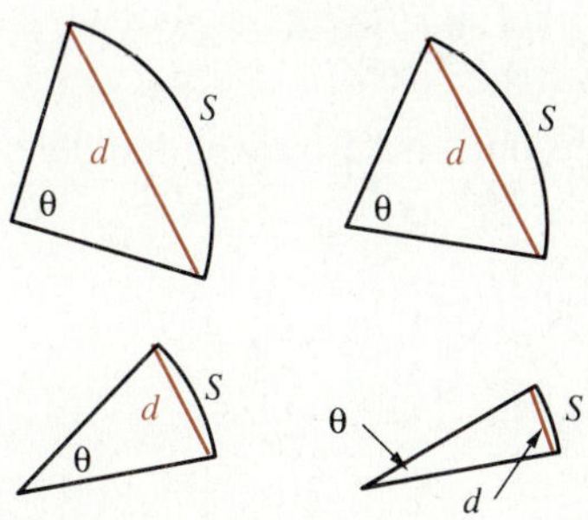

Figure 1.6-2

Small-angle approximation As θ gets smaller and smaller, d and S approach the same value.

> **SMALL-ANGLE APPROXIMATION**
>
> **For very small angle θ,**
>
> $$\theta \approx \frac{d}{r}$$
>
> 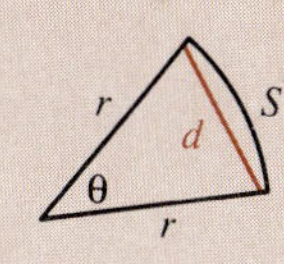
>
>
> **This gives θ in radians.**

▸ **Important.** Notice that this approximation will be true if the *chord d is much smaller than the radius r.*

▸ **How Small Is Small?** As a rule of thumb, we generally use the small-angle approximation if θ is less than about 10° ($\approx \frac{1}{5}$ rad).

Example 37

When you look at your upright thumb at arm's length, what angle does it subtend?

Solution

A typical thumb is about 2 in. long and reaches about 25 in. from the eye (Figure 1.6-3). (These measurements are from the author's body.) We can use the

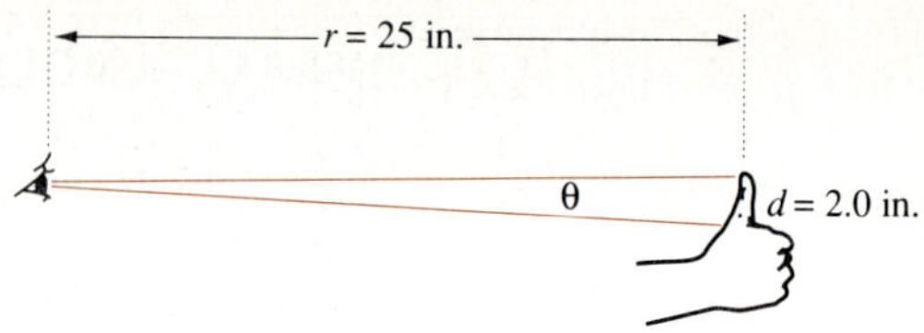

Figure 1.6-3
Thumb viewed at arm's length

small-angle approximation ($\theta \approx d/r$) to calculate θ because d (the thumb's height, 2 in.) is much smaller than r (the arm length, 25 in.). This calculation gives us

$$\theta \approx \frac{d}{r} = \frac{2 \text{ in.}}{25 \text{ in.}} = 0.080 \text{ rad}$$

Converting θ to degrees gives us

$$\theta \approx (0.080 \text{ rad})\left(\frac{180°}{\pi \text{ rad}}\right) \approx 4.6° \quad ■$$

NOW IT'S YOUR TURN

Measure your own thumb and arm lengths and use these measurements to calculate the angle subtended by your thumb at arm's length.
(Answer: Your answer shouldn't be much different from the one we got.)

Example 38
Baseball Bat

At what angle do the sides of a baseball bat slope outward from the handle?

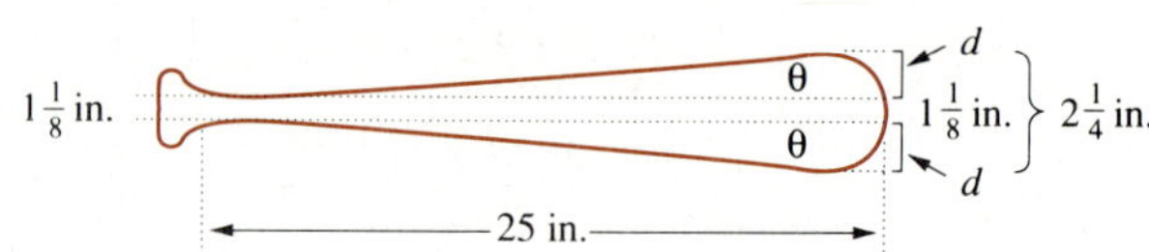

Figure 1.6-4
Baseball bat

Solution

First make a clear sketch of the bat, showing the results of actual measurements (Figure 1.6-4). We want to find the angle θ. From experience (or the actual measurements), we know that θ is quite small, so we can use the small-angle approximation. This approximation tells us that $\theta \approx d/(25 \text{ in.})$. From Figure 1.6-4, we see that

$$\begin{aligned} d + d + 1\tfrac{1}{8} \text{ in.} &= 2\tfrac{1}{4} \text{ in.} \\ 2d &= 2\tfrac{1}{4} \text{ in.} - 1\tfrac{1}{8} \text{ in.} = 1\tfrac{1}{8} \text{ in.} = \tfrac{9}{8} \text{ in.} \\ d &= \tfrac{9}{16} \text{ in.} \end{aligned}$$

Now we can calculate θ.

$$\theta \approx \frac{d}{25\text{ in.}} = \frac{\frac{9}{16}\text{ in.}}{25\text{ in.}} = 0.023\text{ rad}$$

Converting this to degrees:

$$\theta \approx (0.023\text{ rad})\left(\frac{180^\circ}{\pi\text{ rad}}\right) \approx 1.3^\circ \blacksquare$$

NOW IT'S YOUR TURN

If the bat were 3.0 in. wide (instead of $2\frac{1}{4}$ in.) at the widest place, but all other dimensions were unchanged, what would θ be? (Answer: 2.1°)

Example 39

The Size of Retinal Cells

Ever since the 1600s, people have realized that the smallest things that we can distinguish clearly subtend an angle of approximately 1.0′. Since this angle is nearly the same for all humans, it is reasonable to think that it might be related to the receptor cells that make up the retina of the eye.

As light enters the eye, it is focused on the retina (the rear of the eye) by the lens, as shown in Figure 1.6-5. If the two light rays shown are *very* close together, they will both be focused on the same retinal cell, which means that the brain cannot tell these two rays apart. For the brain to be able to distinguish between these two rays, they must fall on different cells. Therefore, the rays must be separated on the retina by a *minimum of one cell width*. Thus rays that are 1.0′ apart must intercept about one cell width on the retina because rays closer together would fall on the same cell (Figure 1.6-6).

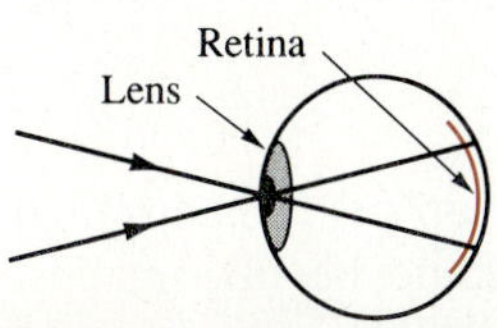

Figure 1.6-5
The eye

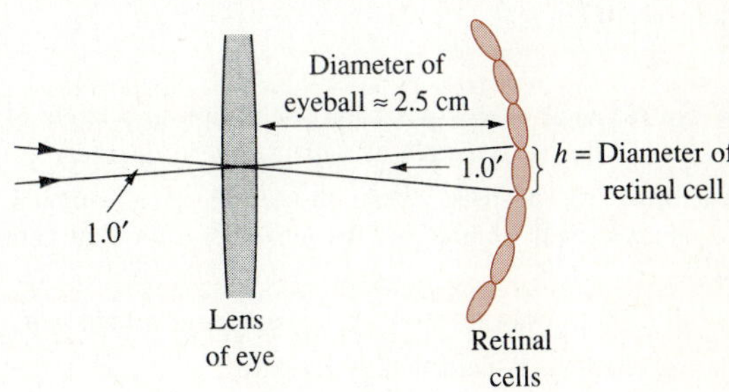

Figure 1.6-6
Rays 1.0′ apart intercept one retinal cell.

Because 1.0′ is much less than 1 rad, we can apply the small-angle approximation in Figure 1.6-6.

$$\theta \approx \frac{d}{r}$$

where r = diameter of eyeball $\approx$ 2.5 cm for typical human eyes, $d = h$, and $\theta = 1.0'$, which we must convert to radians.

$$(1.0')\left(\frac{1^\circ}{60'}\right)\left(\frac{\pi \text{ rad}}{180^\circ}\right) = 2.9 \times 10^{-4} \text{ rad}$$

(See the section on Scientific Notation.) Substituting these numbers and solving for h gives

$$\theta \approx \frac{d}{r}$$

$$2.9 \times 10^{-4} \text{ rad} \approx \frac{h}{2.5 \text{ cm}}$$

$$h \approx (2.5 \text{ cm})(2.9 \times 10^{-4} \text{ rad}) \approx 7.3 \times 10^{-4} \text{ cm} \quad ■$$

Biological research has shown that retinal cells are, in fact, approximately this size. A rough, simple calculation like this is in good agreement with experiment.

NOW IT'S YOUR TURN

Suppose that an animal with eyes the same size eye as human's eyes had retinal cells 9.5×10^{-4} cm in diameter. Approximately what would be the smallest angle that the eyes of this animal could resolve? (Answer: 1.3′)

EXERCISES 1.6

1. If a central angle subtends a chord of 23 mm in a circle of radius 42 cm, what is the angle in radians and in degrees?

2. What angle, in degrees, would the length of a yardstick subtend when viewed from across the length of a football field (100 yd)?

3. If a hill 2500 ft away subtends at 1.5° angle, how high is it?

4. How far away is a meterstick if its length subtends a 2.5° angle?

5. At what distance does a tree 24 ft tall subtend an angle of 10′?

6. How far must you walk from a 100.0-ft high water tower so that its top will appear 49′ above the horizon?

7. A logger first uses a rangefinder to determine that a tree is 275 ft away and then determines that the top of this tree makes an angle of 2°38′ above the horizontal ground. How tall is the tree?

8. What angle will a 120-ft long airliner subtend as it flies over you at an altitude of 30,000.0 ft?

9. An 18-in. mallard duck flies overhead, subtending an angle of 3°15′. How high is this duck flying?

10. What is the angular width of your full hand viewed at arm's length? Use measurements from your own body.

11. At what distance from you will a quarter appear to be the same size as a penny held at arm's length? Make the appropriate measurements to do this calculation.

12. Using only simple distance measurements that you can make with a ruler, how can you measure the angle that the full moon subtends? (This measurement is called the *angular diameter* of the moon and is approximately $\frac{1}{2}^\circ$.) Go outside and make these measurements when there is a full (or nearly full) moon. How close do you come to $\frac{1}{2}^\circ$?

13. If you were on the moon viewing the earth, what would be the angular diameter of our planet? How does this compare to the moon's angular diameter as viewed from earth?

14. The Landsat earth-mapping satellite was 3.0 m long and orbited 917 km above the earth's surface. What angle would it subtend for a tracking station on the ground?

***15.** In a Landsat (see Exercise 14) photograph taken directly over a rectangular forest, the angular dimensions were measured to be 2.0° × 3.2°. What is the *area* of this forest in square km?

16. In a satellite photo, a 100.0-m long river extends over a 1.1° angle. How high was the satellite when it took this photo?

17. Calculate the angle your foot subtends when you glance down at it, using actual measurements from your own body.

18. As you are sitting in your balcony seats at the San Francisco Opera, you notice that you can just cover the 6.0-ft singer on the stage with a $\frac{3}{4}$-in. section of your thumb held at 24-in. arm's length. How far are you from the stage?

19. If a comet, when it is 75 million miles away, has a tail 3 times the angular width of the moon, how long is this tail?

20. If a certain doughnut has a 1.0-in.-diameter hole, how far away must you hold it so that the hole will just frame the full moon?

21. A tract of farmland is an equilateral triangle 6.48 mi on a side. In an aerial photo, the corners of this field are 3.61° apart. How high was the satellite when it took the photo?

***22.** The earth has a nearly circular orbit of radius 93 million miles. Mars' orbit is also nearly circular, but of radius 140 million miles. Mars has a diameter of approximately 4200 mi.

(a) Find the largest angular diameter of Mars as viewed by an astronomer on earth.
(b) If you were on Mars viewing earth, what would be the largest angular diameter of our planet?

23. The Ring Nebula, shown in Figure 1.6-7, is 2150 light-years from us and measures 1.00′ × 1.40′ in angular size. What are its dimensions in light-years and in miles?

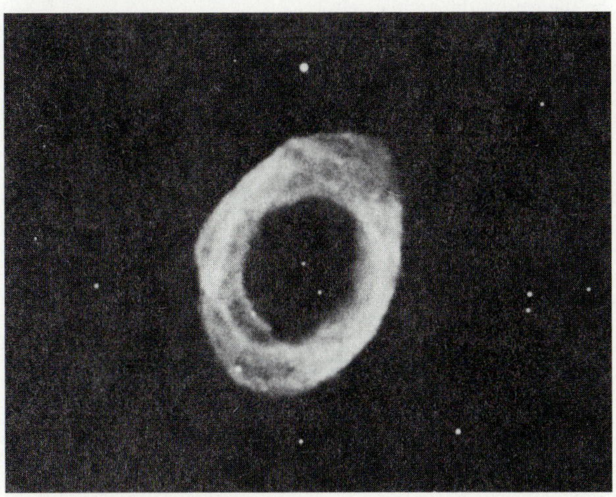

Figure 1.6-7 ***Ring Nebula***

24. Andromeda, our nearest galaxy, is a great spiral about 200,000 light-years across and about 2 million light-years away (see Figure 1.6-8). What angle does Andromeda subtend from earth? If we could see it (which we cannot because it is too dim), how would it compare to the full moon?

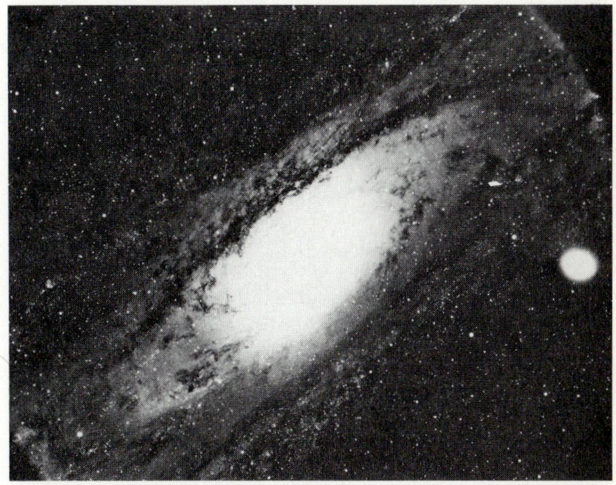

Figure 1.6-8 ***Andromeda Galaxy***

25. In March of 1986, the European spacecraft Giotto came within 324 mi of the nucleus of Halley's comet. In the photos, this nucleus resembles a "burnt black potato" 10.0 mi long by about 6.0 mi wide, which shrinks about 6.0 ft with each visit to the sun. What angles did the width and length of the nucleus present to Giotto's camera. By what angle, as viewed from Giotto's closest distance, does the nucleus shrink with each visit to the sun?

***26.** A spherical water tank of volume 1000.0 m^3 appears to be the same size as a 9.00-in. circumference baseball viewed at 2.00-ft. How far away is the water tank?

***27.** A basketball is 30.0 in. in circumference. How far away should it be so that it will appear to be the same size as the full moon ($\frac{1}{2}°$)?

***28.** You have been asked to buy ribbon to decorate the outside rim of a round window on a building across the street, 45 ft away. You don't have time to go over and measure it, but you do notice that if you hold a ruler at 24-in. arm's length, $\frac{5}{8}$ in. is all it takes to span the width of the window. What length of ribbon should you buy?

***29.** You are viewing a 30.0-ft tall building from 400.0 ft away and notice that you can just see the tip of a 70.0-ft building in line with the top of the 30-ft building. How far away from each other are these two buildings?

***30.** A 1.0-ft long duck flying overhead at 50.0 ft appears to be 3 times as long as a 130-ft airliner also flying overhead. How high is the airliner?

***31.** A certain baseball bat measures 3.93 in. in circumference at its narrowest point and 7.85 in. in circumference at its widest point. The distance between these two points is 26.5 in. At what angle does the bat slope out from its narrowest point?

***32.** Devise a simple way to measure the width of a river by obscuring a building on the other side with your thumb. Which things would you have to measure and/or estimate?

***33.** A tree trunk with constant inward taper is 6.0 ft around at ground level and 5.5 ft around at 7.0 ft above the ground.

(a) How high off the ground should a logger cut this tree to obtain a log 4.0 ft in circumference at that point?
(b) What is the maximum height this tree can reach?

****34. Area of a Circle.** You have all learned that the area of a circle is πr^2, but how do you know this is true? In this exercise and the next one, you will see how to derive this formula. Start with a circle of radius r. It will have a circumference of $2\pi r$, by the *definition* of π. First, cut the circle into four equal quarters, A, B, C, and D, and lay them out next to each other, as shown in Figure 1.6-9.

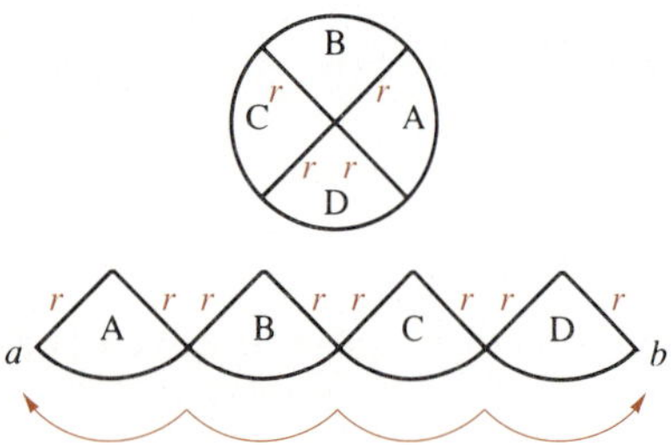

Figure 1.6-9

(a) In the four quarter circles at the bottom of the figure, what is the total arc length ab equal to in relation to the circle?
(b) Now add four identical wedges, as shown in Figure 1.6-10. How does the area of this figure compare to the area of the original circle?
(c) Now repeat the steps in **(a)** and **(b)** by cutting the original circle into more and more wedges. What *simple* shape does the figure in **(b)** approach as the number of wedges gets larger and larger? Use the answer in **(a)** to calculate the area of this simple figure. Then relate this area to that of the original circle to derive $A = \pi r^2$ for the circle.

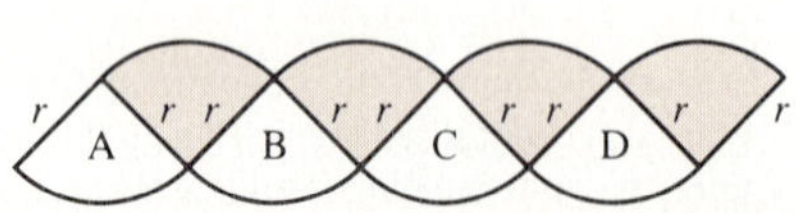

Figure 1.6-10

****35. Area of a Circle.** This exercise is a variation on the method in the previous problem. Start with a circle of radius r and divide it into n equal-size wedges, each of central angle θ, as shown in Figure 1.6-11.

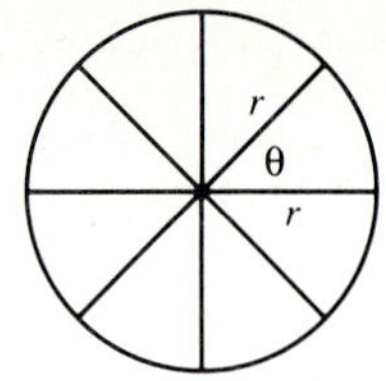

Figure 1.6-11

(a) Express θ in terms of π and n.
(b) If A is the area of the circle, what is the area of each wedge?

Now let the *number of wedges* get larger and larger.

(c) Using the small-angle approximation, what is the area of each wedge?
(d) Using your answer to **(a)**, what is the area of each wedge?
(e) Now add all n wedges to find the total area of the circle, and, in so doing, prove that $A = \pi r^2$.

For Exercises 36–43, see Example 39.

36. If our eyes were 6.0 cm in diameter, what would be the smallest angle we could resolve?

37. If retinal cells were 1.2×10^{-6} m across, what would be the smallest angle we could resolve?

38. If retinal cells were 1.2×10^{-6} m across *and* if our eyes were 6.0 cm in diameter, what would be the smallest angle we could resolve?

39. What would have to be the diameter of our eyes for us to resolve angles of $0.10'$?

40. If the smallest angle we could resolve were $1.8'$, what would be the size of our retinal cells?

41. What is the farthest distance a 6.0-ft person can walk from you and still be barely visible to you?

***42.** Suppose an animal has eyes 5.0 cm in diameter and can resolve angles as small as 0.010°. Calculate the approximate size of this animal's retinal cells.

***43.** A certain creature can just see a 1.0-ft tall plant when it is 150 ft away. It has eyeballs 7.50 cm in diameter. What is the approximate size of this animal's retinal cells?

44. The Very-Long-Baseline Array (VLBA) radiotelescope, expected to be operating by the mid-1990s, should be able to resolve an angle as small as that subtended by a pea in New York viewed from San Francisco! What is this angle in radians and in seconds?*

---

*For a good discussion of this telescope, see "The Very-Long-Baseline Array" by K. I. Kellermann and A. R. Thompson in the January 1988 *Scientific American*.

1.7 TRIGONOMETRY AND SPACE SCIENCE

In this section, we show you how a small amount of mathematics, along with some rather simple observations, can yield far-reaching results about the scope of our universe. The purpose of this section is *not* to teach you astronomy. Rather, it is to show what a powerful tool mathematics can be. To most people, the distances and sizes of astronomical bodies are complete mysteries. Yet, as you will see in this section, these quantities can be calculated to a high degree of precision using the mathematics from just the first chapter of this book!

Since trigonometry was first used in astronomy, we begin with one of these early calculations—Eratosthenes' calculation of the size of the earth.

Example 40
How Large Is the Earth?

In approximately 240 B.C., the Greek scholar Eratosthenes made the following two observations.

1. At noon on the summer solstice (June 21), a vertical pole in Alexandria, Egypt, cast a shadow. By comparing the length of this shadow to the height of the pole, he could determine that the sun's rays hit the pole at approximately a $7\frac{1}{2}°$ angle (in our own notation), as shown in Figure 1.7-1.

2. At the same time in Syene (present-day Aswan), 500 mi south of Alexandria, the sun's rays went straight down a deep, narrow well without hitting the sides, which means that a pole there would cast no shadow.

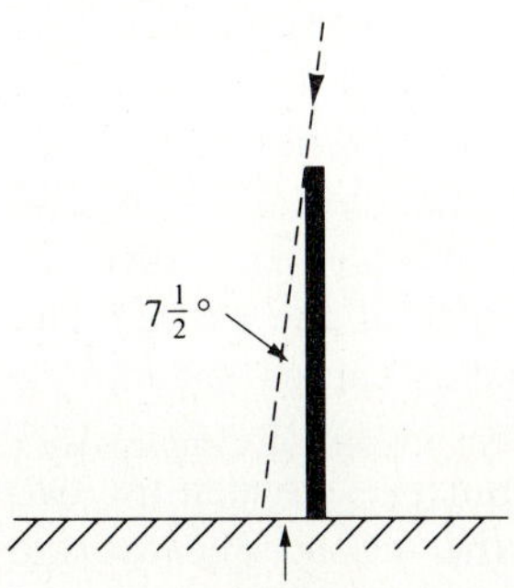

Figure 1.7-1
Shadow cast by sun in Alexandria

Eratosthenes realized that the sun must be so far away that the rays reaching the earth are essentially parallel to each other. According to his reasoning, the ray down the well at Syene must be travelling perpendicular to the surface of the earth because it did not strike the sides of the vertical well. Therefore, this ray, if it continued, would pass through the center of the earth, as shown in Figure 1.7-2. Because the sun's rays are parallel, the angle θ in this figure must be $7\frac{1}{2}°$ because alternate interior angles of two parallel lines are equal.

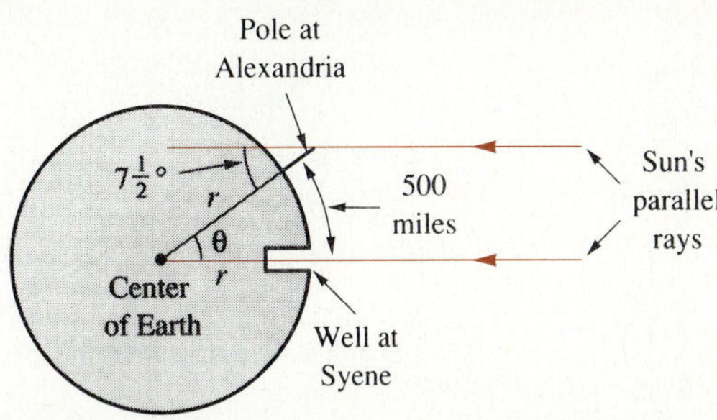

Figure 1.7-2
Eratosthenes' method

Therefore, using the results of Section 1.4, we can calculate the earth's radius r as follows. The distance between the two cities, 500 mi, is the arc length subtended by the angle θ. Therefore, this distance must obey the formula $S = r\theta$, where S is 500 mi and θ is $7\frac{1}{2}°$. Solving for r gives $r = S/\theta$. Before we can use this equation, we must convert the $7\frac{1}{2}°$ to radians.

$$\theta = (7\tfrac{1}{2}^\circ)\left(\frac{\pi \text{ rad}}{180^\circ}\right) = 0.13 \text{ rad}$$

Now we can calculate r.

$$r = \frac{S}{\theta} = \frac{500 \text{ mi}}{0.13 \text{ rad}} = 3850 \text{ mi} \quad ■$$

This result is slightly different from the modern value of 3960 mi because of limitations on the accuracy of Eratosthenes' measurements. (Remember, this was 240 B.C.!) Eratosthenes' method was slightly different from that presented here, but he used the same basic reasoning and calculated the earth's circumference.* Columbus was not the first person to show that the earth is round!

Example 41
How Far Is the Moon?

Suppose two astronomers, A and B, are observing the same point on the moon, as shown in Figure 1.7-3. Because these astronomers are viewing the moon from different positions, they have to point their telescopes in different directions. By comparing the directional settings of their telescopes, they can easily tell that these settings differ by the angle $2p$, as shown in Figure 1.7-3 where p is called the *parallax angle*. In other words, astronomer A would have to move his telescope through angle $2p$ if he went to B's position. (We call this angle $2p$ because astronomers prefer to measure p from the center of the earth.) This type of parallax is called *geocentric parallax* because it is caused by moving around the *earth*.

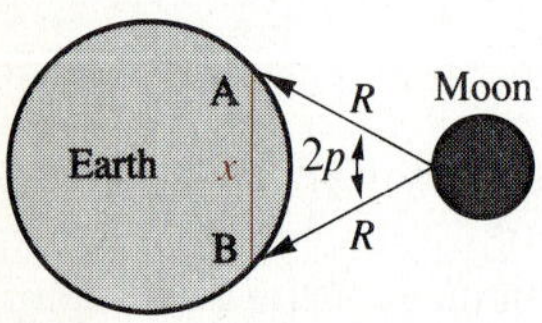

Figure 1.7-3

Distance to the moon The size of p is greatly exaggerated.

The size of the parallax angle p depends on the distance x between the two observers and the distance R between the earth and the moon. Because the distance R to the moon is very much greater than the distance x, we can use the small-angle approximation, which for this figure is $p = x/R$. Solving for R gives $R = x/2p$. The angle p will be greatest when A and B are at opposite ends of the earth, in which case x will be the earth's diameter D. This gives

$$R = \frac{D}{2p}$$

Therefore, we can calculate the moon's distance R by measuring the parallax angle p for two observations made from opposite ends of the earth since we already know the earth's diameter.

---

* Eratosthenes used ratio and proportion to show that

$$\frac{500 \text{ mi}}{7\frac{1}{2}^\circ} = \frac{\text{earth's circumference}}{360^\circ}$$

Therefore, the earth's circumference is equal to $(360^\circ/7\frac{1}{2}^\circ)$ (500 mi) = 24,000 mi. From this we get the earth's radius, using the fact that the circumference is $2\pi r$. (His value for the distance from Alexandria to Syene was highly uncertain.)

Actual measurements of the parallax angle p give a value of 0.95°. Converting this to radians (**why?**) gives

$$p = (0.95°)\left(\frac{\pi \text{ rad}}{180°}\right) = 0.0166 \text{ rad}$$

We have measured the earth's diameter to be 7900 mi, so we can now find R.

$$R = \frac{D}{2p} = \frac{7900 \text{ mi}}{2(0.0166) \text{ rad}} = 240{,}000 \text{ mi}$$

Notice that a parallax angle of 0.95° is large enough to be easily measured using ordinary equipment.

Example 42
How Large Is the Moon?

Now that we know how far the moon is from the earth, we can use this distance to find its size. If we look at the moon, it subtends an angle of approximately $\frac{1}{2}°$,* which allows us to use the small-angle approximation. Applying this to Figure 1.7-4 gives us $\theta = D_M/R$. Solving for the moon's diameter D_M gives us

$$D_M = R\theta$$

We know from the previous example, that $R = 240{,}000$ mi and that $\theta \approx \frac{1}{2}°$, which we must convert to radians.

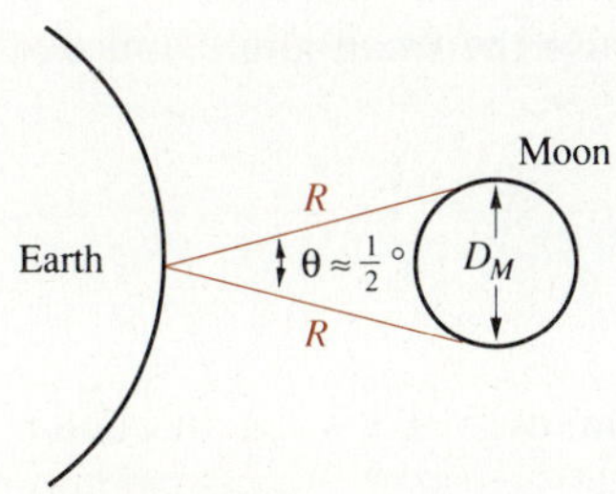

Figure 1.7-4

The size of the moon The size of θ is greatly exaggerated.

$$\theta = \left(\frac{1°}{2}\right)\left(\frac{\pi \text{ rad}}{180°}\right) = 0.0087 \text{ rad}$$

Therefore,

$$D_M = R\theta = (240{,}000 \text{ mi})\ (0.0087 \text{ rad}) = 2100 \text{ mi} \ ■$$

This answer is slightly inaccurate because the $\frac{1}{2}°$ was only approximate. Using more precise figures, the actual diameter is 2160 mi.

We can calculate the sun's distance and size using exactly the same methods we used for the moon.

* See Exercise 12 in Section 1.6 for a way that you yourself can easily measure this angle without any special equipment.

Example 43

How Far Are the Stars?

In principle, geocentric parallax can be used to find the distances to other stars. Attempts to do this, however, have detected no measurable parallax for the stars, which means that the parallax angle is too small to measure with present-day equipment. Therefore, the stars must be extremely far away compared to the diameter of the earth. To increase the parallax angle to a measurable value, we need a greater separation between the two observations. We would have to make A and B, in Figure 1.7-3, farther apart. Although we cannot make the earth larger, we can take advantage of the fact that the earth is in motion about the sun in a nearly circular orbit approximately 93 million miles in radius (Figure 1.7-5). For the nearer stars, this motion produces measurable parallax because of the large size of the orbit. Because this parallax is due to motion around our sun, we call it *heliocentric parallax*.

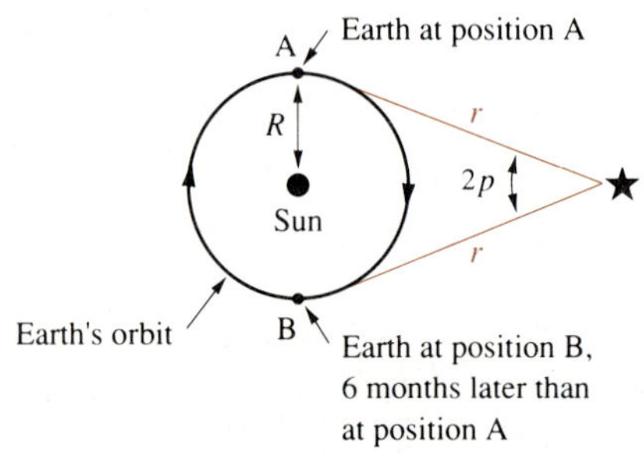

Figure 1.7-5

Distance to the stars The size of p is greatly exaggerated.

Because parallax angles for the stars are *very small* (ask yourself how you could know this from personal experience), we can use the small-angle approximation in Figure 1.7-5 to give us

$$2p = \frac{2R}{r} \rightarrow p = \frac{R}{r}$$

where R is the radius of the earth's orbit (93 million miles $= 9.3 \times 10^7$ mi) and r is the distance to the star.

The measured parallax for our nearest star, Alpha Centauri, is $0.765''$, which we must first convert to radians (**why?**).

$$p = (0.765'')\left(\frac{1'}{60''}\right)\left(\frac{1^\circ}{60'}\right)\left(\frac{\pi \text{ rad}}{180^\circ}\right)$$

$$p = 3.71 \times 10^{-6} \text{ rad}$$

Now we can calculate the distance r to the star.

$$r = \frac{R}{p} = \frac{9.3 \times 10^7 \text{ mi}}{3.71 \times 10^{-6} \text{ rad}} = 2.5 \times 10^{13} \text{ mi}$$

Such distances are so huge that astronomers express them in a different unit, the *light-year*, which is the distance that light travels in one year. This distance turns out to be 5.88×10^{12} mi. Converting our answer to light-years gives

$$d = (2.5 \times 10^{13} \text{ mi})\left(\frac{1 \text{ light-year}}{5.88 \times 10^{12} \text{ mi}}\right) = 4.3 \text{ light-years} \blacksquare$$

You have just seen some samples of the power of a little bit of mathematics coupled with a few measurements. As you go through this book, you will see even more of the value of basic mathematics.

EXERCISES 1.7

1. Suppose that Eratosthenes had mismeasured the angle of the sun with the pole in Alexandria to be 10.0° instead of $7\frac{1}{2}°$. What value would he have calculated for the earth's radius? By what percent would his result be in error?

2. If Alexandria were 1200 mi north of Syene instead of 500 mi, at what angle would Eratosthenes have observed the sunlight striking the vertical pole at noon on June 21? ***Hint:*** Make a clear sketch.

3. If the distance between Alexandria and Syene had been mismeasured to be 400 mi rather than 500 mi, what value would Eratosthenes have found for the earth's radius? By what percent would his result be in error?

4. What is the difference in latitude between Syene and Alexandria?

****5.** It may appear that Eratosthenes was lucky that Syene just happened to lie at a latitude for which the sun's rays would go straight down the well. However, his method works even without this special case. Suppose that at city A the sun's June 21 noon rays hit a vertical pole at 10.0° and at city B, 620 mi to the north, they hit a vertical pole at 19°. Make a clear sketch, like that in Example 40, showing these rays and angles. Then calculate the value these data give for the earth's radius.

6.** If the sun's rays make a 12° angle with a vertical pole at city X, what angle would they make with a similar pole in city Y, 700.0 mi directly north? ***Hint: Make a clear sketch.

7. How Far Away Is the Sun? The geocentric parallax for a point on the sun, such as a sunspot, has been experimentally measured to be 0.147′. How far is the sun from the earth?

8. How Large Is the Sun? The sun subtends an angle of approximately $\frac{1}{2}°$, just as the moon does. How large is the sun?

***9.** The following maximum geocentric parallax angles have been measured for Mars and Jupiter: 8.15×10^{-5} rad for Mars and 1.01×10^{-5} rad for Jupiter. Use this information to calculate the radius of the orbit of each of these planets.

10. From earth, the following maximum angular diameters have been measured for Mars and Jupiter: 38.1″ for Jupiter and 18″ for Mars. Using this information, as well as the results of Exercise 9, find the diameter of each planet.

11. What would be the *geocentric* parallax angle for our nearest star, Alpha Centauri? (See Example 43.)

12. In the future we will no doubt establish observatories on our moon. Then we can use geocentric parallax with a much greater separation than the earth's diameter. For a lunar-based telescope, find the geocentric parallax for our sun and Alpha Centauri.

13. Jupiter (Jove) is 11.2 times the diameter of the earth and 5.19 times as far from the sun. If we lived on this giant planet (a neat trick, since it has no surface!), what would be the "Joviocentric" parallax of our sun? Of Alpha Centauri?

14. Jupiter's satellite Io is 1820 km in radius and orbits 422,000 km from Jupiter, center-to-center. Jupiter's radius is 70,850 km.

(a) Approximately how large, compared to our moon's angular diameter as seen from earth, would Jupiter appear from Io? Is the small-angle approximation very accurate here? Why?

(b) How large would Io appear from Jupiter, compared to our moon's angular diameter from earth? Is the small-angle approximation very accurate here? Why?

15. Mars' satellite Phobos is 140 km in radius and orbits 9380 km from Mars, center-to-center. Mars' radius is 3400 km.

(a) How large would Phobos appear from Mars, compared to our moon's angular diameter from earth? Is the small-angle approximation very accurate here? Why?

(b) What would be the approximate angular diameter of Mars as viewed from Phobos? Is the small-angle approximation very accurate here? Why?

16. The first star whose heliocentric parallax was measured was 61 Cygni in 1838. This was 1.42×10^{-6} rad. How far away is 61 Cygni?

***17.** If a star, as viewed from earth, appears to change its position by 0.22″ over the course of a year, how far away from us in this star?

18. The Parsec. The *parsec*, a distance unit commonly used in astronomy, is defined as the distance a star (or any object) would have to be from us to have a heliocentric parallax angle of 1″. (The name *parsec* (ps) comes from *par*allax of 1 *sec*ond.) How many miles and light-years are in 1 parsec?

19. What is the heliocentric parallax of the star Fomalhaut, 7.0 parsecs away? (See Exercise 18.)

***20.** Suppose that the parsec were defined in its original way, *except* that the parallax was geocentric instead of heliocentric. We could call this new parsec the *geoparsec* for *geo*centric *par*allax of 1 *sec*ond. How many miles would there be in a geoparsec? How many geoparsecs in a parsec?

21. Proper Motion of Stars. Astronomers call the motion of a star across the sky (perpendicular to the line of sight) the *proper motion* and express it in angle per year. (Notice that this motion is just angular speed.) For nearby stars, typical proper motions are a few seconds of arc per year. (These are the motions *after* we correct for the earth's motion around the sun, and they are due to the actual motion of the stars through space.)

(a) Barnard's Star, 5.90 light-years away, has a proper motion of 10.3″/yr. How fast is this star moving in miles per year? In miles per second?
(b) If the red giant Betelgeuse, 520 light-years away, had the same *linear* speed as Barnard's star, what would be its proper motion in seconds per year?
(c) Sirius, 8.6 light-years away, moves 1.32″/yr. How many miles does it move in a year?

***22.** When it is finally launched, the Hubble space telescope should be able to resolve angles as small as 0.0010″ into clear images.

(a) What is the farthest star whose distance this telescope can measure by heliocentric parallax?
(b) How close would a star of our sun's size have to be for this telescope to be able to resolve it into a clear photograph? Are there any stars this close?

23. Comets, such as Halley's comet, are believed to have rocky nuclei about 5 mi across. Could the Hubble space telescope (see Exercise 22) resolve a comet nucleus if the comet came within 10 million miles of earth (close for a comet)?

24. If we put the Hubble space telescope (see Exercise 22) into orbit around Jupiter rather than earth, it would be 5.2 times farther from the sun. What would be the farthest star whose distance it could measure by heliocentric parallax?

25. Using the maximum resolving limit of the Hubble space telescope, (see Exercise 22), what is the farthest star whose distance we could measure by *geocentric* parallax? Are there any stars this close?

26. Figure 1.7-6 shows a 1915 photo of the double star system Kruger-60. The two stars close to each other in the upper left-hand corner are the ones in question. This system is 12.8 light-years from us, and the angular separation of the stars is 6.76×10^{-4} degrees. How far apart are these two stars? Compare this distance with the earth's distance from the sun.

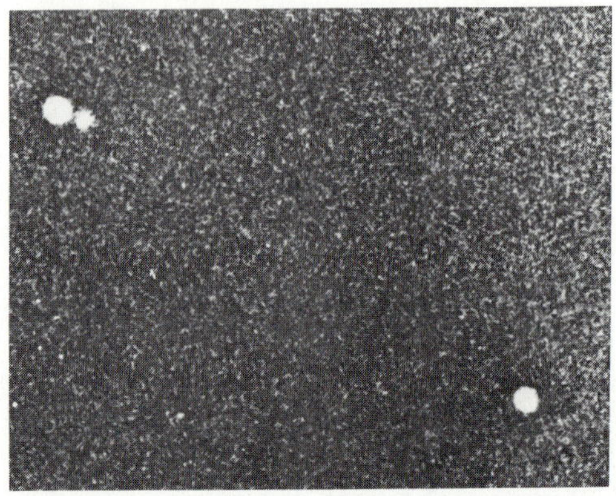

Figure 1.7-6 **Kruger-60**

27. The giant Hercules globular star cluster, M-13, is shown in Figure 1.7-7. It contains more than 100,000 stars, has an angular diameter of 23′, and is 22,200 light-years from us. How many light-years across is it?

Figure 1.7-7 ***Hercules Cluster***

***28.** Astronomers, when comparing photos of the supernova remnant known as the Crab Nebula (see Figure 1.7-8), noticed that between 1936 and 1986 it grew in angular diameter by 9′ of arc. Its spectral lines tell us that it is moving 10,000 km/s. How far away is this nebula?

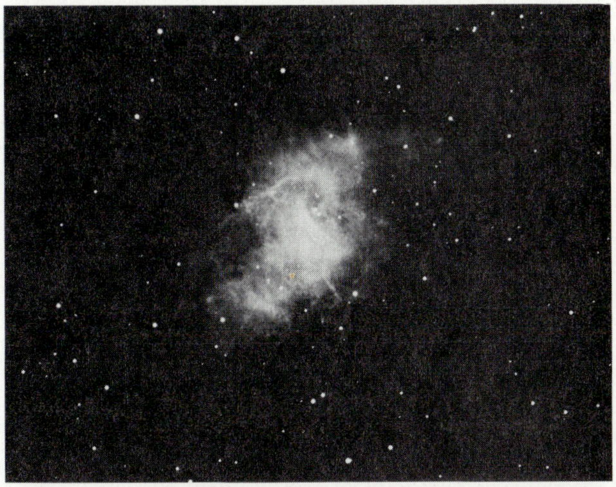

***Figure 1.7-8* Crab Nebula**

***29.** A certain supernova shoots off gas in all directions at a speed of 7250 mi/s. If this exploding star is 511 light-years from us, by what angle will its image grow in 25.0 yr?

30. The United States is planning a huge radiotelescope consisting of 10 antennas located over the whole country. It will be called the Very Long-Base-Line Array (VLBA) and is expected to be ready by the mid-1990s.* It should be able to resolve angles approximately as small as a ten-thousandth of a second of arc (0.0001″). What is the smallest feature on our nearest star, Alpha Centauri, that can be resolved by this telescope? How does this compare to the size of sunspots, which are about 10,000 mi across?

****31.** In 1979 a supernova (exploding star) occurred in the Virgo cluster. Radioastronomy measurements show that this star is growing in size at an angular rate of 0.003″/yr, while optical spectra give its speed of expansion at 11,000 km/s. How many light-years is this supernova from earth?

* For a good discussion of this telescope, see "The Very-Long-Baseline Array" by K. I. Kellermann and A. R. Thompson in the January 1988 *Scientific American*.

1.8 THE PATH OF π

- The circumference of a circle of diameter d is πd.
- The area of a circle of radius r is πr^2.
- The volume of a sphere of radius r is $\frac{4}{3}\pi r^3$.
- The time for one complete swing of a pendulum of length L is equal to $2\pi\sqrt{L/g}$ where g is the acceleration of a freely falling object.
- Rule a large sheet with equidistant parallel lines, a units apart. If you drop a thin needle of length $L \leq a$ at random onto this sheet, the probability p that it will intersect one of the lines is $p = 2L/(\pi a)$.

  And so on, and so on. . . .

It would be impossible to list all the places in mathematics and science where the number π is used, or to overestimate the importance of π in these areas. Few numbers, with the possible exception of $e = 2.718\ldots$ and $i = \sqrt{-1}$, play such an important role in mathematics. (Interestingly, the numbers π, e, and i can be related by the equation $e^{\pi i} + 1 = 0$.)

Although we do not know exactly who first studied π, in this section we attempt to outline several important events in its history.

The people who first studied geometry noticed that the ratio of the circumference of a circle to its diameter does not depend on the size of the circle; it is a constant for all circles. We call this constant π.

$$\pi = \frac{\text{circumference of circle}}{\text{diameter of circle}}$$

In ancient Babylonia (approximately 2,000–1,600 B.C.), people calculated the circumference of a circle by taking three times its diameter. Because π is the ratio of the circumference to the diameter of a circle, they used $\pi = 3$.

A great deal of our knowledge of ancient mathematics, including the number π, comes from the Rhind Papyrus of ancient Egypt. One of the geometric problems in this papyrus is the method the Egyptians used to find the area of a circle:

> Example of a round field of diameter 9 khet. What is its area? Take away $\frac{1}{9}$ of the diameter, namely 1; the remainder is 8. Multiply 8 times 8; it makes 64. Therefore it contains 64 setat of land.

Using modern-day algebraic symbols, we can write this problem as

$$A = \left(d - \frac{d}{9}\right)^2 = \left(\frac{8d}{9}\right)^2 = \left(\frac{64}{81}\right)d^2$$

where A is the area of the circle and d the diameter. But the area of a circle is πr^2 and $r = d/2$, so

$$A = \pi r^2 = \pi\left(\frac{d}{2}\right)^2 = \frac{1}{4}\pi d^2$$

Equating these two expressions for the area gives $\pi = 3.16049\ldots$.

A value for π also appears in the Old Testament, in I Kings 7:23, where the dimensions of a bath in the temple of Solomon are given. The Talmud, the collection of Jewish law and tradition, also gives a value for π.

As surprising at it may seem, even the Great Pyramid enters the story of π. The pyramid mystics have claimed that the Great Pyramid was constructed so that half of the perimeter of its base divided by its height should be equal to π. Using the dimensions of the Great Pyramid gives us

$$\pi = \left(\frac{1}{2}\right)\left(\frac{3023.12 \text{ ft}}{481.2 \text{ ft}}\right) = 3.141230258$$

Although this value is very close to the value of π, it is believed that this is probably accidental. (See Exercise 19 for a reasonable nonmystic explanation.)

The first scientific attempt to compute π seems to have started with the Greeks, especially Archimedes.

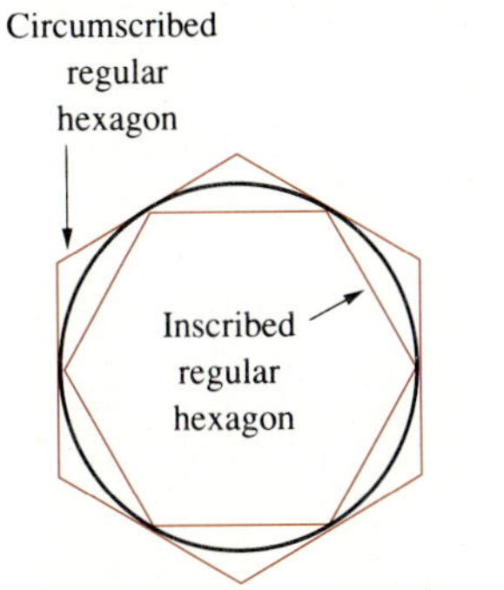

Figure 1.8-1

Inscribed and circumscribed polygons

1. The Early Greeks. **Archimedes of Syracuse (287–212 B.C.).** Greek mathematicians recognized that the ratio of the circumference of a circle to its diameter is the same, regardless of the size of the circle, and they were very interested in determining the value of π.* In his work *Measurement of a Circle*, Archimedes proved that π is between $3\frac{10}{71}$ and $3\frac{1}{7}$. He did this by approximating the circumference of a circle by the perimeters of polygons. It works like this: Choose a circle and construct inscribed and circumscribed regular (having sides of equal length) polygons, as shown in Figure 1.8-1. Since we can compute the perimeters of regular polygons algebraically, we can find *bounds* for π because the circle's circumference lies between the perimeters of these two polygons.

$$\text{perimeter of inner polygon} < 2\pi r < \text{perimeter of outer polygon}$$

At various times, Archimedes used polygons of 6, 12, 24, 48, and 96 sides to obtain bounds of $\frac{223}{71}$ and $\frac{22}{7}$ ($3\frac{10}{71}$ and $3\frac{1}{7}$). By averaging these bounds, he calculated $\pi = 3.14185$. This method, is called the method of exhaustion because, as you take polygons of more and more sides, the area between them gets smaller and smaller, or is exhausted.

Claudius Ptolemy of Alexandria (approximately 85–165 B.C.). The great Greek astronomer Ptolemy, in his famous *Syntaxis Mathematica* (better known by its Arabic title, *Almagest*), gave π the value 3 8′30″, which is *sexagesimal* notation† for 3.14167. Ptolemy used chords of angles, and Book I of the *Almagest* contains a table giving the lengths of the chords of central angles in a circle of radius 60 units, increasing by half a degree at a time from $\frac{1}{2}°$ to 180°. To find the value of π, Ptolemy multiplied the length of the chord of the 1° central angle by 360 and divided this result by the length of the diameter of the circle. Notice that Ptolemy was actually using the small-angle approximation here.

2. Chinese Mathematicians. The Han dynasty in China, from A.D. 206 to 222, saw the revival of literature and the introduction of Buddhism. The post-Han period extended to about A.D. 600. During this time, a number of people calculated π. **Wang Fan**, a general, found an approximation $\pi = \frac{142}{46} = 3.155$. In *Sea Island Mathematical Manual*, the mathematician **Liu Hui**, presented the relationship $3.1410 < \pi < 3.1427$. **Tsu Ch'ung-chih** (A.D. 430–501), an engineer, found the relationship $3.1415926 < \pi < 3.1415927$ and the rational approximation $\frac{355}{113}$, which gives π correct to six decimal places.

3. Indian Mathematicians. Most of the Indian computations for π were done by **Aryabhata (approximately A.D. 476–550), Brahmagupta (approximately A.D.**

---

* Although we use the symbol π here, the Greeks never used such a symbol. It was introduced by the Swiss mathematician Leonhard Euler in 1739.

† In sexagesimal notation, this means $3 + \frac{8}{60} + 30/[(60)(60)]$.

600), and **Bhaskara** (**A.D. 1114–1185**). Aryabhata, an astronomer, computed the value of π using the formula $\pi = [8(100 + 4) + 62{,}000]/20{,}000 = 3.1416$. Brahmagupta gave $\sqrt{10}$ as an approximation for π. Later Bhaskara, a mathematician, gave several values of π: $\pi = 3927/1250$ and $\pi = 754/240$. It is interesting that he also gave $\frac{22}{7}$ as an "inaccurate" value.

In the early twentieth century, the brilliant mathematician **Srinivasa Ramanujan** developed methods of determining π that were later valuable for computer calculations.*

4. The Europeans. **Leonardo Pisano (approximately A.D. 1175–1250).** The Italian mathematician Pisano (better known as Fibonacci) found bounds for the value of π,

$$\frac{1440}{458\frac{4}{9}} < \pi < \frac{1440}{458\frac{1}{5}}$$

Ludolph van Ceulen (1540–1610). In 1610 van Ceulen used Archimedes' method of exhaustion to find π to 33 decimal places using 2^{62} sides for his polygon! This later became known as the Ludolphine number.

The seventeenth century was one of the most prolific periods in mathematics. The development of calculus and analytic geometry made it possible to calculate the area of regions bounded by curves other than the circle. Surprisingly, mathematicians found that these areas could be expressed in terms of π. Thus *π was no longer restricted to the circle.*

John Wallis (1616–1703). Wallis, a British mathematician, tried to determine the value of π by finding an expression for the arc length of a quadrant of the unit circle (of radius 1). He failed because he did not know the general binomial theorem, but instead he derived an *infinite product* expression for π.

$$\frac{4}{\pi} = \frac{3 \times 3 \times 5 \times 5 \times 7 \times 7 \times 9 \times 9 \times \ldots}{2 \times 4 \times 4 \times 6 \times 6 \times 8 \times 8 \times 10 \times 10 \times \ldots}$$

James Gregory (1638–1675). In 1671 this Scottish mathematician found an alternating infinite series for $\frac{\pi}{4}$.

$$\frac{\pi}{4} = 1 - \frac{1}{3} + \frac{1}{5} - \frac{1}{7} + \cdots$$

Gottfried Wilhelm Leibnitz (1646–1716). The great German mathematician Leibnitz, co-inventor of calculus, found the same alternating series for $\frac{\pi}{4}$ as Gregory, but did so independently of Gregory. Isaac Newton, the other creator of calculus, also showed some interest in computing π.

Leonhard Euler (1707–1738). In his great 1748 textbook, *Introductio in Analysin Infinitorum*, Euler *approved* of the Greek letter π to represent the ratio of the circumference to the diameter of a circle, but he was not the first to use

* See "Ramanujan and Pi" in the February 1988 *Scientific American* for a fascinating account of this unconventional mathematician. A shorter article appears in the April 25, 1987 *Science News*.

this symbol. The English writer William Jones used it in 1706, and William Oughtred used it to represent the *circumference* of a circle as early as 1647. (Interestingly, π is the first letter in the Greek word for *periphery*, which may be why people first thought to use it as the ratio of the circumference—periphery—of a circle to its diameter.)

Johann Heinrich Lambert proved in 1767 that π is irrational, and **Adrien-Marie Legendre** showed in 1794 that π^2 is irrational. (A number is irrational if it cannot be expressed as the ratio of two integers.)

George Louis Leclerc, Comte de Buffon presented his famous needle problem in 1777; it used a probability method to compute π. (Exercise 13)

William Rutherford. In 1841 Rutherford used Gregory's series to compute π to 208 decimal places, but he later found that the calculation was correct to only 152 places. In 1853 he computed π again, this time obtaining 400 correct decimal places.

William Shanks, of England, computed π to 707 decimal places in 1873. It took him 15 years of calculation, and later an error was found in the 528th place.

Ferdinand Lindermann proved in 1882 that π is transcendental.* This means that the circle could not be "squared." That is, it is impossible to construct, using only compass and straightedge, a square whose area is *exactly* equal to the area of a given circle. Before this, many people spent many years trying to square the circle!

5. Enter the Computer. In 1949 scientists at the Army Ballistic Research Laboratories in Aberdeen, Maryland, calculated π to 2,036 decimal places using the ENIAC computer. In 1954 they extended this to 3,093 decimal places. In 1958 in Paris **Francois Genuys** computed π to 10,000 decimal places using an IBM-704 computer, and he extended this to 16,167 decimal places the following year. In 1961 **Daniel Shanks** and his colleague **Wrench** used an IBM-7090 computer to calculate π to 100,265 decimal places. Five years later, **M. Jean Guilloud** and his colleagues at the Commissariot a l' Energie Atomique in Paris computed π to 250,000 decimal places using the STRETCH computer, and the following year they extended this to 500,000 places on a CDC-6600 computer. Using the Cray-2 supercomputer in 1986 **David H. Bailey** at the NASA Ames Research Center in California calculated π to 29,360,128 digits. Less than a year later, **Yasumasa Kanada** at the University of Tokyo used an NEC SX-2 supercomputer to calculate π to more than 134 million digits. In 1988, he extended this to 201,326,000 decimal places using a new Hitachi supercomputer, and he hopes to reach 400 million digits† in 1989.‡

---

* A number is called *algebraic* if it is a root of some polynomial having rational coefficients. If a number is not algebraic, it is called *transcendental.*

† You may wonder why anyone needs to know π to so many digits. There are mathematical reasons that we won't go into here, but for a highly fanciful reason, see Carl Sagan's 1985 novel *Contact.*

‡ Two excellent sources of information on the history of π are H. Eves, *Introduction to the History of Mathematics* (1983); New York: Holt, Rinehart, and Winston, and the three-part article by H. C. Schepler, "The Chronology of Pi," *Mathematics Magazine*, January–February, March–April, and May–June, 1950. Both sources, from which we have drawn heavily, require only a basic knowledge of trigonometry. A very thorough but much more advanced treatment is the two-part article by Dario Castellanos, "The Ubiquitous Pi," *Mathematics Magazine*, April and June 1988.

Table 1.8-1

π to 5000 decimal places. **Can anyone spot any errors???**

1415926535	8979323846	2643383279	5028841971	6939937510	5820974944	5923078164	0628620899	8628034825	3421170679
8214808651	3282306647	0938446095	5058223172	5359408128	4811174502	8410270193	8521105559	6446229489	5493038196
4428810975	6659334461	2847564823	3786783165	2712019091	4564856692	3460348610	4543266482	1339360726	0249141273
7245870066	0631558817	4881520920	9628292540	9171536436	7892590360	0113305305	4882046652	1384146951	9415116094
3305727036	5759591953	0921861173	8193261179	3105118548	0744623799	6274956735	1885752724	8912279381	8301194912
9833673362	4406566430	8602139494	6395224737	1907021798	6094370277	0539217176	2931767523	8467481846	7669405132
0005681271	4526356082	7785771342	7577896091	7363717872	1468440901	2249534301	4654958537	1050792279	6892589235
4201995611	2129021960	8640344181	5981362977	4771309960	5187072113	4999999837	2978049951	0597317328	1609631859
5024459455	3469083026	4252230825	3344685035	2619311881	7101000313	7838752886	5875332083	8142061717	7669147303
5982534904	2875546873	1159562863	8823537875	9375195778	1857780532	1712268066	1300192787	6611195909	2164201989
3809525720	1065485863	2788659361	5338182796	8230301952	0353018529	6899577362	2599413891	2497217752	8347913151
5574857242	4541506959	5082953311	6861727855	8890750983	8175463746	4939319255	0604009277	0167113900	9848824012
8583616035	6370766010	4710181942	9555961989	4676783744	9448255379	7747268471	0404753464	6208046684	2590694912
9331367702	8989152104	7521620569	6602405803	8150193511	2533824300	3558764024	7496473263	9141992726	0426992279
6782354781	6360093417	2164121992	4586315030	2861829745	5570674983	8505494588	5869269956	9092721079	7509302955
3211653449	8720275596	0236480665	4991198818	3479775356	6369807426	5425278625	5181841757	4672890977	7727938000
8164706001	6145249192	1732172147	7235014144	1973568548	1613611573	5255213347	5741849468	4385233239	0739414333
4547762416	8625189835	6948556209	9219222184	2725502542	5688767179	0494601653	4668049886	2723279178	6085784383
8279679766	8145410095	3883786360	9506800642	2512520511	7392984896	0841284886	2694560424	1965285022	2106611863
0674427862	2039194945	0471237137	8696095636	4371917287	4677646575	7396241389	0865832645	9958133904	7802759009
9465764078	9512694683	9835259570	9825822620	5224894077	2671947826	8482601476	9909026401	3639443745	5305068203
4962524517	4939965143	1429809190	6592509372	2169646151	5709858387	4105978859	5977297549	8930161753	9284681382
6868386894	2774155991	8559252459	5395943104	9972524680	8459872736	4469584865	3836736222	6260991246	0805124388
4390451244	1365497627	8079771569	1435997700	1296160894	4169486855	5848406353	4220722258	2848864815	8456028506
0168427394	5226746767	8895252138	5225499546	6672782398	6456596116	3548862305	7745649803	5593634568	1743241125
1507606947	9451096596	0940252288	7971089314	5669136867	2287489405	6010150330	8617928680	9208747609	1782493858
9009714909	6759852613	6554978189	3129784821	6829989487	2265880485	7564014270	4775551323	7964145152	3746234364
5428584447	9526586782	1051141354	7357395231	1342716610	2135969536	2314429524	8493718711	0145765403	5902799344
0374200731	0578539062	1983874478	0847848968	3321445713	8687519435	0643021845	3191048481	0053706146	8067491927
8191197939	9520614196	6342875444	0643745123	7181921799	9839101591	9561814675	1426912397	4894090718	6494231961
5679452080	9514655022	5231603881	9301420937	6213785595	6638937787	0830390697	9207734672	2182562599	6615014215
0306803844	7734549202	6054146659	2520149744	2850732518	6660021324	3408819071	0486331734	6496514539	0579626856
1005508106	6587969981	6357473638	4052571459	1028970641	4011097120	6280439039	7595156771	5770042033	7869936007
2305587631	7635942187	3125147120	5329281918	2618612586	7321579198	4148488291	6447060957	5270695722	0917567116
7229109816	9091528017	3506712748	5832228718	3520935396	5725121083	5791513698	8209144421	0067510334	6711031412
6711136990	8658516398	3150197016	5151168517	1437657618	3515565088	4909989859	9823873455	2833163550	7647918535
8932261854	8963213293	3089857064	2046752590	7091548141	6549859461	6371802709	8199430992	4488957571	2828905923
2332609729	9712084433	5732654893	8239119325	9746366730	5836041428	1388303203	8249037589	8524374417	0291327656
1809377344	4030707469	2112019130	2033038019	7621101100	4492932151	6084244485	9637669838	9522868478	3123552658
2131449576	8572624334	4189303968	6426243410	7732269780	2807318915	4411010446	8232527162	0105265227	2111660396
6655730925	4711055785	3763466820	6531098965	2691862056	4769312570	5863566201	8558100729	3606598764	8611791045
3348850346	1136576867	5324944166	8039626579	7877185560	8455296541	2665408530	6143444318	5867697514	5661406800
7002378776	5913440171	2749470420	5622305389	9456131407	1127000407	8547332699	3908145466	4645880797	2708266830
6343285878	5698305235	8089330657	5740679545	7163775254	2021149557	6158140025	0126228594	1302164715	5097925923
0990796547	3761255176	5675135751	7829666454	7791745011	2996148903	0463994713	2962107340	4375189573	5961458901
9389713111	7904297828	5647503203	1986915140	2870808599	0480109412	1472213179	4764777262	2414254854	5403321571
8530614228	8137585043	0633217518	2979866223	7172159160	7716692547	4873898665	4949450114	6540628433	6639379003
9769265672	1463853067	3609657120	9180763832	7166416274	8888007869	2560290228	4721040317	2118608204	1900042296
6171196377	9213375751	1495950156	6049631862	9472654736	4252308177	0367515906	7350235072	8354056704	0386743513
6222247715	8915049530	98444489333	0963408780	7693259939	7805419341	4473774418	4263129860	8099888687	4132604721

EXERCISES 1.8

***1.** The Rhind Papyrus from ancient Egypt is the oldest known mathematical document. It dates from about 2000 B.C. and may be from an even earlier work going back to about 3400 B.C. This papyrus gives a rule for constructing a square with the same area as a given circle: "Cut $\frac{1}{9}$ off the circle's diameter and construct a square on the remainder." Show that this leads to a value for π of $(\frac{16}{9})^2$. How accurate is this value of π?

2. (a) π in the Bible. In the Bible, a passage in I Kings 7:23 states: "And he made a molten sea, 10 cubits from the one brim to the other; it was round all about . . . and a lake of 30 cubits did compass it round about." Essentially the same thing is contained in II Chronicles 4:2.

(b) π in the Talmud. The Talmud states "that which measures 3 units in circumference is one length across." What value of π does each of these passages imply?

***3. Legislating the Value of π.** In 1897 the Indiana State Legislature attempted to legislate the value of π. House bill No. 246 stated, in part: "Be it enacted by the General Assembly of the State of Indiana: It has been found that a circular area is to the square on a line equal to the quadrant of the circumference, as the area of an equilateral rectangle is to the square on one side . . . ." (And you think math books are hard to understand!) This bill was passed by the House, but not by the Senate because of ridicule in the newspapers. Why do you think the newspapers ridiculed it? What value does it legislate for π?

4. Numerical Approximations for π. Many people at various times established approximations to be used for π. For each approximation that follows, express it in decimal form to see to how many figures it is accurate.

(a) Hero of Alexandria (approximately 100 B.C.), a Greek surveyor, used 3 for very rough work and $3\frac{1}{7}$ for practical measurement. Many other people also used this value of $3\frac{1}{7}$, or $\frac{22}{7}$.

(b) Vitruvius (approximately 20 B.C.), a Roman architect and engineer, used $3\frac{1}{8}$.

(c) Ch'ang Höng (approximately A.D. 125), a Chinese court astrologer, used $\sqrt{10}$.

(d) Ptolemy (approximately A.D. 150), a Greek astronomer, used a value in sexagesimal (base 60) form because it was reintroduced to Europe by the Arabs. He used $\pi = 3 + \frac{8}{60} + \frac{30}{60^2}$.

(e) Liu Hui (approximately A.D. 263), a Chinese engineer, used $\frac{22}{7}$ as the "inaccurate" value. For the "accurate" value, he formed the sequence 113355 and then divided the last three digits by the first three to obtain $\pi = \frac{355}{113}$. (It is not known how he arrived at this method.) How accurate were his "inaccurate" and "accurate" values?

(f) Arya-Bhata (approximately A.D. 480), a Hindu mathematician, used Archimedes' polygon method, with 384 sides, to obtain the value $\pi = 62{,}832/20{,}000$. He also used $\pi = 3\ 177/1250$.

(g) Muhammed Ibn al-Khowarizmi (approximately A.D. 825), an Arabian astronomer and mathematician, gave $\frac{22}{7}$ as an approximate value, $\sqrt{10}$ for geometers, and 62,832/20,000 for astronomers. Which of these three got the most accurate value of π?

(h) Nicolaus de Cusa (1464), a German physicist, geometer, astronomer, and theologian, used $\pi = \frac{3}{4}(\sqrt{3} + \sqrt{6})$.

(i) Tycho Brahe (approximately 1530), a Danish astronomer, used a geometric construction to get

$$\pi = \left[9^2 + \left(\frac{19^2}{22}\right)\right]^{1/4}$$

(j) Carl Heisel (1934), an American, proposed $\pi = 3\frac{13}{81}$. Have calculations always improved through the years?

***5. Polygon Methods.** The value of π can be estimated in the following way: Inscribe a regular polygon inside a circle and another regular polygon (with the same number of sides) around the circle. The circumference of this circle can be no less than the perimeter of the inner polygon and no greater than the perimeter of the outer polygon.

$$\text{perimeter of inner polygon} < \pi d < \text{perimeter of outer polygon}$$

where d is the diameter of the circle. This equation leads to upper and lower limits on π. To estimate π from these limits, we could just average the two values. Let's try this using a square as the regular polygon, as shown in Figure 1.8-2.

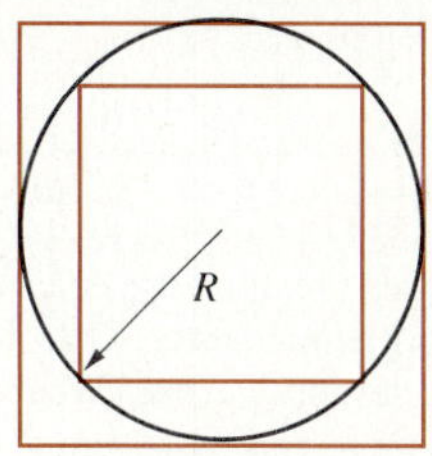

Figure 1.8-2

(a) Show that the outer square has perimeter $8R$.

(b) Show that the inner square has perimeter $4R\sqrt{2}$. (This requires use of the Pythagorean theorem, which we cover in Chapter 2. You may remember this theorem from an earlier math course or you may want to read ahead in Section 2.1.)

(c) Show that the limits on the circumference of the circle lead to the following limits for π: $2\sqrt{2} < \pi < 4$. Average these two values to get π. Notice that even this very crude method leads to a reasonably accurate value for π.

6. (a) Archimedes, the great Greek mathematician, engineer, architect, physicist, . . . , was the first known person to use the method of polygons discussed in Exercise 5. In around 240

B.C. using polygons of 96 sides, he found that $3\frac{1}{7} < \pi < 3\frac{10}{71}$. If you average these limits, what value do you get for π? To how many decimal places was Archimedes accurate?
(b) In around A.D. 1220 the Italian mathematician Leonardo Pisano (Fibonacci), using 96-sided polygons, found

$$\frac{1440}{458\frac{4}{9}} < \pi < \frac{1440}{458\frac{1}{5}}$$

Who got a more accurate value, Archimedes or Fibonacci?

Infinite Series and Expressions. **A number of infinite expressions have been worked out for π, many of which are derived using calculus. Theoretically the answers are exact only when you use an infinite number of terms, which is obviously impossible. But you can often get excellent results with just the first few terms, and the more terms you use, the more accurate the answer becomes. Most people who have calculated π to many decimal places did so using such series.**

***7.** Francois Vieta (approximately 1579), a French mathematician, was the first person to obtain an infinite series for π. He found

$$\frac{2}{\pi} = \sqrt{\tfrac{1}{2}}\,\sqrt{\tfrac{1}{2} + \tfrac{1}{2}\sqrt{\tfrac{1}{2}}}\,\sqrt{\tfrac{1}{2} + \tfrac{1}{2}\sqrt{\tfrac{1}{2} + \tfrac{1}{2}\sqrt{\tfrac{1}{2}}}}\cdots$$

(a) We have shown three factors here (each of the major square roots). What is the next factor?
(b) Evaluate π using: Only the first factor $(\sqrt{1/2})$; the first two factors; the first three factors; the first four factors. How accurate is each of these? If you feel super-energetic, evaluate the fifth factor and use all five to calculate π.

8. John Wallis (approximately 1650), a British mathematician, found the infinite expression

$$\frac{4}{\pi} = \frac{3\cdot 3\cdot 5\cdot 5\cdot 7\cdot 7\cdot 9\cdot 9\cdots}{2\cdot 4\cdot 4\cdot 6\cdot 6\cdot 8\cdot 8\cdot 10\cdots}$$

(a) In evaluating this expression, use equal numbers of terms in the numerator and denominator. Calculate π using: four terms; eight terms; twenty terms. What happens to the accuracy as you use more and more terms?
(b) Show that the preceding expression can be rewritten as

$$\frac{\pi}{2} = \frac{2\cdot 2\cdot 4\cdot 4\cdot 6\cdot 6\cdot 8\cdot 8\cdot 10\cdot 10\cdots}{1\cdot 3\cdot 3\cdot 5\cdot 5\cdot 7\cdot 7\cdot 9\cdot 9\cdot 11\cdot 11\cdots}$$

9. James Gregory (approximately 1668), a Scottish astronomer and mathematician, proved that

$$\frac{\pi}{4} = 1 - \frac{1}{3} + \frac{1}{5} - \frac{1}{7} + \frac{1}{9} - + \cdots$$

(a) Evaluate π by using the first: five terms; ten terms; twenty terms; let your patience be your guide!
(b) This series was shown to be equivalent to the following:

$$\frac{\pi}{4} = 1 - 2\left(\frac{1}{3\times 5} + \frac{1}{7\times 9} + \frac{1}{11\times 13} + \frac{1}{15\times 17} + \cdots\right)$$

Repeat part **(a)** using this equivalent series.

10. Abraham Sharp (approximately 1699), with instructions from Edmund Halley (of comet fame) calculated π to 71 correct decimal places using the series

$$\frac{\pi}{6} = \sqrt{\frac{1}{3}}\left(1 - \frac{1}{3\cdot 3} + \frac{1}{3^2\cdot 5} - \frac{1}{3^3\cdot 7} + \cdots\right)$$

Can you match Sharp's effort? Try using the first five, ten, and then fifteen terms.

***11. Weighing π.** Englishman William Baddeley in 1833 wrote: "From a piece of carefully rolled sheet brass was cut out a circle $1\frac{9}{10}$ inch in diameter, and a square $1\frac{7}{10}$ inch in diameter. On weighing them, they were found to be of exactly the same weight, which proves that, as each are of the same thickness, the surfaces must be precisely similar." What value does this imply for π?

***12.** In 1863 the Englishman S. M. Drach gave this awkward but highly accurate way to find the circumference of a circle: "From 3 diameters, deduct 8/1000 and 7/1,000,000 of a diameter, and add 5% to the result. This gives a length which is smaller than the circumference by about $1\frac{1}{60}$ inch in 14,000 miles." What value of π does this give?

***13. Measuring π by Dropping Sticks.** In 1777 Count de Buffon derived an intriguing way to calculate π: On a sheet of paper draw a series of parallel lines a distance D apart. Now take a straight stick of length L (less than or equal to D) and drop it at random on the paper. Buffon proved that the probability that this stick would lie across one of the lines is $P = 2L/\pi D$. (This proof requires calculus.) When $L = D$, the probability simplifies to $P = \frac{2}{\pi}$. The most accurate known calculation done this way was made by the Italian mathematician Lazzerini in 1901 using 3408 tosses of a needle, which gave a value of $\pi = 3.1415929$. **Your assignment:** Find π this way. Select arbitrary convenient values for D and L and use as many tosses as your patience will allow. (Note that you will have to calculate the probability *directly* using the total number of tosses and the number of tosses in which a line was crossed.) Anyone feel up to breaking Lazzerini's record???!

***14.** In 1868 Rev. Cyrus Pitt Grosvenor of New York published the following rule for finding the area of a circle: "Square the diameter of the circle; multiply the square by 2; extract the square root of the product; from the root subtract the diameter of the circle; square the remainder; multiply this square by $\frac{5}{4}$; subtract the product from the square of the diameter of the circle." Show that this wordy recipe is equivalent to the algebraic formula $A = D^2 - (\frac{5}{4})(\sqrt{2D^2} - D)^2$. Now use this formula to show that Grosvenor's value of π was $\pi = 4 - 5(\sqrt{2} - 1)^2$, and calculate π to see how accurate this approximation is.

****15.** In about A.D. 97 the Roman engineer Frontius stated: "The square digit is greater than the circle digit by $\frac{3}{14}$ of itself; the circle digit is less than the square digit by $\frac{3}{11}$." Interpret this statement.

What value of π does it imply? ***Hint:*** Think of a circle inscribed inside of a square. What does he mean by *digit*?

***16.** In about A.D. 850 the Indian mathematician Mahavira stated formulas for the volume of a sphere: The "approximate" expression was $V = (\frac{9}{2})(\frac{1}{2}d)^3$, whereas the "accurate" expression was $V = (\frac{9}{10})(\frac{9}{2})(\frac{1}{2}d)^3$, where d is the diameter. What value of π does each expression imply? How accurate is each one?

17. Remembering π. A *mnemonic* is any kind of trick to aid the memory. A number of mnemonics have been invented to help remember π to many decimal places. For each of the following mnemonics, see if you can figure out how it works and then write out the value of π that each one gives.

(a) Now I, even I, would celebrate
In rhymes inapt, the great
Immortal Syracusan, rivaled nevermore,
Who is his wondrous lore,
Passed on before
Left men his guidance
How to circles mensurate. (A. C. Orr, American, 1906)

(b) See I have a rhyme assisting my feeble brain its tasks sometimes resisting. (*Scientific American*, March 21, 1914)

(c) Sir, I bear a rhyme excelling
In mystic force and magic spelling
Celestial sprites elucidate
All my own striving can't relate.

(d) How I want a drink, alcoholic of course, after the heavy lectures involving quantum mechanics. (*Scientific American*)

(e) May I have a large container of coffee?

***(f)**
$$\frac{\text{calculator will get fair accuracy}}{\text{but not to } \pi \text{ exact}}$$

***(g)**
$$\frac{\text{dividing top lot through (a nightmare)}}{\text{by number below, you approach } \pi}$$

Do you feel creative enough to invent any mnemonics of your own?

18. Charles Piazzi Smyth (godson of Giuseppe Piazzi, discoverer of the first known asteroid) measured the Great Pyramid (Khufu) and found a niche in the Queen's Chamber that was 185 "pyramid inches" (1.001 times the ordinary inch) long. In his book, *Our Inheritance in the Great Pyramid*, Smyth made the following statement: "The length of the Grand Niche multiplied by 10π is equal to the height of the Great Pyramid."

(a) Test the accuracy of Smyth's statement, using the height of the Great Pyramid given earlier in this section.

(b) Smyth also said: "The square root of ten times the height of the north wall in the Queen's Chamber (which measures 182.4 "pyramid inches") divided by the length of the Grand Niche is equal to π." Check the accuracy of this statement.

****19.** Earlier in this section, we stated that the mystics believed that, for the Great Pyramid (Khufu), *half the perimeter of the base divided by the height was equal to* π. In fact, this belief is true to a high degree of precision. Let's see if there might be a plausible nonmystical reason.

Long horizontal distances cannot be layed out accurately with a rope because it would sag and stretch. But a wheel can be rolled easily. So horizontal distances can be specified as "rolls" of a standard wheel. Each roll, of course, would be one circumference. Vertical distances can be measured with a rope or a pole, and a likely unit is the diameter, or width, of the standard wheel. (Don't be surprised if people use different units for horizontal and vertical distances. We measure ground distances in miles but mountain heights in feet.)

If the Egyptians designed the Great Pyramid to slope so that the four faces rise four of their vertical units (widths) above the square base for every one of their horizontal units (rolls), show that this plan *automatically* makes half the perimeter of the base divided by the height equal to π when we measure in *our* standard units of feet, meters, and so on. Was it necessary for the Egyptians to have any knowledge of π for this result to be true?

Notice that this explanation is a *plausible* one rather than a mystical one. The Great Pyramid is not the only one to obey this "mystical" ratio. Any pyramid that slopes at an angle of 51°52′ will do so, and many of them do.

SUMMARY

1. Trigonometric Angle. For two rays from the same vertex, a trigonometric angle is the amount of rotation needed to turn one ray (the initial side) so that it coincides with the other ray (the terminal side). If the rotation is counterclockwise, the angle is called positive; if it is clockwise, the angle is negative (see Figure S1-1).

2. Coterminal Angles. Two trigonometric angles that have the same initial side and the same terminal side are called coterminal angles.

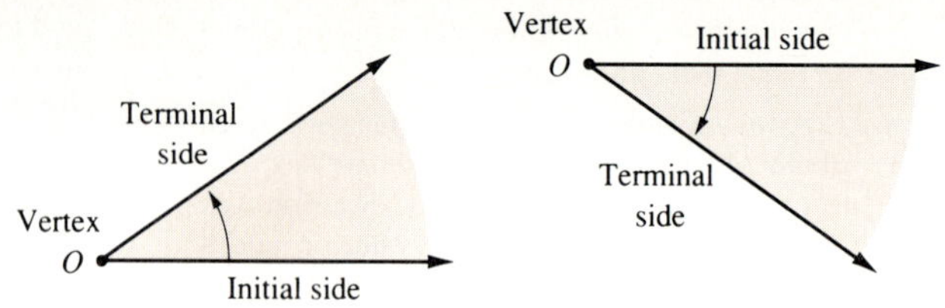

Figure S1-1 Positive angle Negative angle

3. Standard Position. A trigonometric angle is in standard position when it is drawn in a Cartesian coordinate system with its vertex at the origin and its initial side coinciding with the positive x axis.

4. Measure of an Angle. Angles are commonly measured in revolutions, degrees, and radians.

$$1 \text{ revolution} = 360° = 2\pi \text{ radians}$$

A *degree* is $\frac{1}{360}$ of a revolution and is written 1°. A degree can be subdivided into *minutes* (′) and *seconds* (″).

$$1° = 60 \text{ minutes (written } 60')$$

$$1' = 60 \text{ seconds (written } 60'')$$

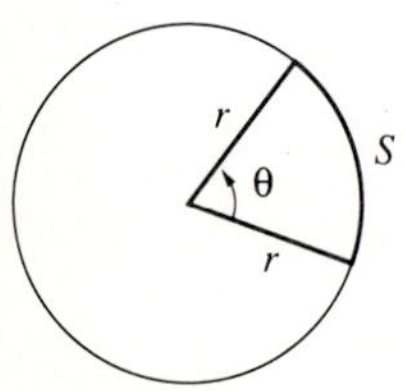

Figure S1-2
An angle in radians

A *radian* is the angle which, if its vertex is placed at the center of a circle, intercepts an arc equal in length to the radius of the circle. If a central angle θ intercepts an arc length S on a circle of radius r (see Figure S1-2), then θ in radians is

$$\theta = \frac{S}{r}$$

5. Degree-Radian Conversions

$$180° = \pi \text{ radians}$$

6. Acute Angle. An acute angle is a positive angle measuring less than 90°.

7. Obtuse Angle. An obtuse angle is a positive angle measuring more than 90° but less than 180°.

8. Complementary Angles. Complementary angles are two positive angles whose sum is 90°.

9. Supplementary Angles. Supplementary angles are two positive angles whose sum is 180°.

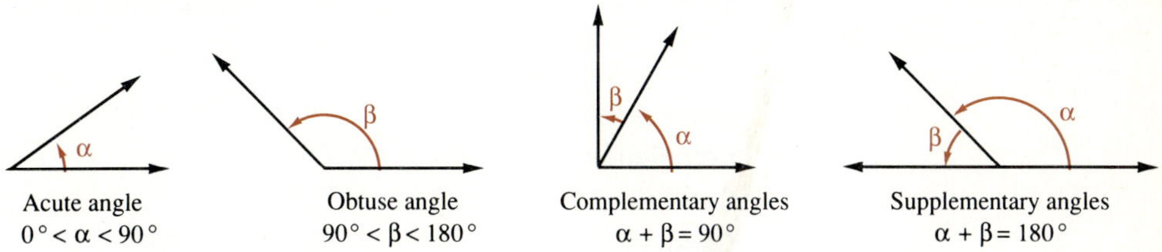

Figure S1-3

10. Formula for Arc Length. If θ is a central angle in a circle of radius r, then the length S of the intercepted (subtended) arc is given by the formula $S = r\theta$. The angle θ must be in radians to use this formula (see Figure S1-2).

11. Area of a Circular Sector. The area A of a sector of a circle is given by $A = \frac{1}{2}r^2\theta$, where r is radius of the circle and θ is the central angle of the sector, measured in *radians* (see Figure S1-4).

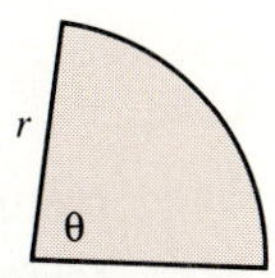

Figure S1-4

12. Linear Speed and Angular Speed

$$\text{linear speed} = \frac{\text{distance travelled}}{\text{time to travel this distance}} \quad \Rightarrow \quad v = \frac{d}{t}$$

$$\text{angular speed} = \frac{\text{angle moved through}}{\text{time to move through this angle}} \quad \Rightarrow \quad \omega = \frac{\theta}{t}$$

13. Relating Angular and Linear Speed. The linear speed v of a moving point on a circle of radius r is given by $v = r\omega$, where ω must be in radian measure (see Figure S1-5).

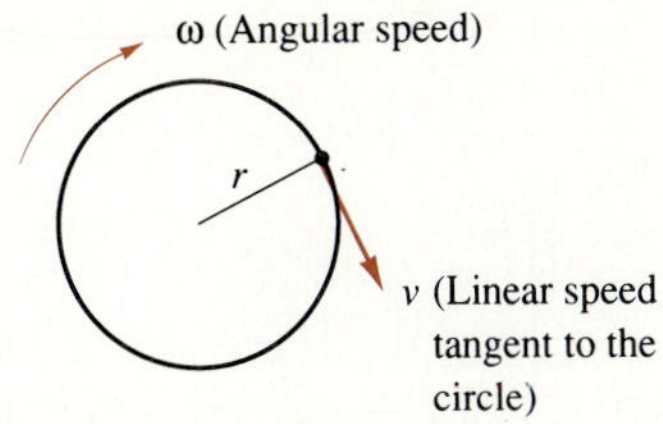

Figure S1-5

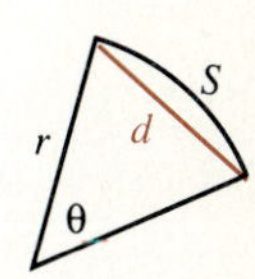

Figure S1-6

14. Small-Angle Approximation. If θ is much less than 1 rad, then $\theta \approx d/r$. This approximation gives θ in radians (see Figure S1-6).

IMPORTANT TERMS

Make sure you understand the meanings of all these terms.

ANGLES

Acute angle
Complementary angles
Coterminal angles
Geometric angle
Initial side of angle
Obtuse angle
Positive and negative angles
Quadrant
Quadrantal angle
Right angle
Rotation
Standard position
Straight angle
Supplementary angles
Terminal side of angle
Trigonometric angle
Vertex of angle

MEASURE OF ANGLES

Degree
Degree mode (for calculator)
Minute
Radian
Radian mode (for calculator)
Revolution
Second

CIRCLES AND ANGLES

Angular speed
Arc length
Central angle
Chord
Circular sector
Circumference
Linear speed
Pi (π)
Radius
Small-angle approximation
Subtend

REVIEW EXERCISES

1. Write the value of the angles (a)–(d) in degrees and in radians.

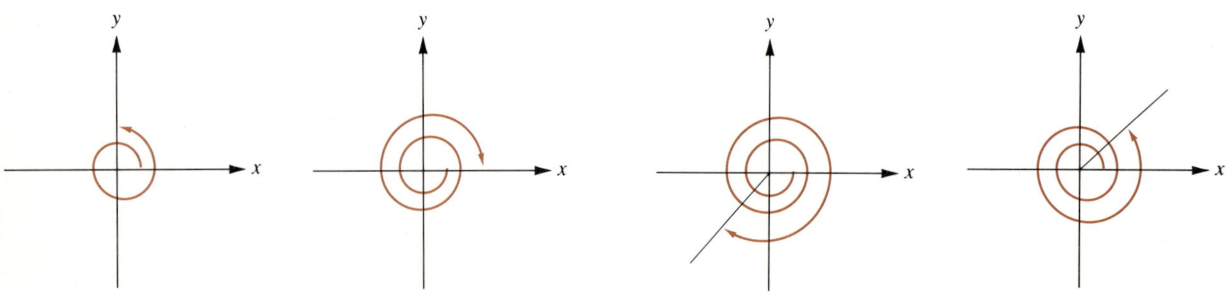

(a) *(b)* *(c)* *(d)*

2. A ferris wheel 75 ft in diameter turns through a 75° angle in 14 s. Through what distance have the riders at the rim of this wheel moved?

3. Find the angle that is

(a) complementary to $\frac{\pi}{6}$
(b) supplementary to $\frac{2\pi}{7}$
(c) supplementary to 105°20′14″
(d) complementary to 79.9°
(e) complementary to 1.20 rad
(f) supplementary to 0.82π
(g) supplementary to 0.20 rev

4. Make the following conversions. Use a calculator as needed, but be sure to write all of your conversion factors.

(a) 42°23′ to decimal degrees
(b) $\frac{5\pi}{6}$ to decimal degrees
(c) −111° to radians
(d) 2.65 rev to radians
(e) 115π to revolutions
(f) 197″ to degree-minute-seconds
(g) 250 rpm to units of radians per second
(h) 4.03π for use in a calculator
(i) 43.7π to revolutions
(j) 18°25′ for use in a calculator
(k) 1″ to radians

5. A bucket of concrete is lowered over a 4.00-cm diameter pulley using an 38.0-cm radius winch as shown.

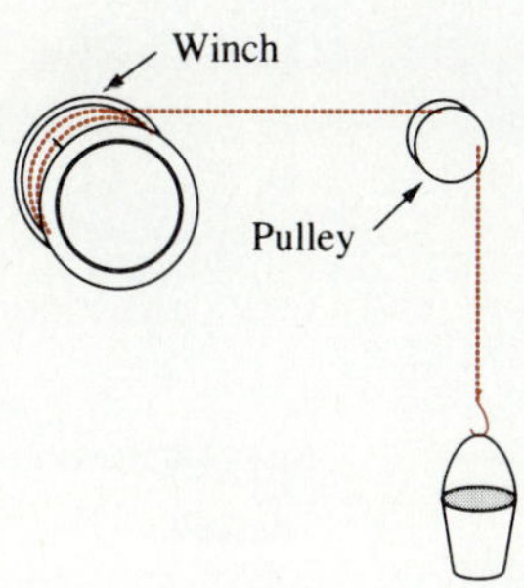

(a) When the bucket has travelled downward by 1.50 m, through what angles have both the winch and the pulley turned?
(b) How far does the bucket rise for each 1.0° turn of the winch?
(c) At what rate in radians per second must the winch turn so that the bucket will move at 6.0 cm/s? At what rate in radians per second will the pulley be turning?

6. Find *two* angles, one positive and one negative, that are coterminal with the given angle. Include a clear sketch.

(a) 65° **(b)** −75.3°
(c) $\frac{4\pi}{3}$ **(d)** -1.20π
(e) 320.7° **(f)** 124°18′

7. A wheel 3.50 ft in diameter rolls a distance of 675 ft in 25.0 s with uniform speed.

(a) Through how many degrees does each of its spokes turn in this time?
(b) What is the angular speed in degrees per second of this wheel?
(c) How long will it take this wheel to turn through 140°?

8. The range of a certain lighthouse beam is 5.0 mi. How many square miles of ocean does this beam cover as it turns through a 75° angle?

9. The sun is 93.0 million miles from earth and subtends an angle of 0.533° as viewed from the earth. Use this information to calculate the approximate diameter of the sun. Include a clear sketch.

10. If a central angle in a 4.50-m diameter circle cuts off an arc 1.37 m long on the rim of that circle, what is the radian measure of this angle?

11. A lot in a housing development is in the shape of the wedge (sector) of a circle. Find the diameter of the circle if the lot contains 810 m^2 and has a central angle of 130°.

2

Right Triangles

The Great Pyramid of Khufu (Cheops) at Gizeh

2.1 SIMILAR TRIANGLES AND RIGHT TRIANGLES

In Chapter 1, we discussed some direct methods for measuring distances and angles. However, many distances are very difficult to measure directly. For example, how would you find the height of a mountain or the distance across a river by direct measurement? To answer such questions, we need to develop an *indirect* method of measurement.

The search for indirect methods has led to great advances in mathematics and science, as well as to many practical applications. For instance, the Egyptians became expert surveyors and geometricians because they needed to reestablish their boundaries after the annual flooding of the Nile.

Greek mathematicians helped to develop many indirect methods of measurement. In about 600 B.C. Thales of Miletus calculated the height of the Great Pyramid by measuring the length of its shadow at the instant when the shadow of a vertical stick next to the pyramid was exactly as long as the stick itself. He reasoned that the height of the pyramid would be equal to the length of its own shadow at that moment (Figure 2.1-1).

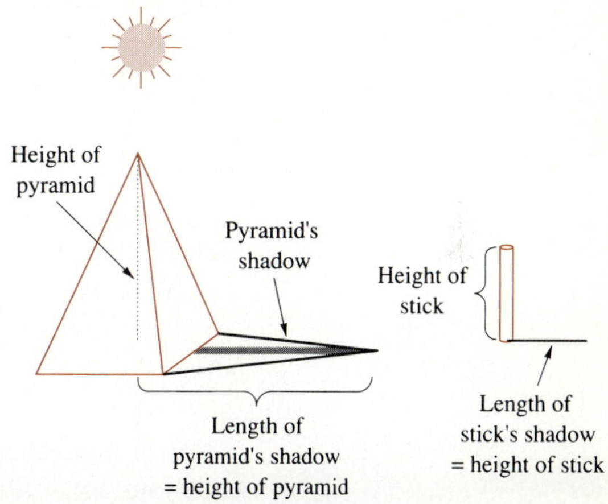

Figure 2.1-1

Thales' method of measuring the height of the great pyramid When the stick's shadow is equal to its height, the length of the pyramid's shadow is equal to its height.

To use this method, the observer must be present at the exact moment when the length of the stick's shadow is equal to the length of the stick. However, according to the Greek biographer Plutarch, Thales also presented a second method that did not require the observer to be present at that exact moment. This second method, which we discuss in Example 2, makes use of similar triangles.

2.1-1 Similar Triangles

Look carefully at the two triangles in Figure 2.1-2. We have drawn these triangles so that they have equal angles but different sizes. Do you notice how similar to each other these triangles look, except for size? For this reason, we call triangles having equal angles **similar triangles**.

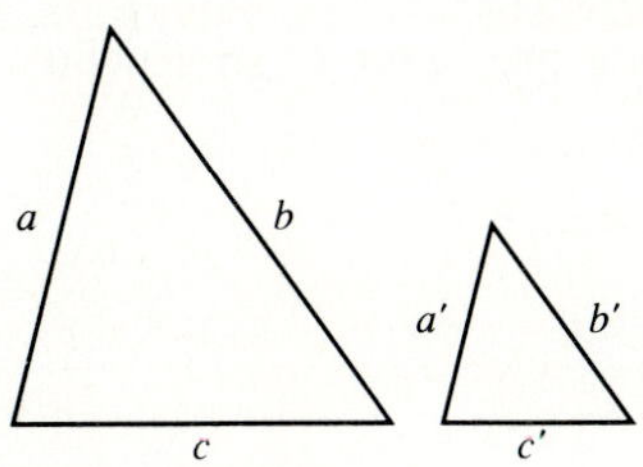

Figure 2.1-2

Similar triangles

DEFINITION OF SIMILAR TRIANGLES

Two triangles are *similar* when the three angles of one triangle are equal to the corresponding three angles of the other triangle.

Now take another look at the triangles in Figure 2.1-2. As you look at them, ask yourself *why* they look so similar. You will notice that the larger triangle is just a scaled up version of the smaller one; that is, the sides of the smaller triangle have all been increased by the *same proportion* to make the larger triangle. Mathematically, this means that a is proportional to a' in the same ratio that b is proportional to b' and c is to c'. We can write this as

$$\frac{a}{a'} = \frac{b}{b'} = \frac{c}{c'}$$

This very important property of similar triangles is called Euclid's theorem. We say that a and a' are *corresponding sides* because *they are opposite the same angle*. Likewise, b and b' are corresponding sides, as are c and c'.

EUCLID'S THEOREM

The corresponding sides of similar triangles are proportional. If a, b, c and a', b', c' are the corresponding sides of two similar triangles, then

$$\frac{a}{a'} = \frac{b}{b'} = \frac{c}{c'}$$

It is important to realize that Euclid's theorem is really three separate equations.

$$\frac{a}{a'} = \frac{b}{b'} \qquad \frac{b}{b'} = \frac{c}{c'} \qquad \text{and} \qquad \frac{a}{a'} = \frac{c}{c'}$$

It is in this latter form that you will make use of the theorem.

Example 1

The triangles in Figure 2.1-3 have the same angles. Find the unknown sides of the larger one.

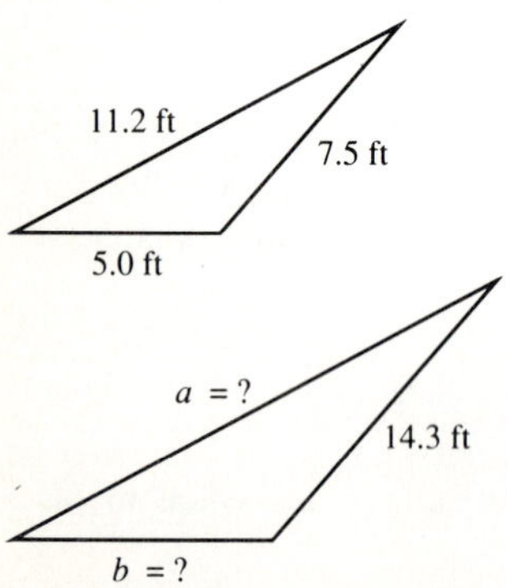

Figure 2.1-3

Solution

Because the two triangles have equal angles, they are similar, so Euclid's theorem applies to them.

$$\frac{a}{a'} = \frac{c}{c'}$$

$$\frac{a}{(11.2 \text{ ft})} = \frac{(14.3 \text{ ft})}{(7.5 \text{ ft})}$$

$$a = \frac{(11.2 \text{ ft})\ (14.3 \text{ ft})}{(7.5 \text{ ft})} = 21 \text{ ft}$$

$$\frac{b}{b'} = \frac{c}{c'}$$

$$\frac{b}{(5.0\text{ ft})} = \frac{(14.3\text{ ft})}{(7.5\text{ ft})}$$

$$b = \frac{(5.0\text{ ft})\ (14.3\text{ ft})}{(7.5\text{ ft})} = 9.5\text{ ft} \ \blacksquare$$

NOW IT'S YOUR TURN

If the 14.3 ft side were 28.6 ft in the larger triangle, what would be the lengths of its other two sides? (Answer: 42 ft, 19 ft)

Example 2 illustrates how Thales used similar triangles to measure the height of the Great Pyramid.

Example 2

Suppose we place a vertical 10.0-ft stick near the pyramid whose height we wish to find. Then we measure the lengths of the shadows of the pyramid and the stick and find that the pyramid's shadow is 250.0 ft long at the moment when the stick's shadow is 20.0 ft long. How tall is the pyramid?

Solution

First make a sketch showing the shadows of the stick and pyramid, calling h the height of the pyramid (Figure 2.1-4). The longest side of each triangle in the sketch is the path of the sun's rays, and the other two sides of each are horizontal and vertical. Therefore, these two triangles have equal angles and are therefore similar. Applying Euclid's theorem gives us

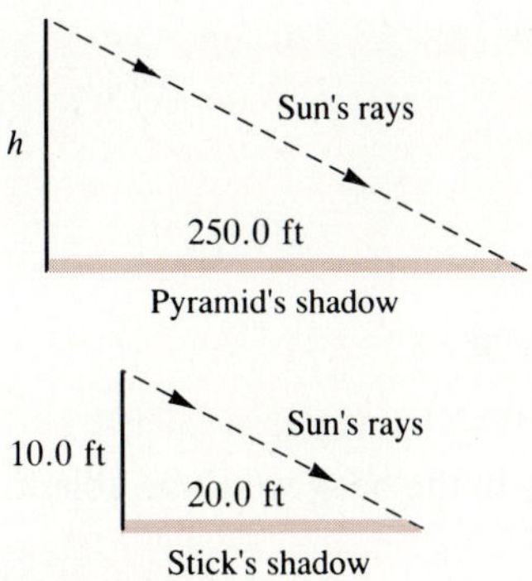

Figure 2.1-4

Measuring the height of a pyramid

$$\frac{h}{10.0\text{ ft}} = \frac{250.0\ \cancel{\text{ft}}}{20.0\ \cancel{\text{ft}}}$$

Solving for h we get

$$h = (10.0\text{ ft})\left(\frac{250.0}{20.0}\right) = 125\text{ ft} \ \blacksquare$$

Thus the pyramid is 125 feet high.

NOW IT'S YOUR TURN

If a 12-ft pole casts a 34-ft shadow at the same time that a pyramid casts a 315-ft shadow, how tall is that pyramid? (Answer: 110 ft)

Example 3
Designing Theater Sets

You are designing sets for a play in a theater that has a stage 30.0 ft deep and an auditorium 80.0 ft deep. The audience is seated with their heads just about level

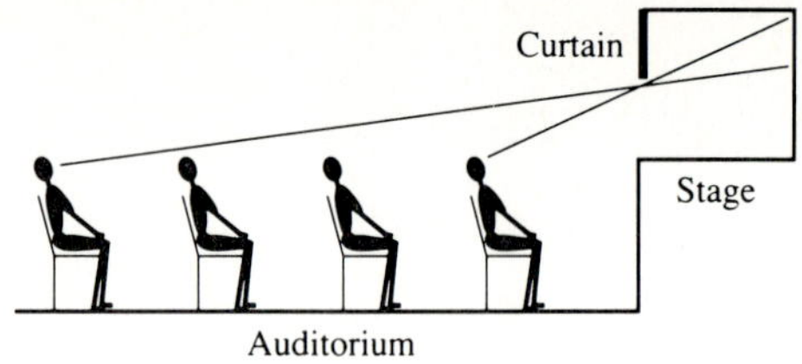

Figure 2.1-5

Audience viewing play

with the stage, as shown in Figure 2.1-5. During the play, the upper part of the stage opening is obscured by the curtain, which leaves an opening 16 ft tall above the stage. What is the tallest set you can design for the rear of the stage which will be completely visible to the entire audience?

Solution

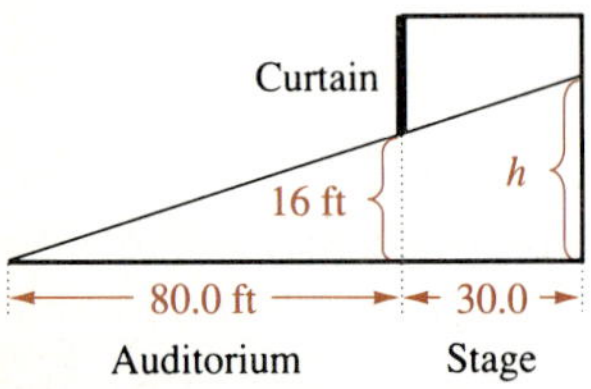

Figure 2.1-6

Theater set design

As you can see by following the lines of sight for the people in the audience, the people in the back have more of their view cut off by the curtain than the people in front. Therefore, if the people in back can see the set, so can everyone else. We can make a simplified drawing of the theater showing the line of sight of only the people in the back (Figure 2.1-6). In this figure, h is the maximum height the scenery can have and still be seen by the people in the back. Here we have two triangles, one with height h and base 110 ft (80.0 ft + 30.0 ft) and one with height 16 ft and base 80.0 ft. These triangles are similar because their angles are equal. Applying Euclid's theorem, we have

$$\frac{h}{110 \text{ ft}} = \frac{16 \cancel{\text{ft}}}{80.0 \cancel{\text{ft}}}$$

Solving for h gives us

$$h = (110 \text{ ft})\left(\frac{16}{80.0}\right) = 22 \text{ ft} \quad \blacksquare$$

Therefore, if the sets are taller than 22 ft, the people in the back won't be able to see the tops of them.

NOW IT'S YOUR TURN

Suppose you have enough material to build sets only 21 ft tall. Find the minimum height of the opening between the stage and curtain that will not obscure the view of the people in the back row? (Answer: 15 ft)

2.1-2 Right Triangles

Perpendicular lines are very common. (Just look around the room to convince yourself of that!) We call any triangle that has two perpendicular sides a ***right***

> A *right triangle* is a triangle having one right angle.
> An *oblique triangle* is a triangle that does not contain a right angle.

triangle because it contains a right angle. We call any other kind of triangle an ***oblique triangle***. Figure 2.1-7 shows both right and oblique triangles.

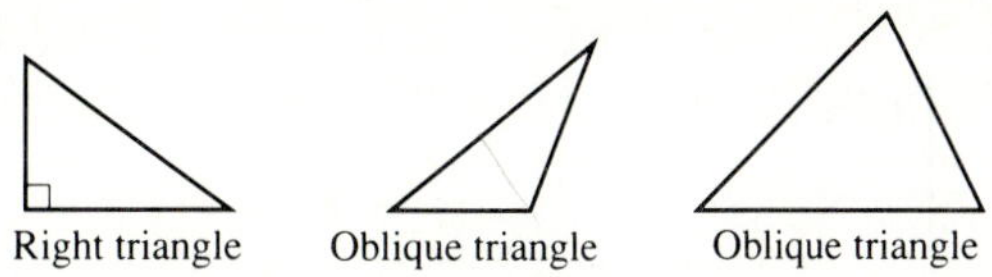

Figure 2.1-7
Types of triangles

If you consider all the possible shapes that a triangle can have, the problem of solving triangles may seem enormous. But fear not! We shall soon see that solving right triangles enables us to solve *all* triangles because we can decompose every triangle into two right triangles, as Figure 2.1-8 shows. Therefore, we first concentrate on solving right triangles.

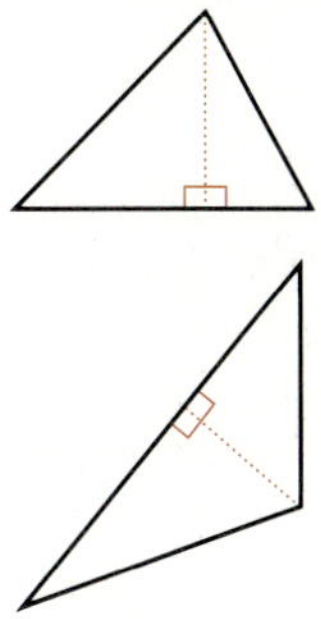

Figure 2.1-8
Oblique triangles decomposed into two right triangles

What do we mean by "solving" a triangle? Every triangle, whether right or oblique, has six parts: three sides and three angles. We call the process of finding the size of these parts solving the triangle. If we know certain parts, we can find the others.

Consider the right triangle ABC in Figure 2.1-9. We usually represent the angles by the capital letters A, B, and C, and the sides by the small letters a, b, and c. Side a is opposite angle A, side b is opposite angle B, and side c is opposite angle C.* In geometry you learned that the sum of the three angles in *any* triangle is 180°. Therefore, because $C = 90°$, the two acute angles (A and B) in a *right* triangle must add up to 90°—they are *complementary*.

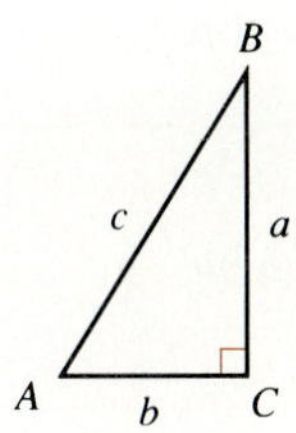

Figure 2.1-9
Right triangle

The Pythagorean Theorem

One of the most important characteristics of any right triangle is that the lengths of its sides obey a famous formula called the Pythagorean theorem. This formula relates the length of the *hypotenuse* (the side opposite the right angle—the longest side) to the lengths of the other two sides (the *legs*). It was probably proved by the Greek mathematician Pythagoras (a student of Thales) in approximately 550 B.C. (see Exercises 2.1-2 for proofs of this famous theorem).

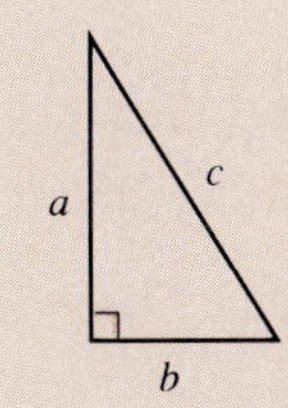

PYTHAGOREAN THEOREM

In a right triangle, the square of the length of the hypotenuse is equal to the sum of the squares of the lengths of the legs. That is, if c is the length of the hypotenuse, and if a and b are the lengths of the legs, then

$$a^2 + b^2 = c^2$$

Example 4

The "size" of a TV screen is really the length of its *diagonal*. If you have a 19-in. TV with a rectangular screen, and if one side is 12 in. long, how long is the other side?

---

* This method of denoting angles and sides was used by Leonhard Euler in the 1700s.

Solution

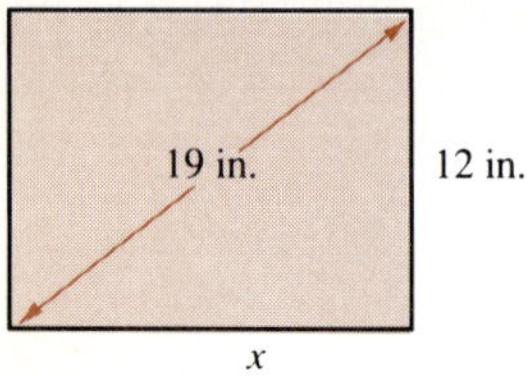

Figure 2.1-10
A 19-in. TV screen

First make a sketch of the screen calling x the length of the unknown side (Figure 2.1-10). Applying the Pythagorean theorem to the right triangle that the diagonal forms with the sides of the screen gives us

$$x^2 + (12 \text{ in.})^2 = (19 \text{ in.})^2$$

Solving for x, we get

$$x^2 = (19 \text{ in.})^2 - (12 \text{ in.})^2 = 361 \text{ in}^2 - 144 \text{ in}^2 = 217 \text{ in}^2$$

$$x = \sqrt{217 \text{ in}^2} = 15 \text{ in.} \ \blacksquare$$

NOW IT'S YOUR TURN

If a TV screen has sides 18 in. and 24 in., what is the size of its screen? (Answer: 30 in.)

The Converse of the Pythagorean Theorem

The Pythagorean theorem says that if a triangle is a right triangle, then its sides a, b, and c obey the formula $a^2 + b^2 = c^2$. But, it also works the other way around: *If the sides of a triangle obey this formula, then the triangle must be a right triangle.* This statement is the converse of the Pythagorean theorem.

> **CONVERSE OF THE PYTHAGOREAN THEOREM**
>
> **If the lengths of the sides of a triangle obey the Pythagorean theorem, then the triangle is a right triangle.**

Example 5

If a triangle has sides of lengths 15, 17, and 8 units, is it a *right* triangle?

Solution

We check to see if the lengths of the sides obey the formula $a^2 + b^2 = c^2$. If so, it must be a right triangle. The hypotenuse c is the longest side, so that must be the 17 unit side. It doesn't matter which sides we call a and b.

$$a^2 + b^2 = 15^2 + 8^2 = 225 + 64 = 289$$

$$c^2 = 17^2 = 289$$

Because the sides obey the Pythagorean theorem ($15^2 + 8^2 = 17^2$), this is a right triangle. ■

NOW IT'S YOUR TURN

Is a triangle with sides of 16, 10, and 20 units a right triangle? (Answer: No)

EXERCISES 2.1-1

In Exercises 1–4, find the sides x for each pair of similar triangles.

1.

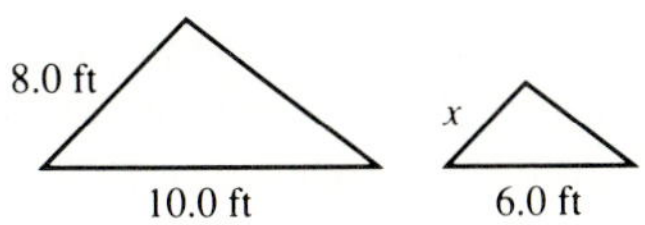

2.

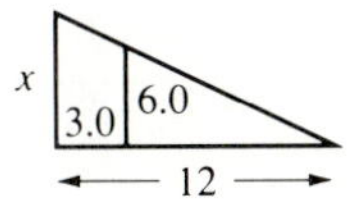

3.

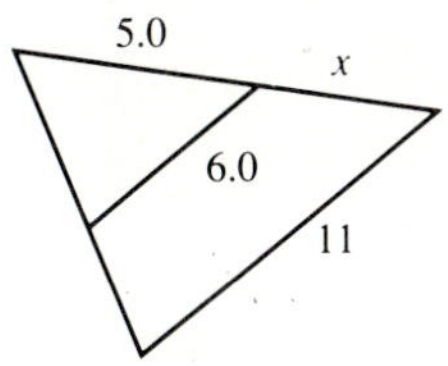

4.

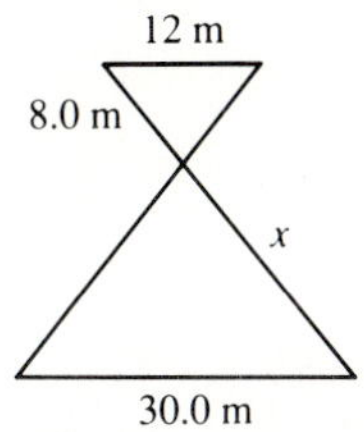

5. If Thales had measured a 6.0-ft pole's shadow to be 8.0 ft at the same time that a pyramid's shadow was 175 ft, how tall would that pyramid be?

6. If a tower casts a 120-ft shadow when a 20.0-ft telephone pole casts a 15-ft shadow, how high is the tower?

7. While you are looking at the top of a 5-story building 60.0 ft away, you notice that the top of a 45-story building is just in line with its top. If an average story is 10.0 ft, how far away from you is the taller building?

8. A roof slopes so that it gains 4.0 in. of vertical elevation (the rise) for every 12 in. of horizontal distance (the run), as shown by the small triangle in Figure 2.1-11. As an architect, you want this roof to extend 30.0 in. horizontally beyond the 8.0-ft tall walls to protect against direct sunlight in summer.

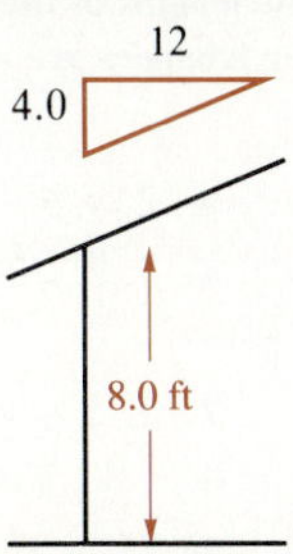

Figure 2.1-11

(a) What is the length of the roof that extends beyond the walls?
(b) How far is the lowest point on the roof above the ground?

9. A 6.0-ft person is 5.0 ft from a 10.0-ft tall street lamp.

(a) How long is this person's shadow?
(b) By how much does the shadow's length change when the person walks 1.0-ft closer to the lamp? When she walks 1.0-ft *farther* from the lamp, starting at her original position?
(c) How far away from the lamp should she be so that her shadow will be 10 times her height?

10. A door 32 in. wide swings on hinges 9.0-in. from a corner. You want to place a light fixture on the adjacent wall, 19 in. from the corner, as shown in Figure 2.1-12. How far can this light fixture protrude from the wall without getting hit by the door when the door is open as far as possible?

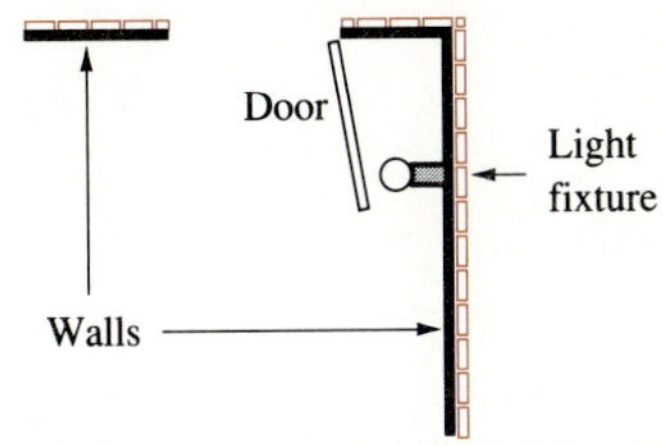

Figure 2.1-12

***11.** A small light is mounted 6.0 in. above a round collar 14 in. in diameter atop an 8.0-ft lamp post, as shown in Figure 2.1-13. Find the *area* of the shadow of this collar on the ground.

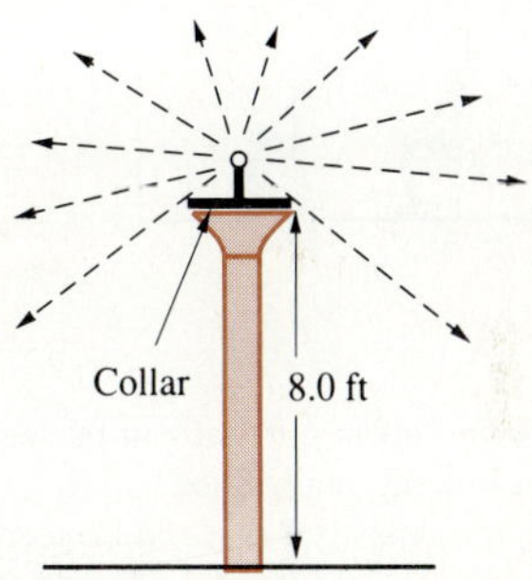

Figure 2.1-13

***12.** You want to add an 8.0-ft high flat-roofed addition to the top of your house, as shown in Figure 2.1-14. For aesthetic reasons, you do not want this addition to be visible to a person on the sidewalk 20.0 ft in front of your house. If you assume a typical height of 6.0 ft for this person, how close can the front of your addition be to the front of your house without being visible to the person on the sidewalk?

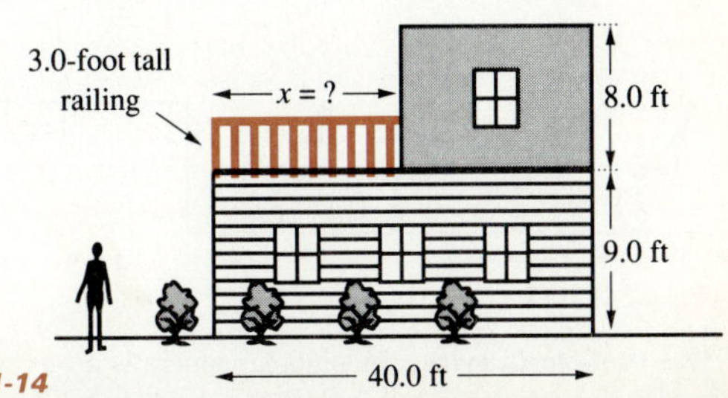

Figure 2.1-14

***13.** In a nuclear-reactor research facility, some radioactive plutonium is on the floor 12 ft from a doorway opposite its center, as shown in Figure 2.1-15. You need to erect a vertical lead shield to protect this doorway completely from the radioactive rays. This shield must allow 4.0 ft of clearance so that people can walk into the room, as shown in the figure. If the doorway is 3.0 ft wide and 7.0 ft high, what must be the minimum *width* and *height* of this lead shield?

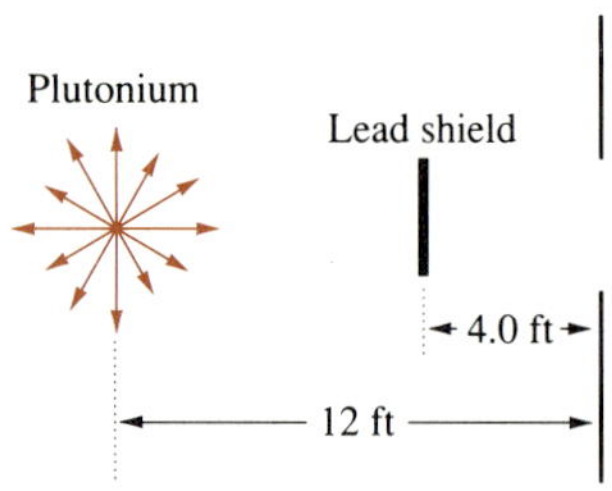

Figure 2.1-15

14. In a certain theater, the stage is 5.0 ft high and 15.0 ft deep, and the seats are arranged so that the heads of the audience are 4.0 ft above the horizontal floor. The first row begins 3.0 ft from the stage, and the last row is 45 ft from the stage (see Figure 2.1-16). You need to design sets for the back of this stage.

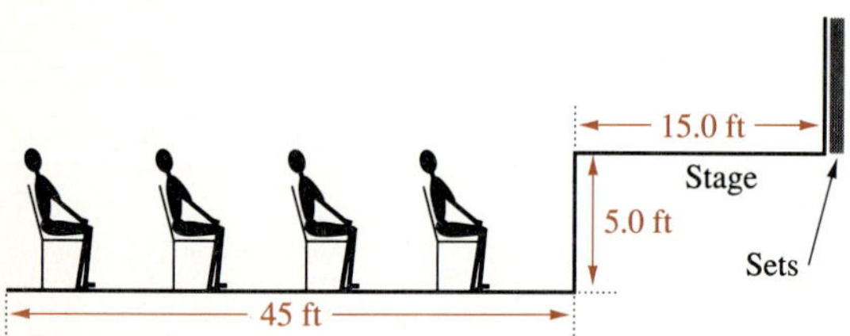

Figure 2.1-16

(a) What is the tallest set that can be seen by everyone in the audience on the main floor?

(b) What is the tallest set that can be seen by people in the front row?

15. The stage in a theater is 15 ft deep and 35 ft high, as shown in Figure 2.1-17. The heads of the audience are at stage level, and the first row starts 5.0 ft from the stage. How far below the ceiling should you lower the curtain so that no one in the audience can see the ceiling of the stage? ***Hint:*** Draw lines of sight.

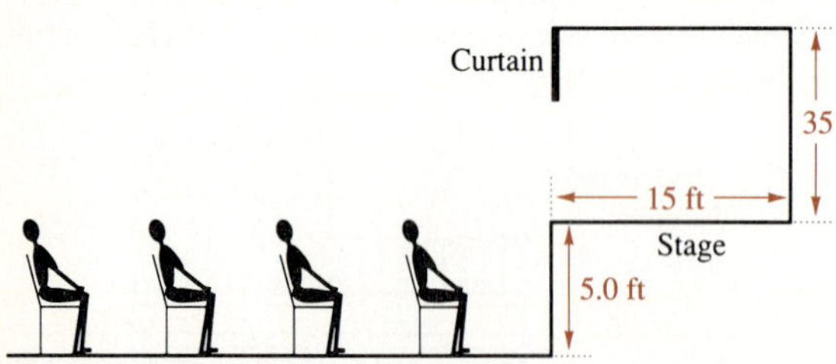

Figure 2.1-17

16. Pinhole Camera. A pinhole camera is a box with a small hole (the pinhole) in the middle of one face and the film on the opposite face. No lens is used. Lght from the object to be photographed passes through the pinhole and strikes the film, as shown in Figure 2.1-18. If your camera has a depth $D = 6.0$ in. and uses $2\frac{1}{4}$- × $2\frac{1}{4}$-in. film:

(a) How close to the front of your camera must a 6.0-ft person be so that his vertical image completely fills the height of the film?

(b) How tall will the image of a 50.0-ft tall tree be on your film if you take the photograph 120 ft away from this tree?

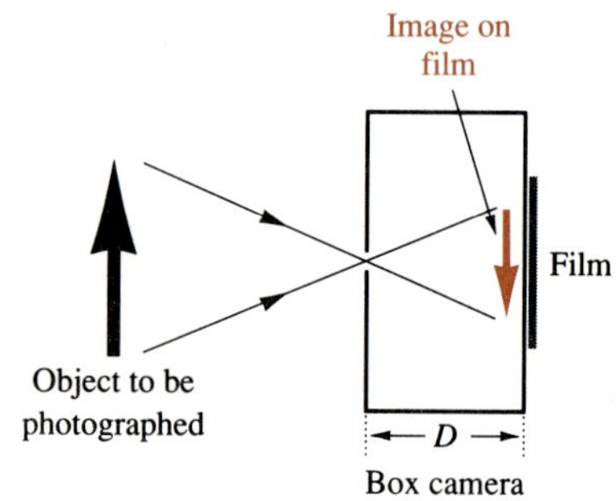

Figure 2.1-18

17.** Prove that the ratios of the lengths of the sides of a triangle do not depend on the *size* of the triangle, provided that the angles remain the same. That is, for the triangle in Figure 2.1–19, prove that BC/AC, AB/BC, AC/AB, and so on do not depend on the size of this triangle. ***Hint: Inside the triangle construct a line parallel to BC, making a smaller triangle that is similar to the original one. Then use the properties of similar triangles.

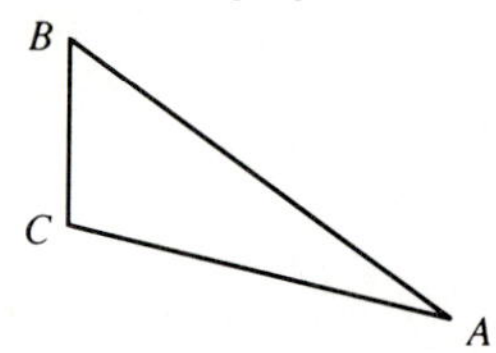

Figure 2.1-19

EXERCISES 2.1–2

In Exercises 1–11, using only the lengths of the sides given, determine which of the following triangles are *right* triangles. Sketch each triangle to scale.

1. 2, 2, 2   **2.** 3, 4, 5

3. 6, 8, 10   **4.** 1, 2, $\sqrt{3}$

5. 1, 2, 4 (Is anything strange here?)

6. 1, 2, $\sqrt{2}$ **7.** 3, 5, $\sqrt{34}$

8. 1, 3, $\sqrt{11}$ **9.** 1.50, 2.80, 3.18

10. 1, 1, 1   **11.** 2.3, 3.7, 5.0

In Exercises 12–24 find the third side of the following right triangle when the other two are given.

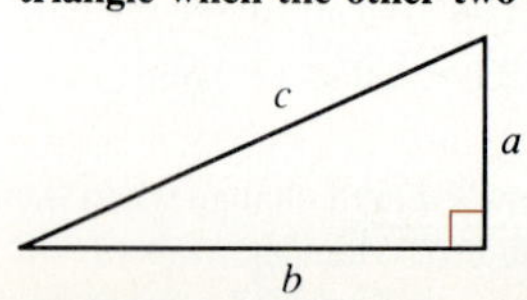

12. $a = 1, b = 2$

13. $a = 2.5, b = 3.1$

14. $a = 2, c = 5$

15. $a = 6.5, c = 9$

16. $b = 65, c = 75$

17. $b = 5.2, c = 4.9$

18. $a = 3, b = 4$

19. $a = 1, b = 1$

20. $b = 1, c = 2$

21. $a = \sqrt{2}, b = \sqrt{7}$

22. $b = \sqrt{5}, c = 4.2$

23. $a = 3, c = \pi$

24. $a = \pi, b = \frac{\pi}{2}$

25. If a square picture must have a diagonal of 30.0 cm, how long must each side be?

26. The picture area of a 35-mm slide is 24.0 × 35.0 mm. How long is the diagonal of the picture?

27. Find the area of a square having a diagonal of 40.0 in.

***28.** If a 28.3-in. TV screen is to have an area of 360.0 in^2, what must be its rectangular dimensions?

***29.** A rectangular swimming pool is to have a diagonal of 24.15 ft and a perimeter of 68.0 ft. Find the length of each side and the area of the surface of this pool.

30. You are doing exercises, like jumping jacks, in which you stand with your feet first together and then 4.0 ft apart. If your legs are 3.0 ft long, by how much does your height change while you do these exercises?

31. Camera as Rangefinder. You can use a manual-focus camera to measure the height of a building as follows. Standing across the street, first focus on the top of the building and read this distance. Then focus on the base of the building and read this distance.

(a) How can you find the building's height from these measurements?

(b) Why can't you use this method to measure the height of a mountain? (If you have very limited experience with a camera, you may not be able to answer this part.)

(c) If you have a manual-focus camera, go out and measure the height of a building using this method.

32. In college football, the playing field is 100.0 yd long and $53\frac{1}{3}$ yd wide; the end zones are 10.0 yd deep, and the kickoff occurs at the center of the 40.0 yd line (see Figure 2.1-20). You are covering an important game for the school newspaper, and you want to photograph the kicker just as he boots the ball. If you are at the 60.0 yd line but 8.00 yd outside the edge of the playing field, at what distance should you prefocus your camera lens to be ready at the instant of kickoff?

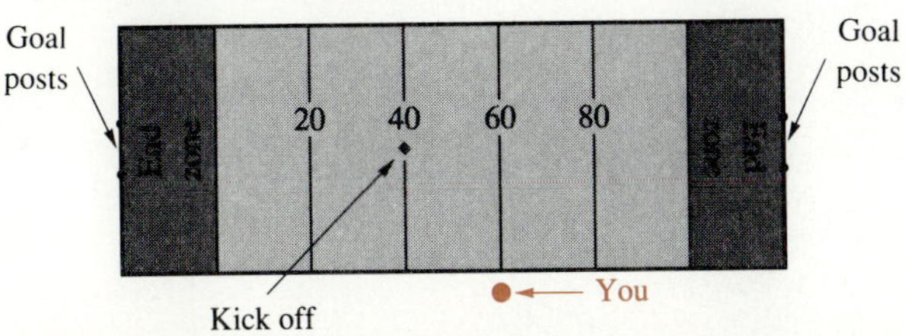

Figure 2.1-20

33. A circle of radius R is inscribed inside of a square.

(a) How far is it from the center of this circle to a corner of the square?

(b) What is the *shortest* distance from a corner of the square to the circle?

***34.** A square is inscribed inside a circle of radius R. What is the shortest distance from the center of this circle to any side of the square?

35. Find the length of the diagonal of any *face* of a cube of side 2.0 ft.

36.** Find the length of the diagonal of a cube 1.0 ft on each side. ***Hint: Reduce this three-dimensional problem to a two-dimensional problem by sketching the cube as in Figure 2.1-21. Notice that we now have a right triangle, with the diagonal as hypotenuse. Apply the method from Exercise 35.

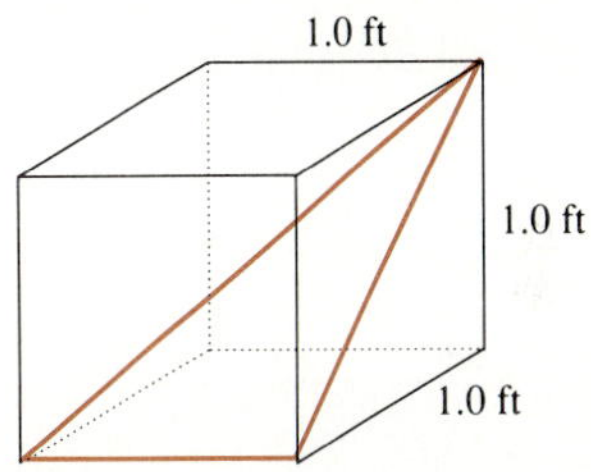

Figure 2.1-21

***37.** A certain room is a cube 12 ft on each side. Find the straight-line distance from the upper northeast corner to the lower southwest corner.

***38.** If a cube has a diagonal 10.0 m long, find its volume.

***39.** If a cube has a diagonal 20.0 ft long, find its total surface area.

***40.** A cube 35 cm on a side has the largest possible sphere inside it.

(a) Find the *shortest* distance from each corner to the surface of the sphere.

(b) What is the total volume outside the sphere but inside the cube?

***41.** A pyramid rises to a vertical height of 120 m above the midpoint of its square base 100.0 m on a side. What is the total area of stone needed to cover the four faces of this pyramid?

42.** A fishbowl has a rectangular base 18 in. × 12 in. and vertical rectangular walls 15 in. high. Find the longest straight-line distance a fish can swim in this bowl. ***Hint: Use the hint from Exercise 36 as a guide.

***43.** You are kneeling at the 70.0-yd line, just about at ground level at the edge of the football field (see Figure 2.1-20), hoping to get a good photo of the ball just as it clears the center of the horizontal bar of the goal posts on your side of the field. This bar is 10.0 ft above the ground. At what distance should you prefocus your camera lens to be ready?

***44.** A jogger runs around the edges of a 10.0-acre field in the shape of a right triangle, the shortest side of which is 800.0 ft long. How far does she run in each complete lap?

The following exercises may involve the use of both similar triangles and the Pythagorean theorem.

45. A bookcase is against a wall near a 30.0 in. wide door, as shown in Figure 2.1-22, and it extends 12.0 in. out from the wall and is placed 6.00 in. from the door's hinge. To plan the arrangement of your other furniture, you need to know how far this door will extend into the room when it is open as far as possible. How far from the wall will it extend?

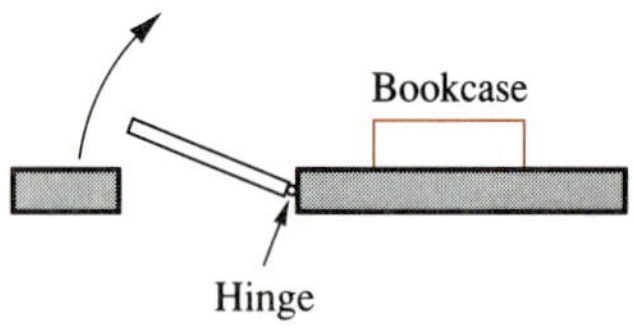

Figure 2.1-22

***46.** Find AB, BD, and EC in Figure 2.1-23.

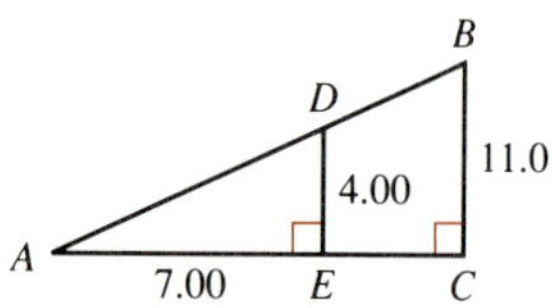

Figure 2.1-23

47. Suppose you bisect any angle of an equilateral triangle of side L, as shown in Figure 2.1-24.

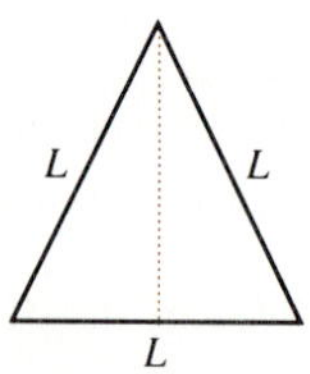

Figure 2.1-24

(a) Prove that the two resulting smaller triangles are (i) similar triangles and (ii) right triangles.

(b) Find the lengths of the sides of the two smaller right triangles that are formed.

(c) What are the two acute angles in each of the smaller triangles?

48. A road has a pitch (or slope) of 2:12 (2/12), which means that it rises two vertical units for every twelve horizontal units. When you have gained 16 mi in elevation, how far have you travelled horizontally?

***49.** A trail has a pitch (or slope) of 4: 12 (4/12) (see Exercise 48). When you have hiked 5.0 mi along this trail, how many feet have you gained in elevation?

***50.** A house 24 ft wide and 36 ft deep has a roof with a pitch (slope) of 3 : 12 (3/12) (see Exercise 48) on both sides. Find the total area of shingles needed to reroof it.

****51.** A building 40.0 ft wide and 32 ft deep with 8.0-ft walls has a roof with two different slopes, 3:12 and 12:6, as shown in Figure 2.1-25 (see Exercise 48). What area of shingles is needed for its roof, and how high is the peak above the ground?

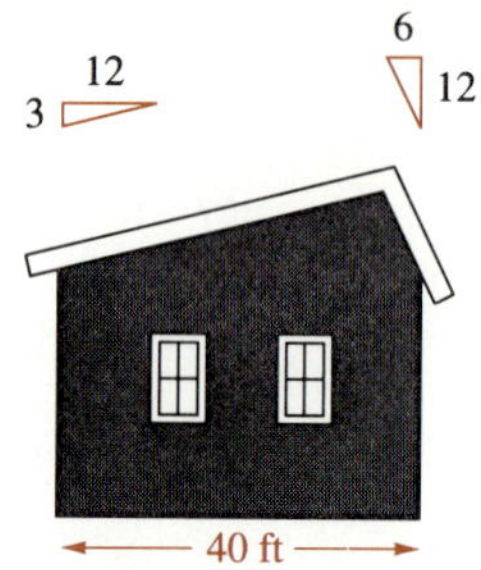

Figure 2.1-25

****52. Proof of the Pythagorean Theorem.** Pythagoras himself is believed to have used the following proof of his famous theorem. Construct the large outer square using four identical triangles of sides a, b, and c, as shown in Figure 2.1-26.

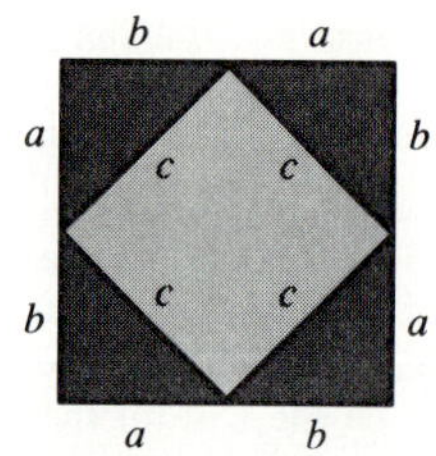

Figure 2.1-26

(a) Prove that the lightly shaded inner quadrilateral is a square *without* using the Pythagorean theorem. ***Hint:*** Think about the *angles* of a square and the sum of the angles of a triangle.

(b) How can you use this figure to prove the Pythagorean theorem? ***Hint:*** Think of *areas*.

(c) Why can't you use the Pythagorean theorem in part (a)?

****53. Another Proof of the Pythagorean Theorem.** Start with a right triangle ABC with acute angles α and β. Divide it into two additional right triangles, as shown in Figure 2.1-27.

(a) Show that we now have three similar triangles: ABC, ACD, and BCD.

(b) Use Euclid's theorem to show that $\overline{AC}^2 = \overline{AB}\,\overline{AD}$ and $\overline{BC}^2 = \overline{AB}\,\overline{DB}$.

(c) Add the results from (b), simplify, and hence prove the Pythagorean theorem!

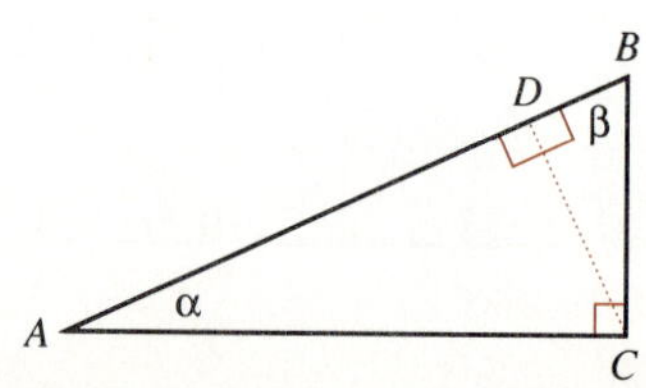

Figure 2.1-27

2.2 THE TRIGONOMETRIC FUNCTIONS FOR A RIGHT TRIANGLE

2.2-1 Definitions of the Trigonometric Functions

In Section 2.1 we discussed two properties of a right triangle, namely:

1. The acute angles of a right triangle are complementary.

$$A + B = 90°$$

2. The lengths of the sides obey the Pythagorean theorem.

$$a^2 + b^2 = c^2$$

The first of these properties relates only the angles, whereas the second relates only the sides. We do not have any relations that tell us the angles if we know the sides. Therefore, to solve right triangles, we must find additional relations that *relate the angles to the lengths of the sides*. The formulation of such relations is the foundation of trigonometry.

To find these relations, first consider the right triangle ABC in Figure 2.2-1a. Suppose we draw any arbitrary line $B'C'$ perpendicular to the base, as shown in Figure 2.2-1b. This line divides the original right triangle into two right triangles, as shown in Figure 2.2-2. Because these triangles have equal angles, they must be similar triangles. Therefore, *the lengths of the corresponding sides are proportional*. For example, $a/a' = c/c'$, which we can rearrange as

$$\frac{a}{c} = \frac{a'}{c'}$$

Now remember, the line $B'C'$ is completely arbitrary. It could be anywhere in the triangle perpendicular to AC, and it would still divide the original triangle

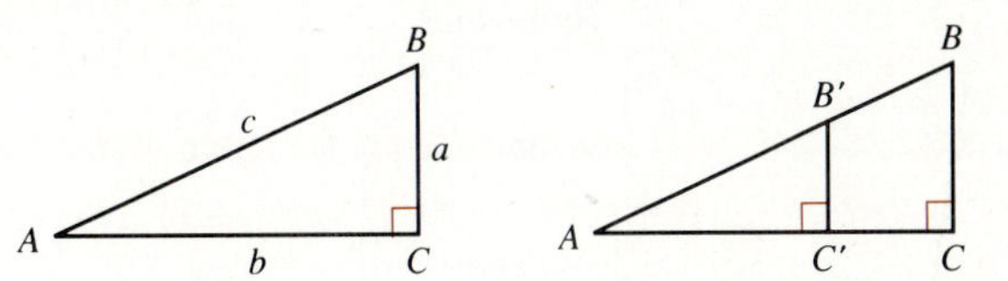

Figure 2.2-1
Right triangles

(a) Right triangle

(b) Right triangle divided into two right triangles

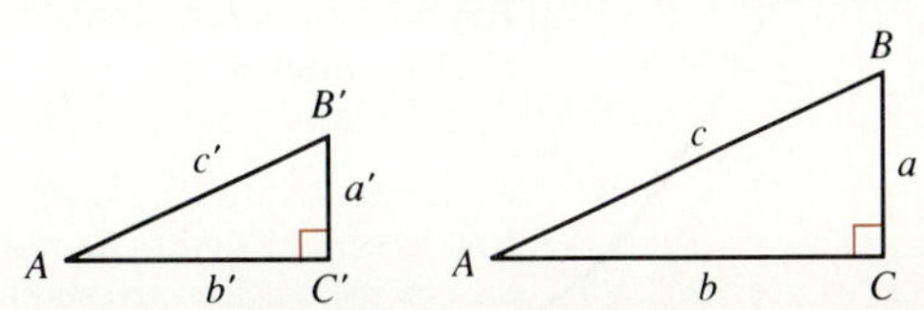

Figure 2.2-2
The two right triangles of Figure 2.2-1b

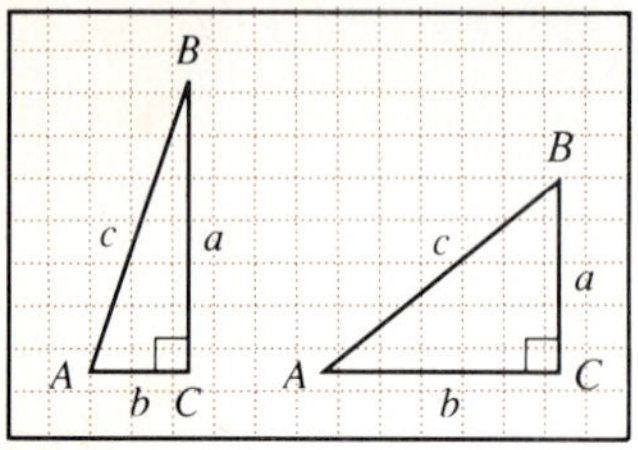

Figure 2.2-3

Right triangles with different angles *A*

> **RATIOS OF THE SIDES OF A RIGHT TRIANGLE**
>
> The *ratios of the lengths of the sides* in a right triangle depend only on the angles of the triangle, not on the size of the triangle.

into different size right triangles. All of these triangles would have equal angles, and they would therefore all be similar triangles. Therefore, the condition $a/c = a'/c'$ would hold *regardless of the size of the triangle*. Consequently, it must be true that the value of the ratio a/c (and a'/c' also) does not depend on the size of the right triangle—it must be true for all sizes, provided they all have the same angles. In other words, *the ratio of the length of the side opposite angle* A *to the length of the hypotenuse in any of the right triangles* ABC *and* AB′C′ *does not depend on the size of the triangle.*

On the other hand, the ratio a/c *does* depend on the size of the angle A, as Figure 2.2-3 shows.

In this figure, we have kept the hypotenuses c the same length, but the lengths of a are different in these two triangles because the angles A are different. Therefore, the ratio a/c will also be different in these two triangles because the angles A are different. The value of the ratio a/c depends *only on the angle* A, so we can regard it as a **function of angle *A***.*

By similar reasoning, we can show that all of the six possible ratios in a right triangle ABC— a/c, b/c, a/b, b/a, c/b, and c/a—depend only on the angle A but not on the size of the triangle. Therefore, they are all functions of the angle A. This is a very important property of right triangles.

We now use these ratios as the formal definitions of the six trigonometric functions of an acute angle A in a right triangle (see Table 2.2-1).

Name of Function	Abbreviation	Definition
sine	sin	$\sin A = \dfrac{\text{length of side opposite angle } A}{\text{length of hypotenuse}}$
cosine	cos	$\cos A = \dfrac{\text{length of side adjacent to angle } A}{\text{length of hypotenuse}}$
tangent	tan	$\tan A = \dfrac{\text{length of side opposite angle } A}{\text{length of side adjacent to angle } A}$
cotangent	cot	$\cot A = \dfrac{\text{length of side adjacent to angle } A}{\text{length of side opposite angle } A}$
secant	sec	$\sec A = \dfrac{\text{length of hypotenuse}}{\text{length of side adjacent to angle } A}$
cosecant	csc	$\csc A = \dfrac{\text{length of hypotenuse}}{\text{length of side opposite angle } A}$

Table 2.2-1

Trigonometric functions of acute angles

*One quantity y is called a *function* of another quantity x if the value of x uniquely determines the value of y. In this case, the value of the angle A determines the value of the ratio a/c.

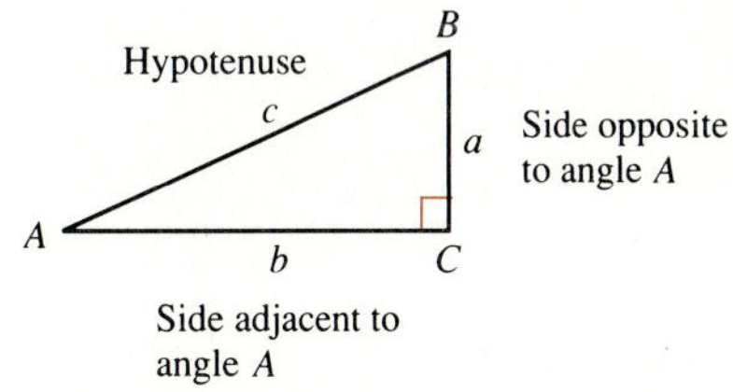

Figure 2.2-4

The sides of a right triangle

If a, b, and c are the lengths of the sides, as shown in Figure 2.2-4, then we can write these trigonometric functions* in Table 2.2-1 more briefly as follows:

DEFINITION OF THE TRIGONOMETRIC FUNCTIONS

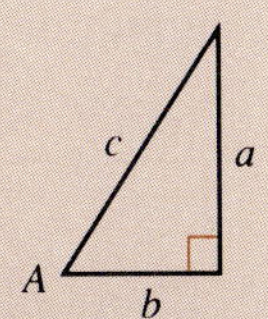

$$\sin A = \frac{a}{c}$$

$$\cos A = \frac{b}{c}$$

$$\tan A = \frac{a}{b}$$

$$\cot A = \frac{b}{a}$$

$$\sec A = \frac{c}{b}$$

$$\csc A = \frac{c}{a}$$

Example 6

Determine the values of all six trigonometric functions of θ to two decimal places for the triangle in Figure 2.2-5.

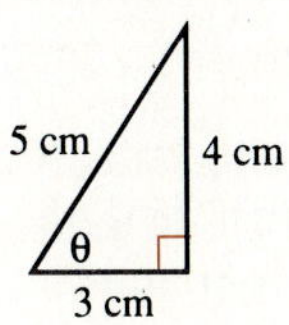

Figure 2.2-5

Solution

We can use the definitions of the trigonometric functions. The side opposite θ is 4 cm long, the side adjacent to θ is 3 cm long, and the hypotenuse is 5 cm long.

$$\sin \theta = \frac{\text{length of opposite side}}{\text{length of hypotenuse}}$$

* The word *sine* comes from the Latin word *sinus*, meaning gulf. Why gulf? The Arabs used *jiba* (from the Hindu word *jiva*, meaning bowstring or cord), but it was later confused with the more common word *jaib*, meaning gulf. In the twelveth century, Plato Tiburtinus translated the work of the Arab Al-Battani into Latin, translating *jaib* into the Latin word for gulf—*sinus*. Danish physician Thomas Finck invented the words *tangent* and *secant* about 1583, although the Arabs had introduced the concepts earlier. The word *cosine* comes from contracting the Latin words *complementis sinus*, meaning the sine of the complementary angle.

$$= \frac{4 \text{ cm}}{5 \text{ cm}} = \frac{4}{5} = 0.80$$

$$\cos \theta = \frac{\text{length of adjacent side}}{\text{length of hypotenuse}}$$

$$= \frac{3 \text{ cm}}{5 \text{ cm}} = \frac{3}{5} = 0.60$$

$$\tan \theta = \frac{\text{length of opposite side}}{\text{length of adjacent side}}$$

$$= \frac{4 \text{ cm}}{3 \text{ cm}} = \frac{4}{3} = 1.33$$

$$\cot \theta = \frac{\text{length of adjacent side}}{\text{length of opposite side}}$$

$$= \frac{3 \text{ cm}}{4 \text{ cm}} = \frac{3}{4} = 0.75$$

$$\sec \theta = \frac{\text{length of hypotenuse}}{\text{length of adjacent side}}$$

$$= \frac{5 \text{ cm}}{3 \text{ cm}} = \frac{5}{3} = 1.67 \blacksquare$$

NOW IT'S YOUR TURN

Find csc θ. (Answer: 1.25)

Example 7

Evaluate all six trigonometric functions of β in Figure 2.2-6, leaving the answers as square roots.

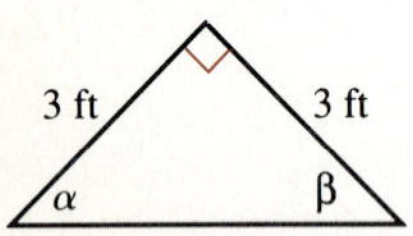

Figure 2.2-6

Solution

To use the definitions of the trigonometric functions, we first must find the length of the hypotenuse in Figure 2.2-6. Let c be the length of the hypotenuse. The Pythagorean theorem gives us

$$c^2 = (3 \text{ ft})^2 + (3 \text{ ft})^2 = 9 \text{ ft}^2 + 9 \text{ ft}^2 = 18 \text{ ft}^2$$

$$c = \sqrt{18 \text{ ft}^2} = \sqrt{9 \times 2 \text{ ft}^2} = 3\sqrt{2} \text{ ft} \blacksquare$$

Now we can apply the definitions of trigonometric functions. The side opposite β is 3 ft long, the side adjacent to β is also 3 ft long, and the hypotenuse is $3\sqrt{2}$ ft long. Therefore,

$$\sin \beta = \frac{\text{length of opposite side}}{\text{length of hypotenuse}}$$

$$= \frac{a}{c} = \frac{3 \text{ ft}}{3\sqrt{2} \text{ ft}} = \frac{1}{\sqrt{2}} = \frac{\sqrt{2}}{2} \quad \blacksquare$$

Notice that we rationalized the denominator in the final answer to keep the square root in the numerator.

$$\cos \beta = \frac{\text{length of adjacent side}}{\text{length of hypotenuse}}$$

$$= \frac{b}{c} = \frac{3 \text{ ft}}{3\sqrt{2} \text{ ft}} = \frac{1}{\sqrt{2}} = \frac{\sqrt{2}}{2}$$

$$\tan \beta = \frac{\text{length of opposite side}}{\text{length of adjacent side}}$$

$$= \frac{a}{b} = \frac{3 \text{ ft}}{3 \text{ ft}} = 1$$

$$\cot \beta = \frac{\text{length of adjacent side}}{\text{length of opposite side}}$$

$$= \frac{b}{a} = \frac{3 \text{ ft}}{3 \text{ ft}} = 1$$

$$\sec \beta = \frac{\text{length of hypotenuse}}{\text{length of adjacent side}}$$

$$= \frac{c}{b} = \frac{3\sqrt{2} \text{ ft}}{3 \text{ ft}} = \sqrt{2}$$

$$\csc \beta = \frac{\text{length of hypotenuse}}{\text{length of opposite side}}$$

$$= \frac{c}{a} = \frac{3\sqrt{2} \text{ ft}}{3 \text{ ft}} = \sqrt{2} \quad \blacksquare$$

NOW IT'S YOUR TURN

Find sin α, cos α, tan α, and sec α for Figure 2.2-6.
(Answer: $\sin \alpha = \sqrt{2}/2$, $\cos \alpha = \sqrt{2}/2$, $\tan \alpha = 1$, and $\sec \alpha = \sqrt{2}$)

Units of the Trigonometric Functions

In Examples 6 and 7, the trigonometric functions came out to be pure numbers because the units of the lengths of the sides (centimeters and feet) cancelled. Because we have defined the trigonometric functions as the *ratios* of the lengths of the sides, the units *always* cancel, leaving pure, dimensionless numbers for the trigonometric functions. If you *ever* get an answer for a trigonometric function that has units such as feet or meters, you have made a mistake!

UNITS OF THE TRIGONOMETRIC FUNCTIONS

The trigonometric functions have no units; they are always pure, dimensionless numbers.

2.2-2 Trigonometric Functions of Special Angles: 30°, 60°, and 45°

In Examples 6 and 7, we knew the lengths of the sides of each right triangle and used them to find the trigonometric functions of their angles. But suppose we know only the angle. How can we find the trigonometric functions of that angle? For example, how can we find the sine of 30°? To see how, we start with some special angles for which we can find relationships between the lengths of the sides of the triangle. Then we can find the values of the trigonometric functions using their definitions.

Functions of 45°

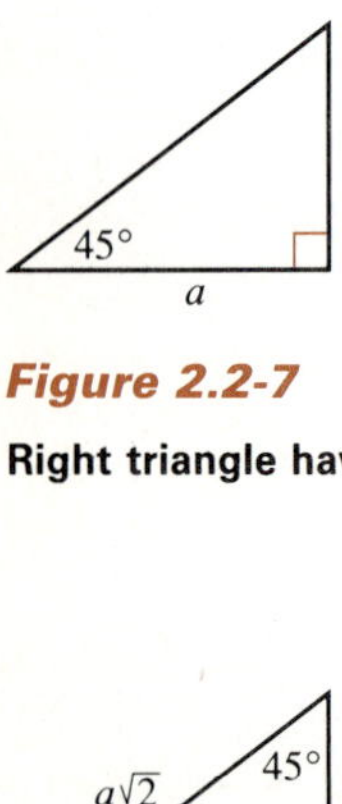

Figure 2.2-7

Right triangle having a 45° angle

Let's look at a right triangle with an acute angle of 45° and one leg having length a, as shown in Figure 2.2-7. Because the acute angles of a right triangle are *complementary*, the other acute angle must also be 45°. Because both acute angles are equal, the sides opposite them must be equal, so the other leg must also have length a. We can now use the Pythagorean theorem to find the length of the hypotenuse.

$$c^2 = a^2 + a^2 = 2a^2$$
$$c = \sqrt{2a^2} = a\sqrt{2}$$

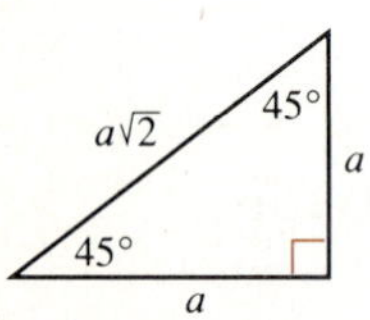

Figure 2.2-8

Sides of a right triangle having 45° angles

We summarize these results in Figure 2.2-8.

Now we can use their definitions to evaluate the trigonometric functions of 45°. Figure 2.2-8 shows us that

$$\sin 45° = \frac{\text{length of opposite side}}{\text{length of hypotenuse}}$$
$$= \frac{a}{a\sqrt{2}} = \frac{1}{\sqrt{2}} = \frac{\sqrt{2}}{2}$$

$$\cos 45^\circ = \frac{\text{length of adjacent side}}{\text{length of hypotenuse}}$$

$$= \frac{a}{a\sqrt{2}} = \frac{1}{\sqrt{2}} = \frac{\sqrt{2}}{2}$$

$$\sec 45^\circ = \frac{\text{length of hypotenuse}}{\text{length of adjacent side}}$$

$$= \frac{a\sqrt{2}}{a} = \sqrt{2}$$

NOW IT'S YOUR TURN

Continue this method to show that $\tan 45^\circ = 1$, $\cot 45^\circ = 1$, and $\csc 45^\circ = \sqrt{2}$.

▸ **Memory Trick.** It is easy to figure out the trigonometric functions of 45° using the triangle in Figure 2.2-9, where $a = 1$. We have a right triangle that is easy to remember, with the legs equal to 1 and the hypotenuse equal to $\sqrt{2}$. By applying the definitions of the trigonometric functions to this triangle, you can easily find their values for 45°. For example, $\sin 45^\circ = 1/\sqrt{2} = \sqrt{2}/2$ and $\tan 45^\circ = 1/1 = 1$. For practice, you should figure out the rest.

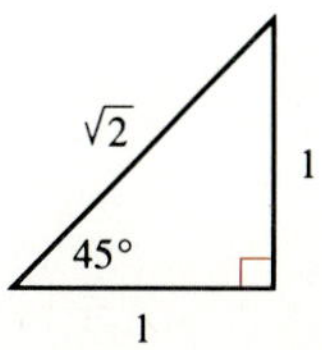

Figure 2.2-9

Triangle for finding the trigonometric functions of 45°

Functions of 30° and 60°

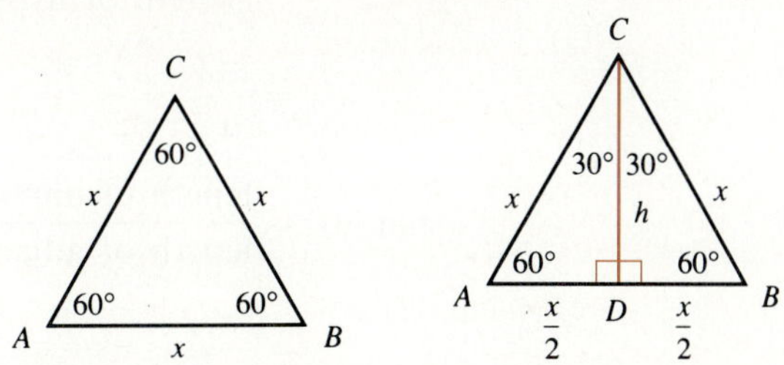

(a) Equilateral triangle ***(b) The angle C is bisected***

Figure 2.2-10

Equilateral triangles and right triangles

We can produce right triangles with acute angles of 30° and 60° by dividing in half an equilateral triangle with sides x. We do this by bisecting one of its angles, as shown in Figure 2.2-10. The acute angles (30° and 60°) in each of the smaller triangles add to 90°, which means that the angles at D must be 90°, or right angles, making the two smaller triangles right triangles. Both of these right triangles have sides h and x, so the third sides of each must be equal (you can prove this using the Pythagorean theorem) and must add to x. Therefore, $AD = x/2$ and $DB = x/2$. We can find h by applying the Pythagorean theorem to either of the small right triangles.

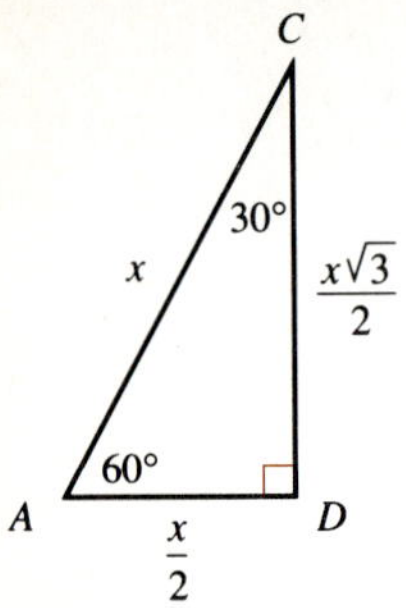

Figure 2.2-11

One of the right triangles isolated from Figure 2.2-10

$$
\begin{aligned}
h^2 + (x/2)^2 &= x^2 \\
h^2 + x^2/4 &= x^2 \\
h^2 &= \tfrac{3}{4}x^2 \\
h &= \sqrt{\tfrac{3}{4}x^2} = x\sqrt{3}/2
\end{aligned}
$$

Isolating triangle ACD in Figure 2.2-10 gives us Figure 2.2-11.

Now we can use the definitions of the trigonometric functions to evaluate them for 30° and 60°. For example, in Figure 2.2-11

$$
\begin{aligned}
\sin 60^\circ &= \frac{\text{length of opposite side}}{\text{length of hypotenuse}} \\
&= \frac{x\sqrt{3}/2}{x} \\
&= \frac{\sqrt{3}}{2} \\
\tan 60^\circ &= \frac{\text{length of opposite side}}{\text{length of adjacent side}} \\
&= \frac{x\sqrt{3}/2}{x/2} = \sqrt{3} \\
\sin 30^\circ &= \frac{\text{length of opposite side}}{\text{length of hypotenuse}} \\
&= \frac{x/2}{x} = \frac{1}{2} \\
\tan 30^\circ &= \frac{\text{length of opposite side}}{\text{length of adjacent side}} \\
&= \frac{x/2}{x\sqrt{3}/2} = \frac{1}{\sqrt{3}} = \frac{\sqrt{3}}{3}
\end{aligned}
$$

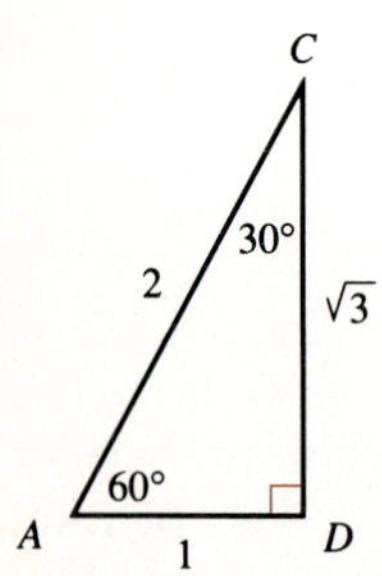

Figure 2.2-12

Triangle to find trigonometric functions of 30° and 60°

NOW IT'S YOUR TURN

Continue the preceding process to find all the other trigonometric functions of 30° and 60°.

(Answer: $\cos 30^\circ = \sqrt{3}/2$, $\cot 30^\circ = \sqrt{3}$, $\sec 30^\circ = 2\sqrt{3}/3$, $\csc 30^\circ = 2$, $\cos 60^\circ = \frac{1}{2}$, $\cot 60^\circ = \sqrt{3}/3$, $\sec 60^\circ = 2$, $\csc 60^\circ = 2\sqrt{3}/3$)

▸ **Memory Trick.** Just as we had a little trick to find the trigonometric functions of 45°, so we have a similar one for 30° and 60°. Letting $x = 2$ in Figure 2.2-11 gives the triangle in Figure 2.2-12. We can use this triangle to find the values of the trigonometric functions of 30° and 60°. For example,

$$\sec 60^\circ = \frac{\text{hypotenuse}}{\text{adjacent}} = \frac{2}{1} = 2$$

$$\tan 30^\circ = \frac{\text{opposite}}{\text{adjacent}} = \frac{1}{\sqrt{3}} = \frac{\sqrt{3}}{3}$$

Do the others yourself to be sure you know how to use this triangle.

Table 2.2-2 will help you organize some of these results. Notice how the progressions inside the square root signs help you to remember this table.

Table 2.2-2

θ	30°	45°	60°
$\sin\theta$	$\frac{1}{2}\sqrt{1}$	$\frac{1}{2}\sqrt{2}$	$\frac{1}{2}\sqrt{3}$
$\cos\theta$	$\frac{1}{2}\sqrt{3}$	$\frac{1}{2}\sqrt{2}$	$\frac{1}{2}\sqrt{1}$

▸ **It is not necessary to memorize any of these trigonometric functions. Of course you need to find their values so you can use them, but it is much more important for you to know how to figure them out than to memorize them!**

Example 8

Find the lengths to three significant figures of the two sides of the right triangle in Figure 2.2-13.

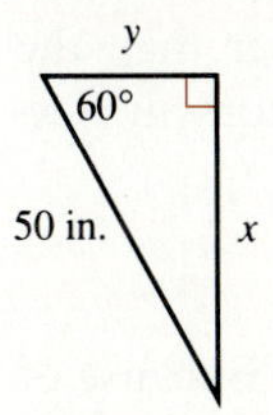

Figure 2.2-13

Solution

Find x: We notice that the side of length x is opposite the 60° angle, and we know the length of the hypotenuse—50 in. The only trigonometric functions involving the opposite side and the hypotenuse are sin 60° and csc 60°. We choose to work with sin 60°. (Why choose the sine? See the explanation after this example.)

$$\sin 60^\circ = \frac{\text{length of opposite side}}{\text{length of hypotenuse}}$$

$$\sin 60^\circ = \frac{x}{50 \text{ in.}}$$

$$x = (50 \text{ in.})(\sin 60^\circ) \qquad \textbf{(solve for } x\textbf{)}$$

$$x = (50 \text{ in.})\left(\frac{\sqrt{3}}{2}\right) = 25\sqrt{3} \text{ in.} \qquad \textbf{(use } \sin 60^\circ = \sqrt{3}/2\textbf{)}$$

$$x = 25\sqrt{3} \text{ in.} = (25)(1.732) \text{ in.} = 43.3 \text{ in.} \qquad \textbf{(convert to decimal form)}$$ ■

NOW IT'S YOUR TURN

Find x using the cosecant of 60° instead of the sine.

Find y: The side of length y is adjacent to the 60° angle. In this case, we want the adjacent side y and know the hypotenuse—50 in.—so the two likely trigonometric functions to use are the cosine and the secant of 60°. We choose to use the cosine. The definition of the cosine gives us

$$\cos 60° = \frac{\text{length of adjacent side}}{\text{length of hypotenuse}}$$

$$\cos 60° = \frac{y}{50 \text{ in.}}$$

$$y = (50 \text{ in.})\,(\cos 60°) \qquad \textbf{(solve for } y\textbf{)}$$

$$y = (50 \text{ in.})(\tfrac{1}{2}) = 25.0 \text{ in.} \qquad \textbf{(use } \cos 60° = \tfrac{1}{2}\textbf{)}$$

NOW IT'S YOUR TURN

Instead of finding y the way we did, find it using *both* of the following methods: (i) the Pythagorean theorem and (ii) the secant of 60°. It is important to be able to find other methods to check your original answer.

Which Trigonometric Function Should We Choose?

In Example 8 we had a choice of trigonometric functions: sin 60° or csc 60° for finding x and cos 60° or sec 60° for finding y. We chose to use the sine and cosine because calculators have keys only for the sine, cosine, and tangent. Therefore, we almost always work with these three functions rather than the cotangent, secant, and cosecant. We explain how to use calculators in Section 2.3.

Example 9

Passive Solar Design

An architect is designing a passive solar house for New Orleans, LA. Because of the hot climate, he wants the eaves of the roof to shade the south-facing walls completely throughout spring and summer. In a solar energy handbook, he finds that the noon sun is at 60° above the horizon on March 21 and September 21, and it is even higher during the summer.* How far should the eaves extend beyond the 9.0-ft high south walls to provide the desired shade?

Solution

First make a clear sketch showing the sun's rays on March 21 and September 21, when the sun is 60° above the horizon (Figure 2.2-14). On these dates, the eaves

* For the appropriate angle in your area, consult a solar energy handbook.

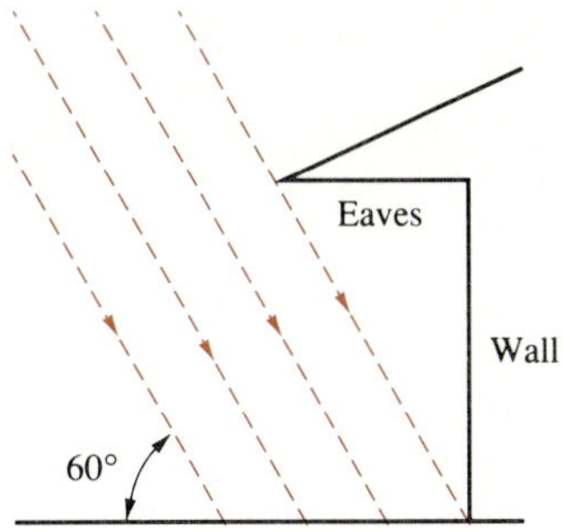

Figure 2.2-14

Sun's rays on March 21 and September 21

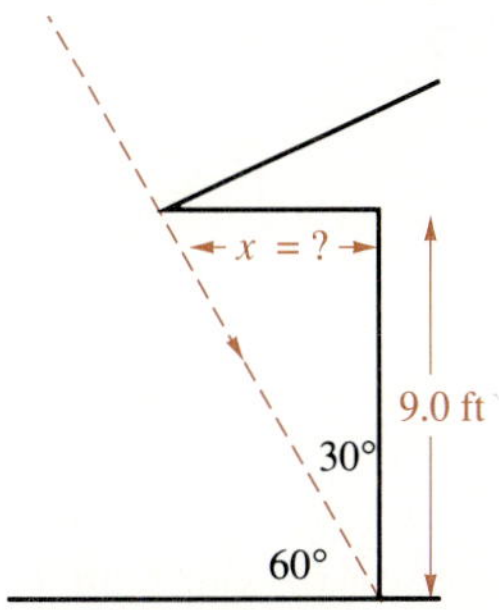

Figure 2.2-15

Simplified sketch of sun's rays, eaves, and wall

will just barely be able to shield the wall, so the rays just reach the bottom of the wall.

Now simplify the sketch, showing the angles and sides we know and what we are looking for, as shown in Figure 2.2-15. Notice that the rays make a 30° angle with the vertical wall and that the eaves and the wall make a right angle.

We know the side adjacent to the 30° angle (9.0 ft), and we want x (the side opposite this angle), so the tangent (or cotangent) is the function to use. Using the tangent gives

$$\tan 30° = \frac{\text{opposite}}{\text{adjacent}} = \frac{x}{9.0 \text{ ft}}$$

$$x = (9.0 \text{ ft}) \tan 30°$$

$$x = (9.0 \text{ ft})\left(\frac{1}{\sqrt{3}}\right) = (9.0 \text{ ft})\left(\frac{1}{1.732}\right) = 5.2 \text{ ft} \quad ■$$

NOW IT'S YOUR TURN

If the eaves can extend 6.0 ft, how high could the walls be and still be shaded as required? (Answer: 10 ft)

Example 10
Architecture

As an architect, you are designing a house for a steeply sloping lot in San Francisco (Figure 2.2-16). The side of the hill slopes at a 30° angle above the horizontal. This house is to be 25 ft tall at entry level and 47 ft deep. How high will be the side of the house above the sloping surface of the hill?

Solution

First look at the architect's plan of the house (Figure 2.2-16). Then make a simplified sketch of it, as shown in Figure 2.2-17, identifying what we know and what we want to find as well as any right triangles that are present. We want to find h, but first we must find y because $h = y + 25$ ft.

The lower part of the building forms a right triangle with a 30° angle (Figure 2.2-17b). We know the side adjacent to this angle (47 ft), and we are look-

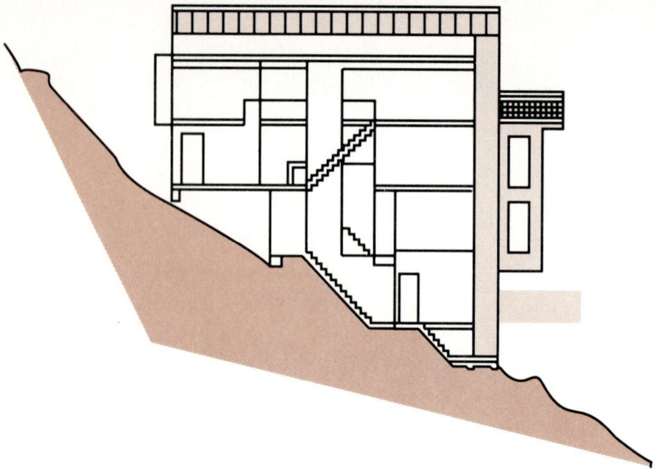

Figure 2.2-16

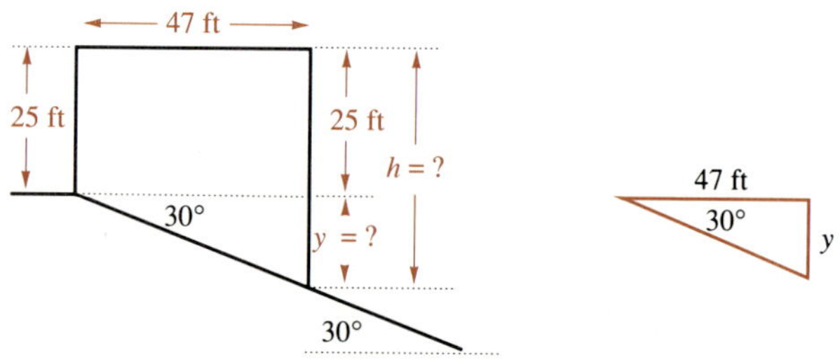

Figure 2.2-17

Simplified sketch of house on a hill

(a) Entire house

(b) Right triangle formed by lower part of house

ing for the side *opposite* (y). The only trigonometric functions involving these quantities are tan 30° and cot 30°. Using the tangent, we have

$$\tan 30° = \frac{\text{opposite}}{\text{adjacent}} = \frac{y}{47 \text{ ft}}$$

$$y = (47 \text{ ft}) \tan 30° = (47 \text{ ft})\left(\frac{1}{\sqrt{3}}\right) = \frac{47 \text{ ft}}{1.732} = 27 \text{ ft}$$

Finally, we can find h.

$$h = y + 25 \text{ ft} = 27 \text{ ft} + 25 \text{ ft} = 52 \text{ ft} \blacksquare$$

NOW IT'S YOUR TURN

If the diagonal base of the house, just along the surface of the hill, must be spanned by a single diagonal metal beam for extra support in case of an earthquake, how long must this beam be? (Answer: 54 ft)

Example 11
Archaeology

An archaeologist has discovered the ruins of the ancient building shown in Figure 2.2-18. At each side, the roof begins 28.0 ft above the ground but slopes upward at 30.0° on one side and at 45.0° on the other. How high was the peak of this building before it was destroyed?

Solution

Figure 2.2-18
Ruins of ancient building

First make a sketch of the building before it fell down, showing the information you know and what you want to find (Figure 2.2-19).

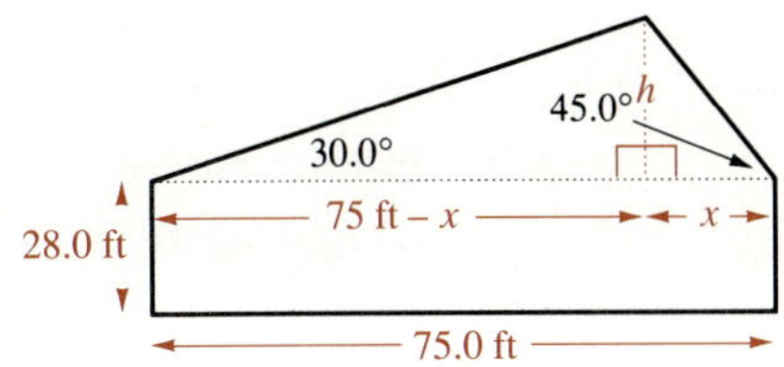

Figure 2.2-19
Original building before it was ruined

We want to find h. Notice that h divides the roof section into two right triangles. In the left-hand triangle,

$$\tan 30.0° = \frac{h}{75.0 \text{ ft} - x}$$

$$\frac{1}{\sqrt{3}} = \frac{h}{75.0 \text{ ft} - x}$$

In the right-hand triangle,

$$\tan 45.0° = \frac{h}{x}$$

$$1 = \frac{h}{x}$$

$$x = h$$

Substituting $x = h$ into the first equation gives us

$$\frac{1}{\sqrt{3}} = \frac{h}{75.0 \text{ ft} - h}$$

$$75.0 \text{ ft} - h = h\sqrt{3} \quad \textbf{(cross multiply)}$$

$$75.0 \text{ ft} = h + h\sqrt{3} = h(1 + \sqrt{3})$$

$$h = \frac{75.0 \text{ ft}}{1 + \sqrt{3}} = \frac{75.0 \text{ ft}}{1 + 1.732} = 27.5 \text{ ft}$$

Therefore, including the walls, the *total* height would have been

$$H_{\text{total}} = 28.0 \text{ ft} + 27.5 \text{ ft} = 55.5 \text{ ft} \quad ■$$

NOW IT'S YOUR TURN

Suppose the two parts of the roof sloped upward at 45.0° and at 60.0° instead but that all else was the same. How high was the original building?
(Answer: 75.5 ft)

2.2-3 Basic Relations Between the Trigonometric Functions

Reciprocal Relations

Take a look at the definitions of the six trigonometric functions that follow.

$$\sin A = \frac{a}{c}$$

$$\cos A = \frac{b}{c}$$

$$\tan A = \frac{a}{b}$$

$$\cot A = \frac{b}{a}$$

$$\sec A = \frac{c}{b}$$

$$\csc A = \frac{c}{a}$$

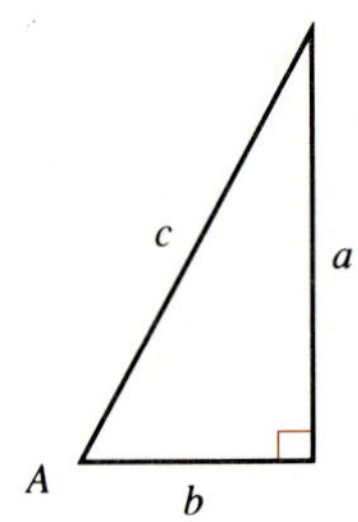

Figure 2.2-20

Notice, for example, that both sin A and csc A involve the same two quantities, a and c, which suggests that these two trigonometric functions are not independent but are related to each other. The same sort of thing is true for cos A—sec A and for tan A—cot A. We can use their definitions to see just how they are related.

From the definitions, we get the following results:

$$\sin A \csc A = \left(\frac{a}{c}\right)\left(\frac{c}{a}\right) = 1$$

$$\cos A \sec A = \left(\frac{b}{c}\right)\left(\frac{c}{b}\right) = 1$$

$$\tan A \cot A = \left(\frac{a}{b}\right)\left(\frac{b}{a}\right) = 1$$

▸ **Notice that these results are true regardless of what value the angle A has. Therefore, they are true for all angles between 0° and 90°.**

The preceding results give us the following relations:

$$\sin A \csc A = 1 \quad \Rightarrow \quad \sin A = \frac{1}{\csc A}$$

$$\cos A \sec A = 1 \quad \Rightarrow \quad \cos A = \frac{1}{\sec A}$$

$$\tan A \cot A = 1 \quad \Rightarrow \quad \tan A = \frac{1}{\cot A}$$

These relations mean that certain trigonometric functions are *reciprocals* of other trigonometric functions, provided of course that the denominator is not zero. (Remember from algebra that division by zero is not possible.)

THE RECIPROCAL PROPERTIES OF THE TRIGONOMETRIC FUNCTIONS

$$\sin A = \frac{1}{\csc A} \qquad \csc A = \frac{1}{\sin A}$$

$$\cos A = \frac{1}{\sec A} \qquad \sec A = \frac{1}{\cos A}$$

$$\tan A = \frac{1}{\cot A} \qquad \cot A = \frac{1}{\tan A}$$

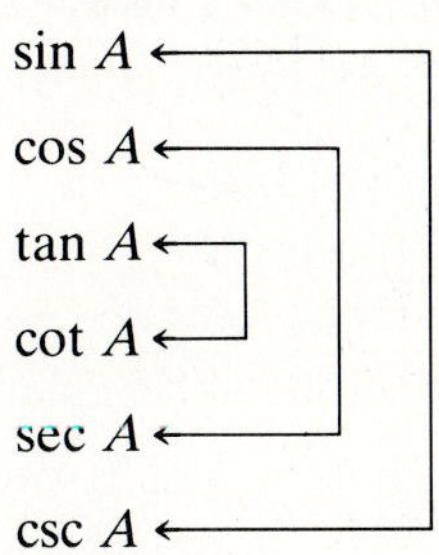

Reciprocals

▸ **The denominators of these reciprocals cannot be zero!**

The scheme shown in the left margin is helpful in remembering the reciprocal properties of trigonometric functions. For example, sec A is the reciprocal of the cos A, sin A is the reciprocal of the csc A, and so on.

▸ **Important Note.** These reciprocal properties are especially helpful when we use calculators to find the values of the trigonometric functions. Most calculators have only sine, cosine, and tangent keys, so we have to use these reciprocal relations to evaluate the other three trigonometric functions.

Quotient Relations

Besides the reciprocals, there are other relations between the trigonometric functions. Figure 2.2-20 tells us that

$$\tan A = \frac{a}{b}$$

If we divide both numerator and denominator by c, we find

$$\tan A = \frac{a/c}{b/c}$$

but Figure 2.2-20 tells us that $\sin A = a/c$ and $\cos A = b/c$. Therefore, we can write $\tan A$ as

$$\boxed{\tan A = \frac{\sin A}{\cos A}}$$

NOW IT'S YOUR TURN

Use the same procedure to prove the second important result.

$$\boxed{\cot A = \frac{\cos A}{\sin A}}$$

There are *two* ways you could prove this result. Can you find *both*? Both of these relations involve quotients of the sine and cosine, so we call them the **quotient relations**.

Example 12

If $\sin \theta = 0.75$ and $\cos \theta = 0.66$, evaluate all the other trigonometric functions of θ to two decimal places.

Solution

We can use the basic relations we just derived.

$$\tan \theta = \frac{\sin \theta}{\cos \theta} = \frac{0.75}{0.66} = 1.14$$

$$\cot \theta = \frac{1}{\tan \theta} = \frac{1}{1.14} = 0.88$$

$$\sec \theta = \frac{1}{\cos \theta} = \frac{1}{0.66} = 1.52$$

$$\csc \theta = \frac{1}{\sin \theta} = \frac{1}{0.75} = 1.33 \quad \blacksquare$$

NOW IT'S YOUR TURN

If $\sin \alpha = 0.42$ and $\tan \alpha = 0.46$, evaluate all the other trigonometric functions of α to two decimal places.

(Answer: $\cos \alpha = 0.91$, $\cot \alpha = 2.17$, $\sec \alpha = 1.10$, $\csc \alpha = 2.38$)

Example 13

If $\sin \theta = \frac{1}{3}$, evaluate all the other trigonometric functions of θ.

Solution

Because $\sin \theta = \frac{1}{3}$, the ratio of the opposite side to the hypotenuse is $\frac{1}{3}$.

$$\frac{\text{opposite}}{\text{hypotenuse}} = \frac{1}{3}$$

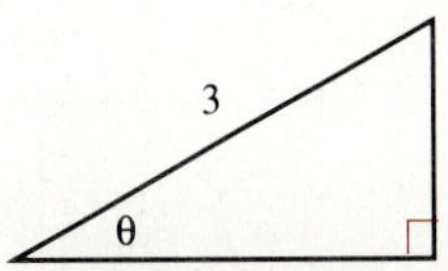

Figure 2.2-21

A right triangle with $\sin \theta = \frac{1}{3}$

Any right triangle with the same angle θ will have these ratios, so we can draw a simple right triangle with these ratios, as shown in Figure 2.2-21. We can get the third side (call it b) by the Pythagorean theorem.

$$3^2 = 1^2 + b^2$$

$$9 = 1 + b^2$$

$$b = \sqrt{8}$$

Therefore,

$$\cos \theta = \frac{\text{adjacent}}{\text{hypotenuse}} = \frac{b}{c} = \frac{\sqrt{8}}{3}$$

Now we can use the basic relations we just found to get the other trigonometric functions.

$$\tan \theta = \frac{\sin \theta}{\cos \theta} = \frac{1/3}{\sqrt{8}/3} = \frac{1}{\sqrt{8}} = \frac{\sqrt{8}}{8}$$

$$\cot \theta = \frac{1}{\tan \theta} = \frac{1}{1/\sqrt{8}} = \sqrt{8}$$

$$\sec \theta = \frac{1}{\cos \theta} = \frac{1}{\sqrt{8}/3} = \frac{3}{\sqrt{8}} = \frac{3\sqrt{8}}{8}$$

$$\csc \theta = \frac{1}{\sin \theta} = \frac{1}{\frac{1}{3}} = 3 \quad \blacksquare$$

NOW IT'S YOUR TURN

If $\sin\theta = \frac{2}{3}$, evaluate all the other trigonometric functions of θ.

(Answer: $\cos\theta = \sqrt{5}/3$, $\tan\theta = 2\sqrt{5}/5$, $\cot\theta = \sqrt{5}/2$, $\sec\theta = 3\sqrt{5}/5$, $\csc\theta = 3/2$)

EXERCISES 2.2-1

In Exercises 1–6, for the angle θ in the right triangles, find the opposite side, the adjacent side, and the ratio of the adjacent side to the hypotenuse. Leave your answers in whole numbers or square roots.

1.
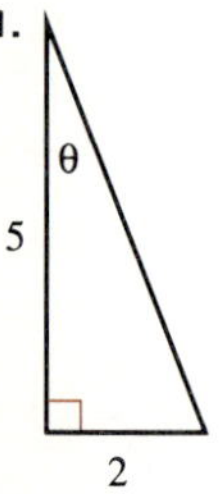

2.
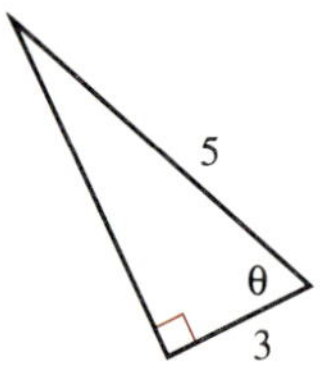

3.
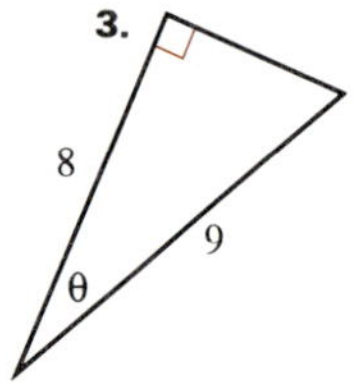

4.
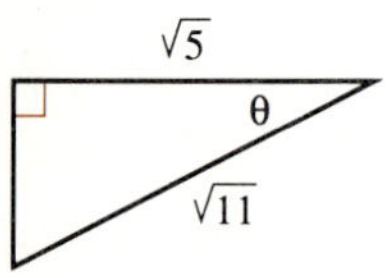

5.
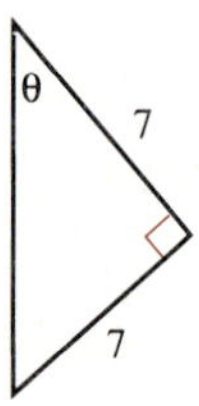

6.
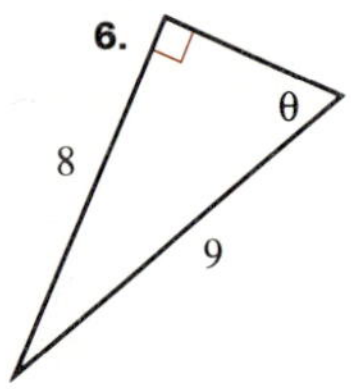

In Exercises 7–21, find the values of the six trigonometric functions of the angle A for the triangle in Figure 2.2-22. Express numerical answers to two decimal places.

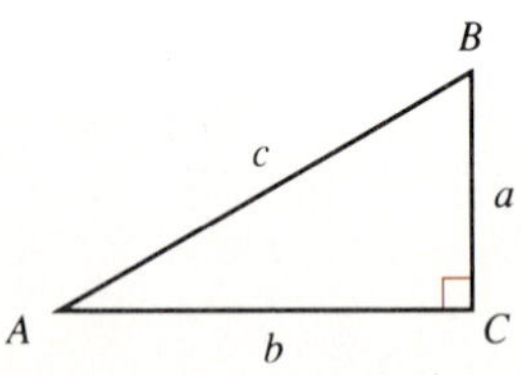

Figure 2.2-22

7. $a = 3, b = 4, c = 5$ **8.** $a = 5, b = 12, c = 13$

9. $a = 8, b = 15, c = 17$ **10.** $a = 9, b = 40, c = 41$

11. $a = 3.9, b = 8, c = 8.9$

12. $a = 1.19, b = 1.20, c = 1.69$

13. $a = 6, b = 8, c = 10$ **14.** $a = 40, b = 9, c = 41$

15. $a = 6, b = 5$ **16.** $a = 9, c = 13$

17. $a = 2, c = 2\sqrt{2}$

***18.** $a = 2p,\ b = p^2 - 1,\ c = p^2 + 1$

***19.** $a = 2pq,\ b = q^2 - p^2,\ c = q^2 + p^2$

***20.** $a = p,\ b = \frac{1}{2}(p^2 - 1),\ c = \frac{1}{2}(p^2 + 1)$

***21.** $a = 2pq/(q - p),\ b = p + q,\ c = (q^2 + p^2)/(q - p)$

In Exercises 22-27, evaluate all six trigonometric functions of the angle θ in the right triangles. Leave your answers in whole numbers or square roots.

22.
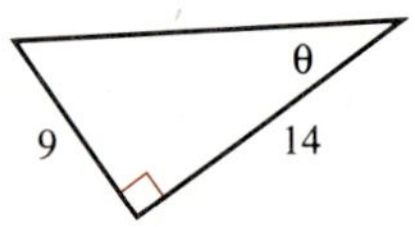

23.
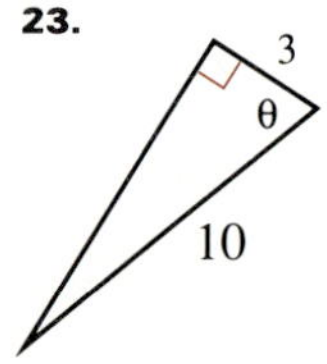

24.
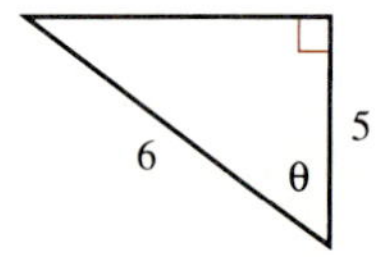

25.
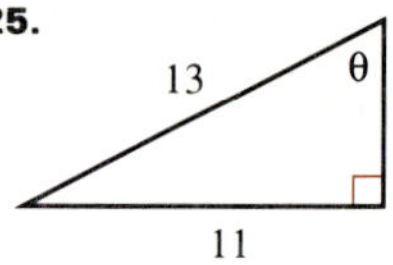

26.
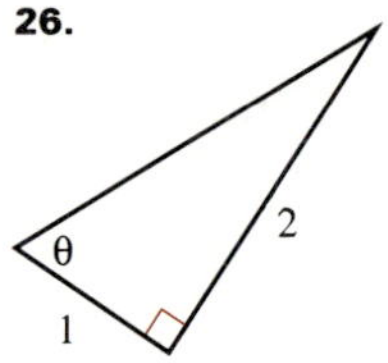

27.
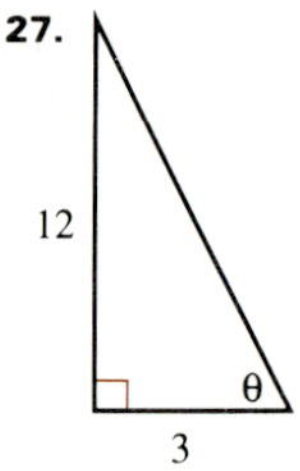

Using Figure 2.2-22, evaluate all six trigonometric functions of A when:

28. $a = b$ **29.** $a = 2b$ **30.** $a = \frac{2}{3}c$

In Exercises 31–37, refer to Figure 2.2-22.

31. Find the length of side a if $\sin A = \frac{3}{5}$ and $c = 20.5$.

32. Find the length of side b if $\cos A = 0.44$ and $c = 3.5$.

33. Find the length of side a if $\tan A = \frac{11}{3}$ and $b = \frac{27}{11}$.

34. Find the length of side b if $\cot A = 4$ and $a = 1700$.

35. Find the length of the hypotenuse if $\sec A = \frac{13}{5}$ and $b = 10.0$.

36. Find the length of the hypotenuse if $\cos A = \frac{8}{15}$ and $b = 16$.

37. Find the length of the hypotenuse if $\csc A = 6.40$ and $a = 35.6$.

In Exercises 38–41, find the missing sides of the triangle in Figure 2.2-22 given that:

***38.** $\tan A = \frac{3}{2}$ and $b = 6$

***39.** $\cos A = \frac{1}{2}$ and $a = 3\frac{1}{2}$

***40.** $\csc A = 1\frac{2}{3}$ and $b = 4$

***41.** $\sin A = 0.60$ and $b = 2.0$

***42. Area of a Triangle.** Prove that the area of the triangle in Figure 2.2-23 is $A = \frac{1}{2}ab \sin \theta$. Prove that it is also given by $A = \frac{1}{2}bc \sin \beta$ and $A = \frac{1}{2}ac \sin \phi$. *Hint*: Think of each side, in turn, as the base and construct the altitude h perpendicular to that base. Then express h in terms of the sides and angles of the given triangle. Finally, use the formula for the area of a triangle.

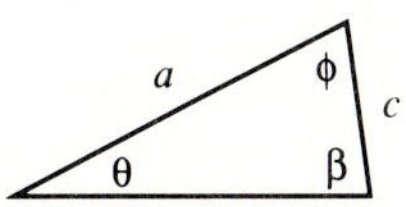

Figure 2.2-23

In Exercises 43–45, find the area of the triangle in Figure 2.2-23 if:

43. $a = 5$ ft, $b = 6$ ft, and $\sin \theta = \frac{3}{5}$

44. $b = 12.0$ cm, $c = 7.5$ cm, and $\sin \beta = \frac{3}{4}$

45. $a = 18.5$ mm, $c = 39.0$ mm, and $\sin \phi = \frac{4}{9}$

46. If the area of the triangle in Figure 2.2-23 is 150 cm^2, $a = 120$ cm, and $b = 25$ cm, find $\sin \theta$.

47. If the area of the triangle in Figure 2.2-23 is 80.0 in^2, $c = 22.5$ in., and $b = 10.2$ in., find $\sin \beta$.

48. If the area of the triangle in Figure 2.2-23 is 125 ft^2, $a = 16.50$ ft, and $c = 22.6$ ft, find $\sin \phi$.

49. Draw a right triangle with sides a, b, and c and an acute angle θ. Use this triangle to show that, *no matter what the angle θ is*, it must be true that $\csc \theta = 1/\sin \theta$, $\sec \theta = 1/\cos \theta$, and $\cot \theta = 1/\tan \theta$.

50. Use the same approach as in Exercise 49 to show that, *no matter what the angle θ is*, $\tan \theta = \sin \theta/\cos \theta$ and $\cot \theta = \cos \theta/\sin \theta$.

51. If two angles θ and α are complementary, how is $\sin \theta$ related to $\cos \alpha$; how is $\csc \theta$ related to $\sec \alpha$; how is $\tan \theta$ related to $\cot \alpha$? ***Hint:*** Sketch a right triangle having α and θ as its acute angles and then interpret this problem geometrically.

****52.** In Figure 2.2-24, the line segment BN is tangent to the circle at B, and MA is tangent to the circle at A. If the circle has radius 1, express the length of each of the following segments in terms of θ: **(a)** OH (Answer: $OH = \cos \theta$), **(b)** BN, **(c)** ON, **(d)** HP, **(e)** AM, **(f)** OM.

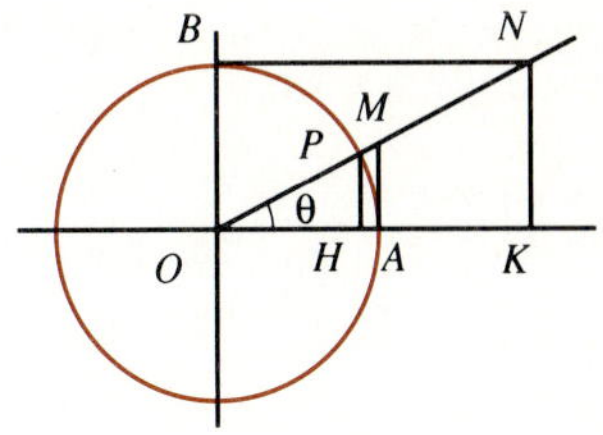

Figure 2.2-24

Exercises 2.2-2

In Exercises 1–11, evaluate each expression without using a calculator, leaving your answers in radical (square root) form.

1. $4 \sin 30° - 3 \tan 45° + \frac{1}{2} \sec 45°$

2. $\sec \frac{\pi}{3} + \frac{\pi}{4}$

3. $\sin 30° \cos 60° + \cos 30° \sin 60°$

4. $\tan 60° - \sec 30°$

5. $\tan 60° - [(\sin 30°)/(1 + \cos 30°)]$

6. $\sec \frac{\pi}{3} \sin \frac{\pi}{3}$

7. $\sin^2 30° + \cos^2 30°$ [$\sin^2 30°$ means $(\sin 30°)^2$]

8. $\sin^2 45° + \cos^2 45°$

9. $\sin^2 \frac{\pi}{3} + \cos^2 \frac{\pi}{3}$

10. $1 + \tan^2 60° - \sec^2 60°$

11. $\sin^2 \frac{\pi}{6} + \sin^2 \frac{\pi}{4} + \sin^2 \frac{\pi}{3}$

In Exercises 12–26, check to see if each equality is true by substituting in the values of the trigonometric functions.

12. $2 \sin 30° - 1 = 0$

13. $\cos 60° = \cos^2 30° - \sin^2 30°$

14. $-\frac{15}{4} + 7 \sin \frac{\pi}{6} + \sin^2 \frac{\pi}{6} = 0$

15. $2 \cos \frac{\pi}{4} \sin \frac{\pi}{4} = 1$

16. $\cos 60° = 2 \cos^2 30° - 1$

17. $\sin \frac{\pi}{6} = \sqrt{(1 - \cos \pi/3)/2}$

18. $\cos 30° = \sqrt{(1 + \cos 60°)/2}$

19. $\tan 30° = (\sin 60°)/(1 + \cos 60°)$

20. $\tan 30° = (1 - \cos 60°)/\sin 60°$

21. $2 \cos^2 \frac{\pi}{6} = 1 + \cos \frac{\pi}{3}$

22. $2 \sin^2 \frac{\pi}{6} = 1 - \cos \frac{\pi}{3}$

23. $\sin 60° = 2 \cos 30° \sin 30°$

24. $\tan 60° = (2 \tan 30°)/(1 - \tan^2 30°)$

25. $1 + \tan^2 \frac{\pi}{4} = \sec^2 \frac{\pi}{4}$

26. $\csc^2 45° = 1 + \cot^2 45°$

In Exercises 27–32, find the acute angle θ (in degrees) that satisfies the given equation. Use Table 2.2-2.

Sample Problem For what acute angle θ is the equation $2 \cos \theta - 1 = 0$ valid? Solution. First solve the equation for $\cos \theta$.

$$2 \cos \theta - 1 = 0 \quad \rightarrow \quad 2 \cos \theta = 1 \quad \rightarrow \quad \cos \theta = \tfrac{1}{2}$$

Therefore, θ must be the angle whose cosine is $\frac{1}{2}$. Using Table 2.2-2, we find that $\theta = 60°$.

27. $2 \cos \theta - \sqrt{2} = 0$

28. $2 \sin \theta - \sqrt{2} = 0$

29. $2 \sin \theta - 1 = 0$

30. $2 \cos \theta - \sqrt{3} = 0$

31. $2 \sin \theta - \sqrt{3} = 0$

***32.** $\sin \theta = \cos \theta$

Area of a Triangle. See Exercise 42 in the Exercises for 2.2-1. In Exercises 33–35, find the area of the triangle in Figure 2.2-23 if:

33. $a = 5.0$ mi, $b = 14$ mi, and $\theta = 30°$

34. $a = 8.0$ mm, $c = 10.0$ mm, and $\phi = \frac{\pi}{3}$

35. $b = 7.88$ ft, $c = 9.54$ ft, and $\beta = \frac{\pi}{4}$

***36.** If the area of the triangle in Figure 2.2-23 is 6.75 ft^2, $a =$ 3.0 ft, and $b = 9.0$ ft, find θ.

***37.** An isosceles triangle has two sides equal to 40.0 cm with a 120° angle between them. Find the length of the third side and the area of this triangle.

38. Find the area of the triangle in Figure 2.2-25.

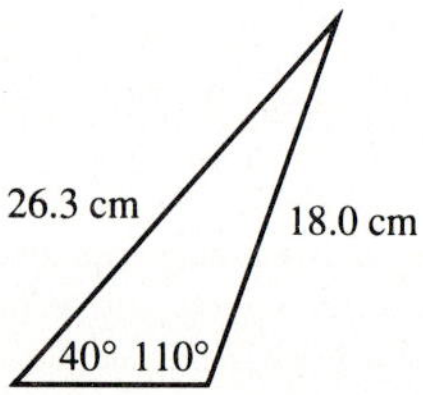

Figure 2.2-25

In Exercises 39–47, find the lengths of all three sides of the right triangle in Figure 2.2-26, using the information given.

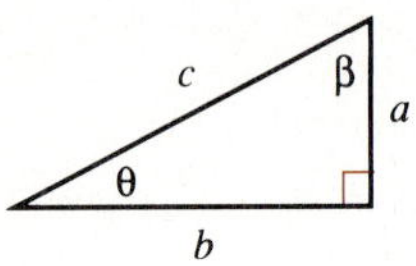

Figure 2.2-26

39. $\theta = 30°$, $c = 6.0$ ft

40. $\beta = \frac{\pi}{6}$, $c = 6.0$ ft

41. $\beta = \frac{\pi}{4}$, $a = 16.0$ cm

42. $\theta = 60°$, $b = 4.0$ m

43. $\beta = \frac{\pi}{3}$, $b = 45$ mm

***44.** $\sin \theta = 0.240$, $c = 8.0$ ft

***45.** $\tan \theta = 2.85$, $b = 1.8$ ft

***46.** $\sec \beta = 6.5$, $a = 10.0$ m

***47.** $\cos \theta = 0.650$, $a = 150.0$ cm

48. An A-frame cabin has walls that slope upward at 60° above the horizontal as shown in Figure 2.2-27. If the base is 35 ft wide, how high is the peak?

49. Some building codes specify that the top of a chimney must be at least 18 in. above the highest point on the roof. For the house in Figure 2.2-28, the roof begins at 8.0 ft above ground level, and both sides slope upward at 45° above the horizontal. How high must the chimney be?

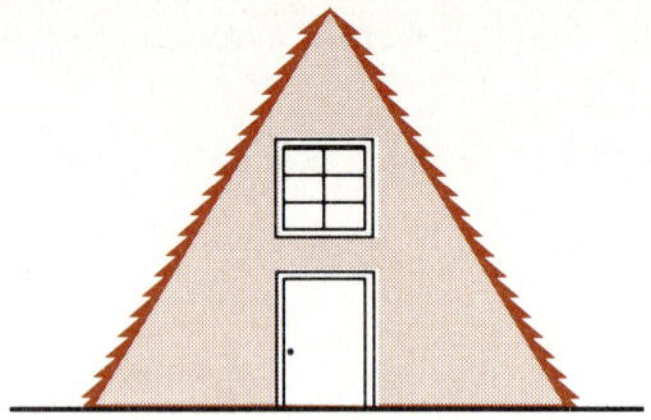

Figure 2.2-27

Figure 2.2-28

***50.** You are designing a two-story, 20.0-ft deep A-frame cabin for the mountains. To prevent the roof's collapse from the weight of snow, the roof must slope upward at 60.0° above the horizontal. If the lower level has a maximum ceiling height of 9.0 ft and a floor area of 500.0 ft^2, find the floor area of the second floor and the maximum height of its ceiling (Figure 2.2-29).

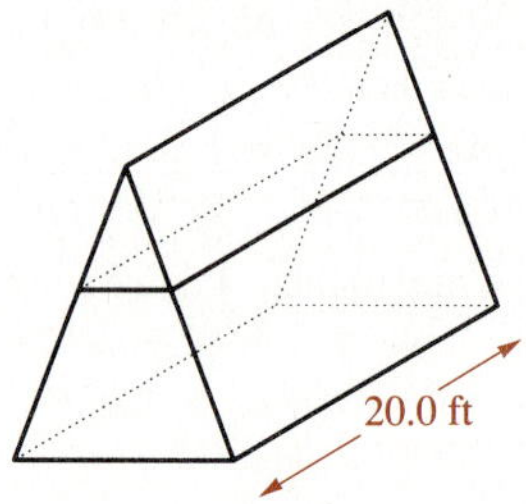

Figure 2.2-29

51. If you are ordering shingles to cover the outside of the slanting sides for the cabin in Figure 2.2-29, what is the total area you must cover?

52. You are adding an attic room, as shown in Figure 2.2-30, with the two side walls each 4.00 ft high. If your roof slopes upward at 30.0°, find the width W of this room. Will it be high enough in the center for a 6.0-ft person to stand?

53. As an archaeologist, you have discovered the ruins of the ancient building in Figure 2.2-31. On each wall, the roof begins 25.6 ft above the ground and slopes upward at 30.0°. How high was the peak of this building before it was destroyed?

54. Near the ruins of the building in Figure 2.2-31, you discover another similar building, also with a crumbled peak. This building was 50.0 ft across at the base, and its roof began 18.0 ft above the ground, but it had two slopes. One side sloped upward at 45° and

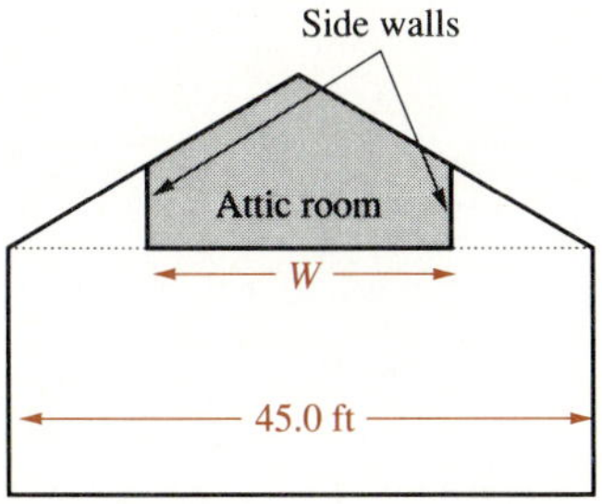

Figure 2.2-30

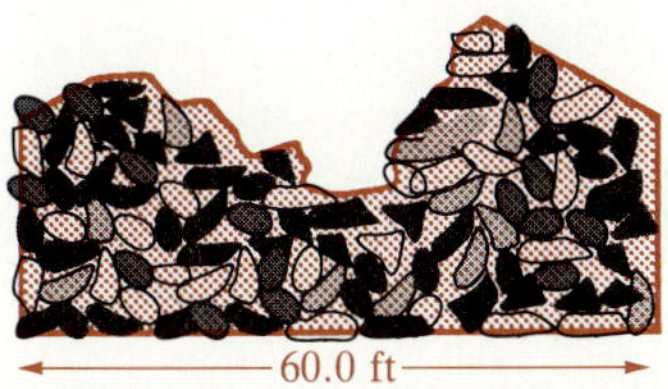

Figure 2.2-31

the other at 60.0°. How tall was the peak of this building? If you stood directly under the peak, how far would you be from the two supporting walls?

55. You live in New Orleans, and because of your skimpy 18-in. roof overhang, too much sun enters your south-facing window. You want to hang a shade from the edge of your roof, as shown in Figure 2.2-32, to block the noon sun completely from that window on March 21st. What is the shortest shade that will do this? (See Example 9 for pertinent information.)

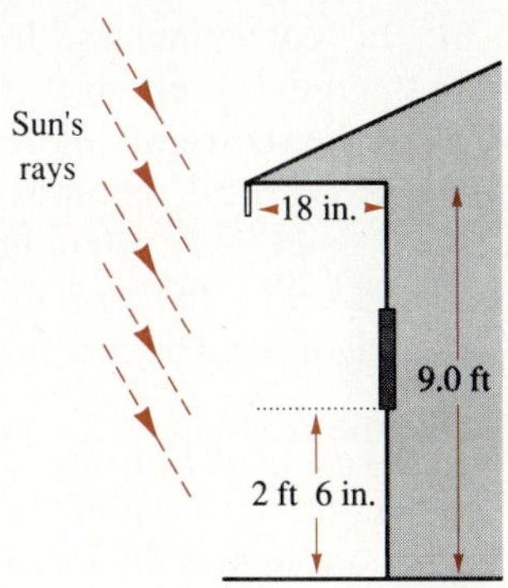

Figure 2.2-32

56. The dump truck in Figure 2.2-33 has a bed 15 ft long that can be raised to a maximum angle of 60° above the horizontal. If the pivot is 4.0 ft above the ground, what vertical clearance is needed so the truck can raise its bed completely?

57. An equilateral triangle is inscribed inside a circle 20 cm in diameter. How far is it from the center of this circle to the midpoint of each side of the triangle?

Figure 2.2-33

58. A circle of radius 12.0 cm is inscribed inside an equilateral triangle. How far is each vertex of this triangle from the center of the circle?

59. An equilateral triangle is 100.0 cm on each side. Find the location of the point inside this triangle that is equidistant from all three vertices. How far is it from this point to each vertex?

***60.** For the triangle in Exercise 59, find the location of the point inside the triangle that is equidistant from all three sides. How far is this point from each side?

61. A regular hexagon is inscribed inside a circle of radius 12 in. How far is each vertex of this hexagon from its nearest neighbor?

62. A circle of radius 12 in. is inscribed inside a regular hexagon. How far is each vertex of this hexagon from its nearest neighbor?

63. Find the area of a regular hexagon 20.0 ft on a side without looking up any special formulas about areas of hexagons.

EXERCISES 2.2-3

In Exercises 1–6, find the values of the indicated functions.

1. $\csc A = \frac{5}{3}$, $\sin A =$ **2.** $\cot B = \frac{3}{2}$, $\tan B =$

3. $\cos\theta = \frac{5}{13}$, $\sec\theta =$ **4.** $\tan\beta = 0.300$, $\cot\beta =$

5. $\sec\alpha = 1.60$, $\cos\alpha =$ **6.** $\sin\phi = 0.61$, $\csc\phi =$

In Exercises 7–10, find the missing angle or trigonometric function.

7. $\cos 40° = 1/\sec\ (?)$ **8.** $\csc 52° = 1/(?)(52°)$

9. $\cot \frac{2\pi}{9} = 1/(?)(\frac{2\pi}{9})$ **10.** $\sin 24°18' = 1/(?)(24°18')$

Without using tables or calculators, find the products in Exercises 11–17.

11. $\tan 20° \cot 20° =$ **12.** $\cos 72° \sec 72° =$

13. $\sin 17°37' \csc 17°37' =$ **14.** $\frac{1}{3} \sec 36° \cos 36° =$

15. $-5 \cot\frac{\pi}{7} \tan\frac{\pi}{7} =$ **16.** $p \cos 1° \sec 1° =$

17. $k \csc 8° \sin 8\,° =$

In Exercises 18–22, evaluate the other four trigonometric functions of θ using the two given functions.

18. $\sin\theta = 0.3746$ and $\cos\theta = 0.9272$

19. $\tan\theta = 1.1918$ and $\cos\theta = 0.6428$

20. $\sec\theta = 1.7434$ and $\tan\theta = 1.4281$

21. $\csc \theta = 1.5557$ and $\sec \theta = 1.3054$

22. $\cot \theta = 3.7323$ and $\sin \theta = 0.2588$

In Exercises 23–31, evaluate the other five trigonometric functions of θ by using the given function.

23. $\sin \theta = \frac{1}{4}$

24. $\cos \theta = \frac{2}{7}$

25. $\tan \theta = \frac{3}{5}$

26. $\sec \theta = 3$

27. $\cot \theta = 1.5$

28. $\sin \theta = 0.650$

29. $\tan \theta = 6.90$

***30.** $\csc \theta = x$

***31.** $\cos \theta = a$

32. Prove that $\tan \theta = \sec \theta / \csc \theta$ and that $\cot \theta = \csc \theta / \sec \theta$.

2.3 FINDING AND USING THE TRIGONOMETRIC FUNCTIONS

2.3-1 Finding the Trigonometric Functions

In Section 2.2-2, we saw how to find the trigonometric functions of 30°, 45°, and 60° using special properties of certain right triangles. Unfortunately, most applications do not involve these angles, so we need to learn how to find the trigonometric functions of all angles. We can use two ways to find them: trigonometric tables or calculators. Tables have been available for a long time, but pocket calculators have virtually replaced them.

Until about 1970, the only good way to find trigonometric functions was to look them up in a table. You could use a slide rule but it wasn't very accurate. This situation changed with the introduction of inexpensive pocket calculators—due to significant developments in microchips design and production.

Pocket calculators make it extremely simple to evaluate trigonometric functions, but we do pay a small price for this convenience. First, calculators usually operate only in radians or decimal degrees, rarely in degrees-minutes-seconds. Second, calculators give only three of the six trigonometric functions—sine, cosine, and tangent. We must use the reciprocal relations developed in Section 2.2-3 to find the other three functions. Here we explain how to find the trigonometric functions using calculators and leave the use of tables as an optional section in the appendix. To find the trigonometric function of an angle:

1. First, set your calculator to the mode—radian (RAD) or degree (DEG)—you want to use for angles. Be sure to set it to the number of decimal places you need in your answer. If your calculator does not allow you to set the number of decimal places, then you must round off the answer yourself.

2. Next, enter the numerical value of the angle. If the angle is in degrees, it should be in decimal form. If it is in radians, it should also be a decimal, not in terms of π.

3. Finally, press the key (sin, cos, tan) of the trigonometric function you want for that angle. If you want the secant, cosecant, or cotangent, you must use the reciprocal relations, as we show in the following examples.

Example 14 Find sin 53.2° to four decimal places.

Solution

1. Set your calculator to four decimal places in the DEG mode.
2. Enter "53.2".
3. Press the SIN key. Your screen should show 0.8007, meaning that sin 53.2° = 0.8007. ■

Example 15 Find tan 29°45′ to six decimal places.

Solution

1. Set your calculator to six decimal places in the DEG mode.
2. Convert 29°45′ to decimal form, which is 29.75°.
3. Enter "29.75".
4. Press the TAN key. Your screen should show 0.571547, meaning that tan 29°45′ = 0.571547. ■

Example 16 Find cos 1.05 to three-decimal accuracy.

Solution

1. Set your calculator to three decimal places in the RAD mode. (How do we know we should use the RAD mode?)
2. Enter "1.05".
3. Press the COS key. Your screen should show 0.498. ■

Example 17 Find cos $\frac{2\pi}{9}$ to four decimal places.

Solution

1. Set the calculator to four decimal places in the RAD mode.
2. Convert $\frac{2\pi}{9}$ to decimal form.

$$\frac{2\pi}{9} = 2(3.1416)/9 = 0.6981$$

Notice that we must use π to four decimal places because we want our answer to four decimal places. After this last calculation, your calculator should show 0.6981. If you erased it, re-enter it.

3. Press the COS key. Your screen should show 0.7660. ■

NOW IT'S YOUR TURN

Use a calculator to find: (a) tan 29.6° to four decimals, (b) sin 19°6′ to four decimals, and (c) tan $\frac{\pi}{5}$ to three decimals. (Answer: 0.5681, 0.3272, 0.727)

Example 18

Find sec 69.8° to five decimal places.

Solution

The calculator does not have a SEC key, but we can use the fact that the secant is the reciprocal of the cosine: $\sec \theta = 1/\cos \theta$. First find cos 69.8°, and then take its reciprocal to get sec 69.8°.

1. Set the calculator to five decimal places in the DEG mode.
2. Enter "69.8".
3. Press the COS key. This gives cos 69.8°, which is 0.34530.
4. Because $\sec \theta = 1/\cos \theta$, press the "$1/x$" (reciprocal) key, which gives you the reciprocal of cos 69.8°. Your screen should show 2.89605, which means that sec 69.8° = 2.89605. ■

Example 19

Find cot 0.62 to five decimal places.

Solution

There is no COT key, but $\cot \theta = 1/\tan \theta$. First find tan 0.62 and then take its reciprocal to get cot 0.62.

1. Set the calculator to five decimal places in the RAD mode.
2. Enter "0.62".
3. Press the TAN key. This gives you tan 0.62 = 0.71391.
4. Press the "$1/x$" (reciprocal) key to get cot 0.62, which should be 1.40074. ■

Example 20

Find csc $\frac{5\pi}{11}$ to four decimal places.

Solution

There is no CSC key, but $\csc \theta = 1/\sin \theta$. First find sin $\frac{5\pi}{11}$ and then take its reciprocal to get the cosecant.

1. Set the calculator to four decimal places in the RAD mode.
2. Convert $\frac{5\pi}{11}$ to a decimal to obtain 1.4280.
3. Press the SIN key to get $\sin \frac{5\pi}{11} = 0.9898$.
4. Press the "$1/x$" key to take the reciprocal of the sine. You should get 1.0103. ■

NOW IT'S YOUR TURN

Use a calculator to find csc 24.9° to four decimal places and cot $\frac{4}{\pi}$ to six decimal places. (Answer: 2.3751, 0.306661)

2.3-2 Using the Trigonometric Functions

Now that we can find the trigonometric functions of *any* acute angle, we are ready to use trigonometry for real-world applications where the angles are rarely 30°, 45°, or 60°. We have selected the following examples from a wide variety of fields.

Example 21

Measuring the Height of a Building

Measuring the height of a building using a ruler or tape measure is likely to be very difficult—and dangerous! Yet people like building inspectors must make such measurements every day to check for compliance with height regulations. How do they make these measurements?

Solution

It is very difficult to make measurements far above the ground, so we prefer to calculate the height from quantities we can measure on the ground. Figure 2.3-1 shows a person looking up at the top of a building, his eyes a distance h above the ground. From the ground, he can easily measure his distance d from the building using a tape measure and the angle above the horizon to the top of the building (its *angle of elevation* θ) using a surveyor's transit. He can then use these measurements to calculate y and then get H.

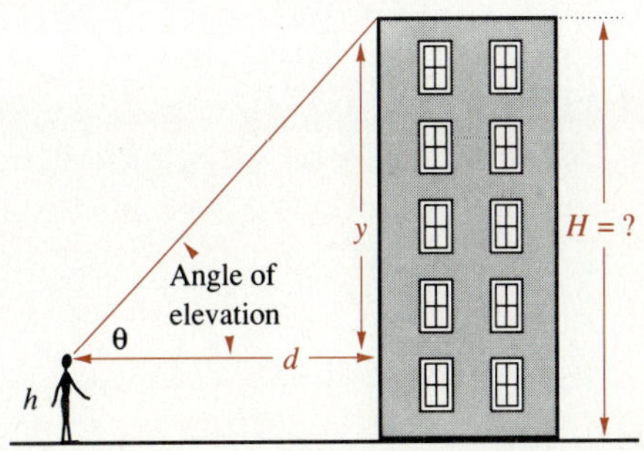

Figure 2.3-1

Measuring the height of a building

Suppose that an inspector who is 36.5 ft from a building measures the angle of elevation of the top to be 52°45′ from a point 5.00 ft above the ground. How tall is this building?

First we find y. In the right triangle in Figure 2.3-1, y is the length of the side opposite θ, and d is the length of the adjacent side. Therefore, we use the tangent.

$$\tan\theta = \frac{\text{opposite}}{\text{adjacent}} = \frac{y}{d}$$

$$y = d\tan\theta$$

Substituting the numbers we get

$$y = d\tan\theta = (36.5\text{ ft})\tan(52°45') = (36.5\text{ ft})(1.315) = 48.0\text{ ft}$$

Therefore, the height H of the building is

$$H = y + h = 48.0 \text{ ft} + 5.0 \text{ ft} = 53.0 \text{ ft} \; \blacksquare$$

which is approximately a five or six story building.

NOW IT'S YOUR TURN

If the inspector measures another building that has the same angle of elevation but is three times as far away, how tall is this building? (Answer: 149 ft)

Example 22

Volcanoes on Mars

Astronomers, geologists, and engineers must often determine the height of a natural feature using only aerial photographs. Because such photos do not show any depth, it is necessary to calculate the feature's height using the length of its shadow and the sun's angle in the photo. The sun's angle is normally recorded for such photos. The photograph in Figure 2.3-2 is a 1972 Mariner-9 image of Mars showing Olympus Mons, the largest known volcano in our solar system.

From the height of the Mariner-9 satellite, astronomers know that the shadow in the upper right-hand corner of the central crater is 10 km wide. At the time this photo was taken, the sun was 82° from the vertical on Mars, coming from the upper right-hand corner. How deep is the central crater below its rim?

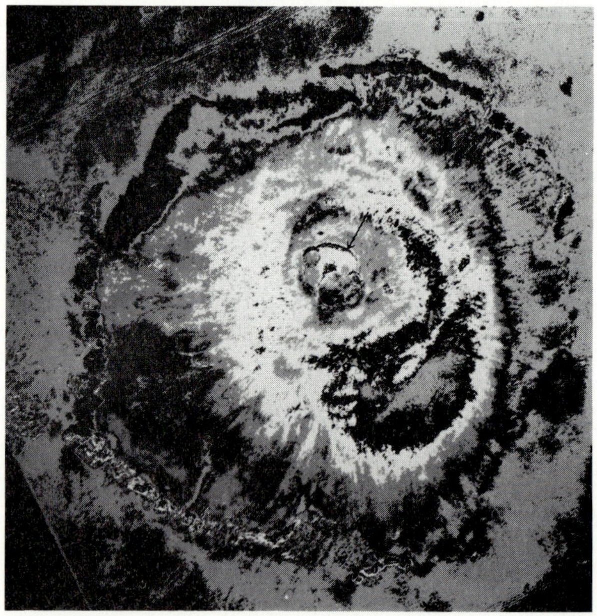

Figure 2.3-2

Olympus Mons volcano on Mars

Solution

First, sketch a simplified cross-sectional view of the walls of this crater (Figure 2.3-3). We know the angle 82° and the side opposite it (10 km), and we want

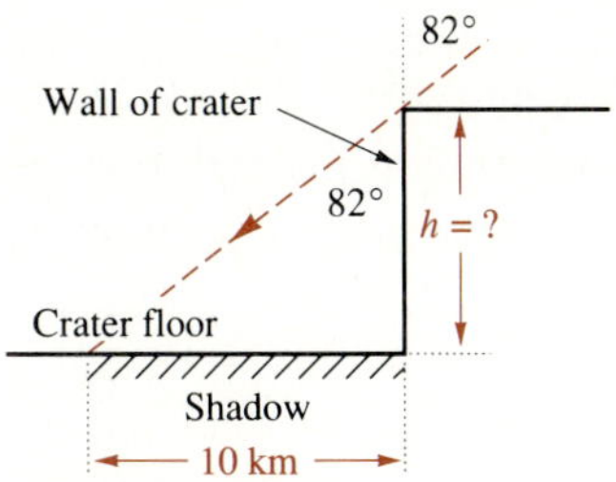

Figure 2.3-3

Cross-sectional view of crater wall

to calculate its adjacent side h. Therefore, the logical trigonometric function to use would be tan 82° or cot 82°. Because calculators have the tangent function, we use that.

$$\tan 82^\circ = \frac{\text{opposite}}{\text{adjacent}} = \frac{10\text{ km}}{h}$$

$$h = \frac{10\text{ km}}{\tan 82^\circ} = \frac{10\text{ km}}{7.115} = 1.4\text{ km} \blacksquare$$

NOW IT'S YOUR TURN

At the top of Figure 2.3-2, a cliff casts a shadow 15 km wide. How tall is this cliff? (Answer: 2.1 km)

Example 23

Water Molecule

In the water (H_2O) molecule in Figure 2.3-4, the hydrogen H atoms are each 0.958 Å from the oxygen O atom, and the H—O—H angle is 104°. (The angstrom unit Å, defined to be 10^{-10} m, is a unit of length frequently used on the atomic scale. It is roughly the diameter of a hydrogen atom.) How far are the hydrogen atoms from each other?

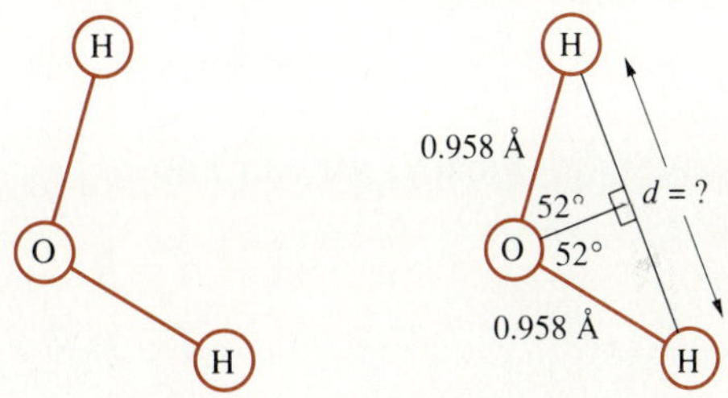

Figure 2.3-4

Water molecule

(a) Water molecule

(b) Water molecule with its dimensions shown

Solution

First sketch this molecule, showing what we know and what we want to find, as shown in Figure 2.3-4b. Consider either of the two identical right triangles. Because we want d and we know the hypotenuse, the sine (or cosecant) is the logical function to use. Because calculators have the sine function, we use that.

$$\sin 52^\circ = \frac{\text{opposite}}{\text{hypotenuse}} = \frac{d/2}{0.958\text{ Å}}$$

$$d = 2(0.958\text{ Å})\sin 52^\circ = 2(0.958)(0.7880) = 1.51\text{ Å} \blacksquare$$

NOW IT'S YOUR TURN

How far apart would the H atoms be if the angle were 121°? (Answer: 1.67 Å)

Example 24

Interior of the Earth

Geologists use earthquake waves to study the earth's interior structure. By monitoring the waves when an earthquake occurs, they find that waves travelling inward at angles less than 57° below the surface move straight through rock as expected. But those directed inward at a greater angle strike a spherical interior core. How far below the surface is this core?

Solution

First sketch the waves, as shown in Figure 2.3-5. Next make a simplified sketch, showing only the wave as it grazes the core, as shown in Figure 2.3-6. We want to find the depth d of the core below the surface. If we can find x, we can get d, because $x + d = R = 3960$ mi. The radius R is perpendicular to the surface, so

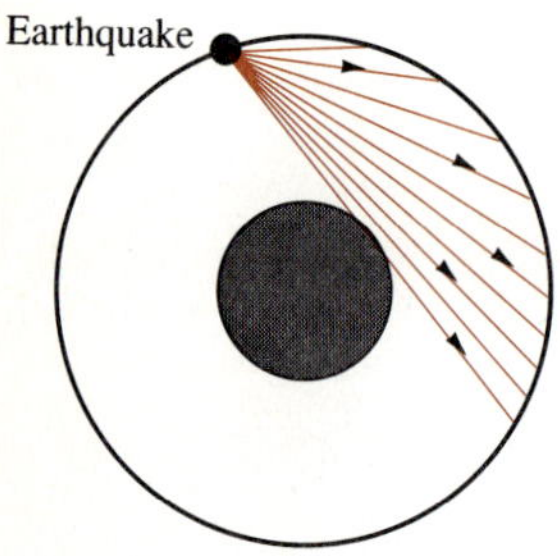

Figure 2.3-5

Earthquake waves travelling through the Earth

$$\alpha + 57° = 90°$$
$$\alpha = 33°$$

Now we can use $\sin \alpha$ to get x since we know R.

$$\sin \alpha = \frac{x}{R}$$
$$x = R \sin \alpha = (3960 \text{ mi})(\sin 33°) = (3960 \text{ mi})(0.5446) = 2160 \text{ mi}$$

Finally, we can find d.

$$x + d = R$$
$$d = R - x = 3960 \text{ mi} - 2160 \text{ mi} = 1800 \text{ mi}$$ ■

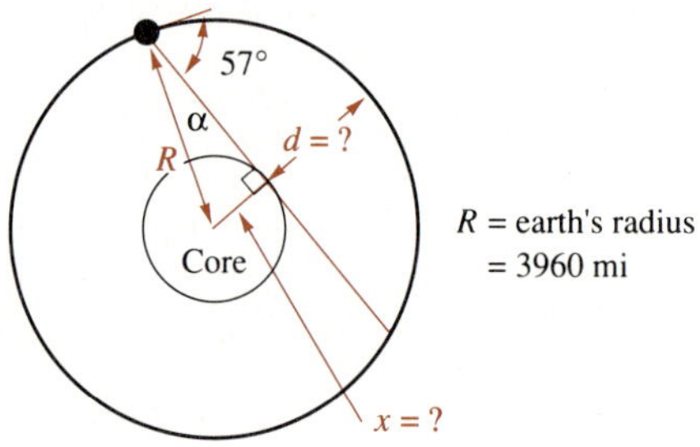

Figure 2.3-6

The innermost wave just grazes the core

NOW IT'S YOUR TURN

Suppose that earthquake waves got bent after going in at a 25° angle rather than a 57° angle. How deep would the core be in this case? (Answer: 370 mi)

EXERCISES 2.3-1

In Exercises 1–30, use a calculator to evaluate each function to four decimal places.

1. $\sin 22°$
2. $\cos 76°$
3. $\tan 8°$
4. $\cos 76.37°$
5. $\tan 0.86°$
6. $\sin 15.23°$
7. $\cos 11.4°$
8. $\sin 19.1°$
9. $\tan 41.9°$
10. $\sin 79.4°$
11. $\tan 11.2°$
12. $\tan 15°12'$
13. $\sin 79°15'$
14. $\cos 1°20'$
15. $\sin 54°41'$
16. $\tan 62°13'$
17. $\cos 80°25'$
18. $\cos 82°5''$
19. $\tan 11°11'$
20. $\sin 32°15'45''$
21. $\cos 57°24'6''$
22. $\tan 9°9'9''$
23. $\cos 71°39°7''$
24. $\sin 17°19''$
25. $\tan 40'14''$
26. $\cos 39'16''$
27. $\sin(18° + 36')$
28. $\tan(1° + 6')$
***29.** $\sin 0.55'$
***30.** $\cos 0.88''$

In Exercises 31–42, use a calculator to evaluate each function to three decimal places.

31. $\sin 0.13$
32. $\tan 0.39$
33. $\cos 1.55$
34. $\tan 0.95$
35. $\cos 0.69$
36. $\tan \frac{\pi}{9}$
37. $\cos \frac{3\pi}{8}$
38. $\sin \frac{\pi}{7}$
39. $\tan \frac{2\pi}{11}$
40. $\cos 0.23\pi$
41. $\sin 0.17\pi$
42. $\tan 0.41\pi$

In Exercises 43–62, use a calculator to evaluate each function to five decimal places.

43. $\cot 50°$
44. $\sec 87°$
45. $\csc 11°$
46. $\sec 29.8°$
47. $\cot 62.87°$
48. $\sec 0.134°$
49. $\cot 19.387°$
50. $\csc 19.19°$
51. $\sec 1°$
52. $\csc 6°15'$
53. $\cot 35°8'$
54. $\sec 13°51'$
55. $\sec 1.12$
56. $\csc 0.873$
57. $\cot 0.950$
58. $\sec \frac{2\pi}{9}$
59. $\cot \frac{4\pi}{15}$
60. $\csc 0.40\pi$
61. $\cot 0.12\pi$
62. $\sec 0.037\pi$

In Exercises 63–81, use a calculator to evaluate each function to three decimal places.

63. $\sin \frac{1}{\pi}$
64. $\cot \frac{3}{\pi}$
65. $\sin(1 + \frac{1}{\pi})$
***66.** $\sin(\sin 1.44)$
***67.** $\tan(\cos 1.12)$
***68.** $\cos(\sec 0.36)$
69. $\text{cost}\, \frac{\pi}{1.4}$
***70.** $\sin(\frac{1}{2}\sqrt{\pi})$
***71.** $\tan \pi^{\frac{1}{4}}$
72. $\sin(\pi - 2.3)$
73. $\tan(\pi - 3)$
***74.** $\sin \frac{1}{\pi}$
75. $\tan[(2.7\pi + 2)/27]$
76. $\cos(0.20 + \frac{\pi}{3})$
77. $\sin(\pi - 2.50)$
78. $\sin(\frac{\pi}{6} + \frac{\pi}{4})$
79. $\tan(\frac{3\pi}{4} - 2)$
***80.** $\cos \frac{1}{\pi^2}$
***81.** $\tan\left(\frac{\pi}{10}^{2}\right)$

In Exercises 82–88, use a calculator to evaluate each function to four decimal places.

***82.** $\sin \pi°$
***83.** $\tan (\frac{2}{\pi})°$
***84.** $\tan \frac{2}{\pi}$
85. $\sin^2 1.13 + \cos^2 1.13 =$
***86.** $\cos(\pi + 1)°$
***87.** $\sin(\pi°\pi'\pi'')$
88. $\sin^2 31.4° + \cos^2 31.4° =$

In Exercises 89–97, find out if each equality is true by evaluating both sides of the equation using a calculator.

89. Is $\sin 1 = \sin 1°$?
***90.** Is $\cos(0.75)^2 = \cos^2(0.75)$?
***91.** Is $\tan(\frac{1}{\pi^2}) = \tan^2(\frac{1}{\pi})$?
***92.** Is $\sin(\frac{3}{\pi}) = 1/\sin(\frac{\pi}{3})$?
***93.** Is $\cos(0.20 + 0.36) = \cos 0.20 + \cos 0.36$?
***94.** Is $\tan[2(18°)] = 2 \tan 18°$?
***95.** Is $\sin[(1.3)(0.86)] = \sin 1.3 \sin 0.86$?
***96.** Is $\tan(\frac{\pi}{5} + \frac{\pi}{4}) = \tan \frac{\pi}{5} + \tan \frac{\pi}{4}$?
***97.** Is $\sin(\frac{\pi}{3}) = \sin(\frac{\pi}{3})°$?

In Exercises 98–107, prove whether each property is true. *Hint:* To prove that something is **not** true, you simply need to find one case that is not true.

***98.** $\sin(A + B) = \sin A + \sin B$
***99.** $\cos(A + B) = \cos A + \cos B$
***100.** $\tan(A + B) = \tan A + \tan B$
***101.** $\sin(2\theta) = 2 \sin \theta$
***102.** $\cos(2\theta) = 2 \cos \theta$
***103.** $\tan(2\theta) = 2 \tan \theta$
***104.** $\sin AB = \sin A \sin B$
***105.** $\cos AB = \cos A \cos B$
***106.** $\sin A/B = \sin A/\sin B$
***107.** $\tan A/B = \tan A/\tan B$

***108.** Use your calculator to make a table showing the angle θ in degrees *and* in radians, $\sin \theta$, and $\tan \theta$, as shown in Table 2.3-1. Give all numbers to six decimal places, and include the following angles: 45°, 35°, 25°, 15°, 10°, 5°, 4°, 3°, 2°, 1°, 0.5°. Now look at your completed table, and compare the values of $\tan \theta$ and $\sin \theta$ and of $\sin \theta$ and θ as θ gets closer and closer to zero. Make a clear

θ Deg	θ Rad	$\sin\theta$	$\tan\theta$
45	0.785398	0.707107	1.0000000
35 etc	etc	etc	etc

Table 2.3-1

word statement about what happens to the values of θ, $\sin\theta$, and $\tan\theta$, as θ approaches zero. Does it matter whether you use θ in degrees or in radians?

EXERCISES 2.3-2

In Exercises 1–10, find all three sides of the triangle in Figure 2.3-7 using the information given.

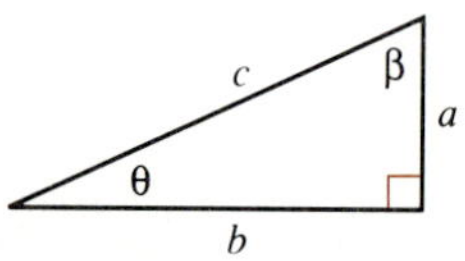

Figure 2.3-7

1. $c = 5.0$ ft, $\theta = 36°$

2. $c = 10.0$ m, $\beta = 60°$

3. $b = 8.5$ ft, $c = 11.8$ ft

4. $a = 60.5$ cm, $\beta = 55.4°$

5. $c = 200.0$ mm, $\theta = 19.2°$

6. $b = 200.0$ in., $\theta = 25°18'$

7. $c = 35.1$ mi, $\beta = 0.660$ rad

8. $a = 160.0$ cm, $\theta = 1.18$

9. $b = 9.7$ ft, $\beta = \frac{2\pi}{7}$

***10.** $\theta = 27°$, $\beta = 63°$

***11.** For the triangle in Figure 2.3-7, draw the altitude from the right angle perpendicular to c. If $b = 10.0$ ft and $\beta = 39°$, find the length of this altitude.

Area of a Triangle. In Exercises 12–14, find the area of the triangle in Figure 2.3-8 using the information given. See Exercise 42 in Section 2.2-1.

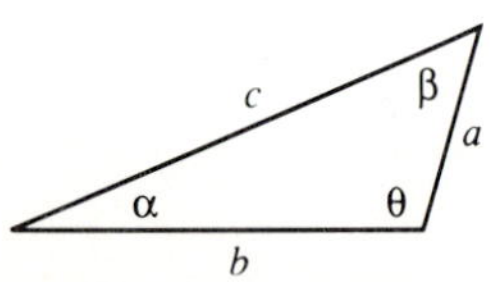

Figure 2.3-8

12. $b = 10.8$ ft, $c = 12.0$ ft, $\alpha = 25.3°$

13. $a = 6.5$ m, $c = 10.7$ m, $\beta = \frac{3\pi}{16}$

14. $a = 21.4$ cm, $c = 37.0$ m, $\alpha = 33.4°$, $\theta = 112°$

Exercises 15–17 refer to Figure 2.3-8.

15. If the area is 150 ft^2, $c = 10.0$ ft, and $\alpha = 52°$, find b and sketch the triangle approximately to scale.

***16.** If $a = 15.0$ cm, $b = 20.0$ cm, $\alpha = 32.0°$, and $\beta = 45.0°$, find c.

****17.** If $\alpha = 25°$, $\beta = 35°$, and $a = 28.0$ in., find b, c, and θ. *Hint:* Express the area in terms of α and then in terms of β.

****18.** For the oblique triangle in Figure 2.3-9, prove that

$$\frac{\sin\alpha}{a} = \frac{\sin\beta}{b} = \frac{\sin\gamma}{c}$$

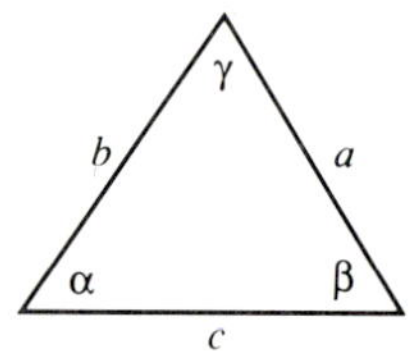

Figure 2.3-9

This relationship is called the *law of sines*. *Hint:* First recognize that this is really three separate equations

$$\frac{\sin\alpha}{a} = \frac{\sin\beta}{b} \qquad \frac{\sin\beta}{b} = \frac{\sin\gamma}{c} \qquad \text{and} \qquad \frac{\sin\alpha}{a} = \frac{\sin\gamma}{c}$$

so you must derive each pair. First express the area of the oblique triangle in Figure 2.3-9 in terms of α, then in terms of β, and finally in terms of γ. Then ask yourself what must be true about these three expressions for the area.

19. In a heavy rainstorm, the wind blows the rain at a 23° angle away from the vertical toward your patio (Figure 2.3-10). If your patio roof is 7.0 ft high, how far inside the patio will the pavement get wet?

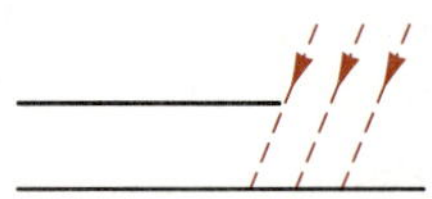

Figure 2.3-10

20. You walk 30.0 ft from a tree and then measure the angle of elevation of its top to be 72°. How tall is this tree?

21. From the fifteenth floor of a building, 150 ft above the ground, you look out the window at a building 360 ft away and find that you must raise your eyes by 56° to see the top. How tall is this other building?

22. Measuring the Angle of Elevation. You can measure the angle of elevation very simply using a protractor, a soda straw, and a weight W hanging from a string, as shown in Figure 2.3-11. This weight determines true vertical.

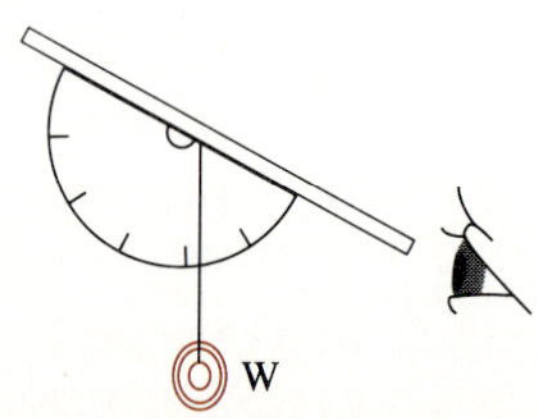

Figure 2.3-11

(a) How would you measure the angle of elevation using this simple device? Be specific!

(b) Make one of these devices and use it to measure the height of some buildings or trees near your home or dormitory.

23. Two buildings that are right next to each other are 100.0 ft away from you. To see the top of building A, you must look up at a 40° angle, and to see the top of building B, you must look up at an 80° angle. Does this mean that building B is twice as tall as building A? Justify your answer numerically.

***24.** You are trying to measure the height of a building using its angle of elevation, but you cannot measure the distance to the building because a river is in the way. However, you can measure that a balcony, which you know is 125 ft above the ground, has an angle of elevation of 41°18′, whereas the top of the building has an angle of elevation of 66°24′. How tall is this building and how far are you from it?

***25.** Using your single-lens reflex camera, you focus on the top of a building and find it is 42 m away. Using the method of Exercise 22, you measure the angle of elevation to be 36.8°. How tall is this building? How far away from its base are you?

26. Angle of Depression. The angle of depression is similar to the angle of elevation, except it is the angle *below* the horizontal of something that is below you. Suppose you are looking across a canyon that a map tells you is 280 ft wide. To see the other rim, you find that you must look 21° *below* the horizontal. This is the angle of depression. How far below your level is the other rim? (See Figure 2.3-12.)

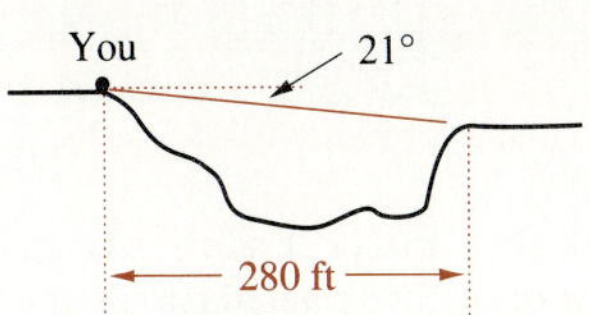

Figure 2.3-12

27. From the top of the Sears Tower in Chicago, which is 1454 ft tall, you measure that a building 575 ft away has an angle of depression of 38.2°. How tall is this other building?

28. You can look at the floor of the Barringer meteor crater in Arizona from its rim, but you are not allowed to go down into the crater itself. You know that this circular crater is 1300 m in diameter, and you measure that the center has an angle of depression of 15°30′ from the rim. How deep is this crater?

***29.** While hiking in the Cascade Mountains, you follow a trail around the rim of a roughly circular volcanic crater (hopefully dormant!). If this trail is 1080 m long, and if the angle of depression of the center of the crater is 38°12′, how deep is this crater?

***30.** You are 300.0 ft from the launching pad of a Saturn rocket. Because of the blinding light at takeoff, you do not see the rocket until it is 24°30′ above the horizon. It soon climbs to an angle of 46°48′. What elevation did it gain between these two observations?

31. In an aerial photo of a canyon, the walls cast shadows 53.8 ft long when the sun is at 65.7° above the horizon. How deep is this canyon?

32. The Grand Canyon is approximately 1.0 mi deep. How long a shadow will its walls cast when the sun is at $\frac{4\pi}{9}$ above the horizon?

33. In an aerial photo of a top secret research lab, a tower casts a shadow 149 ft long when the sun is at 51°18′ above the horizon. How tall is this tower?

***34.** If a tree has a shadow 112 ft long when the sun is at 64° above the horizon, how long will the shadow be when the sun has sunk to 28° above the horizon?

35. If you are 6.0 ft tall, how long will your shadow be when the sun is 80° above the horizon? 40° above the horizon? Did you expect it to be twice as long?

36. Chemistry. Ozone (O_3) is an oxygen molecule consisting of three oxygen atoms, as shown in Figure 2.3-13. The distance between the nearest pairs of oxygen (O) atoms is 1.278 Å. How far apart are the two farthest oxygen atoms?

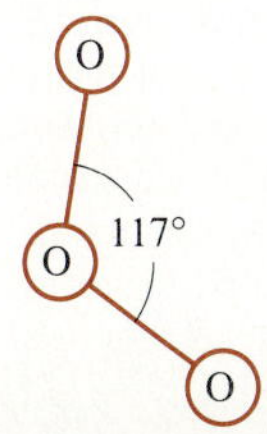

Figure 2.3-13

37. Chemistry. The sulfur dioxide (SO_2) molecule has the shape shown in Figure 2.3-14. The O—S—O angle is 119.536° and the two oxygen (O) atoms are 2.4747 Å apart. How far apart are the oxygen and sulfur (S) atoms?

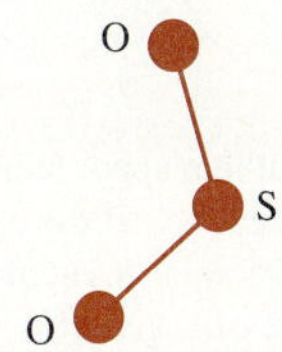

Figure 2.3-14

***38. Chemistry.** Formaldehyde (CH_2O), a common and poisonous lab preservative, has the molecular structure shown in Figure 2.3-15. The H—C—H angle is 125.8°, the C—H distance is 1.060Å, and the C—O distance is 1.230 Å. How far are the hydrogen (H) atoms from each other and from the oxygen (O) atom?

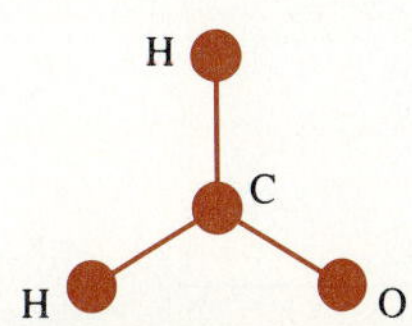

Figure 2.3-15

***39. Crystals.** In a certain type of crystal, called monoclinic, one face has the shape shown in Figure 2.3-16. How far apart are atoms A and D? Atoms B and C?

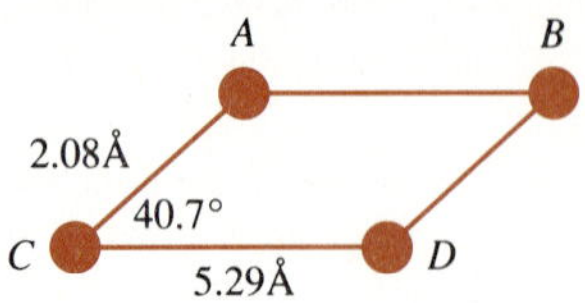

Figure 2.3-16

40. Suppose we measured "marsquake" waves on Mars. The waves that went deeper than 32°36′ below the surface hit a spherical inner core (see Example 24). How deep would this core lie below the surface of Mars? What would be the radius of this core?

41. If our moon had an inner core of diameter 550 mi (see Example 24), what would be the greatest angle below the surface for which "moonquake" waves would not hit this core?

42. Horizon Correction for Telescopes. When you want to locate an object in the sky, such as a star or planet, you point your telescope at a certain angle above the horizon. You can find this angle in an astronomical handbook, but such books assume that you are viewing from sea level, and this horizon is called the *astronomical horizon.* If you are above sea level, the apparent horizon is different from the astronomical horizon, as Figure 2.3-17 shows. To know what angle to use, you must correct for your height above sea level.

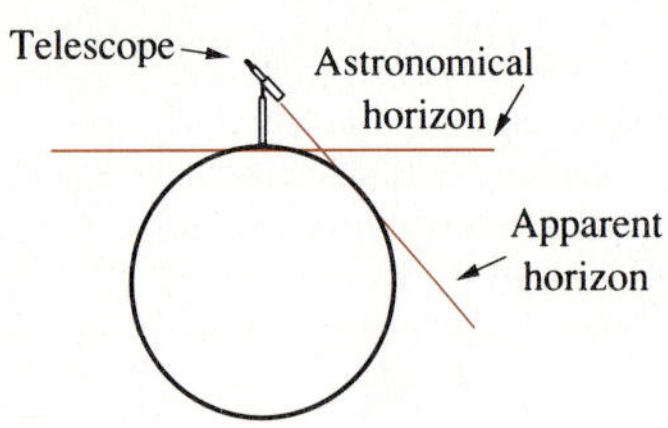

Figure 2.3-17

Suppose you are an astronaut in a space shuttle, orbiting at a typical altitude of 150 mi above sea level. If the astronomical atlas tells you that a certain star's elevation is 28.16° above the horizon, what angle must you add to the one given in the star atlas?

43. A kitchen tray for draining dishes has parallel horizontal rubber-coated support wires running horizontally 4.0 in. apart, 1.0 in. above the rubber pad that catches the water (see Figure 2.3-18). Find the diameter of the smallest circular plates that can be supported by two adjacent support wires and just barely touch the rubber pad.

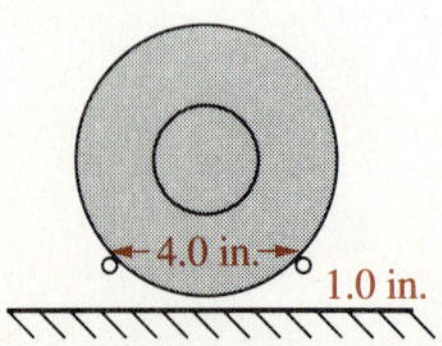

Figure 2.3-18

44. If a staircase is to rise at 33° above the horizontal, what must be the depth of each step if its height is 9.0 in.? (See Figure 2.3-19.)

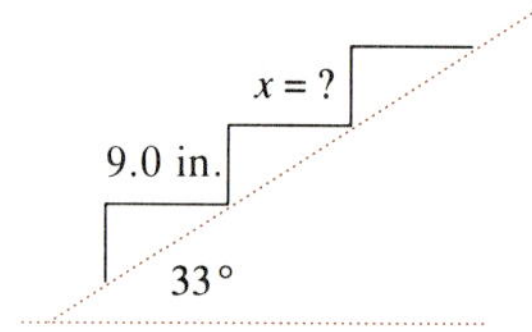

Figure 2.3-19

45. The house in Figure 2.3-20 measures 32 ft wide with outside walls 8 ft 9 in. high. Measure the necessary angles on the figure and calculate the height of the peak.

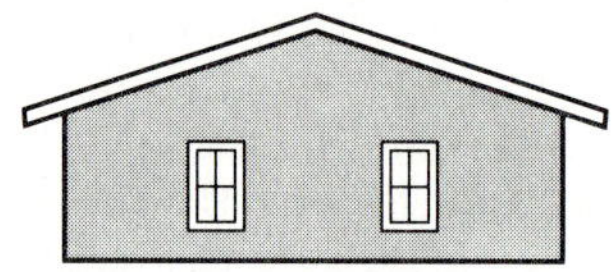

Figure 2.3-20

46. The two segments of the root in Figure 2.3-21 slope upward at angles of 25° and 48° above the horizontal, respectively.

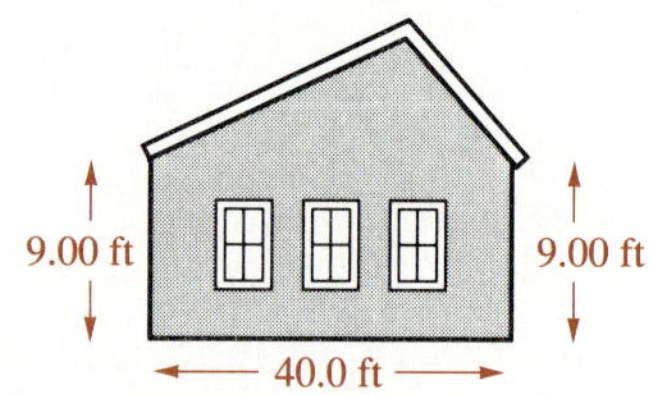

Figure 2.3-21

(a) How high is the peak above the ground?
(b) If each of the two slanted parts of the roof consists of a single beam, how long must each beam be?

***47.** You are building brick facing against a slanting wall, as shown in Figure 2.3-22, using 4.0 in.-high bricks. If this wall had an original height of 8.0 ft *before* it started to slant and has now tilted 26° off vertical, how many bricks will you need to complete the facing?

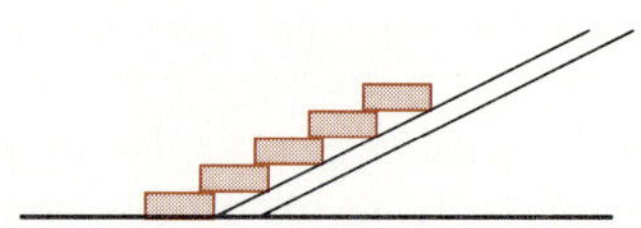

Figure 2.3-22

***48.** You are a solar-conscious architect deciding where to place a window in an 8.0-ft tall south-facing wall in Montreal, Canada, at 46° north latitude. The roof overhang is 3.0 ft. You want to place this window so that it will receive no noon sun on June 21 when the sun is at 68° above the horizon, but so that it will receive all of the December 21 noon sun which is at 20° above the horizon (see Figure 2.3-23).

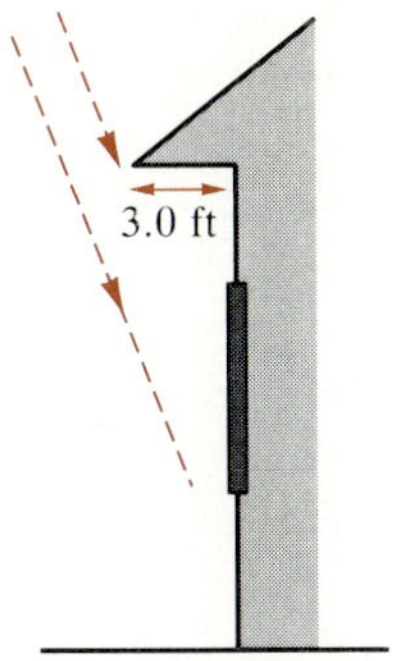

Figure 2.3-23

(a) How close should the bottom of this window be to the ground?

(b) How tall can this window be?

(c) What percent of this window will receive light at noon on March 21st when the sun is 44° above the horizon?

49. Some early civilizations, such as the Anasazi in present-day Mesa Verde National Park, Colorado, built their houses in shallow cliff overhangs on south-facing walls, possibly to make use of solar heat (see Figure 2.3-24). If you were the Anasazi architect:

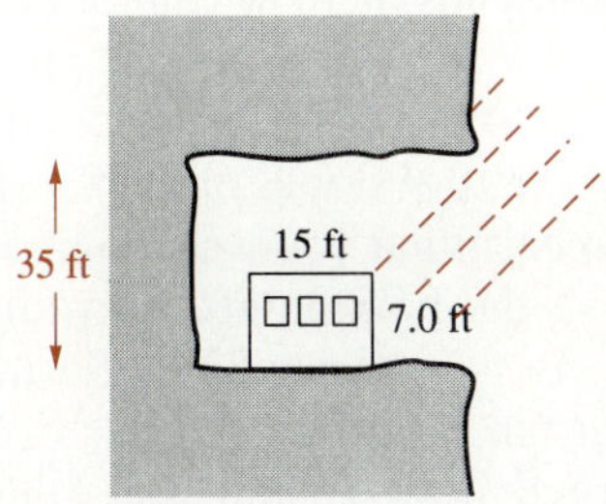

Figure 2.3-24

(a) What is the farthest back from the mouth of the opening that you could build a 7.0-ft high house and still have its front face receive full sun on Dec. 21 at noon when the sun is at 29° above the horizon?

(b) What is the closest you could build the house to the mouth of the opening and still have it completely shaded from the June 21 noon sun at 77° above the horizon?

50. Solar Panel. You are installing a 2.0-ft × 4.0-ft solar panel on the south-facing roof of a house in Denver. This roof rises at 10.0° above the horizontal, and you want the panel to face perpendicular to the sun's noon rays on June 21 when the sun is 74° above the horizon. Figure 2.3-25 shows an edge-on view of the panel on the roof where the support h is perpendicular to the roof. What should h be?

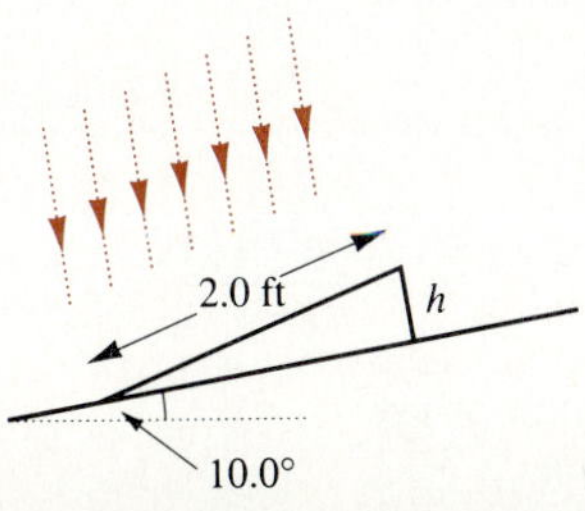

Figure 2.3-25

51. The volcano Skjaldbreidur, in Iceland, has a nearly round base approximately 19 mi in circumference, a flat 1000-ft diameter crater at the top, and sides that rise at approximately $7\frac{1}{2}°$ with respect to the horizontal (Figure 2.3-26). How high is this volcano above its base, assuming it is symmetric?

Figure 2.3-26

***52.** You want to shade your south-facing patio in Denver at 40° north latitude from the noon sun on June 21st. To do this, you want to use a series of equally spaced 6.0-in. wide boards running east-west, tilted upward at a 30° angle above the horizontal, as shown in Figure 2.3-27. How far apart should these boards be so that no direct sun comes through? (See Exercise 50.)

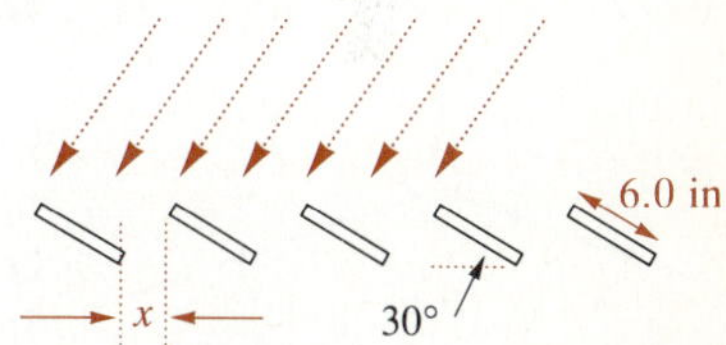

Figure 2.3-27

***53.** As part of a modern sculpture display, you want to balance a heavy triangular sheet of metal horizontally atop a post. For maximum stability, you need to make the point of contact with the post equidistant from all sides of the triangle. If this metal sheet has equal sides and a perimeter of 30.0 ft, where should the balance point be?

54. Mirror Image. Light bounces off a mirror with the same angle that it struck the mirror. The reflected light appears to be coming from a point behind the mirror (the image), as shown in Figure 2.3-28. Prove that the image appears to be just as far behind the mirror as the object is in front of the mirror.

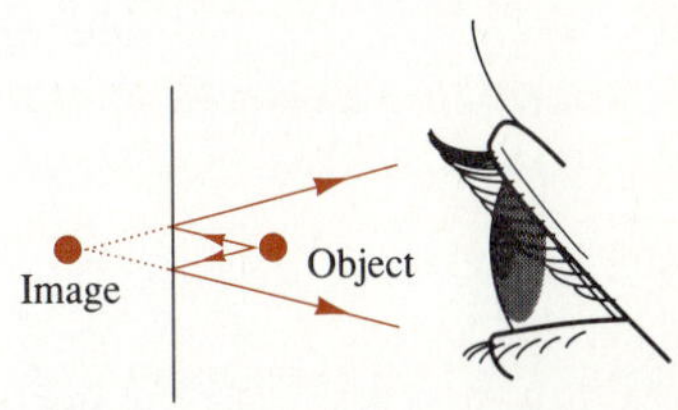

Figure 2.3-28

55. A certain fire truck has a ladder that can be extended up to 95 ft (measured from the back of the truck, which is 3.0 ft above the ground) and raised to a maximum angle of 65° above the horizontal. If you were dispatching fire trucks, what would be the tallest building to which you could send this truck to rescue people safely?

56. You are designing a gazebo with a regular octagonal base framed by eight 10.0-ft planks, as shown in Figure 2.3-29. How far apart are the opposite parallel sides of the floor?

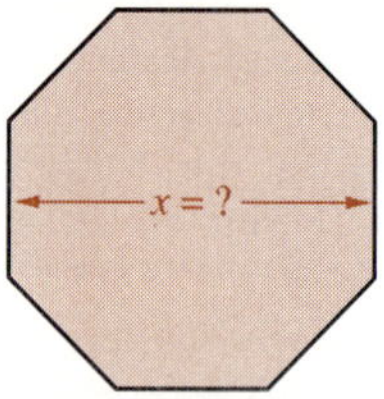

Figure 2.3-29

57. The gazebo in Exercise 56 was so successful that you are building another one, but this time as a regular hexagon, also with 10.0-ft planks as the frame. Find the width of this second gazebo between opposite parallel sides of the floor.

58. You are designing a patio in the shape of a regular hexagon of width 20.0 ft between opposite parallel faces (Figure 2.3-30). How long is each side?

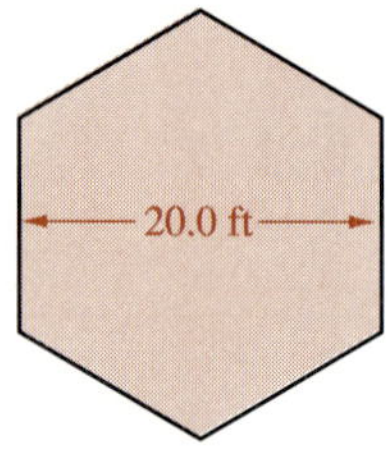

Figure 2.3-30

2.4 FINDING THE ANGLE: AN INTRODUCTION TO INVERSE TRIGONOMETRIC FUNCTIONS

Figure 2.4-1

Slope of a roof

Congratulations! You've won the lottery and now you can build your dream home! Your architect draws the plans and sends them to you for approval. If you consult such plans, you do not normally see any angles for the pitch of the roof. Instead, you find a little right triangle next to the sloping surface, such as the one in Figure 2.4-1. This triangle means that the roof slopes in such a way that it rises 4 in. for every 12 in. of horizontal distance. What you want to know is: At what angle does the roof slope upward above the horizontal? As you can see from Figure 2.4-1, the tangent of this angle is $\frac{4}{12} = \frac{1}{3} = 0.3333$. So the problem is this: How do we find α if we know that $\tan \alpha = 0.3333$?

In the previous sections, we learned how to find the values of the six trigonometric functions if we know the angle. However, as with the slope of the roof, many applications of trigonometry require the reverse problem—finding the angle if you know the values of the trigonometric functions of that angle.

Suppose we know that $\sin \theta = \frac{1}{2}$ and we want to find what θ is. What we are looking for is the *inverse* of the sine; that is, we want to find the angle θ that has a sine equal to $\frac{1}{2}$.

"$\sin \theta = \frac{1}{2}$" means exactly the same thing as "θ is an angle whose sine is $\frac{1}{2}$." We can abbreviate this statement in two ways.

$$\theta \text{ is an angle whose sine } = \tfrac{1}{2}$$

is the same as

$$\theta = \underbrace{\arcsin \tfrac{1}{2}}_{} \quad \text{or} \quad \theta = \underbrace{\sin^{-1}(\tfrac{1}{2})}_{}$$

an angle whose sine is $\frac{1}{2}$

We read these last two as either "θ is the arcsin $\frac{1}{2}$" or "θ is the inverse $\sin(\frac{1}{2})$." Both arcsin and $\sin^{-1}$ mean exactly the same thing, so both can be called either *arcsin* or *inverse sine*.

▸ "Arcsin" can be read as "arcsin" or "inverse sine."
"Sin^{-1}" can be read as "arcsin" or "inverse sine."

▸ **Caution.** When we write something like $\sin^{-1} y$, the $^{-1}$ is not an exponent! It does not mean $\sin y$ to the $^{-1}$ power. That is, $\sin^{-1} y \neq 1/\sin y$.

$\sin^{-1} y$	Here $^{-1}$ means the *inverse* of the sine (or *arcsin*). It does **not** mean $\dfrac{1}{\sin y}$

Notice that

$$\sin\theta = \tfrac{1}{2} \quad \text{means } \theta \text{ is an angle whose sine} = \tfrac{1}{2}$$

$$\begin{bmatrix} \theta = \arcsin\frac{1}{2} \\ \text{or} \\ \theta = \sin^{-1}\frac{1}{2} \end{bmatrix} \quad \text{means } \theta \text{ is an angle whose sine} = \tfrac{1}{2}$$

Therefore,

$$\sin\theta = \tfrac{1}{2} \qquad \text{is equivalent to} \qquad \theta = \arcsin\tfrac{1}{2} \ (\text{or } \theta = \sin^{-1}\tfrac{1}{2})$$

Now let's go back to our problem. If $\sin\theta = \frac{1}{2}$, what is θ? Using the inverse sine, we can write θ as

$$\theta = \sin^{-1}\tfrac{1}{2} = \text{an angle whose sine is } \tfrac{1}{2}$$

In Section 2.2-2, we proved that $\sin 30° = \frac{1}{2}$, which means that 30° is an angle whose sine is $\frac{1}{2}$. Therefore, θ must be 30°. Using the inverse sine, we can write the result as $30° = \sin^{-1}\frac{1}{2}$.

We can now generalize what we have just seen for the inverse sine to the other trigonometric functions.

$$\theta = \sin^{-1} y \qquad \text{is equivalent to} \qquad y = \sin\theta$$

Keeping in mind that $\sin^{-1} y$ and $\arcsin y$ mean exactly the same thing, it is also true that

$$\theta = \arcsin y \qquad \text{is equivalent to} \qquad y = \sin\theta$$

We have similar equivalences for all the trigonometric functions, which we summarize as follows:

$\sin^{-1} y = \theta$	is equivalent to	$y = \sin\theta$
$\cos^{-1} y = \theta$	is equivalent to	$y = \cos\theta$
$\tan^{-1} y = \theta$	is equivalent to	$y = \tan\theta$
$\cot^{-1} y = \theta$	is equivalent to	$y = \cot\theta$
$\sec^{-1} y = \theta$	is equivalent to	$y = \sec\theta$
$\csc^{-1} y = \theta$	is equivalent to	$y = \csc\theta$

In trying to understand the inverse trigonometric functions, it is helpful to remember that $\sin^{-1} y$ is an *angle*—the angle whose sine is equal to y. Similarly, $\cos^{-1} y$ is an *angle*—the angle whose cosine is equal to y. The same thing holds for the other functions.

For the time being we always suppose that $\sin^{-1} y$, $\cos^{-1} y$, and so on are angles between 0° and 90°. In the Chapter 3 we extend these trigonometric functions to larger angles and negative angles.

Example 25

Find $\sin^{-1}(\sqrt{2}/2)$.

Solution

Remember that $\sin^{-1}(\sqrt{2}/2)$ is an *angle*—the angle whose sine is equal to $\sqrt{2}/2$. If we call this angle θ, then we have

$$\theta = \sin^{-1}(\sqrt{2}/2) \qquad \text{is equivalent to} \qquad \sin\theta = \sqrt{2}/2$$

In Section 2.2-2, we saw that $\sin 45° = \sqrt{2}/2$, so θ must be 45°. ■

NOW IT'S YOUR TURN

Find $\sec^{-1}(2/\sqrt{3})$. (Answer: 30°)

The angle in Example 25 was a special one whose trigonometric functions we recognized right away. But most of the time we won't recognize them. (Would you recognize the angle that has a secant equal to 5.0925? Neither would we!) We need calculators to handle such angles.

When you found the trigonometric function of an angle, you first told your calculator the angle, and it told you the trigonometric function of that angle when you pressed the proper key. To find the *inverse* trigonometric functions, you proceed in the reverse way. You tell your calculator the trigonometric function, and it tells you what angle has that trigonometric function when you press the proper key. We summarize the procedure as follows:

FINDING INVERSE TRIGONOMETRIC FUNCTIONS USING YOUR CALCULATOR

1. Set the calculator to the mode (RAD or DEG) you will use for angles.

2. Set the calculator to the number of decimal places you want in your answer (if your calculator allows you to do so).

3. Enter the numerical value of the trigonometric function whose inverse you are trying to find.

4. Find the "inverse" key on your calculator. (It may be marked by a different color or by ARC or by INV or by "$^{-1}$". Consult your calculator's instruction manual if necessary.) Press the "inverse" key.

5. Finally, press the key of the trigonometric function whose inverse you want to find. (If you want the inverse of the cot, sec, or csc, you will have to use the reciprocal relations of the trigonometric functions, which we explain in Example 27.)

Example 26

(a) Find $\tan^{-1}(7.268)$ in degrees to two decimal places.
(b) Find arcsin(0.6901) in radians to three decimal places.
(c) If $\cos\theta = 0.5124$, find θ in degrees to four decimal places.

Solution

(a) **1.** Set the calculator to the DEG mode.
2. Set your calculator to two decimal places if you can.
3. Enter 7.268.
4. Press the "inverse" key.
5. Press the TAN key. Your screen should show 82.17°.

(b) **1.** Set the calculator to the RAD mode.
2. Set the calculator to three decimal places if you can.
3. Enter 0.6901.
4. Press the "inverse" key.
5. Press the SIN key. Your screen should show 0.762.

(c) $\cos\theta = 0.5124$ means that $\theta = \cos^{-1}(0.5124)$.
1. Set the calculator to the DEG mode.
2. Set the calculator to four decimal places if you can.
3. Enter 0.5124.
4. Press the "inverse" key.
5. Press the COS key. Your screen should show 59.1762°. ■

NOW IT'S YOUR TURN

Find each of the following to two decimal places: arctan(1.234) in radians, $\cos^{-1}(0.4231)$ in degrees, $\sin^{-1}(0.8281)$ in degrees.
(Answer: 0.89 rad, 64.97°, 55.90°):

Calculators do not have keys for the cotangent, secant, or cosecant. Therefore, we must use the reciprocal relations from Section 2.2-3 to find the inverses of these functions.

Example 27

(a) Find $\cot^{-1}(1.213)$ in decimal degrees to two decimal places.

(b) Find $\sec^{-1}(6.012)$ in radians to three decimal places.

(c) Find $\text{arccsc}(1.90256)$ in decimal degrees to three decimal places.

Solution

All of these problems involve the three functions that calculators do not have, so we need to use the reciprocal relations we learned in Section 2.2-3.

(a) Call $\theta = \cot^{-1}(1.213)$. It is θ we must find. Remember that $\theta = \cot^{-1}(1.213)$ is equivalent to $\cot\,\theta = 1.213$. The reciprocal relation gives us

$$\tan\,\theta = \frac{1}{\cot\,\theta} = \frac{1}{1.213} = 0.8244$$

$$\theta = \tan^{-1}(0.8244)$$

Now use what you just learned in Example 26 to find

$$\theta = \tan^{-1}(0.8244) = 39.50° \ \blacksquare$$

(b) If we call $\theta = \sec^{-1}(6.012)$, we want to find θ, which is equivalent to $\sec\,\theta = 6.012$. The reciprocal relation gives us

$$\cos\,\theta = \frac{1}{\sec\,\theta} = \frac{1}{6.012} = 0.1663$$

Using the method of Example 26 to find θ gives us

$$\theta = \cos^{-1}(0.1663) = 1.404 \text{ rad}$$

(c) Letting $\theta = \text{arccsc}(1.90256)$, we want to find θ, which is equivalent to $\csc\,\theta = 1.90256$. Using reciprocals gives us

$$\sin\,\theta = \frac{1}{\csc\,\theta} = \frac{1}{1.90256} = 0.5256$$

The method of Example 26 gives us

$$\theta = \arcsin(0.5256) = 31.709° \ \blacksquare$$

NOW IT'S YOUR TURN

Find each of the following to two decimal places: arccot(0.1206) in radians, $\csc^{-1}(1.194)$ in decimal degrees, and $\sec^{-1}(2.5291)$ in decimal degrees. (Answer: 1.45 rad, 56.88°, 66.71°)

We are now ready to solve for the pitch of the roof of your dream home.

Example 28

Slope of a Roof

At what angle does the roof of your dream home slope upward?

Figure 2.4-2

Slope of a roof

Solution

Remember that the little triangle means that the roof slopes in such a way that it rises 4.00 in. for every 12.0 in. of horizontal distance (Figure 2.4-2). We want α, and we know the sides opposite and adjacent to it, so we use the tangent.

$$\tan \alpha = \frac{\text{opposite}}{\text{adjacent}} = \frac{4.00 \text{ in.}}{12.0 \text{ in.}} = \frac{1}{3} = 0.3333$$

$$\alpha = \arctan 0.3333 = 18.4° \quad ■$$

NOW IT'S YOUR TURN

If a roof rises 2.00 in. for every 12.0 in. of horizontal distance, at what angle does it slope upward? (Answer: 9.46°)

Example 29

You are photographing a bird's nest atop a 300-ft tall coast redwood tree, with your camera on a tripod 5.0 ft above the ground and 250 ft from the base of the tree. At what angle above the horizontal must you tilt the camera? (Notice that we are just asking for the angle of elevation of the top of the tree.)

Solution

First sketch the tree showing the camera and bird's nest, including the information you know (Figure 2.4-3).

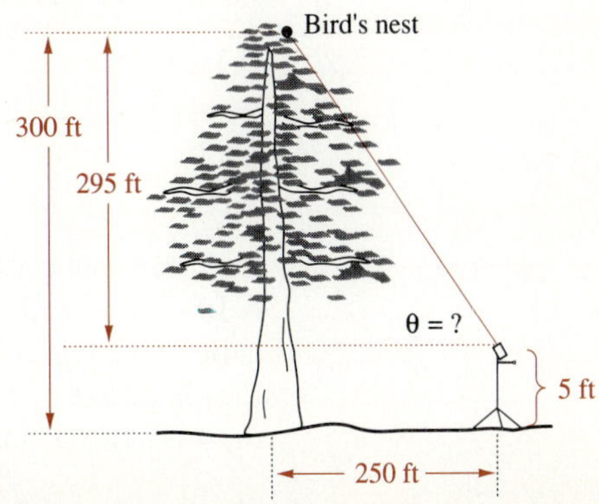

Figure 2.4-3

Photographing a bird's nest atop a coast redwood

$$\tan\theta = \frac{295\text{ ft}}{250\text{ ft}} = 1.18$$

$$\theta = \arctan(1.18) = 50° \blacksquare$$

NOW IT'S YOUR TURN

If you walked so that you were half the distance away, would the angle of elevation double? If not, what would it be? (Answer: No. It would be 67°.)

Example 30
Space Science

The Landsat earth-mapping satellite orbited 570.0 mi above the earth's surface. What angle would the earth subtend in a photograph taken from this satellite?

Solution

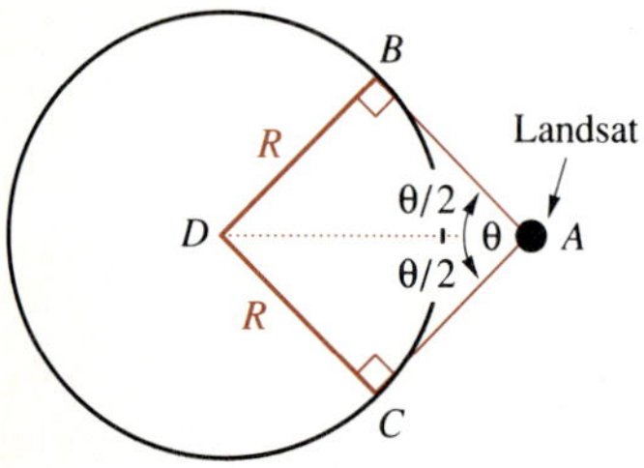

Figure 2.4-4
Landsat earth-mapping satellite

First carefully draw the earth and the Landsat (Figure 2.4-4). The angle subtended by the earth (θ in this figure) is between the lines AB and AC, which are tangent to the earth and therefore perpendicular to the radius R. If we call h the height of Landsat above the surface, then $AD = R + h$. Looking at either right triangle, we know the hypotenuse (AD) and the side opposite $\theta/2$ (R), so we can use the $\sin(\theta/2)$.

$$\sin\left(\frac{\theta}{2}\right) = \frac{R}{AD} = \frac{R}{R+h}$$

where R = radius of earth = 3960 mi and h = 570.0 mi. Therefore,

$$\sin\left(\frac{\theta}{2}\right) = \frac{R}{R+h} = \frac{3960\text{ mi}}{3960\text{ mi} + 570.0\text{ mi}} = 0.8742$$

$$\frac{\theta}{2} = \sin^{-1}(0.8742) = 60.95°$$

$$\theta = 2(60.95°) = 122° \blacksquare$$

NOW IT'S YOUR TURN

If Landsat were mapping our moon, which has a radius of 1080 mi, from the same height (570.0 mi) above its surface, what angle would the moon subtend? (Answer: 81.8°)

Example 31
Crystals △

Gold forms a crystal in the form of a cube 4.07 Å on a side (remember that 1 Å is 10^{-10} m), with gold atoms at the corners and at the center of each face of the cube. This type of crystal is called a *face-centered* cubic. Find the angle between the diagonal of this cube and the edge AC. See Figure 2.4-5, where we have omitted the atom in each face for clarity.

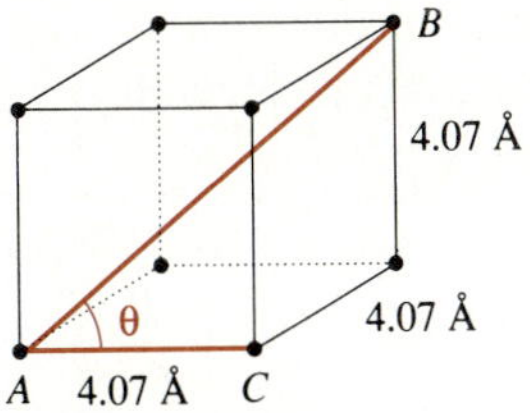

Figure 2.4-5

Gold crystal (Atoms in face are omitted.)

Solution

First carefully draw the cube and its diagonal (Figure 2.4-5). We want to find θ, the angle between AB and AC. This problem is a *three-dimensional* one, and most people have trouble visualizing figures in three dimensions. The secret to such a problem is *to reduce it to a two-dimensional problem* when possible. In this case, we can draw the triangle formed by the lines AC, AB, and BC (the diagonal of the face from B to C) (Figure 2.4-6). The faces of a cube are perpendicular, so the angle C in Figure 2.4-6 is a right angle. The line BC is the diagonal of a face of the cube, so we can find it by the Pythagorean theorem.

$$(BC)^2 = (4.07\ \text{Å})^2 + (4.07\ \text{Å})^2 = 16.6\ \text{Å}^2 + 16.6\ \text{Å}^2 = 33.1\ \text{Å}^2$$

$$BC = \sqrt{33.1\ \text{Å}^2} = 5.756\ \text{Å}$$

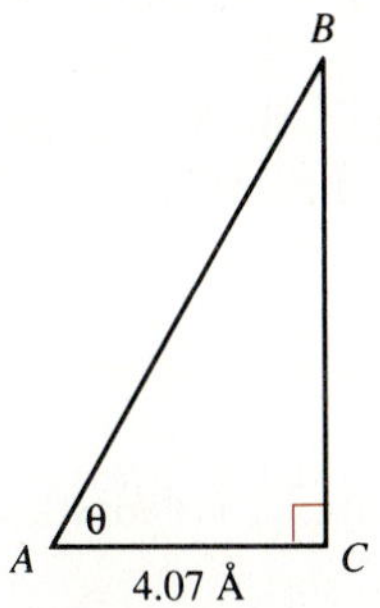

Figure 2.4-6

Reduction of a three-dimensional crystal to a two-dimensional figure

Now we can find θ using $\tan\theta$ because we know the sides opposite and adjacent to θ.

$$\tan\theta = \frac{BC}{AC} = \frac{5.756\ \text{Å}}{4.07\ \text{Å}} = 1.4143$$

$$\theta = \tan^{-1}(1.4143) = 54.7° \quad ■$$

NOW IT'S YOUR TURN

What would the angle be if the cube were twice as large? (Answer: 54.7°)

Example 32

Evaluate sin(arcsec 2).

Solution

First we find the angle arcsec 2 and then the sine of that angle.

Remember that arcsec 2 is an angle having secant 2. Calling this angle θ means θ = arcsec 2, which is equivalent to saying that $\sec\theta = 2$. To find what θ is, sketch a triangle in which $\sec\theta = 2$ (Figure 2.4-7).

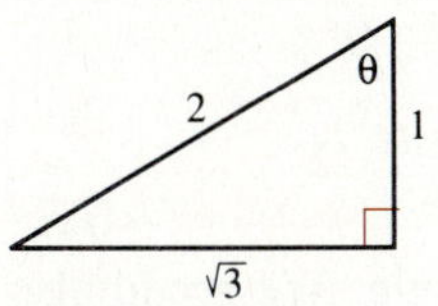

Figure 2.4-7

Triangle with sec θ = 2

$$\sec\theta = \frac{\text{hypotenuse}}{\text{adjacent}} = \frac{2}{1} = 2$$

The Pythagorean theorem tells us that the third side must be $\sqrt{3}$. We recognize this triangle as one of the special one from Section 2.2-2, so $\theta = 60°$. Therefore,

$$\sin(\text{arcsec } 2) = \sin 60° = \sqrt{3}/2 \quad ■$$

In Example 32 θ turned out to be a special angle (60°), but this usually does not happen. When the angle is not a special one, we must use a calculator, as Example 33 illustrates.

Example 33

(a) Evaluate $\tan[\cos^{-1}(0.7547)]$ to four decimal places.

(b) Find $\sin^{-1}(\tan 38.4°)$ to three significant figures.

(c) Evaluate $\sin^{-1}(\cos^{-1} 0.8207)$.

Solution

(a) $\tan[\cos^{-1}(0.7547)]$. Call $\cos^{-1}(0.7547) = \theta$. First find θ and then find its tangent. Because $\cos^{-1}(0.7547) = \theta$, $\cos\theta = 0.7547$. A calculator gives $\theta = 41.0°$. Therefore,

$$\tan[\cos^{-1}(0.7547)] = \tan 41.0° = 0.8693$$

(b) $\sin^{-1}(\tan 38.4°)$. A calculator gives $\tan 38.4° = 0.793$. Using a calculator again gives us

$$\sin^{-1}(\tan 38.4°) = \sin^{-1}(0.793) = 52.5°$$

(c) $\sin^{-1}(\cos^{-1} 0.8207)$. Call $\theta = \sin^{-1}[\cos^{-1}(0.8207)]$. First we use a calculator to find that $\cos^{-1}(0.8207) = 34.8°$, but this would give us

$$\theta = \sin^{-1}[\cos^{-1}(0.8207)] = \sin^{-1}(34.8°)$$

The latter says that $\sin\theta = 34.8°$, which has no meaning because the sine cannot be in degrees; it must be a pure number. Therefore, we must find $\cos^{-1}(0.8207)$ in radians, which is a pure number. A calculator gives $\cos^{-1}(0.8207) = 0.6082$ rad. Therefore,

$$\theta = \sin^{-1}[\cos^{-1}(0.8207)] = \sin^{-1}(0.6082) = 37.5° \quad ■$$

We can use degrees for the final answer because it is an angle. We could also have used radians.

NOW IT'S YOUR TURN

Evaluate $\tan[\sin^{-1}(0.4203)]$, $\cos^{-1}(\sin 61.8°)$, and $\cos^{-1}(\sin^{-1} 0.1234)$.
(Answer: 0.4632, 28.2°, and 82.9°)

EXERCISES 2.4

In Exercises 1–15, find the acute angle satisfying the given condition without using a calculator or a table.

1. $\cos^{-1}(\frac{1}{2})$ **2.** $\arcsin(\sqrt{2}/2)$ **3.** $\tan^{-1}(1)$
4. $\arccos(\sqrt{2}/2)$ **5.** $\cot^{-1}(1)$ **6.** $\operatorname{arcsec}\sqrt{2}$
7. $\csc^{-1}(\sqrt{2})$ **8.** $\cos^{-1}(1/\sqrt{2})$ **9.** $\sec^{-1}(2)$
10. $\operatorname{arccsc} 2$ **11.** $\arccos \frac{1}{2}\sqrt{3}$ **12.** $\sin^{-1}(\sqrt{3}/2)$
13. $\tan^{-1}(\sqrt{3})$ **14.** $\operatorname{arccot}(1/\sqrt{3})$ **15.** $\csc^{-1}(2/\sqrt{3})$

In Exercises 16–36, use a calculator to evaluate each expression to one decimal place, giving each answer in decimal degrees.

16. $\arcsin 0.0611$ **17.** $\arccos 0.9690$ **18.** $\tan^{-1} 1.1918$
19. $\cos^{-1} 0.4263$ **20.** $\arcsin 0.1526$ **21.** $\tan^{-1} 0.1111$
22. $\arctan 11.55$ **23.** $\sin^{-1} 0.9207$ **24.** $\sin^{-1} \frac{3}{4}$
25. $\cos^{-1} \frac{9}{13}$ **26.** $\sin^{-1} 0.9573$ **27.** $\cos^{-1} 0.010$
28. $\cot^{-1} 3.0415$ **29.** $\operatorname{arcsec} 1.0635$ **30.** $\csc^{-1} 44.077$
31. $\operatorname{arccot} 0.4101$ **32.** $\sec^{-1} 44.077$ **33.** $\operatorname{arccsc} 3.707$
34. $\csc^{-1} \frac{9}{7}$ **35.** $\cot^{-1} 0.0020$ **36.** $\sec^{-1} \pi$

In Exercises 37–48, use a calculator to evaluate each expression to three decimal places, giving the answer in radians.

37. $\sin^{-1} 0.9854$ **38.** $\cos^{-1} 0.1771$ **39.** $\arctan 0.8332$
40. $\arctan 1.192$ **41.** $\cos^{-1} 0.9619$ **42.** $\sin^{-1} 0.9924$
43. $\cot^{-1} 5.558$ **44.** $\operatorname{arcsec} 2.967$ **45.** $\csc^{-1} 5.702$
46. $\operatorname{arcsec} 2.0183$ **47.** $\operatorname{arccsc} 5.2650$ **48.** $\operatorname{arccot} 1.6577$

In Exercises 49-80, use a calculator to find θ in degrees accurate to two decimal places.

49. $\sin\theta = 0.9801$ **50.** $\theta = \tan^{-1} 7.892$
51. $\tan\theta = 1.50$ **52.** $\cos\theta = 0.9560$
53. $\theta = \arcsin \frac{19}{26}$ **54.** $\theta = \cos^{-1} 0.8501$
55. $\cos\theta = \frac{7}{8}$ ***56.** $\theta = \cos^{-1} \frac{\pi}{5}$
***57.** $\theta = \tan^{-1} \pi$ ***58.** $\theta = \arcsin \frac{2\pi}{7}$
***59.** $\theta = 3.2 \sin^{-1} 0.5709$ ***60.** $\theta = \frac{2}{3} \arctan 6.091$
***61.** $\frac{3}{4}\theta = \cos^{-1} \frac{3}{4}$ ***62.** $4\theta = \cos^{-1} 0.7878$
***63.** $7.23\theta = \tan^{-1} 0.1234$ ***64.** $\pi\theta = \arcsin 0.6502$
***65.** $\frac{\pi}{4}\theta = 2.57 \arctan \frac{\pi}{4}$ ***66.** $\frac{1}{6}\theta = \sin^{-1} \frac{1}{6}$
67. $\cot\theta = 3.43$ **68.** $\theta = \csc^{-1} \frac{7}{5}$
69. $\sec\theta = 9.999$ **70.** $\sec\theta = 1.25$
***71.** $\cot\theta = \pi$ ***72.** $1.5 \sec\theta = 6.80$
***73.** $\sec\theta = \pi$ ***74.** $\theta = 3.5 \csc^{-1}(3.5)$
***75.** $\sin \frac{\theta}{4} = 0.7604$ ***76.** $\frac{1}{4} \cos \frac{\theta}{4} = 0.1250$
****77.** $\tan \frac{1}{\theta} = 5.75$ ***78.** $\sin(\theta + 25°) = 0.880$
***79.** $\cos(2\theta - 34°) = 0.10$ ****80.** $\sec(3\theta + 50°) = 8.00$

In Exercises 81–92, use a calculator to find θ in radians accurate to three decimal places.

81. $\theta = \operatorname{arccot} 0.4167$ **82.** $\theta = \sec^{-1}(2.941)$
83. $\theta = \csc^{-1}(137.5)$ **84.** $\cot\theta = 9.761$
85. $\sec\theta = 3.000$ **86.** $\csc\theta = \frac{8}{7}$
***87.** $\csc\theta = \pi$ ***88.** $\cot\theta = \frac{2}{\pi}$
***89.** $\tan(2\pi\theta) = 2\pi$ ***90.** $0.0811 \sec\theta = 2.000$
***91.** $\cos(3\theta + 1.5) = 0.8502$ ***92.** $\frac{2}{3} \sin(\frac{2}{3}\theta) = 0.500$

In Exercises 93–106, evaluate each expression without using a calculator or a table.

93. $\sin(\tan^{-1} 1)$ **94.** $\cos[\cos^{-1}(0.4532)]$
95. $\sec(\operatorname{arcsec} 1.532)$ **96.** $\tan(\arcsin \frac{1}{2})$
97. $\tan(\cot^{-1} 1)$ **98.** $\cot[\cos^{-1}(0.5000)]$
***99.** $\cos(\arcsin \frac{7}{25})$ ***100.** $\cos(\arctan \frac{5}{12})$
***101.** $\cot[\sin^{-1}(\sqrt{2}/2)]$ ***102.** $\cos^{-1}(\cos \frac{\pi}{3})$
***103.** $\cot[\cos^{-1}(\sqrt{3}/2)]$ ***104.** $\cos(\operatorname{arccsc} 2)$
***105.** $\sec(\cos^{-1} \frac{5}{13})$ ***106.** $\cos[(\sin^{-1} \frac{1}{2}) + \frac{\pi}{6}]$

In Exercises 107–125, evaluate each expression accurate to two decimal places. When the answer is an angle, express it in radians. Use special angles whenever you can, but use a calculator when necessary.

***107.** $\arcsin(\cos \frac{\pi}{3})$ ***108.** $\tan(\sin \frac{\pi}{3})$
***109.** $\tan^{-1}(\sin \frac{\pi}{3})$ **110.** $\sin^{-1}(\sin \frac{\pi}{4})$
***111.** $\tan(\tan^{-1} 1.2 + \sin^{-1} 0.43)$
***112.** $\sin(\sin 30°)$ ***113.** $\tan[\tan(\tan^{-1} 1.5)]$
***114.** $\tan(2 \sin^{-1} \frac{1}{2})$ ***115.** $\csc(\frac{1}{2} \arctan \sqrt{3})$
116. $\sin^{-1} \frac{1}{2} + \sin^{-1} \sqrt{3}/2$ ***117.** $\sin^{-1}(\sin^{-1} 0.500)$
***118.** $\cos^{-1}(\sin^{-1} 0.25)$ ***119.** $\tan^{-1}(\cos^{-1} 0.7830)$
***120.** $\cos^{-1}(\cos^{-1} \frac{2}{3})$ ***121.** $\csc^{-1}(\sin^{-1} 0.951)$
***122.** $\sin^{-1}(\sin 0.451)$ ***123.** $\cos^{-1}(\tan 19°)$
***124.** $\tan^{-1}(\cos 68°)$ ***125.** $\sin^{-1}(\tan 23°)$

In Exercises 126–131, check to see if each equality is true by evaluating both sides numerically. Use the properties of the special angles whenever they occur.

126. Is $\tan^{-1} 1 + \tan^{-1} \sqrt{3} = \tan^{-1}(1 + \sqrt{3})$?
127. Is $\sin^{-1}(0.10) + \sin^{-1}(0.20) = \sin^{-1}(0.30)$?

128. Is arccos 0.33 + arccos 0.21 = arccos 0.54?

129. Is $\sin^{-1}(0.24 + 0.31) = \sin^{-1}(0.24) + \sin^{-1}(0.31)$?

130. Is arctan(1.52 + 2.83) = arctan 1.52 + arctan 2.83?

131. Is $\cos^{-1}(0.62 + 0.21) = \cos^{-1}(0.62) + \cos^{-1}(0.21)$?

In Exercises 132–136, prove each of the equalities without using a calculator or a table.

****132.** $\tan^{-1}\frac{1}{2} + \tan^{-1}\frac{1}{3} = \frac{\pi}{4}$

****133.** $\sin^{-1}\frac{3}{5} + \sin^{-1}\frac{12}{13} = \sin^{-1}\frac{63}{65}$

****134.** $\tan^{-1}(2 + \sqrt{3}) - \tan^{-1}(2 - \sqrt{3}) = \sec^{-1} 2$

****135.** $\tan^{-1} 1 + \tan^{-1} 2 + \tan^{-1} 3 = \pi$

****136.** $\cos^{-1}(3\sqrt{10}/10 + \sin^{-1}(\sqrt{5}/5) = \frac{\pi}{4}$

In Exercises 137–144, find θ and β in Figure 2.4-8.

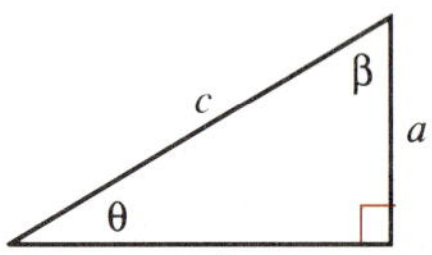

Figure 2.4-8

137. $c = 10.0$ ft, $b = 6.0$ ft

138. $a = 2.5$ m, $c = 3.0$ m

139. $b = 16$ mm, $a = 23$ mm

140. $\sin\theta = 0.7290$

141. $\cot\beta = 4.2303$

142. $\cos\beta = 0.9489$

143. $\sec\beta = 2.6131$

144. $\tan\theta = 1.6709$

Area of a Triangle. In Exercises 145–148, find the area of the triangle in Figure 2.4-8 (see Exercise 42 in Section 2.2-1).

145. $b = 10.0$ ft, $c = 16$ ft

146. $c = 8.82$ m, $\cos\beta = 0.5606$

147. $b = 83.0$ mm, $\tan\beta = 2.5129$

148. $a = 164$ cm, $\sec\theta = 2.7529$

149. At what angle does a roof slope upward from the horizontal if it rises 3.0 in. for every 12 in. of horizontal distance? (Figure 2.4-9)

Figure 2.4-9

150. If a roof rises 3.0 in. for every 12 in. measured along the roof, as shown in Figure 2.4-10, at what angle does this roof rise above the horizontal?

Figure 2.4-10

151. Measuring the Angle of a Roof. Using only a level, a ruler, and a plumb bob, devise a simple procedure to measure, from inside, the angle that a roof slopes upward. (A plumb bob is just a weight hanging from a string; it defines true vertical.) If you have access to such a roof, measure its angle (Figure 2.4-11).

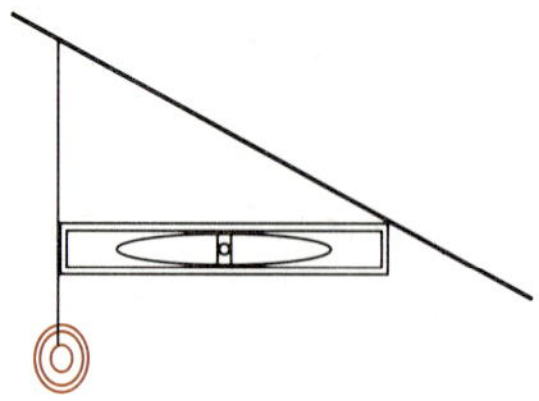

Figure 2.4-11

152. You are adjusting floodlights to illuminate the name on the front of the public library, 35 ft above the ground. If the light is 20.0 ft from the building, at what angle above the ground should you point it so it will illuminate the name?

153. To protect the spotlight in Exercise 152, you decide to put it at the top of a 7.0-ft post 20.0 ft from the building. At what angle should you now point it above the horizontal?

154. According to the *Guinness Book of World Records*, a man in Pakistan is the world's tallest person at 8 ft 3 in. If your eyes are 5.50 ft from the ground, at what angle would you have to raise them to look at the top of his head when you are standing 6.00 ft in front of him?

155. While using binoculars to study a detail at the top of the fifth floor of a building while standing on the ground, you have to lift the binoculars 45.0° above the horizontal. If you are 6.00 ft tall and each story of this building is 10.0 ft:

(a) At what angle above your present viewing angle will you have to raise the binoculars to study a detail at the top of the tenth floor?

(b) How far are you from this building?

156. The great 1982 eruption of the Mexican volcano El Chichon threw up clouds of debris that reached an altitude of about 16 km. If you were viewing these clouds 5.0 mi away from the volcano at its same altitude, at what angle above the horizon would you have to raise your eyes?

***157.** While backpacking in the mountains, you take a short cut and get lost. You consult a topographic map to try to get your bearings by identifying mountains around you. You are in a valley at an elevation of 4,200 ft and can see a ridge that the map says has a crest 6,600 ft high and is 1.25 mi away from you. Beyond this ridge, you can see a peak that you think might be one shown on

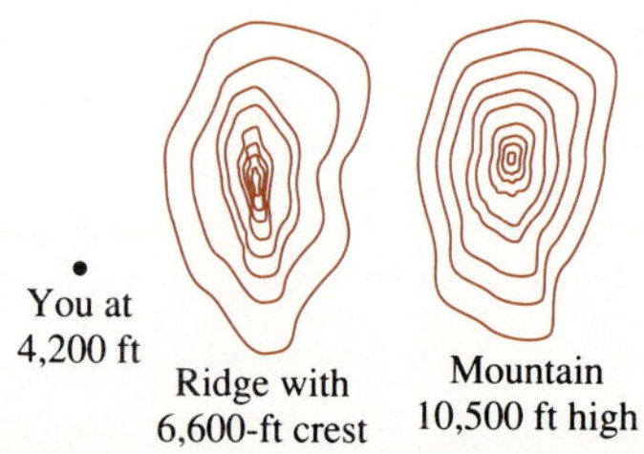

Figure 2.4-12

your map. The map says that this peak is 10,500 ft high and 2.5 mi away from you (see Figure 2.4-12). Can this peak be the mountain shown on your map?

158. Olympus Mons, the largest volcano on Mars and probably in the entire solar system (see Example 22) is about 300 mi across at its base and rises to a maximum height of approximately 15 mi above that base. At the top is a more-or-less flat crater about 40 mi in diameter, as shown in Figure 2.4-13.

Figure 2.4-13

(a) At what angle do the slopes of Olympus Mons rise above the horizontal?

***(b)** Compare this angle with typical earth volcanoes, such as Mt. Hood. Look up the information you need to calculate the angle.

159. Make the appropriate measurements on stairs in your home or college and use these to calculate the angle at which the staircase rises above the horizontal (Figure 2.4-14).

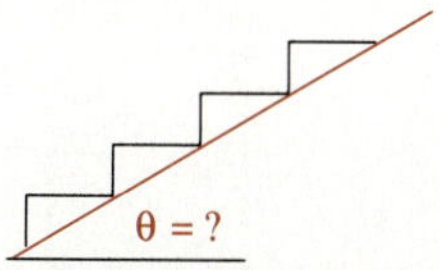

Figure 2.4-14

160. A vertical yardstick placed 1.50 ft from the edge of an 8.0-in. high step casts a 5.0-in. long shadow on the ground at the foot of the step (Figure 2.4-15). What angle do the sun's rays make with the horizontal?

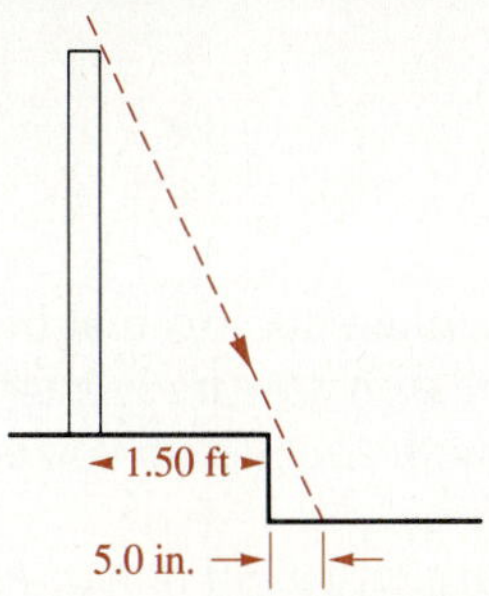

Figure 2.4-15

161. Prove that the diagonal of a square makes a 45° angle with each side.

162. A rectangle is 2.45 m × 8.50 m.

(a) Find the angle between the diagonal and the shorter side.

(b) Find the larger angle formed by the intersection of the two diagonals. What is the smaller angle?

****163.** A certain rectangle has an area of 20.0 cm^2 and a diagonal of 10.0 cm. Find the angle between the diagonal and the longer side.

164. Mariner-9 Mars Orbiter. The Mariner-9 orbiter of Mars, due to its elliptical orbit, varies in altitude from 1,650 km to 17,100 km above the surface of the planet. At each extreme, what angle does Mars subtend in a photo of the planet taken from Mariner-9? ***Hint:*** See Example 30.

***165.** Jupiter is approximately 90,000 mi in diameter. Its inner satellite, Amalthea, is 93 mi in diameter and orbits 110,000 mi from Jupiter's center. What angle would Jupiter subtend in a photo taken from Amalthea? What angle would Amalthea subtend in a photo taken from the "surface" of Jupiter? (We use quotation marks because Jupiter has no solid surface.)

166. The element calcium forms a face-centered cubic crystal, similar to that of gold, except the corners are 5.56 Å apart. Find the angle θ in Figure 2.4-5 for the calcium crystal.

167. For the gold crystal in Example 31, find the angle between the diagonal of the cube (AB) and the diagonal of the lower face (AD) (See Figure 2.4-16).

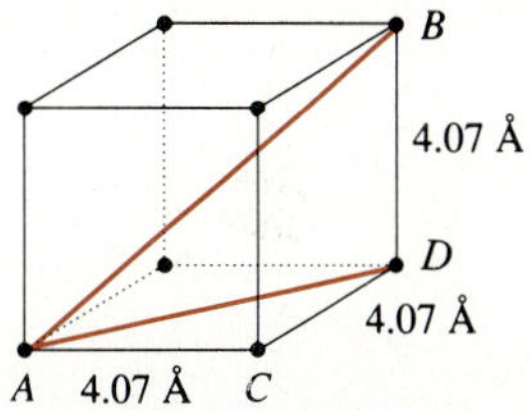

Figure 2.4-16

***168.** In Example 31 we stated that the gold crystal contains atoms centered in each face of the cube as well as at the corners. In Figure 2.4-16, call E the location of the gold atom in the face containing B, C, and D. Find the angle between **(a)** AC and EC and **(b)** AC and AE.

169. For the nitrous oxide (NO_2) molecule shown in Figure 2.4-17, the N—O distance is 1.188 Å and the O—O distance is 2.188 Å. Find the O—N—O angle.

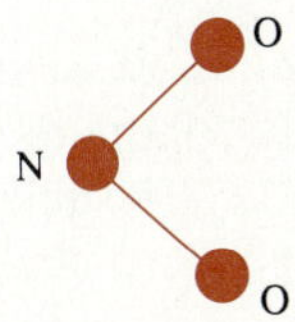

Figure 2.4-17

170. We have three-dimensional vision (or depth perception) partly because our eyes are separated from each other. Because of this separation, our right eye sees things from a slightly different angle than our left eye. Therefore, each eye forms a slightly different image of an object, and our brain interprets the differences in these images to give us depth perception (see Figure 2.4-18, which is a view from above). In the figure, the greater θ, the greater the difference in the two images, and therefore the easier it is for the brain to interpret depth perception.

In a typical human, the eyes are 2.5 in. apart, center-to-center. Calculate the angle θ when the object being viewed is at a distance from you of 1.0 ft, 10 ft, and 100 ft. Use these answers to

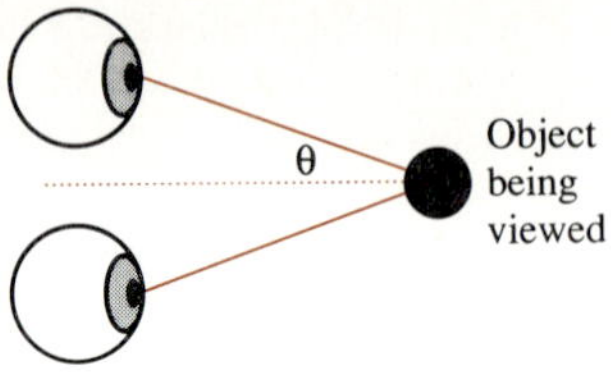

Figure 2.4-18

explain why we see depth more easily for nearby objects than for distant objects.

171. In college football, the goal posts are 23 ft 4 in. apart at the back of an end zone that is 10.0 yd deep (see Figure 2.1-20). If you are trying to kick a field goal from the center of the 75-yd line, which is 25 yd from the end zone, within what angle must you kick the ball so it will not go to either side of the goal posts?

***172. Angles in a Regular Polygon.** A regular polygon with equal angles has n sides.

(a) Prove that the angle between any two adjacent sides is $180° (1 - 2/n)$.

(b) Prove that the sum of the interior angles is $180° (n - 2)$. ***Hint:*** Inscribe the polygon inside of a circle.

(c) Test your formulas for **(a)** and **(b)** for the important special cases of a triangle, a square, a pentagon, and a hexagon.

(d) For a regular polygon of 100 sides, find the sum of the interior angles and the angle between two adjacent sides.

173. You are fitting two roof rafters together, as shown in Figure 2.4-19. You want to cut the ends so that they will meet at the top in a smooth vertical line. At what angle should you cut the ends of these rafters?

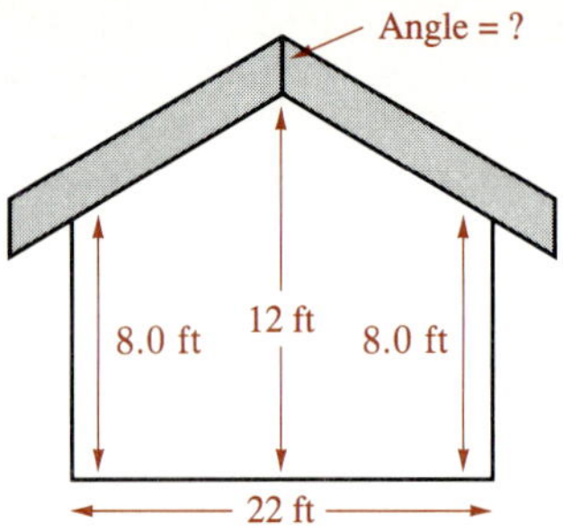

Figure 2.4-19

174. An architect is designing a bay window for the living room of a client's house. The room is 13 ft wide, the front of the bay is 8.0 ft wide, and the bay will be 3.0 ft deep, as shown in Figure 2.4-20. Find the angle between the front of the bay and its diagonal sides and the length of each of the two diagonal sides.

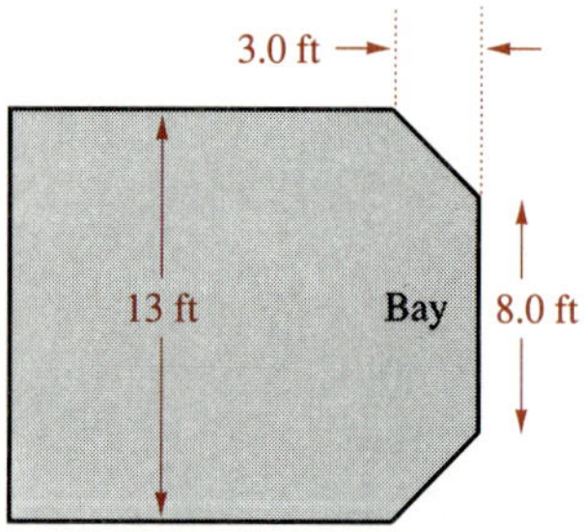

Figure 2.4-20

2.5 PROBLEM SOLVING

Most students have trouble applying their mathematics to real-life situations—the dreaded word problems! Yet the very reason you study trigonometry is to be able to use it for your major or for other math courses, so you must learn how to apply it.

Problem solving is a skill that you can learn, not a gift of nature, although, as with all skills, some people are better at it than others. The best—and probably the only—way to learn problem solving is to study examples and then practice! Here we outline the general approach to solving problems, illustrate this approach with many examples, and finally provide plenty of exercises for your practice.

GENERAL APPROACH TO PROBLEM SOLVING

1. Read the problem carefully (often several times) and make a note of important *quantitative* information, such as angles, lengths, and so on.

2. Decide just what it is that you are trying to find.

3. Always make a clear sketch of the problem. Show the information you are given and what you are trying to find. Often you have to introduce your own symbols for unknown quantities.

4. Look for right triangles and then apply the appropriate mathematical principles to those triangles.

Notice that we have already followed these steps in our examples.

In real-life applications, answers rarely come out to be whole numbers, and often you need to use several principles to solve just one problem. In this section, some of the examples are easy, some are hard, and some are messy. As before, we have chosen them from a wide variety of topics.

Example 34
Wheelchair Ramps

California building codes specify that ramps for wheelchair access to public buildings require handrails for slopes greater than 1 : 15, and in no case can ramps exceed a slope of 1 : 12. (A slope of 1 : 15 means that the surface rises 1 vertical distance unit for every 15 units of horizontal distance.)

(a) What is the maximum angle above the horizontal that a ramp can have?

(b) If an architect is designing a ramp that rises at 3.5° to reach a doorway 4 ft 8 in. above the sidewalk, how far from the doorway must this ramp begin?

Solution

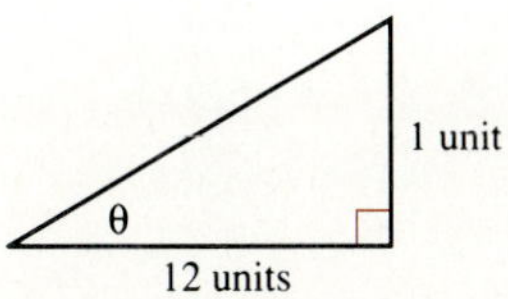

Figure 2.5-1

Steepest possible ramp, having a slope of 1.12

(a) Make a sketch of the steepest possible ramp, which rises 1 vertical unit for each 12 units of horizontal distance (Figure 2.5-1). We want to find θ and we know the sides opposite and adjacent to it, so we use the tangent.

$$\tan\theta = \frac{1 \text{ unit}}{12 \text{ units}} = \frac{1}{12} = 0.0833$$

$$\theta = \arctan(0.0833) = 4.8°$$

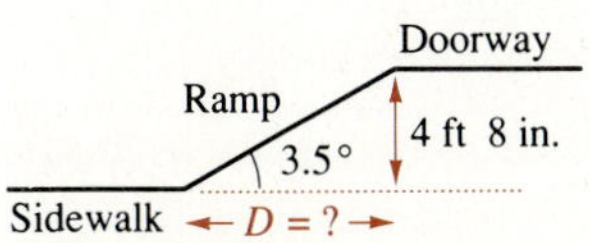

Figure 2.5-2

Wheelchair ramp rising at 3.5°

(b) Sketch the ramp, which must rise to 4 ft 8 in. at a 3.5° angle (Figure 2.5-2). We want to find distance D. We see that tan 3.5° relates sides we know and want, so

$$\tan 3.5° = \frac{4 \text{ ft } 8 \text{ in.}}{D} = \frac{4.67 \text{ ft}}{D}$$

$$D = \frac{4.67 \text{ ft}}{\tan 3.5°} = \frac{4.67 \text{ ft}}{0.0612} = 76 \text{ ft} \ ■$$

NOW IT'S YOUR TURN

If the ramp to the doorway in (b) could begin no more than 50 ft from the doorway, would it be legal? Why or why not?

(Answer: No. It would have to slope up at 5.3°, which is greater than the 4.8° angle we found for the maximum.)

Example 35

35-mm Photography

When a camera focuses on a distant object, the distance between the lens and the film is equal to the focal length (f) of the lens. The picture area on 35-mm film is 24.0 mm high by 35.0 mm wide. (Measure a 35-mm slide to check this.)

(a) Find the horizontal field of view of a 35-mm camera with the standard lens of 50.0-mm focal length.

(b) How far back from the center of a 45.0-ft wide building would you have to stand to get the whole width of the building in a photo with this lens?

Solution

(a) To find the field of view (which is expressed as an angle), sketch the camera as viewed from above, showing light rays that just strike the edges of the film (Figure 2.5-3). (Light through the center of the lens is essentially unbent by the lens.) The angle θ is the field of view. Next concentrate on the light inside the camera (Figure 2.5-4). The film is a distance f (the focal length) from the lens.

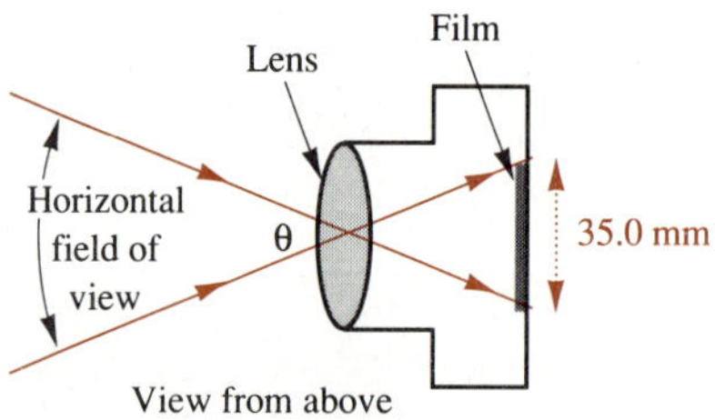

Figure 2.5-3

Horizontal field of view for a camera

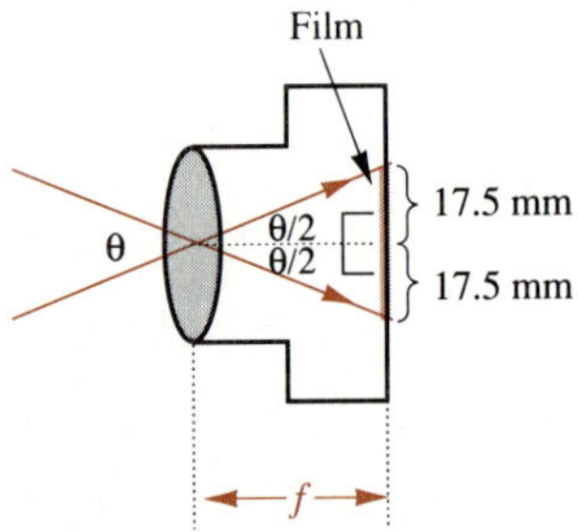

Figure 2.5-4

Light rays inside the camera

We can use $\tan \theta/2$.

$$\tan \frac{\theta}{2} = \frac{17.5 \text{ mm}}{f}$$

For a standard 50-mm lens, $f = 50.0$ mm, so

$$\tan \frac{\theta}{2} = \frac{17.5 \text{ mm}}{50.0 \text{ mm}} = 0.3500$$

$$\frac{\theta}{2} = \tan^{-1}(0.3500) = 19.3°$$

$$\theta = 2(19.3°) = 38.6°$$

(b) To see where to stand to photograph the 45.0-ft wide building, make another sketch (Figure 2.5-5).

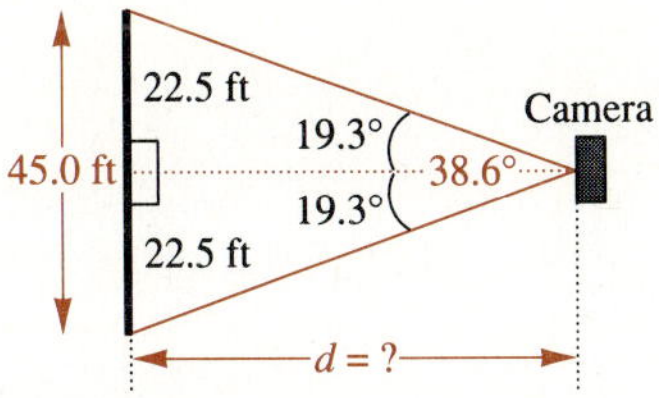

Figure 2.5-5

Camera focused on a building 45.0 ft wide

For the closest possible distance, the building will just fill the field of view. Using either of the two right triangles gives

$$\tan 19.3° = \frac{22.5 \text{ ft}}{d}$$

$$d = \frac{22.5 \text{ ft}}{\tan 19.3°} = \frac{22.5 \text{ ft}}{0.3502} = 64.2 \text{ ft} \blacksquare$$

NOW IT'S YOUR TURN

Suppose you photograph a 50.0-ft wide building with a 28.0-mm wide-angle lens. How close can you stand and still get the whole building in your picture? (Answer: 40.0 ft)

Example 36

Patio Construction

You are building a patio roof, as shown in Figure 2.5-6. What length horizontal beams should you order so they will extend 15.0 ft beyond the walls. How far from the edge of the roof should you attach them?

Solution

We need to find the angle of the roof above the horizontal, and we can find it as we did in Example 28. Using the little triangle in the upper left-hand corner of Figure 2.5-6 gives us

$$\tan \theta = \tfrac{3}{12} = 0.2500$$

$$\theta = \tan^{-1}(0.2500) = 14°$$

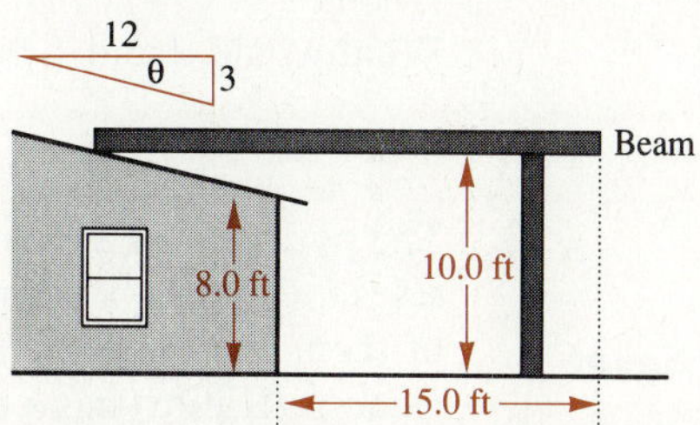

Figure 2.5-6

Patio with overhead beam

Now simplify the drawing, showing what you know and want to find (Figure 2.5-7). Call L the length of the beam and d the distance from the roof's edge.

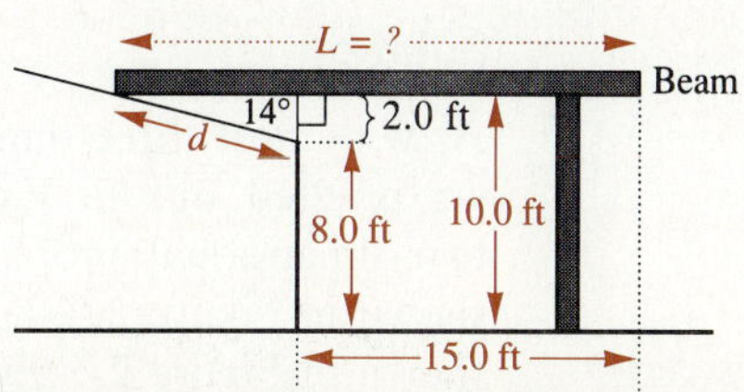

Figure 2.5-7

Simplified sketch of patio

We can find d using the small right triangle.

$$\sin 14^\circ = \frac{2.0 \text{ ft}}{d}$$

$$d = \frac{2.0 \text{ ft}}{\sin 14^\circ} = \frac{2.0 \text{ ft}}{0.2419} = 8.3 \text{ ft}$$

Now notice that

$$L = 15.0 \text{ ft} + (\text{length of side adjacent to } 14^\circ)$$

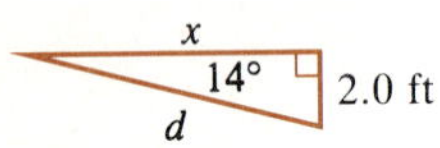

Figure 2.5-8

The small right triangle from Figure 2.5-7

Calling this side x, we have the little triangle in Figure 2.5-8. Since we now know d, we can use *any* trigonometric function of 14° to find x. But let's *not* use d, just in case we made a mistake earlier. Only tan 14° and cot 14° do not involve d, so we'll use the tangent because we can use a calculator.

$$\tan 14^\circ = \frac{2.0 \text{ ft}}{x}$$

$$x = \frac{2.0 \text{ ft}}{\tan 14^\circ} = \frac{2.0 \text{ ft}}{0.2493} = 8.0 \text{ ft}$$

Therefore,

$$L = 15.0 \text{ ft} + 8.0 \text{ ft} = 23 \text{ ft} \blacksquare$$

NOW IT'S YOUR TURN

What would d and L be if your roof had a slope of 4:12? (Answer: 6.0 ft, 21 ft)

Example 37

The Pyramids of Egypt (Archaeology)

The pyramids of ancient Egypt were built on square bases, with the four sloping sides meeting at a point directly above the center of the base. The sides of most of them rise at an angle of 51°52′ above the horizontal. The Great Pyramid of Cheops (Khufu) has a base 3023 ft in perimeter and rises at the typical angle, but part of the top is missing today. As an archaeologist publishing a paper on ancient Egypt, you need to find the original height of this pyramid. How high was it?

Solution

As with most three-dimensional problems, this one is hard to visualize. But, as we pointed out in Example 31, the trick is to reduce this problem to a two-dimensional problem. First sketch the pyramid in three dimensions, as shown in Figure 2.5-9a. Then slice the pyramid in half by a vertical plane, as shown in Figure 2.5-9b. The two-dimensional result is the isosceles triangle in

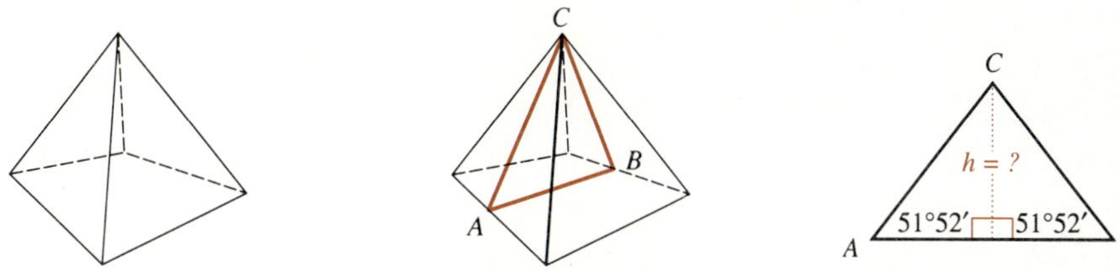

Figure 2.5-9

The Great Pyramid of Cheops

(a) Three-dimensional view of pyramid

(b) Pyramid sliced in half by a vertical plane

(c) Two-dimensional sketch of triangle ABC from part (b)

Figure 2.5-9c, in which $AC = BC$ and h is the original height. AB is the length of one side of the base of the pyramid, which is $\frac{1}{4}$ the perimeter.

$$AB = \tfrac{1}{4}\,(3023 \text{ ft}) = 755.8 \text{ ft}$$

Now we can get h using the tangent.

$$\tan 51°52' = \frac{h}{\frac{1}{2}\,AB} = \frac{h}{\frac{1}{2}\,(755.8 \text{ ft})} = \frac{h}{377.9 \text{ ft}}$$

$$h = (377.9 \text{ ft}) \tan 51°52'$$

To use a calculator, we must convert the angle to decimal degrees.

$$51°52' = 51° + 52' = 51° + (\tfrac{52}{60})° = 51° + 0.87° = 51.87°$$

Finally, we get h.

$$h = (377.9 \text{ ft}) \tan 51.87° = (377.9 \text{ ft})(1.2740) = 481.4 \text{ ft}$$ ■

NOW IT'S YOUR TURN

How far was it originally from the center of any side of the base to the top of Cheops' pyramid? (Answer: 612.0 ft)

Example 38

Latitude and Longitude

Both Denver and Philadelphia are at roughly 40° north latitude, but Philadelphia is at 75° west longitude whereas Denver is at 105° west longitude (see Example 28 in Chapter 1). If you drove straight west from Philadelphia to Denver, how far would you travel?

Solution

First make a sketch showing these cities on the earth (Figure 2.5-10). The distance we want is the circular arc DP connecting the two cities. Because this problem is three-dimensional, we must try to reduce it to a two-dimensional figure. The difference in longitude between the cities of Denver and Philadelphia

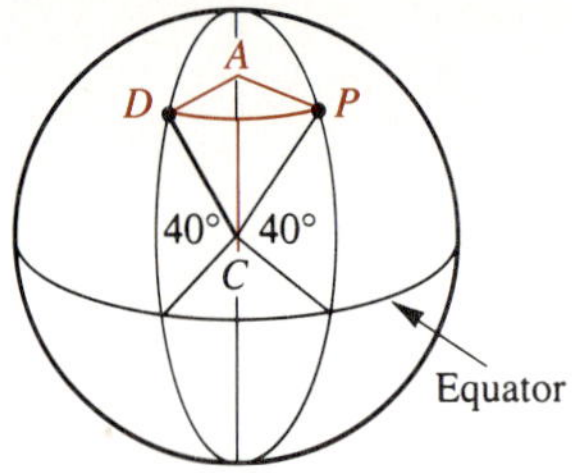

Figure 2.5-10

Map of the earth showing Philadelphia and Denver

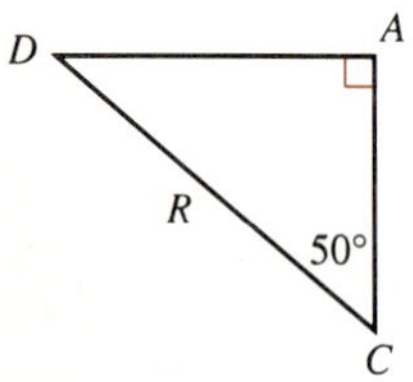

Figure 2.5-11

Triangle *ACD* from Figure 2.5-10

is $105° - 75° = 30°$, the angle between AD and AP in Figure 2.5-10. If we find the lengths AD and AP, we can use the arc length formula ($S = R\theta$) from Section 1.4 to find the arc length DP. Notice that $AD = AP$ because both cities are on the same latitude, but AD and AP are *not* equal to the earth's radius R. But CD and CP *are* equal to R.

Let's zero in on the triangle ACD, in which C is at the earth's center (Figure 2.5-11). The angle at A is a right angle, and the angle at C is 50° because it is complementary to 40°. We can now find AD.

$$\sin 50° = \frac{AD}{R}$$

$$AD = R \sin 50° = (3960 \text{ mi})(\sin 50°) = (3960 \text{ mi})(0.7660) = 3030 \text{ mi}$$

Now we can find the arc DP using $S = r\theta$, where θ must be in radians (Figure 2.5-12). Because $30° = \frac{\pi}{6}$, we have

$$S = r\theta = (3030 \text{ mi})\left(\tfrac{\pi}{6}\right) = 1590 \text{ mi} \ ■$$

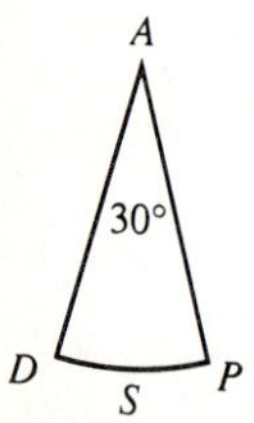

Figure 2.5-12

NOW IT'S YOUR TURN

If Denver were 2000.0 mi west of Philadelphia, by how many degrees would they differ in longitude? (Answer: 37.8°)

Example 39

Heights of Mountains

In Example 21 we could use the method we used for measuring building heights because the tops of buildings are directly above their bases and the distance from the person to the building could be measured. But a natural feature, like a mountain, slopes irregularly to the peak, so it is physically impossible to measure the horizontal distance from the peak. How then, can we measure the heights of mountains?

Recall that as you get closer to a mountain, it appears to get taller, which means that its angle of elevation depends on its distance. We can use this fact to find the height.

Suppose that a National Park Service ranger, mapping a wilderness area, measures a mountain's angle of elevation to be 32°10′. She then walks 1000.0 ft closer to the mountain and measures the angle to be 4°20′ greater. If the ranger is at an elevation of 3200 ft, what should she record on her map for this mountain's height?

Solution

First sketch the mountain, showing the values for the ranger's measurements (Figure 2.5-13). (We did not include the ranger's height because it is insignificant compared to the height of the mountain.) The ranger wants to find the height h. The distance x is not known, nor can it be easily measured.

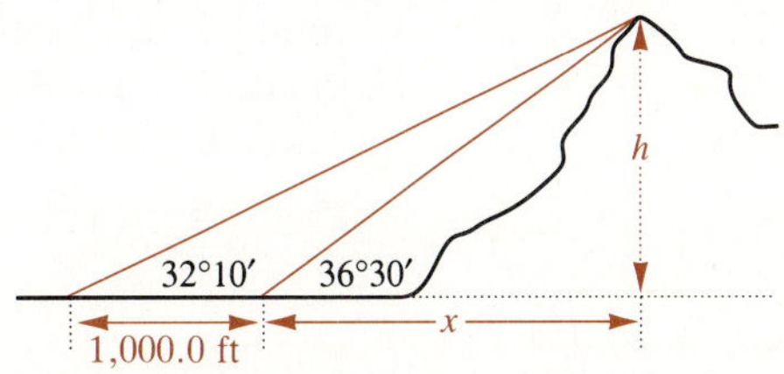

Figure 2.5-13

The two observations of the mountain peak

We have two right triangles here, both involving the sides adjacent to and opposite a known angle, which suggests that we should use the tangent or cotangent. In this case, using the cotangent leads to a much simpler method than using the tangent. (This fact is not obvious at first, but it becomes so after you have started the problem. We suggest that, after you understand our solution, you try doing this example using the tangent so you will see how messy the algebra gets.) Writing the cotangent of each angle, we get

$$\cot 36°30' = \frac{\text{adjacent}}{\text{opposite}} = \frac{x}{h}$$

$$x = h \cot 36°30' \qquad (1)$$

$$\cot 32°10' = \frac{x + 1000.0 \text{ ft}}{h}$$

$$x + 1000.0 \text{ ft} = h \cot 32°10' \qquad (2)$$

Because we do not know x, we must eliminate it between these two equations and solve for h. Substituting equation (1) into eq. (2) gives us

$$h \cot 36°30' + 1000.0 \text{ ft} = h \cot 32°10'$$

$$1000.0 \text{ ft} = h \cot 32°10' - h \cot 36°30'$$

$$= h(\cot 32°10' - \cot 36°30')$$

$$h = \frac{1000.0 \text{ ft}}{\cot 32°10' - \cot 36°30'}$$

Substituting numbers gives us

$$h = \frac{1000.0 \text{ ft}}{1.590 - 1.351} = 4180 \text{ ft}$$

This answer is the height of the mountain above the ranger's elevation. The total height above sea level (which is what a map would show) is 4180 ft + 3200 ft = 7380 ft—the height she should mark on her map. ■

NOW IT'S YOUR TURN

To double check her work, the ranger walks another 500.0 ft closer to the mountain and measures its angle of elevation to be 39°04′. Use this additional information to recalculate the mountain's height. Does it agree with the previous measurement? (Answer: Yes!)

EXERCISES 2.5

For Exercises 1–4, see Example 34.

1. If a wheelchair ramp has a slope of 1:14, at what angle does it rise above the horizontal, and how far in front of a 2.8-ft high doorway must it start?

2. Find the maximum angle above the horizontal for which a wheelchair ramp in California does not require a handrail.

3. An architect is designing a wheelchair ramp for a new state office building in California. This ramp must reach a doorway 2 ft 3 in. above the sidewalk and, for economic reasons, it should not be required to have a handrail. What is the minimum horizontal distance from the doorway that he can start this ramp to comply with the state building codes?

4. A California library has a wheelchair ramp that rises to a doorway 4 ft 9 in. above the sidewalk. The distance from the beginning of the ramp to the doorway is 67.0 ft, measured along the surface of this ramp. Will it be necessary to install a handrail to comply with California state building codes?

5. The Percent Grade of a Slope. The percent grade of a slope, such as the surface of a hill, is the ratio of the vertical elevation gained (the rise) divided by the horizontal distance travelled (the run), expressed as a percent (see Figure 2.5-14).

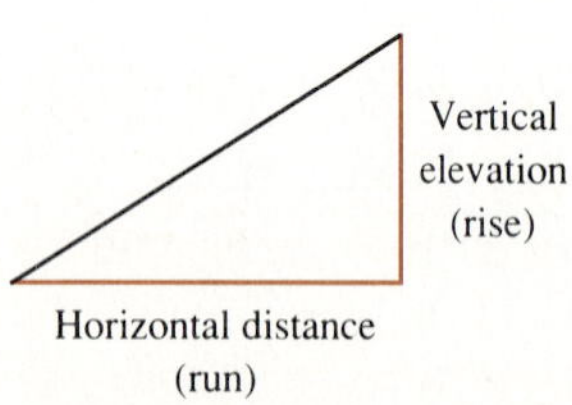

Figure 2.5-14

For example, on a 4% grade, the vertical gain is 4% (0.04) of the horizontal distance, so the grade is

$$\text{grade} = \frac{\text{vertical elevation}}{\text{horizontal distance}} = \frac{0.004 \times \text{horizontal distance}}{\text{horizontal distance}} = 0.04 = 4\%$$

(a) What trigonometric function does the percent grade really express, and what is the angle involved?
(b) If a highway rises 1.50 ft for every 10.0 ft of horizontal distance, what is its percent grade?
(c) If a highway rises 1.50 ft for every 10.0 ft of distance travelled along the surface of the road, what is its percent grade?
(d) A 6.0% highway grade is considered quite steep. At what angle does the surface of such a road slope upward?
(e) While driving on a 3.00% highway grade, how much elevation have you gained when your mileage gauge shows you have travelled 2.00 mi?

6. You are hiking in the Rocky Mountains up a steep trail at a 20% grade. When you have walked 2000 steps at 2.5 ft/step, how high are you above your starting point?

In Exercises 7–14, refer to Example 35.

7. Photography. Find the vertical field of view in a horizontal photo for a 35-mm camera with a standard 50.0-mm lens.

8. Photography. You are using a 135-mm telephoto lens on a 35-mm camera. Calculate the horizontal and vertical fields of view and compare these with those of a 50-mm lens.

9. Photography. As a nature photographer, you are photographing a 10.0-ft long elephant using a 300.0-mm telephoto lens. How close must you get to this animal so that it will completely fill the horizontal field of view?

10. Photography: Wide-Angle Lens. A 50.0 mm lens is considered standard for a 35-mm camera, and a 35-mm lens is considered a wide-angle lens. Compare the fields of view for these two lenses to see why the 35-mm lens is considered to be a wide-angle lens.

11. Photography: Telephoto Lens. A 50.0-mm lens is considered standard for a 35-mm camera, and a 135-mm lens is considered a telephoto lens. Compare their fields of view. If a distant object just fills the frame of a 135-mm lens, will it fill the frame of a 50-mm lens? In which lens will it occupy a larger fraction of the picture area? Do you see why the 135-mm lens is called a telephoto lens?

12. Photography. Which gives a greater magnification, a 135-mm lens or a 250-mm lens? Calculate the field of view of each lens.

13. Photography. What should be the focal length of a lens so that its field of view is $\frac{1}{10}$ that of a standard 50-mm lens? Is this lens a wide-angle or a telephoto lens?

***14. Photography.** Find the height and width of the largest building you can photograph in a horizontal picture with a 150-mm lens on a 35-mm camera, standing 100.0 ft from the center of the building, without tilting the camera upward. Assume that the camera is 5.0 ft above the ground at eye level.

***15.** The Pentagon, in Arlington, VA, is a regular pentagon with a perimeter of approximately 1.0 mi.

- **(a)** What area does the Pentagon cover?
- **(b)** Your construction firm has contracted to build a circular metal fence around the Pentagon. Find the minimum length of this fence.

16. You are in charge of a music program to be held inside a regular pentagon-shaped arena with a perimeter of 500.0 m. You need to locate a loudspeaker in the center of the arena so that it will be equally loud in all five corners, but the nearest electrical outlet is in one of the corners. How long an extension cord will you need?

****17.** As the junior member of an architectural firm, you are asked to design a tentlike building with a regular hexagonal floor and six identical triangular walls that slope upward to a single point. If the floor must have an area of 1500.0 ft^2 and the walls must slope upward at 72.0° above the horizontal:

- **(a)** How high will this building be?
- **(b)** What area of sheetrock should you order to cover the inside of the walls, assuming none is to be wasted?

***18.** The owner of an Atlanta construction company specializing in passive solar homes decides to relocate her company to Chicago. A major feature of her homes is that the roof overhangs completely shade all south-facing windows from the noon sun at the beginning of summer, June 21. Unfortunately for the builder, all of her house plans were developed for Atlanta, where the noon sun is at 80° on June 21; therefore they will not work for Chicago, where it is at 72°. To avoid the expense of designing entirely new house plans, she hires you to figure out how much the horizontal roof overhangs must be changed to shade the southern windows in Chicago. What do you tell her? Should they be increased or decreased, and by what factor? To what should a 3.0-ft overhang be changed?

19. A house has the dimensions shown in Figure 2.5-15. The height of the peak is 12.0 ft above the ground level and the steeper part of the roof slopes up at 11.0° above the horizontal.

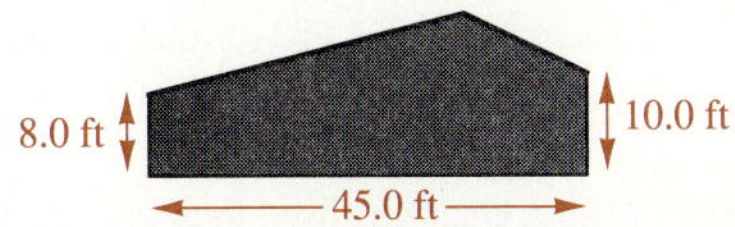

Figure 2.5-15

- **(a)** At what angle does the less steep section of the roof slope upward above the horizontal?
- **(b)** Find the length of each of the two segments of the roof.

***20.** You are designing the house shown in Figure 2.5-16. Room E is storage space, and room D is to have maximum and minimum ceiling heights of 11.0 ft and 6.0 ft.

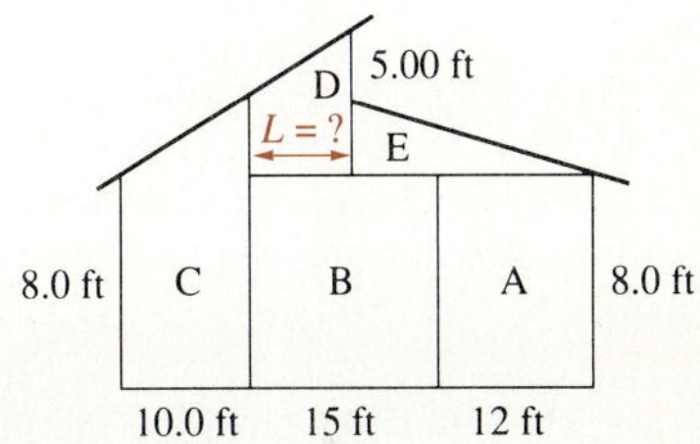

Figure 2.5-16

- **(a)** What is the length L of room D?
- **(b)** How long is the slanted roof covering room E?

21. Your house has a roof that slopes upward at 30.0° to a crest 14.0 ft high, with walls 8.00 ft high. You are adding a room, as shown in Figure 2.5-17, with a crest that is 12.0 ft high and walls 8.00 ft high. Find the length of the crest of the addition and the angle at which the addition's roof slopes upward.

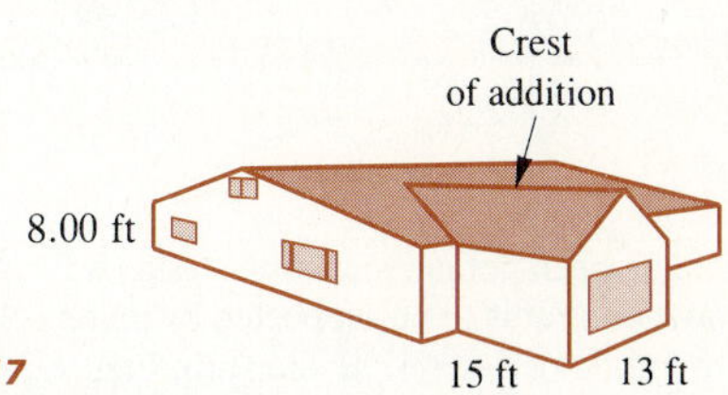

Figure 2.5-17

22. A child swings a ball in a horizontal circle of diameter 2.50 ft by a string 1.00 yd long, as shown in Figure 2.5-18.

- **(a)** What angle does this string make with the vertical?
- **(b)** How far vertically is the ball below the child's hand?

Figure 2.5-18

***23.** How many kilometers above the earth's surface must a satellite orbit so that the earth will subtend an angle of 40.0° in photos taken from this satellite?

24. A drawbridge must be raised to allow a 19-ft high sailboat to go through the middle of the 50.0-ft wide channel shown in Figure 2.5-19. The pivot of this drawbridge is 10.0 ft above water level.

Figure 2.5-19

(a) To what minimum angle must the bridge be raised to allow the boat to pass through?
(b) If this drawbridge can be raised to no more than 40.0°, what is the tallest boat that can go through the middle of this channel?

***25.** A circular swimming pool 20.0 ft in diameter is backed by a straight fence which is 15.0 ft from the rim of the pool at the closest point, as shown in Figure 2.5-20. If you go to the pool's rim at its nearest point to the fence and walk 35.0 ft along the edge of the pool, how far are you from the fence?

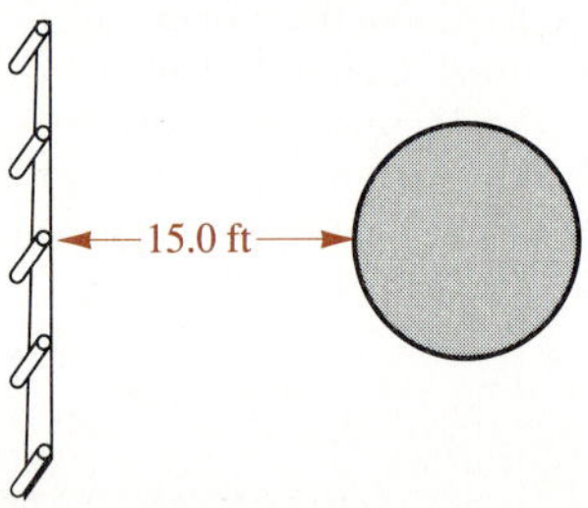

Figure 2.5-20

***26.** You have been commissioned to design a 200.0-ft diameter circular pavilion that is to be supported by seven columns equally spaced around its perimeter, as shown in Figure 2.5-21. A small speakers platform is to be built at column *a* with aisles leading to all the other columns.

(a) How long must each aisle be?
(b) If the entire pavilion is surrounded by a circular fence connecting the columns, how long is the fence segment between columns *a* and *c*?

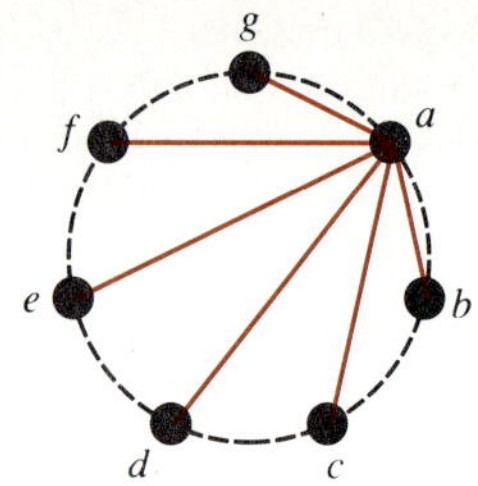

Figure 2.5-21

****27.** A foursided solid (tetrahedron) has faces made up of four equilateral triangles 2.00 m on a side.

(a) Find the shortest distance from any vertex to the opposite face.
(b) Find the angle between any two intersecting faces.

***28.** An archaeologist is trying to understand an ancient civilization's technological capabilities by studying the remains of its architecture. She discovers a pyramid that rises to a point 340 ft above the center of its square base, having walls that slope upward at 58° above the horizontal (see Figure 2.5-22). The outer surface of this pyramid was once covered with cubic blocks of limestone 4 ft × 4 ft × 4 ft, only a few of which remain at the top.

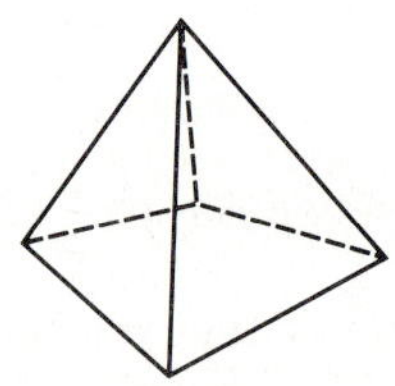

Figure 2.5-22

(a) Find the area of the base of this pyramid.
(b) How many limestone blocks were needed to surface the entire pyramid?

For Exercises 29–32, see Example 37.

***29.** If you started at one corner of the base and climbed directly up to the original top of Cheops' pyramid, what distance would you travel?

30. In your archaeological fieldwork you discovered a pyramid 250 ft high with a square base 2400 ft around. At what angle do the faces of this pyramid slope upward above the horizontal?

***31.** If a pyramid with a square base rises to a maximum height of 340 ft with sides sloping upward at 60°:

(a) What would be the area of the base?
(b) What would be the area of each face?

***32.** Suppose a pyramidlike structure has a rectangular base 40.0 ft × 100.0 ft and rises to a maximum height of 180 ft above the center of the base. At what angle do its four faces slope upward?

33. Benzene Ring. The molecule benzene (C_6H_6) forms a regular hexagonal carbon (C) ring (Figure 2.5-23) with hydrogen (H) atoms pointing directly outward. Adjacent C atoms are 1.395 Å apart, whereas each H atom is 1.084 Å from its nearest C atom.

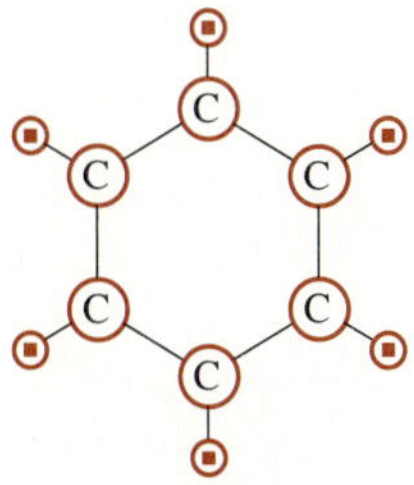

Figure 2.5-23

(a) Find the H—C—C angle.

(b) Find the distance between an H atom and its nearest H neighbor.

34. Phosphorus Crystal. In its solid form, the atom phosphorus (P) forms a simple cubic crystal, in which the P atoms are at the corners of a cube 2.18 Å on each side.

(a) How far apart are the P atoms in opposite corners of the cube?

(b) Find the angle between the diagonal of the cube and the diagonal of one of the faces, where both diagonals start at the same atom.

35. The top of a hill has an angle of elevation of 31°12′. When you walk 42 ft closer to the hill on horizontal ground, its angle of elevation increases by 8°19′. How high above you is the hilltop?

36. If a 165-ft building has an angle of elevation of 38°45′, by how much will this angle increase when you walk 22 ft closer?

****37.** A National Park Service ranger is measuring the altitude of the top of a cliff face, but he has no horizontal ground to work on. Instead, he must work from a slope that rises at 30° above the horizontal, as shown in Figure 2.5-24. From his first observation, he measures the angle of elevation of the top of the cliff face to be 55.6° above the horizontal. He then walks 50.0 yd closer, measured along the surface of the slope, and measures the angle of elevation to be 61.2°, also above the horizontal. How tall is the top of the cliff above his original position?

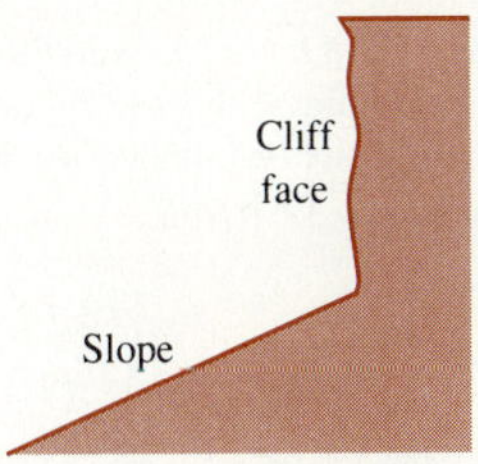

Figure 2.5-24

***38. Truss.** A truss is used to support the roof of a house. For the truss in Figure 2.5-25, the outer framing wood will be 6.0 in. wide, as shown. Find the angles at which this 6.0-in. wide lumber would have to be cut to fit together and find the inside dimensions of the triangular truss.

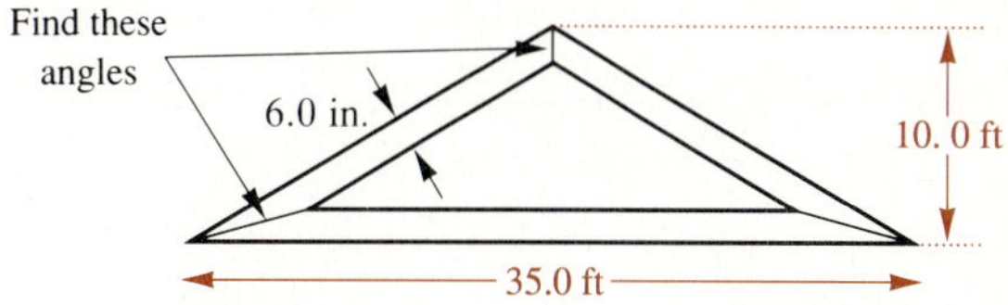

Figure 2.5-25

****39.** A symmetric tent has support rods sloping as shown in Figure 2.5-26. If this tent must be 7.0 ft tall at the center and 10.0 ft wide at the ground, find the length of each support rod. How tall is the tent where the two supports meet at the sides?

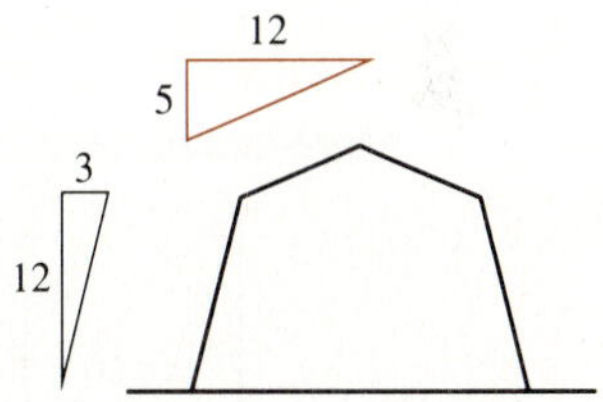

Figure 2.5-26

***40.** An architect is using an unusual window in a very contemporary house design. It is a circle of diameter 24 in. minus a section cut off by the 18 in. window sill (Figure 2.5-27). She needs to know the area of glass in this window to see whether she is complying with the building code governing heat loss. What is the area of the glass?

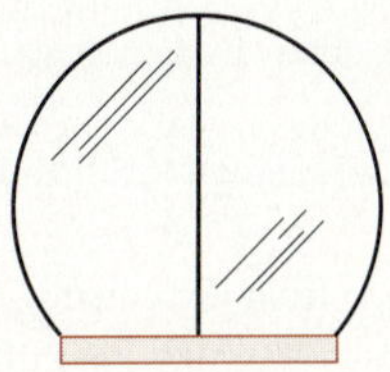

Figure 2.5-27

41. **(a)** If one object is twice as tall as another, will it necessarily have twice the angle of elevation when both are viewed from the same place?

(b) When viewed from the same place, if one object has twice the angle of elevation of another object, is it necessarily twice as tall?

***42.** You are using six equal pieces of 4.0-in. wide wood to frame a regular hexagonal mirror whose parallel sides are 18 in. apart. These pieces will be cut and mitred together where they meet, as shown in Figure 2.5-28.

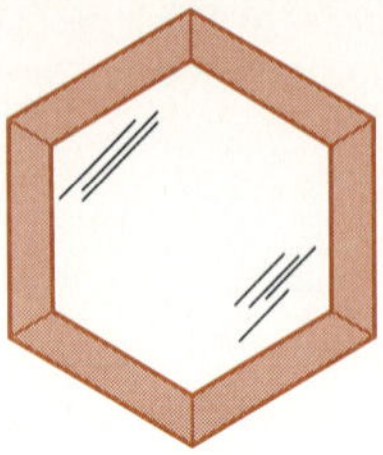

Figure 2.5-28

(a) At what angle should you set your mitre box (or saw) to cut these six pieces so they fit properly?
(b) Find the minimum length that each of the six framing pieces must be before you make any mitre cuts.

***43.** You are designing a horizontal roof for your patio. It is to be made with 12-in. wide lumber. The end of the roof is to extend 15.0 ft beyond the house walls, with an inside vertical clearance of 10.0 ft. Your house walls are 8.0 ft high and your roof has a slope of 5 : 12, as shown in Figure 2.5-29.

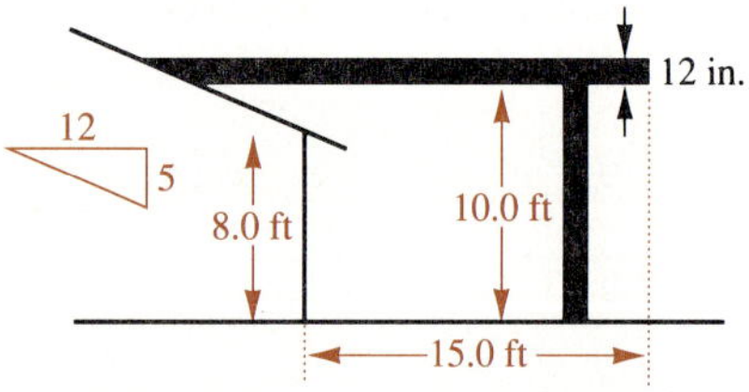

Figure 2.5-29

(a) How long should the 12-in. wide lumber for the roof be?
(b) At what angle should you cut the 12-in. wide lumber so that it will fit snugly against your roof?

For Exercises 44–48, see Example 38.

***44.** If two cities are both at a north latitude of 46°28′ and one is 550 mi directly east of the other, by how much do they differ in longitude?

***45.** Chicago is at a north latitude of 41°50′ and a west longitude of 87°45′.

(a) How far is Chicago from the equator?
(b) How far west is Chicago from the prime meridian?

***46.** Seattle is at a north latitude of 47°40′ and a west longitude of 122°20′. Through what distance does Seattle move in 1 day due to the earth's spin on its axis?

***47.** A ship with a 300.0-ft mast sails directly westward from Singapore, at the equator. How many miles will it have to sail before its mast is no longer visible from the shore due to the curvature of the earth?

****48.** Suppose that the ship in the previous problem sails eastward from New York City at a north latitude of 41°. How many miles will it have to sail until its mast is no longer visible due to the earth's curvature?

***49.** A wire mesh basket used in the kitchen to store fruits and vegetables is 10.0 in. in diameter at the top rim. It is supported by three small chains equally spaced around its rim (Figure 2.5-30). You want to cut the lengths of these chains so that the top rim is 15.0 in. below the hook. How long should each chain be? Find the angle between adjacent chains at the hook.

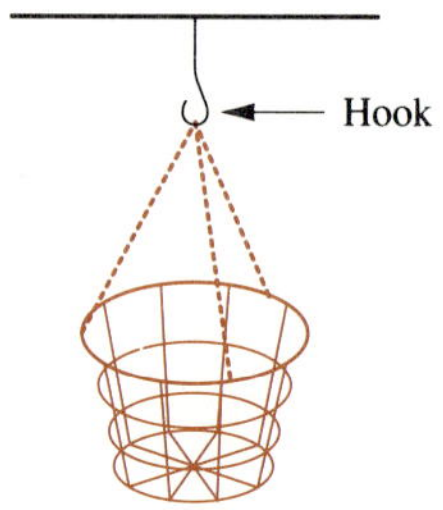

Figure 2.5-30

****50.** Find the total area of glass in the window in Figure 2.5-31 if the circular arch is 1.5 ft high. What percent of the total area of the glass comes from the arched part?

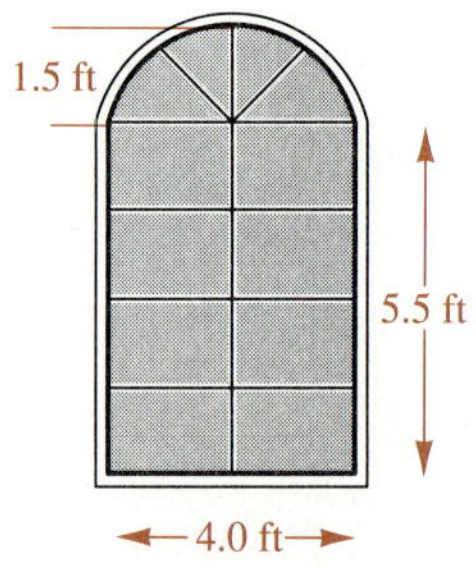

Figure 2.5-31

***51.** A circle of radius 16.0 m has an isosceles triangle cut from it, as shown in Figure 2.5-32. Find the area of each of the three curved, cut-away sections.

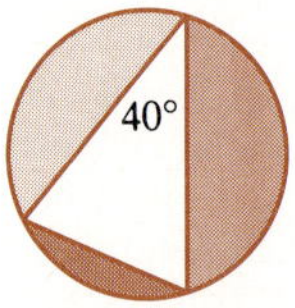

Figure 2.5-32

****52.** A circle of radius 6.00 m has a wedgelike section of angle 30° cut out by two equal-length lines, AB and BC, as shown in Figure 2.5-33. Find the arc length AC subtended by this wedge. Why is this arc length not equal to $(\overline{AB})(\theta)$?

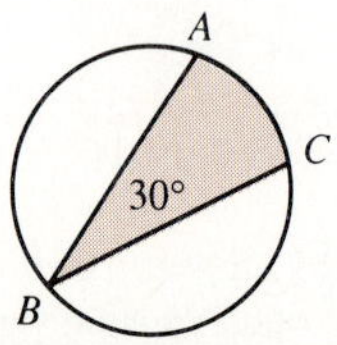

Figure 2.5-33

***53.** A box 40.0 cm wide holds two balls, each 25.0 cm in diameter (Figure 2.5-34). How high is the top of the upper ball above the bottom of the box?

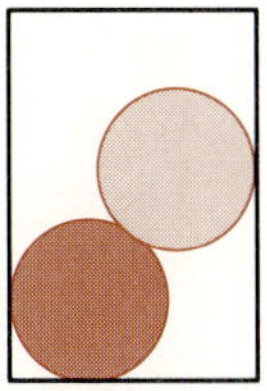

Figure 2.5-34

***54.** A hemispheric fruit bowl 50.0 cm in diameter contains two grapefruit, each 20.0 cm in diameter, as shown in Figure 2.5-35. How high will the tops of this fruit be above the bottom of the bowl?

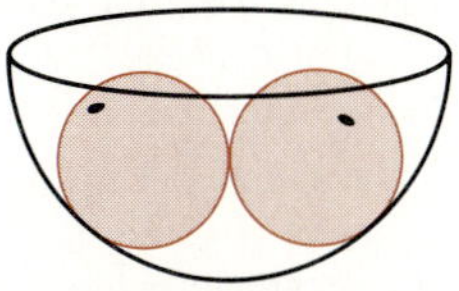

Figure 2.5-35

***55.** Three quarters, each 24.0 mm in diameter, are stacked together as shown in Figure 2.5-36. Find h.

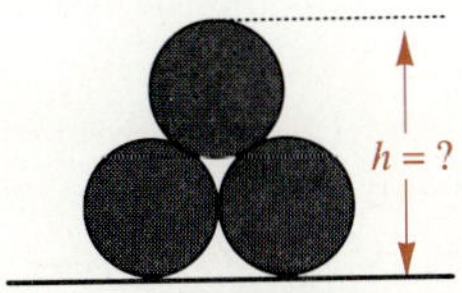

Figure 2.5-36

***56.** Two quarters and a penny are stacked as shown in Figure 2.5-36, with the penny on the top. If the quarters are each 24 mm in diameter and the penny is 19 mm in diameter, find h.

****57.** Three circles are arranged as shown in Figure 2.5-36. If the two lower circles have diameters of 50.0 cm each, find the maximum diameter the upper circle can have such that it does not protrude above the tops of the bottom two circles. ***Hint:*** Use the results of Exercise 88 in this section.

***58.** A sugar cone contains a spherical scoop of rich chocolate ice cream. If the cone has a vertex angle of 20.0 (see Figure 2.5-37), and the scoop of ice cream has a diameter of 12 cm, find the total height of this ice cream cone from its tip to the top of the ice cream before it melts.

Figure 2.5-37

***59.** Two identical balls, each 18 cm in diameter, rest against each other, hanging by identical 45-cm threads from a single hook, as shown in Figure 2.5-38. The threads point directly away from the centers of the balls. Find the angle between these two threads.

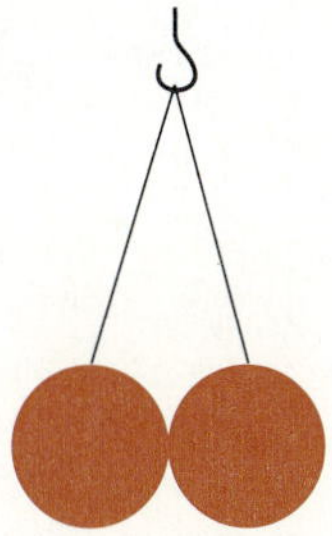

Figure 2.5-38

60. Interpretation of UFO Sightings. A major problem in studying UFO (Unidentified Flying Object) night sightings is that the speeds of the UFOs are not known because their distances are not known. Suppose you watch a light in the sky move through a 40° angle in 15 s, starting directly overhead. Assuming that the object is moving horizontally, find the distance it moved during this time and its *speed* if the object is

(a) 1.0 mi above the ground?

(b) 100 ft above the ground? (Notice that there is *quite* a difference!)

61. A certain 100.0-ft tall redwood tree is growing at a constant rate of 6.0 in./yr. Find the change in the angle of elevation of its top in a century when the tree is viewed from 300.0 ft away from its base. Would the change be the same if the tree were viewed from 100.0 ft away?

62. Take an ordinary ice cream cone (with a pointed tip) and, by making only length measurements on it, calculate the vertex angle at its tip.

***63. Aircraft Tracking.** Two radar stations 150.0 mi apart track an unidentified plane as it passes between them. Station A records its position as 33.4° above the horizon at the same time as station B records it as 48.9°. How high is this aircraft and where is it relative to these two tracking stations?

****64. Satellite Tracking.** Two tracking stations 3010 mi apart track a satellite as it passes between them. Station A records its position at 41.6° above the horizon at the same time that station B also records it as 41.6°. How high is this satellite? ***Hint:*** Apply the ideas of Exercise 63, but notice that you must consider the curvature of the earth.

****65. Canyons on Mars.** Mars contains a great rift valley, Valles Marineris (see photo in Figure 2.5-39). This canyon is nearly 200 km wide and as much as 6.0 km deep. Mars is 6800 km in diameter. If you stood at the center of Valles Marineris, could you see its walls? ***Hint:*** What effect would the curvature of Mars have on your line of sight?

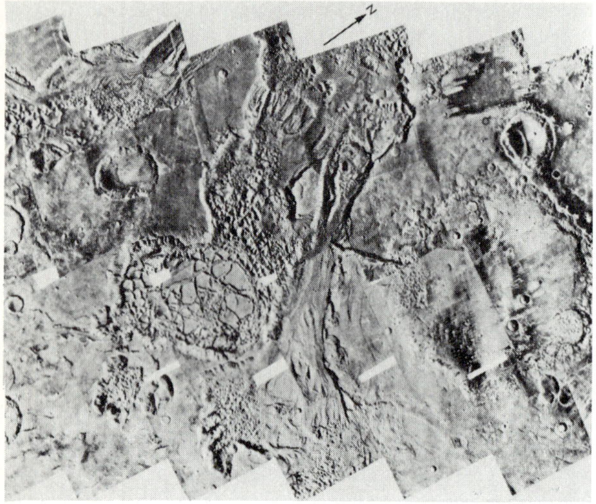

Figure 2.5-39 Valles Marineris on Mars

***66.** **How Far Is the Horizon?** If the earth were flat, the distance to the horizon would be infinite—or at least it would extend to the edge of the earth. But, because the earth is round, there is a limit to how far we can see, as shown in Figure 2.5-40.

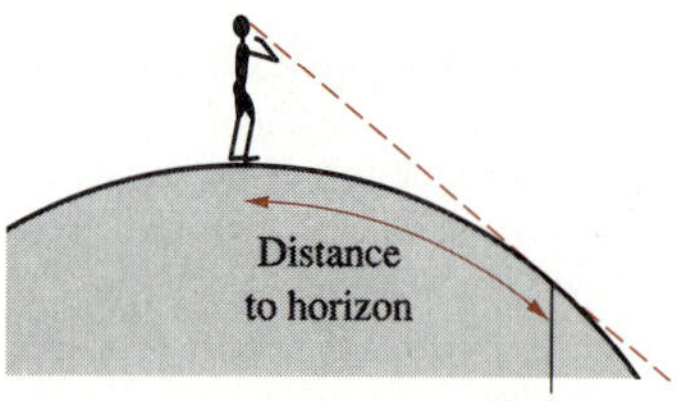

Figure 2.5-40

(a) How far away is the horizon on earth for (i) a 6.0-ft tall person and (ii) a person viewing from the top of a 850-ft hill?
(b) Would these answers also hold on our moon, which has a diameter of 3476 km? If not, find the new horizon distances.

67. To what angle must you raise the hood of your car so that it will cover half of the engine when it is fully raised?

68. **Chemistry.** Formic acid (HCOOH) has the structure shown in Figure 2.5-41. Chemists have measured the angles and distances to be: H—C—O′ = 117.8°, O′—C—O = 124.3°, C—O—H′ = 107.8°, C—O′ = 1.245 Å, C—O = 1.312 Å, C—H = 1.085 Å, and O—H′ = 0.95 Å.
Find the distance between O′ and H′.

****69.** Two circles intersect as shown in Figure 2.5-42. Find the area common to both circles if they have radii of 10.0 ft and 6.00 ft and if the center-to-center distance is 12.0 ft.

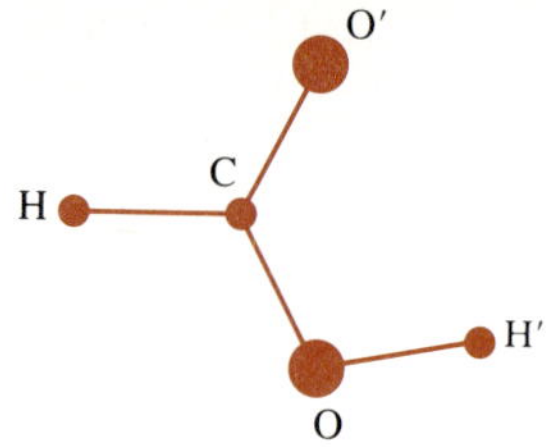

Figure 2.5-41

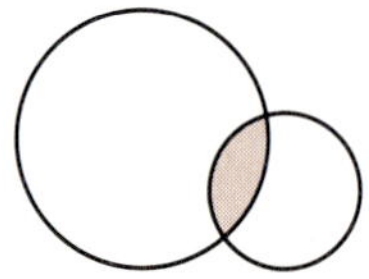

Figure 2.5-42

70. **Basketball.** In professional basketball, the basket rim is 10.0 ft above the floor and the front of the rim is 24.0 in. in front of the backboard. The free throw line is 15.0 ft. from the backboard. If Wilt Chamberlain, who was 7 ft $1\frac{1}{16}$ in. tall, was shooting a free throw, at what angle above the horizontal would he have to raise his eyes to look at the front of the rim, assuming that his eyes were 6.00 in. below the top of his head? ***Hint:*** If you make a careful sketch, this problem is quite easy.

71. **Balance.** A rectangular box will tip over if its center of gravity goes past the pivot point at the corner. The greatest angle it can tip is when the center of gravity is just over the pivot point. If the box in Figure 2.5-43 is 2.5 ft wide and 6.8 ft tall, and if it is uniform so that the center of gravity is in the middle, find the greatest angle θ for which the box will not tip over. ***Hint:*** Construct an appropriate right triangle inside the box using θ.

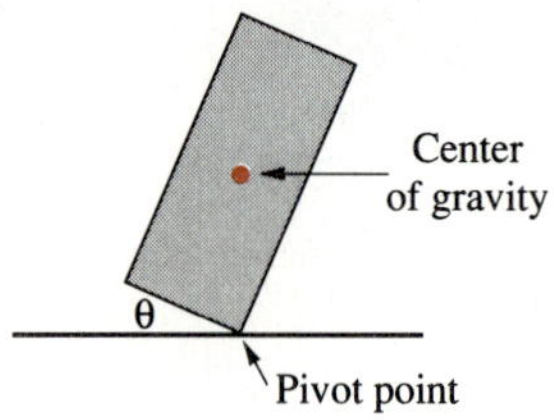

Figure 2.5-43

72. **Balance.** Which is more stable, a tall box or a short box of the same width? To answer this, assign values to the height and calculate the maximum angle that each can be tipped.

73. **Balance.** Suppose that the box in Exercise 71 is partially full, so that the center of gravity is only 2.60 ft above the bottom. Find the maximum angle θ in this case, assuming that the contents do not move inside the box.

***74.** **Balance.** Suppose the uniform box in Exercise 71 is placed on a slanted hill, as shown in Figure 2.5-44. What is the greatest that β can be without causing the box to tip over?

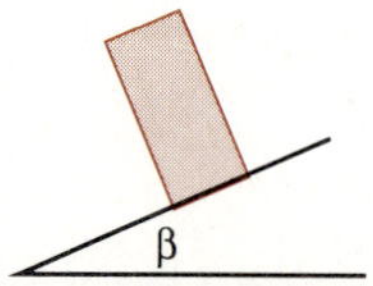

Figure 2.5-44

***75. Balance.** Suppose that the partially filled box in Exercise 73 is placed on the hill in the previous example. What can be the maximum value of β in this case without tipping the box over, assuming the contents don't move inside the box?

***76.** A wedge-shaped garden of radius 12 ft and central angle 30° is completely surrounded by a paved sidewalk 3.0 ft wide, as shown in Figure 2.5-45. Find the area of this sidewalk.

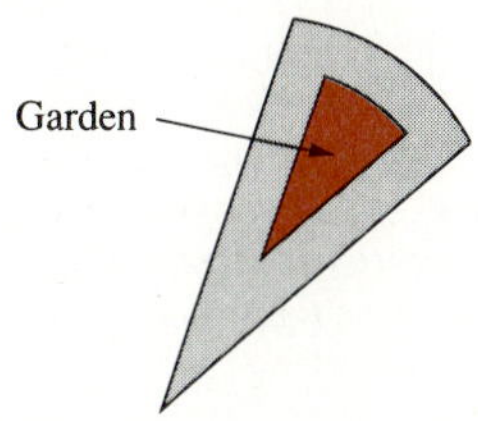

Figure 2.5-45

77. Projections of a Line. We often think of one line as lying partly along another line, as shown in Figure 2.5-46. The line AB lies partly along the line CD. We say that "AB has a *projection* along the line CD." To find the projection of AB along CD, we drop lines from the ends of AB perpendicular to CD. The length intercepted on CD (ab in Figure 2.5-46) is the projection of AB along CD. In a similar way, a line can have projections along the x and y axes, as shown in Figure 2.5-47. The two shaded lines are the projections of AB along the x and y axes.

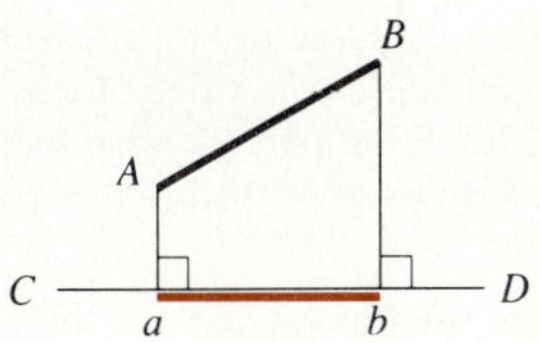

Figure 2.5-46

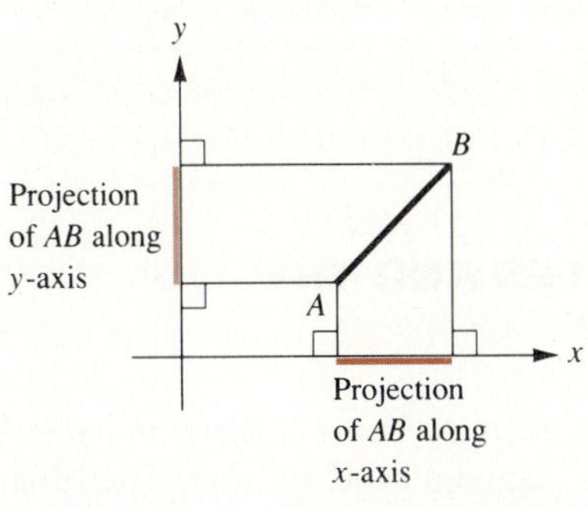

Figure 2.5-47

(a) If a line is 10.0 m long and makes an angle of 30° with the $+x$ axis, find its projections on the x and y axes.
(b) If a line is 58.0 cm long and makes an angle of 51.3° with the $+y$ axis, find its projections along both axes.
(c) If a line has a projection of 15 ft along the $+x$ axis and 20.0 ft along the $+y$ axis, find the length of this line. ***Hint:*** Use the projections to construct a right triangle with this line.
(d) If a line is 25.0 mm long and has a projection on the $+x$ axis of 18.0 mm, find its projection along the y axis and the angle it makes with the x axis.
(e) If a line is 45.0 in. long and has a projection along the $+y$ axis of 32.0 in., find the angle it makes with the $+x$ axis.
(f) If a line has a projection along the $+x$ axis of 1.81 m and a projection along the $+y$ axis of 2.24 m, find the angle it makes with each axis and find its length.

78.** If we draw two lines trisecting the angle θ in Figure 2.5-48, do these lines trisect the side opposite θ? Prove it! ***Hint: To prove that something is **not** true, find just one case in which it is not true.

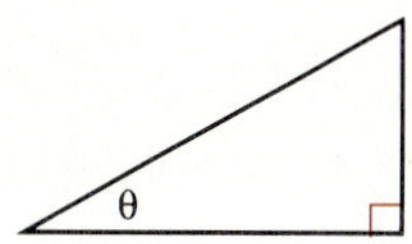

Figure 2.5-48

79. Find the angles α, β, and θ so that the 41.2-ft line in Figure 2.5-49 will be divided into three equal lengths.

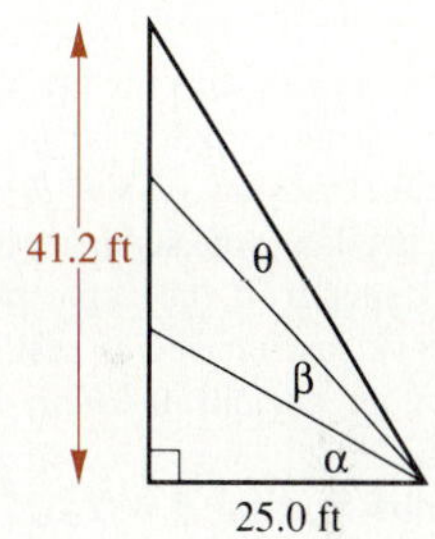

Figure 2.5-49

80. A point in the x-y plane is 14.0 cm from the origin, and a line from the origin to this point makes an angle of +58.0° with the positive x axis. Find the x and y coordinates of this point.

***81.** A point in the x-y plane is 6.75 in. from the origin and a line from the origin to this point makes an angle of 1.84 rad with the $+x$ axis. What are the x and y coordinates of this point?

***82.** A point in the x-y plane is 10.0 mm from the origin and a line from the origin to this point makes an angle of $\frac{9\pi}{8}$ with the $+x$ axis. What are the x and y coordinates of this point?

***83.** A point in the x-y plane is 50.0 m from the origin and a line from the origin to this point makes an angle of 310.0° with the $+x$ axis. What are the x and y coordinates of this point?

84. A point in the x-y plane has coordinates $x = 8.50$ ft and $y = 6.92$ ft.

(a) How far is this point from the origin?
(b) If you draw a line from the origin to this point, what angle does this line make with the $+x$ axis?

***85.** A point in the x-y plane has coordinates $x = -2.56$ cm and $y = +2.00$ cm.

(a) How far is this point from the origin?
(b) If you draw a line from the origin to this point, what angle does this line make with the $+x$ axis?

***86.** A point in the x-y plane has coordinates $x = -6.50$ in. and $y = -4.81$ in.

(a) How far is this point from the origin?
(b) If you draw a line from the origin to this point, what angle does this line make with the $+x$ axis?

***87.** A point in the x-y plane has coordinates $x = +95.60$ mm and $y = -128.1$ mm.

(a) How far is this point from the origin?
(b) If you draw a line from the origin to this point, what angle does this line make with the $+x$ axis?

***88.** **(a)** Pick *any* angle θ and calculate $\sin^2 \theta$, $\cos^2 \theta$, and the sum of $\sin^2 \theta + \cos^2 \theta$.
(b) Now pick five or more different angles and repeat the proceeding calculations. Does it seem that you may have discovered a general law that might be true for *all* angles? What is this law?
(c) To test your hypothesis, pick another angle and *predict* what the sum $\sin^2 \theta + \cos^2 \theta$ should be. Now calculate it. Were you correct?
***(d)** Now let's see if you can *prove* the law you seem to have discovered. Pick any general right triangle with sides a, b, and c. Now calculate the sum of $\sin^2 \theta + \cos^2 \theta$ for this triangle and prove the law you seem to have discovered. Does your answer depend on the value of θ?
(e) The law you have just proved is really just the trigonometric version of an ancient theorem in mathematics. What theorem is this?
(f) Suppose you know that $\sin \theta = 0.451$. Use your result from (b) to find $\cos \theta$ without using trig tables or finding the value of θ.
(g) If the cosine of an angle is 0.200, find the sine of this angle without finding the angle or using tables.
***(h)** Is it also true that $\sin \theta + \cos \theta = 1$? Test this numerically.

89. Is it possible that $\sin \theta + \cos \theta > 1$? Test this numerically.

***90. How Large Can the Sine Be?**

(a) Use your calculator to find the angles having the following sines: (i) 0.50, (ii) 0.75, (iii) 0.89, (iv) 0.95, (v) 0.99, (vi) 0.999, (vii) 0.9999, (viii) 1.00, (ix) 1.01, (x) 1.50, and (xi) 2.53. Make a table of these results.
(b) Can you conclude anything about the maximum value of the sine from what you have just done? What is it?
***(c)** Now let's try to prove the maximum value of the sine in a geometric way. Draw a right triangle of sides a, b, and c, as shown in Figure 2.5-50.

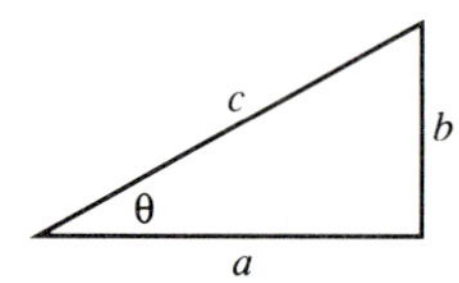

Figure 2.5-50

(i) What is $\sin \theta$ for this triangle? (ii) Which is larger, b or c? Why? Using this result, what can you conclude about the value of $\sin \theta$?
***(d)** Now redraw this triangle several times for values of θ that get closer and closer to 90°. (i) What happens to the value of b compared to the value of c as θ gets closer and closer to 90°? (ii) From part (i), what happens to $\sin \theta$ as θ gets closer and closer to 90°? Does this agree with what you found in part (c)?

***91. How Large Can the Cosine Be?** Using Exercise 90 as a guide, find the maximum value of the cosine. For what angle does the cosine reach its maximum value?

2.6 COMPLEMENTARY ANGLES AND THE COFUNCTION THEOREM

You may have noticed that there seem to be only three *independent* names for the trigonometric functions (sine, tangent, and secant). The others are derived from these three (*co*sine, *co*tangent, and *co*secant). We call trigonometric functions that are related this way ***cofunctions*** of each other. For example, the cosine is the cofunction of the sine. (It is also true that the sine is the cofunction of the cosine.) Let's see what is so special about cofunctions.

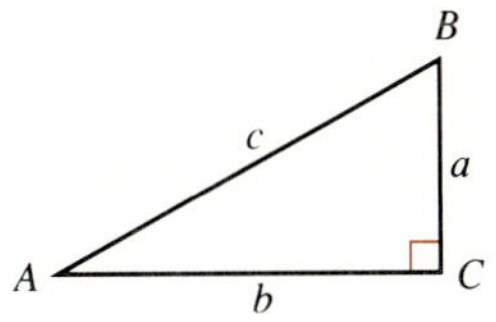

Figure 2.6-1

Consider the right triangle in Figure 2.6-1. Notice that the angles A and B are complementary. Let's investigate each trigonometric function and its cofunction, using this triangle. First, we see that $\sin A = a/c$ and $\cos B = a/c$, which means that

$$\sin A = \cos B$$

Next, we see that $\tan A = a/b$ and $\cot B = a/b$, which means that

$$\tan A = \cot B$$

Finally, $\sec A = c/b$ and $\csc B = c/b$, which means that

$$\sec A = \csc B$$

By the same type of procedure, we can also show that

$$\cos A = \sin B \qquad \cot A = \tan B \qquad \csc A = \sec B$$

Now notice that these six special relations work because the angles A and B are complementary. Therefore, we have just discovered that if A and B are complementary angles, then any trigonometric function of A is equal to its cofunction of B. Also, any trigonometric function of B is equal to its cofunction of A.* We summarize this result as the ***cofunction theorem***.

COFUNCTION THEOREM

If two angles are complementary, then any trigonometric function of one of them is equal to its cofunction of the other, which means that

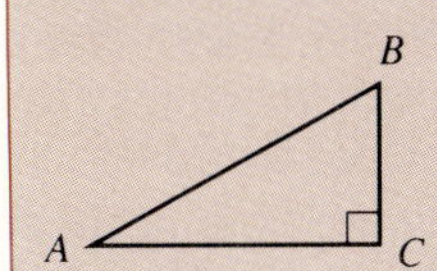

any trigonometric function of A = its cofunction of B

any trigonometric function of B = its cofunction of A

Simply stated, this means that if $A + B = 90°$, then $\sin A = \cos B$, $\cot A = \tan B$, $\sec B = \csc A$, and so on.

---

* We can now see where the names cosine, cotangent, and cosecant came from. The name cosine comes from a contraction of the Latin *complementis sinus*, meaning "sine of the complementary angle." Therefore, $\cos A$ is the sine of the *complement* of A, which is what we just discovered: $\cos A = \sin(\text{complement of } A)$. Likewise, $\cot A$ is the tangent of the complement of A, and $\csc A$ is the secant of the complement of A.

Example 40

If you know that $\sin 52° = 0.7880$, what is $\cos 38°$?

Solution

38° is complementary to 52° because they add to 90°. Therefore, by the cofunction theorem, any trigonometric function of 52° is equal to its cofunction of 38°. Because the cosine is the cofunction of the sine, $\sin 52° = \cos 38°$, so $\cos 38° = 0.7880$. ■

Example 41

If you have only a table of *secants*, find $\csc 15°12'$.

Solution

By the cofunction theorem, $\csc 15°12'$ will be equal to the secant of its complementary angle. The complementary angle of $15°12'$ is

$$90° - 15°12' = 89°60' - 15°12' = 74°48'$$

Therefore,

$$\csc 15°12' = \sec 74°48' \quad ■$$

NOW IT'S YOUR TURN

(a) If $\cot 11° = 5.1446$, find $\tan 79°$.

(b) Using only a table of sines, how would you find $\cos 78°24'$?

(Answer: $\tan 79° = \cot 11° = 5.1446$, $\cos 78°24' = \sin 11°36' = 0.2011$)

If A and B are complementary angles, then the cofunction theorem tells us that $\sin A = \cos B$. But, suppose instead we know that $\sin A = \cos B$. We can work the cofunction theorem backward to conclude that A and B must be complementary, provided they are both acute angles. This is the ***converse of the cofunction theorem***, and we can generalize it for *any* trigonometric function.

> **CONVERSE OF THE COFUNCTION THEOREM**
>
> **If a trigonometric function of A is equal to its cofunction of B, then A and B must be complementary angles, provided they are both acute.**

Simply stated, this means that if $\sin A = \cos B$, then A and B are complementary. Likewise, if $\cot B = \tan A$, or if $\sec B = \csc A$, or if $\cos A = \sin B$, and so on, then A and B must be complementary.

Example 42

If $\sin \theta = 0.200$ and $\cos \alpha = 0.200$, what must be true about θ and α?

Solution

Because $\sin \theta = \cos \alpha$, the converse of the cofunction theorem tells us that α and θ must be complementary because the sine and cosine are cofunctions. So

$$\alpha + \theta = 90° \quad \blacksquare$$

Example 43

If $\sec \theta = \csc \theta$, what is θ?

Solution

Think of this as saying that $\sec A = \csc B$, where $A = B = \theta$ in this case. The converse of the cofunction theorem tells us that A and B must be complementary because secant and cosecant are cofunctions. Thus $A + B = 90°$. Therefore,

$$\theta + \theta = 90° \quad \rightarrow \quad \theta = 45° \quad \blacksquare$$

Example 44

Find an angle θ for which $\cot(\theta + 10°) = \tan \theta$.

Solution

Because the cotangent and tangent are cofunctions, the converse of the cofunction theorem tells us that $(\theta + 10°)$ and θ are complementary if they are acute angles; that is,

$$(\theta + 10°) + \theta = 90°$$

$$2\theta + 10° = 90°$$

$$2\theta = 80° \quad \rightarrow \quad \theta = 40° \quad \blacksquare$$

We must check to see if both θ and $(\theta + 10°)$ are acute angles. Substituting $\theta = 40°$ into each expression, we see that $\theta = 40°$ and $\theta + 10° = 40° + 10° = 50°$. Because both are acute angles, the answer $\theta = 40°$ is the solution.

NOW IT'S YOUR TURN

(a) If $\csc x = 1.892$ and $\sec y = 1.892$, what must be true about x and y?
(b) Find an angle β for which $\sin(\beta + 40°) = \cos \beta$.
(c) If $\sin \alpha = \cos \alpha$, find α.
(Answer: (a) $x + y = 90°$, (b) $\beta = 25°$, (c) $\alpha = 45°$)

EXERCISES 2.6

In Exercises 1–28, express each of the functions as a function of the complementary angle.

1. $\cos 48°$
2. $\tan 32°$
3. $\sin 58°16'$
4. $\sin 55°$
5. $\cot 45°28'$
6. $\cos 40°$
7. $\csc \theta°$
8. $\tan A°$
9. $\csc 24°$
10. $\sec 28°43'$
11. $\cot x°$
12. $\sec 33°$
13. $\sec 72.35°$
14. $\sin 14.43°$
15. $\csc 0.013°$
16. $\cos 78.3°$
17. $\csc 21°$
18. $\sin 1°$
19. $\cos 1$
20. $\tan(0.578 \text{ rad})$
21. $\sin \frac{\pi}{6}$
22. $\cot(1.50 \text{ rad})$
23. $\csc \frac{\pi}{2.5}$
24. $\sec(1.5)$
25. $\cos(0.207)$
26. $\sin 25°18'33''$
27. $\cot 88°59'59''$
28. $\sec 1'1''$

In Exercises 29–34, name the missing trigonometric function.

29. $\sec 40° = \_\_\_\_ 50°$
30. $\cos 25° = \_\_\_\_ 65°$
31. $\cot 38°24' = \_\_\_\_ 51°36'$
32. $\csc 75°15' = \_\_\_\_ 14°45'$
33. $\sin y° = \_\_\_\_ (90° - y°)$
34. $\tan(90° - x°) = \_\_\_\_ x°$

In Exercises 35–50, find the number of degrees in the acute angle θ, if possible.

35. $\cos \theta = \sin \theta$
36. $\cos \theta = \sin 2\theta$
37. $\sec 4\theta = \csc 5\theta$
38. $\tan(\frac{1}{2}\theta) = \cot(3\theta/2)$
39. $\tan 3\theta = \cot 2\theta$
40. $\cot(3\theta + 20°) = \tan(2\theta - 30°)$
41. $\csc 3\theta = \sec(3\theta + 18°)$
42. $\sec \theta = \csc 90°$

*43. $\cot(40°/\theta) = \tan 82°$

*44. $\tan 89° = \cot(1/\theta)$

45. $\sin(\theta + 45°) = \cos(7\theta - 15°)$
46. $\cos(2\theta + 31°) = \sin(31° + 2\theta)$
47. $\cos(\theta + 5°) = \sin(15° - \theta)$
48. $\csc(100° - 4\theta) = \sec 2\theta$

*49. $\cot \theta = \tan n\theta$

*50. $\sin[(p + 1)\theta] = \cos \theta$

*51. If $\sin(x + y) = \cos(2x + 3y)$, express x in terms of y.

*52. If $\tan(x/2 - y) = \cot(x + 2y)$, express y in terms of x.

*53. If $\sec(2x + y + 1) = \csc(x + y)$, express y in terms of x.

In Exercises 54–57, determine whether the given statement is true.

*54. $\cos \theta = \dfrac{1}{\csc(90° - \theta)}$

*55. $\sin \theta = \dfrac{1}{\sec(90° - \theta)}$

*56. $\cot \theta = \dfrac{\cos(90° - \theta)}{\sin(90° - \theta)}$

*57. $\cot \theta = \dfrac{\sin(90° - \theta)}{\cos(90° - \theta)}$

*58. If A, B, and C are angles of any right triangle, is the expression $\sin(\frac{1}{2}A) = \cos[\frac{1}{2}(B + C)]$ true? Explain your answer.

*59. In triangle ABC, if $\sin A = y$ and $\cos B = y$, then what can be said about $A + B$?

2.7 THE SMALL-ANGLE APPROXIMATION REVISITED

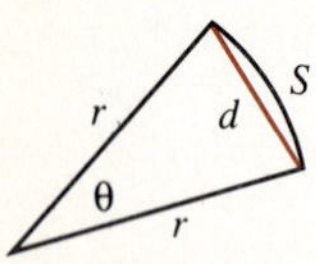

$$\theta \approx \frac{d}{r}$$

Figure 2.7-1
Small-angle approximation

In Section 1.6, we saw that if a central angle is much less than 1 rad, it is approximately equal to the *chord* it subtends (rather than the circular arc) divided by the radius of the circle, as shown in Figure 2.7-1. Now if we consider the right triangle in the circles in Figure 2.7-2, we can extend this approximation to the sine and tangent. As θ gets smaller and smaller in Figure 2.7-2, notice that two things happen.

1. The length of side b gets closer and closer to the length of side r.
2. The length of side a gets closer and closer to the length of the *chord* subtended by θ.

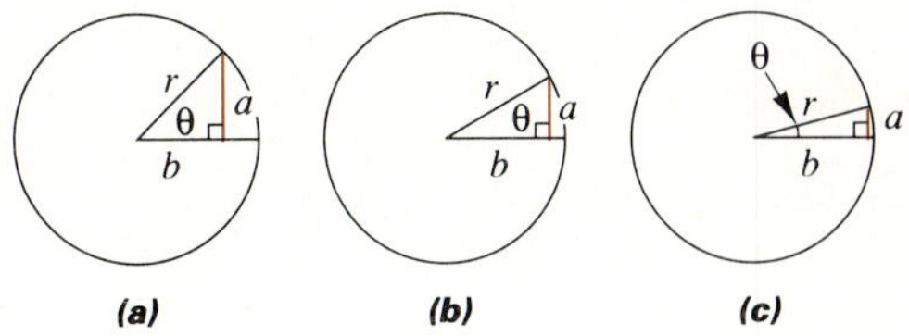

Figure 2.7-2

As θ gets smaller and smaller, r and b are nearly the same length, and a is nearly the length of the chord subtended by θ.

Let's see what effect this has on $\sin\theta$ and $\tan\theta$.

Consider $\tan\theta$: $\tan\theta = a/b$. As θ gets smaller and smaller, b get closer and closer to r. Therefore, $\tan\theta \approx a/r$. But $a/r = \sin\theta$, so for small θ

$$\tan\theta \approx \sin\theta \; \blacksquare$$

Consider $\sin\theta$: $\sin\theta = a/r$. As θ gets smaller and smaller, a gets closer and closer to the length of the *chord* subtended by θ. Thus

$$\sin\theta \approx \frac{\text{chord subtended by } \theta}{r}$$

If we call d the length of this chord, then

$$\sin\theta \approx \frac{d}{r}$$

But Figure 2.7-1 tells us that for small θ, $d/r \approx \theta$, so

$$\sin\theta \approx \theta \; \blacksquare$$

SMALL-ANGLE APPROXIMATION

If $\boldsymbol{\theta}$ is much less than 1 radian,

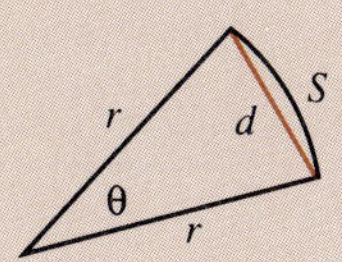

$$\boldsymbol{\theta \approx \frac{d}{r}}$$

$$\boldsymbol{\tan\theta \approx \sin\theta \approx \theta}$$

These approximations do not *replace* the small-angle approximation you learned in Section 1.6; they just *extend it*. You still use that approximation to calculate small angles.

We summarize all of these approximations as the **small-angle approximation***.

▸ **How Small Is Small?** Small-angle approximations can be used whenever θ is much less than 1 rad—but how much less? Although there is no rigid rule, we usually feel comfortable using the small-angle approximation when θ is less than about 10° ($\approx\frac{1}{5}$ rad).

We have space to illustrate only a few examples of the small-angle approximation. However, you should realize that these approximations are used *extensively* in physics, engineering, and calculus.

---

* We have arrived at these approximations *geometrically*, but it is also possible to discover them by making a table of the values of θ, $\sin\theta$, and $\tan\theta$ and noticing what happens as θ gets smaller and smaller. If you have not yet worked it, you may want to go back to Exercise 108 in Section 2.3-1.

Example 45
Hiking

You are out hiking on a trail that rises at a 4% grade. At what angle (in degrees) does it rise, and how far will you have to walk to gain 1000 ft in elevation?

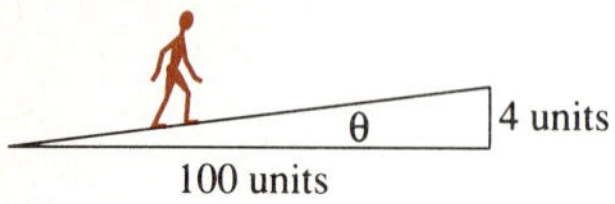

Figure 2.7-3

Hiking up a 4% grade

Solution

A 4% grade means that the trail rises at an angle such that the vertical gain in elevation is 4% (0.04) of the horizontal distance travelled. This means that the trail rises 4 vertical units for every 100 horizontal units walked (Figure 2.7-3). We can use this information to find the angle θ. Figure 2.7-3 tells us that

$$\tan\,\theta = \frac{4 \text{ units}}{100 \text{ units}} = 0.040$$

Because 4 units is so much less than 100 units, θ will be much less than 1 rad. Thus we can use the small-angle approximation, $\tan\,\theta \approx \theta$, which gives us

$$\theta \approx \tan\,\theta = 0.040$$

$$\theta \approx 0.040 \text{ rad}$$

Converting this answer to degrees, we get

$$\theta \approx (0.040 \text{ rad})\left(\frac{180°}{\pi \text{ rad}}\right) \approx 2.3° \blacksquare$$

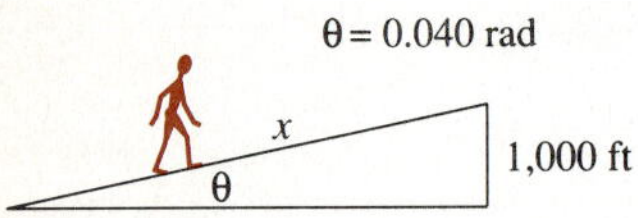

Figure 2.7-4

A 4% grade climbs 1000 ft

Now we can find how far we must hike to gain 1000 ft of elevation (see Figure 2.7-4).

We want the length of the hypotenuse x.

$$\sin\,\theta = \frac{1000 \text{ ft}}{x}$$

Because θ is much less than 1 rad, we can use the small-angle approximation $\sin\,\theta \approx \theta$.

$$\theta \approx \frac{1000 \text{ ft}}{x}$$

$$x \approx \frac{1000 \text{ ft}}{\theta} \approx \frac{1000 \text{ ft}}{0.040} \approx 25{,}000 \text{ ft} \blacksquare$$

or almost 5 mi.

NOW IT'S YOUR TURN

On a trail with a 7% grade, how much elevation would you gain after hiking 10 mi? (Answer: 3700 ft)

Example 46 Snellen Eye Chart

When you have your vision tested, the ophthalmologist usually has you sit 20 ft from a Snellen chart (see Figure 2.7-5) and read down it until you can no longer make out the letters. Based on these tests, the doctor rates your visual acuity as the ratio of two numbers, such as 20/20, 20/35, and so on, always with 20 as the top number. What this ratio means is this: If, for example, your vision is 20/35, the smallest object you can see at 20 ft can be seen by a person with normal vision at 35 ft. Normal vision is 20/20 because you can see at 20 ft what someone with normal vision can see at 20 ft.

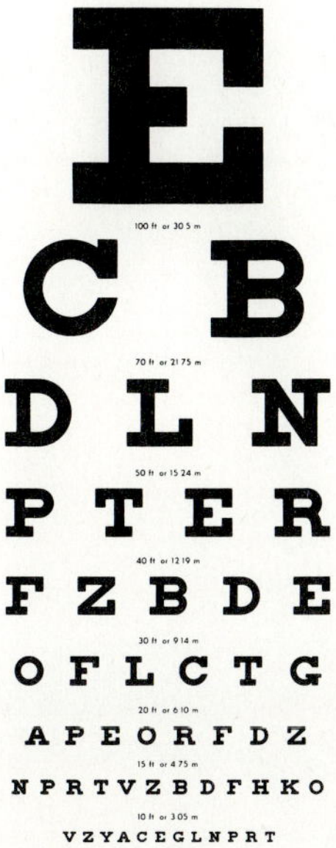

Figure 2.7-5

Snellen eye chart ($\frac{1}{5}$ actual size)

(a) By measuring the necessary quantities on the Snellen Eye Chart in Figure 2.7-5, find the smallest angle in minutes that someone with normal vision can resolve and read comfortably.

(b) If you have 20/30 vision, what is the smallest angle in minutes that you can comfortably read?

Solution

(a) We measured the height of the letters in row 6 (the 20/20 row) to be $\frac{2.5}{32}$ of an inch. This chart is only $\frac{1}{5}$ normal size, so on an actual chart the letters would be five times this height, or $5\left(\frac{2.5}{32}\text{ in.}\right) = 0.39$ in. A normal eye can read this letter at 20 ft (see Figure 2.7-6). Using either right triangle in Figure 2.7-6 gives us

$$\tan\frac{\alpha}{2} = \frac{(0.39\text{ in.})/2}{20\text{ ft}}$$

$$= \frac{(0.39\text{ in.})/2}{240\text{ in.}} = 8.13 \times 10^{-4}$$

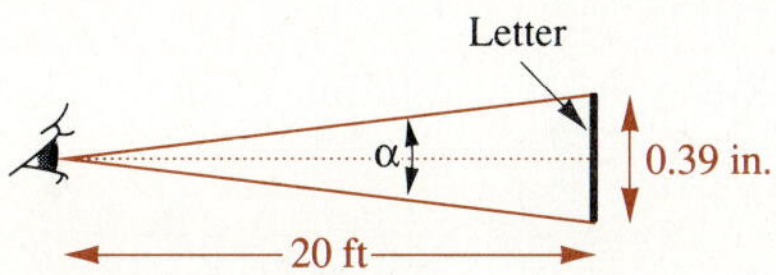

Figure 2.7-6

Normal eye reading its smallest letter at 20 ft

Because α subtends only 0.39 in. at 20 ft, it must be very small. Thus we can use the small-angle approximation $\tan\theta \approx \theta$, where $\theta = \alpha/2$. Thus we get

$$\frac{\alpha}{2} \approx \tan\frac{\alpha}{2} = 8.13 \times 10^{-4}$$

$$\frac{\alpha}{2} \approx 8.13 \times 10^{-4}$$

$$\alpha \approx 1.6 \times 10^{-3}\text{ rad}$$

Converting this to minutes,

$$\alpha \approx (1.6 \times 10^{-3} \text{ rad})\left(\frac{180°}{\pi \text{ rad}}\right)\left(\frac{60'}{1°}\right) \approx 5.5' \blacksquare$$

(b) With 20/30 vision, you can read at 20 ft what a normal eye can read at 30 ft. (Remember, the normal eye can read an angle of 1.6×10^{-3} rad. See Figure 2.7-7.)

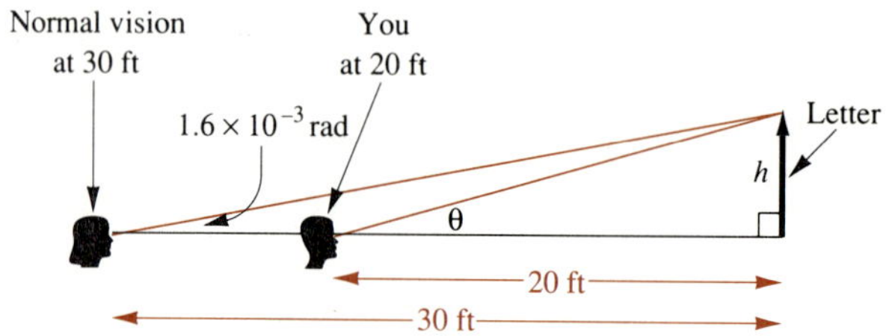

Figure 2.7-7

Someone with 20/30 vision can read at 20 ft what a normal eye can read at 30 ft.

Using the larger triangle to get h,

$$\tan (1.6 \times 10^{-3} \text{ rad}) = \frac{h}{30 \text{ ft}}$$

$$h = (30 \text{ ft}) \tan (1.6 \times 10^{-3} \text{ rad})$$

Using the small-angle approximation gives us

$$h = (30 \text{ ft}) \tan (1.6 \times 10^{-3} \text{ rad}) \approx (30 \text{ ft})(1.6 \times 10^{-3})$$

$$h \approx 4.8 \times 10^{-2} \text{ ft} = 0.048 \text{ ft}$$

Now that we know h, the smaller triangle gives us θ.

$$\tan \theta = \frac{h}{(20 \text{ ft})}$$

$$h = (20 \text{ ft}) \tan \theta$$

The small-angle approximation $\tan \theta \approx \theta$ gives us

$$h \approx (20 \text{ ft})(\theta)$$

$$\theta \approx \frac{h}{20 \text{ ft}} \approx \frac{0.048 \text{ ft}}{20 \text{ ft}} \approx 2.4 \times 10^{-3} \text{ rad}$$

Converting this to minutes, we get

$$\theta \approx (2.4 \times 10^{-3} \text{ rad})\left(\frac{180°}{\pi \text{ rad}}\right)\left(\frac{60'}{1°}\right) \approx 8.25' \blacksquare$$

NOW IT'S YOUR TURN

If you have 20/45 vision, what is the smallest angle in minutes you can read on a Snellen Eye Chart at 20 ft? (Answer: 12.4′. You need glasses!)

EXERCISES 2.7

1. If a lighthouse on an island is 37 ft tall and subtends an angle of 2.3° from shore, how far away is it?

2. A road rises at a 4.8% grade. By what angle does it rise?

3. A mountain trail descends at a 10% grade. At what angle does it descend? When you have hiked a distance of 2.00 mi, how many feet of elevation have you lost?

4. A trail up a volcanic cinder cone rises 600.0 ft in 1.4 mi of trail distance. What is its percent grade and at what angle does it rise?

5. In college football, the goal posts are at the back of the end zone, 10.0 yd beyond the 100.0-yd line, and their horizontal bar is 10 ft high. The kickoff occurs at the 40.0 yd line (see Figure 2.1-20).

(a) Find the angle of elevation of the goal posts' horizontal bar from the kickoff position.

(b) In professional football, the kickoff occurs at the 35-yd line. By how much does this change the angle of elevation in part (a)?

6. In women's softball, the outer edge of the outfield should be a circular arc with a minimum radius of 200.0 ft from home plate (see Figure 2.7-8). The "diamond" is 60.0 ft on each side. If the outfield fence is 15 ft tall, what will be the angular height of its center as viewed by: (a) The batter? (b) The second basewoman? (c) The first basewoman?

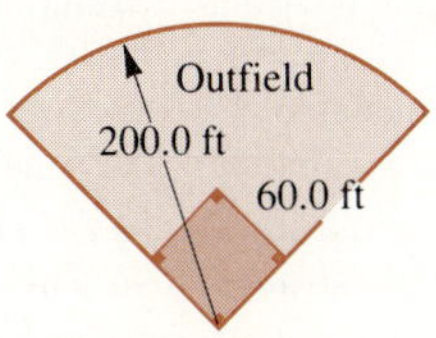

Figure 2.7-8

For Exercises 7–11, see Example 46.

7. If we could comfortably see angles as small as 60.0″, rather than 5.5′, how tall would the letters for 20/20 vision have to be on a Snellen Eye Chart to be viewed from 20.0 ft?

8. If you are designing a Snellen Eye Chart to be viewed from 15 ft, how tall should the letters be for the 20/20 row?

9. If you can read letters down to 0.10 in. tall on a Snellen Eye Chart at 20 ft, what is the smallest angle your eye can comfortably resolve?

***10.** If you have 20/35 vision, what is the smallest angle you can comfortably resolve, and how tall is the smallest letter you can read on a Snellen Eye Chart viewed from 20 ft?

****11.** If you can read letters down to 0.25 in. tall on a Snellen Eye Chart at 20 ft, what is your visual acuity expressed in the conventional way, as $20/x$; that is, what is x?

12. Mount St. Helens Eruption. Mount St. Helens in Washington was originally 2950 m above sea level. After the 1980 eruptions, the angular height of the top, as viewed from a location 10 km (horizontal distance) from the volcano at an altitude of 1000 m, was reduced by 2.2° (see the before-and-after photos in Figure 2.7-9 from the March 1981 *Scientific American*). By how much had these eruptions reduced the volcano's height above sea level?

Figure 2.7-9 Mount St. Helens

13. A radiotransmitter is atop a 460-ft tower 1.33 mi from your home. At what angle above the horizontal must you raise your antenna to point at the transmitter?

14. According to a 1986 newspaper article,* the Environmental Protection Agency predicts that the greenhouse effect, caused by the release of carbon dioxide (CO_2) from burning fossil fuels such as oil and coal, will heat the atmosphere and thus melt polar ice. If the polar ice melts, it will cause sea levels around the United States to rise about 1 ft in the next 30–40 yr, and by 3–5 ft over the next century. Suppose you have a beach front house on land

*"The Tide's Coming In," by Erik Eckholm, in the *Sacramento Bee*, March 29, 1986, from the *New York Times*.

that slopes upward at 2°, which is not uncommon in many areas. How many feet will the ocean advance onto your land in the next 30–40 yr? In the next century, assuming an average 4.0-ft rise?

***15.** In the triangle in Figure 2.7-10, first calculate h using the small-angle approximation without using any trig functions. Then calculate h using the appropriate trig function. Which answer is more exact? By what percent are you in error using the small-angle approximation? By what percent would you be in error if the angle were 15° instead?

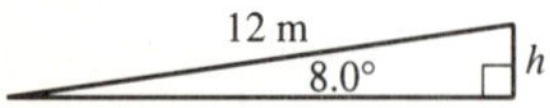

Figure 2.7-10

16. By consulting a map, you find that you are 8.5 horizontal miles away from a mountain peak having an angle of elevation of 9.3°. How high is this peak above your elevation? What error do you make using the small-angle approximation?

17. Halley's Comet. In 1986 the European space probe Giotto photographed the nucleus of Halley's comet from 1703 km. In those photos, the nucleus was a potato-shaped object subtending angles of 0.25° and 0.47°. The Soviet spacecraft Vega 2 photographed the nucleus from 8030 km.

(a) Find the length and width of the nucleus.

(b) What angles did the nucleus subtend in the photos taken by Vega?

18. A spy satellite 25.0 mi directly overhead photographs a building when the sun is at 20.00° above the horizon. The shadow of this building covers an angle of 0.135′ in this photo. How tall is the building?

***19. Moons of Mars.** Mars has two moons, Phobos and Deimos. Phobos is 27 km wide and orbits 9,400 km from the center of Mars, whereas Deimos is 15 km wide and orbits 23,000 km from Mars' center. Mars' diameter is 6,787 km.

(a) Calculate the angular diameter of Mars from each moon. Is it safe to use the small-angle approximation? Why? How do these angular diameters compare with our moon's apparent size as viewed from earth?

(b) Calculate the largest angular diameter of Deimos when viewed from Phobos.

20. Volcanoes on Mars. Use the following table to calculate the angle of the average slope of the sides of the Martian volcanoes given, assuming that they rise symmetrically to a peak.

Volcano	Height (km)	Diameter (km)
Arsia Mons	19	400
Elysium Mons	15	250
Hecates Tholus	7	200

21. Lasers. Commercial lasers for use in schools have typical beam divergence angles of about ±1.3 milliradians. (The prefix *milli* means 10^{-3}.) If such a laser beam has negligible thickness when it leaves the laser, what will be its diameter when it hits the wall 24 ft away?

22. Lasers. If a laser beam starts out with negligible thickness and makes a dot 1.40 mm in diameter on a screen 3.20 m away, what is the divergence of this beam in milliradians? (See Exercise 21.)

***23. Lasers.** If a laser beam with a beam divergence of ±1.10 milliradians has a diameter of 2.00 mm when it leaves the laser, what will be its diameter when it hits a screen 1.50 m away? (See Exercise 21.)

***24. Laser Surgery.** A surgeon is using a laser beam to perform surgery on the retina of the eye. If the beam is 0.00010 mm in diameter when it leaves the laser and must produce a spot no larger than 0.0015 mm on the retina, what must be its maximum divergence in milliradians if the laser is to be used 20 cm from the retina? (See Exercise 21 and Figure 2.7-11.)

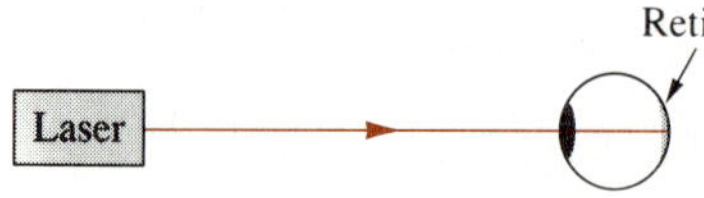

Figure 2.7-11

25. Resolving Power of a Telescope. The opening of a telescope bends light (causes diffraction) and therefore makes the image slightly unclear. The larger the telescope, the less the distortion. The smallest angle θ that can be viewed clearly (or resolved) is given by $\sin \theta = 1.22\ \lambda/d$, where λ is the wavelength of the light and d is the diameter of the telescope lens or mirror.

(a) Using visible light of wavelength 5000.0 Å, calculate the smallest angle that can be resolved by (i) a 6.0-in. (15 cm) diameter telescope and (ii) the 200.0-in. (5.00 m) Hale telescope.

(b) The nearest star is Alpha Centauri, about 4.28 light-years away (1 light-year = 9.46×10^{12} km). If this star is 1.4×10^6 km in diameter, like our sun, could the Hale telescope resolve it to obtain a sharp picture of it?

26. Resolving Power of the Human Eye. The opening of the eye—the pupil—produces distortion just as the opening of a telescope. Therefore, the equation given in Exercise 25 tells us the resolving power of the eye. The pupil's diameter is about 3 mm, and we are most sensitive to light of wavelength 5500 Å. Find the smallest angle our eye can resolve. In view of what we learned in Example 39 in Chapter 1, would there be any advantage to having much larger pupils (greater d) to resolve even smaller angles?

27. Volcanoes on Io. The Voyager I rocket in March 1979 photographed active volcanoes on Jupiter's moon Io (see Figure 2.7-12). In the photo, taken from a distance of 4.5 million kilometers, the volcanic plume subtends an angle of 12.4″. How high did this volcano eject debris in its eruption?

Figure 2.7-12 Volcanoes on Io

28. Solar Prominences. This great eruption (solar prominence) on our sun was photographed on June 4, 1946 (Figure 2.7-13). In this photo, the height of the gas arch subtends an angle of 7.4′ from earth. How high was the arch above the sun?

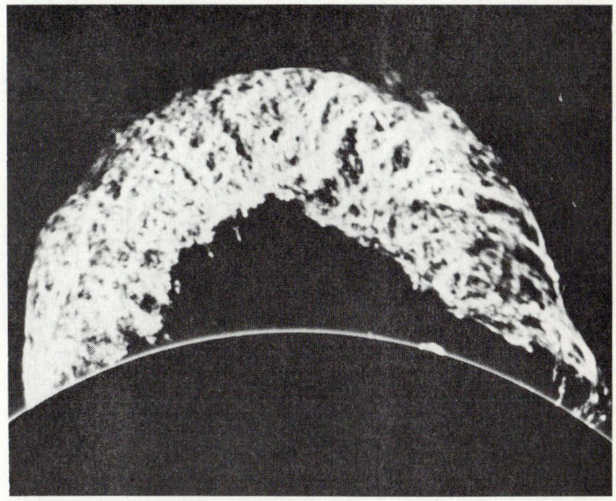

Figure 2.7-13 Solar prominence

SUMMARY

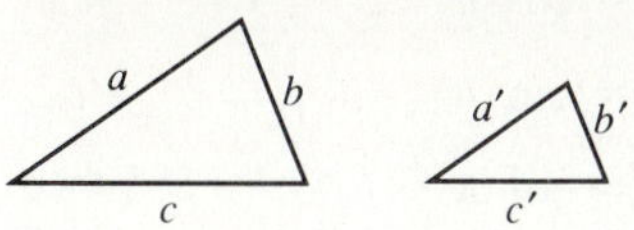

1. Similar Triangles. Two triangles are *similar* when the three angles of one of them are equal to the corresponding three angles of the other. The corresponding sides of similar triangles are proportional.

$$\frac{a}{a'} = \frac{b}{b'} = \frac{c}{c'} \quad \text{(Euclid's theorem)}$$

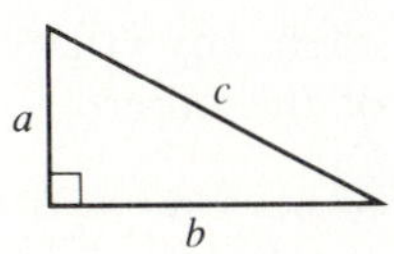

2. Pythagorean Theorem. In a right triangle, the square of the length of the hypotenuse is equal to the sum of the squares of the lengths of the legs.

$$a^2 + b^2 = c^2$$

3. Converse of the Pythagorean Theorem. If the lengths of the sides of a triangle obey the Pythagorean theorem, then it must be a right triangle.

4. Trigonometric Functions for a Right Triangle.

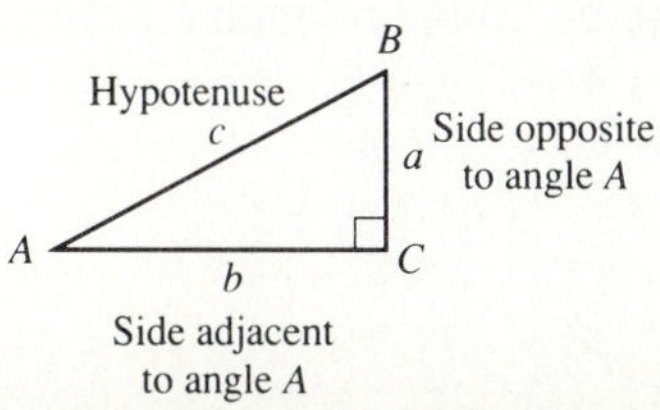

$$\sin A = \frac{a}{c} \qquad \tan A = \frac{a}{b} \qquad \sec A = \frac{c}{b}$$

$$\cos A = \frac{b}{c} \qquad \cot A = \frac{b}{a} \qquad \csc A = \frac{c}{a}$$

5. Special Angles: 30°, 45°, 60°.

θ	30°	45°	60°
$\sin\theta$	$\frac{1}{2}\sqrt{1}$	$\frac{1}{2}\sqrt{2}$	$\frac{1}{2}\sqrt{3}$
$\cos\theta$	$\frac{1}{2}\sqrt{3}$	$\frac{1}{2}\sqrt{2}$	$\frac{1}{2}\sqrt{1}$

6. Reciprocal Properties of the Trigonometric Functions.

$$\csc A = \frac{1}{\sin A} \qquad \sec A = \frac{1}{\cos A} \qquad \cot A = \frac{1}{\tan A}$$

$$\sin A = \frac{1}{\csc A} \qquad \cos A = \frac{1}{\sec A} \qquad \tan A = \frac{1}{\cot A}$$

7. Quotient Relations.

$$\tan A = \frac{\sin A}{\cos A} \qquad \cot A = \frac{\cos A}{\sin A}$$

8. Inverse Trigonometric Functions.

$$\sin^{-1} y = \theta \text{ is equivalent to } y = \sin\theta$$
$$\cos^{-1} y = \theta \text{ is equivalent to } y = \cos\theta$$
$$\tan^{-1} y = \theta \text{ is equivalent to } y = \tan\theta$$
$$\cot^{-1} y = \theta \text{ is equivalent to } y = \cot\theta$$
$$\sec^{-1} y = \theta \text{ is equivalent to } y = \sec\theta$$
$$\csc^{-1} y = \theta \text{ is equivalent to } y = \csc\theta$$

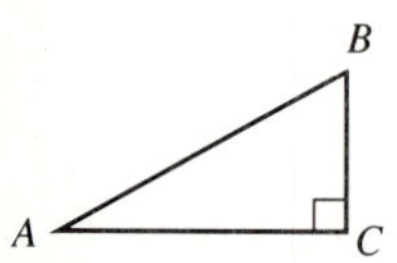

9. Cofunction Theorem. If two angles are complementary, then any trigonometric function of one of them is equal to its cofunction of the other.

$$\text{any trigonometric function of } A = \text{its cofunction of } B$$
$$\text{any trigonometric function of } B = \text{its cofunction of } A$$

10. Converse of the Cofunction Theorem. If a trigonometric function of A is equal to its cofunction of B, then A and B must be complementary angles, provided they are acute.

11. Small-Angle Approximation.

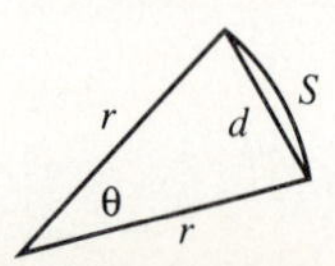

For very small angle θ in *radians*:

$$\theta \approx d/r$$

$$\tan\theta \approx \sin\theta \approx \theta$$

IMPORTANT TERMS

adjacent side
cofunction
converse of the Pythagorean theorem
cosecant
cosine
cotangent
equilateral triangle
Euclid's theorem
hypotenuse
inverse cosecant
inverse cosine
inverse cotangent
inverse secant
inverse sine
inverse tangent
isosceles triangle
oblique triangle
opposite side
Pythagorean theorem
reciprocal
right triangle
secant
similar triangles
slope (of a surface or incline)
sine
tangent

REVIEW EXERCISES

1. Use a calculator to evaluate the six trigonometric functions of θ to four decimal places.

(a) $\theta = 28.7°$ **(b)** $\theta = 80°18'$
(c) $\theta = 56°32'12''$ **(d)** $\theta = 0.812$
(e) $\theta = \frac{\pi}{8}$ **(f)** $\theta = \frac{4}{\pi}$
***(g)** $\theta = \sin 1.27$ ***(h)** $\theta = \cot 61.7°$

2. Find the third side of the given triangle and then find the sine, cosine, and tangent of angles A and B without finding A and B.

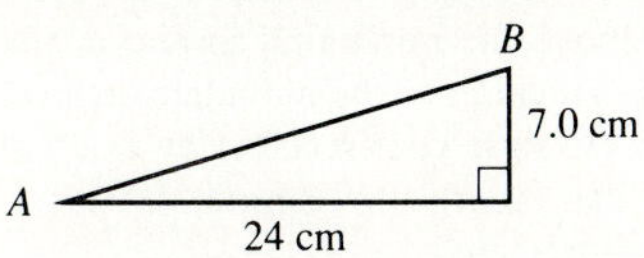

(a)

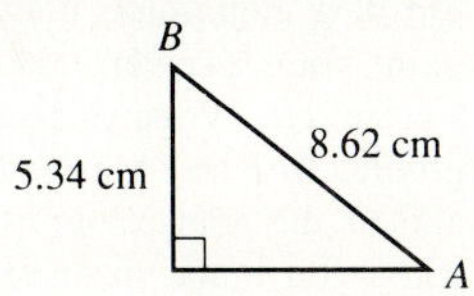

(b)

3. In each case, one trigonometric function of θ is given. Evaluate the other five trigonometric functions of θ without finding θ.

(a) $\cos \theta = \frac{15}{17}$ **(b)** $\csc \theta = \frac{9}{5}$
(c) $\tan \theta = \frac{3}{7}$ **(d)** $\cos \theta = 0.324$

4. Find all the angles and the lengths of the sides in the triangle using the information given.

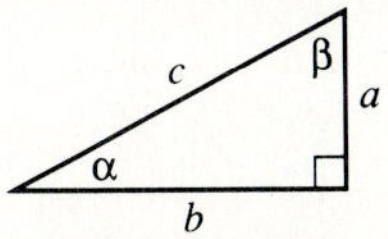

(a) $\beta = 62.4°$, $c = 132$ cm
(b) $\alpha = 1.20$ rad, $c = 2.20$ m
(c) $\alpha = \frac{3\pi}{7}$, $a = 41{,}200$ yd **(d)** $\beta = 16°18'$, $a = 14.0$ ft
(e) $\beta = 67.4°$, $\alpha = 22.6°$ **(f)** $c = 6.25$ m, $b = 4.07$ m

5. Suppose you observed that last May 1 the persimmon tree in your back yard was 12.80 ft tall and cast a shadow 9 ft 7 in. long at noon. This year, at the same time on the same day, the shadow is 4 ft longer. How much did the tree grow during the past year?

6. Evaluate the following without using a calculator or a table.

(a) $\cos(\arcsin \frac{1}{3})$ **(b)** $\tan(\cos^{-1} \frac{2}{3})$
(c) $\sin(\arctan 4)$

7. Evaluate the following in terms of x.

(a) $\cos(\cos^{-1} x)$ **(b)** $\sin(\arccos x)$
(c) $\sin(\tan^{-1} x)$

8. An archaeologist is trying to reconstruct the triangular clay tiles used to pave the courtyard of an ancient building. Unfortunately, these tiles have been broken over the years. From the measurements she has made on the broken tile in the figure, reconstruct the original dimensions of the tile and find its area.

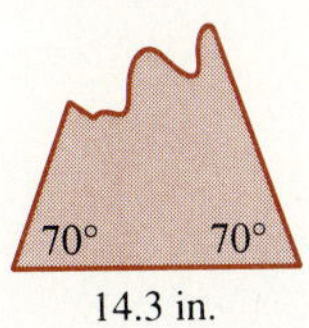

9. In each equation, use a calculator to solve for the acute angle x in radians accurate to two decimal places.

(a) $\sin x = 0.5556$ **(b)** $7 \cos x = 6.8393$
(c) $5.1 - 4.7 \tan x = 3.2$ **(d)** $\cot x = 2.50$
***(e)** $3 + 12 \cos^2 x = 13 \cos x$ ***(f)** $15 \sin^2 x + 1 = 12 \sin x$

10. Sharpen a pencil. Then use only length measurements to calculate the vertex angle α.

11. Use a calculator to find the acute angle θ in decimal degrees accurate to two decimal places.

(a) $\theta = \arccos 0.73373$ **(b)** $\theta = \csc^{-1} 4.091$
(c) $\theta = \frac{3}{4} \tan^{-1} 1.7764$ **(d)** $\cot \theta = 0.68679$
***(e)** $\sin(\theta^2) = 0.83997$ **(f)** $6\ \theta = \sec^{-1} 2.73$

12. Use a calculator to evaluate each acute angle θ in radians accurate to two decimal places.

(a) $\theta = \sin^{-1} 0.10572$ **(b)** $\cot \theta = 0.54980$
(c) $\tan(3.7\ \theta) = 11.869$ **(d)** $\cos(\theta^2) = 0.75601$
(e) $\theta = \csc^{-1} 1.2880$ **(f)** $8.27\ \theta = \text{arcsec } 1.6032$

13. A building having the dimensions shown in the figure has a roof consisting of two sloping segments. If the highest point of the roof is to be 13.0 ft above ground level, find the angle in degrees with which each slopes above the horizontal.

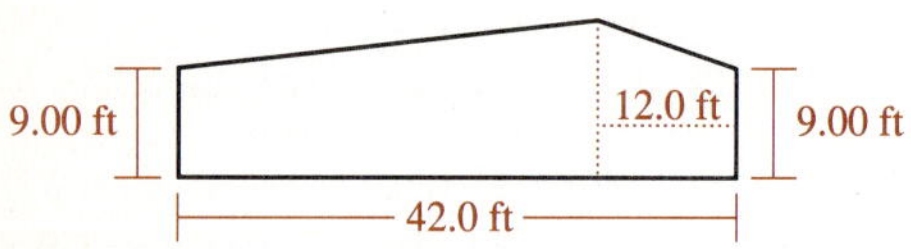

14. A staircase rises at a 35° angle to a floor 12 ft above its base. The handrail for this staircase is 32 in. above the steps. How long is the handrail?

***15.** You are manufacturing canvas tents in the shape shown in the figure. The height of the peak above ground level will be 4.00 ft. Find the total area of canvas you will need to cover the front, back, and sloping walls. The bottom, of course, will not be canvas.

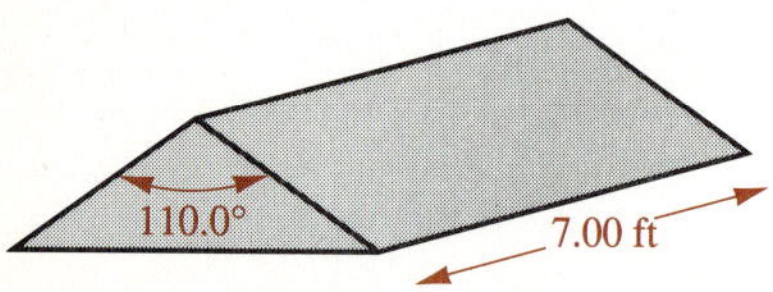

16. Find each answer without using a calculator or a table.

(a) Given that $\sin 61° = 0.875$, find $\cos 29°$.
(b) Given that $\cot 19.4° = 2.840$, find $\tan 70.6°$.
(c) Given that $\sec 11° = 1.019$, find $\csc 79°$.
(d) Given that $\cos \frac{\pi}{9} = 0.940$, find $\sin \frac{7\pi}{18}$.
(e) Given that $\tan 0.810 = 1.050$, find $\cot 0.761$.
(f) Given that $\sin \theta = \frac{3}{4}$, find $\sec(90° - \theta)$.
(g) If $\tan \theta = 1$, what is θ?
(h) Solve for θ in degrees: $\sin(\theta + 20°) = \cos(4\theta + 5°)$.
(i) Solve for θ in radians: $\tan(2\theta + \frac{\pi}{12}) = \cot(3\theta + \frac{\pi}{8})$.
(j) Solve for x: $2 \sin x = \sqrt{3}$.
(k) Find sin(0.524 rad).
***(l)** Solve for x in radians: $2\sqrt{2} \cos^2 x - 2 \cos x + 1 = \sqrt{2} \cos x$.
(m) Evaluate: $\sin^2 \frac{\pi}{3} + \cos^2 \frac{\pi}{4} - 3 \tan^2 \frac{\pi}{6}$.
(n) Solve for ϕ in degrees: $4 \cos^2 \phi = 3$.
(o) Solve for α in radians: $4 + 3 \cot \alpha = 7$.
(p)** Solve for x: $4 \sin^3 x + 1 = 2 \sin x + 2 \sin^2 x$. ***Hint: This can be factored into two factors.

17. Communications Satellites. Communications satellites, such as Telstar, orbit over the earth's equator at an altitude of 22,500 mi. What angle does our planet subtend in a photo taken from Telstar?

18. When you are 50.0 ft from the foot of a 300.0-ft tall coast redwood tree, at what angle must you lift your eyes to see the top of this tree? When you walk 25 ft closer, will the angle be twice as large? Justify your answer numerically. Is it necessary to take your height into consideration? Why?

19. The staircase leading to a large public building has stairs that are 8.0 in. high and 2.0 ft deep. At what angle are you rising above the horizontal as you walk up this staircase?

20. You are tuning your receiver to a transmitter that is at the top of a mountain. You observe that at one location you must point your receiver at 60.7° above the horizontal to aim at the transmitter. When you walk 680 ft closer to the mountain on level ground, you find that you have to raise your receiver by 11.3° to aim at the transmitter. How far was the transmitter from you when you made the first measurement?

21. In major-league baseball, the bases lie at the corners of a square (called the "diamond") 90.0 ft on a side. The pitcher's plate is 60 ft 6 in. from home plate and equidistant from first base and third base. How far is it from the pitcher's plate to second base?

3

Trigonometric Functions of a Single Angle

In Chapter 2 we defined the trigonometric functions of *acute* angles. However, many natural processes are not limited to angles less than 90°. For instance, the planets revolve through more than 90° as they orbit the sun; phonograph records turn through angles larger than 90°; oblique triangles can contain angles greater than 90°; our field of vision extends beyond 90°; and so forth. From these examples, it should be clear that to make trigonometry as useful as possible, the trigonometric functions must not be limited to acute angles. In this chapter, we extend these functions to include all angles: acute, oblique, quadrantal, positive, or negative.

Our approach in this chapter is slightly different from that of Chapters 1 and 2. Because we are generalizing many of the concepts and ideas from those chapters, our approach here is more mathematical. As a result of this generalization, we can, however, introduce many new and diverse applications in later chapters.

3.1 TRIGONOMETRIC FUNCTIONS FOR ANY ANGLE

3.1-1 Generalizing the Trigonometric Functions

In Chapter 2, we defined the trigonometric functions of an acute angle in terms of the sides of a right triangle. If we are now going to consider angles greater than 90°, these definitions are inadequate because we no longer have a right triangle. To see how to proceed for angles larger than 90°, let us use the right triangle as a *starting point*, but not limit the angle θ to less than 90°.

To begin, let us draw a right triangle with θ in standard position in an x-y coordinate system, as shown in Figure 3.1-1. To generate the angles between 0° and 90°, we can keep the hypotenuse at a fixed length (call it r) and swing it around in an arc of a quarter circle of radius r, as shown in Figure 3.1-2. The tip of this hypotenuse is a point $P(x, y)$ having coordinates x and y, moving in a quarter circle. The x coordinate of this point is just the side adjacent to θ, while the y coordinate is the side opposite θ, as shown in Figure 3.1-3.

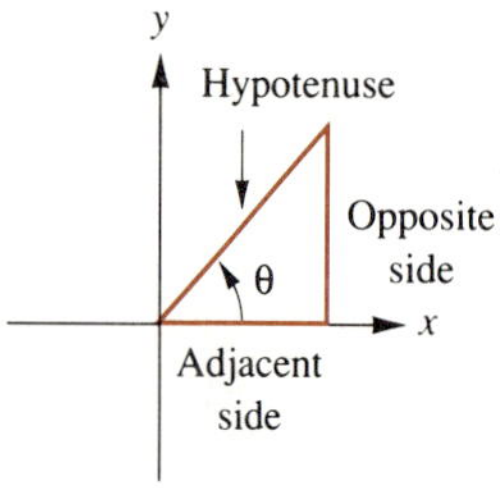

Figure 3.1-1

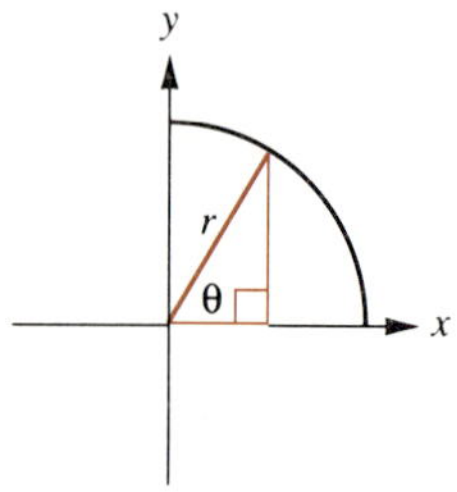

Figure 3.1-2

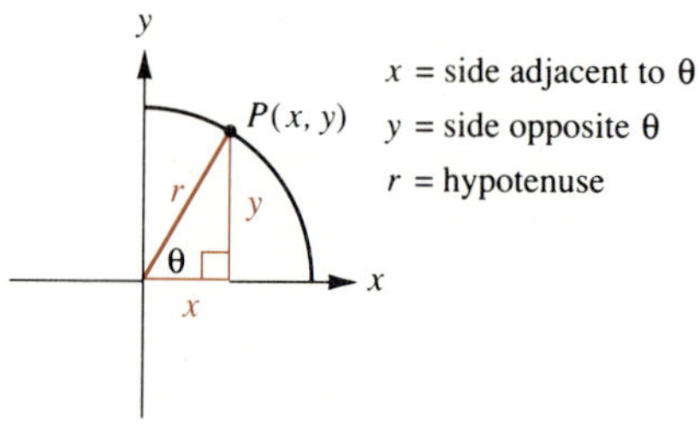

Figure 3.1-3

We can now use the definitions from Chapter 2 to express the trigonometric functions in terms of the coordinates of the point $P(x, y)$ as follows:

$$\sin\theta = \frac{\text{opposite}}{\text{hypotenuse}} = \frac{y \text{ coordinate of } P}{\text{radius}} = \frac{y}{r}$$

$$\cos\theta = \frac{\text{adjacent}}{\text{hypotenuse}} = \frac{x \text{ coordinate of } P}{\text{radius}} = \frac{x}{r}$$

$$\tan\theta = \frac{\text{opposite}}{\text{adjacent}} = \frac{y \text{ coordinate of } P}{x \text{ coordinate of } P} = \frac{y}{x}$$

$$\cot\theta = \frac{\text{adjacent}}{\text{opposite}} = \frac{x \text{ coordinate of } P}{y \text{ coordinate of } P} = \frac{x}{y}$$

$$\sec\theta = \frac{\text{hypotenuse}}{\text{adjacent}} = \frac{\text{radius}}{x \text{ coordinate of } P} = \frac{r}{x}$$

$$\csc\theta = \frac{\text{hypotenuse}}{\text{opposite}} = \frac{\text{radius}}{y \text{ coordinate of } P} = \frac{r}{y}$$

▸ **Notice that we started with the definitions of the trigonometric functions for a right triangle, but now we have expressed them in terms of the coordinates of a point $P(x, y)$ lying on a circle.**

We can use this new expression of the trigonometric functions to extend them to angles greater than 90°.

When θ is in a right triangle, it cannot exceed 90°, but the point $P(x, y)$ on a circle *does not have to stop at 90°*. It can go on to generate any angle, and it can even go in a negative (clockwise) direction. The angle θ is in *standard position* and the point $P(x, y)$ is on its *terminal* side. In Chapter 2 we saw that the ratios of the lengths of the sides of a triangle do not depend on the *size* of the triangle. Therefore, the values of the trigonometric functions do not depend on the radius of the circle that $P(x, y)$ lies on.

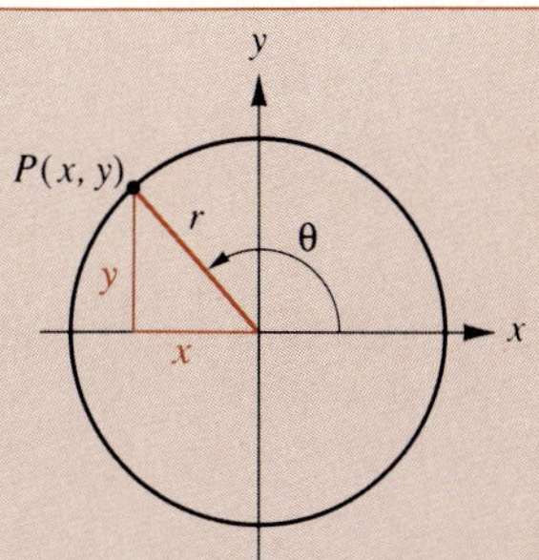

TRIGONOMETRIC FUNCTIONS FOR ANY ANGLE

If $P(x, y)$ is a point other than the origin on a circle of radius r, the radius to P sweeps out an angle θ in standard position. The trigonometric functions of θ are

$$\sin\theta = \frac{y}{r} \qquad \cos\theta = \frac{x}{r} \qquad \tan\theta = \frac{y}{x}$$

$$\csc\theta = \frac{r}{y} \qquad \sec\theta = \frac{r}{x} \qquad \cot\theta = \frac{x}{y}$$

We illustrate several possibilities in Figure 3.1-4. We have given the trigonometric functions their most general definition in terms of the point P on a circle. Therefore, these functions are often called the **circular functions**.

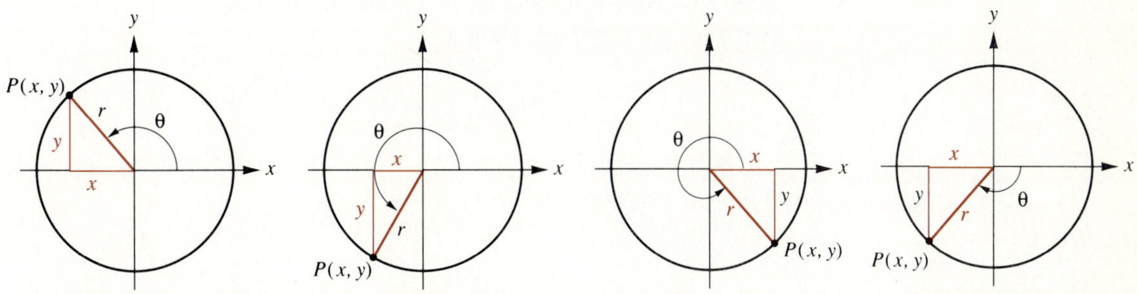

(a) θ a quadrant II angle ***(b) θ a quadrant III angle*** ***(c) θ a quadrant IV angle*** ***(d) θ a negative angle***

Figure 3.1-4

Example 1 A line goes from the origin to the point (5, 3). Evaluate all six trigonometric functions of the angle between this line and the positive x axis.

Solution

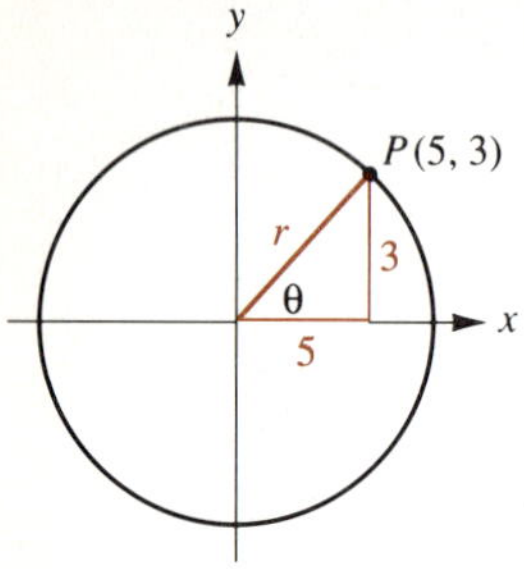

Figure 3.1-5

First make a sketch, thinking of $P(5, 3)$ as lying on a circle. We show the circle in Figure 3.1-5 although it is not necessary to do this each time. The Pythagorean theorem gives us

$$r = \sqrt{x^2 + y^2} = \sqrt{5^2 + 3^2} = \sqrt{34}$$

Using the definitions of the six trigonometric functions of θ, we get

$$\sin\theta = \frac{y}{r} = \frac{3}{\sqrt{34}} = \frac{3\sqrt{34}}{34} \qquad \csc\theta = \frac{r}{y} = \frac{\sqrt{34}}{3}$$

$$\cos\theta = \frac{x}{r} = \frac{5}{\sqrt{34}} = \frac{5\sqrt{34}}{34} \qquad \sec\theta = \frac{r}{x} = \frac{\sqrt{34}}{5}$$

$$\tan\theta = \frac{y}{x} = \frac{3}{5} \qquad \cot\theta = \frac{x}{y} = \frac{5}{3} \quad \blacksquare$$

NOW IT'S YOUR TURN

Find the sine, cosine, and tangent of the angle between the $+x$ axis and a line from the origin to the point $(1, 2)$. Start with a neat sketch!
(Answer: $2\sqrt{5}/5$, $\sqrt{5}/5$, and 2)

Example 2

The terminal side of angle θ in standard position goes through $(-12, -5)$. Evaluate the six trigonometric functions of θ.

Solution

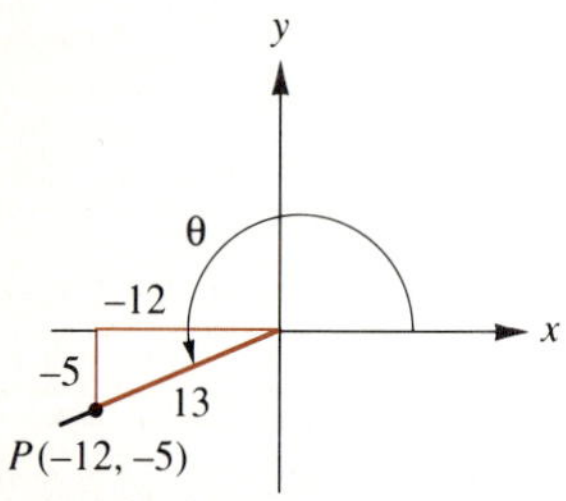

Figure 3.1-6

We can think of the point $P(-12, -5)$ as being on a circle, but we do not have to draw this circle. As shown in Figure 3.1-6, $x = -12$ and $y = -5$. Using the Pythagorean theorem, we get

$$r = \sqrt{x^2 + y^2} = \sqrt{(-12)^2 + (-5)^2} = \sqrt{169} = 13$$

It is **important** for you to notice that x and y may be negative, but r, the radius of the circle, is *always positive*.

Our definitions of the trigonometric functions give us

$$\sin\theta = \frac{y}{r} = \frac{-5}{13} = -\frac{5}{13}$$

$$\cos\theta = \frac{x}{r} = \frac{-12}{13} = -\frac{12}{13}$$

$$\tan\theta = \frac{y}{x} = \frac{-5}{-12} = \frac{5}{12}$$

$$\csc \theta = \frac{r}{y} = \frac{13}{-5} = -\frac{13}{5}$$

$$\sec \theta = \frac{r}{x} = \frac{13}{-12} = -\frac{13}{12}$$

$$\cot \theta = \frac{x}{y} = \frac{-12}{-5} = \frac{12}{5} \quad \blacksquare$$

NOW IT'S YOUR TURN

The terminal side of an angle θ in standard position passes through the point $(3, -6)$. Evaluate all six trigonometric functions of θ to two decimal places.

(Answer: $\sin \theta = -0.89$, $\cos \theta = 0.45$, $\tan \theta = -2.00$, $\cot \theta = -0.50$, $\sec \theta = 2.24$, $\csc \theta = -1.12$)

Example 3

Show that the value of $\sin \theta$ does not depend on the size of the circle, provided the radius is not zero.

Solution

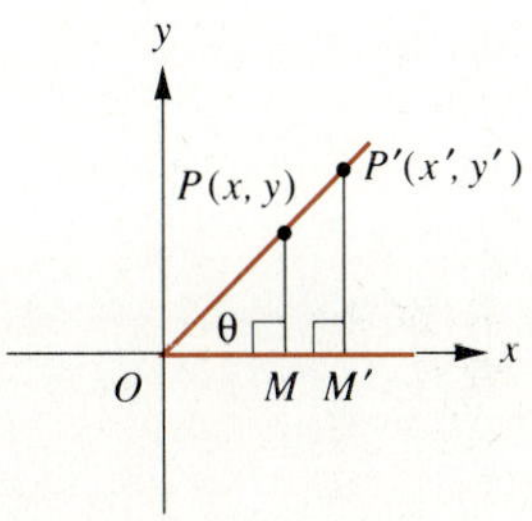

Figure 3.1-7

In Figure 3.1-7, we choose any two distinct points on the terminal side of θ, say P and P'. Each of these points represents a different size circle. The coordinates of P are (x, y) and the coordinates of P' are (x', y'). Let r be the length of the hypotenuse of triangle OPM and r' the hypotenuse of triangle $OP'M'$. Because these two triangles have the same angles, they are similar, so their corresponding sides are proportional; that is,

$$\frac{y}{r} = \frac{y'}{r'}$$

Therefore, the definition of $\sin \theta$ gives us $\sin \theta = y/r = y'/r'$, which means that $\sin \theta$ is the same for both P and P'. This in turn means that $\sin \theta$ does not depend on the size of the circle. $\blacksquare$

NOW IT'S YOUR TURN

Show that the values of the other five trigonometric functions are independent of the point P chosen on the terminal side of the angle.

Example 4

Find $\sin 135°$, $\cos 135°$, and $\tan 135°$.

Solution

First sketch this angle, noticing that it lies in quadrant II. Because the values of the trigonometric functions do not depend on the size of the circle, we can arbitrarily use $r = 1$ for simplicity. We must first find the x and y coordinates of $P(x, y)$, and then we can use them in the definition of the trigonometric functions.

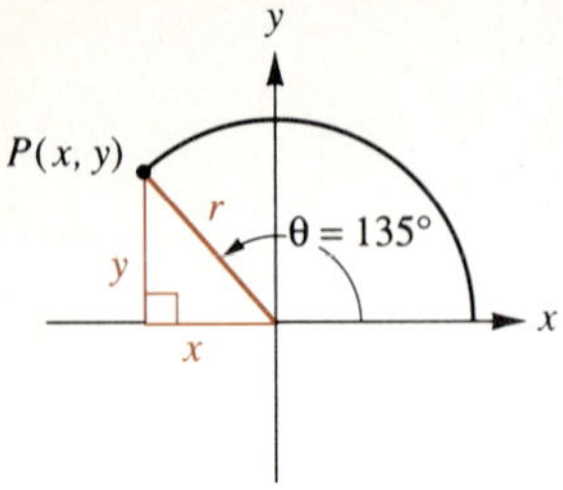

Figure 3.1-8

To find x and y: In the right triangle in Figure 3.1-8, the acute angle between r and the x-axis must be 45° because it is supplementary to 135°. Applying what we learned in Chapter 2 to this right triangle gives us

$$\sin 45° = \frac{y}{r}$$

$$y = r \sin 45° = (1) \sin 45° = \frac{\sqrt{2}}{2}$$

$$\cos 45° = \frac{x}{r}$$

$$x = r \cos 45° = (1) \cos 45° = \frac{\sqrt{2}}{2}$$

However, we are not yet finished. Notice that these values of x and y are just the *lengths of the sides of the right triangle*, and we need the *coordinates* of $P(x, y)$. Because P lies in the second quadrant, its x coordinate lies to the left of the origin and is therefore negative, so $x = -\sqrt{2}/2$. The y coordinate of P is positive because it is above the origin, so $y = \sqrt{2}/2$.

Now we apply the definitions of the trigonometric functions.

$$\sin 135° = \frac{y}{r} = \frac{\sqrt{2}/2}{1} = \frac{\sqrt{2}}{2}$$

$$\cos 135° = \frac{x}{r} = \frac{-\sqrt{2}/2}{1} = -\frac{\sqrt{2}}{2}$$

$$\tan 135° = \frac{y}{x} = \frac{\sqrt{2}/2}{-\sqrt{2}/2} = -1 \quad ■$$

NOW IT'S YOUR TURN

Find cot 135°, sec 135°, and csc 135° by using a circle of radius 2 instead of 1.
(Answer: -1, $-\sqrt{2}$, and $\sqrt{2}$)

Remember that in Chapter 1, we said that if the terminal side of an angle in standard position ends on either the $\pm x$ axis or the $\pm y$ axis, the angle is a *quadrantal angle*. The trigonometric functions of the quadrantal angles are especially easy to find using the generalized definitions we have given in this section.

Example 5
Quadrantal Angles

Evaluate the trigonometric functions of 90°.

Solution

We begin by placing the angle in standard position on a circle of radius R, as shown in Figure 3.1-9. The point $P(x, y)$ on this circle is on the y axis, and so its x

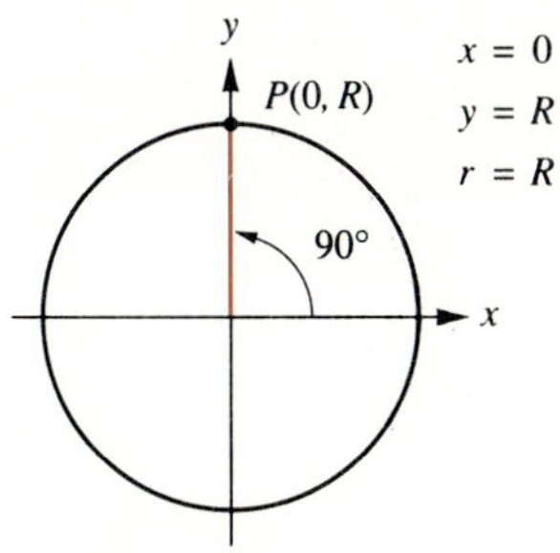

Figure 3.1-9

coordinate is 0, its y coordinate is R, and $r = R$: $x = 0$, $y = R$, $r = R$. Now we can apply our generalized definitions of the trigonometric functions.

$$\sin 90^\circ = \frac{y}{r} = \frac{R}{R} = 1$$

$$\cos 90^\circ = \frac{x}{r} = \frac{0}{R} = 0$$

$$\tan 90^\circ = \frac{y}{x} = \frac{R}{0}$$

Remember that division by zero is not defined, so

$$\tan 90^\circ = \text{undefined}$$

$$\cot 90^\circ = \frac{x}{y} = \frac{0}{R} = 0$$

$$\sec 90^\circ = \frac{r}{x} = \frac{R}{0} = \text{undefined}$$

$$\csc 90^\circ = \frac{r}{y} = \frac{R}{R} = 1 \quad ■$$

NOW IT'S YOUR TURN

Evaluate the trigonometric functions of 0°.

(Answer: $\sin 0^\circ = 0$, $\cos 0^\circ = 1$, $\tan 0^\circ = 0$, $\cot 0^\circ =$ undefined, $\sec 0^\circ = 1$, $\csc 0^\circ =$ undefined)

Example 6

Quadrantal Angles

Evaluate the trigonometric functions of π.

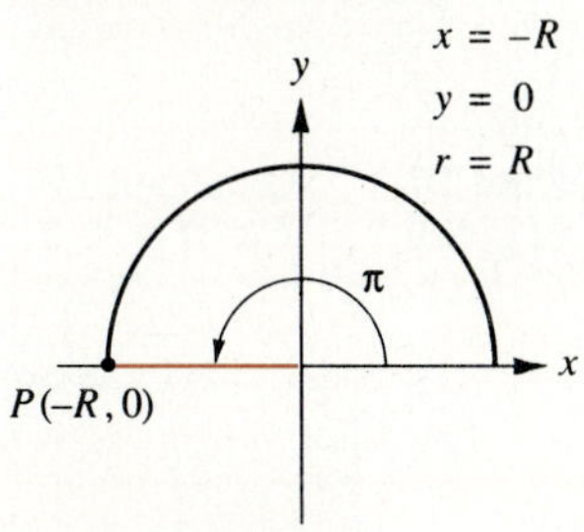

Figure 3.1-10

Solution

Placing the angle π in standard position on a circle of radius R, we see from Figure 3.1-10 that the point P is on the negative x axis, a distance R to the left of the origin. Therefore, its x coordinate is $-R$, its y coordinate is 0, and $r = R$. Using our generalized definitions gives us

$$\sin \pi = \frac{y}{r} = \frac{0}{R} = 0$$

$$\cos \pi = \frac{x}{r} = \frac{-R}{R} = -1$$

$$\tan \pi = \frac{y}{x} = \frac{0}{-R} = 0$$

$$\cot \pi = \frac{x}{y} = \frac{-R}{0} = \text{undefined}$$

$$\sec \pi = \frac{r}{x} = \frac{R}{-R} = -1$$

$$\csc \pi = \frac{r}{y} = \frac{R}{0} = \text{undefined} \quad ■$$

NOW IT'S YOUR TURN

Evaluate the trigonometric functions of $\frac{3\pi}{2}$.

(Answer: $\sin \frac{3\pi}{2} = -1$, $\cos \frac{3\pi}{2} = 0$, $\tan \frac{3\pi}{2} =$ undefined, $\cot \frac{3\pi}{2} = 0$, $\sec \frac{3\pi}{2} =$ undefined, $\csc \frac{3\pi}{2} = -1$)

We can combine the results of Example 5 with the results we got for the special angles in Chapter 2 in Table 3.1-1. Notice the progression of the integers inside the square roots.

Table 3.1-1

	0°	30°	45°	60°	90°
$\sin \theta$	$\frac{1}{2}\sqrt{0}$	$\frac{1}{2}\sqrt{1}$	$\frac{1}{2}\sqrt{2}$	$\frac{1}{2}\sqrt{3}$	$\frac{1}{2}\sqrt{4}$
$\cos \theta$	$\frac{1}{2}\sqrt{4}$	$\frac{1}{2}\sqrt{3}$	$\frac{1}{2}\sqrt{2}$	$\frac{1}{2}\sqrt{1}$	$\frac{1}{2}\sqrt{0}$

3.1-2 Signs of the Trigonometric Functions

As we saw in our examples, the trigonometric functions can be positive or negative depending on which quadrant the angle is in.

Using our generalized definition, we have $\sin \theta = y/r$. Because the radius r is always positive, $\sin \theta$ will have the same algebraic sign as y. Therefore, $\sin \theta$ is positive when y is positive, which is when the angle θ is in quadrants I or II. Sin θ will be negative when y is negative, which is when θ is in quadrants III or IV.

Table 3.1-2

	Quadrant I $x > 0, y > 0$	**Quadrant II** $x < 0, y > 0$	**Quadrant III** $x < 0, y < 0$	**Quadrant IV** $x > 0, y < 0$
$\sin \theta = y/r$	+	+	−	−
$\cos \theta = x/r$	+	−	−	+
$\tan \theta = y/x$	+	−	+	−
$\cot \theta = x/y$	+	−	+	−
$\sec \theta = r/x$	+	−	−	+
$\csc \theta = r/y$	+	+	−	−

In a similar way, we can find the algebraic signs of the remaining functions in all four quadrants and summarize them as shown in Table 3.1-2. Do *not* memorize this table! Instead, you should be able to figure out the sign of each trigonometric function by looking at the signs of x and y when θ is in the various quadrants. Use this table only as a guide in checking your answers.

The sine, cosine, and tangent are the three most frequently used trigonometric functions, so it is especially important to know their signs. Figure 3.1-11 will help you remember them.

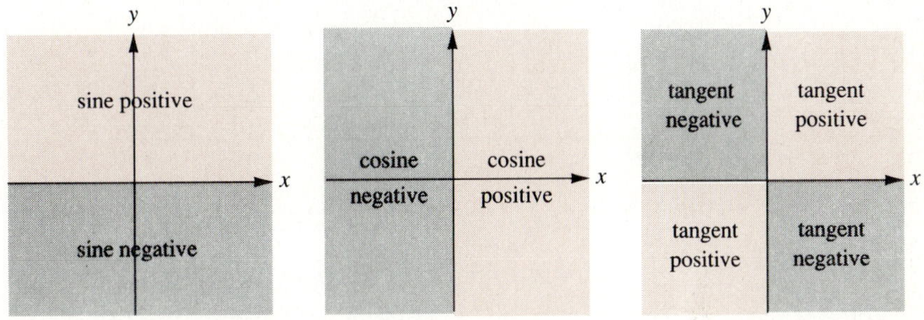

Figure 3.1-11

The signs of the sine, cosine, and tangent in the four quadrants

Example 7

If θ is an angle in quadrant III having $\sin \theta = -\frac{3}{7}$, evaluate the other five trigonometric functions of θ.

Solution

To use the definitions of the trigonometric functions, we must find x, y, and r. By definition, $\sin \theta = y/r$, so $y/r = -\frac{3}{7}$. The values of the trigonometric functions do not depend on the size of the circle, so we can let r have any positive value we want. We can use $r = 7$, which means that y must be -3. We show these values in Figure 3.1-12.

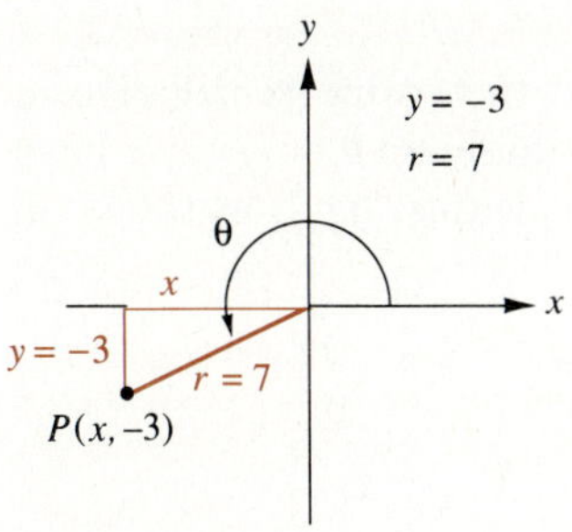

Figure 3.1-12

To find x, we use the Pythagorean theorem.

$$x^2 + y^2 = r^2$$
$$x^2 = r^2 - y^2 = 7^2 - (-3)^2 = 49 - 9 = 40$$
$$x = \pm\sqrt{40} = \pm 2\sqrt{10}$$

The two roots are $x = +2\sqrt{10}$ and $x = -2\sqrt{10}$. The angle θ is in quadrant III, so x must be negative because it is to the left of the origin. Consequently we must take the negative square root, so $x = -2\sqrt{10}$. Therefore, we have

$$x = -2\sqrt{10} \qquad y = -3 \qquad r = 7$$

Using the definitions of the trigonometric functions, we get

$$\sin \theta = \frac{y}{r} = \frac{-3}{7}$$

$$\cos\theta = \frac{x}{r} = \frac{-2\sqrt{10}}{7}$$

$$\tan\theta = \frac{y}{x} = \frac{-3}{-2\sqrt{10}} = \frac{3}{2\sqrt{10}} = \frac{3\sqrt{10}}{20}$$

$$\cot\theta = \frac{x}{y} = \frac{-2\sqrt{10}}{-3} = \frac{2\sqrt{10}}{3}$$

$$\csc\theta = \frac{r}{y} = \frac{7}{-3} = \frac{-7}{3}$$

$$\sec\theta = \frac{r}{x} = \frac{7}{-2\sqrt{10}} = \frac{-7\sqrt{10}}{-20} \quad ■$$

NOW IT'S YOUR TURN

If θ is a quadrant II angle having $\cos\theta = -\frac{3}{4}$, find $\sin\theta$, $\tan\theta$, and $\sec\theta$. (Answer: $\sqrt{7}/4$, $-\sqrt{7}/3$, $-4/3$)

Example 8

If $\csc\theta = \frac{3}{2}$ and $\cos\theta$ is negative, evaluate the other trigonometric functions of θ.

Solution

The cosecant is positive in quadrants I and II, and the cosine is negative in quadrants II and III, so θ must lie in quadrant II. Because $\csc\theta = r/y$, we have $r/y = \frac{3}{2}$. We can choose r to be 3, which means that y must be 2, as shown in Figure 3.1-13.

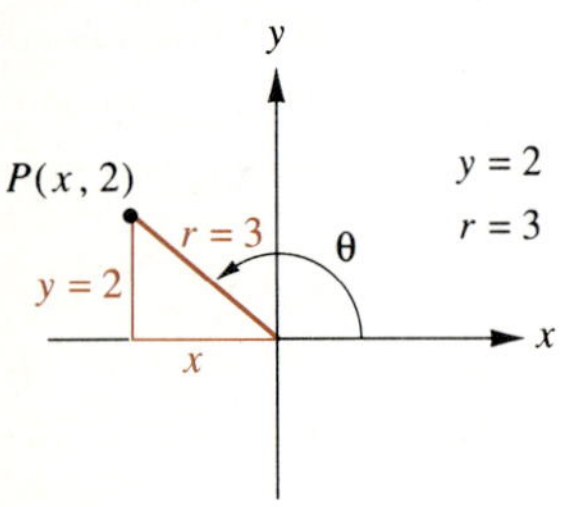

Figure 3.1-13

Using the Pythagorean theorem to find x,

$$x^2 + y^2 = r^2$$
$$x^2 = r^2 - y^2 = 3^2 - 2^2 = 9 - 4 = 5$$
$$x = \pm\sqrt{5}$$

Because θ is in quadrant II, x must be negative, so we must use the negative square root. Therefore, for this angle

$$x = -\sqrt{5} \qquad y = 2 \qquad r = 3$$

The definitions of the trigonometric functions give us

$$\sin\theta = \frac{y}{r} = \frac{2}{3}$$

$$\cos\theta = \frac{x}{r} = \frac{-\sqrt{5}}{3}$$

$$\tan\theta = \frac{y}{x} = \frac{2}{-\sqrt{5}} = \frac{2\sqrt{5}}{-5}$$

$$\cot\theta = \frac{x}{y} = \frac{-\sqrt{5}}{2}$$

$$\sec\theta = \frac{r}{x} = \frac{3}{-\sqrt{5}} = \frac{3\sqrt{5}}{-5}$$

$$\csc\theta = \frac{r}{y} = \frac{3}{2}$$ ■

NOW IT'S YOUR TURN

If $\sin\theta = -\frac{1}{3}$ and $\cos\theta$ is positive, find $\cos\theta$, $\tan\theta$, and $\csc\theta$.
(Answer: $\sqrt{8}/3$, $-\sqrt{8}/8$, -3)

Example 9

If $\tan\theta = -\frac{4}{3}$ and θ is between 0° and 360°, find the other trigonometric functions of θ.

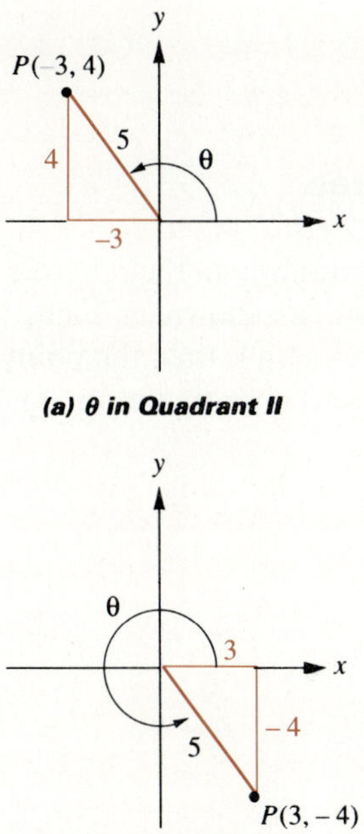

(a) θ in Quadrant II

(b) θ in Quadrant IV

Figure 3.1-14

Solution

The tangent is negative in quadrants II and IV, so there are *two* possible angles.

1. If θ is in quadrant II, $\tan\theta = -\frac{4}{3}$. By the definition of the tangent, $\tan\theta = y/x$, so

$$\frac{y}{x} = -\frac{4}{3}$$

Which is negative, x or y? Because θ is in quadrant II, x will be negative and y will be positive. Thus we can use $x = -3$ and $y = 4$. The Pythagorean theorem gives us

$$r = \sqrt{x^2 + y^2} = \sqrt{(-3)^2 + (4)^2} = \sqrt{25} = 5$$

Therefore, $x = -3$, $y = 4$, and $r = 5$.

2. If θ is in quadrant IV, $\tan\theta = -\frac{4}{3}$. By the definition of the tangent, $\tan\theta = y/x$, so $y/x = -\frac{4}{3}$. In this case, θ is in quadrant IV, so x is positive and y is negative. Thus we can use $x = 3$, $y = -4$. As in the last solution, $r = 5$.

We illustrate both possibilities in Figure 3.1-14. Applying the definitions of the trigonometric functions gives us the following:

For θ in Quadrant II: $x = -3, y = 4, r = 5$	**For θ in Quadrant IV:** $x = 3, y = -4, r = 5$
$\sin\theta = \frac{y}{r} = \frac{4}{5}$	$\sin\theta = \frac{y}{r} = \frac{-4}{5}$
$\cos\theta = \frac{x}{r} = \frac{-3}{5}$	$\cos\theta = \frac{x}{r} = \frac{3}{5}$
$\tan\theta = \frac{y}{x} = \frac{4}{-3}$	$\tan\theta = \frac{y}{x} = \frac{-4}{3}$
$\cot\theta = \frac{x}{y} = \frac{-3}{4}$	$\cot\theta = \frac{x}{y} = \frac{3}{-4}$
$\sec\theta = \frac{r}{x} = \frac{5}{-3}$	$\sec\theta = \frac{r}{x} = \frac{5}{3}$
$\csc\theta = \frac{r}{y} = \frac{5}{4}$	$\csc\theta = \frac{r}{y} = \frac{5}{-4}$

NOW IT'S YOUR TURN

If $\cos\theta = \frac{1}{4}$ and θ is between 0 and 2π, find $\sin\theta$, $\tan\theta$, and $\sec\theta$.
(Answer: $\sqrt{15}/4$, $\sqrt{15}$, 4 and $-\sqrt{15}/4$, $-\sqrt{15}$, 4)

3.1-3 Trigonometric Functions of Negative Angles

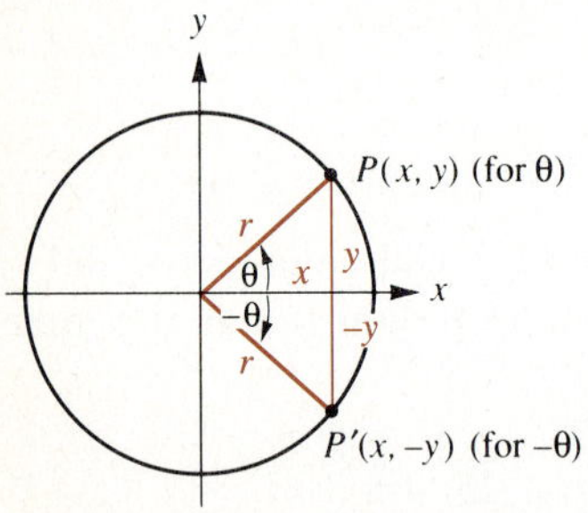

Figure 3.1-15

If θ is any angle, then $-\theta$ represents the same amount of rotation in the *opposite* direction. If we place both angles in standard position, as shown in Figure 3.1-15, we can see that if the point P for θ has coordinates (x, y), then the point P' for $-\theta$ will have the same x coordinate but a *reversed* y coordinate, or $-y$.

The trigonometric functions of $-\theta$ will be

$$\sin(-\theta) = \frac{-y}{r} = -\frac{y}{r}$$

$$\cos(-\theta) = \frac{x}{r}$$

$$\tan(-\theta) = \frac{-y}{x} = -\frac{y}{x}$$

$$\cot(-\theta) = \frac{x}{-y} = -\frac{x}{y}$$

$$\sec(-\theta) = \frac{r}{x}$$

$$\csc(-\theta) = \frac{r}{-y} = -\frac{r}{y}$$

Applying these same definitions to θ, we get

$$\sin\theta = \frac{y}{r} \qquad \cos\theta = \frac{x}{r} \qquad \tan\theta = \frac{y}{x}$$

$$\cot\theta = \frac{x}{y} \qquad \sec\theta = \frac{r}{x} \qquad \csc\theta = \frac{r}{y}$$

Therefore, as you can readily see, the trigonometric functions of $-\theta$ are related to the trigonometric functions of θ as follows:

$$\sin(-\theta) = \frac{-y}{r} = -\sin\theta \qquad \left(\textbf{because } \frac{y}{r} = \sin\boldsymbol{\theta}\right)$$

$$\cos(-\theta) = \frac{x}{r} = \cos\theta \qquad \left(\textbf{because } \frac{x}{r} = \cos\boldsymbol{\theta}\right)$$

$$\tan(-\theta) = \frac{-y}{x} = -\tan\theta \qquad \left(\textbf{because } \frac{y}{x} = \tan\boldsymbol{\theta}\right)$$

$$\cot(-\theta) = \frac{-x}{y} = -\cot\theta \qquad \left(\textbf{because } \frac{x}{y} = \cot\boldsymbol{\theta}\right)$$

$$\sec(-\theta) = \frac{r}{x} = \sec\theta \qquad \left(\textbf{because } \frac{r}{x} = \sec\boldsymbol{\theta}\right)$$

$$\csc(-\theta) = \frac{-r}{y} = -\csc\theta \qquad \left(\textbf{because } \frac{r}{y} = \csc\boldsymbol{\theta}\right)$$

We have proven these results only when θ is an acute angle, but as we shall see, they hold no matter what quadrant θ is in.

TRIGONOMETRIC FUNCTIONS OF NEGATIVE ANGLES

$\sin(-\theta) = -\sin\theta \qquad \csc(-\theta) = -\csc\theta$

$\cos(-\theta) = \cos\theta \qquad \sec(-\theta) = \sec\theta$

$\tan(-\theta) = -\tan\theta \qquad \cot(-\theta) = -\cot\theta$

Example 10

Evaluate the trigonometric functions of $-30°$.

Solution

Using the values from Table 3.1-1 and the preceding results for negative angles, we have

$$\sin(-30°) = -\sin 30° = -\tfrac{1}{2}$$

$$\cos(-30°) = \cos 30° = \frac{\sqrt{3}}{2}$$

$$\tan(-30°) = -\tan 30° = \frac{-\sqrt{3}}{3}$$

$$\cot(-30°) = -\cot 30° = -\sqrt{3}$$

$$\sec(-30°) = \sec 30° = \frac{2\sqrt{3}}{3}$$

$$\csc(-30°) = -\csc 30° = -2 \; ■$$

NOW IT'S YOUR TURN

Compute $\sin(-60°)$, $\cos(-60°)$, and $\tan(-60°)$.
(Answer: $-\sqrt{3}/2$, $\frac{1}{2}$, and $-\sqrt{3}$)

Example 11

Prove that $3 \sin \theta + 3 \sin(-\theta) + 5 \cos(-\theta) = 5 \cos \theta$.

Solution

Using the results that $\sin(-\theta) = -\sin \theta$ and $\cos(-\theta) = \cos \theta$, we have

$$\begin{aligned} 3 \sin \theta + 3 \sin(-\theta) + 5 \cos(-\theta) &= 3 \sin \theta + 3(-\sin \theta) + 5 \cos \theta \\ &= 3 \sin \theta - 3 \sin \theta + 5 \cos \theta \\ &= 5 \cos \theta \; ■ \end{aligned}$$

NOW IT'S YOUR TURN

Evaluate the expression $\cos \theta \sin \theta + \cos(-\theta) \sin(-\theta)$. (Answer: 0)

EXERCISES 3.1-1

In Exercises 1–18, evaluate all six trigonometric functions of the angle between the $+x$ axis and a line from the origin passing through the given point. Make a clear sketch for each one.

1. $(-3, 4)$ **2.** $(4, -3)$ **3.** $(-3, -4)$
4. $(3, 4)$ **5.** $(-1, -1)$ **6.** $(-3, -3)$
7. $(-1, 2)$ **8.** $(3, 1)$ **9.** $(2, -2)$
10. $(2, 0)$ **11.** $(0, -6)$ **12.** $(-1, \sqrt{3})$
13. $(1.6, -3.7)$ **14.** $(-4.9, 6.2)$ **15.** $(-9.3, -8.5)$
16. $(2.5, -\sqrt{7})$ **17.** $(-5, 0)$ **18.** $(\pi, -\pi/2)$

In Exercises 19–30, use Example 4 as a guide to find the following trigonometric functions without using a calculator or a table.

***19.** $\sin 225°$, $\cos 225°$, and $\tan 225°$
***20.** $\sec \frac{7\pi}{4}$, $\csc \frac{7\pi}{4}$, and $\cot \frac{7\pi}{4}$
***21.** $\cos 150°$, $\sec 150°$, and $\cot 150°$
***22.** $\sin \frac{2\pi}{3}$, $\csc \frac{2\pi}{3}$, and $\tan \frac{2\pi}{3}$
***23.** $\sin 210°$, $\cos 210°$, and $\tan 210°$
***24.** $\sec 300°$, $\csc 300°$, and $\cot 300°$
***25.** $\sin \frac{4\pi}{3}$, $\sec \frac{4\pi}{3}$, and $\tan \frac{4\pi}{3}$
***26.** $\csc 315°$, $\cot 315°$, and $\cos 315°$
***27.** $\sin 390°$ and $\cos 390°$
***28.** $\tan \frac{11\pi}{4}$ and $\sec \frac{11\pi}{4}$
***29.** $\sin 570°$ and $\cot 570°$
***30.** $\sec 660°$ and $\csc 660°$

In Exercises 31–38, find each function without using a calculator or a table. Use Examples 5 and 6 as guides.

31. Find $\sin \frac{\pi}{2}$, $\cos \frac{\pi}{2}$, and $\tan \frac{\pi}{2}$.
32. Find $\sec 180°$, $\csc 180°$, and $\cot 180°$.
33. Find $\sin 270°$ and $\cot 270°$.
34. Find the trigonometric functions of 2π.
***35.** Find $\sin 3\pi$ and $\cos 3\pi$.
***36.** Find $\tan 450°$ and $\sec 450°$.
***37.** Find $\sec 11\pi$ and $\csc 11\pi$.
***38.** Find $\sin 990°$ and $\cot 990°$.

In Exercises 39–48, evaluate each expression without using a calculator or a table.

39. $\sin 90° + 2\cos 90°$ **40.** $3\tan 0° - 5\cos \frac{\pi}{2}$
41. $2\cos \pi - 3\sec \frac{3\pi}{2}$ **42.** $2\csc 270° + 3\sin 180°$
43. $3\cos \pi + 2\sin^2 \frac{\pi}{2} + \tan 2\pi$
44. $-4\sec 0° + 2\cot^2 90° + \cos 360°$
45. $\sin^2 270° + \cos^2 270°$ **46.** $\cos^2 2\pi + \sin^2 2\pi$
47. $1 + \tan^2 180° - \sec^2 180°$ **48.** $1 - \csc^2 \frac{\pi}{2} + \cot^2 \frac{\pi}{2}$

***49.** When we generalized the trigonometric functions in this section, we said that $P(x, y)$ was any point except the origin on a circle of radius r. Why can't we use the origin, $P(0, 0)$?

EXERCISES 3.1-2

For Exercises 1–10, in what quadrants can θ lie if:

1. both $\sin \theta$ and $\cos \theta$ are positive?
2. $\cos \theta$ is negative but $\sin \theta$ is positive?
3. both $\sec \theta$ and $\csc \theta$ are negative?
4. $\sin \theta$ is negative but $\cot \theta$ is positive?
5. $\cos \theta$ is negative but $\tan \theta$ is positive?
***6.** both $\tan \theta$ and $\cot \theta$ are negative?
***7.** both $\sin \theta$ and $\csc \theta$ are negative?
***8.** both $\cot \theta$ and $\tan \theta$ are positive?
***9.** both $\cos \theta$ and $\sec \theta$ are positive?
***10.** $\tan \theta$ is positive but $\cot \theta$ is negative?

11. Are there any quadrants in which the trigonometric functions are **(a)** all positive or **(b)** all negative?

12. In the following table, fill in the missing signs of the trigonometric functions and tell which quadrant the angle is in. Figure out each one starting with a clear sketch.

$\sin \theta$	$\cos \theta$	$\tan \theta$	$\cot \theta$	$\sec \theta$	$\csc \theta$	Quadrant
+			−			
			+		−	
−	+					
+		−				
	−			+		
−				+		
	−	+				
−	−					
				−	+	

In Exercises 13–29, use the given conditions to evaluate the six trigonometric functions of θ *without finding* θ. (Use Examples 7 and 8 as guides.)

13. $\sin\theta = \frac{3}{4}$ and θ lies in quadrant I.

14. $\cos\theta = -\frac{1}{2}$ and θ lies in quadrant III.

15. $\sin\theta = \frac{3}{5}$ and θ lies in quadrant I.

16. $\tan\theta = \frac{4}{3}$ and θ lies in quadrant I.

17. $\tan\theta = -\frac{5}{12}$ and θ lies in quadrant II.

18. $\sec\theta = 2$ and θ lies in quadrant IV.

19. $\sin\theta = -\frac{4}{5}$ and $\cos\theta$ is negative.

***20.** $\cos\theta = -\frac{2}{3}$ and θ is in the quadrant IV.

21. $\cot\theta = \frac{3}{4}$ and $\sin\theta$ is positive.

22. $\tan\theta = \sqrt{3}$ and $\sin\theta$ is negative.

23. $\cot\theta = -3$ and $\sin\theta$ is positive.

***24.** $\sin\theta = -0.632$ and $\cos\theta$ is positive.

***25.** $\tan\theta = -6.52$ and θ is in quadrant III.

***26.** $\csc\theta = -2.6$ and $\tan\theta$ is positive.

***27.** $\cos\theta = -0.754$ and $\sin\theta$ is negative.

***28.** $\cot\theta = 1.6$ and θ is in quadrant II.

***29.** $\sec\theta = -2.42$ and $\csc\theta$ is positive.

In Exercises 30–44, evaluate the six trigonometric functions of θ where θ is a positive angle $\le 360°$ without finding θ. (Use Example 9 as a guide.)

30. $\csc\theta = 3$

31. $\csc\theta = -3$

32. $\sin\theta = -\frac{1}{2}$

33. $\tan\theta = -1$

34. $\sin\theta = -1$

35. $\sec\theta = -\frac{3}{2}$

36. $\cos\theta = \frac{5}{12}$

37. $\cot\theta = 5$

***38.** $\sin\theta = -0.42$

***39.** $\cot\theta = 1.64$

***40.** $\sec\theta = -1.72$

***41.** $\cos\theta = 0.35$

***42.** $\csc\theta = -1.3$

***43.** $\cot\theta = p\ (p > 0)$

***44.** $\cos\theta = -b/2\ (b > 0)$

In Exercises 45–49, evaluate each expression if $0 \le \theta \le \pi$.

45. $(1 + \cot\theta)/(\cos\theta + \sin\theta)$ if $\tan\theta = -\frac{4}{3}$.

46. $\tan\theta - \sin\theta\sec\theta$ if $\sec\theta = -\frac{13}{5}$.

47. $[\sin\theta\cos^2\theta/(\cos\theta - \sin\theta)] + 2/\tan\theta$ if $\sec\theta = \frac{18}{7}$.

48. $(\cos^2\theta + \sin^2\theta)/(1 - \sec^2\theta)$ if $\sec\theta = -\frac{11}{4}$.

49. $\tan\theta(1 - \sin^2\theta)/(\cos\theta$ if $\sec\theta = -\frac{13}{9}$.

***50. Unit Circle.** The *unit circle* has its center at the origin and a radius of one unit.

(a) Does the unit circle have unit area?
(b) Pick any point P on the unit circle. The radius to P makes an angle θ with the $+x$ axis. Show that the x and y coordinates of P are $\cos\theta$ and $\sin\theta$, respectively.

***51. (a)** Does the point $P(0.60, -0.80)$ lie on the unit circle? Sketch this point and the circle.
(b) Does the point $P(-0.30, 0.90)$ lie on the unit circle? (See Exercise 50).

***52.** If a point P has an x coordinate of -0.40, what must be its y coordinate if it lies on the unit circle? (See Exercise 50.)

EXERCISES 3.1-3

In Exercises 1–18, evaluate each function without using a calculator or a table.

1. $\cos(-\frac{\pi}{4})$

2. $\cos(-120°)$

3. $\sin(-120°)$

4. $\sec(-\frac{5\pi}{4})$

5. $\sec(-300°)$

6. $\sin(-\frac{3\pi}{4})$ and $\cos(-\frac{3\pi}{4})$

7. $\sin(-\frac{\pi}{3})$

8. $\cos(-\frac{\pi}{6})$

9. $\cot[3(-20°)]$

10. $\tan[2(-15°)]$

11. $\sin(-\frac{\pi}{2})$

12. $\sec(-\frac{3\pi}{2})$

13. $\cot(-270°)$

14. $\tan(-90°)$

15. $\cos(-\pi)$

***16.** $\cos(-3\pi)$

***17.** $\csc(-4\pi)$

***18.** $\sin(-\frac{5\pi}{2})$

Without using a calculator or a table, prove or disprove the statements in Exercises 19–22.

19. $\tan(-38°)\cot(38°) = -1$

20. $\sec(-110°)\cos(110°) = -1$

21. $\dfrac{\cos\theta}{\sin(-\theta)} = \cot(-\theta)$

22. $\dfrac{\cos(-\theta)}{\sin(-\theta)} = -\cot\theta$

***23.** Determine the sign of $\sin(-\theta)\cos(-\theta)\tan(-\theta)$ if θ is in quadrant II.

***24.** Determine the sign of $\csc(-\theta)\sec(-\theta)\cot(-\theta)$ if θ is in quadrant III.

25. Simplify: $\dfrac{\sin(-\theta) + \cos(-\theta)}{\sin(-\theta) - \cos(-\theta)}$.

26. Simplify: $(1 + \sin\theta)[1 + \sin(-\theta)]$.

3.2 REFERENCE ANGLES

3.2-1 Definition of the Reference Angle

> **DEFINITION OF THE REFERENCE ANGLE**
>
> For an angle θ in standard position, the *reference angle* is the positive acute angle between the terminal side of θ and the x axis.

In Chapter 2 we learned how to find the trigonometric functions of angles between 0° and 90° (0 and $\frac{\pi}{2}$). In this section, we see how to find the trigonometric functions of all angles, not just acute angles.

Look back at Examples 2, 4, 7, 8, and 9 in Section 3.1. In each of these, we dealt with angles greater than 90°, but to find the trigonometric functions, we had to construct a right triangle by dropping a line from $P(x, y)$ perpendicular to the x axis. Figure 3.2-1 shows two examples of this. We then used the sides of this right triangle to find the trigonometric functions of the angle θ, always being careful to notice that x and y could be positive or negative, depending on which quadrant θ was in. We call this right triangle the **reference triangle**; the acute angle its hypotenuse makes with the x axis we call the **reference angle**. Figure 3.2-2 shows several examples of reference angles and reference triangles.

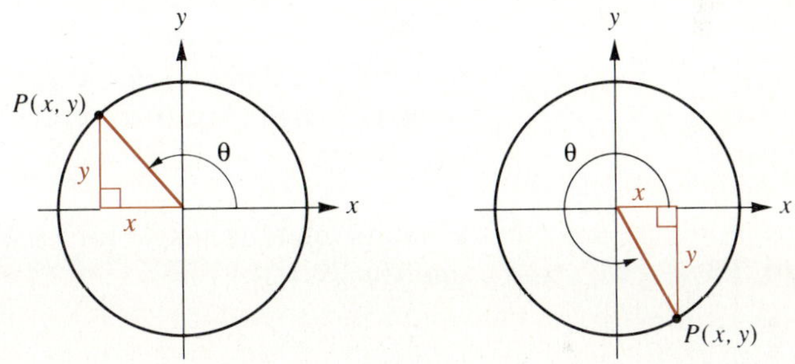

Figure 3.2-1

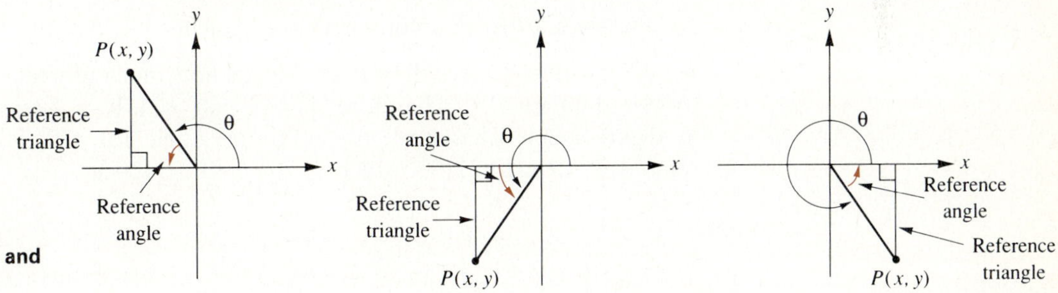

Figure 3.2-2
Reference triangles and reference angles

Example 12

Find the reference angle for each of the following angles: (a) 60°, (b) 145°, (c) 230°, and (d) 315°.

Solution

When finding reference angles, it is always helpful to sketch the angle first (Figure 3.2-3).

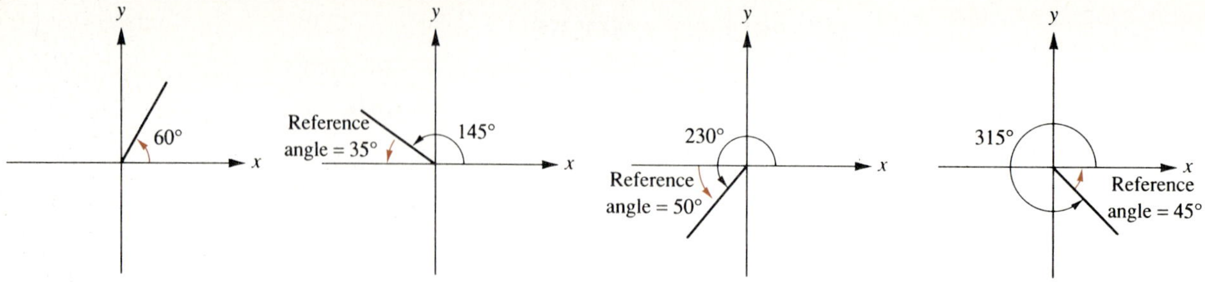

(a) Reference angle of 60° is 60°.

(b) Reference angle of 145° is 180° − 145° = 35°.

(c) Reference angle of 230° is 230° − 180° = 50°.

(d) Reference angle of 315° is 360° − 315° = 45°.

Figure 3.2-3

NOW IT'S YOUR TURN

Find the reference angle of each of the following: 290°, 11°, 92°, and 192°. (Answer: 70°, 11°, 88°, and 12°)

What about negative angles, angles in degrees and minutes, angles greater than 360°, and angles in radians? As the following examples show, we find the reference angles for these in the same way as in Example 12.

Example 13

Find the reference angle for (a) 470°, (b) 1130°, (c) −155°, (d) 196°11′, (e) 1.9π, and (f) $\frac{33\pi}{8}$.

Solution

As always, start each solution with a sketch.

(a) 470°: 470° is greater than 1 rev. To see how much greater, first subtract 360° from it to obtain 110°. Therefore, 470° = 360° + 110° = 1 rev + 110°, as shown in Figure 3.2-4. Therefore, the reference angle for 470° is the same as the reference angle for 110°, which is 70° (180° − 110°).

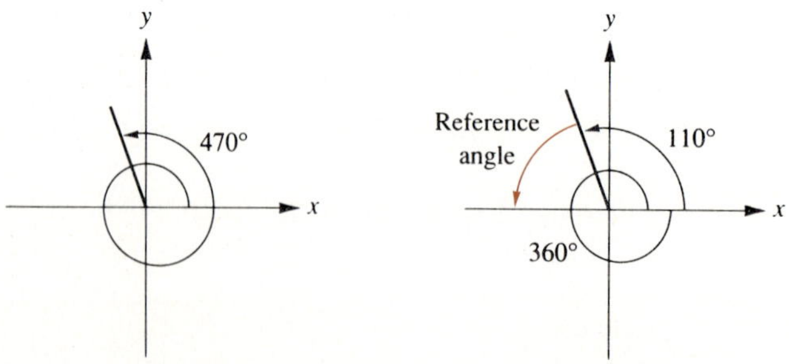

(a) ***(b)***

Figure 3.2-4

Reference angle for 470°

(b) 1130°: This angle is also greater than 1 rev. As in part (a), we must first find out how much greater. The easiest way to do this is to keep subtracting 360° until we end up with a number less than 360°.

$$1130° - 360° = 770°$$
$$1130° - 360° - 360° = 410°$$
$$1130° - 360° - 360° - 360° = 50°$$

We can stop subtracting now because 50° is less than 360°. Therefore, 1130° can be expressed as

$$1130° = \underbrace{360° + 360° + 360°}_{3\text{ rev}} + 50°$$

As you can see in Figure 3.2-5, 1130° is equal to 3 rev plus 50°, so the reference angle for 1130° is the same as for 50°, which is 50°.

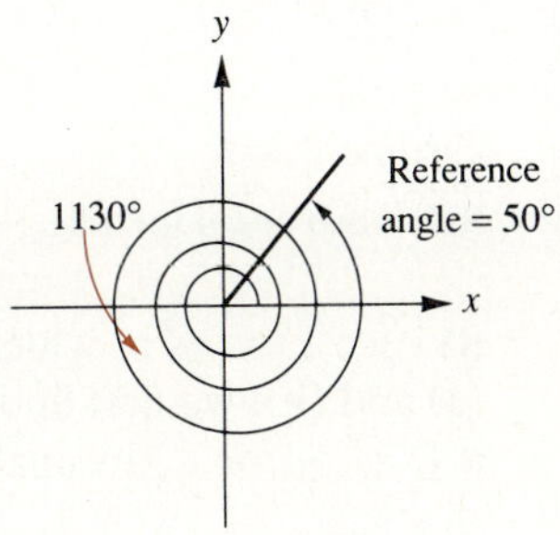

Figure 3.2-5

Reference angle for 1130°

(c) −155°: As Figure 3.2-6 shows, the positive acute angle between the terminal side of −155° and the x axis is 25° (180° − 155°), so the reference angle of −155° is 25°.

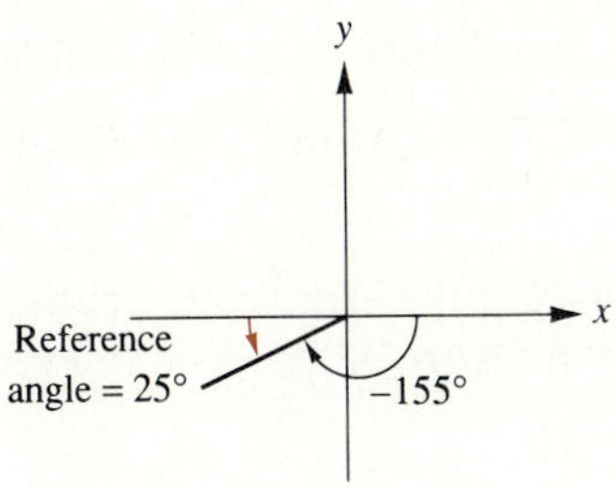

Figure 3.2-6

Reference angle for −155°

(d) 196°11′: As Figure 3.2-7 shows, the reference angle is

$$196°11' - 180° = 16°11'$$

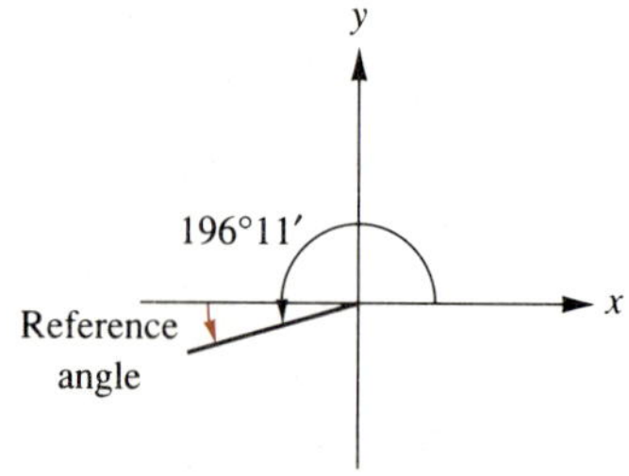

Figure 3.2-7

Reference angle for 196°11′

(e) 1.9π: 1.9π is less than 2π but greater than $\frac{3\pi}{2}$ (1.5π), so it lies in quadrant IV (see Figure 3.2-8). The reference angle is $2\pi - 1.9\pi = 0.1\pi$.

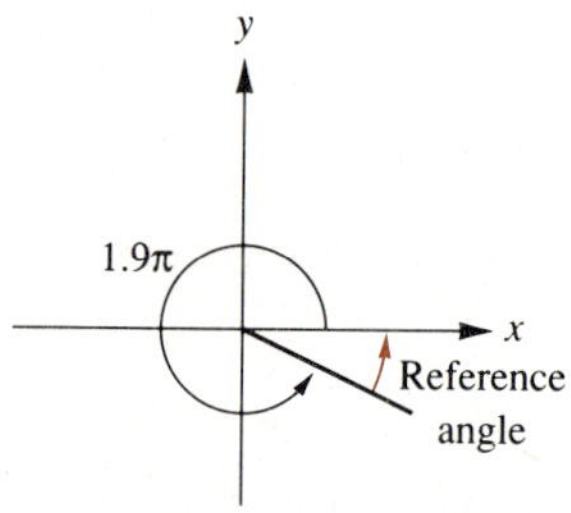

Figure 3.2-8

Reference angle for 1.9π

(f) $\frac{33\pi}{8}$: $\frac{33\pi}{8} = \frac{33}{8}\pi$, which is greater than 2π and thus more than 1 rev. As in parts (a) and (b), we first find out how much more. To do this we can treat $\frac{33}{8}$ either as a fraction or a decimal.

Treating $\frac{33}{8}$ as a decimal,

$$\frac{33}{8}\pi = 4.13\pi = 4\pi + 0.13\pi = 2(2\pi) + 0.13\pi = 2 \text{ rev} + 0.13\pi$$

Treating $\frac{33}{8}$ as a fraction,

$$\frac{33}{8}\pi = \frac{(32 + 1)}{8\pi} = (4 + \tfrac{1}{8})\pi = 4\pi + \frac{\pi}{8} = 2 \text{ rev} + \frac{\pi}{8}$$

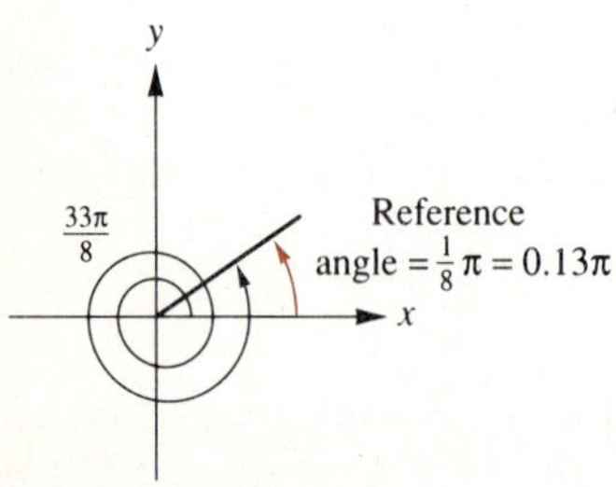

Figure 3.2-9

Reference angle for $\frac{33\pi}{8}$

Because $\frac{1}{8} = 0.13$, we get the same result either way. Now we sketch the angle using these results (see Figure 3.2-9). The reference angle is obviously $\frac{1}{8}\pi$, which is also 0.13π. ■

NOW IT'S YOUR TURN

Find the reference angle for: (a) 160°, (b) 294°, (c) 600°, (d) −97°, (e) 295°45′, (f) $\frac{4\pi}{5}$, and (g) $\frac{65\pi}{7}$. (Answer: 20°, 66°, 60°, 83°, 64°15′, $\frac{\pi}{5}$, and $\frac{2}{7}\pi = 0.29\pi$)

3.2-2 Finding Trigonometric Functions Using the Reference Angle

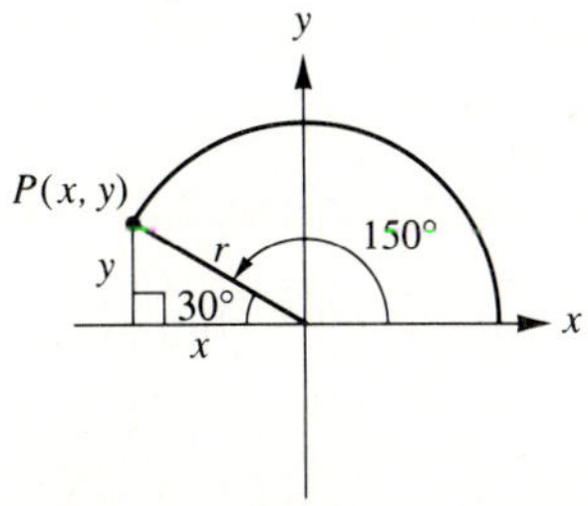

Figure 3.2-10

Reference triangle and reference angle for 150°

Remember that we introduced the reference angle because we were trying to find the trigonometric functions for angles greater than 90°. Let's see how to do it.

Consider, for example, the angle 150°. Its reference angle is 30° and its reference triangle is shown in Figure 3.2-10. Using the generalized definitions of the trigonometric functions from Section 3.1, we have

$$\sin 150° = \frac{y}{r} \text{ and is positive}$$

$$\cos 150° = \frac{x}{r} \text{ and is negative}$$

$$\tan 150° = \frac{y}{x} \text{ and is negative}$$

$$\cot 150° = \frac{x}{y} \text{ and is negative}$$

$$\sec 150° = \frac{r}{x} \text{ and is negative}$$

$$\csc 150° = \frac{r}{y} \text{ and is positive}$$

Now look at the reference triangle in Figure 3.2-10.

$$\sin 150° = \frac{y}{r} = \sin 30°$$

$$\cos 150 = \frac{x}{r} = -\cos 30° \text{ (cosine is negative in quadrant II)}$$

$$\tan 150° = \frac{y}{x} = -\tan 30° \text{ (tangent is negative in quadrant II)}$$

In a similar way, we also obtain

$$\cot 150° = \frac{x}{y} = -\cot 30°$$

$$\sec 150° = \frac{r}{x} = -\sec 30°$$

$$\csc 150° = \frac{r}{y} = \csc 30°$$

REFERENCE ANGLE THEOREM

The trigonometric function of any angle θ is equal to the same trigonometric function of the *reference angle of* θ, except for possible differences in sign. The quadrant in which the terminal side of θ lies determines the sign of the function.

Remembering that 30° is the reference angle of 150°, we have shown the following:

▸ **The trigonometric functions of 150° are equal to the trigonometric functions of its reference angle, except for possible differences in sign.**

We can find the sign of each trigonometric function by looking at the quadrant the angle is in, as we learned in Section 3.1-2.

We have only shown this property for a quadrant II angle (150°), but as we see in the examples and exercises, it can be generalized for any angle whatsoever.

This result is so important that we call it a theorem—the **reference angle theorem**.

To find the trigonometric function of any angle, first find the trigonometric function of its reference angle. Then determine the sign of the function by looking at the quadrant the angle is in, as in Section 3.1-2.

Example 14

Find all six trigonometric functions of 120°.

Solution

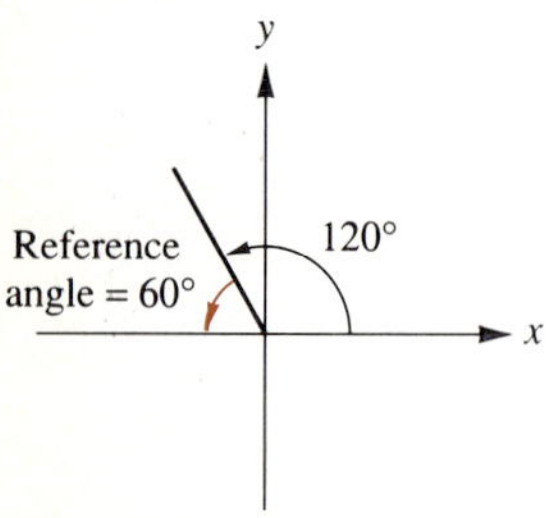

Figure 3.2-11

The angle 120° and its reference angle

First sketch the angle in standard position and find its reference angle, which is 60° (Figure 3.2-11). The reference angle theorem tells us that the trigonometric functions of 120° are equal to those of its reference angle except for possible differences in sign. Because 120° is in quadrant II, only the sine and cosecant are positive, whereas the other functions are negative. Using the reference angle theorem, we have

$$\sin 120^\circ = \sin 60^\circ = \frac{\sqrt{3}}{2}$$

$$\cos 120^\circ = -\cos 60^\circ = -\tfrac{1}{2}$$

$$\tan 120^\circ = -\tan 60^\circ = -\sqrt{3}$$

$$\cot 120^\circ = -\cot 60^\circ = \frac{-\sqrt{3}}{3}$$

$$\sec 120^\circ = -\sec 60^\circ = -2$$

$$\csc 120^\circ = \csc 60^\circ = \frac{2\sqrt{3}}{3} \quad ■$$

NOW IT'S YOUR TURN

Find sin 150°, cos 150°, and tan 150°. (Answer: $\frac{1}{2}$, $-\sqrt{3}/2$, and $-\sqrt{3}/3$)

We use the same approach for finding angles in other quadrants, angles greater than 360°, and negative angles: First find the reference angle, then use the reference angle theorem to find the magnitude of the trigonometric function and, finally, determine the sign of the function by considering what quadrant the angle is in. The following examples illustrate this approach.

Example 15

Find $\tan \frac{11\pi}{6}$ and $\sec \frac{11\pi}{6}$.

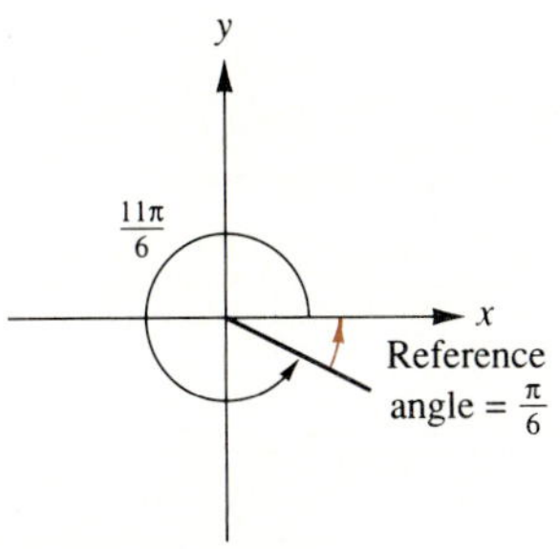

Figure 3.2-12

$\frac{11\pi}{6}$ lies in quadrant IV and has reference angle $\frac{\pi}{6}$.

Solution

First sketch the angle in standard position. $\frac{11\pi}{6} = (12\pi - \pi)/6 = 2\pi - \frac{\pi}{6}$, so this angle lies in quadrant IV. Its reference angle is $(2\pi - \frac{11\pi}{6}) = \frac{\pi}{6}$ (Figure 3.2-12). Because $\frac{11\pi}{6}$ is in quadrant IV, the tangent is negative but the secant is positive. The reference angle theorem gives us

$$\tan \frac{11\pi}{6} = -\tan \frac{\pi}{6} = \frac{-\sqrt{3}}{3}$$

$$\sec \frac{11\pi}{6} = \sec \frac{\pi}{6} = \frac{2\sqrt{3}}{3} \quad ■$$

NOW IT'S YOUR TURN

Find $\sin \frac{5\pi}{3}$ and $\cot \frac{5\pi}{3}$. (Answer: $-\sqrt{3}/2$ and $-\sqrt{3}/3$)

Example 16

Find $\sin(-135°)$ and $\tan(-135°)$.

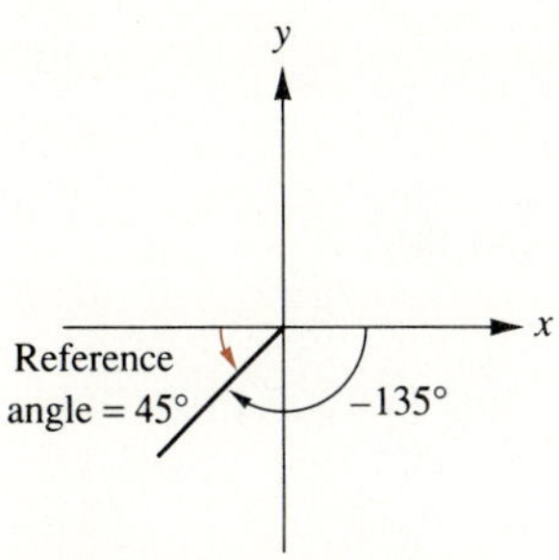

Figure 3.2-13

The reference angle for −135° is 45°.

Solution

We treat negative angles the same as positive angles. First sketch the angle in standard position (Figure 3.2-13). You can readily see that the reference angle is 45° (180° − 135°). The angle −135° lies in quadrant III, where the sine is negative but the tangent is positive. The reference angle theorem gives us

$$\sin(-135°) = -\sin 45° = \frac{-\sqrt{2}}{2}$$

$$\tan(-135°) = \tan 45° = 1 \quad ■$$

NOW IT'S YOUR TURN

Find $\csc(-225°)$ and $\cot(-225°)$. (Answer: $\sqrt{2}$ and -1)

Most angles are not special angles, as the following examples show.

Example 17

Find (a) sin 263° and (b) csc 502°.

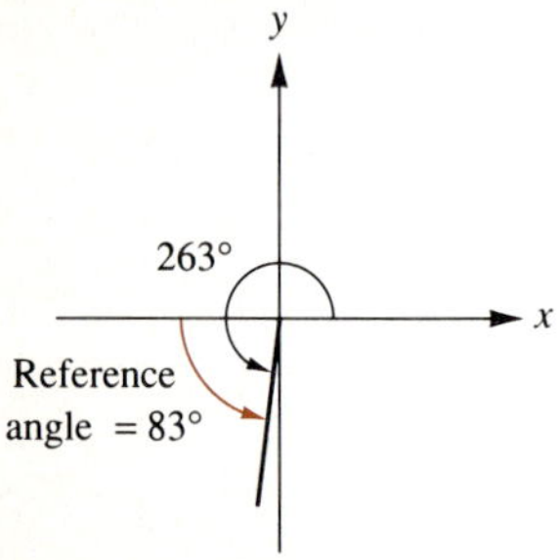

Figure 3.2-14

Solution

(a) sin 263°: Because 263° is in quadrant III, the sine is negative and its reference angle is 83° (263° − 180°) (Figure 3.2-14). Using the reference angle theorem, we get

$$\sin 263^\circ = -\sin 83^\circ = -0.9925$$

We used a calculator to get sin 83°.

(b) csc 502°: 502° is greater than 1 rev, so we handle it as in Example 13a: 502° = 360° + 142° = 1 rev + 142°. Now we sketch the angle (Figure 3.2-15).

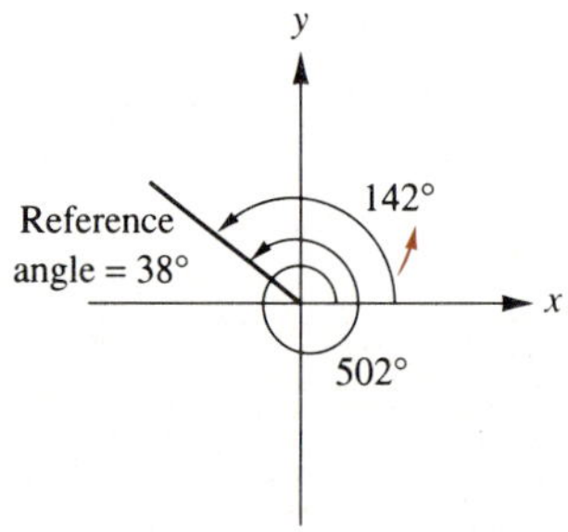

Figure 3.2-15

As you can see from Figure 3.2-15, the reference angle for 502° is the same as for 142°, which is 38° (180° − 142°). Because 502° is in quadrant II, the cosecant is positive. The reference angle theorem gives us

$$\csc 502^\circ = \csc 38^\circ = 1.6243$$

Again, we used a calculator to get csc 38°. ■

NOW IT'S YOUR TURN

Find cos 219° and cot 133°. (Answer: −0.7771 and −0.9325)

When angles are expressed in radians, we use the same general procedure.

Example 18

Find: (a) cos(4.20 rad) and (b) sin(1.1π).

Solution

(a) cos(4.20 rad): 4.20 rad is greater than π (3.14) but less than $\frac{3\pi}{2}$ (4.71), so it lies in quadrant III (Figure 3.2-16). The reference angle is $4.20 - \pi = 4.20 - 3.14 = 1.06$ rad. Because the cosine is negative in quadrant

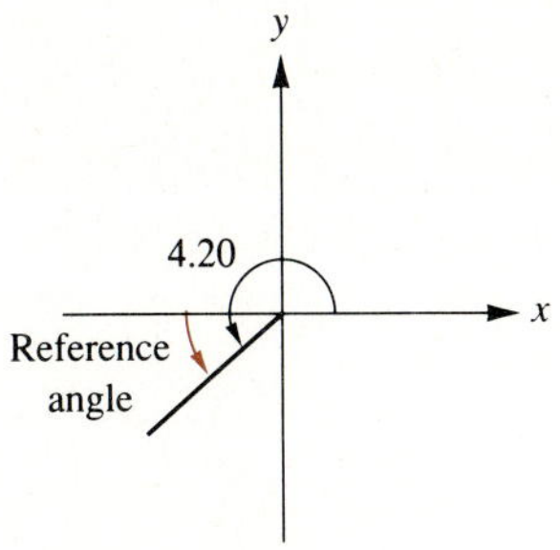

Figure 3.2-16

III, the reference angle theorem gives us

$$\cos(4.20) = -\cos(1.06) = -0.4889$$

(b) $\sin(1.1\pi)$: The angle 1.1π is greater than π but less than $\frac{3\pi}{2}$ (1.5π), so it is in quadrant III. As Figure 3.2-17 shows, the reference angle is

$$1.1\pi - \pi = 1.1\pi - 1.0\pi = 0.1\pi = (0.1)(3.14) = 0.314$$

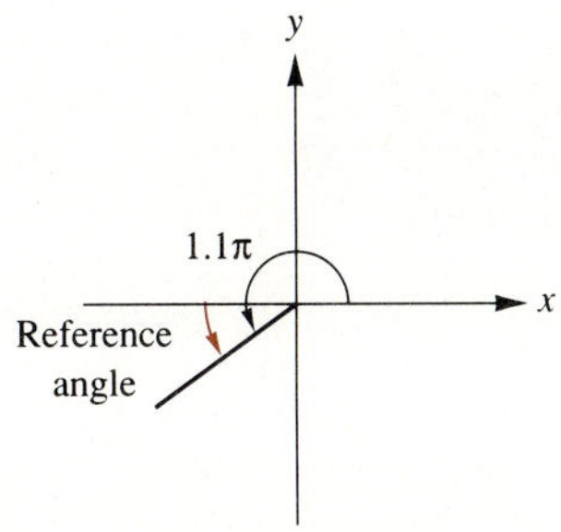

Figure 3.2-17

The sine is negative in quadrant III, so

$$\sin(1.1\pi) = -\sin 0.314 = -0.3089$$ ■

NOW IT'S YOUR TURN

Using $\pi = 3.14$, find $\tan(6.00)$ and $\cos(-20\pi/11)$.
(Answer: −0.2876 and 0.8414)

3.2-3 Calculators and Reference Angles

In the previous examples, we used a calculator only to find the trigonometric function of the reference angle. However, calculators can compute the trigonometric functions of any angle directly. There is no need to find the reference angle. Just enter the angle and find its trigonometric function in the same way you learned in Section 2.3-1.

Example 19

Use a calculator to find each trigonometric function to four decimal places without finding the reference angle. (a) cos 251°, (b) cot 820°, (c) tan(5.21 rad), and (d) $\sin(-\frac{6\pi}{7})$.

Solution

(a) cos 251°:

1. Set the calculator to the DEG mode.
2. Enter 251.

3. Press COS. Your screen should show -0.3256.

(b) cot 820°:

Remember that $\cot 820° = 1/(\tan 820°)$.

1. Set the calculator to the DEG mode.
2. Enter 820.
3. Press TAN. This gives you tan 820°.
4. Press $1/x$. Your screen should show -0.1763.

(c) tan(5.21 rad):

1. Set the calculator to the RAD mode.
2. Enter 5.21.
3. Press TAN. Your screen should show -1.8409.

(d) $\sin(-\frac{6\pi}{7})$:

1. Calculate the angle in decimal form:

$$-\tfrac{6\pi}{7} = \frac{-6(3.14)}{7} = -2.69$$

2. Set the calculator to the RAD mode.
3. Enter -2.69.
4. Press SIN. Your screen should show -0.4364. ■

NOW IT'S YOUR TURN

Using $\pi = 3.14$, use a calculator to find each of the following to four decimal places without finding the reference angle: sin 139°, sec 951°, cos $(\frac{-4\pi}{7})$, and tan(4.03 rad). (Answer: 0.6561, −1.5890, −0.2216, and 1.2306)

NOTE ON CALCULATORS

You may think that the first eight examples were a waste of time because a calculator will find any trigonometric function without using the reference angle. However, in the next section we learn how to find the inverse functions of angles greater than 90°. You will then see that it is necessary to use reference angles even when you use a calculator. The calculator will be able to tell you only the reference angle. Then you must use the reference angle theorem to find the angle.

▸ **Your calculator cannot think for you!**

EXERCISES 3.2

In Exercises 1–62, find the reference angle of each angle. Always start with a clear sketch.

1. 102°
2. 215°
3. 157°
4. 550°
5. 203°
6. −147°
7. 260°
8. −10°
9. 316°
10. 17°
11. −278°
12. −390°
13. 477°
14. 525°
15. −620°
16. −708°
17. 110°10′
18. 431°20′

19. $-14°30'$ **20.** $-246°50'$ **21.** $105°42'16''$

22. $308°11''$ **23.** $221°40'46''$ **24.** $169.3°$

25. $271.2°$ **26.** $-253.6°$ **27.** $358.4°$

28. $-400.4°$ **29.** $985.8°$ **30.** 2.50 rad

31. 1.4 rad **32.** 5.36 **33.** -2.61

34. 3.34 **35.** 7.11 **36.** -7.11

37. 20.0 **38.** -15.4 **39.** 1.4π

40. -1.36π **41.** π **42.** -1.1π

43. -0.69π **44.** 0.82π **45.** 1.70π

46. $\frac{10\pi}{9}$ **47.** $\frac{20\pi}{13}$ **48.** $\frac{-5\pi}{4}$

49. $\frac{7\pi}{8}$ **50.** $\frac{16\pi}{9}$ **51.** 2.3π

52. $\frac{19\pi}{6}$ **53.** $\frac{30\pi}{7}$ **54.** 11.3π

55. 169.4π ***56.** π^2 ***57.** $\pi°$

***58.** $\pi + \frac{1}{\pi}$ ***59.** -9.2π ***60.** $\frac{-17\pi}{5}$

***61.** $\pi + 2$ ***62.** $(\pi + 2)°$

In Exercises 63–82, use the reference angle theorem but no calculators or tables to evaluate each function.

63. $\sec \frac{5\pi}{6}$ and $\cot \frac{5\pi}{6}$ **64.** $\cos 330°$ and $\csc 330°$

65. $\sin 210°$ and $\tan 210°$ **66.** $\sec \frac{4\pi}{3}$ and $\cot \frac{4\pi}{3}$

67. $\sin \frac{5\pi}{3}$ and $\cot \frac{5\pi}{3}$ **68.** $\sec 135°$ and $\sin 135°$

69. $\sin \frac{5\pi}{4}$ and $\cos \frac{5\pi}{4}$ **70.** $\tan \frac{7\pi}{4}$ and $\csc \frac{7\pi}{4}$

71. $\cos(-150°)$ and $\csc(-150°)$

72. $\sin(-\frac{2\pi}{3})$ and $\tan(-\frac{2\pi}{3})$

73. $\tan(-225°)$ and $\sin(-225°)$

74. $\sec(-\frac{11\pi}{6})$ and $\sin(-\frac{11\pi}{6})$

75. $\sin 405°$ and $\cos 405°$

76. $\tan 495°$ and $\sec 495°$

77. $\cot \frac{13\pi}{4}$ and $\cos \frac{13\pi}{4}$

78. $\sin 570°$ and $\cot 570°$

79. $\sec \frac{11\pi}{3}$ and $\csc \frac{11\pi}{3}$

80. $\sin(-480°)$ and $\cos(-480°)$

81. $\tan(-\frac{13\pi}{4})$ and $\sec(-\frac{13\pi}{4})$

82. $\sin(-\frac{11\pi}{3})$ and $\sec(-\frac{11\pi}{3})$

In Exercises 83–135, use the reference angle theorem and a calculator to evaluate each function to three decimal places. Use $\pi \approx 3.14$ when it is needed.

83. $\sin 320°$ **84.** $\cos 152°$ **85.** $\csc 200°$

86. $\cos 290°$ **87.** $\sin 217°$ **88.** $\tan 245°$

89. $\sec 460°$ **90.** $\cot 1220°$ **91.** $\csc 3890°$

92. $\sin(-1190°)$ **93.** $\tan(-915°)$ **94.** $\sin 131°42'$

95. $\tan 214°24'$ **96.** $\cos 305°36'$ **97.** $\sin(-16°18')$

98. $\tan(-201°36')$ **99.** $\tan(-75°54')$ **100.** $\sec 99.9°$

101. $\cos 243.2°$ **102.** $\tan 308.7°$ **103.** $\cot 143.4°$

104. $\sin 269.9°$ **105.** $\csc 419.1°$ **106.** $\cos 461.2°$

107. $\csc(-18.9°)$ **108.** $\cot(-95.4°)$ **109.** $\sin(-409.3°)$

110. $\cos(-529.9°)$ **111.** $\cos 1.92$ **112.** $\tan 2.49$

113. $\cot 3.81$ **114.** $\sec 4.00$ **115.** $\sin 5.00$

116. $\csc 6.00$ **117.** $\cos 8.30$ **118.** $\tan 11.86$

119. $\sin 22.84$ **120.** $\cos 29.12$ **121.** $\sec(-1.62)$

122. $\cot(-5.04)$ **123.** $\sin(-4.66)$ **124.** $\cos(-0.11)$

125. $\sin \frac{11\pi}{7}$ **126.** $\tan \frac{8\pi}{5}$ **127.** $\sec \frac{17\pi}{8}$

128. $\cot(-\frac{21\pi}{11})$ **129.** $\csc 1.3\pi$ **130.** $\cos 0.97\pi$

131. $\sin(-1.81\pi)$ **132.** $\sin^3 4$ **133.** $\cot(\pi + 1)$

134. $\csc[3(\pi - 1.2)]$ **135.** $2 \cos^2 2$

In Exercises 136–174, use a calculator to evaluate each function without using the reference angle theorem. Use $\pi \approx 3.14$ as needed.

136. $\tan 143.7°$ **137.** $\cos 187.4°$

138. $\sin 257.5°$ **139.** $\csc 285.9°$

140. $\sec 341.6°$ **141.** $\cot 95.3°$

142. $\cos(-39.6°)$ **143.** $\cot(-103.7°)$

144. $\sin(-306.8°)$ ***145.** $\sin[\cos(-214.3°)]$

146. $\tan(-443.5°)$ **147.** $\csc(-651.9°)$

148. $\cot(-205°45')$ **149.** $\sin(141°25')$

150. $\sin(-7.88)$ **151.** $\tan[\sec(476°20')]$

***152.** $\sin(\sin 1.82)$ **153.** $\cos(-4.03)$

***154.** $\cot(\tan 6.11)$ ***155.** $\tan(\tan 91.43°)$

156. $\sin \frac{4\pi}{5}$ **157.** $\cos \frac{11\pi}{7}$

158. $\tan 1.37\pi$ **159.** $\sec 1.42\pi$

160. $\cot 2.3\pi$ **161.** $\sec 2.04$

162. $\csc 4.19$ **163.** $\cot(-5.162)$

164. $\sec(-6.91)$ **165.** $\csc(-3.00)$

166. $\tan^2 2.1\pi$ **167.** $\sin(2\pi - 1)$

168. $\cos(\pi + 2)$ **169.** $\sin^2 0.90\pi$

170. $\sin(0.90\pi)^2$ **171.** $\tan[\frac{1}{2}(\pi + 2)]$

172. $\cot^2 4.01$ **173.** $\sec^2 2.17$

174. $\csc^2 4.95$

In Exercises 175–186, evaluate each expression without using a calculator.

175. $\cos 135° \tan 150° \sin 120°$

176. $\sec 150° \cos 300° \sin 210°$

177. $\cot 300° \csc 120° \sec 240°$

178. $\cot 120° \cos 240° \tan 330°$

***179.** $3 \sin[3\pi(\cos^2 \frac{5\pi}{4})]$

180. $2 \sin(-60°) \cos(-60°)$

***181.** $6 \tan^3((\frac{\pi}{2} \sin \frac{7\pi}{6}))$

***182.** $\dfrac{\sin 120° - \cos 240°}{\tan 225° + \sec^2 45°}$

183. $\cot 30° + \cos 210° + \tan 330° - \cot 300°$

184. $\tan(-315°) + \cos(-45°) + \sin 135°$

***185.** $\dfrac{\csc 135° \sin 990° + \cos 810°}{\sin(-45°) + \cot 240° \tan 540°}$

***186.** $(m^2 + 2mn + n^2) \sin 90° + (m^2 - 2mn + n^2) \cos 180° - 4mn \tan 225°$

In Exercises 187–189, test to see if each is true without using a calculator.

***187.** $\sin 825° \cos(-15°) + \cos 75° \sin(-555°) + \tan 155° \tan 245° = 0$

***188.** $[\sin(-328°)\sin 958°]/\cot 572° - [\cos(-508°)\cos(-1022°)]/\tan(-212°) = -1$

***189.** $[\cos^2 696° + \tan(-260°)\tan 530° - \cos^2 156°]/(\tan^2 252° + \cot^2 342°) = \frac{1}{2} \tan^2 18°$

****190.** Prove that the reference angle theorem is true for angles in all quadrants and for negative angles.

In Exercises 191–196, a point *P* lies on the unit circle (see Exercises 50–52 for Section 3.1-2). If the radius to *P* makes the given angle with the +*x* axis, find the *x* and *y* coordinates of *P*.

191. 42° **192.** $\frac{6\pi}{7}$ **193.** 203°

194. 1.40π **195.** 6.03 **196.** −115°

***197.** Use the unit circle (see Exercises 50–52 for Section 3.1-2) to prove that $\sin^2 \theta + \cos^2 \theta = 1$ for all θ.

In Exercises 198–206, the point *P* is a distance *r* from the origin and *r* makes an angle *θ* with the +*x* axis. Find the *x* and *y* coordinates of *P*.

198. $r = 6.5, \theta = 40°$ **199.** $r = 1.4, \theta = \frac{3\pi}{5}$

200. $r = 0.30, \theta = 318°$ **201.** $r = 2.65, \theta = 1.21\pi$

202. $r = 1.75, \theta = -1.3\pi$ **203.** $r = 6.40, \theta = \pi$

204. $r = 1.86, \theta = 3P/2$ **205.** $r = 0.82, \theta = 90°$

206. $r = 5.5, \theta = 720°$

****207. Curved Space.** In Chapter 1 we defined π as the ratio of the circumference of a circle to its diameter. If the circle is on a flat surface, like a sheet of paper, π has the value 3.14159 . . . no matter what the size of the circle. In other words, in flat space $\pi = 3.14159\ldots$, and its value does not depend on the *size* of the circle. Will π have the same properties in curved space, such as on the surface of the earth or a balloon? To investigate this, draw a circle on the surface of a sphere of radius R, as shown in Figure 3.2-18.

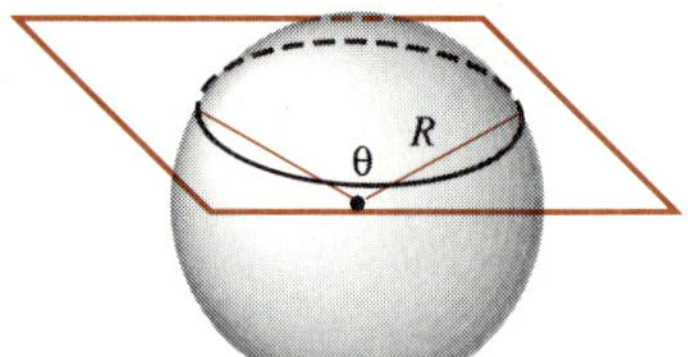

Figure 3.2-18

The plane intersects the sphere, and the curve formed where they meet is a circle. This circle is actually the boundary of two circular areas: the area of the circle on the curved surface of the sphere and the area of the flat circle lying in the plane inside the sphere. Call θ the central angle intercepted by the diameter of this circle, as shown in Figure 3.2-18.

(a) Show that the diameter of the circle on the *curved* surface of the sphere is $d_c = R\theta$, whereas the diameter of the *flat* circle is $d_f = 2R \sin \theta/2$.

(b) Show that the circumference of this circle is $c = 2\pi R \sin(\theta/2)$, as measured for either the flat or the curved circle.

(c) Use these results to show that π on the curved surface (call it π_c) is

$$\pi_c = \pi_f \frac{\sin \theta/2}{\theta/2}$$

where $\pi_f = 3.14159\ldots$ is the flat-space value of π. ***Hint:*** Use the definition of π on the curved surface.

(d) Evaluate π_c when $\theta = 90°$ and $\theta = 180°$, and draw the curved circle in each case. Is π_c a constant independent of the size of the circle on the curved space?

(e) In the formula you just derived in (c), why must θ be restricted to be between 0° and 180° when in fact it could actually go beyond that on the figure? ***Hint:*** Suppose that $\theta > 90°$ and show that this could lead to two different values of π_c.

(f) Express π_c in terms of the radius r_c of the curved circle and show that this is $\pi_c = \pi_f R\,[\sin(r_c/R)]/r_c$. Notice that this result tells you that if you measured both the radius of a circle r_c and the value of π_c for that circle, you could use these to calculate the value of R, the curvature of the curved space. In principle, astronomers could make such measurements on very large circles in space to determine the curvature of the universe. These measurements would in turn tell us whether the universe will expand forever or eventually collapse in on itself.

3.3 FINDING THE ANGLE: INVERSE TRIGONOMETRIC FUNCTIONS REVISITED

3.3-1 Positive Angles Less Than One Revolution

Using the Reference Angle

In Section 2.4 we learned how to find the angle when we knew a trigonometric function of that angle. In that section we were restricted to positive angles less than 90°, and now we want to extend those ideas to cover all angles.

Let's start with a concrete example. Suppose we know that $\sin \theta = \frac{1}{2}$ and we want to find out what θ is. In Chapter 2 we would have concluded that $\theta = 30°$ because $\sin 30° = \frac{1}{2}$, and that would have been the end of it. But, from our work in the present chapter, we may wonder if there are any other angles whose sine is $\frac{1}{2}$.

To answer this, remember two things: (a) The reference angle theorem tells us that if two angles have the same reference angle, their trigonometric functions will be the same, except for possible differences in sign, and (b) the sine is positive in quadrants I and II. These two statements tell us that any angle in the first two quadrants having a reference angle of 30° will have a sine of $\frac{1}{2}$. In quadrant I, such an angle is 30°. In quadrant II, 150° has a reference angle of 30° and therefore also has a sine of $\frac{1}{2}$, as you can see in Figure 3.3-1. The sine is negative in quadrants III and IV so no angle there can have a sine of $\frac{1}{2}$ even if it has a reference angle of 30°. Therefore, if $\sin \theta = \frac{1}{2}$, θ has two possible values: 30° and 150°.*

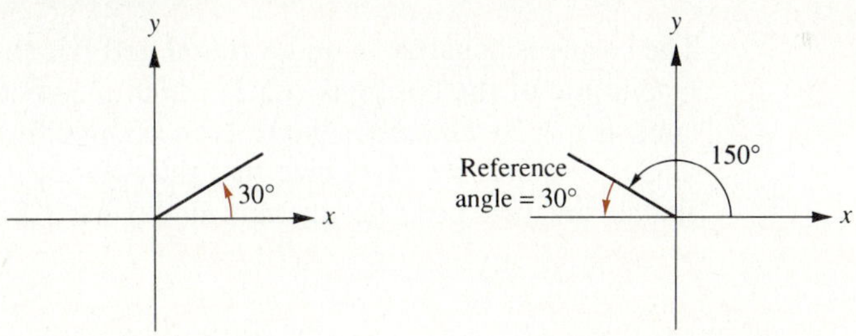

Figure 3.3-1

Both 30° and 150° have a reference angle of 30° and a sine of $\frac{1}{2}$.

(a) Quadrant I angle ***(b) Quadrant II angle***

Before we generalize this process for finding the angles, let's work a few examples to discover the general procedure.

Example 20

Find all the positive values of θ less than 360° for which $\sin \theta = -\frac{1}{2}$.

Solution

Because $\sin \theta$ is negative, we know that θ can lie in only quadrants III or IV. The fact that the magnitude of $\sin \theta$ is $\frac{1}{2}$ means that the sine of its reference angle

---

* In fact, if we allow θ to be more than 1 rev or to be negative, there can be infinitely many solutions. We shall investigate such cases shortly.

must be $\frac{1}{2}$. Therefore, the reference angle is 30°. We sketch the two possibilities in Figure 3.3-2.

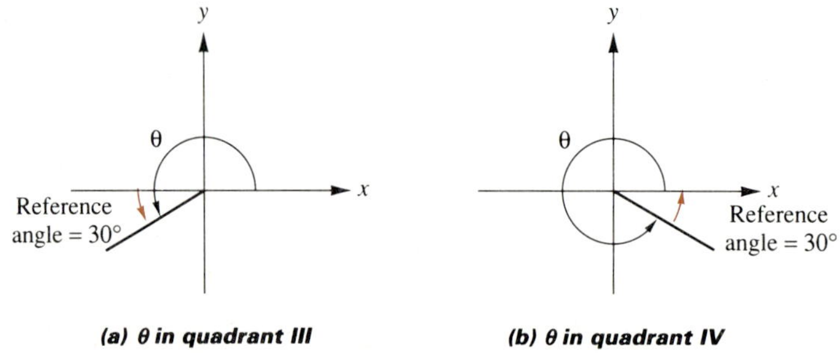

Figure 3.3-2

For both cases shown, sin $\theta = -\frac{1}{2}$ and the reference angle is 30°.

As you can see from Figure 3.3-2, the two possibilities for θ are

In quadrant III: $\theta = 180° + 30° = 210°$

In quadrant IV: $\theta = 360° - 30° = 330°$

Therefore, if $\sin \theta = -\frac{1}{2}$, $\theta = 210°$ or $330°$. ■

Example 21

If $\cos \theta = -\sqrt{2}/2$, find all the values that θ can have between 0 and 2π.

Solution

The cosine is negative in quadrants II and III, so θ can be in either of these. The magnitude of the cosine is $\sqrt{2}/2$, which means that the cosine of the reference angle is $\sqrt{2}/2$ (remember the reference angle theorem!). Because cos (reference angle) = $\sqrt{2}/2$, we recognize that the reference angle is a special angle, $\frac{\pi}{4}$. We sketch the two possible answers in Figure 3.3-3.

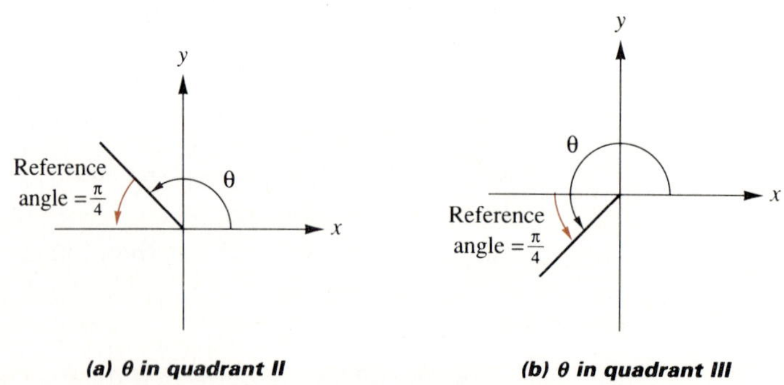

Figure 3.3-3

For both cases, the reference angle of θ is $\frac{\pi}{4}$ and $\cos \theta = -\sqrt{2}/2$.

As you can see from Figure 3.3-3, the two possible answers for θ are

In quadrant II: $\theta = \pi - \frac{\pi}{4} = \frac{3\pi}{4}$

In quadrant III: $\theta = \pi + \frac{\pi}{4} = \frac{5\pi}{4}$ ■

NOW IT'S YOUR TURN

Find all the positive values of θ less than 2π for which $\cos\theta = \sqrt{3}/2$. (Answer: $\frac{\pi}{6}$, $\frac{11\pi}{6}$)

Now let us generalize what we have done by recognizing patterns in the solution procedure. First, we use the *sign* of the trigonometric function to find which *quadrants* the angle can be in. Second, recognize that the *magnitude* of the trigonometric function ($\frac{1}{2}$ and $\sqrt{2}/2$ in these examples) gives us the *reference angle*. Finally, we make a sketch showing the possible angles and read the answers from this sketch. This general procedure works for *all* angles so we summarize it.

FINDING INVERSE FUNCTIONS* FOR ANY ANGLE:

1. **Use the sign of the trigonometric function to find which quadrants the angle can be in.**
2. **Use the magnitude of the trigonometric function to find the reference angle.**
3. **Sketch the possible angles, showing the reference angle.**
4. **Read the values of the angles from your sketch.**

Example 22

Find all possible positive angles θ less than 360° for which $\theta = \tan^{-1}(-1)$.

Solution

$\theta = \tan^{-1}(-1)$ means that θ is an angle whose tangent is -1, or $\tan\theta = -1$. Following the steps we just outlined:

1. The tangent is negative in quadrants II and IV, so θ can be in either of these quadrants.

2. The magnitude of $\tan\theta$ is 1, which means that the tangent of the reference angle is 1, so the reference angle is 45°.

3. Sketch the possible angles (Figure 3.3-4).

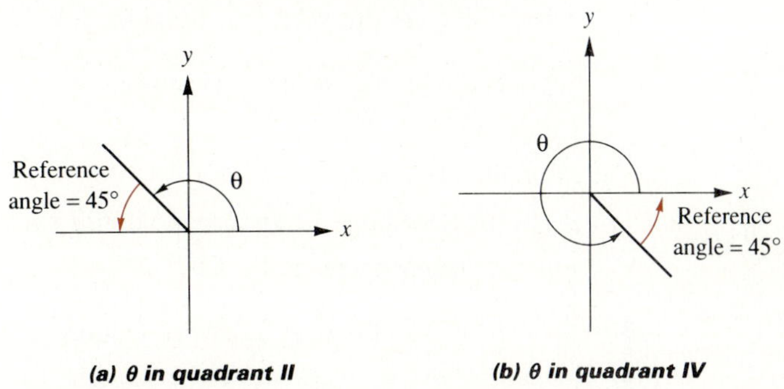

Figure 3.3-4

Tan $\theta = -1$ for both angles θ.

---

* Strictly speaking, we are finding the inverse trigonometric *relationships* rather than *functions* because more than one angle can have the same value for its trigonometric function. For example, both $\frac{\pi}{4}$ and both $\frac{5\pi}{4}$ have tangent equal to 1. We discuss functions in greater depth in Chapter 8.

4. Read the two answers from the sketch.

If θ is in quadrant II: $\theta = 180° - 45° = 135°$

If θ is in quadrant IV: $\theta = 360° - 45° = 315°$ ■

NOW IT'S YOUR TURN

Find all the positive angles θ less than 360° for which $\theta = \tan^{-1} 1$.
(Answer: 45°, 225°)

Using Calculators

So far the angles have been special ones, but in practice, they rarely are. For other angles, we follow the same general procedure, but we use a calculator to find the reference angle.

As we saw in the previous examples, when we deal with angles greater than 90°, *two* angles can have the same trigonometric function. For example, both 30° and 150° have a sine of $\frac{1}{2}$. But your calculator can give you only *one* answer.

You may want to check this statement. With your calculator set to the DEG mode, enter $\frac{1}{2}$. Then find the inverse sine of this number. Your calculator displays 30, meaning 30°. It does not give you 150°. Therefore, *to find inverse trigonometric functions with your calculator, you will have to make use of reference angles.*

Example 23

Use a calculator to find all the positive values of $\tan^{-1}(-1.89)$ less than 360°, accurate to two decimal places.

Solution

Calling θ the unknown angle, we have $\theta = \tan^{-1}(-1.89)$, which means that tan $\theta = -1.89$. Following the general procedure gives us

1. The negative tangent means that θ can be in quadrants II or IV.

2. The tangent of the reference angle is 1.89. We use a calculator to find the reference angle as described in Section 2.4:

Set the calculator to the DEG mode.

Enter 1.89.

Press the INV key.

Press the TAN key. Your screen should show 62.12.

The reference angle is 62.12°.

3. Sketch the two possible answers (Figure 3.3-5).

4. Read the two answers from the sketch.

In quadrant II: $\theta = 180° - 62.12° = 117.88°$

In quadrant IV: $\theta = 360° - 62.12° = 297.88°$ ■

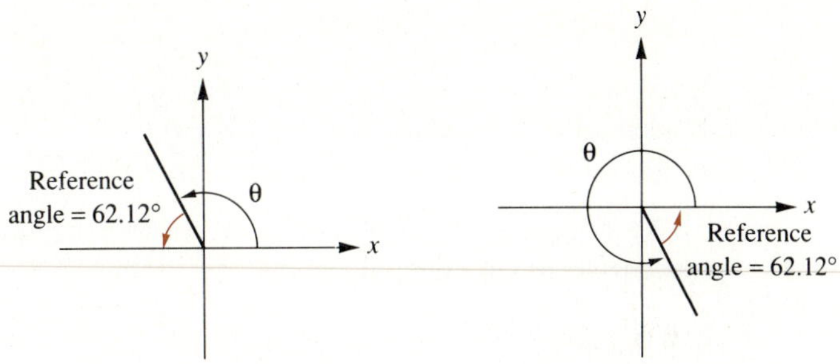

Figure 3.3-5 *(a) θ in quadrant II* *(b) θ in quadrant IV*

NOW IT'S YOUR TURN

Use a calculator to find all the positive values of $\sin^{-1}(-0.614)$ less than 360° accurate to two decimal places. (Answer: 217.88°, 322.12°)

Example 24

Use a calculator to find all the positive values of $\cos^{-1}(-0.253)$ that are less than 2π, accurate to two decimal places.

Solution

Calling the unknown angle θ gives $\theta = \cos^{-1}(-0.253)$, which means that $\cos\theta = -0.253$. Following the general procedure gives us

1. The cosine is negative in quadrants II and III, so θ can be in either of these.
2. The cosine of the reference angle is 0.253, so the reference angle = $\cos^{-1}(0.253)$. A calculator, as described in Section 2.4, tells us that the reference angle is 1.32 rad.
3. Sketch the two possible answers (Figure 3.3-6).

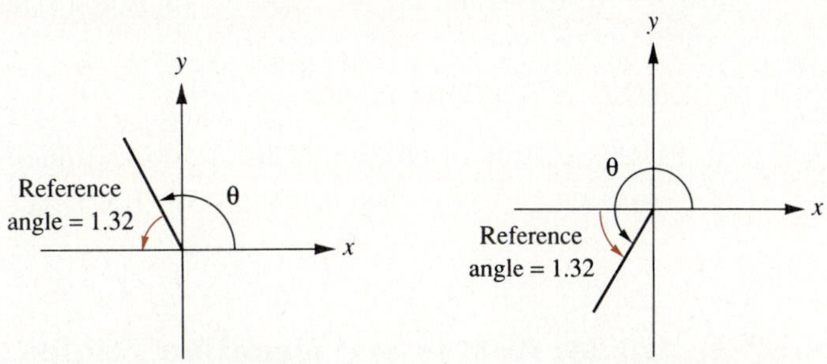

Figure 3.3-6 *(a) θ in quadrant II* *(b) θ in quadrant III*

4. Read the two answers from the sketch.

In quadrant II: $\theta = \pi - 1.32 = 3.14 - 1.32 = 1.82$

In quadrant III: $\theta = \pi + 1.32 = 3.14 + 1.32 = 4.46$ ■

NOW IT'S YOUR TURN

Use a calculator to find all the positive values of arctan(3.65) less than 2π to two decimal places. (Answer: 1.30, 4.44)

Example 25

Use a calculator to find all the positive values of $\sec^{-1}(5.03)$ less than 360° accurate to one decimal place.

Solution

Letting $\theta = \sec^{-1} 5.03$ means that $\sec\theta = 5.03$. Now use the general procedure.

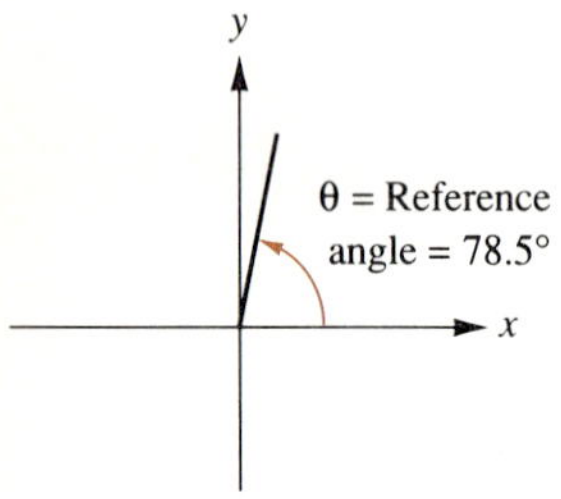

(a) θ in quadrant I

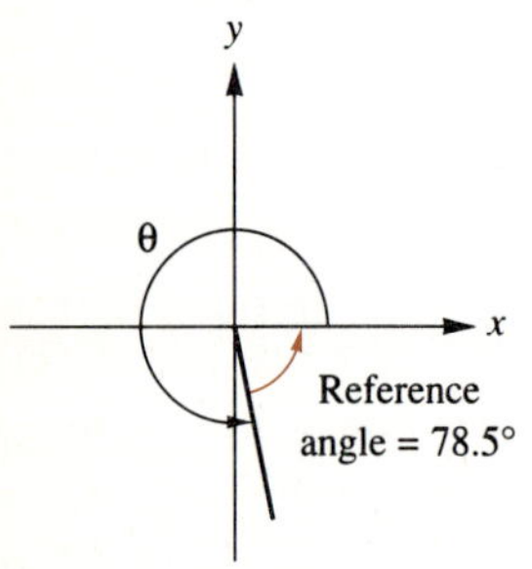

(b) θ in quadrant IV

Figure 3.3-7

1. The secant is positive in quadrants I and IV, so θ can lie in either of these.
2. To find the reference angle, use $\sec\theta = 1/\cos\theta$. The secant of the reference angle is 5.03, so

$$\sec(\text{ref angle}) = 5.03 = \frac{1}{\cos(\text{ref angle})}$$

$$\cos(\text{ref angle}) = \frac{1}{5.03} = 0.1988$$

Now use the calculator to find the reference angle.

$$\text{ref angle} = \cos^{-1}(0.1988) = 78.5^\circ$$

3. Sketch the two possible answers (Figure 3.3-7).
4. Read the answers from the sketch.

In quadrant I: $\theta = 78.5^\circ$

In quadrant IV: $\theta = 360^\circ - 78.5^\circ = 281.5^\circ$ ■

NOW IT'S YOUR TURN

Use a calculator to find all the positive values of $\csc^{-1}(-1.50)$ less than 360° accurate to three significant figures. (Answer: 222°, 318°)

3.3-2 Larger Angles and Negative Angles

Angles Greater Than One Revolution

So far we have limited ourselves to angles less than 1 rev. To deal with larger angles, we use the same general approach as before, except that the angle can include *multiple revolutions*.

Example 26

Find all the positive values of $\sin^{-1} \frac{1}{2}$ less than 2 rev in units of degrees.

Solution

Calling $\theta = \sin^{-1} \frac{1}{2}$ means that $\sin \theta = \frac{1}{2}$. As we saw at the beginning of this section, the solutions to this are $\theta = 30°$ and $150°$, as shown in Figure 3.3-8. In that case, θ was limited to be less than 360°, but now we can include larger angles. As you can see in Figure 3.3-9, angles of 30° + 1 rev and 150° + 1 rev have the same terminal sides as 30° and 150°. This means that these larger angles must have the same trigonometric functions as the smaller ones, so they must also be solutions. The *additional* solutions are

In quadrant I: $\theta = 30° + 1 \text{ rev} = 30° + 360° = 390°$

In quadrant II: $\theta = 150° + 1 \text{ rev} = 150° + 360° = 510°$

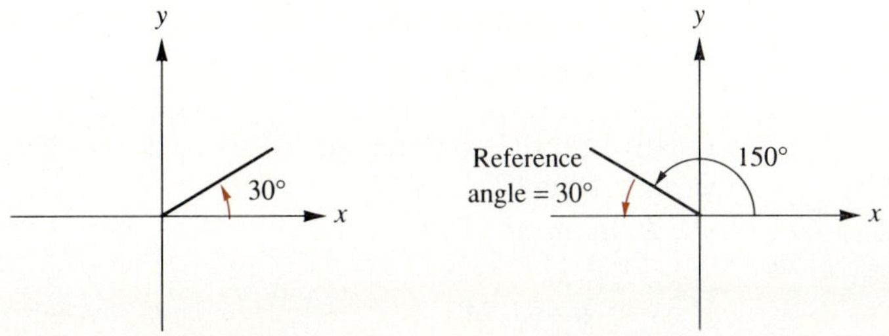

Figure 3.3-8
Both 30° and 150° have sine equal to $\frac{1}{2}$.

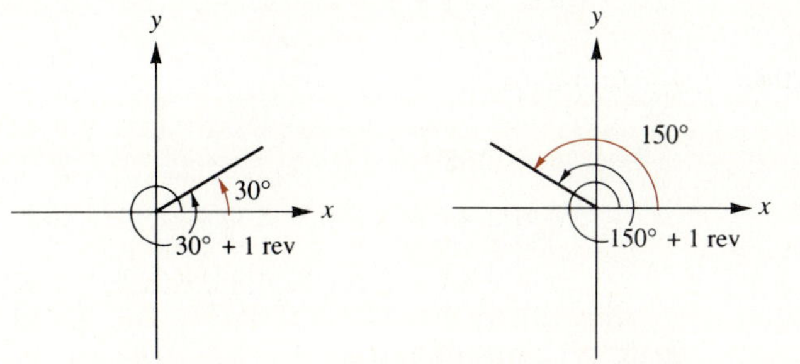

(a) 30° + 1 rev ***(b) 150° + 1 rev***

Figure 3.3-9
These angles also have sine equal to $\frac{1}{2}$.

The complete set of solutions is $\theta = 30°, 150°, 390°, 510°$ ■

NOW IT'S YOUR TURN

If $\tan \theta = 1$, find all the positive values of θ (in degrees) less than 2 rev. (Answer: 45°, 225°, 405°, 585°)

Example 27

Use a calculator to find all the positive values in radians of arctan(−5.90) less than 3 rev.

Solution

First we find the possible angles less than 1 rev, as in Example 26. Then we get all the solutions by adding 1 rev and 2 rev to these angles.

Calling $\theta = \arctan(-5.90)$ means $\tan\theta = -5.90$. Now follow the usual steps.

1. The minus sign means that θ must be in quadrant II or IV, where the tangent is negative.
2. We get the reference angle using tan(ref angle) = 5.90. Using a calculator, we get 1.40 rad.
3. Sketch both possible answers for θ less than 2π (Figure 3.3-10).
4. Read the two answers from the sketch.

In the quadrant II: $\theta = \pi - 1.40 = 3.14 - 1.40 = 1.74$

In quadrant IV: $\theta = 2\pi - 1.40 = 6.28 - 1.40 = 4.88$

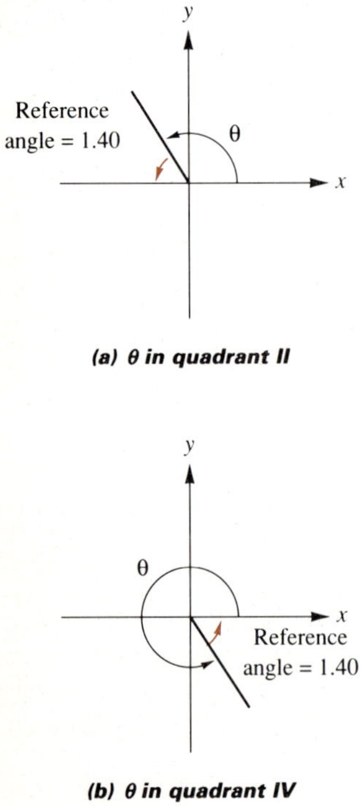

Figure 3.3-10

The possible solutions less than 1 revolution

Finally, get the complete set of six answers by adding 1 and 2 rev to the answers we just found.

In quadrant II:

$$\theta = 1.74$$

$$\theta = 1.74 + 1 \text{ rev} = 1.74 + 2\pi = 1.74 + 6.28 = 8.02$$

$$\theta = 1.74 + 2 \text{ rev} = 1.74 + 2(2\pi) = 1.74 + 12.56 = 14.30$$

In quadrant IV:

$$\theta = 4.88$$

$$\theta = 4.88 + 1 \text{ rev} = 4.88 + 2\pi = 4.88 + 6.28 = 11.16$$

$$\theta = 4.88 + 2 \text{ rev} = 4.88 + 2(2\pi) = 4.88 + 12.56 = 17.44$$ ■

NOW IT'S YOUR TURN

Use a calculator to find all of the positive values in radians of arcsin(−0.465) less than 3 rev. (Answer: 3.62, 5.80, 9.90, 12.08, 16.18, 18.36)

Negative Angles

Up to now we have considered only positive angles, but negative angles frequently occur. We handle them in the same general way as positive angles, as the following examples show.

Example 28

Find all the angles θ between 0° and −360° for which $\sin\theta = \frac{1}{2}$.

Solution

Because $\sin\theta = \frac{1}{2}$, θ must have a reference angle of 30° and a terminal side in either quadrant I or II. Sketch these two possible answers (Figure 3.3-11).

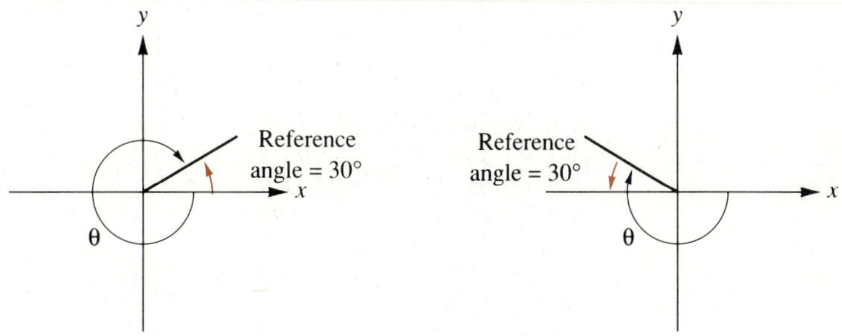

(a) Quadrant I solution   *(b) Quadrant II solution*

Figure 3.3-11

For both values of θ, $\sin\theta = \frac{1}{2}$ and the reference angle is 30°. The rotation of θ is clockwise because θ is negative.

Using Figure 3.3-11, we can see that

For quadrant I,

$$\theta = 360° - 30° = 330°$$

but in a *negative* direction, so $\theta = -330°$.

For quadrant II,

$$\theta = 180° + 30° = 210°$$

but in a *negative* direction, so $\theta = -210°$. ■

NOW IT'S YOUR TURN

Find all angles between 0° and −360° for which $\cos\theta = \frac{1}{2}$.
(Answer: −60°, −300°)

Example 29

Find all the angles θ between 0 and -2π for which $\theta = \cos^{-1}(-0.4447)$.

Solution

1. The angle can lie in quadrants II or III because the cosine is negative.

2. cos(ref angle) = 0.4447. A calculator in the RAD mode tells us that the reference angle is 1.11 rad.

3. Sketch the two possible answers (Figure 3.3-12).

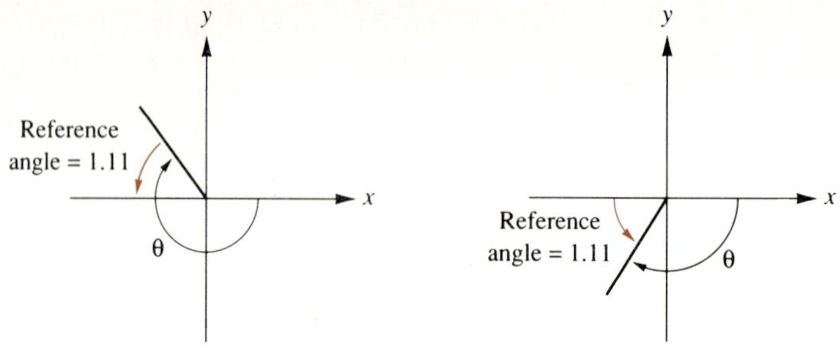

Figure 3.3-12 **(a) Quadrant II angle** **(b) Quadrant III angle**

4. Read the answers from the sketch.

In quadrant II,

$$\theta = \pi + 1.11 = 3.14 + 1.11 = 4.25$$

but in a negative direction, so $\theta = -4.25$.

In quadrant III,

$$\theta = \pi - 1.11 = 3.14 - 1.11 = 2.03$$

but in a negative direction, so $\theta = -2.03$. ■

NOW IT'S YOUR TURN

Find all the angles θ between 0 and -2π to two decimal places for which $\theta = \tan^{-1}(0.4228)$. (Answer: −2.74, −5.88)

EXERCISES 3.3

In Exercises 1–12, find all the possible positive values of θ less than 1 rev in degrees without using a calculator or a table.

1. $\sin\theta = \sqrt{3}/2$

2. $\sin\theta = -\sqrt{3}/2$

3. $\csc\theta = 2$

4. $\cot\theta = -1$

5. $\theta = \arctan(\sqrt{3})$

6. $\sin\theta = -\sqrt{2}/2$

7. $\sin\theta = 0$

8. $\sec\theta = -2$

9. $\theta = \cos^{-1}(-1)$

10. $\sec\theta = 1$

11. $\theta = \cos^{-1}(\sqrt{2}/2)$

12. $\cos\theta = 0$

In Exercises 13–25, find all the possible positive values of θ less than 1 rev in radians without using a calculator or a table.

13. $\theta = \csc^{-1}(-2\sqrt{3}/3)$

14. $\theta = \text{arcsec}(-\sqrt{2})$

15. $\cos\theta = 0.50$

16. $\theta = \arctan 0$

17. $\theta = \cos^{-1}(1)$

18. $\csc\theta = -1$

19. $\theta = \cot^{-1}(-\sqrt{3})$

20. $\theta = \text{arccot}(\text{undefined})$

21. $\theta = \sin^{-1}(-1)$

22. $\sec\theta = \text{undefined}$

23. $\theta = \tan^{-1}(\text{undefined})$

24. $\theta = \cot^{-1}(0)$

25. $\theta = \csc^{-1}(\text{undefined})$

In Exercises 26–44, use a calculator to find all the possible positive values of θ less than 1 rev in degrees accurate to two decimal places.

26. $\sin\theta = 0.2571$
27. $\tan\theta = 0.1228$
28. $\cos\theta = -0.4801$
29. $\sin\theta = -0.9527$
30. $\cot\theta = -1.212$
31. $\theta = \csc^{-1}(44.077)$
32. $\sec\theta = 2.760$
33. $\theta = \cos^{-1} 0.1840$
34. $\theta = \arctan(-0.1717)$
35. $\theta = \cot^{-1}(1.9458)$
***36.** $\theta = \sin^{-1}(-\frac{7}{8})$
37. $\sec\theta = -5.5396$
***38.** $\sin\theta = \frac{\pi}{4}$
***39.** $1 + \sin\theta = 1.2790$
40. $\operatorname{arccsc}(-1.011) = \theta$
***41.** $2\cos\theta = 1.7676$
***42.** $3\tan\theta + 4 = 7$
****43.** $\theta = \frac{\pi}{4} + \cos^{-1}(0.6967)$
***44.** $\theta - 11° = \arcsin(0.5736)$

In Exercises 45–62, use a calculator to find all the possible positive values of θ less than 1 rev in radians accurate to two decimal places.

45. $\sin\theta = -0.6210$
46. $\cos\theta = 0.540$
47. $\tan\theta = -7.8$
48. $\theta = \arctan(-1.69)$
49. $\theta = \sin^{-1}(-0.98)$
50. $\theta = \cos^{-1}(0.402)$
51. $\sec\theta = 1.65$
52. $\cot\theta = -2.03$
53. $\theta = \csc^{-1}(3.13)$
54. $\sec\theta = \pi$
55. $\cos\theta = 3 - \pi$
***56.** $1 + \sin\theta = 0.62$
***57.** $\sin(1 + \theta) = 0.62$
***58.** $4\tan\theta = 7$
***59.** $3\cos\theta = -2.2$
***60.** $4\sec\theta = 9.10$
***61.** $\sin(2\theta) = 0.84$
***62.** $2.81\sin(\theta + 0.3) = 1.50$

In Exercises 63–66, use a calculator to find all the possible positive values of $\theta \le 3$ rev in degrees accurate to one decimal place.

63. $\cos\theta = 0.8192$
64. $\sin\theta = -0.7833$
65. $\tan\theta = -1.5399$
66. $\cos\theta = -0.9553$

In Exercises 67–72, use a calculator to find all the possible positive values of $\theta \le 3$ rev in radians accurate to one decimal place.

67. $\sin\theta = -0.3518$
68. $\tan\theta = 1.116$
69. $\theta = \arctan 1.8040$
70. $\csc\theta = 1.771$
71. $\theta = \sec^{-1}(-2.5593)$
72. $\cot\theta = 3.916$

In Exercises 73–76, use a calculator to find all the possible values of θ between ± 1 rev in degrees accurate to two decimal places.

73. $\cos\theta = 0.9063$
74. $\tan\theta = 0.2027$
75. $\theta = \sin^{-1}(-0.9272)$
76. $\sin\theta = -0.8912$

In Exercises 77–82, use a calculator to find all possible values of θ between ± 1 rev in radians accurate to two decimal places.

77. $\theta = \arctan(-5.6713)$
78. $\cos\theta = -0.1700$
79. $\csc\theta = 8.2055$
80. $\sec^{-1}(-3.171) = \theta$
81. $\theta = \arcsin(-0.9593)$
82. $\theta = \cot^{-1}(-6.617)$

In Exercises 83–88, use a calculator to find all the possible values of θ between ± 3 rev accurate to two decimal places. Express the answers to the first three in radians and the last three in degrees.

83. $\tan\theta = 0.2309$
84. $\theta = \arcsin 0.8912$
85. $\cos\theta = -0.4695$
86. $\theta = \arctan(-0.5463)$
87. $\theta = \csc^{-1}(2.8555)$
88. $\cot\theta = -9.967$

In Exercises 89–94, use a calculator or your knowledge of special angles to find only the positive solutions in degrees to two decimal places.

***89.** If $\sin 2x = \frac{1}{2}$, find all $x \le 360°$.
***90.** If $\cos 3x = 0$, find all $x \le 2\pi$.
***91.** If $\tan 2y = -1$, find all $y \le 360°$.
***92.** If $6\cos 2x = 3$, find all $x \le 2\pi$.
***93.** If $9\sin 4z = -4.5$, find all $z \le 360°$.
***94.** If $8 + \tan\alpha = 3\tan\alpha$, find all $\alpha \le 2\pi$.

In Exercises 95–101, use a calculator or your knowledge of special angles to find only the positive solutions in radians to two decimal places.

***95.** If $\cos(\theta + 15.2°) = 0.4679$, find all $\theta \le 360°$.
****96.** If $\cot^2 x = 3$, find all $x \le 2\pi$.
***97.** If $4\tan(2\theta + 40°) = 1.8652$, find all $\theta \le 720°$.
****98.** If $\sin^2\theta = \frac{1}{2}$, find all $\theta \le 2\pi$.
***99.** If $3\sec(t + 50°) = 3.3102$, find all $t \le 360°$.
****100.** If $\cos^2\theta = \frac{2}{3}$, find all $\theta \le 3\pi$.
****101.** If $\tan^2\theta + \tan\theta = 6$, find all $\theta \le 2\pi$. ***Hint:*** Use the quadratic formula to solve for $\tan\theta$.

Use radian measure in Exercises 102–105. In each case, write a formula in terms of an integer n and generate all the solutions by letting $n = 0, 1, 2, 3, \ldots$. (Equations like these often occur in engineering and physics in the study of vibrating strings and surfaces, transmission lines, the structure of solids, and quantum mechanics.)

****102.** If $\sin kx = 0$, find all the positive values of x in terms of k.
****103.** If $\cos bx = 0$, find all the positive values of x in terms of b.
****104.** If $\sin^2(A\theta) = 1$, find all the positive values of θ in terms of A.
****105.** If $\sin 3x \cos 8t = 0$, find all the positive values of x and t.

In Exercises 106–113, a point P on the unit circle (see Exercise 50 in Section 3.1-2) has the x or y coordinates given. Find the other coordinate of P and the angle between the radius to P and the $+x$ axis.

106. $x = 0.20$, P is in quadrant I.

107. $x = 0.60$, P is in quadrant IV.

108. $y = 0.71$, P is in quadrant II.

109. $x = -0.25$, P is in quadrant II.

110. $y = -0.40$, P is in quadrant III.

***111.** $x = -0.75$, y is in any possible quadrant.

***112.** $y = 0.80$, x is in any possible quadrant.

***113.** $x = -1.40$, y is in any possible quadrant.

In Exercises 114–117, the point P has coordinates x and y. Find the distance of P from the origin and the angle between the radius to P and the $+x$ axis.

114. $x = 4.0$, $y = 5.0$

115. $x = -1.4$, $y = -2.3$

116. $x = -6.3$, $y = 8.8$

117. $x = 0.72$, $y = -1.10$

3.4 SOME IMPORTANT PROPERTIES OF THE TRIGONOMETRIC FUNCTIONS

In this chapter, we have been generalizing the trigonometric functions to all possible angles. As part of this process, we want to look at several important properties that are obeyed by the trigonometric functions of all angles.

When a trigonometric function obeys a certain relationship, *no matter what the value of the angle*, this relationship is *identically true for all angles*. We call such a relationship an **identity**. An example of an identity is the relationship $\sin\theta = y/r$. This relationship is true no matter what θ is, so it is true for all angles. On the other hand, the relationship $\sin\theta = \frac{1}{2}$ is not an identity because it is true only for specific angles. For example, it is true for 30° and 150°, but not for 45°.

NOW IT'S YOUR TURN

Which of the following are identities and which are not? (a) $\cos\theta = \frac{2}{3}$, (b) $\tan\theta = 2.1$, (c) $\sec\theta =$ undefined, (d) $\cot\theta = x/y$.
(Answer: Only (d) is an identity.)

3.4-1 Reciprocal and Quotient Identities

In Section 2.2-3, we proved some important fundamental relationships between the trigonometric functions of *acute* angles, such as $\sin\theta = 1/\csc\theta$, $\tan\theta = \sin\theta/\cos\theta$, and so on. We now wish to extend these to all angles, not just acute ones.

In Section 3.1, we extended the definitions of the trigonometric functions of an angle θ by expressing them in terms of the coordinates (x, y) of a point $P(x, y)$ (Figure 3.4-1) and obtained

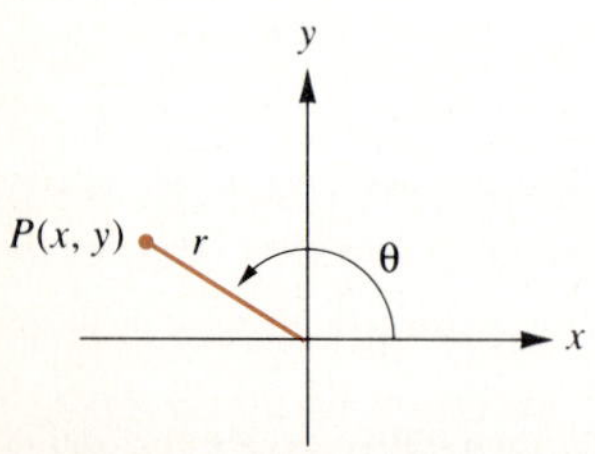

Figure 3.4-1

$$\sin\theta = \frac{y}{r} \qquad \csc\theta = \frac{r}{y}$$

$$\cos\theta = \frac{x}{r} \qquad \sec\theta = \frac{r}{x}$$

$$\tan \theta = \frac{y}{x} \qquad \cot \theta = \frac{x}{y}$$

From these definitions, we have

$$\sin \theta \csc \theta = \left(\frac{y}{r}\right)\left(\frac{r}{y}\right) = 1$$

$$\cos \theta \sec \theta = \left(\frac{x}{r}\right)\left(\frac{r}{x}\right) = 1$$

$$\tan \theta \cot \theta = \left(\frac{y}{x}\right)\left(\frac{x}{y}\right) = 1$$

Therefore,

$$\sin \theta \csc \theta = 1 \qquad \cos \theta \sec \theta = 1 \qquad \text{and} \qquad \tan \theta \cot \theta = 1$$

We can solve each of these last three equations for one trigonometric function in terms of the other and obtain $\sin \theta = 1/\csc \theta$ or $\csc \theta = 1/\sin \theta$; $\cos \theta = 1/\sec \theta$ or $\sec \theta = 1/\cos \theta$; and $\tan \theta = 1/\cot \theta$ or $\cot \theta = 1/\tan \theta$. We call these identities the **reciprocal identities** because certain trigonometric functions are reciprocals of others.

RECIPROCAL IDENTITIES

$$\sin \theta = \frac{1}{\csc \theta} \qquad \csc \theta = \frac{1}{\sin \theta}$$

$$\cos \theta = \frac{1}{\sec \theta} \qquad \sec \theta = \frac{1}{\cos \theta}$$

$$\tan \theta = \frac{1}{\cot \theta} \qquad \cot \theta = \frac{1}{\tan \theta}$$

where θ is any angle for which the functions are defined.

Notice that we have called these relationships identities because they are true for all allowed values of θ, not just specific ones.

Example 30

(a) Find $\sin \theta$ if $\csc \theta = \sqrt{5}/2$.

(b) Find $\sec \theta$ if $\cos \theta = -\frac{12}{13}$.

(c) Find $\cot 116°$ from $\tan 116°$.

(d) Find $\cos \frac{\pi}{2}$ from $\sec \frac{\pi}{2}$.

Solution

Using the reciprocal identities, we get

(a) $\sin \theta = \dfrac{1}{\csc \theta} = \dfrac{1}{\sqrt{5}/2} = 2/\sqrt{5} = 2\sqrt{5}/5$

(b) $\sec \theta = \dfrac{1}{\cos \theta} = \dfrac{1}{-12/13} = -13/12$

(c) $\cot 116° = \dfrac{1}{\tan 116°} = \dfrac{1}{-2.0503} = -0.4877$
where we used a calculator to find tan 116°.

(d) We might expect to use the reciprocal identity $\cos \frac{\pi}{2} = \dfrac{1}{\sec \frac{\pi}{2}}$. But $\sec \frac{\pi}{2}$ is undefined, so $\frac{\pi}{2}$ is not an allowed value of the angle and we cannot use this identity for this angle. Notice that it would lead to an incorrect result: $\sec \frac{\pi}{2}$ is undefined, so $1/(\sec \frac{\pi}{2})$ is also undefined. The identity would say that $\cos \frac{\pi}{2}$ is also undefined, yet we know that $\cos \frac{\pi}{2} = 0$. ■

NOW IT'S YOUR TURN

(a) Use the reciprocal identities to find (i) $\csc \theta$ if $\sin \theta = -\frac{1}{4}$ and (ii) $\cot \theta$ if $\tan \theta = 0.1234$.

(b) Find sin 0° from csc 0°.

(c) Find sec 228.1° from cos 228.1°.

(Answer: (a) −4, 8.104, (b) not possible, (c) −1.497).

In Section 2.2-3, we also found relationships relating the tangent and cotangent to the sine and cosine. We now extend these to cover all angles.

Using the definition of the sine and cosine, we have

$$\frac{\sin \theta}{\cos \theta} = \frac{\frac{y}{r}}{\frac{x}{r}} = \left(\frac{y}{r}\right)\left(\frac{r}{x}\right) = \frac{y}{x} = \tan \theta \quad ■$$

> **QUOTIENT IDENTITIES**
>
> $$\tan \theta = \frac{\sin \theta}{\cos \theta}$$
>
> $$\cot \theta = \frac{\cos \theta}{\sin \theta}$$
>
> where θ is any angle for which the functions are defined.

Combining this result with the reciprocal identity for the tangent and cotangent, we get

$$\cot \theta = \frac{1}{\tan \theta} = \frac{1}{\frac{\sin \theta}{\cos \theta}} = \frac{\cos \theta}{\sin \theta} \quad ■$$

These last two results involve quotients of sines and cosines, so we call them the **quotient identities**.

These relationships are identities because they are true for all values of θ.

Example 31

If $\sin \theta = \frac{1}{2}$ and $\cos \theta = -\sqrt{3}/2$, find the values of the other four trigonometric functions.

Solution

Using the quotient identities, we have

$$\tan\ \theta = \frac{\sin\ \theta}{\cos\ \theta} = \frac{\frac{1}{2}}{-\sqrt{3}/2} = \frac{1}{-\sqrt{3}} = \frac{-\sqrt{3}}{3}$$

Now we use the reciprocal identities to find

$$\cot\ \theta = \frac{1}{\tan\ \theta} = \frac{1}{-\sqrt{3}/3} = \frac{-3}{\sqrt{3}} = -3\left(\frac{\sqrt{3}}{3}\right) = -\sqrt{3}$$

$$\sec\ \theta = \frac{1}{\cos\ \theta} = \frac{1}{-\sqrt{3}/2} = \frac{-2}{\sqrt{3}} = \frac{-2\sqrt{3}}{3}$$

$$\csc\ \theta = \frac{1}{\sin\ \theta} = \frac{1}{\frac{1}{2}} = 2$$ ■

NOW IT'S YOUR TURN

If $\sin\theta = -\frac{1}{3}$ and $\cos\theta = -\sqrt{8}/3$, evaluate the other trigonometric functions of θ.
(Answers: $\tan\theta = \sqrt{8}/8$, $\cot\theta = \sqrt{8}$, $\sec\theta = -3\sqrt{8}/8$, $\csc\theta = -3$)

Notice from Examples 30 and 31 that we can use the values of the sine and cosine to evaluate all the other trigonometric functions.

3.4-2 The Pythagorean Identities

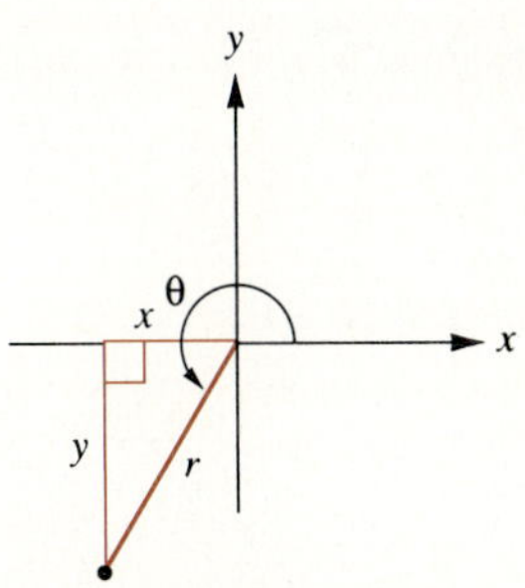

Figure 3.4-2

In Section 3.1 we generalized the definitions of the trigonometric functions by expressing them in terms of the coordinates (x, y) of a point $P(x, y)$. The magnitudes of x and y are also the sides of the reference triangle, which is a right triangle (Figure 3.4-2). Therefore, they obey the Pythagorean theorem. Even though x and y may be negative, they will obey this formula because the square of a negative number is a positive number.

$$x^2 + y^2 = r^2$$

$$\frac{x^2 + y^2}{r^2} = \frac{r^2}{r^2} = 1 \quad \text{(divide by } r^2\text{)}$$

$$\frac{x^2}{r^2} + \frac{y^2}{r^2} = 1$$

$$\left(\frac{x}{r}\right)^2 + \left(\frac{y}{r}\right)^2 = 1$$

From the definitions of the trigonometric functions, we recognize that $x/r = \cos\theta$ and $y/r = \sin\theta$. Therefore, the last equation becomes

$$\sin^2\theta + \cos^2\theta = 1$$

Similarly, if we divide $x^2 + y^2 = r^2$ by x^2, we get

$$\frac{x^2 + y^2}{x^2} = \frac{r^2}{x^2}$$

$$1 + \frac{y^2}{x^2} = \frac{r^2}{x^2}$$

$$1 + \left(\frac{y}{x}\right)^2 = \left(\frac{r}{x}\right)^2$$

From the definition of the trigonometric functions, we see that $y/x = \tan\theta$ and $r/x = \sec\theta$, so the last equation becomes

$$1 + \tan^2\theta = \sec^2\theta$$

NOW IT'S YOUR TURN

Use a similar procedure to prove that

$$1 + \cot^2\theta = \csc^2\theta$$

These three results all came from the Pythagorean theorem, so we call them the **Pythagorean identities.**

PYTHAGOREAN IDENTITIES

$$\sin^2\theta + \cos^2\theta = 1$$

$$1 + \tan^2\theta = \sec^2\theta$$

$$1 + \cot^2\theta = \csc^2\theta$$

for any angle θ for which the functions are defined.

These results are valid no matter how the angle is expressed. For example,

$$\sin^2\left(\tfrac{3\pi}{2}\right) + \cos^2\left(\tfrac{3\pi}{2}\right) = 1$$

$$1 + \tan^2(-1.83) = \sec^2(-1.83)$$

$$1 + \cot^2(x + 5) = \csc^2(x + 5)$$

We can combine these identities to find the values of the other trigonometric functions when just one of them is known, as Example 32 illustrates.

Example 32

If θ is a quadrant II angle having $\tan\theta = -\frac{5}{12}$, find the other five trigonometric functions of θ.

Solution

1. Starting with the Pythagorean identity $1 + \tan^2\theta = \sec^2\theta$, we solve for $\sec\theta$:

$$\sec\theta = \pm\sqrt{1 + \tan^2\theta} = \pm\sqrt{1 + \left(-\tfrac{5}{12}\right)^2} = \pm\sqrt{1 + \tfrac{25}{144}} = \pm\sqrt{\tfrac{169}{144}}$$

$$\sec\theta = \pm\tfrac{13}{12}$$

Because θ is in quadrant II, the secant is negative, so we must use the negative square root: $\sec\theta = -\frac{13}{12}$.

2. Using the reciprocal identities, we get

$$\cot\theta = \frac{1}{\tan\theta} = \frac{1}{-\frac{5}{12}} = -\tfrac{12}{5}$$

$$\cos\theta = \frac{1}{\sec\theta} = \frac{1}{-\frac{13}{12}} = -\tfrac{12}{13}$$

3. Solving the quotient identity $\tan\theta = \dfrac{\sin\theta}{\cos\theta}$ for $\sin\theta$ gives us

$$\sin\theta = \tan\theta\cos\theta = \left(-\tfrac{5}{12}\right)\left(-\tfrac{12}{13}\right) = \tfrac{5}{13}$$

4. Finally, a reciprocal identity gives

$$\csc\theta = \frac{1}{\sin\theta} = \frac{1}{\frac{5}{13}} = \tfrac{13}{5} \quad ■$$

NOW IT'S YOUR TURN

If θ is a quadrant IV angle having $\cos\theta = \frac{1}{4}$, find the other trigonometric functions of θ.

(Answer: $\sin\theta = -\sqrt{15}/4$, $\tan\theta = -\sqrt{15}$, $\cot\theta = -\sqrt{15}/15$, $\sec\theta = 4$, $\csc\theta = -4\sqrt{15}/15$)

Example 33

If θ is a positive angle less than 360° having $\sin\theta = 0.426$, find all the other trigonometric functions of θ.

Solution

Because $\sin \theta$ is positive, θ can lie in quadrants I or II. In quadrant I, all the trigonometric functions are positive, but in quadrant II only the sine and cosecant are positive and the rest are negative.

1. First use the Pythagorean identity to find $\cos \theta$.

$$\sin^2 \theta + \cos^2 \theta = 1$$

$$\cos^2 \theta = 1 - \sin^2 \theta = 1 - (0.426)^2 = 0.819$$

$$\cos \theta = \pm\sqrt{0.819} = \pm 0.905$$

Because θ can lie in the quadrants I and II, $\cos \theta$ can be positive and negative, so we need to use *both* values of $\cos \theta$.

2. Now use a quotient identity to find $\tan \theta$.

$$\tan \theta = \frac{\sin \theta}{\cos \theta} = \frac{0.425}{\pm 0.905} = \pm 0.470$$

As with $\cos \theta$, $\tan \theta$ can be positive and negative, so we need both values.

3. Finally, the reciprocal identities give us the other functions.

$$\cot \theta = \frac{1}{\tan \theta} = \frac{1}{\pm 0.471} = \pm 2.13$$

As with $\tan \theta$, we need both values of $\cot \theta$.

$$\sec \theta = \frac{1}{\cos \theta} = \frac{1}{\pm 0.905} = \pm 1.105$$

where we again need both values because $\sec \theta$ is positive in the quadrant I but negative quadrant II.

$$\csc \theta = \frac{1}{\sin \theta} = \frac{1}{0.426} = 2.347$$

We have only one value for $\csc \theta$ because it is positive in both quadrants I and II.

In summary, the trigonometric functions can have the following values: $\sin \theta = 0.426$ (given), $\cos \theta = \pm 0.905$, $\tan \theta = \pm 0.470$, $\cot \theta = \pm 2.13$, $\sec \theta = \pm 1.105$, and $\csc \theta = 2.347$. ■

NOW IT'S YOUR TURN

If θ is a positive angle less than 360° having $\cos \theta = 0.440$, find the other trigonometric functions of θ.

(Answer: $\sin \theta = \pm 0.898$, $\tan \theta = \pm 2.041$, $\cot \theta = \pm 0.490$, $\sec \theta = 2.273$, $\csc \theta = \pm 1.114$)

These identities are often especially useful for simplifying complex expressions by reducing them to a single trigonometric function, as the following examples illustrate.

Example 34

Simplify the expression $(\sin^3 \theta + \sin \theta \cos^2 \theta) \cot \theta$.

Solution

$$
\begin{aligned}
(\sin^3 \theta + \sin \theta \cos^2 \theta)\cot \theta &= \sin \theta(\sin^2 \theta + \cos^2 \theta)\cot \theta && \textbf{(factor out } \sin \boldsymbol{\theta}\textbf{)} \\
&= \sin \theta \cot \theta && \textbf{(use } \sin^2 \boldsymbol{\theta} + \cos^2 \boldsymbol{\theta} = 1\textbf{)} \\
&= \sin \theta \frac{\cos \theta}{\sin \theta} && \textbf{(use } \cot \boldsymbol{\theta} = \cos \boldsymbol{\theta}/\sin \boldsymbol{\theta}\textbf{)} \\
&= \cos \theta && \textbf{(cancel } \sin \boldsymbol{\theta}\textbf{)}
\end{aligned}
$$

NOW IT'S YOUR TURN

Simplify the expression $(\cos^4 \theta + \sin^2 \theta \cos^2 \theta) \tan \theta$. (Answer: $\sin \theta \cos \theta$)

Example 35

Simplify the following expression by expressing it in terms of $\cos x$.

$$\frac{\sin x \cot x}{\csc^2 x - \cos^2 x \csc^2 x}$$

Solution

It is not always obvious where to begin on a complicated expression. If in doubt, start by expressing things in terms of sines and cosines.

$$
\begin{aligned}
\frac{\sin x \cot x}{\csc^2 x - \cos^2 x \csc^2 x} &= \frac{\sin x\,(\cos x/\sin x)}{\csc^2 x - \cos^2 x \csc^2 x} && \textbf{(use } \cot x = \cos x/\sin x\textbf{)} \\
&= \frac{\cos x}{\csc^2 x - \cos^2 x \csc^2 x} && \textbf{(cancel } \sin x\textbf{)} \\
&= \frac{\cos x}{\csc^2 x\,(1 - \cos^2 x)} && \textbf{(factor out } \csc^2 x \textbf{ in the denominator)} \\
&= \frac{\cos x}{\csc^2 x \sin^2 x} && \textbf{(use Pythagorean identify } \sin^2 x = 1 - \cos^2 x\textbf{)} \\
&= \frac{\cos x}{(1/\sin^2 x) \sin^2 x} && \textbf{(use } \csc x = 1/\sin x\textbf{)} \\
&= \cos x && \textbf{(cancel } \sin^2 x\textbf{)}
\end{aligned}
$$

NOW IT'S YOUR TURN

Simplify the following expression by expressing it in terms of cos θ.

$$\frac{\sin\theta\cos^2\theta + \cos^2\theta\sin\theta\tan^2\theta}{\tan\theta}$$

(Answer: cos θ)

3.4-3 An Introduction to Periodic Functions

> **PERIODIC FUNCTION**
>
> **A *periodic function* repeats its values at regular intervals. The *period* is the smallest interval over which a periodic function repeats itself.**

Many things in nature occur with repeating regularity: for example, the periodic outbreaks of flu each winter, the periodic ebb and flow of the tides, the periodic menstruation of women, and the periodic cycles of the moon. As you can see, we call such repeating things periodic. In mathematics there are many *functions* that repeat their values regularly, or *periodically*. We call them **periodic functions**, and the *smallest interval* over which they repeat their values we call their **period**.

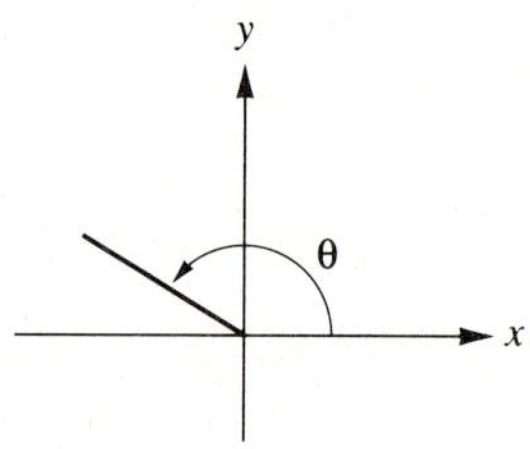

Figure 3.4-3

The trigonometric functions are all periodic functions. To see why, first sketch any arbitrary angle θ, as shown in Figure 3.4-3. Now sketch $\theta + 1$ rev and $\theta - 1$ rev, as shown in Figure 3.4-4. As you can see from these figures, the angles θ, $\theta + 1$ rev, and $\theta - 1$ rev all lie in the same quadrant and all have exactly the same reference angle. Therefore, the trigonometric functions of all three angles must be exactly the same. In fact, if we draw these figures without the arrows, as shown in Figure 3.4-5, you cannot tell if the angle represents θ, $\theta \pm 1$ rev, $\theta \pm 2$ rev, $\theta \pm 3$ rev, and so on. Therefore, the trigonometric functions of all these angles must be the same. This means that the trigonometric functions of *any* angle θ repeat in value every time that θ increases or decreases by 1 rev. Therefore, the trigonometric functions are periodic and repeat every time θ changes by 1 rev (2π or 360°).

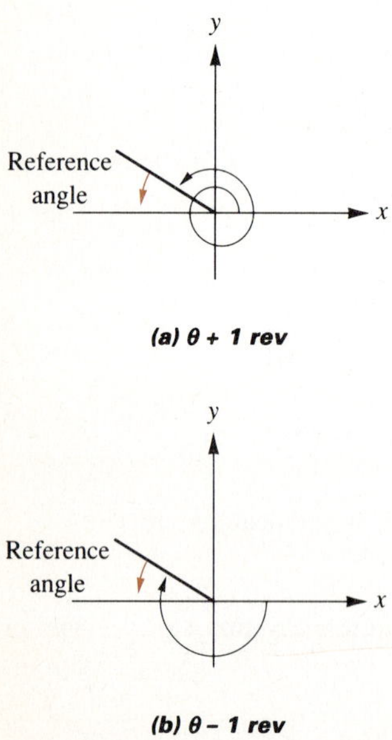

Figure 3.4-4

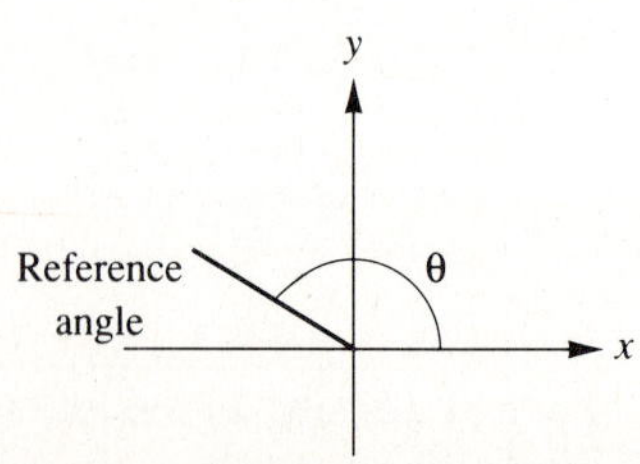

Figure 3.4-5

Periods of the Trigonometric Functions

We have just seen that the trigonometric functions are *periodic* and that they repeat in value every time the angle increases or decreases by 1 rev (2π or 360°). It might seem as if this means that they all have a period of 2π (360°), but remember that the period is the *smallest* interval over which a periodic function repeats itself. The sine, cosine, secant, and cosecant do, in fact, require a full revolution to go through all their possible positive and negative values, and thus they have a *period of 2π or 360°*. The tangent and the cotangent, however, repeat more rapidly. They go through all of their positive and negative values in just 180°, and then they repeat again in the next 180°. Therefore, the tangent and the cotangent have a period of π or 180°. Summarizing these results, we have

PERIODS OF THE TRIGONOMETRIC FUNCTIONS

The sine, cosine, secant, and cosecant have a period of 2π (360°).

$$\sin(\theta \pm 2\pi) = \sin\theta \qquad \cos(\theta \pm 2\pi) = \cos\theta$$

$$\sec(\theta \pm 2\pi) = \sec\theta \qquad \csc(\theta \pm 2\pi) = \csc\theta$$

The tangent and cotangent have a period of π (180°).

$$\tan(\theta \pm \pi) = \tan\theta \qquad \cot(\theta \pm \pi) = \cot\theta$$

Notice that these results are identities because they are true for all allowed values of θ.

▸ **The periodic nature of the trigonometric functions is one of their most characteristic and important properties.**

Example 36

Use the periodicity of the trigonometric functions to find the value of (a) tan 210° and (b) $\sin\frac{9\pi}{4}$.

Solution

(a) tan 210° = tan(180° + 30°). Because the tangent has a period of 180°, it follows that tan(180° + 30°) = tan 30°. Thus

$$\tan 210° = \tan(180° + 30°) = \tan 30° = \sqrt{3}/3$$

(b) $\sin\frac{9\pi}{4} = \sin[(8+1)\frac{\pi}{4}] = \sin(2\pi + \frac{\pi}{4})$. Because the sine has a period of 2π, it must be true that $\sin(2\pi + \frac{\pi}{4}) = \sin\frac{\pi}{4}$. Thus

$$\sin\frac{9\pi}{4} = \sin(2\pi + \frac{\pi}{4}) = \sin\frac{\pi}{4} = \sqrt{2}/2 \ ■$$

NOW IT'S YOUR TURN

Use the periodicity of the trigonometric functions to find cos 405° and cot $\frac{7\pi}{4}$.
(Answer: $\sqrt{2}/2$ and -1)

Graphs of Periodic Functions

A periodic function repeats its values over finite intervals. Therefore, the graph of a periodic function repeats itself over the same finite intervals. The *smallest* such interval is the *period*.

Example 37

Which of the graphs in Figures 3.4-6 and 3.4-7 represents a periodic function, and what is its period?

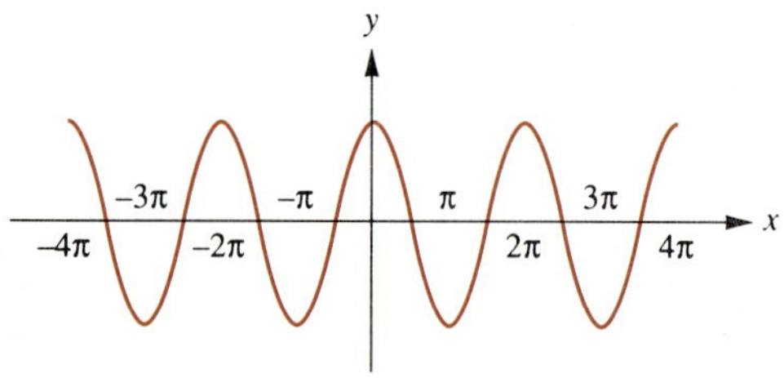

Figure 3.4-6

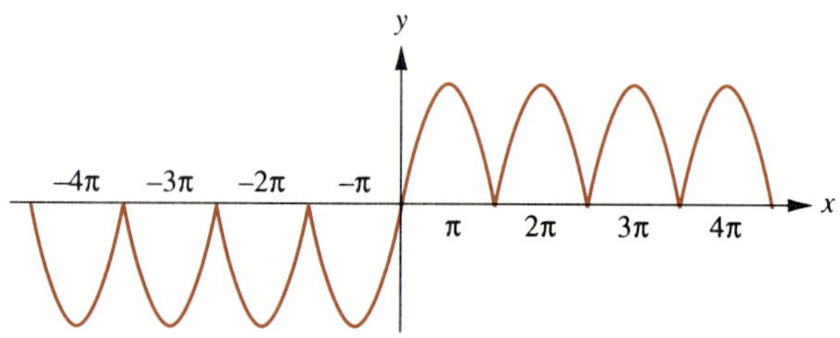

Figure 3.4-7

Solution

As you can see from Figure 3.4-8, this graph repeats itself every time that x changes by 2π. Therefore, Figure 3.4-6 represents a periodic function, and the period is 2π.

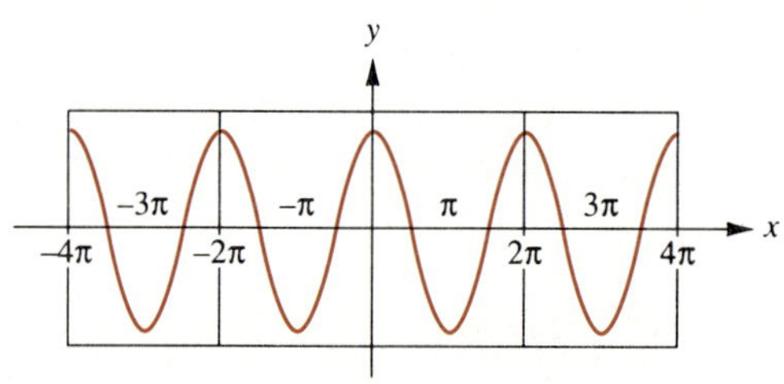

Figure 3.4-8

Although the graph in Figure 3.4-7 repeats its general pattern, it does not repeat its values, because y has positive values when x is positive and negative values when x is negative. Therefore, it does not represent a periodic function.

NOW IT'S YOUR TURN

Which of the graphs in Figure 3.4-9, 3.4-10, and 3.4-11 represent a periodic function? What is the period for those that do?

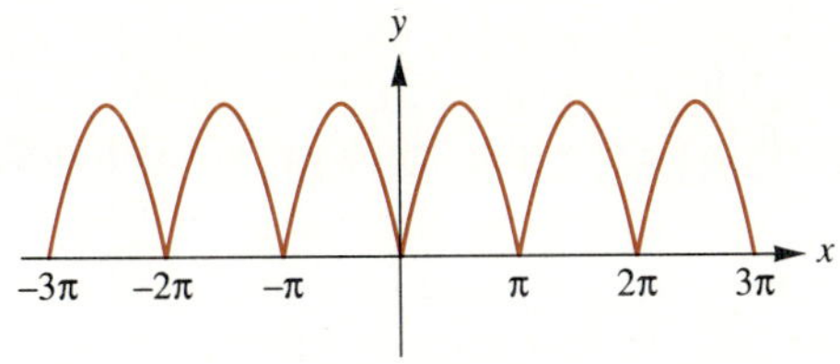

Figure 3.4-9

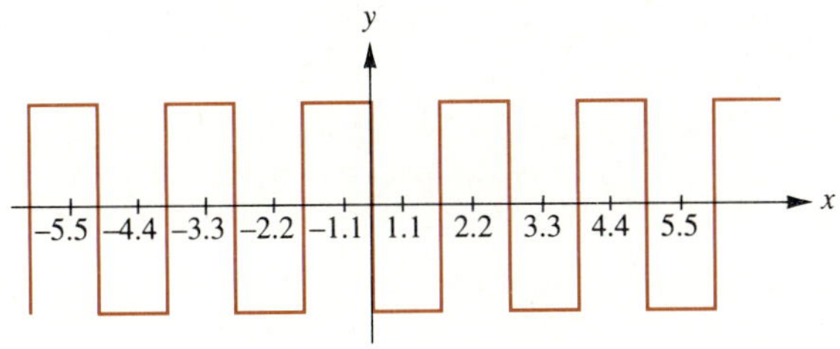

Figure 3.4-10

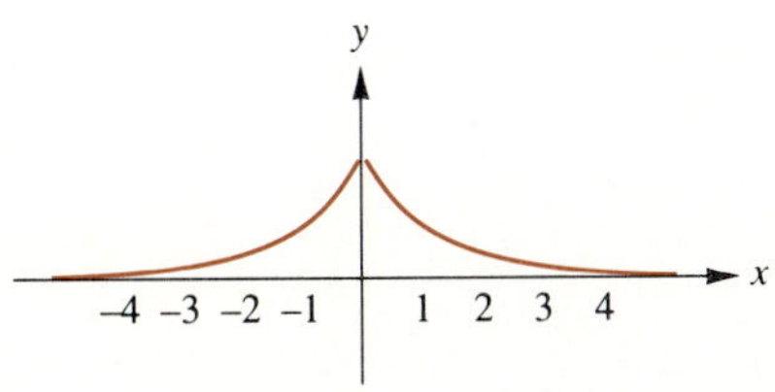

Figure 3.4-11

(Answer: The graph of Figure 3.4-9 is periodic with period π. The graph of Figure 3.4-10 is periodic with period 2.75. The graph of Figure 3.4-11 is not periodic.)

In most applications, the angle in the trigonometric function has a number in front of it, such as $2x$ or 3.5θ, or a number added to it, such as $\theta + 4$. Sometimes the trigonometric function itself may have a number in front of it, such as $5 \sin \theta$. As Example 38 shows, some of these things can affect the period.

Example 38

Find the period of (a) $6 \cos \theta$, (b) $\sin(3\theta)$, and (c) $\cot(\theta + 5)$.

Solution

By the period, we mean the interval of the angle θ for which the function repeats itself.

(a) $6 \cos \theta$: We know that $\cos \theta$ repeats whenever θ changes by 2π (360°). Therefore, $6[\cos(\theta + 2\pi)] = 6 \cos \theta$, which means that the period of $6 \cos \theta$ is the same as that of $\cos \theta$. Therefore, multiplying a trigonometric function by a number does not affect the period of that function.

(b) $\sin(3\theta)$: We know that $\sin \theta$ has a period of 2π (360°). To see the effect of the 3, think of 3θ as a single unit and call it α: $\sin(3\theta) = \sin \alpha$. We know that $\sin \alpha$ has period 2π, so when α changes from 0 to 2π, $\sin \alpha$ repeats itself. This means that when 3θ changes from 0 to 2π, $\sin(3\theta)$ repeats itself. Therefore, $\sin(3\theta)$ will repeat itself when $3\theta = 2\pi$, which gives $\theta = 2\pi/3(120°)$. The new period is $2\pi/3(120°)$. Notice that because θ was *multiplied* by 3, the period was *divided* by 3.

(c) $\cot(\theta + 5)$: Because 5 is a pure number, θ must be a pure number also, so it must be in radians. As in (b), think of $(\theta + 5)$ as a single unit α: $\cot(\theta + 5) = \cot \alpha$. We know that $\cot \alpha$ has a period of π, which means that it repeats its values whenever α changes by π. Therefore,

$$\cot(\alpha + \pi) = \cot \alpha$$

$$\cot(\theta + 5 + \pi) = \cot(\theta + 5) \quad \textbf{(since } \boldsymbol{\alpha = \theta + 5}\textbf{)}$$

This means that the period of $\cot(\theta + 5)$ is π. Notice that adding a number to θ does not affect the period of the trigonometric function of θ. ■

NOW IT'S YOUR TURN

Find the period of (a) $\cos(x + 7)$, (b) $\tan 4x$, and (c) $8 \cot x$.
(Answer: 2π, $\frac{\pi}{4}$, and π)

3.4-4 Maximum and Minimum Values of the Sine and Cosine

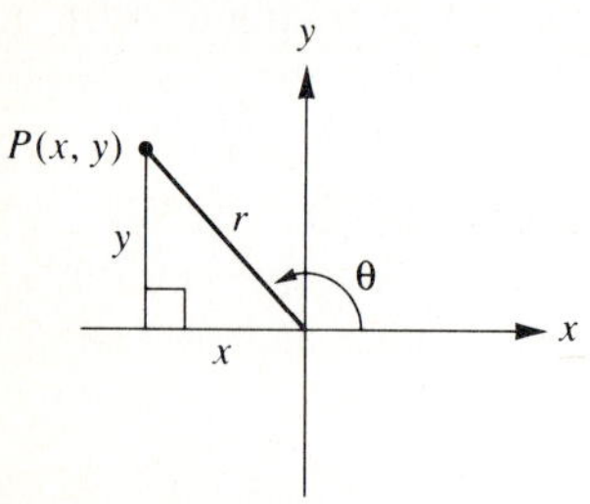

Figure 3.4-12

You may have noticed that in all the examples and exercises you have done, it has never turned out that the sine or the cosine was greater than +1 or less than −1, whereas the other trigonometric functions have all had such values. To see if this has been just a coincidence, enter a number greater than 1 in your calculator and find the inverse sine and cosine of that number. Repeat this for a number less than −1. What do you get? Apparently, the sine and the cosine must be between ±1. To see why, look at Figure 3.4-12 and apply the definition of these two functions.

$$\sin \theta = \frac{y}{r}, \qquad \cos \theta = \frac{x}{r}$$

LIMITS ON THE SINE AND COSINE

$$-1 \le \sin \boldsymbol{\theta} \le +1$$

$$-1 \le \cos \boldsymbol{\theta} \le +1$$

The legs of the reference triangle are x and y, and neither of these can be longer than the hypotenuse r. They can be equal to r only if θ is a quadrantal angle. Keeping in mind that both x and y (but not r) can be both positive and negative, it follows that x and y can be no greater than r but no less than $-r$. Therefore, x/r and y/r can be no greater than +1 and no less than −1. Because $x/r = \cos \theta$ and $y/r = \sin \theta$, $\sin \theta$ and $\cos \theta$ must both lie between +1 and −1 no matter what value θ has.

The values of these functions vary between ±1 as θ ranges between 0° and 360° (0 and 2π). Using their values for the quadrantal angles, we can summarize these results in Table 3.4-1.

Table 3.4-1

When θ increases	$\sin \theta$	$\cos \theta$
from 0° to 90°	increases from 0 to 1	decreases from 1 to 0
from 90° to 180°	decreases from 1 to 0	decreases from 0 to −1
from 180° to 270°	decreases from 0 to −1	increases from −1 to 0
from 270° to 360°	increases from −1 to 0	increases from 0 to 1

We express this graphically in Figure 3.4-13.

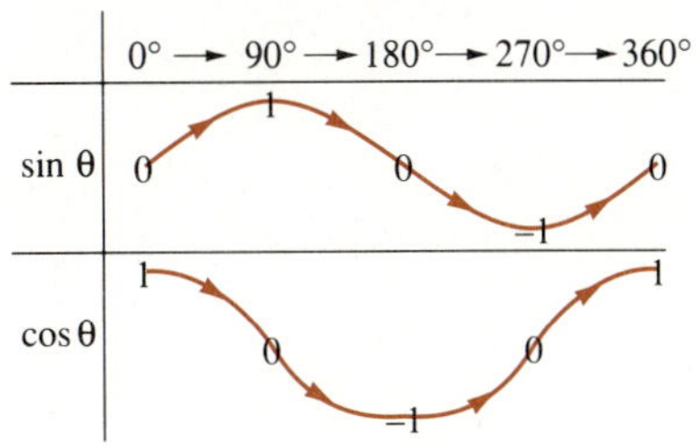

Figure 3.4-13

Example 39

What are the maximum and minimum values of (a) $4.8 \cos \theta$, and (b) $\sin(8.4\theta)$?

Solution

(a) The maximum and minimum values of $\cos \theta$ are ± 1. Therefore, the maximum and minimum values of $4.8 \cos \theta$ are $4.8(\pm 1) = \pm 4.8$.

(b) Think of 8.4θ as a unit and call it α: $\sin(8.4\theta) = \sin \alpha$. No matter what α is, $\sin \alpha$ is between ± 1, so it follows that $\sin(8.4\theta)$ is between ± 1. Therefore, the maximum and minimum values of $\sin(8.4\theta)$ are ± 1. ■

NOW IT'S YOUR TURN

Find the maximum and minimum values of (a) $6.11 \sin \theta$, (b) $\cos 3\theta$, and (c) $5 \sin 9\theta$. (Answer: (a) ± 6.11, (b) ± 1, and (c) ± 5).

EXERCISES 3.4-1

In Exercises 1–8, use the reciprocal identities and the quotient identities to write each expression as a single trigonometric function of an angle.

1. $\sin 45°/\cos 45°$
2. $\cos 3\pi/\sin 3\pi$
3. $1/\sec 10°$
4. $8/\csc 3$
5. $1/\sin 4$
6. $1/\tan 3B$
7. $6/\cot 400°$
8. $2/\cos A$

In Exercises 9–16, find the value of the requested function without finding θ.

9. Find $\cos \theta$ if $\sec \theta = -\frac{4}{3}$.
10. Find $\tan \theta$ if $\cot \theta = \frac{9}{10}$.
11. Find $\csc \theta$ if $\sin \theta = \frac{2}{7}$.
12. Find $\sec \theta$ if $\cos \theta = -0.822$.
13. Find $\cot \theta$ if $\tan \theta = 1.839$.

14. Find $\sin\theta$ if $\csc\theta = \frac{\pi}{2}$.

***15.** Find $\tan(-\theta)$ if $\cot\theta = 8.45$.

***16.** Find $\csc(-\theta)$ if $\sin\theta = -\frac{2}{3}$.

In Exercises 17–24, use the reciprocal identities and the quotient identities to find the other four trigonometric functions without finding the angle, given the following:

17. $\sin\theta = \frac{3}{5}$ and $\cos\theta = \frac{4}{5}$.

18. $\sin\alpha = 1$ and $\cos\alpha = 0$

19. $\sin\beta = 2/\sqrt{5}$ and $\cos\beta = 1/\sqrt{5}$

20. $\sin\delta = 3/\sqrt{13}$ and $\cos\delta = -2/\sqrt{13}$

21. $\sec\theta = -7/3$ and $\csc\theta = 7/\sqrt{40}$

22. $\tan\theta = \frac{1}{3}$ and $\cos\theta = -\sqrt{8}/3$

***23.** $\sin\phi = -0.610$ and $\cos\phi = -0.792$

***24.** $\tan x = 2.47$ and $\sin x = -0.929$

EXERCISES 3.4-2

In Exercises 1–14, find the other five trigonometric functions of θ without finding θ.

1. $\sin\theta = \frac{15}{17}$, θ in quadrant II

2. $\cos\theta = -\frac{4}{5}$, θ in quadrant III

3. $\tan\theta = \frac{4}{3}$, θ in quadrant III

4. $\cot\theta = -\sqrt{3}/3$, θ in quadrant IV

5. $\sin\theta = -0.425$, θ in quadrant III

6. $\tan\theta = -1.84$, θ in quadrant IV

7. $\cot\theta = -4$, θ in quadrant II

8. $\csc\theta = 3.7$, θ in quadrant II

9. $\tan(-\theta) = 0.25$, θ in quadrant IV

10. $\sec\theta = \frac{4}{3}$, $\sin\theta < 0$

11. $\sin\theta = 0.20$, $\tan\theta < 0$

12. $\tan\theta = -4.6$, $\cos\theta < 0$

***13.** $\sin\theta = a/b$, where $0 < a < b$ and θ in quadrant II

***14.** $\tan\theta = a^2/b^2$ and $\cos\theta < 0$

In Exercises 15–24, find all possible values of the other five trigonometric functions of θ, without finding θ, given the following:

15. $\sin\theta = \frac{1}{3}$

16. $\cos\theta = -\frac{2}{3}$

17. $\tan\theta = \frac{5}{2}$

18. $\cot\theta = \frac{4}{7}$

19. $\sec\theta = -\frac{8}{5}$

20. $\sin\theta = -0.62$

21. $\cos\theta = 0.884$

22. $\sec\theta = -2.43$

23. $\cot\theta = 1.9$

24. $\csc\theta = 2.6$

In Exercises 25–33, reduce each expression to an equivalent expression involving only sin *x*.

25. $\cot x \cos x$

26. $\dfrac{\tan x - 1}{1 - \cot x}$

27. $\dfrac{\sec x}{\tan x + \cot x}$

28. $\dfrac{\tan x - \cot x}{\tan x + \cot x}$

29. $\dfrac{\tan x - \sin x}{1 - \cos x}$

30. $\cot^2 x - \cos^2 x$

31. $\tan^2 x + \sec^2 x$

***32.** $\dfrac{\sin x \sec x}{1 + \sec x} + \dfrac{1 + \sec x}{\sin x \sec x}$

***33.** $\dfrac{1 + \tan x}{\tan x\,(\sin x + \cos x)}$

In Exercises 34–48, use the identities in Sections 3.4-1 and 3.4-2 to simplify each expression as much as possible (see Examples 34 and 35).

34. $\sin x\,(\sin x - 1) + \cos^2 x - 1$

35. $\sin x + \cos^2 x - \sin x\,(1 - \sin x)$

36. $\dfrac{1 - \sin^2 B}{\cos B}$

37. $\dfrac{\sin^2 B}{1 + \cos B}$

38. $\dfrac{\cos^2\theta}{1 - \sin^2\theta}$

39. $\cos^2 y + 1 - \sin^2 y$

40. $\cos^2\phi + 1 - \sin^2\phi$

41. $\dfrac{(\cos x + \sin x)^2 - 2\sin x\cos x}{\sin x}$

42. $\dfrac{\sin x\cos x}{(1 - \sin x)\tan x}$

43. $\dfrac{\tan^2 x + 1}{\sec^3 x}$

***44.** $\left(\dfrac{\sin x + \sin x\cos x}{1 - \sin x}\right)\left(\dfrac{\cos^2 x}{1 + \cos x}\csc x\right)$

***45.** $\dfrac{\sin x}{\cot x\,(1 - \cos x) - 1}$

***46.** $\cot x\cos x\sin x + \sin^2 x$

***47.** $\dfrac{\sec x}{\sin x\cot x}$

****48.** $\left[\dfrac{\csc x\,(1 + \tan x)}{\sec x}\csc x - \sec x\right]\sin x$

In Exercises 49–53, evaluate each expression without finding the angle.

***49.** $(\sec^2 x + 1)\cot x$, if $\tan x = 2$

***50.** $(\cot^2 \theta + \tan^2 \theta)(\csc^2 \theta + \sec^2 \theta)$, if $\sec \theta = 2$

***51.** $(1 - \csc^2 y)\tan y$, if $\cot y = 3$

****52.** $(1 - \cot^2 x)\tan x$, if $\sin x = -\frac{1}{3}$ and $\cos \theta < 0$

****53.**

$$\tan x/(\sec x\sqrt{2} + \cot x), \text{ if } \sin x = -\tfrac{3}{4}$$

In Exercises 54–64, which are identities (x, y, r, and θ refer to the quantities in Figure 3.1-3)?

54. $\cot \theta = 1.60$

55. $\cos \theta = x/r$

56. $\cot \theta \sin \theta = 1$

57. $\dfrac{\sin \theta}{\cos \theta} = \tan \theta$

58. $\sin^2 \theta + \cos^2 \theta = 1$

59. $\cot^2 \theta = \csc^2 \theta - 1$

***60.** $\cos \theta = \sin \theta \cot \theta$

61. $1 + \tan^2 \theta = \sec^2 60°$

62. $\cos^2 \theta = 1 - \sin^2 \theta$

63. $\sec \theta \cos \theta = 1$

64. $\sin \theta = \frac{1}{4}$

***65.** Write in terms of $\sin \theta$ and $\cos \theta$ and simplify as much as you can: $\tan(-\theta) + \cot(-\theta)$.

66.** Is $\sin \theta + \cos \theta = 1$ an identity? ***Hint: To prove that something is not true, all you have to do is find one example in which it is false.

***67.** Is $\sin^2(-\theta) + \cos^2(-\theta) = 1$ an identity?

***68.** Use the Pythagorean identity involving the sine and cosine to derive the other two Pythagorean identities.

****69.** If $\sin^2 \alpha = 3\cos^2 \alpha$, find all positive values of $\alpha < 360°$.

****70.** If $\sin \theta = \cos \theta$, find all positive values of $\theta < 2\pi$.

71. If $\cos x = -\sin x$, find all positive values of $x < 360°$.

EXERCISES 3.4-3

In Exercises 1–8, determine the period of the function represented by the given graph if it is periodic.

1.

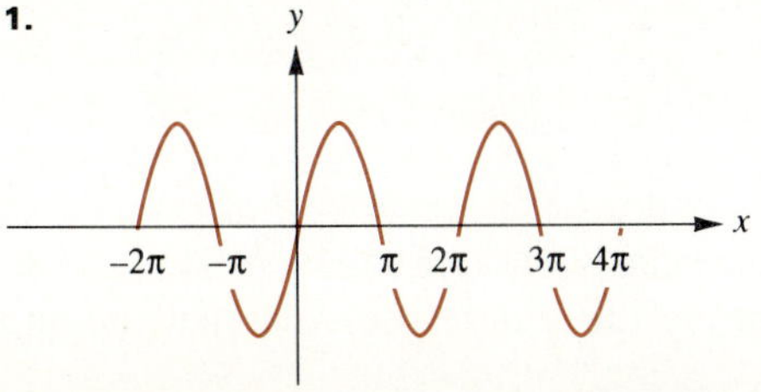

2.

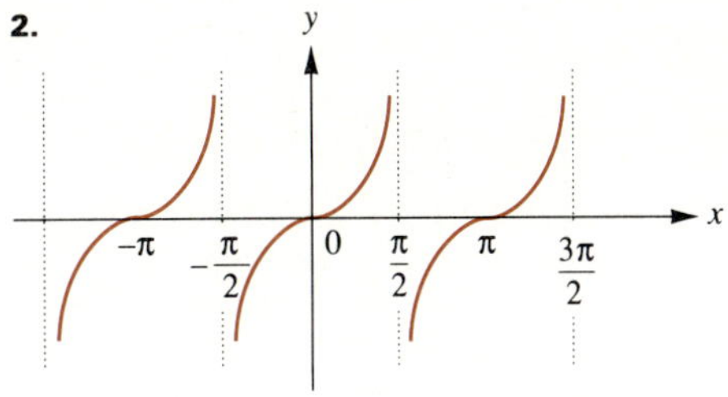

3.

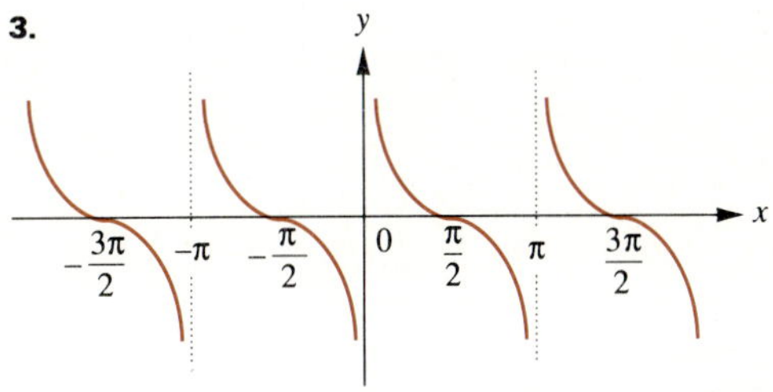

4.

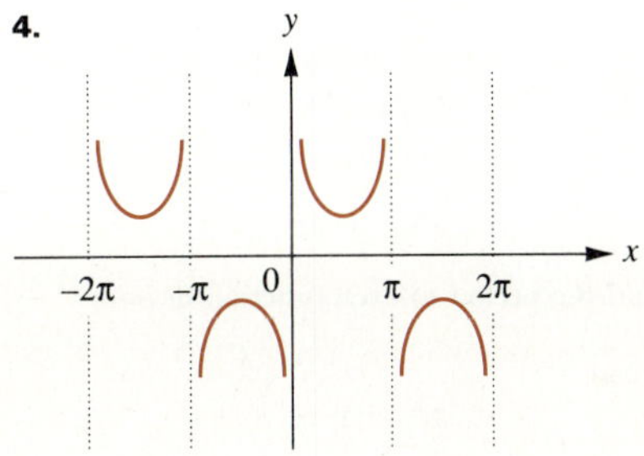

5.

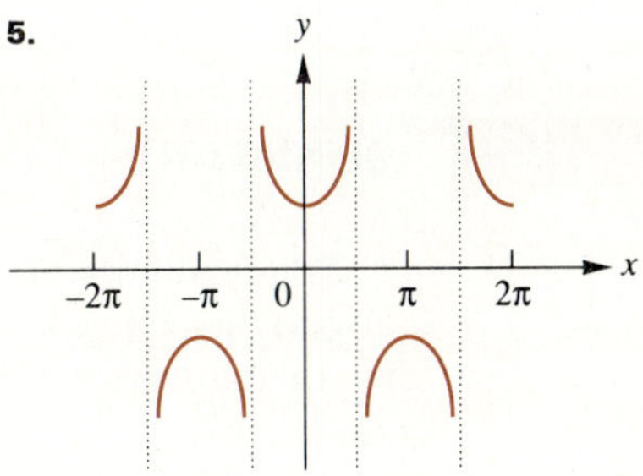

6.

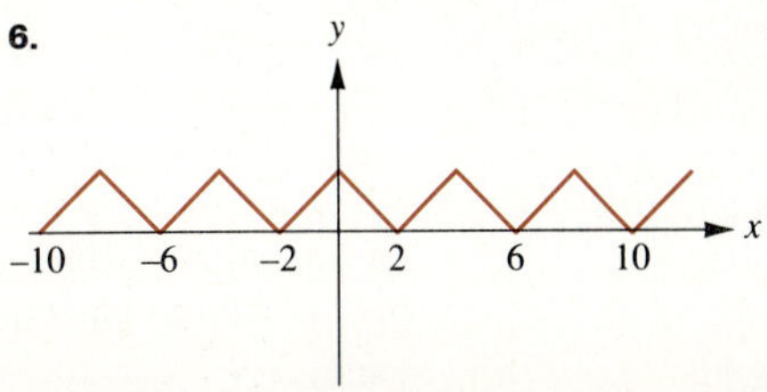

***7.**

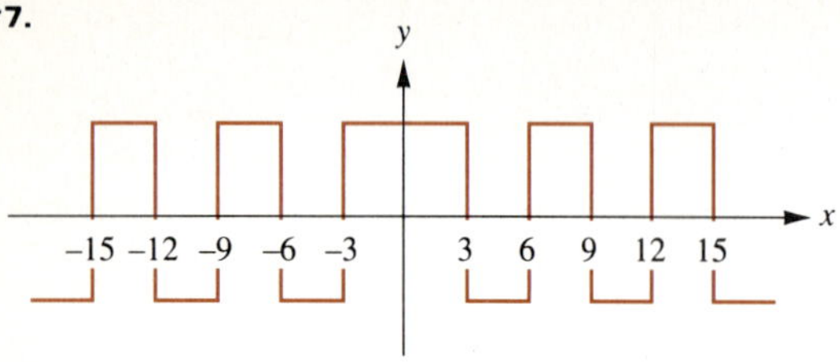

***8.**

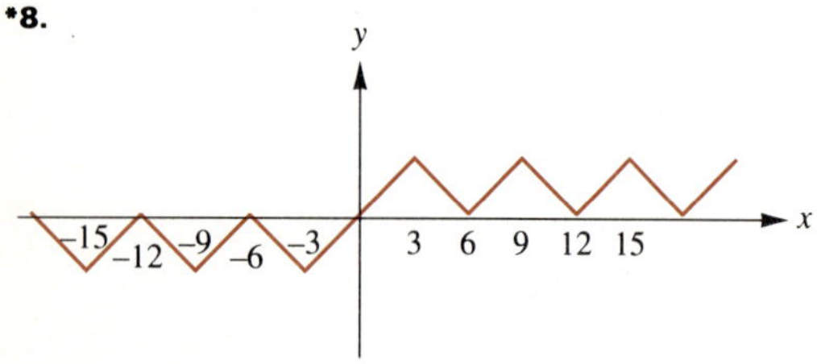

In Exercises 9–18, use the periodicity of the trigonometric functions to find each trigonometric function.

9. $\cos 390°$ **10.** $\sin 450°$

11. $\cot 240°$ **12.** $\sec \frac{9\pi}{4}$

13. $\csc 765°$ **14.** $\cos 810°$

15. $\tan 630°$ **16.** $\sin \frac{13\pi}{4}$

17. $\cos \frac{7\pi}{6}$ **18.** $\cos \frac{13\pi}{2}$

In Exercises 19–30, find the period of each function in both degrees and radians.

19. $\sin 2x$ **20.** $\tan 6x$

21. $\cos(1.8\theta)$ **22.** $\sec(\theta + 12°)$

23. $3 \cos(x + 9.11)$ **24.** $\sin(-x)$

25. $\cos(-3x + 2.4)$ **26.** $6 \cot \theta$

27. $9 \tan(11° - x)$ **28.** $-7 \cos(x + 44°)$

29. $\sec(\theta - 2.1)$ **30.** $\sin[3(x + 25°)]$

****31.** What is the period of $\sin^2 x$? Of $\cos^2 x$?

****32.** What is the period of $\tan^2 \theta$? Of $\cot^2 \theta$?

EXERCISES 3.4-4

In Exercises 1–12, find the maximum and minimum values of y.

1. $y = 63.8 \cos \phi$ **2.** $y = 1.2 \sin \theta$

3. $y = -3 \cos x$ **4.** $y = 6 \cos(2\theta)$

5. $y = 2.7 \sin(x + 35°)$ ***6.** $y = 1.64 \sin(3\alpha - 7.5)$

***7.** $y = \sin^2 x$ ***8.** $y = 8 \cos^2 x$

***9.** $y = \sin^3 x$ ***10.** $y = \sin^4(3\theta)$

***11.** $y = \cos(\theta^2)$ ***12.** $y = -3 \sin^2 \phi$

****13.** Prove that $\sec \theta$ can never be between ± 1.

****14.** Prove that $\csc \theta$ can never be between ± 1.

****15.** Prove that there are no limits to the value of $\tan \theta$.

****16.** Prove that there are no limits to the value of $\cot \theta$.

SUMMARY

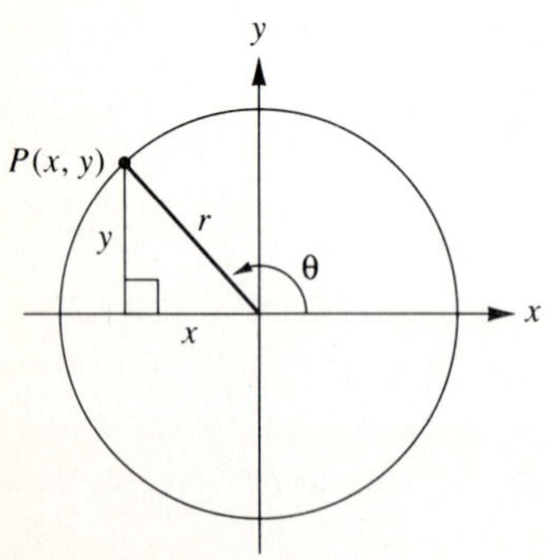

Figure S3-1

1. Trigonometric Functions for All Angles. If $P(x, y)$ is a point (except the origin) on a circle of radius r (Figure S3-1), the radius to P sweeps out an angle θ in standard position. The trigonometric functions of θ are

$$\sin \theta = \frac{y}{r} \qquad \cos \theta = \frac{x}{r} \qquad \tan \theta = \frac{y}{x}$$

$$\csc \theta = \frac{r}{y} \qquad \sec \theta = \frac{r}{x} \qquad \cot \theta = \frac{x}{y}$$

2. The sine, cosine, and tangent of some special angles are shown in Table S3-1. Notice the progression of the integers inside the square roots for the sine and cosine.

Table S3-1

	0°	30°	45°	60°	90°
$\sin\theta$	$\frac{1}{2}\sqrt{0}$	$\frac{1}{2}\sqrt{1}$	$\frac{1}{2}\sqrt{2}$	$\frac{1}{2}\sqrt{3}$	$\frac{1}{2}\sqrt{4}$
$\cos\theta$	$\frac{1}{2}\sqrt{4}$	$\frac{1}{2}\sqrt{3}$	$\frac{1}{2}\sqrt{2}$	$\frac{1}{2}\sqrt{1}$	$\frac{1}{2}\sqrt{0}$
$\tan\theta$	0	$\sqrt{3}/3$	1	$\sqrt{3}$	undefined

3. Signs of the Trigonometric Functions.

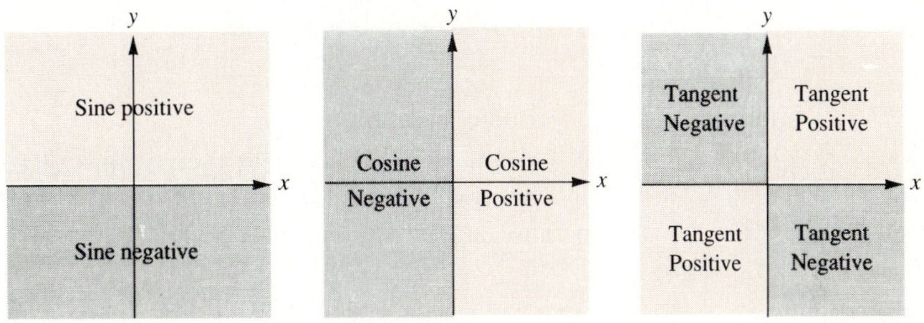

Figure S3-2

The signs of the sine, cosine, and tangent in the four quadrants.

The signs of the other trigonometric functions can be found using the reciprocal identities.

4. Trigonometric Functions of Negative Angles

$$\sin(-\theta) = -\sin\theta \qquad \csc(-\theta) = -\csc\theta$$
$$\cos(-\theta) = \cos\theta \qquad \sec(-\theta) = \sec\theta$$
$$\tan(-\theta) = -\tan\theta \qquad \cot(-\theta) = -\cot\theta$$

5. Reference Angle. For any angle θ in standard position, the reference angle is the positive acute angle between the terminal side of θ and the x axis.

6. Reference Angle Theorem. The trigonometric function of any angle θ is equal to the same trigonometric function of the reference angle of θ except for possible differences in sign. The quadrant in which the terminal side of θ lies determines the sign of the function.

7. Some Important Identities. A trigonometric **identity** is a relationship which is true for all allowed angles.

Reciprocal Identities:

$$\sin\theta = \frac{1}{\csc\theta} \qquad \csc\theta = \frac{1}{\sin\theta}$$
$$\cos\theta = \frac{1}{\sec\theta} \qquad \sec\theta = \frac{1}{\cos\theta}$$
$$\tan\theta = \frac{1}{\cot\theta} \qquad \cot\theta = \frac{1}{\tan\theta}$$

Quotient Identities:

$$\tan \theta = \frac{\sin \theta}{\cos \theta} \qquad \cot \theta = \frac{\cos \theta}{\sin \theta}$$

Pythagorean Identities:

$$\sin^2 \theta + \cos^2 \theta = 1$$
$$1 + \tan^2 \theta = \sec^2 \theta$$
$$1 + \cot^2 \theta = \csc^2 \theta$$

8. Periodic Functions

(a) Periodic functions repeat their values at regular intervals.
(b) The smallest interval over which they repeat their values is called the period.
(c) The tangent and cotangent have period π (180°).

$$\tan(\theta \pm \pi) = \tan \theta \qquad \tan(\theta \pm 180°) = \tan \theta$$
$$\cot(\theta \pm \pi) = \cot \theta \qquad \cot(\theta \pm 180°) = \cot \theta$$

(d) The other four trigonometric functions have period 2π (360°).

$$\sin(\theta \pm 2\pi) = \sin \theta \qquad \sin(\theta \pm 360°) = \sin \theta$$
$$\cos(\theta \pm 2\pi) = \cos \theta \qquad \cos(\theta \pm 360°) = \cos \theta$$
$$\sec(\theta \pm 2\pi) = \sec \theta \qquad \sec(\theta \pm 360°) = \sec \theta$$
$$\csc(\theta \pm 2\pi) = \csc \theta \qquad \csc(\theta \pm 360°) = \csc \theta$$

9. Maximum and Minimum Values of Sine and Cosine. Neither sine nor cosine can be greater than +1 nor less than −1.

$$-1 \leq \sin \theta \leq +1 \qquad \text{and} \qquad -1 \leq \cos \theta \leq +1.$$

IMPORTANT TERMS

identity
period
periodic function
Pythagorean identities
quotient identities
reciprocal
reference angle
reference triangle

REVIEW EXERCISES

1. The angle θ is in standard position and its terminal side passes through the given point. Evaluate all six trigonometric functions of θ, leaving your answers as whole numbers and/or square roots.

(a) $(5, -12)$ **(b)** $(-8, 15)$
(c) $(-\sqrt{2}/2, 0)$ **(d)** $(-\sqrt{3}, -\sqrt{2})$
(e) $(\frac{3\pi}{2}, \frac{\pi}{4})$

2. Evaluate the six trigonometric functions of θ accurate to three decimal places without finding θ.

(a) $\sin\theta = \frac{3}{4}$ and θ lies in quadrant II.
(b) $\cos\theta = -0.50$ and θ lies in quadrant III.
(c) $\cot\theta = -\frac{7}{5}$ and $\sin\theta < 0$.
(d) $\csc\theta = -3.5714$ and $\sec\theta < 0$.
(e) $\cos\theta = \frac{2}{3}$—include all possibilities.
(f) $\cot\theta = -2.00$—include all possibilities.

3. Use a calculator to evaluate each function to four decimal places.

(a) $\cos(-136°17')$ **(b)** $\tan(-342.43°)$
(c) $\cot 25.75°$ **(d)** $\sec 188.6°$
(e) $\sin(\sqrt{7})$ **(f)** $\csc(-41°)$
(g) $\sin(-421°)$ **(h)** $\tan\frac{11\pi}{12}$
(i) $\sec 3.547$ **(j)** $\tan(-5.61)$
(k) $\csc(-1.215)$ **(l)** $\sin 1.2\pi$
(m) $\sin(\cos 3.25)$ **(n)** $\sin(\cos 121°)$
(o) $\cos(\sin 1.1\pi)$ **(p)** $\cos^2(-4.3\pi)$
(q) $\cot^3[-\cos(-228.3°)]$ **(r)** $\sin(\tan 55°)$
(s) $\tan(\sin 55°)$

4. Evaluate the following without a calculator.

(a) $\sin(-\frac{14\pi}{3}) + \cot^2\frac{29\pi}{4} - \tan^2\frac{3\pi}{4} + 1$
(b) $2\cos(-\frac{23\pi}{6}) + \cot\frac{23\pi}{4} - \tan^2\frac{16\pi}{3} - 1$

5. Evaluate the following using a calculator.

(a) $\dfrac{(\cot 44° + \tan 226°)\cos 406°}{\cos 316°} - \cot 72° \cot 18°$

(b) $\dfrac{\cos^2 197° + \cos^2 287°}{1 - \sin^2 217°} - \dfrac{\sin^2 323°}{\cos^2 37°}$

6. Use a calculator to find all possible values of θ less than 1 rev. Express each answer to two decimal places in the units specified.

(a) $\theta = \cos^{-1} 0.86603$ (deg)
(b) $\theta = \sin^{-1}(-0.360)$ (rad)
(c) $\tan\theta = 0.516$ (rad)
(d) $\sec\theta = 0.27778$ (deg)
(e) $\cos 2\theta = -0.39526$ (deg)
(f) $2.7\cot\theta = -3.1$ (deg)
(g) $5\tan\theta = 21.965$ (rad)
(h) $-\frac{1}{2}\cos\theta = 0.456$ (rad)
(i) $3\sin 2\theta = 2.52$ (rad)
(j) $8\cos(\theta + 1.3) = 6.125$ (rad)

7. Use inverse trigonometric functions to express x in terms of y.

(a) $y = \cos 2x$ **(b)** $x = 4$ arccot $(y/2)$
(c) $y = 3\sin 4x$ **(d)** $y = 5\tan(x/2)$
(e) $y = n\cos mx$ **(f)** $y = p\sec nx$

8. Solve for y in terms of x.

(a) $2x = \sin^{-1} y$ **(b)** $x = 4$ arccot $(y/2)$
(c) $x - 2 = 5\tan^{-1}(y/2)$ **(d)** $x + 7 = 5\cos^{-1} 4y$
(e) $3x - 8 = \frac{1}{2}$ arccsc $(y/7)$

9. If $\sec\theta = \frac{13}{12}$ and $\csc\theta = -\frac{13}{5}$, use the reciprocal and quotient identities to find the following functions without finding θ.

(a) $\sin\theta$ **(b)** $\cos\theta$ **(c)** $\cot\theta$

10. If $\cos\theta = 0.2588$ and $\sin\theta < 0$, use trigonometric identities to find the other trigonometric functions of θ without finding θ.

11. Use the identities from this chapter to simplify each expression as much as possible.

(a) $(\sin\theta + \cos\theta)^2 - 2\sin\theta\cos\theta$
(b) $\cos\theta(\sec\theta + \cos\theta\csc^2\theta)$
(c) $\cos\theta(\sec\theta + \tan\theta)$
(d) $(1 - \cos^2\theta)/\cos^2\theta$
(e) $(\sec\theta/\sin\theta) - (\sec\theta/\csc\theta)$
(f) $1/(\sec\theta + \tan\theta) + 1/(\sec\theta - \tan\theta)$

12. Determine the period of each function and the maximum and minimum values of y if appropriate.

(a) $y = -7.8\sin x$ **(b)** $y = \cos 2x$
(c) $y = 2\sin 4x$ **(d)** $y = \cot 2x$
(e) $y = -3\sin(x/2)$ **(f)** $y = -\cos(3x - \pi/2)$
(g) $y = \sec\pi x$ **(h)** $y = 6\csc 5x$

13. Evaluate each function without using a calculator or a table.

(a) $\sin 225°$ **(b)** $\cos 240°$
(c) $\tan\frac{13\pi}{6}$ **(d)** $\cot(-120°)$
(e) $\sec(-\frac{3\pi}{2})$ **(f)** $\cot\pi$
(g) $\sin(-\frac{5\pi}{4})$ **(h)** $\cos(-\frac{25\pi}{6})$
(i) $\tan 1110°$ **(j)** $\sec 120°$
(k) $\cot(\frac{3}{8}$ rev)

14. Find all possible positive $\theta < 2\pi$ to two decimal places without using a calculator or a table.

(a) $\sqrt{2}\sin\theta + 1 = 0$
(b) $\frac{\pi}{6} + \theta = \sin^{-1}\sqrt{3}/2$
(c) $2\cos\theta = -2.50$
(d) $3\theta = \arctan(-1)$
***(e)** $-2\theta = \cos^{-1}(-\frac{1}{2})$
***(f)** $2\sin^2\theta + 4 + 3\sin\theta = 3$

4

Oblique Triangles

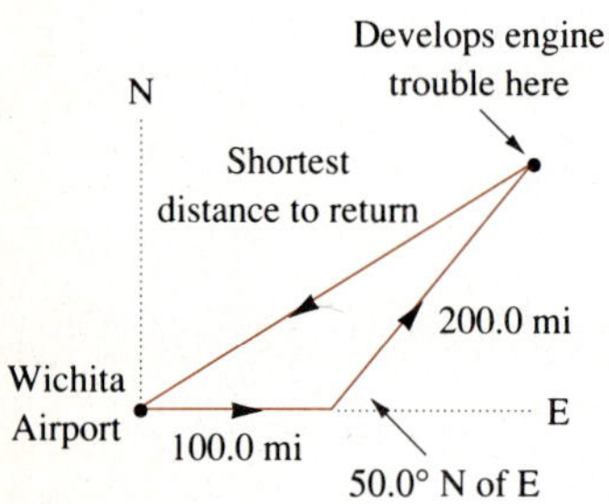

Figure 4.0-1

Path of plane

A plane takes off from the Wichita airport and travels eastward for 100.0 mi. It then changes course and heads 50.0° north of east.* After travelling 200.0 mi in this direction, it suddenly develops engine trouble. The navigator calculates that there is enough fuel left to fly only 285 mi. Can this plane return to the original airport, or must the pilot seek a closer one for an emergency landing? If it can return, in what direction must it head for the shortest return?

Figure 4.0-1 shows a sketch of the plane's path. As you can see, the triangle in this figure does not contain any right angles, so it is an *oblique triangle*. Therefore, to solve this problem, we need to learn how to solve oblique triangles.

We first must explain the notation we use. The six parts of a triangle are its three sides and its three angles, as shown in Figure 4.0-2. We shall call the angles A, B, C, and the lengths of the sides opposite those angles a, b, and c, respectively. That is, the side of length a is opposite angle A, b is opposite angle B, and c is opposite angle C.

Figure 4.0-2

Oblique triangles

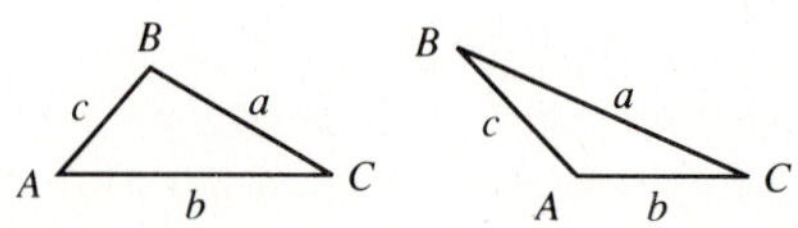

In Chapter 2 we learned how to use the known parts of a right triangle to find the unknown parts. This process is called solving the triangle. In this chapter, we learn how to do the same thing for oblique triangles. To do this, we

---

* The notation 50° north of east (or 50° N of E) means that the direction of the plane's path points 50° away from the east towards the north, as the 200 mi line in Figure 4.0-1. Similarly, 50° E of N would point 50° away from the north towards the east.

need to find relationships between the angles and the sides, just as we did for a right triangle in Chapter 2.

The first of these relationships, which you already know from geometry, is that the sum of the angles in any triangle is 180°: $A + B + C = 180°$. This property enables us to find the third angle of a triangle if we know the other two.

The new relationships relate the sides to the angles; they are the *law of cosines* and the *law of sines*. We derive these relationships in the next two sections.

4.1 LAW OF COSINES

The Pythagorean theorem, which we studied in Chapter 2, relates the lengths of the sides of a right triangle. In this section, we derive a more general law relating the angles and the lengths of the sides of *any* triangle.

To begin, we start with the general oblique triangle in Figure 4.1-1a. Now we construct the altitude h perpendicular to the base (Figure 4.1-1b), which produces a figure having two right triangles, one with sides b, h, m and one with sides a, h, n.

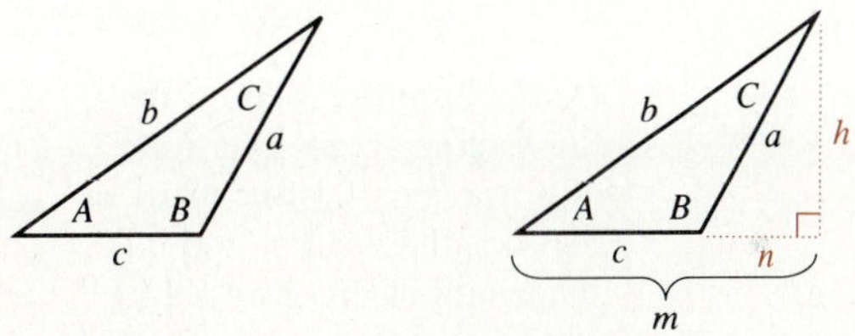

Figure 4.1-1

Law of cosines ***(a) Oblique triangle*** ***(b) Oblique triangle with altitude h***

Applying the Pythagorean theorem to the smaller of these right triangles gives us

$$a^2 = h^2 + n^2 \qquad (1)$$

Using the larger triangle, we find that

$$\sin A = \frac{h}{b} \quad \rightarrow \quad h = b \sin A \qquad (2)$$

and that

$$\cos A = \frac{m}{b} \quad \rightarrow \quad m = b \cos A$$

From this figure, we can see that $m = n + c$, so $n = m - c$. Substituting $m = b \cos A$ into this last result gives us

$$n = m - c = b \cos A - c \tag{3}$$

Now substituting eqs. (2) and (3) into eq. (1) gives us

$$\begin{aligned} a^2 &= h^2 + n^2 = (b \sin A)^2 + (b \cos A - c)^2 \\ &= b^2 \sin^2 A + b^2 \cos^2 A - 2bc \cos A + c^2 \\ &= b^2(\sin^2 A + \cos^2 A) + c^2 - 2bc \cos A \end{aligned}$$

We remember from the Pythagorean identity in Chapter 3 that $\sin^2 A + \cos^2 A = 1$, so our equation becomes

$$a^2 = b^2 + c^2 - 2bc \cos A$$

This result is the **law of cosines**. In this law, b and c are *any* two sides of an oblique triangle, A is the angle *between those sides*, and a is the length of the side opposite A.

LAW OF COSINES

$a^2 = b^2 + c^2 - 2bc \cos A$

In this formula, b and c are the lengths of any two sides of a triangle, A is the angle between those sides, and a is the length of the third side, opposite angle A.

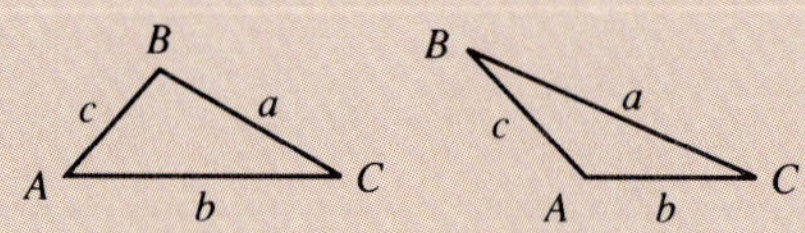

We have proved the law of cosines only for a triangle having an obtuse angle A, but, as the exercises will show, it is true for any triangle.*

The law of cosines is especially useful when we know either two sides and the included angle or three sides of a triangle.

Example 1

Find the third side of the triangle in Figure 4.1-2.

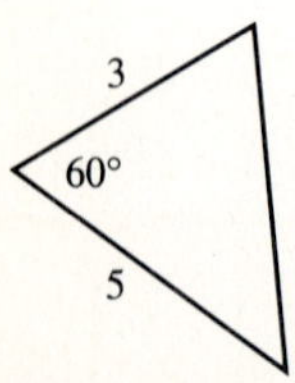

Figure 4.1-2

Solution

We know two sides and the included angle. If we let the unknown side be a, then $b = 3$, $c = 5$, and $A = 60°$. Applying the law of cosines gives us

$$a^2 = b^2 + c^2 - 2\ bc \cos A$$

$$a^2 = 3^2 + 5^2 - 2(3)(5) \cos 60° = 9 + 25 - (30)(\tfrac{1}{2}) = 19$$

---

* The discovery of the law of cosines seems to be due to the French mathematician Francis Vieta in the late 1500s.

Therefore, $a = \pm\sqrt{19}$. Because a represents a length, it cannot be negative, so we use the positive square root: $a = \sqrt{19}$. ■

NOW IT'S YOUR TURN

Find the third side of a triangle having sides of length 3.0 and 6.5 units with a 50.0° angle between these two sides. (Answer: 5.1 units)

Example 2

Find the angle θ in the triangle in Figure 4.1-3.

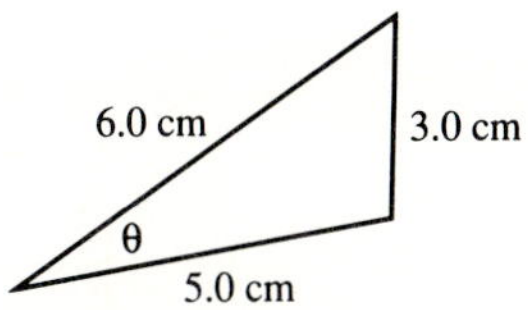

Figure 4.1-3

Solution

Because we know all three sides, we can use the law of cosines to find cos θ. Let $a = 3.0$ cm, $b = 5.0$ cm, and $c = 6.0$ cm. The angle opposite a is θ, so $A = \theta$. The law of cosines gives us

$$a^2 = b^2 + c^2 - 2\ bc \cos A$$

$$(3.0 \text{ cm})^2 = (5.0 \text{ cm})^2 + (6.0 \text{ cm})^2 - 2(5.0 \text{ cm})(6.0 \text{ cm}) \cos \theta$$

$$9.0 \text{ cm}^2 = 25 \text{ cm}^2 + 36 \text{ cm}^2 - 60 \text{ cm}^2 \cos \theta$$

$$-52 \text{ cm}^2 = -60 \text{ cm}^2 \cos \theta$$

$$\cos \theta = \frac{-52 \text{ cm}^2}{-60 \text{ cm}^2} = +0.87$$

Using a calculator gives us $\theta = \cos^{-1}(0.87) = 30°$. ■

NOW IT'S YOUR TURN

For the given triangle, find the angle opposite the 5-unit side.
(Answer: 56°)

Example 3

We can now use the law of cosines to solve part of the introductory problem about the plane with engine trouble. Find the distance back to the airport.

Solution

First, re-draw Figure 4.0-1, as shown in Figure 4.1-4b. In Figure 4.1-4b, we can see that $\theta + 50.0° = 180°$ because there are 180° in a straight angle. Thus $\theta = 180° - 50.0° = 130°$. In the oblique triangle in this figure, we know two sides and the included angle, and we want the third side, so we can use the law of cosines.

$$a^2 = b^2 + c^2 - 2\ bc \cos A$$

In this case, $b = 100.0$ mi, $c = 200.0$ mi, $A = \theta = 130°$, and $a = d$.

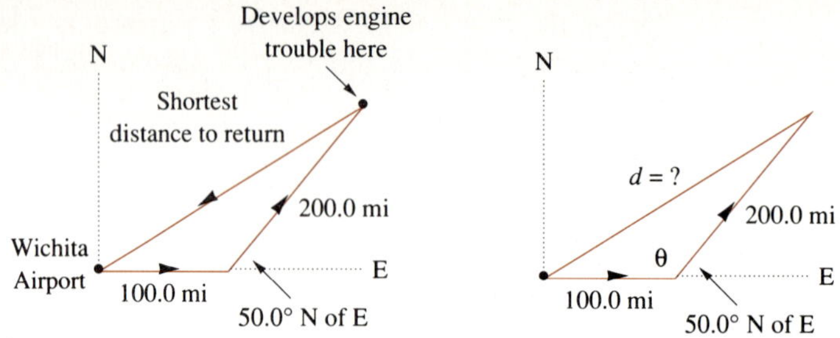

(a) Original figure (b) Simplified sketch

Figure 4.1-4

Plane with engine trouble

$$d^2 = (100.0 \text{ mi})^2 + (200.0 \text{ mi})^2 - 2(100.0 \text{ mi})(200.0 \text{ mi}) \cos 130°$$
$$d^2 = 75{,}700 \text{ mi}^2$$
$$d = \sqrt{75{,}700 \text{ mi}^2} = 275 \text{ mi}$$

Because the pilot has enough fuel left to fly 285 mi, she can just make it back. ■

NOW IT'S YOUR TURN

If the navigator found that by mistake the plane had travelled for 200.0 mi at 35.0° north of east rather than 50.0°, would there be enough fuel for the plane to return to Wichita?

(Answer: Look for another airport! There is only enough fuel for 285 mi and the distance is 288 mi.)

Example 4

Astronomy

The star Vega is 8.1 ps (the parsec—ps—is about 3 light years) from earth and the star Ross 154 is 2.9 ps from earth. As we look at these stars, they are separated by an angle of 62°8′ in the sky. How far are they from each other?

Solution

Sketch these stars as seen from earth (Figure 4.1-5), letting d be the distance between them. We know the two sides and included angle of this oblique triangle. so we can use the law of cosines to find d.

$$a^2 = b^2 + c^2 - 2bc \cos A$$

where $b = 8.1$ ps, $c = 2.9$ ps, $A = 62°8'$, and $a = d$.

$$d^2 = (8.1 \text{ ps})^2 + (2.9 \text{ ps})^2 - 2(8.1 \text{ ps})(2.9 \text{ ps}) \cos(62°8')$$
$$d^2 = 65.6 \text{ ps}^2 + 8.41 \text{ ps}^2 - 22.0 \text{ ps}^2 = 52.0 \text{ ps}^2$$
$$d = \sqrt{52.0 \text{ ps}^2} = 7.2 \text{ ps} \; ■$$

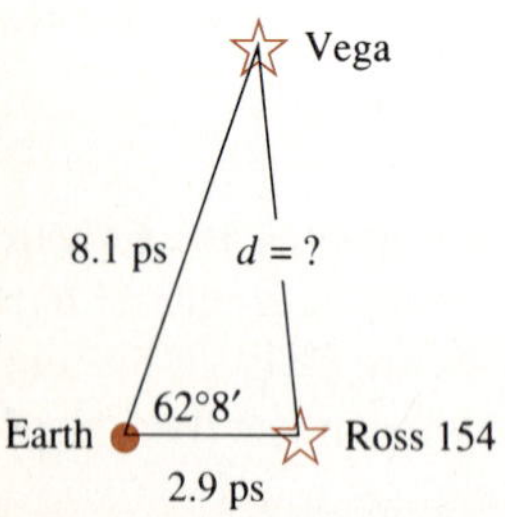

Figure 4.1-5

The stars Vega and Ross 154

NOW IT'S YOUR TURN

If an error in measurement were discovered and if Ross 154 were really 7.4 ps from earth, how far would it be from Vega? (Answer: 8.0 ps)

Example 5

The Scissors Truss (Architecture)

Architects use trusses made of wood or steel to support the roof of a building. The scissors truss shown in Figure 4.1-6 is used to support arched ceilings (such as those found in some churches) by hanging the ceiling from the truss.

In the plans for a certain building, the architect has specified a scissors truss to span a width of 48.0 ft. The top chords, which will have the same length, will rise with a slope of 8 : 12 and be met $\frac{1}{3}$ of the way from the top by the bottom chords. As the carpenter on this job, what length should you cut the lumber for the top chords and the bottom chords?

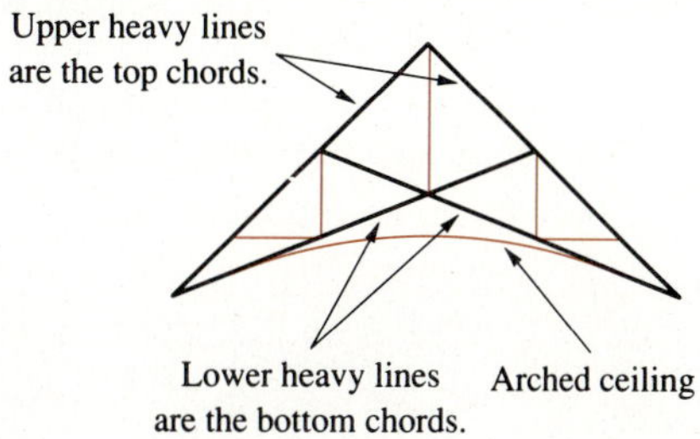

Figure 4.1-6

Scissors truss

Solution

First sketch the truss, simplifying it and showing what you know (Figure 4.1-7). From this figure, we see that if we can find the angle θ, we can use the law of cosines in triangle ABC to find the lengths of the top chords AC and BC. Then we can use the law of cosines again, this time in triangle ACE, to find the lengths of the bottom chords AE and BD.

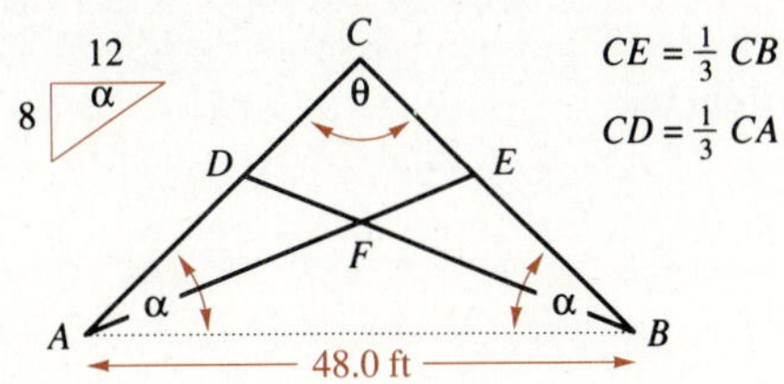

Figure 4.1-7

Diagram of scissors truss

Use the slope to find α.

$$\tan \alpha = \tfrac{8}{12} = 0.667$$

$$\alpha = \arctan 0.667 = 33.7°$$

In triangle ABC, $\theta + 2\alpha = 180°$ because a triangle has 180°. This gives us

$$\theta = 180° - 2\alpha = 180° - 2(33.7°) = 113°$$

Let AC and $BC = x$, and apply the law of cosines to triangle ABC.

$$c^2 = a^2 + b^2 - 2ab \cos C$$
$$(48.0 \text{ ft})^2 = x^2 + x^2 - 2(x)(x) \cos 113°$$
$$2304 \text{ ft}^2 = 2\,x^2 - 2\,x^2 \cos 113°$$

Dividing by 2, factoring out the x^2, and simplifying gives us

$$1152 \text{ ft}^2 = x^2(1 - \cos 113°) = x^2[1 - (-0.391)] = x^2(1.39)$$
$$x^2 = \frac{1152 \text{ ft}^2}{1.39} = 828.8 \text{ ft}^2 \quad \textbf{(solve for } x^2\textbf{)}$$
$$x = \sqrt{828.8 \text{ ft}^2} = 28.8 \text{ ft} \quad \textbf{(take square roots)}$$

Therefore, the top chords should each be 28.8 ft long.

Now apply the law of cosines to triangle ACE. It often helps to simplify your sketch, as in Figure 4.1-8.

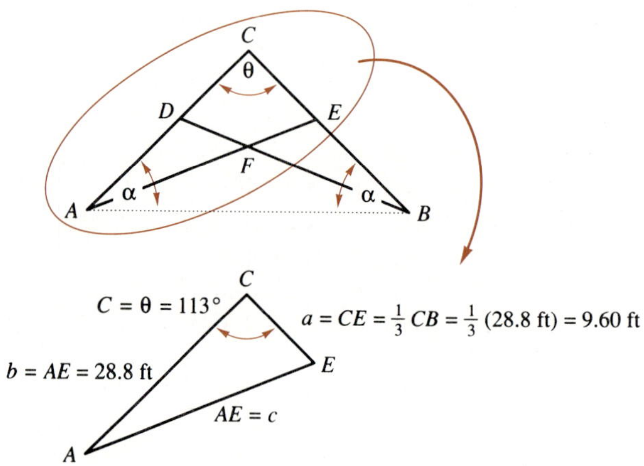

Figure 4.1-8

Isolate triangle *ACE* from the original figure

$$c^2 = a^2 + b^2 - 2ab \cos C$$

where $a = CE = \frac{1}{3}CB = \frac{1}{3}(28.8 \text{ ft}) = 9.60$ ft, $b = AC = 28.8$ ft, and $C = \theta = 113°$. Let c be the length of the bottom chord AE.

$$c^2 = (9.60 \text{ ft})^2 + (28.8 \text{ ft})^2 - 2(9.60 \text{ ft})\,(28.8 \text{ ft})\cos 113°$$
$$c^2 = 92.2 \text{ ft}^2 + 829 \text{ ft}^2 - (553 \text{ ft}^2)\,(-0.391)$$
$$c^2 = 92.2 \text{ ft}^2 + 829 \text{ ft}^2 + 216 \text{ ft}^2 = 1140 \text{ ft}^2$$
$$c = \sqrt{1140 \text{ ft}^2} = 33.8 \text{ ft}$$

The bottom chords must each be 33.8 ft long. ■

NOW IT'S YOUR TURN

If you need a scissors truss to span 60.0 ft having the same slope as this one, what length should you cut for the top and bottom chords?
(Answer: 36.0 ft, 42.2 ft)

Example 6

Organic Molecules (Biochemistry)

Most organic molecules have complex three-dimensional structures. Chemists and biologists need to understand such structures to explain the chemical activity of these molecules.

One of the simplest such molecules is methane, or natural gas (CH_4). The four hydrogen atoms form a regular tetrahedron (a three-dimensional solid figure having four identical faces), with the carbon atom inside (Figure 4.1-9). Measurements reveal that the C—H distances are all 1.091 Å and that the H—C—H angles are all 109°28′. (Remember that the angstrom unit—Å—is 10^{-10} m.) How far apart are the neighboring hydrogen atoms?

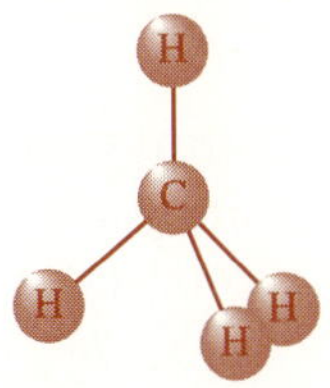

Figure 4.1-9

Methane molecule

Solution

As with most three-dimensional problems, we try to reduce this to a two-dimensional figure. To do this, look at the triangle formed by any two hydrogen atoms and the carbon atom (Figure 4.1-10).

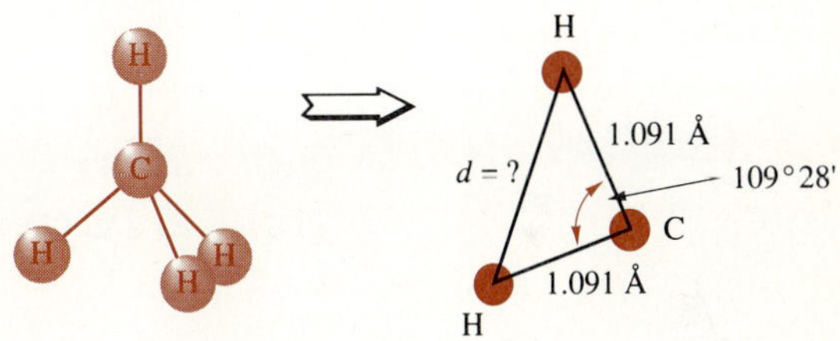

Figure 4.1-10

Reduction of three-dimensional figure to two dimensions

We want to find d in this figure. Applying the law of cosines gives us

$$c^2 = a^2 + b^2 - 2ab\cos C$$

$$d^2 = (1.091 \text{ Å})^2 + (1.091 \text{ Å})^2 - 2(1.091 \text{ Å})(1.091 \text{ Å})\cos(109°28')$$

Collecting terms and converting the angle to decimal form (for use in a calculator) gives us

$$d^2 = 2(1.091 \text{ Å})^2 - 2(1.09 \text{ Å})^2 \cos 109.47°$$

$$d^2 = 2(1.091 \text{ Å})^2(1 - \cos 109.47°) = 2(1.091 \text{ Å})^2(1 + 0.333)$$

$$d^2 = 3.174 \text{ Å}^2$$

$$d = \sqrt{3.174 \text{ A}^2} = 1.782 \text{ Å} \quad \blacksquare$$

NOW IT'S YOUR TURN

Consider the triangle formed by any three hydrogen atoms in methane. Find the angles of that triangle. (Answer: 60°)

Example 7

Interior of the Earth (Geology)

In Example 24 in Chapter 2, we saw how geologists learn that the earth has interior zones or layers. One way that geologists learn about the conditions of these zones is by measuring the speed of earthquake waves as the waves travel through the zones. The greater the density of the material, the faster the waves travel.

An earthquake occurs in El Salvador at 14° north latitude and 88° west longitude; 5.6 min later the first p waves (p for pressure) arrive at a seismic station in Chicago, which is at the same longitude but at 42° north latitude. Knowing that the earth's radius is 6370 km, and assuming that the p waves travel in straight lines at constant speed (which is not quite true), find the speed of these p waves in kilometers per second.

Solution

Sketch the earth, showing both Chicago and El Salvador (Figure 4.1-11). Now simplify this three-dimensional problem by reducing it to a two-dimensional figure. To do this, sketch the circular sector from Figure 4.1-11 connecting Chicago, El Salvador, and the earth's center (Figure 4.1-12).

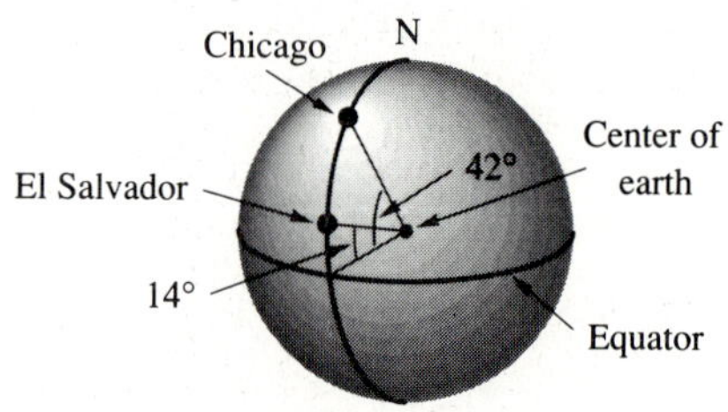

Figure 4.1-11
El Salvador and Chicago

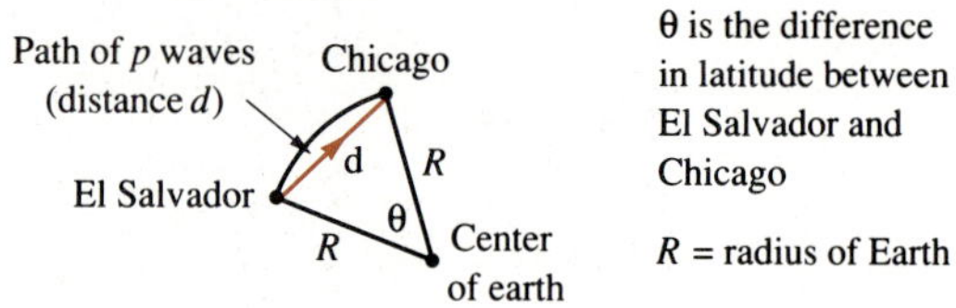

Figure 4.1-12
Reduction of three-dimensional figure to a simplified two-dimensional figure

By the definition of speed from Section 1.5, $v = d/t$. Thus, if we can find d, we can get v because we know that $t = 5.6$ min.

In Figure 4.1-12, θ is the difference in latitude between El Salvador and Chicago, which is $\theta = 42° - 14° = 28°$. Now use the law of cosines in Figure 4.1-12, where $d = a$, $b = c = R$, and $A = \theta = 28°$.

$$a^2 = b^2 + c^2 - 2bc\cos A$$

$$d^2 = R^2 + R^2 - 2RR\cos 28°$$

$$d^2 = 2R^2 - 2R^2\cos 28° = 2R^2(1 - \cos 28°)$$

$$d^2 = 2(6370\text{ km})^2(1 - 0.883) = 9.50 \times 10^6\text{ km}^2$$

$$d = \sqrt{9.50 \times 10^6\text{ km}^2} = 3080\text{ km}$$

Now we can find the speed using $v = d/t$. We want the speed in kilometers per second, so we must convert the 5.8 min to seconds.

$$t = (5.8 \text{ min}) \times \left(\frac{60 \text{ s}}{1 \text{ min}}\right) = 350 \text{ s}$$

Therefore, the speed is

$$v = \frac{d}{t} = \frac{3080 \text{ km}}{350 \text{ s}} = 8.8 \text{ km/s}$$

This speed is nearly twice the speed of p waves in the earth's crust which tells us that the interior is much denser than the crust. ■

NOW IT'S YOUR TURN

If an earthquake occurred in the ocean floor at 88° west longitude and 5° north latitude, how long in minutes would it take the p waves to reach Chicago if they had the speed we just found? (Answer: 7.7 min)

EXERCISES 4.1

In Exercises 1–10, use the following figure to find the indicated length or angle to one decimal place.

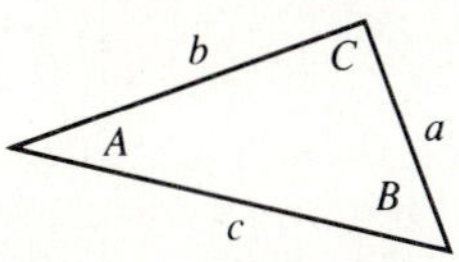

1. $b = 4.0$, $c = 3.0$, $A = 45°$; find a.
2. $C = 120°$, $a = 3.0$ cm, $b = 2.0$ cm; find c.
3. $a = 6.5$ ft, $b = 7.3$ ft, $c = 4.1$ ft; find C in degrees.
4. $a = 3.9$ mm, $b = 7.75$ mm, $c = 5.2$ mm; find B in radians.
5. $a = 10.0$ m, $B = 51.3°$, $c = 8.1$ m; find b.
6. $a = 13$ ft, $b = 8\sqrt{3}$ ft, $C = \frac{\pi}{6}$; find c.
7. $a = 5.0$, $b = 9.8$, $c = 7.3$; find A, B, and C in degrees.
8. $a = 66.2$ in, $b = 58.4$ in, $c = 27.1$ in; find B in radians.
9. $b = 11$, $c = 8$, $A = 2.00$; find a.
10. $b = 0.385$ cm, $c = 0.612$ cm, $A = 118°30'$; find a.
11. If a triangle has sides of length 23 ft, 19 ft, and 18 ft, find its largest angle.
12. If a triangle has sides of length 37 cm, 43 cm, and 55 cm, find the smallest angle.
13. Two adjacent sides of a parallelogram are 6.0 cm and 10.0 cm long, and the angle between them measures $\frac{2\pi}{3}$. Find the lengths of both diagonals.

*14. The diagonals of a parallelogram are 14 in. and 16 in. and they intersect at a smaller angle of $\frac{\pi}{3}$. Find the lengths of both sides of this parallelogram. ***Hint:*** The diagonals of a parallelogram bisect each other.

15. Without using the sine, find the length of the chord of a central angle of 55.8° on a circle 15.6 ft in diameter.

16. A central angle subtends a chord 18.4 mm long on a circle of diameter 40.7 mm. Find this angle without using the sine. Why are there two possible answers?

*17. A central angle subtends an arc length of 23.4 cm on a circle of radius 18.70 cm. Find the length of the chord subtended by this angle without using the sine.

*18. A central angle on a circle of diameter 75.0 m subtends a chord 65.3 m long. Find the arc length this angle subtends without using the sine. Why are there two possible answers?

*19. A central angle in a circle of radius 4.08 m subtends a chord 6.40 m long. Find the area of the smaller wedge cut out by this angle without using the sine.

20. To what famous law does the law of cosines reduce when the included angle is a right angle?

21. An air traffic controller at New York's Kennedy Airport sees two jets on the radar, both at the same altitude. One is 750 m

away from the control tower and the other is 875 m away, with a 7.5° angle between them. How far are these jets from each other?

22. A 22-ft long beam must be supported at an angle of 54° above the horizontal using a cable attached 12 ft above the base of the beam, as shown in Figure 4.1-13. How long must the cable be?

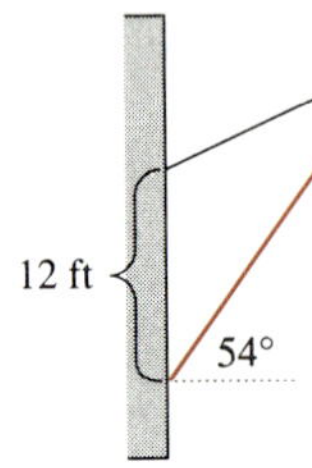

Figure 4.1-13

23. If the cable in Exercise 22 is 19.0 ft long and is attached to the same places as the cable in that problem, at what angle above the horizontal will it support the beam?

24. The plans for a building call for a series of slanted decorative beams on an open ceiling, as shown in Figure 4.1-14. The sides of the ceiling meet at a 150° angle, and each beam is to be nailed 2.50 ft from the peak on one side and 4.75 ft from the peak on the other side. How long should the carpenter cut each beam?

Figure 4.1-14

25. In applying for a patent on your new Super-Comfo chair, you must specify the angle between the back and the seat. It is difficult to measure this angle directly, but you measure the lengths shown in Figure 4.1-15. What is the angle θ?

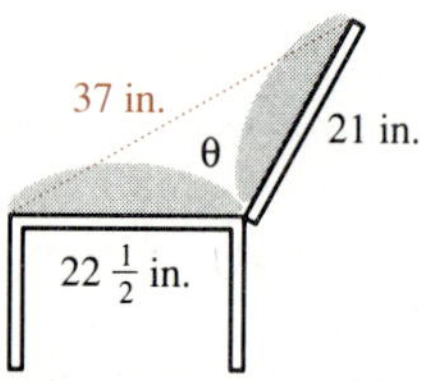

Figure 4.1-15

26. Measurements on a certain lawnmower are shown in Figure 4.1-16. At what angle does the handle slope above the horizontal?

27. Using the previous two problems as a guide, make the appropriate measurements on a chair in your home to calculate the angle that its back slopes above the horizontal.

28. A brace connects the midpoints of the back and the armrest of the chair in Figure 4.1-17. How long is this brace?

29. Make the necessary length measurements to calculate the angle your leg moves through as you take a typical step while walking. Then look at this angle as you take a step (perhaps with the help of a friend) and see if your answer seems reasonable.

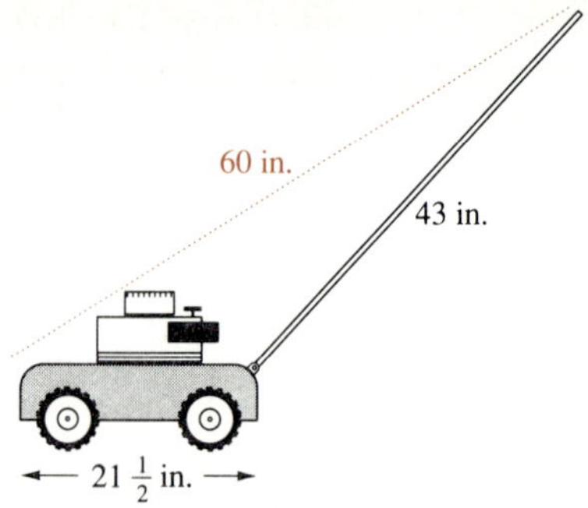

Figure 4.1-16

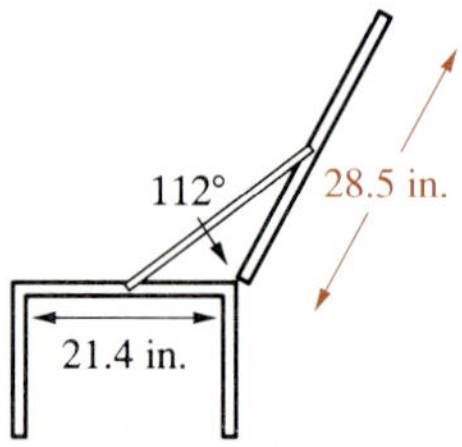

Figure 4.1-17

30. A 500-lb, 24.0-ft long beam is to be supported over the edge of a flat roof at an angle of 32.0° above the horizontal, as shown in Figure 4.1-18. If you attach the cable to a hook 11.3 ft from the base of this beam, how long must the cable be?

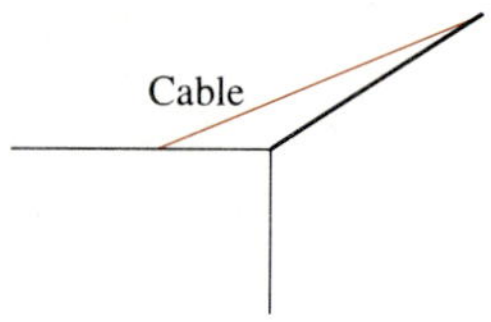

Figure 4.1-18

***31.** A TV antenna is to be mounted on the peak of the roof of the house in Figure 4.1-19. It will be supported by two guy wires attached to the midpoint of each side of the roof and attached to the antenna 12.0 ft above its base. This roof if 32.0 ft wide, and each side rises at an angle of 20.5° above the horizontal. How long must each guy wire be?

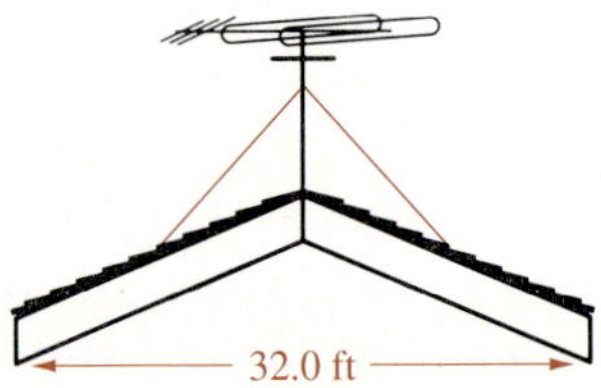

Figure 4.1-19

***32.** A surveyor is trying to determine the width of a river by measuring the distances between three prominent trees along its banks, as shown in Figure 4.1-20. How wide is this river?

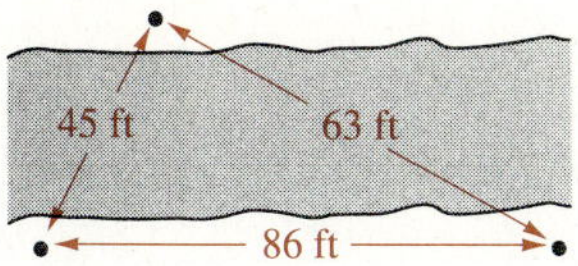

Figure 4.1-20

***33.** The chair in Figure 4.1-21 has gone undusted for far too long. If a spider spins a web from points A to B, what is the minimum length of silk he will need?

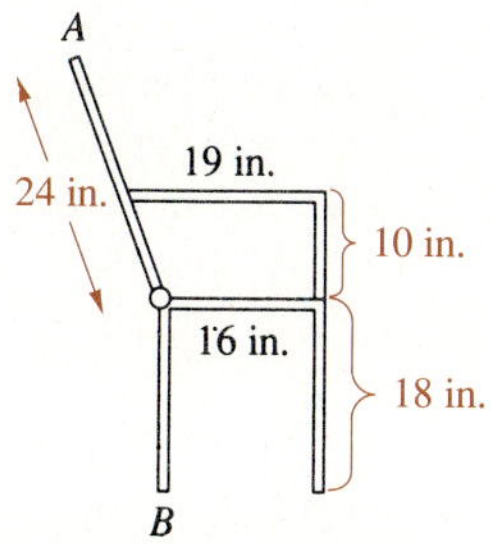

Figure 4.1-21

34. Two jets, a 747 and a DC-10, take off from the Dallas-Fort Worth airport at the same time and fly at the same altitude. The 747 flies at 23.0° north of east at 500.0 mph, whereas the DC-10 flies at 15.6° west of south at 700.0 mph. How far apart are they at the end of 2.00 hr?

***35.** The situation is same as it is in the previous problem, except that the 747 takes off $\frac{1}{2}$ hr after the DC-10. How far apart are they when the DC-10 has flown for 2.00 hr?

***36. Space Colonies.** Advocates of space colonization have suggested that we place populated satellites at two special points in the moon's orbit, called the Lagrangian points L_4 and L_5. (See *The High Frontier* by Gerard O'Neil, Bantam Books, New York, 1978 or *Colonies in Space* by T. A. Heppenheimer, Warner Books, New York, 1977.) These points lie in the moon's orbit at 60° in front of the moon and 60° behind it. If someone in L_4 sends a radiowave message to L_5, how far must it travel and how long will it take to get there, given that radiowaves travel at the speed of light?

37. The tone arm on the author's turntable is 8.5 in. long and pivots about a point at its end (see Figure 4.1-22). Through what angle does this tone arm pivot while playing an LP record for which the outer radius is 6.0 in. and the inner radius of the playing area is 2.5 in.?

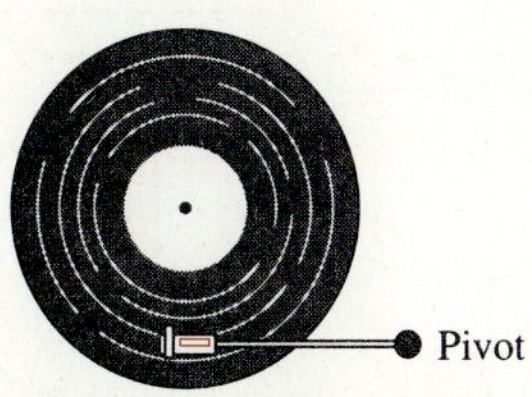

Figure 4.1-22

38. In the nitrosyl bromide molecule (NOBr) in Figure 4.1-23, the O—N distance is 1.15 Å, the N—Br distance is 2.14 Å, and the Br—N—O angle is 117°. What is the distance between the Br and O atoms?

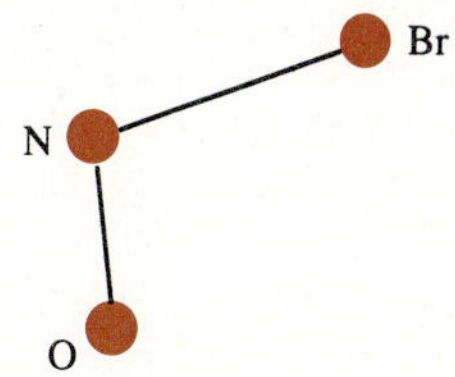

Figure 4.1-23

39. The nitric acid (HNO_3) molecule shown in Figure 4.1-24 has a symmetric structure. Find the distance between **(a)** the two right-hand O atoms, **(b)** the O atoms farthest from each other and **(c)** the H atom and the most distant O atom.

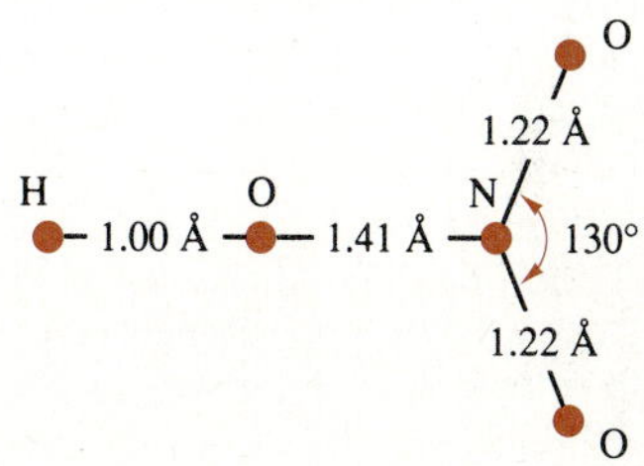

Figure 4.1-24

40. The methanol molecule, CH_3OH, which is very poisonous wood alcohol and may be able to replace gasoline in cars, has the structure shown in Figure 4.1-25. The C—O—H′ angle is 108.9°. How far apart are the C and H′ atoms?

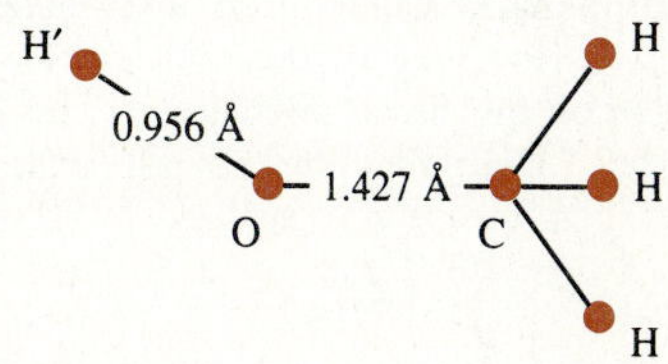

Figure 4.1-25

41. The dimethyl ether molecule is shown in Figure 4.1-26. Each CH_3 group is 1.43 Å from the O atom. How far apart are the CH_3 groups from each other?

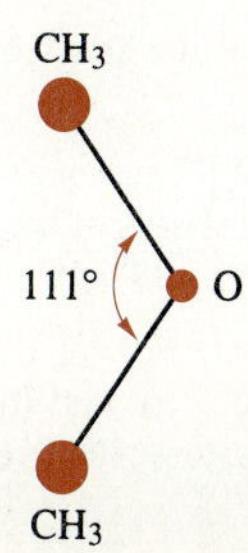

Figure 4.1-26

****42.** In the ammonia molecule (NH_3), the three hydrogen atoms form a plane and the nitrogen atom is outside of that plane. The N—H distance is 1.008 Å and the H—N—H angle is 107.3°.

(a) Carefully sketch this molecule showing its three dimensional structure.
(b) How far apart are the H atoms from each other?
(c) How far is the N atom from the plane formed by the three H atoms?

***43.** Carbon tetrachloride (CCl_4) has the same structure as methane, with the carbon inside the regular tetrahedron (see Example 6). The C—Cl distance is 1.766 Å and the Cl—Cl distance is 2.887 Å.

(a) Carefully sketch this molecule to show its three-dimensional structure.
(b) Find the Cl—Cl—C angle and the Cl—C—Cl angle.

44. A scissors truss is to support an arched roof over walls 18.0 m wide; it will rise at a 40.0° angle above the horizontal. As in Example 5, the bottom chords will meet the top chords $\frac{1}{3}$ of the way from the top. How long must its top chords and bottom chords be?

45. If an earthquake occurs at Santiago, Chile, at a south latitude of 33°, how long does it take the earthquake p waves to reach Santo Domingo, Dominican Republic, at the same longitude but at a north latitude of 18°20′, if they travel at the speed found in Example 7?

***46.** Suppose an earthquake measuring 8.3 on the Richter scale occurs in Los Angeles (34° north latitude and 118°15′ west longitude). How long does it take the p waves, travelling at the speed found in Example 7, to reach a seismic station in Columbia, South Carolina (34° north latitude and 81° west longitude)?

47.** Two lines of equal length start at the same point on the rim of a circle of radius 42 cm and form a 40° angle with each other (Figure 4.1-27). How long are these lines, and what arc length do they intercept on the circle? (Be careful. The 40° angle is not a central angle!) ***Hint: Find the central angle this arc length subtends.

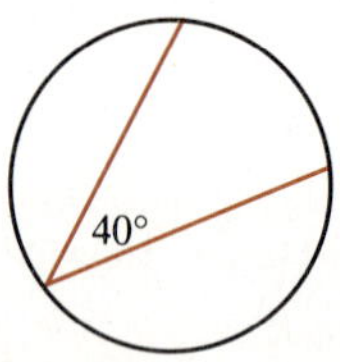

Figure 4.1-27

***48.** Find the area bounded by the two lines and the circle in the Exercise 47.

49.** Two circles, each of radius 64.0 cm, have their centers 100.0 cm apart (Figure 4.1-28). Find the length of the chord Z connecting the points of intersection. ***Hint: First find the angle subtended by this chord.

***50.** Two circles, one of radius 80.0 cm and the other of radius 50.0 cm, intersect with their centers 100.0 cm apart (Figure 4.1-29). Find the length of the chord connecting the points of intersection and the central angle subtended by the chord in each circle.

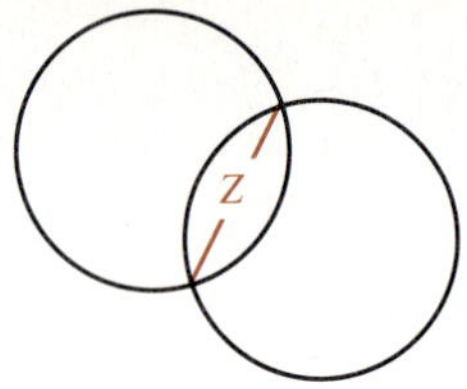

Figure 4.1-28

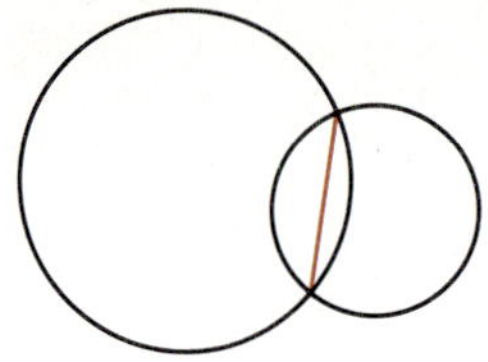

Figure 4.1-29

****51.** Two circles, each of radius 50.0 ft, have their centers 80.0 ft apart (Figure 4.1-30). Find the area common to both circles.

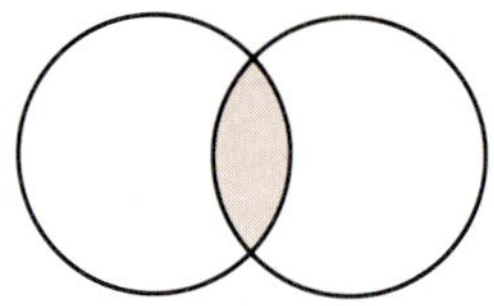

Figure 4.1-30

****52.** The situation is the same as it is in the previous problem, except one circle has a radius of 40.0 ft. Find the area common to both circles.

***53.** Two circular garden paths of diameters 10.0 ft and 20.0 ft are tangent to each other (Figure 4.1-31). As you walk along the larger path, where will you be equidistant from the center of each path?

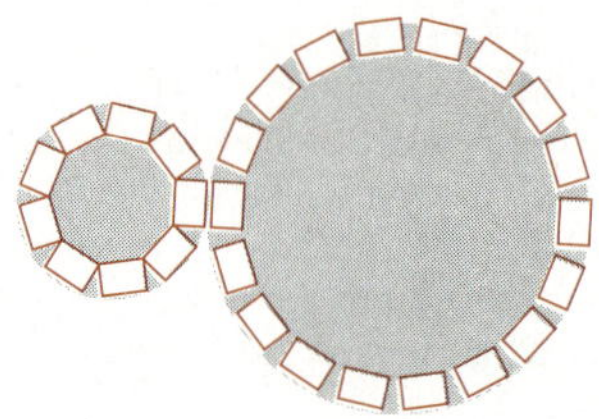

Figure 4.1-31

***54.** For the cube shown in Figure 4.1-32, E is the midpoint of face $BDFG$. Find the angle between AE and AB and between AE and AD.

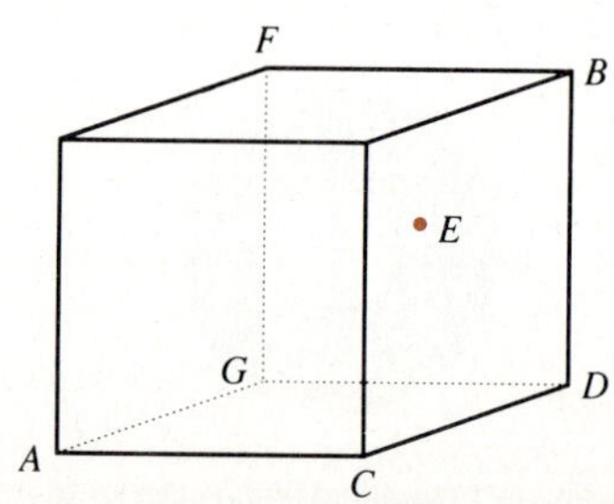

Figure 4.1-32

55. You are seated at a football game on the 40-yd line, ready to take a picture of the kickoff. Your seats are 45.0 ft up in the stands which rise at 25° and begin 10.0 yd from the edge of the field (Figure 4.1-33). In college football, the kickoff occurs at the center of the 40-yard line, and the field is 160.0 ft wide. At what distance should you prefocus your camera to be ready to snap the picture at the instant of kickoff?

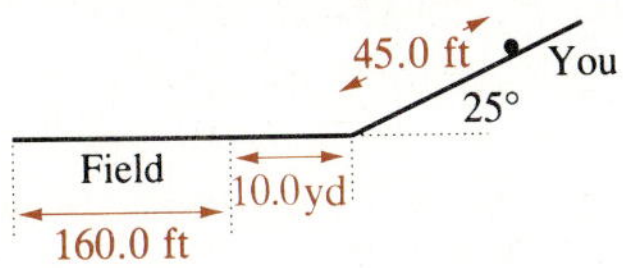

Figure 4.1-33

***56.** Suppose that the measurements are the same as in Exercise 55 except that your seats are at the 50 yd line. At what distance should you prefocus your camera to be ready for the kickoff?

4.2 THE LAW OF SINES

In Section 4.1, we saw that the law of cosines relates the sides and angles of an oblique triangle. However, this law by itself is not always sufficient to solve every oblique triangle. For example, if we knew the angles A and B and the length of the included side c in Figure 4.2-1, we could not find the lengths a and b using the law of cosines. Another law that is helpful in such cases is the law of sines. This law also relates the sides and angles of an oblique triangle.

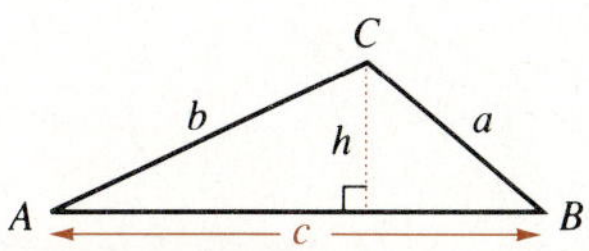

Figure 4.2-1

Oblique triangle

As you recall from plane geometry, the area of any triangle is equal to the product of $\frac{1}{2}$ its base times its altitude. For the triangle in Figure 4.2-1, the base is c and the altitude is h, so its area is *area* $= \frac{1}{2}ch$. But, from the left-hand right triangle in this figure, we can see that $\sin A = h/b$, so $h = b \sin A$. Therefore, the area of the oblique triangle is area $= \frac{1}{2}ch = \frac{1}{2}c(b \sin A) = \frac{1}{2}bc \sin A$.

$$\text{area} = \tfrac{1}{2}bc \sin A$$

Notice that A is the angle between b and c.

In the same way, we can treat b as the base and construct the altitude perpendicular to b to obtain

$$\text{area} = \tfrac{1}{2}ab \sin C$$

or we can treat a as the base and obtain

$$\text{area} = \tfrac{1}{2}ac \sin B$$

Remember that each of these three expressions represents the area of the same oblique triangle. Consequently, they must be equal; that is,

$$\tfrac{1}{2}bc \sin A = \tfrac{1}{2}ac \sin B = \tfrac{1}{2}ab \sin C$$

Dividing each term in this expression by $\frac{1}{2}abc$ we get

$$\frac{\frac{1}{2}bc \sin A}{\frac{1}{2}abc} = \frac{\frac{1}{2}ac \sin B}{\frac{1}{2}abc} = \frac{\frac{1}{2}ab \sin C}{\frac{1}{2}abc}$$

Cancelling the factors common to the numerators and denominators, we get

$$\frac{\sin A}{a} = \frac{\sin B}{b} = \frac{\sin C}{c}$$

This set of equations is the **law of sines**. Because there was nothing special about the triangle we used, this law applies to all triangles.*

LAW OF SINES

$$\frac{\sin A}{a} = \frac{\sin B}{b} = \frac{\sin C}{c}$$

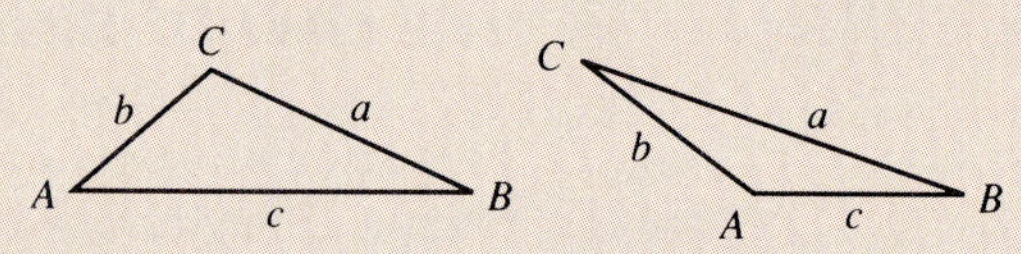

We usually write the law of sines in this compact form because it is easy to remember, but it is important to realize that this form really stands for *three separate equations*.

$$\frac{\sin A}{a} = \frac{\sin B}{b} = \frac{\sin C}{c}$$

When you use the law of sines, you must always use it in one of these three forms, never with the three terms together.

▸ **Notice that the denominator is always the length of the side opposite the angle in the numerator.**

Example 8

For the triangle in Figure 4.2-2a, find all the unknown sides and angles.

Solution

First redraw the given triangle and introduce symbols for the unknown sides and angles (Figure 4.2-2b). Because the sum of the angles in any triangle is 180°, we have

$$A + 30.0^\circ + 45.0^\circ = 180^\circ \quad \rightarrow \quad A = 105^\circ$$

* The first statement and proof of the law of sines appears to have been given in a treatise on trigonometry by the Persian Nasir-Eddin in the 1200s. He was the first person known to treat trigonometry as a subject separate from astronomy.

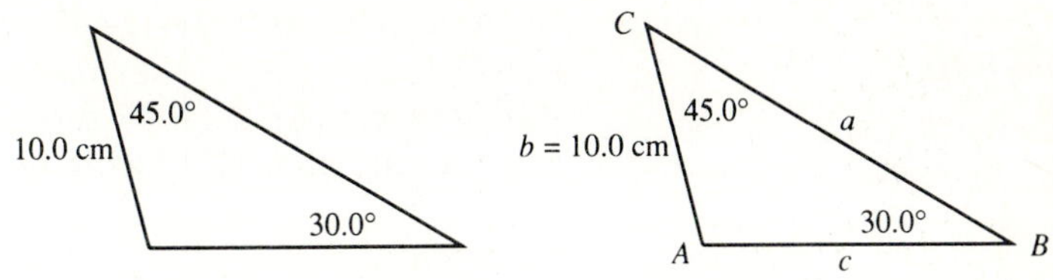

Figure 4.2-2 *(a) The given triangle* *(b) The triangle with symbols introduced*

In Figure 4.2-2b, we know all three angles and one side, so we can use the law of sines to find the other two sides. If we start with sides a and c, we have two unknowns (a and c), but if we start with sides b and a, we have only one unknown (a).

$$\frac{\sin B}{b} = \frac{\sin A}{a}$$

$$\frac{\sin 30^\circ}{10.0 \text{ cm}} = \frac{\sin 105^\circ}{a}$$

$$a \sin 30^\circ = (10.0 \text{ cm}) \sin 105^\circ$$

$$a = \frac{(10.0 \text{ cm})(\sin 105^\circ)}{\sin 30^\circ}$$

$$a = \frac{(10.0 \text{ cm})(0.966)}{0.500} = 19.3 \text{ cm}$$

Applying the law of sines again, this time using sides b and c, we get

$$\frac{\sin B}{b} = \frac{\sin C}{c}$$

$$\frac{\sin 30^\circ}{10.0 \text{ cm}} = \frac{\sin 45^\circ}{c}$$

$$c = \frac{(10.0 \text{ cm})(\sin 45^\circ)}{\sin 30^\circ} = \frac{(10.0 \text{ cm})(\sqrt{2}/2)}{\frac{1}{2}} = 14.1 \text{ cm} \quad ■$$

NOW IT'S YOUR TURN

For Figure 4.2-2b, if $C = 52.0^\circ$, $b = 43.0$ cm, and $c = 65.0$ cm, find the other angles and side. (Answer: 96.6°, 31.4°, 81.9 cm)

Example 9

Now we can finish solving the navigation problem of the plane with engine trouble that we used to introduce this chapter. In Example 3, we found that the plane has enough fuel to fly the 275 mi to the Wichita airport, but we still need to find the direction in which it must head.

Solution

First sketch the plane's path (Figure 4.2-3). In this figure, the angle at which the plane must head, in a direction south of west, is shown as θ. But, θ and A are alternate interior angles of two parallel lines, so they must be equal. Thus, we need to find the angle A.

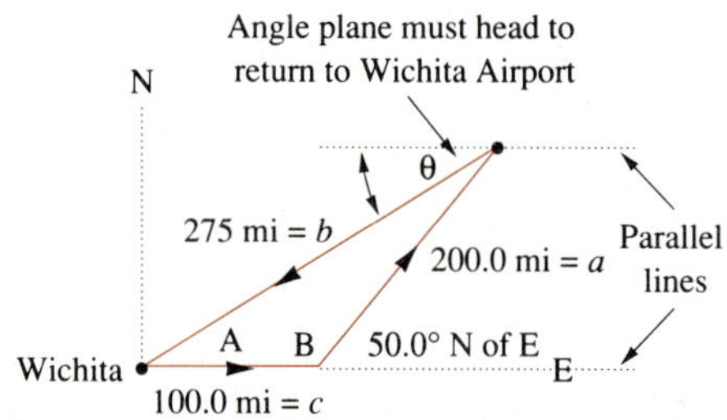

Figure 4.2-3

Path of the plane returning to Wichita

There are 180° in a straight angle, so $B + 50° = 180°$, which means that $B = 130°$. Now use the law of sines to find A.

$$\frac{\sin A}{a} = \frac{\sin B}{b}$$

$$\frac{\sin A}{200.0 \text{ mi}} = \frac{\sin 130°}{275 \text{ mi}}$$

$$\sin A = \frac{(200.0 \text{ mi})(\sin 130°)}{275 \text{ mi}} = 0.5571$$

$$A = \sin^{-1}(0.5571) = 33.9°$$

This direction is 33.9° away from the west toward the south, so the pilot must head 33.9° south of west to return to the Wichita Airport. ■

NOW IT'S YOUR TURN

Use the law of sines to find the third angle in the oblique triangle of Figure 4.2-3. (Answer: 16.2°)

Example 10
(Geometry)

In a certain isosceles triangle, the angle between the two equal sides is 40.0° and the side opposite that angle is 50.0 cm. Find the lengths of the two equal sides and the area of the triangle.

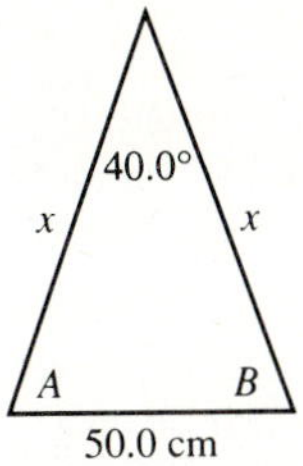

Figure 4.2-4

Isosceles triangle

Solution

First sketch the triangle, showing what you know and what you want to find (Figure 4.2-4). Call x the length of the unknown sides.

Because a triangle has 180°, $A + B + 40° = 180°$. But, because the triangle is isosceles, $A = B$. Therefore,

$$A + A + 40° = 180°$$
$$2A = 140°$$
$$A = 70.0°$$

Now we can apply the law of sines to find x.

$$\frac{\sin A}{x} = \frac{\sin 40.0°}{50.0 \text{ cm}}$$

Using $A = 70.0°$, we get

$$x = \frac{(50.0 \text{ cm})(\sin 70.0°)}{\sin 40.0°} = 73.1 \text{ cm}$$

To find the area of this triangle, we can use the area formulas from which we derived the law of sines at the beginning of this section.

$$\text{area} = \tfrac{1}{2}bc \sin A$$

where b and c are the lengths of any two sides and A is the angle between those sides. Although we can use any two sides, for illustration we use angle A, which is 70°, in Figure 4.2-4. Therefore, $b = 50.0$ cm and $c = x = 73.1$ cm. This gives us

$$\text{area} = \tfrac{1}{2}bc \sin A = \tfrac{1}{2}(50.0 \text{ cm})(73.1 \text{ cm}) \sin 70.0° = 1720 \text{ cm}^2 \ ■$$

NOW IT'S YOUR TURN

If an isosceles triangle has a 50.0° angle between the equal sides, and if the side opposite that angle is 60.0 cm long, find the lengths of the other two sides and the area of the triangle. (Answer: 71.0 cm, 1930 cm^2)

Example 11

Rock Art Preservation (Archaeology)

A ranger for the National Park Service in Canyonlands National Park, Utah, must erect a horizontal shade to protect some prehistoric rock art from the noon sun. This art is on a rock face 9.60 ft long sloping inward from the vertical cliff at 20.0°, as shown in Figure 4.2-5. At the latitude of the park, the noon sun varies between 28.0° (on December 21) and 76.0° (on June 21) above the horizon over the course of a year. What should be the minimum length of this shade so that it will protect the rock art at noon throughout the year?

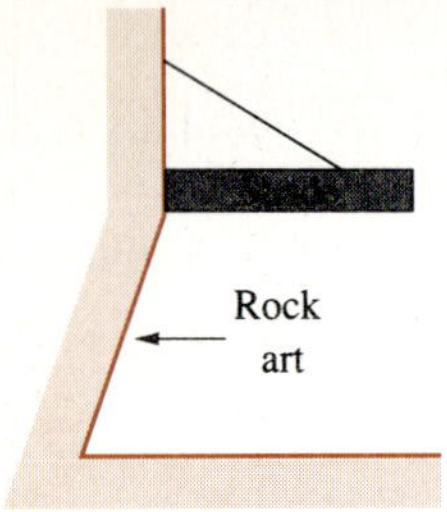

Figure 4.2-5

Rock art preservation

Solution

First of all, the shade would have to be longest when the sun is lowest in the sky. Therefore, if the ranger protects the art on December 21, it will be safe throughout the rest of the year.

Sketch the cliff face on December 21 (Figure 4.2-6). We know that α must be 28.0° because the opposite angles formed by two intersecting straight lines are equal. Vertical is perpendicular to horizontal, so $28.0° + \beta + 20.0° = 90.0°$, which gives $\beta = 90.0° - 20.0° - 28.0° = 42.0°$. We can now use the law of sines in the oblique triangle.

$$\frac{\sin \beta}{x} = \frac{\sin \alpha}{9.60 \text{ ft}}$$

$$x = \frac{(9.60 \text{ ft})\ (\sin \beta)}{\sin \alpha}$$

$$x = \frac{(9.60 \text{ ft})\ (\sin 42.0°)}{(\sin 28.0°)} = 13.7 \text{ ft} \quad ■$$

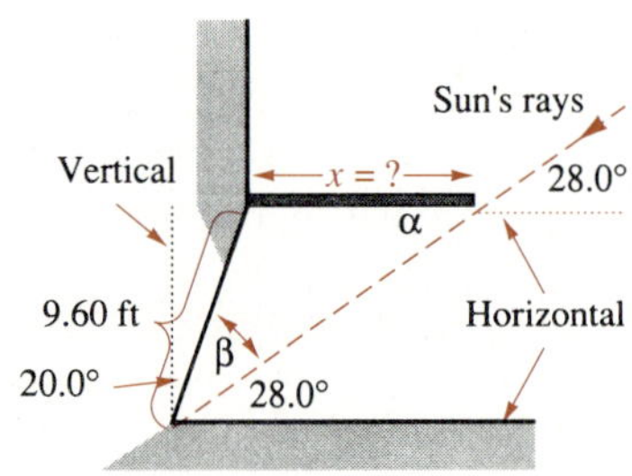

Figure 4.2-6

Rock art on December 21 at noon

NOW IT'S YOUR TURN

If a similar cliff nearby has the same 9.60-ft length but slopes inward at 30.0° instead of 20.0°, how long must the shade be? (Answer: 10.8 ft)

To solve the introductory example about the plane with engine trouble completely, we used both the law of cosines and the law of sines. It is frequently necessary to use both laws, as the following examples illustrate.

Example 12

Solve the triangle in Figure 4.2-7.

Solution

The law of cosines gives us d.

$$d^2 = (18 \text{ ft})^2 + (24 \text{ ft})^2 - 2(18 \text{ ft})\ (24 \text{ ft}) \cos 35° = 192 \text{ ft}^2$$

$$d = 14 \text{ ft}$$

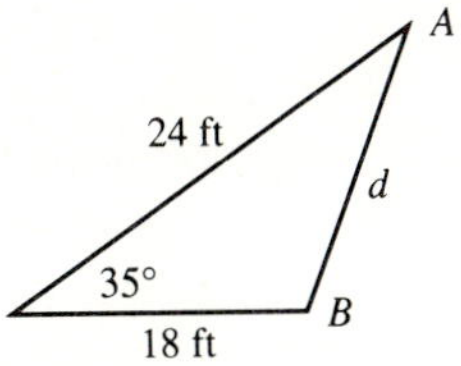

Figure 4.2-7

Now use this answer in the law of sines to get A (or B).

$$\frac{\sin A}{18\text{ ft}} = \frac{\sin 35°}{14\text{ ft}}$$

$$\sin A = \frac{(18\text{ ft})\,(\sin 35°)}{14\text{ ft}} = 0.737$$

$$A = \arcsin(0.737) = 48°$$

Finally, because a triangle has 180°,

$$A + B + 35° = 180°$$

$$48° + B + 35° = 180°$$

$$B = 180° - 35° - 48° = 97° \blacksquare$$

NOW IT'S YOUR TURN

If, in Figure 4.2-7, the 35° angle is changed to 130° but the given lengths are not changed, solve the new triangle. (Answer: 38 ft, $A = 21°$, $B = 29°$)

Example 13
Baseball

In professional baseball, the "diamond" is a square 90.0 ft on each side. The pitcher's plate is 60 ft 6 in. (60.5 ft) from home plate along a line from home plate to second base. If the pitcher throws a ball to the first baseman to pick off a runner on first base:

(a) How far must he throw the ball?

(b) At what angle, relative to the first base line, should the first baseman be prepared for the ball to come from the pitcher?

Solution

Start with a sketch showing what we know and what we want to find (Figure 4.2-8).

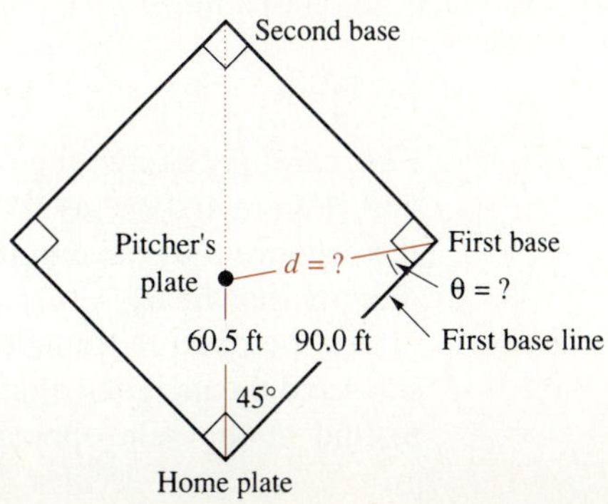

Figure 4.2-8
The baseball diamond

(a) Notice that the line from home plate to the pitcher's plate makes a 45° angle with the first base line. (Why?) Applying the law of cosines to the oblique triangle formed by home plate, the pitcher's plate, and first base, we have

$$a^2 = b^2 + c^2 - 2\,bc \cos A$$

$$d^2 = (60.5 \text{ ft})^2 + (90.0 \text{ ft})^2 - 2(60.5 \text{ ft})(90.0 \text{ ft})\cos 45^\circ$$

$$d^2 = 3660 \text{ ft}^2 + 8100 \text{ ft}^2 - 7700 \text{ ft}^2 = 4060 \text{ ft}^2$$

$$d = \sqrt{4060 \text{ ft}^2} = 63.7 \text{ ft}$$

(b) Now use this answer in the law of sines to find θ.

$$\frac{\sin \theta}{60.5 \text{ ft}} = \frac{\sin 45^\circ}{63.7 \text{ ft}}$$

$$\sin \theta = \frac{(60.5 \text{ ft})\,(\sin 45^\circ)}{63.7 \text{ ft}} = 0.672$$

$$\theta = \sin^{-1}(0.672) = 42.2^\circ \quad ■$$

NOW IT'S YOUR TURN

If the pitcher's plate were misplaced at 50.0 ft from home plate but everything else remained the same, what are the revised answers to this example?

(Answer: 65.1 ft, 32.9°)

Example 14
Space Science

NASA (National Aeronautics and Space Administration) communicates with satellites using radiowaves that travel in straight lines at the speed of light (3.00×10^8 m/s). NASA scientists must send a command to the Pioneer Venus orbiter when Venus is 125° ahead of the earth in its orbit.

(a) How far will the radiowaves have to travel?

(b) How many minutes will it take this message to reach Pioneer?

(c) At what angle relative to the earth-sun line must they aim the radiowaves?

The radius of Venus' orbit is 1.08×10^{11} m, whereas that of the earth is 1.50×10^{11} m.

Solution

First carefully sketch the problem, showing what we know and what we want to find (Figure 4.2-9). (a) We know the speed of the radiowaves, so if we can find the distance they have to travel to reach Venus, we can find how long it will take them to get there.

In the oblique triangle formed by the earth, sun, and Venus, we know two sides and the included angle (R_E, R_V, and 125°), so we can use the law of cosines to find d, the side opposite the 125° angle.

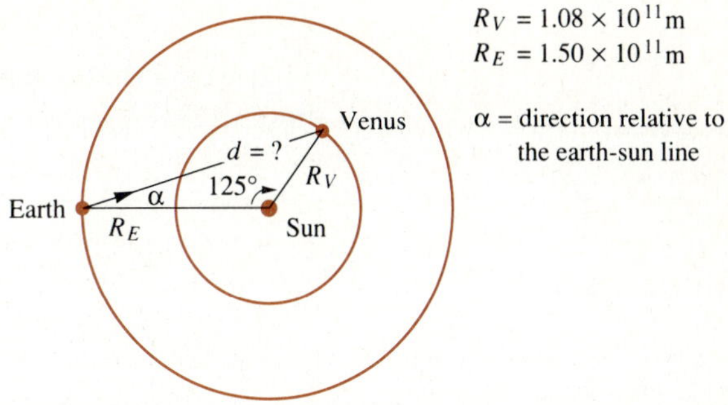

Figure 4.2-9

Earth, Venus, and the Sun

$$a^2 = b^2 + c^2 - 2\,bc \cos A$$
$$d^2 = R_V^{\,2} + R_E^{\,2} - 2\,R_V\,R_E \cos 125°$$
$$d^2 = (1.08 \times 10^{11}\text{ m})^2 + (1.50 \times 10^{11}\text{ m})^2 - 2(1.08 \times 10^{11}\text{ m})\,(1.50 \times 10^{11}\text{ m})\,(-0.574)$$
$$d^2 = 1.17 \times 10^{22}\text{ m}^2 + 2.25 \times 10^{22}\text{ m}^2 + 1.86 \times 10^{22}\text{ m}^2$$
$$d^2 = 5.28 \times 10^{22}\text{ m}^2$$
$$d = \sqrt{5.28 \times 10^{22}\text{ m}^2} = 2.30 \times 10^{11}\text{ m}$$

This is the distance the radiowaves must travel, moving at the speed of light.

(b) To find the time for this message to reach Venus, we recall from Section 1.5 that $v = d/t$ if the speed is constant, which it is here. Solving for the time t gives us

$$t = \frac{d}{v} = \frac{2.30 \times 10^{11}\text{ m}}{3.00 \times 10^{8}\text{ m/s}} = 767\text{ s}$$

Converting this answer to minutes gives us

$$t = (767\text{ s})\left(\frac{1\text{ min}}{60\text{ s}}\right) = 12.8\text{ min}$$

(c) To find the direction in which they must aim the radiowaves, we can use the law of sines to find α.

$$\frac{\sin\alpha}{R_V} = \frac{\sin 125°}{d}$$
$$\sin\alpha = \frac{R_V \sin 125°}{d}$$
$$\sin\alpha = \frac{(1.08 \times 10^{11}\text{ m})\,(0.819)}{2.30 \times 10^{11}\text{ m}} = 0.385$$
$$\alpha = \sin^{-1}(0.385) = 22.6° \quad ■$$

NOW IT'S YOUR TURN

What would the direction and the distance be if Venus were only 41.0° ahead of the earth? (Answer: 46.0°, 9.85×10^{10} m)

EXERCISES 4.2

In Exercises 1–21, solve the triangle *ABC* in Figure 4.2-10 given the following information.

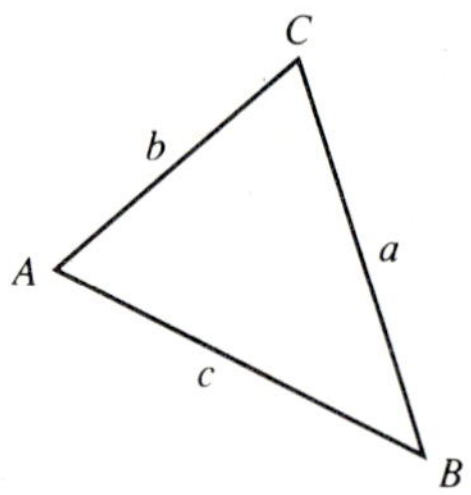

Figure 4.2-10

1. $A = 42°$, $B = 78°$, $c = 25.5$.

2. $A = 52.5°$, $B = 78.2°$, $c = 300.5$ cm.

3. $B = 26°50'$, $C = 88°10'$, $c = 615.3$ ft.

4. $A = \frac{\pi}{4}$, $C = \frac{2\pi}{3}$, $c = 18.2$ m.

5. $A = 10°12'$, $B = 46°36'$, $a = 500.0$ mm.

6. $B = 13°57'$, $C = 57°13'$, $b = 13.57$ mi.

7. $A = \frac{3\pi}{8}$, $C = \frac{6\pi}{11}$, $b = 0.0873$ in.

8. $B = 1.00$, $C = 1.50$, $b = 9.86$ Å.

***9.** $B = 41.3°$, $a = 207$ m, $c = 148$ m.

***10.** $A = 104.2°$, $b = 31.43$ km, $c = 82.60$ km.

***11.** $B = 30.67°$, $C = 127.0°$, $a = 95.0$ ft.

***12.** $C = 29.6°$, $a = 20.4$ mi, $b = 48.3$ mi.

***13.** $A = 37°10'30''$, $C = 78°22'10''$, $b = 88.8$ m.

***14.** $a = 4.0$ cm, $b = 6.0$ cm, $c = 8.0$ cm.

***15.** $A = 24.0°$, $B = 111.8°$.

16.** $A = 30°$, $a = 40.0$ m, $b = 60.0$ m. ***Hint: There are two possibilities!

17.** $a = 19.0$ ft, $b = 28.0$ ft, $A = 50°$. ***Hint: Make a careful sketch of this triangle to interpret your answer.

18.** $A = \frac{\pi}{6}$, $b = 45$ cm, $a = 18$ cm. ***Hint: Make a careful sketch of this triangle to interpret your answer.

19. $A = \frac{\pi}{4}$, $b = 75.0$ m, $a = 85.0$ m.

***20.** $a = 15.0$ ft, $b = 25.0$ ft, $c = 9.0$ ft.

***21.** $a = 12$ m, $b = 18$ m, $c = 32$ m.

In Exercises 22–32, compute the area of triangle *ABC* in Figure 4.2-10 using the following information.

22. $B = 40°$, $a = 27$ cm, $c = 32$ cm.

23. $a = 35$ in., $c = 43$ in., $B = 37°$.

24. $a = 4.8$ m, $c = 5.3$ m, $B = 39°27'$.

***25.** $A = 42.7°$, $B = 31.4°$, $a = 18.10$ m.

***26.** $a = 17.3$ m, $b = 19.4$ m, $B = 56.25°$.

***27.** $a = 9.800$ ft, $c = 12.6$ ft, $C = 68.5°$.

28. $b = 127.8$ ft, $c = 168.5$ ft, $A = 1.27$ rad.

***29.** $a = 423.9$ yd, $b = 503$ yd, $A = 68°27'$.

***30.** $a = 15$ cm, $c = 21$ cm, $A = 21.4°$.

***31.** $a = 11$ m, $b = 15$ m, $c = 21$ m.

32.** $A = 34.0°$, $a = 95.0$ cm, $b = 156$ cm. ***Hint: There are two possibilities. Sketch both possible triangles.

***33.** Find the area of the quadrilateral *ABCD*, given that $AB = 3.0$ cm, $AC = 4.2$ cm, $AD = 3.8$ cm, angle $BAD = 88°10'$, and angle $BAC = 36°20'$.

****34.** Prove that the area of a parallelogram is equal to the product of the base, the diagonal, and the sine of the angle between them.

****35.** In the pentagon *ABCDE*, given that $AC = 6.00$ in., $A = 110°50'$, $B = 106°30'$, $E = 104°10'$, angle $BAC = 30°$, angle $DAE = 34°56'$, and angle $ADC = 52°30'$, find *C*, *D*, and lengths of the sides of the pentagon.

***36.** A museum exhibitor wants to hang a 29.6-in. sword as shown in Figure 4.2-11 by attaching two strings to its ends. How long must each string be?

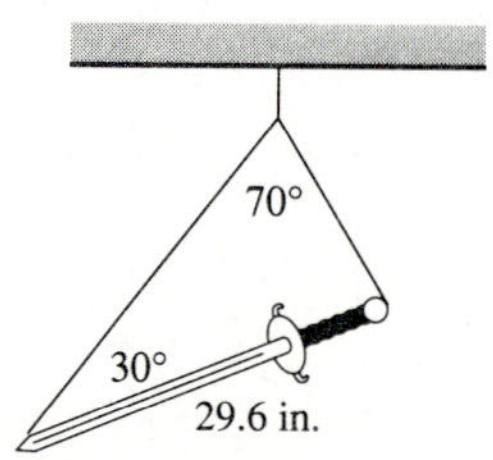

Figure 4.2-11

37. A 20.0-ft, 3,000-lb beam is hinged to a wall and supported by a thin cable attached to the beam 4.00 ft from its free end (Figure

4.2-12). The cable makes a 40° angle with the beam and must hold the beam up at a 30° angle above the horizontal. How long must this cable be, and how high above the hinge must it be fastened to the wall?

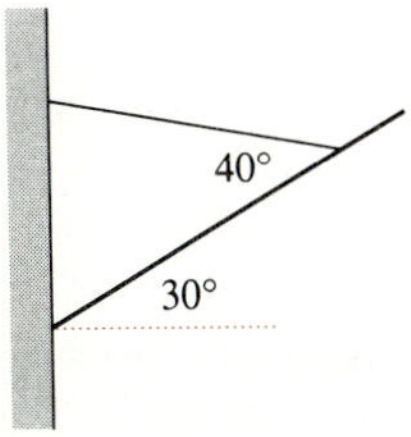

Figure 4.2-12

38. A 115-ft ponderosa pine leans 24° off of vertical toward the north (Figure 4.2-13). How long will the shadow of this tree be when the noon sun is 68° above the horizon?

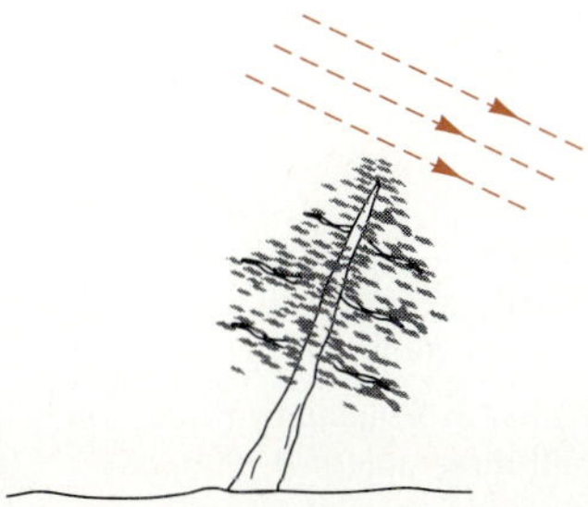

Figure 4.2-13

39. You are building a gazebo with a floor in the shape of a regular hexagon 10.0 ft on each side. This floor will be supported underneath by six beams, all of which meet under the midpoint of the floor, as shown in Figure 4.2-14. How long should you cut each support beam?

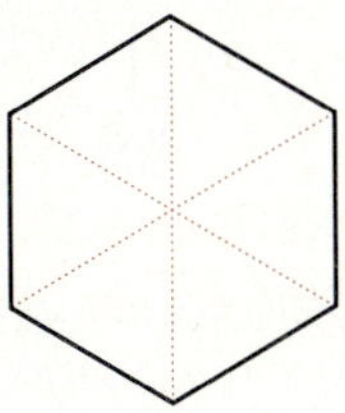

Figure 4.2-14

***40.** Suppose that you want to design another gazebo, similar to the one in Exercise 39, except that you want it to be a regular octagon 10.0 ft on each side. This time the floor will be supported by eight equal beams meeting in the center. How long should each beam be?

***41.** The staircase to a public building slopes upward at 22.4° above the horizontal. At the top, the building rises 5.40 m above the top of the stairs. When the sun is at 65.0° above the horizontal and is coming from behind the building, how long will the building's shadow be on the staircase? (Figure 4.2-15)

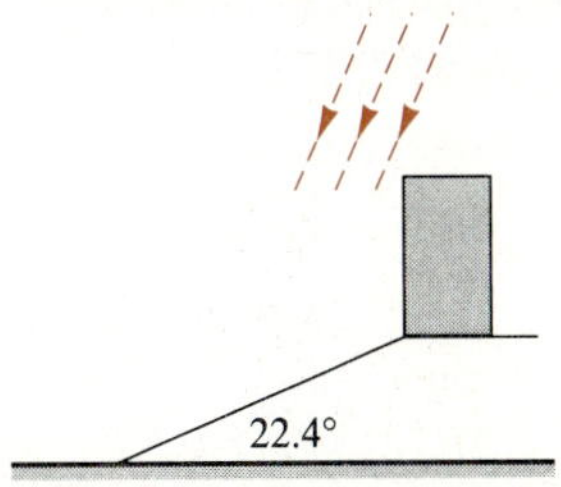

Figure 4.2-15

***42.** As an environmentally aware architect, you are designing a passive solar house for Los Angeles, where the June 21 sun is at 80° above the horizon and the December 21 sun is at 32° above the horizon. This home has a south-facing window that you want to shield completely from the sun on June 21 but have totally illuminated on December 21. To do so, you allow 4.00 ft of the sloping roof to extend beyond the wall (Figure 4.2-16). What should be the distance between the roof and the top of the window (y in the figure), and what is the maximum vertical height this window can have (z in the figure)?

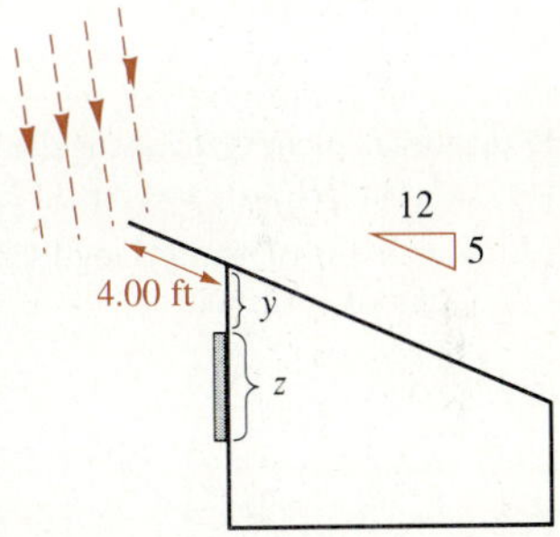

Figure 4.2-16

43. A cliff 3.20 m long slopes outward 11.0° from the vertical (Figure 4.2-17). An archaeologist wants to build an overhang to protect some petroglyphs carved on this cliff from the sun when it is 75.0° above the horizon. How far beyond the cliff should this overhang extend?

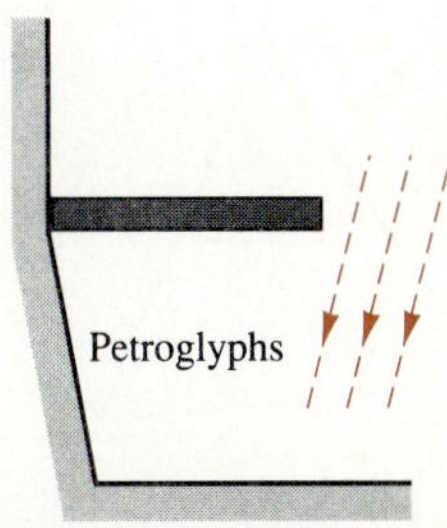

Figure 4.2-17

****44.** In a certain baseball field, the "diamond" is a 90.0-ft square, the pitcher's plate is 60 ft 6 in. from home plate, and the outfield is bounded by the foul lines and the arc of a circle of radius 100.0 ft centered on the pitcher's plate (Figure 4.2-18). You are buying sod to plant grass in the outfield. What area will you need to cover with sod?

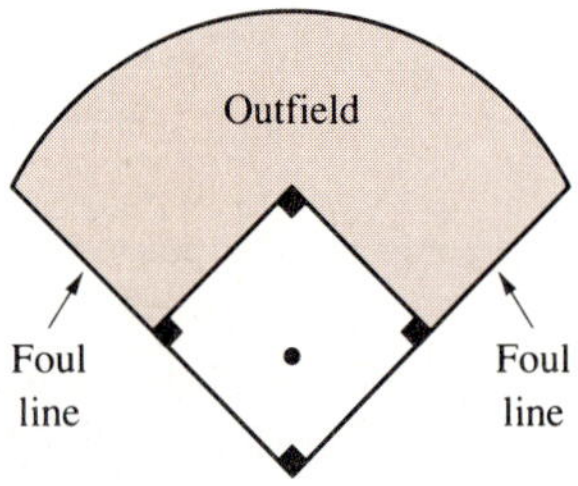

Figure 4.2-18

***45.** Is there a "law of cosines" as follows

$$\frac{\cos A}{a} = \frac{\cos B}{b} = \frac{\cos C}{c}$$

that is analogous to the law of sines? ***Hint***: To prove that something is not true, all you have to do is find one case in which it does not hold.

****46.** Derive the law of sines from the law of cosines. ***Hint***: Use the law of cosines to find expressions for $\cos A$, $\cos B$, and $\cos C$. Then use the Pythagorean identity involving the sine and cosine to compute $(\sin^2 A)/a^2$, $(\sin^2 B)/b^2$, and $(\sin^2 C)/c^2$.

***47.** Obtain formulas expressing the lengths a, b, c, and d in Figure 4.2-19 in terms of x, θ, and ϕ.

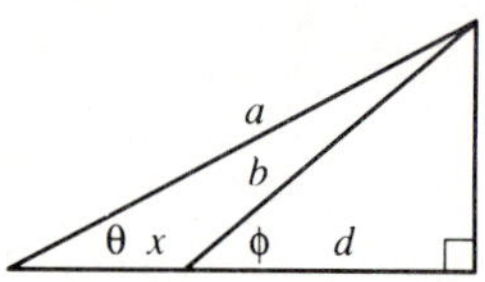

Figure 4.2-19

***48.** Show that the radius r of the circle circumscribed about the equilateral triangle ABC in Figure 4.2-20 is $r = a/(2 \sin \theta)$.

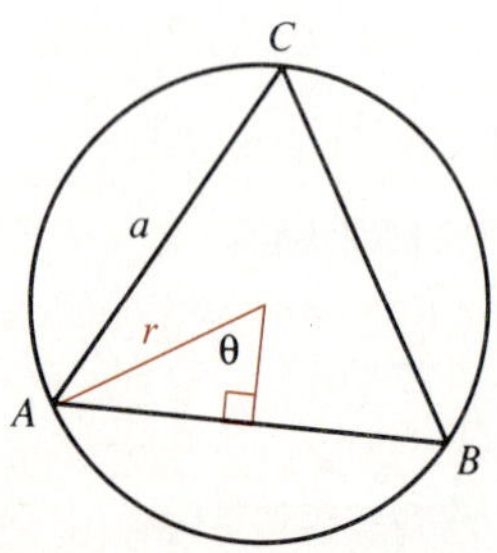

Figure 4.2-20

The following exercises may involve the use of both the law of sines and the law of cosines.

****49.** In Figure 4.1-6 of Section 4.1, showing the scissors truss, there are three thin vertical and two thin horizontal lines. These lines represent supports called *members*. How long are the members in Example 5 of that section?

50. When a runner is midway between third base and home plate, how far is he from second base?

51. A runner is midway between first base and second base. How far is he from the pitcher's plate and from third base?

52. For the letter A shown, use only length measurements made on the letter to calculate the three angles in the upper part of the letter. Then measure these angles using a protractor to check your calculations.

***53.** A farmer's triangular field measures 2000.0 ft × 3000.0 ft × 4000.0 ft. If his soil yields an average of 15.0 bushels of corn per acre, how much corn will this field produce?

54. An ornament is hanging by three strings, as shown in Figure 4.2-21. Find all three angles α, β, and ϕ.

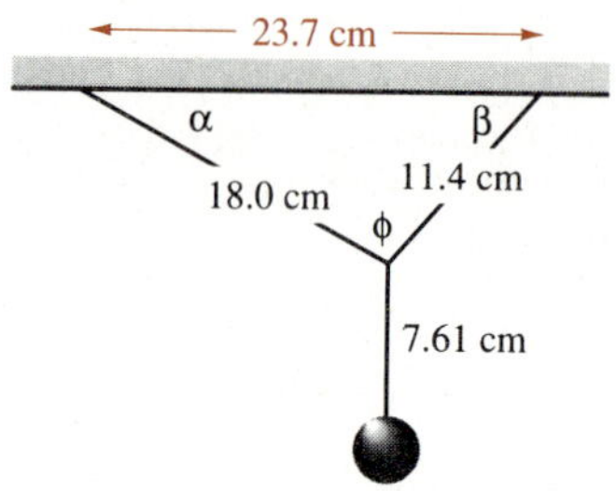

Figure 4.2-21

***55.** A metal plate is to be attached to strengthen the wooden supports (called members) of a truss, as shown in Figure 4.2-22. The five members divide the top chords into equal-length segments. Find the angles between the spokes of this plate.

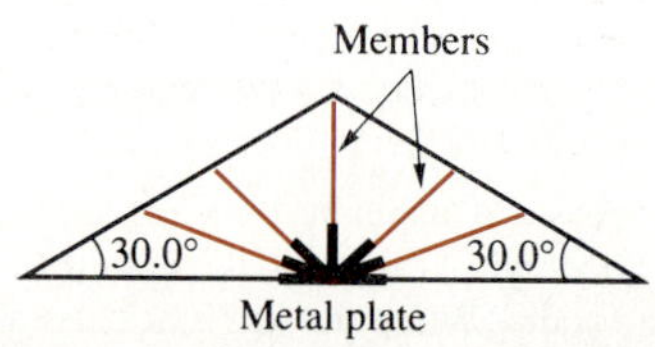

Figure 4.2-22

56. You are designing a stepladder with a step portion 8.0 ft long that is to stand at angles of 50.0° and 75°, as shown in Figure 4.2-23. A horizontal brace (Z) is needed 4.0 ft above the ground. Find the length of the other side (L), the length of the brace (Z), and the width (W) this ladder will span on the ground.

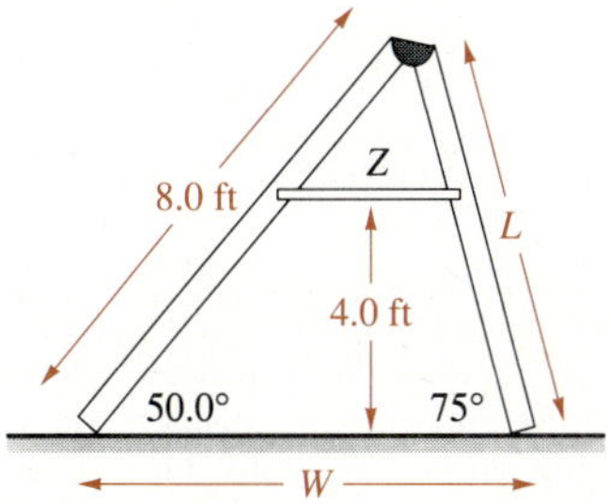

Figure 4.2-23

57. The atoms of formic acid (HCOOH) lie in a plane as shown in Figure 4.2-24. Chemists have measured the following angles and distances: H—C—O′ = 117.8°, O′—C—O = 124.3°, C—O—H′ = 107.8°, C—O′ = 1.245 Å, C—O = 1.312 Å, C—H = 1.085 Å, and O—H′ = 0.9500 Å. Find the distances O′—O and O—H.

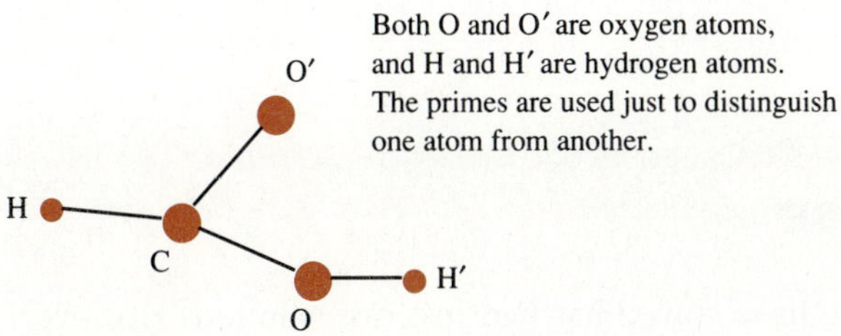

Figure 4.2-24

58. Benzene (C_6H_6)forms a regular hexagon called the benzene ring (Figure 4.2-25). This molecule is perfectly symmetric, so the H—C line bisects the larger angle formed by C—C—C. Chemists have measured the C—C distance to be 1.395 Å and the C—H distance to be 1.084 Å. Find a, b, d, and the angle between a and b.

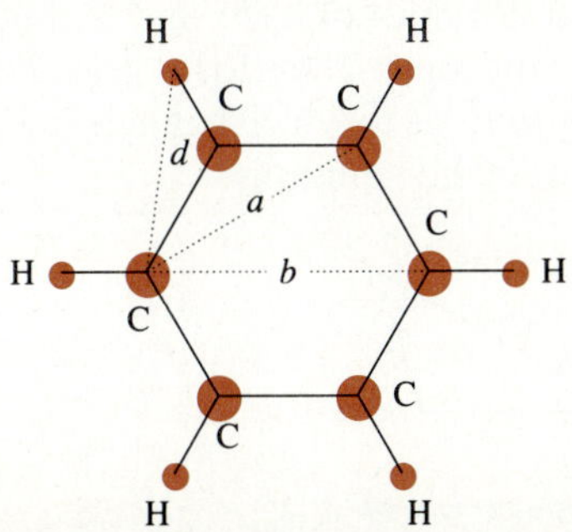

Figure 4.2-25

****59.** A parallelopiped is formed by sliding one face of a 2.00-cm cube parallel to itself to form the solid in Figure 4.2-26. You now form a triangle by connecting corner A to the midpoints of faces $BCDE$ and $DEFG$ and then connecting those midpoints with each other. Find the angles of this triangle.

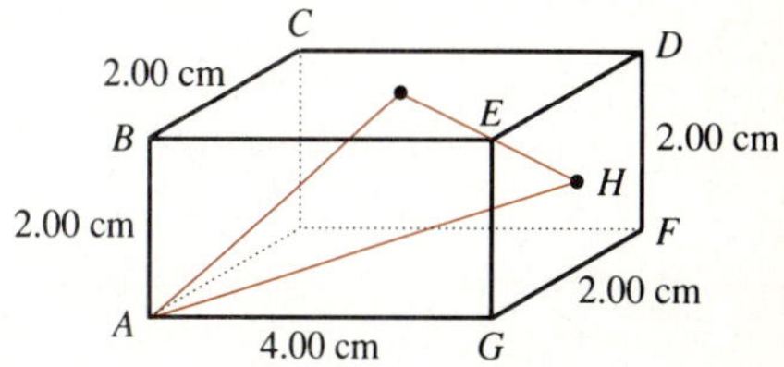

Figure 4.2-26

***60.** In the parallelopiped of Figure 4.2-26, find the angle between AH and AD.

61. Sailboat Racing. The Seawanhaka rule for rating a sailboat in a race is

$$\text{rating of sailboat} = \tfrac{1}{2}\left(\text{length of boat} + \sqrt{\text{area of sail}}\right)$$

This rule was used until around 1905. Modern formulas are much more complicated, but they still make use of the area of the sail. Suppose you enter a race with a boat 30.0 ft long with a sail that measures 10.0 ft × 20.0 ft × 25.0 ft. What is the rating of your boat by the Seawanhaka rule?

***62.** When Mercury is 152° ahead of the earth in its orbit and Venus is 98.0° behind earth:

(a) How far apart are Mercury and Venus?
(b) What is the angular separation of Mercury and Venus as viewed from earth?

Hint: Make a careful sketch showing all three planets and the sun. Mercury is 57.9×10^6 km from the sun, whereas Venus is 108.2×10^6 km from the sun.

****63.** When Mercury and Venus are farthest apart in the sky as viewed from earth:

(a) What is their angular separation as viewed from earth?
(b) How many kilometers apart are they?

Hint: Start by making a careful sketch of all three planets and the sun. See Exercise 62 for useful data.

****64.** A chord of a circle is 8.00 m long and the circle is 11.0 m in radius. This chord makes a 118° angle with another chord, as shown in Figure 4.2-27. Find the length of the second chord.

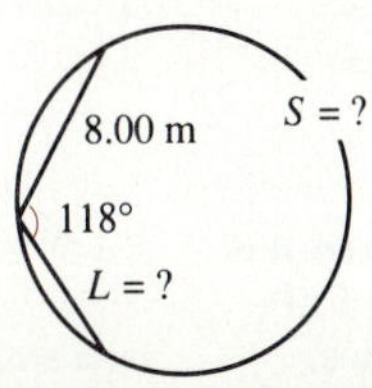

Figure 4.2-27

****65.** Find the length of the arc S subtended by the two chords in Exercise 64.

****66.** Find the area bounded between the two chords and the circle in Exercise 64.

****67.** In Figure 4.2-28, find the angle between the two chords if the circle has a radius of 50.0 cm.

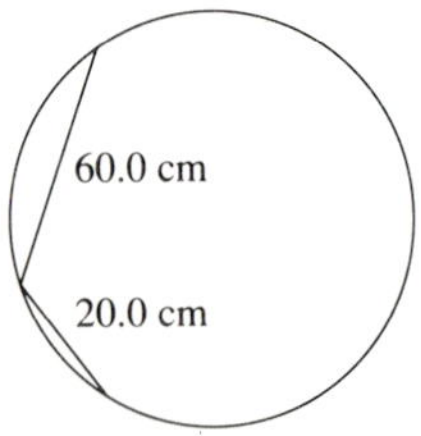

Figure 4.2-28

****68.** As a newly hired civil engineer, you are asked to determine the area of the irregular lot in Figure 4.2-29. What is the area?

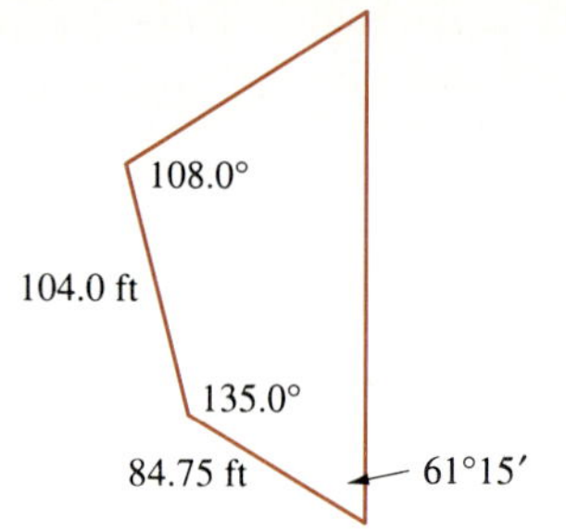

Figure 4.2-29

***69.** You have been asked to build the bicycle shown in Figure 4.2-30. The frame of this bike is made of two triangles. If side a is 28.0 in., measure the necessary angles on the figure and calculate what the lengths of the other four sides of this frame should be.

Figure 4.2-30

4.3 THE AMBIGUOUS CASE

Up to now, every triangle we have solved has had just one solution. However, when dealing with oblique triangles, we find that if we know two sides of a triangle and an angle not included between those sides, there can be two possible solutions or even none at all. Because of the possibility of two solutions or no solutions, we often call this the **ambiguous case**.

Let's start by sketching an oblique triangle for which we know two sides and an acute angle not included between those sides, as shown in Figure 4.3-1a. It is possible, however, to have another triangle with exactly the same values of a, b, and A, with the side a sloping toward side b as shown in Figure 4.3-1b. Both triangles in Figure 4.3-1 have the same values of a, b, and A, but two different possible values for the other angles and side. Therefore, if we knew only a, b, and A, it would be unclear, or *ambiguous*, as to which triangle in Figure 4.3-1 we should use. In fact, both triangles are valid solutions.

(a) a slopes away from b *(b) a slopes toward b*

Figure 4.3-1

The Ambiguous Case when *A* is an acute angle. *A*, *a*, and *b* are the same for both triangles.

Example 15

Solve the triangle having sides of 7.0 and 10.0 units with a 30° angle adjacent to the 10.0-unit side.

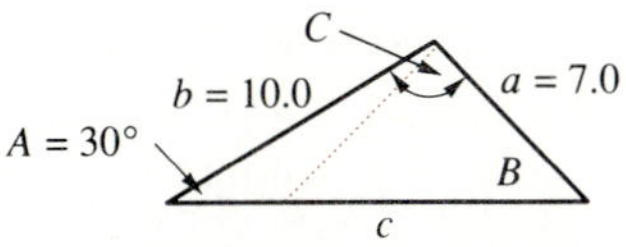

(a) a slopes away from b

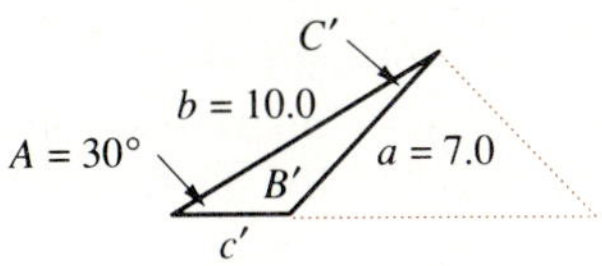

(b) a slopes toward b

Figure 4.3-2

Both possibilities for the given triangle

Solution

Because we know two sides and a nonincluded angle, this case is ambiguous. First sketch the triangle, showing both possibilities (Figure 4.3-2).

Let's start with the triangle in Figure 4.3-2a. Applying the law of sines gives us

$$\frac{\sin B}{b} = \frac{\sin A}{a}$$

$$\frac{\sin B}{10.0} = \frac{\sin 30°}{7.0}$$

$$\sin B = \frac{(10.0)(\sin 30°)}{(7.0)} = 0.714$$

$$B = \arcsin(0.714) = 46°$$

A triangle contains 180°, so

$$A + B + C = 180°$$

$$C = 180° - A - B = 180° - 30° - 46° = 104°$$

THE AMBIGUOUS CASE

The *ambiguous case* occurs when we know two sides of a triangle and an angle not included between those sides.

Now we can find c using either the law of cosines or the law of sines. We use the law of sines because it is a little easier.

$$\frac{\sin C}{c} = \frac{\sin A}{a}$$

$$\frac{\sin 104°}{c} = \frac{\sin 30°}{7.0}$$

$$c = \frac{(7.0)(\sin 104°)}{\sin 30°} = 14$$

Now let's try to solve the triangle in Figure 4.3-2b. Because we know the same information as for the previous triangle, we can proceed in exactly the same way. Applying the law of sines gives us

$$\frac{\sin B'}{b} = \frac{\sin A}{a}$$

$$\frac{\sin B'}{10.0} = \frac{\sin 30°}{7.0}$$

$$\sin B' = \frac{(10.0)(\sin 30°)}{7.0} = 0.714$$

$$B' = \arcsin(0.714)$$

But this answer is exactly the same one we just got for the triangle in Figure 4.3-2a when we were solving for B. Therefore, we might expect to get the same answer, namely $B' = B = 46°$. Yet by looking at these two triangles, it is obvious that B' is not the same as B. B was acute whereas $B' > 90°$. Because of these conflicting results, we must interpret our answers to understand their meaning. To do this, let's superimpose the two triangles on top of each other (Figure 4.3-3).

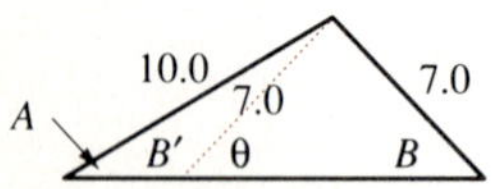

Figure 4.3-3

The two triangles superimposed on one another

In Figure 4.3-3, the angle θ must be equal to B because both are at the base of equal legs of an isosceles triangle having equal sides 7.0 units long. Also, θ is supplementary to B' because there are 180° in a straight angle. Therefore, because $B = \theta$ and θ is supplementary to B', B must be supplementary to B': $B + B' = 180°$. Now we can find B' since we just found that $B = 46°$.

$$B + B' = 180°$$

$$B' = 180° - B = 180° - 46° = 134°$$

The reason that the law of sines gave us the same answer for both $\sin B$ and $\sin B'$ is this: Since $B + B' = 180°$, B is the reference angle for B'. Now remember from Chapter 3 that the sine of an angle and the sine of its reference angle are the *same* if the angle is less than 180°. Therefore, because $\sin B' = \sin B$, it does *not* follow that $B = B'$. It just means that B and B' are supplementary angles.

This is a very important result. In Figure 4.3-2, the angles B and B' in the two possible triangles are supplementary angles. Once we know B, we can easily use this fact to find B'.

Now that we know B' we can complete the solution for the triangle in Figure 4.3-2b.

$$A + B' + C' = 180°$$

$$C' = 180° - A - B'$$

$$= 180° - 30° - 134° = 16°$$

The law of sines gives us

$$\frac{\sin C'}{c'} = \frac{\sin A}{a}$$

$$\frac{\sin 16°}{c'} = \frac{\sin 30°}{7.0}$$

$$c' = \frac{(7.0)(\sin 16°)}{\sin 30°} = \frac{(7.0)(0.28)}{0.50} = 3.9 \text{ units}$$ ■

NOW IT'S YOUR TURN

Solve the same triangle if $A = 60°$, $a = 19$ units, and $b = 21$ units. (Answer: 73°, 47°, 16 units and 107°, 13°, 4.9 units)

Example 16

If a triangle has sides of 15.0 ft and 25.0 ft and a 20.0° angle adjacent to the 25.0-ft side, find all possible solutions for this triangle.

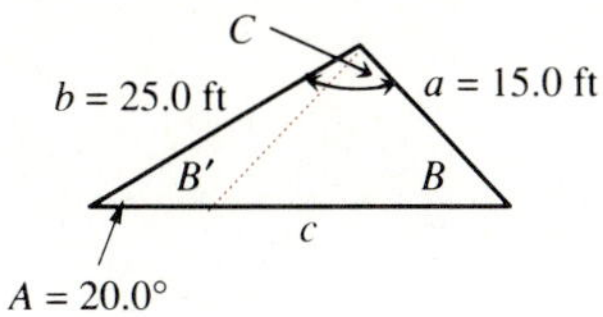

(a)

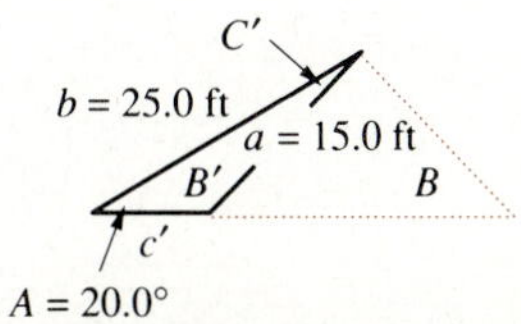

(b)

Figure 4.3-4
The two possible solutions

Solution

We know two sides and a nonincluded angle, so this will be the ambiguous case. First sketch the triangle showing both possible solutions (Figure 4.3-4). The first solution is the heavy triangle in Figure 4.3-4a. We can use the law of sines to find B.

$$\frac{\sin B}{b} = \frac{\sin A}{a}$$

$$\frac{\sin B}{25.0 \text{ ft}} = \frac{\sin 20.0°}{15.0 \text{ ft}}$$

$$\sin B = \frac{(25.0 \text{ ft})(\sin 20.0°)}{15.0 \text{ ft}} = 0.570$$

$$B = \sin^{-1}(0.570) = 34.8°$$

Now find C.

$$A + B + C = 180°$$

$$C = 180° - 34.8° - 20.0° = 125° \quad \text{(to three-figure accuracy)}$$

Now use the law of sines to find c.

$$\frac{\sin C}{c} = \frac{\sin A}{a}$$

$$\frac{\sin 125°}{c} = \frac{\sin 20.0°}{15.0 \text{ ft}}$$

$$c = \frac{(15.0 \text{ ft})(\sin 125°)}{\sin 20.0°} = 35.9 \text{ ft}$$

The second solution is the heavy triangle in Figure 4.3-4b. As we showed in the previous example, B and B' are supplementary, so

$$B + B' = 180°$$

$$B' = 180° - B = 180° - 34.8° = 145° \quad \text{(to three-figure accuracy)}$$

Now we can get C'.

$$A + B' + C' = 180^\circ$$
$$C' = 180^\circ - A - B'$$
$$= 180^\circ - 20.0^\circ - 145^\circ = 15.0^\circ$$

Now use the law of sines to get c'.

$$\frac{\sin A}{a} = \frac{\sin C'}{c'}$$

$$\frac{\sin 20.0^\circ}{15.0 \text{ ft}} = \frac{\sin 15.0^\circ}{c'}$$

$$c' = \frac{(15.0 \text{ ft})(\sin 15.0^\circ)}{\sin 20.0^\circ} = 11.4 \text{ ft} \blacksquare$$

NOW IT'S YOUR TURN

If the original triangle is the same except that the given angle is 32.0° rather than 20.0°, find all possible solutions.

(Answer: 62.0°, 86.0°, 28.2 ft and 118°, 30.0°, 14.2 ft)

FOOD FOR THOUGHT

What are the possible solutions if B turns out to be 90°?

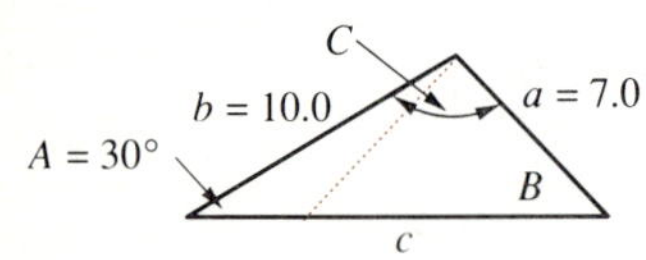

(a) a slopes away from b

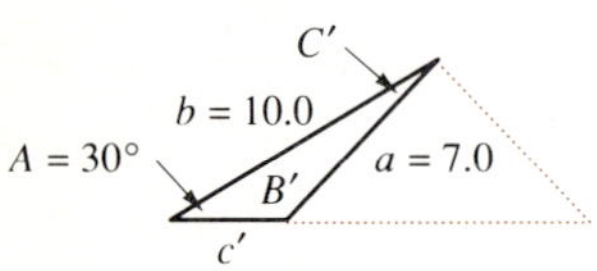

(b) a slopes toward b

Figure 4.3-5

Both possibilities for the given triangle

At the beginning of this section, we said that sometimes there are *no solutions* to an oblique triangle if we know only two sides and a nonincluded angle. Let us see geometrically how this is possible.

Consider the triangles in Figure 4.3-5, where A is an acute angle. There are two possible solutions in this case because the side a is long enough that it can slope either toward b or away from b.

Suppose that instead the side a is just long enough to reach the base (Figure 4.3-6a). Then a would have to be perpendicular to the base and there would be just one solution. The triangle in Figure 4.3-6a is a right triangle, so

$$\sin A = \frac{a}{b} \quad \rightarrow \quad a = b \sin A$$

Therefore, if $a = b \sin A$, there is only one solution and the triangle is a right triangle.

Suppose that a is not even long enough to reach the base, as shown in Figure 4.3-6b. Then there is no possible solution.

The question to ask is: How do we recognize that there are no solutions? Drawing the figure to scale is not accurate enough. One way is to compute $b \sin A$. If a is less than $b \sin A$, then it is not long enough to reach the base, so there is no solution. Another way is to apply the law of sines, as shown in Example 17.

Figure 4.3-6
Triangles having one solution and no solutions

(a) Side a is just long enough to reach the base. There is only one solution to this triangle.

(b) Side a is not long enough to reach the base. There are no solutions to this triangle.

Example 17

A triangle has two sides of length 11 cm and 24 cm with a 41° angle adjacent to the 24-cm side. Find the other side and angles.

Solution

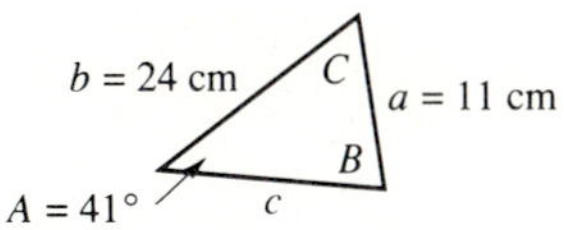

Figure 4.3-7

First sketch the triangle and label the unknown side and angles, as in Figure 4.3-7. Without calculating the angles, we do not know how the 11-cm side should slope, so we have arbitrarily drawn it sloping away from the 24-cm side. First, just as in the previous examples, we apply the law of sines to find B.

$$\frac{\sin B}{24\text{ cm}} = \frac{\sin 41°}{11\text{ cm}}$$

$$\sin B = \frac{(24\text{ cm})(\sin 41°)}{11\text{ cm}} = 1.4$$

It is impossible for $\sin B$ to be 1.4 because the sine can never be greater than 1, as we learned in Chapter 3. Therefore, there is no angle B that satisfies this condition. Consequently there is *no solution* to this triangle; it is not possible to have a triangle with the given dimensions.

As a check, we can see if a is less than $b \sin A$.

$$b \sin A = (24\text{ cm})(\sin 41°) = 15.7\text{ cm}$$

We find that $b \sin A$ is greater than a, so a is not long enough to reach the base of the triangle. Therefore, this triangle is impossible. ■

▸ **This example illustrates an important point. If, while solving a triangle, we get an impossible result (such as a sine greater than 1), then the triangle is impossible and thus has no solution.**

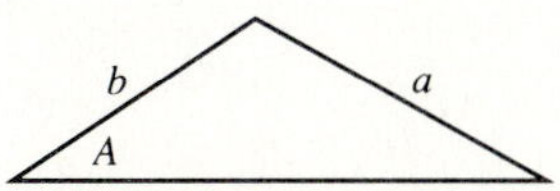

Figure 4.3-8
a* > *b

NOW IT'S YOUR TURN

How many solutions are there if $a > b$ in Figure 4.3-8?

(Answer: One solution. To see why, make a careful sketch and try to find two possibilities.)

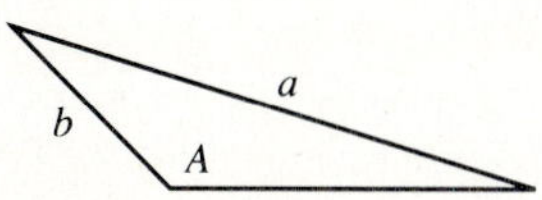

Figure 4.3-9
***A* > 90°**

In everything we have done so far in this section, A has been an acute angle. It is certainly possible, however, for A to be obtuse (>90°). To see what happens then, look at Figure 4.3-9. The first thing to notice is that side a must be long enough to reach the base and yet keep A inside the triangle. Therefore, *a must be greater than b*. The second thing to note is that *a can slope only one way*. If a slopes the other way, A will be outside the triangle (see Figure 4.3-10).

Figure 4.3-10

The side *a* can slope only one way and still keep the angle *A* inside the triangle.

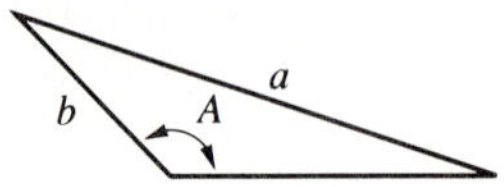

(a) A is inside the triangle.

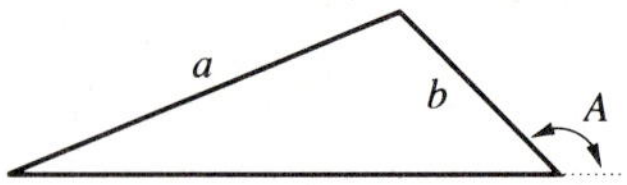

(b) If a slopes the other way, A is outside the triangle.

We can now summarize these results using Figure 4.3-11. *When A is > 90°:*

1. If $a \leq b$, there is no solution.
2. If $a > b$, there is only one solution.

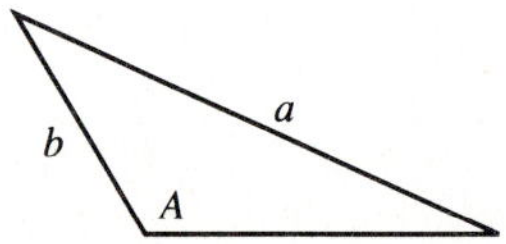

Figure 4.3-11

Example 18

Find all possible solutions to the triangle in Figure 4.3-12.

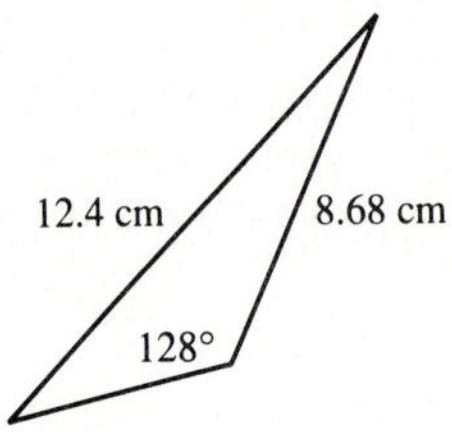

Figure 4.3-12

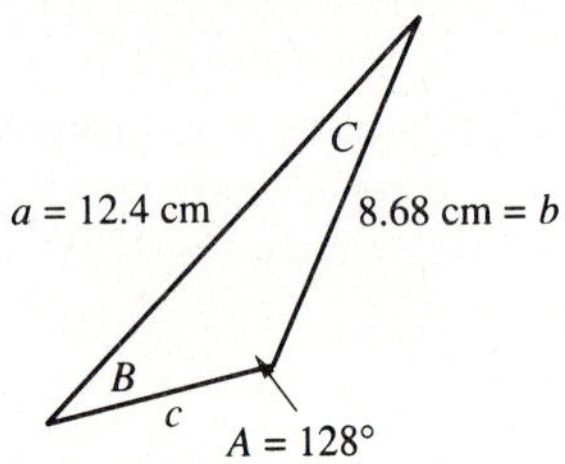

Figure 4.3-13

Solution

First redraw the triangle, introducing symbols as needed (Figure 4.3-13). Before doing a lot of work, we should check to make sure there is a solution. Because $a > b$ (12.4 cm > 8.68 cm), there will be a solution. But because $A > 90°$, there will be only one solution.

We find B by the law of sines.

$$\frac{\sin B}{b} = \frac{\sin A}{a}$$

$$\frac{\sin B}{8.68\text{ cm}} = \frac{\sin 128°}{12.4\text{ cm}}$$

$$\sin B = \frac{(8.68 \text{ cm})(\sin 128°)}{12.4 \text{ cm}} = 0.552$$

$$B = \sin^{-1}(0.552) = 33.5°$$

Now find C:

$$A + B + C = 180°$$
$$C = 180° - A - B$$
$$= 180° - 128° - 33.5° = 18.5°$$

Now use the law of sines to find c.

$$\frac{\sin C}{c} = \frac{\sin A}{a}$$

$$\frac{\sin 18.5°}{c} = \frac{\sin 128°}{12.4 \text{ cm}}$$

$$c = \frac{(12.4 \text{ cm})(\sin 18.5°)}{\sin 128°} = 4.99 \text{ cm}$$ ■

NOW IT'S YOUR TURN

If a triangle has two sides of length 965 in. and 2080 in. with a 110° angle adjacent to the 965-in. side, find all possible solutions to this triangle.
(Answer: 25.8°, 44.2°, 1540 in.)

Example 19

If a triangle has two sides of length 15 m and 18 m with a 131° angle adjacent to the 18-m side, find all possible solutions.

Solution

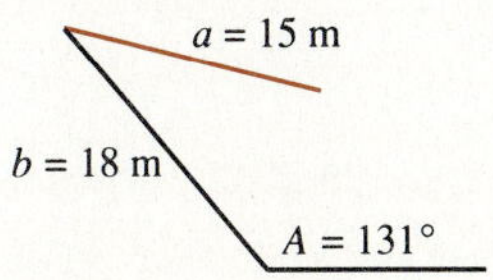

Figure 4.3-14

Sketch the triangle first (Figure 4.3-14). We notice that $a < b$ (15 m < 18 m), so there are no solutions. ■

NOW IT'S YOUR TURN

In Example 18, suppose that the 128° angle were adjacent to the 12.4-m side instead of the 8.68-m side. Find all possible solutions.
(Answer: Surprise! There are none.)

We have covered many possible cases in this section, and you may find them a bit confusing at first. We summarize the basic concepts to help you remember them. Do not try to memorize all these special cases. Rather, just be sure you can figure them out by applying the laws of trigonometry to the triangles involved. Use this summary only as a reference in case you need to check yourself.

SUMMARY OF AMBIGUOUS CASE

1. If A is $< 90°$:

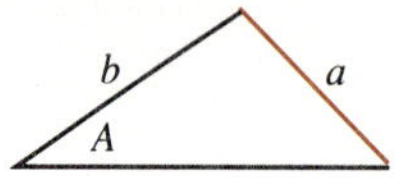

(a) If $a = b \sin A$, there is only *one solution* and the triangle is a *right triangle*.

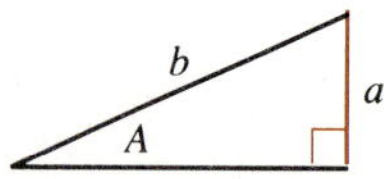

(b) If $a < b \sin A$, there are *no solutions*.

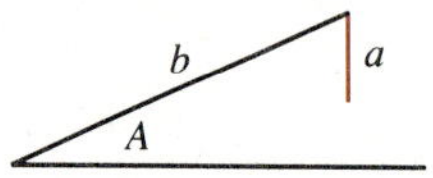

(c) If $a > b \sin A$ but less than b, there are *two solutions*.

(d) If $a > b$, there is only *one solution*.

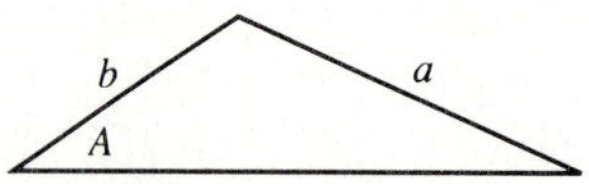

2. If $A > 90°$:

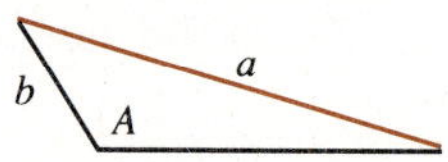

(a) If $a \leq b$, there are *no solutions*.

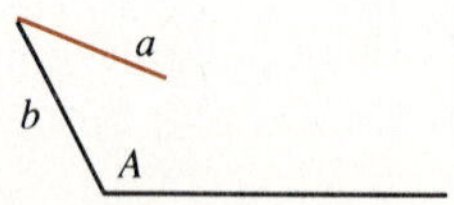

(b) If $a > b$, there is only *one solution*.

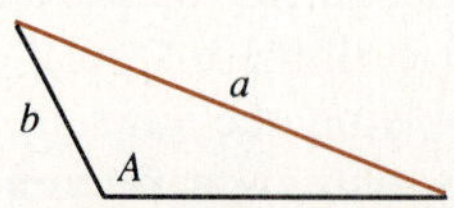

EXERCISES 4.3

In Exercises 1–10, find the number of possible triangles that satisfy the given conditions. Do not solve these triangles; just find out how many there are.

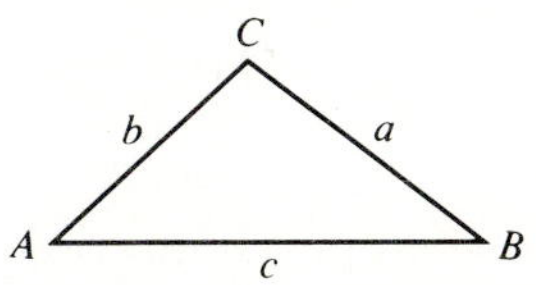

1. $A = 0.73$, $a = 16$ km, $b = 20$ km.
2. $C = 41°13'$, $a = 32.56$ in., $c = 26.49$ in.
3. $A = 73°2'$, $a = 14.12$ mm, $c = 23.57$ mm.
4. $A = 49.30°$, $a = 44.58$ cm, $b = 58.80$ cm.
5. $C = 78°19.4'$, $b = 4931.4$ ft, $c = 6623.2$ ft.
6. $A = \frac{3\pi}{11}$, $a = 109$ m, $b = 128$ m.
7. $A = 109°13.8'$, $a = 673.56$ yd, $c = 786.49$ yd.
8. $A = 2.93$, $b = 75.1$ m, $a = 98.2$ m.
9. $A = 60°58.1'$, $a = 3706.1$ km, $b = 4238.7$ km.
10. $B = \frac{2\pi}{9}$, $b = 0.527$ Å, $c = 0.886$ Å.

In Exercises 11–22, first determine the number of triangles satisfying the given conditions. Then solve each possible triangle completely.

11. $A = 30.0°$, $a = 80.0$ mi, $b = 100.0$ mi.
12. $A = \frac{\pi}{6}$, $a = 50.0$ km, $b = 100.0$ km.
13. $A = 0.524$, $a = 40.0$ m, $b = 100.0$ m.
14. $A = 30°$, $a = 1.000$ m, $b = 100.0$ cm.
15. $A = 1.350$, $a = 13.4$ ft, $b = 11.46$ ft.
16. $A = \frac{\pi}{3}$, $a = 70.0$ light-years, $b = 75.0$ light-years.
17. $B = 52°9'$, $a = 134.16$ Å, $b = 84.54$ Å.
18. $A = \frac{\pi}{4}$, $a = 200.0$ nautical miles, $b = 100.0$ nautical miles.
19. $a = 150.0$ units, $b = 50.0$ units, $A = 130.0°$.
20. $A = \frac{5\pi}{7}$, $a = 75.0$ mm, $b = 86.0$ mm.
21. $a = 809$ cm, $b = 948$ cm, $A = 58.6°$.
22. $A = \frac{\pi}{8}$, $a = 34.6$ m, $b = 45.8$ m.

In Exercises 23–28, find the area of all possible triangles satisfying the given conditions.

23. $A = 2.00$, $b = 15.0$ km, $a = 24.0$ km.
24. $C = 53.1°$, $c = 390.0$ yd, $b = 425$ yd.
25. $b = 28.6$ m, $c = 24.3$ cm, $C = \frac{7\pi}{8}$.
26. $B = 38.72°$, $b = 53.55$ cm, $c = 85.61$ cm.
27. $A = 1.00$, $b = 68.3$ ft, $a = 42.42$ ft.
28. $B = 45°45'$, $b = 850.0$ light-years, $a = 1{,}000.0$ light-years.

*29. If $B = 16°15'$, $a = 75.0$ in., and $b = 29.0$ in., find the difference between the areas of the two triangles satisfying these conditions.

30. As the personnel manager for the Sure Way Survey Co., you are testing a job applicant. In surveying a triangular field, he produces the following measurements: two sides measure 145.0 ft and 94.2 ft and the angle opposite the 94.2-ft side is 42.1°. Should you recommend this person be hired? Back up your recommendation with *quantitative* reasons.

31. The next applicant to the Sure Way Survey Co. (see Exercise 30) surveys a triangular field and produces the following measurements for its sides: 122.0 ft, 153.2 ft, and 24.07 ft. Should you recommend this person be hired? Give a *quantitative* reason for your decision.

32. As a civil engineer, you must determine the amount of concrete needed to form the border of a triangular platform. The surveyor gives you only the following measurements: two sides measure 20.7 m and 25.0 m, and the angle opposite the 20.7-m side is 52°18′. What are the possibilities for the length of this concrete border?

33. As an inspector for the U.S. Patent Office, you can award a patent only if the description is completely clear and unambiguous. A budding inventor submits a patent application for a Power Pyramid and gives the following specifications for one of its triangular faces: Two of its sides are 45.4 in. and 64.1 in. long, and the angle opposite the shorter of these two sides is 28°41′. Does this information specify the lengths and angles of this triangular face unambiguously? If so, what are these dimensions? If not, prove it. Should you award a patent for the Power Pyramid?

34. The next inventor to visit the patent office (see Exercise 33) submits plans for a perpetual-motion machine having a triangular part with the following specifications: Two sides are 36.80 cm and 9.403 cm long, and the angle opposite the shorter of these two sides is 18.08°. Should this person receive a patent? Why? (Justify your answer quantitatively.)

35. You are ordering ceramic tile for a triangular table top. The only measurements you have are that one side measures 85 cm, another measures 71 cm, and the angle opposite the 71-cm side is 49°. What is the area of tile you will possibly need?

36. A client has contracted your jewelry store to gold-plate a triangular metal medallion on both sides. He gives you the following measurements: Two sides measure 67.0 mm and 58.2 mm, and the angle opposite the longer of these two sides is 140°34′. What area of gold should you buy?

4.4 VECTORS

4.4-1 What Is a Vector?

If you told someone that your age is 19 yr southeast, you would invite stares because age has only *one* property—its *magnitude* (or scale). Direction does not apply to it! Other similar quantities include height, time of day, temperature, and so on. But suppose you are planning a trigonometry study session at your house, and you need to give directions to your friends. If you only tell them to go 1.5 mi from campus, you will probably study alone! It is obvious that in this case *two* properties are essential: *magnitude* and *direction*. Other familiar quantities for which direction is essential include *force* (pushing a refrigerator with a 50-lb force eastward is different from pushing it northward) and *velocity* (driving 55 mph northeast is different from driving 55 mph northwest).

As we have just seen, there are two different types of quantities. One has only *magnitude* (scale) but no direction, whereas the other has both *magnitude and direction*. We call the first type a *scalar* quantity and the second type a *vector* quantity.*

DEFINITION OF SCALARS AND VECTORS

A *Scalar* is any quantity that has only *magnitude*.

A *Vector* is any quantity that has both *magnitude and direction*.

Examples of *scalars* include age, temperature, time, the number of pages in a book, distance, speed, and so on.

Examples of *vectors* include displacement (directed distance), force, and velocity (speed with direction). For example, a displacement of 25 mi southwest would have a magnitude of 25 mi and a direction southwest. A force of 80 lb at 30° south of east would have a magnitude of 80 lb and a direction 30° south of east.

Writing and Drawing Vectors

Because scalars and vectors are quite different from each other, we must write them differently. We write scalars in the ordinary way. But we usually indicate a vector by putting an arrow over it (or sometimes by underlining it). In books or on computers, arrows are not always easy to use, so we usually use **boldface** type for vectors. Thus we can represent a force vector as

$$\vec{F} \quad \text{or} \quad \mathbf{F}$$

* Vectors first entered mathematics in the early 1800s, mostly as a geometric interpretation of complex numbers. They became especially prominent in applied mathematics due to the work in electromagnetism by the Scottish physicist James Clerk Maxwell in the 1860s and 1870s.

When we want to represent only the *magnitude* of a vector, we usually do it either by putting absolute value signs around the vector or by writing the vector without the arrow or without boldface. For example, to represent a force vector of 65 lb directed southeast, we could represent the *vector* as

$$\vec{F} = 65 \text{ lb southeast} \quad \text{or} \quad \mathbf{F} = 65 \text{ lb southeast}$$

and the *magnitude* of this vector as

$$|\vec{F}| = 65 \text{ lb} \quad \text{or} \quad |F| = 65 \text{ lb} \quad \text{or} \quad F = 65 \text{ lb}$$

Figure 4.4-1

A directed line segment represents a vector.

It is very helpful to represent vectors by sketching them. We do this by letting a *directed line segment* (arrow) represent the vector. We call the front tip of this arrow the head and the back the tail (see Figure 4.4-1). The arrow points in the *direction* of the vector, and its length represents the vector's *magnitude*.

DRAWING VECTORS

A directed line segment represents a vector. The *direction* of the line segment represents the *direction* of the vector. The *length* of the line segment represents (is proportional to) the *magnitude* of the vector.

Example 20

If **v** is a velocity vector of 40 mph at 30° north of east, and **w** is velocity vector of 80 mph at 45° southwest, draw these two vectors (Figure 4.4-2).

Solution

Notice that the length of the line segment representing **w** is twice as long as the one representing **v** because the magnitude of **w** is twice that of **v**. ■

NOW IT'S YOUR TURN

Sketch the vectors **A** = 50 km/hr at 40° south of east and **B** = 75 km/hr at 20° east of north. (Answer: See Figure 4.4-3)

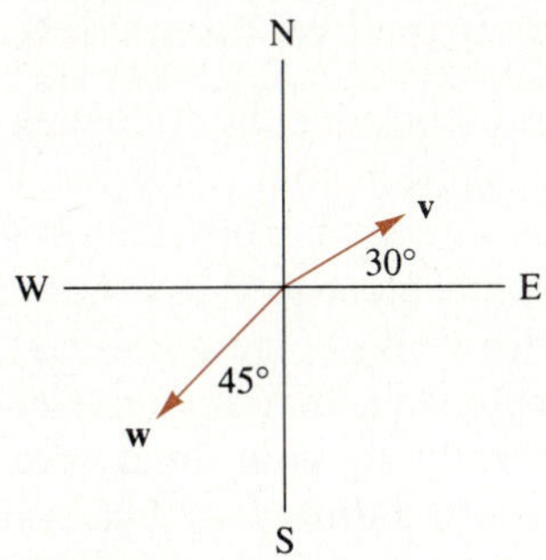

Figure 4.4-2

Vectors v and w

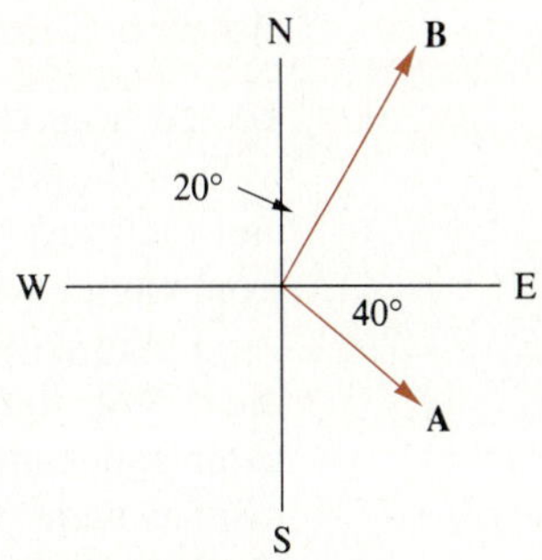

Figure 4.4-3

Vectors A and B

Moving Vectors

A vector has two properties—magnitude and direction. Therefore, if two vectors have the same magnitude and direction, they must be the *same* vector—*even if they are drawn at different places*. For example, in Figure 4.4-4, all three line segments represent the same vector because all three have the same magnitude and direction. This very important fact means that *we can move vectors around provided that we do not change their magnitude and direction* while doing so.

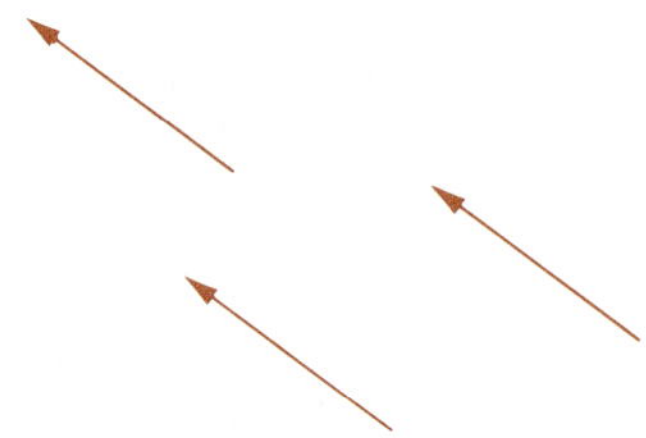

Figure 4.4-4

All three line segments represent the same vector because they have the same magnitude and direction.

4.4-2 Adding Vectors

Usually several vectors act together. For example, two people can push on the same box, or you can walk on a moving escalator, thus adding your velocity to that of the escalator. In these cases, the vectors are combining—or adding—so that they are *equivalent to a single vector*. We call this single vector the **resultant** or sum of the individual vectors.

DEFINITION OF RESULTANT

The *resultant* (or sum) of several vectors is the single vector which, by itself, is completely equivalent to the individual vectors acting together.

How to Find the Resultant of Two Vectors

To see how vectors add, consider the following situation. Suppose you take a walk in the woods starting from your campsite. You first walk for 2 mi northeast and then turn and walk directly northward for 3 mi. What single walk, starting from camp, would be equivalent to these two walks?

First, sketch the two walks (Figure 4.4-5a). You can readily see from Figure 4.4-5b that the single walk that is equivalent to these two walks would be from the campsite directly to your final location. This means that, if the displacement vectors representing your two walks are drawn head-to-tail, the single vector that is *equivalent* to these two vectors is drawn from the tail of the first vector to the head of the second one. Therefore, this vector by itself is equivalent to the two vectors acting together, so it must be the *resultant* of these two vectors (Figure 4.4-6).

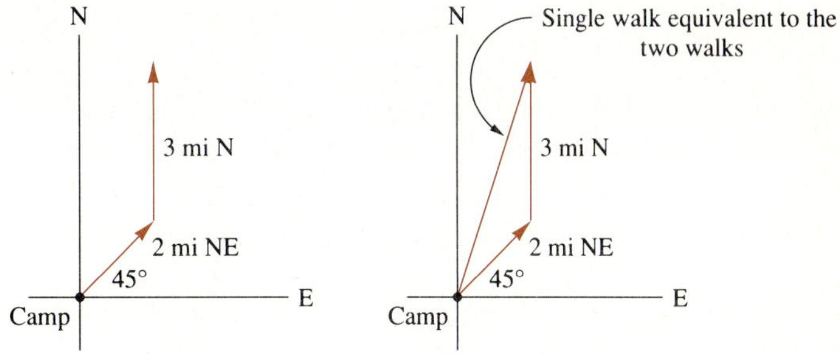

Figure 4.4-5 *(a) Two walks alone* *(b) Two walks and the equivalent walk*

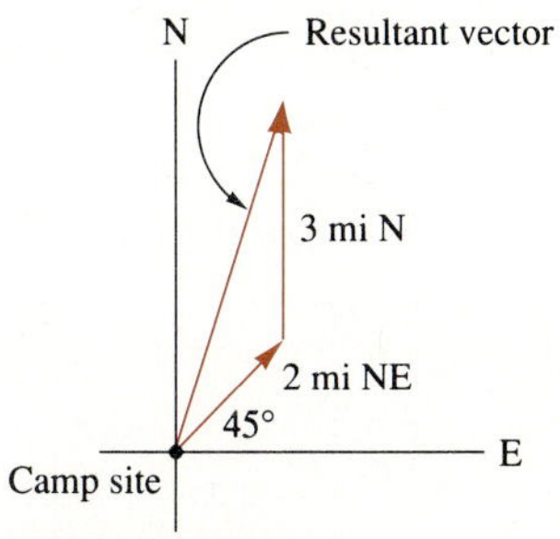

Figure 4.4-6

Resultant vector of the two displacement vectors

We have used displacement vectors to discover how to add vectors, but the same procedure applies to *all* vectors. We can now generalize the method we have learned. Because the two vectors and their resultant form a closed polygon, we often call this the *polygon method*.

ADDING TWO VECTORS

Polygon Method

1. If the vectors are not already head-to-tail, put them that way by keeping one (it doesn't matter which one) fixed and moving the other one *without changing its magnitude or direction*. This leaves one free head and one free tail.

2. The resultant of these two vectors is *from* the free *tail* of the vector you kept fixed *to* the free *head* of the vector you moved.

Example 21

Find the resultant of the vectors **A** and **B** in Figure 4.4-7.

Figure 4.4-7

Solution

Let's *choose* to keep **A** fixed. First put these vectors head-to-tail by moving **B** so that its tail is at the head of **A**, as shown in Figure 4.4-8a. Now draw the resultant **R** from the free tail of **A** to the free head of **B**, as shown in Figure 4.4-8b. ■

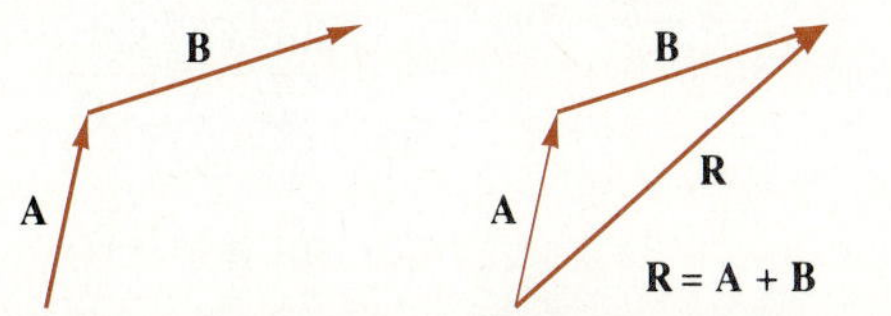

Figure 4.4-8
Adding A and B *(a) A and B drawn head to tail* *(b) The resultant R of A and B*

NOW IT'S YOUR TURN

We kept **A** fixed. Now add these two vectors by keeping **B** fixed and moving **A**. Do this carefully to see that you get the same resultant no matter which vector you keep fixed. This fact is very important. It means that you can add vectors in *any order*.

Example 22

Let's solve the example we used to discover how to add vectors, in which you took a walk in the woods. Starting from your campsite, you walk 2.0 mi northeast, then turn and walk 3.0 mi north. What single walk would be equivalent to these two?

Solution

As we saw in Figure 4.4-6, the single walk is the resultant of these two displacement vectors. We need to find the magnitude and direction of this resultant vector. First sketch it (Figure 4.4-9). In the figure, $\beta = 45°$ and $\theta + \beta = 180°$. Thus $\theta + 45° = 180°$, which gives us $\theta = 135°$. Now we can use the law of cosines.

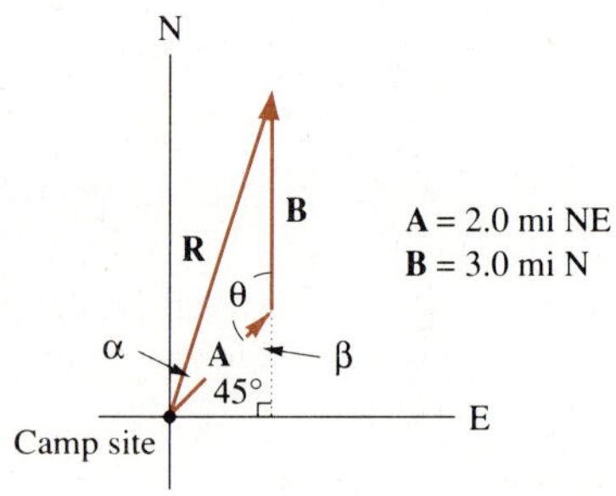

Figure 4.4-9
The resultant of A and B

$$R^2 = A^2 + B^2 - 2AB\cos\theta$$
$$R^2 = (2.0\text{ mi})^2 + (3.0\text{ mi})^2 - 2(2.0\text{ mi})(3.0\text{ mi})\cos 135°$$
$$R^2 = 4.0\text{ mi}^2 + 9.0\text{ mi}^2 - (12.0\text{ mi}^2)(-\sqrt{2}/2) = 21.5\text{ mi}^2$$
$$R = \sqrt{21.5\text{ mi}^2} = 4.6\text{ mi}$$

We need α to find the direction of **R**. The law of sines is useful here.

$$\frac{\sin\alpha}{B} = \frac{\sin\theta}{R}$$
$$\sin\alpha = \frac{B(\sin\theta)}{R} = \frac{(3.0\text{ mi})(\sin 135°)}{4.6\text{ mi}} = 0.46$$
$$\alpha = \sin^{-1}(0.46) = 27°$$

The direction of **R** is therefore

$$45° + \alpha = 45° + 27° = 72° \text{ north of east.}$$

Therefore, if you walk 4.6 mi at 72° north of east, you wind up in the same place as you would when you take the two walks. ■

NOW IT'S YOUR TURN

If you walk 4.0 km at 60° south of east and then walk directly eastward for another 6.0 km, find the magnitude and direction of your resultant displacement. (Answer: 8.7 km at 23° south of east)

Parallelogram Method for Adding Vectors

Although you can always use the polygon method to add two vectors, there is a short cut that is often useful. In Fig. 4.4-10a and 4.4-10b, we have added two vectors by the polygon method. Now notice that the resultant **R** is the diagonal of a *parallelogram* having **A** and **B** as two sides, as shown in Figure 4.4-10c. Therefore, if **A** and **B** are tail-to-tail, just complete the parallelogram. Then draw the resultant from the corner where their tails meet to the opposite corner of the parallelogram (Figure 4.4-11). This method for adding vectors is called the *parallelogram method*.

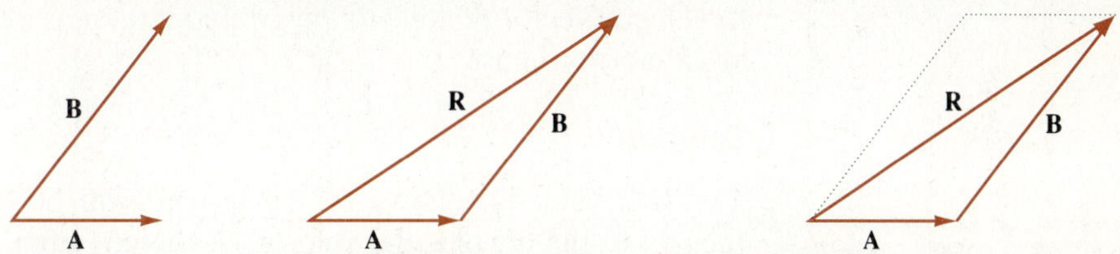

(a) The original vectors

(b) Adding the vectors A and B

(c) R is the diagonal of a parallelogram formed by the vectors when they are tail to tail.

Figure 4.4-10
Vector addition

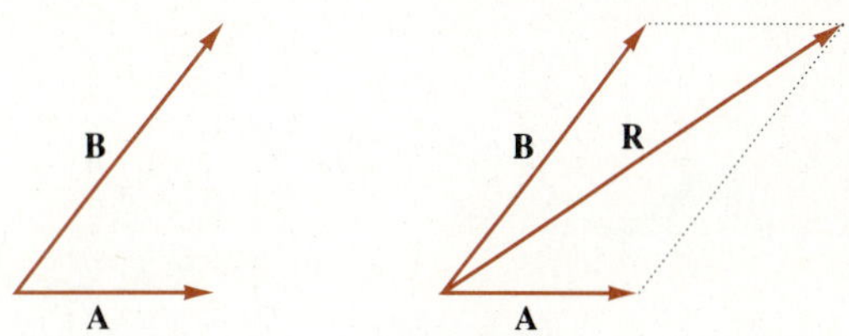

(a) The original vectors tail to tail

(b) Complete the parallelogram and draw R from the corner where the tails meet to the opposite corner.

Figure 4.4-11
Adding vectors by the parallelogram method

Example 23

Solve Example 21 by the parallelogram method.

Solution

First put the two vectors tail-to-tail (Figure 4.4-12b). Then draw the resultant **R** from the corner where the tails meet to the opposite corner (Figure 4.4-12c). ■

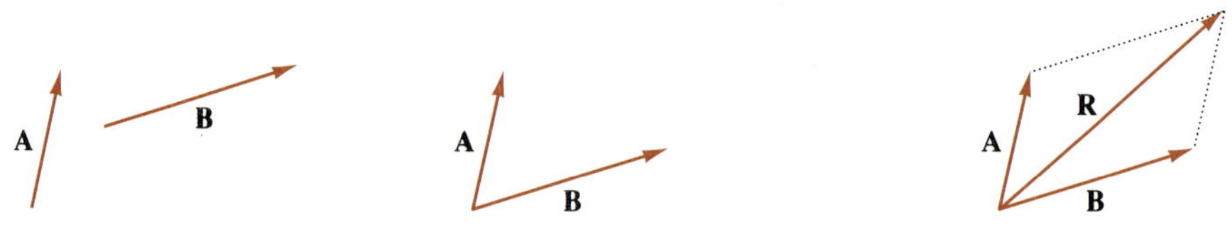

Figure 4.4-12
Adding A and B by the parallelogram method

(a) The original vectors *(b) Put the vectors tail to tail* *(c) Complete the parallelogram and draw in the resultant R*

NOW IT'S YOUR TURN

Solve Example 22 by the parallelogram method. ***Hint:*** Put the vectors tail-to-tail.

Example 24
Airplane Navigation

An airplane gets its lift due to its speed *relative to the air* (called *air speed*). If the air is moving (which it usually is), the plane's velocity relative to the ground will be the resultant of the velocity of the air and the plane's velocity relative to the air.

If a jet has an air speed of 600 mph and is heading at 40° north of west, find its velocity vector relative to the ground if there is a steady 70 mph wind blowing at 25° north of east.

Solution

The jet's velocity relative to the ground will be the resultant of its velocity relative to the air plus the velocity of the air. First make a sketch showing the plane and the wind (Figure 4.4-13a), and then sketch the vectors representing

Figure 4.4-13
Velocities of plane and air

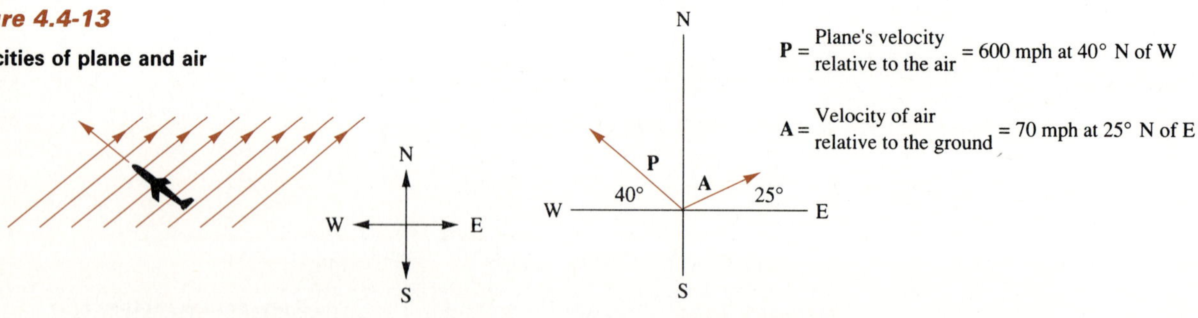

(a) Motion of plane and air *(b) Velocity vectors of plane and air*

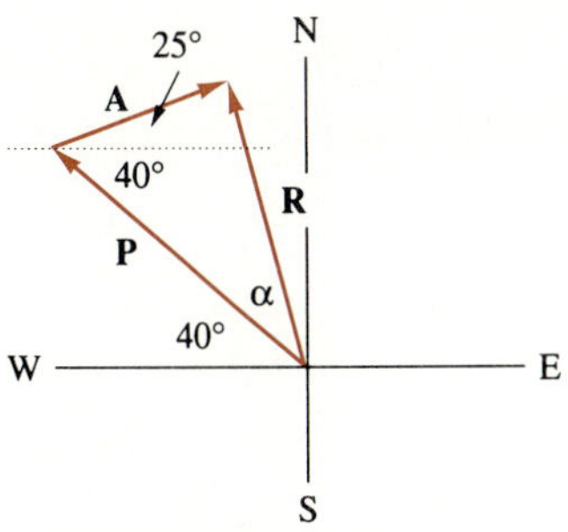

Figure 4.4-14

Adding the two velocities

these velocities (Figure 4.4-13b). Now add these two velocities, keeping **P** fixed and calling **R** the resultant (Figure 4.4-14).

We can use the laws of geometry and trigonometry to find the lengths of the sides of this oblique triangle. But remember that when we draw a vector, the length of the directed line segment is proportional to the vector's magnitude. Therefore, these lengths actually represent *speeds*.

We can use the law of cosines to find R.

$$R^2 = A^2 + P^2 - 2AP\cos 65^\circ$$

$$R^2 = (70\text{ mph})^2 + (600\text{ mph})^2 - 2(70\text{ mph})(600\text{ mph})\cos 65^\circ$$

$$R^2 = 329{,}000\text{ (mph)}^2$$

$$R = \sqrt{329{,}000\text{ (mph)}^2} = 570\text{ mph}$$

This is the speed relative to the ground, often called the *ground speed*.

The direction of **R** is $(40^\circ + \alpha)$ north of west, so we need to find α. The law of sines gives us

$$\frac{\sin\alpha}{A} = \frac{\sin 65^\circ}{R}$$

$$\sin\alpha = \frac{A(\sin 65^\circ)}{R}$$

$$\sin\alpha = \frac{(70\text{ mph})(\sin 65^\circ)}{570\text{ mph}} = 0.111$$

$$\alpha = \sin^{-1}(0.111) = 6.4^\circ$$

Therefore, the direction of **R** is $40^\circ + 6.4^\circ = 46^\circ$ north of west. ■

NOW IT'S YOUR TURN

If the wind changed to 150 mph at 35° north of east, what would be the jet's ground speed? (Answer: 580 mph at 57° north of west)

Example 25

Nursing

A patient must have his broken leg put in traction, as shown in Figure 4.4-15a. The physician specifies that the resultant traction force must be 10.0 lb and must be directed *along the leg* while the leg is supported at 25° above the horizontal. If the nurse attaches a 5.0 lb weight to pull horizontally on the leg, what must the other weight be and at what angle must it be adjusted to pull?

Solution

First sketch the traction arrangement and call W the unknown weight (Figure 4.4-15a). The two weights produce force vectors **W** and **F** on the leg, as shown in Figure 4.4-15b. The angle we want to find is β.

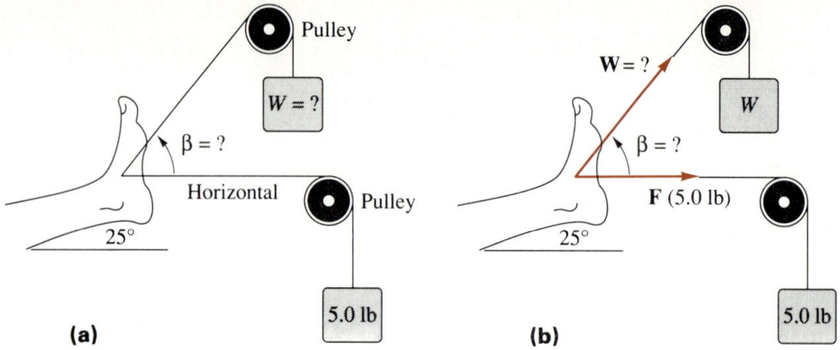

Figure 4.4-15

This problem is different from the previous examples because here *we know the resultant*—it must be 10.0 lb directed along the leg, so it must be at 25° above the horizontal (Figure 4.4-16). Now add **W** and **F** to obtain **R**. We choose to keep **F** fixed and move **W** using the polygon method (Figure 4.4-17).

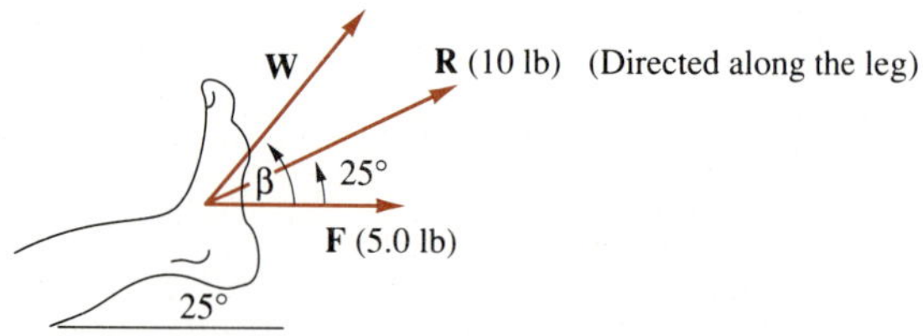

Figure 4.4-16

Force vectors and resultant vector acting on leg

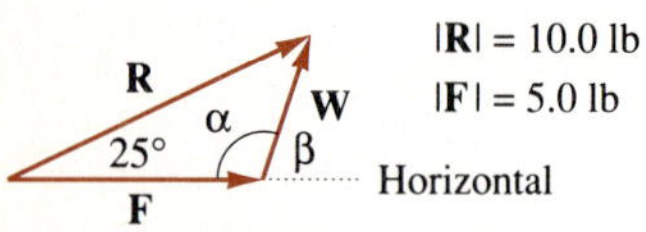

Figure 4.4-17

Adding the vectors

Because we know two sides and the included angle in this oblique triangle, we can use the law of cosines to find W.

$$W^2 = (5.00 \text{ lb})^2 + (10.0 \text{ lb})^2 - 2(5.00 \text{ lb})(10.0 \text{ lb})\cos 25° = 34.4 \text{ lb}^2$$

$$W = \sqrt{34.4 \text{ lb}^2} = 5.9 \text{ lb}$$

Now we can use the law of cosines* to find α.

$$R^2 = W^2 + F^2 - 2WF \cos \alpha$$

$$(10.0 \text{ lb})^2 = (5.9 \text{ lb})^2 + (5.0 \text{ lb})^2 - 2(5.9 \text{ lb})(5.0 \text{ lb}) \cos \alpha$$

$$40.2 \text{ lb}^2 = -59 \text{ lb}^2 \cos \alpha$$

$$\cos \alpha = \frac{40.2 \text{ lb}^2}{-59 \text{ lb}^2} = -0.681$$

$$\alpha = \cos^{-1}(-0.681) = 133°$$

Now we can find β: $\beta + \alpha = 180°$, so

$$\beta = 180° - \alpha = 180° - 133° = 47° \text{ above horizontal} \blacksquare$$

*If you are wondering why we didn't use the law of sines instead, try it and resolve the ambiguous answer it gives.

NOW IT'S YOUR TURN

Suppose that the physician decides that the leg must be kept at the same angle but the traction force must be increased to 12.0 lb. If the 5.0-lb vector is kept the same, find the new value of the weight W and the new angle to which it must be set. (Answer: 7.8 lb at 42° above the horizontal)

EXERCISES 4.4

In Exercises 1–4, carefully sketch each of the vectors and identify its magnitude and direction.

1. A displacement of 60 m at 25° east of north.

2. A velocity of 55 mph at $\frac{\pi}{3}$ south of west.

3. A force of 100 lb at 40° north of west and a force of 30 lb at 20° south of east.

4. A displacement of 50 m at $\frac{\pi}{6}$ west of south and a displacement of 75 m at 1.0 rad south of west.

In Exercises 5–12, make measurements on the vectors in the figures and then use the distance scale shown to find the magnitude and direction of each vector.

5.

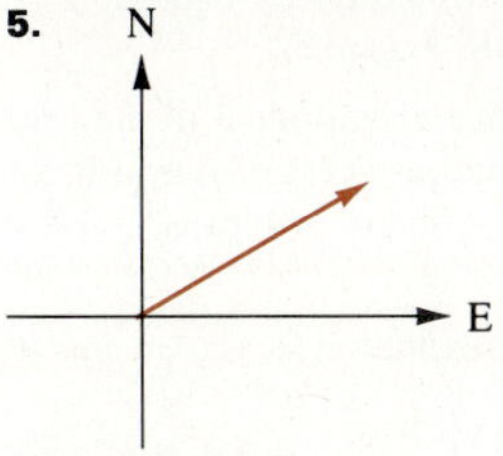

1.0 cm = 2.0 lb

6.

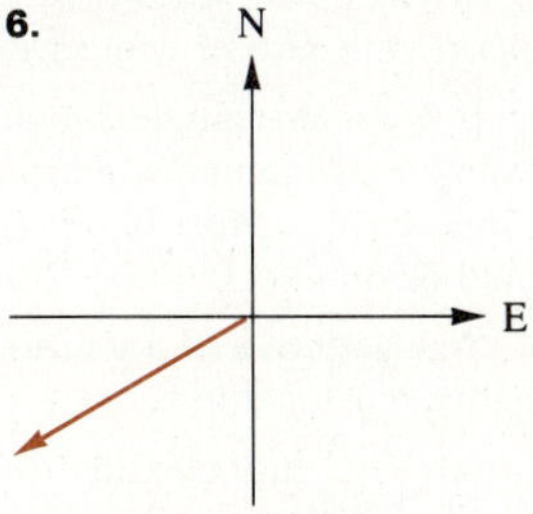

1.0 cm = 3.0 ft

7.

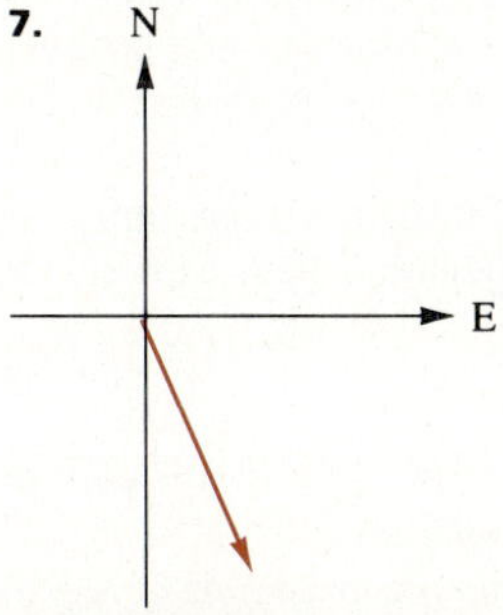

2.0 cm = 3.0 lb

8.

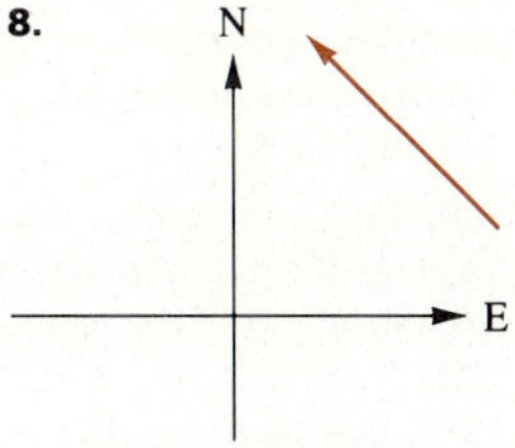

1.0 cm = 4.0 m/s

9.

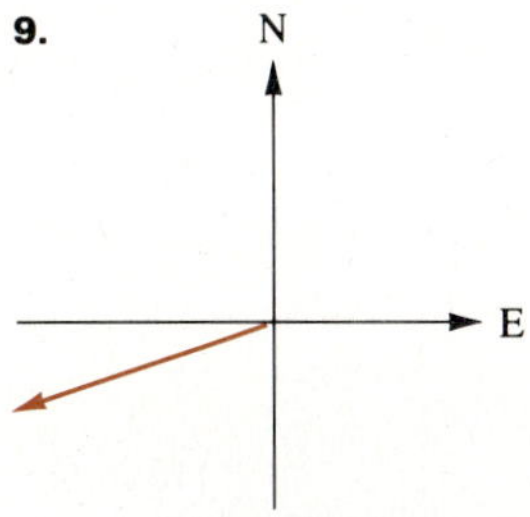

1.5 in. = 3.6 lb

10.

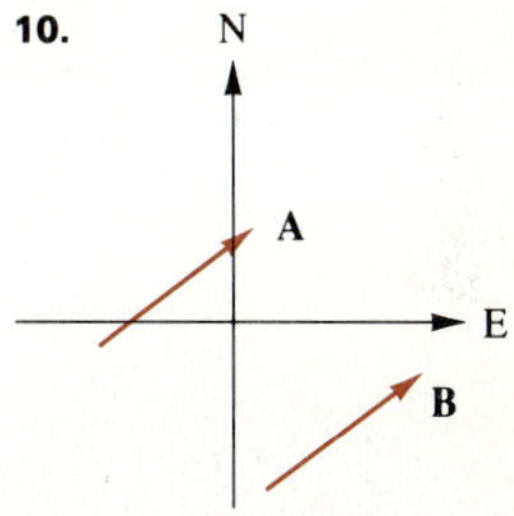

2.0 cm = 8.0 mph

11.

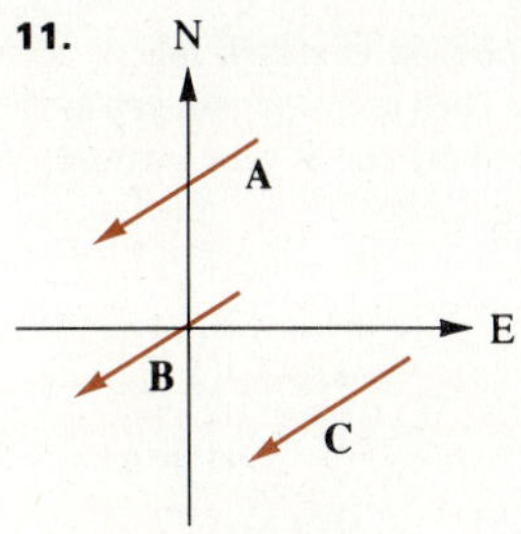

2.0 cm = 5.0 lb

12.

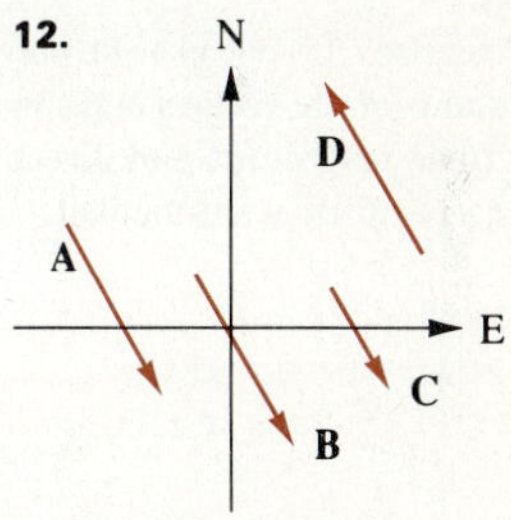

3.0 cm = 5.0 km/s

In Exercises 13–18, draw the resultants of the vectors in the figures graphically using the polygon method.

13.

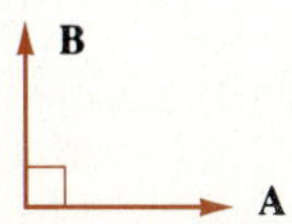

14.

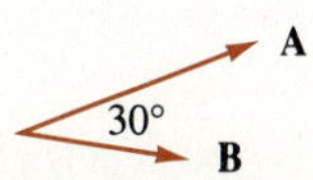

15.

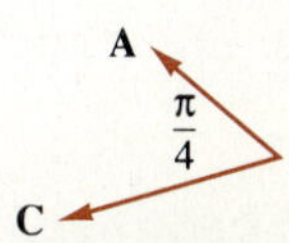

16.

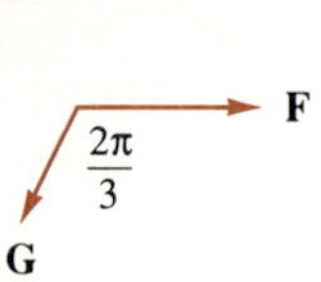

17.

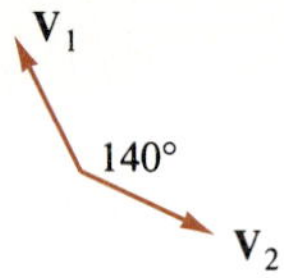

18.

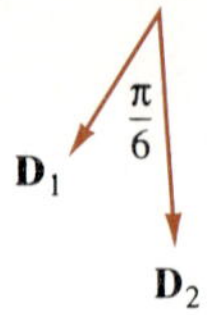

In Exercises 19–24, find each resultant in the figures graphically using the polygon method. Then use trigonometry to find the magnitude and direction of the resultant. (Express the direction by giving the angle between the resultant and the direction of the smaller of the given vectors.) Finally, check each answer by adding the vectors by the parallelogram method.

19.

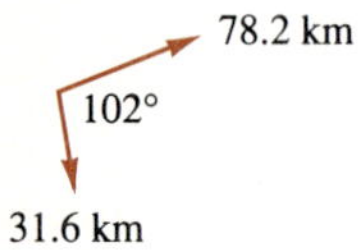

20.

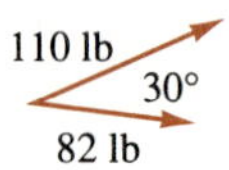

21.

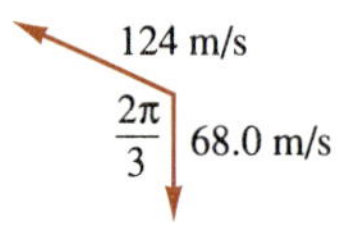

22.

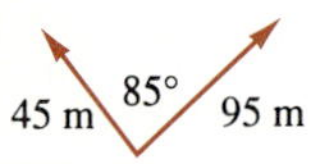

23.

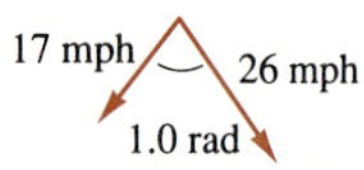

24.

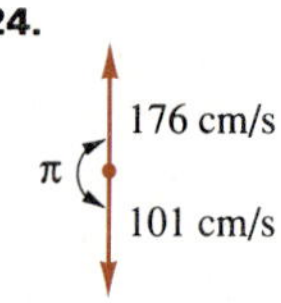

In Exercises 25–30, use the polygon method to sketch the resultants of the vectors in the figures. Then use trigonometry to find their magnitude and direction. Finally check your answer by the parallelogram method.

25.

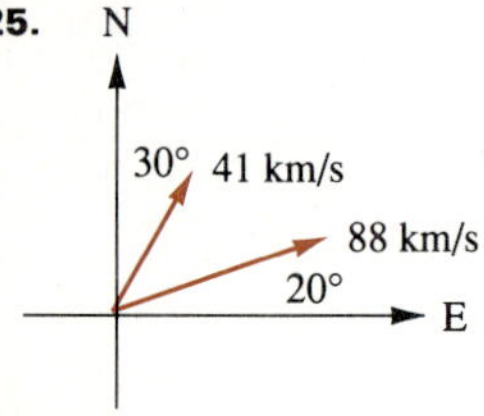

26.

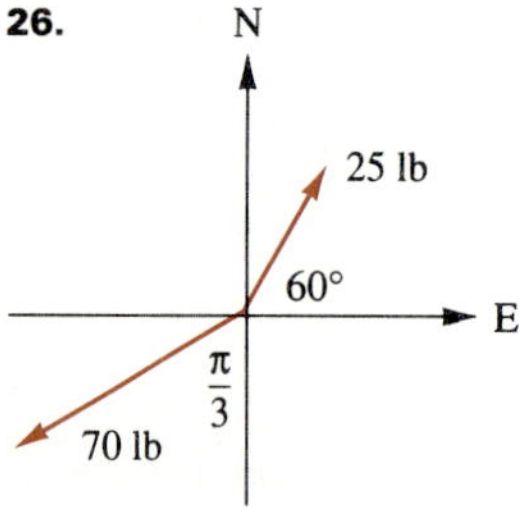

27.

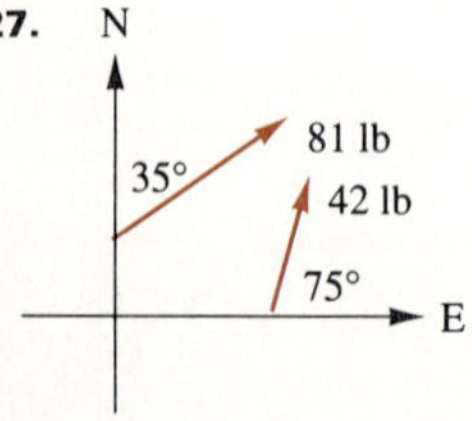

28.

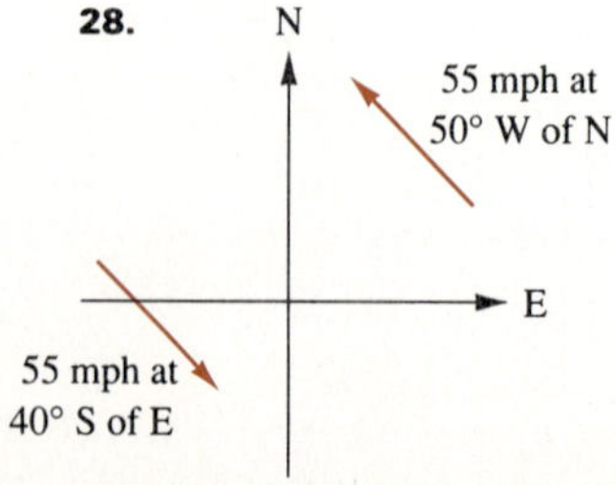

***29.**

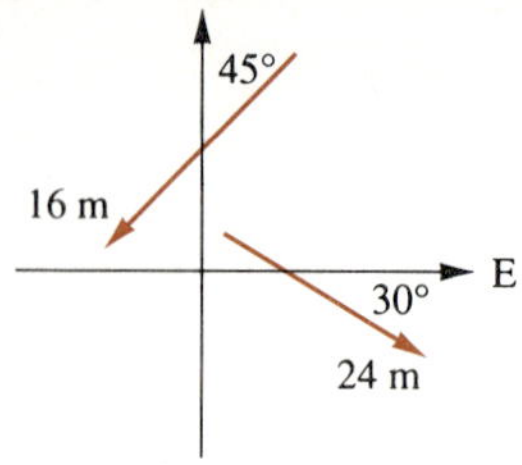

***30.**

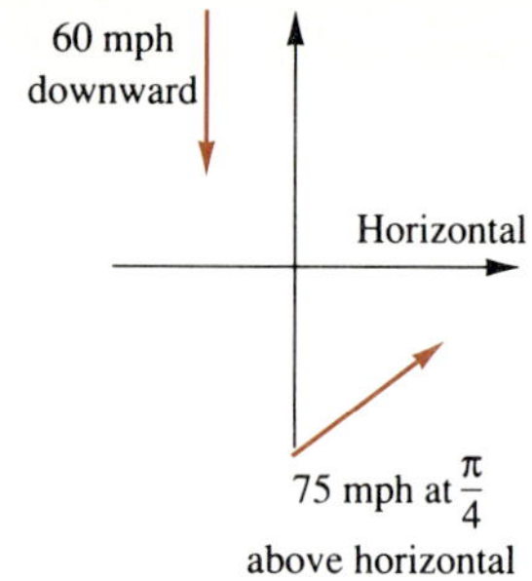

In Exercises 31–36, what vector B must you add to the vector A so the resultant will be as given? Find the magnitude and direction of B. Check each answer with a graphic sum.

31. **A** = 46 lb at 30° north of east, **R** = 60 lb northward.

32. **A** = 50 m/s at $\frac{\pi}{4}$ south of west, **R** = 50 m/s directly southward.

33. **A** = 100 ft at 1.00 rad south of east, **R** = 80 ft directly northward.

34. **A** = 650 km at 15° west of north, **R** = 440 km at 25° south of east.

35. **A** = 61.0 mph at $\frac{\pi}{6}$ south of east, **R** = 125 mph at $\frac{\pi}{10}$ south of east.

36. **A** = 14.7 lb at 21.6° north of west, **R** = **O**.

37. When adding vectors, is $\overrightarrow{1 \text{ ft}} + \overrightarrow{1 \text{ ft}}$ *necessarily* equal to $\overrightarrow{2 \text{ ft}}$? Support your answer with several good examples.

38. If **A** has magnitude 2 lb and **B** has magnitude 3 lb, find the maximum and minimum magnitude the resultant of **A** and **B** can possibly have. ***Hint:*** Interpret this problem *graphically* and it should be quite easy.

39. The Negative of a Vector. By definition, a vector plus its negative must add to zero.

(a) Use this definition to interpret the negative of a vector graphically.
(b) If **A** is 50 lb north, what is −**A**?
(c) If **B** = 66 mph at 20° north of west, find the magnitude and direction of −**B**.

***40. Subtraction of Vectors.** We know how to find **A** + **B**, but how do we find **A** − **B**?

(a) First, recognize that **A** − **B** is the same thing as **A** + (−**B**). For the vectors in Figure 4.4-18, express −**B** graphically (see Exercise 39).
(b) Now, using the fact that **A** − **B** = **A** + (−**B**), add −**B** to **A** to obtain **A** − **B**.
(c) If **A** = 50 lb eastward and *B* = 75 lb at 60° north of west, find **A** + **B** and **A** − **B** graphically.
(d) Use trigonometry to calculate the magnitude of **A** − **B** in part (c).

Figure 4.4-18

41. Subtraction of Vectors. (See Exercise 39 and 40.) For each pair of vectors **A** and **B**, find **A** − **B** graphically. Then find its magnitude and direction using trigonometry.

(a) **A** = 88 ft/s northward, **B** = 40 ft/s westward

(b) **A** = 70.0 lb at $\frac{\pi}{6}$ south of west, **B** = 100.0 lb at $\frac{\pi}{4}$ east of south

(c) **A** = 240 mi at $\frac{\pi}{12}$ north of west, **B** = 300 mi at 1.0 rad north of east.

(d) **A** = 55 mph at 40° west of north, **B** = 105 mph at 15° south of east.

42. Components of a Vector. Mathematicians, scientists, and engineers usually work with vectors in a plane using their projection along *the x and y axes*. They call these projections the components of the vector as follows:

x component of **A** = A_x = projection of **A** along the x axis
y component of **A** = A_y = projection of **A** along the y axis

(See Figure 4.4-19.) Show that the x and y components of **A** are

$$A_x = A\cos\theta \qquad A_y = A\sin\theta$$

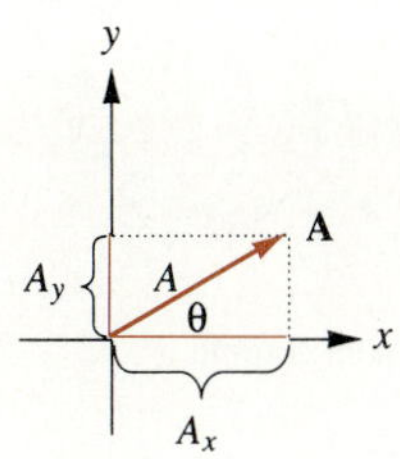

Figure 4.4-19

43. Vector Components. (See Exercise 42.) Find the x and y components of each of the following vectors, where θ is the angle with the $+x$ axis and A is the vector's magnitude.

(a) $A = 15$ m, $\theta = 60°$

(b) $A = 25.0$ m/s, $\theta = 23.4°$

(c) $A = 65$ mph, $\theta = 340°$

(d) $A = 36.7$ lb, $\theta = 1.2\pi$

(e) $A = 100.0$ cm/hr, $\theta = 105.1°$

(f) $A = 145$ m/s, $\theta = -\frac{5\pi}{6}$

44. Vector Components. (See Exercises 42 and 43.) Show that the magnitude and direction of the vector **A** are related to its components by the formulas

$$A = \sqrt{A_x^2 + A_y^2} \qquad \tan\theta = \frac{A_y}{A_x}$$

where θ is defined in Figure 4.4-19.

45. Vector Components. (See Exercise 44.) Find the magnitude and direction of each of the vectors having the components given below. In each case, sketch the vector clearly.

(a) $A_x = 4.0$ ft, $A_y = 3.0$ ft

(b) $A_x = -6.0$ m/s, $A_y = 5.0$ m/s

(c) $B_x = 12.5$ lb, $B_y = -10.0$ lb

(d) $F_x = -16.0$ lb, $F_y = -11.4$ lb

(e) $V_x = -46$ km/hr, $V_y = 0$

(f) $D_x = 0$, $D_y = -18.2$ cm

46. Two bulldozers are pushing a huge boulder at the same time. If one exerts 950 lb at 40.0° north of east and the other exerts 1200 lb at 20.0° east of north, find the magnitude and direction of the resultant force they exert on the boulder.

47. A pilot files the following flight plan at an airport. He will fly 175 mi at 31.2° north of west and then fly 152 mi at 19.6° east of south. Find the magnitude and direction of the resultant displacement of this plane after it makes this flight.

48. A plane leaves Seattle and flies 85.0 mi at 22.0° north of east and then changes direction to 48.0° south of east. After flying 115 mi in this new direction, he must make an emergency landing on a field. The Seattle Airport dispatches a rescue crew. In what direction and how far should this crew fly to go directly to this field?

49. A leg in traction has two weights attached to it, as shown in Figure 4.4-20. Find the resultant force these weights exert on the leg.

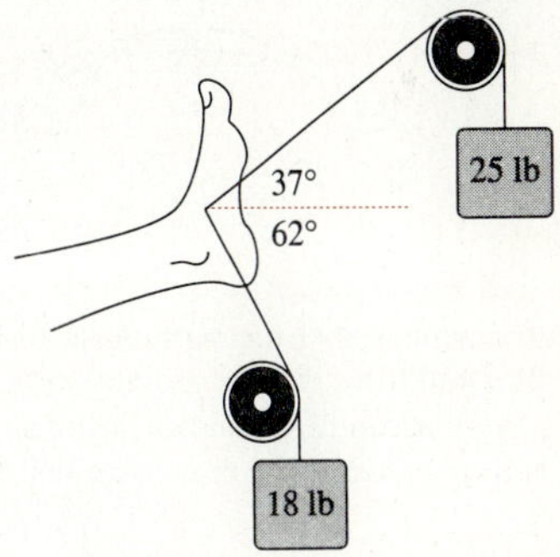

Figure 4.4-20

50. A rocket fires two engines simultaneously. One produces a thrust of 568 lb directly forward, whereas the other gives a 418-lb thrust at 32.4° above the forward direction. In Figure 4.4-21, the thrust will be *opposite* to the direction of the exhaust. Find the magnitude and direction (relative to the forward direction) of the resultant force that these engines exert on the rocket.

Figure 4.4-21

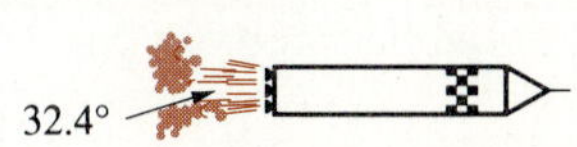

51. A 25-mph wind is blowing out of the northeast at 60.0 west of south. A duck is flying in this wind with a speed of 12 mph relative to the air heading directly westward. How fast and in what direction is this duck moving relative to the ground?

52. A ship leaves from Honolulu and sails 285 km at 50.0° west of north. In what direction must it then head, and how far must it sail, so that its resultant displacement will be 115 km directly west of Honolulu?

***53.** A 150-lb patient is to be suspended by two slings in a hospital so that he does not touch the floor, as shown in Figure 4.4-22. What force vector **F** must the other sling exert on this patient?

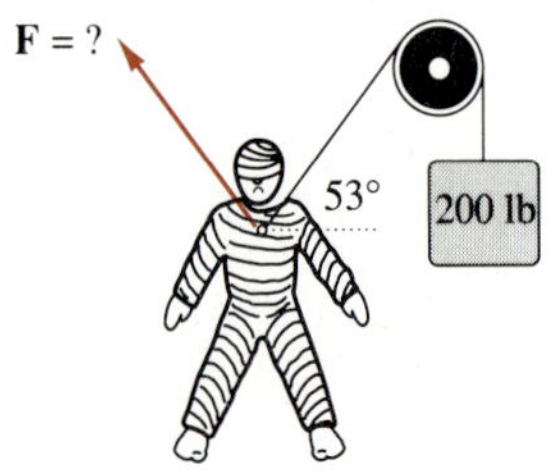

Figure 4.4-22

***54.** The wind is blowing steadily from west to east at 45 mph. In what direction and with what *air speed* should a plane fly so it has a ground speed of 65 mph at 50.0° north of east?

55. A dentist puts the string as shown to pull inward on a protruding incisor (Figure 4.4-23). If the string must produce a force of 10.0 oz directly inward, to what tension in ounces must she adjust the string?

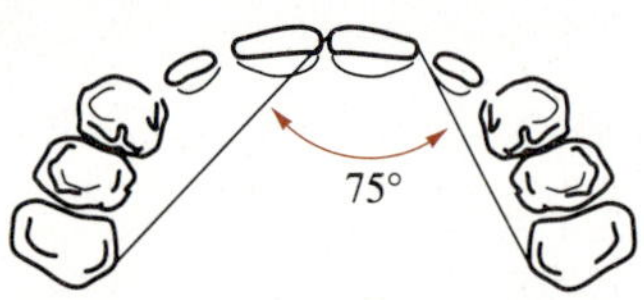

Figure 4.4-23

56. A 240-lb lineman is at the scrimmage line in his crouched position. If the front force that the ground exerts on him is 65 lb at 60.0° above the horizontal, as shown in Figure 4.4-24, find the magnitude and direction of the rear force that the ground exerts on him.

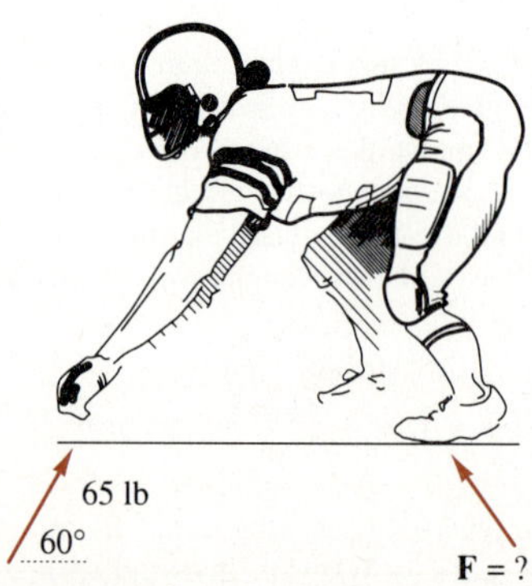

Figure 4.4-24

****57.** A 125-lb crate is resting on a ramp that slants above the horizontal at 35.0°, as shown in Figure 4.4-25. Besides its weight, which is vertical, this crate feels a friction force up the ramp directed parallel to its surface and a contact force perpendicular to the ramp's surface. Find the magnitude of both the friction force and the contact force. ***Hint:*** Use components perpendicular to and parallel to the surface. See Exercises 42–45.

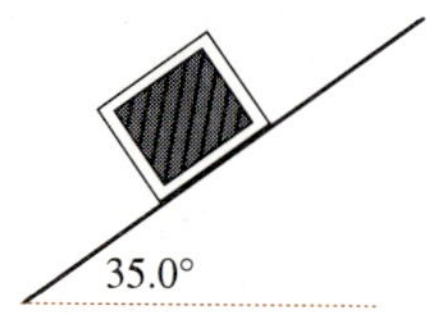

Figure 4.4-25

***58.** A student intently reading *Applied Trigonometry* has his head inclined at 60.0° above the horizontal, as shown in Figure 4.4-26. Besides its weight, his head feels a pull due to the neck muscles directed along the neck, and a force of contact due to his spinal column. If his head weighs 25 lb (mostly brains!) and the pull due to the neck muscles is 29.0 lb, find the magnitude and direction of the force due to the spinal column.

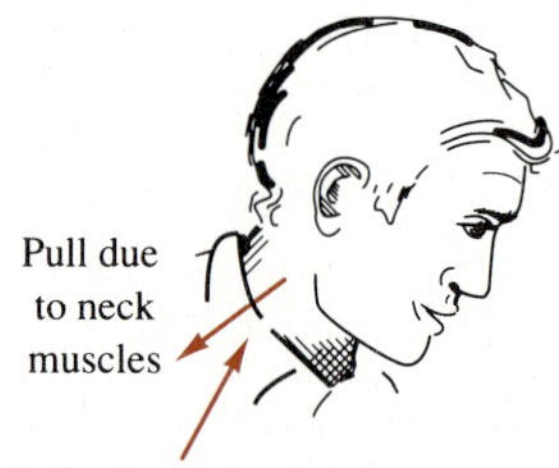

Figure 4.4-26

***59.** A weight lifter has a forearm that weighs 5.0 lb and is lifting a 22-lb weight. The only other significant forces on his forearm come from the biceps muscle, which acts perpendicular to the forearm, and the force at the elbow. If the biceps produce a pull of 68 lb when the forearm is raised 41° above the horizontal, find the magnitude and direction for the force that the elbow exerts on the forearm (Figure 4.4-27).

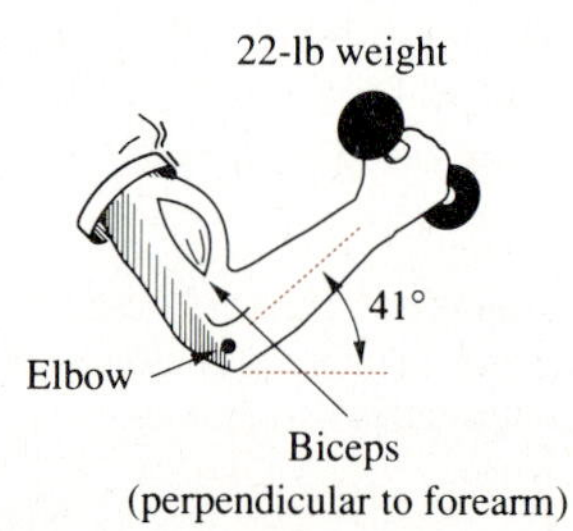

Figure 4.4-27

***60.** Three weights are hanging stationary, as shown in Figure 4.4-28. Find W and θ.

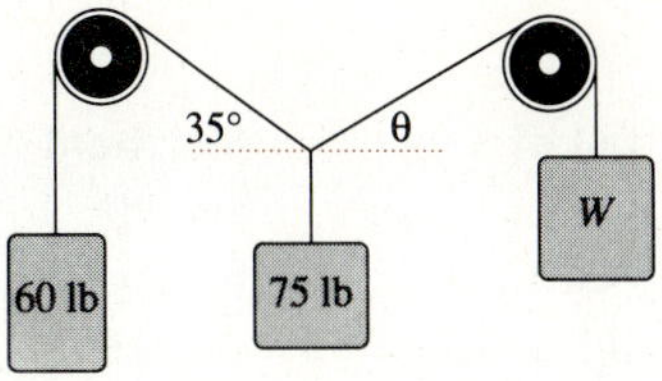

Figure 4.4-28

***61.** A 46.0-lb shelf is supported horizontally by a hinge and a cable, as shown in Figure 4.4-29. If the tension force in the cable is 31.8 lb, find the magnitude and direction of the force that the hinge exerts on the shelf.

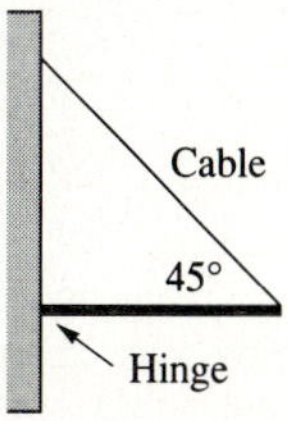

Figure 4.4-29

***62.** A 124-lb beam is hanging as shown in Figure 4.4-30. Find the magnitude and direction of **F**, the force that the floor exerts on this beam.

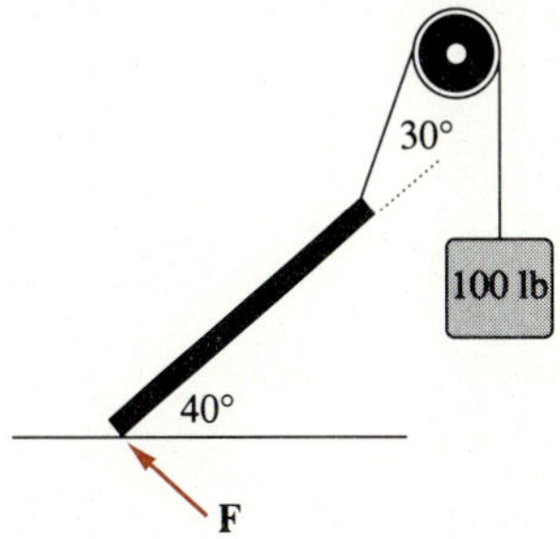

Figure 4.4-30

SUMMARY

1. Law of Cosines

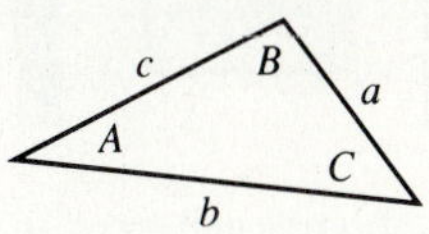

$$a^2 = b^2 + c^2 - 2bc \cos A$$

where b and c are *any* two sides of a triangle, A is the angle between those sides, and a is the third side.

2. Law of Sines

$$\frac{\sin A}{a} = \frac{\sin B}{b} = \frac{\sin C}{c}$$

3. Area of Triangle. The area of a triangle is one-half the product of the lengths of any two sides times the sine of the angle between them.

$$\text{area} = \tfrac{1}{2}ab \sin C$$

4. Ambiguous Case. If we know two sides and an angle opposite one of them, the triangle may have one, two, or no solutions.

5. Vectors. A vector is a quantity having both magnitude and direction.

6. Addition of Vectors. Find the *resultant* (or sum) of two vectors using the polygon method, which is equivalent to the parallelogram method.

IMPORTANT TERMS

Ambiguous case
Law of cosines
Law of sines
Oblique triangle
Vectors
Direction of a vector
Head and tail of a vector
Magnitude of a vector
Parallelogram method of vector addition
Polygon method of vector addition
Resultant of two vectors
Scalar
Vector

REVIEW EXERCISES

1. Use the methods discussed in this chapter to solve completely (find all angles and sides) the following oblique triangle for the cases given. As you do each solution, sketch the triangle approximately *to scale*. If there is more than one solution, find both.

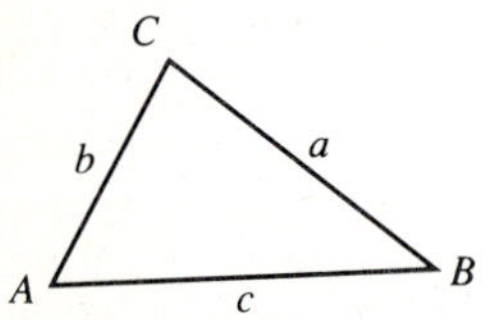

(a) $b = 7.00$ ft, $c = 10.0$ ft, $A = 49.1°$
(b) $a = 65$ m, $b = 87$ m, $c = 44$ m
(c) $C = 73.4°$, $B = 43.4°$, $b = 122$ cm
(d) $A = 64.8°$, $a = 52.1$ m, $c = 45.3$ m
(e) $c = 21.0$ ft, $A = \frac{\pi}{6}$, $B = \frac{\pi}{5}$
(f) $a = 15.0$ m, $b = 6.50$ m, $B = 0.559$ rad

2. Use the triangle in Exercise 1 to determine the number of possible triangles that satisfy the given conditions. Sketch each possible one, but do not solve it.

(a) $b = 50.39$ m, $A = 26°50'$, $B = 57°41'$
(b) $a = 14.0$ ft, $b = 11.6$ ft, $c = 18.18$ ft
(c) $a = 840$ in., $b = 485$ in., $A = 21.5°$
(d) $b = 150$ cm, $c = 130$ cm, $C = 125.1°$
(e) $a = 25.7$ m, $b = 18.2$ m, $c = 6.20$ m
(f) $a = 34$ mi, $b = 22$ mi, $B = \frac{\pi}{6}$
(g) $a = 11.3$ km, $c = 12.8$ km, $A = 54.4°$
(h) $a = 11.2$ cm, $b = 6.30$ cm, $c = 21.3$ cm.

3. Find the area of the triangle satisfying the given conditions, using the figure in Exercise 1. If there is more than one possibility, find *both*. Always include a clear sketch.

(a) $a = 8.0$ in., $c = 6.0$ in., $B = \frac{5\pi}{6}$
(b) $a = 11.0$ mm, $b = 16.0$ mm, $c = 25.0$ mm
(c) $a = 40.0$ m, $b = 30.0$ m, $B = \frac{\pi}{6}$
(d) $b = 74.5$ ft, $a = 59.4$ ft, $A = 64.4°$

4. A 1500-lb sign must be hung using two wires. If one can support 850 lb as shown, what force must the other wire support and in what direction must it pull on the sign?

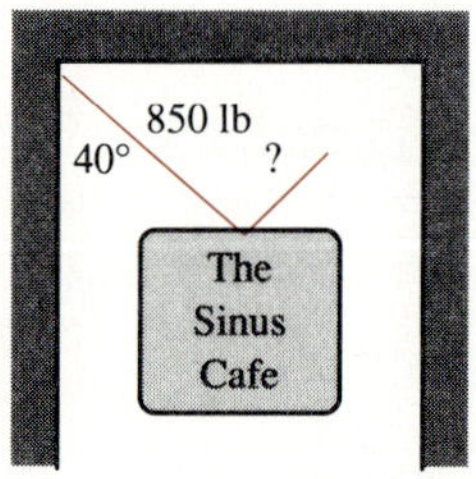

***5.** In a baseball game, to catch the base runner in a rundown, the pitcher must throw the ball to the third baseman who is midway between third base and home plate. How far must the pitcher throw the ball? (See Example 13 for pertinent information.)

6. Your construction firm is given the plan shown in the following figure for a triangular patio. This plan is not to scale; it is just a rough sketch showing only certain dimensions for this patio. If the patio is to be covered with square tiles 25 cm on each side, what is the maximum number of tiles needed, assuming no waste? Would you have any reason to call this architect to demand some further explanation of his plans?

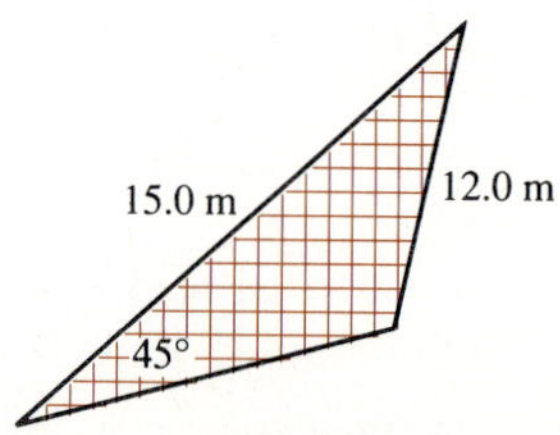

7. Show that the area of the shaded region in the following figure is given by $A = \frac{1}{2}r^2(\theta - \sin\theta)$.

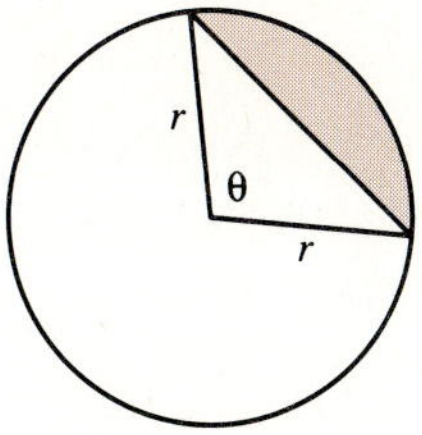

8. For the parallelogram in the following figure find its area and the lengths of both diagonals.

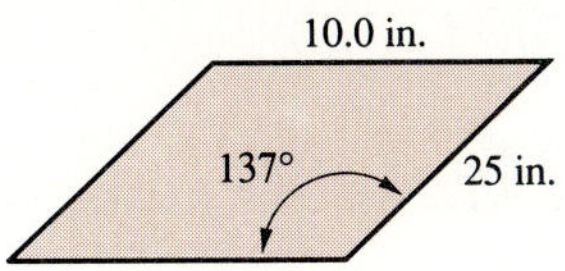

9. For the formic acid molecule in Fig. 4.2-24, find **(a)** the C—H′ distance and ***(b)** the O′—H′ distance.
(See Exercise 57 of Section 4.2 for pertinent information.)

10. Two pickup trucks roped to a stump are pulling horizontally, trying to dislodge it. One pulls with 150 lb at 45° north of east while the other pulls with 340 lb at 30° west of south. Find the magnitude and direction of the single pull that is equivalent to these two.

11. If two vectors have magnitudes of 75 lb and 35 lb, find the largest and smallest that their resultant can be. In each case, sketch the orientation of the vectors.

12. Two wires are attached to the same point on a tree. One pulls with 12.0 lb and the other pulls with 15.0 lb. It the equivalent pull is 21.6 lb, find the angle between them.

5

The Graphs of the Trigonometric Functions

5.1 PERIODIC FUNCTIONS

One of the reasons that trigonometry is so important is that the trigonometric functions are *periodic*—they repeat themselves at regular intervals. Many natural processes do the same thing. Sound and light waves, breathing movements, heart beats, the motion of the tides, the seasons of the year, and the vibrations of a piano string all have periodic or repeating patterns called *cycles*. The periodicity of the trigonometric functions makes them especially useful for describing such processes.

First of all, remember how we defined **periodic functions** in Section 3.4-3. We suggest you reread that section now.

> **PERIODIC FUNCTIONS**
>
> A *periodic function* repeats its values at regular intervals.
> The *period* is the *smallest interval* over which a periodic function repeats itself.
> One *cycle* occurs in each period.

Example 1
Hands of a Clock

The position of the minute hand on a clock is periodic because it keeps repeating itself (Figure 5.1-1). It takes this hand 60 min to make one cycle, so the period is 60 min. The position of the hour hand is also periodic with a period of 12 hr. ■

Figure 5.1-1

Clock showing hour and minute hands

NOW IT'S YOUR TURN

What is the period of the second hand? (Answer: 1 min)

Example 2

Motion of the Earth

The earth's position relative to the sun is periodic because it keeps repeating (see Figure 5.1-2). Each cycle is one complete revolution about the sun and takes approximately 365 days. Therefore, the period is 365 days, which means that 365 days from today the earth will be in the same position as it is now relative to the sun. The earth's position will also be the same 2 × 365 days from today, 3 × 365 days from today and so forth. ■

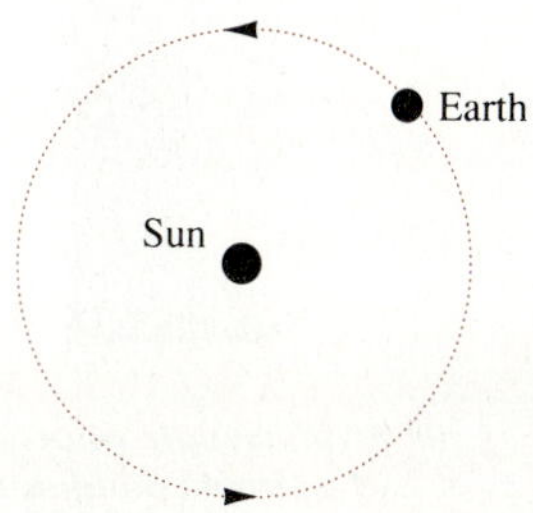

Figure 5.1-2

Earth's orbit around the sun

NOW IT'S YOUR TURN

What is the period of the position of your town or city on the earth due to our planet's spin? (Answer: 24 hr)

Example 3

Blood Pressure

The pulse beat is the regular throbbing of the arteries caused by the successive contractions of the heart. The *systolic pressure* is the maximum pressure (when the heart has contracted), and the *diastolic pressure* is the minimum pressure (when the heart has relaxed). The length of time between successive systolic peaks (or diastolic valleys) is the period (see Figure 5.1-3). For the blood pressure in Figure 5.1-3, the period is 0.9 s because this is the time needed for

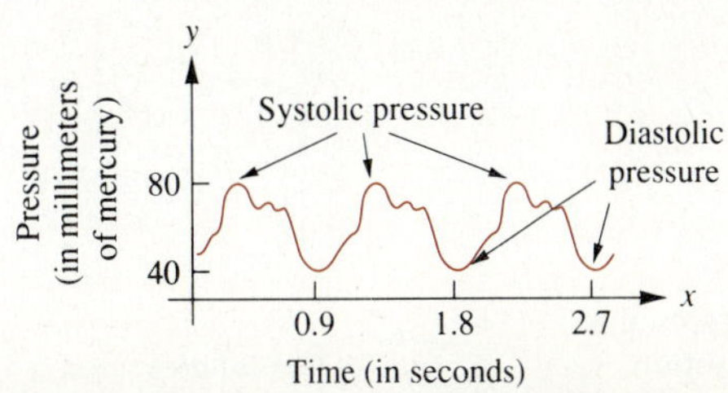

Figure 5.1-3

Graph of blood pressure as a function of time

the cycle to repeat itself. The maximum value of the function (the systolic pressure) is 80 mm of mercury, and the minimum value (the diastolic pressure) is 40 mm of mercury. ■

NOW IT'S YOUR TURN

Carefully sketch a graph for someone whose pulse has a period of 1.2 s.

(Answer: It would have the same shape as Figure 5.1-3 except the time interval between systolic peaks would be 1.2 s instead of 0.9 s.)

Example 4
Pendulum

A simple pendulum is a small object swinging by a string or wire from a stationary hook, as shown in Figure 5.1-4. The string swings between $+\theta$ and $-\theta$ from its neutral position. In one cycle, the pendulum swings from A to B and back to A, and it would continue to do so if there were no friction. The period of a pendulum depends on the length of the string.

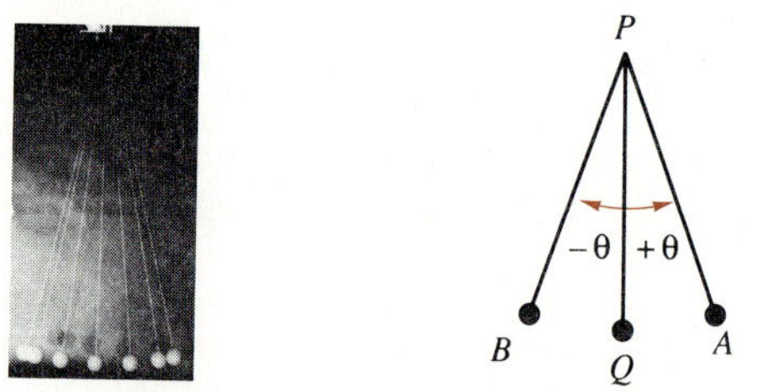

(a) Stroboscopic photograph of a simple pendulum

(b) Pendulum in the neutral position (Q), at $+\theta$ (A) and at $-\theta$ (B)

Figure 5.1-4
Simple pendulum

We can produce a graph of the pendulum's motion by attaching a felt pen to its lower end and letting it trace a curve on a strip of paper moving at constant speed underneath it, as shown in Figure 5.1-5. (Chart recorders and seismographs produce graphs by moving paper in this way.) If the pendulum is not moving, the pen draws a straight line (Figure 5.1-5a), but if the pendulum is swinging the pen draws a periodic curve (Figure 5.1-5b). If we straighten out the graph paper from Figure 5.1-5b, the graph on it looks like the one shown in

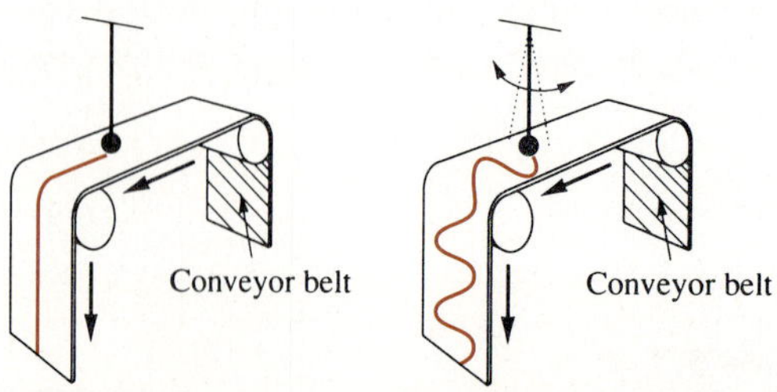

(a) Pendulum at rest

(b) Swinging pendulum

Figure 5.1-5
Making the graph of a pendulum's motion

Figure 5.1-6. The y axis represents the distance of the pendulum from its neutral position, and the x axis represents time. We call such a graph, in which the x axis is time, a *time graph*. ■

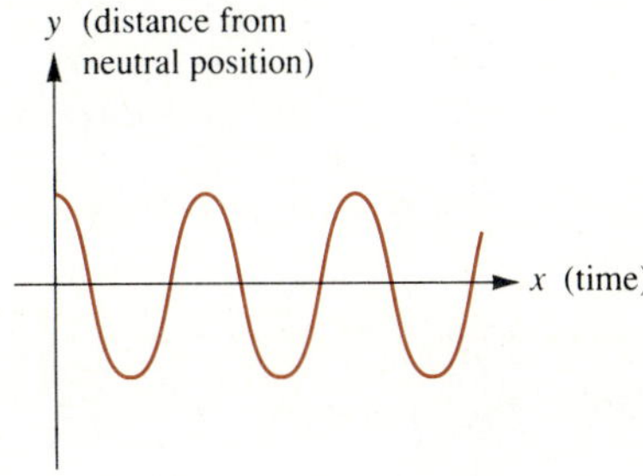

Figure 5.1-6

Graph of the pendulum's motion. The y-axis gives the distance of the pendulum from its neutral position, and the x-axis gives the time.

NOW IT'S YOUR TURN

Use ordinary objects to make a pendulum with an adjustable length. Then measure the period and use it to sketch a time graph of the pendulum's motion. If you are inquisitive, measure the period for several different lengths to see if the period does, in fact, depend upon the length.

EXERCISES 5.1

In Exercises 1–4, each graph represents a function of x. Determine the period, maximum value, and minimum value of each function.

1.

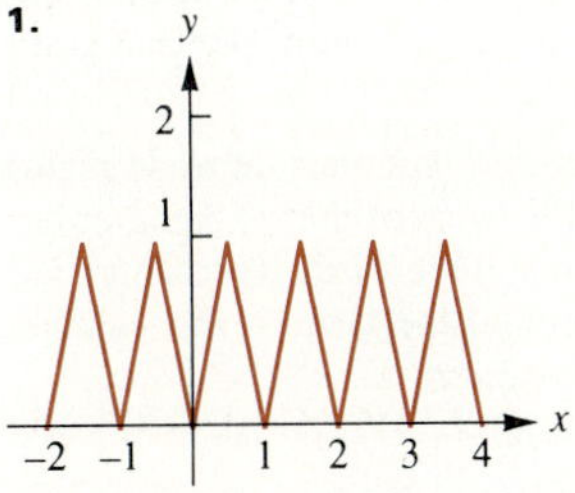

2.

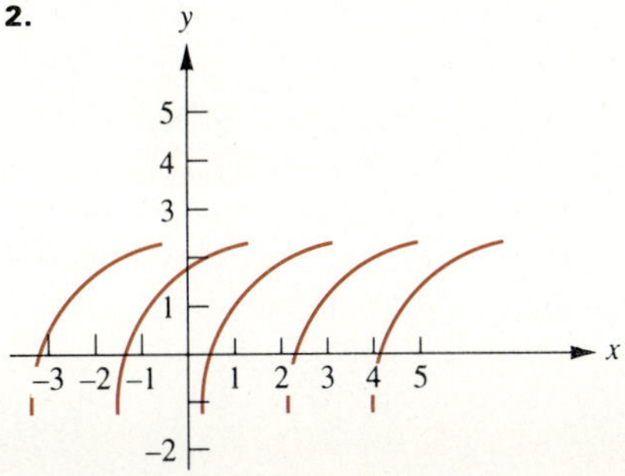

3.

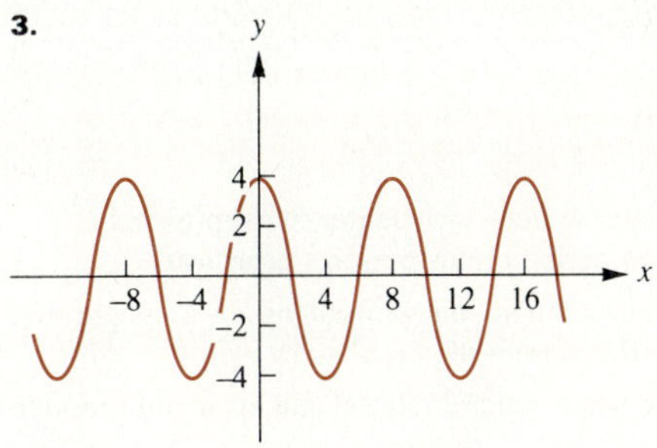

4.

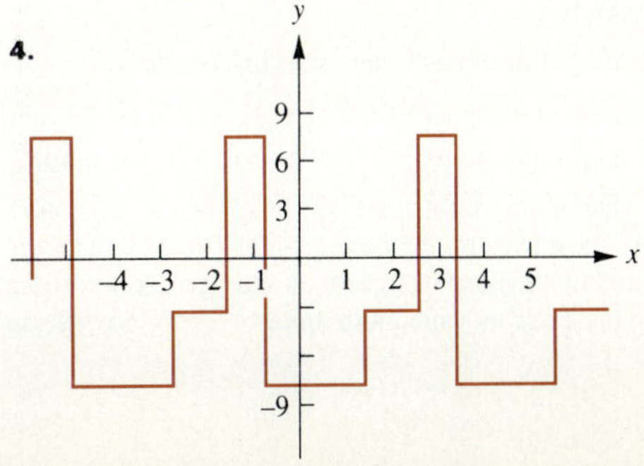

5. Breathing. The graph in Figure 5.1-7 represents a person's breathing. It was made by strapping a pressure-sensitive detector around this person's chest. As he breathed, the pressure changed due to the expansion and contraction of his chest. The pressure reading was fed into a chart recorder (very similar to that shown in Figure 5.1-5) to make the graph. Therefore, the y axis represents pressure whereas the x axis represents time. The paper moved through the chart recorder at 1.0 cm/s, so each centimeter on the x axis represents 1.0 s. Is this graph exactly periodic? Would you expect it to be? Why? Find the average period of this person's breathing.

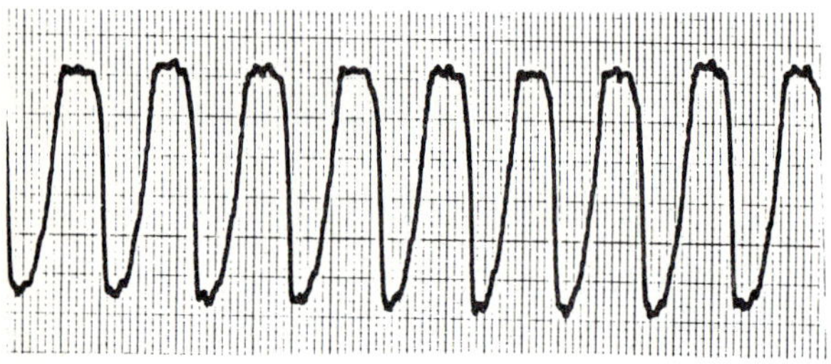

Figure 5.1-7

6. Electrocardiogram (ECG). The graph in Figure 5.1-8 is an electrocardiogram of a person's heart. The hills and valleys are electrical currents caused by the contraction and relaxation of the various parts of the heart (the atria and the ventricles). These electrical currents were fed into a chart recorder with the paper moving at 3.0 cm/s.

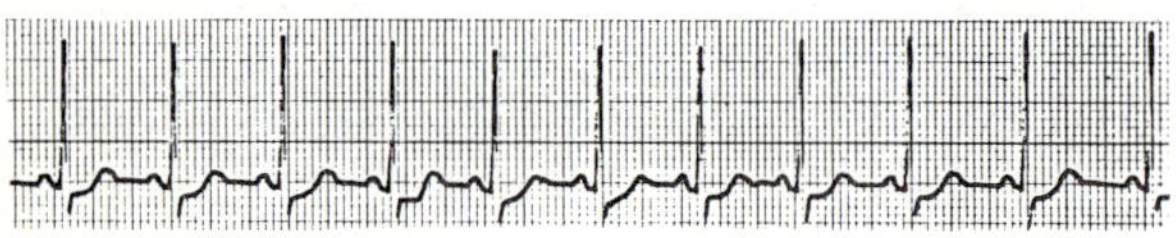

Figure 5.1-8

(a) How much time does each centimeter represent?
(b) What is the period of this person's heartbeat?

7. Make actual observations on yourself or on a cooperative friend to measure the periods of:

(a) your pulse when you are relaxed and again after moderate exercise.
(b) your breathing when you are relaxed and again after moderate exercise.

Then make a time-graph of each one, showing the period.

8. What is the period of the seasons?

9. What is the approximate period of the phases of the moon?

10. Sunspot Cycle. Sunspots are dark spots on the sun's surface; they are dark because they are cooler than the rest of the surface. The number of sunspots increases and decreases with an 11-yr period. This cycle is called the *sunspot cycle*. Sketch an approximate numerical graph of the number of sunspots as a function of time over a century.

11. Tree Rings. Tree rings periodically repeat their pattern (see Figure 5.1-9). What would you expect their period to be? Why?

Figure 5.1-9

12. Architecture. Skyscrapers are designed so that they will sway with the wind or in earthquakes. The period, which may be several seconds, depends on the structure of the building. The distance moved may be several feet for a tall building such as the Sears Tower in Chicago. If a building takes 4.5 s for one cycle and moves ±5.0 ft from its neutral (vertical) position, sketch a graph of its position as a function of time.

13. The graph of a certain pendulum's motion is shown in Figure 5.1-10. The y axis represents the distance of the swinging object from its neutral (vertical) position, and the x axis represents time. Find the period of this pendulum and the farthest the swinging object ever gets from its neutral position.

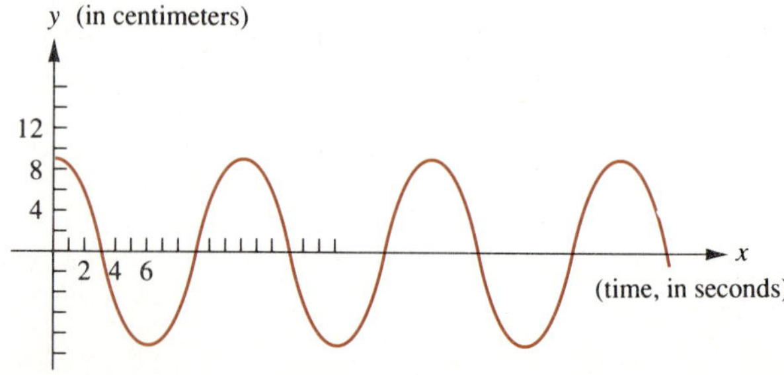

Figure 5.1-10

***14.** As the hands of a clock move, the tip moves to the right and left of the clock's center. If a clock has a minute hand 6.0 in. long, sketch the graph of the position of the tip of this hand to the right and left of the clock's center, calling right positive and left negative. Start the graph at midnight and let it cover several hours. Label the numerical values of the points where the graph crosses the x and y axes.

***15.** Make the same type of graph as in Exercise 14, except graph the position of the tip of the minute hand above and below the clock's center. Call the positions above the center positive and those below the center negative.

***16.** The graph in Figure 5.1-11 represents the position to the right and left of the clock's center of one of the hands on the clock. Which hand does it represent, and how long is this hand?

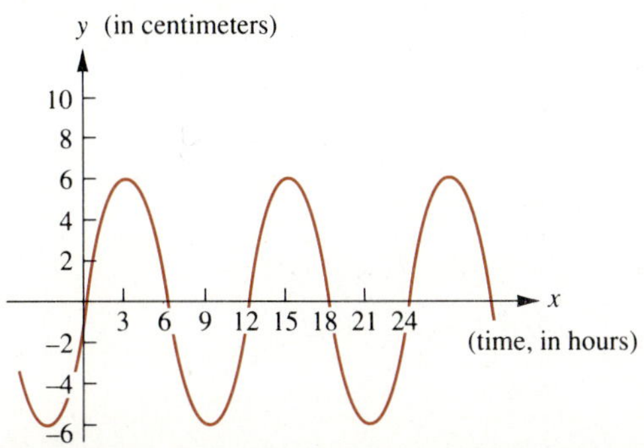

Figure 5.1-11

***17. Cycloid.** Draw a point on the rim of a circle. Now roll the circle along. Sketch the path this point traces, and show that it is the same as in Figure 5.1-12. This path is called a *cycloid.* Express the period of the cycloid in terms of the radius of the circle. Is each cycle of the cycloid a semicircle?

Figure 5.1-12

***18. Frequency.** The frequency f of a graph or process is the number of cycles per second.

(a) If a saxophone reed is vibrating at 100 cycles/s, how long does each cycle take? How long does each one take if it is vibrating at 1000 cycles/s? What is the name for the times you just found?

(b) What does $1/f$ represent? Relate $1/f$ to the period. *Hint:* f = number of cycles/s, so what are the units of $1/f$? Interpret what this means.

For Exercises 19–24, refer to Exercise 18.

19. What are the frequency and period of:

(a) the second hand of a clock?

(b) the earth as it spins on its axis?

(c) the earth as it goes around the sun?

20. What are the period and frequency of an LP record and a 45-rpm record?

21. What is the frequency of the pulse beat in Example 3?

22. Fig. 5.1-13 shows an electrocardiogram (top) and a stethoscope graph (bottom) (see Exercise 6) recorded simultaneously for the same person. The paper was moving through the chart recorder at 10 cm/s. (In case you don't remember, a stethoscope is the instrument an M.D. puts on your chest to listen to your heart.) Make measurements on both of these graphs to calculate the period (in seconds) and the frequency (in beats per second *and* in beats per minute) of this person's heart. Compare the frequencies of the two graphs. Is it a coincidence that they are related? Why?

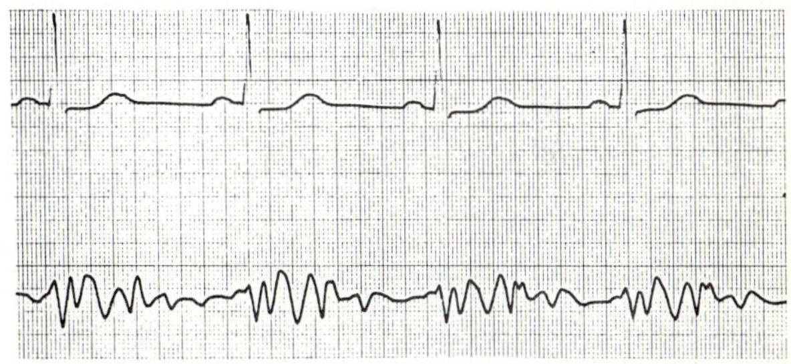

Figure 5.1-13

23. Seismograph. A seismograph measures movement of the earth's surface. When earthquake waves hit the instrument, they cause a pen attached to a spring to oscillate back and forth. As it does so, it draws a line on a rotating cylinder. The stronger the earthquake, the larger the oscillations. The pattern in Figure 5.1-14 is a test pattern to see if the seismograph is operating properly. The places where the straight lines are offset are time calibration markers, and they occur once every minute. Using measurements made on this pattern, calculate the frequency and the period with which the pen oscillates back and forth. (This demands a keen eye!)

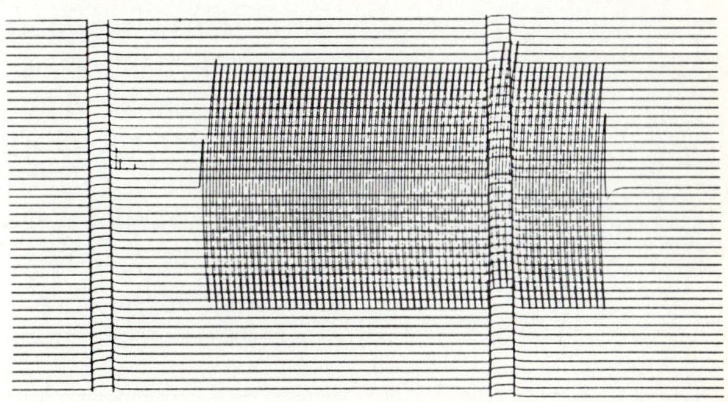

Figure 5.1-14 Seismograph test pattern

24. Underground Nuclear Test. Figure 5.1-15 shows a seismograph pattern for an underground nuclear test done at the Nevada test site on April 18, 1987 (see Exercise 23). Using measurements made on this figure, calculate the frequency and the period with which the pen oscillates back and forth. Is it the same frequency as for the test pattern? (This demands a keen eye!)

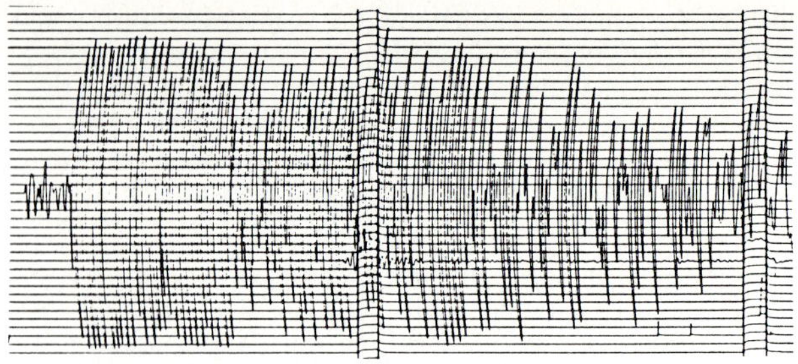

Figure 5.1-15

5.2 THE GRAPHS OF THE TRIGONOMETRIC FUNCTIONS

In Section 3.4-3, we found that the trigonometric functions all have periods of 2π (360°), except for the tangent and cotangent, which have periods of π (180°). This periodicity is very helpful in finding the graphs of these functions. The graphic representation of trigonometric functions is a vivid and convenient way to understand many of their properties.

PERIODS OF THE TRIGONOMETRIC FUNCTIONS

The sine, cosine, secant, and cosecant all have a period of 2π (360°).

$$\sin(\theta \pm 2\pi) = \sin\theta \qquad \cos(\theta \pm 2\pi) = \cos\theta$$

$$\sec(\theta \pm 2\pi) = \sec\theta \qquad \csc(\theta \pm 2\pi) = \csc\theta$$

The tangent and cotangent have a period of π (180°).

$$\tan(\theta \pm \pi) = \tan\theta \qquad \cot(\theta \pm \pi) = \cot\theta$$

5.2-1 The Graphs of the Sine and Cosine

The Graph of $y = \sin x$

To obtain this graph as easily as possible, we use the fact that the sine function has a period of 2π. If we find the graph when x is between 0 and 2π, we can get the complete graph by repeating this pattern every 2π.

First, we construct a table of values of $y = \sin x$, where the angle x is measured in radians.* For clarity we make the first graph in four steps, each one $\frac{1}{4}$ of a period, although you usually would not break it up this way (see Figures 5.2-1 through 5.2-4).

After we have one complete cycle, we can use periodicity to extend the graph to other values of x. Because the sine has period 2π, the values of $\sin x$ repeat as x varies from 2π to 4π, 4π to 6π, and so on. Similarly, the values of $\sin x$ repeat as x varies from 0 to -2π, from -2π to -4π, and so on. Figure 5.2-5 shows a more extended graph of $y = \sin x$. From the graph it is easy to see that

* We could work in degrees, but in most applications it is necessary to use radians. Also, notice that we are using x to represent the angle, rather than θ as we did earlier, because we usually plot the angle on the x axis.

Step 1: x goes from 0 to $\frac{\pi}{2}$:

x(rad)	$y = \sin x$
0	0.00
$\frac{\pi}{6}$	0.50
$\frac{\pi}{4}$	0.71
$\frac{\pi}{3}$	0.87
$\frac{\pi}{2}$	1.00

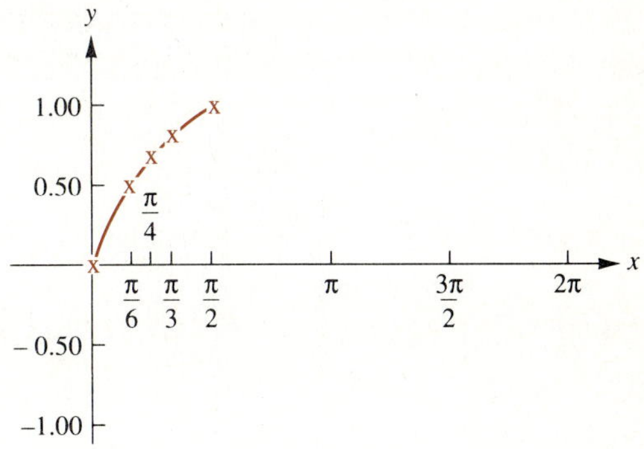

Figure 5.2-1

$y = \sin x$ increases from 0 to +1 as x goes from 0 to $\frac{\pi}{2}$.

Step 2: x goes from $\frac{\pi}{2}$ to π:

x(rad)	$y = \sin x$
$\frac{\pi}{2}$	1.00
$\frac{2\pi}{3}$	0.87
$\frac{3\pi}{4}$	0.71
$\frac{5\pi}{6}$	0.50
π	0.00

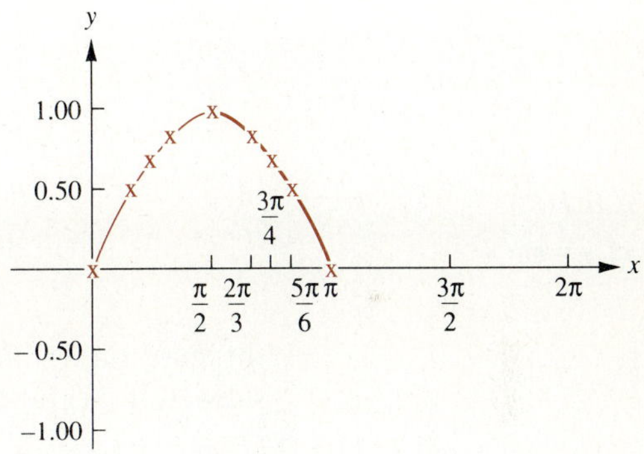

Figure 5.2-2

$y = \sin x$ decreases from 1 to 0 as x goes from $\frac{\pi}{2}$ to π.

Step 3: x goes from π to $\frac{3\pi}{2}$:

x(rad)	$y = \sin x$
π	0.00
$\frac{7\pi}{6}$	−0.50
$\frac{5\pi}{4}$	−0.71
$\frac{4\pi}{3}$	−0.87
$\frac{3\pi}{2}$	−1.00

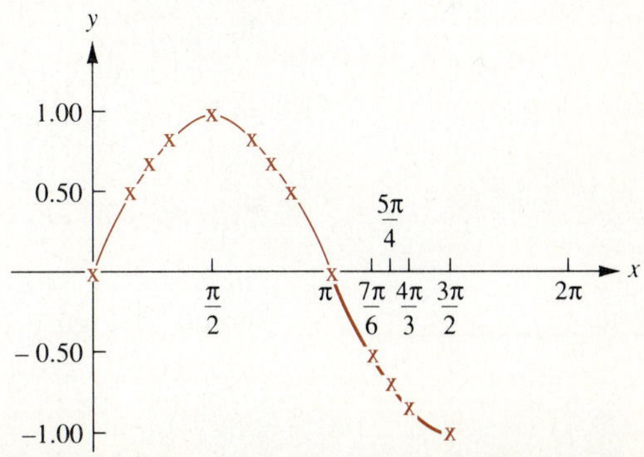

Figure 5.2-3

$y = \sin x$ decreases from 0 to −1 as x goes from π to $\frac{3\pi}{2}$.

Step 4: x goes from $\frac{3\pi}{2}$ to 2π:

x(rad)	$y = \sin x$
$\frac{3\pi}{2}$	−1.00
$\frac{5\pi}{3}$	−0.87
$\frac{7\pi}{4}$	−0.71
$\frac{11\pi}{6}$	−0.50
2π	0.00

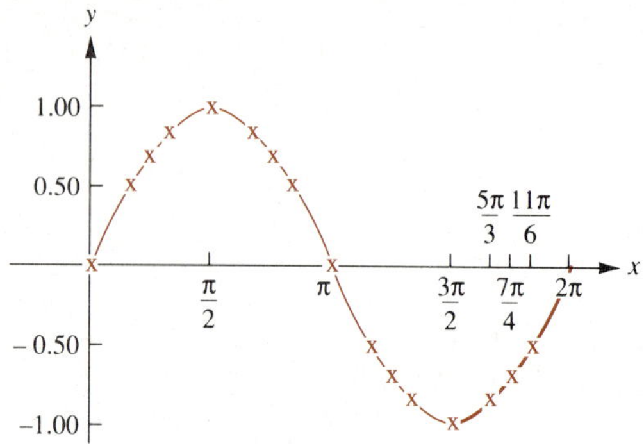

Figure 5.2-4

$y = \sin x$ increases from −1 to 0 as x goes from $\frac{3\pi}{2}$ to 2π.

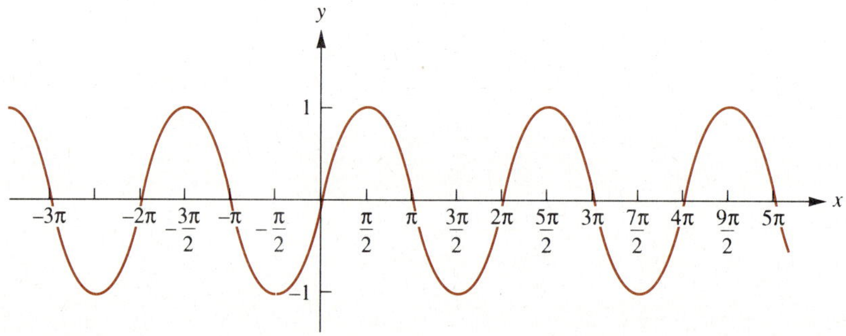

Figure 5.2-5

Graph of $y = \sin x$

the maximum value of $\sin x$ is +1 and the minimum value is −1, which we showed in Section 3.4-4. Thus, no matter what the value of x,

$$-1 \leq \sin x \leq 1$$

NOW IT'S YOUR TURN

Construct a table that gives the values of $\cos x$ as x varies from 0 to 2π and use these values to graph $y = \cos x$. Then use the periodicity of the cosine to extend this graph for all values of x. (To obtain the preceding graphs, we used only special angles. We suggest, however, that you use your calculator to obtain additional points to plot a clear smooth curve.) (Answer: See Figure 5.2-6.)

The Graph of $y = \cos x$

In the work you just did, you should have found the graph of $y = \cos x$ as shown in Figure 5.2-6. Notice that, no matter what value x has,

$$-1 \leq \cos x \leq 1$$

as we showed in Section 3.4-4.

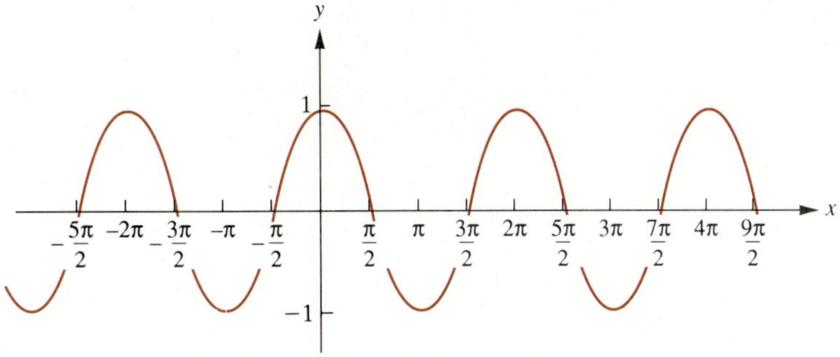

Figure 5.2-6

Graph of $y = \cos x$

5.2-2 The Graph of $y = \tan x$

The tangent is a periodic function with a period of π. But remember that tan x is undefined when $x = \pm\frac{\pi}{2}$. To avoid dealing with this point (for now), we construct a table of values of tan x with x between $-\frac{\pi}{2}$ and $\frac{\pi}{2}$. Table 5.2-1 still covers one full period, and it shows such values (from a calculator) rounded off to two decimal places. We have used these values to sketch the graph of $y = \tan x$ for x between $-\frac{\pi}{2}$ and $\frac{\pi}{2}$ (see Figure 5.2-7).

x	tan x
$-\frac{\pi}{2}$	undefined
$-\frac{5\pi}{12}$	−3.73
$-\frac{\pi}{3}$	−1.73
$-\frac{\pi}{4}$	−1.00
$-\frac{\pi}{6}$	−0.58
$-\frac{\pi}{12}$	−0.27
0	0.00
$\frac{\pi}{12}$	0.27
$\frac{\pi}{6}$	0.58
$\frac{\pi}{4}$	1.00
$\frac{\pi}{3}$	1.73
$\frac{5\pi}{12}$	3.73
$\frac{\pi}{2}$	undefined

Table 5.2-1

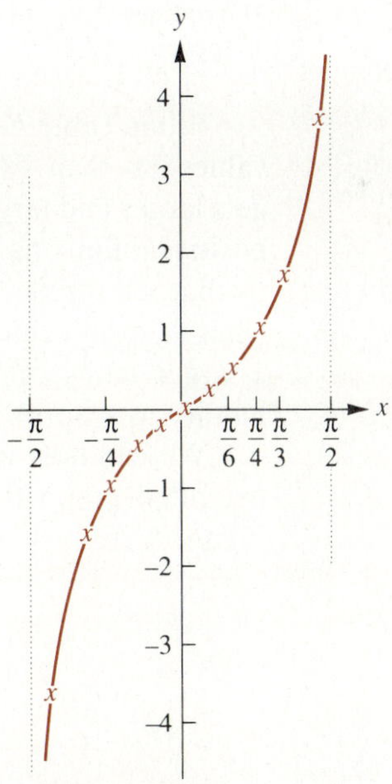

Figure 5.2-7

Graph of $y = \tan x$

The graph we obtained is fairly accurate for values of x between $-\frac{5\pi}{12}$ and $\frac{5\pi}{12}$. We do, however, need a more detailed investigation of the values of tan x as x gets closer to $\frac{\pi}{2}$ and $-\frac{\pi}{2}$. To do so, we use a calculator to construct more detailed tables of $\tan x$ (Tables 5.2-2 and 5.2-3). Remember that $\frac{\pi}{2}$ is approximately 1.571 rad and $\tan(-x) = -\tan x$.

x(rad)	$y = \tan x$
1.47	9.89
1.48	10.98
1.49	12.35
1.50	14.10
1.51	16.43
1.52	19.67
1.53	24.50
1.54	32.46
1.55	48.08
1.56	92.62
1.57	1255.77
1.5705	3374.65

Table 5.2-2

The values of $y = \tan x$ as x approaches $\frac{\pi}{2}$

x(rad)	$y = \tan x$
−1.47	−9.89
−1.48	−10.98
−1.49	−12.35
−1.50	−14.10
−1.51	−16.43
−1.52	−19.67
−1.53	−24.50
−1.54	−32.46
−1.55	−48.08
−1.56	−92.62
−1.57	−1255.77
−1.5705	−3374.65

Table 5.2-3

The values of $y = \tan x$ as x approaches $-\frac{\pi}{2}$

From Table 5.2-2 we see that, as x gets closer and closer to $\frac{\pi}{2}(1.571)$ from values less than $\frac{\pi}{2}$ (that is, values to the left of $\frac{\pi}{2}$ on the x axis), the value of $\tan x$ gets larger and larger. We describe this behavior by saying that $\tan x$ approaches positive infinity as x approaches $\frac{\pi}{2}$ from the *left*. In a similar way, Table 5.2-3 tells us that, as x gets closer and closer to $-\frac{\pi}{2}(-1.571)$ from values greater than $-\frac{\pi}{2}$ (that is, from values to the right of $-\frac{\pi}{2}$ on the x axis), the value of $\tan x$ gets to be a larger and larger negative number. We say that $\tan x$ approaches negative infinity as x approaches $-\frac{\pi}{2}$ from the *right*.

We can now use the fact that the tangent has a period of π to give a more extended graph of $y = \tan x$ by repeating the pattern we just obtained (see Figure 5.2-8).

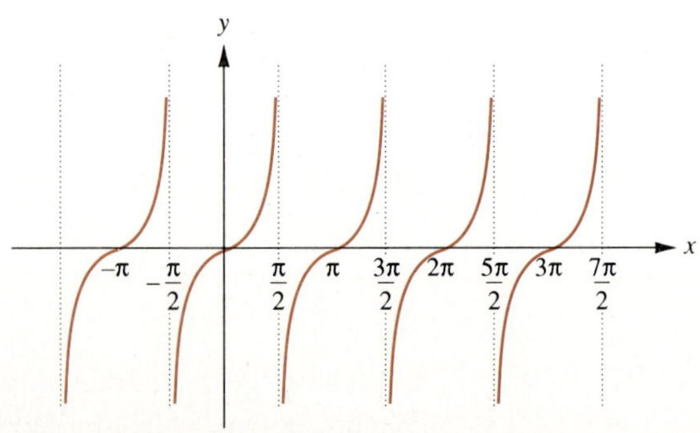

Figure 5.2-8

Graph of $y = \tan x$

What It Means When the Tangent Is Undefined

We have already seen that the tangent is *undefined* at $x = \frac{\pi}{2}$ because division by zero is undefined. The graph in Figure 5.2-8 can also help us understand just what this means. This graph shows that as x gets closer and closer to $\frac{\pi}{2}$ from the left, $\tan x$ gets more and more *positive*. But as x gets closer and closer to $\frac{\pi}{2}$ from the right, $\tan x$ gets more and more *negative*. Therefore, $\tan x$ does not approach a *definite value* as x approaches $\frac{\pi}{2}$. The value of $\tan x$ is *different*, depending on the direction from which x approaches $\frac{\pi}{2}$. The same behavior occurs when $x = -\frac{\pi}{2}, \pm\frac{3\pi}{2}, \pm\frac{5\pi}{2}$, and so on; thus $\tan x$ is also undefined at these points.

5.2-3 The Graphs of Cotangent, Secant, and Cosecant

We have discussed the graphs of the sine, cosine, and tangent functions in considerable detail because these three trigonometric functions are used most often. The remaining three trigonometric functions occur much less frequently, although we do sometimes encounter their graphs. We present the graphs of the cotangent, secant, and cosecant functions so you will be familiar with them, but we leave the details of how to obtain these graphs for the exercises (see Figures 5.2-9 through 5.2-11).

Figure 5.2-9
Graph of $y = \cot x$

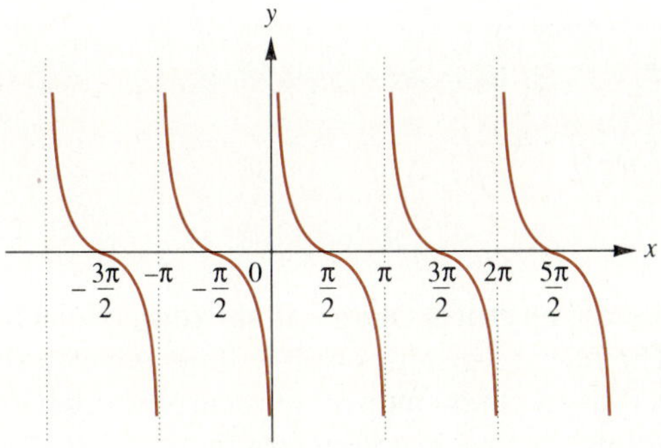

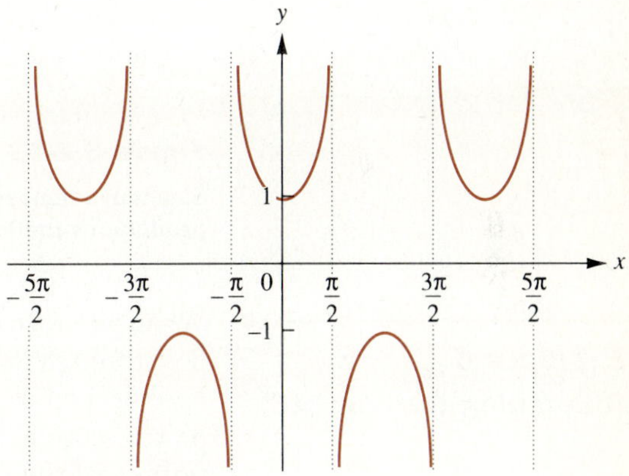

Figure 5.2-10
Graph of $y = \sec x$

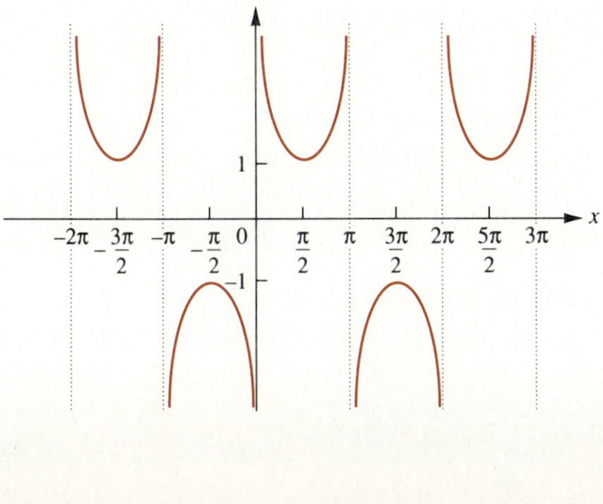

Figure 5.2-11
Graph of $y = \csc x$

5.2-4 Sine Waves

Figure 5.2-12

Sine wave

In Section 5.1, we looked at a number of natural periodic processes, such as a pendulum. We also pointed out that because the trigonometric functions are periodic, they are especially useful for describing such periodic processes. But natural processes do not become infinite. (For example, a building with an infinite amount of vibration would fall down!) Therefore, only the trigonometric functions that remain *finite* are useful for describing most natural periodic processes. These two functions are the sine and the cosine because all the others approach $\pm\infty$ at some points in their cycles. For this reason, *periodic behavior is nearly always described using sines and cosines.* The graphs of such processes are sine and cosine graphs, and we often call such graphs **sine waves** (see Figure 5.2-12). Any process that can be represented by a sine wave is called **sinusoidal**.

Example 5
Simple Pendulum

In Section 5.1, we showed that the swinging motion of a simple pendulum is periodic, and in Figure 5.1-5, we saw how to obtain a time graph of the pendulum's motion. This graph is a sine wave in which the y axis can represent the distance the pendulum swings. But usually we let the y axis represent the angle θ of its swing, while the x axis represents time (Figure 5.2-13). ■

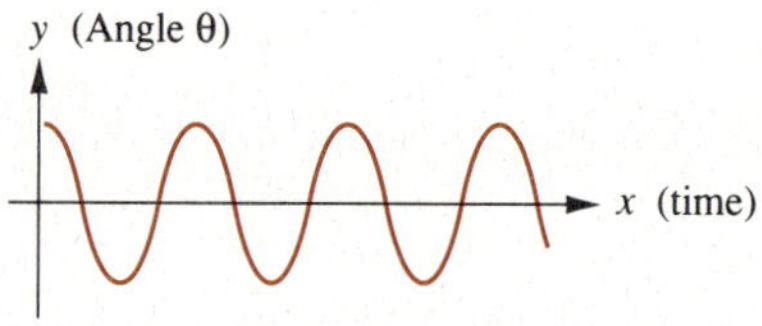

Figure 5.2-13

Sine wave representing pendulum's motion

Example 6
Alternating Current (AC)

When you turn on your stereo, electrical charges in the form of electrons start to flow through the wires. The *rate* at which this charge flows is called the electrical *current*, and it is measured in amperes (amps). This current is generated at the power station by wire loops turning in strong magnetic fields. The periodic turning of these loops causes the electrons to reverse direction periodically (or alternate), completing 60 cycles each second. Therefore, each cycle takes $\frac{1}{60}$ s, so the period is $\frac{1}{60}$ s. Figure 5.2-14 is a photograph of an oscilloscope trace showing such an alternating current. As you can see, the current is a *sine wave*. The y axis represents current in amperes, and the x axis is time. The graph alternates between positive and negative as the electrons reverse direction. ■

NOW IT'S YOUR TURN

Sketch a time graph for current that has 120 cycles/s.

(Answer: It is the same as Figure 5.2-14 except the period is 1/120 s).

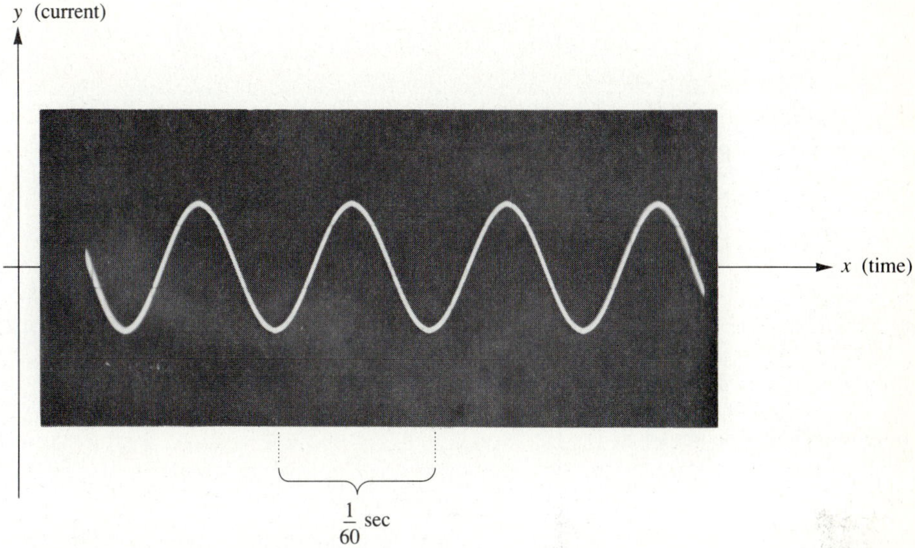

Figure 5.2-14

Alternating current This photograph shows an oscilloscope trace. The period is $\frac{1}{60}$ s. The y axis is current and the x axis is time.

Example 7

Sound

A vibrating object, such as a guitar string, a bell, or the reed of a clarinet, periodically bunches up the molecules of the air and causes a pressure wave (Figure 5.2-15). We experience this pressure wave as sound. Figure 5.2-16 is a photograph of an oscilloscope trace showing a sound wave for the note middle C. The y axis represents pressure and the x axis time. For this note, the sound makes 262 cycles every second, so each cycle takes $\frac{1}{262}$ s $= 3.82 \times 10^{-3}$ s. Therefore, the period is 3.82×10^{-3} s. ■

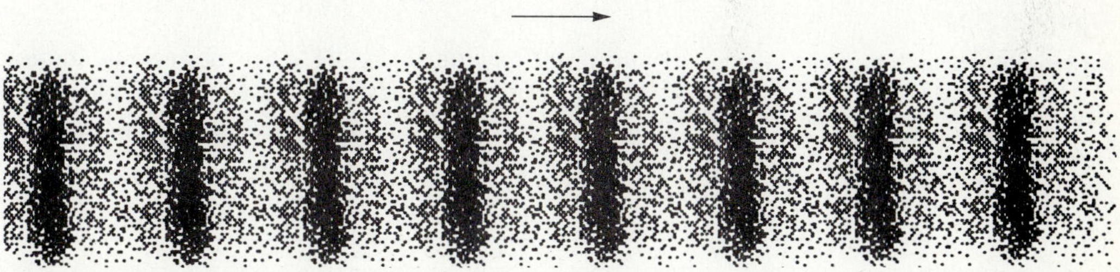

Figure 5.2-15

A travelling sound wave The molecules are periodically bunched up by a vibrating object, such as a guitar string.

NOW IT'S YOUR TURN

Sketch the sine wave time graph representing the note A, which has 440 cycles each second.

(Answer: The same as Figure 5.2-16, except the period is 2.27×10^{-3} s.)

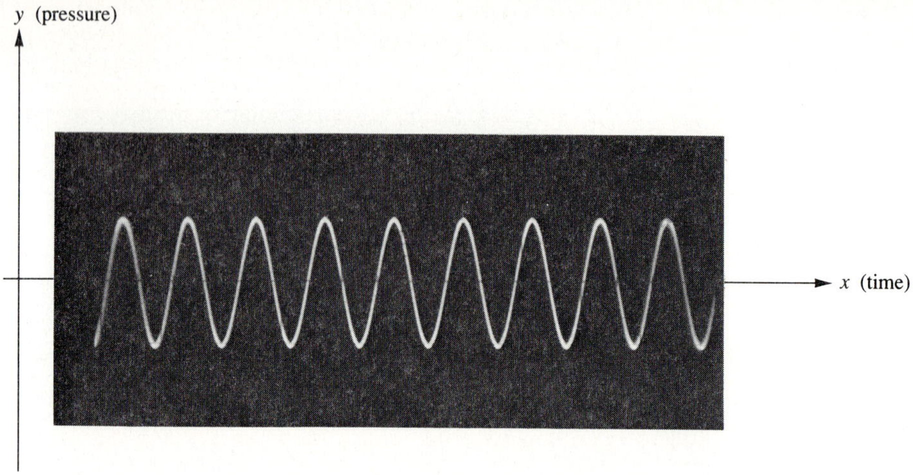

Figure 5.2-16

Photograph of an oscilloscope trace of the sound wave for the note middle C The period is 3.82×10^{-3} s.

Example 8

Musical Instruments

Any musical instrument has some part that vibrates back and forth, usually several hundred times per second. This part may be a string, a reed, a larynx (for a singer), or the lips of the players (for brass instruments). Figure 5.2-17 shows a vibrating string. As you can see, the string is in the shape of a sine wave. The y axis represents the distance the string has vibrated away from its neutral position, whereas the x axis represents distance measured along the length of the string. This is not a time graph because the x axis represents *distance*, not time. Vibrations such as these produce the pressure waves we experience as sound (Example 7). ■

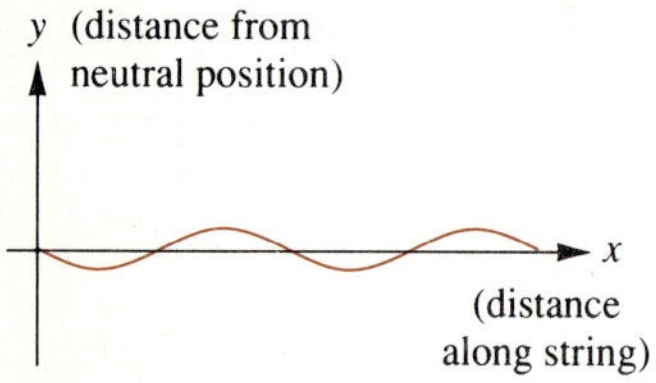

Figure 5.2-17

A vibrating string

By now you may have the idea that all periodic motion in nature is a sine wave, but it is not. Nevertheless, even when periodic motion is *not* sinusoidal, we can always construct it by *adding together a combination of sines and cosines.* We discuss this further in Section 5.4.

EXERCISES 5.2-1

1. How is the graph of $y = \sin(x + 2\pi)$ related to the graph of $y = \sin x$? Why?

2. How is the graph of $y = \cos(x - 2\pi)$ related to the graph of $y = \cos x$? Why?

3. Find the values of x for which the graph of $y = \sin x$ crosses the x axis. What is the value of y when it crosses the y axis?

4. Find the values of x for which the graph of $y = \cos x$ crosses the x axis. What is the value of y when it crosses the y axis?

5. At what values of x does the graph of $y = \sin x$ reach its maximum value? What is this maximum value?

6. At what values of x does the graph of $y = \cos x$ reach its maximum value? What is this maximum value?

7. Carefully sketch the graphs of $y = \sin x$ and $y = \cos x$ on the *same* set of axes.

***8.** On a *single* set of axes, sketch the graphs of $y = \sin x$ and $y = -\sin x$. (Do not do this by plotting points, but rather start with the known graph of $y = \sin x$ and ask yourself how the minus sign changes this graph.) What effect does the minus sign have on the period of the sine graph? On the maximum distance of the sine graph from the x axis? Just what effect does this minus sign have on the sine graph?

***9.** Repeat Exercise 8, except for $y = \cos x$ and $y = -\cos x$.

***10.** Prepare a table of values using calculator and plot each pair of graphs on a single set of axes, covering several cycles of the graph. For each of these graphs, what is its maximum distance from the x axis and what is its period? Explain how the coefficient of $\sin x$ alters the basic $\sin x$ graph.

(a) $y = \sin x$ and $y = 2 \sin x$
(b) $y = \sin x$ and $y = 3 \sin x$
(c) $y = \sin x$ and $y = 4 \sin x$

***11.** Repeat Exercise 10, except for the cosine instead of the sine.

***12.** Making use of what you have learned in Exercises 10 and 11, sketch each of the following without preparing a table of values.

(a) $y = 5 \sin x$ **(b)** $y = 6 \cos x$.

***13.** Use a calculator to prepare a table of values and plot each pair of graphs on a single set of axes, covering several cycles. For each of these graphs, what is its maximum distance from the x axis and what is its period? Explain how the coefficient of x affects the basic $\sin x$ graph.

(a) $y = \sin x$ and $y = \sin 2x$
(b) $y = \sin x$ and $y = \sin 3x$
(c) $y = \sin x$ and $y = \sin \frac{x}{2}$

***14.** Repeat Exercise 13 except for the cosine instead of the sine.

***15.** Prepare a table of values using a calculator and plot each pair of functions on a single set of axes, covering several cycles. What is the maximum distance of each graph from the x axis? What is the period of each graph? Interpret the effect of the added angles $(\pm\frac{\pi}{4}, \pm\frac{\pi}{2})$ in terms of shifting the basic graph of $y = \sin x$.

(a) $y = \sin x$ and $y = \sin(x - \frac{\pi}{4})$
(b) $y = \sin x$ and $y = \sin(x + \frac{\pi}{4})$
(c) $y = \sin x$ and $y = \sin(x - \frac{\pi}{2})$
(d) $y = \sin x$ and $y = \sin(x + \frac{\pi}{2})$

***16.** Repeat Exercise 15, except for the cosine instead of the sine.

***17. $\sin^2 x$ and $\cos^2 x$.** Without calculating any values or making a graph, make a plausible argument to show why the period of $y = \sin^2 x$ and $y = \cos^2 x$ is π, rather than 2π.

***18. $\sin^2 x$.** Sketch the graph of $y = \sin^2 x$ and find its period from the graph. Do this by plotting as few points as possible. (You should be able to do this simply by plotting the maximum and minimum values and finding the points where the graph crosses the x axis.)

***19. $\cos^2 x$.** Using Exercise 18 as a guide, sketch the graph of $y = \cos^2 x$ and find its period from the graph.

20. $\sin^2 x$. On a single set of axes, sketch the graphs of $y = \sin x$ and $y = \sin^2 x$.

21. $\cos^2 x$. On a single set of axes, sketch the graphs of $y = \cos x$ and $y = \cos^2 x$.

****22.** Make a table of values and plot the graph of $y = \sin x^2$. Is it periodic? Is it the same as the graph of $y = \sin^2 x$?

***23.** Make a table of values and plot the graph of $y = 1 + \sin x$. Is this graph periodic? If so, what is its period?

***24.** Make a table of values and plot the graph of $y = \sin x + \cos x$ covering several cycles. Is it periodic? What is the period?

***25.** Identify each of the following graphs. They *could* include $y = \pm\sin x, y = \pm 2 \sin x, y = \pm 4 \sin x, y = \pm\cos x, y = \pm 2 \cos x$, and $y = \pm 4 \cos x$ but no others.

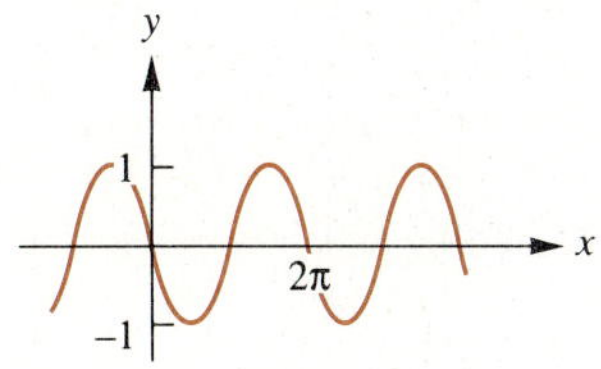

(a)

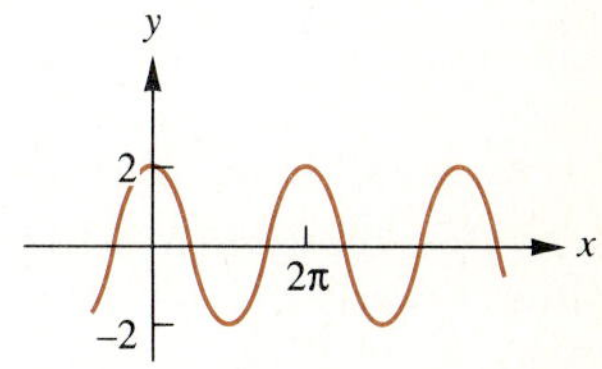

(b)

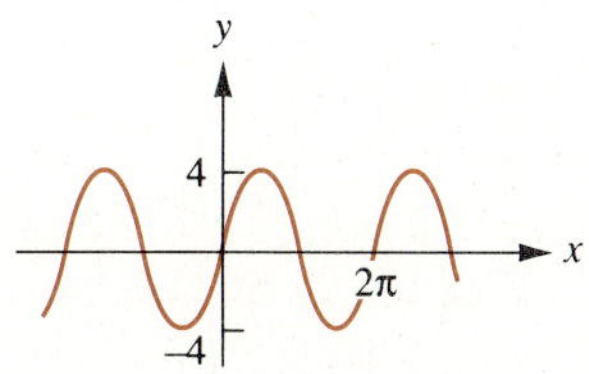

(c)

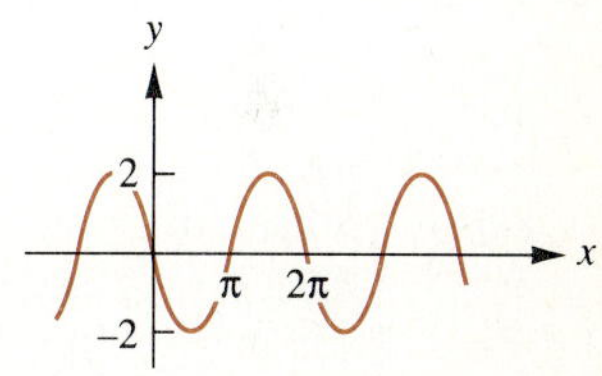

(d)

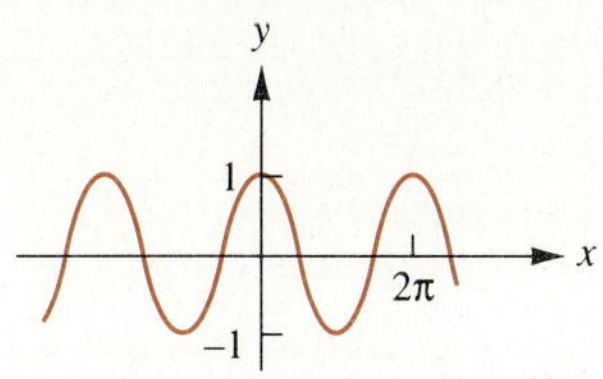

(e)

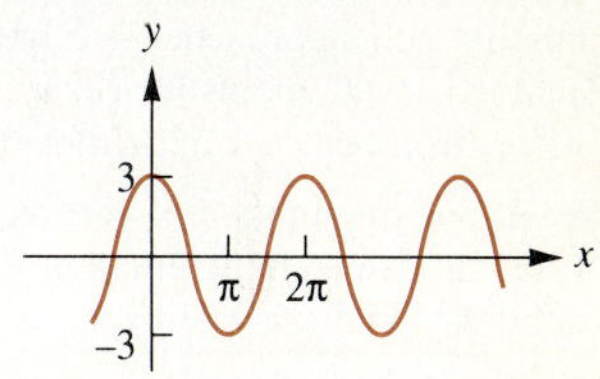

(f)

EXERCISES 5.2-2

1. How is the graph of $y = \tan(x + \pi)$ related to the graph of $y = \tan x$? Why?

2. Find the values of x for which the graph of $y = \tan x$ crosses the x axis. What is the value of y when it crosses the y axis?

3. At what values of x is the graph of $y = \tan x$ not defined?

4. Carefully sketch the graphs of $y = \sin x$ and $y = \tan x$ on the *same* set of axes.

5. Carefully sketch the graphs of $y = \cos x$ and $y = \tan x$ on the *same* set of axes.

***6.** Carefully sketch the graph of $y = -\tan x$. Do not plot points, but rather start with the known graph of $y = \tan x$ and then see how the minus sign affects this graph.

***7.** Carefully sketch the graph of $y = 3 \tan x$. Do not plot points, but rather start with the known graph of $y = \tan x$ and then see how the coefficient 3 affects this graph.

***8.** Carefully sketch the graph of $y = -2 \tan x$. Do not plot points, but rather start with the known graph of $y = \tan x$ and then see how the coefficient -2 affects this graph.

***9.** Make a table of values and use them to sketch a careful graph of $y = \tan 2x$. What is the period of this function? How is it related to the period of $y = \tan x$?

***10.** Repeat Exercise 9 except for $y = \tan 3x$. How does the 3 affect the period compared to the period of $y = \tan x$? Compare your result with that of Exercise 9.

***11.** Using the results from Exercises 9 and 10, predict the period of each function: (a) $y = \tan 6x$, (b) $y = \tan \frac{x}{2}$, (c) $y = \tan \frac{\pi x}{4}$.

EXERCISES 5.2-3

1. Carefully sketch each pair of graphs on a single set of axes.

- **(a)** $y = \sin x$ and $y = \csc x$
- **(b)** $y = \cos x$ and $y = \sec x$
- **(c)** $y = \tan x$ and $y = \cot x$
- **(d)** $y = \sec x$ and $y = \csc x$
- **(e)** $y = \sin x$ and $y = \sec x$
- **(f)** $y = \cos x$ and $y = \csc x$

2. Use the fact that $\cot x = 1/\tan x$ to show that $\cot x$ must have a period of π.

3. In Section 5.2-2, we found that $\tan x$ approaches positive infinity as x approaches $\frac{\pi}{2}$ from the left. Make the appropriate analogous statements for $\cot x$ as x approaches $0, \pm\pi, \pm 2\pi$, and so on from the *left* and from the *right*.

***4.** Using the method of Section 5.2-2 as a guide, make a table of values and sketch the graph of $y = \cot x$. Investigate its behavior near the points $x = 0, \pm\pi, \pm 2\pi$, and so on.

****5. Getting the Cotangent Graph from the Tangent.** Carefully sketch the graph of $y = \tan x$ for several cycles. Now use the fact that $\cot x = 1/\tan x$ to sketch carefully the graph of $y = \cot x$ on the *same* set of axes. (For example, you should use the fact that the magnitude of $\cot x$ gets small as the magnitude of $\tan x$ gets large, and vice versa.)

6. Without plotting points, show that the graph of $y = \sec x$ can never be closer than one unit to the x axis.

***7.** Using the method of Section 5.2-2 as a guide, make a table of values and sketch the graph of $y = \sec x$. Investigate and clearly explain its behavior near the points $x = \pm\frac{\pi}{2}, \pm\frac{3\pi}{2}, \pm\frac{5\pi}{2}$, and so on.

8. Rework Exercise 3 except make the statement for $\sec x$ as x approaches $\pm\frac{\pi}{2}, \pm\frac{3\pi}{2}, \pm\frac{5\pi}{2}$, and so on from both the *right* and the *left*.

****9. Getting the Secant Graph from the Cosine.** Sketch the graph of $y = \cos x$. Now use the fact that $\sec x = 1/\cos x$ to sketch the graph of $y = \sec x$. ***Hint:*** See Exercise 5.

10. Using the fact that $\csc x = 1/\sin x$, show that the graph of $y = \csc x$ has period 2π.

11. Without plotting points, show that the graph of $y = \csc x$ never gets closer than one unit to the x axis.

12. Rework Exercise 3 except make the statement for $\csc x$ as x approaches $0, \pm\pi, \pm 2\pi, \pm 3\pi$, and so on from both the *right* and the *left*.

****13. Getting the Cosecant Graph from the Sine.** Sketch the graph of $y = \sin x$. Now use the fact that $\csc x = 1/\sin x$ to sketch the graph of $y = \csc x$. ***Hint:*** See Exercise 5.

14. Identify which of the six trigonometric functions each graph in the following figures represents.

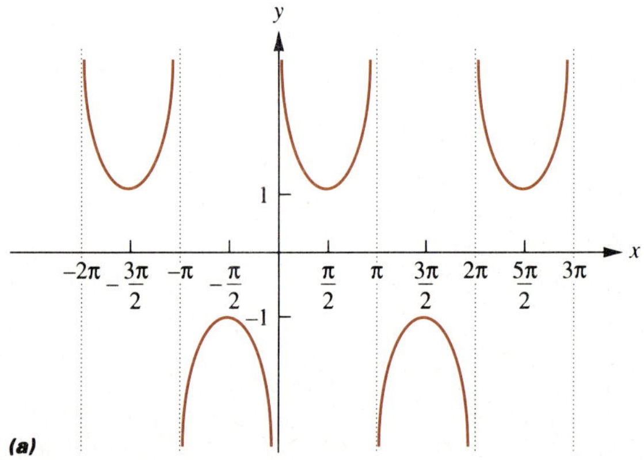

(a)

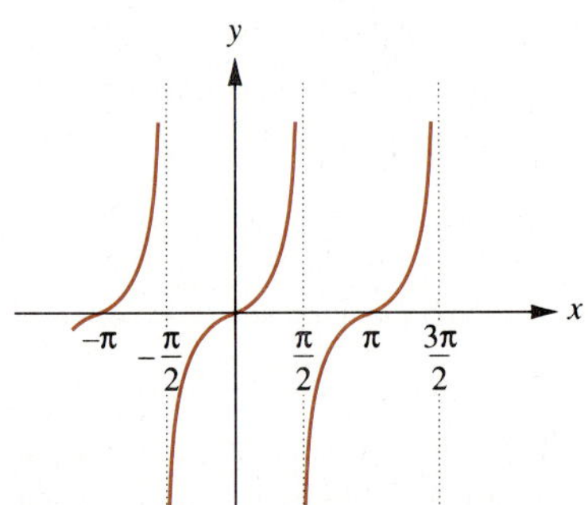

(b)

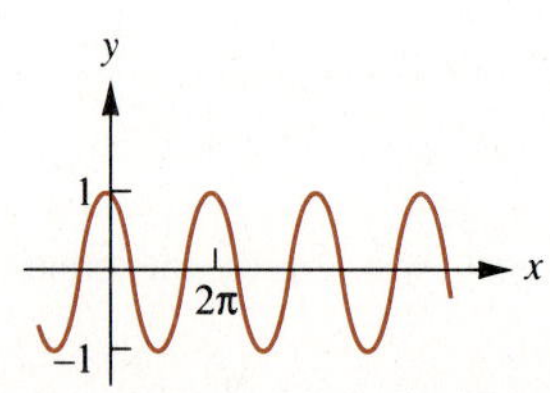

(c)

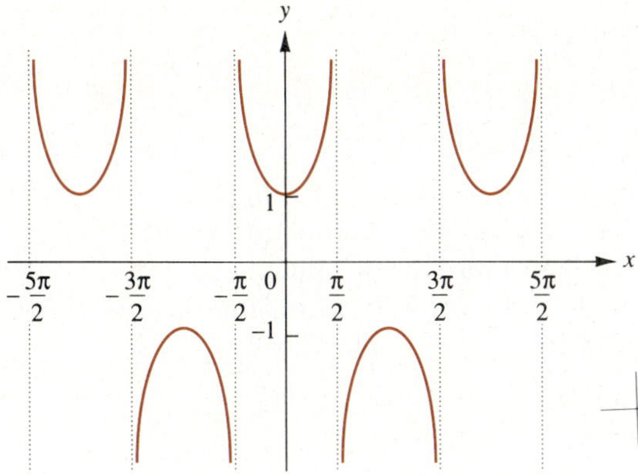

(d)

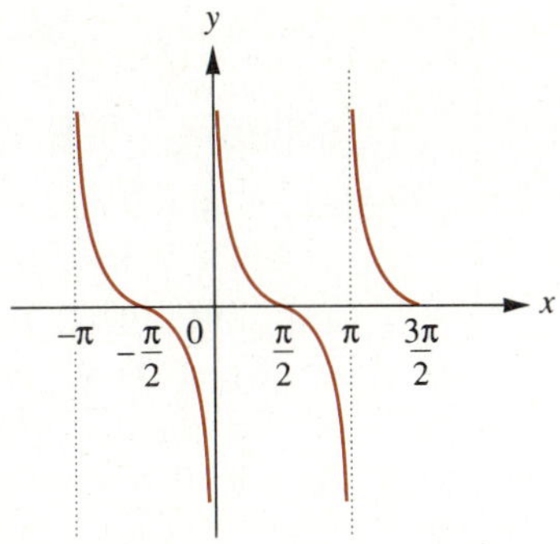

(f)

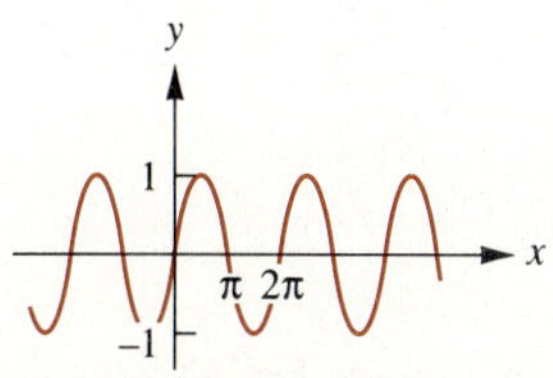

(e)

EXERCISES 5.2-4

1. Some parts of the world use AC current having 50 (instead of 60) cycles/s. Sketch a time graph of this current and find its period.

2. The note D# has 311.2 vibrations/s. Sketch the sound wave of this note and find its period.

3. Is the graph of the cosine a sine wave? Explain why.

5.3 GRAPHS OF $y = A\sin(Bx + C)$ and $y = A\cos(Bx + C)$

In the previous section we looked at the graphs of $y = \sin x$ and $y = \cos x$. We now investigate the effect on these graphs if we:

1. Multiply the function by a constant A; that is, $y = A\sin x$ and $y = A\cos x$.
2. Multiply the angle by a constant B; that is $y = \sin Bx$ and $y = \cos Bx$.
3. Add a constant C to the angle; that is, $y = \sin(x + C)$ and $y = \cos(x + C)$.

We first look at these effects one at a time, and then combine what we have learned. We can then quickly sketch the graph of any function of the form $y = A\sin(Bx + C)$ or $y = A\cos(Bx + C)$.

5.3-1 Graphs of $y = A\sin x$ and $y = A\cos x$

As we saw in Section 5.2, the largest that $\sin x$ and $\cos x$ can be is $+1$ and the smallest is -1. Now let us see what happens if we multiply either $\sin x$ or $\cos x$ by a constant A. We begin with concrete examples.

Example 9

Sketch the graph of $y = 3 \sin x$.

Solution

We could simply construct a table of values for $y = 3 \sin x$ and graph these numbers. There is, however, a much simpler way. By multiplying $\sin x$ by 3, we are multiplying each value of $\sin x$ by 3. This multiplication has the effect of *stretching* the graph in the y direction so that the maximum and minimum values are $+3$ and -3 rather than $+1$ and -1. Figure 5.3-1 illustrates this stretching; we have also included the graph of $y = \sin x$ for comparison.

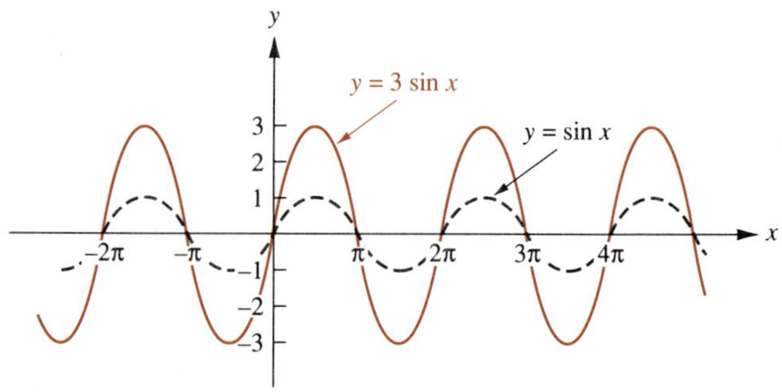

Figure 5.3-1
Graph of $y = 3 \sin x$

Notice two things:

1. Multiplying by 3 has *no effect on the period*, which remains 2π.
2. Because $\sin x$ is multiplied by 3, the graph can get no more than 3 units from the x axis. ■

NOW IT'S YOUR TURN

Sketch the graph of $y = 4.5 \sin x$.

(Answer: It is the same as the graph in Figure 5.3-1 except the maximum and minimum values of y are ± 4.5 instead of ± 3.)

Example 10

Sketch the graph of $y = -2 \cos x$.

Solution

Notice that in this case two things happen:

1. The minus sign turns the positive values of y negative and the negative values of y positive, which is the same thing as reflecting the cosine graph in the x axis, or turning the cosine graph upside down.
2. The factor of 2 multiplies all the values of $\cos x$ by 2, which makes the maximum and minimum values ± 2 instead of ± 1. This stretches the graph in the y direction. Notice that it is the *magnitude* (or absolute value) of the -2 that stretches the graph in the y direction.

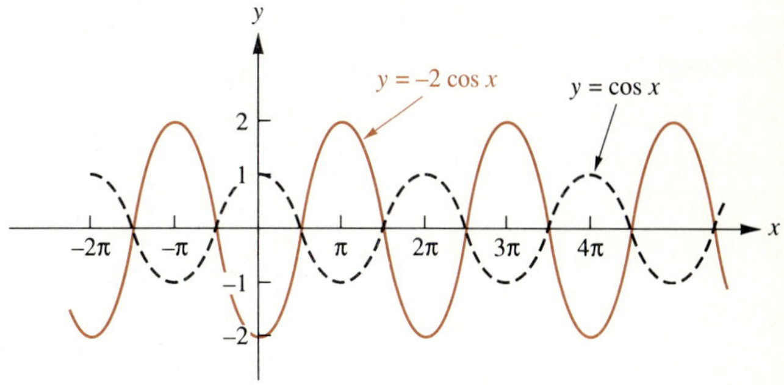

Figure 5.3-2

Graph of $y = -2\cos x$

In Figure 5.3-2 we show the results of these two effects, and we also show the graph of $y = \cos x$ for comparison. As in Example 9, notice that the multiplication by -2 has no effect on the period, which remains 2π. This multiplication does, however, limit the graph so it can get no more than two units from the x axis. ■

NOW IT'S YOUR TURN

Sketch the graph of $y = -1.5\cos x$.

(Answer: It is the same as the graph in Figure 5.3-2 except the maximum and minimum values of y are ± 1.5 instead of ± 2.)

Now let us generalize what we have learned from these two examples. Both are of the form $y = A\sin x$ or $y = A\cos x$. In both cases, the maximum distance that the graph can get from the x axis is equal to the magnitude of A. Therefore, we call the magnitude of A—which is its *absolute value* $|A|$—the **amplitude** of the graph.

AMPLITUDE

The *amplitude* of $y = A\sin x$ or $y = A\cos x$ is the maximum distance the graph can get from the x axis. This amplitude is equal to the absolute value of A, $|A|$.

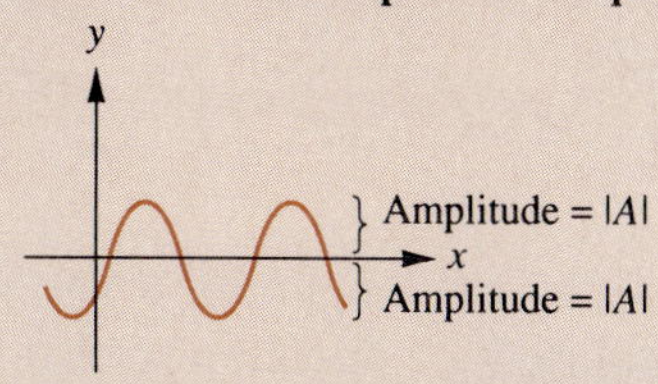

If A is negative, the sine or cosine graph is reflected in the x axis.

The other four trigonometric functions all approach $\pm\infty$, so there is no limit to how far they can get from the x axis. Therefore, it makes no sense to define amplitude for them. Thus we define the amplitude only for the sine and cosine.

Example 11

Without making a table of values, find the amplitude and sketch the graph of each of the following: (a) $y = 5.2 \sin x$, (b) $y = -6 \cos x$, and (c) $y = \frac{1}{3} \cos x$.

Solution

(a) $y = 5.2 \sin x$: Amplitude $= |A| = |5.2| = 5.2$. The graph is the same as that of $y = \sin x$, except it will be stretched in the y direction so its maximum distance from the x axis is 5.2 (see Figure 5.3-3).

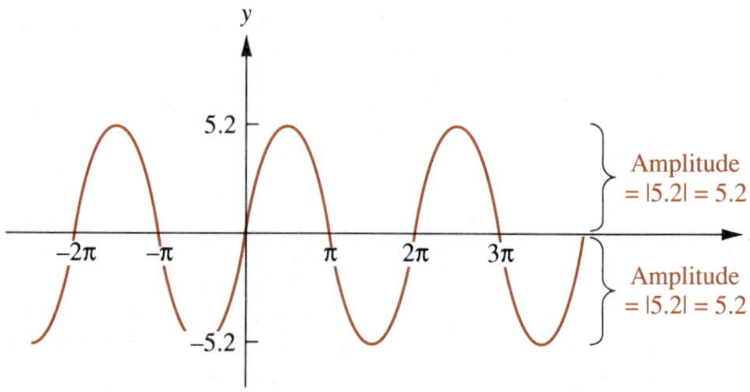

Figure 5.3-3

Graph of $y = 5.2 \sin x$

(b) $y = -6 \cos x$: Amplitude $= |A| = |-6| = 6$. The minus sign means that the graph is a cosine graph reflected in the x axis, or turned upside down. This graph is stretched in the y direction so that its maximum distance from the x axis is 6 (see Figure 5.3-4).

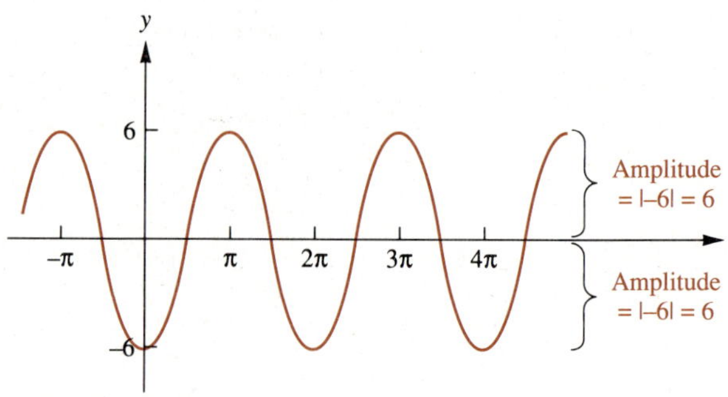

Figure 5.3-4

Graph of $y = -6 \cos x$

(c) $y = \frac{1}{3} \cos x$: Amplitude $= |A| = |\frac{1}{3}| = \frac{1}{3}$. The graph is a cosine graph compressed in the y direction so that its maximum distance from the x axis is $\frac{1}{3}$ unit (see Figure 5.3-5). ■

NOW IT'S YOUR TURN

Find the amplitude and sketch the graph of $y = -7.5 \sin x$.

(Answer: Amplitude $= 7.5$, see Figure 5.3-6.)

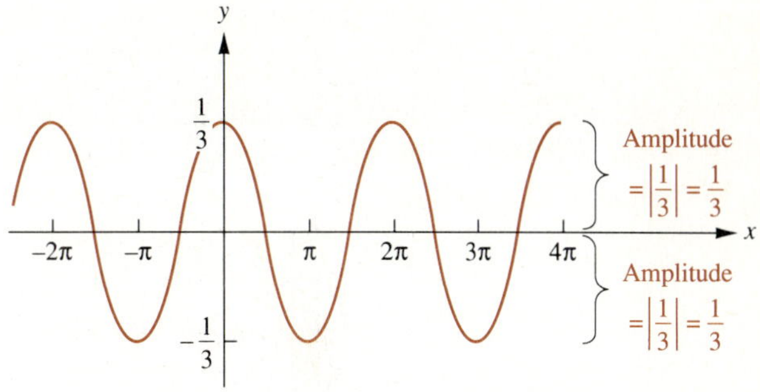

Figure 5.3-5

Graph of $y = \frac{1}{3}\cos x$

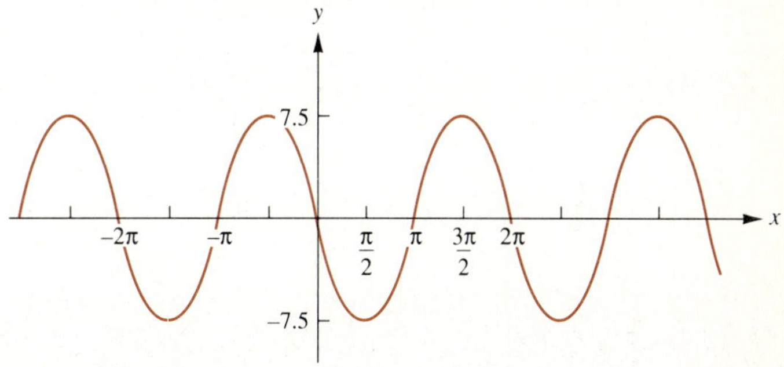

Figure 5.3-6

Graph of $y = -7.5\sin x$

The Physical Meaning of Amplitude

Mathematically the amplitude of a sine wave is the greatest distance that the graph gets from the x axis. This distance is the greatest value that the quantity on the y axis can have. This may represent an actual distance, but it doesn't always do so, as the following examples show.

Example 12

Pendulum

In Example 5 we saw how a pendulum traces a sine wave time graph. When graphing a pendulum's motion, we usually let the y axis be its angle (θ) with the vertical and plot time on the x axis. The amplitude of this sine wave is the maximum *angle* the pendulum swings, which is not a distance (see Figure 5.3-7). ■

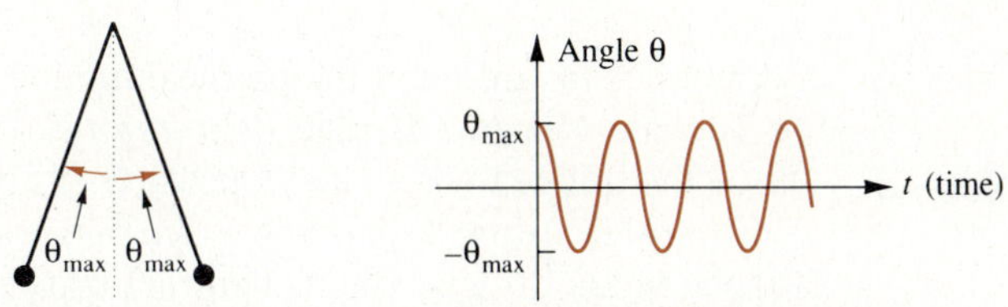

Figure 5.3-7

Amplitude of a pendulum's swing

(a) The pendulum at maximum angles

(b) Graph of the pendulum's motion (The amplitude represents the maximum angle of swing.)

NOW IT'S YOUR TURN

Sketch a careful time graph for a pendulum that swings back and forth between $\pm 15°$ and takes 2.0 s for each complete cycle. (Answer: See Figure 5.3-8)

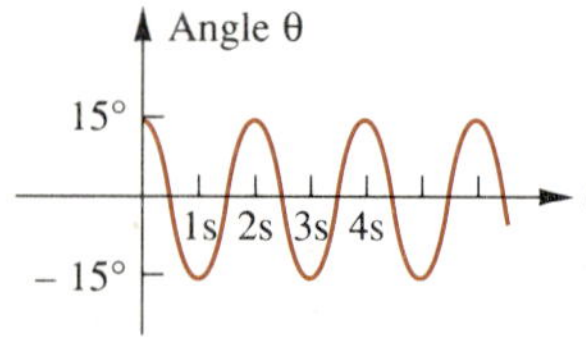

Figure 5.3-8

Time graph for a pendulum that swings between ±15° in 2.0 s

Example 13
Sound

In Example 7 we saw that sound is a pressure wave that can be represented by a sine wave time graph with pressure on the y axis. In this case, *the amplitude represents the maximum pressure*, which we experience as *loudness*. Figure 5.3-9 is an oscilloscope photograph showing two sound waves of different amplitudes. The wave with the greater amplitude is the louder sound. ■

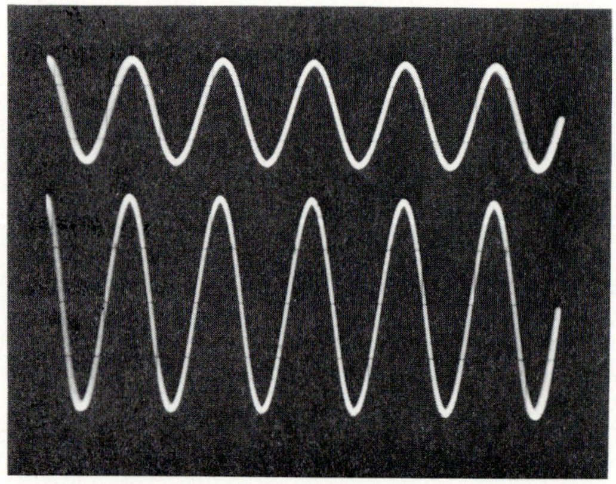

Figure 5.3-9

Oscilloscope trace showing sound waves of different amplitude. We experience the amplitude of such a wave as *loudness*. The bottom wave has a greater amplitude and therefore is the louder sound.

NOW IT'S YOUR TURN

For the vibrating string in Example 8, interpret the meaning of the amplitude of that sine wave.

(Answer: The amplitude is the maximum amount the string stretches from its neutral position.)

Example 14
Collapse of the Tacoma Narrows Bridge

A dramatic illustration of the physical significance of amplitude is the collapse of the Tacoma Narrows Bridge near Seattle on November 10, 1940. The winds caused the bridge to oscillate violently (see Figure 5.3-10a). Notice that the edge of the bridge forms a sine wave. In this case the amplitude is the maximum distance of the bridge from its neutral position. Figure 5.3-10b shows the drastic results of an amplitude that was too great!* ■

* There were no human injuries, but a dog that refused to leave an abandoned car went down with the bridge.

Figure 5.3-10

The Tacoma Narrows Bridge (a) The bridge vibrating in a sine wave. The amplitude is the maximum distance of the bridge from its neutral position. (b) The result of too much amplitude!

5.3-2 Graphs of $y = \sin Bx$ and $y = \cos Bx$

Changes in the Period

We now look at how a constant B in $y = \sin Bx$ or $y = \cos Bx$ affects the graphs, starting with three numerical examples.

Example 15

Graph $y = \sin 2x$ for $0 \le x \le 2\pi$.

Solution

We start by making a table of values of $\sin 2x$ and using these to sketch the graph (see Figure 5.3-11). As you can see in this figure, $y = \sin x$ completes one cycle as x increases from 0 to 2π. But $y = \sin 2x$ completes *two* cycles in this interval. If we think of the cycle as starting when $2x = 0$, it repeats when $2x = 2\pi$, which is when $x = \frac{2\pi}{2} = \pi$. Therefore, the period is $P = \pi$. ■
Notice that the period is $P = \frac{2\pi}{2} = 2\pi/(\text{coefficient of } x)$.

x	$2x$	$\sin 2x$
0	0	0.00
$\frac{\pi}{4}$	$\frac{\pi}{2}$	1.00
$\frac{\pi}{2}$	π	0.00
$\frac{3\pi}{4}$	$\frac{3\pi}{2}$	−1.00
π	2π	0.00
$\frac{5\pi}{4}$	$\frac{5\pi}{2}$	1.00
$\frac{3\pi}{2}$	3π	0.00
$\frac{7\pi}{4}$	$\frac{7\pi}{2}$	−1.00
2π	4π	0.00

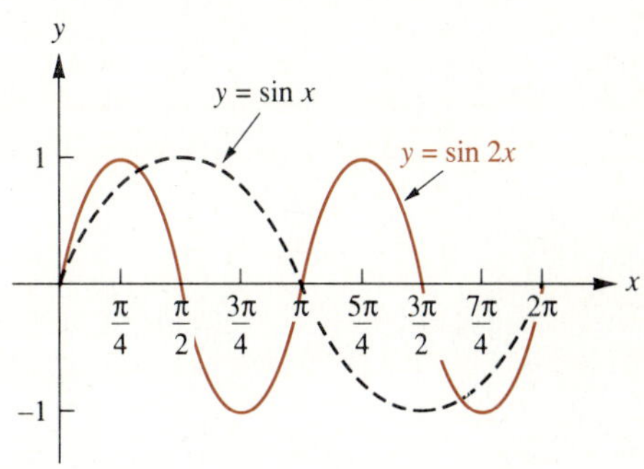

Figure 5.3-11

Table of values and the graph of $y = \sin 2x$

Example 16

Graph $y = \cos(\frac{1}{2}x)$.

Solution

We first construct a table of values and use these to graph the function (see Figure 5.3-12). As you can see from the graph, $y = \cos \frac{x}{2}$ completes one cycle when x increases from 0 to 4π, but $y = \cos x$ completes two cycles in this same interval. To see why this is true, we think of $x/2$ as a unit, as in Example 15. Then $\cos(x/2)$ repeats its values when $x/2$ increases from 0 to 2π. The end of this cycle occurs when $x/2 = 2\pi$, which is when $x = 2\pi/\frac{1}{2} = 4\pi$. Thus the period must be $P = 2\pi/(\frac{1}{2}) = 4\pi$. ■

Notice that, just as in Example 15, the period is $P = 2\pi/(\frac{1}{2})$ $= 2\pi/(\text{coefficient of } x)$.

x	$\frac{x}{2}$	$\cos \frac{x}{2}$
0	0	1.00
$\frac{\pi}{2}$	$\frac{\pi}{4}$	0.71
π	$\frac{\pi}{2}$	0.00
$\frac{3\pi}{2}$	$\frac{3\pi}{4}$	−0.71
2π	π	−1.00
$\frac{5\pi}{2}$	$\frac{5\pi}{4}$	−0.71
3π	$\frac{3\pi}{2}$	0.00
$\frac{7\pi}{2}$	$\frac{7\pi}{4}$	0.71
4π	2π	1.00

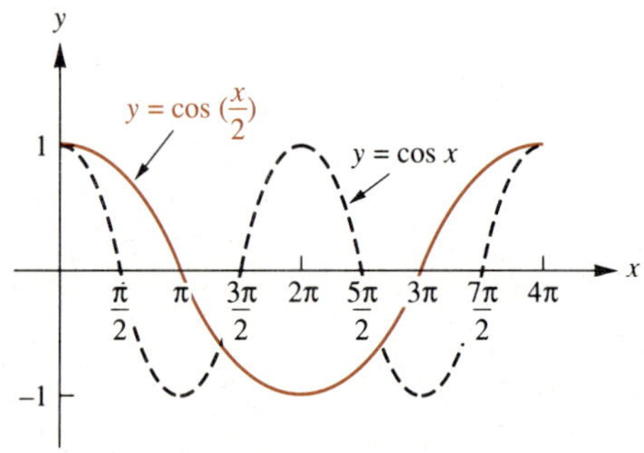

Figure 5.3-12

Table of values and graph of $y = \cos x/2$

NOW IT'S YOUR TURN

By making tables of data, graph each function and find its period: (a) $y = \sin 4x$, (b) $y = \cos x/3$.

(Answer: (a) Period $= \frac{\pi}{2}$, (b) Period $= 6\pi$. See Figure 5.3-13 for the graphs.)

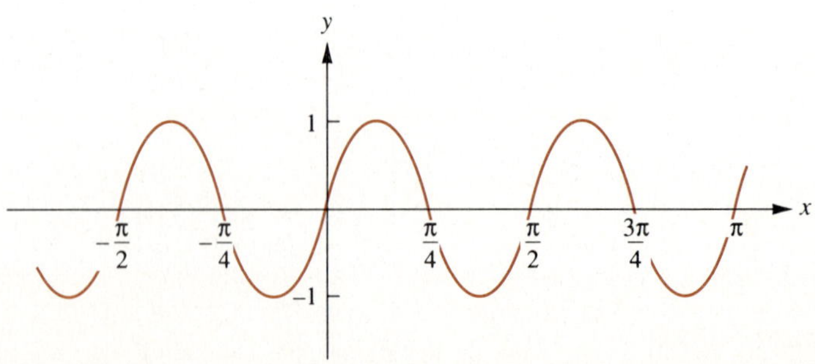

Figure 5.3-13

(a) Graph of $y = \sin 4x$

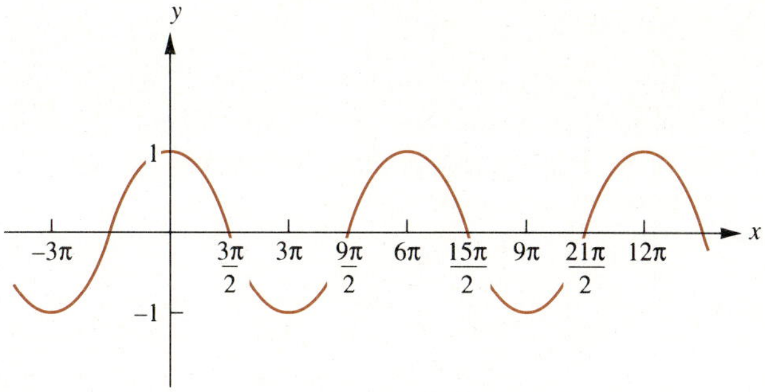

***Figure 5.3-13** (Continued)*

***(b) Graph of** $y = \cos \frac{x}{3}$*

Example 17

Find the period of $y = \sin(-2x)$.

Solution

We can rewrite $\sin(-2x)$ using the fact that $\sin(-\theta) = -\sin \theta$, which we proved in Section 3.1-3. Therefore,

$$\sin(-2x) = -\sin 2x$$

This result means that the period of $\sin(-2x)$ must be the same as the period of $-\sin 2x$. As we saw in Section 5.3-1, the -1 in front of $\sin 2x$ has no effect on the period. Therefore the *period* of $\sin(-2x)$ is the *same* as the period of $\sin 2x$, which we saw was π in Example 15. The minus sign does affect the graph by reflecting it in the x axis, but it has *no effect on the period.* ■

NOW IT'S YOUR TURN

Find the period of $\cos(-2x)$ using the same procedure as the one used in this example. Does the minus sign affect the *period*? Will it affect the *graph*? (Answer: π, no, no)

> **PERIOD**
>
> The period of the graph of $y = \sin Bx$ and $y = \cos Bx$ is $2\pi/|B|$.

Now we can generalize what we have just seen in these examples; $y = \sin Bx$ (or $y = \cos Bx$) completes one cycle as Bx increases from 0 to 2π. Therefore, the value of x for which the graph completes one cycle is $Bx = 2\pi$, which gives us $x = 2\pi/B$, so the period is $2\pi/B$. If B is negative, the minus sign has no effect on the period. Therefore, to include both positive and negative values of B, we can express the period as $P = 2\pi/|B|$. ■

Example 18

Find the period and sketch the graph of each function: (a) $y = \sin 3x$, (b) $y = \cos(0.64\, x)$, and (c) $y = \sin(-x/4)$.

Solution

(a) $y = \sin 3x$: Period $= 2\pi/|B| = 2\pi/3$. The graph goes through one cycle as x goes from 0 to $2\pi/3$. First draw one cycle between $x = 0$ and $x = 2\pi/3$ along the x axis. Then repeat the pattern at intervals of $2\pi/3$ (see Figure 5.3–14).

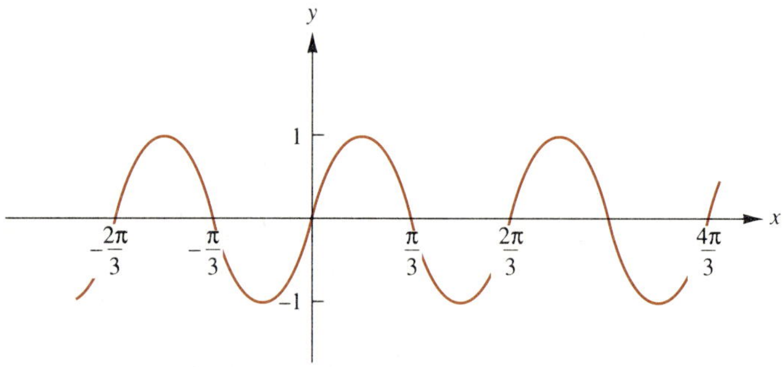

Figure 5.3-14

Graph of $y = \sin 3x$

(b) $y = \cos(0.64x)$: Period $= 2\pi/|B| = 2\pi/0.64 = 9.8$. First sketch one cycle between $x = 0$ and $x = 9.8$ and then repeat that pattern every 9.8 units on the x axis (see Figure 5.3-15).

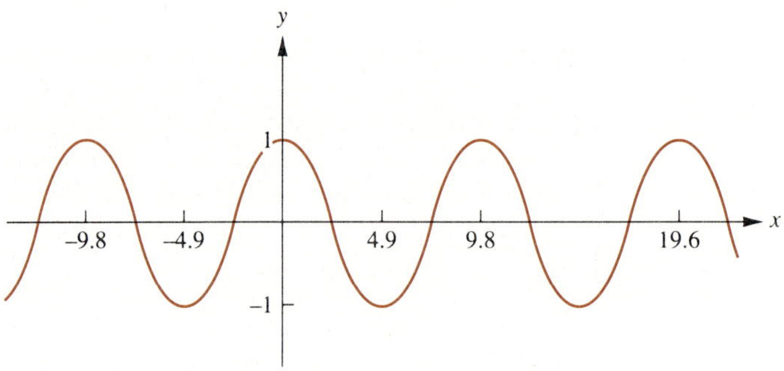

Figure 5.3-15

Graph of $y = \cos(0.64x)$

(c) $y = \sin(-x/4)$: Period $= 2\pi/|B| = 2\pi/|-\frac{1}{4}| = 2\pi/\frac{1}{4} = 8\pi$. Now use the fact that $\sin(-\theta) = -\sin\theta$: $y = \sin(-x/4) = -\sin(x/4)$. Therefore, the graph of $y = \sin(-x/4)$ is the same as the graph of $y = -\sin(x/4)$. The minus sign reflects the graph of $y = \sin(x/4)$ in the x axis. First sketch one cycle between $x = 0$ and $x = 8\pi$, and then repeat the pattern every 8π units along the x axis (see Figure 5.3-16). ■

NOW IT'S YOUR TURN

Find the period and sketch the graph of: (a) $y = \sin(4.5x)$ and (b) $y \cos(-0.64x)$.

(Answer: (a) $P = 1.4$, see Figure 5.3-17 for the graph. (b) $P = 9.8$, same graph as Figure 5.3-15.)

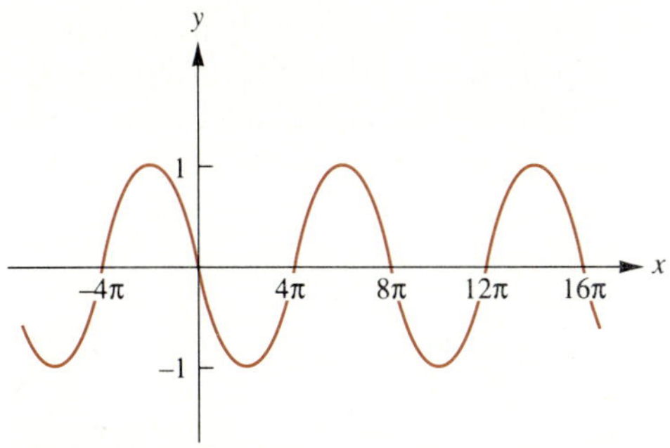

Figure 5.3-16

Graph of $y = \sin(-x/4)$

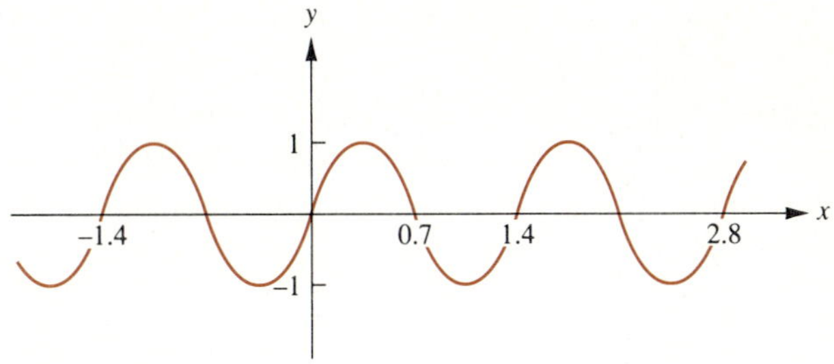

Figure 5.3-17

Graph of $y = \sin(4.5x)$

Combining Changes in the Amplitude and the Period

So far we have dealt with changes in the amplitude and changes in the period separately, but often we have to graph equations of the form $y = A \sin Bx$ or $y = A \cos Bx$. In these cases, *both* the amplitude and period are affected. Such graphs are not difficult to do because we can look at the effects on the amplitude and period *separately*, and then combine the results to make the graph.

Example 19

For each of the following functions, find the amplitude and the period and sketch its graph. (a) $y = -2 \sin x/3$, (b) $y = \frac{1}{2} \cos \pi x$, and (c) $y = 4 \cos(-3x)$.

Solution

As we have seen, *A has no effect on the period and B does not affect the amplitude*. Therefore, we can find the amplitude and the period just as we did before.

(a) $y = -2 \sin x/3$: $A = -2$ and $B = \frac{1}{3}$. Amplitude $= |A| = |-2| = 2$. Period $= 2\pi/|B| = 2\pi/(\frac{1}{3}) = 6\pi$. The minus sign in A reflects the graph of $y = 2 \sin x/3$ through the x axis. First sketch one cycle between $x = 0$ and 6π, and then repeat the pattern as x changes by 6π (see Figure 5.3-18).

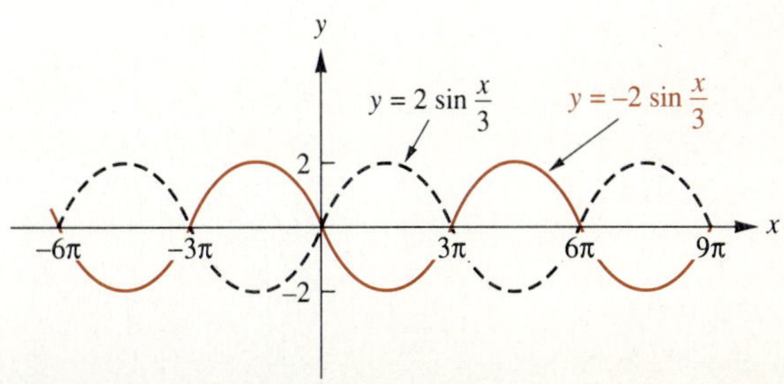

Figure 5.3-18

Graph of $y = -2 \sin x/3$

(b) $y = \frac{1}{2}\cos \pi x$: $A = \frac{1}{2}$ and $B = \pi$. As in part (a), we can find the amplitude and period independently because they do not affect each other. Amplitude $= |A| = |\frac{1}{2}| = \frac{1}{2}$. Period $= 2\pi/|B| = 2\pi/\pi = 2$. First sketch one cycle, then repeat the pattern every two units along the x axis (see Figure 5.3-19).

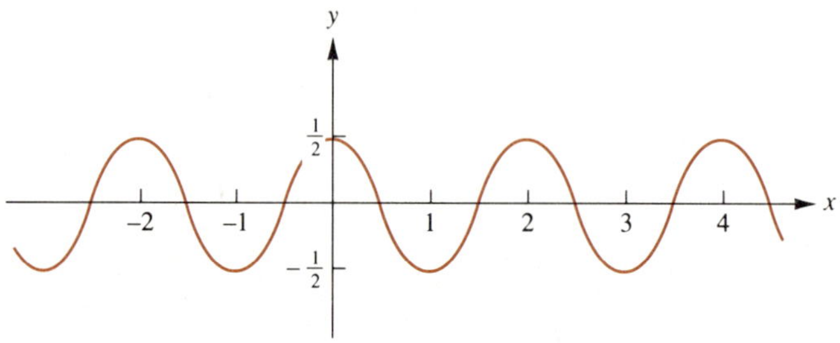

Figure 5.3-19
Graph of $y = \frac{1}{2}\cos \pi x$

(c) $y = 4\cos(-3x)$: $A = 4$, $B = -3$. Amplitude $= |A| = 4$. Period $= P = 2\pi/|B| = 2\pi/|-3| = 2\pi/3$. From Section 3.1-3, we know that $\cos(-\theta) = \cos\theta$, which gives us $y = 4\cos(-3x) = 4\cos(3x)$. This means that the minus sign in B will *not* reflect the graph in the x axis. In this case, the fact that B is negative has no effect whatsoever on the graph. Therefore, the graph of $y = 4\cos(-3x)$ is exactly the same as the graph of $y = 4\cos 3x$ (see Figure 5.3-20). ■

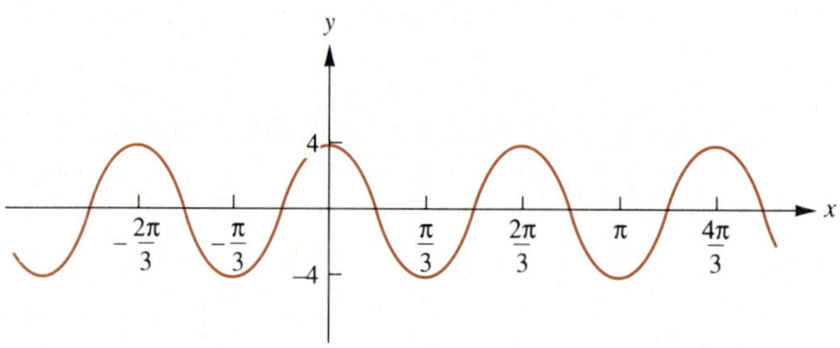

Figure 5.3-20
Graph of $y = 4\cos(-3x)$

NOW IT'S YOUR TURN

Determine the amplitudes and periods and sketch the graphs of (a) $y = -3\sin x/2$ and (b) $y = -\sin(-2x)$.

(Answer: (a) $P = 4\pi$, amplitude $= 3$, see Figure 5.3-21 for the graph. (b) $P = \pi$, amplitude $= 1$, the graph is the same as in Figure 5.3-11.)

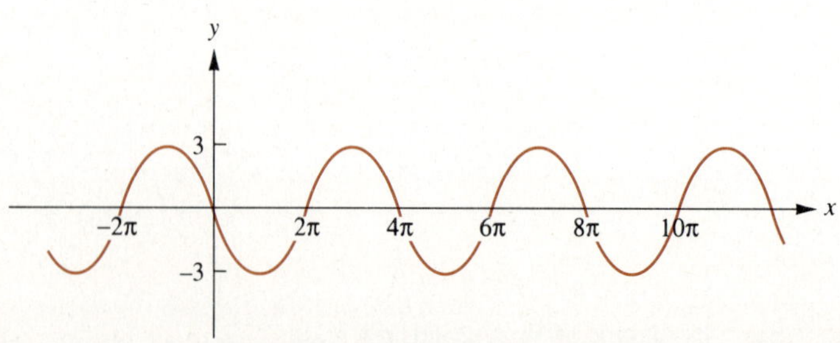

Figure 5.3-21
Graph of $y = -3\sin x/2$

The Physical Meaning of *B* for Time Graphs

We have seen a number of examples of sine waves in which the x axis represents time (the pendulum, sound waves, AC current). Remember that we call such graphs *time graphs*. When we write a sine wave such as $y = A \sin Bt$, what is the physical meaning of B? To find out, remember that the period P of the sine wave is $P = 2\pi/|B|$. Solving for $|B|$ gives $|B| = 2\pi/P$. When describing such waves, we usually refer to their *frequency* f, which is the *number of cycles per second*. The *reciprocal* of the frequency is the *number of seconds per cycle*. But the number of seconds per cycle is just the *time for one cycle*, which is the period. Therefore, *the period is the reciprocal of the frequency*: $P = 1/f$. We can also write this as $f = 1/P$.

FREQUENCY

frequency = number of cycles per second

$f = 1/P$ and $P = 1/f$

We usually measure frequencies in *cycles per second* (cps or cycle/s), which we call the *hertz* (*Hz*). A cycle is not a physical quantity like a length or a time, so it has no units. Therefore, the cycle per second is the same thing as $1/\text{s} = \text{s}^{-1}$.

UNITS OF FREQUENCY

$1 \text{ Hz} = 1 \text{ cycle/s} = 1/\text{s} = \text{s}^{-1}$

Now we can find the physical meaning of B.

$$|B| = \frac{2\pi}{P} = 2\pi\frac{1}{P} = 2\pi(f) = 2\pi f$$

Therefore, *$|B|$ is 2π times the frequency of the wave.*

Example 20
Sound

In Example 7, we saw that sound can be represented by a sine wave in which the y axis is pressure and the x axis time. The equation of this time graph is $y = A \sin Bt$, where $|B|$ is 2π times the frequency of the sound, or $|B| = 2\pi f$. Audible sounds have frequencies between 20 Hz and 20,000 Hz. We experience these frequencies as *pitch*. A sound with a *high pitch* has a *high frequency* (a *short* period), whereas a sound with a *low pitch* has a *low frequency* (a *long* period). Thus the constant $|B| = 2\pi f$ represents the *pitch* of the sound.

Figure 5.3-22 is a photograph of an oscilloscope trace of two sound waves. They have different periods, and therefore different frequencies and pitches.

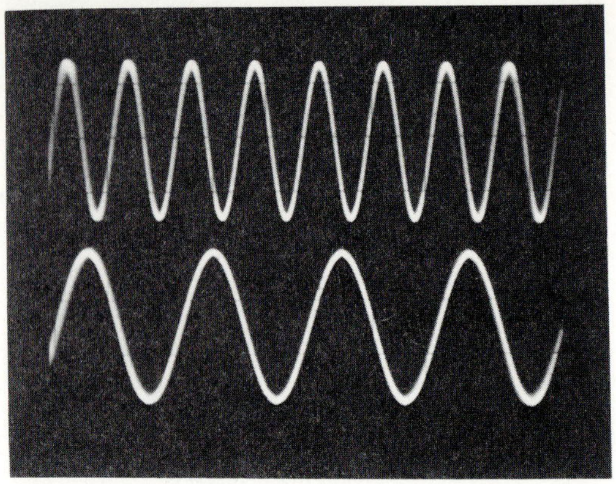

Figure 5.3-22

Photograph of an oscilloscope trace of two sound waves The upper sound has a shorter period and therefore a higher frequency and pitch than the lower wave.

The note concert A has 440 vibrations/s. Each vibration is one cycle, so its frequency is f = 440 cycles/s = 440 Hz. Therefore, for this note, $|B|$ is

$$|B| = 2\pi f = (2\pi)(440 \text{ Hz}) = (2\pi)(440 \text{ s}^{-1}) = 2800 \text{ s}^{-1}$$

Its period is

$$P = \frac{1}{f} = \frac{1}{440 \text{ Hz}} = \frac{1}{440 \text{ s}^{-1}} = 2.3 \times 10^{-3} \text{ s}$$

Thus each vibration, or cycle, takes 2.3×10^{-3} s. The sine wave for this note is $y = \sin Bt = \sin(2800 \text{ s}^{-1}\, t)$, and Figure 5.3-23 shows its graph.* Although it is technically correct to write the s^{-1} units in the parentheses, we usually do not do so because it is a bit awkward. We usually write this wave as $y = \sin(2800\, t)$, where it is understood that the 2800 had units of s^{-1}. ■

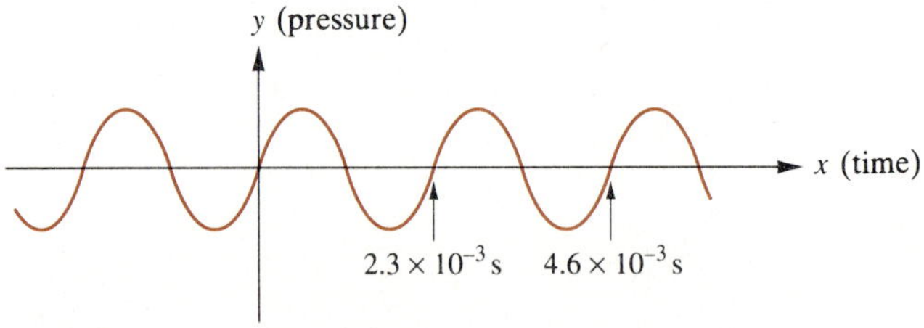

Figure 5.3-23

The sine wave representing the note A The amplitude represents loudness and the frequency represents pitch.

NOW IT'S YOUR TURN

The note $G\#$ in the chromatic scale has a frequency of 415 Hz. Find the period of this note, express it as a sine wave of the form $y = \sin Bt$, and sketch its time graph.

---

* We have let the amplitude be 1, meaning one unit of pressure.

(Answer: $P = 2.41 \times 10^{-3}$ s, $y = \sin(2610t)$, where 2810 has units of s^{-1}; see Figure 5.3-24 for the graph.)

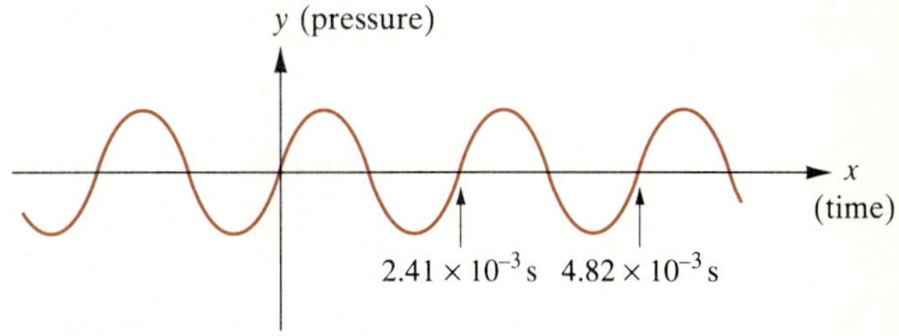

Figure 5.3-24

Sine wave representing the note G♯

5.3-3 Graphs of $y = \sin(x + C)$ and $y = \cos(x + C)$

Phase Shifts

To see how C affects the graph, let's start with two examples.

Example 21

Graph (a) $y = \cos(x + \frac{\pi}{2})$ and (b) $y = \sin(x - \frac{\pi}{4})$.

Solution

(a) $y = \cos(x + \frac{\pi}{2})$: Notice that $C = +\frac{\pi}{2}$. We start by making a table of values and graphing them (see Figure 5.3-25). Notice this important result: The graph of $y = \cos(x + \frac{\pi}{2})$ is the same as the graph of $y = \cos x$, except it is shifted to the left by $\frac{\pi}{2}$ units. We say that the graph of $y = \cos(x + \frac{\pi}{2})$ is *out of phase* with the graph of $y = \cos x$ by an amount $\frac{\pi}{2}$ along the x axis. Therefore, we call $\frac{\pi}{2}$ a *phase shift*. The *amount* of phase shift is $\frac{\pi}{2}$ and the *direction* is to the left. Notice that *when C is positive* the graph of $y = \cos x$ is shifted to the *left*. Also notice that the phase shift has no effect on the amplitude or the period.

x	$x + \frac{\pi}{2}$	$\cos(x + \frac{\pi}{2})$
$-\frac{\pi}{2}$	0	1.00
0	$\frac{\pi}{2}$	0.00
$\frac{\pi}{2}$	π	−1.00
π	$\frac{3\pi}{2}$	0.00
$\frac{3\pi}{2}$	2π	1.00

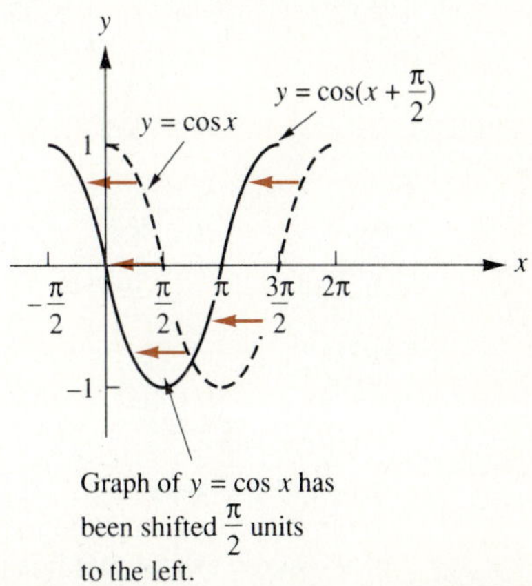

Figure 5.3-25

Graph of $y = \cos(x + \frac{\pi}{2})$
Notice that this graph is the same as the graph of $y = \cos x$ shifted to the left by $\frac{\pi}{2}$ units along the x axis.

(b) $y = \sin(x - \frac{\pi}{4})$: Notice that $C = -\frac{\pi}{4}$. We begin by making a table of values and graphing this function (see Figure 5.3-26). Notice that the graph of $y = \sin(x - \frac{\pi}{4})$ is the same as the graph of $y = \sin x$, except it is shifted to the right by $\frac{\pi}{4}$ units. As in part (a), we say that the graph of $y = \sin(x - \frac{\pi}{4})$ is out of phase with the graph of $y = \sin x$ by $\frac{\pi}{4}$ units along the x axis. As in that example, we call $-\frac{\pi}{4}$ a phase shift. The amount of phase shift is $\frac{\pi}{4}$, and the direction is to the right. Notice that when C *is negative*, the graph of sin x is shifted to the *right*. Also note that the phase shift has *no effect* on the amplitude or the period. ■

x	$x - \frac{\pi}{4}$	$\sin(x - \frac{\pi}{4})$
$\frac{\pi}{4}$	0	0.00
$\frac{\pi}{2}$	$\frac{\pi}{4}$	0.71
$\frac{3\pi}{4}$	$\frac{\pi}{2}$	1.00
π	$\frac{3\pi}{4}$	0.71
$\frac{5\pi}{4}$	π	0.00
$\frac{3\pi}{2}$	$\frac{5\pi}{4}$	−0.71
$\frac{7\pi}{4}$	$\frac{3\pi}{2}$	−1.00
2π	$\frac{7\pi}{4}$	−0.71
$\frac{9\pi}{4}$	2π	0.00

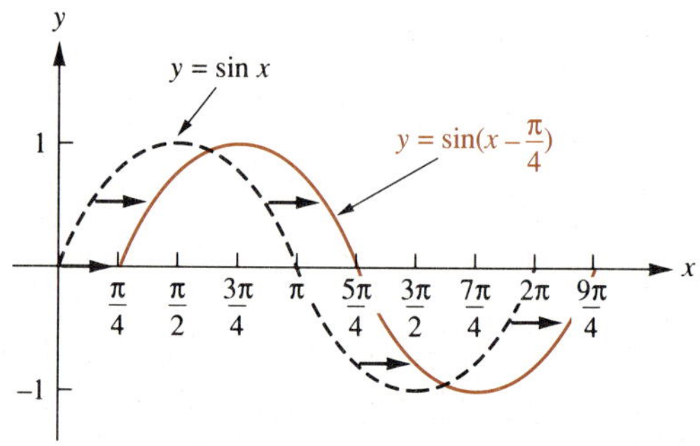

Figure 5.3-26

Table of values and graph of $y = \sin(x - \frac{\pi}{4})$ The graph of $\sin(x - \frac{\pi}{4})$ is the same as the graph of sin x, except it is shifted to the right by $\frac{\pi}{4}$ units along the x axis.

NOW IT'S YOUR TURN

Make a table of values to graph each function and find the amount and direction of the phase shift. (a) $y = \sin(x - \frac{\pi}{2})$ and (b) $y = \cos(x + \frac{\pi}{4})$.

(Answer: The phase shifts are $\frac{\pi}{2}$ to the right for (a) and $\frac{\pi}{4}$ to the left for (b). For the graphs, see Figures 5.3-27 and 5.3-28.)

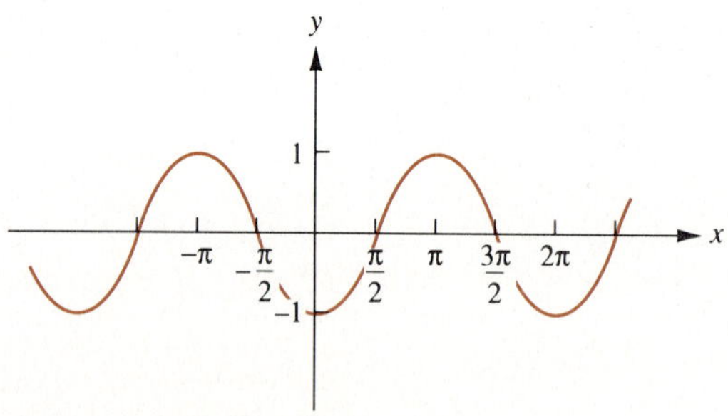

Figure 5.3-27

Graph of $y = \sin(x - \frac{\pi}{2})$. This graph is the same as the graph of $y = \sin x$, except that it is shifted to the right by $\frac{\pi}{2}$ units.

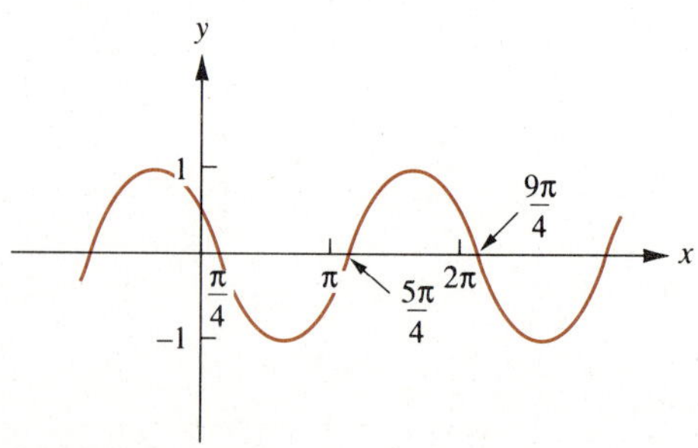

Figure 5.3-28

Graph of $y = \cos(x + \frac{\pi}{4})$ This graph is the same as the graph of $y = \cos x$, except that it is shifted to the left by $\frac{\pi}{4}$ units.

From the results of the last two examples, we can see the meaning of the C in $y = \sin(x + C)$ or $y = \cos(x + C)$. Both examples show that C is a *phase shift*—the distance that the graph is shifted along the x axis. The *magnitude* of C tells us the *amount* of phase shift, while the *sign* of C tells us its *direction*. If C is *positive*, the phase shift is to the *left*. If C is *negative*, the phase shift is to the *right*. At first glance these results may appear to contradict our usual convention of the right being positive and the left negative, but they are perfectly consistent with our notions of signed numbers. To see why, consider $y = \sin(x + C)$. This function goes through one cycle as $x + C$ increases from 0 to 2π. This cycle begins when $x + C = 0$, which is when $x = -C$. If C is positive, the cycle now begins at $x = -C$ instead of $x = 0$, meaning the graph is shifted C units to the *left*. If C is negative, then the cycle also begins when $x = -C$, but $-C$ is now a *positive* number because C itself is negative. We can write this positive number as $|C|$. Therefore, the cycle now begins at $x = -C = |C|$ instead of $x = 0$, which is $|C|$ units to the right, meaning the graph is shifted to the *right* by $|C|$ units.

PHASE SHIFT

$$y = \sin(x + C) \qquad \text{and} \qquad y = \cos(x + C)$$

Phase Shift The *phase shift* is the *distance* that the graph of $\sin x$ or $\cos x$ is shifted to the right or left along the x axis.

Amount of Phase Shift The *magnitude* (absolute value) of C is the *amount of the phase shift*.

Direction of Phase Shift If C is *positive*, the graph is shifted to the *left*. If C is *negative*, the graph is shifted to the *right*.

Example 22

For each function, find its phase shift and sketch its graph. (a) $y = \sin(x - 2)$ and (b) $y = \cos(x + \pi)$.

Solution

(a) $y = \sin(x - 2)$: $C = -2$, so the phase shift is $|C| = |-2| = 2$. Because C is negative, the graph of $\sin x$ is shifted to the right by two units (see Figure 5.3-29).

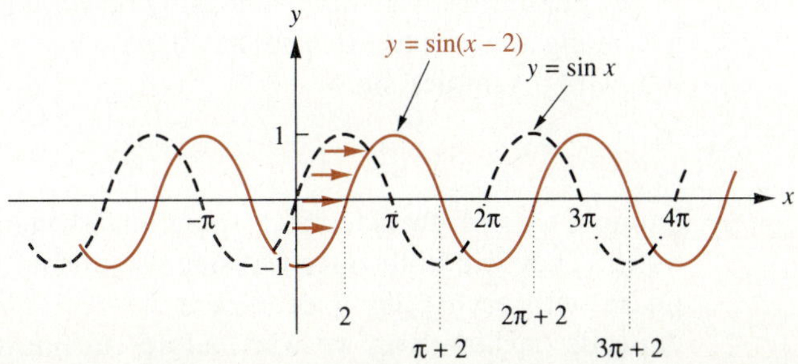

Figure 5.3-29

Graph of $y = \sin(x - 2)$ This is the same as the graph of $y = \sin x$ shifted two units to the right.

(b) $y = \cos(x + \pi)$: $C = \pi$, so the phase shift is $|C| = |\pi| = \pi$. Because C is positive, the graph of $y = \cos x$ is shifted to the left by π units (see Figure 5.3-30). ■

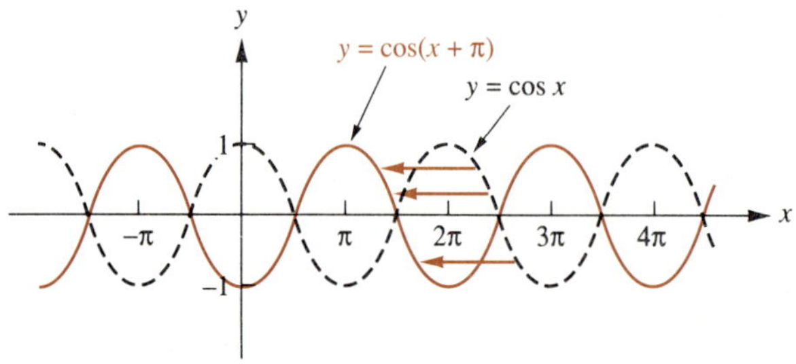

Figure 5.3-30

Graph of $y = \cos(x + \pi)$ This is the same as the graph of $y = \cos x$ shifted by π units to the left.

NOW IT'S YOUR TURN

Find the phase shift for each function and sketch its graph. (a) $y = \sin(x + 1)$ and (b) $y = \cos(x - \frac{\pi}{3})$.

(Answer: (a) One unit to the left, (b) $\frac{\pi}{3}$ units to the right. See Figure 5.3-31 for the graphs.)

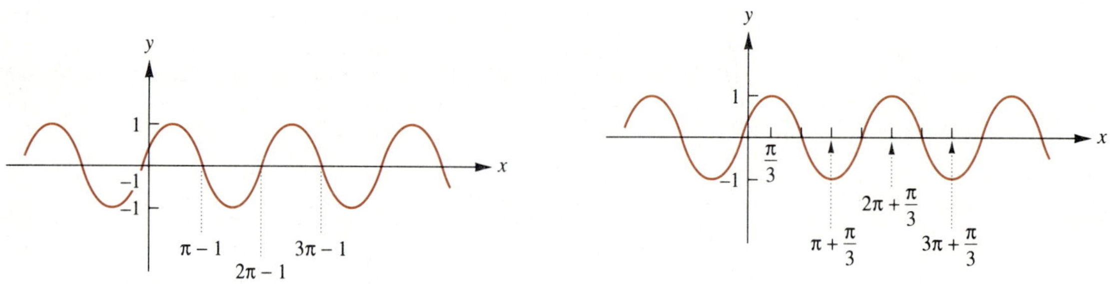

(a) Graph of $y = \sin(x + 1)$

(b) Graph of $y = \cos(x - \frac{\pi}{3})$

Figure 5.3-31

A phase shift has *graphic* meaning because it shifts the graph to the right or left along the x axis. It also has a *physical* meaning in applications, as the following examples show.

Example 23
Pendulum

Suppose we start two identical pendulums swinging, but we begin one *before* the other. They will both trace out sine waves, but these sine waves will be out of phase with each other (see Figure 5.3-32). The amount of phase difference depends on how long we wait before starting the second pendulum. ■

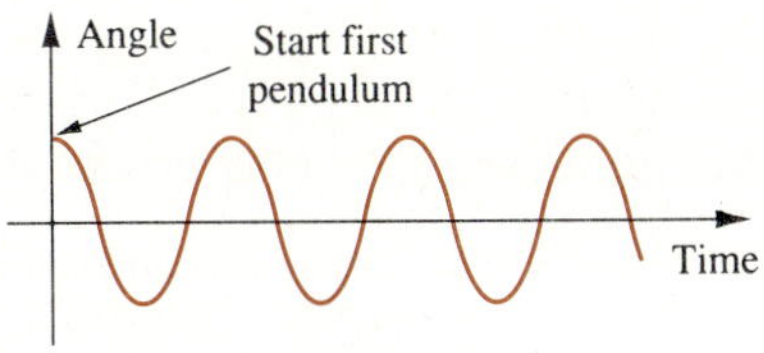

(a) Graph of the first pendulum

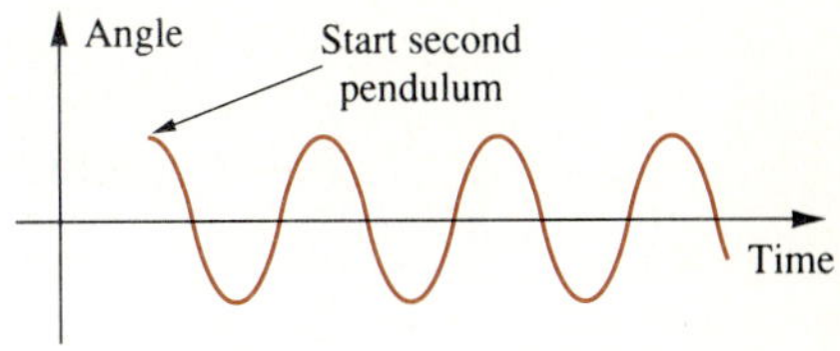

(b) Graph of the second pendulum

Figure 5.3-32

Sine waves of two pendulums that began swinging at different times These sine waves are out of phase because they started at different times.

Example 24
Echo

If you shout in a canyon, you often hear your own voice coming back—an echo. The echo is just your voice bouncing off the canyon's walls. Because both the original sound and the echo came from you, they are the same sine wave *except* that they are out of phase because of the time delay of the echo. Therefore, in this case the phase difference is a measure of the distance to the reflecting wall (see Figure 5.3-33). ■

Figure 5.3-33

Echo The original sound travels directly to your ears, but the echo is delayed because it travels an extra distance. Therefore, the echo is out of phase with the original sound when it reaches your ears. These two sound waves resemble those shown in Figure 5.3-35.

Example 25
Directional Hearing

One reason we can tell what direction a sound comes from is that our ears are very sensitive to the small phase difference between waves arriving at our right and left ears (see Figure 5.3-34). This phase difference is due to the fact that the sound must travel different distances to reach each ear. When the source of sound is directly in front of us, the phase difference is zero. (Why?) The farther to the right or left the source is, the greater the phase difference.

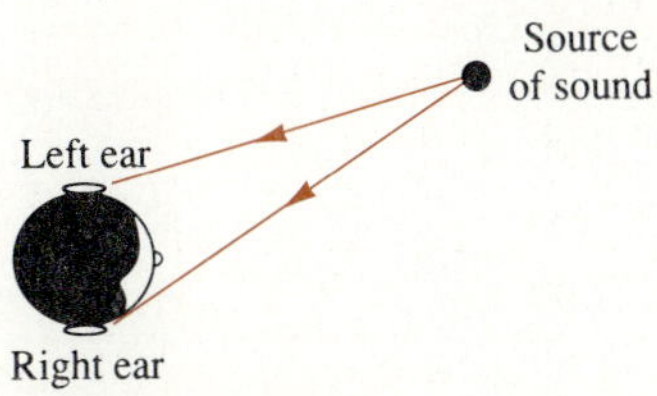

Figure 5.3-34

Directional hearing The sound must travel a greater distance to reach the right ear than the left ear, so the two waves are out of phase when they reach our ears. Our brain interprets these phase differences to tell what direction the sound came from.

Figure 5.3-35 is a photograph of an oscilloscope trace showing two sound waves that are slightly out of phase. Our brain responds to these phase differences to enable us to tell the direction the sound is coming from. ■

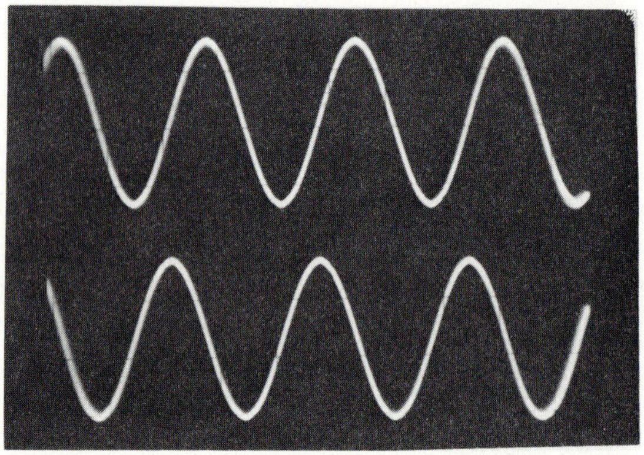

Figure 5.3-35

Oscilloscope trace of two sound waves slightly out of phase

Graphing $y = A \sin(Bx + C)$ and $y = A \cos(Bx + C)$

We are now ready to graph equations that combine all three effects: differences in amplitude, period, and phase. These equations are of the form $y = A \sin(Bx + C)$ and $y = A \cos(Bx + C)$. We start with a concrete example.

Example 26

Graph $y = \sin(2x + \pi)$.

Solution

This equation is of the form $y = \sin(Bx + C)$. Earlier in this section, we saw how B affects the period. The period of this graph is $P = 2\pi/|B| = \frac{2\pi}{2} = \pi$. At first glance, we might think that the phase shift is π units to the left. With the help of a table, we can see whether this is correct (see Figure 5.3-36).

x	$2x + \pi$	$\sin(2x + \pi)$
$-\frac{\pi}{2}$	0	0.00
$-\frac{\pi}{4}$	$\frac{\pi}{2}$	1.00
0	π	0.00
$\frac{\pi}{4}$	$\frac{3\pi}{2}$	−1.00
$\frac{\pi}{2}$	2π	0.00

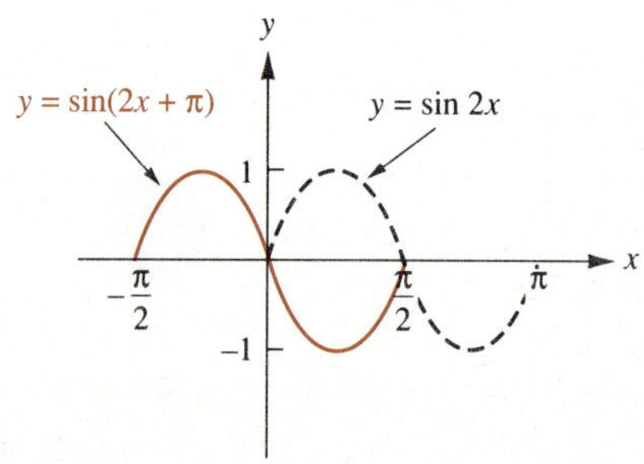

Figure 5.3-36

Table of values and graph of $y = \sin(2x + \pi)$

As we can see from Figure 5.3-36, the graph of $y = \sin(2x + \pi)$ is the same as the graph of $y = \sin 2x$, except it is shifted to the left by $\frac{\pi}{2}$ units. Therefore, the phase shift is $\frac{\pi}{2}$, not π as we first suspected. We can also see this by treating $(2x + \pi)$ as a unit. A cycle begins when $2x + \pi = 0$, which is when $x = -\frac{\pi}{2}$. Because this cycle begins at $x = -\frac{\pi}{2}$ instead of at $x = 0$, the graph of $y = \sin 2x$ is shifted $\frac{\pi}{2}$ units to the left. ■

We can generalize this result to include negative values of C and B: The graph of $y = A\sin(Bx + C)$ begins a cycle when $Bx + C = 0$, which is when $x = -C/B$. Because both B and C could be negative, the magnitude of this phase shift is $|C/B|$. If C/B is positive, the phase shift is to the left, whereas if C/B is negative, the phase shift is to the right.

▶ **Thus the graph of $y = A\sin(Bx + C)$ is the same as the graph of $y = A\sin Bx$, except shifted $|C/B|$ units to the right or left, depending on the sign of C/B.**

THE GRAPHS OF $y = A\sin(Bx + C)$ and $y = A\cos(Bx + C)$

Amplitude Amplitude $= |A|$. If A is negative, the graph will be reflected about the x axis.

Period Period $= 2\pi/|B|$.

Phase Shift Phase shift $= |C/B|$. It is to the left if C/B is positive and to the right if C/B is negative. This phase shift is the distance that the graph of $y = A\sin Bx$ is shifted along the x axis.

Example 27

Graph $y = 1.5\cos(2\pi x - \frac{\pi}{3})$.

Solution

For this graph, $A = 1.5$, $B = 2\pi$, and $C = -\frac{\pi}{3}$. Therefore,

$$\text{amplitude} = |A| = |1.5| = 1.5$$

$$\text{period} = P = \frac{2\pi}{|B|} = \frac{2\pi}{2\pi} = 1$$

$$\text{phase shift} = \left|\frac{C}{B}\right| = \left|\frac{-\frac{\pi}{3}}{2\pi}\right| = \left|\frac{\frac{\pi}{3}}{2\pi}\right| = \frac{1}{6}$$

C/B is negative, so the shift will be to the right by $\frac{1}{6}$ unit.

This graph will be the same as the graph of $y = 1.5\cos 2\pi x$, except it will be shifted $\frac{1}{6}$ unit to the right. First sketch the graph of $y = 1.5\cos 2\pi x$ and then shift it $\frac{1}{6}$ unit to the right (see Figure 5.3-37). ■

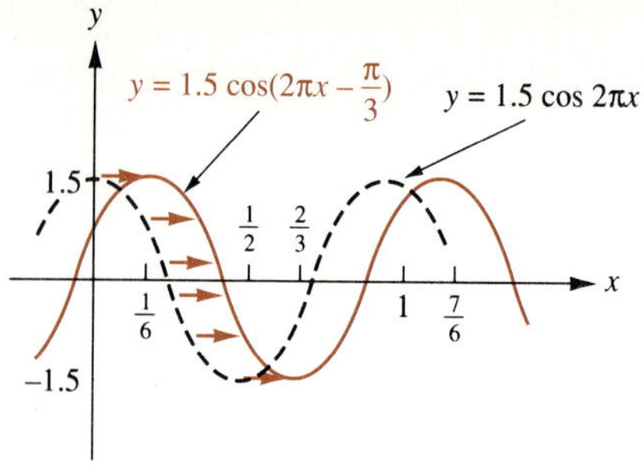

Figure 5.3-37

Graph of $y = 1.5\cos(2\pi x - \frac{\pi}{3})$ Amplitude = 1.5, period = 1, phase shift = $\frac{1}{6}$. This is the same as the graph of $y = 1.5\cos 2\pi x$, except it is shifted $\frac{1}{6}$ unit to the right.

Example 28

Graph $y = -2\sin(3x + \frac{\pi}{2})$.

Solution

For this equation, $A = -2$, $B = 3$, and $C = \frac{\pi}{2}$.

$$\text{amplitude} = |A| = |-2| = 2$$

$$\text{period} = \frac{2\pi}{|B|} = \frac{2\pi}{3}$$

$$\text{phase shift} = \left|\frac{C}{B}\right| = \left|\frac{\frac{\pi}{2}}{3}\right| = \frac{\pi}{6}$$

Because C/B is positive, the shift will be to the left $\frac{\pi}{6}$ units.

This graph is the same as the graph of $y = -2\sin 3x$, except that it is shifted $\frac{\pi}{6}$ units to the left. First sketch the graph of $y = -2\sin 3x$, and then shift it $\frac{\pi}{6}$ units to the left (see Figure 5.3-38). ■

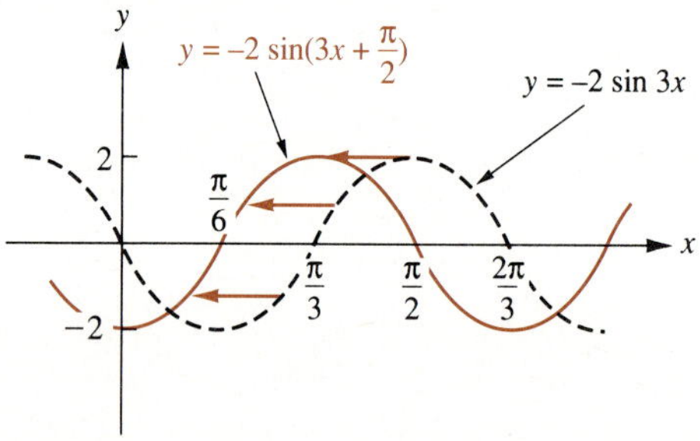

Figure 5.3-38

Graph of $y = -2\sin(3x + \frac{\pi}{2})$ The amplitude is 2, the period is $\frac{2\pi}{3}$, and the phase shift is $\frac{\pi}{6}$ to the left.

Example 29

Sketch the graph of $y = 4\sin(-2x + \pi)$.

Solution

For this equation, $A = 4$, $B = -2$, and $C = \pi$.

$$\text{amplitude} = |A| = 4$$

$$\text{period} = P = \frac{2\pi}{|B|} = \frac{2\pi}{|-2|} = \frac{2\pi}{2} = \pi$$

$$\text{phase shift} = \left|\frac{C}{B}\right| = \left|\frac{\pi}{-2}\right| = \frac{\pi}{2}$$

and is to the right because C/B is negative.

This graph is the same as the graph of $y = 4\sin(-2x)$, except that it is shifted to the right by $\frac{\pi}{2}$ units. First sketch the graph of $y = \sin(-2x)$ and then shift it $\frac{\pi}{2}$ units to the right (see Figure 5.3-39). ■

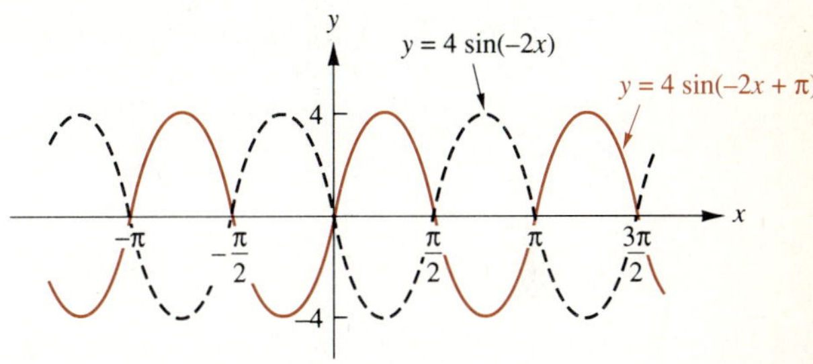

Figure 5.3-39

Graph of $y = 4\sin(-2x + \pi)$
This is the same as the graph of $y = 4\sin(-2x)$ except it is shifted $\frac{\pi}{2}$ units to the right.

NOW IT'S YOUR TURN

Find the amplitude, period, and phase shift and sketch the graph of (a) $y = 6.0\sin(\pi x/2 - \pi/4)$ and (b) $y = 2\sin(-4x - \pi)$.

(Answer: (a) Amplitude = 6.0, period = 4, phase shift = $\frac{1}{2}$ to the right; see Figure 5.3-40 for the graph. (b) Amplitude = 2, period = $\frac{\pi}{2}$, phase shift = $\frac{\pi}{4}$ units to the left; see Figure 5.3-41 for the graph.)

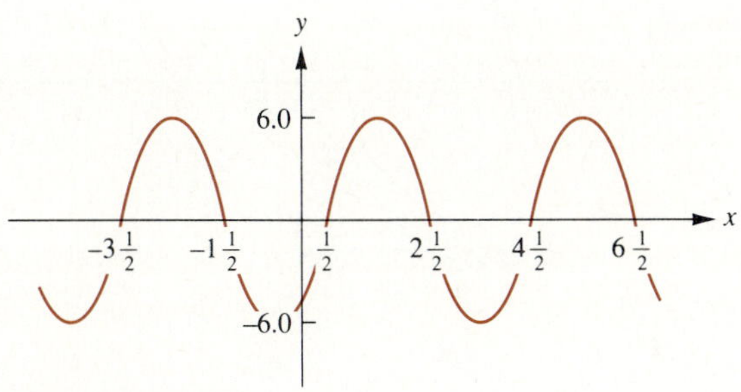

Figure 5.3-40

Graph of $y = 6\sin(\pi x/2 - \pi/4)$
Amplitude = 6.0, period = 4, phase shift = $\frac{1}{2}$ to the right.

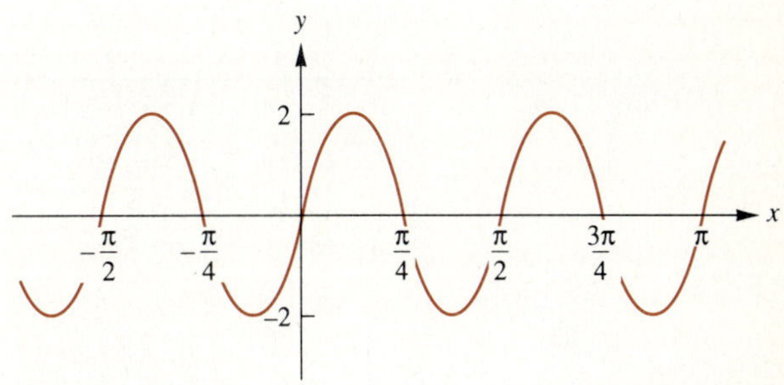

Figure 5.3-41

Graph of $y = 2\sin(-4x - \pi)$

EXERCISES 5.3-1

In Exercises 1–10, find the amplitude and sketch the graph of each function without plotting points.

1. $y = 2 \sin x$

2. $y = 3 \cos x$

3. $y = -\cos x$

4. $y = -3 \sin x$

5. $y = \frac{1}{4} \sin x$

6. $y = 0.60 \cos x$

7. $y = -\frac{1}{3} \cos x$

8. $y = \pi \sin x$

9. $y = -3.4 \cos x$

10. $y = \frac{1}{\pi} \sin x$

In Exercises 11–20, use the method developed in Section 5.3-1 to sketch each graph without plotting points.

11. $y = 5 \tan x$

12. $y = -3 \tan x$

13. $y = -6 \tan x$

14. $y = \frac{1}{2} \tan x$

15. $y = 3 \sec x$

16. $y = \frac{1}{4} \csc x$

17. $y = -\csc x$

18. $y = -\cot x$

19. $y = 4 \cot x$

20. $y = -\frac{1}{3} \sec x$

In Exercises 21–26, identify each graph as a sine or cosine, find A and the amplitude. Then write the numerical equation of the graph, such as $y = \sin 2x$, for example.

21.

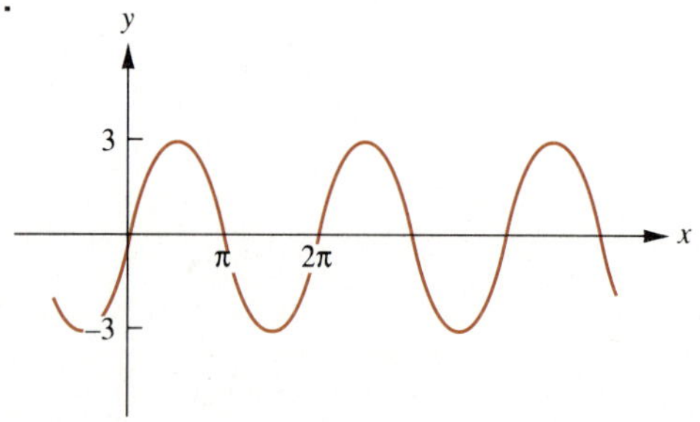

22.

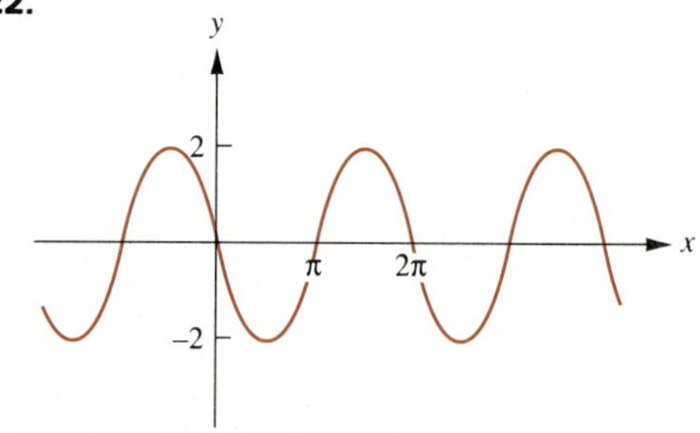

23.

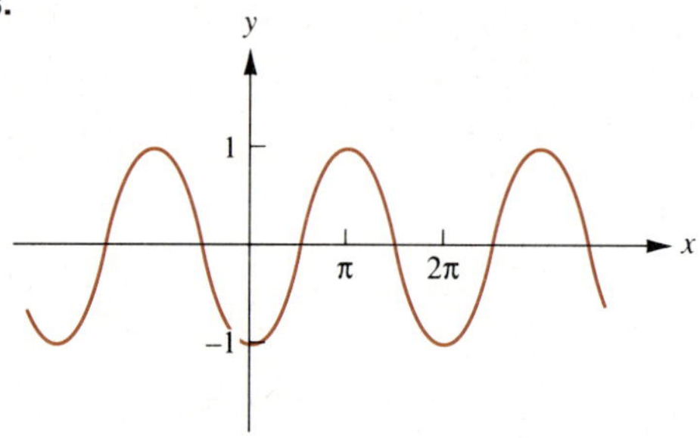

24.

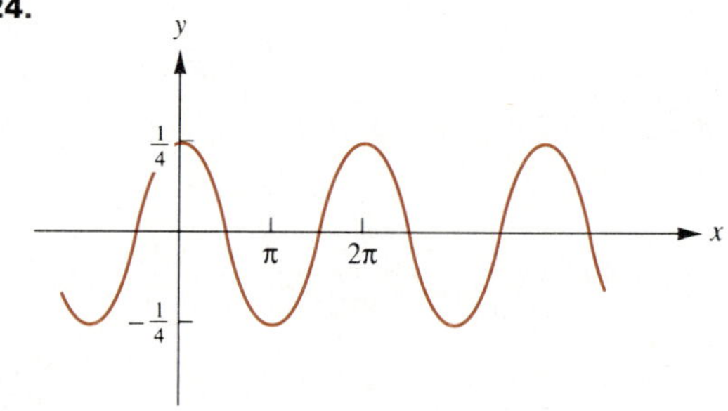

25.

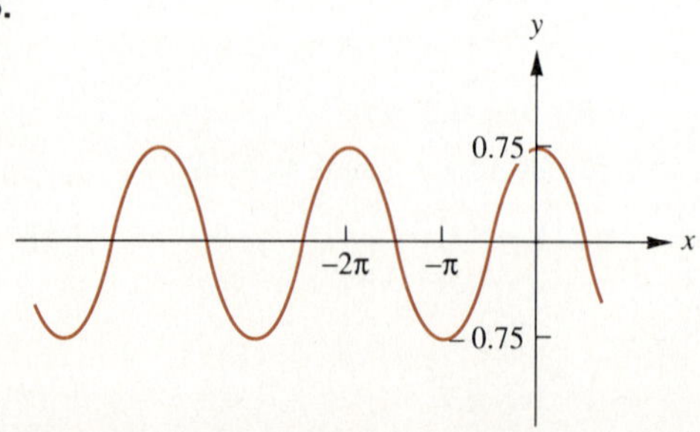

26.

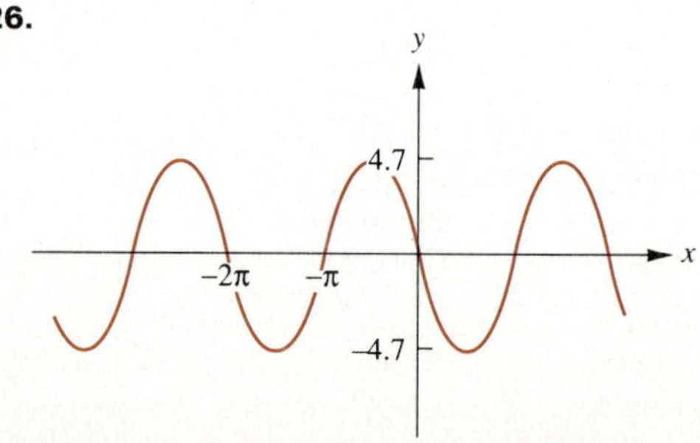

EXERCISES 5.3-2

In Exercises 1–16, find the period and sketch the graph of the function without plotting points.

1. $y = \sin 2x$

2. $y = \cos 3x$

3. $y = \sin(x/2)$

4. $y = \cos(x/4)$

5. $y = \sin 2\pi x$

6. $y = \cos(\pi x/3)$

7. $y = \sin 9.74x$

8. $y = \cos 1.2x$

9. $y = \sin 0.28x$

***10.** $y = \cos(x/\pi)$

11. $y = \sin(-3x)$

12. $y = \cos(-\pi x/4)$

13. $y = \cos(-2x/3)$

14. $y = \sin(-\pi x/2)$

15. $y = \sin(-2.4x)$

16. $y = \cos(-0.35x)$

In Exercises 17–31, find the amplitude and period of each function, and sketch its graph without plotting points.

17. $y = 2\sin 2x$

18. $y = \frac{1}{3}\cos 4x$

19. $y = 2\sin\frac{1}{3}x$

20. $y = 4\cos(\pi x/4)$

21. $y = -\sin 5x$

22. $y = -3\cos 6x$

23. $y = 0.45\sin \pi x$

24. $y = -2.4\cos \pi x$

25. $y = \pi\sin(3x/2)$

26. $y = -2\pi\cos 2\pi x$

27. $y = 4\sin(-x/3)$

28. $y = \cos(-6x/5)$

29. $y = -1.5\sin(-\pi x/2)$

30. $y = -2\cos(-2\pi x/3)$

31. $y = \frac{1}{4}\sin(-1.7x)$

In Exercises 32–39, each graph is of $y = \sin Bx$ or $y = \cos Bx$. From the graph, identify which function (sine or cosine) it is, find B, and write the numerical equation of the graph, such as $y = \sin 3x$, for example. In a few cases, it is not possible to determine the sign of B; be sure to indicate when this occurs.

32.

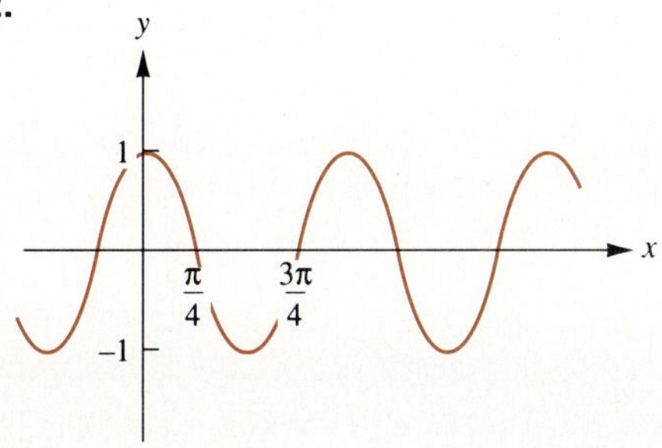

33.

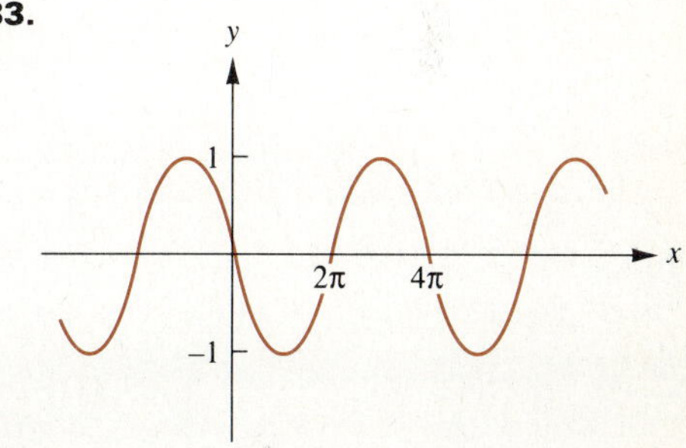

34.

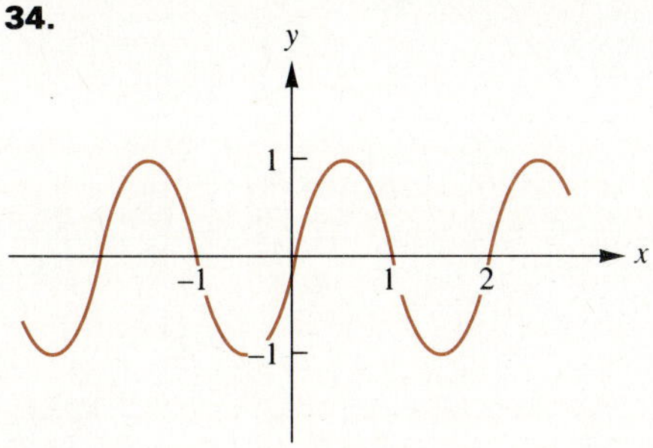

35.

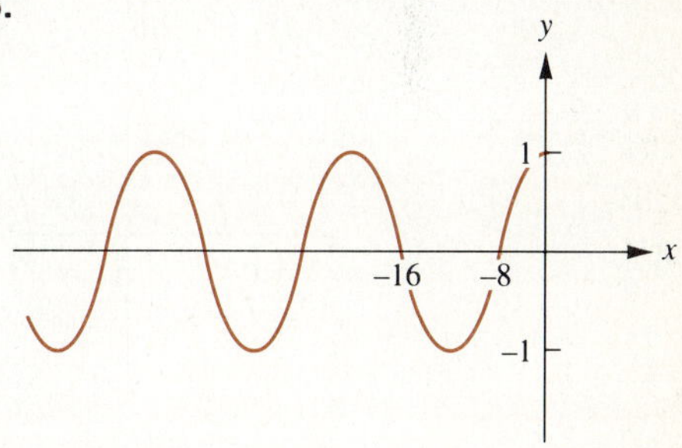

36.

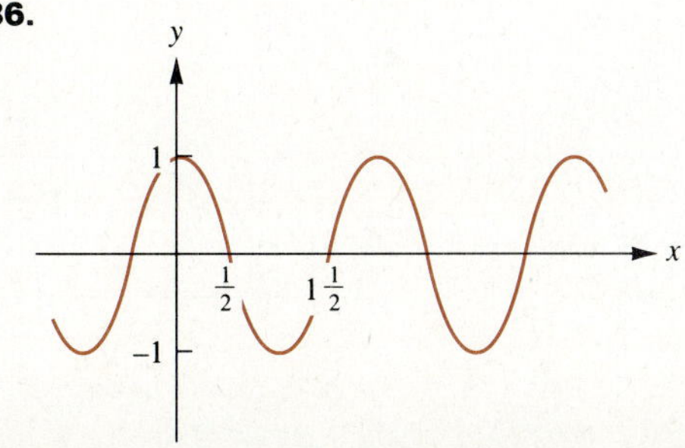

37.

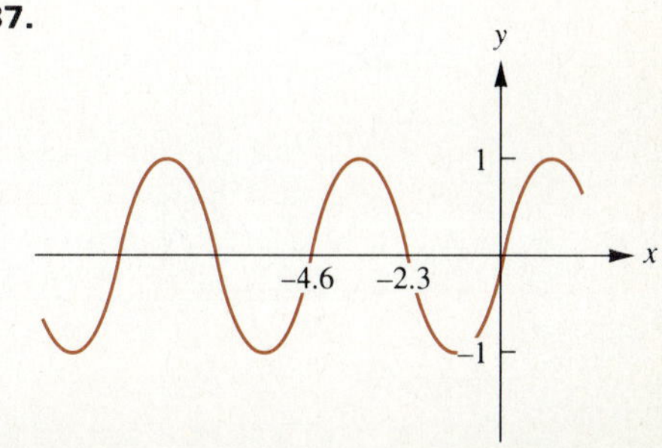

38.

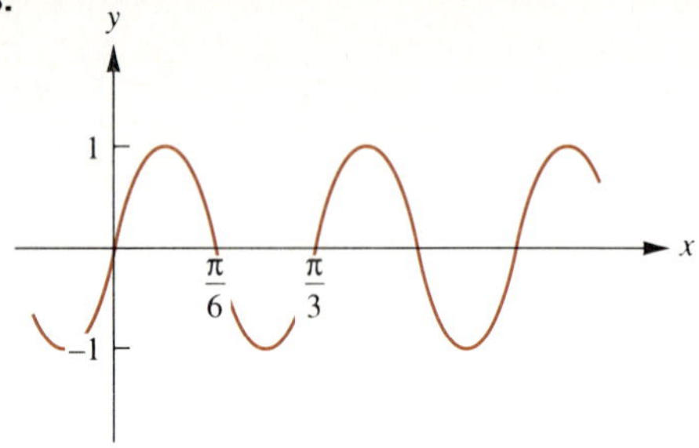

39.

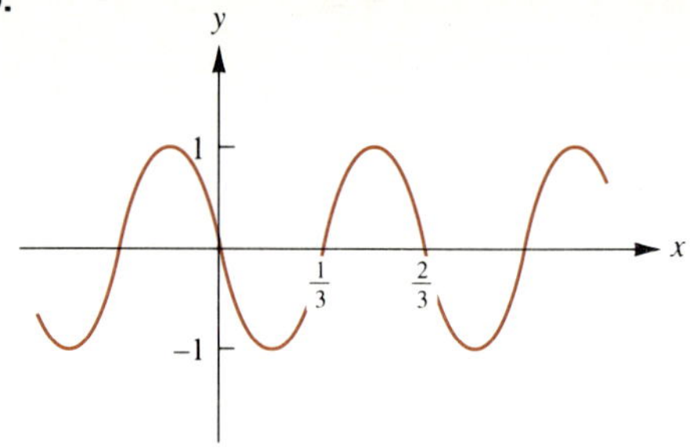

In Exercises 40–47, each graph is of $y = A \sin Bx$ or $y = A \cos Bx$. From the graph, decide which function (sine or cosine) it is, find A and the amplitude, and find B, assuming that $B > 0$. Finally, write the numerical equation for the graph, such as $y = -2 \sin 3x$, for example.

40.

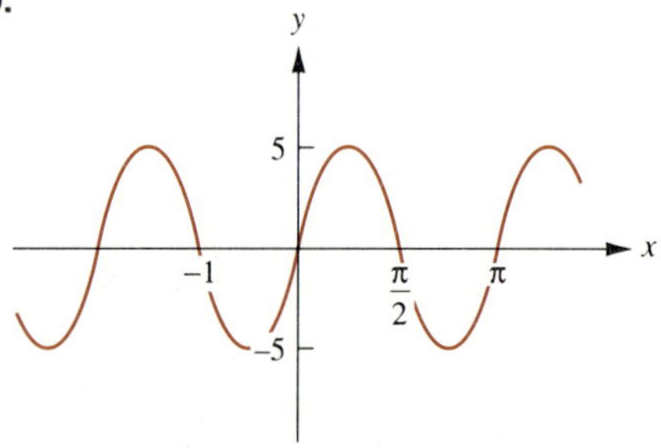

41.

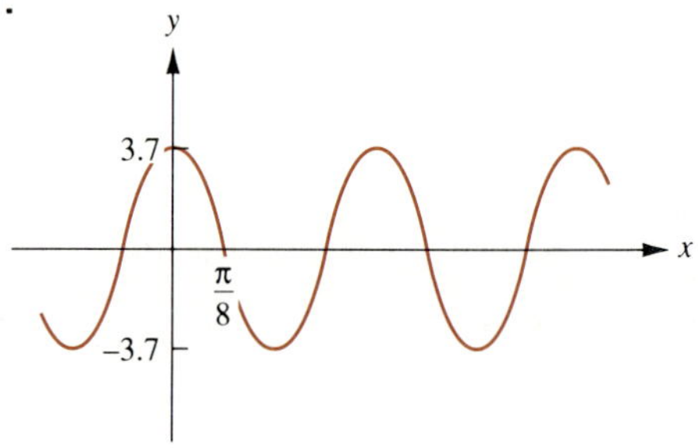

42.

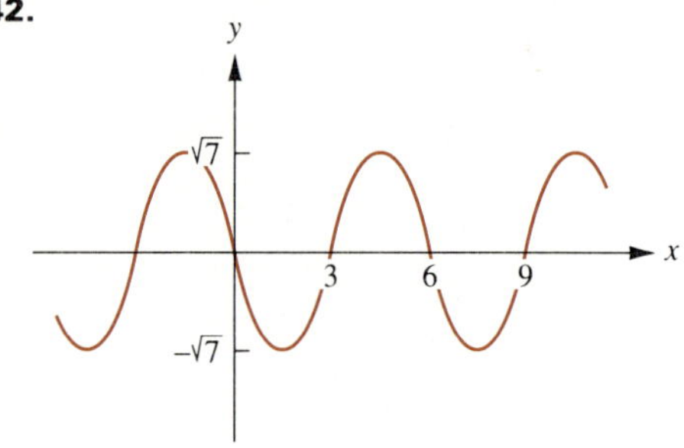

43.

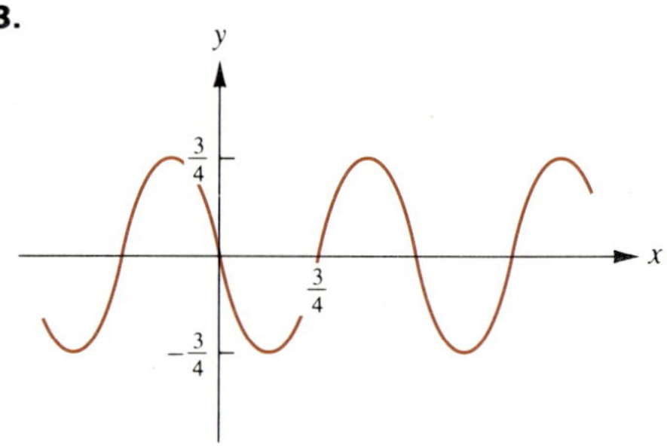

44.

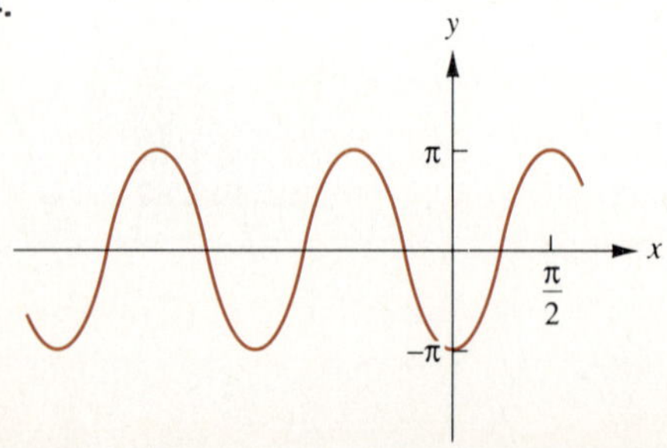

45.

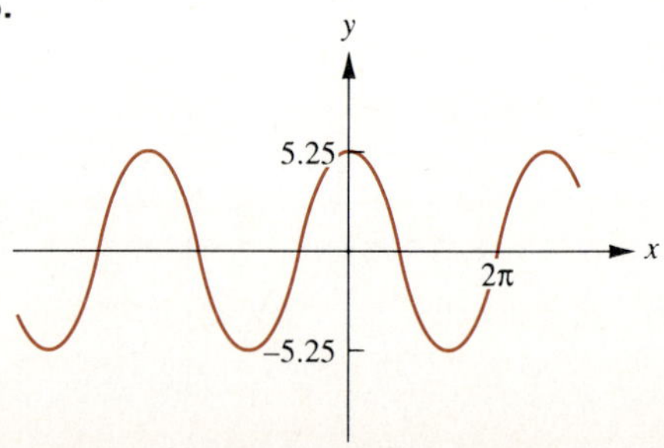

46.

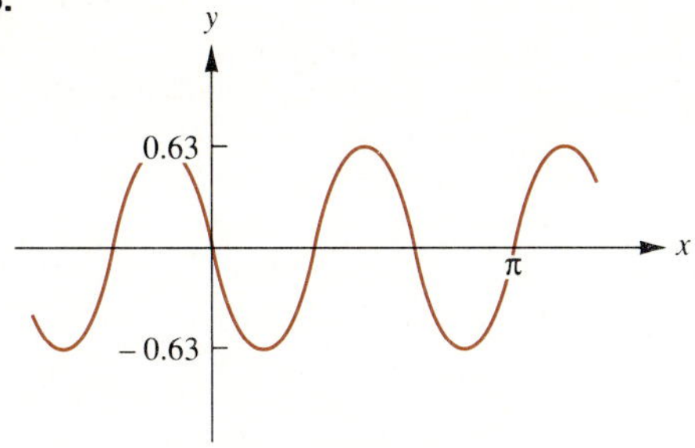

47.

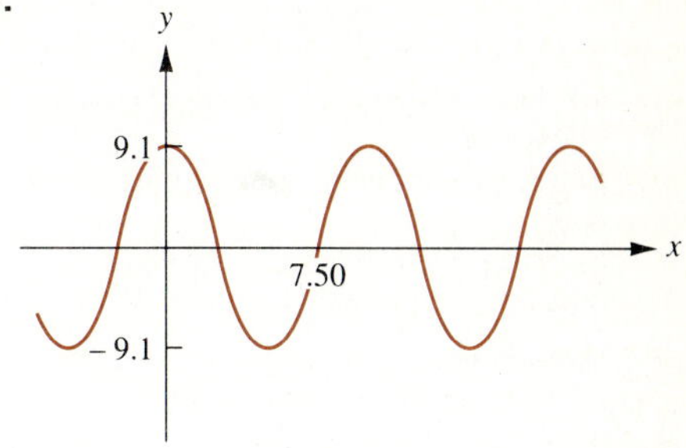

Exercises 48–55 refer to the graphs for Exercises 40–47, except now assume that $A > 0$ but B can be positive or negative. In each case, identify the function (sine or cosine), find A, the amplitude, and B. Finally, write the numerical equation for the graph such as $y = 3.1 \cos 5\pi x$, for example. Be sure to indicate those cases where it is not possible to determine the sign of B.

48. The graph for Exercise 40.

49. The graph for Exercise 41.

50. The graph for Exercise 42.

51. The graph for Exercise 43.

***52.** The graph for Exercise 44.

53. The graph for Exercise 45.

54. The graph for Exercise 46.

55. The graph for Exercise 47.

Exercises 56–63 refer to the graphs for Exercises 40–47, except now assume that *both* A and B can be positive or negative. In each case, identify the function (sine or cosine), find A and the amplitude, and find B. There are several cases where there are two possibilities, such as positive A and negative B or negative A and positive B. Be sure to include *both* possibilities. In addition, indicate when it is not possible to determine the sign of B.

***56.** The graph for Exercise 40.

***57.** The graph for Exercise 41.

***58.** The graph for Exercise 42.

***59.** The graph for Exercise 43.

***60.** The graph for Exercise 44.

***61.** The graph for Exercise 45.

***62.** The graph for Exercise 46.

***63.** The graph for Exercise 47.

In Exercises 64–74, remember that the tangent has period π and completes one cycle between $\pm\frac{\pi}{2}$. Use the methods developed in Section 5.3-2 to find the period of each function and sketch its graph without plotting points.

***64.** $y = \tan 2x$ ***65.** $y = \tan(x/3)$

***66.** $y = \tan(-x)$ ***67.** $y = \tan(\pi x/2)$

***68.** $y = \tan(2.37x)$ ***69.** $y = \tan(-\pi x/4)$

***70.** $y = -\tan 4x$ ***71.** $y = 5 \tan 3x$

***72.** $y = -7 \tan(-\pi x)$ ***73.** $y = -3 \tan(\pi x/2)$

***74.** $y = 3.5 \tan(4.82x)$

In Exercises 75–83, use the methods developed in Section 5.3-2 to find the period and sketch the graph of each function without plotting points.

***75.** $y = \cot 2x$ ***76.** $y = \sec 3x$

***77.** $y = \csc \pi x$ ***78.** $y = \cot(-x)$

***79.** $y = 3 \cot(x/2)$ ***80.** $y = 6 \sec(-x/3)$

***81.** $y = -5 \sec(\pi x/2)$ ***82.** $y = 4 \csc(-3x/4)$

***83.** $y = -2 \csc 4x$

84. What are the frequency in hertz and the period in seconds of a 78-rpm record?

85. The note B^b has 466.2 vibrations/s. Find the frequency in hertz and the period in seconds of this sound and sketch the sound wave of this note.

86. The Octave. Musical scales are separated by the octave. If note 2 is one octave higher than note 1, note 2 has 2 times the frequency of note 1. For example, concert A has a frequency of 440 Hz. The note A one octave higher has a frequency of 2×440 Hz = 880 Hz, whereas the note A one octave lower has a frequency of $\frac{1}{2} \times 440$ Hz = 220 Hz. High C for a tenor is one octave above middle C (frequency 262 Hz), whereas for a soprano it is two octaves above middle C.

(a) Find the frequency of a tenor's high C *and* a soprano's high C.

(b) Sketch a time graph of all three Cs on a *single set of axes*.

87. Range of the Piano. The lowest note on a piano is A three octaves below concert A (440 Hz), whereas the highest note is C four octaves above middle C (262 Hz) (see Exercise 86).

(a) Find the frequency and period of the highest and lowest notes on a piano.

(b) Sketch *separate* time graphs of the sounds these keys produce.

(c) If we express these sound waves as $y = \sin Bt$, find B for these two extremes.

88. Range of Hearing. Humans can hear sounds having frequencies between approximately 20 Hz and 20,000 Hz.

(a) Find the time for one sound cycle for each of these two frequencies.
(b) Sketch separate time graphs of the sounds of these limits.
(c) If we express these sound waves as $y = \sin Bt$, find B for these two extremes.

89. Ultrasound. Ultrasound is any sound having a frequency greater than 20,000 Hz (beyond human hearing). Such sound is often used in medicine, much like x rays. In obstetrics, ultrasound of frequency approximately 2 megahertz (mega $= 10^6$) is used to get a picture of the fetus of a pregnant woman. Find the period of this ultrasound and sketch its time graph. If we express the sound wave as $y = \cos Bt$, find B.

90. Color. Light is a sine wave consisting of fluctuating electric and magnetic fields. Color is our perception of light waves of different frequencies. The visible spectrum ranges approximately from a frequency of 4.3×10^{14} Hz (red) to 7.5×10^{14} Hz (violet). Find the periods of each of these two colors and sketch their sine waves on the same set of axes.

91. Pendulum. If a pendulum takes 5.60 s to swing back to its original position:

(a) Find its frequency and period.
(b) Sketch the time graph of the angle it makes with the vertical.
(c) If we express the equation of this angle in the form $\theta = \cos Bt$ with $B > 0$, find B.

92. Atomic Clocks: Modern-Day Pendulums. Years ago people actually used pendulums to keep careful time—remember the grandfather's clock. Today we use vibrations of atoms and molecules (atomic clocks) because they are much more regular. The modern standard for the second is the atom cesium, which makes 9,192,631,770 vibrations each second. A time graph of its motion is—you guessed it!—a sine wave.

(a) Find the frequency and period of the cesium atom.
(b) Sketch a time graph of its position (y) showing the period.
(c) If $y = \sin Bt$ with $B > 0$ represents the motion of the cesium clock, find B.

EXERCISES 5.3-3

In Exercises 1–10, find the phase shift, decide if it is to the right or the left, and sketch the graph of the function without plotting points.

1. $y = \sin(x + \frac{\pi}{2})$ **2.** $y = \cos(x - \frac{\pi}{4})$
3. $y = \sin(x - \frac{\pi}{6})$ **4.** $y = \cos(x - \pi)$
5. $y = \sin(x + \frac{\pi}{3})$ **6.** $y = \cos(x + 2)$
7. $y = \sin(x - 4)$ **8.** $y = \cos(x - 0.50)$
9. $y = \sin(x + 2.5)$ **10.** $y = \cos(x - 2\pi)$

In Exercises 11–32, find the amplitude, period, and phase shift (decide if it is to the right or left), and then sketch the graph of each function without plotting points.

11. $y = \sin(2x - \pi)$ **12.** $y = \sin(2x + \frac{\pi}{4})$
13. $y = \cos(2x + \pi)$ **14.** $y = -\cos(x - \frac{\pi}{5})$
15. $y = -\sin(x - \frac{\pi}{4})$ **16.** $y = \sin(4x + \pi)$
17. $y = \sin(-x - \frac{\pi}{2})$ **18.** $y = -\cos(2x - \frac{\pi}{2})$
19. $y = \sin(-3x + \pi)$ **20.** $y = \sin(x/3 - \frac{\pi}{2})$
21. $y = \cos(x/2 - 2\pi/3)$ **22.** $y = 2\sin(2x + 2)$
23. $y = \pi\cos(-\pi x + \pi)$ **24.** $y = 3\cos(3x/2 - 3\pi/8)$
25. $y = 4\cos(2\pi x - \frac{\pi}{4})$ **26.** $y = -2\sin(\pi x - 4.0)$
27. $y = \sin(-2x + \frac{\pi}{2})$ **28.** $y = \cos(-3x/4 - \pi/4)$
29. $y = -3\cos(-x/2 - 1)$ **30.** $y = -3\cos(x/2 + 1)$
31. $y = -2.5\sin(-\pi x/3 - \frac{\pi}{2})$ **32.** $y = 2.5\sin(\pi x/3 + \pi/2)$

In Exercises 33–46, find the phase shift (is it to the right or left) and period of each function and sketch its graph. Do not plot points, but rather use the methods we developed for the sine and cosine in Section 5.3-3, making use of the fact that the period of $y = \tan x$ is π.

***33.** $y = \tan(x - \pi/4)$ ***34.** $y = \tan(x + \pi/2)$
***35.** $y = \tan(x + \pi)$ ***36.** $y = \tan(x - 2)$
***37.** $y = \tan(-x - 1.50)$ ***38.** $y = 2\tan(-x + \pi/4)$
***39.** $y = -\tan(2x + \pi)$ ***40.** $y = \tan(3x - 3)$
***41.** $y = -4\tan(x/2 - \pi/8)$ ***42.** $y = \tan(-4x + \pi)$
***43.** $y = \tan(-\pi x - \pi/2)$ ***44.** $y = -\tan(\pi x + \pi/2)$
***45.** $y = \tan(-\pi x/8 + \pi/32)$ ***46.** $y = \tan(-2x + 2)$

In Exercises 47–62, find the phase shift (is it to the right or left) and period of each function and sketch its graph. Do not plot points, but use the methods we developed for the sine and cosine in Section 5.3-3, making use of the fact that the period of $y = \sec x$ and $y = \csc x$ is 2π, whereas that of $y = \cot x$ is π.

***47.** $y = \cot(x - \frac{\pi}{2})$ ***48.** $y = \sec(x + 6)$
***49.** $y = \csc(x - \frac{\pi}{4})$ ***50.** $y = \cot(-x + \frac{\pi}{4})$
***51.** $y = \sec(-x + \pi)$ ***52.** $y = \csc(-x + \frac{\pi}{2})$
***53.** $y = 4\cot(3x - \pi)$ ***54.** $y = -2\sec(x + \frac{\pi}{6})$
***55.** $y = 6\csc(x/2 - \pi/2)$ ***56.** $y = \sec(-4x + \pi)$
***57.** $y = \cot(-\pi x/3 - 1)$ ***58.** $y = \sec(4x - \pi)$
***59.** $y = -3\csc(-6x + 2\pi)$ ***60.** $y = -\cot(\pi x/3 + 1)$
***61.** $y = -3\sec(-2x + 1)$ ***62.** $y = 9\csc(-x - \frac{\pi}{4})$

In Exercises 63–69, which ones are true and which are not? For those that are true, are they identities? Why?

63. $\sin(x + \frac{\pi}{2}) = \cos x$ **64.** $\cos(x + \frac{\pi}{2}) = \sin x$

***65.** $\sin(x \pm \pi) = -\sin x$

***66.** $\cos(x \pm \pi) = -\cos x$

67. $\sin(x \pm 2\pi) = \sin x$

***68.** $\tan(x - \frac{\pi}{2}) = -\cot x$

69. $\tan(x \pm \pi) = \tan x$

***70.** Show that a phase shift of π to the right for the graphs of $y = \sin x$ and $y = \cos x$ is the same thing as multiplying $\sin x$ or $\cos x$ by -1.

The graphs in Exercises 71–82 represent the equation $y = A\sin(Bx + C)$. By interpreting each graph, find A, B, and C (with $|C| < \pi$ and $B > 0$), and decide if the phase shift is to the right or left. Then write the numerical equation for each graph, such as $y = 3\sin(2x - \pi)$, for example.

71.

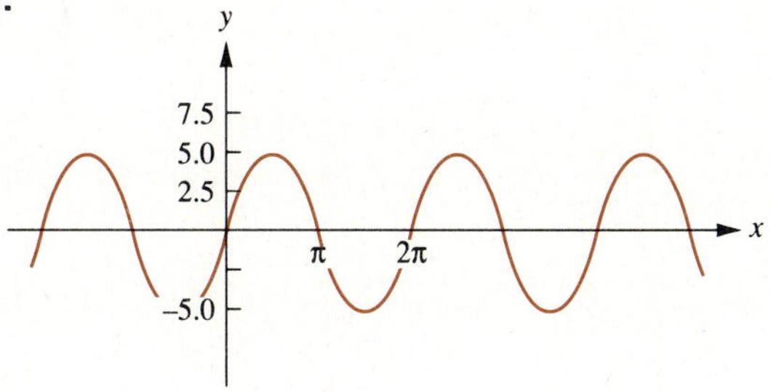

72.

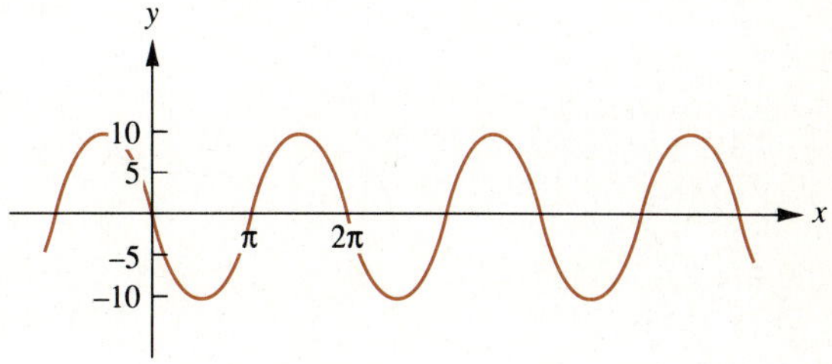

***73.**

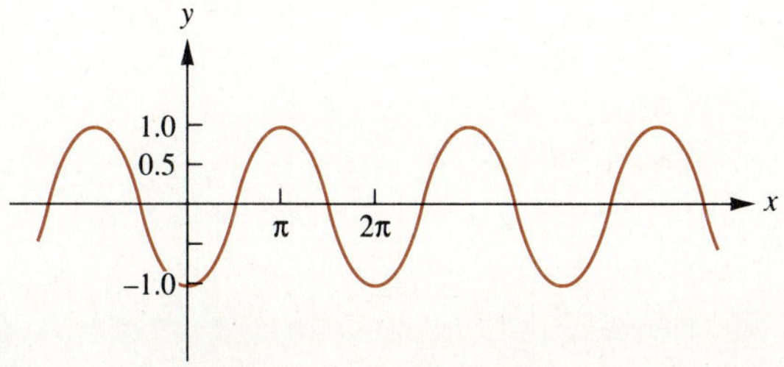

***74.**

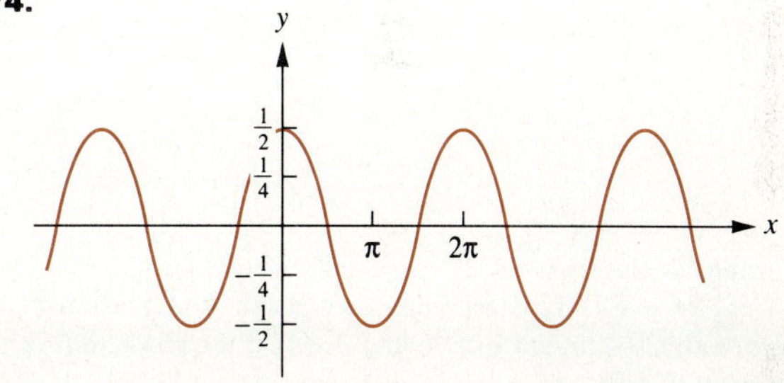

***75.**

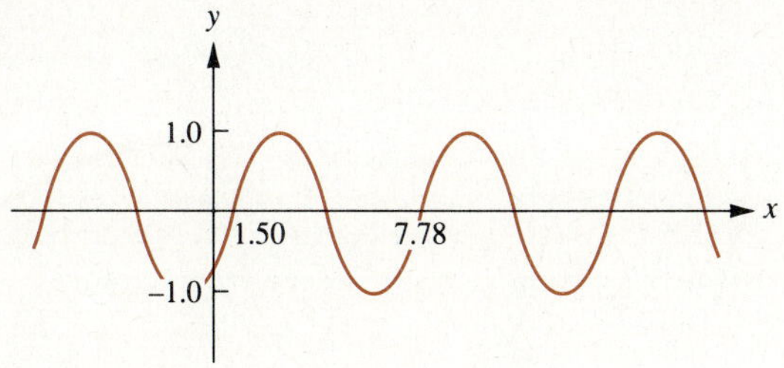

***76.**

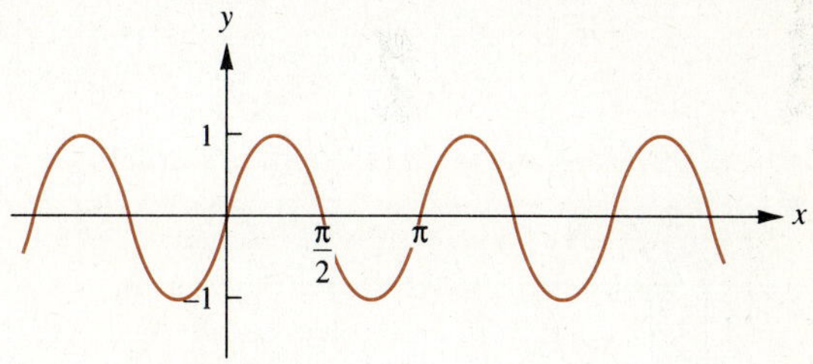

***77.**

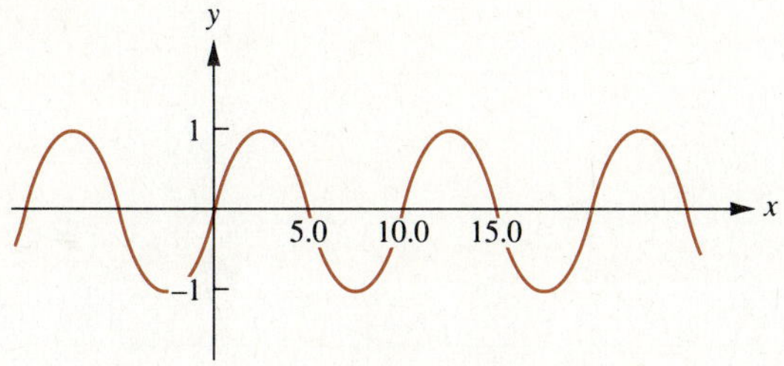

***78.**

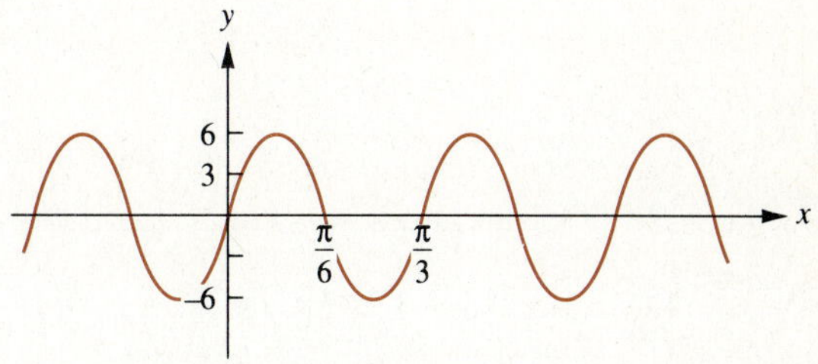

***79.**

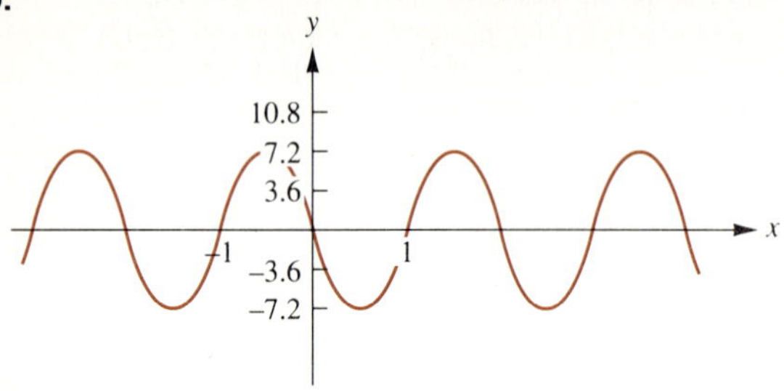

****80.**

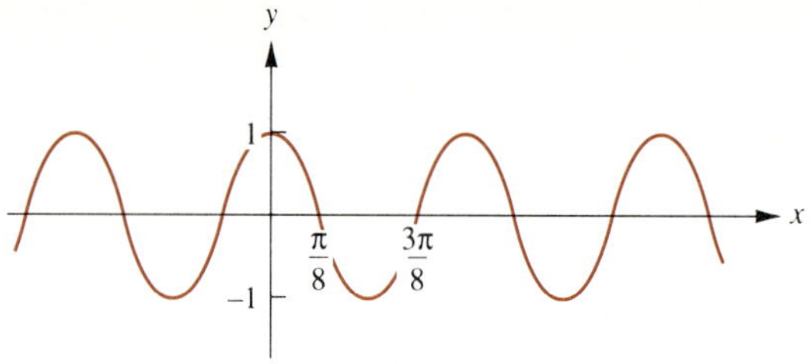

****81.**

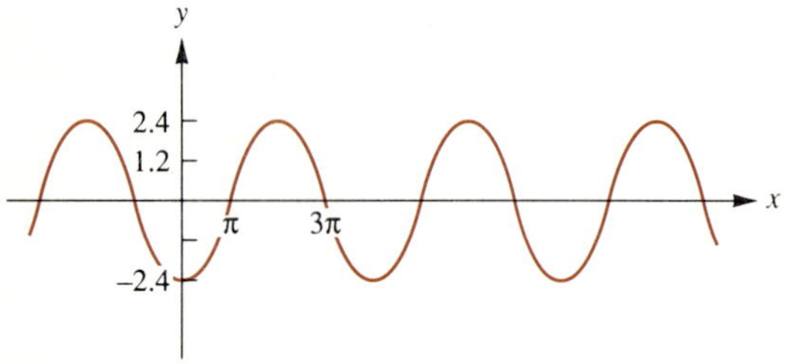

****82.**

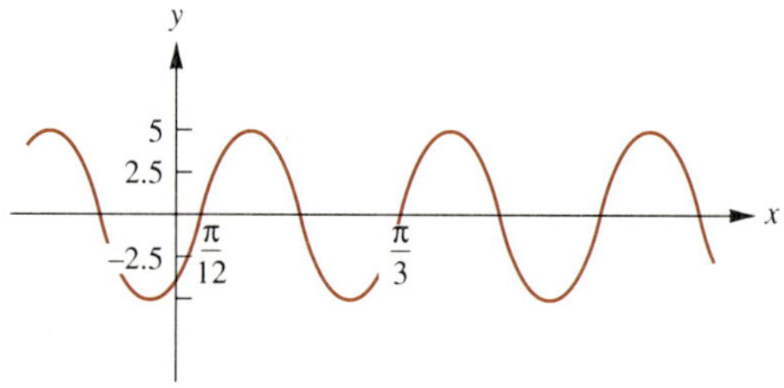

The graphs in Exercises 83–88 represent the equation $y = A\cos(Bx + C)$. By interpreting each graph, find A, B, and C (such that $|C| < \pi$ and $B > 0$) and decide if the phase shift is to the right or left.

83.

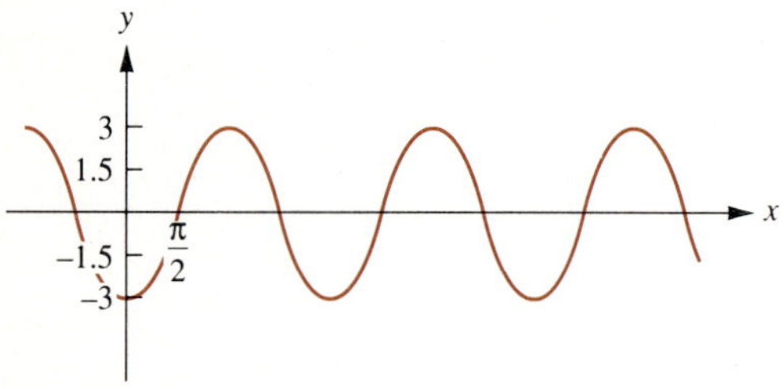

***84.**

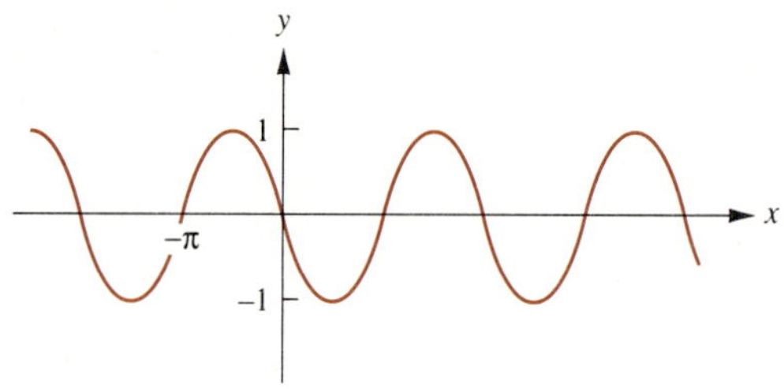

***85.**

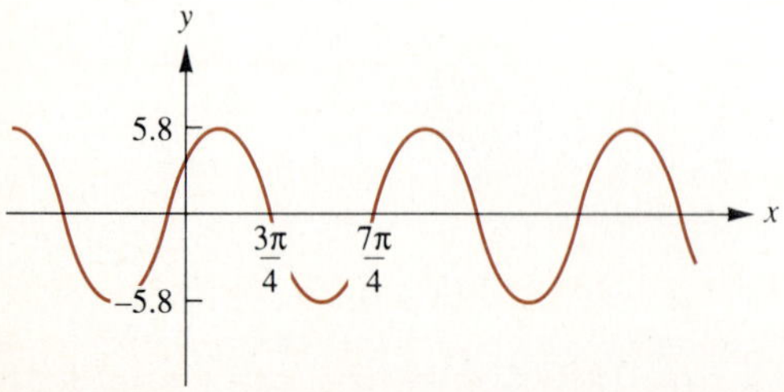

***86.**

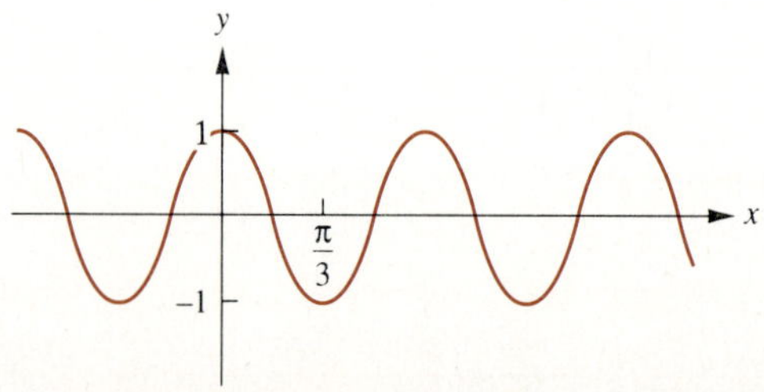

***87.**

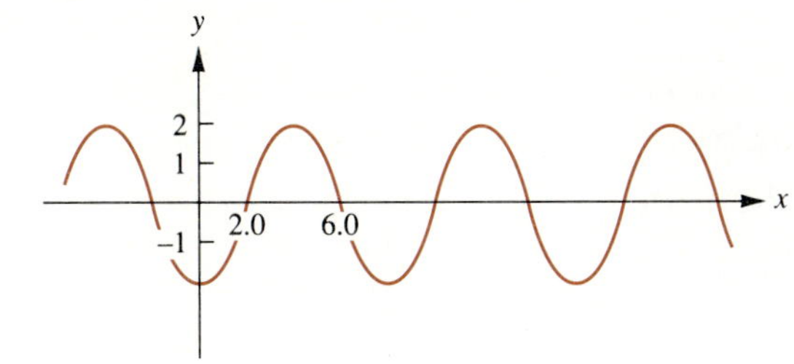

***88.**

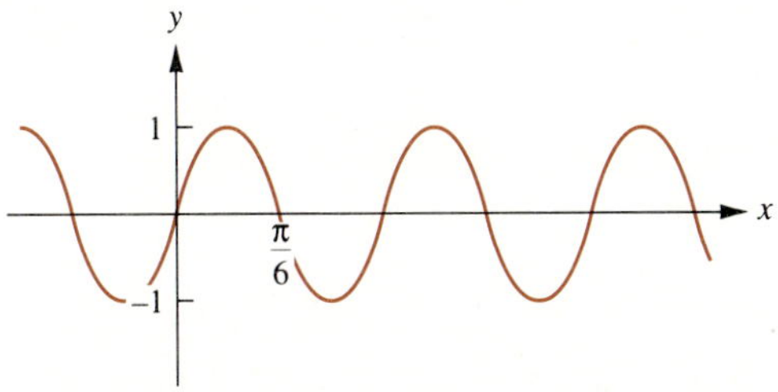

****89. Directional Hearing.** You hear a 1000 Hz sound coming from a distant place 60° to the right of straight ahead.

(a) If your head is 23 cm wide at the ears, find the phase difference between the sound waves arriving at your ears. Sound travels at 330 m/s in air. ***Hint:*** The phase difference is due to the extra distance the sound must travel to reach your left ear. What fraction of the period is this sound delayed due to this extra distance? Convert this fraction of a period to a phase difference. If, for example, the extra time is $\frac{1}{4}$ of a period, then the phase difference is $\frac{1}{4} \times 2\pi = \frac{\pi}{2}$

(b) On the same set of axes, sketch the graphs of both sound waves as they arrive at your ears.

90. The sine wave in Figure 5.3-42 represents a sound wave.

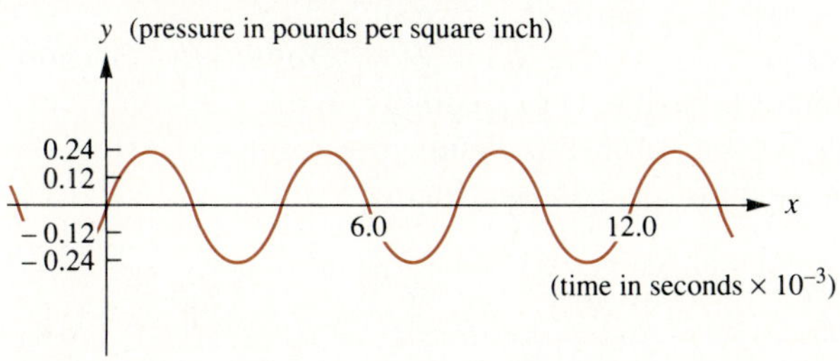

Figure 5.3-42

(a) Find the period and frequency of this sound.

(b) If we write its equation in the form $y = A\sin(Bt + C)$, find A, B, and C (with $A > 0$ and $B > 0$).

91. The sine wave in Figure 5.3-43 represents the angle of swing of a pendulum.

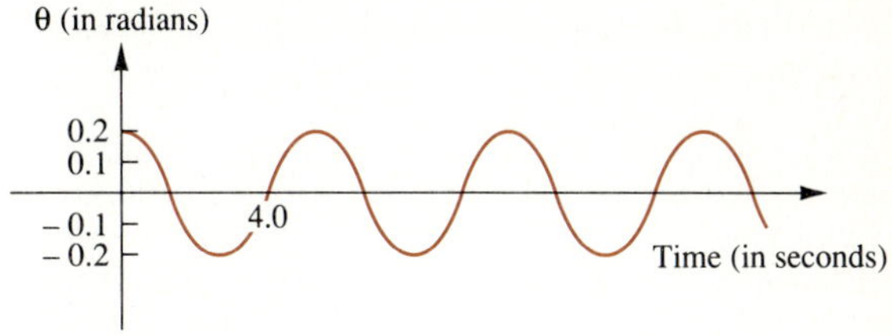

Figure 5.3-43

(a) Find the maximum angle the pendulum makes with the vertical.

(b) Find the period and frequency of the pendulum's motion.

(c) If we express the equation of this pendulum's motion in the form $\theta = A\cos(Bt + C)$ with $A > 0$ and $B > 0$, find A, B, and C from the graph.

92. Amount of Daylight. You have noticed that the number of daylight hours varies periodically over the course of a year. This variation is due to the fact that the earth spins on a tilted axis. During part of the year (summer), our hemisphere is pointed toward the sun, and we get more daylight. The other part of the year (winter), we are pointed away from the sun, and we get less daylight. On the equinoxes (March 21 and Sept. 21), day and night have equal length, so we have 12 hr of daylight. Figure 5.3-44 is a graph of the number of hours of daylight at a north latitude of 40° (roughly Denver's location). The x axis represents time, and the y axis represents the number of hours of daylight above (or below) the 12-hr day of the equinoxes. (Actually, the y axis gives only the number of hours the sun is above the horizon; daylight lasts longer than this.)

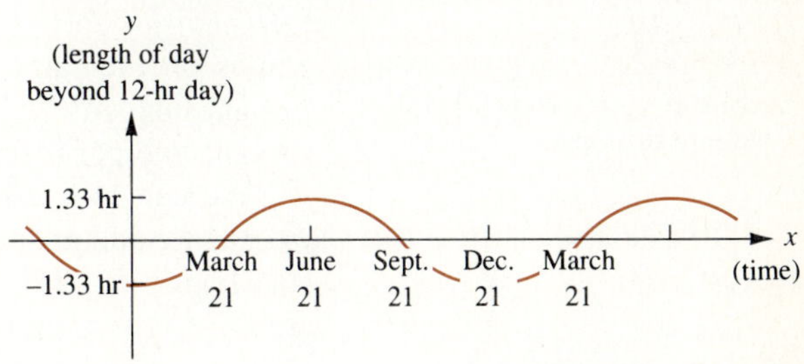

Figure 5.3-44

(a) Find the period in years, the frequency in cycles per year, and the amplitude of this graph.

(b) If we use $y = A\cos(Bt + C)$ to represent this graph, find A, B, and C using $|C| \le \pi$, $A > 0$, and $B > 0$.

(c) For how many hours is the sun above the horizon on the longest (June 21) and shortest (December 21) days of the year at Denver?

93. You run a pipe manufacturing company and get an order for 100,000 cylindrical hollow pipes with very thin metal walls. Each pipe is to have a radius R and will have its top sloping upward at an angle θ above the base as shown in Figure 5.3-45. You want to manufacture these pipes by stamping the pattern out of a flat sheet of thin metal and then wrapping each pattern into a cylinder so that its sides meet at the shortest length of the pipe. What should be the shape of the flat pattern so that it forms the proper slant at the top when it is folded in a cylinder? That is, determine the equation of the part shown in Figure 5.3-46. ***Hint:* Imagine unwrapping a properly made pipe into a flat sheet. To test your answer, actually make one of these patterns out of a sheet of paper and wrap it to make a pipe 5.0 cm in radius sloping up at a 30° angle. (This exercise is a real brain twister!)

Figure 5.3-45

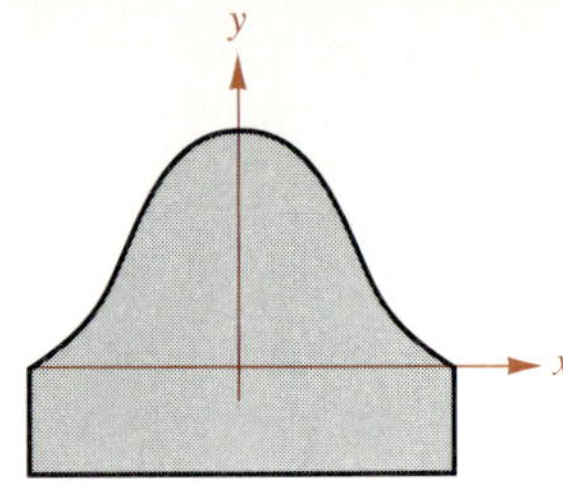

Figure 5.3-46

5.4 ADDITION OF GRAPHS

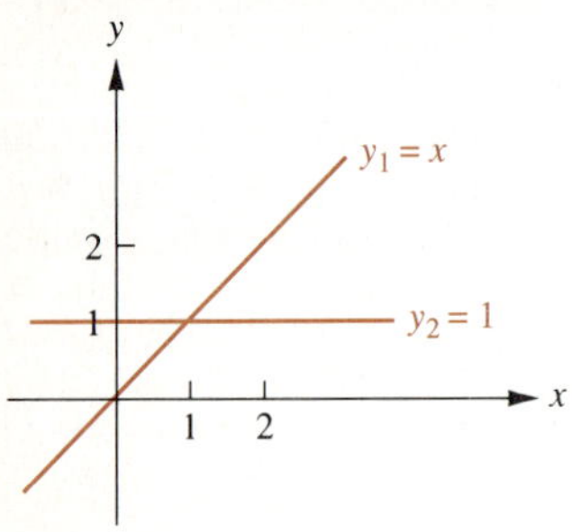

Figure 5.4-1

Graphs of $y_1 = x$ and $y_2 = 1$ on a single set of axes

So far in this chapter, we have learned how to graph the trigonometric functions one at a time. Yet in most actual applications, we receive several waves at once. For example, the sound from a rock band is the sum of many sine waves. Therefore, we need to learn how to add graphs of the trigonometric functions, especially the sine and the cosine.

What does it mean to add two graphs, such as $y = \sin x + \sin 2x$? To answer that question, let us consider a simple case first. Suppose we want to graph $y = x + 1$. This can be viewed as the sum of *two* graphs: $y_1 = x$ and $y_2 = 1$. At any point x, the value of y at that point is equal to the sum of $y_1 + y_2$ *both evaluated for the same x*. To interpret this sum graphically, let us draw both graphs on the same set of axes (see Figure 5.4-1). The graph of $y_1 = x$ is a straight line with a slope of 1 going through the origin. The graph of $y_2 = 1$ is a straight line parallel to the x axis and one unit above it.

We can now add the value of y_2 to the value of y_1 *graphically*. Because $y_2 = 1$, we add the same number (which is 1) to all points on the graph of $y_1 = x$ (see Figure 5.4-2a). As you can see from this figure, the result is to shift the graph of $y_1 = x$ up one unit along the y axis (see Figure 5.4-2b).

NOW IT'S YOUR TURN

Graph $y = x + 2$ by adding two graphs. (Answer: See Figure 5.4-3.)

Notice that the effect of adding a *constant* term K to a graph is to shift that graph up the y axis by K units. If K is negative, it shifts the graph down the y axis by $|K|$ units. This important result applies to any graph.

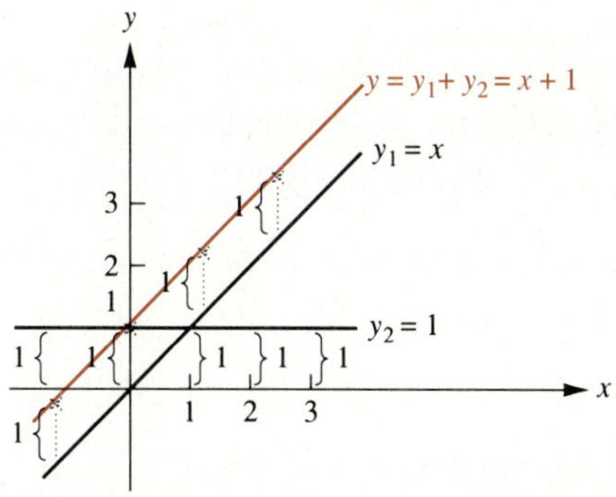

(a) For each point x, we add the value of y_2 to the value of y_1. The effect of this is to shift the graph of $y_1 = x$ one unit up along the y-axis.

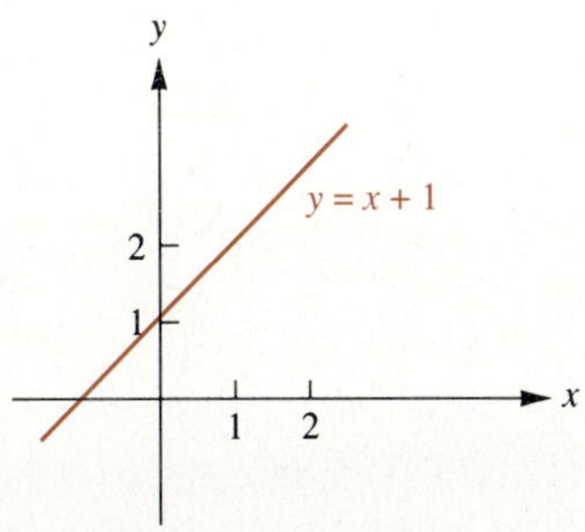

Figure 5.4-2

Graphs of $y = x + 1$

(b) The final graph of $y = x + 1$

Figure 5.4-3

Graph of $y = x + 2$ The effect of the constant 2 is to shift the graph of $y = x$ up the y axis by 2 units.

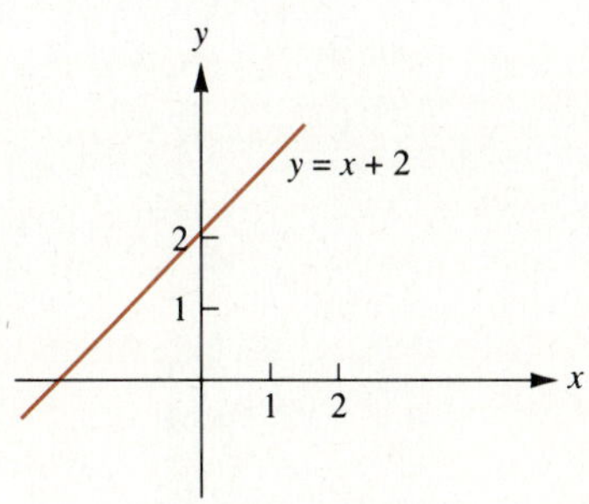

ADDING A CONSTANT TO A GRAPH

Adding a constant K to a graph shifts the entire graph up the y axis by K units if K is positive, and down the y axis by $|K|$ units if K is negative.

Using what we have just done as a guide, we are ready to apply this method to the trigonometric functions. Throughout this section, keep in mind that we want to add graphs by plotting the *minimum* number of points. Whenever possible, we use the general properties of the trigonometric functions to sketch the sum of graphs. When we must plot points, we try to choose the easiest ones, especially the points where the sine or cosine are equal to zero or their extreme values of ± 1.

Example 30

Sketch the graph of $y = 2 + \sin x$.

Solution

As before, think of this graph as the sum of two graphs, in this case $y_1 = 2$ and $y_2 = \sin x$: $y = y_1 + y_2$. The effect of adding 2 to $\sin x$ is to add 2 to every value of $\sin x$. As we just saw, this addition shifts the graph of $y_2 = \sin x$ up two units along the y axis (see Figure 5.4-4). ■

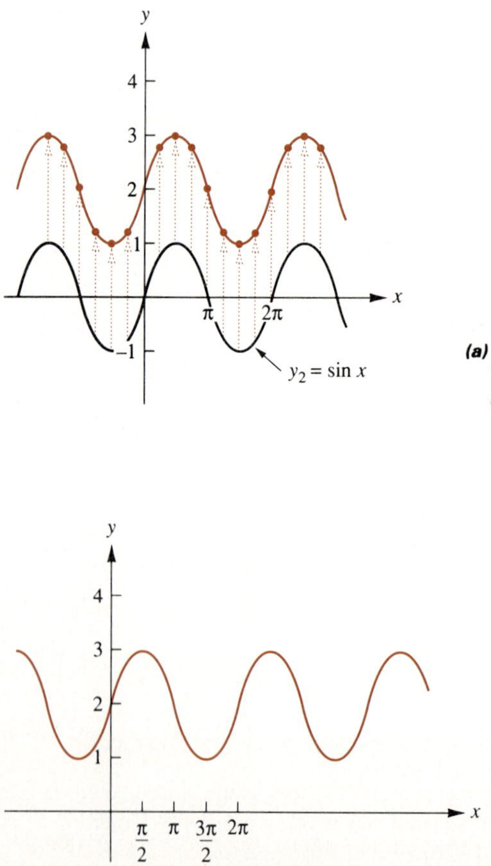

(a) Adding 2 to each value of $y_2 = \sin x$. This shifts the graph of $y_2 = \sin x$ up two units along the y-axis.

(b) The graph of $y = 2 + \sin x$

Figure 5.4-4

Graphing $y = 2 + \sin x$

NOW IT'S YOUR TURN

Sketch the graph of $y = 3 + \cos x$. (Answer: See Figure 5.4-5)

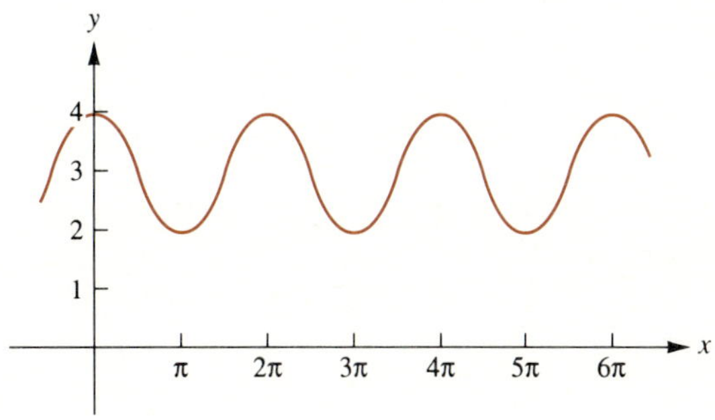

Figure 5.4-5

Graph of $y = 3 + \cos x$

Example 31

Sketch the graph of $y = x + \sin x$.

Solution

In this case, we want to add the graph of $y_2 = \sin x$ to the graph of $y_1 = x$. We start by sketching both graphs on the same set of axes (see Figure 5.4-6). Now we must add the value of $\sin x$ to the value of x to get the value of $y = x + \sin x$. This addition is more complicated than it was in the previous example because we are not always adding the same value to the graph. However, the same general method applies. The easiest way to do it is to use the values of x for which $\sin x$ is equal to zero and its extreme values ± 1. At these points, we add the value of $\sin x$ to the graph of $y = x$ (see Figure 5.4-7a). We know the graph must fit *smoothly* between these points, so we can carefully sketch the rest of the graph (see Figure 5.4-7b). We show the final result in Figure 5.4-7c). ■

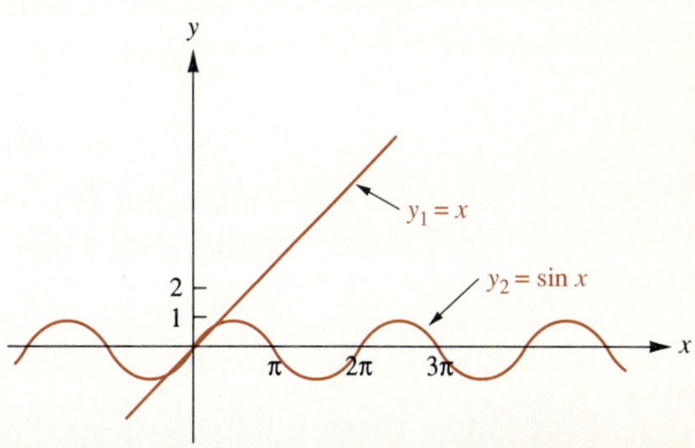

Figure 5.4-6

Graphs of $y_1 = x$ and $y_2 = \sin x$

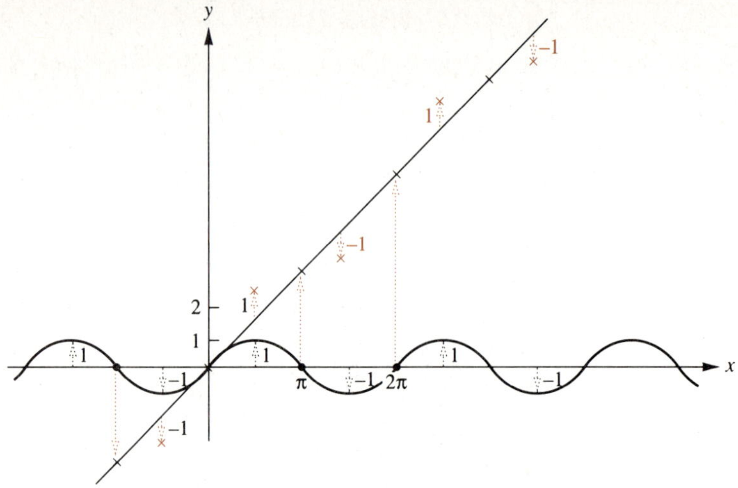

(a) Adding the values of sin x to the graph of x. For simplicity, we have picked the values of x for which sin x = 0 and ± 1.

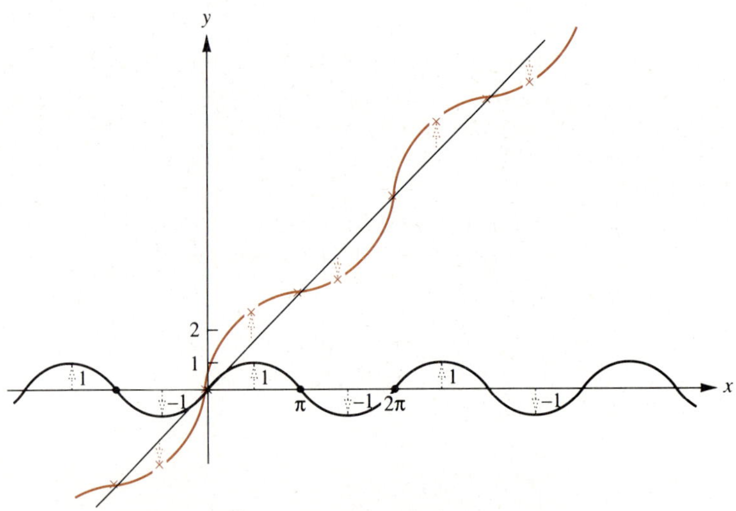

(b) Smoothly connect the points marked by x in (a).

Figure 5.4-7

Adding $y = x + \sin x$

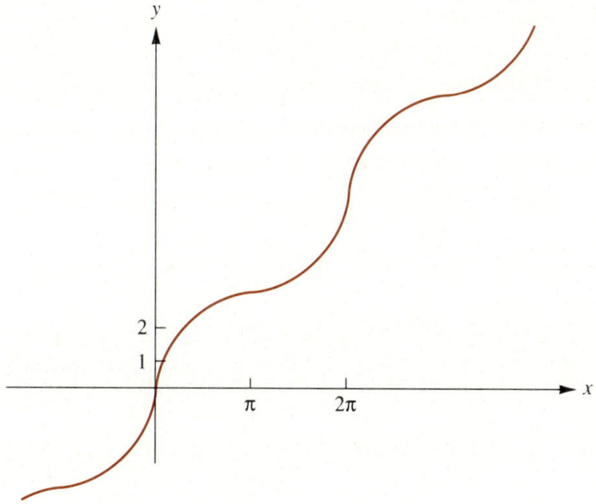

(c) The final result: the graph of y = x + sin x.

Figure 5.4-7 *(Continued)*

NOW IT'S YOUR TURN

Sketch the graph of $y = x + \cos x$. (Answer: See Figure 5.4-8.)

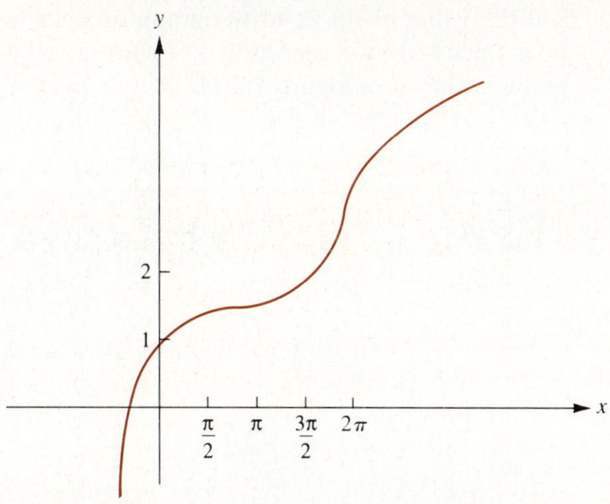

Figure 5.4-8

Graph of $y = x + \cos x$

Example 32

Sketch the graph of $y = \sin x + \sin 2x$.

Solution

We use the same procedure as in Example 31. First sketch the two graphs ($\sin x$ and $\sin 2x$) on the same set of axes; $y = \sin x$ has period 2π and $\sin 2x$ has period π (see Figure 5.4-9).

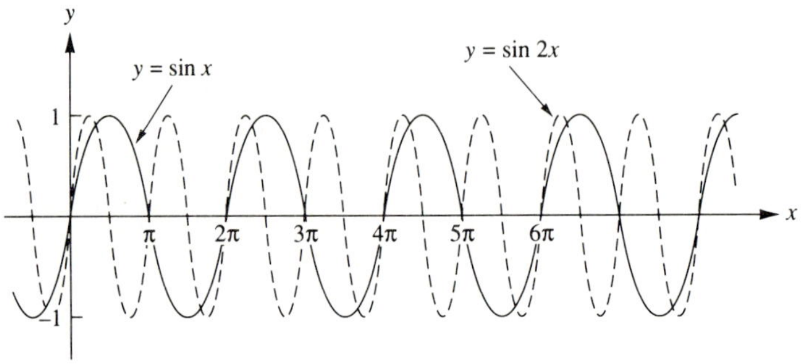

Figure 5.4-9

Graphs of sin *x* and sin 2*x*

Now look at the points at which the *more rapidly fluctuating* graph (which is $\sin 2x$ in this case) is equal to zero and its extreme values ± 1. At these points, add the value of $\sin 2x$ to the graph of $\sin x$ (see Figure 5.4-10). Connect the points marked by $\times$ smoothly in Figure 5.4-10a (see Figure 5.4-10b). The final graph is shown in Figure 5.4-11. ■

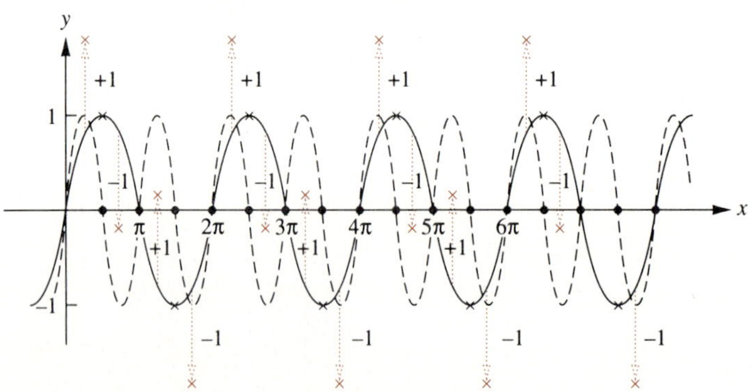

(a) At the points for which sin 2x = 0 and ±1, add its value to the graph of sin x.

Figure 5.4-10

Adding sin 2*x* to sin *x*

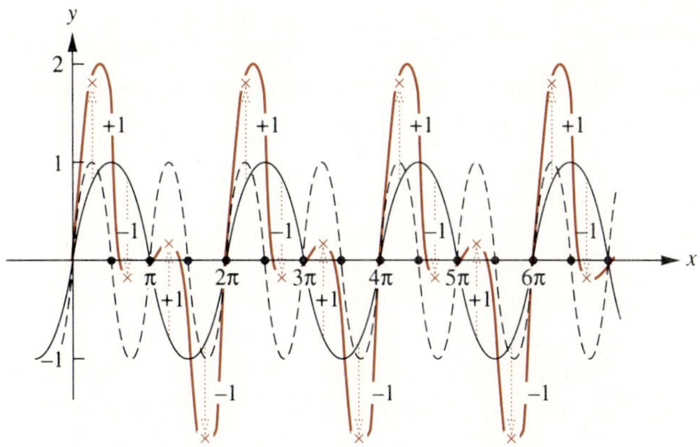

(b) Smoothly connect the points marked x in (a).

Figure 5.4-10 *(Continued)*

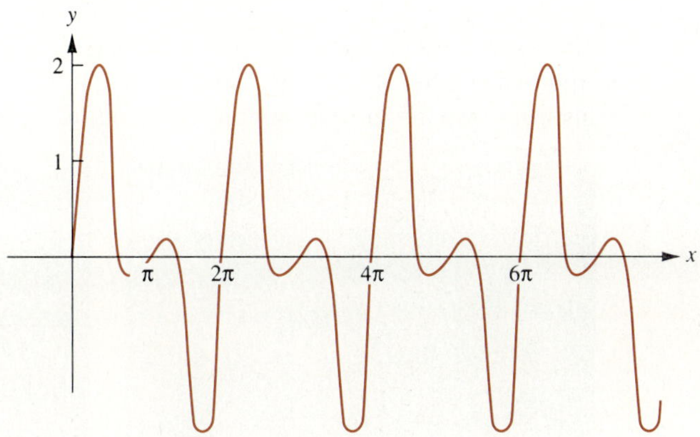

Figure 5.4-11

Graph of $y = \sin x + \sin 2x$

NOW IT'S YOUR TURN

Sketch the graph of $y = \cos x + \cos 2x$. (Answer: Figure 5.4-12)

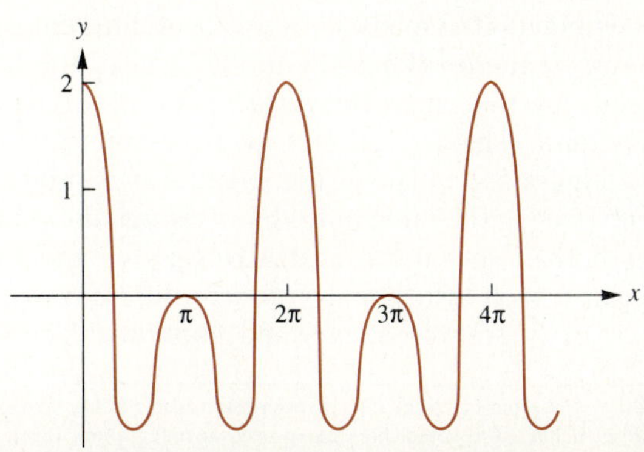

Figure 5.4-12

Graph of $y = \cos x + \cos 2x$

Here we find something that you may not have expected. When we add two sine (or cosine) graphs of different periods (such as $\sin x + \sin 2x$), we get a graph that doesn't look anything like a sine or a cosine. We still get a *periodic graph*—made up of sines or cosines—but its shape is very different from either of the graphs that combine to make it. If we were to vary the amplitude and the period (such as $y = 5 \sin 4x + 2 \cos 3x$), we could produce an even wider variety of shapes. And, if we included sines *and* cosines *together*, we could produce the widest possible variety of shapes for periodic graphs. This situation suggests that we should be able to duplicate any periodic graph, if we add together enough sine and cosine terms of differing amplitudes and periods. Such sums are called *Fourier series* and have been studied since the 1700s*. If you go on in mathematics, physics, or engineering, you will certainly encounter Fourier series.

There are many applications in which complex waves are formed by adding sines and cosines. We discuss only a few here.

Example 33
Sound Quality

Earlier in this chapter we saw that sound is a pressure wave that we can represent as a sine wave. However, only an exceptionally pure sound, such as one from a good tuning fork or an electronically produced sound, is a perfect sine wave. Nearly all the sounds we hear, including singers and musical instruments, are not so simple (see Figure 5.4-13).

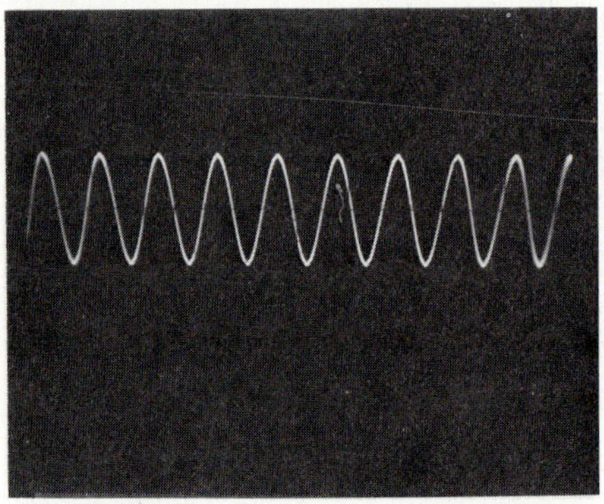

(a) Pure tone

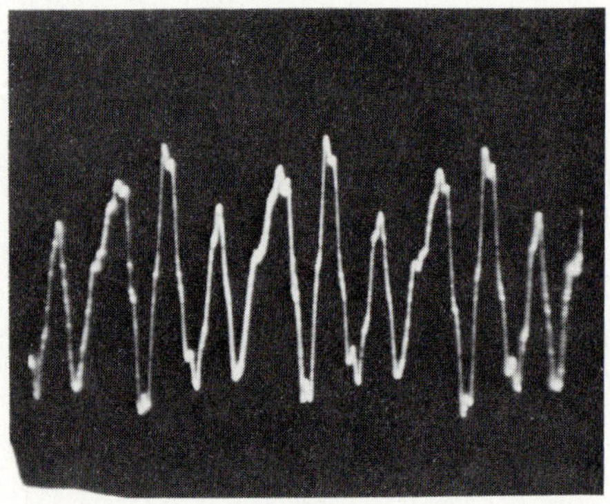

(b) A more typical sound

Figure 5.4-13
Oscilloscope traces showing sound waves

When an instrument, such as a guitar, plays a note, the sound wave is really a combination of many sines waves of different frequencies and amplitudes. The *lowest* frequency (longest period) is called the *fundamental* frequency. It is this frequency that gives the *pitch* to the note. The higher frequencies are called *overtones*. These overtones give the sound its characteristic *quality*.

Figure 5.4-14 shows the sound waves produced by a clarinet and an oboe, both playing the same note. Even though the fundamental frequency is the same for both notes, the sound waves look (and therefore *sound*) very different because each instrument produces different overtones. Each *overtone* is a *sine wave* of different frequency and amplitude. The sound quality produced by these

* But *not* by Fourier! Although these series are named after him, French mathematician and physicist Joseph Fourier neither discovered them nor proved anything about them. He just used them in his work.

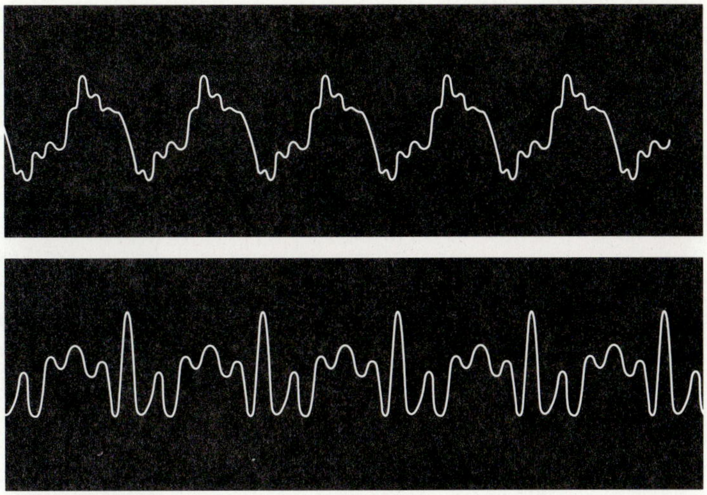

Figure 5.4-14

Sound waves produced by a clarinet (top) and an oboe (bottom), both playing the same note Both patterns have the same period but different overtones. These overtones enable us to distinguish between the two instruments.

overtones enables us to tell the difference between a clarinet and an oboe, even when they are playing the same note. If it were not for these overtones, we could not tell one from the other. ■

Example 34
Voice Prints

Just as musical instruments produce overtones (Example 33), so do human vocal chords. These sine wave overtones add to produce the pattern of a person's voice.

Figure 5.4-15 shows oscilloscope traces of the voices of two people. Notice how different the patterns look because of the addition of different overtones. The overtones are due to the structure of the person's larynx, so they are unique to each person since we all have different vocal chords. No two people have identical overtones. The pattern of overtones is called the *voice print* of the person. Because this pattern is unique to each person, voice prints can be (and are) used to identify people in much the same way as fingerprints. These overtones enable you to recognize the voices of your friends or of different singers. ■

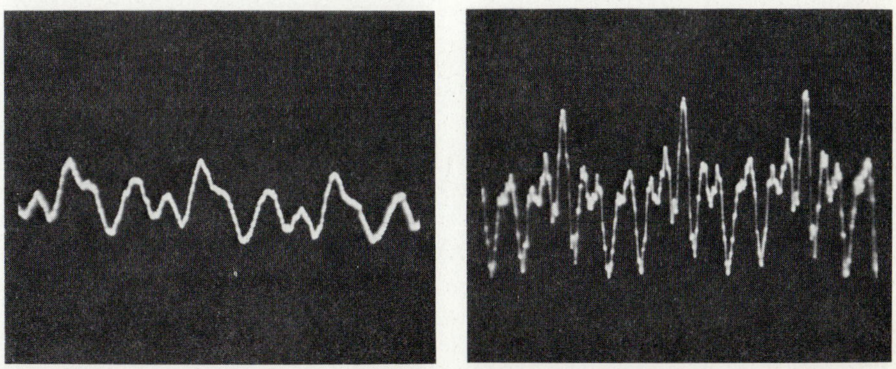

Figure 5.4-15

Voice prints of two different people The overtones, which are sine waves, give the patterns their shapes. Because no two people have exactly the same overtones, the voice print provides a positive identification of each person.

Example 35

Electronics

In electronics labs it is often necessary to generate a type of electrical voltage wave called a *square wave*, as shown in Figure 5.4-16. Good electronic generators (called *function generators*) produce square waves by adding together many different sine waves. Figure 5.4-17 shows how this can be done using four waves of different amplitudes and periods. As you can see, the more waves you use, the better the results. ■

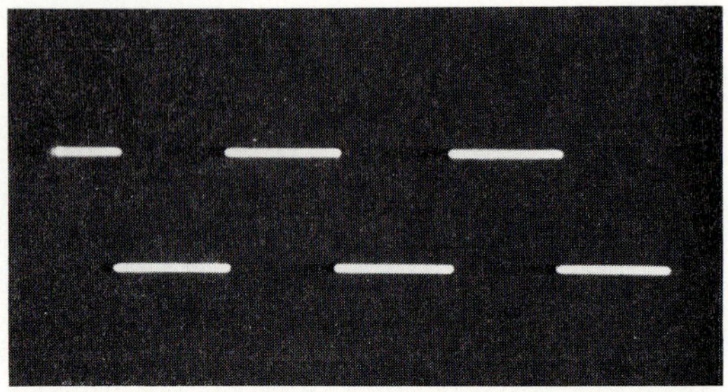

Figure 5.4-16

Square wave Electrical voltage is on the y axis and time is on the x axis.

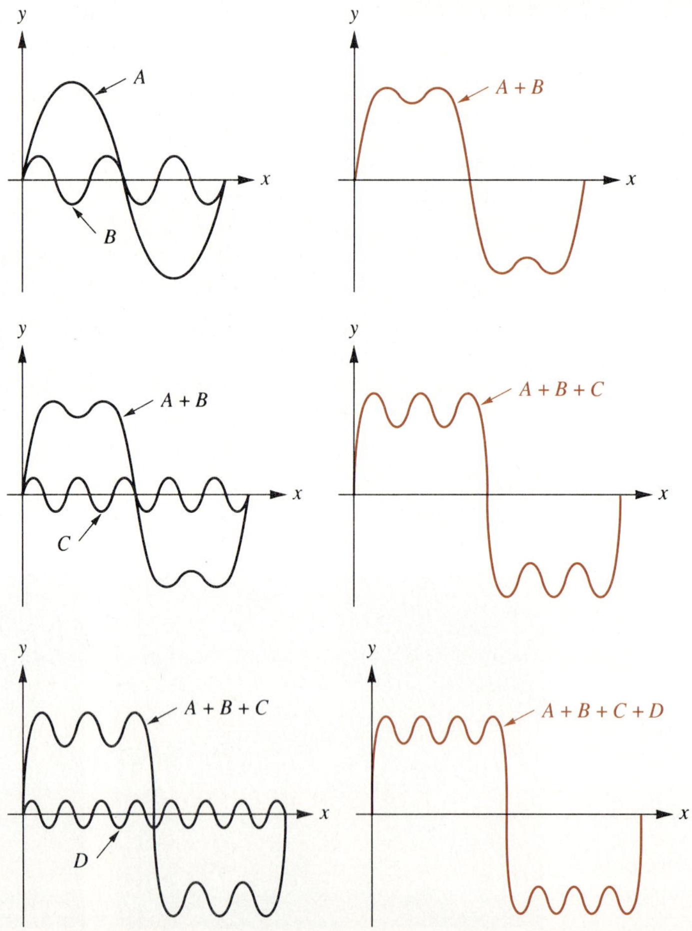

Figure 5.4-17

Adding sine waves to make a square wave Adding more and more waves gives a better and better result.

EXERCISES 5.4

In Exercises 1–10, sketch the graph of y as a function of x by adding the two graphs, not by plotting points.

1. $y = x + 3$

2. $y = x - 1$

3. $y = x + 1.75$

4. $y = 2x + 1$

5. $y = x/2 + 1$

6. $y = 3x - 6$

7. $y = 2 - x$

8. $y = 5 - x/5$

9. $y = -3 - 2x$

10. $y = \pi x - \pi$

In Exercises 11–20, sketch the graph of y as a function of x by *adding* the two graphs. You may want to plot just a few points.

***11.** $y = x^2 + 1$

***12.** $y = 2 - x^2$

***13.** $y = x^3 + 4$

***14.** $y = 1 + \sqrt{x}$

***15.** $y = 2 - \sqrt{x}$

***16.** $y = x + x^2$

***17.** $y = x^2 - 3x$

***18.** $y = x + \sqrt{x}$

19.** $y = 2x^2 + x - 1$ ***Hint: Think of $x - 1$ as a single graph.

***20.** $y = x + 1/x$

In Exercises 21–51, sketch the graph of y as a function of x by adding the two graphs. You may need to plot just a few points to help you, as we did in the examples. Include several complete cycles for each graph.

21. $y = 1 + \sin x$

22. $y = \cos x + 2$

23. $y = 0.50 + \sin x$

24. $y = \sin x - 1$

25. $y = \cos x - 2$

***26.** $y = 1 - \sin x$

27. $y = 4 + \sin x$

***28.** $y = 4 - \sin x$

29. $y = 1 + \cos 2x$

30. $y = 2 + \cos 3x$

31. $y = 2 \sin x - 1$

32. $y = 3 \cos \pi x + 4$

33. $y = 1 + \cos(x - \frac{\pi}{4})$

***34.** $y = 2 + 3 \sin(\pi x/2 - 1)$

***35.** $y = \pi + \pi \cos(\pi x + \pi)$

36. $y = 2x + \sin x$

37. $y = \sin x - x$

38. $y = x + 2 \cos x$

39. $y = 1 + x + \sin x$ ***Hint:*** Think of $1 + x$ as a single graph.

40. $y = x - \cos x$

***41.** $y = x^2 + \sin x$

***42.** $y = x^3 + \cos x$

43. $y = \sin x + \sin 3x$

44. $y = \cos x + \cos 3x$

45. $y = \sin x + \cos x$

46. $y = 2 \cos x + \cos 2x$

47. $y = \sin x - \cos x$

48. $y = \sin x + \cos 2x$

49. $y = \sin x + \frac{1}{2} \sin 2x$

50. $y = \cos x + \frac{1}{3} \cos 3x$

***51.** $y = \sin x + \frac{1}{2} \sin 2x + \frac{1}{3} \sin 3x$

52. Sound. An electronic synthesizer produces the notes C (262 Hz) and C# (277 Hz) at the same time. If these two notes are equally loud, sketch the sound wave that will be produced. ***Hint:*** Sketch the sine waves for both notes on the same axes and add them.

53. Sound. A guitarist simultaneously plays the notes G# (415 Hz) and G# one octave higher (see Exercise 86 in Section 5.3-2). Sketch the sound wave produced if both notes are equally loud. See the hint in Exercise 52.

54. Sound. The photograph in Figure 5.4-18 shows an oscilloscope trace of the sound waves of two notes.

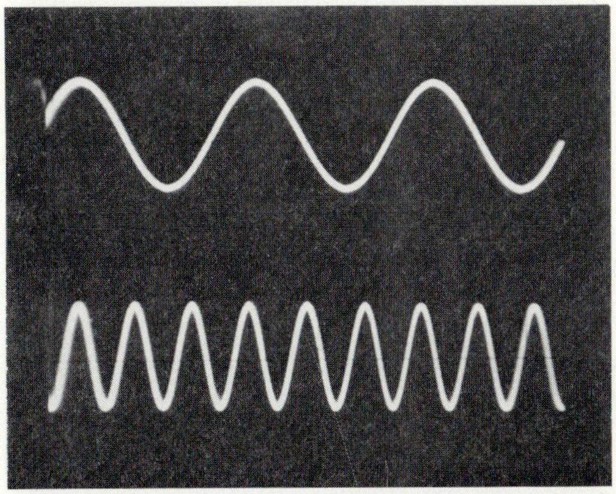

Figure 5.4-18

(a) How many octaves separate these notes? (see Exercise 86 in Section 5.3-2.)

(b) Add these waves graphically to find the pattern of the two notes sounded together.

***55. Sound.** While analyzing the sound pattern shown in Figure 5.4-19, the analyzer believes that this sound is a combination of two waves of frequencies 1 and 2 kilohertz. Is this possible?

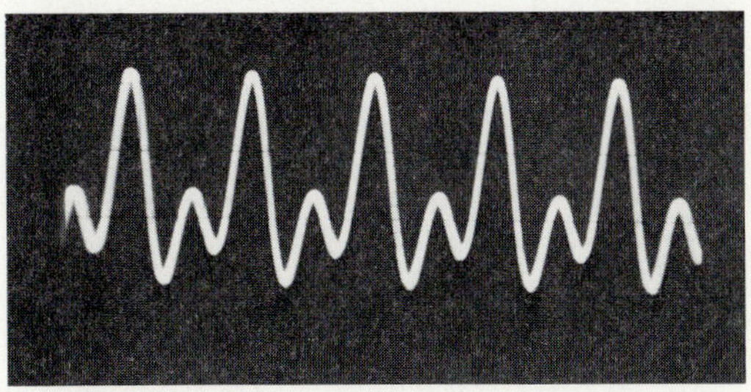

Figure 5.4-19

56. Voice Prints. Two singers are singing the *same* note A (440 Hz). This note is the *fundamental frequency*. One singer has an overtone of 880 Hz with $\frac{1}{2}$ the amplitude of the fundamental, whereas the other singer has an overtone of 880 Hz at $\frac{1}{4}$ the amplitude of the fundamental. Sketch the voice prints of the two singers.

***57. Voice Prints.** A singer singing F# (370 Hz, the fundamental) has one overtone of frequency 740 Hz with $\frac{1}{2}$ the amplitude of the fundamental and another overtone of frequency 1110 Hz at $\frac{1}{3}$ the amplitude of the fundamental. Draw the voice print of this singer.

Exercises 58–60 illustrate examples from electronics. They also show examples of the use of Fourier series to produce periodic functions by adding more and more sine and/or cosine terms. To do these problems correctly, you should work very carefully and use graph paper.

***58.** Show that the square wave in Example 35 can be produced by adding more and more terms of the sum

$$y = \left(\frac{12}{\pi}\right)\left[\sin\left(\frac{\pi x}{4}\right) + \frac{1}{3}\sin\left(\frac{3\pi x}{4}\right) + \frac{1}{5}\sin\left(\frac{5\pi x}{4}\right) + \cdots\right]$$

What are the period and the amplitude of this square wave? ***Hint:*** First sketch the first term in the sum. Then add the second term to it. Then add the third term to that sum. Do as many terms (at least three) as you have patience for. Notice that as you add more and more terms, the sum becomes a better and better approximation to a square wave.

***59.** Figure 5.4-20 shows a photo of a *sawtooth* wave often used in electronics. Using Exercise 58 as a guide, show that this sawtooth wave can be produced by adding more and more terms of the sum

$$y = \frac{\pi}{2} - \left(\frac{4}{\pi}\right)\left[\cos x + \frac{\cos 3x}{3^2} + \frac{\cos 5x}{5^2} + \cdots\right]$$

What is the period of this wave?

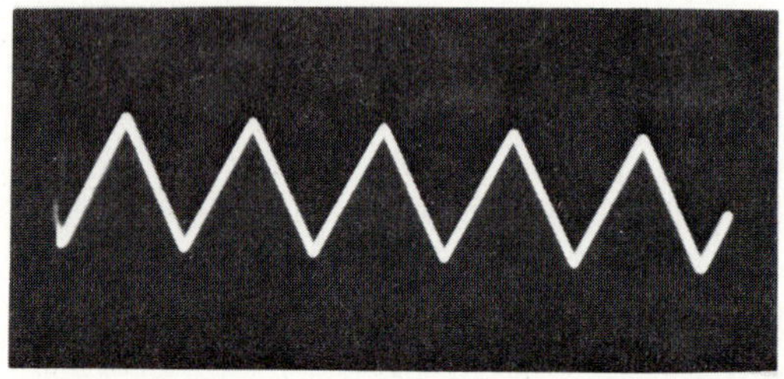

Figure 5.4-20

***60.** What wave does the following sum represent?

$$y = \left(\tfrac{8}{\pi}\right)\left(\sin x + \tfrac{1}{3}\sin 3x + \tfrac{1}{5}\sin 5x + \ldots\right)$$

Find its period.

SUMMARY

1. Periodic Function. Any function that repeats its values at regular intervals is a *periodic function.* All of the trigonometric functions are periodic.

2. Period. The *period P* is the shortest distance along the *x* axis over which the function repeats itself.

3. Frequency. The *frequency f* of a time graph is the number of cycles per second. The frequency and period are reciprocals of each other.

$$f = \frac{1}{P} \quad \textit{and} \quad P = \frac{1}{f}$$

4. Graphs of the Trigonometric Functions.

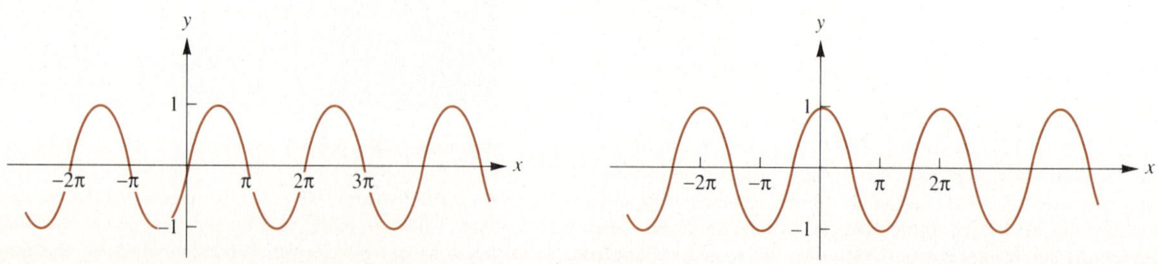

***(a)** $y = \sin x$: amplitude = 1, period = 2π, $-1 \le \sin x \le 1$*

***(b)** $y = \cos x$: amplitude = 1, period = 2π, $-1 \le \cos x \le 1$*

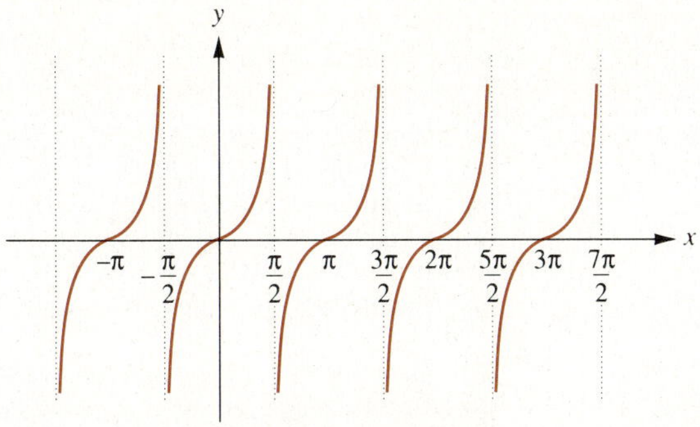

(c) ***y = tan x: period = π***

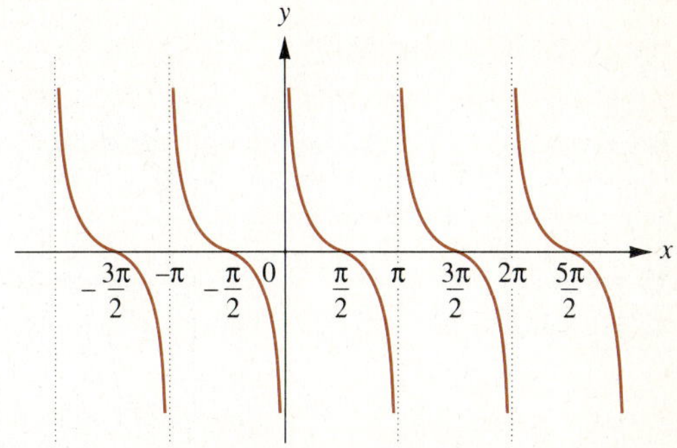

(d) ***y = cot x: period = π***

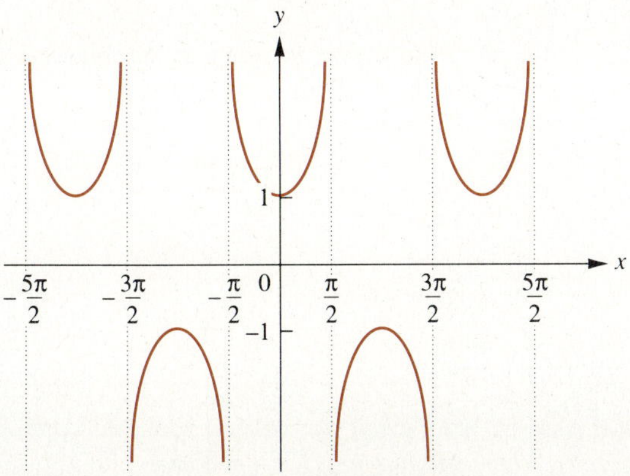

(e) ***y = sec x: period = 2π***

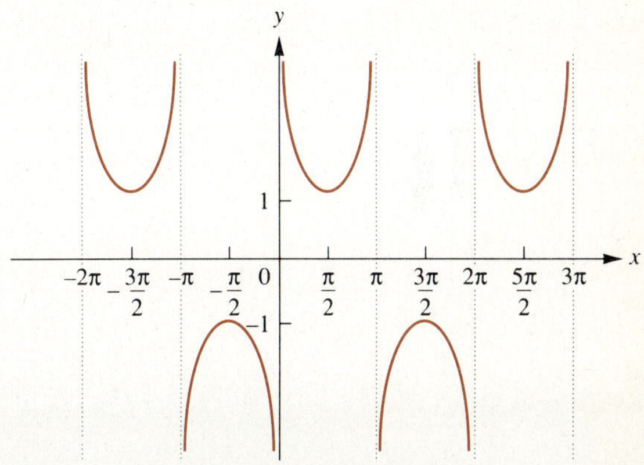

(f) ***y = csc x: period = 2π***

5. **Phase Shift.** The *phase shift* is the distance that the graph is shifted to the right or the left along the x axis.

6. **The Graphs of $y = A\sin(Bx + C)$ and $y = A\cos(Bx + C)$.** These graphs have:
   (a) Amplitude $= |A|$. If A is negative, the graph is reflected in the x axis. The amplitude is the greatest distance that the graph gets from the x axis.
   (b) Period $= 2\pi/|B|$. For a time graph, $|B| = 2\pi f$.
   (c) Phase shift $= |C/B|$. The graph is shifted to the right if C/B is negative and to the left if C/B is positive.

7. **Sine Wave.** If a graph has the shape of the sine graph, we call that graph a sine wave.

IMPORTANT TERMS

amplitude	**period**	**sine wave**
cycle	**periodic function**	**sinusoidal**
frequency	**phase shift**	**time graph**
hertz (Hz)		

REVIEW EXERCISES

In Exercises 1–14, find the amplitude (if applicable), period, and phase shift (including its direction), and then sketch the graph of each function without plotting points.

1. $y = \sin 3x$

2. $y = -3\cos 2x$

3. $y = \cos(x - \frac{\pi}{2})$

4. $y = \frac{1}{2}\sin(-3x - \pi)$

5. $y = 2\sin(-x + 1)$

6. $y = -3\cos(x - \frac{\pi}{6})$

7. $y = -\tan(x - \frac{\pi}{2})$

8. $y = 3\tan(2x + \pi)$

9. $y = -\cos(-\pi x + \frac{\pi}{2})$

10. $y = -4\cos(-\pi x/2)$

11. $y = -2\sin(0.42x - 1.3)$

***12.** $y = -2\sec(2x + 6)$

***13.** $y = 1.5\cot(\pi x + \frac{\pi}{4})$

***14.** $y = 5\csc(2\pi x + 1)$

15. Write the equation of $y = A\sin(Bx + C)$ with $A > 0$ and $B > 0$ satisfying the following conditions. Then sketch each graph.

(a) Amplitude = 3, period = 2π, no phase shift
(b) Amplitude = 1, period = π, no phase shift
(c) Amplitude = 2, period = $\frac{\pi}{2}$, no phase shift
(d) Amplitude = 4, period = 2π, phase shift = $\frac{\pi}{4}$ to the right
(e) Amplitude = $\frac{1}{2}$, period = 4, phase shift = $\frac{\pi}{3}$ to the right
(f) Amplitude = $\frac{1}{4}$, period = $\frac{2}{3}$, phase shift = $\frac{\pi}{6}$ to the left
(g) Amplitude = $\frac{2}{3}$, period = 1, phase shift = $\frac{1}{6}$ to the left

16. A sound wave has a frequency of 1200 Hz. Find its period, write its equation in the form $y = \sin Bt$, and sketch its time graph.

17. Find the period for each of the following graphs that is periodic.

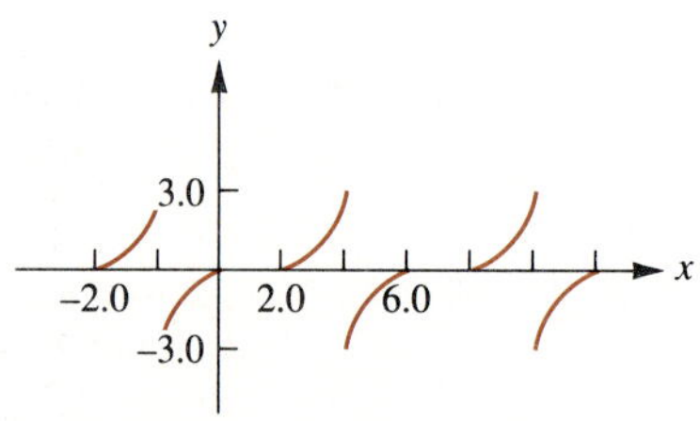

(a)

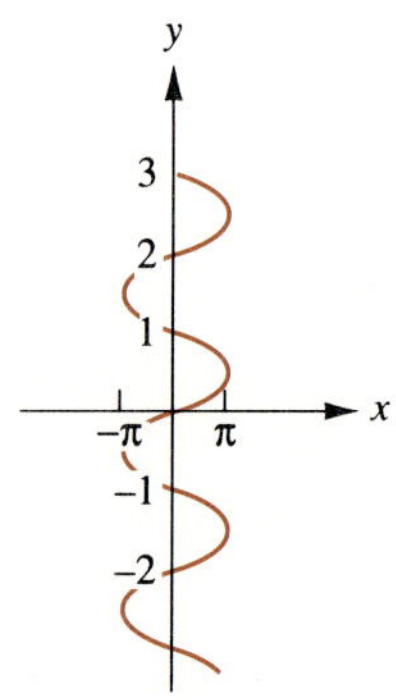

(b)

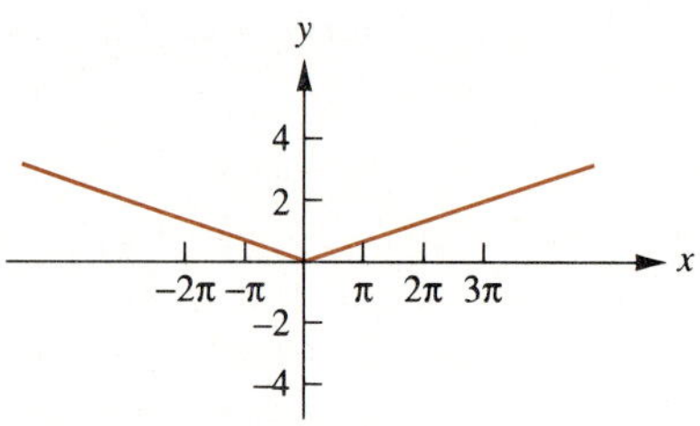

(c)

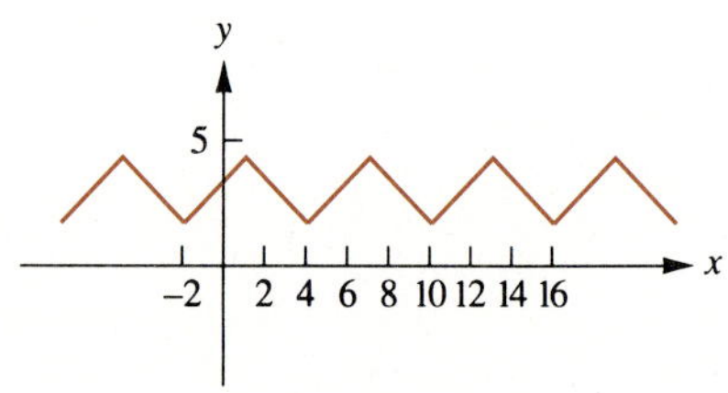

(d)

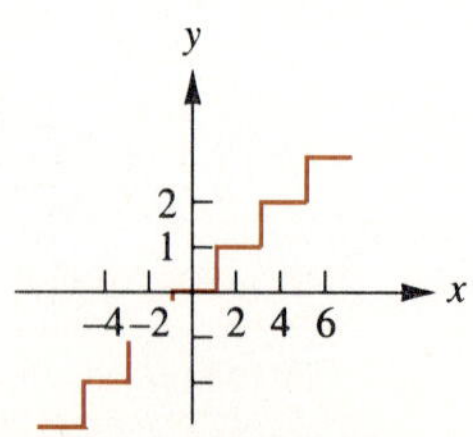

(e)

18. Each of the following graphs is either of the equation $y = A \sin Bx$ or $y = A \cos Bx$, where $B > 0$. For each one, find the amplitude and the period and write its numerical equation.

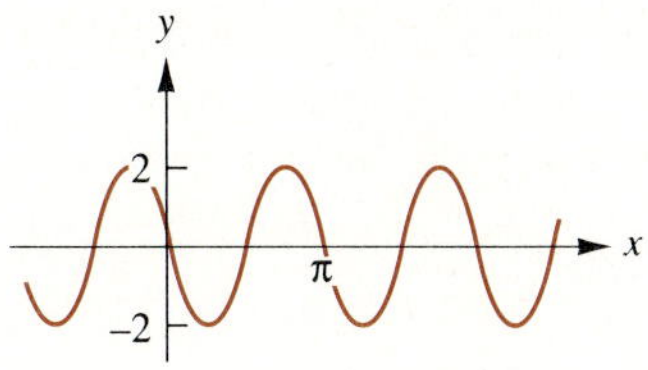

(a)

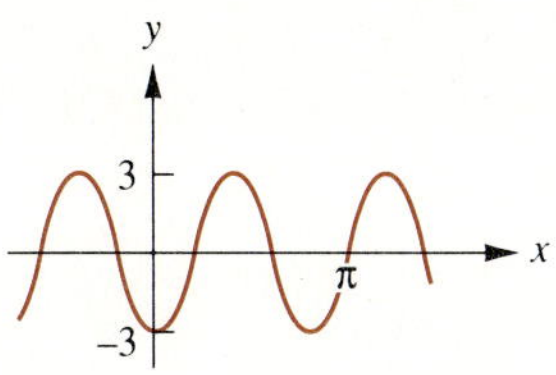

(b)

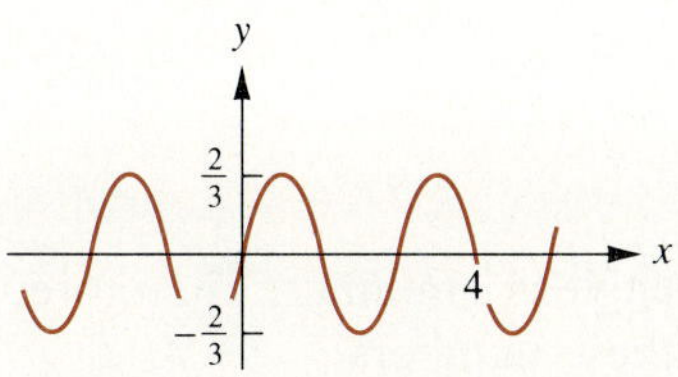

(c)

19. Each of the following graphs is of the form $y = A \sin(x + C)$ where $|C| < \pi$. Find the amplitude and phase shift of each and write its numerical equation.

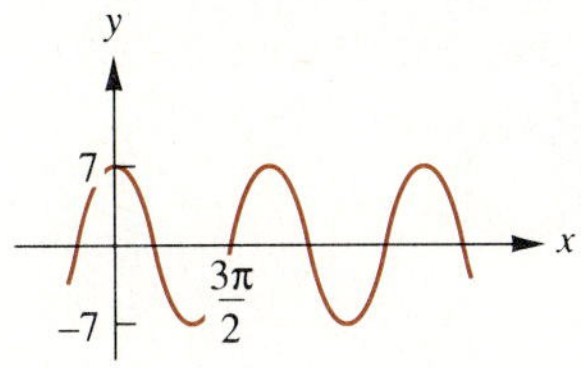

(a)

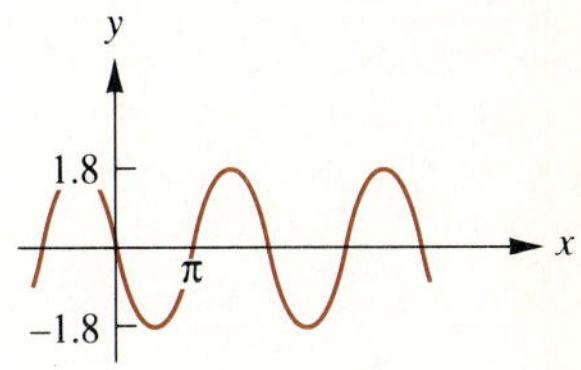

(b)

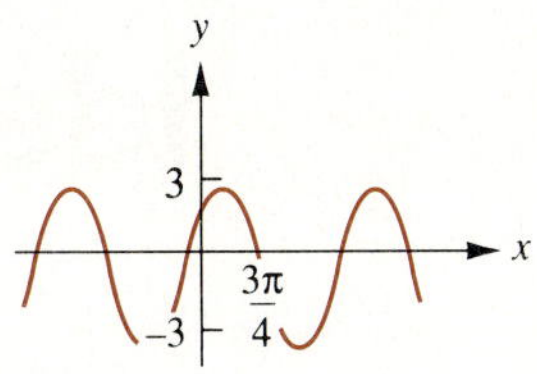

(c)

20. Graph the function $y = x - \cos 2x$ by adding two graphs, as discussed in Section 5.4.

6

Identities

6.1 ALGEBRAIC OPERATIONS WITH THE TRIGONOMETRIC FUNCTIONS

Remember that each trigonometric function of an angle is a number. Therefore, the operations of algebra can be performed with these numbers.

Just as $12 \times 12 = 12^2$, we also have $(\sin \theta)(\sin \theta) = (\sin \theta)^2$, which we usually write as $\sin^2 \theta$. This is *not* the same thing as $\sin \theta^2$. The term $\sin \theta^2$ means that we square the angle first, then take the sine of that product. However, $\sin^2 \theta$ means that we find the value of $\sin \theta$ first and then square that number. Similarly, $\cos^3 \theta$ means $(\cos \theta)^3$, $\tan^4 \theta$ means $(\tan \theta)^4$, and so on.

Example 1

Evaluate (a) $\cos^3 135°$, (b) $\sin^2 \pi$, and (c) $\sin \pi^2$.

Solution

(a) $\cos^3 135° = (\cos 135°)^3 = (-\sqrt{2}/2)^3 = -\dfrac{2\sqrt{2}}{8} = -\dfrac{\sqrt{2}}{4}$

(b) $\sin^2 \pi = (\sin \pi)^2 = (0)^2 = 0$

(c) $\sin \pi^2 = \sin(3.14)^2 = \sin 9.86 = -0.422$

Notice that if the angle is taken to a power, we must express it in radian measure to find its trigonometric function. For example, if we want to evaluate

$\sin x^2$ when $x = 180°$, we must first convert x to radians, then square that number, and finally find its sine.

$$\sin(180°)^2 = \sin \pi^2 = \sin(3.14)^2 = \sin 9.86 = -0.422$$

To see why we must use radian measure, notice that $(180°)^2 = (180°)\,(180°) = 32{,}400\ (°)^2$. The "square degree," $(°)^2$, is not an angular unit. But because the radian is a pure number, the "square radian" (rad^2) is also a pure number and is, therefore, a proper unit for angles. ■

NOW IT'S YOUR TURN

Find $\tan^2 \frac{\pi}{4}$ and $\cos\left(\frac{\pi}{2}\right)^2$.
(Answer: $\tan^2 \frac{\pi}{4} = 1$, $\cos\left(\frac{\pi}{2}\right)^2 = \cos 2.46 = -0.777$)

CAUTION

Does $\sin^{-1}\theta = (\sin\theta)^{-1} = \dfrac{1}{\sin\theta}$? (Answer: No!)
Remember that in trigonometry the superscript $^{-1}$ after a trigonometric function is not an exponent and therefore does *not* mean the reciprocal of that function. As we saw in Chapter 2, the $^{-1}$ means the *inverse* of that function, so that $\sin^{-1}\theta = \arcsin\theta$. To indicate the *reciprocal* of $\sin\theta$, we write $(\sin\theta)^{-1}$ or $\dfrac{1}{\sin\theta}$.

The following examples illustrate that many operations from arithmetic and algebra also hold for the trigonometric functions because these functions represent numbers.

Example 2

Find the product $(\cos\theta + \tan\theta)\,(\cos\theta - \tan\theta)$.

Solution

Because $(a + b)\,(a - b) = a^2 - b^2$, we have

$$(\cos\theta + \tan\theta)\,(\cos\theta - \tan\theta) = \cos^2\theta - \tan^2\theta$$ ■

Example 3

Expand $(\sec x + \sin x)^2$.

Solution

Because $(a + b)^2 = a^2 + 2ab + b^2$, we have

$$(\sec x + \sin x)^2 = \sec^2 x + 2\sec x \sin x + \sin^2 x$$ ■

Example 4

Factor $\cot^3 y - \csc^3 y$.

Solution

Because $a^3 - b^3 = (a - b)(a^2 + ab + b^2)$, we have

$$\cot^3 y - \csc^3 y = (\cot y - \csc y)(\cot^2 y + \cot y \csc y + \csc^2 y) \quad ■$$

NOW IT'S YOUR TURN

Expand $(\sin x + \cos x)^2$ and factor $\sec^2 x - \cot^2 x$.
(Answer: $\sin^2 x + 2 \sin x \cos x + \cos^2 x$, $(\sec x - \cot x)(\sec x + \cot x)$.)

We can use these algebraic operations to solve equations involving the trigonometric functions, as the next example shows.

Example 5

Find the values of $\sin x$ that satisfy the equation $2 \sin^2 x = 1 + \sin x$.

Solution

First bring everything to one side of the equation.

$$2 \sin^2 x - \sin x - 1 = 0$$

Now factor the left-hand side.

$$(2 \sin x + 1)(\sin x - 1) = 0$$

Therefore, $2 \sin x + 1 = 0$ or $\sin x - 1 = 0$, which gives us

$$\sin x = -\tfrac{1}{2} \qquad \text{or} \qquad \sin x = 1 \quad ■$$

NOW IT'S YOUR TURN

Find the values of $\sec x$ that satisfy the equation $2 \sec^2 x = 4 - 2 \sec x$.
(Answer: $\sec x = 1, -2$)

Example 6

Simplify:

$$\frac{\cot x + \dfrac{1}{\cot x}}{1 + \dfrac{1}{\cot^2 x}}$$

Solution

$$\frac{\cot x + \dfrac{1}{\cot x}}{1 + \dfrac{1}{\cot^2 x}} = \frac{\dfrac{\cot^2 x + 1}{\cot x}}{\dfrac{\cot^2 x + 1}{\cot^2 x}} \qquad \text{(find common denominators)}$$

$$= \frac{\cot^2 x + 1}{\cot x} \cdot \frac{\cot^2 x}{\cot^2 x + 1} \qquad \text{(divide denominator into numerator)}$$

$$= \frac{\cot^2 x}{\cot x} \qquad \text{(cancel common terms)}$$

$$= \cot x \; ■ \qquad \text{(cancel cot } x\text{)}$$

NOW IT'S YOUR TURN

Simplify:

$$\frac{\dfrac{1}{\sin x} - \sin x}{1 - \dfrac{1}{\sin^2 x}}$$

(Answer: $-\sin x$)

EXERCISES 6.1

In Exercises 1–7, evaluate each expression numerically.

1. $\sin^2 60°$

2. $\tan^3 \frac{\pi}{6}$

3. **(a)** $\cos^2 24°$, **(b)** $(\cos 24°)^2$

4. $0.241 - \sin^4 11.2°$

5. $(\sin 141°)^2$

6. $\dfrac{\sin^2 15°}{\tan^3 (-24°)}$

7. $\cos^3 101°$

In Exercises 8–15, evaluate each expression for the value of the angle given.

8. $\tan x^3$ where $x = 2.4$ rad

9. When $x = 32°$, evaluate **(a)** $\cos^2 x$, **(b)** $\cos x^2$, and **(c)** $(\cos x)^2$

10. $\sin \phi^2 + \sin^2 \phi$, where $\phi = 30.0°$

11. $\cos x^2 + \cos^2 x$, where $x = 1.20$ rad

12. $\sin^2 \theta + \cos^2 \theta$, where $\theta = 23°$

13. $\sin x^2 + \cos x^2$, where $x = 0.85$ rad

14. $1 + \cot^2 11.4°$

15. $\tan^2 \frac{\pi}{7} - \sec^2 \frac{\pi}{7}$

In Exercises 16–30, perform the indicated operations and simplify.

16. $(\cos x - 3 \sin x)(\cos x + 3 \sin x)$

17. $(4 \tan y - 3)^2$

18. $(\sec \beta + 2 \cos \beta)^2$

19. $2 \cot \phi(1 + 5 \cot^2 \phi)$

20. $(2 \tan x + 3 \sec x)(2 \tan x - 3 \sec x)$

***21.** $(\sin \theta + \cos \theta)^3$

***22.** $(\sin \alpha - \tan \alpha)(\sin^2 \alpha + \sin \alpha \tan \alpha + \tan^2 \alpha)$

23. $\dfrac{7\sin^5 x - 4\sin^3 x}{\sin^2 x}$

24. $\dfrac{-\sec x \tan x - \sec x \cot x}{\sin x \cot x + \sin x \tan x}$

25. $\dfrac{\csc^2 x - 1}{\csc x + 1}$

*26. $\dfrac{\sin^3 x + \cos^3 x}{\sin x + \cos x}$

27. $\dfrac{1 - \dfrac{1}{\tan^2 x}}{\dfrac{1}{\tan^2 x} + \dfrac{1}{\tan^3 x}}$

28. $\dfrac{\dfrac{1}{\sin x} - \dfrac{1}{\sin^3 x}}{\sin x - 1}$

29. $\dfrac{\cos x + \dfrac{1}{\cos^2 x}}{\cos x - 1 + \dfrac{1}{\cos x}}$

30. $\dfrac{\sec x + \dfrac{1}{\sec x}}{\dfrac{1}{\sec^2 x} + 1}$

In Exercises 31–41, find the values of the trigonometric function that satisfy the given equation.

31. $\csc^2 x + 3\csc x = 0$
32. $2\cos^2 x - \cos x = 1$
33. $\tan^2 x + 4\tan x - 5 = 0$
34. $6\sec^2 x - 2 = \sec x$
35. $16\cot^2 x - 1 = 0$
36. $\sin^3 x - 9\sin x = 0$
37. $\csc x - 4\csc^3 x = 0$
38. $\tan^2 x - 3 = 0$

*39. $\sin^2 x + 3\sin x + 1 = 0$ *Hint:* Try the quadratic formula.

*40. $2\cos^2 x + 5\cos x = 2$

*41. $6\tan^3 x - 2\tan^2 x - 3\tan x + 1 = 0$

In Exercises 42–51, find the possible values of x that satisfy each equation. Limit x to the range $0 \le x \le 2\pi$. You may sometimes need to use the quadratic formula.

42. $2\cos^2 x - \cos x = 1$
43. $\sin^2 x + 2\sin x = 3$
44. $\sin x \tan x - \sin x = 0$
45. $1 - \sin^2 x = \frac{1}{4}$
46. $\cos^2 x + \cos x + 1 = 4$
47. $8\sin^2 x + 2\sin x = 3$
48. $4 + \tan^2 x = 3$

*49. $\cos^3 x + \cos x \sin^2 x = 0.200$

*50. $\sin^2 x - 2\sin x \cos x + \cos^2 x = 0$

**51. $(\sin x - 2)(\sin^2 x + 2\sin x + 4) = -7.50$ *Hint:* See Example 4.

6.2 REVIEW OF THE FUNDAMENTAL TRIGONOMETRIC RELATIONS

In mathematics, the sciences, and other fields we often need to simplify trigonometric expressions, solve equations, and derive formulas. Such work relies heavily on the relationships between the trigonometric functions. These relationships allow us to express trigonometric expressions in many different yet *equivalent* forms.

In this section, we first review the fundamental relations from Sections 2.2 and 3.4, and then we see how to use them to simplify complicated expressions.

In applications, the sine and cosine occur most often. Also, because these two are the *only* trigonometric functions defined for *all* angles, it is often useful to reduce many trigonometric expressions to sines and cosines.

Example 7

Express $\dfrac{\sec x}{\cot x + \tan x}$ in terms of $\sin x$.

Solution

From the fundamental relations, we need to use $\sec x = \dfrac{1}{\cos x}$, $\cot x = \dfrac{\cos x}{\sin x}$,

FUNDAMENTAL TRIGONOMETRIC RELATIONS

Reciprocal Identities

$$\csc\theta = \frac{1}{\sin\theta} \quad \text{or} \quad \sin\theta = \frac{1}{\csc\theta}$$

$$\sec\theta = \frac{1}{\cos\theta} \quad \text{or} \quad \cos\theta = \frac{1}{\sec\theta}$$

$$\cot\theta = \frac{1}{\tan\theta} \quad \text{or} \quad \tan\theta = \frac{1}{\cot\theta}$$

Quotient Identities

$$\tan\theta = \frac{\sin\theta}{\cos\theta} \quad \text{and} \quad \cot\theta = \frac{\cos\theta}{\sin\theta}$$

Pythagorean Identities

$$\sin^2\theta + \cos^2\theta = 1 \qquad \tan^2\theta + 1 = \sec^2\theta \qquad \cot^2\theta + 1 = \csc^2\theta$$

and $\tan x = \dfrac{\sin x}{\cos x}$.

$$\frac{\sec x}{\cot x + \tan x} = \frac{\frac{1}{\cos x}}{\frac{\cos x}{\sin x} + \frac{\sin x}{\cos x}} \quad \text{(use the three fundamental relations above)}$$

$$= \frac{\frac{1}{\cos x}}{\frac{\cos^2 x + \sin^2 x}{\sin x \cos x}} \quad \text{(find common denominator in the denominator)}$$

$$= \frac{\frac{1}{\cos x}}{\frac{1}{\sin x \cos x}} \quad \text{(use Pythagorean identity } \sin^2 x + \cos^2 x = 1\text{)}$$

$$= \frac{\sin x \cos x}{\cos x} \quad \text{(divide numerator by denominator)}$$

$$= \sin x \quad \text{(cancel } \cos x\text{)}$$

Therefore,

$$\frac{\sec x}{\cot x + \tan x} = \sin x \ ■$$

NOW IT'S YOUR TURN

Express $\dfrac{\cot^2 x}{\csc^2 x + \sec^2 x}$ in terms of $\cos x$. (Answer: $\cos^4 x$)

Example 8

Express $1 - 2\sin^2 x + \dfrac{\tan^4 x}{(1 + \tan^2 x)^2}$ in terms of $\cos x$.

Solution

Using the relations $\tan x = \dfrac{\sin x}{\cos x}$, $\sec x = \dfrac{1}{\cos x}$ and $1 + \tan^2 x = \sec^2 x$, we get

$$\begin{aligned}
1 - 2\sin^2 x + \frac{\tan^4 x}{(1 + \tan^2 x)^2} &= 1 - 2\sin^2 x + \frac{\tan^4 x}{(\sec^2 x)^2} && (\text{use } \sec^2 x = 1 + \tan^2 x)\\
&= 1 - 2\sin^2 x + \frac{\left(\dfrac{\sin x}{\cos x}\right)^4}{\sec^4 x} && \left(\text{use } \tan x = \frac{\sin x}{\cos x}\right)\\
&= 1 - 2\sin^2 x + \frac{\left(\dfrac{\sin x}{\cos x}\right)^4}{\dfrac{1}{\cos^4 x}} && \left(\text{use } \sec x = \frac{1}{\cos x}\right)\\
&= 1 - 2\sin^2 x + \sin^4 x && (\text{invert denominator and cancel } \cos^4 x)\\
&= (1 - \sin^2 x)^2 && (\text{factor})\\
&= (\cos^2 x)^2 && (\text{use } 1 - \sin^2 x = \cos^2 x)\\
&= \cos^4 x \quad \blacksquare
\end{aligned}$$

NOW IT'S YOUR TURN

Express

$$\frac{(1 - \cot^4 x)\sin^2 x}{1 - 2\sin^2 x} + 4\csc^2 x$$

in terms of $\sin x$. (Answer: $3/\sin^2 x$)

Example 9

Simplify:

$$\sec x \csc x (\sin^2 x - \cos^2 x) + \cot x$$

Solution

When you have a complicated expression such as this and do not recognize any obvious relationships, it is usually best to express everything in terms of the sine and cosine. Using the relations

$\sec x = \dfrac{1}{\cos x}$, $\csc x = \dfrac{1}{\sin x}$, and $\cot x = \dfrac{\cos x}{\sin x}$, we get

$$\begin{aligned}
\sec x \csc x\,(\sin^2 x - \cos^2 x) + \cot x &= \frac{1}{\cos x}\cdot\frac{1}{\sin x}(\sin^2 x - \cos^2 x) + \frac{\cos x}{\sin x}\\
&= \frac{\sin^2 x - \cos^2 x}{\cos x \sin x} + \frac{\cos x}{\sin x}\\
&= \frac{\sin^2 x - \cos^2 x + \cos^2 x}{\cos x \sin x} && \text{(find common denominator)}\\
&= \frac{\sin^2 x}{\cos x \sin x} && \text{(cancel } \cos^2 x\text{)}\\
&= \frac{\sin x}{\cos x} && \text{(cancel } \sin x\text{)}\\
&= \tan x \;\blacksquare && \left(\text{use } \tan x = \frac{\sin x}{\cos x}\right)
\end{aligned}$$

NOW IT'S YOUR TURN

Simplify

$$\frac{\sec x + \csc x}{\tan x + \cot x}$$

by reducing it to sines and cosines. (Answer: $\sin x + \cos x$)

EXERCISES 6.2

In Exercises 1–15, reduce each expression to an equivalent expression involving only sin *x*, and simplify it as much as possible.

1. $(1 - \tan x)\cos x$

2. $\dfrac{\csc x}{\tan x + \cot x}$

3. $\dfrac{\sec^2 x + \csc^2 x}{\sec x \csc x}$

4. $\dfrac{\tan^2 x}{1 + \tan^2 x}$

5. $\dfrac{\sin x + \cos x}{\sec x + \csc x}$

6. $\sec^2 x + \csc^2 x$

7. $\sec x - \tan x$

8. $\sin^3 x + \cos^2 x \sin x$

9. $\dfrac{\sin x \cos x}{(1 - \sin x)\tan x}$

10. $\left(\dfrac{\sin x + \sin x \cos x}{1 - \sin x}\right)\left(\dfrac{\cos^2 x}{1 + \cos x} \cdot \csc x\right)$

*11. $\dfrac{\cos^2 x}{\sin^2 x - \sin^2 x \cos x} \div \dfrac{1 + \sin x}{(1 - \cos x)^2(1 + \cos x)}$

*12. $\dfrac{\sin x\,(\sin x - \cos x \tan x \cot^2 x)}{\cos x(\tan x + 1)} + \cos x$

*13. $\cos^4 x + \sin^4 x + 2\cos^2 x \sin^2 x - \cos^2 x$

*14. $\dfrac{\sin^2 x - 1}{\cos^2 x - 1} + \cot x \tan x$

*15. $(\cot x + 1)^2 + (\cot x - 1)^2$

In Exercises 16–25, write each expression as an equivalent expression involving only sin x and/or cos x, and simplify it as much as possible.

16. $\dfrac{1 - \sin^2 x}{\cos x}$

17. $\dfrac{\sin^2 x}{1 + \cos x}$

18. $\tan x \csc x$

19. $\tan x + \cot x$

20. $\dfrac{\tan^2 x + 1}{\sec^3 x}$

21. $\dfrac{2\sin^2 x}{\sin x \cos x} - \tan x$

*22. $\dfrac{\sec x - \tan x}{\csc x - \cot x}$

*23. $\dfrac{\cot^2 x - \cos^2 x}{\tan^2 x - \sin^2 x}$

*24. $\left(\dfrac{\cos x}{\sin x - 1} + \dfrac{\sin x}{\cos x}\right) \div \dfrac{1}{\sin x - 1}$

*25. $\cos^4 x - \sin^4 x$

6.3 PROVING IDENTITIES

In algebra you studied equations such as

$$(x + 1)^2 = x^2 + 2x + 1 \qquad (1)$$

$$x^2 - 3^2 = (x + 3)(x - 3) \qquad (2)$$

$$x + 4 = 3 \qquad (3)$$

$$x^2 - 2x - 15 = 0 \qquad (4)$$

Equations (1) and (2) are true no matter what values we assign to x. Equations such as these two, which are true for *all values* of the variable, are called *identities*. Equations (3) and (4), on the other hand, are true *only* for certain values of the variable x. Equation (3) is true only for $x = -1$, and eq. (4) is true only for $x = -3$ and $x = 5$. An equation that is true only for *certain values* of the variable is called a *conditional equation*.

TRIGONOMETRIC IDENTITY

A *trigonometric identity* is an equation involving the trigonometric functions that is true for *all allowed values of the angle.*

In trigonometry we also have identities and conditional equations, as we saw in Section 3.4-2. For example, the equation $\sin^2 x + \cos^2 x = 1$ is true for *any* value of the angle x, and therefore it is an *identity*. However, the equation $\sin x = \frac{1}{2}$ is true *only* when $x = 30°, 150°, 390°, 510°$, and so on, so it is a *conditional equation*. In trigonometry as in algebra, we *prove* identities, whereas we *solve* conditional equations. In Chapter 8 we discuss how to solve such equations, but in this chapter we concentrate on proving identities.

All of the fundamental trigonometric relationships listed at the beginning of Section 6.2 are identities. That is, $\csc x = \dfrac{1}{\sin x}$, $\sec x = \dfrac{1}{\cos x}$, $\cot x = \dfrac{1}{\tan x}$, $\tan x = \dfrac{\sin x}{\cos x}$, $\cot x = \dfrac{\cos x}{\sin x}$, $\sin^2 x + \cos^2 x = 1$, $\tan^2 x + 1 = \sec^2 x$, and $\cot^2 x + 1 = \csc^2 x$ are all identities. Therefore, they are true for all *allowed* values of the angle x.

Example 10

What values of x are not allowed for the identity $\cot x = 1/\tan x$?

Solution

Any value of x is not allowed if it makes *any term* in the identity *undefined*. In this case, the right-hand side ($1/\tan x$) is undefined whenever the denominator is zero *and* when $\tan x$ is undefined. Therefore, all values of x for which $\tan x = 0$ or $\tan x$ is undefined are not allowed values.

$$\tan x = 0 \text{ when } x = 0, \pm\pi, \pm 2\pi, \pm 3\pi, \ldots$$

$$\tan x = \text{undefined when } x = \pm\tfrac{\pi}{2}, \pm\tfrac{3\pi}{2}, \pm\tfrac{5\pi}{2}, \ldots$$

The left-hand side ($\cot x$) is undefined when $x = 0, \pm\pi, \pm 2\pi, \pm 3\pi$, and so on, but these values are the same ones we just found. Therefore, the values of x that are not allowed for this identity are

$$x = 0, \pm\tfrac{\pi}{2}, \pm\pi, \pm\tfrac{3\pi}{2}, \pm 2\pi, \pm\tfrac{5\pi}{2}, \pm 3\pi, \ldots \quad ■$$

To see what it means to say that these values of x are not allowed, let us try to use this identity when $x = \frac{\pi}{2}$, which we just saw is not allowed.

$$\cot x = \frac{1}{\tan x}$$

$$\cot \frac{\pi}{2} = \frac{1}{\tan(\pi/2)}$$

We know that $\cot \frac{\pi}{2} = 0$. But $\tan \frac{\pi}{2}$ is undefined, which means that $1/\tan(\frac{\pi}{2})$ must also be undefined. Therefore, this identity would give us

$$\underbrace{\cot \tfrac{\pi}{2}}_{0} = \frac{1}{\underbrace{\tan \tfrac{\pi}{2}}_{\text{undefined}}} \quad \Rightarrow \quad 0 = \text{undefined}$$

This result is obviously wrong because 0 is defined. Thus the identity gave us an incorrect result when we tried to use a value of x that is not allowed.

NOW IT'S YOUR TURN

What values of x are not allowed for the identity $\sin x = \dfrac{1}{\csc x}$?

(Answer: $x = 0, \pm\pi, \pm 2\pi, \pm 3\pi, \ldots$)

Example 11

Show that $\cos 2x = 2 \cos x$ is *not* an identity.

Solution

To show that a trigonometric equation is *not* an identity, all we have to do is find *one allowed* value of the angle for which the equation does *not* hold. For the equation $\cos 2x = 2\cos x$, there are no unallowed values of x, because the cosine is defined for all values of x. We can try $x = \frac{\pi}{4}$, which gives us

$$\cos 2x = \cos[2(\tfrac{\pi}{4})] = \cos\tfrac{\pi}{2} = 0$$
$$2\cos x = 2\cos\tfrac{\pi}{4} = 2(\sqrt{2}/2) = \sqrt{2}$$

Therefore, $\cos 2x \neq 2\cos x$ because $0 \neq \sqrt{2}$, which means that $\cos 2x = 2\cos x$ is a *conditional* equation rather than an identity. ■

NOW IT'S YOUR TURN

Is $\tan x/3 = \frac{1}{3}\tan x$ an identity?
(Answer: No. Try any allowed value of x except $0, 3\pi, 6\pi, 9\pi, \ldots$.)

How to Prove Identities

To prove an identity, we must prove that *both* sides of the equation are the same for *all* allowed values of the angle. To do this, we use trigonometric substitutions and algebraic operations to transform one or both sides of the equation until both sides are *exactly the same*, or *identical*. It is usually best to work with the more complicated side and simplify it. Sometimes it is helpful to do so by reducing all the trigonometric functions to sines and cosines.

▸ **Caution.** Proving identities is not like solving conditional equations. We work with only one side of the equation at a time.

▸ **Because we can work with only one side of the equation at a time, we cannot multiply (or divide) both sides of the equation by the same number or function, nor can we add (or subtract) a number or function to both sides.**

There are no general rules that always work for all identities, but we do offer the following general suggestions:

TIPS FOR PROVING IDENTITIES

1. Transform only one side of the equation at a time until it is exactly the same as the other side.

2. Transform the more complicated side by simplifying it.

3. If you do not see any obvious way to simplify a side, start by reducing all of its trigonometric functions to sines and cosines.

4. Not all equations are identities! Before you spend a lot of time trying to prove an equation is an identity, first try substituting a few *allowed* numerical values of the angle to see if the equation holds. It is best to avoid 0, $\frac{\pi}{2}$, π, $\frac{3\pi}{2}$, and 2π because many trigonometric functions are 0 or undefined for these angles. If it does not hold, the equation is not an identity, as we saw in Example 11. However, if the equation does hold, it may or may not be an identity. The only way to be sure that it is an identity is to continue with the proof and transform one side until it is identical to the other side.

The best way to learn to prove identities is to follow some examples and then try lots of them on your own. The following examples illustrate some useful techniques, and the exercises at the end of the section provide practice. Proving identities can be an exciting challenge! If you view them as a game or puzzle, they can actually be stimulating and fun!

Example 12

Is the following equation an identity?

$$\frac{1 - \tan^2 x}{1 + \tan^2 x} = \cos^2 x - \sin^2 x$$

Solution

First try a few allowed numerical values to find out if the equation is not an identity. We leave that to you.

Remember that we must work with the two sides of the equation independently. To remember to do so, we separate the two sides with a vertical line. In this case, the left-hand side is more complicated, so we shall simplify it.

$$\frac{1 - \tan^2 x}{1 + \tan^2 x} \stackrel{?}{=} \cos^2 x - \sin^2 x$$

(use $\tan x = \dfrac{\sin x}{\cos x}$)
$$\frac{1 - \dfrac{\sin^2 x}{\cos^2 x}}{1 + \dfrac{\sin^2 x}{\cos^2 x}}$$

(find common denominator)
$$\frac{\dfrac{\cos^2 x - \sin^2 x}{\cos^2 x}}{\dfrac{\cos^2 x + \sin^2 x}{\cos^2 x}}$$

(invert denominator and cancel $\cos^2 x$)
$$\frac{\cos^2 x - \sin^2 x}{\cos^2 x + \sin^2 x}$$

(use $\sin^2 x + \cos^2 x = 1$)
$$\frac{\cos^2 x - \sin^2 x}{1}$$

$$\cos^2 x - \sin^2 x \quad \Big| \quad \cos^2 x - \sin^2 x$$

The two sides are now exactly the same, so this equation is an identity. ■

NOW IT'S YOUR TURN

Prove the identity in this example by using $\sec^2 x = 1 + \tan^2 x$ as your first step.

Example 13

Is the equation $\sin^2 x = 1 - \cot^2 x \sin^2 x$ an identity?

Solution

After checking several numerical values (we leave this up to you), we try to transform the more complicated right-hand side so it is the same as the left-hand side.

$$\sin^2 x \stackrel{?}{=} 1 - \cot^2 x \sin^2 x$$

$$\begin{array}{|ll} 1 - \dfrac{\cos^2 x}{\sin^2 x}\sin^2 x & \left(\text{use } \cot x = \dfrac{\cos x}{\sin x}\right) \\ 1 - \cos^2 x & (\text{cancel } \sin^2 x) \\ \sin^2 x & (\text{use } \sin^2 x = 1 - \cos^2 x) \end{array}$$

The two sides are now exactly the same, so this equation is an identity. ■

NOW IT'S YOUR TURN

Prove that the equation $1 - \tan^2 x \cos^2 x = \cos^2 x$ is an identity.

Example 14

Is the equation $\sec x \csc x\,(1 + \tan^2 x) = \sin^2 x$ an identity?

Solution

First check a few allowed numerical values. Trying $x = \frac{\pi}{4}$ gives

$$\sec\tfrac{\pi}{4}\csc\tfrac{\pi}{4}\left(1 + \tan^2\tfrac{\pi}{4}\right) = \sec^2\tfrac{\pi}{4}$$

$$\sqrt{2}\,\sqrt{2}\,(1 + 1^2) = (\sqrt{2}/2)^2$$

$$(2)(2) = \tfrac{1}{2}$$

$$4 = \tfrac{1}{2}$$

Obviously $4 \neq \frac{1}{2}$, so this equation does not hold for $x = \frac{\pi}{4}$. Because $\frac{\pi}{4}$ is an allowed value of x, this equation cannot be an identity. ■

NOW IT'S YOUR TURN

Would it have been any good to try $x = \frac{\pi}{2}$ instead of $\frac{\pi}{4}$? Why?

(Answer: No, because $\tan x$ and $\sec x$ are not defined when $x = \frac{\pi}{2}$. Thus $\frac{\pi}{2}$ is not an allowed value. Even identities do not necessarily hold at *nonallowed* values of the angle, so trying $\frac{\pi}{2}$ would prove nothing.)

Example 15

Is the following equation an identity?

$$\sin x\left(\frac{1}{\tan x} + \tan x\right) = \sec x$$

Solution

After trying some allowed numerical values, we start with the more complicated side of the equation and try to transform it so that it is exactly the same as the simpler side.

$$\sin x\left(\frac{1}{\tan x} + \tan x\right) \stackrel{?}{=} \sec x$$

(combine into a fraction with common denominator tan x) $\quad \sin x\left(\dfrac{1 + \tan^2 x}{\tan x}\right)$

(use $\sec^2 x = 1 + \tan^2 x$) $\quad \sin x\left(\dfrac{\sec^2 x}{\tan x}\right)$

$\left(\text{substitute } \dfrac{1}{\cos x} = \sec x \text{ and } \tan x = \dfrac{\sin x}{\cos x}\right)$ $\quad \sin x\left(\dfrac{\dfrac{1}{\cos^2 x}}{\dfrac{\sin x}{\cos x}}\right)$

(divide fractions) $\quad \sin x\left(\dfrac{1}{\sin x \cos x}\right)$

(cancel $\sin x$) $\quad \dfrac{1}{\cos x}$

$\left(\text{use } \sec x = \dfrac{1}{\cos x}\right)$ $\quad \sec x$

The two sides are now exactly equal, so this equation is an identity. ■

NOW IT'S YOUR TURN

Prove that the equation $\cos x\left(\cot x + \dfrac{1}{\cot x}\right) = \csc x$ is an identity.

Sometimes we can prove an identity by reducing *both* sides to the same expression. But remember we work with only *one side at a time.*

Example 16

Prove that the equation $\frac{1}{\sin x} - \sin x = \cos x \cot x$ is an identity by transforming *both* sides of the equation into the same expression.

Solution

$$\frac{1}{\sin x} - \sin x \stackrel{?}{=} \cos x \cot x$$

	Left side	Right side	
(common denominator)	$\frac{1 - \sin^2 x}{\sin x}$	$\cos x \frac{\cos x}{\sin x}$	$\left(\text{use } \cot x = \frac{\cos x}{\sin x}\right)$
(use $\cos^2 x = 1 - \sin^2 x$)	$\frac{\cos^2 x}{\sin x}$	$\frac{\cos^2 x}{\sin x}$	(multiply)

Because we have reduced both sides to the same expression, we have proven that the equation is an identity. ■

NOW IT'S YOUR TURN

Prove that the equation $\sec x - \frac{1}{\sec x} = \tan x \sin x$ is an identity by transforming both sides into the same expression.

Example 17

Prove the identity

$$\frac{\sin^4 x - \cos^4 x}{\cos^2 x} = \tan^2 x - 1$$

Solution

We work with the left side because it is more complex. To factor the numerator, we can use the algebraic identity $a^2 - b^2 = (a + b)(a - b)$ where, in this case, $a = \sin^2 x$ and $b = \cos^2 x$.

$$\frac{\sin^4 x - \cos^4 x}{\cos^2 x} \stackrel{?}{=} \tan^2 x - 1$$

(use the algebraic identity to factor numerator)	$\frac{(\sin^2 x + \cos^2 x)(\sin^2 x - \cos^2 x)}{\cos^2 x}$
(use $\sin^2 x + \cos^2 x = 1$)	$\frac{(\sin^2 x - \cos^2 x)}{\cos^2 x}$
(divide each term in numerator by $\cos^2 x$)	$\frac{\sin^2 x}{\cos^2 x} - \frac{\cos^2 x}{\cos^2 x}$
$\left(\text{use } \tan x = \frac{\sin x}{\cos x}\right)$	$\tan^2 x - 1$

The two sides are now identical, so this equation is an identity. ■

NOW IT'S YOUR TURN

Prove that the equation $\sin^4 x + 2\sin^2 x \cos^2 x + \cos^4 x = 1$ is an identity.

Example 18

Prove that the following equation is an identity.

$$\frac{1}{\sec x + \tan x} = \frac{1 - \sin x}{\cos x}$$

Solution

We begin by multiplying the numerator *and* the denominator on the left-hand side by $(\sec x - \tan x)$.* This multiplication makes the denominator equal to $\sec^2 x - \tan^2 x$, which we recognize as being equal to 1 because $\sec^2 x = 1 + \tan^2 x$.

$$\frac{1}{\sec x + \tan x} \stackrel{?}{=} \frac{1 - \sin x}{\cos x}$$

(multiply numerator and denominator by $\sec x - \tan x$)
$$\frac{\sec x - \tan x}{\sec x - \tan x} \cdot \frac{1}{\sec x + \tan x}$$

(multiply denominators)
$$\frac{\sec x - \tan x}{\sec^2 x - \tan^2 x}$$

(use $\sec^2 x = 1 + \tan^2 x$)
$$\sec x - \tan x$$

$\left(\text{use } \sec x = \dfrac{1}{\cos x} \text{ and } \tan x = \dfrac{\sin x}{\cos x}\right)$
$$\frac{1}{\cos x} - \frac{\sin x}{\cos x}$$

(common denominator)
$$\frac{1 - \sin x}{\cos x}$$

The two sides are now identical, so this equation must be an identity. ■

---

* When we multiply by the fraction $(\sec x - \tan x)/(\sec x - \tan x)$, we *may* introduce new values for which x is not allowed. In such a case, this fraction is equal to 1 provided that $\sec x$ and $\tan x$ are *both* defined and also provided that the denominator is not 0. Both $\sec x$ and $\tan x$ are undefined only when $x = \pm\frac{\pi}{2}, \pm\frac{3\pi}{2}, \pm\frac{5\pi}{2}, \ldots$, so these are not allowed values. The denominator is 0 only when $\sec x = \tan x$, which also means that $\sec^2 x = \tan^2 x$. But the Pythagorean identity $\sec^2 x = 1 + \tan^2 x$ tells us that $\sec^2 x$ is always 1 unit greater than $\tan^2 x$, so the denominator in this fraction cannot ever be 0. Therefore, the only nonallowed values of x we introduce when we multiply by this fraction are $x = \pm\frac{\pi}{2}, \pm\frac{3\pi}{2}$, and so on. But these values were already nonallowed values of x because the original identity is not defined for these values of x. Therefore, we have introduced no *new* nonallowed values of x in this case. However, this will not always be true, so you must check each time.

NOW IT'S YOUR TURN

Prove that the following equation is an identity.

$$\frac{1}{\sin x - \cos x} = \frac{\sin x + \cos x}{\sin^2 x - \cos^2 x}$$

Example 19

Prove that the equation $\sin^6 x + \cos^6 x = 1 - 3\sin^2 x \cos^2 x$ is an identity. ***Hint:*** In "Now It's Your Turn" in Example 17 you proved that $\sin^4 x + 2\sin^2 x \cos^2 x + \cos^4 x = 1$.

Solution

From algebra we know that $a^3 + b^3 = (a + b)(a^2 - ab + b^2)$. Because $\sin^6 x = (\sin^2 x)^3$ and $\cos^6 x = (\cos^2 x)^3$, we can let $a = \sin^2 x$ and $b = \cos^2 x$ and then write $\sin^6 x + \cos^6 x = (\sin^2 x)^3 + (\cos^2 x)^3$. Next we can use the algebraic result to factor the trigonometric expression.

$$\sin^6 x + \cos^6 x \stackrel{?}{=} 1 - 3\sin^2 x \cos^2 x$$

	$(\sin^2 x)^3 + (\cos^2 x)^3$
(algebraic identity)	$(\sin^2 x + \cos^2 x)(\sin^4 x - \sin^2 x \cos^2 x + \cos^4 x)$
(use $\sin^2 x + \cos^2 x = 1$)	$\sin^4 x - \sin^2 x \cos^2 x + \cos^4 x$
(solve for $\sin^4 x + \cos^4 x$ in hint)	$1 - 2\sin^2 x \cos^2 x - \sin^2 x \cos^2 x$
(combine like terms)	$1 - 3\sin^2 x \cos^2 x$

The two sides are identical, so this equation is an identity. ■

NOW IT'S YOUR TURN

Prove that the equation $\sin^3 x + \cos^3 x = (\sin x + \cos x)(1 - \sin x \cos x)$ is an identity.

Remember that when proving identities, we can only manipulate one side at a time. We cannot use steps from algebra that involve both sides of the identity at once. Operations such as adding, subtracting, multiplying, and dividing both sides by the same expression, or squaring both sides, must be avoided. The following example illustrates what can happen if you are careless about this.

Example 20

An Example of an Incorrect Proof

Is the equation $\cos x = -\sqrt{1 - \sin^2 x}$ is an identity?

Solution

Starting with $\cos x = -\sqrt{1 - \sin^2 x}$, we first square both sides, giving

$$\cos^2 x = 1 - \sin^2 x$$

Adding $\sin^2 x$ to both sides gives

$$\sin^2 x + \cos^2 x = 1$$

This result is the basic Pythagorean identity, so

$$1 = 1$$

You might think that we have proven the identity. However, if we try $x = 0$, we have $\cos 0 = 1$ and $-\sqrt{1 - \sin^2 0} = -\sqrt{1 - 0} = -\sqrt{1} = -1$. But $1 \neq -1$, so $\cos 0 \neq -\sqrt{1 - \sin^2 0}$, which means this equation *cannot* be an identity. *Where did we go wrong?* When we squared both sides, we operated on *both* sides of the equation at the same time. In doing so, we lost the minus sign in front of the square root and therefore changed the original equation into a *new* equation ($\cos^2 x = 1 - \sin^2 x$). This *new* equation is an identity, but the *original* equation was not. ■

EXERCISES 6.3

In Exercises 1–56, prove that each equation is an identity, if it is an identity.

1. $1 - \sin x = \cos x$

2. $1 - \sin^2 x - \cos^2 x = 0$

3. $\sin x \cos x = \tan x$

4. $\sin x \cos x = \sec x$

5. $\tan x \csc x = \sec x$

6. $2 \sin x \cos x (\tan x + \cot x) = 2$

7. $\tan x \sec x (\csc x - \sin x) = 1$

8. $\cos^2 x - (1 - \sin x)(1 + \sin x) = \sin x$

9. $2 \cos^2 x - 1 = \cos^2 x - \sin^2 x$

10. $\sin x(1 - \sin x) - \cos^2 x + 1 = \sin x$

11. $(1 - \sin x)^2 = \cos^2 x$

12. $\cos^2 x - \sin^2 x = 1 - 2 \sin^2 x$

13. $\dfrac{\sin^3 x}{1 - \cos^2 x} = \sin x$

14. $\dfrac{\cos^2 x - 1}{\sin x} = \cos x$

15. $\dfrac{\sin^2 x}{1 - \cos x} = 1 + \cos x$

16. $\dfrac{\cos^2 x}{1 - \sin x} = 1 + \sin x$

17. $\sec^2 x - \csc^2 x = \tan^2 x - \cot^2 x$

18. $\dfrac{\csc x}{\sin x} - \dfrac{\sec x}{\cos x} = \cot^2 x - \tan^2 x$

19. $(\tan x - 1)^2 = \sec^2 x - 2 \tan x$

20. $(\sin x - \cos x)^2 = 1 - 2 \sin x \cos x$

21. $\sin x^2 + \cos x^2 = 1$

22. $\cot^2 x - \cos^2 x = \cot^2 x \cos^2 x$

23. $\sec^2 x - \sec^2 x \sin^2 x = 1$

24. $\sin^4 x - \cos^4 x = \sin^2 x - \cos^2 x$

25. $\csc^4 x - \cot^4 x = \csc^2 x + \cot^2 x$

26. $\dfrac{\cos x}{\sin x} + \dfrac{\csc x}{\sec x} = 2 \cot x$

27. $\dfrac{\csc x}{\cot x} - \dfrac{\cot x}{\csc x} = \tan x \sin x$

28. $\dfrac{\cot x}{\cos x} + \dfrac{\sec x}{\cot x} = \sec^2 x \csc x$

29. $\dfrac{\sec x}{\sin x} - \dfrac{2 \sin x}{\cos x} = \cot x - \tan x$

30. $\dfrac{1 - \cos x}{1 + \cos x} = (\cot x - \csc x)^2$

31. $\dfrac{1 + \cos x}{1 - \cos x} = \dfrac{(1 + \cos x)^2}{\sin^2 x}$

32. $\dfrac{1 + \tan x}{1 - \tan x} = \sec^2 x$

33. $\dfrac{1}{1 - \cos x} + \dfrac{1}{1 + \cos x} = 2 \csc^2 x$

34. $\dfrac{1 - \sin x}{1 + \sin x} - (1 + \sin x)(1 - \sin x) = -4 \sec x \tan x$

35. $\dfrac{\cos x}{\tan x + \sec x} - \dfrac{\cos x}{\tan x - \sec x} = 2$

36. $\dfrac{1}{\sec x + \tan x} = \sec x - \tan x$

37. $\dfrac{\cos x}{1 - \sin x} - \dfrac{1 - \sin x}{\cos x} = 2 \tan x$

38. $\dfrac{\sin x - \cos x}{\cos x} + \dfrac{\sin x + \cos x}{\sin x} = \csc x \sec x$

*39. $\dfrac{\sin^3 x + \cos^3 x}{\sin x + \cos x} = 1 - \sin x \cos x$

*40. $\dfrac{\tan^3 x + \cot^3 x}{\tan x + \cot x} = \tan^2 x + \cot^2 x - 1$

*41. $\dfrac{\sin^3 x + \csc^3 x}{\sin x + \csc x} = \csc^2 x - \cos^2 x$

*42. $\dfrac{\tan^3 x - \cot^3 x}{\tan x - \cot x} = \tan^2 x + \csc^2 x$

*43. $\sin^6 x + \cos^6 x = 1 - 3 \sin^2 x + 3 \sin^4 x$

*44. $\sec^6 x - \tan^6 x = 1 + 3 \sin^2 x \sec^4 x$

*45. $\dfrac{\cos^3 x - \sin^3 x}{\cos^2 x - \sin^2 x} = \dfrac{1 + \sin x \cos x}{\sin x + \cos x}$

*46. $\dfrac{\sin^3 x + \cos^3 x}{2 \sin^2 x - 1} = \dfrac{\sec x - \sin x}{\tan x - 1}$

**47. $\dfrac{1 + \sin x - \cos x}{1 + \sin x + \cos x} = \dfrac{\sin x}{1 + \cos x}$

**48. $\dfrac{\cos x + \sin x + 1}{\cos x - \sin x - 1} = \dfrac{1 + \cos x}{-\sin x}$

**49. $\dfrac{\tan x + \sec x + 1}{\tan x + \sec x - 1} = \dfrac{1 + \cos x}{\sin x}$

**50. $\dfrac{\csc x + 1 + \cot x}{\csc x + 1 - \cot x} = \dfrac{1 + \cos x}{\sin x}$

*51. $(\sin x \cos y + \cos x \sin y)^2 + (\cos x \cos y - \sin x \sin y)^2 = 1$

*52. $(\sin x \cos y - \cos x \sin y)^2 + (\cos x \cos y + \sin x \sin y)^2 = 1$

*53. $\dfrac{\sin x \cos y - \cos x \sin y}{\cos x \cos y + \sin x \sin y} = \dfrac{\tan x - \tan y}{1 + \tan x \tan y}$

*54. $\dfrac{\cos x \cos y - \sin x \sin y}{\sin x \cos y + \cos x \sin y} = \dfrac{\cot x \cot y - 1}{\cot x + \cot y}$

*55. $\sqrt{\dfrac{1 - \sin x}{1 + \sin x}} = \dfrac{1 - \sin x}{\cos x}$, for $0 < x < \frac{\pi}{2}$

*56. $\sqrt{\dfrac{1 - \sin x}{1 + \sin x}} = \sec x - \tan x$, for $0 < x < \frac{\pi}{2}$

*57. Show that if $\alpha + \beta = \frac{\pi}{2}$, then **(a)** $\cos^2 \alpha + \cos^2 \beta = 1$ and **(b)** $\tan \alpha \tan \beta = 1$.

*58. If $x = r \cos \theta$ and $y = r \sin \theta$, show that $x^2 + y^2 = r^2$.

*59. If $x = a \cos \theta$ and $y = b \sin \theta$, show that $\dfrac{x^2}{a^2} + \dfrac{y^2}{b^2} = 1$.

*60. If $x = a \sec \theta$ and $y = b \tan \theta$, show that $\dfrac{x^2}{a^2} - \dfrac{y^2}{b^2} = 1$.

*61. If $x = r \sin \theta \cos \varphi$, $y = r \sin \theta \sin \varphi$, and $z = r \cos \theta$, show that $x^2 + y^2 + z^2 = r^2$.

In Exercises 62–83, give a counter example to show that each equation is not an identity.

62. $\sin 4x = 4 \sin x$

63. $\sin(x + y) = \sin x + \sin y$

64. $\cos(x + y) = \cos x + \cos y$

65. $\tan(x + y) = \tan x + \tan y$

66. $\sin(x - y) = \sin x - \sin y$

67. $\cos(x - y) = \cos x - \cos y$

68. $\tan(x - y) = \tan x - \tan y$

69. $\sin(x/2) = \frac{1}{2} \sin x$

70. $\cos(x/2) = \frac{1}{2} \cos x$

71. $\sin(xy) = \sin x \sin y$

72. $\cos(xy) = \cos x \cos y$

73. $\cos(x + \pi) = \cos x$

74. $\sqrt{1 + \sin^2 x} = 1 + \sin x$

75. $\cos x^2 = \cos^2 x$

76. $(\sin x + \cos x)^2 = 1$

77. $\sin(x/y) = \dfrac{\sin x}{\sin y}$

78. $\sin^{-1} x = \dfrac{1}{\sin x}$

79. $\cos(x/y) = \dfrac{\cos x}{\cos y}$

80. $1 + \tan x^2 = \sec x^2$

81. $1 + \cot x^2 = \csc x^2$

*82. $\log(\cos x) = \cos(\log x)$

*83. $\sin(\cos x) = \sin x \cos x$

In Exercises 84–88, use identities to simplify the expression enough to solve for x. Limit your answers to $0 \le x \le 2\pi$.

84. $\cos^2 x + \tan x \sin^2 x = 1$

85. $\cos x \sin x + \dfrac{\cos^3 x}{\sin x} = 1$

86. $\sin x \tan^2 x - \sec^2 x \sin^2 x = 0.70$

87. $(1 - \cos x)(1 + \cos x) = 0.400$

*88. $\sec^2 x \sin^2 x - \tan^2 x \sin^2 x + \sec^2 x \cos^2 x - \tan^2 x \cos^2 x = 2.40 \cos x$

SUMMARY

1. **Conditional Equation.** An equation that is true only for certain values of the variable is a *conditional equation*.

2. **Identity.** An equation that is true for *all* values of the variable is an *identity*.

3. **Trigonometric Identity.** An equation involving the trigonometric functions that is true for *all allowed* values of the angle is a *trigonometric identity*.

4. **Proving Trigonometric Identities.** To prove a trigonometric identity, use trigonometric substitutions and algebraic operations to transform one or both sides of the equation until they are *exactly the same*. Work with only one side at a time.

IMPORTANT TERMS

Allowed angles	**Identity**	**Solving an equation**
Conditional equation	**Proving an identity**	**Trigonometric identity**

REVIEW EXERCISES

In Exercises 1–7, simplify each expression as much as possible.

1. $\dfrac{\sin^2 x}{\cos x - 1}$

2. $\dfrac{\tan x}{1 - \tan^2 x} \cdot \dfrac{\cot^2 x - 1}{\cot x}$

3. $1 + \cos^2 x - \sin^2 x$

4. $\sin^4 x + 2\sin^2 x \cos^2 x + \cos^4 x$

5. $\sin^4 x - \cos^4 x - \sin^2 x + \cos^2 x$

6. $(\tan x + \cot x)^2 - (\tan x - \cot x)^2$

7. $(\tan x + 1)^2 - (\tan x - 1)^2$

In Exercises 8–32, prove that each equation is an identity, if it is one.

8. $(1 + \tan^2 x)\cos^2 x = 1$

9. $\sin x \cos x = 1 + \cos x - \sin x$

10. $(1 - \cos^2 x)(1 + \tan^2 x) = \tan^2 x$

11. $\dfrac{\cos^2 x - 1}{\sin^2 x - 1} = \tan^2 x$

12. $\dfrac{1}{\sin^2 x} - \dfrac{1}{\tan^2 x} = 1$

13. $\dfrac{\sin x}{\cos x \tan^2 x} = \cot x$

14. $\dfrac{1 - 2\sin^2 x}{2\cos^2 x - 1} = 1$

15. $\dfrac{\sin^3 x - \cos^3 x}{1 + \sin x \cos x} = \sin x - \cos x$

16. $\dfrac{\sin x - \cos x}{\sin x + \cos x} = \dfrac{\tan x - 1}{\tan x + 1}$

17. $\dfrac{\sin^2 x - \cos^2 x}{\sin x \cos x} = \dfrac{\tan^2 x - 1}{\tan x}$

18. $\dfrac{\cos x + \tan x}{\cot x + \sec x} = \sin x$

19. $\dfrac{\cos x + \cot x}{\cot x} = 1 + \sin x$

20. $\dfrac{\sin x + \tan x}{\tan x} = 1 + \cos x$

21. $\dfrac{\tan x}{\tan x + \cot x} = \sin^2 x$

22. $\dfrac{\cot x}{\tan x + \cot x} = \cos^2 x$

23. $\sin^4 x + \sin^2 x \cos^2 x + \cos^2 x = 1$

24. $\sin^2 x + \sin^2 x \cos^2 x + \cos^4 x = 1$

25. $\dfrac{1 - \tan^2 x}{1 + \tan^2 x} = \cos^2 x - \sin^2 x$

26. $\dfrac{\sin^2 x}{\sec^2 x - 1} + \dfrac{\cos^2 x}{\csc^2 x - 1} = 1$

27. $\tan x + \cot x = \sec x \csc x$

28. $\csc x - \cos x \cot x = \sin x$

29. $\dfrac{\tan x + \sec x - 1}{\tan x - \sec x + 1} = (1 + \sin x) \sec x$

30. $\dfrac{\tan x + \tan y}{\cot x + \cot y} + \dfrac{1 - \tan x \tan y}{1 - \cot x \cot y} = 0$

31. $\dfrac{\sin x + \cos y}{\sin x - \cos y} - \dfrac{\sec y + \csc x}{\sec y - \csc x} = 0$

32. $\sin(\cos x) = \cos(\sin x)$

7

Trigonometric Functions of More Than One Angle

7.1 sin(A ± B), cos(A ± B), and tan(A ± B)

So far in our study of trigonometry, we have limited ourselves to trigonometric functions of a single angle. Yet consider the following cases.

1. A tree or building on a hill casts a longer shadow than one on level ground. If the hill rises above the horizontal at an angle α and the sun's rays are at an angle θ above the horizontal, the length of the shadow will depend on the angle $\theta - \alpha$, as you can see in Figure 7.1-1.

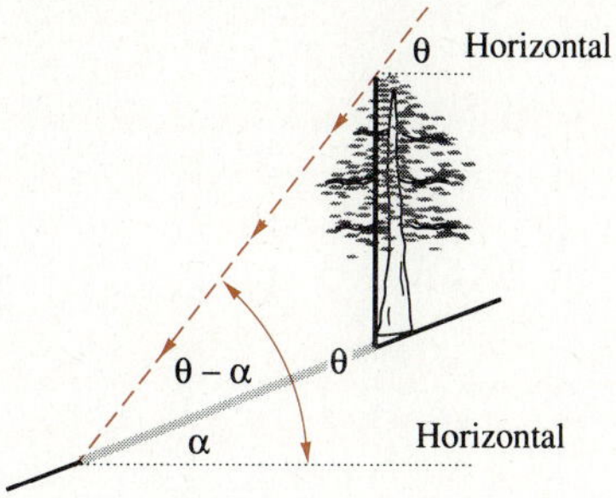

Figure 7.1-1

Shadow cast by a tree on a slanted hill The shadow's length depends on the angle $\theta - \alpha$.

2. If a plane flying at an angle θ above the horizontal launches a rocket at an angle α measured with respect to the plane's path, the path this rocket follows depends on the angle $\theta + \alpha$ (Figure 7.1-2).

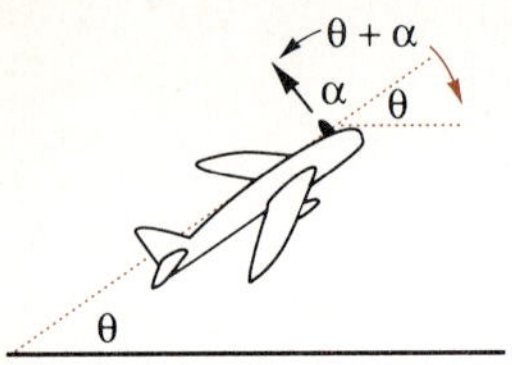

Figure 7.1-2

A plane flying at an angle θ above the horizontal launches a rocket at an angle α with respect to the plane's path. The path the rocket follows depends on the angle $\theta + \alpha$.

As you can see, the quantities we need to find often depend on the *sum* or *difference* of two angles. Therefore, we must extend our knowledge of trigonometry to cover the trigonometric functions of the sum and difference of two angles.

The obvious first question is: Is $\sin(A \pm B) = \sin A \pm \sin B$? What about $\cos(A \pm B)$ and $\tan(A \pm B)$? To find out, let us try some numerical values. First we look at the sine. Let $A = 60°$ and $B = 30°$. Then

$$\sin(60° + 30°) = \sin 90° = 1$$

$$\sin 60° + \sin 30° = \frac{\sqrt{3}}{2} + \tfrac{1}{2} = \tfrac{1}{2}(\sqrt{3} + 1)$$

Obviously, $\sin(60° + 30°)$ is not equal to $\sin 60° + \sin 30°$, so $\sin(A + B)$ *cannot* be identical to $\sin A + \sin B$.

NOW IT'S YOUR TURN

Try numerical values to show that $\cos(A + B)$ is not identical to $\cos A + \cos B$ and $\tan(A \pm B)$ is not equal to $\tan A \pm \tan B$.

We can see now that we must derive formulas to find $\sin(A \pm B)$, $\cos(A \pm B)$ and $\tan(A \pm B)$. We shall begin with $\cos(A \pm B)$.

7.1-1 cos(*A* + *B*) and cos(*A* − *B*)

We start by drawing a circle of radius R centered at the origin, showing two arbitrary adjacent central angles A and B (see Figure 7.1-3). We can use the law of cosines to find the length of the line segment $\overline{P_1P_2}$ in terms of $A + B$.

$$\overline{P_1P_2}^2 = R^2 + R^2 - 2\,RR\cos(A + B)$$

$$\overline{P_1P_2}^2 = 2R^2[1 - \cos(A + B)]$$

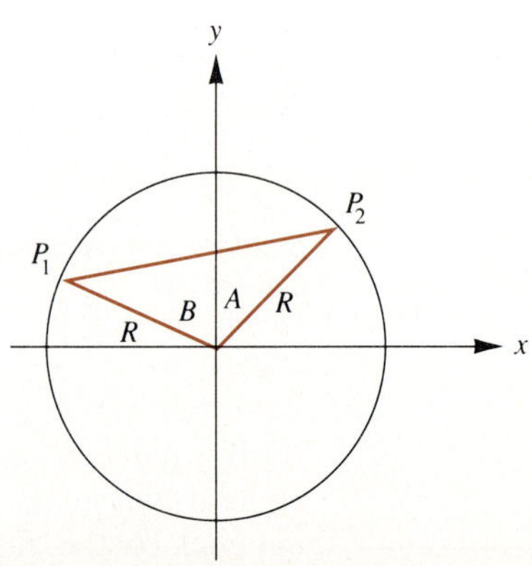

Figure 7.1-3

Two adjacent central angles *A* and *B* of a circle of radius *R* centered at the origin

Now we want to find the length of $\overline{P_1P_2}$ in terms of the individual angles A and B. To do this, first construct a right angle $P_1P_2P_3$ with legs parallel to the x and y axes and with $\overline{P_1P_2}$ as its hypotenuse (see Figure 7.1-4). The Pythagorean theorem tells us that

$$\overline{P_1P_2^2} = \overline{P_1P_3^2} + \overline{P_2P_3^2}$$

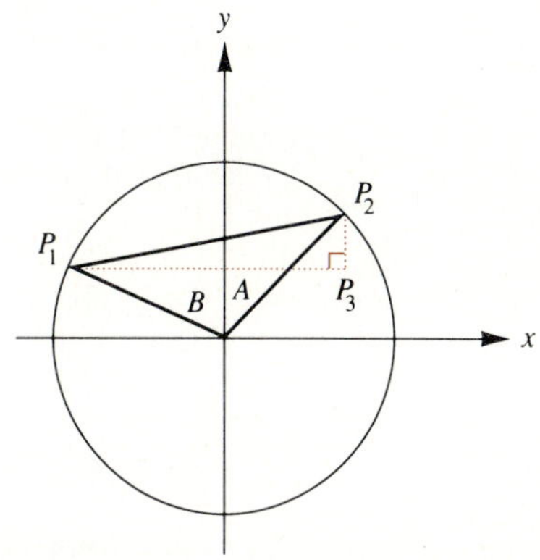

Figure 7.1-4

P_1P_2 is the hypotenuse of the right triangle $P_1P_2P_3$ having legs parallel to the x and y axes.

Now we can use trigonometry on this figure to find the lengths of the legs of this right triangle. First we find the distance of points P_1 and P_2 from the x and y axes (see Figure 7.1-5).

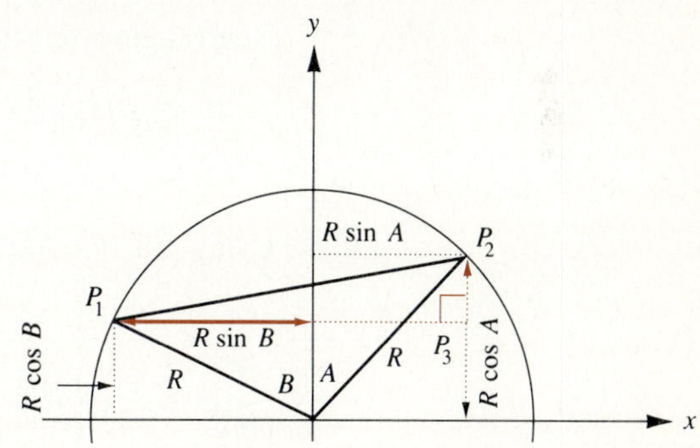

Figure 7.1-5

The distances of P_1 and P_2 from the x and y axes

We can now combine the results of Figures 7.1-4 and 7.1-5 to get the sides of the right triangle $P_1P_2P_3$ (see Figure 7.1-6). From this figure we can see that

$$\overline{P_1P_3} = R \sin A + R \sin B = R(\sin A + \sin B)$$

$$\overline{P_2P_3} = R \cos A - R \cos B = R(\cos A - \cos B)$$

In Figure 7.1-6, A and B are acute angles, but that does not always have to be true. To keep the distances $\overline{P_1P_3}$ and $\overline{P_2P_3}$ from coming out negative, we must use the absolute values in the last two equations.

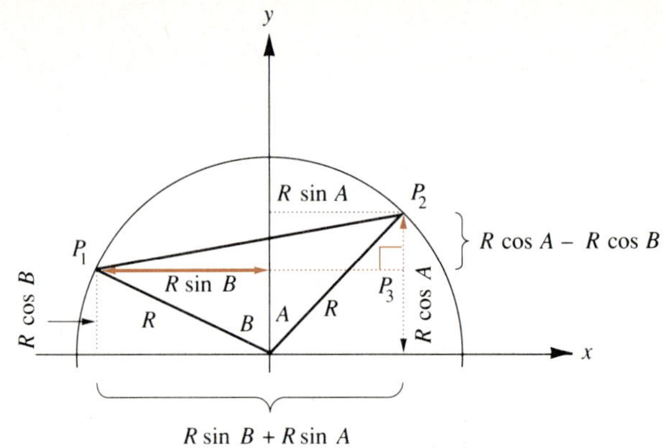

Figure 7.1-6

Combine the results of Figures 7.1-4 and 7.1-5 to get the lengths of the sides P_1P_3 and P_2P_3.

$$\overline{P_1P_3} = |R \sin A + R \sin B| = |R(\sin A + \sin B)|$$
$$\overline{P_2P_3} = |R \cos A - R \cos B| = |R(\cos A - \cos B)|$$

Now apply the Pythagorean theorem to the triangle $P_1P_2P_3$.

$$\overline{P_1P_2}^2 = \overline{P_1P_3}^2 + \overline{P_2P_3}^2$$
$$\overline{P_1P_2}^2 = |R(\sin A + \sin B)|^2 + |R(\cos A - \cos B)|^2$$
$$= R^2(\sin^2 A + 2 \sin A \sin B + \sin^2 B) + R^2(\cos^2 A - 2 \cos A \cos B + \cos^2 B)$$

Rearranging terms,

$$\overline{P_1P_2}^2 = R^2[(\sin^2 A + \cos^2 A) + (\sin^2 B + \cos^2 B) + 2 \sin A \sin B - 2 \cos A \cos B]$$

Using $\sin^2 \theta + \cos^2 \theta = 1$, this equation simplifies to

$$\overline{P_1P_2}^2 = R^2[1 + 1 + 2 \sin A \sin B - 2 \cos A \cos B]$$
$$\overline{P_1P_2}^2 = 2R^2(1 + \sin A \sin B - \cos A \cos B)$$

We have now found two different expressions for $\overline{P_1P_2}^2$, the first involving $\cos(A + B)$ and the second involving sines and cosines of A and B separately.

$$\overline{P_1P_2}^2 = 2R^2[1 - \cos(A + B)]$$
$$\overline{P_1P_2}^2 = 2R^2(1 + \sin A \sin B - \cos A \cos B)$$

Because both expressions are for the same line segment $\overline{P_1P_2}$, they must be equal to each other, so we can equate them.

$$2R^2[1 - \cos(A + B)] = 2R^2(1 + \sin A \sin B - \cos A \cos B)$$

Dividing by $2R^2$, subtracting 1 from both sides of the equation, and solving for $\cos(A + B)$, we get

$$\boxed{\cos(A + B) = \cos A \cos B - \sin A \sin B} \qquad (1)$$

This proof does not depend on the magnitudes or signs of A and B, so eq. (1) holds for all values of A and B, both positive and negative. Therefore, this equation is an identity.

It took a lot of work to prove this result, but all the other equations follow from it with much less effort. We can extend this equation to $\cos(A - B)$ by using the fact that $A - B = A + (-B)$, as follows.

$$\begin{aligned} \cos(A - B) &= \cos[A + (-B)] \\ &= \cos A \cos(-B) - \sin A \sin(-B) && \text{(eq. (1))} \\ &= \cos A \cos B - \sin A(-\sin B) && \text{(because } \cos(-B) = \cos B \text{ and } \sin(-B) = -\sin B) \\ &= \cos A \cos B + \sin A \sin B \end{aligned}$$

$$\boxed{\cos(A - B) = \cos A \cos B + \sin A \sin B} \qquad (2)$$

Example 1

Using only trigonometric functions of 30° and 45°, evaluate (a) cos 75° and (b) cos 15°.

Solution

(a) We can express 75° as the sum of these two angles (75° = 30° + 45°) and then apply eq. (1) for $\cos(A + B)$. Because we know the trigonometric functions of the special angles, we can find cos 75°.

$$\begin{aligned} \cos 75^\circ &= \cos(30^\circ + 45^\circ) \\ &= \cos 30^\circ \cos 45^\circ - \sin 30^\circ \sin 45^\circ && \text{(eq. (1))} \\ &= \left(\frac{\sqrt{3}}{2}\right)\left(\frac{\sqrt{2}}{2}\right) - \left(\frac{1}{2}\right)\left(\frac{\sqrt{2}}{2}\right) \\ &= (\sqrt{6} - \sqrt{2})/4 \end{aligned}$$

(b) We can express 15° as the difference of the two angles (15° = 45° − 30°) and then apply eq. (2).

$$\begin{aligned} \cos 15^\circ &= \cos(45^\circ - 30^\circ) \\ &= \cos 45^\circ \cos 30^\circ + \sin 45^\circ \sin 30^\circ && \text{(eq. (2))} \end{aligned}$$

$$= \left(\frac{\sqrt{2}}{2}\right)\left(\frac{\sqrt{3}}{2}\right) + \left(\frac{\sqrt{2}}{2}\right)\left(\frac{1}{2}\right)$$

$$= \sqrt{6}/4 + \sqrt{2}/4 = (\sqrt{6} + \sqrt{2})/4 \quad ■$$

NOW IT'S YOUR TURN

Use only trigonometric functions of 45° and 60°, to find cos 105°.
(Answer: $(\sqrt{2} - \sqrt{6})/4$)

Example 2 Use the identities for $\cos(A \pm B)$ to show that (a) $\cos(90° - x) = \sin x$ and (b) $\sin(90° - x) = \cos x$.

Solution

(a) $\cos(90° - x) = \sin x$:

$$\cos(90° - x) = \cos 90° \cos x + \sin 90° \sin x$$

$$= (0)\cos x + (1)\sin x = \sin x \quad \textbf{(eq. (2))}$$

(b) $\sin(90° - x) = \cos x$:
We can use the result from part (a), $\sin x = \cos(90° - x)$. Because this result is true for any value of x, it must also be true if we replace x by $(90° - x)$. All we have to do is write $(90° - x)$ wherever there was an x in the result from (a).

$$\sin \textcircled{x} = \cos(90° - \textcircled{x})$$

replace x by $(90° - x)$

which gives us

$$\sin(90° - x) = \cos[90° - (90° - x)]$$

Simplifying gives us

$$\sin(90° - x) = \cos[90° - 90° + x] = \cos x$$

Notice that this example illustrates the cofunction theorem that we introduced in Section 2.6. ■

NOW IT'S YOUR TURN

Show that $\cos(90° + x) = -\sin x$. Then use this result to show that $\sin(x - 90°) = -\cos x$.

7.1-2 sin(*A* + *B*) and sin(*A* − *B*)

We can now use the results of Example 2 and the eqs. (1) and (2) for $\cos(A \pm B)$ to derive formulas for $\sin(A + B)$ and $\sin(A - B)$.

From Example 2*a* we know that $\sin x = \cos(90° - x)$. We can let $x = A + B$ and apply this formula.

$$\sin(A + B) = \cos[90° - (A + B)]$$

$$= \cos[(90° - A) - B] \qquad \textbf{(rearrange)}$$

$$= \cos(90° - A)\cos B + \sin(90° - A)\sin B \qquad \textbf{(eq. (2))}$$

$$\sin(A + B) = \sin A \cos B + \cos A \sin B \qquad \textbf{(use Examples 2a and 2b)}$$

$$\boxed{\sin(A + B) = \sin A \cos B + \cos A \sin B} \qquad (3)$$

NOW IT'S YOUR TURN

Extend eq. formula (3) to $\sin(A - B)$. (See the answer in the following box.) ***Hint:*** Use the same approach we used to extend eq. (1) to eq. (2) in Section 7.1-1.

$$\boxed{\sin(A - B) = \sin A \cos B - \cos A \sin B} \qquad (4)$$

Example 3

If α and β are quadrant I angles for which $\cos \alpha = \frac{1}{3}$ and $\cos \beta = \frac{1}{2}$, find $\sin(\alpha - \beta)$ without finding α or β.

Solution

Because we know (or can find) the trigonometric functions of α and of β, we can use eq. (4) to express $\sin(\alpha - \beta)$ in terms of these functions.

$$\sin(\alpha - \beta) = \sin \alpha \cos \beta - \cos \alpha \sin \beta$$

We can see that we need $\sin \alpha$ and $\sin \beta$. To find them, first sketch the reference triangles for α and β and use these to find $\sin \alpha$ and $\sin \beta$ (see Figure 7.1-7). We have used the Pythagorean theorem to find the third side of each triangle. Figure 7.1-7a tells us that $\sin \alpha = \sqrt{8}/3$, and Figure 7.1-7b tells us that $\sin \beta = \sqrt{3}/2$. Therefore,

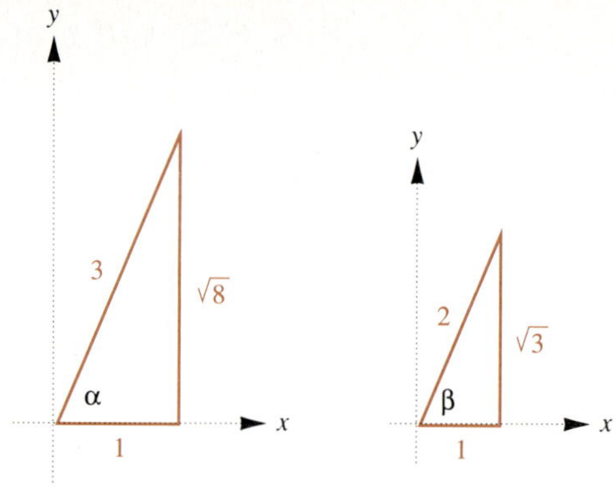

Figure 7.1-7
Reference triangles for α and β

(a) Reference triangle for α. We know that $\cos\alpha = \frac{1}{3}$.

(b) Reference triangle for β. We know that $\cos\beta = \frac{1}{2}$.

$$\sin(\alpha - \beta) = \sin\alpha\cos\beta - \cos\alpha\sin\beta$$

$$= \left(\frac{\sqrt{8}}{3}\right)\left(\frac{1}{2}\right) - \left(\frac{1}{3}\right)\left(\frac{\sqrt{3}}{2}\right) = (\sqrt{8} - \sqrt{3})/6 \quad ■$$

NOW IT'S YOUR TURN

If θ and ϕ are quadrant I angles for which $\sin\theta = \frac{3}{4}$ and $\sin\phi = \frac{1}{3}$, find $\sin(\theta + \phi)$ without finding θ or ϕ. (Answer: $(\sqrt{7} + 3\sqrt{8})/12$)

7.1-3 tan(*A* + *B*) and tan(*A* − *B*)

To derive identities for $\tan(A \pm B)$, we use the identity $\tan x = \sin x/\cos x$ and substitute $A \pm B$ for x.

$$\tan(A + B) = \frac{\sin(A + B)}{\cos(A + B)}$$

$$= \frac{\sin A\cos B + \cos A\sin B}{\cos A\cos B - \sin A\sin B} \quad \text{(expand numerator and denominator)}$$

$$= \frac{\dfrac{\sin A\cos B}{\cos A\cos B} + \dfrac{\cos A\sin B}{\cos A\cos B}}{\dfrac{\cos A\cos B}{\cos A\cos B} - \dfrac{\sin A\sin B}{\cos A\cos B}} \quad \text{(divide numerator and denominator by } \cos A\cos B\text{)}$$

$$= \frac{\dfrac{\sin A}{\cos A} + \dfrac{\sin B}{\cos B}}{1 - \dfrac{\sin A}{\cos A}\dfrac{\sin B}{\cos B}} \qquad \text{(simplify)}$$

$$= \frac{\tan A + \tan B}{1 - \tan A \tan B} \qquad (\text{use } \tan\theta = \sin\theta/\cos\theta)$$

$$\boxed{\tan(A + B) = \frac{\tan A + \tan B}{1 - \tan A \tan B}} \qquad (5)$$

NOW IT'S YOUR TURN

Derive a similar identity for $\tan(A - B)$ using the same method we used to extend eq. (1) to eq. (2) and eq. (3) to eq. (4). (Answer: See the following box.)

$$\boxed{\tan(A - B) = \frac{\tan A - \tan B}{1 + \tan A \tan B}} \qquad (6)$$

Example 4

If α is a quadrant II angle for which $\sin \alpha = \frac{3}{4}$ and β is a quadrant I angle for which $\cos \beta = \frac{1}{3}$, find $\tan(\alpha + \beta)$ to four decimal places without finding α or β.

Solution

We can use Eq. 5 for $\tan(A + B)$, where $A = \alpha$ and $B = \beta$. We need $\tan \alpha$ and $\tan \beta$. To find them, first construct the reference triangles for α and β (see Figure 7.1-8), as we did in Example 3. Because α is in quadrant II, $\tan \alpha$ is negative and is equal to

$$\tan \alpha = \frac{3}{(-\sqrt{7})} = \frac{-3}{(2.6458)} = -1.1339$$

Because β is in quadrant I, $\tan \beta$ is positive and equal to

$$\tan \beta = \sqrt{8}/1 = \sqrt{8} = 2.8284$$

Eq. (5) for $\tan(A + B)$ gives:

$$\tan(\alpha + \beta) = \frac{\tan \alpha + \tan \beta}{1 - \tan \alpha \tan \beta} = \frac{-1.1339 + 2.8284}{1 - (-1.1339)(2.8284)}$$

$$= \frac{1.6945}{4.2071} = 0.4028 \quad ■$$

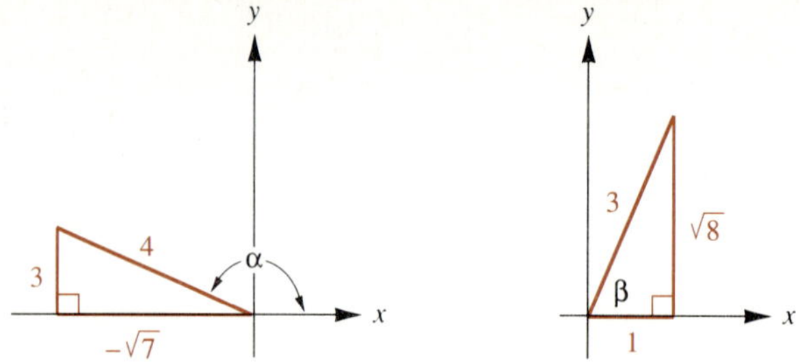

Figure 7.1-8
Reference triangles for α and β

(a) Reference triangle for α.   *(b) Reference triangle for β.*

NOW IT'S YOUR TURN

If α is a quadrant I angle for which $\cos \alpha = \frac{2}{3}$ and β is a quadrant II angle for which $\sin \beta = \frac{1}{4}$, find $\tan(\alpha + \beta)$ to four decimal places without finding α or β.
(Answer: 0.6672)

7.1-4 Some Additional Examples

Example 5

If angle C in the triangle ABC is the right angle, show that $\sin(A - B) = (a^2 - b^2)/(a^2 + b^2)$.

Solution

We begin by drawing the triangle ABC (see Figure 7.1-9). From this figure, we see that $\sin A = a/c$, $\cos A = b/c$, $\sin B = b/c$, $\cos B = a/c$, and the Pythagorean theorem tells us that $c^2 = a^2 + b^2$. Therefore,

$$\begin{aligned}\sin(A - B) &= \sin A \cos B - \cos A \sin B \\ &= \left(\frac{a}{c}\right)\left(\frac{a}{c}\right) - \left(\frac{b}{c}\right)\left(\frac{b}{c}\right) \\ &= \frac{a^2}{c^2} - \frac{b^2}{c^2} = \frac{a^2 - b^2}{c^2} = \frac{a^2 - b^2}{a^2 + b^2} \quad ■\end{aligned}$$

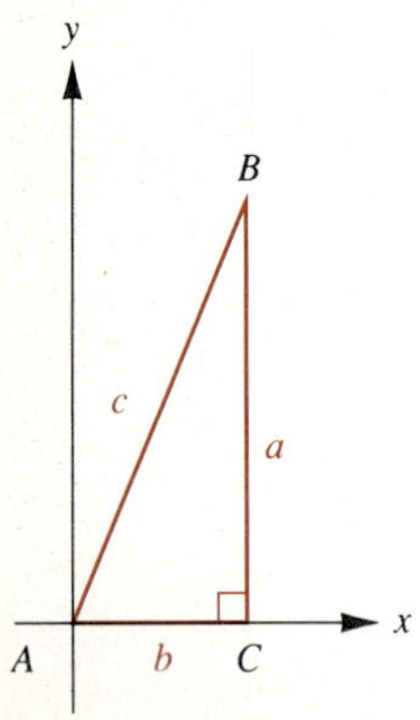

Figure 7.1-9
Triangle *ABC*

NOW IT'S YOUR TURN

For the triangle in this example, show that $\cos(A - B) = 2ab/c^2$.

Example 6

Express $\cos 8x \cos 5x + \sin 8x \sin 5x$ as a single function of one angle. (This procedure is especially helpful for solving trigonometric equations.)

Solution

Using the identity $\cos(A - B) = \cos A \cos B + \sin A \sin B$ with $A = 8x$ and $B = 5x$, we get

$$\cos 8x \cos 5x + \sin 8x \sin 5x = \cos(8x - 5x) = \cos 3x \quad ■$$

NOW IT'S YOUR TURN

Express $\sin 2x \cos 7x + \cos 2x \sin 7x$ as a single function of one angle. (Answer: $\sin 9x$)

Example 7

Find all values of x between 0° and 360° that satisfy the equation $\sin 3x \cos x - \cos 3x \sin x = \frac{1}{2}$.

Solution

First use the method of Example 6 to reduce this equation to one trigonometric function of a single angle. We have $\sin(A - B) = \sin A \cos B - \cos A \sin B$, where $A = 3x$ and $B = x$. This gives us

$$\sin 3x \cos x - \cos 3x \sin x = \sin(3x - x) = \sin 2x$$

We have reduced this equation to $\sin 2x = \frac{1}{2}$, which tells us that $2x = \arcsin \frac{1}{2}$. Because x is between 0° and 360°, $2x$ can be between 0° and 720°. The first two solutions are $2x = 30°$ and $2x = 150°$. To find the solutions greater than 360°, we just add 360° to the first two solutions. Thus

$$2x = \arcsin \tfrac{1}{2} = \begin{cases} 30° \\ 150° \\ 30° + 360° = 390° \\ 150° + 360° = 510° \end{cases}$$

Dividing by 2 to solve for x gives us

$$x = \tfrac{1}{2}(30°) = 15°$$
$$x = \tfrac{1}{2}(150°) = 75°$$
$$x = \tfrac{1}{2}(390°) = 195°$$
$$x = \tfrac{1}{2}(510°) = 255° \quad ■$$

NOW IT'S YOUR TURN

Find all values of x between 0° and 360° which satisfy the equation $\cos 4x \cos 2x + \sin 4x \sin 2x = \sqrt{2}/2$. (Answer: 22.5°, 157.5°, 202.5°, and 337.5°)

Example 8

A tree on a hill casts a shadow that is 90.0% as long as the tree's height. If this hill slopes upward at 24.0° above the horizontal, find the angle of the sun above the horizontal.

Solution

First make a sketch of the tree and its shadow on the hill. Call θ the sun's angle above the horizontal, h the height of the tree, and L the length of the tree's shadow (see Figure 7.1-10).

First we must realize that the tree will grow vertically, which is perpendicular to horizontal. Then we can modify the previous figure, showing the relationships between the angles (see Figure 7.1-11).

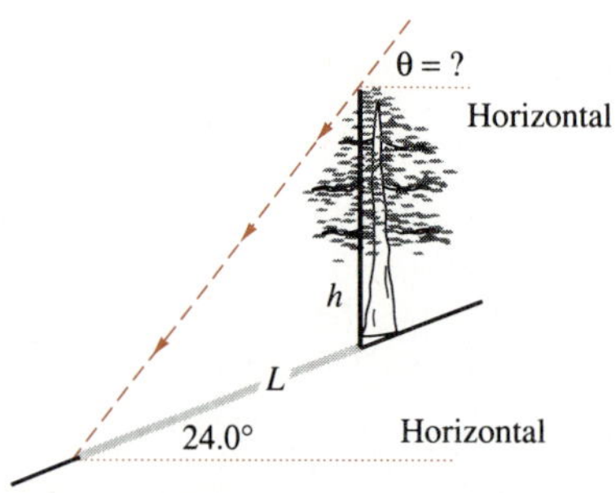

Figure 7.1-10

A tree casts a shadow on a hill's sloping surface.

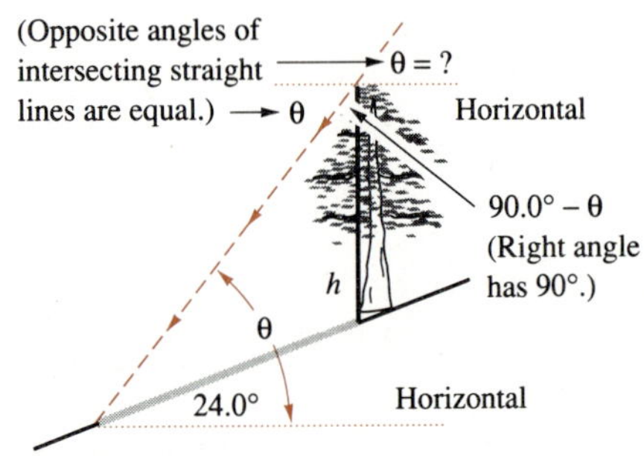

Figure 7.1-11

Relationships between angles

Next we simplify this figure, showing the quantities necessary to find θ (see Figure 7.1-12). Applying the law of sines to the oblique triangle in Figure 7.1-12 gives us

$$\frac{\sin(\theta - 24°)}{h} = \frac{\sin(90° - \theta)}{L}$$

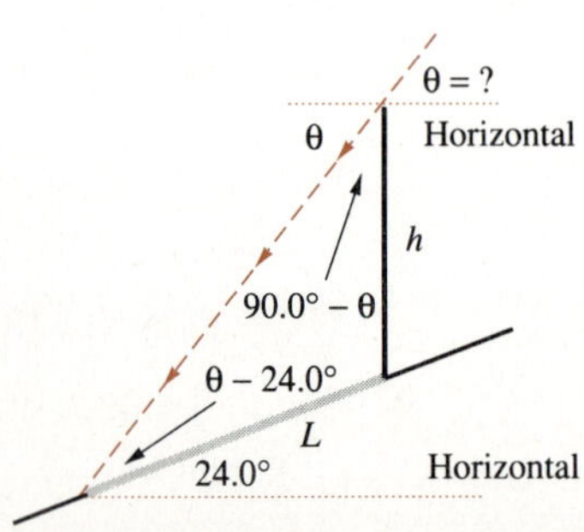

Figure 7.1-12

Simplification of Figure 7.1-11

To solve for θ, we must be able to isolate a single trigonometric function of θ. We begin by expanding the numerators using the formula for $\sin(A - B)$.

$$\frac{\sin \theta \cos 24° - \cos \theta \sin 24°}{h} = \frac{\sin 90° \cos \theta - \cos 90° \sin \theta}{L}$$

Substituting numerical values gives us

$$\frac{0.9135 \sin \theta - 0.4067 \cos \theta}{h} = \frac{\cos \theta}{L}$$

Because the length of the shadow is 90.0% of the tree's height,

$$L = 90.0\% \text{ of } h = 0.900h.$$

Putting this into the last equation gives us

$$\frac{0.9135 \sin \theta - 0.4067 \cos \theta}{h} = \frac{\cos \theta}{0.900h}$$

Multiplying both sides by $0.900h$ gives us

$$\begin{aligned}(0.900)(0.9135 \sin \theta - 0.4067 \cos \theta) &= \cos \theta \\ 0.8222 \sin \theta - 0.3660 \cos \theta &= \cos \theta \\ 0.8222 \sin \theta &= 1.3660 \cos \theta \\ \sin \theta = \frac{1.3660 \cos \theta}{0.8222} &= 1.6614 \cos \theta\end{aligned}$$

Dividing by $\cos \theta$ and using $\dfrac{\sin \theta}{\cos \theta} = \tan \theta$ gives us

$$\begin{aligned}\tan \theta &= 1.6614 \\ \theta &= \tan^{-1} 1.6614 = 59.0° \ ■\end{aligned}$$

NOW IT'S YOUR TURN

If the hill slopes up at a 30.0° angle, and if the length of the shadow is 1.10 times the tree's height, what is the sun's angle above the horizontal?
(Answer: 58.4°)

We close this section with a summary of the six major identities. You can reduce these to only three formulas if you just remember the formulas for $A + B$. You can then use these formulas to derive the formulas for $A - B$ by replacing B by $-B$, which is just how we derived them.

$$\sin(A + B) = \sin A \cos B + \cos A \sin B$$
$$\sin(A - B) = \sin A \cos B - \cos A \sin B$$
$$\cos(A + B) = \cos A \cos B - \sin A \sin B$$
$$\cos(A - B) = \cos A \cos B + \sin A \sin B$$
$$\tan(A + B) = \frac{\tan A + \tan B}{1 - \tan A \tan B}$$
$$\tan(A - B) = \frac{\tan A - \tan B}{1 + \tan A \tan B}$$

EXERCISES 7.1-1

In Exercises 1–10, find the cosine of each of the following angles without using a calculator or a table.

1. $-15°$ **2.** $165°$ **3.** $-75°$
4. $-105°$ **5.** $195°$ **6.** $405°$
7. $315°$ **8.** $\frac{\pi}{12}$ **9.** $\frac{7\pi}{12}$
10. $-345°$

In Exercises 11–16, evaluate the indicated functions without finding x and y.

11. Both x and y are quadrant I angles for which $\sin x = \frac{3}{5}$, $\cos x = \frac{4}{5}$, $\sin y = \frac{5}{13}$, and $\cos y = \frac{12}{13}$. Find

(a) $\cos(x + y)$ **(b)** $\cos(x - y)$

***12.** Both x and y are quadrant I angles for which $\sin x = \frac{1}{3}$ and $\cos y = \frac{3}{4}$. Find

(a) $\cos(x + y)$ **(b)** $\cos(x - y)$

***13.** Both x and y are quadrant I angles for which $\cos x = 0.581$ and $\sin y = 0.213$. Find

(a) $\cos(x + y)$ **(b)** $\cos(x - y)$

***14.** If x is a quadrant I angle for which $\sin x = \frac{2}{3}$ and y is a quadrant II angle for which $\sin y = \frac{1}{4}$, find

(a) $\cos(x + y)$ **(b)** $\cos(x - y)$

***15.** If x is a quadrant III angle for which $\cos x = -0.525$ and y is a quadrant IV angle for which $\cos y = 0.125$, find

(a) $\cos(x + y)$ **(b)** $\cos(x - y)$

***16.** If x is a quadrant IV angle for which $\tan x = -\frac{2}{3}$ and y is a quadrant II angle for which $\cot y = -\frac{4}{5}$, find

(a) $\cos(x + y)$ **(b)** $\cos(x - y)$

EXERCISES 7.1-2

In Exercises 1–10, find the sine of each of the following angles without using a calculator or a table.

1. $-15°$ **2.** $165°$ **3.** $-75°$
4. $-105°$ **5.** $195°$ **6.** $405°$
7. $315°$ **8.** $\frac{\pi}{12}$ **9.** $\frac{7\pi}{12}$
10. $-345°$

In Exercises 11–16, evaluate the indicated functions without finding x and y.

11. Both x and y are quadrant I angles for which $\cos x = \frac{3}{5}$, $\sin x = \frac{4}{5}$, $\cos y = \frac{5}{13}$, and $\sin y = \frac{12}{13}$. Find

(a) $\sin(x + y)$ **(b)** $\sin(x - y)$

***12.** Both x and y are quadrant I angles for which $\sin x = \frac{2}{5}$ and $\cos y = \frac{3}{4}$. Find

(a) $\sin(x + y)$ **(b)** $\sin(x - y)$

***13.** Both x and y are quadrant I angles for which $\cos x = 0.172$ and $\sin y = 0.608$. Find

(a) $\sin(x + y)$ **(b)** $\sin(x - y)$

***14.** If x is a quadrant II angle for which $\sin x = \frac{1}{2}$ and y is a quadrant IV angle for which $\cos y = \frac{2}{3}$, find

(a) $\sin(x + y)$ **(b)** $\sin(x - y)$

***15.** If x is a quadrant III angle for which $\cos x = -0.666$ and y is a quadrant I angle for which $\sin y = 0.425$, find

(a) $\sin(x + y)$ **(b)** $\sin(x - y)$

***16.** If x is a quadrant III angle for which $\sin x = -\frac{1}{5}$ and y is a quadrant II angle for which $\tan y = -\frac{2}{3}$, find

(a) $\sin(x + y)$ **(b)** $\sin(x - y)$

EXERCISES 7.1-3

In Exercises 1–10, find the tangent of each of the following angles without using a calculator or a table.

1. $-15°$ **2.** $165°$ **3.** $-75°$

4. $-105°$ **5.** $195°$ **6.** $405°$

7. $315°$ **8.** $\frac{\pi}{12}$ **9.** $\frac{7\pi}{12}$

10. $-345°$

In Exercises 11–16, evaluate the indicated functions without finding x or y.

11. Both x and y are quadrant I angles for which $\sin x = \frac{3}{5}$, $\cos x = \frac{4}{5}$, $\sin y = \frac{5}{13}$, and $\cos y = \frac{12}{13}$. Find

(a) $\tan(x + y)$ **(b)** $\tan(x - y)$

12. Both x and y are quadrant I angles for which $\tan x = \frac{2}{5}$ and $\tan y = \frac{4}{7}$. Find

(a) $\tan(x + y)$ **(b)** $\tan(x - y)$

13. Both x and y are quadrant I angles for which $\cos x = 0.124$ and $\sin y = 0.657$. Find

(a) $\tan(x + y)$ **(b)** $\tan(x - y)$

14. If x is a quadrant I angle for which $\tan x = \frac{2}{3}$ and y is a quadrant II angle for which $\sin y = \frac{1}{4}$, find

(a) $\tan(x + y)$ **(b)** $\tan(x - y)$

***15.** If x is a quadrant II angle for which $\cos x = -0.885$ and y is a quadrant IV angle for which $\cos y = 0.772$, find

(a) $\tan(x + y)$ **(b)** $\tan(x - y)$

***16.** If x is a quadrant IV angle for which $\sec x = \frac{3}{2}$ and y is a quadrant II angle for which $\csc y = \frac{5}{4}$, find

(a) $\tan(x + y)$ **(b)** $\tan(x - y)$

EXERCISES 7.1-4

In Exercises 1–6, evaluate $\sin(A + B)$, $\sin(A - B)$, $\cos(A + B)$, and $\cos(A - B)$ without finding A or B.

1. A is a quadrant I angle with $\sin A = \frac{12}{13}$, and B is a quadrant II angle with $\cos B = -\frac{4}{5}$.

2. A and B are quadrant I angles with $\sin A = \frac{7}{25}$ and $\sin B = \frac{15}{17}$.

3. A and B are both acute, $\tan A = \frac{1}{2}$, and $\cot B = 3$.

4. A and B are in the same quadrant with $\sin A = \frac{3}{5}$ and $\cos B = -\frac{12}{13}$.

5. A and B are in the same quadrant with $\sin A = \frac{2}{5}$ and $\cos B = \frac{3}{5}$.

***6.** $\cos B > 0$, $\csc A = -1$, and $\csc B = -2$.

***7.** Find $\sin x$, $\cos x$, and $\tan x$ if x and y are positive acute angles satisfying the conditions $\cos(x + y) = \frac{3}{5}$ and $\sin y = \frac{1}{5}$.

***8.** Find $\sin x$, $\cos x$, and $\tan x$ if x and y are positive acute angles satisfying the conditions $\cos(x - y) = \frac{4}{5}$ and $\sin y = \frac{1}{5}$.

In Exercises 9–14, use the formulas of this section to express each function in terms functions of x alone. (Example: $\sin(x + \frac{\pi}{2}) = \sin x \cos \frac{\pi}{2} + \cos x \sin \frac{\pi}{2} = (0) \sin x + (1) \cos x = \cos x$.)

9. $\sin(x + \pi)$ **10.** $\cos(x - \pi)$ **11.** $\tan(x + \frac{\pi}{2})$

12. $\sin(x - \frac{\pi}{2})$ **13.** $\cos(\frac{\pi}{2} - x)$ **14.** $\sin(\pi - x)$

15. Use the formulas of this section to prove that: **(a)** $\sin \theta$ has period 2π. *Hint:* If you expand $\sin(\theta + 2\pi)$, what do you get? **(b)** $\cos \theta$ has period 360°. **(c)** $\tan \theta$ has period π.

In Exercises 16–36, express each quantity as a single function of one angle.

16. $\sin 30° \cos 20° + \cos 30° \sin 20°$

17. $\sin 50° \cos 30° - \cos 50° \sin 30°$

18. $\cos \frac{\pi}{4} \cos \frac{3\pi}{4} - \sin \frac{\pi}{4} \sin \frac{3\pi}{4}$

19. $\cos 40° \cos 21° + \sin 40° \sin 21°$

20. $\dfrac{\tan 23° + \tan 22°}{1 - \tan 22° \tan 23°}$

21. $\dfrac{\tan \frac{\pi}{3} - \tan \frac{\pi}{4}}{1 + \tan \frac{\pi}{3} \tan \frac{\pi}{4}}$

22. $\sin 4x \cos 3x + \cos 4x \sin 3x$

23. $\dfrac{\tan 3\theta + \tan 2\theta}{1 - \tan 3\theta \tan 2\theta}$

24. $\sin 5x \cos 5x + \cos 5x \sin 5x$

25. $\cos(3x/5) \cos(2x/5) - \sin(3x/5) \sin(2x/5)$

***26.** $\dfrac{\sqrt{3}}{2} \sin x + \dfrac{1}{2} \cos x$

Hint: Do you recognize $\sqrt{3}/2$ and $\frac{1}{2}$ as the sine and cosine of any special angle?

***27.** $\dfrac{\sqrt{2}}{2} (\sin x - \cos x)$

***28.** $\dfrac{\sqrt{2}}{2} (\cos x + \sin x)$

***29.** $\dfrac{1}{2} \cos x - \dfrac{\sqrt{3}}{2} \sin x$

***30.** $8\sqrt{3} \cos x - 8 \sin x$

***31.** $7.4 \sin x + 7.4\sqrt{3} \cos x$

***32.** $\sin x + \cos x$

***33.** $0.9336 \sin x + 0.3584 \cos x$

***34.** $0.2588 \cos x + 0.9659 \sin x$

***35.** $\cos(x + y) \cos y + \sin(x + y) \sin y$

***36.** $\sin(x + y) \cos y - \cos(x + y) \sin y$

In Exercises 37–46, put each expression in the form $\tan(x \pm A)$, where $|A| \le 90°$.

***37.** $\dfrac{1 + \tan x}{1 - \tan x}$ ***38.** $\dfrac{1 - \tan x}{1 + \tan x}$

*39. $\dfrac{\frac{\sqrt{3}}{3} + \tan x}{1 - \frac{\sqrt{3}}{3}\tan x}$

*40. $\dfrac{\tan x - \frac{\sqrt{3}}{3}}{1 + \frac{\sqrt{3}}{3}\tan x}$

*41. $\dfrac{\sqrt{3} + \tan x}{1 - \sqrt{3}\tan x}$

*42. $\dfrac{\tan x - \sqrt{3}}{1 + \sqrt{3}\tan x}$

*43. $\dfrac{1.1918 + \tan x}{1 - 1.1918\tan x}$

*44. $\dfrac{\tan x - 0.8391}{1 + 0.8391\tan x}$

**45. $\dfrac{3.2610 + 6.40\tan x}{1 - 0.5095\tan x}$

**46. $\dfrac{0.214\tan x - 0.6215}{1 + 2.9042\tan x}$

In Exercises 47–61, solve for all values of x between 0° and 360°.

47. $\cos 2x \cos x + \sin 2x \sin x = \frac{1}{2}$

48. $\sin x \cos 2x - \cos x \sin 2x = \sqrt{3}/2$

49. $\cos 6x \cos 4x + \sin 6x \sin 4x = 1$

50. $\cos 7x \sin 4x - \sin 7x \cos 4x = \sqrt{2}/2$

51. $\sin 3x \cos x - \cos 3x \sin x = 0.40674$

52. $\cos 5x \cos 3x + \sin 5x \sin 3x = -0.32447$

*53. $2 \sin x \cos x = \frac{1}{2}$ *Hint:* $2 \sin x \cos x = \sin x \cos x + \cos x \sin x$.

*54. $\cos^2 x - \sin^2 x = \sqrt{2}/2$ *Hint:* $\cos^2 x - \sin^2 x = \cos x \cos x - \sin x \sin x$.

55. $\sin(x - \frac{\pi}{2}) - \cos x = 1$

56. $\cos(\frac{\pi}{2} - x) + 2 \sin x - 1 = 0$

*57. $\sin(x + \frac{\pi}{3}) + \sin(x - \frac{\pi}{3}) = 0.642788$

*58. $\sin(x + 30°) + \cos(x + 60°) = -\frac{1}{2}$

*59. $\sin(x + 20.0°) + \cos(x + 70.0°) = 0.66410$

*60. $\sin(x + 50.0°) = \frac{1}{2} + \cos(x + 40.0°)$

*61. $\sin x + \cos x = 1$

In Exercises 62–78, prove that each equation is an identity.

62. $\cos(x + 30°) - \cos(x - 30°) = -\sin x$

63. $\cos(x + \frac{\pi}{4}) + \cos(x - \frac{\pi}{4}) = \sqrt{2}\cos x$

64. $\sin(\frac{\pi}{6} + x) + \cos(\frac{\pi}{3} + x) = \cos x$

65. $\sin(30° + x) - \cos(60° + x) = \sqrt{3}\sin x$

66. $\sin(x + 25°) + \cos(x + 65°) = 2 \cos x \sin 25°$

67. $\cos(x + y) + \cos(x - y) = 2 \cos x \cos y$

68. $\sin(x + y) - \sin(x - y) = 2 \cos x \sin y$

69. $\sin(x + y) + \sin(x - y) = 2 \sin x \cos y$

70. $\cos(x + y) - \cos(x - y) = -2 \sin x \cos y$

71. $\cos(x + y)\cos(x - y) = \cos^2 x - \sin^2 y$

72. $\sin(x + y)\sin(x - y) = \sin^2 x - \sin^2 y$

73. $\sin(x + y)\cos y - \cos(x + y)\sin y = \sin x$

*74. $\dfrac{\sin(x + y) + \sin(x - y)}{\cos(x + y) + \cos(x - y)} = \tan x$

*75. $\sin(x + y + z) = \sin x \cos y \cos z + \cos x \sin y \cos z + \cos x \cos y \sin z - \sin x \sin y \sin z$

Hint: Write $x + y + z$ as $(x + y) + z$

*76. $\cos(x + y + z) = \cos x \cos y \cos z - \sin x \cos y \sin z - \cos x \sin y \sin z - \sin x \sin y \cos z$

*77. $\sec(A + B) = \dfrac{\sec A \sec B}{1 - \tan A \tan B}$

*78. $\csc(A + B) = \dfrac{\csc A \sec B}{1 + \cot A \tan B}$

*79. Derive a formula for $\cot(A + B)$ in terms of $\cot A$ and $\cot B$, similar to the formula we derived in this section for $\tan(A + B)$.

80. Use the results of Exercise 79 to derive a formula for $\cot(A - B)$ in terms of $\cot A$ and $\cot B$.

81. Use the identities in Exercises 77 and 78 to prove that

(a) $\sec(A - B) = \dfrac{\sec A \sec B}{1 + \tan A \tan B}$

(b) $\csc(A - B) = \dfrac{\csc A \sec B}{1 - \cot A \tan B}$

*82. Show that in any triangle ABC, $\sin A \cos B + \cos A \sin B = \sin C$.

*83. Show that in any triangle ABC, $\cos A \cos B - \sin A \sin B = -\cos C$.

*84. Apply the results of Exercises 82 and 83 to a right triangle with C being the right angle, and explain why the results are reasonable.

In Exercises 85–88, the right angle is C.

*85. Prove that in any right triangle

$$\tan(B - A) = \frac{(b + a)(b - a)}{2ab}$$

*86. By expanding $\sin(A + B)$, prove that in any right triangle ABC, $\sin(A + B) = 1$. Could you have guessed this without expanding the sine? Why?

*87. By expanding $\cos(A + B)$, prove that in any right triangle ABC, $\cos(A + B) = 0$. Could you have guessed this without expanding the cosine?

*88. By expanding $\tan(A + B)$, prove that for any right triangle ABC, $\tan(A + B)$ is undefined.

*89. If ABC is not a right triangle, prove that

$$\tan A + \tan B + \tan C = \tan A \tan B \tan C$$

Hint: Use the formula for $\tan(x - y)$

*90. If $\tan x = \frac{1}{3}$ and $x - y = \frac{\pi}{4}$, find $\tan y$ without finding x or y.

***91.** If $\tan y = 2$ and $x + y = 135°$, find $\tan x$ without finding x or y.

***92.** Prove the identity

$$\sin(A + B)\sin(A - B) = (\sin A + \sin B)(\sin A - \sin B)$$

Does this mean that $\sin(A + B) = \sin A + \sin B$ and $\sin(A - B) = \sin A - \sin B$? Why?

***93.** Prove that $y = A\cos(x + \sigma)$ and $y = a\sin x + b\cos x$ are identical if $a = -A\sin\sigma$ and $b = A\cos\sigma$.

***94.** Prove that $y = B\sin(x + \phi)$ and $y = C\sin x + D\cos x$ are identical if $C = B\cos\phi$ and $D = B\sin\phi$.

****95.** Express $y = 2\sin x + 3\cos x$ in the form $y = A\cos(x + \sigma)$, $-\frac{\pi}{2} \le \sigma \frac{\pi}{2}$. ***Hint:*** See Exercise 93 and solve for A and σ.

****96.** Express $y = \sin x - 2\cos x$ in the form $y = B\sin(x + \phi)$, $-\frac{\pi}{2} \le \phi \le \frac{\pi}{_2}$. ***Hint:*** See Exercise 94 and solve for B and ϕ.

In Exercises 97–102, derive each equation.

97. $\sin 2A = 2\sin A\cos A$ ***Hint:*** $2A = A + A$.

98. $\cos 2A = \cos^2 A - \sin^2 A$

99. $\cos 2A = 1 - 2\sin^2 A = 2\cos^2 A - 1$ ***Hint:*** Use the results of Exercise 98.

100. $\tan 2A = \dfrac{2\tan A}{1 - \tan^2 A}$

***101.** **(a)** $\cos\dfrac{\theta}{2} = \pm\sqrt{\dfrac{1 + \cos\theta}{2}}$

(b) $\sin\dfrac{\theta}{2} = \pm\sqrt{\dfrac{1 - \cos\theta}{2}}$

Hint: In the appropriate form of Exercise 99, let $2A = \theta$.

***102.** $\tan\dfrac{\theta}{2} = \pm\sqrt{\dfrac{1 - \cos\theta}{1 + \cos\theta}}$

In Exercises 103–106, prove each identity.

103. $\sin A\cos B = \frac{1}{2}[\sin(A + B) + \sin(A - B)]$

104. $\cos A\sin B = \frac{1}{2}[\sin(A + B) - \sin(A - B)]$

105. $\cos A\cos B = \frac{1}{2}[\cos(A + B) + \cos(A - B)]$

106. $\sin A\sin B = \frac{1}{2}[\cos(A - B) - \cos(A - B)]$

107. A solar panel has a slope of 2:12 measured with respect to a sloping roof, which itself has a slope of 4:12 (see Figure 7.1-13). At what angle does this panel slope upward above the horizontal, and what is this slope in the standard form of xx:12?

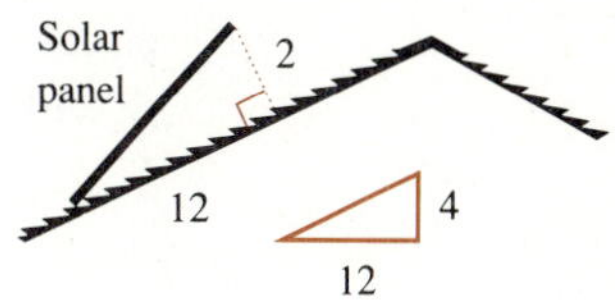

Figure 7.1-13

****108.** A ladder is leaning against a wall. When it makes an angle θ with the floor, it reaches to a height h on the wall (see Figure 7.1-14). If you slide the ladder up the wall so that θ increases by 21.0°, you find that h increases by 20.0%. What was the original angle θ?

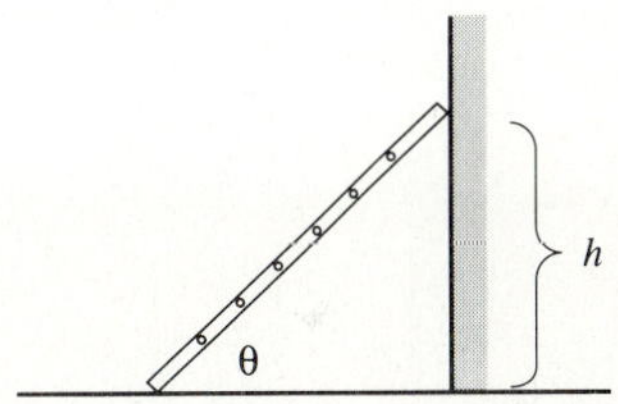

Figure 7.1-14

****109.** A ladder is leaning as shown in Figure 7.1-14. If it slips so that θ decreases by 15.0°, causing h to decrease by 10.0%, what was the original angle?

7.2 DOUBLE-ANGLE AND HALF-ANGLE FORMULAS

When solving equations or deriving formulas, it is often necessary to express a trigonometric function of 2θ or $\theta/2$ in terms of functions of just θ. We can do this by using the results of Section 7.1 to find formulas for $\sin 2A$, $\cos 2A$, and $\tan 2A$ (the double-angle formulas), and we can then use these formulas to get formulas for $\sin A/2$, $\cos A/2$, and $\tan A/2$ (the half-angle formulas).

7.2-1 Double-Angle Formulas

How is $\sin 2A$ related to $\sin A$? You might expect that $\sin 2A = 2\sin A$. To check whether this is so, try $A = 45°$.

$$\sin 2A = \sin[2(45°)] = \sin 90° = 1$$
$$2 \sin A = 2 \sin 45° = 2\sqrt{2}/2 = \sqrt{2}$$

Obviously sin $2A$ is not identical to 2 sin A, so we need to find a formula for sin $2A$.

NOW IT'S YOUR TURN

Is cos $2A$ identical to 2 cos A? What about tan $2A$? (Answer: No!)

To derive formulas for sin $2A$, cos $2A$, and tan $2A$, we can apply the equations $\sin(A + B)$, $\cos(A + B)$ and $\tan(A + B)$ from Section 7.1. If we let A and B be equal in these equations, we get

$$\sin 2A = \sin(A + A) = \sin A \cos A + \cos A \sin A = 2 \sin A \cos A$$

$$\tan 2A = \tan(A + A) = \frac{\tan A + \tan A}{1 - \tan A \tan A} = \frac{2 \tan A}{1 - \tan^2 A}$$

NOW IT'S YOUR TURN

Show that $\cos 2A = \cos^2 A - \sin^2 A$.

Before going on to the examples, we summarize these important results.

DOUBLE-ANGLE FORMULAS

$$\sin 2A = 2 \sin A \cos A$$
$$\cos 2A = \cos^2 A - \sin^2 A$$
$$\tan 2A = \frac{2 \tan A}{1 - \tan^2 A}$$

We recommend that you do *not* memorize these formulas. Instead, start with the formulas for the trigonometric functions of $A + B$ and make sure you can figure them out, just as we did in deriving them. Six months (days?) from now, you will not remember these formulas, but if you understand how to figure them out, you can easily derive them without a book.

Example 9

Use the double-angle formulas to find $\sin \frac{2\pi}{3}$, $\cos \frac{2\pi}{3}$, and $\tan \frac{2\pi}{3}$.

Solution

$$\sin 2A = 2 \sin A \cos A$$
$$\sin \frac{2\pi}{3} = \sin\left[2\left(\frac{\pi}{3}\right)\right] = 2 \sin \frac{\pi}{3} \cos \frac{\pi}{3}$$
$$= 2\frac{\sqrt{3}}{2}\frac{1}{2} = \frac{\sqrt{3}}{2}$$

In a similar way, we get

$$\cos\frac{2\pi}{3} = \cos\left[2\left(\frac{\pi}{3}\right)\right] = \cos^2\frac{\pi}{3} - \sin^2\frac{\pi}{3}$$
$$= \left(\frac{1}{2}\right)^2 - \left(\frac{\sqrt{3}}{2}\right)^2 = \frac{1}{4} - \frac{3}{4} = -\frac{1}{2}$$

$$\tan\frac{2\pi}{3} = \tan\left[2\left(\frac{\pi}{3}\right)\right] = \frac{2\tan\frac{\pi}{3}}{1-\tan^2\frac{\pi}{3}} = \frac{2\sqrt{3}}{1-(\sqrt{3})^2} = \frac{2\sqrt{3}}{1-3} = -\sqrt{3}$$ ■

NOW IT'S YOUR TURN

Use the double-angle formulas to find $\sin\frac{\pi}{2}$, $\cos\frac{\pi}{2}$, and $\tan\frac{\pi}{2}$.
(Answer: By this time you don't need answers to these!)

Example 10

If θ is a quadrant III angle for which $\cos\theta = -4/5$, use the double-angle formulas to find $\sin 2\theta$, $\cos 2\theta$, and $\tan 2\theta$.

Solution

We need $\sin\theta$ and $\tan\theta$. To find these, first sketch θ with its reference triangle (see Figure 7.2-1). Because $\cos\theta = -\frac{4}{5} = x/r$, then $x = -4$ and $r = 5$. We can use the Pythagorean theorem to find y.

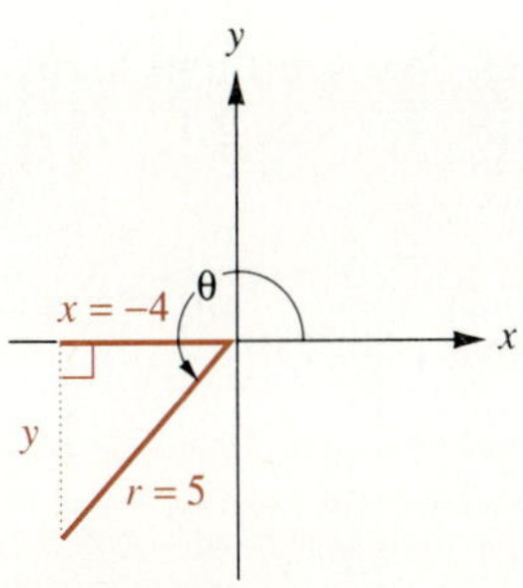

Figure 7.2-1
Reference triangle for θ

$$x^2 + y^2 = r^2$$
$$(-4)^2 + y^2 = 5^2$$
$$y^2 = 5^2 - (-4)^2 = 25 - 16 = 9$$
$$y = \pm\sqrt{9} = \pm 3$$

Because θ is in quadrant III, y is negative, so we must use the negative sign. Therefore, we have $x = -4$, $y = -3$, and $r = 5$. This result gives us $\sin\theta = y/r = -\frac{3}{5}$, $\cos\theta = x/r = -\frac{4}{5}$, and $\tan\theta = y/x = (-3)/(-4) = \frac{3}{4}$. Now we can use the double-angle formulas.

$$\sin 2\theta = 2\sin\theta\cos\theta = 2(-\tfrac{3}{5})(-\tfrac{4}{5}) = \tfrac{24}{25}$$
$$\cos 2\theta = \cos^2\theta - \sin^2\theta = (-\tfrac{4}{5})^2 - (-\tfrac{3}{5})^2 = \tfrac{16}{25} - \tfrac{9}{25} = \tfrac{7}{25}$$
$$\tan 2\theta = \frac{2\tan\theta}{1-\tan^2\theta} = \frac{2(\frac{3}{4})}{1-(\frac{3}{4})^2} = \frac{\frac{3}{2}}{1-\frac{9}{16}} = \frac{\frac{3}{2}}{\frac{7}{16}} = \frac{24}{7}$$

As a check, we can use

$$\tan 2\theta = \frac{\sin 2\theta}{\cos 2\theta} = \frac{\frac{24}{25}}{\frac{7}{25}} = \frac{24}{7}$$ ■

NOW IT'S YOUR TURN

If θ is a quadrant IV angle for which $\tan\theta = -\frac{1}{2}$, use the double-angle formulas to find $\sin 2\theta$, $\cos 2\theta$, and $\tan 2\theta$.
(Answer: $\sin 2\theta = -\frac{4}{5}$, $\cos 2\theta = \frac{3}{5}$, and $\tan 2\theta = -\frac{4}{3}$.)

The major value of the double-angle formulas is to simplify expressions and prove other identities. The formulas are especially useful in solving trigonometric equations. The following examples illustrate several of these uses.

Example 11

Express $\cos 2A$ in terms of only $\sin A$.

Solution

$$\begin{aligned}\cos 2A &= \cos^2 A - \sin^2 A\\ &= (1 - \sin^2 A) - \sin^2 A \qquad (\text{use } \sin^2 A + \cos^2 A = 1)\\ &= 1 - 2\sin^2 A \;\blacksquare\end{aligned}$$

NOW IT'S YOUR TURN

Express $\cos 2A$ in terms of only $\cos A$. (Answer: $\cos 2A = 2\cos^2 A - 1$)

The two identities in this example are very useful ones. We see them many times in this chapter. Along with the previous identity for $\cos 2A$, these identities allow us to express $\cos 2A$ in three ways.

THREE FORMS OF COS 2A

$$\cos 2A = \begin{cases}\cos^2 A - \sin^2 A\\ 1 - 2\sin^2 A\\ 2\cos^2 A - 1\end{cases}$$

Example 12

Simplify (a) $4\sin x\cos x\cos 2x$, and (b) $7\cos^2 5x - 7\sin^2 5x$.

Solution

(a) $4\sin x\cos x\cos 2x$: Whenever we see $\sin x\cos x$, we should think of $\sin 2x$ because $\sin 2x = 2\sin x\cos x$. Therefore, we can write the given expression as

$$\begin{aligned}4\sin x\cos x\cos 2x &= 2\cdot\underbrace{2\sin x\cos x}_{\sin 2x}\cos 2x\\ &= \underbrace{2\sin 2x\cos 2x}_{\sin[2(2x)]} = \sin 4x\end{aligned}$$

(b) $7 \cos^2 5x - 7 \sin^2 5x$: Whenever we see $\cos^2 \theta - \sin^2 \theta$, we should think of $\cos 2\theta$. In this case, $\theta = 5x$.

$$7 \cos^2 \theta = 7 \sin^2 \theta = 7(\cos^2 \theta - \sin^2 \theta) = 7 \cos 2\theta$$

$$7 \cos^2 5x - 7 \sin^2 5x = 7\underbrace{(\cos^2 5x - \sin^2 5x)}_{\cos[2(5x)]} = 7 \cos 10x \quad ■$$

NOW IT'S YOUR TURN

Simplify (a) $3.4 \sin^2 6x - 3.4 \cos^2 6x$ and (b) $\cos 2x \cos 4x \sin 2x$.
(Answer: (a) $-3.4 \cos 12x$; (b) $\frac{1}{4} \sin 8x$)

Example 13

Prove the identity

$$\tan^2 x = \frac{\sec 2x - 1}{\sec 2x + 1}$$

Solution

We simplify the more complicated right-hand side.

$$\tan^2 x \stackrel{?}{=} \frac{\sec 2x - 1}{\sec 2x + 1}$$

$$\frac{\dfrac{1}{\cos 2x} - 1}{\dfrac{1}{\cos 2x} + 1} \qquad \left(\text{use } \sec A = \frac{1}{\cos A}\right)$$

$$\frac{\dfrac{1 - \cos 2x}{\cos 2x}}{\dfrac{1 + \cos 2x}{\cos 2x}} \qquad \text{(find common denominator)}$$

$$\frac{1 - \cos 2x}{1 + \cos 2x} \qquad \text{(divide fractions)}$$

$$\frac{1 - (\cos^2 x - \sin^2 x)}{1 + (\cos^2 x - \sin^2 x)} \qquad \text{(use double-angle formula for } \cos 2x\text{)}$$

$$\frac{(1 - \cos^2 x) + \sin^2 x}{(1 - \sin^2 x) + \cos^2 x} \qquad \text{(rearrange and group with parentheses)}$$

$$\frac{2 \sin^2 x}{2 \cos^2 x} \qquad \text{(use identity } \sin^2 x + \cos^2 x = 1 \text{ in numerator and denominator)}$$

$$\frac{\sin^2 x}{\cos^2 x} \qquad \text{(divide by 2)}$$

$$\tan^2 x \qquad \text{(identity } \tan x = \sin x/\cos x\text{)} \quad ■$$

NOW IT'S YOUR TURN

Prove that the following equation is an identity.

$$\cot^2 x = \frac{\sec 2x + 1}{\sec 2x - 1}$$

Example 14

Find all values of x between 0° and 360° that satisfy the equation $\sin^2 x - \cos^2 x = \frac{1}{2}$.

Solution

In solving this equation, our goal is to reduce it to a single trigonometric function of a single angle. Remember that whenever we see $\cos^2 A - \sin^2 A$, we should think of $\cos 2A$.

$$\sin^2 x - \cos^2 x = \tfrac{1}{2}$$

$$-(\cos^2 x - \sin^2 x) = \tfrac{1}{2} \qquad \textbf{(factor } -1\textbf{)}$$

$$-\cos 2x = \tfrac{1}{2} \qquad \textbf{(use identity } \cos^2 x - \sin^2 x = \cos 2x\textbf{)}$$

$$\cos 2x = -\tfrac{1}{2} \qquad \textbf{(multiply both sides by } -1\textbf{)}$$

$$2x = \cos^{-1}(-\tfrac{1}{2})$$

Because x can be between 0° and 360°, $2x$ can be between 0° and 720°. The cosine is negative in quadrants II and III, so $2x$ can lie in either of these quadrants. Because $\cos(2x) = -\frac{1}{2}$, the reference angle theorem from Chapter 3 tells us that the reference angle for $2x$ is 60° (see Figure 7.2-2).

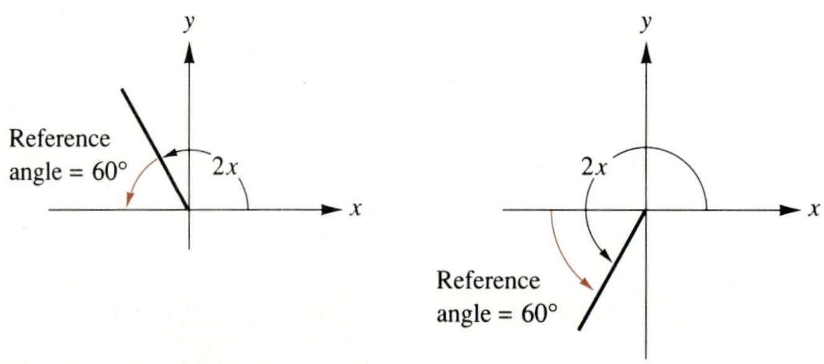

(a) If 2x is in quadrant II ***(b) If 2x is in quadrant III***

Figure 7.2-2
Reference angle for 2*x*

As you can see from Figure 7.2-2, if $2x$ is in quadrant II, then

$$2x = 180° - 60° = 120°$$

If $2x$ is in quadrant III, then

$$2x = 180° + 60° = 240°$$

To get the solutions when $2x$ is between 360° and 720°, we just add 360° to the solutions we just found. Therefore, the solutions for $2x$ are

$$2x = \begin{cases} 120^\circ \\ 240^\circ \\ 120^\circ + 360^\circ = 480^\circ \\ 240^\circ + 360^\circ = 600^\circ \end{cases}$$

Dividing by 2 to get the values of x gives us

$$x = 60^\circ,\ 120^\circ,\ 240^\circ,\ 300^\circ \quad ■$$

NOW IT'S YOUR TURN

Find all values of θ between 0° and 360° that satisfy the equation $\sin x \cos x = \frac{1}{2}$.
(Answer: 45°, 225°)

Example 15

To design an energy-efficient house, an architect needs to know the angle of the noon sun in a certain city on December 21 and March 21. All she knows is that noon shadows on December 21 are 2.5 times as long as they are on March 21, whereas the sun is twice as high in the sky on March 21 as it is on December 21. Find the angle of the noon sun above the horizon on both these days.

Solution

First sketch a typical shadow on December 21 and March 21 (see Figure 7.2-3). The sun is at an angle θ on December 21 and at an angle 2θ (twice as high in the sky) on March 21. Let us call S_D the shadow's length on December 21, S_M the shadow's length on March 21, and h the height of the object casting the shadow.

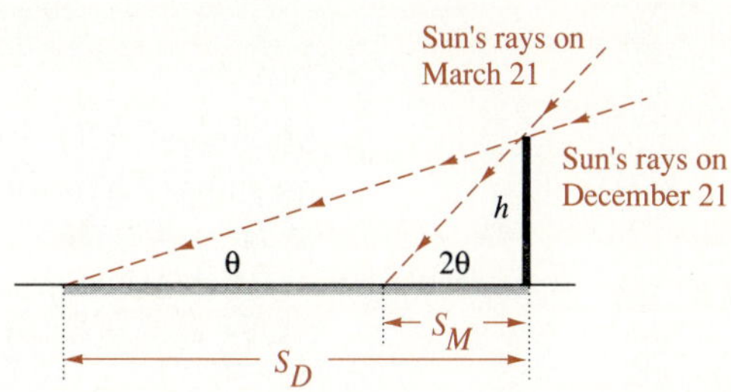

Figure 7.2-3

Shadows on December 21 and March 21

We know that $S_D = 2.5\, S_M$ and want to use this information to find θ and 2θ. Figure 7.2-3 tells us that

$$\tan\theta = \frac{h}{S_D} \rightarrow S_D = \frac{h}{\tan\theta}$$

$$\tan 2\theta = \frac{h}{S_M} \rightarrow S_M = \frac{h}{\tan 2\theta}$$

Using the fact that $S_D = 2.5\,S_M$ gives us

$$\frac{h}{\tan\,\theta} = 2.5\left(\frac{h}{\tan\,2\theta}\right)$$

Dividing by h and cross-multiplying gives us

$$\tan\,2\theta = 2.5\,\tan\,\theta$$

To solve this equation, we must reduce it to a single function of a single angle. We can do this using the double-angle formula for $\tan\,2\theta$.

$$\frac{2\,\tan\,\theta}{1 - \tan^2\,\theta} = 2.5\,\tan\,\theta$$

Because the noon sun is not at the horizon on these days, θ cannot be 0 or π, so $\tan\,\theta$ cannot be 0. Therefore, we can divide both sides of this equation by $\tan\,\theta$. (Remember that division by zero is not allowed.)

$$\frac{2}{1 - \tan^2\,\theta} = 2.5$$

Multiplying by $1 - \tan^2\,\theta$ and solving for $\tan^2\,\theta$ gives us

$$2 = 2.5(1 - \tan^2\,\theta) = 2.5 - 2.5\,\tan^2\,\theta$$

$$\tan^2\,\theta = \frac{2.5 - 2}{2.5} = \frac{0.5}{2.5} = 0.20$$

$$\tan\,\theta = \sqrt{0.20} = 0.4472$$

$$\theta = \tan^{-1}(0.4472) = 24°$$

Mathematically, θ could also be in quadrant III, but this solution has no physical meaning. Therefore, on December 21 the sun is at $\theta = 24°$, and on March 21 it is at $2\theta = 48°$. ■

NOW IT'S YOUR TURN

If the shadows were 3.0 times as long, rather than 2.5 times, but everything else was the same, what would be the sun's angle on December 21 and March 21? (Answer: 30°, 60°)

Example 16

Range of a Rocket

An important problem in physics and aeronautics is to find how far a projectile, such as a rocket, will go if we launch it at an angle θ above the horizontal and let it coast without power after that. This horizontal distance is called the range (see Figure 7.2-4). Although you will have to study physics to see where this formula comes from, we can state the result here. If v is the speed with which you initially

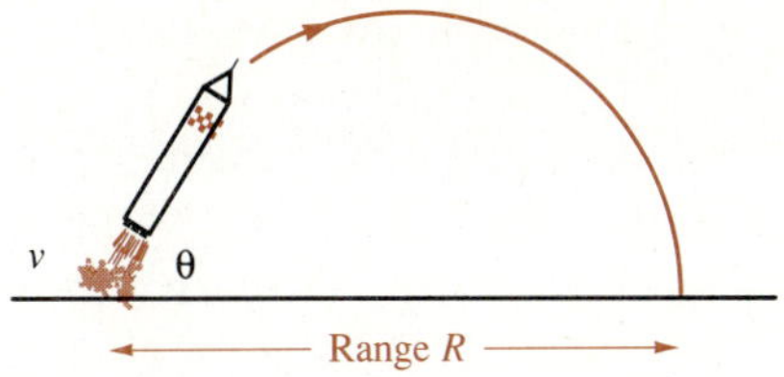

Figure 7.2-4
Range of a rocket

launch the rocket, then the range R is given by

$$R = \frac{2v^2}{g} \sin\theta \cos\theta$$

where g is the acceleration of any freely falling object. Now we can ask: What angle θ will give the maximum range?

Solution

For any angle θ, the range is

$$R = \frac{2v^2}{g} \sin\theta \cos\theta$$

We recognize $\sin 2\theta$ lurking here because of the $\sin\theta\cos\theta$. Thus we can rewrite the range as

$$R = \frac{v^2}{g} \underbrace{2 \sin\theta \cos\theta}_{\sin 2\theta}$$

$$R = \frac{v^2}{g} \sin 2\theta$$

The range will be a maximum when $\sin 2\theta$ is a maximum, and that occurs when $2\theta = 90°$, which is when $\theta = 45°$. This same result would apply to a stone that you throw. Test it the next time you are at a still lake where you can see how far the stone goes. ■

7.2-2 Half-Angle Formulas

We can now use the double-angle formulas to derive equations for $\sin A/2$, $\cos A/2$, and $\tan A/2$. These formulas are called the half-angle formulas. We begin with the identity

$$\cos 2x = \cos^2 x - \sin^2 x$$

Expressing this identity in terms of only $\sin x$, as in Example 11, gives us

$$\cos 2x = 1 - 2\sin^2 x$$

Solving for $\sin x$, we get

$$2\sin^2 x = 1 - \cos 2x$$

$$\sin^2 x = \frac{1 - \cos 2x}{2} \qquad \textbf{[divide by 2]}$$

$$\sin x = \pm\sqrt{\frac{1 - \cos 2x}{2}} \qquad \textbf{[take square roots]}$$

Now we let $2x = A$, so $x = A/2$. Substituting this into the last equation for $\sin x$ gives us

$$\sin\frac{A}{2} = \pm\sqrt{\frac{1 - \cos A}{2}}$$

The $\pm$ sign in this result means that we can use *either* the plus sign *or* the minus sign, depending on which quadrant $A/2$ lies in. It does *not* mean that we use both signs. If $A/2$ is in quadrants I or II where $\sin A/2$ is positive, then we use the plus sign. If $A/2$ is in quadrants III or IV where $\sin A/2$ is negative, we use the negative sign.

NOW IT'S YOUR TURN

Derive the half-angle formula for $\cos A/2$. ***Hint:*** Start by expressing $\cos 2x$ in terms of only $\cos x$ and follow the same general procedure we just used. Then use the formulas for $\sin A/2$ and $\cos A/2$ to derive the formula for $\tan A/2$.

(Answer: $\cos\frac{A}{2} = \pm\sqrt{\frac{1 + \cos A}{2}}$ and $\tan\frac{A}{2} = \pm\sqrt{\frac{1 - \cos A}{1 + \cos A}}$)

Summarizing these results, we get the following formulas.

HALF-ANGLE FORMULAS

$$\sin\frac{A}{2} = \pm\sqrt{\frac{1 - \cos A}{2}}$$

$$\cos\frac{A}{2} = \pm\sqrt{\frac{1 + \cos A}{2}}$$

$$\tan\frac{A}{2} = \pm\sqrt{\frac{1 - \cos A}{1 + \cos A}}$$

In each case, the correct sign depends on the quadrant in which $A/2$ lies.

Example 17

Use the half-angle formulas to find sin 22.5°, cos 22.5°, and tan 22.5° to three decimal places.

Solution

Because $22.5° = 45°/2$, we can use the half-angle formulas with $A = 45°$. Because 22.5° is in quadrant I, all the trigonometric functions are positive.

$$\sin 22.5° = \sin\frac{45°}{2} = \sqrt{\frac{1-\cos 45°}{2}} = \sqrt{\frac{1-\dfrac{\sqrt{2}}{2}}{2}} = \sqrt{\frac{2-\sqrt{2}}{4}} = \frac{\sqrt{2-\sqrt{2}}}{2}$$

$$= \frac{\sqrt{2-1.414}}{2} = \frac{\sqrt{0.586}}{2} = 0.383$$

$$\cos 22.5° = \cos\frac{45°}{2} = \sqrt{\frac{1+\cos 45°}{2}} = \sqrt{\frac{1+\dfrac{\sqrt{2}}{2}}{2}} = \sqrt{\frac{2+\sqrt{2}}{4}} = \frac{\sqrt{2+\sqrt{2}}}{2}$$

$$= \frac{\sqrt{2+1.414}}{2} = \frac{\sqrt{3.414}}{2} = 0.924$$

$$\tan 22.5° = \tan\frac{45°}{2} = \sqrt{\frac{1-\cos 45°}{1+\cos 45°}} = \sqrt{\frac{1-\dfrac{\sqrt{2}}{2}}{1+\dfrac{\sqrt{2}}{2}}} = \sqrt{\frac{\left(\dfrac{2-\sqrt{2}}{2}\right)}{\left(\dfrac{2+\sqrt{2}}{2}\right)}}$$

$$= \sqrt{\frac{2-\sqrt{2}}{2+\sqrt{2}}} = \sqrt{\frac{2-1.414}{2+1.414}} = \sqrt{0.1716} = 0.414 \quad ■$$

NOW IT'S YOUR TURN

Use the half-angle formulas to find sin 15°, cos 15°, and tan 15° to three decimal places. Check your answers with a calculator. (Answer: Ask your calculator!)

Example 18

Use the half-angle formulas to find $\cos\frac{7\pi}{12}$ to three decimal places.

Solution

In this case, $A/2 = \frac{7\pi}{12}$, so $A = \frac{7\pi}{6}$. Because $A/2 = \frac{7\pi}{12}$ is in quadrant II, $\cos\frac{7\pi}{12}$ is negative. Using the half-angle formula we get

$$\cos\frac{7\pi}{12} = \cos\left(\frac{1}{2}\frac{7\pi}{6}\right) = -\sqrt{\frac{1+\cos\dfrac{7\pi}{6}}{2}} = -\sqrt{\frac{1-\dfrac{\sqrt{3}}{2}}{2}} = -\frac{\sqrt{2-\sqrt{3}}}{2}$$

$$= -\frac{\sqrt{2-1.732}}{2} = -0.259 \quad ■$$

NOW IT'S YOUR TURN

Use the half-angle formulas to find $\sin \frac{7\pi}{12}$ to three decimal places, and check your answer with a calculator. (Answer: Ask your calculator!)

Example 19

If θ is a quadrant II angle for which $\csc \theta = \frac{13}{12}$, find $\sin \theta/2$ and $\cos \theta/2$ using the half-angle formulas.

Solution

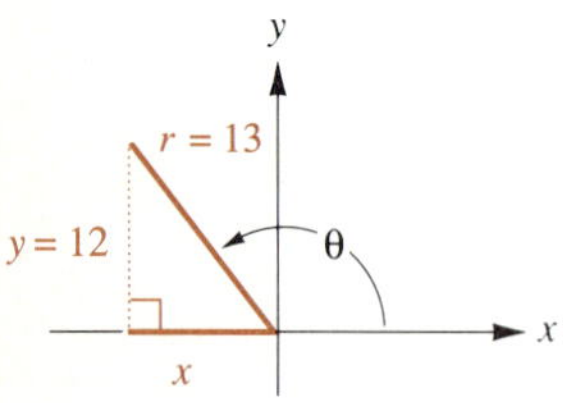

Figure 7.2-5

Reference triangle for θ

To use the half-angle formulas, we need $\cos \theta$. Figure 7.2-5 shows its reference triangle. Because $\csc \theta = \frac{13}{12}$ and θ is in quadrant II, then $r = 13$, $y = 12$, and x is negative. The Pythagorean theorem tells us that $x = \pm 5$. Because x is negative in quadrant II, we must use the negative value $x = -5$. Therefore, $\cos \theta = x/r = -\frac{5}{13}$.

Because θ is in quadrant II, it is between 90° and 180°. Therefore, $\theta/2$ must be between 45° and 90° which means that $\theta/2$ is in quadrant I, so its sine and cosine are both positive. Now we can find $\sin \theta/2$ and $\cos \theta/2$.

$$\sin \frac{\theta}{2} = +\sqrt{\frac{1 - \cos \theta}{2}} = \sqrt{\frac{1 - \left(\frac{-5}{13}\right)}{2}} = \sqrt{\frac{\frac{18}{13}}{2}} = \sqrt{\frac{18}{26}}$$

$$= \sqrt{\frac{9}{13}} = \frac{3}{\sqrt{13}} = \frac{3\sqrt{13}}{13}$$

$$\cos \frac{\theta}{2} = +\sqrt{\frac{1 + \cos \theta}{2}} = \sqrt{\frac{1 + \left(\frac{-5}{13}\right)}{2}} = \sqrt{\frac{8}{26}} = \sqrt{\frac{4}{13}} = \frac{2}{\sqrt{13}} = \frac{2\sqrt{13}}{13}$$ ■

NOW IT'S YOUR TURN

If θ is a quadrant II angle for which $\sec \theta = -4$, use the half-angle formulas to find $\sin \theta/2$ and $\cos \theta/2$. (Answer: $\sin \theta/2 = \sqrt{10}/4$, $\cos \theta/2 = \sqrt{6}/4$.)

Example 20

Evaluate $\sin\left(\frac{1}{2} \cos^{-1} \frac{3}{5}\right)$ assuming quadrant I angles.

Solution

First find $\cos^{-1} \frac{3}{5}$ by letting $\theta = \cos^{-1} \frac{3}{5}$, which means that $\cos \theta = \frac{3}{5}$. Now apply the half-angle formula.

$$\sin\left(\underbrace{\frac{1}{2} \cos^{-1} \frac{3}{5}}_{\theta}\right) = \sin \frac{\theta}{2} = \sqrt{\frac{1 - \cos \theta}{2}} = \sqrt{\frac{1 - \frac{3}{5}}{2}} = \sqrt{\frac{\frac{2}{5}}{2}} = \sqrt{\frac{1}{5}} = \frac{1}{\sqrt{5}} = \frac{\sqrt{5}}{5}$$ ■

NOW IT'S YOUR TURN

Evaluate $\cos(\frac{1}{2}\cos^{-1}\frac{1}{4})$ assuming quadrant I angles. (Answer: $\sqrt{10}/4$)

The half-angle formulas are especially helpful for solving trigonometric equations because we can use them to reduce an equation involving general trigonometric functions to one involving only a single function of one angle. The following example illustrates this reduction.

Example 21

While camping out, you use a rectangular sheet of plastic as a lean-to to protect your campsite from rain. You normally keep it raised (Figure 7.2-6a), but you can lower it quickly if a sudden rain occurs. You want to find what θ should be so that if you lower the lean-to to $\theta/2$, the area sheltered from vertical rain is three times greater than the area sheltered when the lean-to is raised (Figure 7.2-6b).

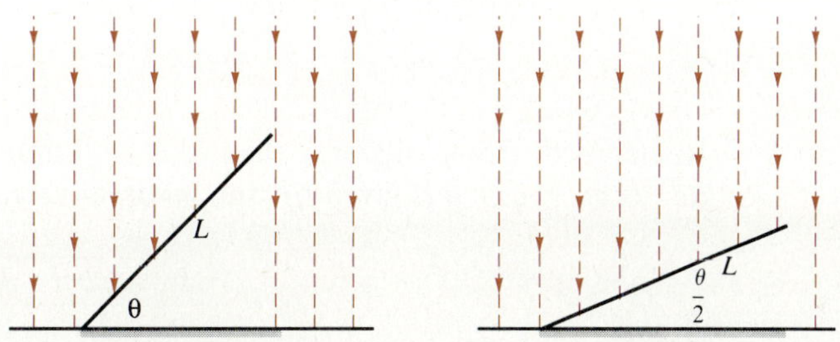

(a) Lean-to raised to angle θ. ***(b) Lean-to lowered to angle θ/2.***

Figure 7.2-6

Lean-to providing shelter from vertical rain

Solution

In Figure 7.2-6, the shaded region is protected from vertical rain. If L is the length of the lean-to, this region has a length $L\cos\theta$ with the lean-to raised, and a length $L\cos(\theta/2)$ with the lean-to lowered. We want to find θ so that the lowered protected length is three times the raised protected length. Mathematically this means that

$$L\cos\frac{\theta}{2} = 3L\cos\theta$$

Cancelling the L gives

$$\cos\frac{\theta}{2} = 3\cos\theta$$

As it stands, we cannot solve this equation because it involves θ and $\theta/2$. However, we can use the half-angle formula for $\cos \theta/2$ to reduce this equation to one involving only θ.

$$\cos \frac{\theta}{2} = 3 \cos \theta$$

$$\pm\sqrt{\frac{1 + \cos \theta}{2}} = 3 \cos \theta \qquad \textbf{(use half-angle formula for } \cos \boldsymbol{\theta}/2\textbf{)}$$

$$\frac{1 + \cos \theta}{2} = 9 \cos^2 \theta \qquad \textbf{(square)}$$

$$1 + \cos \theta = 18 \cos^2 \theta \qquad \textbf{(multiply by 2)}$$

$$18 \cos^2 \theta - \cos \theta - 1 = 0 \qquad \textbf{(rearrange)}$$

This is a *quadratic equation* in which the unknown quantity is $\cos \theta$. If we write x for $\cos \theta$, this equation may look more familiar to you.

$$18x^2 - x - 1 = 0$$

Remember from algebra that the solution of the quadratic equation $ax^2 + bx + c = 0$ is given by the *quadratic formula*

$$x = \frac{-b \pm \sqrt{b^2 - 4ac}}{2a}$$

In this case, $a = 18$, $b = -1$, and $c = -1$. The solution is then

$$x = \frac{-(-1) \pm \sqrt{(-1)^2 - 4(18)(-1)}}{2(18)}$$

$$x = \frac{1 \pm \sqrt{73}}{36} = \frac{1 \pm 8.54}{36} = \begin{cases} \dfrac{9.54}{36} = 0.265 \\[2ex] \dfrac{-7.54}{36} = -0.209 \end{cases}$$

Because x represents $\cos \theta$, $\cos \theta = 0.265$ and $\cos \theta = -0.209$. The negative answer would mean that θ is in quadrants II or III ($90° \le \theta \le 270°$), which makes no sense for a lean-to. Therefore, the only answer that has *physical* meaning is the positive answer, $\cos \theta = 0.265$. Finding θ,

$$\theta = \cos^{-1}(0.265) = 74.6° \approx 75°$$

To check this answer, we see if $L \cos \theta/2 = 3\ L \cos \theta$.

$$L\cos\frac{\theta}{2} = L\cos\frac{74.6^\circ}{2} = L\cos 37.3^\circ = 0.795\ L$$

$$3L\cos\theta = 3L\cos 74.6^\circ = 3L(0.266) = 0.797\ L$$

These answers agree. The slight discrepancy in the third decimal place is from rounding off. ■

NOW IT'S YOUR TURN

What should θ be so that the protected area will be four times the original area when θ is reduced to $\theta/2$? Be sure to check your answer! (Answer: 78.9°)

Another important use of the half-angle formulas is in proving identities, as Example 22 shows.

Example 22

Prove that $\tan(x/2) + \cot(x/2) = 2\csc x$.

Solution

Working on the more complicated left-hand side gives us

$$\tan\frac{x}{2} + \cot\frac{x}{2} \stackrel{?}{=} 2\csc x$$

(use $\tan A = \dfrac{\sin A}{\cos A}$ and $\cot A = \dfrac{\cos A}{\sin A}$)	$\dfrac{\sin\frac{x}{2}}{\cos\frac{x}{2}} + \dfrac{\cos\frac{x}{2}}{\sin\frac{x}{2}}$
(find common denominator)	$\dfrac{\sin^2\frac{x}{2} + \cos^2\frac{x}{2}}{\sin\frac{x}{2}\cos\frac{x}{2}}$
(use $\sin^2 A + \cos^2 A = 1$)	$\dfrac{1}{\sin\frac{x}{2}\cos\frac{x}{2}}$
(multiply numerator and denominator by 2)	$\dfrac{2}{2\sin\frac{x}{2}\cos\frac{x}{2}}$
(double-angle identity $\sin 2A = 2\sin A\cos A$, with $A = x/2$)	$\dfrac{2}{\sin x}$
(use $\csc A = 1/\sin A$)	$2\csc x$ ■

NOW IT'S YOUR TURN

Prove the identity $\tan(x/2) - \cot(x/2) = -2\cot x$.

Historical Note

Nowadays the double- and half-angle formulas, plus those we found in Section 7.1, are most useful for deriving formulas and solving equations in calculus, physics, and engineering. For numerical work, it is easier to use a calculator. However, at one time these formulas were very helpful for compiling trigonometric tables. For example, we can get the trigonometric functions of 30°, 45°, and 60° using some basic properties of geometry, as we did in Chapter 2. We can then use the half-angle formulas to get the trigonometric functions of 15° and 22.5°, and then repeat the process to get those of 7.5°, 3.75°, . . ., and 11.25°, 5.625°, ... We can then use the equations from Section 7.1 to find trigonometric functions of other angles. For example, we can find sin 67.5° using $\sin 67.5° = \sin(60° + 7.5°) = \sin 60° \cos 7.5° + \cos 60° \sin 7.5°$. In this way, with a lot of patience, mathematicians compiled extensive tables of the values of the trigonometric functions for practical use. As late as the early 1970s, people still used such tables regularly for numerical work. *Thank goodness for the pocket calculator!*

EXERCISES 7.2-1

In Exercises 1–10, find $\sin 2\theta$, $\cos 2\theta$, and $\tan 2\theta$ to three decimal places, using only trigonometric functions of θ. Check each answer by computing the trigonometric functions of 2θ directly using a calculator.

1. $\theta = 11.0°$ **2.** $\theta = 39.6°$ **3.** $\theta = 48°$

4. $\theta = 160°$ **5.** $\theta = \frac{\pi}{3}$ **6.** $\theta = -\frac{\pi}{6}$

7. $\theta = 195°$ **8.** $\theta = \frac{8\pi}{5}$ **9.** $\theta = -216°$

10. $\theta = \frac{13\pi}{6}$

11. **(a)** Use only trigonometric functions of 90° to find $\sin \pi$ and $\cos \pi$.

(b) Can you use tan 90° to get $\tan \pi$? Why?

12. For each of the following triangles, find $\sin 2\theta$, $\cos 2\theta$, and $\tan 2\theta$ by using only trigonometric functions of θ without finding θ.

(a)

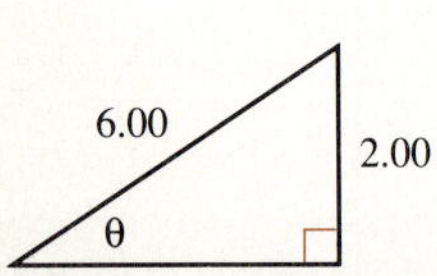

(b)

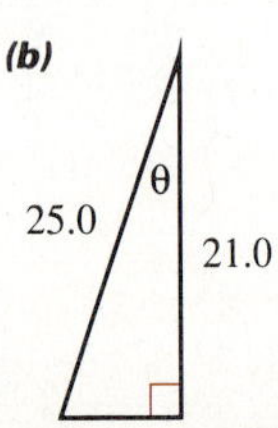

(c)

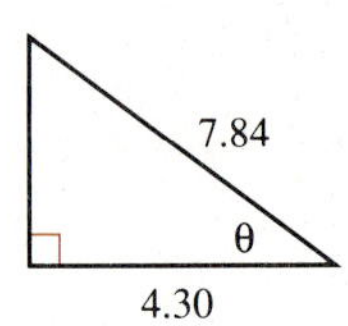

(d)

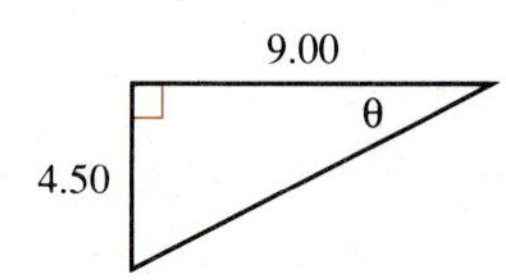

(e)

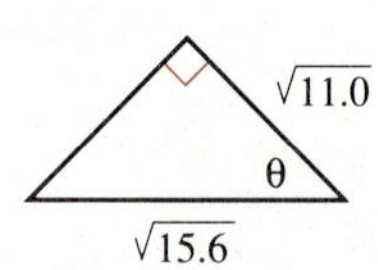

(f)

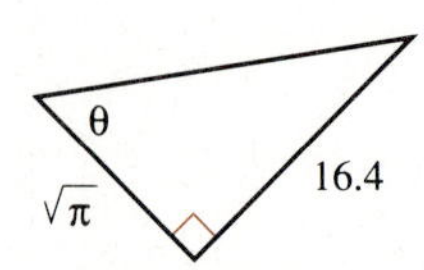

In Exercises 13–24, write each expression in terms of functions of half the angle. For example, $\sin 10x = \sin[2(5x)] = 2 \sin 5x \cos 5x$.

13. $\sin 4x$ **14.** $\sin 3x$ **15.** $\cos 6x$

16. $\cos(x/2)$ **17.** $\tan 4x$ **18.** $\sin 120°$

19. $\cos 35°$ **20.** $\cos 90°$ **21.** $\tan (x/3)$

22. $\cos 220°$ **23.** $\sin(90° + 2x)$ **24.** $\cos(60° + x)$

In Exercises 25–34, evaluate each trigonometric function without finding any angles.

25. $\sin(2 \sin^{-1} \frac{1}{3})$ *Hint:* let $\theta = \sin^{-1} \frac{1}{3}$.

26. $\cos(2 \cos^{-1} \frac{2}{3})$

27. $\sin(2 \cos^{-1} \frac{3}{4})$ **28.** $\cos(2 \sin^{-1} 0.420)$

29. $\tan(2 \tan^{-1} 3)$ **30.** $\tan(2 \sin^{-1} \frac{1}{4})$

31. $\sin(2 \csc^{-1} 3)$ **32.** $\cos(2 \csc^{-1} 4)$

33. $\tan(2 \cot^{-1} \frac{1}{3})$ **34.** $\tan(2 \sec^{-1} 1.5)$

In Exercises 35–43, evaluate each trigonometric function in terms of *a*.

***35.** $\sin(2 \tan^{-1} a)$ ***36.** $\tan(2 \tan^{-1} a)$

***37.** $\tan(2 \sin^{-1} a)$ ***38.** $\cos(2 \csc^{-1} a)$

***39.** $\tan(2 \sec^{-1} a)$ ***40.** $\sin(2 \sin^{-1} a)$

***41.** $\cos(2 \cos^{-1} a)$ ***42.** $\cos(2 \sin^{-1} a)$

***43.** $\sin(2 \cos^{-1} a)$

In Exercises 44–51, find $\sin 2\theta$, $\cos 2\theta$, and $\tan 2\theta$ without finding θ, given:

44. $\sin \theta = \frac{1}{2}$ and θ is in quadrant II.

45. $\cos \theta = \frac{3}{4}$ and θ is in quadrant IV.

46. $\tan \theta = -\frac{2}{3}$ and θ is in quadrant IV.

47. $\tan \theta = 4$ and θ is in quadrant III.

48. $\cot \theta = -\frac{10}{4}$ and $\sin \theta > 0$.

49. $\sec \theta = -3$ and θ is in quadrant III.

***50.** $\csc \theta = \frac{13}{12}$ and $450° < \theta < 540°$.

51. $\cos \theta = -1$ and $0 < \theta < 2\pi$.

In Exercises 52–66, write each expression as a single term. For example, 2 sin 15° cos 15° = sin[2(15°)] = sin 30°.

52. $2 \sin 40° \cos 40°$ **53.** $1 - 2 \sin^2 35°$

54. $2 \cos^2 48° - 1$ **55.** $2 \sin 32° \cos 32°$

56. $\dfrac{2 \tan 25°}{1 - \tan^2 25°}$ **57.** $\cos^2(x/2) - \sin^2(x/2)$

58. $4 \sin 4x \cos 4x$ **59.** $\sin 5x \cos 5x$

60. $2 - 4 \sin^2(7x)$ **61.** $4 \sin^2(3x) \cos^2(3x)$

62. $\dfrac{\tan \frac{x}{2}}{\frac{1}{2} - \frac{1}{2} \tan^2 \frac{x}{2}}$ ***63.** $\dfrac{2 \tan x}{2 - \sec^2 x}$

***64.** $\cos^4(5x) - \sin^4(5x)$ ***65.** $\frac{1}{2} \sec 3x \csc 3x$

***66.** $\dfrac{1 - \tan^2 6x}{2 \tan 6x}$

In Exercises 67–73, express each trigonometric function in terms of trigonometric functions of only θ. Simplify your answers as much as possible.

67. $\sin 4\theta$ *Hint:* $4\theta = 2\theta + 2\theta$

68. $\cos 4\theta$ **69.** $\sin 3\theta$ **70.** $\cos 3\theta$

***71.** $\tan 4\theta$ ***72.** $\sin 6\theta$ ***73.** $\cos 6\theta$.

In Exercises 74–83, solve each equation for x ($0 \le x \le 360°$).

74. $2 \sin x \cos x = 1$ **75.** $\sin x \cos x = -0.500$

76. $\cos^2 x - \sin^2 x = \frac{1}{2}$ **77.** $1 = \dfrac{2 \tan x}{1 - \tan^2 x}$

78. $\sin 2x - \sin x = 0$ **79.** $\cos 2x + \sin^2 x = \frac{3}{4}$

80. $\sin 2x = \sin x \cos x$

***81.** $\cos 4x - \cos^2 2x + \frac{1}{3} = 0$

***82.** $\cos^2 x - 2 \sin x \cos x - \sin^2 x = 0$

***83.** $\sin 2x + \cos 2x + 2 \sin^2 x = 2$

84. Range of a Rocket. (See Example 16)

(a) Show that the maximum range of a rocket is v^2/g.
(b) Suppose that two rockets are launched with the same initial speeds v. One is launched at an angle of $45° + \phi$ and the other at $45° - \phi$. Show that these rockets have the same range of $(v^2/g) \cos 2\phi$. Sketch the paths of these two rockets.
(c) For the rockets in (b), what angle ϕ gives the maximum range?

***85.** In right triangle ABC ($C = 90°$), the bisector of angle B divides the opposite side into two segments. If one segment is 2.4 times longer than the other, what are the angles of this triangle?

86. A ladder leaning against a wall makes an angle θ with the floor, and its top end is 5.00 ft above the floor. When the ladder is moved so that its angle with the floor is doubled, its top end is now 7.50 ft above the floor. How long is this ladder, and what angle did it originally make with the floor?

87. Suppose you cut two triangular pieces off the corners of a square sheet of paper. Both triangles have a hypotenuse of 12.0 cm, but one makes an angle twice that of the other with the original edge of the paper (see Figure 7.2-7). Find θ so that the areas of these two triangles will be equal, and find what that area is.

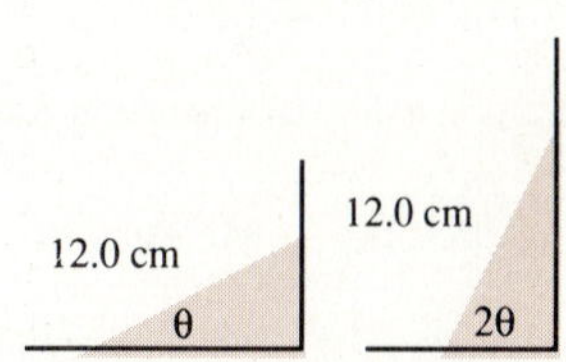

Figure 7.2-7

88. This Exercise is the same as Exercise 87 except that you are to find θ so that the area with angle 2θ is 1.50 times the area with angle θ.

***89.** For the oblique triangle in Figure 7.2-8, prove that $\cos\theta = a/2b$ and

$$\cos 2\theta = \frac{a^2 - b^2}{2b^2}$$

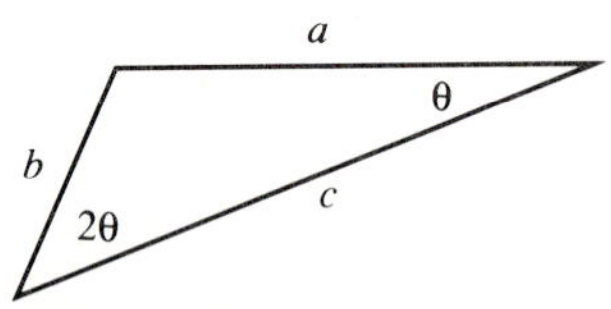

Figure 7.2-8

90. Suppose that while you are walking, the *length* of your stride increases by 60% when you double the angle of your stride. What are the angles of these two strides?

91. A ground-based microwave receiver is supposed to pick up signals from the tops of two transmitting antennas that are right next to each other. One is 50.0 ft tall and other is 140.0 ft tall (see Figure 7.2-9). How far from these antennas should the receiver be located so that the angle of elevation of the top of the taller antenna is twice that of the top of the shorter one? What is the angle of elevation of each one?

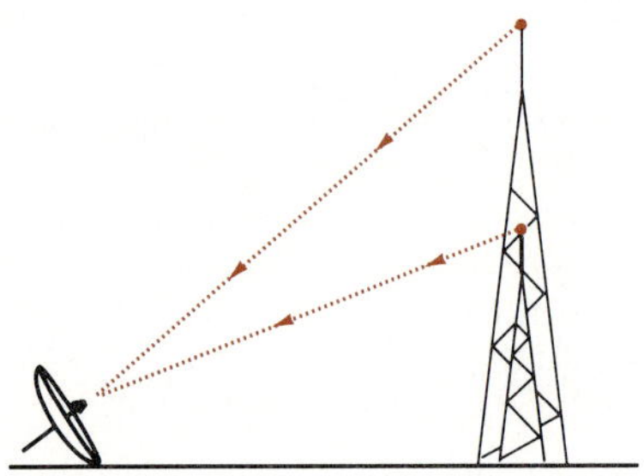

Figure 7.2-9

***92.** Suppose that in the previous problem, the taller antenna was only 90.0 ft tall, but everything else was the same. How far should the receiver be in this case so that the taller antenna has twice the angle of elevation of the shorter antenna? Interpret the meaning of your answer. Justify its significance on geometric grounds.

In Exercises 93–104, prove that each equation is an identity.

93. $\cos^2 x = \sin^2 x + \cos 2x$

94. $\tan x = \dfrac{\sin 2x}{1 + \cos 2x}$

95. $\cos^4 x - \sin^4 x = \cos 2x$

96. $2 \csc 2x = \sec x \csc x$

97. $\tan 2x = \dfrac{2 \sin x}{2 \cos x - \sec x}$

98. $2 \sin x \csc 2x = \sec x$

99. $\cos 2x + 2 \sin^2 x = 1$

100. $\dfrac{\cot^2 x - 1}{\csc^2 x} = \cos 2x$

101. $\dfrac{1 - \sin 2x}{\cos 2x} = \dfrac{1 - \tan x}{1 + \tan x}$

****102.** $\dfrac{\sec x + 1}{\sec x - 1} = \csc^2\left(\dfrac{x}{2}\right) - 1$

***103.** $4(\sin^6 x + \cos^6 x) = 4 - 3 \sin^2 2x$

***104.** $2(\sin 4x \cos 2x - \cos 4x \sin 2x)^2 = 1 - \cos 4x$

In Exercises 105–109, *ABC* is a right triangle ($C = 90°$) having sides *a*, *b*, and *c*. Prove each of the following:

105. $\sin 2A = \sin 2B$

106. $\cos 2A + \cos 2B = 0$

107. $\cos 2A = \dfrac{b^2 - a^2}{c^2}$

108. $\tan 2A = \dfrac{2ab}{b^2 - a^2}$

***109.** $\sin 3A = \dfrac{3ab^2 - a^3}{c^3}$

EXERCISES 7.2-2

In Exercises 1–11, use only trigonometric functions of θ to find $\sin(\theta/2)$, $\cos(\theta/2)$, and $\tan(\theta/2)$ to three decimal places. Be sure to figure out the correct sign of each. Check your answers by finding the functions of $\theta/2$ directly.

1. $\theta = 18.7°$ **2.** $\theta = \frac{\pi}{7}$ **3.** $\theta = \pi$

4. $\theta = 162°$ **5.** $\theta = 190°$ **6.** $\theta = 1.2\pi$

7. $\theta = 280°$ **8.** $\theta = \frac{20\pi}{11}$ **9.** $\theta = -174°$

10. $\theta = -\frac{\pi}{3}$ **11.** $\theta = 400°$

12. Can you use the half-angle formula for the tangent to find $\tan\frac{\pi}{4}$ from $\tan\frac{\pi}{2}$? Why?

13. For each triangle in Exercise 12 of Exercises 7.2-1, find $\sin(\theta/2)$, $\cos(\theta/2)$, and $\tan(\theta/2)$ using only trigonometric functions of θ without finding θ.

14. If $\cos 2\theta = \frac{1}{5}$ and θ is in quadrant I, find $\sin\theta$ and $\cos\theta$ without finding θ.

15. If $\cos\theta = 0.600$ and θ is in quadrant I, find $\sin(\theta/2)$ and $\cos(\theta/2)$ without finding θ.

***16.** Find $\sin(\theta/2)$, $\cos(\theta/2)$, and $\tan(\theta/2)$ without finding θ:

(a) if $\cos\theta = -\frac{12}{13}$ and θ is a positive angle in quadrant II.
(b) if $\cos\theta = -\frac{12}{13}$ and θ is a negative angle in quadrant III.

***17.** Find $\sin(\theta/2)$, $\cos(\theta/2)$, and $\tan(\theta/2)$ without finding θ:

(a) if $\sin\theta = \frac{24}{25}$ and θ is a negative angle in quadrant I.
(b) if $\sin\theta = \frac{24}{25}$ and θ is a positive angle in quadrant II.

In Exercises 18–23, evaluate each trigonometric function without finding any angles or using tables. *Hint:* See Example 20.

18. $\cos(\frac{1}{2}\cos^{-1}\frac{2}{3})$

19. $\sin(\frac{1}{2}\sin^{-1}\frac{1}{3})$

20. $\sin(\frac{1}{2}\tan^{-1}\frac{7}{24})$

21. $\cos(\frac{1}{2}\sin^{-1}\frac{7}{25})$

***22.** $\sin(\pi + \frac{1}{2}\sin^{-1}\frac{7}{25})$

***23.** $\cos(90° - \frac{1}{2}\sin^{-1}\frac{1}{4})$

In Exercises 24–32, prove that each equation is an identity.

24. $2\sin^2\left(\frac{x}{2}\right) = \frac{\sin^2 x}{1 + \cos x}$

25. $2\cos^2\left(\frac{x}{2}\right) = \frac{\sin^2 x}{1 - \cos x}$

26. $\sin^2\left(\frac{x}{2}\right)\cos^2\left(\frac{x}{2}\right) = \frac{\sin^2 x}{4}$

27. $\cos^2\left(\frac{x}{2}\right)\tan\left(\frac{x}{2}\right) = \frac{\sin x}{2}$

***28.** $$\frac{\cos\left(\frac{x}{2}\right) + \sin\left(\frac{x}{2}\right)}{\cos\left(\frac{x}{2}\right) - \sin\left(\frac{x}{2}\right)} = \frac{1 + \sin x}{\cos x}$$

29. $$\sin A + \sin B = 2\sin\left(\frac{A + B}{2}\right)\cos\left(\frac{A - B}{2}\right)$$

30. $$\sin A - \sin B = 2\cos\left(\frac{A + B}{2}\right)\sin\left(\frac{A - B}{2}\right)$$

31. $$\cos A + \cos B = 2\cos\left(\frac{A + B}{2}\right)\cos\left(\frac{A - B}{2}\right)$$

32. $$\cos A - \cos B = -2\sin\left(\frac{A + B}{2}\right)\sin\left(\frac{A - B}{2}\right)$$

***33.** A 25.0-ft long drawbridge is raised to an angle θ above the horizontal (see Figure 7.2-10). If you lower it to $\theta/2$, the vertical clearance under the center of the bridge decreases by $33\frac{1}{3}\%$. Find θ and the center clearance at both angles.

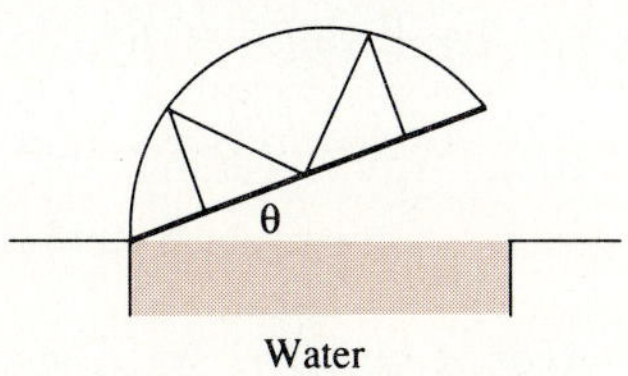

Figure 7.2-10

***34.** An 8.0-ft stepladder reaches 1.6 times higher at angle θ as it does at angle $\theta/2$ (see Figure 7.2-11). Find θ and the ladder's height at these angles.

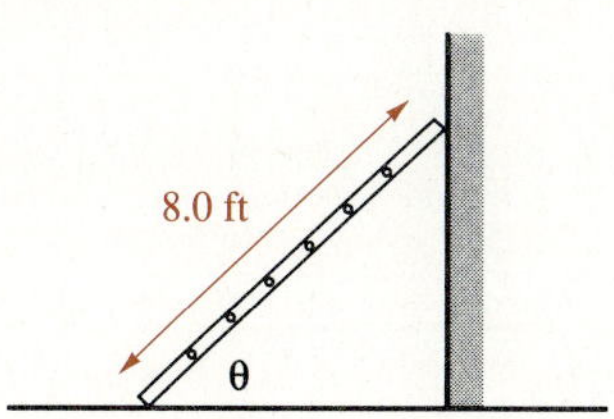

Figure 7.2-11

In Exercises 35–40, *A*, *B*, and *C* are the angles of a triangle. Show that:

***35.** $4\cos(A/2)\cos(B/2)\cos(C/2) = \sin A + \sin B + \sin C$

***36.** $4\sin(A/2)\sin(B/2)\cos(C/2) = \sin A + \sin B - \sin C$

***37.** $4\cos(A/2)\cos(B/2)\sin(C/2) - 1 = \cos A + \cos B - \cos C$

****38.** $\tan(A/2)\tan(B/2) + \tan(A/2)\tan(C/2) + \tan(B/2)\tan(C/2) = 1$

****39.** $\sin^2(A/2) + \sin^2(B/2) + \sin^2(C/2) = 1 - 2\sin(A/2)\sin(B/2)\sin(C/2)$

****40.** $1 + 4\sin(A/2)\sin(B/2)\sin(C/2) = \cos A + \cos B + \cos C$

****41.** Show that the roots of $x^2 - (2\csc\theta)x + 1 = 0$ are $x = \tan(\theta/2)$ and $x = \cot(\theta/2)$.

In Exercises 42–44, derive formulas for each and simplify your answers as much as possible.

42.** Derive a formula for $\cos(\theta/4)$ in terms of functions of θ. ***Hint: $\cos(\theta/4) = \cos[\frac{1}{2}(\theta/2)]$.

***43.** Derive a formula for $\sin(\theta/4)$ in terms of functions of θ.

***44.** Derive a formula for $\tan(\theta/4)$ in terms of functions of θ.

45. Lost on a Desert Island.** Suppose you found yourself stranded on the proverbial desert island with plenty of paper and pencils and a calculator that performs only the four basic operations of addition, subtraction, multiplication, and division. (Alas! No trig functions!) For amusement (you're really getting bored!), you decide to construct a trigonometry table of the values of the sine, cosine, and tangent for angles between 0° and 90°. To compile your table, you may use only the values of the trigonometric functions that you can figure out by algebra and geometry (0°, 30°, 45°, 60°, and 90°), plus the formulas from Sections 7.1 and 7.2. Show clearly how you would find the value for each of the following trigonometric functions. It is not necessary to compute the value unless you feel very ambitious; just show how you would get them. ***Hint: Rather than trying hit-or-miss, develop a strategy to evaluate the trigonometric functions. Then apply your strategy to the angles asked for.

(a) sin 7.5° (This one is a give-away just to get you started!)

(b) cos 18.75°

(c) sin 18.75°

(d) cos 73.125°

(e) cos 89.0625°

(f) sin 4.6875°

(g) cos 78.7500°

7.3 PRODUCT-TO-SUM AND SUM-TO-PRODUCT FORMULAS

In this section we develop formulas that are useful in mathematics, especially calculus and other applied areas. Although rarely used for numerical computation, these formulas are useful for solving equations or for deriving other formulas.

7.3-1 Product-to-Sum Formulas

$$\sin A \cos B = \tfrac{1}{2}[\sin(A + B) + \sin(A - B)] \quad (1)$$

$$\cos A \sin B = \tfrac{1}{2}[\sin(A + B) - \sin(A - B)] \quad (2)$$

$$\cos A \cos B = \tfrac{1}{2}[\cos(A + B) + \cos(A - B)] \quad (3)$$

$$\sin A \sin B = \tfrac{1}{2}[\cos(A - B) - \cos(A + B)] \quad (4)$$

We obtain eq. (1) by adding the identities for $\sin(A \pm B)$ from Section 7.1.

$$\begin{aligned} \sin(A + B) + \sin(A - B) &= (\sin A \cos B + \cos A \sin B) \\ &\quad + (\sin A \cos B - \cos A \sin B) \\ &= 2 \sin A \cos B \end{aligned}$$

Dividing by 2 gives us

$$\sin A \cos B = \tfrac{1}{2}[\sin(A + B) + \sin(A - B)]$$

We can derive eq. (3) in a similar way. Adding the identities for $\cos(A \pm B)$ from Section 7.1 gives us

$$\begin{aligned} \cos(A + B) + \cos(A - B) &= (\cos A \cos B - \sin A \sin B) \\ &\quad + (\cos A \cos B + \sin A \sin B) \\ &= 2 \cos A \cos B \end{aligned}$$

Dividing by 2 gives us

$$\cos A \cos B = \tfrac{1}{2}[\cos(A + B) + \cos(A - B)]$$

NOW IT'S YOUR TURN

Use the $\sin(A \pm B)$ identities to derive eq. (2), and use the $\cos(A \pm B)$ identities to derive eq. (4).

Example 23

Use a product-to-sum formula to find the value of $\sin 67.5° \sin 22.5°$.

Solution

Using eq. (4) with $A = 67.5°$ and $B = 22.5°$ gives us

$$\begin{aligned}\sin 67.5° \sin 22.5° &= \tfrac{1}{2}[\cos(67.5° - 22.5°) - \cos(67.5° + 22.5°)]\\ &= \tfrac{1}{2}[\cos 45° - \cos 90°] = \tfrac{1}{2}[\sqrt{2}/2 - 0] = \sqrt{2}/4\end{aligned}$$ ■

NOW IT'S YOUR TURN

Use a product-to-sum formula to find $\cos 45° \sin 15°$. (Answer: $(\sqrt{3} - 1)/4$)

Example 24

Prove the identity

$$\cos 3x \sin x = \sin x \cos x\,(4\cos^2 x - 3)$$

Solution

Working with the left-hand side gives us

$$\begin{aligned}\cos 3x \sin x &= \tfrac{1}{2}[\sin(3x + x) - \sin(3x - x)] && \text{(eq. (2))}\\ &= \tfrac{1}{2}[\sin 4x - \sin 2x]\\ &= \tfrac{1}{2}\{\sin[2(2x)] - \sin 2x\}\\ &= \tfrac{1}{2}[2\sin 2x \cos 2x - 2\sin x \cos x] && \text{(double-angle formula for } \sin 4x \text{ and } \sin 2x)\\ &= \sin 2x \cos 2x - \sin x \cos x\\ &= (2\sin x \cos x)(2\cos^2 x - 1) - \sin x \cos x && \text{(double-angle formulas for } \sin 2x \text{ and } \cos 2x)\\ &= \sin x \cos x\,[2(2\cos^2 x - 1) - 1] && \text{(factor } \sin x \cos x)\\ &= \sin x \cos x\,(4\cos^2 x - 3)\end{aligned}$$

which is identically equal to the right-hand side. Thus we have proved this identity. ■

NOW IT'S YOUR TURN

Prove that the following equation is an identity: $\sin 3x \cos x = \sin x \cos x\,(4\cos^2 x - 1)$.

7.3-2 Sum-to-Product Formulas

$$\sin A + \sin B = 2\sin\left(\frac{A+B}{2}\right)\cos\left(\frac{A-B}{2}\right) \quad (5)$$

$$\sin A - \sin B = 2\cos\left(\frac{A+B}{2}\right)\sin\left(\frac{A-B}{2}\right) \quad (6)$$

$$\cos A + \cos B = 2\cos\left(\frac{A+B}{2}\right)\cos\left(\frac{A-B}{2}\right) \quad (7)$$

$$\cos A - \cos B = -2\sin\left(\frac{A+B}{2}\right)\sin\left(\frac{A-B}{2}\right) \quad (8)$$

To derive eq. (5), we start with product-to-sum eq. (1) written in terms of x and y instead of A and B.

$$\sin x \cos y = \tfrac{1}{2}[\sin(x+y) + \sin(x-y)]$$

If we let $x + y = A$ and $x - y = B$, we can solve for A and B by first adding and then subtracting these two equations.

Adding:

$$\begin{aligned} x + y &= A \\ x - y &= B \\ \hline 2x &= A + B \\ x &= \frac{A+B}{2} \end{aligned}$$

Subtracting:

$$\begin{aligned} x + y &= A \\ -(x - y) &= -B \\ \hline 2y &= A - B \\ y &= \frac{A-B}{2} \end{aligned}$$

Now substituting these values of x and y into product-to-sum eq. (1) gives us

$$\sin x \cos y = \tfrac{1}{2}[\sin(x+y) + \sin(x+y)]$$

$$\sin\left(\frac{A+B}{2}\right)\cos\left(\frac{A-B}{2}\right) = \frac{1}{2}[\sin A + \sin B]$$

$$\sin A + \sin B = 2\sin\left(\frac{A+B}{2}\right)\cos\left(\frac{A-B}{2}\right) \quad ■$$

NOW IT'S YOUR TURN

Use similar substitutions to prove eqs. (6) through (8).

Example 25

Use a sum-to-product formula to find the value of $\cos 75° + \cos 15°$.

Solution

Letting $A = 75°$ and $B = 15°$ in eq. (7), we get

$$\cos 75° + \cos 15° = 2 \cos\left(\frac{75° + 15°}{2}\right)\cos\left(\frac{75° - 15°}{2}\right) = 2 \cos 45° \cos 30°$$

$$= 2 \frac{\sqrt{2}}{2} \frac{\sqrt{3}}{2} = \frac{\sqrt{6}}{2} \quad ■$$

NOW IT'S YOUR TURN

Use a product-to-sum formula to find $\sin 105° - \sin 15°$. (Answer: $\sqrt{2}/2$)

Example 26

Prove the identity

$$\tan x = \frac{\cos 3x - \cos 5x}{\sin 3x + \sin 5x}$$

Solution

Simplifying the right-hand side, we get

$$\frac{\cos 3x - \cos 5x}{\sin 3x + \sin 5x} = \frac{-2 \sin 4x \sin(-x)}{2 \sin 4x \cos(-x)} \qquad \text{(eqs. (8) \& (5))}$$

$$= \frac{(-2 \sin 4x)(-\sin x)}{2 \sin 4x \cos x} \qquad \text{(use } \sin(-x) = -\sin x \text{ and } \cos(-x) = \cos x\text{)}$$

$$= \frac{(2 \sin 4x)(\sin x)}{2 \sin 4x \cos x} \qquad \text{(multiply)}$$

$$= \frac{\sin x}{\cos x} \qquad \text{(divide)}$$

$$= \tan x \quad ■$$

Example 27

Solve the equation $\cos 3\theta - \cos \theta + \sin 2\theta = 0$ for all nonnegative values of θ less than 180° ($0 \le \theta < 180°$).

Solution

$$\underbrace{\cos 3\theta - \cos \theta}_{-2 \sin(2\theta) \sin \theta} + \sin 2\theta = 0 \qquad \text{(eq. (8))}$$

$$-2 \sin 2\theta \sin \theta + \sin 2\theta = 0$$

$$(\sin 2\theta)(-2 \sin \theta + 1) = 0 \quad \textbf{(factor } \sin 2\theta\textbf{)}$$

To find the solutions, we set each factor equal to zero. The first term gives us

$$\sin 2\theta = 0 \rightarrow 2\theta = \sin^{-1} 0 \rightarrow \begin{cases} 2\theta = 0° \rightarrow \theta = 0° \\ 2\theta = 180° \rightarrow \theta = 90° \end{cases}$$

The second term gives us

$$-2 \sin \theta + 1 = 0 \rightarrow \sin \theta = \tfrac{1}{2} \rightarrow \theta = \sin^{-1} \tfrac{1}{2} \rightarrow \theta = \begin{cases} 30° \\ 150° \end{cases}$$

Therefore, the solutions are $\theta = 0°, 30°, 90°, 150°$. Check these answers by substituting them into the original equation.
$\theta = 0°$: $\cos[3(0°)] - \cos 0° + \sin[2(0°)] = 1 - 1 + 0 = 0\surd$
$\theta = 30°$: $\cos[3(30°)] - \cos 30° + \sin[2(30°)] = 0 - \sqrt{3}/2 + \sqrt{3}/2 = 0\surd$
You check the rest! ■

NOW IT'S YOUR TURN

Find all nonnegative values of x less than 180° that satisfy the equation $\sin 3\theta - \sin \theta = \cos 2\theta$. Check all your answers! (Answer: 30°, 45°, 135°, 150°)

EXERCISES 7.3

In Exercises 1–10, express the sums and differences of functions as products.

1. $\sin 70° + \sin 50°$
2. $\sin 70° - \sin 50°$
3. $\cos 140° + \cos 80°$
4. $\cos \frac{3\pi}{2} + \cos \frac{\pi}{2}$
5. $\sin 3x + \sin x$
6. $\cos 4x - \cos 2x$
7. $\sin(x + y) + \sin(x - y)$
8. $\cos(x + y) - \cos(x - y)$

*9. $\cos 4kt - \cos(10kt - \frac{7\pi}{6})$

*10. $\cos(x + y + \frac{3\pi}{2}) - \cos(x - y - \frac{7\pi}{2})$

In Exercises 11–23, express each sum as a product and simplify it.

11. $\sin 75° + \sin 15°$
12. $\sin 75° - \sin 15°$
13. $\cos \frac{5\pi}{6} + \cos \frac{\pi}{6}$
14. $\cos \frac{5\pi}{6} - \cos \frac{\pi}{6}$
15. $\cos 80° - \cos 40°$
16. $\sin 50° + \sin 70°$
17. $\dfrac{\sin 50° - \sin 30°}{\cos 50° + \cos 30°}$
18. $\dfrac{\sin 70° + \sin 50°}{\cos 70° - \cos 50°}$

*19. $\dfrac{\cos x - \cos y}{\sin x + \sin y}$

*20. $\dfrac{\sin x + \sin y}{\cos x + \cos y}$

*21. $\dfrac{\sin x + \sin y}{\cos x - \cos y}$

*22. $\dfrac{\cos x + \cos y}{\sin x - \sin y}$

*23. $\sin(2\pi f_1 t) + \sin(2\pi f_2 t)$

In Exercises 24–31, prove each identity.

24. $\dfrac{\cos 3x - \cos x}{\cos 3x + \cos x} = -\tan 2x \tan x$

25. $\dfrac{\sin 5x - \sin 3x}{\sin 5x + \sin 3x} = \tan x \cot 4x$

26. $\dfrac{\cos 3x + \cos x}{\sin 3x + \sin x} = \cot 2x$

27. $\dfrac{\cos 3x - \cos x}{\sin 3x - \sin x} = \dfrac{2 \tan x}{\tan^2 x - 1}$

*28. $\dfrac{\cos 2x + \cos 7x}{\cos 3x + \cos 6x} + \dfrac{\cos 6x - \cos x}{\cos 5x - \cos 2x} = \dfrac{2 \sin 4x}{\sin 3x}$

***29.** $\dfrac{\sin 10x + \sin 6x}{\sin 12x + \sin 4x} - \dfrac{\sin 5x - \sin x}{\sin 7x + \sin x} = \dfrac{2 \sin 2x}{\sin 8x}$

****30.** $\sin 2x + \sin 4x + \sin 6x = 4 \cos x \cos 2x \sin 3x$ *Hint:* Use the sum-to-product formula on six $2x + \sin 4x$, and use the double-angle formula on $\sin 6x = \sin(2(3x))$.

****31.** $\cos 2x + \cos 4x + \cos 6x = 4 \cos x \cos 2x \cos 3x - 1$

32. Solve $\sin 2\theta + \sin \theta = 0$ for all nonnegative values of $\theta < 360°$.

33. Solve $\cos 3\theta + \cos \theta = 0$ for all nonnegative values of $\theta < 360°$.

34. Solve $\sin 3\theta = \sin \theta$ for all nonnegative values of $\theta < 360°$.

***35.** Solve $\cos 3\theta + \cos \theta - \sin 2\theta = 0$ for all nonnegative values of $\theta < 360°$.

SUMMARY

1. Addition Formulas

$$\sin(A + B) = \sin A \cos B + \cos A \sin B$$

$$\sin(A - B) = \sin A \cos B - \cos A \sin B$$

$$\cos(A + B) = \cos A \cos B - \sin A \sin B$$

$$\cos(A - B) = \cos A \cos B + \sin A \sin B$$

$$\tan(A + B) = \frac{\tan A + \tan B}{1 - \tan A \tan B}$$

$$\tan(A - B) = \frac{\tan A - \tan B}{1 + \tan A \tan B}$$

2. Double-Angle Formulas

$$\sin 2A = 2 \sin A \cos A$$

$$\cos 2A = \begin{cases} \cos^2 A - \sin^2 A \\ 1 - 2 \sin^2 A \\ 2 \cos^2 A - 1 \end{cases}$$

$$\tan 2A = \frac{2 \tan A}{1 - \tan^2 A}$$

3. Half-Angle Formulas

$$\sin \frac{A}{2} = \pm \sqrt{\frac{1 - \cos A}{2}}$$

$$\cos \frac{A}{2} = \pm \sqrt{\frac{1 + \cos A}{2}}$$

$$\tan \frac{A}{2} = \pm \sqrt{\frac{1 - \cos A}{1 + \cos A}}$$

In each case, the correct sign depends on the quadrant in which $A/2$ lies.

4. Product-to-Sum Formulas

$$\sin A \cos B = \tfrac{1}{2}[\sin(A + B) + \sin(A - B)]$$
$$\cos A \sin B = \tfrac{1}{2}[\sin(A + B) - \sin(A - B)]$$
$$\cos A \cos B = \tfrac{1}{2}[\cos(A + B) + \cos(A - B)]$$
$$\sin A \sin B = \tfrac{1}{2}[\cos(A - B) - \cos(A + B)]$$

5. Sum-to-Product Formulas

$$\sin A + \sin B = 2 \sin\left(\frac{A + B}{2}\right)\cos\left(\frac{A - B}{2}\right)$$
$$\sin A - \sin B = 2 \cos\left(\frac{A + B}{2}\right)\sin\left(\frac{A - B}{2}\right)$$
$$\cos A + \cos B = 2 \cos\left(\frac{A + B}{2}\right)\cos\left(\frac{A - B}{2}\right)$$
$$\cos A - \cos B = -2 \sin\left(\frac{A + B}{2}\right)\sin\left(\frac{A - B}{2}\right)$$

REVIEW EXERCISES

1. Find the sine, cosine, and tangent of each angle without using a calculator or a table.

(a) 255° (use 30° and 225°) **(b)** 285° (use part (a))

(c) −195°

2. Find the sine, cosine, and tangent of each angle without using a calculator or a table.

(a) $\frac{11\pi}{12}$ (use $\frac{\pi}{6}$ and $\frac{3\pi}{4}$) **(b)** $-\frac{13\pi}{12}$

(c) $\frac{19\pi}{12}$ **(d)** $\frac{23\pi}{12}$

3. Find sin $(A + B)$ and cos $(A + B)$ without finding A or B.

(a) $\sin A = \frac{2}{3}$, $\sin B = \frac{3}{4}$, A and B in quadrant I
(b) $\tan A = \frac{2}{3}$, $\cot B = 2$, A and B in quadrant III
(c) $\sin A = 0.850$, $\cos B = -0.652$, A in quadrant I, B in quadrant III

4. Find sin $(A - B)$ and cos $(A - B)$ without finding A or B.

(a) $\cos A = -\frac{1}{4}$, $\tan B = 2$, A in quadrant II, B in quadrant III
(b) $\sin A = \frac{4}{5}$, $\cos B = -\frac{4}{5}$, A and B in quadrant II
(c) $\cos A = -0.408$, $\sin B = 0.621$, A in quadrant III, B in quadrant I

Evaluate each expression in Exercises 5–13 without using a calculator or a table.

5. $\sin 17° \cos 13° + \cos 17° \sin 13°$

6. $\sin 57° \cos 12° - \cos 57° \sin 12°$

7. $\dfrac{\tan 74° - \tan 14°}{1 + \tan 74° \tan 14°}$

8. $\cos 163° \cos 13° + \sin 163° \sin 13°$

9. $\dfrac{\tan 200° + \tan 10°}{1 - \tan 200° \tan 10°}$

***10.** $\cos(-53°) \sin(-337°) + \sin 307° \sin 113°$

11. $\cos 73° \cos 17° - \sin 73° \sin 17°$

***12.** $\dfrac{\sin(7\pi/30) \cos(4\pi/15) + \cos(7\pi/30) \sin(4\pi/15)}{\sin(7\pi/12) \cos(5\pi/12) - \cos(7\pi/12) \sin(5\pi/12)}$

***13.** $[\sin(4\pi/9) \cos(\pi/9) - \sin(\pi/18) \cos(7\pi/18)] \times \dfrac{[1 + \tan(5\pi/12) \tan(\pi/12)]}{\tan(5\pi/12) - \tan(\pi/12)}$

In Exercises 14–21, solve for all positive values of $x < 360°$.

14. $\sin(35° + x) + \sin(35° - x) = 0$

15. $\sin(m + x) - \sin(m - x) = 0$

16. $\cos(m + x) - \cos(m - x) = 0$

17. $1 + \sin x \sin 2x = \cos x \cos 2x$

18. $\tan 20° - \tan x = 1 + \tan 20° \tan x$

19. $\cos 3x + \sin x \sin 2x = 0$ ***Hint***: $3x = x + 2x$.

20. $\sin 5x - \sin 2x \cos 3x = 0$ ***Hint***: Does the hint in Exercise 19 give you any ideas?

21. $\cos(90° + x)\cos(90° - x) = -\frac{1}{2}$

In Exercises 22–26, prove that each equation is an identity if it is one.

22. **(a)** $\sin(45° + x) = \cos(45° - x)$
(b) $\cos(45° + x) = \sin(45° - x)$

23. $\cos(\frac{\pi}{4} + x)\cos(\frac{\pi}{4} - x) - \sin(\frac{\pi}{4} + x)\sin(\frac{\pi}{4} - x) = 0$

24. **(a)** $\sin^2(\frac{\pi}{6} + x) + \sin^2(\frac{\pi}{6} - x) - \sin^2 x = \frac{1}{2}$
(b) $\cos^2(x + 30°) + \cos^2(x - 30°) + \sin^2 x = \frac{3}{2}$

25. **(a)** $\sin 2x + \cos 2x \cot x = \cot x$
(b) $\sin 2x - \cos 2x \tan x = \tan x$

26. $\dfrac{\tan^2 x - \tan^2 y}{1 - \tan^2 x \tan^2 y} = \tan(x + y)\tan(x - y)$

***27.** The sines of two angles of an oblique triangle are equal to $\frac{5}{13}$ and $\frac{99}{101}$. Find the cosine of the third angle without finding any of the angles.

***28.** Evaluate the following without using a calculator or a table.

(a) $\tan(\arctan 2 + \arctan \frac{1}{2})$
(b) $\sin(\arcsin \frac{3}{5} + \arcsin \frac{4}{5})$
(c) $\cos[\sin^{-1}(-\frac{8}{17}) - \cos^{-1}(-\frac{3}{5})]$

In Exercises 29–33, find $\sin 2\theta$ without finding θ.

29. $\sin \theta = \frac{3}{5}$ and θ is in quadrant I

30. $\sec \theta = -\frac{13}{5}$ and θ is in quadrant III

31. $\csc \theta = \frac{25}{7}$ and θ is in quadrant II

32. $\tan \theta = -3$ and θ is in quadrant IV

33. $\cot \theta = 2$ and θ is in quadrant III

***34.** Compute $\sin x$ and $\cos x$ without finding x.

(a) $\cos(x/2) = -\frac{15}{17}$ and $\sin(x/2) > 0$
(b) $\tan(x/2) = -\frac{5}{12}$ and x is in quadrant IV

35. Evaluate each expression without finding x.

(a) $\dfrac{\sin x}{3 - 2\cos x}$ if $\tan \dfrac{x}{2} = 2$

(b) $\dfrac{\tan x + \sin x}{\tan x - \sin x}$ if $\tan \dfrac{x}{2} = \dfrac{2}{15}$

***36.** If $\sin x + \cos x = \frac{5}{4}$, find $\sin 2x$, $\cos 2x$, and $\tan 2x$ without finding x.

In Exercises 37–40, prove that each equation is an identity if it is one.

37. $1 + \cos 2x + 2\sin^2 x = 2$

38. $\dfrac{\tan \frac{x}{2}}{1 + \tan \frac{x}{2}} + \dfrac{\tan \frac{x}{2}}{1 - \tan \frac{x}{2}} = \tan x$

39. **(a)** $\sin 2x - \tan x = \cos 2x \tan x$
(b) $\sin 2x - \cot x = -\cos 2x \cot x$

40. $4\sin^4 x + \sin^2(2x) = 4\sin^2 x$

In Exercises 41–45, compute $\sin x/2$, $\cos x/2$, and $\tan x/2$ without finding x.

41. $\sin x = \frac{3}{5}$ and x is in quadrant I

42. $\tan x = \frac{3}{4}$ and x is in quadrant III

43. $\tan x = \frac{7}{24}$ and x is in quadrant I

44. $\sin x = -0.80$ and $\frac{3\pi}{2} < x < 2\pi$

***45.** $\cot x = -\frac{4}{3}$

In Exercises 46–49, solve for x in radians, $0 \le x < 2\pi$.

46. **(a)** $\sin x \cos x = \sqrt{3}/4$
(b) $\sin^2 x - \cos^2 x = 0.500$

47. **(a)** $\sin x = \sin 2x$
(b) $\cos 2x = \cos x$

48. $2\sin^2 x = \cos 2x$

***49.** $\tan 2x = 3\tan x$

50. Evaluate without using a calculator or a table.

(a) $\cos(\frac{1}{2} \arccos \frac{4}{5})$
(b) $2\cos^2(\frac{1}{2}\cos^{-1}(5/8))$
(c) $1 - 2\sin^2(\frac{1}{2}\cos^{-1}\frac{5}{13})$
(d) $\cos^2(\frac{1}{2} \arccos \frac{4}{5}) - \sin^2(\frac{1}{2}\cos^{-1}\frac{4}{5})$

In Exercises 51–54, solve for all x ($0 \le x < 2\pi$).

***51.** $\sin 4x \cos 2x = \sin 5x \cos x$

***52.** $\cos 5x \cos x = \cos 6x \cos 2x$

***53.** $\sin 2x \sin 6x = \sin 3x \sin 5x$

***54.** $\cos 3x = \sin 2x$

8

Trigonometric Equations and Inverse Trigonometric Functions

8.1 SOLVING TRIGONOMETRIC EQUATIONS

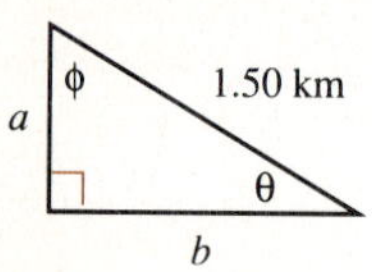

Figure 8.1-1

Jogging track

An urban planner is hired by the Parks and Recreation Department of Los Angeles to design a park with a jogging track 3.50 km long. Because of the space limitations of this park, the track must be a right triangle with a 1.50 km hypotenuse. In the plans for this park, what angles must the planner specify for the jogging track triangle?

To solve this problem, first sketch the jogging track (see Figure 8.1-1). It will be enough to find θ; we can subtract to get ϕ because a triangle has 180°. The definitions of the sine and cosine tell us that

$$\sin\theta = \frac{a}{1.50\text{ km}} \quad\rightarrow\quad a = 1.50\text{ km}\sin\theta$$

$$\cos\theta = \frac{b}{1.50\text{ km}} \quad\rightarrow\quad b = 1.50\text{ km}\cos\theta$$

This track must be 3.50 km long, so

$$a + b + 1.50\text{ km} = 3.50\text{ km}$$

$$1.50\text{ km}\sin\theta + 1.50\text{ km}\cos\theta + 1.50\text{ km} = 3.50\text{ km}$$

$$(1.50\text{ km})(\sin\theta + \cos\theta) = 3.50\text{ km} - 1.50\text{ km} = 2.0\text{ km}$$

Dividing by 1.50 km and using the fact that $\frac{2.0}{1.5} = 1.33 = \frac{4}{3}$ gives

$$\sin\theta + \cos\theta = \tfrac{4}{3} \quad ■$$

To complete this problem, we must solve the equation $\sin\theta + \cos\theta = \frac{4}{3}$. This equation is not an identity because it is not true for all allowed values of θ. (Check this yourself. Put in several values of θ and see that this equation does not hold for all of them.) Because this equation is true only for certain values of θ, we call it a *conditional equation* or simply an equation. In this chapter, we discuss how to solve conditional trigonometric equations, starting with the simpler ones first. The equation we have just found, $\sin\theta + \cos\theta = \frac{4}{3}$, is a bit harder to solve, so we postpone it.

The process of finding all the values of the angle that satisfy a conditional equation is called *solving the equation*. To solve a trigonometric equation, we use the principles of both algebra and trigonometry. There is no single procedure for solving all trigonometric equations, but always keep in mind that the main objective is to reduce the equation so that it involves only one trigonometric function of only one angle. Then it is usually a simple matter to solve for that trigonometric function and then find its inverse to get the angle. The following examples illustrate many useful procedures. (This is a good time for you to review Section 3.3.)

▶ **If the equation involves only one trigonometric function of a single angle, then solve algebraically for the values of that function.**

The first six examples illustrate different aspects of this method.

Example 1

Solve the equation $2\cos x - 1 = 0$ for all values of x between 0 and 360°.

Solution

$$\begin{aligned} 2\cos x - 1 &= 0 \\ 2\cos x &= 1 && \text{(add 1 to both sides)} \\ \cos x &= \tfrac{1}{2} && \text{(divide both sides by 2)} \\ x &= \cos^{-1}(\tfrac{1}{2}) \end{aligned}$$

The cosine is positive in quadrants I and IV, so x can lie in either of these quadrants. The reference angle is $\cos^{-1}(\frac{1}{2}) = 60°$. Therefore, in quadrant I, $x = 60°$. If x is in quadrant IV, $x = 360° - 60° = 300°$. Thus the solutions are $x = 60°$ and $x = 300°$. ■

NOW IT'S YOUR TURN

Solve the equation $\sqrt{2} - 2\sin x = 0$ for all positive values of $x < 360°$.
(Answer: 45°, 135°)

Example 2

Solve $\sin^2 x = 2 \sin x$ for all non-negative values of x less than 2π.

Solution

$$\sin^2 x = 2 \sin x \quad \text{(why can't we divide both sides by } \sin x\text{?)}$$

$$\sin^2 x - 2 \sin x = 0 \quad \text{(subtract 2 } \sin x \text{ from both sides)}$$

$$\sin x(\sin x - 2) = 0 \quad \text{(factor)}$$

We obtain the solutions by setting each factor equal to zero.

$$\sin x = 0 \quad \rightarrow \quad x = \sin^{-1}(0) \quad \rightarrow \quad x = 0, \pi$$

$$\sin x - 2 = 0 \quad \rightarrow \quad \sin x = 2$$

Because $\sin x$ can never be greater than 1, this last equation has no solutions. Therefore, the solutions to this equation are $x = 0, \pi$. ■

NOW IT'S YOUR TURN

Find all values of x ($0 \leq x < 2\pi$) that satisfy the equation $2 \cos^2 x = \cos x$.
(Answer: $\frac{\pi}{3}, \frac{\pi}{2}, \frac{3\pi}{2}, \frac{5\pi}{3}$)

Example 3

Find all values of θ ($0 \leq \theta < 2\pi$) for which $4.50 \cos^2 \theta - 3.66 = 0$.

Solution

$$4.50 \cos^2 \theta - 3.66 = 0$$

$$4.50 \cos^2 \theta = 3.66 \quad \text{(add 3.66 to both sides)}$$

$$\cos^2 \theta = \frac{3.66}{4.50} = 0.813 \quad \text{(divide both sides by 4.50)}$$

$$\cos \theta = \pm \sqrt{0.813} = \pm 0.902 \quad \text{(take square roots of both sides)}$$

Using the positive root: $\cos \theta = 0.902$. The reference angle is $\cos^{-1}(0.902) = 0.447$ rad and θ can be in quadrants I or IV (positive cosine). The two solutions are

$$\theta = 0.447 \quad \text{(if } \theta \text{ is in quadrant I)}$$

$$\theta = 2\pi - 0.447 = 5.83 \quad \text{(if } \theta \text{ is in quadrant IV)}$$

Using the negative root: $\cos \theta = -0.902$. The reference angle is still 0.447 rad, but this time θ can lie in quadrants II or III (negative cosine). These solutions are

$$\theta = \pi - 0.447 = 2.69 \quad \text{(if } \theta \text{ is in quadrant II)}$$

$$\theta = \pi + 0.447 = 3.59 \quad \text{(if } \theta \text{ is in quadrant III)}$$

Therefore, the full set of solutions is

$$x = 0.447, 2.69, 3.59, 5.83 \quad ■$$

NOW IT'S YOUR TURN

Find all θ ($0 \le \theta < 2\pi$) for which $3.00 \sin^2 \theta - 1.00 = 0$.
(Answer: 0.615, 2.53, 3.76, 5.67)

Example 4

Solve $\cos 3x = \sqrt{2}/2$ for all x ($0 \le x \le 2\pi$).

Solution

If $\cos 3x = \sqrt{2}/2$, then $3x = \cos^{-1}(\sqrt{2}/2)$, which means that $3x = \frac{\pi}{4}$ and $3x = \frac{7\pi}{4}$. But there are more values! Because $0 \le x \le 2\pi$, the smallest that $3x$ can be is $3(0) = 0$, and the largest it can be is $3(2\pi) = 6\pi$. Therefore, $3x$ is between 0 and 6π: $0 \le 3x \le 6\pi$. To find all the values of x, we use the fact that the cosine has a period of 2π, so we add multiples of 2π to the two smallest ones.

$$3x = \begin{cases} \frac{\pi}{4} \\ \frac{\pi}{4} + 2\pi = \frac{9\pi}{4} \\ \frac{\pi}{4} + 4\pi = \frac{17\pi}{4} \end{cases} \quad \text{and} \quad 3x = \begin{cases} \frac{7\pi}{4} \\ \frac{7\pi}{4} + 2\pi = \frac{15\pi}{4} \\ \frac{7\pi}{4} + 4\pi = \frac{23\pi}{4} \end{cases}$$

$$3x = \frac{\pi}{4}, \frac{7\pi}{4}, \frac{9\pi}{4}, \frac{15\pi}{4}, \frac{17\pi}{4}, \frac{23\pi}{4}.$$

Dividing by 3 to solve for x gives us

$$x = \frac{\pi}{12}, \frac{7\pi}{12}, \frac{3\pi}{4}, \frac{5\pi}{4}, \frac{17\pi}{12}, \frac{23\pi}{12}. \quad ■$$

NOW IT'S YOUR TURN

Find all x ($0 \le x \le 2\pi$) for which $2 \sin 2x = 1$. (Answer: $\frac{\pi}{12}, \frac{5\pi}{12}, \frac{13\pi}{12}, \frac{17\pi}{12}$)

Example 5

Find *all* values of x in radians for which $8 \tan x + 8 = 0$.

Solution

Solving for $\tan x$ gives us

$$8 \tan x + 8 = 0$$

$$8 \tan x = -8 \quad \text{(subtract 8 from both sides)}$$

$$\tan x = -1 \quad \text{(divide by 8)}$$

The reference angle is $\frac{\pi}{4}$, and the tangent is negative in quadrant II, so $x = \frac{3\pi}{4}$, but this solution is not the only one. The tangent has period π, so if we add $\pm\pi$ to this solution, its tangent will also be -1. Therefore, we can get all the solutions by adding whole number multiples of π to our answer.

$$x = \begin{cases} \frac{3\pi}{4} \\ \frac{3\pi}{4} \pm \pi \\ \frac{3\pi}{4} \pm 2\pi \\ \frac{3\pi}{4} \pm 3\pi \\ \vdots \end{cases}$$

We can express this equation in a more compact form by writing it as

$$x = \tfrac{3\pi}{4} \pm n\pi$$

where n represents an integer (whole number) that can take on all integral values starting from 0; that is, $n = 0, 1, 2, 3, 4, \ldots$. We usually write this solution as

$$x = \tfrac{3\pi}{4} \pm n\pi \; (n = 0, 1, 2, 3, \ldots) \; ■$$

Expressing angles in this form is very common in the study of mathematics, acoustics, waves, transmission lines, vibrations, and so on.

NOW IT'S YOUR TURN

Find all values of x that satisfy the equation $7.8 \tan x - 7.8 = 0$.
(Answer: $\frac{\pi}{4} \pm n\pi$, $n = 0, 1, 2, 3, \ldots$)

Example 6

Solve the equation $6 \cos^2 \theta - \cos \theta - 3 = 0$ for $0 \leq \theta < 360°$.

Solution

Notice that this equation contains only one trigonometric function ($\cos \theta$), but the equation is a quadratic equation in which the variable is $\cos \theta$. If we let $x = \cos \theta$, then this equation may look more familiar to you.

$$6x^2 - x - 3 = 0$$

The first step is to solve for x ($\cos \theta$) and then use this result to solve for θ.

This quadratic equation is not easily factorable, so we use the quadratic formula.

$$x = \frac{-b \pm \sqrt{b^2 - 4ac}}{2a}$$

In this case, $a = 6$, $b = -1$, and $c = -3$, which gives us

$$x = \frac{-(-1) \pm \sqrt{(-1)^2 - 4(6)(-3)}}{2(6)} = \frac{1 \pm \sqrt{1 + 72}}{12} = \frac{1 \pm \sqrt{73}}{12}$$

The two solutions are

$$x = \frac{1 \pm \sqrt{73}}{12} = \begin{cases} \dfrac{1 + \sqrt{73}}{12} = 0.795 \\ \dfrac{1 - \sqrt{73}}{12} = 0.629 \end{cases}$$

Now we can use the fact that $x = \cos\theta$ to find θ.

Using the upper solution: $\cos\theta = 0.795 \quad \rightarrow \quad \theta = \cos^{-1}(0.795)$. The reference angle is 37.3° (from a calculator), and θ can be in quadrants I or IV (positive cosine).

$$\theta = \begin{cases} 37.3° & \textbf{(in quadrant I)} \\ 360° - 37.3° = 323° & \textbf{(in quadrant IV)} \end{cases}$$

Using the lower solution: $\cos\theta = -0.629 \quad \rightarrow \quad \theta = \cos^{-1}(-0.629)$. The reference angle is 51.0° (from a calculator), and θ can lie in quadrants II or III (negative cosine).

$$\theta = \begin{cases} 180° - 51.0° = 129° & \textbf{(in quadrant II)} \\ 180° + 51.0° = 231° & \textbf{(in quadrant III)} \end{cases}$$

Therefore, the solutions are $\theta = 37.3°, 129°, 231°, 323°$. ■

NOW IT'S YOUR TURN

Find all θ $(0 \le \theta < 360°)$ to three-figure accuracy for which $12\sin^2\theta - \sin\theta = 1$. (Answer: 19.5°, 161°, 194°, 346°)

▸ **If an equation involves several trigonometric functions of a single angle, then use the fundamental identities to write the equation in terms of only one trigonometric function.**

The following examples illustrate this procedure.

Example 7

Solve the equation $\sin x = \csc x$ for $0 \le x < 2\pi$.

Solution

Using the fundamental identity $\csc x = \dfrac{1}{\sin x}$, this equation becomes

$$\sin x = \csc x$$

$$\sin x = \frac{1}{\sin x} \qquad \left(\textbf{use fundamental identity } \csc x = \frac{1}{\sin x}\textbf{, provided } \sin x \neq 0\right)$$

$$\sin^2 x = 1 \qquad \textbf{(multiply both sides by } \sin x\textbf{)}$$

$$\sin x = \pm 1 \qquad \textbf{(take square roots)}$$

The positive root gives

$$\sin x = 1 \qquad \rightarrow \qquad x = \sin^{-1}(1) = \tfrac{\pi}{2}$$

The negative root gives

$$\sin x = -1 \qquad \rightarrow \qquad x = \sin^{-1}(-1) = \tfrac{3\pi}{2}$$

Therefore, the solutions are $x = \frac{\pi}{2}$ and $\frac{3\pi}{2}$. ■

NOW IT'S YOUR TURN

Find all x $(0 \leq x < 2\pi)$ for which $\cos x = \sec x$. (Answer: $0, \pi$)

Example 8

Now we can solve the equation we encountered at the beginning of this section, namely $\sin\theta + \cos\theta = \frac{4}{3}$.

Solution

Because this equation contains two trigonometric functions, we first want to reduce these to only *one* such function. We can do this using the Pythagorean identity $\sin^2\theta + \cos^2\theta = 1$. This identity gives us

$$\cos^2\theta = 1 - \sin^2\theta \qquad \rightarrow \qquad \cos\theta = \pm\sqrt{1 - \sin^2\theta}$$

Substituting this equation into the original equation gives us

$$\begin{aligned} \sin\theta + \cos\theta &= \tfrac{4}{3} \\ \sin\theta \pm \sqrt{1 - \sin^2\theta} &= \tfrac{4}{3} \end{aligned}$$

To solve for $\sin\theta$, we must get rid of the square root. To do so, first isolate it on one side of the equation by subtracting $\sin\theta$ from both sides. This gives us

$$\pm\sqrt{1 - \sin^2\theta} = \tfrac{4}{3} - \sin\theta$$

$$1 - \sin^2\theta = (\tfrac{4}{3} - \sin\theta)^2 = \tfrac{16}{9} - \tfrac{8}{3}\sin\theta + \sin^2\theta \quad \textbf{(square)}$$

$$2\sin^2\theta - \tfrac{8}{3}\sin\theta + \tfrac{7}{9} = 0 \quad \textbf{(combine terms)}$$

This problem is the same kind we encountered in Example 6. It is a quadratic equation in which the unknown is $\sin\theta$. Applying the quadratic formula ($a = 2$, $b = -\frac{8}{3}$, $c = \frac{7}{9}$) gives us

$$\sin\theta = \frac{-(-\frac{8}{3}) \pm \sqrt{(-\frac{8}{3})^2 - 4(2)(\frac{7}{9})}}{2(2)} = \frac{\frac{8}{3} \pm \sqrt{\frac{8}{9}}}{4} = \frac{\frac{8}{3} \pm \frac{1}{3}\sqrt{8}}{4} = \frac{2}{3} \pm \frac{\sqrt{8}}{12}$$

The two solutions are

$$\sin\theta = \begin{cases} \dfrac{2}{3} + \dfrac{\sqrt{8}}{12} = 0.667 + 0.236 = 0.903 \\[2ex] \dfrac{2}{3} - \dfrac{\sqrt{8}}{12} = 0.667 - 0.236 = 0.431 \end{cases}$$

The first solution gives us $\sin\theta = 0.903$. The reference angle is 64.6°, and θ can be in quadrants I or II (positive sine).

$$\theta = \sin^{-1}(0.903) = \begin{cases} 64.6° & \textbf{(quadrant I)} \\ 180° - 64.6° = 115° & \textbf{(quadrant II)} \end{cases}$$

The second solution gives us $\sin\theta = 0.431$. The reference angle is 25.5°, and θ can be in quadrants I or II (positive sine).

$$\theta = \sin^{-1}(0.431) = \begin{cases} 25.5° & \textbf{(quadrant I)} \\ 180° - 25.5° = 155° & \textbf{(quadrant II)} \end{cases}$$

Before accepting these answers as final, we must check something. While solving this problem, *we squared both sides and took the square root with the quadratic formula*. As you learned in algebra, this method can introduce extra solutions (called *extraneous roots*) that are *not* solutions to the original equation. Therefore, we must check each answer to see if it satisfies the original equation, which was $\sin\theta + \cos\theta = \frac{4}{3} = 1.33$.

$$\sin 64.6° + \cos 64.6° = 1.33 \quad \checkmark$$

$$\sin 115° + \cos 115° = 0.484 \quad \text{X}$$

$$\sin 25.5° + \cos 25.5° = 1.33 \quad \checkmark$$

$$\sin 155° + \cos 155° = -0.484 \quad \text{X}$$

Only the solutions 64.6° and 25.5° are valid; the other two are extraneous roots. ■

NOW IT'S YOUR TURN

Solve the equation $\sin\theta + \cos\theta = 1$ and check your answers.
(Answer: $\theta = 0$ and $\frac{\pi}{2}$; $\theta = \pi$ is an extraneous root.)

▶ **When an equation involves trigonometric functions of more than one angle, we can often solve it by reducing it to an equation involving trigonometric functions of a single angle.**

The double-angle and half-angle formulas are especially useful for this reduction, as the following examples show.

Example 9

Solve $\cos 2x + 3 \sin x = 2$ for $0 \le x < 360°$.

Solution

Remember that our goal is to reduce this equation so that it involves a single trigonometric function of a single angle. We begin by expressing $\cos 2x$ in terms of $\sin x$ using the double-angle formulas from Chapter 7. There are three forms for $\cos 2x$.

$$\cos 2x = \cos^2 x - \sin^2 x = 1 - 2\sin^2 x = 2\cos^2 x - 1$$

Because the original equation involves $\sin x$, we should use the identity $\cos 2x = 1 - 2\sin^2 x$. Using this identity allows us to express the original equation in terms of only $\sin x$.

$$\begin{aligned} \cos 2x + 3\sin x &= 2 \\ (1 - 2\sin^2 x) + 3\sin x &= 2 \quad \textbf{(use } \cos 2x = 1 - 2\sin^2 x\textbf{)} \\ 2\sin^2 x - 3\sin x + 1 &= 0 \quad \textbf{(simplify)} \\ (2\sin x - 1)(\sin x - 1) &= 0 \quad \textbf{(factor)} \end{aligned}$$

(If this last equation did not factor easily, we would use the quadratic formula at this point.) Equating each factor to zero gives us the solutions.

$$\begin{aligned} 2\sin x - 1 = 0 &\quad\rightarrow\quad \sin x = \tfrac{1}{2} \quad\rightarrow\quad x = 30° \text{ and } 150° \\ \sin x - 1 = 0 &\quad\rightarrow\quad \sin x = 1 \quad\rightarrow\quad x = 90° \end{aligned}$$

Therefore, the solutions are $x = 30°, 90°, 150°$. ■

NOW IT'S YOUR TURN

Solve $\cos 2x = -2 - 3\cos x$ for $0 \le x < 2\pi$. (Answer: $\frac{2\pi}{3}, \pi, \frac{4\pi}{3}$)

Example 10

Solve $6 \sin x \cos x = \sqrt{2}$ for $0 \le x < 2\pi$.

Solution

We could use the Pythagorean identity as in Example 8. Even though that method would work, it would be fairly messy. A much easier approach is to recognize that the product $\sin x \cos x$ suggests using the double-angle formula for $\sin 2x$, namely $\sin 2x = 2 \sin x \cos x$.

$$6 \sin x \cos x = \sqrt{2}$$

$$3\underbrace{(2 \sin x \cos x)}_{\sin 2x} = \sqrt{2} \qquad \textbf{(use double-angle formula)}$$

$$3 \sin 2x = \sqrt{2}$$

$$\sin 2x = \sqrt{2}/3 \qquad \textbf{(divide by 3)}$$

Don't be deceived by $\sqrt{2}/3$; this is *not* a special angle! We must divide this number out as a decimal.

$$\sin 2x = \sqrt{2}/3 = \frac{1.414}{3} = 0.471 \qquad \rightarrow \qquad 2x = \sin^{-1}(0.471)$$

The reference angle is $\sin^{-1}(0.471) = 0.490$ rad, and $2x$ and be in quadrants I or II. Therefore,

$$2x = \begin{cases} 0.490 & \textbf{(quadrant I)} \\ \pi - 0.490 = 2.65 & \textbf{(quadrant II)} \end{cases}$$

Because x is between 0 and 2π, $2x$ is between 0 and 4π. To get the values of $2x > 2\pi$, we add 2π to the values we just found.

$$2x = \begin{cases} 0.490 + 2\pi = 6.77 \\ 2.65 + 2\pi = 8.93 \end{cases}$$

Therefore, the full set of values for $2x$ is

$$2x = 0.490, 2.65, 6.77, 8.93$$

Dividing by 2 to solve for x gives us

$$x = 0.245, 1.33, 3.39, 4.47 \; \blacksquare$$

NOW IT'S YOUR TURN

Solve $4 \sin x \cos x = \sqrt{2}$ for $0 \le x < 2\pi$. (Answer: $\frac{\pi}{8}, \frac{3\pi}{8}, \frac{9\pi}{8}, \frac{11\pi}{8}$)

Example 11

Solve the equation $\sin^2(x/2) + \cos x = \frac{1}{2}$ for $0 \le x < 360°$.

Solution

We must reduce this equation to a single trigonometric function of a single angle. Using the half-angle formula

$$\sin \frac{x}{2} = \pm\sqrt{\frac{1 - \cos x}{2}}$$

we can express this equation in terms of only $\cos x$.

$$\sin^2\left(\frac{x}{2}\right) + \cos x = \frac{1}{2}$$

$$\left(\pm\sqrt{\frac{1-\cos x}{2}}\right)^2 + \cos x = \frac{1}{2} \quad \text{(use half-angle formula)}$$

$$\frac{1-\cos x}{2} + \cos x = \frac{1}{2} \quad \text{(square)}$$

$$\frac{1}{2} - \frac{\cos x}{2} + \cos x = \frac{1}{2} \quad \text{(simplify)}$$

$$\frac{1}{2} + \frac{\cos x}{2} = \frac{1}{2} \quad \text{(simplify)}$$

$$\cos x = 0 \quad \text{(solve for } \cos x\text{)}$$

$$x = \cos^{-1}(0) = 90^\circ \text{ and } 270^\circ$$ ■

NOW IT'S YOUR TURN

Solve the equation $\cos^2(x/2) = \frac{7}{6} - \cos x$ for $0 \le x < 360^\circ$.
(Answer: $x = 63.6^\circ, 296.4^\circ$)

Example 12

Solve the equation $4 \sin x \cos x - 2 \cos x = 2 \sin x - 1$, for $0 \le x < 2\pi$.

Solution

Transposing all terms to one side, we get

$$4 \sin x \cos x - 2 \cos x - 2 \sin x + 1 = 0$$

$$(2 \cos x)(2 \sin x - 1) - (2 \sin x - 1) = 0 \quad \text{(group factoring)}$$

$$(2 \sin x - 1)(2 \cos x - 1) = 0 \quad \text{(more factoring)}$$

Equating each factor to zero, we get

$$2 \sin x - 1 = 0 \quad \rightarrow \quad \sin x = \tfrac{1}{2} \quad \rightarrow \quad x = \tfrac{\pi}{6} \text{ and } \tfrac{5\pi}{6}$$

$$2 \cos x - 1 = 0 \quad \rightarrow \quad \cos x = \tfrac{1}{2} \quad \rightarrow \quad x = \tfrac{\pi}{3} \text{ and } \tfrac{5\pi}{3}$$

The solutions are $x = \frac{\pi}{6}, \frac{\pi}{3}, \frac{5\pi}{3}$, and $\frac{5\pi}{6}$. ■

NOW IT'S YOUR TURN

Solve the equation $2 \tan x \sin x + \tan x = 2 \sin x + 1$, for $0 \le x < 2\pi$.
(Answer: $\frac{\pi}{4}, \frac{7\pi}{6}, \frac{5\pi}{4}, \frac{11\pi}{6}$)

Example 13

Find the values of x $(0 \le x < 2\pi)$ for which $\sin(\frac{3x}{2} + \frac{\pi}{4})$ is a minimum.

Solution

Because $-1 \le \sin\theta \le 1$, the minimum value of the sine is -1. Therefore, we must solve the equation $\sin(\frac{3x}{2} + \frac{\pi}{4}) = -1$. Solving gives us

$$\frac{3x}{2} + \frac{\pi}{4} = \sin^{-1}(-1) = \frac{3\pi}{2}$$

$$\frac{3x}{2} + \frac{\pi}{4} = \frac{3\pi}{2}$$

$$\frac{3x}{2} = \frac{3\pi}{2} - \frac{\pi}{4} = \frac{5\pi}{4}$$

Solving for x gives us

$$x = \frac{\frac{5\pi}{4}}{\frac{3}{2}} = \left(\frac{5\pi}{4}\right)\left(\frac{2}{3}\right) = \frac{5\pi}{6} \quad ■$$

NOW IT'S YOUR TURN

Find all values of x $(0 \le x < 2\pi)$ for which $\sin(\frac{x}{3} + \frac{\pi}{6})$ is a maximum.
(Answer: $x = \pi$)

EXERCISES 8.1

In Exercises 1–86, solve for all values of x $(0 \le x < 2\pi)$, giving answers in radians. Use a calculator for angles that are not special angles, giving answers accurate to two decimal places.

1. $\cos x = 1$

2. $2\sin x + 1 = 0$

3. $\cot x = 0$

4. $4.00 \sin x - 3.50 = 0$

5. $1.54 + \tan x = \pi$

6. $5.00 - 1.80 \cos x = 6.58$

7. $2\sin x = -1.7642$

8. $-1.2 \sin x - 0.8445 = 0$

9. $\cos^2 x = \frac{1}{2}$

10. $\sin^2 x = \frac{3}{4}$

11. $\tan^2 x = 1$

12. $\tan^2 x + 6.32 = \sqrt{23}$

13. $\cos^2 x - 0.4017 = 0$

14. $\sin^2 x + 0.4361 = 0$

15. $\sin 2x = 0.3264$

16. $\cos(x/2) = -0.3984$

17. $(\tan x + 1)\cos x = 0$

18. $(2\cos x + 1)\sin x = 0$

19. $(2\cos x + 1)(\sin x - 1) = 0$

20. $(3\tan^2 x - 1)(\sin x - 1) = 0$

21. $(\sec x - 2)(2\cos x + \sqrt{3}) = 0$

22. $2\sin^2 x + \sin x = 1$

23. $3(1 - \sin x) = 2\cos^2 x$

24. $\sqrt{2}\sin x \cos x = \cos x$

***25.** $\cot^2 x + 2\csc^2 x - 3 = 0$

26. $\tan x = \cot x$

27. $\sec x + 1 = 2\cos x$

28. $\sqrt{4 - \csc^2 x} = \csc x$

29. $\sin^2 x = 1 - \cos^2 x$

30. $3\tan^2 x + \csc^2 x = 5$

31. $\cos^2 x = 2\sin^4 x$

32. $\csc x \tan x = \tan x$

33. $\sec x - \tan x = 2\cos x$

34. $\sec^2 x - 1 = \tan^2 x$

***35.** $\tan x + \sec x = 1$

36. $\sqrt{3}\cot x - \csc x = 1$

37. $8\cos^2 x \sin x + 4\cos^2 x = 6\sin x + 3$

38. $2\sin^4 x - \sin^2 x - 1 = 0$ ***Hint:*** First treat $\sin^2 x$ as a unit and solve for it; then find $\sin x$.

39. $12 \cos^4 x = 7 \cos^2 x - 1$

40. $\sin^2 x \cos^2 x = \frac{1}{2} - \sin^2 x$

41. $1.18 \sin^2 x - 0.280 \sin x = 0.587$

42. $3 \sin^2 x + 5 \sin x = 2$

***43.** $\cos^2 x = 6 + \cos x$

44. $8 \sin^6 x - 9 \sin^3 x + 1 = 0$ *Hint:* First treat $\sin^3 x$ as a unit.

45. $\csc^2 x - \tan^2 x = 1$

46. $\sqrt{3} \tan x = \sec x + 1$

47. $\sin 2x = 1$

48. $\sin 2x = -1$

49. $\cos 3x - 1 = 0$

50. $\cos 3x + 1 = 0$

51. $\sin 2x = 0$

52. $\cos 5x = 0$

53. $\tan 2x = 1$

54. $(2 \sin 2x + 1)\cos x = 0$

55. $(\tan 2x - \sqrt{3})\cos x = 0$

56. $\sin 2x = \sin x$

57. $\sin 2x + \sqrt{2} \sin x = 0$

58. $\sin 2x = \sqrt{2} \sin x$

59. $\sin 2x + \cos x = 0$

60. $\cos 2x + \sin x = 0$

61. $\tan 2x = \tan x$

62. $\tan 2x = 3 \tan x$

63. $\cot 2x + \cot x = 0$

***64.** $\csc 2x = 2\sqrt{3}/5$

65. $\sin^2 (x/2) = \frac{1}{2}$

66. $\cos 2x - 1 = \sin x$

67. $2 + \cos 2x = 3 \cos x$

68. $\cot 2x = \tan x$

69. $3 \cos 2x = 1 - \sin x$

***70.** $\sin 4x + \sin 2x = 0$

***71.** $\cos 4x = \sin 2x$

72. $\cos (x/2) = 1 + \cos x$

73. $24 \cos^2 x = 17 - 13 \sin x$

74. $2 \sin^2 x + 2(\sin x/\csc x) = 4 - 4 \cos^2 x$

75. $\cos (x/2) \cos x - \sin (x/2) \sin x = 0.7142$

76. $\sin x \cos 2x \tan 3x = 0$

***77.** $2 \sin x - 3 \cos x + 2 = 0$

78. $3.52 \sin 2x - 4.84 \cos 2x = 13.42$

79. $\sin x \cos x = \frac{1}{2}$

80. $4 \sin^2 (x/2) = 1$

81. $4 \cos x \sin x = \sqrt{5}$

***82.** $\cos 3x \cos x = \sin^2 2x$ *Hint:* $3x = x + 2x$

****83.** $\sin 3x + \sin x = 0$

****84.** $\sin 3x + \sin x = \cos x$

****85.** $2 \sin 3x + 2 \sin x = \tan x$

****86.** $\cos 3x - \cos x = \sqrt{3} \sin x$

In Exercises 87–91, for which values of x ($0 \le x < 2\pi$) is each expression (a) a maximum, (b) a minimum, or (c) zero?

87. $2 \sin(x - \frac{\pi}{4})$

88. $3 \sin(\frac{3x}{4} + \frac{\pi}{6})$

89. $-5 \cos(\frac{\pi}{3} - 2x)$

****90.** $-2 \sin^2(3 - \frac{x}{4})$

****91.** $7 \cos\left(\frac{x}{3} - \pi\right)$

In Exercises 92–99, for what values of x ($0 \le x < 2\pi$) is each expression (a) zero or (b) undefined?

92. $6.8 \tan(4x - \frac{\pi}{3})$

93. $-8.95 \cot(\frac{3x}{\pi} - 5)$

94. $\sec(2x - \frac{\pi}{4})$

95. $\csc(\frac{-x}{4} + \frac{3}{8})$

***96.** $\sin(\frac{2}{x} + \frac{3}{7})$

***97.** $-\cos[(3 - \frac{\pi}{2})/x]$

***98.** $\cos[\pi/(x - 1)]$

****99.** $\sin[\pi/(x^2 - 4)]$

100. For what values of $x (0 \le x < 2\pi)$ is y (i) a maximum or (ii) a minimum? In each case, find the numeric value of the maximum and minimum value of y.

(a) $y = 3 \cos x \sin x$
(b) $y + 3 + 8 \sin^2 x = 8 \cos^2 x$

In Exercises 101–112, find an expression for *all* values of $x(-\infty < x < \infty)$ for which the given condition is true. Write each answer in terms of an integer n (where n may be 0, 1, 2, 3, . . .).

101. $\sin 2x = 0$

102. $\cos 5x = 0$

103. $\cos (2x/3) = 1$

104. $\sin \pi x = 1$

105. $\sin \pi x = -1$

106. $\cos 2\pi x = -1$

107. $\tan 4x = 1$

108. $\tan 4x = -1$

***109.** $|\cos \pi x| = 1$

***110.** $\cos|\pi x| = 1$

***111.** $\sin^2 3x = 1$

***112.** $\sin[(3x)^2] = 1$

113. In the study of vibrating strings (such as violin strings), the equation $\sin kL = 0$, where L is the length of the string, arises. For what values of k does this equation hold?

In Exercises 114–125, for what range of values of b will each equation have solutions?

114. $\sin x = 1 - b$

115. $\cos x = 1 + b$

116. $\sin x = 9 - b^2$

117. $\sin^2 x = b + 3$

118. $4 \cos x = 1 + b$

***119.** $3|\sin x| = b + 7$

120. $\tan x = b + 8$

121. $\sin 3x = b^2 + 1$

122. $b \cos x = 7$

***123.** $\sin 8x = 1 + 1/b$

124. $b \sec x = 8$

125. $5 \csc^2 x = \frac{1}{2}b^2$

***126.** Solve for x in terms of a: $\tan x + \sec x = a$

***127.** Solve for x in terms of a: $\cos 2x = a(1 - \cos x)$

128. A rectangle is to be deformed to a parallelogram so that the area of the parallelogram is one-half the original area of the rectangle, as shown in Figure 8.1-2. What must θ be?

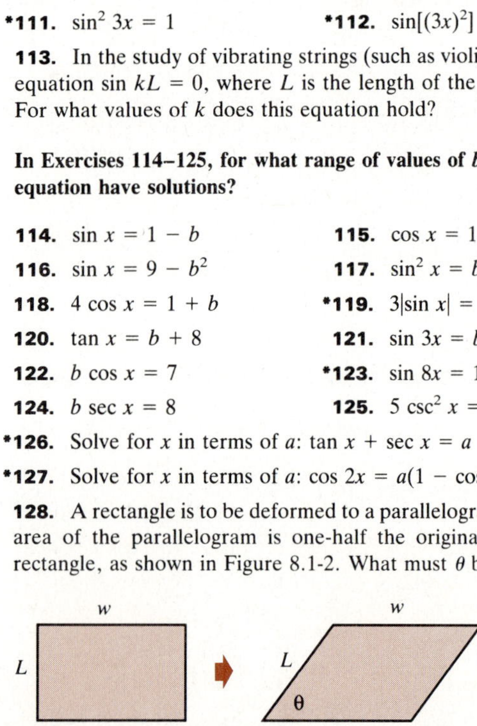

Figure 8.1-2

129. You are designing a triangular path for a public park; the path will have sides 3.0 km and 4.0 km long, as shown in Figure 8.1-3. If the total path must be 13.0 km long, what should θ be?

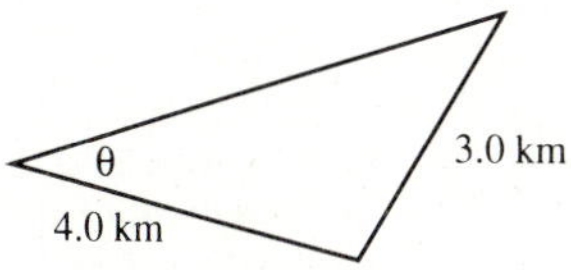

Figure 8.1-3

130. The triangular garden in Figure 8.1-4 is to have 4530 m^2 of planting space. What must θ be?

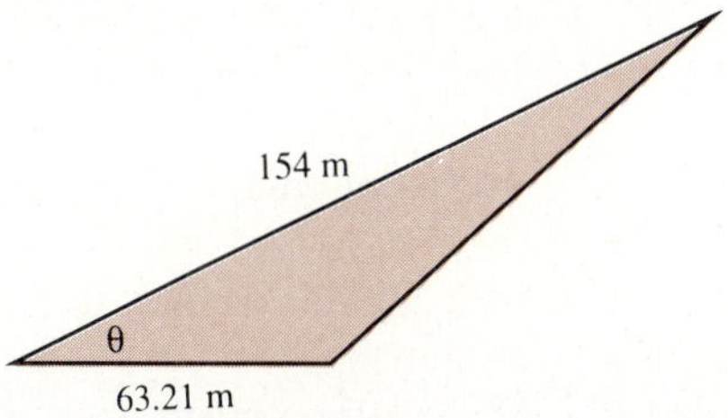

Figure 8.1-4

131. A painter is at the top of a ladder that suddenly starts slipping. At what angle with the floor will the painter be three times as far from the foot of the ladder as she is above the floor? (See Figure 8.1-5.)

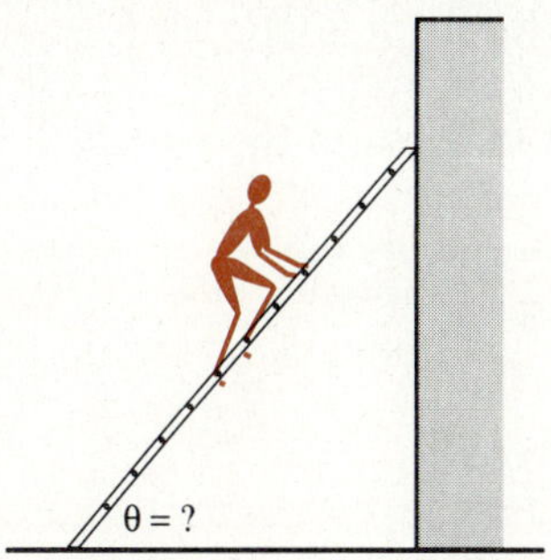

Figure 8.1-5

***132.** A right triangle with a hypotenuse of 10.0 m has an area of 5.00 m^2. Find its acute angles.

***133.** A brick wall encloses a right triangular space with a perimeter of 68.6 m and a hypotenuse 28.9 m long. Find the acute angles of this triangle.

***134.** At point P_2 the angle of elevation of a building is twice as great as at point P_1. P_2 is 150.0 ft from the building, and P_1 is 3.40 times as far from the building as is P_2. Find the building's angle of elevation from both points and find the building's height.

135. A wedge-shaped sector of land of radius 1000.0 m is cut into two segments by connecting the chord of the arc, as shown in Figure 8.1-6. What must be the central angle θ of this sector so that the triangular segment will have an area of 223,000 m^2?

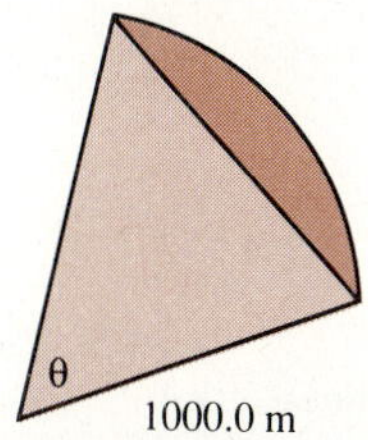

Figure 8.1-6

****136.** At a certain location, a building casts a shadow at noon on day A. On day B, when the sun is 10° lower in the sky than on day A, the building's noon shadow is 50% longer than on day A. Find the angle of the noon sun above the horizon on both days.

***137.** For unknown reasons, the rotation rate of our sun is different at different latitudes on the sun. The approximate formula, which was experimentally measured, is

$$\omega \approx \tfrac{1}{26}\left(1 - \tfrac{1}{8}\sin^2\theta - \tfrac{1}{6}\sin^4\theta\right) \text{ rotations/day}$$

where θ is the angle of latitude above (or below) the sun's equator. (See NASA Publication SP-303, *The Quiet Sun*, by Edward Gibson, p. 31.)

(a) Find the sun's rotation rate at the poles and at the equator.

(b) At what angle of latitude is the rotation rate midway between its values at the poles and the equator?

8.2 GRAPHIC SOLUTION OF TRIGONOMETRIC EQUATIONS

A building contractor wants to divide a large circular plot of land into wedge-shaped lots (see Figure 8.2-1a). He further wants to divide each wedge into two equal-sized lots by running a fence along the chord of the circular part of the wedge (Figure 8.2-1b). What should be the central angle θ of each wedge?

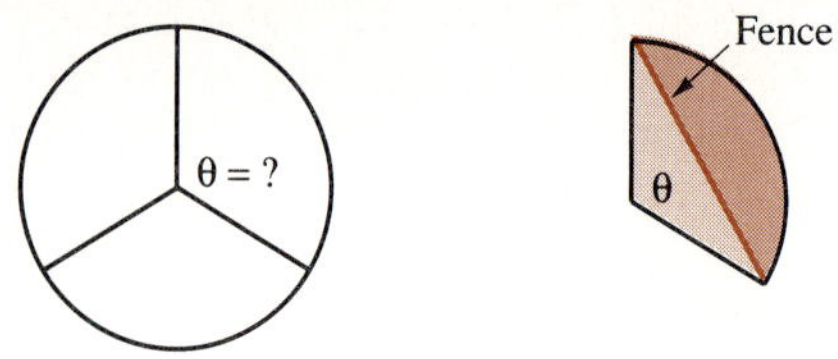

Figure 8.2-1

Building contractor's lots

(a) Wedge-shaped lots

(b) Wedge-shaped lot divided into two equal-sized lots

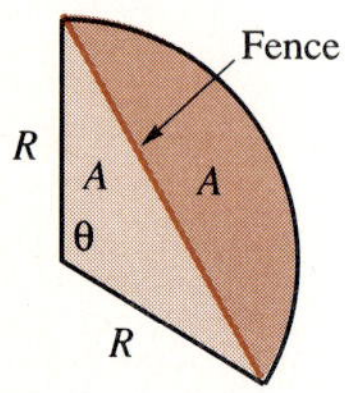

Figure 8.2-2

Wedge-shaped lot divided into two equal-sized lots

First sketch a wedge divided into two equal-sized lots (see Figure 8.2-2), calling R the radius of the circle and A the area of each half. Because the two areas are equal, each one is half the total area of the wedge. Working with the triangular half because we know how to compute its area gives us

$$\text{area of triangle} = \tfrac{1}{2}(\text{area of wedge})$$

In Section 1.4 we showed that the area of a circular wedge is

$$\text{area of wedge} = \tfrac{1}{2}R^2\theta \quad (\boldsymbol{\theta} \textbf{ in radians})$$

In Section 4.2 we showed that the area of any triangle is equal to half the product of the lengths of any two sides times the sine of the angle between them. This gives us

$$\text{area of triangle} = \tfrac{1}{2}RR\sin\theta = \tfrac{1}{2}R^2\sin\theta$$

Because the area of the triangle is half the area of the wedge,

$$\underbrace{\tfrac{1}{2}R^2\sin\theta}_{\text{triangle}} = \tfrac{1}{2}\underbrace{(\tfrac{1}{2}R^2\theta)}_{\text{wedge}}$$

Multiplying by 2 and dividing by R^2 gives us

$$\sin\theta = \frac{\theta}{2} \quad ■$$

To complete the solution, we must solve this last equation. Yet this equation is quite different from the ones we encountered in Section 8.1 because the unknown (θ in this case) occurs both within a trigonometric function and *by itself*. None of the previous methods work on this equation. In fact, no algebraic/trigonometric methods work on it! The only way to solve this type of equation is *graphically* or using a *computer* (more about that later).

To solve this equation graphically, think of it as *two* equations: $y_1 = \sin\theta$ and $y_2 = \theta/2$. If you graph these two equations on the *same* set of axes, the points where the two graphs intersect are the points where $y_1 = y_2$; that is, where $\sin\theta = \theta/2$.

Example 14

Let's solve this building contractor's problem for him graphically:

$$\sin\theta = \frac{\theta}{2}$$

Solution

Carefully graph $y_1 = \sin\theta$ and $y_2 = \theta/2$ on the same set of axes and then find the points of intersection. y_2 is a straight line with slope $\frac{1}{2}$ (see Figure 8.2-3). As you can see from this figure, the two graphs intersect when $\theta \approx -1.85$, 0, and 1.85 rad.

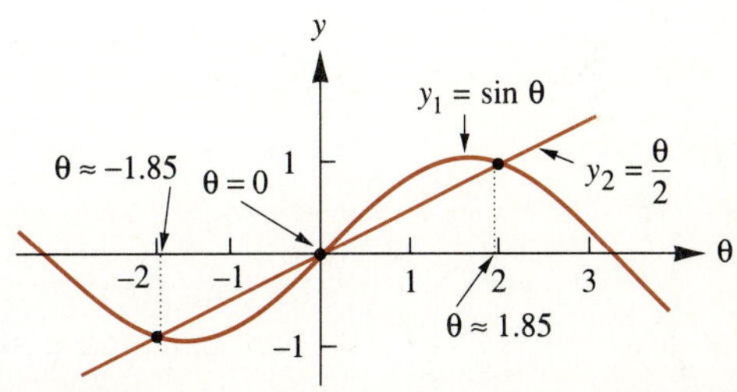

Figure 8.2-3

Graphs of $y_1 = \sin\theta$ and $y_2 = \theta/2$ on the same axes

You can also readily see that there are no more solutions, because the graphs will not intersect at any other points.

We should check these values in the original equation $\sin\theta = \theta/2$, keeping in mind that graphical methods do not give exact answers.

$$\sin(-1.85) = -\tfrac{1.85}{2} \quad \rightarrow \quad -0.93 \approx -0.96 \quad \surd$$

$$\sin 0 = \tfrac{0}{2} \quad \rightarrow \quad 0 = 0 \quad \surd$$

$$\sin 1.85 = \tfrac{1.85}{2} \quad \rightarrow \quad 0.93 \approx 0.96 \quad \surd$$

These solutions satisfy the original equation within the limited accuracy of the graph. In this particular problem, only the value $\theta \approx 1.85$ has physical meaning because the lots cannot have zero or negative area. Thus the building contractor should divide his lots into wedges of central angle 1.85 rad ($\approx 106°$). ■

NOW IT'S YOUR TURN

Find all the solutions to the equation $\cos\theta = \theta$. Check each answer.
(Answer: $\theta \approx 0.75$ rad)

In this discussion we said that graphic solutions are necessary when the unknown occurs both in a trigonometric function and by itself. Such methods are also necessary in many other cases. Almost any time an equation involves a *combination* of trigonometric functions and other functions, that equation cannot be solved by algebraic/trigonometric methods. Such cases include equations like

$$\cos x + \log x = 1.5$$
$$x \sin x = 1 + \cos x$$
$$10^x = \sin x$$
$$\cos x + \sin x = e^x$$

We give you a chance to try some of these in the exercises.

Yet these cases are not the only ones in which graphical solutions can be useful. Sometimes an equation contains only trigonometric functions, but the usual methods produce a complicated mess that either takes too long or is impossible to solve algebraically. For example, consider the innocent-looking equation

$$1 + \sin x = \tan x$$

Reducing this to a single trigonometric function ($\sin x$ or $\cos x$) produces a very complicated equation. Try it! Yet a graphic solution is quite straight forward.

Example 15

Solve the equation $1 + \sin x = \tan x$ graphically.

Solution

As before, we think of $y_1 = 1 + \sin x$ and $y_2 = \tan x$ as two functions and graph both on the same set of axes (see Figure 8.2-4). We can read the solutions from the points where the two graphs intersect: $x \approx 1.1, 3.6, 7.35, \ldots$.

You can also see that there are *infinitely* many solutions to this equation because these graphs will never stop intersecting.

We should check our answers in the original equation, but we leave that up to you. ■

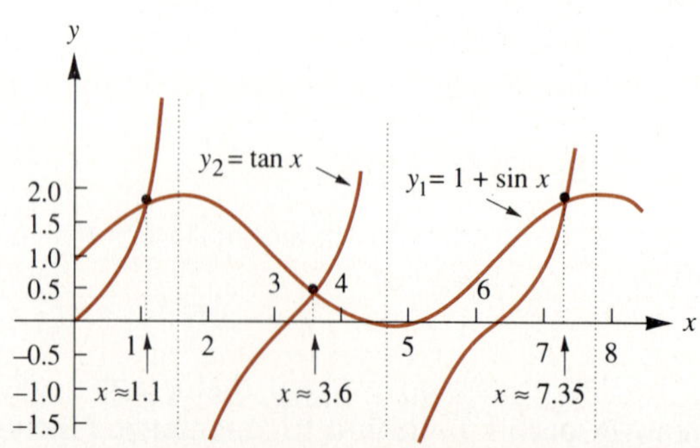

Figure 8.2-4
Graphs of $y_1 = 1 + \sin x$ and $y_2 = \tan x$

NOW IT'S YOUR TURN

Find the solutions to $\tan x = \cos x + 1$ graphically. Are there a finite or infinite number of solutions?

(Answer: 1.0, ... *You* decide if your additional answers are correct by checking them in the original equation. There are infinitely many solutions.)

Solving Equations by Computer

The graphical method gives you an approximate solution to an equation, which in many cases may be more than adequate. If, however, you need a high degree of precision, you must use a computer. Remember that a computer is a number-cruncher; it can perform complex calculations very quickly. Therefore, if you have an equation to solve, such as $\sin \theta = \theta/2$, you can ask (program) the computer to start calculating numerical values for both sides and then compare those values. When the values get close enough to each other (how close depends on the accuracy you need), then the computer tells you it has found a solution and prints it out. It then goes on to find the next solution, if there is one.*

EXERCISES 8.2

In Exercises 1–28, solve for x graphically. In each case, check your answers by substituting them back into the original equation. In cases where there are infinitely many solutions, find the first few positive and negative ones.

1. $\sin x = \frac{2}{3}$

2. $\cos x = -\frac{3}{4}$

3. $\sin x = x$

4. $1 + \sin x = x$

5. $1 + \sin x = -x$

6. $\cos x = x + 1$

7. $\tan x = x$ (This equation occurs in optics.)

8. $3 \sin x = x$

9. $\tan x = 3 - x$

10. $\sin x = x^2$

11. $\cos x = x^3$

***12.** $x^2 - 4 = \cos x$

***13.** $x^2 + 4 = \sin x$

***14.** $\sin x = 1/x$

15. $\sin 2x = x$

16. $\cos (x/2) = -x$

17. $\sin x = \cos x$

18. $2 \sin x = \cos x$

***19.** $3 + \cos x = \sin x$

20. $1 - \cos x = \csc x$

21. $\csc x = 1 + \tan x$

***22.** $2 \sec x = \sin x$

***23.** $\sin x = 10^x$

***24.** $\log x = \tan x$

***25.** $\cos x = e^x$

***26.** $4 \sin x = \ln x$

***27.** $\tan x = 10^{-x}$

***28.** $\sin x = e^{-x}$

29. In Exercise 3 you graphed x and $\sin x$ on the same set of axes. Explain how this graph supports the small-angle approximation, $\sin x \approx x$, for x much less than 1 rad. (See Section 2.7 if you have forgotten this approximation.)

In Exercises 30–41, decide if each equation has a finite or an infinite number of solutions by graphic means. Using only a rough sketch of the graphs; do not plot points and do not find the actual solutions.

30. $\sin x = \frac{1}{3}$

31. $\tan x = \frac{1}{3}$

32. $\tan x = -x$

33. $\sin x = -x$

34. $\cos x = x$

35. $\cos x = x^2$

***36.** $\tan x = x^2$

***37.** $\cos x = \log x$

***38.** $\ln x = \cot x$

***39.** $e^x = \sin x$

***40.** $e^x = \tan x$

41. $6 + \sin x = \cos x$

42. A circular wedge of radius 100 m is divided into two areas by a chord of the circle, as shown in Figure 8.2-5. What must the

* Another obvious method is to have the computer graph both functions on the same set of axes as we did in Examples 14 and 15. Although this method gives limited accuracy, it can be useful and fast on a microcomputer.

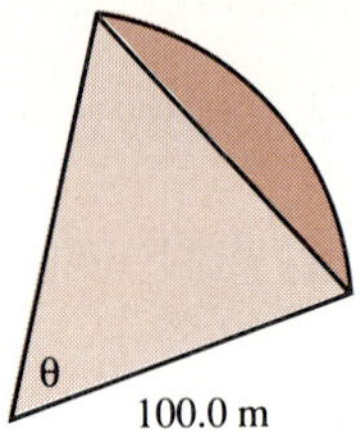

Figure 8.2-5

central angle θ be so that the triangular part is 30% of the area of the total wedge, and what are these two areas?

43. During the first half of its cycle, a pendulum swings through an angle θ from A to B, as shown in Figure 8.2-6. What must θ be so that the horizontal distance from A to B is three-fourths the total distance the pendulum bob swings through from A to B?

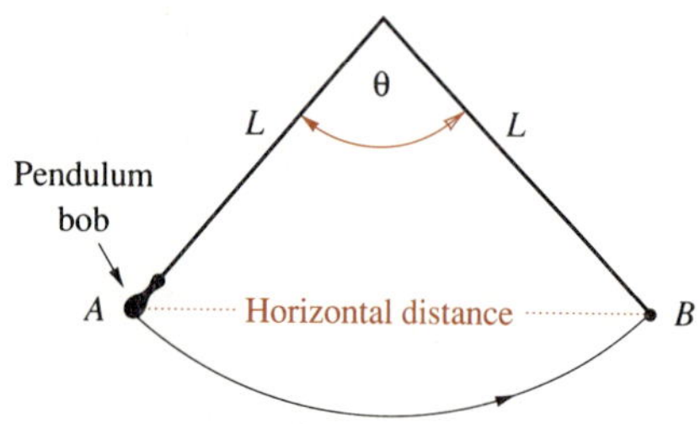

Figure 8.2-6

***44.** A path surrounds a circular, wedge-shaped garden. This path also has a shortcut along a chord of the circle, as shown in Figure 8.2-7. What must θ be so that the round-trip path from the center A by way of the shortcut is 90% as long as the round trip taking the curved part of the path, again starting at A?

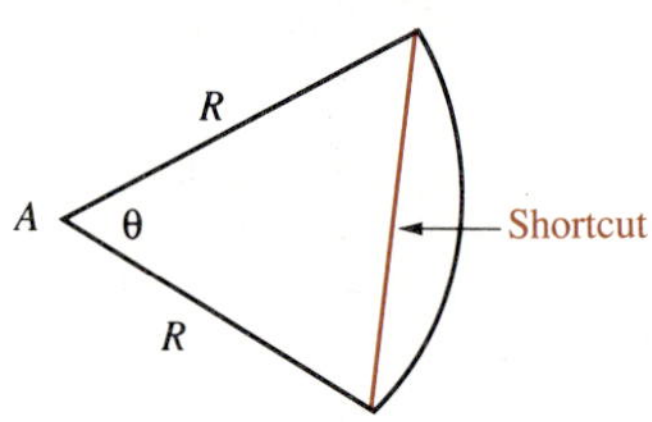

Figure 8.2-7

***45.** A circular sector is divided into two sections by a line from one corner perpendicular to the other radius, as shown in Figure 8.2-8. Find the central angle θ for which these two sections have equal areas.

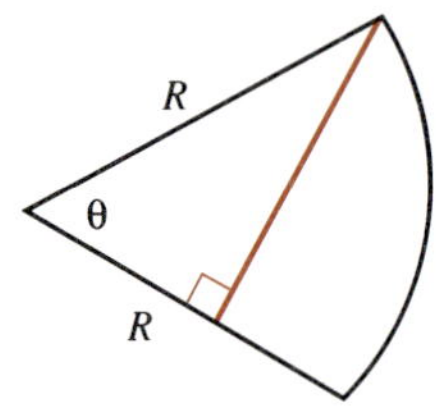

Figure 8.2-8

****46.** Suppose you dug a straight mine shaft through the earth from A to B, as shown in Figure 8.2-9. How long would that tunnel have to be so that it was only 70% of the distance from A to B along the surface of the earth? ***Hint:*** Can you relate these quantities to the central angle of the arc AB?

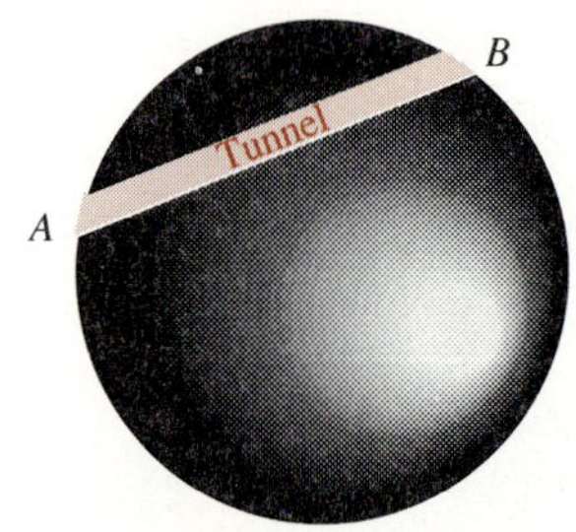

Figure 8.2-9

***47. Curved Space.** (See Exercise 207 in Section 3.2.)

(a) Show that, when there is very little curvature to a surface, or when we are looking at only a very small part of it, then $\pi_c \approx 3.14159\ldots$. ***Hint:*** Interpret "very little curvature" in terms of θ in Figure 3.2-18. Then apply the small-angle approximation to the formula in Exercise 207c in Section. 3.2.

(b) Use the formula from Exercise 207c to find the maximum and minimum values of π on the curved surface.

(c) For what angle θ will the curved-space value of π be equal to 3.0?

8.3 INVERSE TRIGONOMETRIC FUNCTIONS AND THEIR GRAPHS

Throughout most of this book, we have dealt with the trigonometric functions by saying that the value of the trigonometric function depends on (is a function of) the angle. For example, when we write $y = \sin x$, we mean that the value of y depends on (is a function of) the value of the angle x. If we know the angle, then we can compute the value of the sine of that angle. However, we can also think of this process in the *inverse* way: If we know the value of the *sine* of the angle, then we can compute the value of the angle. Likewise, if we know the value of the cosine or tangent of the angle, we can compute the value of the angle. This process involves the *inverse trigonometric functions*, which we studied in Sections 2.4 and 3.3. Now we think of the angle as depending on the value of the trigonometric function. Thus we write these equations as follows.

$$y = \sin^{-1} x = \arcsin x$$

means that y is the angle whose sine is x. Therefore, $\sin y = x$.

$$y = \cos^{-1} x = \arccos x$$

means that y is the angle whose cosine is x. Therefore, $\cos y = x$.

$$y = \tan^{-1} x = \arctan x$$

means that y is the angle whose tangent is x. Therefore, $\tan y = x$.

To graph $y = \sin^{-1} x$, we make a table of values and plot them. Remember, y is the angle in this case and this is equivalent to $x = \sin y$ (see Figure 8.3-1).

x	$y = \sin^{-1} x$
0	-2π
1	$-\frac{3\pi}{2}$
0	$-\pi$
-1	$-\frac{\pi}{2}$
0	0
1	$\frac{\pi}{2}$
0	π
-1	$\frac{3\pi}{2}$
0	2π

NOW IT'S YOUR TURN

Make a table of values and graph $y = \cos^{-1} x$ for $-2\pi \le y \le 2\pi$.
(Answer: See Figure 8.3-2)

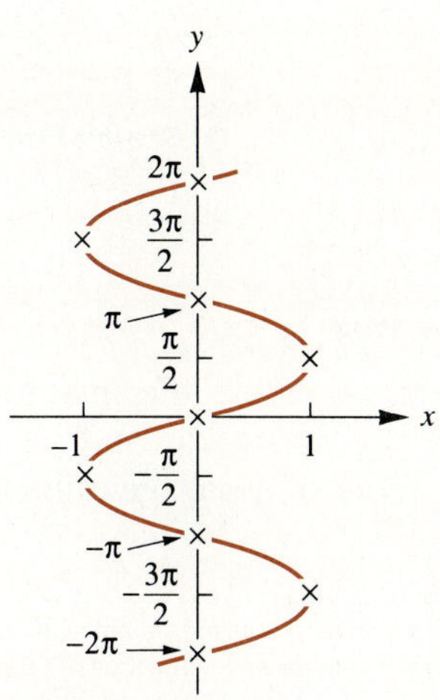

Figure 8.3-1
Graph of $y = \sin^{-1} x$

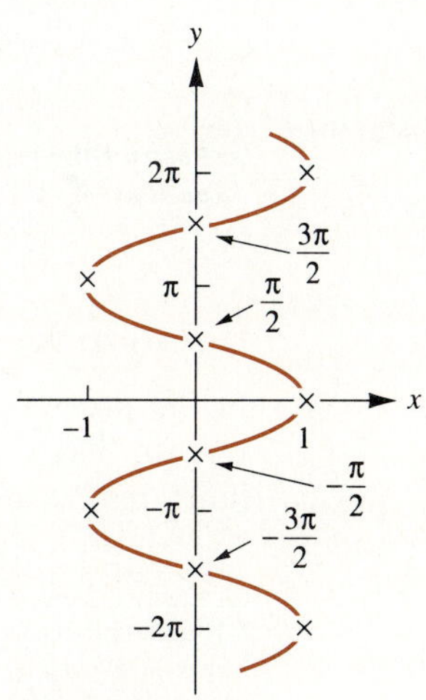

Figure 8.3-2
Graph of $y = \cos^{-1} x$

There is still one very important point we must clear up before we have a complete presentation of the inverse trigonometric functions. Notice that we have used the word *functions*. The value of y is a *function* of the value of x because y depends on x. We mean that *the value of x uniquely determines the value of y*. We use the word *unique* because any value of x can give only *one* value of y. If a value of x gives more than one value of y, we call this a *relationship* but not a function.*

Now look at the graphs of the inverse sine and cosine (Figures 8.3-1 and 8.3-2). Obviously, one value of x gives *many* (in fact, infinitely many) values of y. For example, in Figure 8.3-1, when $x = 0$, $y = 0, \pm\pi, \pm 2\pi, \pm 3\pi, \ldots = \pm k\pi$ $(k = 0, 1, 2, 3, \ldots)$. Therefore, these graphs cannot represent functions; they represent *relationships*. To make them functions, we must restrict the range of y. In Figure 8.3-1, we can restrict y to be between $\frac{\pi}{2}$ and $-\frac{\pi}{2}$, and then there will be only one value of y for each value of x and the graph will represent a function (see Figure 8.3-3).

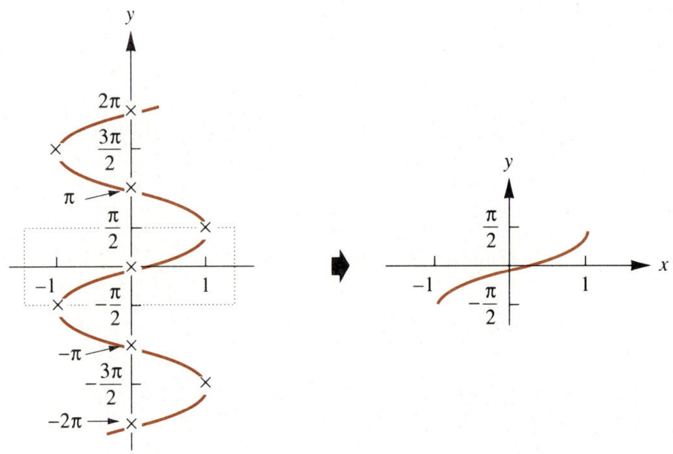

(a) Restrict the graph of $y = \sin^{-1} x$ **so that** $-\frac{\pi}{2} \le y \le \frac{\pi}{2}$.

(b) Graph of the function $y = \text{Sin}^{-1} x$.

Figure 8.3-3

Converting the graph of $y = \sin^{-1} x$ into the graph of the *function* $y = \text{Sin}^{-1} x$ by restricting the range of y

Notation for the Inverse Trigonometric Functions

In the previous chapters, we wrote the inverse trigonometric *relationships* as $y = \sin^{-1} x$, $y = \cos^{-1} x$, $y = \tan^{-1} x$, and so on. These relationships are not functions because one value of x can give many values of y. For example, if

* The reason we make this requirement for a function should be clear from the following. (a) If y represents your height and x represents your age, then your height is a function of your age. At any particular age you can have only one height—one value of y. You may have the same height at several different ages, but you cannot have two different heights at one age. (b) If y represents the temperature and x is the time of day, the temperature is a function of the time of day. At any one time, there can be only *one* temperature—one value of y. You can have the same temperature at different times of day but not different temperatures at the same time of day.

$x = 1$, $y = \tan^{-1}(1) = \frac{\pi}{4}, \frac{5\pi}{4}, \frac{9\pi}{4}, \ldots$ When we want to represent an inverse trigonometric *function*, we begin it with a *capital letter*. For example, $y = \sin^{-1}x$ (or $y = \arcsin x$) represents a relationship, but $y = \text{Sin}^{-1} x$ (or $y = \text{Arcsin}\, x$) is a *function*. Likewise for the other trigonometric functions, $y = \cos^{-1} x$ (or $y = \arccos x$) is a *relationship*, but $y = \text{Cos}^{-1} x$ (or $y = \text{Arccos}\, x$) is a *function*, and so on.

Now we can convert the inverse trigonometric relationships to trigonometric functions by restricting the range of y for each one, as we did with $y = \text{Sin}^{-1} x$ in Figure 8.3-3.

NOW IT'S YOUR TURN

What restrictions must we place on y in the graph of $y = \cos^{-1} x$ (Figure 8.3-2) to make it a function? Sketch the graph of this function.
(Answer: $0 \leq y \leq \pi$; see Figure 8.3-4.)

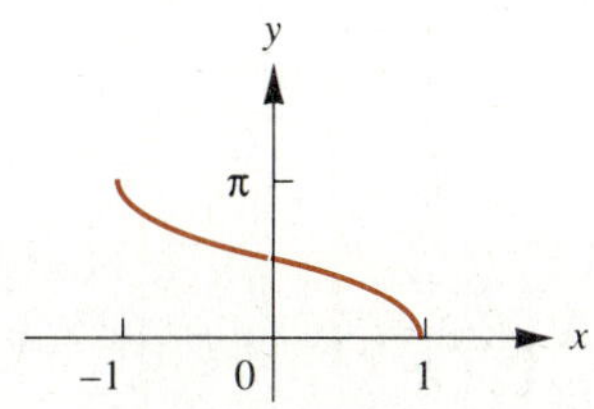

Figure 8.3-4

Graph of the *function* $y = \text{Cos}^{-1} x$

Figure 8.3-5 shows the graph of the function $y = \text{Tan}^{-1} x$. We leave the details of obtaining this graph and the limits on y for the exercises.

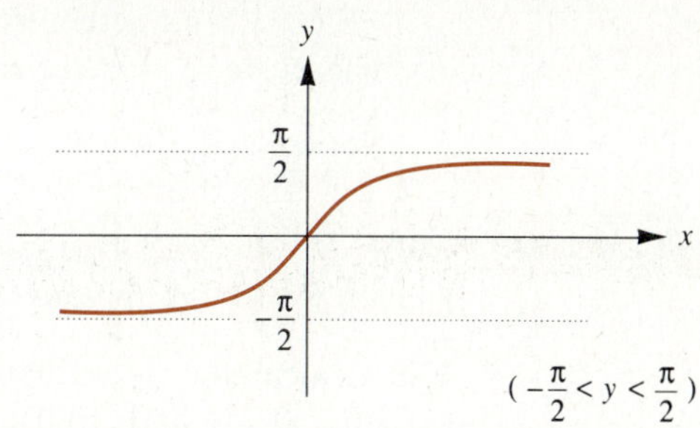

Figure 8.3-5

Graph of the *function* $y = \text{Tan}^{-1} x$

Example 16

Evaluate the function $y = \text{Sin}^{-1}(-\sqrt{2}/2)$.

Solution

$y = \text{Sin}^{-1}(-\sqrt{2}/2)$ means that $\sin y = -\sqrt{2}/2$. Because this equation is a function, y must be in the range $-\frac{\pi}{2} \leq y \leq \frac{\pi}{2}$, which means that y can lie in only quadrants I or IV. But $\sin y$ is positive in quadrant I and negative in quadrant IV. Because $\sin y$ is negative, y must be in quadrant IV. Therefore,

$$\sin(\text{ref angle}) = |-\sqrt{2}/2| = \sqrt{2}/2$$

so the reference angle of y is $\frac{\pi}{4}$. Because $-\frac{\pi}{2} \le y \le \frac{\pi}{2}$, y must be $-\frac{\pi}{4}$. Notice that y is in quadrant IV; it cannot be $\frac{7\pi}{4}$ because y cannot be greater than $\frac{\pi}{2}$. ■

NOW IT'S YOUR TURN

Without using a calculator, evaluate (a) $\text{Cos}^{-1}(-\sqrt{3}/2)$ and (b) $\text{Tan}^{-1}(-1)$. (Answer: $\frac{5\pi}{6}$; $-\frac{\pi}{4}$)

Example 17

Use a calculator to evaluate each function in degrees. (a) $y = \text{Sin}^{-1}(-0.412)$ and (b) $y = \text{Arccos}(-0.683)$.

Solution

Scientific calculators are programmed to handle the restrictions on the inverse functions, so they automatically give you the angle in the proper range.

(a) $y = \text{Sin}^{-1}(-0.412)$:

1. Set your calculator to the DEG mode.
2. Enter -0.412.
3. Press the INV key and then the SIN key to get the angle whose sine is -0.412. Your screen should show -24.3, meaning $-24.3°$.

Therefore, $y = \text{Sin}^{-1}(-0.412) = -24.3°$.

(b) $y = \text{Arccos}(-0.683)$:

1. Set your calculator to the DEG mode.
2. Enter -0.683.
3. Press the INV key and then the COS key to get the angle whose cosine is -0.683. Your screen should show 133, meaning 133°.

Therefore, $y = \text{Arccos}(-0.683) = 133°$.

Notice that the calculator gave the answer for $y = \text{Sin}^{-1} x$ in the range $-\frac{\pi}{2} \le y \le \frac{\pi}{2}$, but for $y = \text{Arccos}\, x$ the answer was in the range $0 \le y \le \pi$. ■

NOW IT'S YOUR TURN

Use a calculator to evaluate each of the following functions in radians: (a) $\text{Sin}^{-1}\, 0.754$ and (b) $\text{Tan}^{-1}(-3.167)$. (Answer: 0.854; −1.265)

Example 18

Find an algebraic expression for $\tan(\text{Arcsin}\, x)$.

Solution

Let $\theta = \text{Arcsin}\, x$. Then $\sin\theta = x$. Now draw a right triangle with acute angle θ (see Figure 8.3-6). Because $\sin\theta = \dfrac{\text{opposite side}}{\text{hypotenuse}}$ and $\sin\theta = x = x/1$, the opposite side $= x$ and the hypotenuse $= 1$. The Pythagorean theorem tells us that the third side must be $\sqrt{1-x^2}$. Therefore, $\tan\theta = \dfrac{\text{opposite side}}{\text{adjacent side}} =$

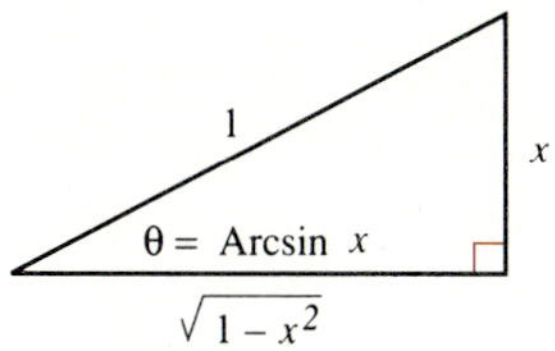

Figure 8.3-6
Triangle for which sin θ = x

$\frac{x}{\sqrt{1-x^2}}$. Because $\theta = \text{Arcsin } x$, this gives us

$$\tan(\text{Arcsin } x) = \frac{x}{\sqrt{1-x^2}}$$

This answer works for both positive and negative values of x. When x is positive, Arcsin x is between 0 and $\frac{\pi}{2}$ so tan(Arcsin x) is positive. When x is negative, Arcsin x is between 0 and $-\frac{\pi}{2}$, so tan(Arcsin x) is negative. ■

NOW IT'S YOUR TURN

Find an algebraic expression for $\sin(\text{Cos}^{-1} x)$. (Answer: $\sqrt{1-x^2}$)

Example 19

Solve the equation $\text{Sin}^{-1} x + \text{Sin}^{-1} 2x = \frac{\pi}{2}$ for x.

Solution

Let $\theta = \text{Sin}^{-1} x$ and $\phi = \text{Sin}^{-1} 2x$; which means that $\sin\theta = x$ and $\sin\phi = 2x$. The original equation becomes

$$\underbrace{\text{Sin}^{-1} x}_{\theta} + \underbrace{\text{Sin}^{-1} 2x}_{\phi} = \frac{\pi}{2}$$

$$\theta + \phi = \frac{\pi}{2}$$

Now take the cosine of each side. (Why the cosine? Because $\cos\frac{\pi}{2} = 0$, this will simplify our work.)

$$\cos(\theta + \phi) = \frac{\pi}{2}$$

$$\cos\theta\cos\phi - \sin\theta\sin\phi = 0 \quad \textbf{(expand } \cos(\boldsymbol{\theta} + \boldsymbol{\phi})\textbf{)}$$

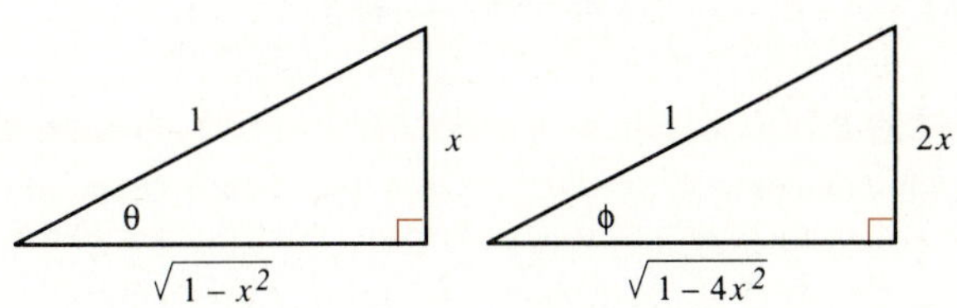

Figure 8.3-7
Right triangles showing θ and ϕ

To figure out $\cos\theta$ and $\cos\phi$, sketch right triangles showing θ and ϕ (see Figure 8.3-7). This figure tells us that $\sin\theta = x$, $\cos\theta = \sqrt{1-x^2}$, $\sin\phi = 2x$, and $\cos\phi = \sqrt{1-4x^2}$. Substituting these expressions for $\cos\theta$, $\cos\phi$, $\sin\theta$, and $\sin\phi$ gives us

$$\cos\theta\cos\phi - \sin\theta\sin\phi = 0$$

$$\sqrt{1-x^2}\,\sqrt{1-4x^2} - (x)(2x) = 0$$

$$\sqrt{1-x^2}\,\sqrt{1-4x^2} = 2x^2$$

$$(1-x^2)(1-4x^2) = 4x^4 \quad \textbf{(square both sides)}$$

$$1 - 5x^2 + 4x^4 = 4x^4 \quad \textbf{(multiply)}$$

$$5x^2 = 1 \quad \textbf{(simplify)}$$

$$x^2 = \tfrac{1}{5} \quad \rightarrow \quad x = \pm\sqrt{5}/5$$

Checking both values, we find that $-\sqrt{5}/5$ does not satisfy the original equation because $\mathrm{Sin}^{-1}(-\sqrt{5}/5)$ and $\mathrm{Sin}^{-1}(-2\sqrt{5}/5)$ are both negative, so their sum cannot possibly be positive $\frac{\pi}{2}$. Therefore, the only root is $x = \sqrt{5}/5$.
(Why do we get an extraneous root?) (Question: Would $\sin^{-1}(-\sqrt{5}/5)$ and $\sin^{-1}(-2\sqrt{5}/5)$ necessarily both be negative? Why?) ■

NOW IT'S YOUR TURN

Solve for x: $\mathrm{Cos}^{-1}x - \mathrm{Sin}^{-1}x = \pi$.

(Answer: $-\sqrt{2}/2$; $+\sqrt{2}/2$ is an extraneous root.)

The inverse cotangent, secant, and cosecant are seldom used, so we just state their definitions and graphs (see Figure 8.3-8 through 8.3-10).

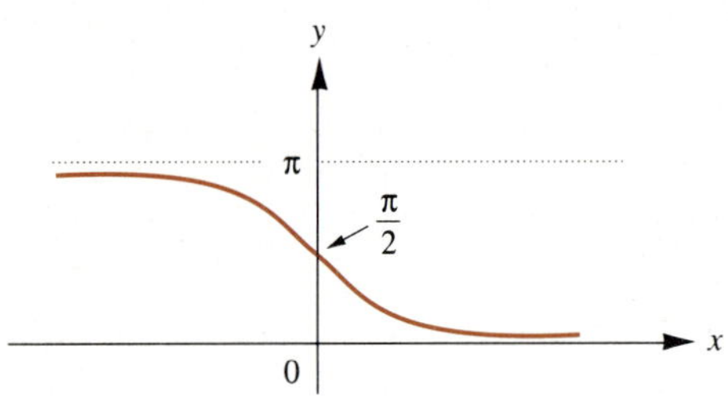

Figure 8.3-8

Graph of the function
$y = \mathrm{Cot}^{-1}x \quad 0 < y < \pi$.

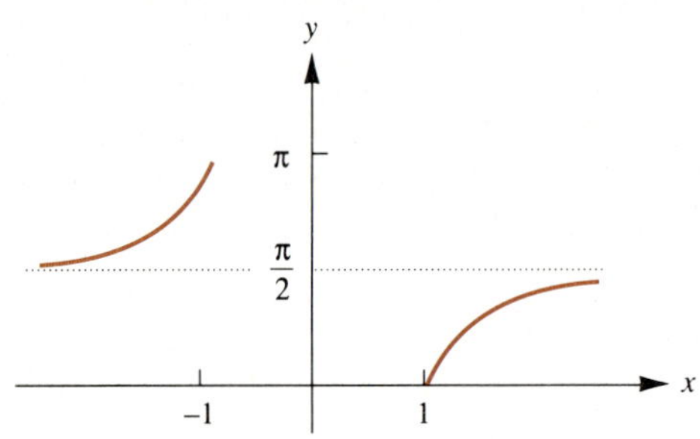

Figure 8.3-9

Graph of the function
$y = \mathrm{Sec}^{-1}x \quad 0 \le y \le \pi,\ y \ne \pi/2$.

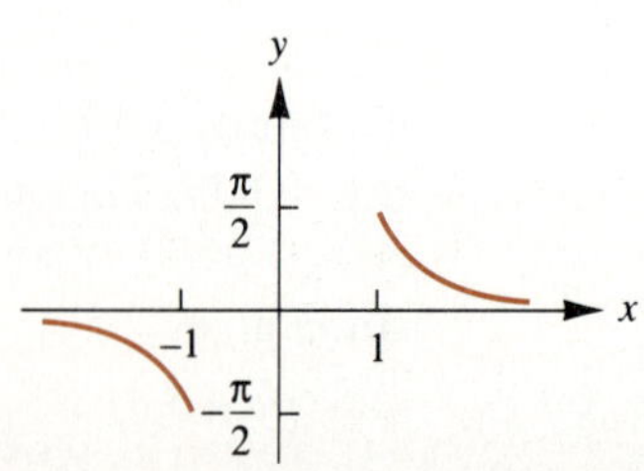

Figure 8.3-10

Graph of the function
$y = \mathrm{Csc}^{-1}x \quad -\frac{\pi}{2} \le y \le \frac{\pi}{2},\ y \ne 0$.

EXERCISES 8.3

▸ **Throughout these exercises, take note of the capital letters and remember what they mean.**

In Exercises 1–24, evaluate each quantity that is defined without using a calculator or a table.

1. $\text{Cos}^{-1}(-\frac{1}{2})$
2. $\text{Sin}^{-1}(-\sqrt{3}/2)$
3. $\sin^{-1}(-\sqrt{3}/2)$
4. $\text{Tan}^{-1}(1)$
5. $\tan^{-1} 1$
6. $\text{Cos}^{-1}(-\sqrt{3}/2)$
7. $\text{Arcsin}(\frac{1}{2})$
8. $\text{Arccos}(0)$
9. $\text{Arcsin}(-\frac{1}{2})$
10. $\text{Arccos}(-1)$
11. $\text{Arcsin}(-1)$
12. $\text{Arccos}(1/\sqrt{2})$
13. $\text{Tan}^{-1}(0)$
14. $\text{Sin}^{-1}(0)$
15. $\text{Cos}^{-1}(\sqrt{3})$
16. $\text{Sin}^{-1}(\sin \frac{\pi}{9})$
17. $\text{Cos}^{-1}[\cos(-\frac{3\pi}{10})]$
18. $\text{Tan}^{-1}(\tan \frac{\pi}{13})$
19. $\sin(\text{Arcsin } 3)$
20. $\cos(\text{Arccos } 117°)$
21. $\tan[\text{Arctan}(-43°)]$
***22.** $\text{Arcsec}\left[\sec\left(\frac{x}{x+1}\right)\right]$
***23.** $\text{Csc}^{-1}(\csc\sqrt{u})$
***24.** $\cot(\text{Arccot}\sqrt{a^2 - b^2})$

In Exercises 25–51, evaluate the given quantities without using a calculator or a table.

25. $\cos[\text{Sin}^{-1}(-\sqrt{2}/2)]$
26. $\sin[\text{Cos}^{-1}(-\sqrt{2}/2)]$
27. $\sin[\text{Sin}^{-1}(-\frac{3}{8})]$
28. $\sin[\text{Arccos}(-\frac{1}{2})]$
29. $\tan[\text{Arcsin}(-\sqrt{5}/2)]$
30. $\cot[\text{Cos}^{-1}(-\sqrt{2}/3)]$
31. $\sin[\text{Tan}^{-1}(-2)]$
32. $\sec(\text{Arcsec } 2.8)$
33. $\text{Sin}^{-1}(\tan 45°)$
34. $\text{Arccos}(\sin \frac{\pi}{4})$
35. $\cot(\text{Cot}^{-1}(-9.68)$
36. $\text{Tan}^{-1}(\cos \frac{\pi}{2})$
37. $(\frac{3}{5})[\text{Arcsin}(1) - \text{Arctan}(-1)]$
38. $\text{Tan}^{-1}(-\sqrt{3}) - \text{Cos}^{-1}(0)$
39. $\text{Sin}^{-1}(-\sqrt{3}/2) - \text{Cos}^{-1}(-1)$
40. $\text{Arctan}(-1) + \text{Tan}^{-1}(1)$
41.** $\sin[2\,\text{Sin}^{-1}(\sqrt{2}/10)]$ ***Hint: Think of this as $\sin 2\theta$.
***42.** $\cos[2\,\text{Cos}^{-1}(\frac{1}{3})]$
***43.** $\sin[\frac{1}{2}\,\text{Sin}^{-1}(\frac{3}{4})]$
***44.** $\cos[\frac{1}{2}\,\text{Sin}^{-1}(\frac{2}{3})]$
***45.** $\tan(2\,\text{Tan}^{-1}\sqrt{7})$
***46.** $\sin[2\,\text{Cos}^{-1}(\sqrt{2}/3)]$
47.** $\sin(\text{Sin}^{-1}\frac{1}{2} + \text{Cos}^{-1}\frac{1}{3})$ ***Hint: Let $\theta = \text{Sin}^{-1}\frac{1}{2}$ and $\phi = \text{Cos}^{-1}\frac{1}{3}$. Then use $\sin(\theta + \phi)$.
***48.** $\cos(\text{Tan}^{-1} 4 + \text{Cot}^{-1} 2)$
***49.** $\sin(\text{Sin}^{-1}\frac{2}{3} + \text{Sin}^{-1}\frac{2}{3})$
****50.** $\tan(\text{Sin}^{-1}\frac{3}{5} + 2\,\text{Cos}^{-1}\frac{2}{3})$
****51.** $\tan(\text{Cos}^{-1}\frac{1}{6} - 2\,\text{Tan}^{-1} 3)$

In Exercises 52–76, use a calculator to evaluate each expression, giving your answers in radians.

52. $\text{Arctan}(3.14)$
53. $\text{Cos}^{-1}(0.936)$
54. $\text{Sin}^{-1}(-0.167)$
55. $\text{Arcsin}(-\sqrt{3})$
56. $\text{Arccos}(0.785)$
57. $\text{Tan}^{-1}\sqrt{2}$
58. $\cos[\text{Tan}^{-1}(-3.64)]$
59. $\tan(\text{Sin}^{-1} 0.375)$
60. $\sin[\text{Arccos}(-0.172)]$
61. $\cos(\text{Arcsin } 0.247)$
62. **(a)** $\sin(\text{Arccos } 0.431)$ **(b)** $\cos(\text{Arcsin } 0.431)$
63. $\text{Cos}^{-1}(0.20) - \text{Cos}^{-1}(0.60)$
64. $\text{Sin}^{-1}(-0.40) - \text{Sin}^{-1}(0.60)$
65. $\text{Arctan}(\sqrt{3}/4) - \text{Tan}^{-1}(\sqrt{3})$
66. $\text{Cos}^{-1}(3/\sqrt{10}) + \text{Cos}^{-1}(2/\sqrt{5})$
67. $\text{Sin}^{-1}\frac{8}{17} + \text{Sin}^{-1}\frac{75}{85}$ (Do you see why you get this answer?)
68. $\text{Sin}^{-1}\frac{63}{65} - \text{Sin}^{-1}\frac{5}{13} - \text{Cos}^{-1}\frac{3}{5}$
69. $\cos(2\,\text{Cos}^{-1} 0.6731)$
70. $\text{Sin}^{-1}(2\,\text{Cos}^{-1} 0.2211)$
71. $\sin[\frac{1}{2}\,\text{Cos}^{-1}(-0.5557)]$
72. $\text{Cos}^{-1}(\frac{1}{2}\,\text{Sin}^{-1} 0.1984)$
73. $\sin(5\,\text{Tan}^{-1} 4.23)$
74. $\text{Tan}^{-1}(3.7\,\text{Cos}^{-1} 0.6701)$
75. $\cos[\frac{2}{3}\,\text{Tan}^{-1}(-0.851)]$
76. $\sin(\frac{1}{10}\,\text{Tan}^{-1} 9.31)$

In Exercises 77–93, write each expression as an algebraic expression involving *x*.

77. $\cos(\text{Sin}^{-1} x)$
78. $\sec(\text{Cos}^{-1} x)$
79. $\cot(\text{Tan}^{-1} x)$
80. $\sin(\text{Tan}^{-1} x)$
***81.** $\cos(2\,\text{Cos}^{-1} x)$
***82.** $\sin(2\,\text{Sin}^{-1} x)$
***83.** $\tan(2\,\text{Tan}^{-1} x)$
***84.** $\sin(\frac{1}{2}\,\text{Cos}^{-1} x)$
***85.** $\cos(\frac{1}{2}\,\text{Sin}^{-1} x)$
86.** $\sin(\text{Cos}^{-1} x + \text{Sin}^{-1} 2x)$ ***Hint: See Exercise 47 for a hint.
***87.** $\cos[\text{Sin}^{-1} x - \text{Sin}^{-1}(x/2)]$
***88.** $\sin[\text{Tan}^{-1} x - \text{Cos}^{-1}(x/3)]$
***89.** $\tan[\text{Tan}^{-1}(x/2) + \text{Sec}^{-1}(x/3)]$
***90.** $\cos(\text{Sin}^{-1}\sqrt{x+1} + \text{Cos}^{-1} x)$
****91.** $\sin(3\,\text{Sin}^{-1} x)$ ***Hint:*** $3 = 2 + 1!$
****92.** $\cos(3\,\text{Cos}^{-1} x)$
93.** $\sin(\frac{\pi}{2} - \text{Cos}^{-1} x)$ ***Hint: Let $\theta = \text{Cos}^{-1} x$
***94.** Prove that if $0 \le x \le 1$, then $\text{Sin}^{-1}(-x) = -\text{Sin}^{-1} x$.
***95.** Prove that if $0 \le x \le 1$, then $\text{Cos}^{-1}(-x) = \pi - \text{Cos}^{-1} x$.
***96.** Prove that $\sin(\text{Cos}^{-1} a + \text{Sin}^{-1} b) = ab + \sqrt{1-a^2}\,\sqrt{1-b^2}$.
***97.** Prove that $\text{Sin}^{-1} x + \text{Cos}^{-1} x = \frac{\pi}{2}$, $(0 \le x \le 1)$.
***98.** Solve for x: $\text{Cos}^{-1} x + \text{Cos}^{-1} 2x = \frac{\pi}{2}$.

**99. Prove that $\text{Tan}^{-1} x + \text{Cot}^{-1} x = \frac{\pi}{2}, x \geq 0$. *Hint:* Take the tangent of both sides and use the fact that $\text{Tan}^{-1}(\text{undefined}) = \frac{\pi}{2}$.

*100. Show that $\text{Tan}^{-1} 3 + \text{Tan}^{-1} 2 + \text{Tan}^{-1} 1 = \pi$.

**101. Given that $\text{Cos}^{-1} A + \text{Cos}^{-1} B + \text{Cos}^{-1} C = \pi$, show that $A^2 + B^2 + C^2 + 2ABC = 1$.

102. Solve the equation $\cos(\frac{\pi}{3} - 3x) = \frac{2}{3}$ for x.

**103. Solve for x: $a\cos(bx + c) = d$, (a, b, c, and d are constants).

**104. Solve for x: $a\sin(bx + c) = d$ (a, b, c, and d are constants).

*105. We have seen that $\sin(A + B) \neq \sin A + \sin B$. But is it true that $\text{Sin}^{-1}(a + b) = \text{Sin}^{-1} a + \text{Sin}^{-1} b$? *Hint:* If you can find at least one set of values for a and b for which this is not true, then you have proven that it is not true.

*106. Is $\text{Cos}^{-1}(a + b) = \text{Cos}^{-1} a + \text{Cos}^{-1} b$? Prove your answer. *Hint:* See Exercise 105.

*107. Prove that $\text{Sin}^{-1}(\sin x) = x$. *Hint:* Let $y = \sin x$. Then show that this leads to $\text{Sin}^{-1} y = \text{Sin}^{-1}(\sin x)$. Next use the fact that $y = \sin x$ means that $x = \text{Sin}^{-1} y$, and then combine your results.

*108. Prove that **(a)** $\text{Cos}^{-1}(\cos x) = x$ and **(b)** $\text{Tan}^{-1}(\tan x) = x$. *Hint:* See Exercise 107.

109. Evaluate each expression without using a calculator or a table. *Hint:* See Exercises 107 and 108.

(a) $\text{Cos}^{-1}(\cos \frac{\pi}{7})$ **(b)** $\text{Tan}^{-1}(\tan 0.818)$
(c) $\text{Sin}^{-1}(\sin 1.4)$ ***(d)** $\text{Cos}^{-1}(\cos 60°)$
***(e)** $\text{Sin}^{-1}(\sin 45°)$

In Exercises 110–121, which graphs represent functions? Explain your reasoning!

110.
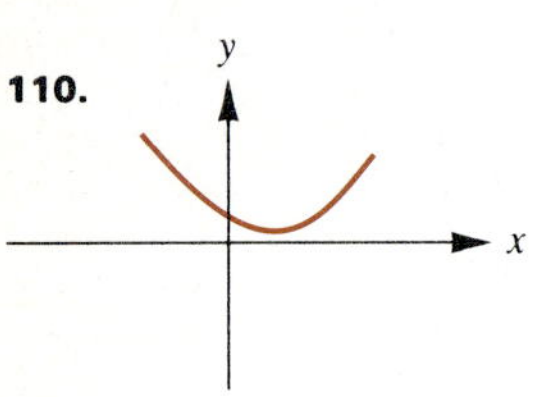

111.
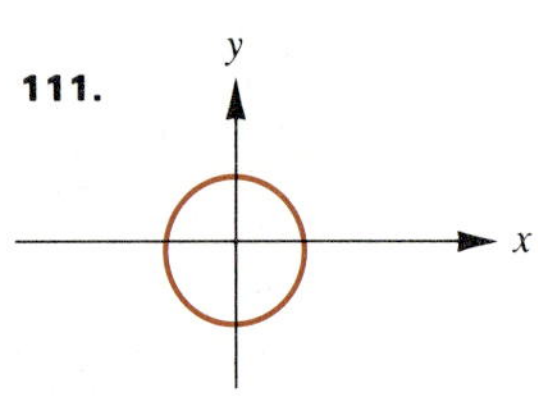

112.
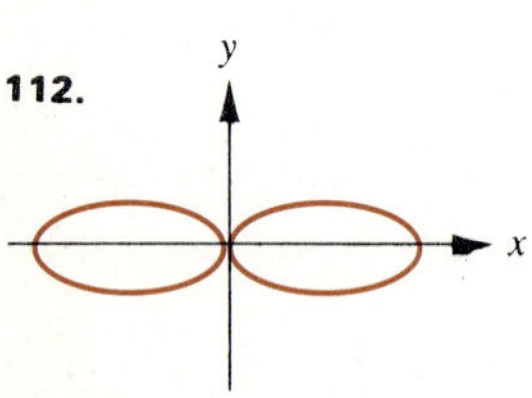

113.
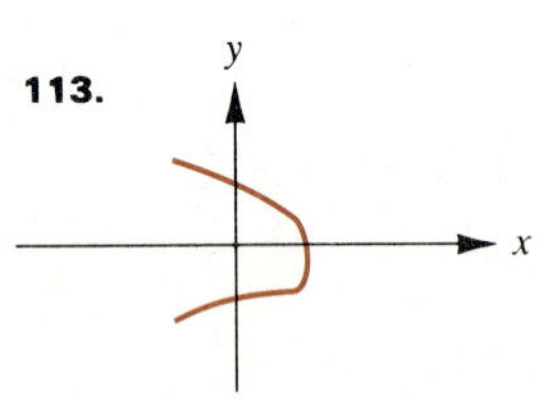

114.
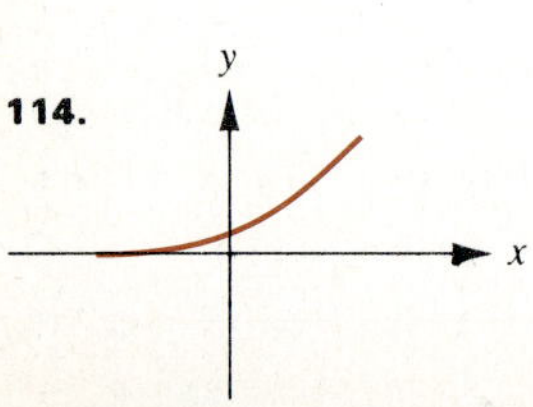

115.
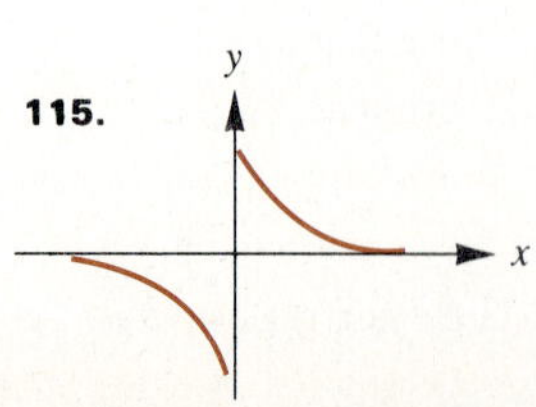

116.
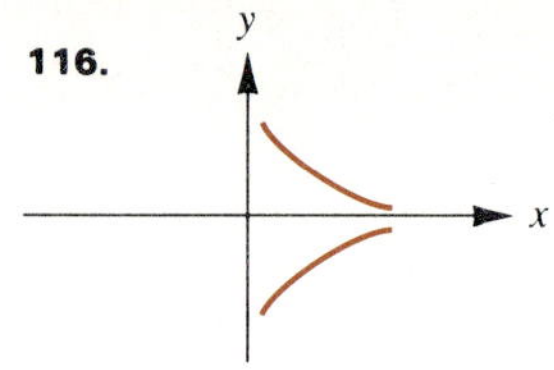

117.
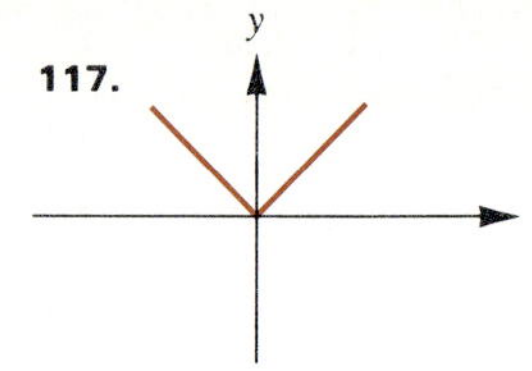

118.
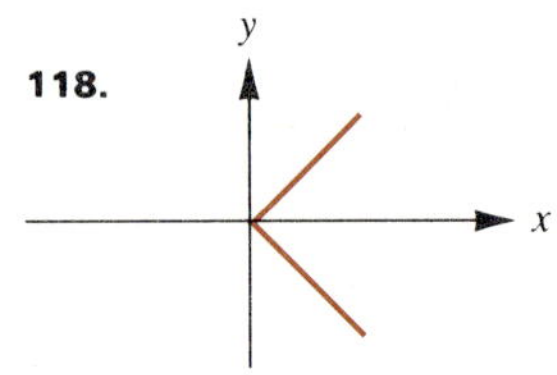

119.
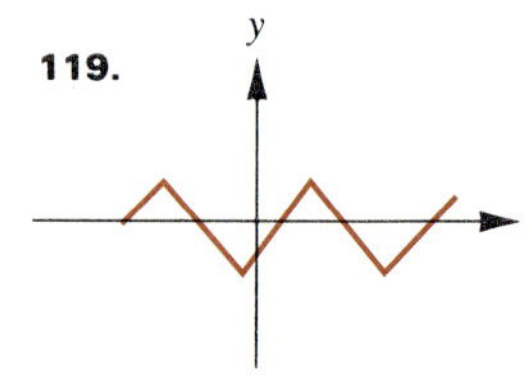

120.
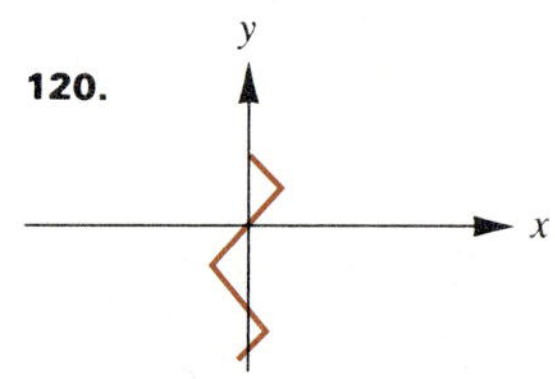

121.
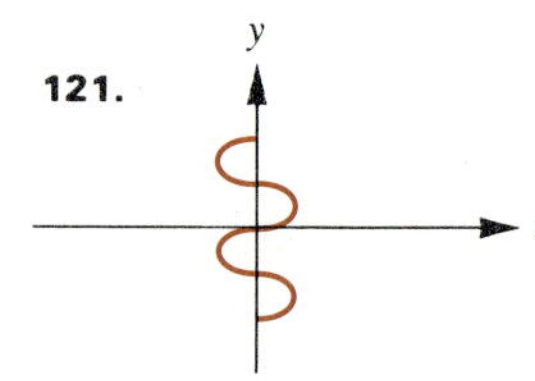

122. For each graph in Exercises 110–121 that does *not* represent a function, restrict the range of y so that it will represent a function. No numerical values are needed; just sketch the part (or parts) of the graph that you need to include to make it represent a function.

In Exercises 123–130, which equations represent functions? Clearly explain why. *Hint:* Try to decide without making a graph by looking at the properties of each equation. If you cannot tell, try graphing the equation as a last resort.

*123. $y = x^2$ *124. $y^2 = x$
*125. $y^2 = 1 - x^2$ *126. $y = |x|$
*127. $|y| = x$ *128. $y^2 = \sin x$
*129. $y = \sin^2 x$
*130. $y^2 = x^2$ (Is this the same as $y = x$?)

*131. **(a)** Make a table of values and use these values to sketch the graph of $y = \tan^{-1} x$ without limiting y. Is this a function? Why?
(b) By looking at your graph, decide what limits should be placed on y so that $y = \text{Tan}^{-1} x$ will be a function.

*132. **(a)** Make a table of values and use these values to sketch the graph of $y = \cot^{-1} x$ without limiting y. Is this a function? Why?
(b) By looking at your graph, decide what limits should be placed on y so that $y = \text{Cot}^{-1} x$ will be a function.

***133.** **(a)** Make a table of values and use these values to sketch the graph of $y = \sec^{-1} x$ without limiting y. Is this a function? Why?
(b) By looking at your graph, decide what limits should be placed on y so that $y = \text{Sec}^{-1} x$ will be a function.

***134.** **(a)** Make a table of values and use these values to sketch the graph of $y = \csc^{-1} x$ without limiting y. Is this a function? Why?
(b) By looking at your graph, decide what limits should be placed on y so that $y = \text{Csc}^{-1} x$ will be a function.

135. **(a)** Is it possible that $\sin^{-1} x$ is negative but x is positive? Justify your answer with a few numerical examples.
(b) Is it possible that $\text{Sin}^{-1} x$ is negative but x is positive? Justify your answer with a few numerical examples.

136. **(a)** Is it possible that $\cos^{-1} x$ is negative but x is positive? Justify your answer with a few numerical examples.
(b) Is it possible that $\text{Cos}^{-1} x$ is negative but x is positive? Justify your answer with a few numerical examples.

137. **(a)** Is it possible that $\tan^{-1} x$ is negative but x is positive? Justify your answer with a few numerical examples.
(b) Is it possible that $\text{Tan}^{-1} x$ is negative but x is positive? Justify your answer with a few numerical examples.

SUMMARY

1. Inverse Sine (Arcsin)

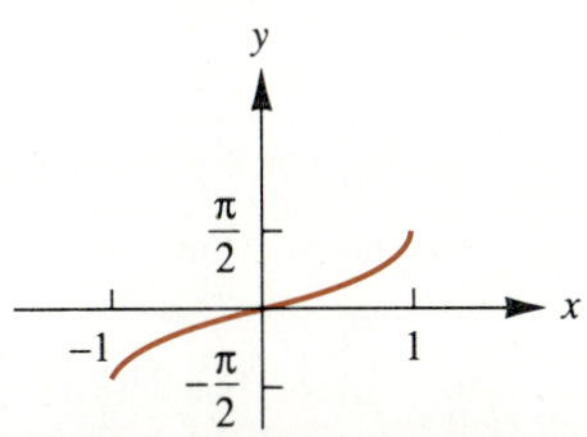

$$y = \text{Sin}^{-1}x \iff x = \sin y$$
$$-\frac{\pi}{2} \le y \le \frac{\pi}{2}$$

2. Inverse Cosine (Arccos)

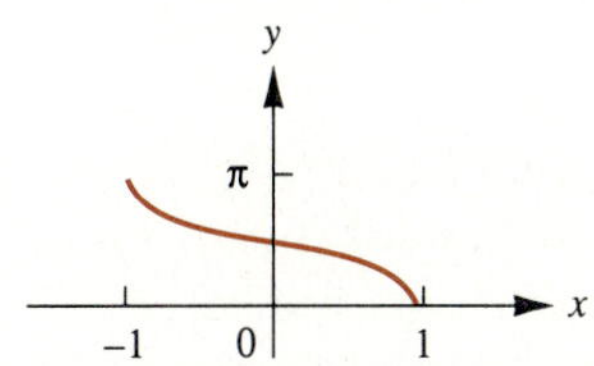

$$y = \text{Cos}^{-1}x \iff x = \cos y$$
$$0 \le x \le \pi$$

3. Inverse Tangent (Arctan)

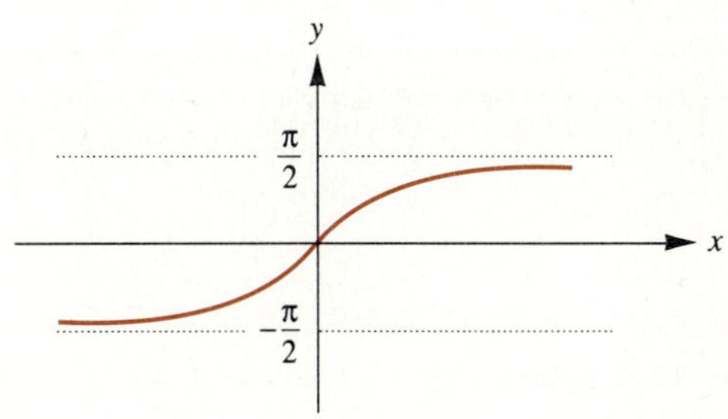

$$y = \text{Tan}^{-1}x \iff x = \tan y$$
$$-\frac{\pi}{2} < y < \frac{\pi}{2}$$

IMPORTANT TERMS

conditional equation
mathematical function
mathematical identity
mathematical relationship

REVIEW EXERCISES

In Exercises 1–48, solve for all values of x ($0 \le x < 2\pi$). When you need to use a calculator, express answers to two decimal places.

1. $\cos x = \frac{1}{2} \sin 2x$
2. $\sin x = 1 - \cos x$
3. $3 \cos x - 1 = -4 \cos 2x$
4. $2 - \tan x \tan 2x = 0$
5. $\cos x = -\cos 2x$
6. $\cot x \tan 2x = 3$
7. $3 \sin^2 x - \cos^2 x = \sin 2x$
8. $2 \sin x = \cos x$
9. $\tan x \sec x = \sqrt{2}$
***10.** $8 \cos^3 x + \cos 3x = 0$
11. $\tan x + \sec x = 3$
12. $\sin^{-1} (x/2) = \frac{\pi}{6}$
13. $\csc x - \tan x = \sec x - \cot x$
14. $\tan^2 x + \cot^2 x = \frac{10}{3}$
15. $6 \sin x + 4 \cos 2x = 5$
16. $\sin^2 2x + 2 \sin^2 x = 2$
***17.** $\sin^2 x + \sin 2x = 1$
18. $\sec 2x = \cot x \tan 2x$
19. $\sin 2x \cot x - \sin^2 x = \frac{1}{2}$
20. $2 \cos x = 1 + \sec 2x$
21. $\csc x - \cot x = \sqrt{3}$
22. $\cos 5x + \cos 3x = -\cos x$
23. $2 \cos x = 1 - \sin x$
24. $\cos x + \sin x = \sec x$
25. $\cos 3x + \cos x = 2 \cos^2 x$
26. $\sin^{-1} x + 2 \cos^{-1} x = \frac{2\pi}{3}$
27. $\dfrac{1 - \tan x}{1 + \tan x} = \cos 2x$
28. $\sqrt{1 + \sin x} - \sqrt{1 - \sin x} = 2 \cos x$
***29.** $(1 - \tan x) \cos 2x = 1 + \tan x$
30. $\sec(x + 120°) + \sec(x - 120°) = 2 \cos x$
31. $\sin x + \sin 2x + \sin 3x = 0$
32. $\tan (\frac{\pi}{3} + x) \tan (\frac{\pi}{3} - x) + 2 = 0$
33. $\sin(x + 120°) + \sin(x + 60°) = 1.50$
34. $\cos 3x = \sin 2x$
35. $\sin^4 x - \cos^4 x = 0.28$
36. $\sec x - 2 \tan x = 0.250$
37. $\cos 2x - 1 = \sin 3x$
***38.** $\cos x + \cos 3x + \cos 5x + \cos 7x = 0$
39. $\arcsin x + \arcsin (x/2) = \frac{2\pi}{3}$
40. $\arctan(x + 1) + \arctan(x - 1) = \arctan 2x$
41. $\cos 2x \sec x + \sec x + 1 = 0$
42. $\tan(x + \frac{\pi}{4}) = 8 \tan x$
43. $\cos x - \cos 3x = \sin x + \sin 3x$
***44.** $\tan 3x = \tan x + \tan 2x$
***45.** $\tan 2x - 2 \sin^2 x = \sin 2x$
46. $\dfrac{\cos 2x}{1 - \tan x} = 0$
****47.** $\sin(\pi \cos x) - \cos(\pi \sin x) = 0$
48. $2 \sin x \cos \dfrac{x}{2} = \sqrt{\dfrac{1 + \cos x}{2}}$

49. For what values of x ($0 \le x < 2\pi$) is y a minimum; for which values is it a maximum.

(a) $y = -2 \sin \left(\dfrac{\pi x}{4} + 3\right)$
(b) $y = -\cos(\frac{\pi}{3} - 2x)$
****(c)** $y = \sin \left(\dfrac{x}{4} - 3\right)$
(d) $y = 1.9 \cos \left(8.0 + \dfrac{\pi x}{3}\right)$

In Exercises 50–53, find all possible values of θ between $\pm 2\pi$.

50. (a) $\theta = \sin^{-1} \frac{1}{2}$ **(b)** $\theta = \text{Sin}^{-1} \frac{1}{2}$
51. (a) $\theta = \text{Cos}^{-1} 0$ **(b)** $\theta = \cos^{-1} 0$
52. (a) $\theta = \tan^{-1}(-1)$ **(b)** $\theta = \text{Tan}^{-1} (-1)$
53. (a) $\theta = \text{Sin}^{-1}(-0.213)$ **(b)** $\theta = \sin^{-1} (-0.213)$

In Exercises 54–73, evaluate each quantity. Pay heed to capital letters.

54. $\text{Arcsec} - 2$
55. $\text{Arctan } 1 + \text{Arctan } \sqrt{3}$
56. $\text{Arccos } 1 + 3 \text{ Arcsin } \frac{1}{2}$
57. $\text{Arccot } 1 - \frac{\pi}{4}$
58. $\text{Tan}^{-1} (-1) - \frac{3\pi}{4}$
59. $\text{Sin}^{-1}\frac{1}{2} + \text{Cos}^{-1}\frac{1}{2}$
60. $\text{Cos}^{-1} (-\frac{1}{2}) - \text{Cos}^{-1} (\sqrt{3}/2)$
61. $\text{Cos}^{-1} (-1) + \pi$
62. $\frac{1}{3} \text{Tan}^{-1} \sqrt{3} + \frac{1}{2} \text{Cot}^{-1}(-\sqrt{3})$
63. $\text{Arcsin}(-\sqrt{3}/2) + \text{Arccos}(-\sqrt{3}/2)$
64. $\cos(2 \text{ Arcsin } \sqrt{2}/2)$
65. $\tan(\text{Arccos } \frac{1}{2})$
66. $\sin(\text{Cot}^{-1}\sqrt{3})$
67. $\cos[2 \text{ Arcsin}(-\sqrt{2}/2)]$
68. $\tan(\text{Arcsin } \frac{8}{17})$
69. $\sin[\text{Arcsin}(-\frac{1}{2}) + \text{Arccos}(-\frac{1}{2})]$
70. $\cot[\text{Arccot}(-0.2239)]$
71. $\tan(\text{Arcsin } 0.912)$
72. $\cos[\text{Arccot}(-0.2839)]$
73. $\cos[\text{Arccos}(-\sqrt{3}/2) + \frac{1}{2} \text{ Arcsin}(\sqrt{3}/2)]$

74. **(a)** Express x as a function of y.

(i) $y = \sin 2x$ **(ii)** $y = \tan 3x$
(iii) $y = 4 \cos 3x$ **(iv)** $3y = 5 \cot (x/2)$
(v) $y = m \sin nx$

(b) Express y as a function of x.

(i) $x = 4 \text{ Arctan } (y/2)$ **(ii)** $x = 6 \arccos 2y$
(iii) $x - 5 = 3 \sin^{-1} y$ **(iv)** $x + 3 = 4 \text{ Tan}^{-1} 5y$
(v) $2x - 4 = \frac{1}{2} \cos^{-1} (y/2)$

In Exercises 75–79, prove each identity if it is one.

***75.** $\text{Arcsin} \sqrt{1 - x^2} = \text{Arctan} \left(\dfrac{\sqrt{1 - x^2}}{|x|} \right)$

***76.** $\text{Arcsin} \left(\dfrac{1}{\sqrt{x^2 + 1}} \right) = \text{Arcsec} \left(\dfrac{\sqrt{x^2 + 1}}{|x|} \right)$

***77.** $\text{Tan}^{-1} \left(\dfrac{1}{\sqrt{x^2 - 1}} \right) = \text{Cos}^{-1} \left(\dfrac{\sqrt{x^2 - 1}}{|x|} \right)$

78. $\sin(\text{Cos}^{-1} x) = \cos(\text{Sin}^{-1} x)$

79. $\tan(\text{Cot}^{-1} x) = \cot(\text{Tan}^{-1} x)$

80. Which of the graphs represents a mathematical function? Explain your reasoning. For those that do not, restrict the range of y to make them functions. No numbers are necessary; do it graphically.

(a)

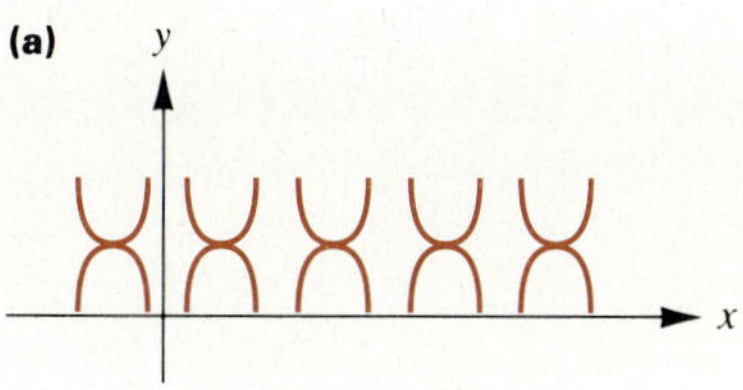

(b)

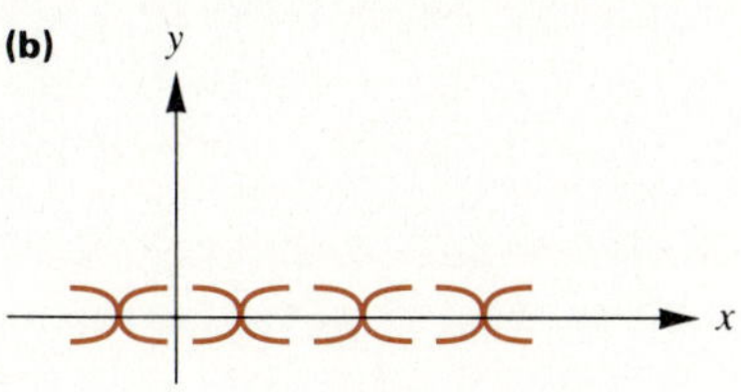

(c)

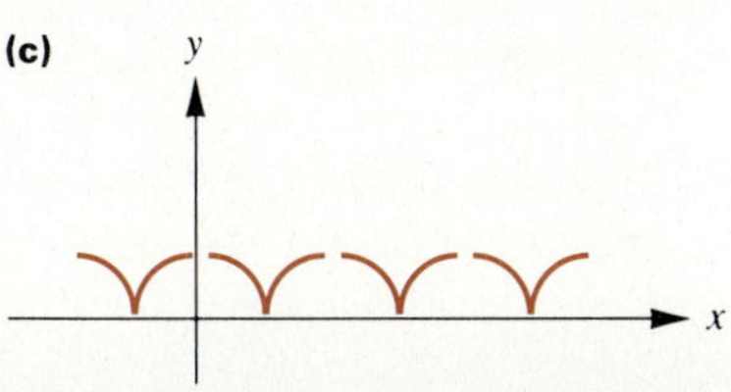

(d)

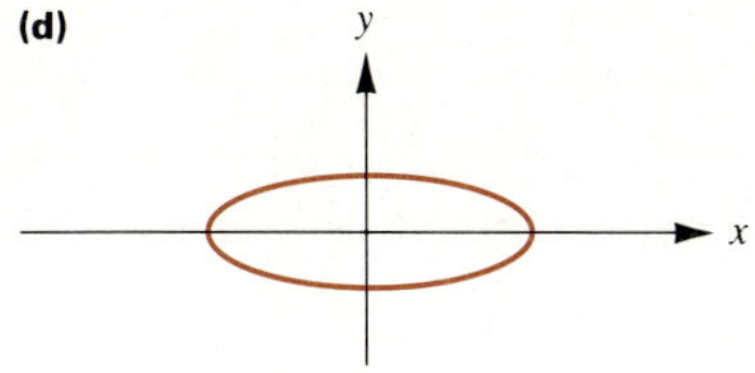

(e)

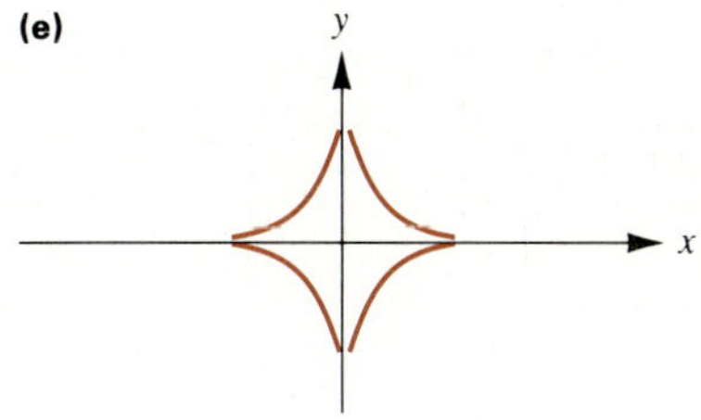

(f)

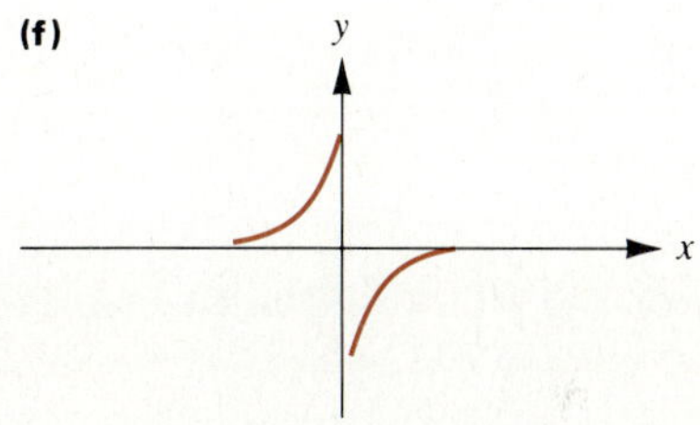

(g)

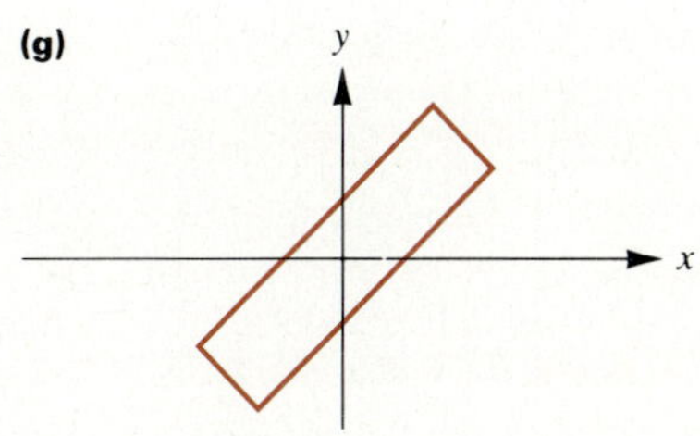

(h)

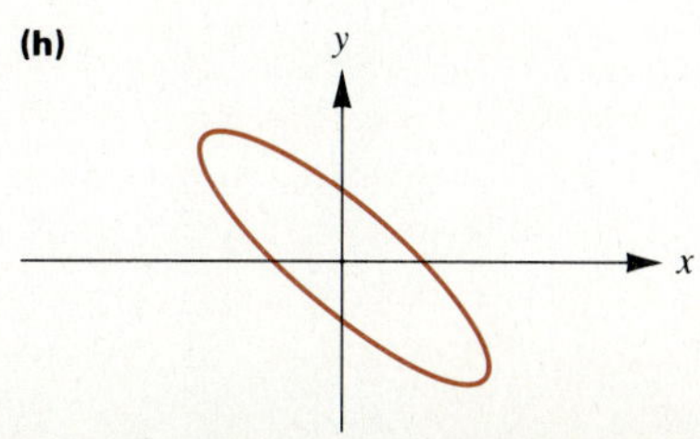

81. If you deform a square to make a parallelogram, find θ so that the new area will be one-third the original area of the square.

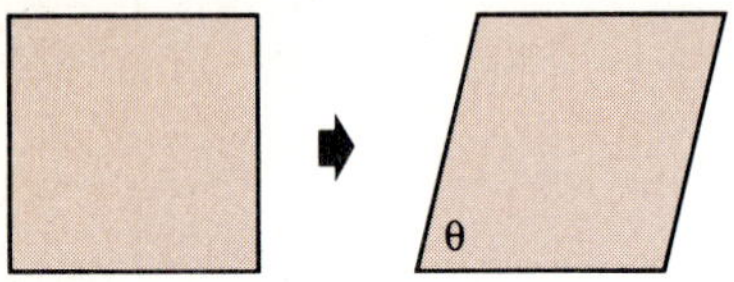

82. A plot of land must be subdivided into lots in the shape of a parallelogram. Find θ so that the area of each plot will be 5800 m^2.

83. A park is to be layed out in the shape of parallelogram $ABCD$ as shown. A nature trail is to go from $A \rightarrow B \rightarrow C \rightarrow A$ and must be 1000 yd long.

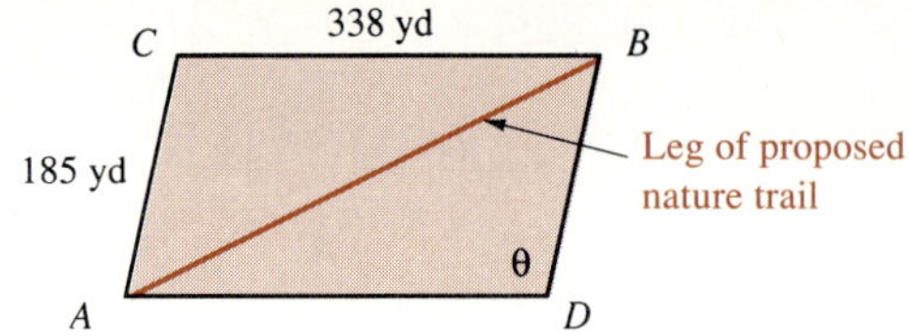

(a) What must be the angle θ?

(b) What would be the length of a nature trail going from $A \rightarrow C \rightarrow D \rightarrow A$?

In Exercises 84–88, solve each question graphically. If there are infinitely many solutions, find the first few positive and negative ones.

84. $\sin x = 1 + 2x$

85. $\tan x = \cos 2x$

86. $\sin^2 x = x/8$

87. $\cos x = \sqrt{x}$

88. $\cos x = |x|$

9

Polar Coordinates and Complex Numbers

9.1 POLAR COORDINATES

9.1-1 Definition of Polar Coordinates

If you were telling a city friend how to reach your favorite pizza parlor, you would probably say something like, "from my house, go 6 blocks east and 9 blocks north." By giving directions this way, you are really giving the x and y coordinates of the pizza parlor, treating your house as the origin (see Figure 9.1-1). The square blocks of most cities function like large x-y coordinate

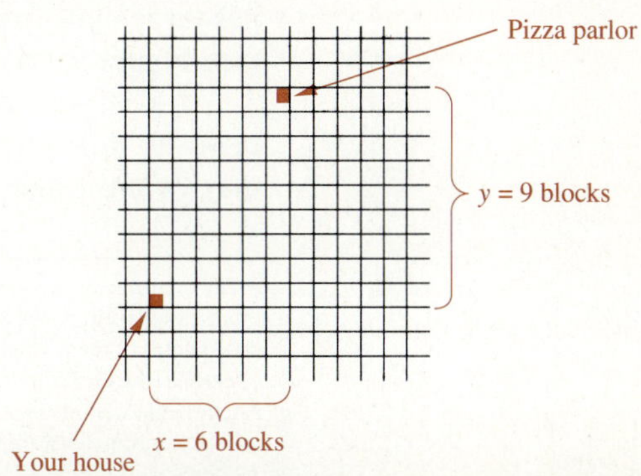

Figure 9.1-1

Location of pizza parlor These city blocks form a rectangular (x, y) coordinate system.

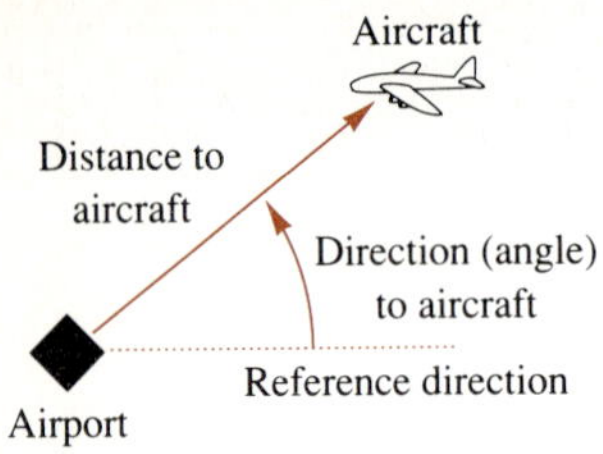

Figure 9.1-2

Air traffic controller screen The location of the plane is given by its distance from the airport and its direction with respect to a fixed reference direction.

systems. We call such an x-y coordinate system a *rectangular* or a *cartesian** coordinate system. Up to now, we have worked only with rectangular coordinates.

Suppose, however, that you are an air traffic controller at a busy airport. The information on your radar screen gives you the location of aircraft but in a very different way. The radar tells how *far* each plane is from the airport and along what *direction* (angle) it lies (see Figure 9.1-2). This information locates the airplane without using rectangular (x, y) coordinates.

This way of locating a point, using its distance from a fixed point and an angle relative to a reference direction, is a new type of coordinate system, which we call a *polar coordinate system.*†

To define a polar coordinate system, we need two things: a fixed point O (called the *pole* or *origin*) from which to measure distances, and a fixed reference line (called the *polar axis*) from which to measure angles (see Figure 9.1-3a). To locate a point P in a plane, we give the *directed* distance OP from the pole O to P (we call this r) and the angle that OP makes with the polar axis (we call this angle θ), as shown in Figure 9.1-3b. We call these numbers (r and θ) the *polar coordinates* of the point P, and we write them as an *ordered pair* (r, θ), just as we write the rectangular coordinates (x, y) as an ordered pair. The first number is always r and the second is always θ. We often write this pair as $P(r, \theta)$, meaning that P is the point having polar coordinates (r, θ). For example, $P(3, 60°)$ is a point for which $r = 3$ and $\theta = 60°$. We use the usual sign convention for θ: counterclockwise is positive, clockwise negative.

Figure 9.1-3

Polar coordinates

(a) Pole (origin) and polar axis ***(b) Polar coordinates (r,θ) of the point P***

To locate the polar coordinates of a point, we construct a grid consisting of concentric circles, each of which has the pole O as its center with rays coming out from the pole. (Better yet, buy polar graph paper.) (See Figure 9.1-4.) The concentric circles allow us to plot the distance r from the pole, and the rays allow us to locate angles θ relative to the polar axis. For example, the point $P(3, 60°)$ lies three units from the pole along the ray making a 60° angle with the polar axis (see Figure 9.1-4).

If r is negative, we measure $|r|$ units from the pole, but along the ray opposite θ (see Figure 9.1-5). If $r = 0$, the point is located at the pole O, so θ may be any angle.

---

* The coordinates are called "cartesian" after the French mathematician-philosopher Rene Descartes, who used them extensively when he founded analytic geometry in 1637. Descartes formulated geometry in terms of the coordinates of the points on lines and curves, but he did not invent this coordinate system. Its roots go back at least as early as the Greek mathematician Appollonius of Perga in around 200 B.C. Appollonius lived at the same time as Euclid and Archimedes.

† Polar coordinates were first introduced by Isaac Newton in around 1671 in his book *Method of Fluxions*, in which he developed the calculus. They were greatly expanded by Euler in 1748. However, *bees* knew about them much sooner than this! (See Example 5.)

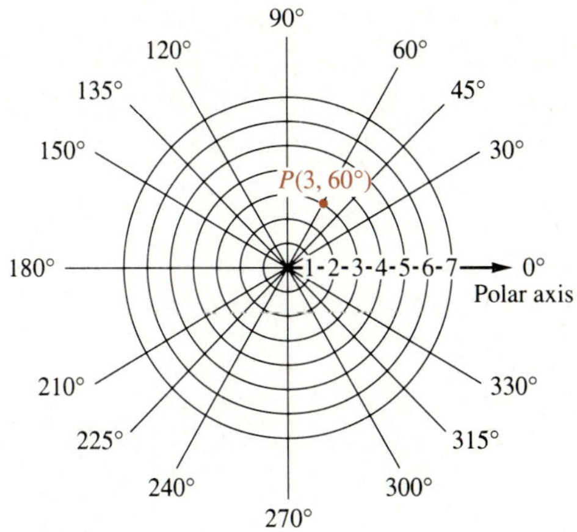

Figure 9.1-4

Polar graph paper, with the point *P*(3, 60°) plotted on it

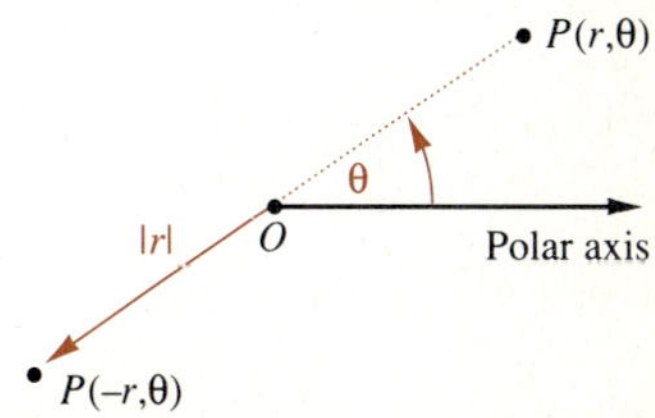

Figure 9.1-5

Plotting a point if *r* is negative

Example 1

Figure 9.1-6 shows an illustration of an air traffic controller's radar screen, each dot being an aircraft. The pole is the airport, the polar axis is from west to east, and r is in kilometers. What are the polar coordinates of each aircraft if $r > 0$ and $0 \le \theta < 2\pi$?

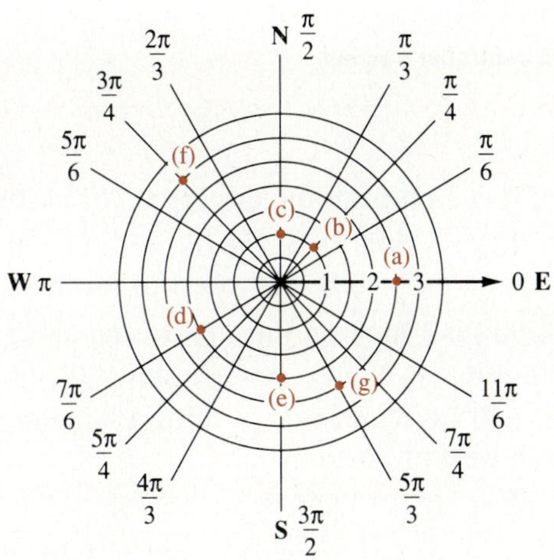

Figure 9.1-6

Aircraft on an air traffic controller's radar screen

Solution

We just read r and θ from the graph. (a) (2.5 km, 0), (b) (1 km, $\frac{\pi}{4}$), (c) (1 km, $\frac{\pi}{2}$), and (d) (2 km, $\frac{7\pi}{6}$) ■

NOW IT'S YOUR TURN

Find (e), (f), and (g).

(Answer: (e) (2 km, $\frac{3\pi}{2}$), (f) (3 km, $\frac{3\pi}{4}$), and (g) (2.5 km, $\frac{5\pi}{3}$))

Example 2

An air traffic controller's screen shows planes at the following polar locations, where the pole is the airport, the polar axis is a line from west to east, and r is in miles. (a) (3, 45°), (b) (5, −120°), (c) (−4, 90°), (d) (−6, −30°), (e) (0, 420°), (f) (3, 0°), and (g) (4, 270°). Plot what this screen shows.

Solution

See Figure 9.1-7.

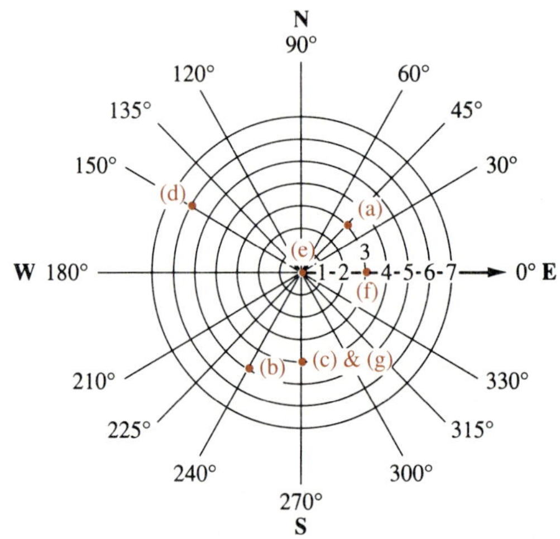

Figure 9.1-7

Air traffic controller's screen

(a) (3, 45°) is 3 mi from the airport at 45° north of east.

(b) (5, −120°) is 5 mi from the airport along a ray making an angle of −120° with the polar axis, so it is 30° west of south.

(c) (−4, 90°) is 4 mi from the airport, but along the ray that is opposite the 90° ray. Therefore, it is 4 mi directly south of the airport.

(d) (−6, −30°) is 6 mi from the airport but along the ray opposite the −30° ray, so it is 60° west of north.

(e) The plane at (0, 420°) is at the pole, so it is at the airport.

(f) (3, 0) is 3 mi directly east of the airport along the polar axis.

(g) The plane at (4, 270°) is 4 mi directly south of the airport. Notice that this position is the same point as (−4, 90°). We have potential trouble here! ■

NOW IT'S YOUR TURN

If planes are at the following points, plot the radar screen: (a) (2.5, 135°), (b) (−3, 180°), (c) (0, 50°), (d) (2.5, −225°), (e) (−2, 0°), and (f) (1, −45°). (Answer: See Figure 9.1-8)

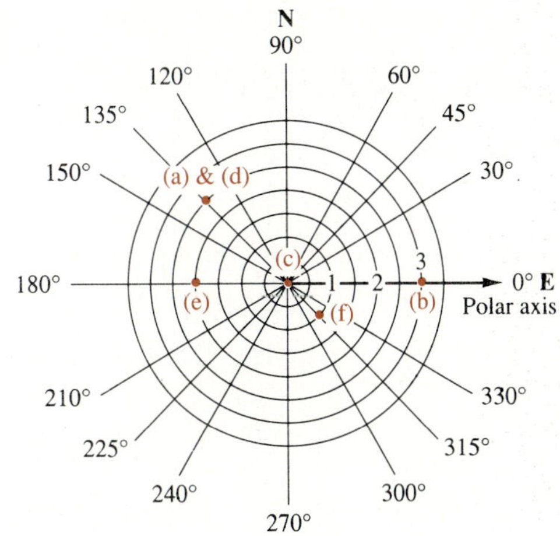

Figure 9.1-8

In Example 2 we discovered something surprising. It is possible for a point to have more than one set of polar coordinates. For example, we saw that $(-4, 90°)$ and $(4, 270°)$ both represent one point and that $(2.5, \frac{3\pi}{4})$ and $(2.5, -\frac{5\pi}{4})$ both represent one point.

NOW IT'S YOUR TURN

Find at least three additional polar coordinates for the point $(1, \frac{\pi}{2})$, at least one of which includes a negative value of r. (Answer: $(-1, \frac{3\pi}{2})$, $(1, \frac{5\pi}{2})$, $(-1, -\frac{\pi}{2})$. There are *infinitely* many more possibilities!)

In rectangular coordinates, every point has one and only one representation. In polar coordinates, however, a point can have many different representations. In fact, it has *infinitely* many because the point repeats whenever θ changes by multiples of 2π.

Example 3

Give polar representations of the point $(3.4, -\frac{\pi}{6})$ that satisfy each of the following conditions.

(a) $r > 0,\ 0 \le \theta < 2\pi$ (b) $r > 0,\ -2\pi < \theta \le 0$
(c) $r < 0,\ 0 \le \theta < 2\pi$ (d) $r < 0,\ -2\pi < \theta \le 0$

Solution

(a) $(3.4, -\frac{\pi}{6}) = (3.4, -\frac{\pi}{6} + 2\pi) = (3.4, \frac{11\pi}{6})$

(b) $(3.4, -\frac{\pi}{6})$ satisfies this condition.

(c) $(3.4, -\frac{\pi}{6}) = (-3.4, -\frac{\pi}{6} + \pi) = (-3.4, \frac{5\pi}{6})$

(d) $(3.4, -\frac{\pi}{6}) = (-3.4, \frac{5\pi}{6}) = (-3.4, \frac{5\pi}{6} - 2\pi) = (-3.4, -\frac{7\pi}{6})$ ■

NOW IT'S YOUR TURN

Give polar representations of the polar point $(1.8, \frac{\pi}{4})$ that satisfy each condition.

(a) $r > 0$, $2\pi \le \theta \le 4\pi$ (b) $r < 0$, $0 < \theta < 4\pi$

(Answer: (a) $(1.8, \frac{9\pi}{4})$, (b) $(-1.8, \frac{5\pi}{4})$ and $(-1.8, \frac{13\pi}{4})$)

9.1-2 Transformations Between Polar and Rectangular Coordinates

When we know the coordinates of a point in one coordinate system, we often want to find the coordinates of that same point in the other coordinate system. This process of changing from one coordinate system to another we call *transformation of coordinates.*

Converting Polar Coordinates to Rectangular Coordinates

We know r and θ for the point $P(r, \theta)$ and want to find x and y for that same point. We start by superimposing the polar and rectangular coordinate systems, placing the pole at the origin of the rectangular system and the polar axis along the positive x axis (see Figure 9.1-9).

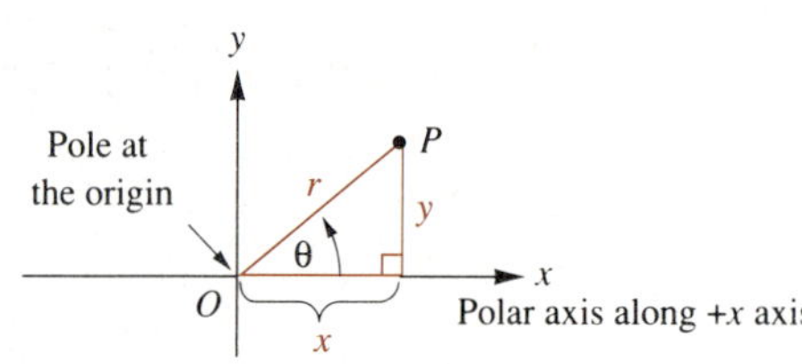

Figure 9.1-9

Polar and rectangular coordinate systems superimposed on each other, with the pole at the origin and the polar axis along the +*x* axis.

Solving the right triangle in this figure for x and y, we get

$$\cos\theta = \frac{x}{r} \quad \rightarrow \quad x = r\cos\theta$$

$$\sin\theta = \frac{y}{r} \quad \rightarrow \quad y = r\sin\theta$$

Therefore, if we know r and θ, we can easily calculate x and y.

TRANSFORMING FROM POLAR TO RECTANGULAR COORDINATES

$$x = r\cos\theta$$

$$y = r\sin\theta$$

Although we only showed this result when P is in quadrant I, it is true for all quadrants.

Example 4

Convert the following polar coordinates to rectangular coordinates. Double check the signs of the answers by plotting each point. (a) $(3, 60°)$, (b) $(7, -\frac{5\pi}{6})$, (c) $(-2.000, 135.6°)$

Solution

Using the equations $x = r\cos\theta$ and $y = r\sin\theta$ gives us

(a) $(3, 60°)$: $r = 3$ and $\theta = 60°$.

$$x = r\cos\theta = 3\cos 60° = 3(\tfrac{1}{2}) = \tfrac{3}{2}$$
$$y = r\sin\theta = 3\sin 60° = 3(\sqrt{3}/2) = 3\sqrt{3}/2$$

(b) $(7, -\frac{5\pi}{6})$: $r = 7$ and $\theta = -\frac{5\pi}{6}$.

$$x = r\cos\theta = 7\cos(-\tfrac{5\pi}{6}) = 7(-\sqrt{3}/2) = -7\sqrt{3}/2$$
$$y = r\sin\theta = 7\sin(-\tfrac{5\pi}{6}) = 7(-\tfrac{1}{2}) = -\tfrac{7}{2}$$

(c) $(-2.000, 135.6°)$: $r = -2$ and $\theta = 135.6°$.

$$x = r\cos\theta = (-2.000)\cos 135.6° = (-2.000)(-0.7145) = 1.429$$
$$y = r\sin\theta = (-2.000)\sin 135.6° = (-2.000)(0.6997) = -1.399$$

We sketch each point to double check the signs (see Figure 9.1-10). ■

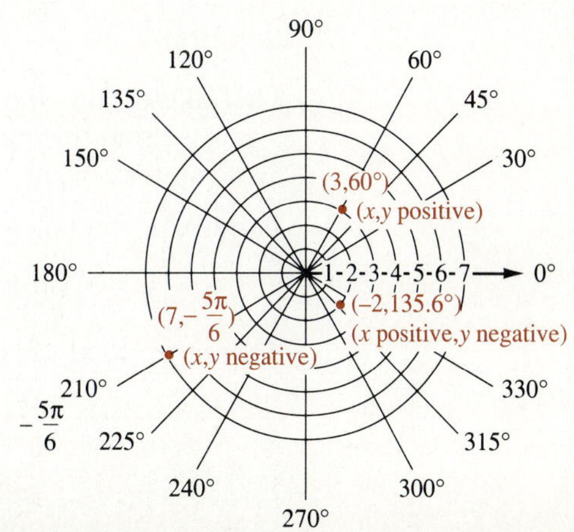

Figure 9.1-10

Double checking the signs of Example 4

NOW IT'S YOUR TURN

Find the rectangular coordinates of the polar points (a) (4.20, 125°) and (b) (−6.40, 3.2 rad). Check the signs by plotting each point.
(Answer: (a) (−2.41, 3.44), (b) (6.4, 0.37))

Example 5

The Tail-Wagging Dance of the Honeybee

When a honeybee discovers a source of food, she returns to the hive and does a figure-eight dance to tell the other bees where the food is.* This dance gives two pieces of information about the food: its *direction* and its *distance* from the hive. The direction is indicated by the *angle* θ that the straight part of the figure-eight makes with a horizontal line from the hive to the direction of the sun, and the distance is given by the *rate* at which the bee dances around the figure-eight (the farther the food, the slower the dance)† (see Figure 9.1-11). You have no doubt recognized that the bee's dance really gives the *polar coordinates* of the food, using the hive as the pole and the hive-sun line as the polar axis.

Suppose that we observe the dances of two bees when the sun is directly east of the hive, and we find that food source A is located 1400 m from the hive at an angle of 72° with the sun, whereas food B is 2400 m from the hive at an angle of 230° with the sun. If we are marking these food sources on a map, we need to use their distances along the cardinal directions (N,S,E,W) from the hive. What would the distances in these directions be?

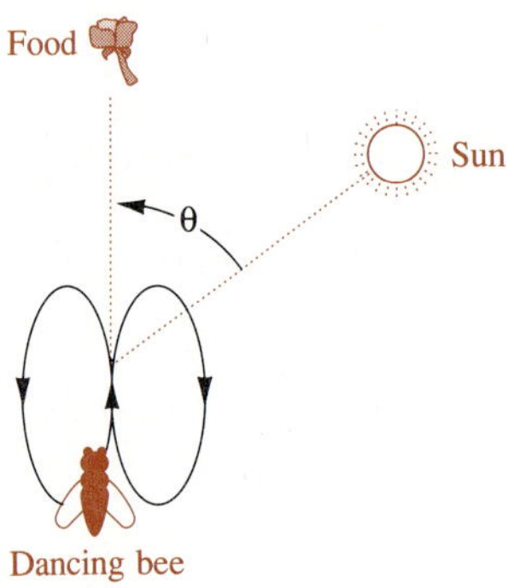

Figure 9.1-11

The meaning of the honeybee's tail-wagging dance The rate of the dance tells the other bees how far the food is, and the angle between the straight part of the figure-eight and a horizontal line to the direction of the sun tells them its direction.

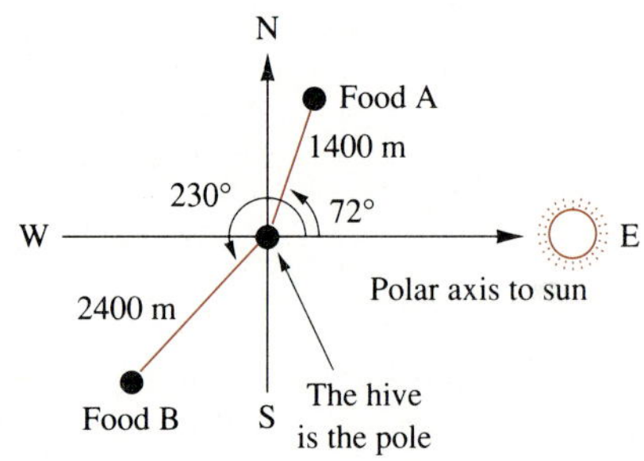

Figure 9.1-12 **Food sources *A* and *B***

Solution

First sketch the food sources and show the cardinal directions (see Figure 9.1-12). From this figure you can see that the cardinal directions are the x and y axes, and we want the x and y coordinates of the food.

$$\text{Food A: } x = r\cos\theta = (1400\text{ m})\cos 72° = (1400\text{ m})(0.309) = 430\text{ m}$$

$$y = r\sin\theta = (1400\text{ m})\sin 72° = (1400\text{ m})(0.951) = 1300\text{ m}$$

Therefore, food A is 430 m east and 1300 m north of the hive. ■

---

*See Karl von Frisch, *The Dance Language and Orientation of Bees*, The Belknap Press (Harvard University Press), 1967, Karl von Frisch, "Decoding the Language of Bees," *Science*, August 23, 1974; and Harald Esch, "The Evolution of Bee Language," *Scientific American*, April 1967, for a more complete discussion.

† Besides bees, birds also navigate by the sun. See John T. Emlen and Richard L. Penney, "The Navigation of Penguins," Scientific American, Oct. 1966, p. 105.

NOW IT'S YOUR TURN

What should be the distances of food B along the cardinal directions on this map? (Answer: 1500 m west and 1800 m south of the hive)

Converting Rectangular Coordinates to Polar Coordinates

We know the rectangular (x, y) coordinates of a point and want to find its polar (r, θ) coordinates from them. Using the equations $x = r\cos\theta$ and $y = r\sin\theta$, we get

$$\begin{aligned} x^2 + y^2 &= (r\cos\theta)^2 + (r\sin\theta)^2 = r^2\cos^2\theta + r^2\sin^2\theta \\ &= r^2\underbrace{(\cos^2\theta + \sin^2\theta)}_{1} = r^2 \end{aligned}$$

Taking square roots gives us

$$r = \sqrt{x^2 + y^2}$$

To find θ, we first find its reference angle and then figure which quadrant it is in by looking at the signs of x and y. Figure 9.1-9 tells us than tan(ref angle of θ) = $|y/x|$.

> **TRANSFORMING FROM RECTANGULAR TO POLAR COORDINATES**
>
> $$r = \sqrt{x^2 + y^2} \qquad \tan(\text{ref angle } \theta) = \left|\frac{y}{x}\right|$$
>
> Use the signs of x and y to find what quadrant θ is in.

As we saw earlier, a point P may have infinitely many polar representations, so the equations we have just found do not determine r and θ *uniquely* from x and y. They will give us r and θ in the range $r \geq 0$ and $0 \leq \theta < 2\pi$.

Example 6

Find the polar coordinates in the range $r \geq 0$, $0 \leq \theta < 2\pi$, for each point in rectangular coordinates. (a) (3, 5) (find θ in radians) and (b) $(-\sqrt{2}, -4)$ (find θ in degrees).

Solution

(a) (3, 5): $x = 3$, $y = 5$.

$$r = \sqrt{x^2 + y^2} = \sqrt{3^2 + 5^2} = \sqrt{34}$$

$$\tan(\text{ref angle } \theta) = |y/x| = |\tfrac{5}{3}| = \tfrac{5}{3}$$

$$\text{ref angle } \theta = \tan^{-1}\tfrac{5}{3} = 1.03 \text{ rad}$$

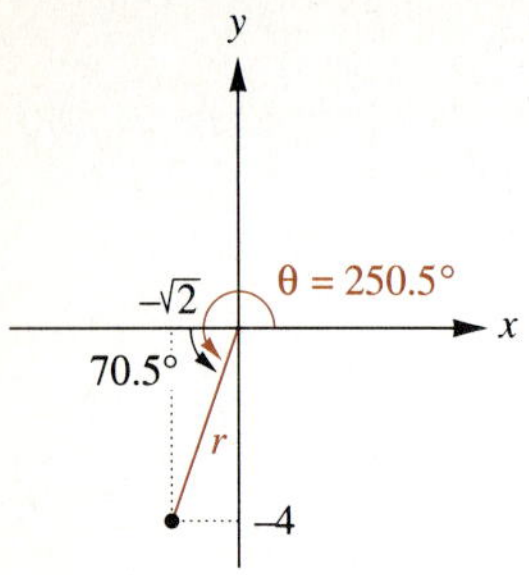

Figure 9.1-13

Both x and y are negative, so θ is in quadrant III.

Because both x and y are positive, θ is in quadrant I, so it is equal to its reference angle. Therefore, $\theta = 1.03$ rad. Thus the equivalent polar coordinates for the point (3, 5) are $(\sqrt{34},\ 1.03)$.

(b) $(-\sqrt{2},\ -4)$: $x = -\sqrt{2}$, $y = -4$.

$$
\begin{aligned}
r &= \sqrt{x^2 + y^2} = \sqrt{(-\sqrt{2})^2 + (-4)^2} = \sqrt{18} = \sqrt{(9)(2)} = 3\sqrt{2} \\
\tan(\text{ref angle } \theta) &= |y/x| = |(-4)/(-\sqrt{2})| = 4/\sqrt{2} = 2\sqrt{2} \\
\text{ref angle } \theta &= \tan^{-1}(2\sqrt{2}) = 70.5^\circ
\end{aligned}
$$

Because x and y are both negative, θ is in quadrant III, as Figure 9.1-13 shows. Thus $\theta = 70.5^\circ + 180^\circ = 250.5^\circ$. Therefore, the rectangular point $(-\sqrt{2}, -4)$ is equivalent to the polar coordinates $(3\sqrt{2},\ 250.5^\circ)$. ■

NOW IT'S YOUR TURN

Find the polar coordinates $(0 \le \theta < 2\pi,\ r > 0)$ of the point having rectangular coordinates $(8.50,\ -10.3)$. (Answer: (13.4, 5.40 rad))

9.1-3 Converting Equations

In many applications and in later mathematics courses, it is often necessary to convert equations from rectangular to polar coordinates and vice versa. The transformations between (x, y) and (r, θ) that we have just derived allow us to make these conversions, as the following examples show.

Example 7

Find the polar equation for (a) $x^2 + y^2 = 36$ and (b) $y^2 = 5x$.

Solution

We know that $x = r\cos\theta$ and $y = r\sin\theta$.

(a) Substituting for x and y gives us

$$
\begin{aligned}
x^2 + y^2 &= 36 \\
x^2 + y^2 &= (r\cos\theta)^2 + (r\sin\theta)^2 = r^2\cos^2\theta + r^2\sin^2\theta \\
&= r^2\underbrace{(\cos^2\theta + \sin^2\theta)}_{1} = r^2
\end{aligned}
$$

The equation in polar coordinates is $r^2 = 36$.

Notice that this equation does not contain the variable θ, which means that it must be true no matter what θ is. In other words, $r^2 = 36$ for *all* values of θ. On a piece of polar graph paper, try to sketch the *graph* of this equation.

(b) $y^2 = 5x$. Substituting for x and y gives us

$$(r \sin \theta)^2 = 5(r \cos \theta)$$

$$r^2 \sin^2 \theta = 5r \cos \theta$$

$$r \sin^2 \theta = 5 \cos \theta \qquad \textbf{(divide by } r\textbf{)}$$

$$r = \frac{5 \cos \theta}{\sin^2 \theta} \quad ■$$

NOW IT'S YOUR TURN

Express the equation $xy = 4$ in polar coordinates. $\left(\text{Answer: } r^2 = \dfrac{4}{\sin \theta \cos \theta}\right)$

Example 8

Express the following polar equations in cartesian coordinates: (a) $r(4 \cos \theta - 13 \sin \theta) = 1$ and (b) $r = 5 \tan \theta \sec \theta$.

Solution

In converting polar equations to rectangular form, we look for the expressions $r \cos \theta$ and $r \sin \theta$ because these expressions are equal to x and y.

(a) $r(4 \cos \theta - 13 \sin \theta) = 1$. Multiplying through by r gives

$$4\underbrace{(r \cos \theta)}_{x} - 13\underbrace{(r \sin \theta)}_{y} = 1$$

$$4x - 13y = 1$$

(b) $r = 5 \tan \theta \sec \theta$.

$$r = 5 \frac{\sin \theta}{\cos \theta} \frac{1}{\cos \theta} \qquad \left(\textbf{use } \tan \boldsymbol{\theta} = \frac{\sin \boldsymbol{\theta}}{\cos \boldsymbol{\theta}} \textbf{ and } \sec \boldsymbol{\theta} = \frac{1}{\cos \boldsymbol{\theta}}\right)$$

$$r \cos^2 \theta = 5 \sin \theta \qquad \textbf{(multiply by } \cos^2 \boldsymbol{\theta}\textbf{, for } \cos \boldsymbol{\theta} \neq 0\textbf{)}$$

$$r^2 \cos^2 \theta = 5\, r \sin \theta \qquad \textbf{(multiply by } r\textbf{)}$$

$$(r \cos \theta)^2 = 5\, r \sin \theta$$

$$x^2 = 5y \qquad \textbf{(use } x = r \cos \boldsymbol{\theta} \textbf{ and } y = r \sin \boldsymbol{\theta}\textbf{)} \quad ■$$

NOW IT'S YOUR TURN

Express the polar equation $r \sin \theta = -\cot \theta$ in cartesian coordinates. (Answer: $y^2 + x = 0$)

EXERCISES 9.1-1

1. Write the polar coordinates (r, θ) of each point in the graph in Figure 9.1-14, using $r \geq 0$ and $0° \leq \theta < 360°$.

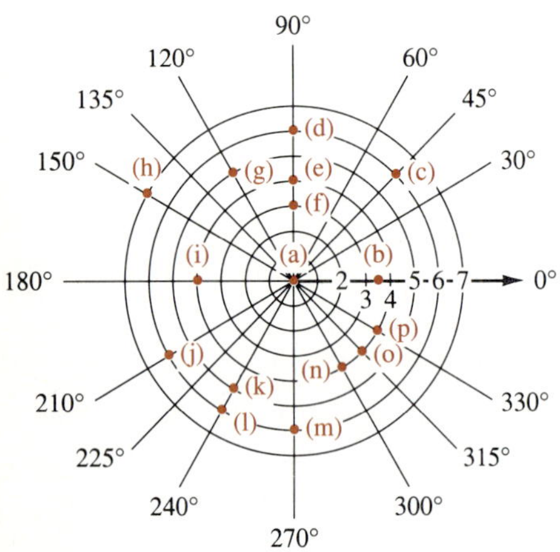

Figure 9.1-14

In Exercises 2–23, plot each point in the polar coordinate system.

2. (3, 45°) **3.** (6, 30°) **4.** (2, 150°)

5. $(2, \frac{11\pi}{6})$ **6.** (1, 0) **7.** $(-3, \frac{\pi}{2})$

8. (−5, 270°) **9.** (3, −180°) **10.** $(-2, -\frac{5\pi}{6})$

11. $(\frac{1}{2}, -\frac{2\pi}{3})$ **12.** (2, 170°) **13.** (4, 430°)

14. (−3, 65°) **15.** (2, −250°) **16.** $(-2.5, -\frac{\pi}{6})$

17. $(-3.5, -\pi)$ **18.** (0, −110°) **19.** (−1.63, 0°)

20. (−1, −135°) **21.** (0, 155°) **22.** $(-\frac{1}{4}, -\frac{3\pi}{2})$

23. $(-\frac{\pi}{2}, -\frac{\pi}{2})$

In Exercises 24–33, each point is given in polar coordinates. Find all possible additional polar representations for that point subject to the restrictions given. ***Hint:*** **Plot each point!**

24. $(3, 0)$, $r \geq 0$, $0 \leq \theta < 2\pi$

25. $(3, 0)$, $r \geq 0$, $0 \leq \theta \leq 2\pi$

26. $(2, \frac{\pi}{4})$, $r \geq 0$, $-2\pi \leq \theta < 2\pi$

27. $(2, \frac{\pi}{4})$, $0 \leq \theta < 2\pi$

28. $(4, \frac{3\pi}{2})$, $r < 0$, $0 \leq \theta < 2\pi$

29. $(-1, \frac{\pi}{6})$, $r \geq 0$, $0 \leq \theta < 4\pi$

***30.** $(6, \frac{\pi}{3})$, $r \geq 0$, $\theta \geq 0$

***31.** $(5, 50°)$, $r \geq 0$

***32.** $(7, \frac{\pi}{2})$, $r \geq 0$

***33.** $(8, \pi)$, no restrictions

34. Water Molecule. A water molecule (H_2O) is placed in a polar coordinate system so that the oxygen atom is at the pole, as shown in Figure 9.1-15. What are the polar coordinates of all three atoms?

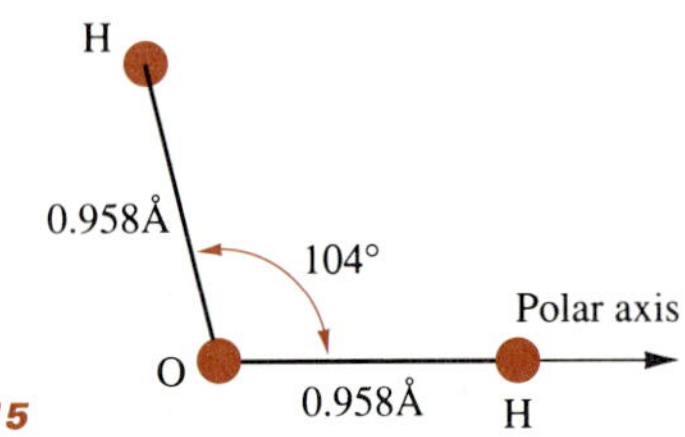

Figure 9.1-15

35. The hour hand on a certain clock is 9.5 cm long. Taking the pole at the center of the clock's face and the polar axis to be the direction this hand points at 12:00 midnight, find the polar coordinates of the tip of the hour hand at the following times of day. Use $r \geq 0$ and $0 \leq \theta < 2\pi$.

(a) 3:00 P.M. **(c)** 8:00 A.M.
(b) 6:00 P.M. **(d)** 1:30 P.M.

36. Using the sun as the pole and the line from the sun to the earth on January 1 as the polar axis (see Figure 9.1-16), find the polar coordinates of the earth on the following dates, assuming its orbit is circular. Use $r \geq 0$, $0 \leq \theta < 2\pi$. **(a)** January 1, **(b)** March 1, **(c)** July 1, **(d)** November 1, **(e)** March 1 one year after the date in (b).

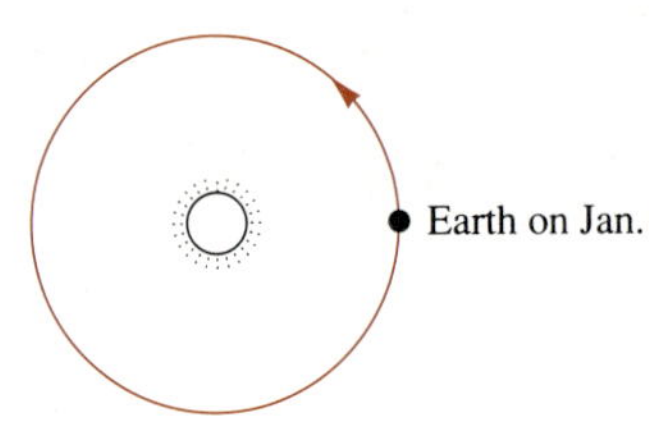

Figure 9.1-16

EXERCISES 9.1-2

In Exercises 1–14, each point is in polar coordinates. Find its rectangular coordinates.

1. $(2\sqrt{2}, \frac{\pi}{4})$ **2.** (2, 150°)

3. (−6, 45°) **4.** (−10, 120°)

5. $(3, \frac{3\pi}{4})$ **6.** $(-2\sqrt{2}, 120°)$

7. (−5, 53°) **8.** (13, 157°)

9. $(5, \tan^{-1} \frac{3}{4})$ **10.** $(2, -\frac{\pi}{6})$

11. (1.365, −103.6°) **12.** (0, 1)

13. (1, 0) **14.** (−0.736, −0.736)

15. Find the rectangular coordinates of each point in Figure 9.1-17.

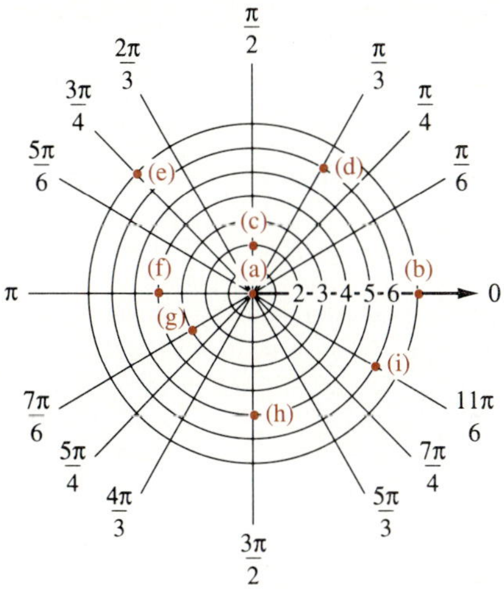

Figure 9.1-17

In Exercises 16–30, each point is in rectangular coordinates. Find the polar coordinates of each point, using $r \geq 0$ and $0 \leq \theta < 2\pi$.

16. $(1, \sqrt{3})$	**17.** $(-1, 1)$	**18.** $(-\sqrt{3}, -1)$
19. $(2, -2)$	**20.** $(2, 0)$	**21.** $(0, 3)$
22. $(-3, 0)$	**23.** $(0, -4)$	**24.** $(2, 3)$
25. $(-4, 3)$	**26.** $(-5, -12)$	**27.** $(-3, -4)$
28. $(4, -3)$	**29.** (π, π)	**30.** $(-\pi, -\pi)$

31. Find the polar coordinates of each point on the cartesian graph in Figure 9.1-18, using $r \geq 0$, $0 \leq \theta < 2\pi$.

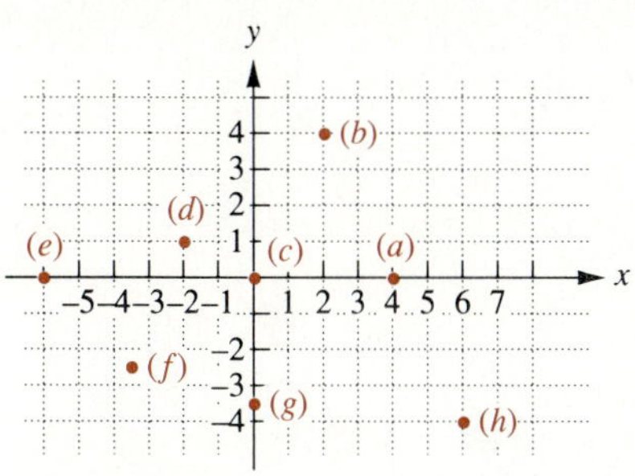

Figure 9.1-18

32. Find the rectangular coordinates of the tip of the hour hand of the clock in Exercise 35 in Section 9.1-1, calling the x axis the polar axis, the y axis from the center through 9 : 00 o'clock, and the origin the center of the clock's face.

33. We began this chapter with the directions to your favorite pizza parlor. Using your home as the pole and the W → E direction as the polar axis, find the polar coordinates of this pizza parlor.

34. The expression "as the crow flies" means the straight-line distance because crows do not worry about x-y coordinate systems. If the distance from your house to the library is 5.40 km as the crow flies, and if the crow must fly at 24.0° south of west to get there, how far east (or west) and how far north (or south) is the library from your home?

35. If the malt shop is 2410 km north and 1720 km west of your house, how far from your house is it as the crow flies (see Exercise 34), and what are its polar coordinates using the same convention as in Exercise 33?

36. In the tail-wagging bee dance (see Example 5), the sun is directly east of the hive, and the rate of the dance indicates that the food is 1500 m away (see Figure 9.1-19). How far north (or south) and east (or west) is the food from this hive?

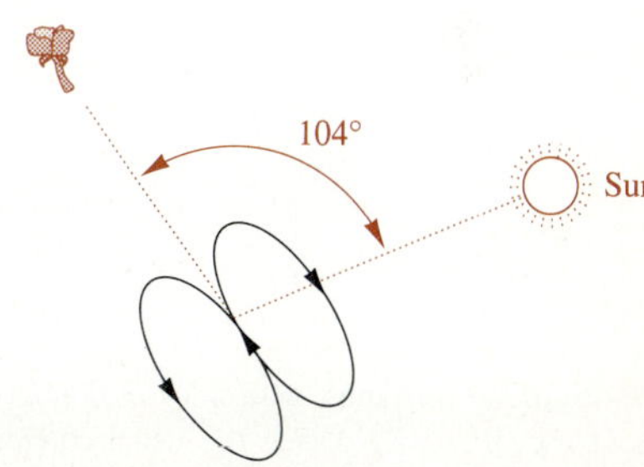

Figure 9.1-19

***37.** A source of bee food is located 450 m north and 1200 m east of a hive. A bee, having located this tasty treat, returns to the hive when the sun is 40° south of east. What angle and what distance will this bee indicate in her tail-wagging dance? (See Example 5) *Hint:* A clear sketch should help considerably!

38. Water Molecule. Find the rectangular coordinates of each atom in the water molecule in Figure 9.1-15 if the x axis is the polar axis and the y axis runs upward through the O atom.

***39.** The circle in Figure 9.1-20 has a diameter of four units. Find the polar coordinates of the points A, B, C, and D.

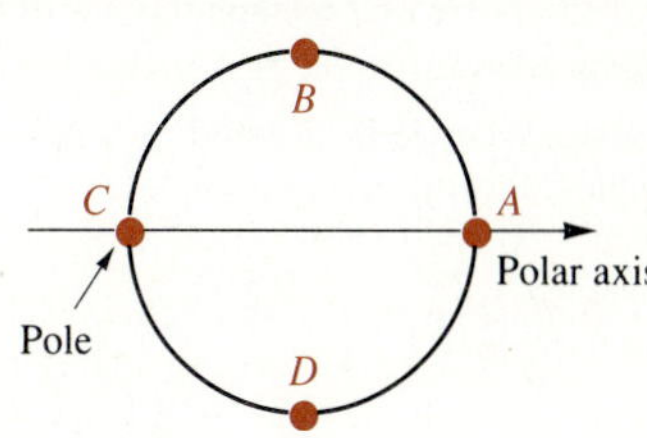

Figure 9.1-20

40. For the circle in Exercise 39, find the rectangular coordinates of the center *and* the points A, B, C, and D, using the pole as the origin, the polar axis the x axis, and the $D \to B$ direction the y axis.

41. **(a)** Find the cartesian coordinates of the four vertex points of the rectangle in Figure 9.1-21.
(b) Find the polar coordinates of the four vertex points of this rectangle, using the usual convention on axes.

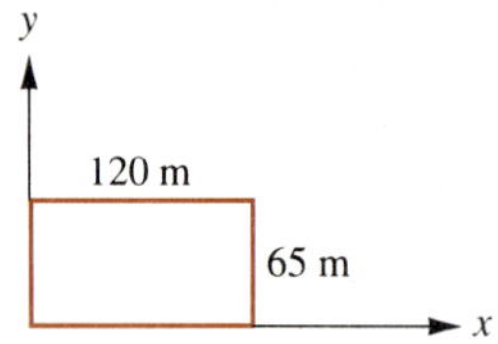

Figure 9.1-21

***42.** A square 6.00 cm on each side is oriented as in Figure 9.1-22. Find the cartesian *and* polar coordinates of the midpoints of its four sides.

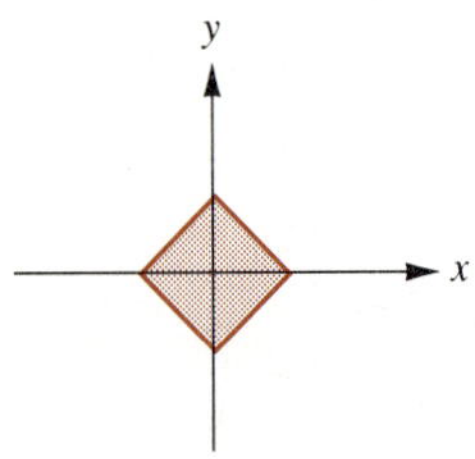

Figure 9.1-22

***43.** An equilateral triangle with sides 22.0 cm long is oriented as shown in Figure 9.1-23. Using the usual convention on coordinates:

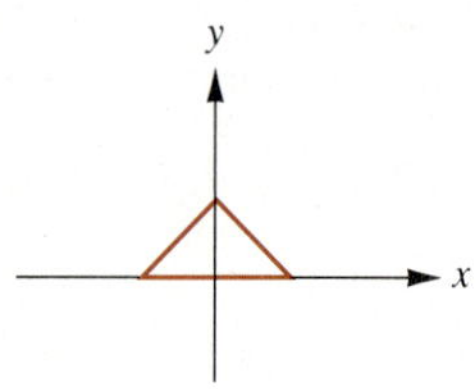

Figure 9.1-23

(a) Find the polar *and* rectangular coordinates of each vertex.
(b) Find the polar *and* rectangular coordinates of the midpoint of each side.

***44.** An isosceles triangle is oriented as shown in Figure 9.1-24. Using the usual convention on coordinates:

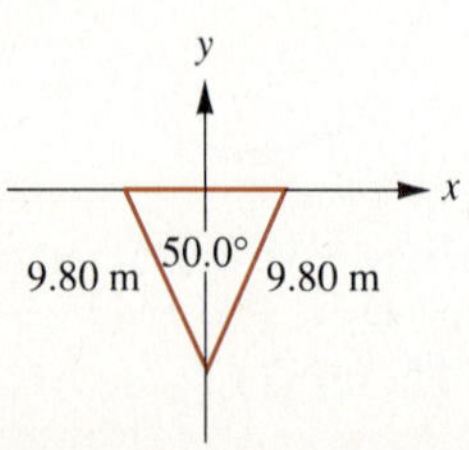

Figure 9.1-24

(a) Find the polar *and* rectangular coordinates of each vertex.
***(b)** Find the polar *and* rectangular coordinates of the midpoint of each side.

45. If the baseball diamond (see Example 13 in Section 4.2) is aligned with a polar coordinate system so that the pole is home plate and the first-base line is the polar axis, find the polar coordinates of **(a)** first base, **(b)** second base, and **(c)** the pitcher's plate.

***46.** For the baseball diamond in Exercise 45, find the polar coordinates of all four bases and the pitcher's plate if the pole is at the pitcher's plate with the polar axis running toward second base.

47. Nitric Acid. The nitric acid ($HN0_3$) in Figure 9.1-25 is aligned along a polar coordinate system with the N atom at the pole and the polar axis toward the H atom. Find the polar coordinates *and* the rectangular coordinates of each atom in this molecule.

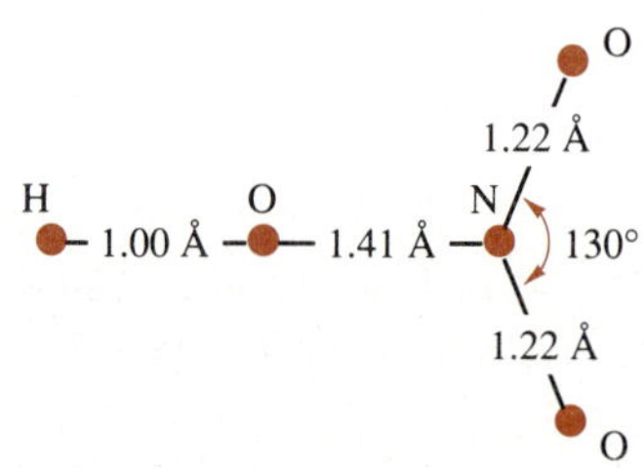

Figure 9.1-25

***48. Benzene Ring.** Figure 9.1-26 shows a benzene ring molecule, C_6H_6. The C atoms form a regular hexagon, and the H atoms point directly outward from its center. Chemists have measured the C—C distance to be 1.395 Å and the C—H distance to be 1.084 Å. If we take the pole at the center of the hexagon with the polar axis running to the right through the C—H pair, find the rectangular *and* polar coordinates of **(a)** each carbon atom C and **(b)** each hydrogen atom H.

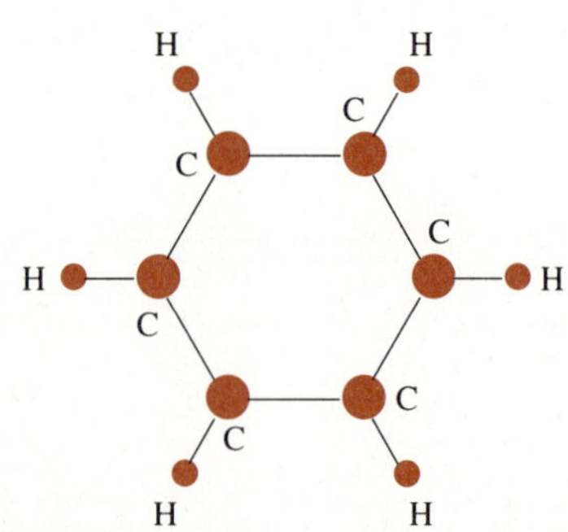

Figure 9.1-26

****49. Benzene Ring.** Suppose that, for the benzene molecule in Exercise 48, we take the pole at the extreme left C atom but keep the same polar axis. Find the rectangular *and* the polar coordinates of **(a)** each C atom and **(b)** each H atom.

****50. Formic Acid.** For the formic acid (HCOOH) molecule in Figure 9.1-27, chemists have measured the following angles and distances: H—C—O′ = 117.8°, O′—C—O = 124.3°, C—O—H′ = 107.8°, C—O′ = 1.245 Å, C—O = 1.312 Å, C—H = 1.085 Å, and O—H′ = 0.95 Å. Taking the pole at the C atom with the polar axis running from C to O (*not* O′), find the polar *and* rectangular coordinates of each atom in this molecule.

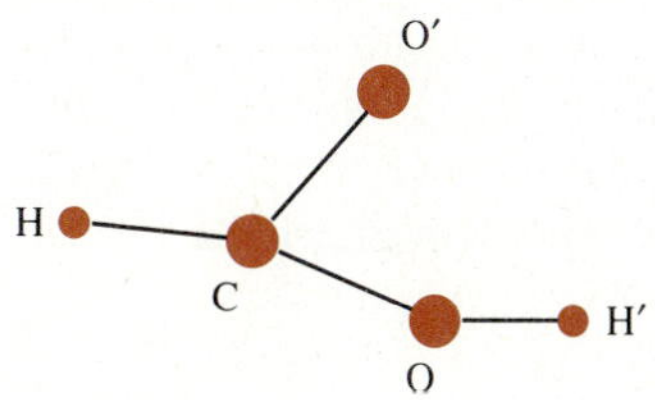

O and O′ are both oxygen atoms, and H and H′ are both hydrogen atoms. Primes are used to distinguish one atom from another.

Figure 9.1-27

51. Show that, in Figure 9.1-28, the distance between the points $P_1(r_1, \theta_1)$ and $P_2(r_2, \theta_2)$ in the polar coordinate system is given by $r_1^2 + r_2^2 - 2r_1r_2\cos(\theta_2 - \theta_1)$.

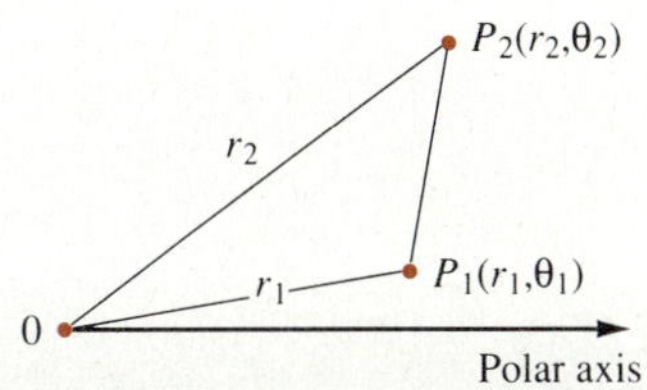

Figure 9.1-28

52.** If the rectangular coordinates of a point P are (x, y) and its polar coordinates are (r, θ), then find the polar coordinates of **(a)** $(-x, y)$, **(b)** $(x, -y)$, and **(c)** $(-x, -y)$, where $r > 0$ and $0 \le \theta < 2\pi$. ***Hint: Make clear sketches!

53.** If the polar coordinates of point P are (r, θ) and its rectangular coordinates are (x, y), then find the rectangular coordinates of **(a)** $(r, -\theta)$, **(b)** $(-r, \theta)$ and **(c)** $(-r, -\theta)$. ***Hint: Make clear sketches!

EXERCISES 9.1-3

In Exercises 1–13, convert each rectangular equation to an equivalent polar equation.

1. $y = 4$
2. $x = -2$
3. $2x - 4y = 1$
4. $5y + 7x = 8$
5. $x^2 = 3y$
6. $y^2 = 9x$
7. $x^2 + y^2 = 144$
8. $3x^2 - 2y^2 = y$
9. $(x - 1)^2 + (y - 2)^2 = 4$
***10.** $x^4 - y^4 = 7xy$
***11.** $x^2 - x + y^2 + y = 0$
***12.** $(x^2 + y^2)^2 = x(x^2 - 3y^2)$
***13.** $(x^2 + y^2 + ax^2)^2 = a^2(x^2 + y^2)$

In Exercises 14–31, convert each polar equation to an equivalent rectangular equation.

14. $r \sin\theta = 4$
15. $-2 = r\cos\theta$
16. $r = 2\sin\theta$
17. $r = -5\cos\theta$
18. $r = \csc\theta$
19. $r = 3\sec\theta$
20. $r = 2\cos\theta + 3\sin\theta$
21. $r = 5$
22. $r = a$
***23.** $\theta = \frac{\pi}{4}$
24. $\tan\theta = \frac{7}{11}$
***25.** $5 = r^2 \sin 2\theta$
***26.** $r\cos(\theta - \frac{\pi}{6}) = 2$
27. $\sin\theta = \frac{1}{2}$
28. $r = \cos\theta\tan\theta$
***29.** $r = \sin\theta\tan\theta$
***30.** $r = 1 + \sin\theta$
31.** $r^2(\cos^4\theta + \sin^4\theta) = \sin 2\theta$ ***Hint: Multiply both sides by r^2 and use the double-angle form for $\sin 2\theta$.

9.2 POLAR GRAPHS

9.2-1 Graphs in Polar Coordinates

Just as we graph equations in cartesian coordinates, we also graph them in polar coordinates. The procedure is basically the same: Make a table of values and plot these points. When graphing trigonometric functions, it is easiest to pick special angles as often as possible. As we shall see, some graphs are especially easy to express in polar coordinates, whereas others are more easily expressed in rectangular coordinates.

Example 9

In Example 7 of Section 9.1, we showed that the equation $x^2 + y^2 = 36$ can be expressed in polar form as $r^2 = 36$, which is equivalent to $r = \pm 6$. Let us graph this equation in polar coordinates.

Solution

Notice that θ does not occur in this equation, which means that for all values of θ, $r = \pm 6$. Plotting some points in Figure 9.2-1 shows that this curve is a circle of radius 6 centered at the pole. Here is a case where the polar form of the equation ($r^2 = 36$) is much simpler than the rectangular form ($x^2 + y^2 = 36$). It is also much easier to plot this polar graph than the cartesian graph. ■

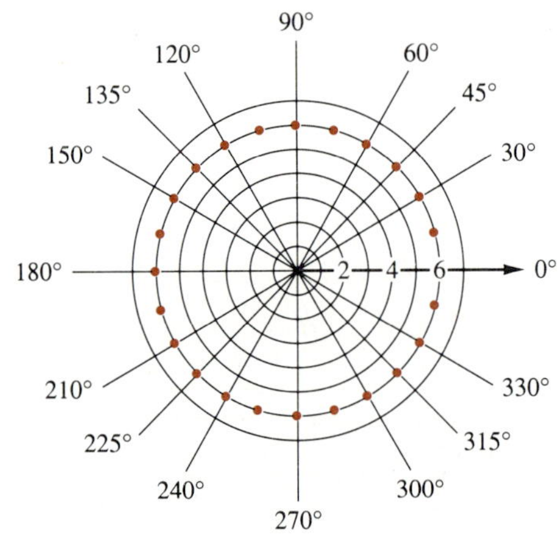

Figure 9.2-1

Graph of $r = \pm 6$ For all values of θ, r has the values ± 6. This curve is a circle of radius 6 centered on the pole.

NOW IT'S YOUR TURN

Draw the graph of $r^2 = 64$.
(Answer: A circle of radius 8 centered on the pole)

Example 10

Graph $r = \cos\theta$.

Solution

Make a table of values covering 1 rev of θ and then plot these values on polar graph paper, remembering to use special angles as often as possible (see Figure 9.2-2). Notice that after 180° the points just repeat. As you can see, this graph is a circle of radius $\frac{1}{2}$ centered at $(\frac{1}{2}, 0°)$, which is the cartesian point $(\frac{1}{2}, 0)$. ■

NOW IT'S YOUR TURN

Graph $r = \sin\theta$.
(Answer: A circle of radius $\frac{1}{2}$ centered at the polar point $(\frac{1}{2}, 90°)$)

θ	$r = \cos\,\theta$
0°	1.00
15°	0.97
30°	$\sqrt{3}/2 = 0.87$
45°	$\sqrt{2}/2 = 0.71$
60°	$\frac{1}{2} = 0.50$
75°	0.26
90°	0.00
105°	−0.26
120°	−0.50
135°	−0.71
150°	−0.87
165°	−0.97
180°	−1.00
210°	−0.87
225°	−0.71
240°	−0.50
270°	0.00
300°	0.50
315°	0.71
330°	0.87
360°	1.00

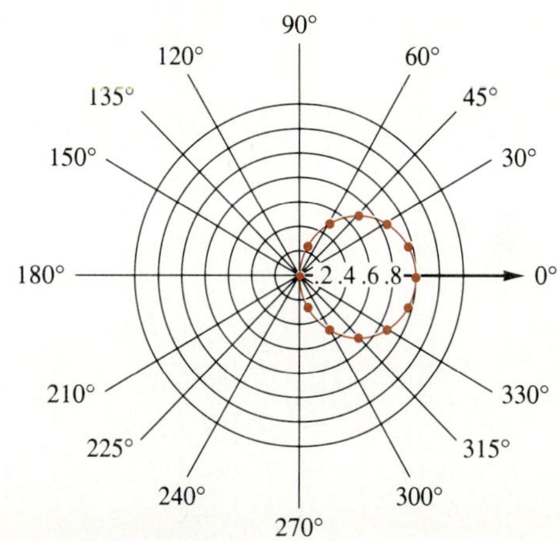

Figure 9.2-2

Graph of $r = \cos\,\theta$

Example 11

Graph $r = 2 + \sin\,\theta$.

Solution

As before, we make a table of values covering 1 rev of θ and then plot these points (see Figure 9.2-3). (As you get more practice, you won't need to plot as many points.) This curve is called a limaçon.* ■

NOW IT'S YOUR TURN

Graph $r = 2 + \cos\,\theta$.

(Answer: The same shape as in Figure 9.2-3 except rotated 90° clockwise about the pole.)

---

* Limaçon means snail in French. This curve was exhaustively studied around 1600 by Etienne Pascal, father of the great mathematician Blaise Pascal.

θ	$\sin \theta$	$r = 2 + \sin \theta$
0°	0.00	2.00
30°	0.50	2.50
45°	0.71	2.71
60°	0.87	2.87
90°	1.00	3.00
120°	0.87	2.87
135°	0.71	2.71
150°	0.50	2.50
180°	0.00	2.00
210°	−0.50	1.50
225°	−0.71	1.29
240°	−0.87	1.13
270°	−1.00	1.00
300°	−0.87	1.13
315°	−0.71	1.29
330°	−0.50	1.50
360°	0.00	2.00

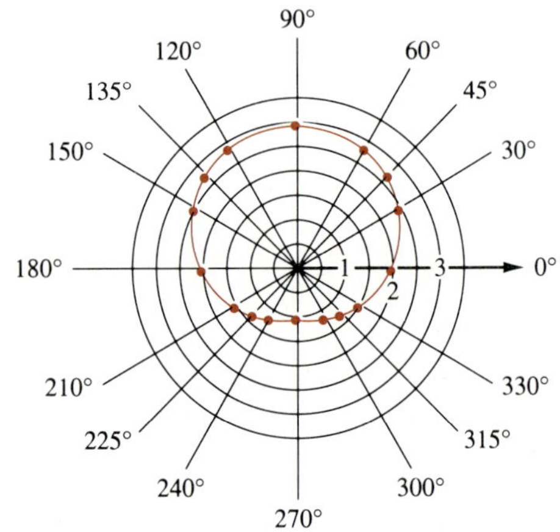

Figure 9.2-3

Graph of $r = 2 + \sin \theta$. This is a limaçon.

Example 12

Graph $r = \sin 2\theta$.

Solution

Again plot points, letting θ go through 1 rev. This time we use θ in radians. To make 2θ the special angle $\frac{\pi}{4}$, we need to use $\theta = \frac{\pi}{8}$ (see Figure 9.2-4). Because this graph resembles a flower with four leaves, it is called a rosette or four-leaved rose. ■

NOW IT'S YOUR TURN

Draw the graph of $r = \cos 2\theta$.

(Answer: A four-leaved rose obtained by rotating the rose in Figure 9.2-4 by $\frac{\pi}{4}$ about the pole.)

9.2-2 Spirals

We have seen how important sine waves are in nature. As you can see from Figure 9.2-5, spirals also occur *very* often in nature and art. In addition to the examples shown in Figure 9.2-5, additional spiral shapes include coiled ropes,

θ	2θ	$r = \sin 2\theta$
0	0	0.00
$\frac{\pi}{8}$	$\frac{\pi}{4}$	0.71
$\frac{\pi}{4}$	$\frac{\pi}{2}$	1.00
$\frac{3\pi}{8}$	$\frac{3\pi}{4}$	0.71
$\frac{\pi}{2}$	π	0.00
$\frac{5\pi}{8}$	$\frac{5\pi}{4}$	−0.71
$\frac{3\pi}{4}$	$\frac{3\pi}{2}$	−1.00
$\frac{7\pi}{8}$	$\frac{7\pi}{4}$	−0.71
π	2π	0.00
$\frac{9\pi}{8}$	$\frac{9\pi}{4}$	0.71
$\frac{5\pi}{4}$	$\frac{5\pi}{2}$	1.00
$\frac{11\pi}{8}$	$\frac{11\pi}{8}$	0.71
$\frac{3\pi}{2}$	3π	0.00
$\frac{13\pi}{8}$	$\frac{13\pi}{4}$	−0.71
$\frac{7\pi}{4}$	$\frac{7\pi}{2}$	−1.00
$\frac{15\pi}{8}$	$\frac{15\pi}{4}$	−0.71
2π	4π	0.00

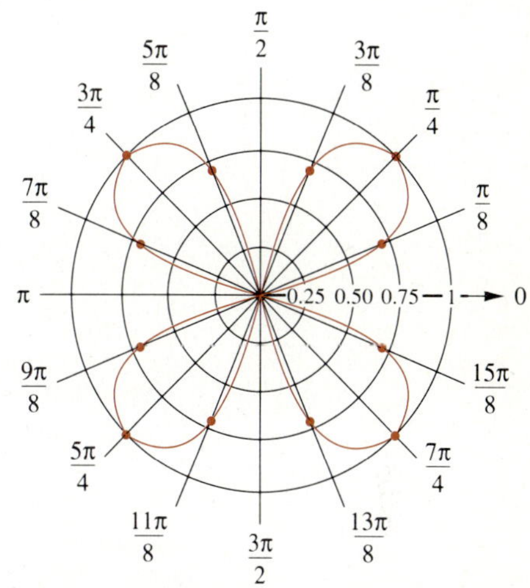

Figure 9.2-4

Graph of $r = \sin 2\theta$, sometimes called a four-leaved rose

(b) Spiral from art

(a) Spirals from nature

Figure 9.2-5

the grooves on LP records, watch springs, the arrangement of seeds on pinecones and sunflowers, plus many more.* Spirals are especially easy to represent in polar coordinates.

---

* For an interesting and very thorough discussion on spirals, see *The Curves of Life*, by Theodore A. Cook, Dover Publications, 1979 (originally published in 1914).

Example 13
The Spiral of Archimedes

The spiral of Archimedes* is approximately the shape of a coiled rattlesnake or rope, and it can be used to approximate many other spirals. Its equation is $r = k\theta$, where k is a constant. Let us plot the graph of $r = 2\theta$.

Figure 9.2-6
Spiral of Archimedes in nature

Solution

Because $r = 2\theta$, both r and θ must have the same units. This can happen only if both are pure numbers, which means that θ is in radians. As usual, we start with a table of values. Notice that the graph will *not* repeat itself as θ sweeps through multiples of 2π because r keeps getting larger and larger as θ increases (see Figure 9.2-7). ■

θ	$r = 2\theta$
0	0.00
$\frac{\pi}{6} = 0.52$	1.0
$\frac{\pi}{4} = 0.79$	1.6
$\frac{\pi}{3} = 1.05$	2.1
$\frac{\pi}{2} = 1.6$	3.2
$\frac{3\pi}{4} = 2.4$	4.8
$\pi = 3.14$	6.3
$\frac{3\pi}{2} = 4.7$	9.4
$2\pi = 6.28$	12.6
$\frac{5\pi}{2} = 7.85$	15.7
$3\pi = 9.4$	18.8
$\frac{7\pi}{2} = 11.0$	22.0
$4\pi = 12.6$	25.2

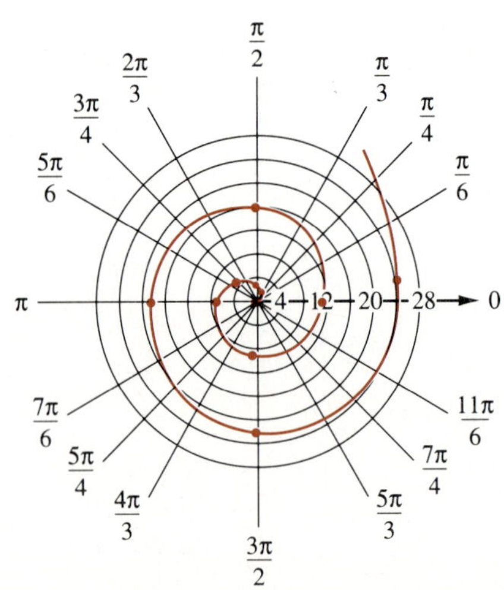

Figure 9.2-7
Graph of $r = 2\theta$ This spiral is called the spiral of Archimedes.

* Although Archimedes studied spirals extensively, his friend Conon of Alexandria actually studied this spiral before him.

NOW IT'S YOUR TURN

Graph $r = -2\theta$. (Answer: See Figure 9.2-8.)

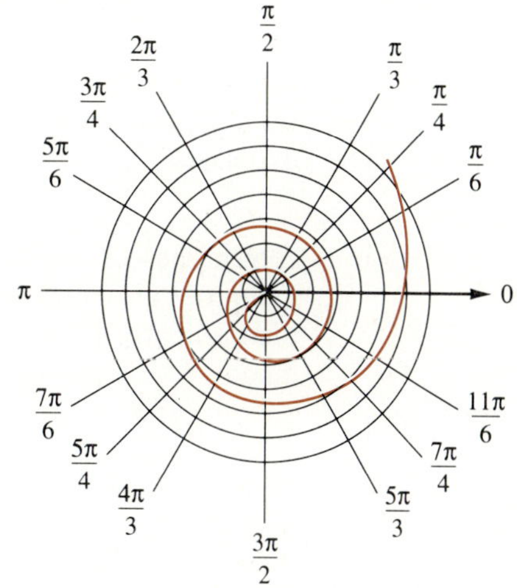

Figure 9.2-8

Archimedes spiral $r = -2\theta$

9.2-3 An Excursion into Three Dimensions: Cylindrical Coordinates

Before leaving polar coordinates, we want to extend them into three dimensions. To see how to do this, think about the discussion of the air traffic controller that opened this chapter. The radar screen gives the polar coordinates (r, θ) of each aircraft, but there is one more essential piece of information needed: the *altitude* of the airplanes. These three numbers—the horizontal distance from the airport, the angle with respect to a fixed direction, and the altitude of the aircraft—tell the controller exactly where the plane is located.

These three quantities are the basis of *cylindrical coordinates*. The first two are the polar coordinates (r and θ) which lie in a plane—the x-y plane. The third dimension, which we call z, is the perpendicular distance above or below this plane. Notice that this z is exactly the same z as used in three-dimensional rectangular coordinates (x, y, z) (see Figure 9.2-9). To write the cylindrical

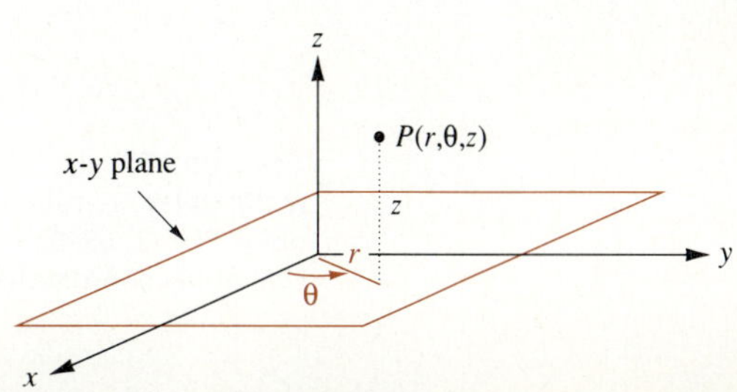

Figure 9.2-9

Cylindrical coordinates The point P has cylindrical coordinates (r, θ, z); r and θ are the usual polar coordinates in the x-y plane, and z is the perpendicular distance of P above or below this plane.

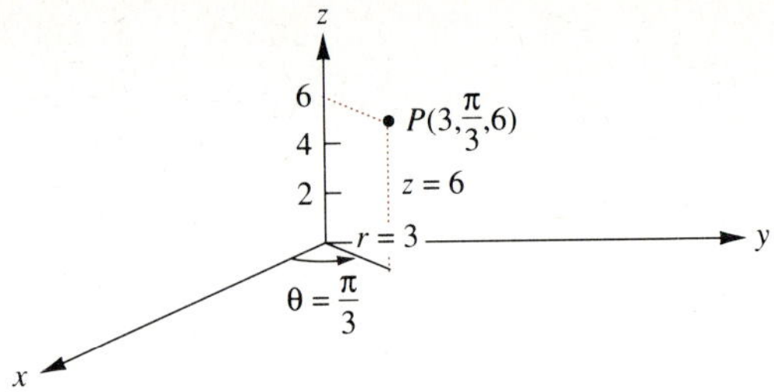

Figure 9.2-10

The point $P(3, \frac{\pi}{3}, 6)$ in cylindrical coordinates

coordinates of a point P, we write an *ordered triple* (r, θ, z). The first two quantities are the polar coordinates (r and θ) in the x-y plane, and the third is the z coordinate. For example, the point $P(3, \frac{\pi}{3}, 6)$ would have cylindrical coordinates $r = 3$, $\theta = \frac{\pi}{3}$, and $z = 6$ (see Figure 9.2-10).

Example 14

If $P(2, 4, -3)$ is in rectangular coordinates, express it in cylindrical coordinates and plot it.

Solution

$P(2, 4, -3)$: $x = 2$, $y = 4$, $z = -3$. First find the polar coordinates.

$$r = \sqrt{x^2 + y^2} = \sqrt{2^2 + 4^2} = \sqrt{20} = 4.47$$

$$\tan(\text{ref angle } \theta) = \left|\frac{y}{x}\right| = \left|\frac{4}{2}\right| = 2$$

Because x and y are both positive, θ is in quadrant I, so it is equal to its reference angle.

$$\theta = \tan^{-1} 2 = 63.4°$$

Therefore, in cylindrical coordinates, this point is $P(4.47, 63.4°, -3)$. ■

We plot it in Figure 9.2-11.

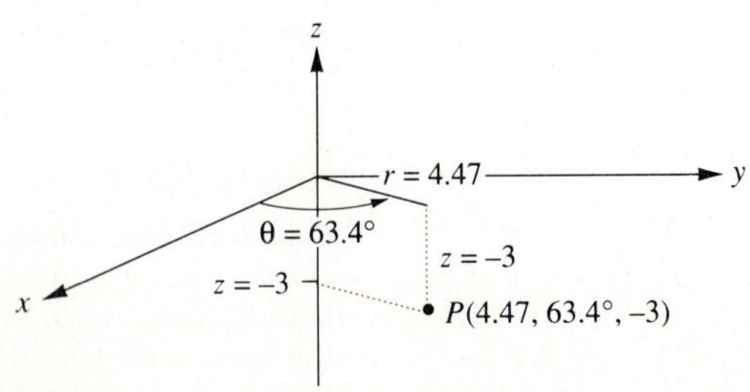

Figure 9.2-11

The point $P(2, 4, -3)$ in rectangular coordinates is the point $P(4.47, 63.4°, -3)$ in cylindrical coordinates.

NOW IT'S YOUR TURN

If the point $P(3, 7, 8)$ is in rectangular coordinates, express it in cylindrical coordinates and plot it. (Answer: 7.62, 66.8°, 8)

Example 15

Sketch the graph of the following figure in cylindrical coordinates.

$$\begin{cases} r = 1 \\ z = \theta \end{cases}$$

Solution

Because $r = 1$, this graph is always one unit from the z axis; because $z = \theta$, z starts at zero when $\theta = 0$ and increases as θ increases. Therefore, this curve is like the path traced by a bug crawling up and around the outside surface of a cylindrical pipe (Figure 9.2-12). We call such a curve a *helix*. The DNA molecule, which carries our genetic code, is a helix (actually it is a *double helix* formed by two long strands of molecules), and many other helices are found in art and nature (see Figure 9.2-13). ■

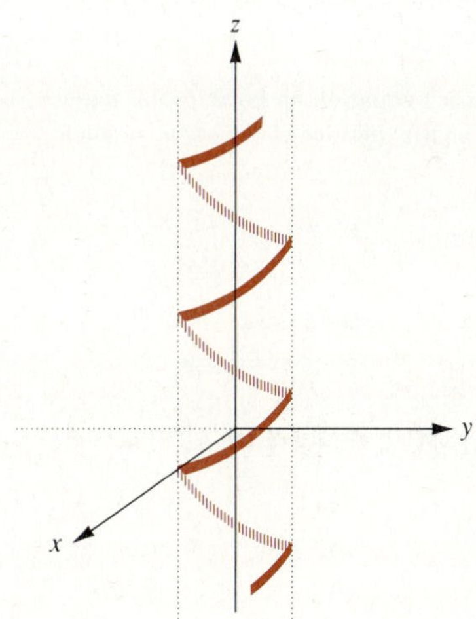

Figure 9.2-12

Graph of $\begin{cases} r = 1 \\ z = \theta \end{cases}$ in cylindrical coordinates is a helix.

NOW IT'S YOUR TURN

Sketch the graph of $r = 3$, $z = 2\theta$ in cylindrical coordinates.

(Answer: A helix, as in Figure 9.2-12, except having three times the radius and rising in the z direction twice as steeply.)

(a) Helices in nature

(b) Helices in art

Figure 9.2-13

EXERCISES 9.2-1

In Exercises 1–25, graph each equation on polar graph paper. Plot enough points to get a clear picture of the shape of each graph.

1. **(a)** $r = 1$ **(b)** $r = -1$

2. $r^2 = 4$

3. $\theta = \frac{\pi}{3}$

4. $\tan \theta = \frac{3}{4}$

5. $r \cos \theta = 3$

6. $r \sin \theta = -2$

7. $r(3 \cos \theta - 2 \sin \theta) = 1$

8. $r = 4 \cos \theta$

9. $r = -9 \sin \theta$

10. **(a)** $r = 1 + \sin \theta$ **(b)** $r = 1 - \sin \theta$

11. $r = 1 + 2 \cos \theta$

12. $r = 3 - 2 \cos \theta$

13. $r = 4 - 2 \sin \theta$

14. $r = 2 + 3 \sin \theta$

15. $r = 4 + 4 \sin \theta$

16. $r = 2(1 - \cos \theta)$

17. $r = 2 \sin 3\theta$

18. $r = \cos 4\theta$

***19.** $r = 3/(1 + \frac{1}{2} \cos \theta)$

***20.** $r = 2/(3 - \sin \theta)$

***21.** $r^2 = 9 \cos \theta$

***22.** $r^2 = -4 \sin \theta$

***23.** $r = \tan \theta$

24. $r = \sin \frac{\theta}{2}$

25. $r = 4 \cos \frac{\theta}{2}$

26. Halley's Comet. From telescopic observations, the polar equation of Halley's comet is approximately

$$r = \frac{1.07 \times 10^8 \text{ mi}}{1 + 0.967 \cos \theta} \qquad \text{with the sun at the pole.}$$

(a) How close does this famous comet get to the sun at its nearest point (perihelion) and farthest point (aphelion)?

(b) To find out what this orbit looks like, plot it in polar coordinates. Look at the range of distances you have to cover before you start plotting!

27. Halley's Comet. To see how the orbit of Halley's comet compares with those of the other planets, on a single set of polar axes plot the orbits of the earth (approximate equation is $r = 9.3 \times 10^7$ mi), Jupiter (approximate equation is $r = 5.0 \times 10^8$ mi), and Halley's comet (see Exercise 26). Look at the range of distances before you start plotting!

EXERCISES 9.2-2

In Exercises 1–10, graph each equation on polar graph paper. Plot enough points to get a clear picture of the shape of each graph. Use $\theta \geq 0$ unless directed otherwise.

1. $r = 3\theta$

2. $r = -\theta$

3. $r = 4 + \theta$, $(\theta \geq -4)$

4. **(a)** $r^2 = \theta$ **(b)** $\theta^2 = r$

5. $r\theta = \pi$

6. **(a)** $r = 10^\theta$ **(b)** $r = e^\theta$

7. **(a)** $r = 4\theta^3$ **(b)** $r = -4\theta^3$

8. **(a)** $r = 3\theta^2$ ***(b)** $r = 3\theta^{-2}$

9. $\ln r = \theta$ (Remember that $\ln \equiv \log_e$ is the natural or napierien log.)

10. $r = \sqrt{\theta}$ **11.** $r = \log \theta$

12. Figure 9.2-14 shows a section of the shell of *Nautilus pompilius*. By graphing $\log r = \theta$, show that this curve is a good approximation to the spiral shape of the shell. This curve is known as the logarithmic spiral.

Figure 9.2-14

EXERCISES 9.2-3

In Exercises 1–6, each point is given in cartesian coordinates. Express it in cylindrical coordinates. Then carefully plot it and show its cartesian *and* cylindrical coordinates on the figure.

1. $P(1, 2, 3)$ **2.** $P(-1, 3, -4)$ **3.** $P(-5, -8, 3)$

4. $P(0, 0, 2)$ **5.** $P(5, -6, 0)$ **6.** $P(-2, -1, -7)$

In Exercises 7–11, each point is given in cylindrical coordinates. Express it in rectangular coordinates. Then carefully plot it and show its cartesian *and* cylindrical coordinates on the figure.

7. $P(4, 45°, 5)$ **8.** $P(3, \frac{\pi}{2}, -2)$ **9.** $P(6, \frac{5\pi}{6}, 5)$

10. $P(2, 300°, -1)$ **11.** $P(5, 240°, -6)$

12. On an air traffic controller's radar screen, a 747 jumbo jet is 4.8 km from the airport at an angle of 118° with the polar axis at an altitude of 2100 m. If the origin is the airport:

(a) Find the cylindrical coordinates of this jet.
(b) Find the three cartesian coordinates of this jet.

13. A DC-10 is 5.60 km south and 3.96 km west of the airport, at an altitude of 1500 m. If the origin is at the airport and the polar axis is from west to east, find the cylindrical coordinates of this point.

***14.** A cube 12 cm on each side is located in the first octant of a rectangular coordinate system with one vertex at the origin, as shown in Figure 9.2-15. (Just as we divide the x-y plane into four quadrants, so we divide three-dimensional space into eight regions called *octants*. The first octant is the region where x, y, and z are all positive.) Find the cylindrical coordinates of all eight vertices of this cube.

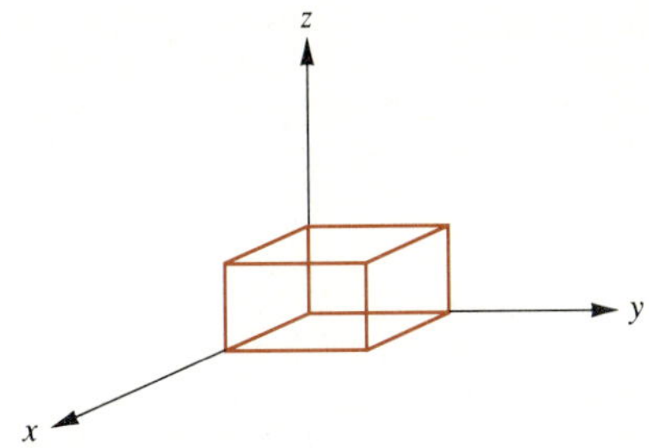

Figure 9.2-15

***15.** A pyramid having a square base 100.0 m on a side rises to a maximum height of 75.0 m directly over the midpoint of the base. If this pyramid is in the first octant (see Exercise 14) as shown in Figure 9.2-16, find the cylindrical coordinates of the highest point.

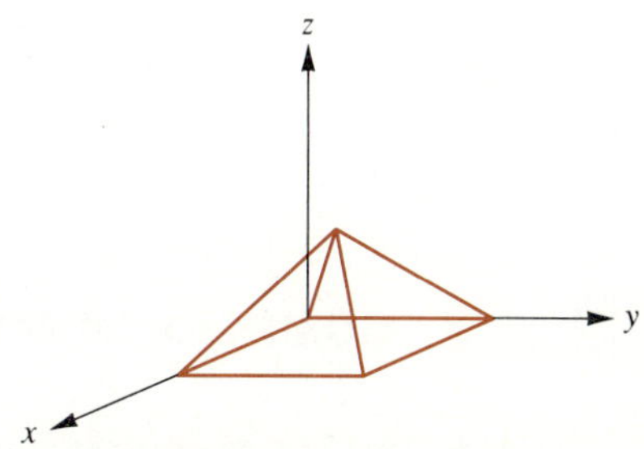

Figure 9.2-16

16. Imagine cartesian and cylindrical coordinate systems embedded in the earth, the pole (origin) at its center. Find the cylindrical coordinates of all the points at which the x, y, and z axes intersect our planet's surface, expressing r in kilometers.

****17.** A sphere of radius 20.0 cm is located in the first octant (see Exercise 14) so that it is tangent to the x-y plane, the x-z plane, and the y-z plane. (These three planes are called the coordinate planes.) Find the cartesian and cylindrical coordinates of:

(a) The center of this sphere.
(b) The three points where it touches the coordinate planes.
(c) The point farthest from (i) the x-y plane, (ii) the y-z plane, and (iii) the x-z plane. ***Hint:*** Make three drawings of the sphere, as viewed from a direction along the x axis, then a direction along the y axis, and finally a direction along the z axis. Remember what we said earlier in this book about reducing three-dimensional problems to two-dimensional figures.

****18.** Imagine a cylindrical coordinate system embedded in the earth, as in Exercise 16. It is oriented so that the x axis comes out at the equator at the same longitude as Greenwich, England, and the z axis comes out at the north pole (see Chapters 1 and 2 where we discussed latitude and longitude). Find the cylindrical coordinates of Seattle, which is at a north latitude of 48° and a west longitude of 122°.

****19. Gold Crystal.** Gold forms a face-centered cubic crystal with gold atoms at the corners of a cube *and* centered in each face (see Example 31 in Section 2.4). This cube is 4.07 Å on each side. Take the x-y plane to be the bottom face of the cube and the z axis to run through the center of the cube through two of the face-centered atoms.

- ***(a)** Find the rectangular coordinates of each gold atom.
- ****(b)** Find the cylindrical coordinates of each gold atom.

****20. Methane Molecule.** In the methane (CH_4) molecule, the four hydrogen atoms form a regular tetrahedron (a three-dimensional solid figure having four identical faces) with the carbon atom inside (see Figure 9.2-17). Measurements reveal that the C—H distances are all 1.091 Å and that the H—C—H angles are all 109°28′ (see Example 6 in Chapter 4). Let the three lower H atoms be in the x-y plane with the z axis running upward through the C atom.

- **(a)** Find the cylindrical coordinates of each atom in methane.
- **(b)** Find the rectangular coordinates of each atom in methane.

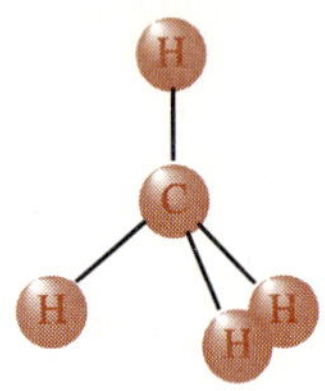

Figure 9.2-17

In Exercises 21–27, sketch the graph of each equation(s) in cylindrical coordinates. Do not plot points (except maybe one or two easy ones to get started), but rather use the general characteristics of the equations to determine the shape of the graph. If the resulting figures are familiar surfaces or curves, identify them by name.

21. $r = 2,\ z = -\theta$

22. $z = 5$

23. $\theta = \frac{\pi}{6}$

***24.** $\theta = \frac{\pi}{4},\ z = 1$

25. $r = 2$

26. $r = \theta,\ z = \theta$

***27.** $z = 2,\ r = 3$

9.3 COMPLEX NUMBERS

9.3-1 Basic Definitions

Around A.D. 75, Heron of Alexandria encountered the number $\sqrt{81 - 144} = \sqrt{-63}$. This number is the first square root of a negative number that we know of. Then, around A.D. 250, Diophantus of Alexandria,* in his *Stereometrica* found $\sqrt{1849 - 2016} = \sqrt{-167}$ as a possible root of a quadratic equation.

We do not know what these great mathematicians thought about such numbers, but they may well have felt that the numbers were impossible or had no meaning.

In his 1545 book *Ars Magna*, the Italian mathematician Girolamo Cardano (also known as Jerome Cardan) proposed the following problem: Divide 10 into two parts whose product is 40. This problem is a straight-forward algebra problem today. Letting x be one part, then $10 - x$ is the other part. The product of these is 40, so

$$x(10 - x) = 40$$
$$10x - x^2 = 40$$
$$x^2 - 10x + 40 = 0$$

* Alexandria, in present-day Egypt, keeps popping up in early mathematics. It was a center of Greek mathematics from the time of Euclid (around 300 B.C.) to past A.D. 400—an incredible period of more than seven centuries!

Solving this by the quadratic formula gives us

$$x = \frac{-b \pm \sqrt{b^2 - 4ac}}{2a} = \frac{-(-10) \pm \sqrt{(-10)^2 - 4(1)(40)}}{2(1)} = \frac{10 \pm \sqrt{-60}}{2}$$

$$= \frac{10 \pm \sqrt{(4)(-15)}}{2} = \frac{10 \pm 2\sqrt{-15}}{2} = 5 \pm \sqrt{-15}$$

Cardano actually obtained these answers. In recognizing the impossibility of finding the square root of the negative number, Cardano called such quantities "truly sophisticated" and wrote that it would be "as subtle as it would be useless" to keep working with them.* He was right about their subtlety, but wrong about their uselessness! Although Cardano realized that quantities involving the square root of negative numbers might be important, mathematicians showed little interest in them until the 1700s, which was one of the greatest periods of mathematical activity in history. Since that time the square roots of negative numbers have become central to mathematics. In the nineteenth and twentieth centuries, these quantities, which we now call *complex numbers*, have found many applications, especially in physics and engineering.

The reason that mathematicians resisted the square roots of negative numbers for so long is that it is impossible to find the square root of a negative number using the ordinary numbers such as positive and negative integers, fractions, and decimals. There is no such number which, when multiplied by itself, gives a negative number. Therefore, we must *define* such a number, and we do it in this way. We *define* the number i to be the square root of -1. This definition gives i the property that its square is -1.

DEFINITION OF *i*

$$i = \sqrt{-1} \quad \Rightarrow \quad i^2 = -1$$

The number i is an *imaginary number* (that's why we use i). A record could not cost $\$i$, nor could you get a score of $i\%$ on a trigonometry exam. (Try to take $\sqrt{-1}$ on your calculator and see what happens.) All the other numbers we have dealt with so far, such as 8, $\frac{1}{4}$, π, $\sqrt{2}$, 6.025, and so on are called *real numbers*.

We can express the square root of any negative number in terms of i by factoring out a -1 inside the square root. For example, Heron's number becomes

$$\sqrt{81 - 144} = \sqrt{-63} = \sqrt{(-1)(63)}$$

It is the -1 inside the square root that prevents this number from being a real number. Thus we can write the number in terms of i as

$$\sqrt{(-1)(63)} = i\sqrt{63}$$

* See Carl B. Boyer, *A History of Mathematics*, Princeton University Press, 1968, and David Wells, *Curious and Interesting Numbers*, Penguin Books, 1986.

NOW IT'S YOUR TURN

Express Diophantus' number in terms of i. (Answer: $i\sqrt{167}$)

We can generalize for any negative number $\sqrt{-a}$ where a is a positive real number.

$$\sqrt{-a} = i\sqrt{a}$$

if a is a positive real number.

Cardano's "truly sophisticated" numbers, $5 \pm \sqrt{-15}$, are a bit more complex. To see why, we express them in terms of i.

$$5 \pm \sqrt{-15} = 5 \pm i\sqrt{15}$$

Notice that each of these numbers is the sum of two numbers, one real (5) and the other imaginary ($\pm\, i\sqrt{15}$). We call such a number a *complex number.** Any complex number has *two parts:* the *real* part and the *imaginary* part. We define the imaginary part as the *coefficient* of i. The real part is the number that does *not* multiply i. For example, in Cardano's numbers $5 \pm i\sqrt{15}$, the real part is 5 and the imaginary parts are $\pm\sqrt{15}$, whereas in Heron's number $i\sqrt{63}$, the real part is zero and the imaginary part is $\sqrt{63}$. Be careful:

▸ **The imaginary part is a real number because it is the coefficient of i.**

NOW IT'S YOUR TURN

What are the real and imaginary parts of the complex number $-6.1 + 3.4i$?
(Answer: -6.1 is the real part and 3.4 the imaginary part.)

The term *complex* number does not mean a *complicated* number. It just means a number with a real part and an imaginary part.

We usually use the letter z to represent a complex number, and we often use a for its real part and b for its imaginary part, where both a and b are real numbers. Thus we write $z = a + bi$.

COMPLEX NUMBER

$$z = a + bi$$

a = real part of z and b = imaginary part of z.

* The use of i for $\sqrt{-1}$ was introduced by Leonhard Euler in 1777 and the term *complex number* by Gauss around 1800.

(We also use x and y instead of a and b; in that case we write z as $z = x + yi$.)

As we have seen, complex numbers first arose when people were trying to solve quadratic equations.

Example 16

Solve the quadratic equations

(a) $x^2 + 4 = 0$ and (b) $4.6x^2 + 1.3x + 2.4 = 0$.

Solution

(a) We can solve this equation simply by taking square roots without using the quadratic formula.

$$x^2 + 4 = 0 \quad \rightarrow \quad x^2 = -4$$

Taking square roots gives us

$$x = \pm\sqrt{-4} = \pm 2i$$

(b) This equation does not factor, so we need the quadratic formula with

$a = 4.6$, $b = 1.3$, and $c = 2.4$.

$$x = \frac{-b \pm \sqrt{b^2 - 4ac}}{2a} = \frac{-1.3 \pm \sqrt{(1.3)^2 - 4(4.6)(2.4)}}{2(4.6)}$$

$$x = \frac{-1.3 \pm \sqrt{-42.47}}{9.2} = \frac{-1.3 \pm i\sqrt{42.47}}{9.2} = \frac{-1.3}{9.2} \pm \frac{6.52i}{9.2} = 0.14 \pm 0.71i$$

$$x = \begin{cases} 0.14 + 0.71i \\ 0.14 - 0.71i \end{cases}$$ ■

NOW IT'S YOUR TURN

Solve for x. (a) $4x^2 + 20 = 0$ and (b) $x^2 + x + 3 = 0$.
(Answer: (a) $x = \pm 2.24i$, (b) $x = -0.50 \pm 1.66i$)

In Example 16(b), the solutions $0.14 + 0.71i$ and $0.14 - 17i$ are pairs of complex numbers which differ only in the sign of the imaginary parts. We call such pairs of complex numbers *conjugates* of one another. *If* $z = a + bi$, then its conjugate (which we write as $\bar{z}$) is $\bar{z} = a - bi$.

COMPLEX CONJUGATE

If $z = a + bi$, then the *conjugate* of z is $\bar{z} = a - bi$

Notice that to find the conjugate of a complex number, you just change the sign of its imaginary part.

Example 17

Find the conjugate of (a) $z = -3 + 5i$, (b) $z = -9i$, and (c) $z = 10$.

Solution

We just change the sign of the imaginary part. (a) $\bar{z} = -3 - 5i$, (b) $\bar{z} = -(-9i) = 9i$, and (c) $\bar{z} = 10$ because its imaginary part is zero. ■

NOW IT'S YOUR TURN

Find the conjugate of $z_1 = -6.3 - 7.4i$, $z_2 = -4.9$, and $z_3 = 2.1i$.
(Answer: $\bar{z}_1 = -6.3 + 7.4i$, $\bar{z}_2 = -4.9$, and $\bar{z}_3 = -2.1i$)

9.3-2 The Algebra of Complex Numbers

Adding Complex Numbers

To add two complex numbers, we add the real parts and the imaginary parts separately.

$$\begin{aligned}(a + bi) + (c + di) &= a + bi + c + di = a + c + bi + di \\ &= (a + c) + (b + d)i\end{aligned}$$

Subtracting Complex Numbers

To subtract two complex numbers, we subtract the real parts and the imaginary parts separately.

$$\begin{aligned}(a + bi) - (c + di) &= a + bi - c - di = a - c + bi - di \\ &= (a - c) + (b - d)i\end{aligned}$$

Example 18

If $z_1 = 4 - 3i$ and $z_2 = -2 + 8i$, find $z_1 + z_2$ and $z_1 - z_2$.

Solution

$$\begin{aligned}z_1 + z_2 &= (4 - 3i) + (-2 + 8i) = 4 - 3i - 2 + 8i \\ &= 4 - 2 - 3i + 8i = 2 + 5i \\ z_1 - z_2 &= (4 - 3i) - (-2 + 8i) = 4 - 3i + 2 - 8i = 4 + 2 - 3i - 8i \\ &= 6 - 11i\end{aligned}$$

■

NOW IT'S YOUR TURN

For z_1 and z_2 in this example, find (a) $z_1 - \overline{z}_1$ and (b) $\overline{z}_2 + z_1$.
(Answer: (a) $-6i$, (b) $2 - 11i$)

Multiplying Complex Numbers

To multiply two complex numbers, multiply as you would any two binomial expressions and use the fact that $i^2 = -1$.

$$(a + bi)(c + di) = ac + adi + bci + bdi^2 = ac + adi + bci + bd(-1)$$
$$= (ac - bd) + (ad + bc)i$$

Example 19

(a) If $z_1 = 4 - 7i$ and $z_2 = -1 + i$, find $z_1 z_2$. (b) If $z = a + bi$ is any complex number, find $z\overline{z}$.

Solution

(a) $$z_1 z_2 = (4 - 7i)(-1 + i) = (4)(-1) + (4)(i) + (-7i)(-1) + (-7i)(i)$$
$$= -4 + 4i + 7i - 7i^2 = -4 + 4i + 7i - 7(-1)$$
$$= -4 + 7 + 4i + 7i = 3 + 11i$$

(b) $$z\overline{z} = (a + bi)(a - bi) = a^2 - abi + bai - b^2i^2 = a^2 - b^2i^2$$
$$= a^2 - b^2(-1) = a^2 + b^2$$

Notice that this number is always *a real number* no matter what z is. ■

NOW IT'S YOUR TURN

If $z_1 = 3 + i$ and $z_2 = -4 + 2i$, find (a) $z_1 z_2$ and (b) $z_2\overline{z}_2$.
(Answer: (a) $-14 + 2i$, (b) 20)

Dividing Complex Numbers

As with multiplication, addition, and subtraction of complex numbers, the goal of division is to get the answer in the form $a + bi$. To accomplish this in division, we get rid of the complex number in the denominator by multiplying by its *conjugate*. This multiplication always makes the denominator a positive real number, as we just saw in Example 19b. Of course, if we multiply the denominator by its conjugate, we must also multiply the numerator by the same number. Therefore,

$$\frac{z_1}{z_2} = \frac{z_1 \times \overline{z}_2}{z_2 \times \overline{z}_2}$$

If $z_1 = a + bi$, and $z_2 = c + di$, this equation becomes

$$\frac{z_1}{z_2} = \frac{z_1 \times \bar{z}_2}{z_2 \times \bar{z}_2} = \frac{(a + bi)(c - di)}{(c + di)(c - di)} = \frac{(a + bi)(c - di)}{c^2 + d^2}$$

We now multiply out the numerator by the method we just learned and then divide this result by the positive real number $c^2 + d^2$.

Example 20

If $z_1 = 2 - 3i$ and $z_2 = 1 + 4i$, find $\frac{z_1}{z_2}$.

Solution

We multiply both the numerator and the denominator by the conjugate of the denominator.

$$\frac{z_1}{z_2} = \frac{2 - 3i}{1 + 4i} = \frac{(2 - 3i)(1 - 4i)}{(1 + 4i)(1 - 4i)} = \frac{2 - 8i - 3i + 12i^2}{1^2 + 4^2} = \frac{-10 - 11i}{17} = -\frac{10}{17} - \frac{11}{17}i \quad ■$$

NOW IT'S YOUR TURN

Find

$$\frac{6 + 3i}{5 - i}$$

(Answer: $\frac{27}{26} + \frac{21}{26}i$)

EXERCISES 9.3-1

In Exercises 1–10, rewrite each expression in terms of *i*.

1. $\sqrt{-16}$ **2.** $\sqrt{-9}$ **3.** $8\sqrt{-7}$
4. $2.5\sqrt{-13}$ **5.** $1 + 7\sqrt{-12}$ **6.** $-8 - 5\sqrt{-47}$
7. $4 - \sqrt{-\frac{5}{9}}$ **8.** $9.54 + \sqrt{-\frac{3}{25}}$ **9.** $\sqrt{-0.04}$
10. $1 + 3\sqrt{-1.93}$

In Exercises 11–20, perform the indicated operations and simplify, putting your final answers in terms of *i* if the number is imaginary.

11. $\sqrt{-1}\sqrt{-1}$ **12.** $\sqrt{-5}\sqrt{-10}$ **13.** $7\sqrt{-3}\sqrt{-20}$
14. $(\sqrt{-1})^3$ **15.** $(\sqrt{-1})^4$ **16.** $\sqrt{-7})^2$
***17.** $(5\sqrt{-2})^2 - 5\sqrt{(-2)^2}$ ***18.** $3\sqrt{(-7)^2} + (3\sqrt{-7})^2$
19. $\sqrt{-x^2y^2}$ **20.** $2y\sqrt{-xy^3}$

21. Show that **(a)** $i^2 = -1$, **(b)** $i^3 = -i$ and **(c)** $i^4 = 1$.
▸ **[These results are important. Be sure you understand how to do them.]**

22. Use the results of Exercise 21 to reduce each expression to $\pm i$ or ± 1. (For example, $i^5 = ii^4 = i(1) = i$.)

(a) i^6 **(b)** i^9 **(c)** i^{10}
(d) i^{15} **(e)** i^{16} **(f)** $-i^{25}$
(g) $-i^{28}$ **(h)** i^{50} **(i)** i^{75}
(j) i^{100} **(k)** i^{200}

In Exercises 23–38, identify the real and imaginary parts of each complex number *and* find its complex conjugate.

23. $2 - 3i$ **24.** $-8 + i$
25. $-2i$ **26.** $\sqrt{7}$
27. $6i - 7$ **28.** $-i + 4$
29. $16i$ **30.** -9.1
31. $\sqrt{-1}$ **32.** $5\sqrt{-1}$
33. $-4 + \sqrt{-9}$ **34.** $13 + 11\sqrt{-25}$
***35.** $12i^2 + 2i$ ***36.** $8 + i\sqrt{-4}$
***37.** $3 + i + \sqrt{-16} - i^2$ ***38.** $1 + i + i^2 + i^3 + i^4 + i^5$

In Exercises 39–48, the complex conjugate of a complex number z is given. Find the complex number z.

39. $\bar{z} = 4 + 7i$ **40.** $\bar{z} = -3 + 6i$

41. $\bar{z} = 8 - i$ **42.** $\bar{z} = -16i$

43. $\bar{z} = -8.90$ **44.** $\bar{z} = -10 - 7i$

45. $\bar{z} = 5 - \sqrt{-36}$ **46.** $\bar{z} = 38 + 4i\sqrt{-49}$

47. $\bar{z} = i^2$ ***48.** $\bar{z} = i - i^3 + i^5$

49. Prove that $\overline{(\bar{z})} = z$ is an identity.

***50.** What is wrong with the following "proof" that $-2 = 2$?

$$\sqrt{-2}\sqrt{-2} = \sqrt{(-2)(-2)} = \sqrt{+4} = 2$$
$$\sqrt{-2}\sqrt{-2} = (i\sqrt{2})(i\sqrt{2}) = i^2\sqrt{4} = (-1)\sqrt{4} = -2$$
$$\therefore \quad -2 = 2$$

51. Use the *incorrect method* of Exercise 50 to incorrectly "prove" that $3 = -3$.

52. A complex number that has only a real part is called a *pure real*. One having only an imaginary part is called a *pure imaginary*. What must a and b be so that $z = a + bi$ will be **(a)** pure real and **(b)** pure imaginary?

53. Which of the complex numbers in Exercises 23–38 are pure real and which are pure imaginary? (See Exercise 52.)

In Exercises 54–60, solve for x.

54. $x^2 + 10 = 1$ **55.** $-2x^2 + x = 1$

56. $3x^2 - 5x + 4 = 0$ **57.** $6(x^2 + 1) = 9x + 2$

58. $\frac{1}{3}x^2 - \frac{1}{2}x + \frac{3}{4} = 0$ **59.** $10.7x - 4.22 = 9.16x^2$

60. $4.66x^2 + 7.50x + 3.85 = 0$

EXERCISES 9.3-2

In Exercises 1–28, perform the indicated operations and express your final answers in the form $a + bi$.

1. $(4 - 3i) + (2 - 4i)$ **2.** $(12 - 5i) + (12 + 5i)$

3. $(6 - 3i) - (7 + i)$ **4.** $(5 - 4i) - (-1 + 3i)$

5. $(\frac{1}{3} - \frac{1}{6}i) - (\frac{5}{6} + \frac{2}{3}i)$

6. $(0.4 + 0.3i) - (0.1 - 0.5i)$

7. $(\frac{1}{2} + \frac{1}{3}i) + (-\frac{1}{3} + \frac{1}{2}i)$

8. $(0.25 - 0.25i) + (0.75 + 0.75i)$

9. $(3 + 4i)(5 + 2i)$ **10.** $(-7 + 3i)(-3 + 2i)$

11. $(\frac{1}{2} + \frac{1}{3}i)(\frac{1}{2} - \frac{1}{3}i)$ **12.** $(4 + 2i)^2$

13. $(1 - i)^2$ ***14.** $(1 + i)^3$

***15.** $(1 - i)^4$ **16.** $\dfrac{1 - i}{1 + i}$

17. $\dfrac{6}{1 - 2i}$ **18.** $\dfrac{1}{i}$

19. $\dfrac{4i}{2 - 3i}$ **20.** $\dfrac{1 + i\sqrt{3}}{1 - i\sqrt{3}}$

21. $\dfrac{4 - i\sqrt{2}}{1 + i\sqrt{2}}$ **22.** $\dfrac{-3\sqrt{2} + i}{1 + 3i\sqrt{2}}$

***23.** $\dfrac{1}{1 - i} + \dfrac{1}{1 + i}$ **24.** $\dfrac{i}{2i - 3} - \dfrac{6i + 2}{3 + 4i}$

***25.** $\dfrac{2i + 3}{i + 5} - \dfrac{3i - 4}{2i - 1}$

***26.** $(2 + i)^2(1 + i) - \dfrac{3(i - 2)}{1 + i}$ ***27.** $\dfrac{c + i\sqrt{d}}{c - i\sqrt{d}}$

***28.** $\dfrac{c + di}{d - ci}$

Two complex numbers are equal if and only if their real parts are equal to each other *and* their imaginary parts are also equal to each other. For example, $3 + xi = y - 4i$ if and only if $y = 3$ and $x = -4$. Use this condition to solve for x and y in Exercises 29–35.

29. $x + \sqrt{-64} = 14 + yi$ **30.** $x + yi = \sqrt{-81}$

31. $2x + 4i = 6 - 2yi$ **32.** $-5 + 3yi = 5x - 12i$

33. $(1 - i)x + (1 + i)y = 1 - 3i$

***34.** $(2 + 3i)x^2 - (3 - 2i)y = 2x - 3y + 15i$

****35.** $(4x^2 + 3xy) + (2xy - 3x^2)i = 4y^2 - \frac{1}{2}x^2 + (3xy - 2y^2)i$

In Exercises 36–39, $z_1 = 1 + 2i$, $z_2 = 3i - 4$, $\bar{z}_3 = z_1 + z_2$, $\bar{z}_4 = z_1 z_2$, and $\bar{z}_5 = z_1/z_2$. Evaluate each expression and put it in the form $a + bi$.

36. z_3 **37.** z_4

38. z_5 **39.** $\bar{z}_3\,\bar{z}_3$

40.** Is $\overline{(z_1 + z_2)} = \bar{z}_1 + \bar{z}_2$? Prove it! ***Hint: Let $z_1 = a + bi$ and $z_2 = c + di$. Then evaluate both sides of the equation and see if they are always equal no matter what a, b, c, and d are.

Use the hint in Exercise 40 for Exercises 41–44.

***41.** Is $\overline{(z_1 - z_2)} = \bar{z}_1 - \bar{z}_2$? Prove it.

***42.** Is $\overline{z_1 z_2} = \bar{z}_1\bar{z}_2$? Prove it.

***43.** Is $\overline{z_1/z_2} = \bar{z}_1/\bar{z}_2$? Prove it.

***44.** Is $\bar{z}^2 = \overline{z^2}$? Prove it.

45. If $z = a + bi$, which of the following are pure real, pure imaginary, or neither pure real nor pure imaginary? (See Exercise 52 in Section 9.3-1.)

(a) $z + \bar{z}$ **(b)** $z - \bar{z}$

(c) z^2 **(d)** $z\bar{z}$

(e) $\dfrac{z}{\bar{z}}$ **(f)** $\dfrac{\bar{z}}{z}$

(g) $\dfrac{z + \bar{z}}{2i}$ **(h)** $\dfrac{z - \bar{z}}{2i}$

In Exercises 46–51, $z_1 = a + bi$ and $z_2 = c + di$. (See Exercise 52 in Section 9.3-1.) What must be the conditions on a, b, c, and d so that:

***46.** $z_1 + z_2$ is pure real

***47.** $z_1 + z_2$ is pure imaginary

***48.** $z_1 - z_2$ is pure real

***49.** $z_1 - z_2$ is pure imaginary

***50.** $z_1 z_2$ is pure real

***51.** $z_1 z_2$ is pure imaginary

***52.** If a quadratic equation has *nonreal* roots, prove that there are *always two distinct roots* and that these roots are *conjugates* of each other. Does this result hold true if the roots are real? Why?

In Exercises 53–57, suppose that each number is the root of some quadratic equation, but you do not know the equation. Find the *other* root of this quadratic equation. *Hint:* See Exercise 52.

53. $x = -3 + 2i$

54. $x = 8.9i + 6.4$

55. $x = -i$

56. $x = 5(3i - 7)$

57. $x = 9.1(2.4i - 3.7)$

58.** Prove that the real part of any complex number z is equal to $(z + \bar{z})/2$. ***Hint: Let $z = a + bi$

***59.** Prove that the imaginary part of any complex number z is equal to $(z - \bar{z})/2$. See Exercise 58 for a hint.

****60.** Let z be a complex number. Prove that if $z^2 = \bar{z}^2$, then z is either real or pure imaginary.

***61.** If $z = a + bi$ obeys the equation $z^2 + \bar{z}^2 = c^2$, where c is a constant, find the equation that a and b must obey.

62. In algebra you learned that you cannot factor the sum of two squares, $a^2 + b^2$. However, show that by using complex numbers you can factor this as $a^2 + b^2 = (a + bi)(a - bi)$.

9.4 GEOMETRIC REPRESENTATION OF COMPLEX NUMBERS

9.4-1 Rectangular Form of a Complex Number

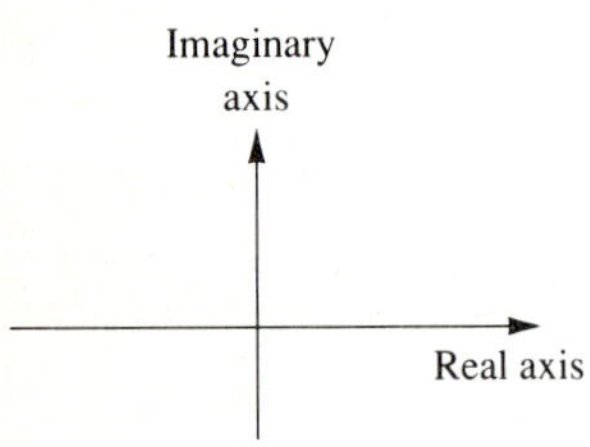

Figure 9.4-1

As you learned in algebra, we can plot all the real numbers on a straight line—the real number line. But we cannot put complex numbers on this line because i is not a real number. To plot complex numbers, we add a line for the imaginary numbers and orient it perpendicular to the real number line. These two lines (or axes) form a plane, which we call the *complex plane** (see Figure 9.4-1).

A complex number z represents a point in this complex plane. We plot the real part of z on the real axis and the imaginary part on the imaginary axis, just as we plot the x and y coordinates of a point in the x-y plane. Therefore, we can express a complex number as an *ordered pair*† of numbers, the first being

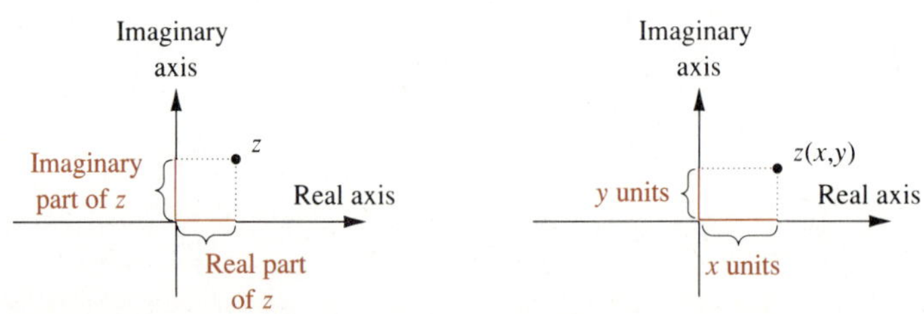

(a) The complex number z plotted in the complex plane

(b) The complex number z = x + yi plotted in the complex plane

Figure 9.4-2

* The complex plane is sometimes called the *Argand plane*. French mathematician Jean R. Argand presented his paper on the complex plane in 1806. This idea had been suggested nine years earlier to the Danish Academy of Sciences by Norwegian mathematician and surveyor Caspar Wessel.

† The idea of using an ordered pair to represent a complex number was developed independently and almost simultaneously by Karl F. Gauss and William R. Hamilton in the early part of the nineteenth century.

the real part of z and the second the imaginary part (see Figure 9.4-2a). If $z = x + yi$, we write $z = (x, y)$ and plot it as shown in Figure 9.4-2b. (Notice that this plot is just like plotting points in the x-y plane, where the real part of z is equivalent to the x axis and the imaginary part is equivalent to the y axis.)

Example 21

Plot the complex numbers (a) $-2 - 3i$, (b) 5, and (c) $2i$.

Solution

(a) $z = -2 - 3i$, so $x = -2$, $y = -3$.

(b) $z = 5 = 5 + 0i$, so $x = 5$, $y = 0$.

(c) $z = 2i = 0 + 2i$, so $x = 0$, $y = 2$ (see Figure 9.4-3). ■

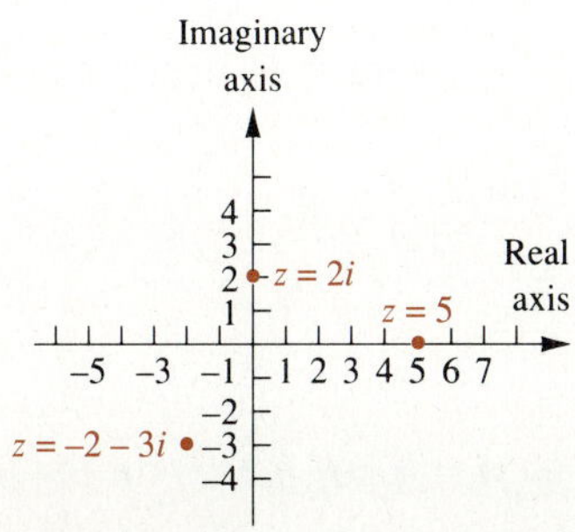

Figure 9.4-3

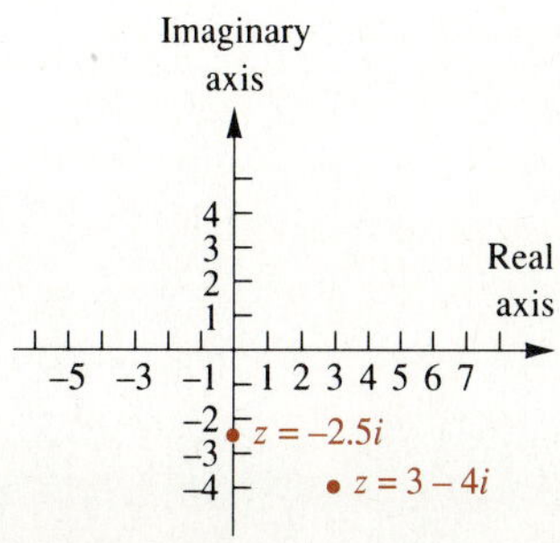

Figure 9.4-4

NOW IT'S YOUR TURN

Plot $z = 3 - 4i$ and $z = -2.5i$ in the complex plane.

(Answer: Figure 9.4-4)

9.4-2 Polar Form of a Complex Number

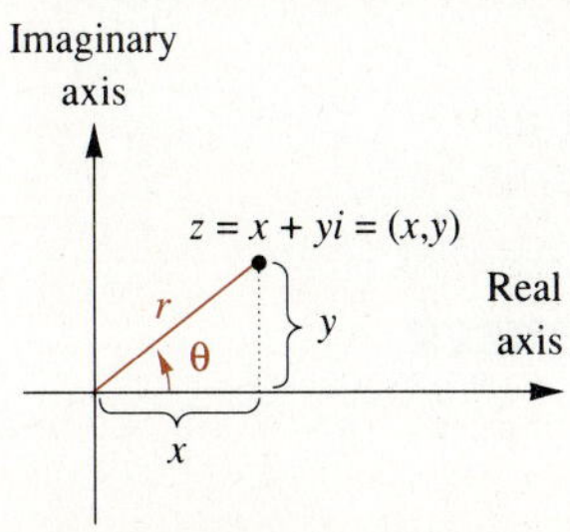

Figure 9.4-5

We have just seen how to plot a complex number in the complex plane as an ordered pair (x, y) in rectangular coordinates. We can also represent the same complex number in polar form, just as we could represent any point in the x-y plane by its polar coordinates in Section 9.1. By analogy with polar coordinates, if $z = x + yi$, then r is the distance from the origin to the complex number $z(x, y)$, and θ is the angle that r makes with the positive real axis. Both r and θ are always real numbers (see Figure 9.4-5).

When working with complex numbers, we usually call r the *modulus* (or *absolute value* or *magnitude*) of z and write it as $r = |z|$. We call θ the *argument* of z and write it as $\theta = \arg z$.

ARGUMENT AND MODULUS OF A COMPLEX NUMBER

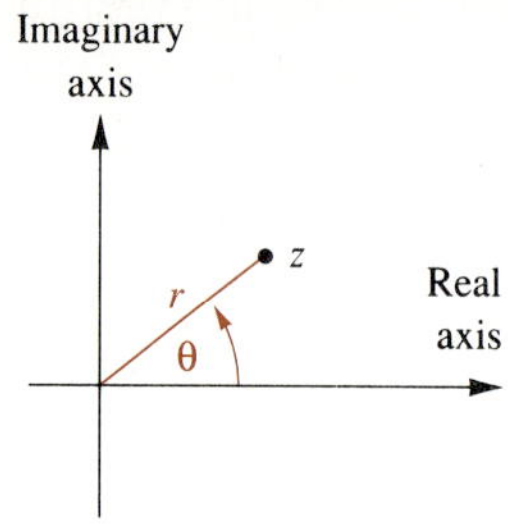

argument of z = arg $z = \theta$

modulus of $z = |z| = r$

We can now express a complex number in polar form. Figure 9.4-5 tells us that

$$\cos\theta = \frac{x}{r} \quad \rightarrow \quad x = r\cos\theta$$

$$\sin\theta = \frac{y}{r} \quad \rightarrow \quad y = r\sin\theta$$

Therefore,

$$z = x + yi = r\cos\theta + ir\sin\theta = r(\cos\theta + i\sin\theta)$$
$$z = r(\cos\theta + i\sin\theta)$$

The expression $\cos\theta + i\sin\theta$ is often abbreviated cis θ (the "cis" means "cos + i sin"). Using this abbreviation, we can write the polar form of z as $z = r \operatorname{cis}\theta$. The value of this abbreviation is that it saves a lot of writing.

As in Section 9.1, we can relate the polar and rectangular coordinates of z. Figure 9.4-5 tells us that

$$r^2 = x^2 + y^2 \quad \rightarrow \quad r = |z| = \sqrt{x^2 + y^2}$$

$$\tan\theta = \frac{y}{x} \quad \rightarrow \quad \theta = \arg z = \tan^{-1}\frac{y}{x}$$

The signs of x and y determine which quadrant θ is in.

POLAR FORM OF A COMPLEX NUMBER

In rectangular form:

$$z = x + yi$$

In polar form:

$$z = r(\cos\theta + i\sin\theta) = r \text{ cis } \theta$$
$$r = |z| = \sqrt{x^2 + y^2}$$
$$\tan\theta = \frac{y}{x}$$

The signs of x and y determine which quadrant θ is in.

Example 22

Express $z = 2 - 2\sqrt{3}\,i$ in polar form.

Solution

Because $z = 2 - 2\sqrt{3}\,i$, then $x = 2$ and $y = -2\sqrt{3}$.

Find r:

$$r = |z| = \sqrt{x^2 + y^2} = \sqrt{(2)^2 + (-2\sqrt{3})^2} = \sqrt{4 + 12} = \sqrt{16} = 4$$

Find θ:

$$\tan(\text{reference angle}) = \left|\frac{y}{x}\right| = \left|\frac{-2\sqrt{3}}{2}\right| = \sqrt{3}$$

$$\text{reference angle} = \tan^{-1}\sqrt{3} = \tfrac{\pi}{3}$$

Because x is positive and y is negative, θ lies in quadrant IV with reference angle $\frac{\pi}{3}$. Thus

$$\theta = \arg z = 2\pi - \tfrac{\pi}{3} = \tfrac{5\pi}{3}$$

Writing z in polar form, we have

$$z = r(\cos\theta + i\sin\theta)$$
$$z = 4(\cos\tfrac{5\pi}{3} + i\sin\tfrac{5\pi}{3}) = 4 \text{ cis } \tfrac{5\pi}{3} \quad ■$$

NOW IT'S YOUR TURN

Express $z = -4\sqrt{3} + 4i$ in polar form. (Answer: $z = 8 \text{ cis } \frac{5\pi}{6}$)

Example 23

Express the complex number $z = 6 \text{ cis } 143°$ in rectangular form.

Solution

$r = 6.00$ and $\theta = 143°$. Thus

$$x = r \cos \theta = 6.00 \cos 143° = (6.00)(-0.799) = -4.79$$

$$y = r \sin \theta = 6.00 \sin 143° = (6.00)(0.602) = 3.61$$

In rectangular form,

$$z = x + yi = -4.79 + 3.61i$$ ■

NOW IT'S YOUR TURN

Express $z = 3.45 (\cos 311° + i \sin 311°)$ in rectangular form.
(Answer: $z = 2.26 - 2.60i$)

9.4-3 Multiplication and Division in Polar Form

Addition and subtraction of complex numbers are easiest in rectangular form. But, as we shall see, multiplication and division are much easier when the numbers are expressed in polar form.

If $z_1 = r_1(\cos \theta_1 + i \sin \theta_1)$ and $z_2 = r_2(\cos \theta_2 + i \sin \theta_2)$, then

$$z_1 z_2 = [r_1(\cos \theta_1 + i \sin \theta_1)][r_2(\cos \theta_2 + i \sin \theta_2)]$$

$$= r_1 r_2(\cos \theta_1 \cos \theta_2 + i \cos \theta_1 \sin \theta_2 + i \sin \theta_1 \cos \theta_2 + i^2 \sin \theta_1 \sin \theta_2)$$

$$z_1 z_2 = r_1 r_2[\underbrace{(\cos \theta_1 \cos \theta_2 - \sin \theta_1 \sin \theta_2)}_{\cos(\theta_1 + \theta_2)} + i \underbrace{(\sin \theta_1 \cos \theta_2 + \cos \theta_1 \sin \theta_2)}_{\sin(\theta_1 + \theta_2)}]$$

$$z_1 z_2 = r_1 r_2[\cos(\theta_1 + \theta_2) + i \sin(\theta_1 + \theta_2)] = r_1 r_2 \text{ cis}(\theta_1 + \theta_2)$$

NOW IT'S YOUR TURN

Show that, if $z_2 \neq 0$, then

$$\frac{z_1}{z_2} = \left(\frac{r_1}{r_2}\right) [\cos(\theta_1 - \theta_2) + i \sin(\theta_1 - \theta_2)] = \left(\frac{r_1}{r_2}\right) \text{cis}(\theta_1 - \theta_2)$$

Hint: **Express z_1 and z_2 in polar form, then form $\frac{z_1}{z_2}$ and multiply numerator and denominator by $\bar{z}_2$, using $\bar{z} = r \text{ cis}(-\theta)$ (see Exercise 49 for Section 9.4-2).**

Summarizing our results:

MULTIPLICATION AND DIVISION IN POLAR FORM

$$z_1 = r_1(\cos\theta_1 + i\sin\theta_1) \quad \text{and} \quad z_2 = r_2(\cos\theta_2 + i\sin\theta_2)$$

$$z_1z_2 = r_1r_2[\cos(\theta_1 + \theta_2) + i\sin(\theta_1 + \theta_2)]$$

$$\frac{z_1}{z_2} = \left(\frac{r_1}{r_2}\right)[\cos(\theta_1 - \theta_2) + i\sin(\theta_1 - \theta_2)] \quad (z_2 \neq 0)$$

Now it should be clear why multiplication and division of complex numbers is especially easy in polar form. To multiply two complex numbers, all we do is *multiply their absolute values* and *add their arguments*. To divide them, we *divide their absolute values* and *subtract their arguments*.

Example 24

If $z_1 = 5.00$ cis 15.0° and $z_2 = 6.00(\cos 20.0° + i\sin 20.0°)$, find z_1z_2 and z_1/z_2. Express the answers in both polar form and rectangular form.

Solution

$$z_1z_2 = r_1r_2[\cos(\theta_1 + \theta_2) + i\sin(\theta_1 + \theta_2)]$$

$$\begin{aligned} z_1z_2 &= (5)(6)[\cos(15° + 20°) + i\sin(15° + 20°)] \\ &= 30(\cos 35° + i\sin 35°) = 30 \text{ cis } 35° \quad \textbf{(in polar form)} \\ &= 30(0.8192 + i0.5736) \\ &= 24.6 + 17.2i \quad \textbf{(in rectangular form)} \end{aligned}$$

$$\begin{aligned} \frac{z_1}{z_2} &= \left(\frac{r_1}{r_2}\right)\text{cis}(\theta_1 - \theta_2) \\ &= (\tfrac{5}{6})\text{cis}(15° - 20°) \\ &= (\tfrac{5}{6})\text{cis}(-5°) = (\tfrac{5}{6})[\cos(-5°) + i\sin(-5°)] \\ &= (\tfrac{5}{6})[0.9962 + i(-0.0872)] \\ &= 0.830 - 0.0726i \quad \textbf{(in rectangular form)} \end{aligned}$$

In this example, both complex numbers were given in polar form. If one or both were not given in polar form, you would first have to convert from rectangular form to polar form, as in Example 22. ■

NOW IT'S YOUR TURN

If $z_1 = 16.0(\cos 147° + i\sin 147°)$ and $z_2 = 8.00$ cis 22.0°, find z_1z_2 and z_1/z_2. Express your answers in both polar form and rectangular form.

(Answer: $z_1z_2 = 128$ cis 169° $= -126 + 24.4i$ and $z_1/z_2 = 2$ cis 125° $= -1.15 + 1.64i$)

Who Uses Complex Numbers?

You may think that because complex numbers are not real they have no physical meaning. Not so! It is true that a complex number cannot represent a *quantity* of something; you cannot walk $2 + 3i$ miles or buy $4 - i$ books. Anything that we can *measure* or *observe* must be a *real* quantity. Therefore, in applying complex numbers, we must ultimately deal with the parts that are real. Remember that the real part, the imaginary part, the modulus $|z|$, and the argument of a complex number are all real numbers. Therefore, it is these quantities that have meaning in physical applications, as the following examples illustrate.

In AC circuits, electrical engineers use complex numbers to represent the currents and voltages. When plotted in the complex plane, the angles between these complex numbers represent *phase differences* between current and voltage in various parts of the circuits. The real part represents the current in one part of a circuit, whereas the imaginary part is the current in another part.

Quantum mechanics is the study of the laws of physics at the very small scale of atoms and molecules. In quantum mechanics, a complex quantity called a *wave function* describes the behavior of a particle, such as an electron in an atom. The observable part, which must be real, is the *square* of the modulus of the wave function. This quantity represents the probability of finding the electron in a small volume of space.

These few examples—there are many more—should help you see that complex numbers are very useful to engineers and scientists. Complex numbers are often used instead of real numbers because it is often much easier to do the mathematics with complex numbers (this may come as a surprise to you) and then zero in at the end on the parts that are real.

Mathematicians make extensive use of complex numbers and are usually not even concerned with their reality. The study of complex numbers is a very important part of *analysis*, a part of higher mathematics that deals with functions. If you go on in mathematics, physics, or engineering, you will certainly make use of complex numbers.

EXERCISES 9.4-1

In Exercises 1–10, plot each complex number and its conjugate in the complex plane.

1. $5 + 2i$ **2.** $-4 + 3i$ **3.** $-3 - 2i$

4. $6 - 3i$ **5.** 7 **6.** $5i$

7. $-4i$ **8.** -6 **9.** $-4 + 4i$

10. $1 + i$

11. Write each complex number (A, B, C, ...) in Figure 9.4-6 as $a + bi$.

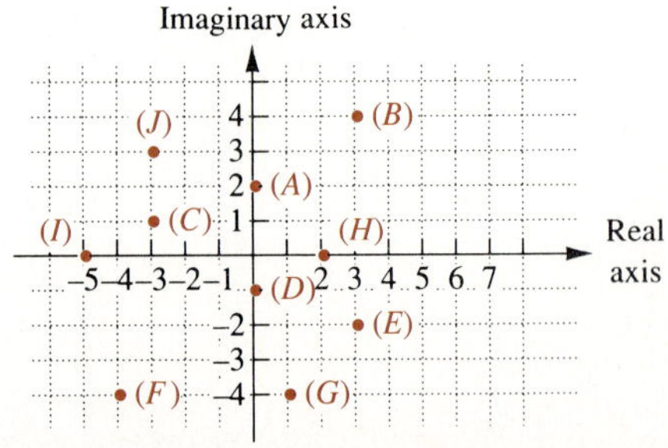

Figure 9.4-6

12.** If z is a complex number in the complex plane, give the geometric interpretation of: **(a)** $-z$, **(b)** $\bar{z}$, **(c)** $-\bar{z}$. ***Hint: Sketch each of these on the same graph with z to see how they relate to the original complex number z.

13.** Show that multiplying any complex number z by i is equivalent to rotating z by $+90°$. ***Hint: If $z = x + yi$, graphically interpret iz by plotting both z and iz on the same axes.

***14.** If $z = x + yi$ obeys the condition that the imaginary part of $(z - i)$ is equal to the real part of $(z + 1)$, find the equation that x and y must obey.

****15. Buried Treasure.** You discover an ancient manuscript that gives the following directions to a fabulous buried treasure on an island: "On this island there are only two palm trees (A and B) and the remains of a gallows. Start at the gallows and count the steps to walk in a straight line to tree A. At the tree turn 90° to the left and then walk forward the same number of steps. At the point where you stop, drive a spike into the ground. Now return to the gallows and walk in a straight line towards tree B, counting your steps. When you reach the tree, turn 90° to the right and take the same number of steps forward, placing another spike at the point where you stop. Dig at the point exactly halfway between the spikes and you will find the buried treasure."*

You save your money and sail to the island, but when you arrive you find that, alas, the trees are there but the gallows has rotted completely away, leaving no trace whatever. However, having just studied the geometric representation of complex numbers, you apply this knowledge to search for the treasure. *Where is the treasure?* (For the solution, see Martin Gardner's "Mathematical Games" section of the September 1979 *Scientific American* or Chapter 2 of George Gamow's book *One, Two, Three, . . . Infinity.*)

EXERCISES 9.4-2

In Exercises 1–16, find the modulus and the argument of each complex number.

1. $4(\cos 25° + i \sin 25°)$

2. $8 \text{ cis } 139°$

3. $\sqrt{8} \text{ cis } \frac{7\pi}{4}$

4. **(a)** i **(b)** $-i$

5. **(a)** $+1$ **(b)** -1

6. $2 + i$

7. $2i - 3$

8. $\sqrt{7} - \sqrt{6}i$

9. $6.84 - 9.17i$

10. $20i - 10$

11. $-3 - 4i$

12. $16 - 14i - 23 + 6i$

***13.** $5(\cos 30° + i \sin 60°)$

***14.** $8(\cos \frac{\pi}{6} - i \sin \frac{\pi}{4})$

***15.** $4(-\cos 50° + i \sin 50°)$

***16.** $9(-\cos 250° - i \sin 250°)$

* Our idea for this problem comes from Martin Gardner's fascinating article "The Imaginableness of Imaginary Numbers" in the "Mathematical Games" section of the August 1979 *Scientific American*.

In Exercises 17–30, write each complex number in polar form and find its modulus and argument.

17. $1 + i$

18. $-1 + i$

19. $1 - i$

20. $-\sqrt{3} + 3i$

21. $4i$

22. 4

23. -8

24. $-6i$

25. $-3 - 3i$

26. $3\sqrt{3} + 3i$

27. $\sqrt{2} - \sqrt{2}i$

28. $-4.23 + 6.57i$

29. $0.52i + 1.3$

30. $2.36 - 3.17i$

In Exercises 31–48, express each complex number in rectangular form.

31. $\sqrt{2}\,(\cos \frac{\pi}{4} + i \sin \frac{\pi}{4})$

32. $8 \text{ cis } \frac{\pi}{3}$

33. $6(\cos 150° + i \sin 150°)$

34. $3\sqrt{2} \cos 315° + i3\sqrt{2} \sin 315°$

35. $2 \text{ cis } \frac{7\pi}{6}$

36. $\frac{5}{2}(\cos \frac{2\pi}{3} + i \sin \frac{2\pi}{3})$

37. $5 \text{ cis } 540°$

38. $2(\cos 750° + i \sin 750°)$

39. $1/\sqrt{2}(\cos \frac{7\pi}{4} + i \sin \frac{7\pi}{4})$

40. $5 \text{ cis } 217°$

41. $3 \text{ cis}(-\frac{11\pi}{6})$

42. $7[\cos(-\frac{\pi}{3}) + i \sin(-\frac{\pi}{3})]$

43. $15[\cos(-\frac{15\pi}{4}) + i \sin(-\frac{15\pi}{4})]$

***44.** $7(\cos\frac{\pi}{6} + i \sin \frac{\pi}{4})$

***45.** $30\,(\cos 45° + i \sin 30°)$

***46.** $3.4(-\cos 120° + i \sin 120°)$

***47.** $6\,(\sin 60° + i \cos 60°)$

48. $-10 \text{ cis } \frac{3\pi}{4}$

49. If $z - r \text{ cis } \theta$, show that $\bar{z} = r \text{ cis}(-\theta)$.

***50.** Describe the curve in the complex plane satisfying the equation $|z| = c$, where c is a constant.

EXERCISES 9.4-3

In Exercises 1–21, perform the indicated operations and express your answers in polar form.

1. $[3(\cos 15° + i \sin 15°)]\,[6(\cos 75° + i \sin 75°)]$

2. $[6(\cos \frac{\pi}{3} + i \sin \frac{\pi}{3})]\,[2(\cos \frac{11\pi}{6} + i \sin \frac{11\pi}{6})]$

3. $[2(\cos \frac{\pi}{4} + i \sin \frac{\pi}{4})][2(\cos 225° + i \sin 225°)]$

4. $(3 \text{ cis } 127°)(4 \text{ cis } 203°)$

5. $(2 \text{ cis } \frac{2\pi}{15})(5 \text{ cis } \frac{\pi}{5})$

6. $[4(\cos \frac{5\pi}{12} + i \sin \frac{5\pi}{12})](2 \text{ cis } \frac{\pi}{12})$

***7.** $[3(\cos 25° + i \sin 25°)][5(\cos 15° + i \sin 20°)]$

***8.** $[9\,(\text{cis } \frac{\pi}{3})][7(\sin 30° + i \sin 60°)]$

***9.** $[8(\cos 45° - i \sin 45°)][13(-\cos 60° + i \sin 60°)]$

10. $\dfrac{8(\cos 135° + i \sin 135°)}{2(\cos 105° + i \sin 105°)}$

11. $\dfrac{4(\cos 47° + i \sin 47°)}{2(\cos 17° + i \sin 17°)}$

12. $\dfrac{18 \text{ cis } 42°}{9 \text{ cis } 7°}$

13. $\dfrac{5 \text{ cis } 38°}{7 \text{ cis } 73°}$

14. $\dfrac{9.3(\cos 91° + i \sin 91°)}{4.0 \text{ cis } 11°}$

15. $\dfrac{\sqrt{7}\left(\cos\dfrac{3\pi}{5} + i \sin\dfrac{3\pi}{5}\right)}{3.2\left(\cos\dfrac{3\pi}{4} + i \sin\dfrac{3\pi}{4}\right)}$

16. $\dfrac{6\left(\cos\dfrac{\pi}{9} + i \sin\dfrac{\pi}{9}\right)}{4 \text{ cis } 10°}$

***17.** $\dfrac{9(\cos 60° + i \sin 45°)}{3(\cos 60° + i \sin 60°)}$

***18.** $\dfrac{24(\sin 120° - i \cos 120°)}{8 \text{ cis } 20°}$

***19.** $\dfrac{49(\sin 50° + i \sin 50°)}{7(\sin 25° + i \cos 25°)}$

***20.** $\dfrac{(5 \text{ cis } 132°)(\sqrt{7} \text{ cis } 15°)(3 \text{ cis } 254°)}{9 \text{ cis } 72°}$

***21.** $\dfrac{(2 \text{ cis } 150°)\,[3 \text{ cis } (-20°)]\,(\sqrt{3} \text{ cis } 210°)}{6 \text{ cis } 51°}$

In Exercises 22–30, perform the indicated operations in polar form. Express your answers in both polar and rectangular form.

22. $(4 + 2i)(3 - i)$

23. $(-3 + 2i)(-i - 1)$

24. $(2 - 5i)(i)$

25. $(2 + 6i)^2$ ***Hint:*** $z^2 = zz$

26. $\dfrac{3 - 6i}{9 + 7i}$

27. $\dfrac{i - 4}{2i + 1}$

28. $\dfrac{7 + 6i}{9 - 5i}$

29. $\dfrac{9i - 7}{i}$

30. $\dfrac{i}{2 - i}$

***31.** Show that the reciprocal of $r(\cos\theta + i \sin\theta)$ is

$$\frac{1}{r}(\cos\theta - i \sin\theta)$$

****32.** Under what conditions are the conjugate and the reciprocal of a complex number equal?

***33.** Find **(a)** $[r(\cos\theta + i \sin\theta)]^2$ and **(b)** $[r(\cos\theta + i \sin\theta)]^3$. Can you generalize parts (a) and (b) to derive a formula for $[r(\cos\theta + i \sin\theta)]^n$, where n is a positive integer?

34. Show that $z\bar{z} = |z|^2$. Prove this in two ways. Use the polar form of z and the rectangular form of z.

***35.** If $z = r \text{ cis } \theta$ obeys the condition that $z^2 + \bar{z}^2 = c^2$, where c is a constant, determine the equation that r and θ obey.

36.** Is $\bar{z}^2 = \overline{z^2}$? ***Hint: Use z in its polar form $z = r \text{ cis } \theta$. Compare this proof with that of Exercise 44 in Section 9.3-2. Notice how much easier it is in polar form!

37.** Show that multiplication of any complex number z by i is equivalent to rotating z by $+90°$. ***Hint: Let $z = r \text{ cis } \theta$ and graphically interpret iz.

***38.** Continue with the idea introduced in Exercise 37 to interpret multiplication of z by $i^2, i^3, i^4, \ldots$.

39. Use the ideas of Exercises 37 and 38 to find the argument of z without actually carrying out any multiplication:

(a) $z = i(5 \text{ cis } 34°)$ **(b)** $z = -(18 \text{ cis } 112°)$
(c) $z = i^3(10 \text{ cis } 221°)$ **(d)** $z = i^4(9 \text{ cis } 305°)$

***40. Euler Identities.** In around 1748, Leonhard Euler introduced the identity $e^{i\theta} = \cos\theta + i \sin\theta$ as an important tool in mathematics although it was actually known to some mathematicians before this time. Use the Euler identity to show that:

(a) $e^{-i\theta} = \cos\theta - i \sin\theta$ (This is the second Euler identity.)

(b) $e^{\pi i} = -1$, $e^{\pi i/2} = i$, $e^{-\pi i} = -1$, and $e^0 = 1$

(c) $\sin\theta = \dfrac{e^{i\theta} - e^{-i\theta}}{2i}$

(d) $\cos\theta = \dfrac{e^{i\theta} + e^{-i\theta}}{2}$

41. The most important numbers in mathematics are probably 1, 0, π, e, and i. Use the Euler identity (see Exercise 40) to show that these five numbers are related by the equation $e^{\pi i} + 1 = 0$.

42. Show that we can express any complex number in terms of the Euler identity (see Exercise 40) as $z = re^{i\theta}$. Also show that the conjugate of a complex number can be expressed as $\bar{z} = re^{-i\theta}$.

***43.** Use the Euler identity (see Exercises 40 and 42) to derive the laws for multiplication and division of complex numbers in polar form.

$$z_1z_2 = r_1r_2 \text{ cis } (\theta_1 + \theta_2) \quad \text{and} \quad \frac{z_1}{z_2} = \frac{r_1}{r_2} \text{cis}(\theta_1 - \theta_2)$$

****44.** Use the Euler identity (see Exercises 40 and 42) to show that if n is any integer (positive, negative, or zero), then $z^n = r^n(\cos n\theta + i \sin n\theta)$. ***Hint:*** $z^n = (r\,e^{i\theta})^n$. Apply this formula for $n = 0, 1, 2, 3$, and 4, and then check your answer by multiplying z^n in the ordinary polar way.

45. Use the results of Exercise 44 to find $(2 - i)^7$ in polar form.

46. Use the Euler identity (see Exercises 40 and 42) to show that $z\bar{z} = r^2$.

****47.** By using the Euler formulas (see Exercise 40), it is possible to derive the formulas for both $\sin(A + B)$ and $\cos(A + B)$ in one proof. Can you figure out how? ***Hint:*** Remember that when multiplying we add exponents, so $e^{iA}e^{iB} = e^{i(A+B)}$. Now use the Euler formulas on *both* sides of this equation. Also recall that if two complex numbers are equal, their real parts must be equal to each other and their imaginary parts must be equal to each other.

****48.** Apply the method of Exercise 47 to derive the formulas for both $\sin(A - B)$ and $\cos(A - B)$ in one proof.

****49.** Apply the method of Exercise 47 to derive the double-angle formulas for both $\sin 2\theta$ and $\cos 2\theta$ in one proof.

9.5 POWERS AND ROOTS OF COMPLEX NUMBERS

9.5-1 Powers of Complex Numbers

In Section 9.4 we showed that if $z_1(r_1, \theta_1)$ and $z_2(r_2, \theta_2)$ are two complex numbers, then their product is

$$z_1z_2 = r_1r_2[\cos(\theta_1 + \theta_2) + i \sin(\theta_1 + \theta_2)] = r_1r_2 \operatorname{cis}(\theta_1 + \theta_2)$$

We can use this result to find integer powers of z by letting $z_1 = z_2 = z$, where $z = r \operatorname{cis} \theta$.

$$z^2 = zz = (r \operatorname{cis} \theta)(r \operatorname{cis} \theta) = r^2 \operatorname{cis}(\theta + \theta) = r^2 \operatorname{cis} 2\theta$$
$$z^3 = zz^2 = (r \operatorname{cis} \theta)(r^2 \operatorname{cis} 2\theta) = r^3 \operatorname{cis}(\theta + 2\theta) = r^3 \operatorname{cis} 3\theta$$
$$z^4 = zz^3 = (r \operatorname{cis} \theta)(r^3 \operatorname{cis} 3\theta) = r^4 \operatorname{cis}(\theta + 3\theta) = r^4 \operatorname{cis} 4\theta$$

Continuing this process for positive integers n suggests the more general statement

$$z^n = r^n \operatorname{cis} n\theta = r^n (\cos n\theta + i \sin n\theta)$$

If we define $z^0 = 1$ and $z^{-n} = 1/z^n$, we can extend this result to include negative integers and $n = 0$ (provided that $z \neq 0$). To do this, let $z = r(\cos \theta + i \sin \theta)$. Then

$$z^{-n} = \frac{1}{z^n} = \frac{1}{r^n(\cos n\theta + i \sin n\theta)}$$

but we can write the numerator as $1 = \cos 0 + i \sin 0$. Writing this expression for 1 and using polar division of complex numbers, this equation becomes

$$z^{-n} = \frac{\cos 0 + i \sin 0}{r^n(\cos n\theta + i \sin n\theta)} = r^{-n}[\cos (0 - n\theta) + i \sin(0 - n\theta)]$$
$$z^{-n} = r^{-n}[\cos(-n\theta) + i \sin(-n\theta)] = r^{-n} \operatorname{cis}(-n\theta)$$

DE MOIVRE'S THEOREM

If $z = r(\cos \theta + i \sin \theta)$ and n is any positive or negative rational number or zero, then

$$z^n = r^n (\cos n\theta + i \sin n\theta)$$

Thus we see that this result holds for *all integers*, positive, negative, and zero. In fact, it even holds if n is *any rational number.** We call this important result *De Moivre's Theorem.*†

* Remember that a rational number is any number that can be expressed as the ratio of two integers.

† This theorem was first proved, but only for positive integers, in around 1707 by Abraham De Moivre, a French mathematician who lived in England to escape religious persecution. Although mainly remembered for this theorem, De Moivre's chief interest was the theory of probability.

Example 25

Find $(-2 + 2i)^5$ and express it in both polar and rectangular form.

Solution

We first put the number in polar form to be able to use De Moivre's theorem; $z = -2 + 2i$, so $x = -2$ and $y = 2$.

$$r = \sqrt{x^2 + y^2} = \sqrt{(-2)^2 + (2)^2} = 2\sqrt{2}$$

Because x is negative and y is positive, θ must be in quadrant II.

$$\theta = \tan^{-1}\left(\frac{y}{x}\right) = \tan^{-1}(-1) = 135° \quad \textbf{(because } \boldsymbol{\theta} \textbf{ is in quadrant II)}$$

Thus the polar form of $z = -2 + 2i$ is $z = 2\sqrt{2}(\cos 135° + i \sin 135°)$. Using De Moivre's theorem with $r = 2\sqrt{2}$, $\theta = 135°$, and $n = 5$ gives us

$$\begin{aligned} z^n &= r^n(\cos n\theta + i \sin n\theta) \\ (-2 + 2i)^5 &= [2\sqrt{2}(\cos 135° + i \sin 135°)]^5 \\ &= (2\sqrt{2})^5[\cos(5 \times 135°) + i \sin(5 \times 135°)] \\ &= 128\sqrt{2}(\cos 675° + i \sin 675°) \end{aligned}$$

Reducing the angle to less than 360° gives us

$$\begin{aligned} (-2 + 2i)^5 &= 128\sqrt{2}(\cos 315° + i \sin 315°) \quad \textbf{(in polar form)} \\ &= 128\sqrt{2}[\sqrt{2}/2 + i(-\sqrt{2}/2)] \\ &= 128 - 128i \quad \textbf{(in rectangular form)} \end{aligned}$$ ■

NOW IT'S YOUR TURN

Find $(1 + i)^3$ and express the answer in both polar and rectangular form. (Answer: $2\sqrt{2}$ cis 135° = $-2 + 2i$)

Example 26

Express $[4(\cos 30° + i \sin 30°)]^{-2}$ in both polar and rectangular form.

Solution

De Moivre's theorem, with $r = 4$, $\theta = 30°$, and $n = -2$ gives us

$$\begin{aligned} z^n &= r^n(\cos n\theta + i \sin n\theta) \\ [4(\cos 30° + i \sin 30°)]^{-2} &= (4)^{-2}[\cos(-2 \times 30°) + i \sin(-2 \times 30°)] \\ &= (\tfrac{1}{16})[\cos(-60°) + i \sin(-60°)] \quad \textbf{(in polar form)} \\ &= \tfrac{1}{16}(\cos 60° - i \sin 60°) \quad \textbf{(use } \cos(-\theta) = \cos\theta, \ \sin(-\theta) = -\sin\theta\textbf{)} \\ &= \tfrac{1}{16}\left(\tfrac{1}{2} - \frac{i\sqrt{3}}{2}\right) = \tfrac{1}{32} - \frac{i\sqrt{3}}{32} \quad \textbf{(in rectangular form)} \end{aligned}$$ ■

NOW IT'S YOUR TURN

Find $(3.6 \text{ cis } 22°)^{-3}$ in both polar and rectangular form.
(Answer: $0.021 \text{ cis}(-66°) = 0.0085 - 0.019i$)

9.5-2 Roots of Complex Numbers

In algebra you learned how to express the root of a number as an exponent. For example, the square root of x is $\sqrt{x} = x^{1/2}$, the cube root is $\sqrt[3]{x} = x^{1/3}$, the fourth root is $\sqrt[4]{x} = x^{1/4}$, . . . , and the nth root is $\sqrt[n]{x} = x^{1/n}$. For a complex number z, the nth root is simply $z^{1/n}$.

Earlier in this section, we stated (but did not prove) that De Moivre's theorem holds for all rational powers of z. Therefore, it holds for rational roots like $\frac{1}{2}, \frac{1}{3}, \frac{1}{4}, \ldots, 1/n$. We must, however, be careful, as the following example shows.

Let us find the square root of -1 using De Moivre's theorem. First write -1 in polar form: $-1 = 1 \text{ cis } 180°$. Now apply De Moivre's theorem where the exponent is $\frac{1}{n} = \frac{1}{2}$ for the square root; $z = -1$, $r = 1$, and $\theta = 180°$.

$$\begin{aligned} z^{1/n} &= r^{1/n}(\cos \tfrac{\theta}{n} + i \sin \tfrac{\theta}{n}) \\ \sqrt{-1} &= (-1)^{1/2} = 1^{1/2}[\cos(\tfrac{1}{2} \times 180°) + i \sin(\tfrac{1}{2} \times 180°)] \\ &= 1(\cos 90° + i \sin 90°) = 0 + i = i \end{aligned}$$

This answer should not surprise us because we *defined* i to be $\sqrt{-1}$. What should surprise us is that *we missed a root*! We know that $-i$ should also be an answer because

$$(-i)^2 = [(-1)(i)]^2 = (-1)^2(i)^2 = i^2 = -1$$

Why did we miss this second root? We missed it because we overlooked the fact that a complex number has many representations in polar form. We wrote -1 as $-1 = \cos 180° + i \sin 180°$, but we could also write -1 as

$$\begin{aligned} -1 &= \cos(180° + 360°) + i \sin(180° + 360°) \\ -1 &= \cos(180° + 2 \times 360°) + i \sin(180° + 2 \times 360°) \\ -1 &= \cos(180° + 3 \times 360°) + i \sin(180° + 3 \times 180°) \\ &\vdots \\ -1 &= \cos(180° + k \times 360°) + i \sin(180° + k \times 360°) \end{aligned}$$

where $k = 0, \pm 1, \pm 2, \pm 3, \pm 4, \ldots$.

We just found $\sqrt{-1}$ using only the $k = 0$ representation for -1. Let us redo this square root using the $k = 1$ representation.

$$-1 = \cos(180° + 360°) + i \sin(180° + 360°)$$

$$\sqrt{-1} = (1)^{1/2}\left[\cos\left(\frac{180° + 360°}{2}\right) + i \sin\left(\frac{180° + 360°}{2}\right)\right]$$

$$= \cos(90° + 180°) + i \sin(90° + 180°)$$

$$\sqrt{-1} = \cos 270° + i \sin 270° = 0 - i = -i$$

This result gives us the missing root, $\sqrt{-1} = -i$.

You might think that we could go on to generate an infinity of roots, but we cannot. Let's try $k = 2$.

$$-1 = \cos(180° + 2 \times 360°)$$

Repeating the preceding steps gives us

$$\sqrt{-1} = (1)^{1/2}\left[\cos\left(\frac{180° + 2 \times 360°}{2}\right) + i \sin\left(\frac{180° + 2 \times 360°}{2}\right)\right]$$

$$= \underbrace{\cos(90° + 360°)}_{\cos 90°} + i \underbrace{\sin(90° + 360°)}_{\sin 90°}$$

$$\sqrt{-1} = \cos 90° + i \sin 90° = 0 + i = i$$

Here we get the first root again. If we keep this up, we keep getting the same roots over and over again.

NOW IT'S YOUR TURN

Try $k = 3$ and show that it repeats the root $-i$.

How do we know when to stop? Notice that in this example, $n = 2$, so we stopped when $k = 1$, which is $n - 1$. Once k reaches or exceeds 2, the roots begin to repeat. In general, for the nth root, we must use all values of k from 0 up to and including $n - 1$ to get all the roots.

We can now generalize this result when n is any integer. If $z = r(\cos\theta + i\sin\theta)$, then the nth roots of z, $z^{1/n}$ are given by

$$z^{1/n} = r^{1/n}\left\{\cos\left[\frac{\theta + k(360°)}{n}\right] + i \sin\left[\frac{\theta + k(360°)}{n}\right]\right\}$$

where $k = 0, 1, 2, 3, \ldots, n - 1$. Each value of k is a different root, so we often put a subscript k on z to indicate which value of k gave that root. For example, z_3 means that $k = 3$ gives that root. Notice that k starts at zero and ends up at $n - 1$, so there are n values of k, which means that there are n different roots. For example, $z^{1/8}$ has eight roots, $z^{1/3}$ has three roots, and so on.

ROOTS OF A COMPLEX NUMBER

$z = r(\cos \theta + i \sin \theta)$

The nth root of z is $z^{1/n}$. These n roots are given by

$$z_k^{1/n} = r^{1/n}\left\{\cos\left[\frac{\theta + k(360°)}{n}\right] + i \sin\left[\frac{\theta + k(360°)}{n}\right]\right\}$$

where $k = 0, 1, 2, \ldots, n - 1$.

If we used θ in radians, we would replace 360° by 2π in this formula for $z_k^{1/n}$.

Example 27

Find the cube roots of 27 and plot them in the complex plane.

Solution

Because $z = 27 = 27 + 0i$, we have $r = 27$ and $\theta = 0°$. The cube root is $z^{1/3}$, so $n = 3$, which means there are three roots. Substituting these values in the formula for $z_k^{1/n}$ gives us

$$z_k^{1/n} = r^{1/n}\left\{\cos\left[\frac{\theta + k(360°)}{n}\right] + i \sin\left[\frac{\theta + k(360°)}{n}\right]\right\}$$

$$27_k^{1/3} = 27^{1/3}\left\{\cos\left[\frac{0° + k(360°)}{3}\right] + i \sin\left[\frac{0° + k(360°)}{3}\right]\right\}$$

where $k = 0, 1, 2$.

For $k = 0$:

$$27_0^{1/3} = 3(\cos 0° + i \sin 0°) = 3(1 + 0) = 3$$

For $k = 1$:

$$27_1^{1/3} = 3\left\{\cos\left[\frac{0° + 1(360°)}{3}\right] + i \sin\left[\frac{0° + 1(360°)}{3}\right]\right\}$$

$$= 3(\cos 120° + i \sin 120°) = 3(-\tfrac{1}{2} + \sqrt{3}/2\, i)$$

For $k = 2$:

$$27_2^{1/3} = 3\left\{\cos\left[\frac{0° + 2(360°)}{3}\right] + i \sin\left[\frac{0° + 2(360°)}{3}\right]\right\}$$

$$= 3[\cos(\tfrac{720}{3})° + i \sin(\tfrac{720}{3})°]$$

$$= 3(\cos 240° + i \sin 240°) = 3(-\tfrac{1}{2} - \sqrt{3}/2\, i)$$

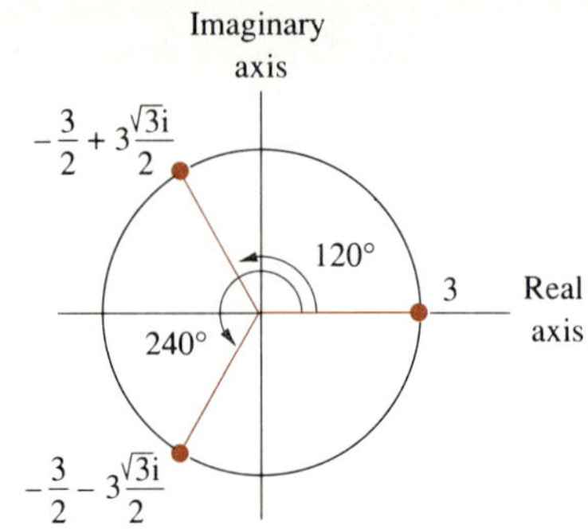

Figure 9.5-1

The three cube roots of 27

Graphing these three roots in the complex plane (Figure 9.5-1), we can see that they are equally spaced around a circle (because they all have the same absolute value) of radius 3 centered at the origin. ■

> **The n roots of a nonzero complex number $z = r(\cos\theta + i\sin\theta)$ are equally spaced around a circle centered at the origin having a radius $r^{1/n}$.**

NOW IT'S YOUR TURN

Find the cube roots of −8 and plot them in the complex plane. (Answer: 2 cis 60° = $1 + i\sqrt{3}$, 2 cis 180° = −2, 2 cis 300° = $1 - \sqrt{3}$. Notice that these roots are equally spaced around a circle of radius 2. See Figure 9.5-2.)

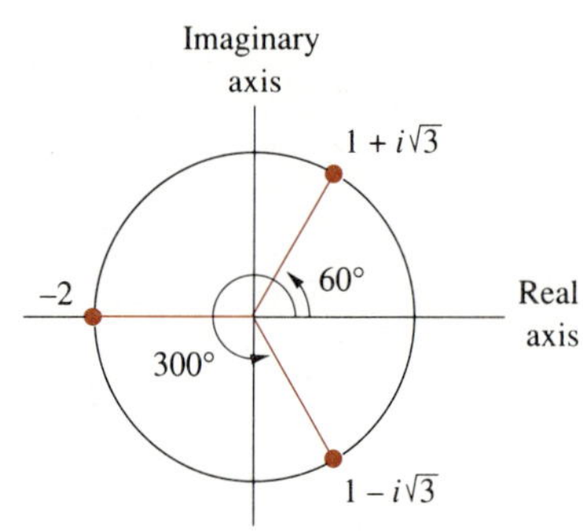

Figure 9.5-2

The three cube roots of −8

The result we saw in Example 27, about the roots being equally spaced about a circle, is a general one.

Example 28

Find the roots $(2 - 2\sqrt{3}i)^{1/4}$ and plot them in the complex plane.

Solution

First we note that the number $2 - 2\sqrt{3}\,i$ is in quadrant IV because x is positive and y is negative. Putting the number in polar form gives us

$$r = \sqrt{(2)^2 + (-2\sqrt{3})^2} = 4 \qquad \text{and} \qquad \theta = \tan^{-1}\left(\frac{-2\sqrt{3}}{2}\right) = \tan^{-1}(-\sqrt{3}) = \frac{5\pi}{3}$$

Therefore,

$$2 - 2\sqrt{3}i = 4(\cos \tfrac{5\pi}{3} + i \sin \tfrac{5\pi}{3})$$

Substituting $r = 4$, $\theta = \frac{5\pi}{3}$, and $n = 4$ into the formula for $z_k^{1/n}$ gives us

$$z_k^{1/n} = r^{1/n}\left\{\cos\left[\frac{\theta + k(360°)}{n}\right] + i \sin\left[\frac{\theta + k(360°)}{n}\right]\right\}$$

$$(2 - 2\sqrt{3}i)^{1/4} = (4)^{1/4}\left[\cos\left(\frac{\frac{5\pi}{3} + 2\pi k}{4}\right) + i \sin\left(\frac{\frac{5\pi}{3} + 2\pi k}{4}\right)\right]$$

$$= \sqrt{2}\,[\cos(\tfrac{5\pi}{12} + k\tfrac{\pi}{2}) + i \sin(\tfrac{5\pi}{12} + k\tfrac{\pi}{2})]$$

where $k = 0, 1, 2, 3$.

For $k = 0$:

$$z_0^{1/4} = \sqrt{2}(\cos \tfrac{5\pi}{12} + i \sin \tfrac{5\pi}{12})$$
$$= 0.3660 + 1.366\, i$$

For $k = 1$:

$$z_1^{1/4} = \sqrt{2}[\cos(\tfrac{5\pi}{12} + \tfrac{\pi}{2}) + i \sin(\tfrac{5\pi}{12} + \tfrac{\pi}{2})]$$
$$= -1.3660 + 0.366i$$

For $k = 2$:

$$z_2^{1/4} = \sqrt{2}[\cos(\tfrac{5\pi}{12} + \pi) + i \sin(\tfrac{5\pi}{12} + \pi)]$$
$$= -0.3660 - 1.366i$$

For $k = 3$:

$$z_3^{1/4} = \sqrt{2}[\cos(\tfrac{5\pi}{12} + \tfrac{3\pi}{2}) + i \sin(\tfrac{5\pi}{12} + \tfrac{3\pi}{2})]$$
$$= 1.366 - 0.3660i$$

These four points are shown in Figure 9.5-3. As you can see, they are equally spaced around a circle of radius $\sqrt{2}$ centered at the origin. ■

NOW IT'S YOUR TURN

Find $\sqrt{1 + i}$ and sketch these roots in the complex plane.

(Answer: $2^{1/4} \text{ cis } \frac{\pi}{8} = 1.1 + 0.46i$ and $2^{1/4} \text{ cis } \frac{9\pi}{8} = -1.1 - 0.46i$. Also see Figure 9.5-4.)

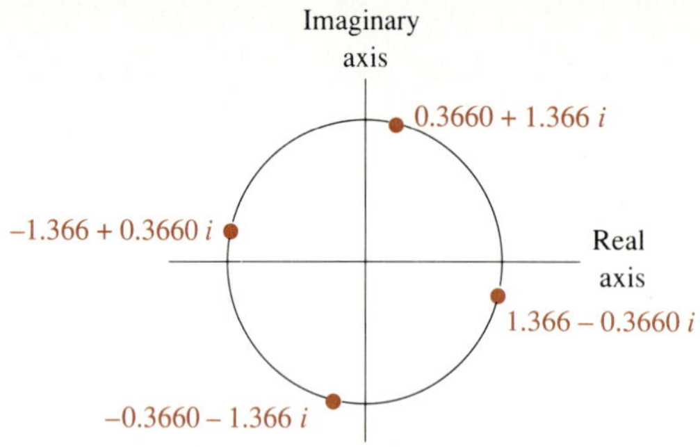

Figure 9.5-3

The four roots of $(2 - 2\sqrt{3}\,i)^{1/4}$

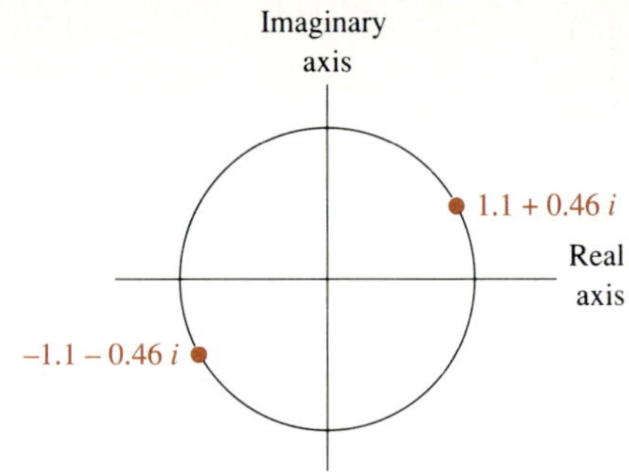

Figure 9.5-4

The two values of $\sqrt{1+i}$

EXERCISES 9.5-1

In Exercises 1–25, use De Moivre's theorem to find the indicated power and express the results in both polar and rectangular form.

1. $[2(\cos 10° + i \sin 10°)]^2$

2. $[3(\cos 15° + i \sin 15°)]^3$

3. $[4(\cos 30° + i \sin 30°)]^6$

4. $[2(\cos 40° + i \sin 40°)]^{10}$

5. $[1.2\,(\cos 50° + i \sin 50°)]^4$

6. $[3(\cos \frac{\pi}{3} + i \sin \frac{\pi}{3})]^4$

7. $[\frac{1}{3}(\cos \frac{2\pi}{3} + i \sin \frac{2\pi}{3})]^3$

8. $[\frac{1}{2}(\cos \frac{5\pi}{4} + i \sin \frac{5\pi}{4})]^4$

9. $[\frac{2}{3}(\cos \frac{7\pi}{4} + i \sin \frac{7\pi}{4})]^4$

10. $[3.7(\cos \frac{11\pi}{6} + i \sin \frac{11\pi}{6})]^5$

11. $(1 - i)^3$

12. $(1 + i)^4$

13. $(\sqrt{3} - i)^3$

14. $(-2\sqrt{3} - 2i)^2$

15. $(\sqrt{3}/2 + i/2)^9$

16. $(\sqrt{2}/2 + i\sqrt{2}/2)^4$

17. $(\sqrt{3}/2 - i/2)^5$

18. $(\frac{1}{2} - i\sqrt{3}/2)^4$

19. $(\frac{1}{2} + i\sqrt{3}/2)^{12}$

20. $(\sqrt{2}/2 - i\sqrt{2}/2)^{100}$

21. $(1 - i)^{-4}$

22. $(-\sqrt{3} + i)^{-3}$

23. $(2 + 3i)^{-3}$

24. $60/(1 + 2i)^5$

25. $32/(1 + i)^6$

EXERCISES 9.5-2

In Exercises 1–22, find the indicated roots. Express each answer in both polar and rectangular form and sketch each one clearly in the complex plane.

1. $\sqrt{16(\cos 120° + i \sin 120°)}$

2. The fourth roots of $36(\cos 240° + i \sin 240°)$.

3. $(\cos 354° + i \sin 354°)^{1/6}$

4. The ninth roots of $\cos 225° + i \sin 225°$

5. $(64 \text{ cis } \frac{6\pi}{7})^{1/3}$

6. $1^{1/3}$

7. $\sqrt[6]{1}$

8. $\sqrt[3]{i}$

9. $\sqrt{i}$

10. The cube roots of 216

11. $\sqrt{-i}$

12. $(-i)^{1/5}$

13. The cube roots of -1

14. $\sqrt{1 - i}$

15. $\sqrt{i - 1}$

16. The square roots of $(1 - \sqrt{3}i)$

17. $\left(\frac{1}{2} - \frac{\sqrt{3}}{2}i\right)^{1/4}$

18. $\sqrt[5]{-4(1 + i)}$

19. The sixth roots of $[\frac{1}{2}(-1 + \sqrt{3}i)]$

20. $\sqrt{\sqrt{1 + i}}$

****21.** $\sqrt{4 - \sqrt{2i - 3}}$

****22.** $\sqrt{\sqrt{3 - i} + 2i}$

In Exercises 23–31, find all the roots of each equation, expressing each root in rectangular form.

23. $z^6 + 1 = 0$

24. $z^6 + i = 0$

25. $z^4 = -i$

26. $z^4 - 256 = 0$

27. $z^6 + 64 = 0$

28. $z^3 = 27i$

29. $z^4 + 2 + 2i\sqrt{3} = 0$

30. $z^4 - 81 = 0$

31. $z^3 + 1 = 0$

***32.** Show that the product of the roots of $z^3 = 1 + i$ equals $1 + i$.

***33.** Show that the product of the roots of $z^4 = -1$ equals 1.

****34.** Show that the product of the roots of $z^2 = r \text{ cis } \theta$ equals $-z$.

In Exercises 35–39, find all possible values of the indicated power of the complex number. When you use a calculator, express answers to four decimal places.

35.** $[16(\cos 160° + i \sin 160°)]^{3/4}$ ***Hint: $z^{3/4} = (z^{1/4})^3$

***36.** $(-1 + i\sqrt{3})^{2/3}$

***37.** $(2 - 2i)^{3/5}$

***38.** $\left(\dfrac{1 - 4i}{6 + i}\right)^{2/3}$

Hint: Multiply both the numerator and the denominator of the fraction inside the parentheses by the conjugate of the denominator.

***39.** $(3 - 4i)^{-3/4}$

***40.** The three cube roots of 1 are often denoted by 1, $\omega = -\frac{1}{2} + i\sqrt{3}/2$, and $\omega^2 = -\frac{1}{2} - i\sqrt{3}/2$. Show that, for any x and y,

$$(x + y)(x + \omega y)(x + \omega^2 y) = x^3 + y^3$$

***41.** (See Exercise 40.) Show that, for any x, y, and z,

$$(x + y + z)(x + \omega y + \omega^2 z)(x + \omega^2 y + \omega z) = x^3 + y^3 + z^3 - 3xyz$$

****42.** Use the Euler identity (see Section 9.4-3, Exercise 40) to derive De Moivre's theorem $z^n = r^n \text{ cis } n\theta$.

SUMMARY

1. Polar Coordinates. The polar coordinates of a point $P(r, \theta)$ are r(distance from the pole) and θ(angle with the polar axis).

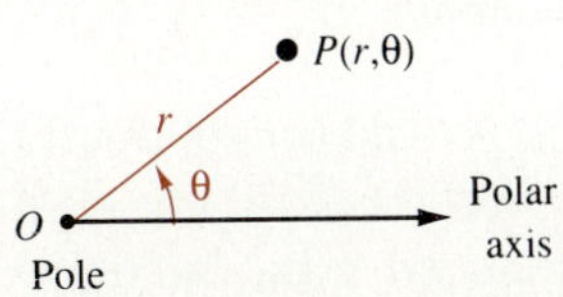

2. Transformation of Coordinates

Polar to rectangular: $x = r \cos \theta$, $y = r \sin \theta$.

Rectangular to polar: $r = \sqrt{x^2 + y^2}$ $\quad \tan(\text{ref } \theta) = \left|\dfrac{y}{x}\right|$.

Use the signs of x and y to find which quadrant θ is in.

3. Complex Numbers

$$i = \sqrt{-1} \quad \rightarrow \quad i^2 = -1$$

$$\sqrt{-a} = i\sqrt{a}$$

where a is a positive real number or zero.
A complex number has a real part and an imaginary part.

In rectangular form: $z = x + yi$; x is the real part of z and y is the imaginary part of z.

In polar form: $z = r(\cos \theta + i \sin \theta) = r \text{ cis } \theta$; $r = |z|$ is the modulus (or magnitude or absolute value) of z and θ is the argument of z.

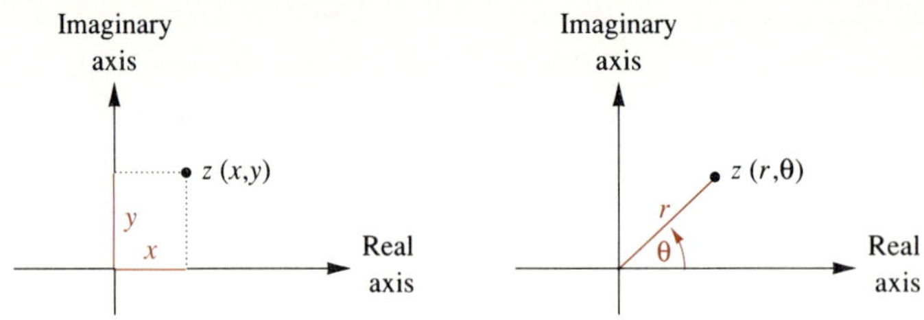

(a) Rectangular form ***(b) Polar form***

Relating the polar and rectangular forms:

$$x = r \cos \theta \qquad y = r \sin \theta$$

$$r = \sqrt{x^2 + y^2} \qquad \tan \theta = \frac{y}{x}$$

The signs of x and y determine the quadrant θ is in.
The complex conjugate of z is $\bar{z}$: If $z = x + yi$, then $\bar{z} = x - yi$.

4. Algebra of Complex Numbers

Adding and Subtracting: Add (or subtract) the real parts and the imaginary parts separately.

Multiplying: In *rectangular* form, multiply like binomial expressions and use $i^2 = -1$ to simplify to the form $z = x + yi$. In *polar* form, multiply the magnitudes and add the arguments: $z_1z_2 = r_1r_2[\cos(\theta_1 + \theta_2) + i \sin(\theta_1 + \theta_2)]$

Dividing: In *rectangular* form, multiply the numerator and the denominator by the conjugate of the denominator

$$\frac{z_1}{z_2} = \frac{z_1 \times \bar{z}_2}{z_1 \times \bar{z}_2}$$

and then simplify to the form $z = x + yi$.
In *polar* form, divide the magnitudes and subtract the arguments: $z_1/z_2 = (r_1/r_2)[\cos(\theta_1 - \theta_2) + i \sin(\theta_1 - \theta_2)]$.

5. Powers of Complex Numbers

$$z^n = r^n(\cos n\theta + i \sin n\theta) = r^n \text{ cis } n\theta$$

where n is any positive or negative rational number or zero (De Moivre's theorem).

6. Roots of Complex Numbers. If $z = r(\cos \theta + i \sin \theta)$, the nth roots of z are given by the formula

$$z_k^{1/n} = r^{1/n}\left\{\cos\left[\frac{\theta + k(360°)}{n}\right] + i\sin\left[\frac{\theta + k(360°)}{n}\right]\right\}$$

where $k = 0, 1, 2, \ldots, n - 1$, and n is an integer. If θ is in radians, replace the 360° by 2π in this formula.

The n roots of a nonzero complex number $z = r \text{ cis } \theta$ are equally spaced around a circle centered at the origin having a radius of $r^{1/n}$.

IMPORTANT TERMS

POLAR COORDINATES

cartesian coordinates
cylindrical coordinates
ordered pair
ordered triple
polar coordinates
polar axis
pole
rectangular coordinates
spiral
transformation of coordinates

COMPLEX NUMBERS

argument of a complex number
complex conjugate
complex number
complex plane
De Moivre's theorem
imaginary number
imaginary part of a complex number
magnitude of a complex number
modulus of a complex number
polar form of a complex number
real number
real part of a complex number
rectangular form of a complex number
roots of a complex number

REVIEW EXERCISES

1. The following figure shows several points (A, B, C, . . .) in a cartesian (rectangular) coordinate system. Find the polar coordinates of each point, using $r \geq 0$ and $0 \leq \theta < 2\pi$.

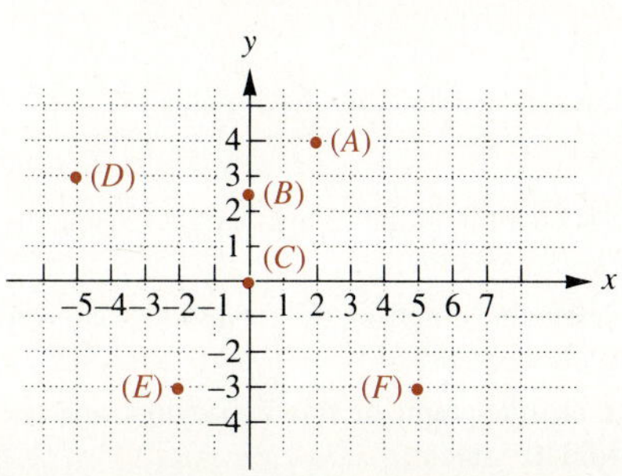

In Exercises 2–9, each point is given in rectangular coordinates. Express it in polar coordinates using $r \geq 0$ and $0 \leq \theta < 360°$.

2. $(-1, -\sqrt{3})$ **3.** $(-2.0, 5.0)$ **4.** $(\frac{2}{3}, -\frac{1}{3})$
5. $(-16, 0)$ **6.** $(0, -6.9)$ **7.** $(3.2, 8.1)$
8. $(0, 0)$ **9.** $(\pi, -\pi)$

10. The following figure shows several points (A, B, C, . . .) in a polar coordinate system. Find the rectangular coordinates of each point.

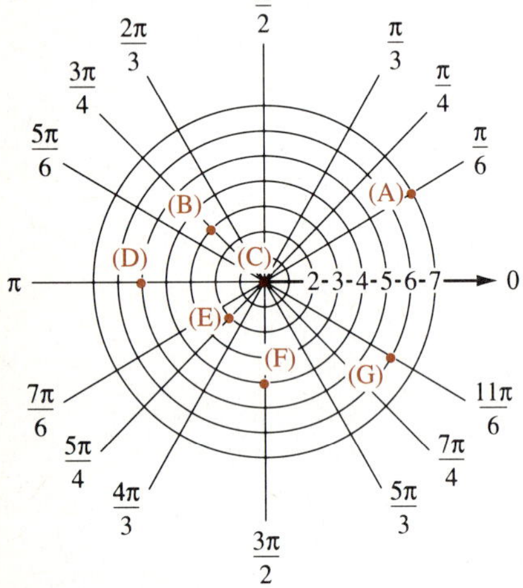

In Exercises 11–17, each point is given in polar coordinates. Find the rectangular coordinates of each point.

11. $(8, \frac{\pi}{4})$ **12.** $(2, \frac{5\pi}{6})$

13. $(-6, 45°)$ **14.** $(10, -\frac{2\pi}{3})$

15. $(-3.4, -135°)$ **16.** $(\pi, -\pi)$

17. **(a)** $(4, 0)$ **(b)** $(0, 4)$

18. An isosceles triangle is in quadrant I, as shown in the following figure. With the x axis as the polar axis and the origin the pole, find both the polar and rectangular coordinates of the triangle's vertices.

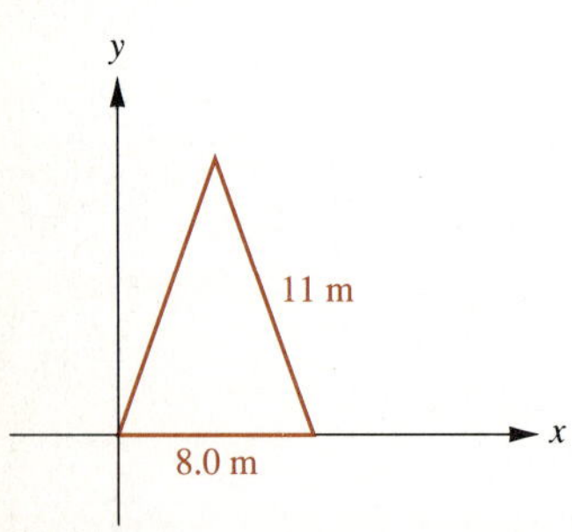

19. A parallelogram is in quadrant II, as shown in the following figure. Find both the polar and rectangular coordinates of its four corners, $r \geq 0$, $0 \leq \theta < 360°$.

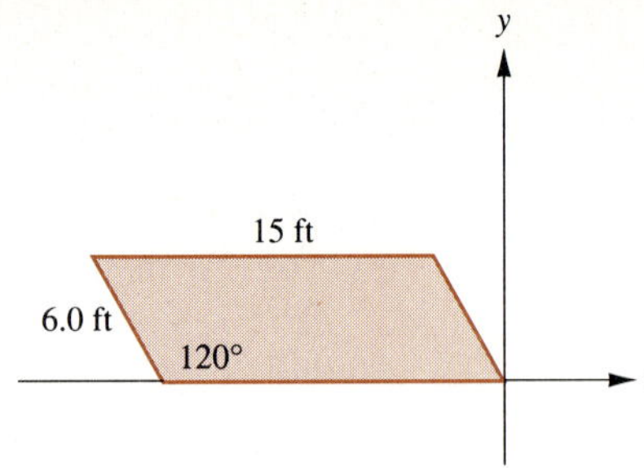

20. A circle of diameter 20.0 cm is in quadrant IV, as shown in the following figure. Find both the polar and rectangular coordinates of points A, B, C (the center), D, and E, with $r \geq 0$ and $0 \leq \theta < 2\pi$.

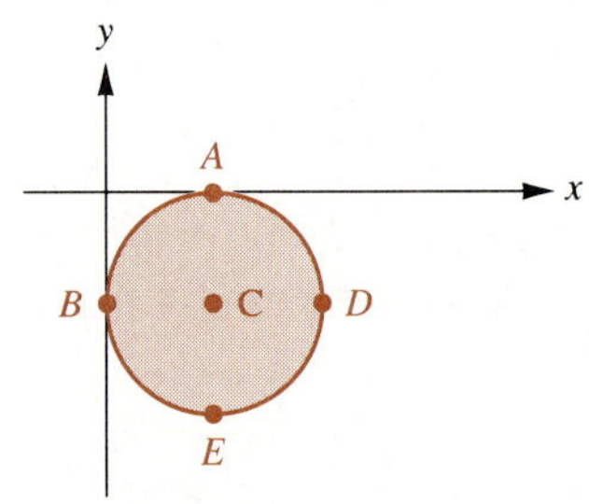

21. A methanol molecule (CH_3OH) is placed in a polar coordinate system with the pole at the carbon atom and the polar axis running to the right from carbon through the central hydrogen (see the following figure). Find both the polar and rectangular coordinates of the C, O, and H′ atom.

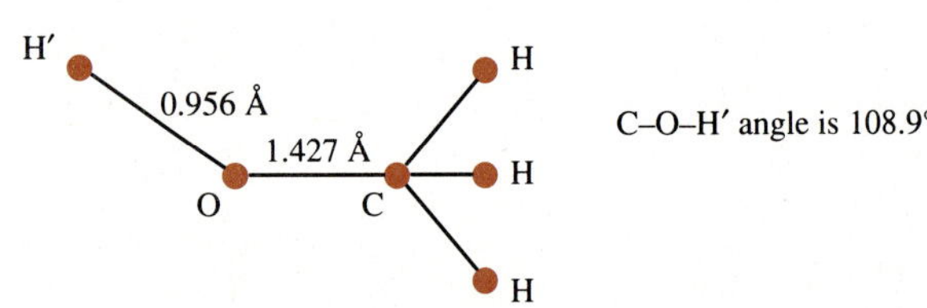

In Exercises 22–27, convert each equation from polar to rectangular form.

22. $r \cos \theta = 3$. Also sketch the graph of this equation using the more convenient coordinate system.

23. $r^2 = 49$. Also sketch the graph of this equation using the more convenient coordinate system.

24. $|r| = 4$. Also sketch the graph of this equation using the more convenient coordinate system.

25. $\theta = -\frac{\pi}{6}$. Also sketch the graph of this equation using the more convenient coordinate system.

26. $\sin\theta = \frac{1}{2}$ (careful). Also sketch the graph of this equation using the more convenient coordinate system.

27. $r^2 = 2(1 + r^2\cos^2\theta)$

In Exercises 28–31, convert each equation from rectangular to polar form.

28. $x^2 + y^2 = 81$. Also sketch the graph of this equation using the more convenient coordinate system.

29. $x = y$. Also sketch the graph of this equation using the more convenient coordinate system.

30. $y = -2$. Also sketch the graph of this equation using the more convenient coordinate system.

31. $y^2 = 4x$

In Exercises 32–46, graph each equation on polar graph paper.

32. **(a)** $r = 4$, **(b)** $r = -4$

33. $r\cos\theta = \frac{1}{2}$

34. $r\sin\theta = -4$

35. $r = 2\sin\theta$

36. $r = -\frac{3}{2}\cos\theta$

37. $r = 1 + \frac{6\theta}{\pi}$ $(\theta > -\frac{\pi}{6})$

38. $r = 2^\theta\,(-\infty < \theta < \infty)$

39. $\theta = r^2$

40. **(a)** $r = 2\sin 3\theta$ **(b)** $r = 3\cos 3\theta$

41. **(a)** $r = 3\cos 2\theta$ **(b)** $r = 4\sin 2\theta$

42. **(a)** $r^2 = 4\cos 2\theta$ **(b)** $r^2 = 9\sin 2\theta$

43. $r^2 = -2\cos\theta$

44. **(a)** $r = 3 + 2\cos\theta$, **(b)** $r = 3 - 2\cos\theta$

45. **(a)** $r = 2 + 2\sin\theta$, **(b)** $r = 2 - 2\sin\theta$

46. **(a)** $\theta = \frac{\pi}{9}$, **(b)** $\theta = -50°$

47. Each point is in rectangular (cartesian) coordinates. Find its cylindrical coordinates and plot the point $(r \geq 0,\ 0 \leq \theta < 2\pi)$.

(a) $(0, 1, 4)$ **(b)** $(0, -2, -3)$
(c) $(1, -3, 5)$ **(d)** $(-6, 5, 2)$
(e) $(-8, -6, 1)$

48. Each point is given in cylindrical coordinates. Find its rectangular (cartesian) coordinates and plot the point.

(a) $(6, \frac{2\pi}{3}, -2)$ **(b)** $(1, \frac{11\pi}{6}, 4)$
(c) $(4, \frac{\pi}{4}, 2)$ **(d)** $(8, -\frac{2\pi}{3}, 3)$

***49.** A face-centered gold crystal is placed in the first octant, as shown in Figure 9.2-15. Find the cylindrical and rectangular coordinates of the atoms at its corners and in its faces. (For pertinent information, see Exercise 19 in Section 9.2-3).

In Exercises 50–54, write each complex number in polar form using $r \geq 0$ and $0 \leq \theta < 2\pi$. Also, find its modulus, argument, magnitude, and complex conjugate.

50. **(a)** -4, **(b)** $+3$

51. **(a)** $5i$ **(b)** $-5i$

52. $-2\sqrt{3} - 2i$

53. **(a)** $\sqrt{2} - \sqrt{2}i$ **(b)** $\sqrt{2} + \sqrt{2}i$

54. $-5.3 + 7i$

In Exercises 55–61, express each complex number in rectangular form. Also, find its argument, modulus, and magnitude.

55. $7(\cos 0° + i\sin 0°)$

56. $3(\cos\frac{\pi}{2} + i\sin\frac{\pi}{2})$

57. $6(\cos\frac{\pi}{3} + i\sin\frac{\pi}{3})$

58. $\cos 150° + i\sin 150°$

59. $3(\cos\frac{5\pi}{4} + i\sin\frac{5\pi}{4})$

60. $5.3\ \text{cis}\ 1.7\pi$

61. $84\ \text{cis}\ 248°$

62. The following figure shows several complex numbers plotted in the complex plane in rectangular form.

(a) Write each one in the rectangular form of $z = x + yi$.
(b) Write each one in polar form $(r \geq 0,\ 0 \leq \theta < 2\pi)$.
(c) Find the complex conjugate of each and write it in both polar and rectangular form.
(d) Find its argument, modulus, and magnitude.

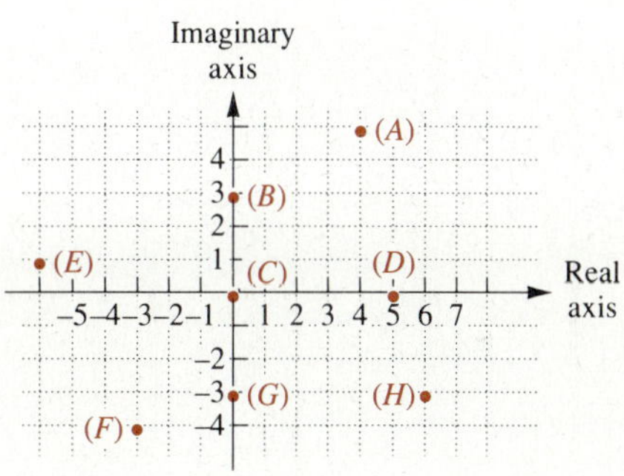

In Exercises 63–70, perform the indicated operations and simplify to the form $a + bi$.

63. **(a)** $\sqrt{-3}\sqrt{-2}$
(b) $\sqrt{(-3)(-2)}$ Are (a) and (b) the *same*?

64. $\sqrt{-12}\sqrt{3}$

65. $(3\sqrt{-2})^2$

66. $(-2\sqrt{-5})^2$

67. $(2 - \sqrt{-3})^2$

68. $(1 - \sqrt{-2})(2 + \sqrt{-3})$

69. $\sqrt{-6}\sqrt{-8}\sqrt{-12}$

70. $(\sqrt{-2}\sqrt{-2})^2$

71. Express each number in terms of 1, i, or a combination of the two.

(a) i^{17} **(b)** i^{56}
(c) $i^{12} - i^{40}$ **(d)** $i^{30} + i^{31} + i^{32}$

In Exercises 72–82, perform the indicated operations using only the rectangular form of the complex numbers and express your final answers in the form $a + bi$.

72. $(6 + 5i) + (3 - 2i)$

73. $(-6 + 3i)(5 + 8i)$

74. $(6i^3)(3i^2)$

75. $(3 - \sqrt{-25})(4 + \sqrt{-9})$

76. $(2 + 3i)(2 - 3i)$

77. $1/(2 - 3i)$

78. $3/(-5 + 2i)$

79. $\dfrac{3 - \sqrt{-25}}{\sqrt{-4}}$

80. $\dfrac{2 + \sqrt{-9}}{3 - \sqrt{-16}}$

81. $\dfrac{-5i + 1}{5i - 3}$

82. $\dfrac{5 - 3i}{1 + 2i} \cdot \dfrac{2 - 3i}{4 + i}$

83. Show that if

$$x = \frac{-1 - \sqrt{3}i}{2} \quad \text{and} \quad y = \frac{-1 + \sqrt{3}i}{2}$$

then:

(a) $y^2 = x$ **(b)** $x^2 = y$

(c) $x^3 = 1$ **(d)** $y^3 = 1$

In Exercises 84–102, perform the indicated operations in polar form and leave your answers in polar form.

84. $[6(\cos 20° + i \sin 20°)]\,[3(\cos 25° + i \sin 25°)]$

85. $[\text{cis}(-18°)]\,(3 \text{ cis } 123°)$

86. $[3(\cos \frac{\pi}{12} + i \sin \frac{\pi}{12})]\,(5 \text{ cis } \frac{5\pi}{12})$

87. $(12 \text{ cis } 232°) \div (3 \text{ cis } 82°)$

88. $(8 \text{ cis } 73°) \div [2(\cos 118° + i \sin 118°)]$

89. $[3(\cos \frac{\pi}{3} + i \sin \frac{\pi}{3})]^6$

90. $(-\cos \frac{\pi}{4} + i \sin \frac{\pi}{4})^8$

91. $(-2 - 2i)^7$

92. $4(-1 + i)^{-3}$

93. $[2(\cos 14° + i \sin 14°)]^{-5}$

94. $\dfrac{60}{(1 + 2i)^5}$

95. $\sqrt{4 + 4i}$

96. $\sqrt[3]{-1 - \sqrt{3}i}$

97. $[16(\cos 240° + i \sin 240°)]^{1/4}$

98. $\sqrt[6]{\cos \frac{4\pi}{3} + i \sin \frac{4\pi}{3}}$

***99.** $(8 \text{ cis } \frac{\pi}{6})^{2/3}$

100. $i^{1/5}$

101. $\sqrt[4]{-1}$

***102.** $\sqrt{2 + i\sqrt{2 + i}}$

APPENDIX A

Use of Trigonometry Tables

A.1 HOW TO FIND TRIGONOMETRIC FUNCTIONS

A typical trigonometry table gives the value of all six trigonometric functions for angles between 0° and 90°.* The angles can be expressed in degree-minute form, decimal degrees, or radians. To use the table, first find the angle and then read the value of its trigonometric function from the appropriate column. For angles greater than 90°, the table gives the reference angle; it is then necessary to use the reference angle theorem (see Chapter 3) to find the functions of the angle.

A.1-1 Angles in Radians

In Table 1 of Appendix B, the left-hand column lists angles in radians, and the other columns give the trigonometric functions of each angle. To find a trigonometric function of an angle, use the following steps:

Step 1: Go down the left-hand column until you find the angle whose function you want. You may have to round off the angle to the same number of decimal places the table uses.

Step 2: Move across the table to the right to the column for the trigonometric function you want and read its value for that angle.

* The present form of these tables goes back to the Germans, who began computing very exact tables soon after the invention of the printing press in about 1450.

Example 1

Use Table 1 to find (a) cos 0.163 and sec 0.163 and (b) tan $\frac{\pi}{15}$.

Solution

(a) cos 0.163 and sec 0.163: In Table 1, angles are expressed to two decimal places, so we must round the angle 0.163 to 0.16.

Figure A.1
A portion of Table 1 in Appendix B.

Table 1 Trigonometric Functions—Radians or Real Numbers

	sin	cos	tan	cot	sec	csc
0.15	0.1494	0.9888	0.1511	6.617	1.011	6.692
0.16	0.1593	0.9872	0.1614	6.197	1.013	6.277
0.17	0.1692	0.9856	0.1717	5.826	1.015	5.911
0.18	0.1790	0.9838	0.1820	5.495	1.016	5.586
0.19	0.1889	0.9820	0.1923	5.200	1.018	5.295
0.20	0.1987	0.9801	0.2027	4.933	1.020	5.033
0.21	0.2085	0.9780	0.2131	4.692	1.022	4.797
0.22	0.2182	0.9759	0.2236	4.472	1.025	4.582
0.23	0.2280	0.9737	0.2341	4.271	1.027	4.386
0.24	0.2377	0.9713	0.2447	4.086	1.030	4.207

Step 1: Locate 0.16 in the left-hand column of Table 1 in Appendix B (see Figure A.1).

Step 2: Go across to the "cos" and "sec" columns and read the answers: cos 0.16 = 0.9872 and sec 0.16 = 1.013.

(b) tan $\frac{\pi}{15}$: To use Table 1 in Appendix B, the angle must be in decimal form to two decimal accuracy.

Step 1: Convert the angle to a decimal: $\frac{\pi}{15} = (3.14)/15 = 0.21$.

Step 2: Locate 0.21 in the left-hand column of the table (see Figure A.1).

Step 3: Go across to the "tan" column and read the answer: tan $\frac{\pi}{15}$ = tan 0.21 = 0.2131.

NOW IT'S YOUR TURN

Use Table 1 in Appendix B to find sec 0.192, cot 0.23, and csc $\frac{\pi}{13}$.

(Answer: 1.018, 4.271, 4.207)

A.1-2 Angles in Degrees

Tables in degrees usually have angles ≤45° on the left-hand side and angles ≥45° on the right-hand side. The function headings on the top of the table are for angles ≤45°, whereas those on the bottom are for angles ≥45°. The basic procedure is the same as for angles in radians.

Example 2

Use Table 2 in Appendix B to find (a) cos 43.7° and (b) sec 45°24′.

Solution

(a) cos 43.7°.

Step 1: Because 43.7° < 45°, we use angles on the left-hand side of Table 2 (see Figure A.2). Go down the decimal-degree column and locate 43.7°.

Step 2: Using the headings at the top of the table, go over to the "cos θ" column and read cos 43.7° = 0.7230.

Table 2 **Trigonometric Functions, Degrees**

θ deg	deg min	sin θ	csc θ	tan θ	cot θ	sec θ	cos θ		
43.6	43 36	0.6896	1.4501	0.9523	1.0501	1.3809	0.7242	46 24	46.4
43.7	43 42	0.6909	1.4474	0.9556	1.0464	1.3832	0.7230	46 18	46.3
43.8	43 48	0.6921	1.4448	0.9590	1.0428	1.3855	0.7218	46 12	46.2
43.9	43 54	0.6934	1.4422	0.9623	1.0392	1.3878	0.7206	46 6	46.1
44.0	44 0	0.6947	1.4396	0.9657	1.0355	1.3902	0.7193	46 0	46.0
44.1	44 6	0.6959	1.4370	0.9691	1.0319	1.3925	0.7181	45 54	45.9
44.2	44 12	0.6972	1.4344	0.9725	1.0283	1.3949	0.7169	45 48	45.8
44.3	44 18	0.6984	1.4318	0.9759	1.0247	1.3972	0.7157	45 42	45.7
44.4	44 24	0.6997	1.4293	0.9793	1.0212	1.3996	0.7145	45 36	45.6
44.5	44 30	0.7009	1.4267	0.9827	1.0176	1.4020	0.7133	45 30	45.5
44.6	44 36	0.7022	1.4242	0.9861	1.0141	1.4044	0.7120	45 24	45.4
44.7	44 42	0.7034	1.4217	0.9896	1.0105	1.4069	0.7108	45 18	45.3
44.8	44 48	0.7046	1.4192	0.9930	1.0070	1.4093	0.7096	45 12	45.2
44.9	44 54	0.7059	1.4167	0.9965	1.0035	1.4118	0.7083	45 6	45.1
45.0	45 0	0.7071	1.4142	1.0000	1.0000	1.4142	0.7071	45 0	45.0
		cos θ	sec θ	cot θ	tan θ	csc θ	sin θ	deg min	θ deg

Figure A.2 **A portion of Table 2 in Appendix B**

(b) sec 45° 24′.

Step 1: Use the angles on the the right-hand side of Table 2 because 45°24′ > 45°. Locate 45°24′ in the "deg min" column (see Figure A.2).

Step 2: Using the headings at the bottom of the table, go over to the "sec θ" column and read sec 45°24′ = 1.4242.

NOW IT'S YOUR TURN

Use Table 2 to find (a) sin 17°18′, (b) cot 37.2°, (c) csc 72.1°, (d) sec 64°48′.
(Answer: (a) 0.2974, (b) 1.3175, (c) 1.0509, (d) 2.3486)

A.2 FINDING ANGLES—INVERSE TRIGONOMETRIC FUNCTIONS

To find inverse trigonometric functions (angles), we work in the opposite way we do to find the functions of an angle. First find the value of the trigonometric function in the appropriate column ("sin," "cos," and so on) and then read the inverse (the angle) from the angle column.

Example 3

Find arctan 0.1923 in radians.

Solution

We must use Table 1 in Appendix B because we want the angle in radians.

Figure A.3
A portion of Table 1 in Appendix B

Table 1 Trigonometric Functions, Radians

	sin	cos	tan	cot	sec	csc
0.17	0.1692	0.9856	0.1717	5.826	1.015	5.911
0.18	0.1790	0.9838	0.1820	5.495	1.016	5.586
(0.19)	0.1889	0.9820	(0.1923)	5.200	1.018	5.295
0.20	0.1987	0.9801	0.2027	4.933	1.020	5.033
0.21	0.2085	0.9780	0.2131	4.692	1.022	4.797
0.22	0.2182	0.9759	0.2236	4.472	1.025	4.582
0.23	0.2280	0.9737	0.2341	4.271	1.027	4.386
0.24	0.2377	0.9713	0.2447	4.086	1.030	4.207

Step 1: In Table 1, locate 0.1923 in the "tan" column (see Figure A.3).

Step 2: Read the angle in the left-hand column: arctan 0.1923 = 0.19 radian.

Example 4

Find (a) $\csc^{-1} 1.4192$ in degree-minute form and (b) arcsin 0.7133 in decimal degrees.

Solution

We must use Table 2 because we want the angles in degrees.

(a) $\csc^{-1} 1.4192$.

Table 2 Trigonometric Functions, Degrees

θ deg	deg min	sin θ	csc θ	tan θ	cot θ	sec θ	cos θ		
44.2	44 12	0.6972	1.4344	0.9725	1.0283	1.3949	0.7169	45 48	45.8
44.3	44 18	0.6984	1.4318	0.9759	1.0247	1.3972	0.7157	45 42	45.7
44.4	44 24	0.6997	1.4293	0.9793	1.0212	1.3996	0.7145	45 36	45.6
44.5	44 30	0.7009	1.4267	0.9827	1.0176	1.4020	0.7133	45 30	45.5
44.6	44 36	0.7022	1.4242	0.9861	1.0141	1.4044	0.7120	45 24	45.4
44.7	44 42	0.7034	1.4217	0.9896	1.0105	1.4069	0.7108	45 18	45.3
44.8	44 48	0.7046	1.4192	0.9930	1.0070	1.4093	0.7096	45 12	45.2
44.9	44 54	0.7059	1.4167	0.9965	1.0035	1.4118	0.7083	45 6	45.1
45.0	45 0	0.7071	1.4142	1.0000	1.0000	1.4142	0.7071	45 0	45.0
		cos θ	sec θ	cot θ	tan θ	csc θ	sin θ	deg min	θ deg

Figure A.4 **A portion of Table 2 in Appendix B**

Step 1: Start at the top of the "csc θ" column in Table 2 (see Figure A.4) and go down this column until you find 1.4192. Because we found this number under the top headings, the angle must be less than 45° so we use angles on the left-hand side of the table.

Step 2: Reading the angle in the "deg min" column on the left side of the table, we find that $\csc^{-1} 1.4192 = 44°48'$.

(b) arcsin 0.7133.

Step 1: Start at the top of the "sin θ" column (see Figure A.4). As we go down this column, the numbers increase but do not reach 0.7133, which means that the angle must be greater than 45°. Therefore, we must continue up the "sin" column, but for angles greater than 45°. To do so, we must switch columns: Go to the "sin θ" column on the bottom and continue up this column until we come to 0.7133.

Step 2: Read the angle on the right-hand side of the table in the "deg" column: arcsin 0.7133 = 45.5°

NOW IT'S YOUR TURN

Use Table 2 to find (a) $\cos^{-1} 0.0209$ in decimal degrees, (b) arccot 3.0595 in degree-minute form, (c) arcsec 1.0201 in decimal degrees, and (d) $\csc^{-1} 1.2299$ in degree-minute form. (Answer: (a) 88.8°, (b) 18°6', (c) 11.4°, and (d) 54°24'.)

APPENDIX B

Trigonometry Tables

Table 1
Trigonometric Functions—Radians or Real Numbers*

↓→	sin	cos	tan	cot	sec	csc
0.00	0.0000	1.0000	0.0000	—	1.000	—
0.01	0.0100	1.0000	0.0100	99.997	1.000	100.00
0.02	0.0200	0.9998	0.0200	49.993	1.000	50.00
0.03	0.0300	0.9996	0.0300	33.323	1.000	33.34
0.04	0.0400	0.9992	0.0400	24.987	1.001	25.01
0.05	0.0500	0.9988	0.0500	19.983	1.001	20.01
0.06	0.0600	0.9982	0.0601	16.647	1.002	16.68
0.07	0.0699	0.9976	0.0701	14.262	1.002	14.30
0.08	0.0799	0.9968	0.0802	12.473	1.003	12.51
0.09	0.0899	0.9960	0.0902	11.081	1.004	11.13
0.10	0.0998	0.9950	0.1003	9.967	1.005	10.02
0.11	0.1098	0.9940	0.1104	9.054	1.006	9.109
0.12	0.1197	0.9928	0.1206	8.293	1.007	8.353
0.13	0.1296	0.9916	0.1307	7.649	1.009	7.714
0.14	0.1395	0.9902	0.1409	7.096	1.010	7.166
0.15	0.1494	0.9888	0.1511	6.617	1.011	6.692
0.16	0.1593	0.9872	0.1614	6.197	1.013	6.277
0.17	0.1692	0.9856	0.1717	5.826	1.015	5.911
0.18	0.1790	0.9838	0.1820	5.495	1.016	5.586
0.19	0.1889	0.9820	0.1923	5.200	1.018	5.295
	sin	cos	tan	cot	sec	csc

Table 1
(continued)

	sin	cos	tan	cot	sec	csc
0.20	0.1987	0.9801	0.2027	4.933	1.020	5.033
0.21	0.2085	0.9780	0.2131	4.692	1.022	4.797
0.22	0.2182	0.9759	0.2236	4.472	1.025	4.582
0.23	0.2280	0.9737	0.2341	4.271	1.027	4.386
0.24	0.2377	0.9713	0.2447	4.086	1.030	4.207
0.25	0.2474	0.9689	0.2553	3.916	1.032	4.042
0.26	0.2571	0.9664	0.2660	3.759	1.035	3.890
0.27	0.2667	0.9638	0.2768	3.613	1.038	3.749
0.28	0.2764	0.9611	0.2876	3.478	1.041	3.619
0.29	0.2860	0.9582	0.2984	3.351	1.044	3.497
0.30	0.2955	0.9553	0.3093	3.233	1.047	3.384
0.31	0.3051	0.9523	0.3203	3.122	1.050	3.278
0.32	0.3146	0.9492	0.3314	3.018	1.053	3.179
0.33	0.3240	0.9460	0.3425	2.920	1.057	3.086
0.34	0.3335	0.9428	0.3537	2.827	1.061	2.999
0.35	0.3429	0.9394	0.3650	2.740	1.065	2.916
0.36	0.3523	0.9359	0.3764	2.657	1.068	2.839
0.37	0.3616	0.9323	0.3879	2.578	1.073	2.765
0.38	0.3709	0.9287	0.3994	2.504	1.077	2.696
0.39	0.3802	0.9249	0.4111	2.433	1.081	2.630
0.40	0.3894	0.9211	0.4228	2.365	1.086	2.568
0.41	0.3986	0.9171	0.4346	2.301	1.090	2.509
0.42	0.4078	0.9131	0.4466	2.239	1.095	2.452
0.43	0.4169	0.9090	0.4586	2.180	1.100	2.399
0.44	0.4259	0.9048	0.4708	2.124	1.105	2.348
0.45	0.4350	0.9004	0.4831	2.070	1.111	2.299
0.46	0.4439	0.8961	0.4954	2.018	1.116	2.253
0.47	0.4529	0.8916	0.5080	1.969	1.122	2.208
0.48	0.4618	0.8870	0.5206	1.921	1.127	2.166
0.49	0.4706	0.8823	0.5334	1.875	1.133	2.125
0.50	0.4794	0.8776	0.5463	1.830	1.139	2.086
0.51	0.4882	0.8727	0.5594	1.788	1.146	2.048
0.52	0.4969	0.8678	0.5726	1.747	1.152	2.013
0.53	0.5055	0.8628	0.5859	1.707	1.159	1.978
0.54	0.5141	0.8577	0.5994	1.668	1.166	1.945
0.55	0.5227	0.8525	0.6131	1.631	1.173	1.913
0.56	0.5312	0.8473	0.6269	1.595	1.180	1.883
0.57	0.5396	0.8419	0.6410	1.560	1.188	1.853
0.58	0.5480	0.8365	0.6552	1.526	1.196	1.825
0.59	0.5564	0.8309	0.6696	1.494	1.203	1.797
	sin	cos	tan	cot	sec	csc

Table 1
(continued)

↓→	sin	cos	tan	cot	sec	csc
0.60	0.5646	0.8253	0.6841	1.462	1.212	1.771
0.61	0.5729	0.8196	0.6989	1.431	1.220	1.746
0.62	0.5810	0.8139	0.7139	1.401	1.229	1.721
0.63	0.5891	0.8080	0.7291	1.372	1.238	1.697
0.64	0.5972	0.8021	0.7445	1.343	1.247	1.674
0.65	0.6052	0.7961	0.7602	1.315	1.256	1.652
0.66	0.6131	0.7900	0.7761	1.288	1.266	1.631
0.67	0.6210	0.7838	0.7923	1.262	1.276	1.610
0.68	0.6288	0.7776	0.8087	1.237	1.286	1.590
0.69	0.6365	0.7712	0.8253	1.212	1.297	1.571
0.70	0.6442	0.7648	0.8423	1.187	1.307	1.552
0.71	0.6518	0.7584	0.8595	1.163	1.319	1.534
0.72	0.6594	0.7518	0.8771	1.140	1.330	1.517
0.73	0.6669	0.7452	0.8949	1.117	1.342	1.500
0.74	0.6743	0.7385	0.9131	1.095	1.354	1.483
0.75	0.6816	0.7317	0.9316	1.073	1.367	1.467
0.76	0.6889	0.7248	0.9505	1.052	1.380	1.452
0.77	0.6961	0.7179	0.9697	1.031	1.393	1.437
0.78	0.7033	0.7109	0.9893	1.011	1.407	1.422
0.79	0.7104	0.7038	1.009	0.9908	1.421	1.408
0.80	0.7174	0.6967	1.030	0.9712	1.435	1.394
0.81	0.7243	0.6895	1.050	0.9520	1.450	1.381
0.82	0.7311	0.6822	1.072	0.9331	1.466	1.368
0.83	0.7379	0.6749	1.093	0.9146	1.482	1.355
0.84	0.7446	0.6675	1.116	0.8964	1.498	1.343
0.85	0.7513	0.6600	1.138	0.8785	1.515	1.331
0.86	0.7578	0.6524	1.162	0.8609	1.533	1.320
0.87	0.7643	0.6448	1.185	0.8437	1.551	1.308
0.88	0.7707	0.6372	1.210	0.8267	1.569	1.297
0.89	0.7771	0.6294	1.235	0.8100	1.589	1.287
0.90	0.7833	0.6216	1.260	0.7936	1.609	1.277
0.91	0.7895	0.6137	1.286	0.7774	1.629	1.267
0.92	0.7956	0.6058	1.313	0.7615	1.651	1.257
0.93	0.8016	0.5978	1.341	0.7458	1.673	1.247
0.94	0.8076	0.5898	1.369	0.7303	1.696	1.238
0.95	0.8134	0.5817	1.398	0.7151	1.719	1.229
0.96	0.8192	0.5735	1.428	0.7001	1.744	1.221
0.97	0.8249	0.5653	1.459	0.6853	1.769	1.212
0.98	0.8305	0.5570	1.491	0.6707	1.795	1.204
0.99	0.8360	0.5487	1.524	0.6563	1.823	1.196
	sin	cos	tan	cot	sec	csc

Table 1
(continued)

↓→	sin	cos	tan	cot	sec	csc
1.00	0.8415	0.5403	1.557	0.6421	1.851	1.188
1.01	0.8468	0.5319	1.592	0.6281	1.880	1.181
1.02	0.8521	0.5234	1.628	0.6142	1.911	1.174
1.03	0.8573	0.5148	1.665	0.6005	1.942	1.166
1.04	0.8624	0.5062	1.704	0.5870	1.975	1.160
1.05	0.8674	0.4976	1.743	0.5736	2.010	1.153
1.06	0.8724	0.4889	1.784	0.5604	2.046	1.146
1.07	0.8772	0.4801	1.827	0.5473	2.083	1.140
1.08	0.8820	0.4713	1.871	0.5344	2.122	1.134
1.09	0.8866	0.4625	1.917	0.5216	2.162	1.128
1.10	0.8912	0.4536	1.965	0.5090	2.205	1.122
1.11	0.8957	0.4447	2.014	0.4964	2.249	1.116
1.12	0.9001	0.4357	2.066	0.4840	2.295	1.111
1.13	0.9044	0.4267	2.120	0.4718	2.344	1.106
1.14	0.9086	0.4176	2.176	0.4596	2.395	1.101
1.15	0.9128	0.4085	2.234	0.4475	2.448	1.096
1.16	0.9168	0.3993	2.296	0.4356	2.504	1.091
1.17	0.9208	0.3902	2.360	0.4237	2.563	1.086
1.18	0.9246	0.3809	2.427	0.4120	2.625	1.082
1.19	0.9284	0.3717	2.498	0.4003	2.691	1.077
1.20	0.9320	0.3624	2.572	0.3888	2.760	1.073
1.21	0.9356	0.3530	2.650	0.3773	2.833	1.069
1.22	0.9391	0.3436	2.733	0.3659	2.910	1.065
1.23	0.9425	0.3342	2.820	0.3546	2.992	1.061
1.24	0.9458	0.3248	2.912	0.3434	3.079	1.057
1.25	0.9490	0.3153	3.010	0.3323	3.171	1.054
1.26	0.9521	0.3058	3.113	0.3212	3.270	1.050
1.27	0.9551	0.2963	3.224	0.3102	3.375	1.047
1.28	0.9580	0.2867	3.341	0.2993	3.488	1.044
1.29	0.9608	0.2771	3.467	0.2884	3.609	1.041
1.30	0.9636	0.2675	3.602	0.2776	3.738	1.038
1.31	0.9662	0.2579	3.747	0.2669	3.878	1.035
1.32	0.9687	0.2482	3.903	0.2562	4.029	1.032
1.33	0.9711	0.2385	4.072	0.2456	4.193	1.030
1.34	0.9735	0.2288	4.256	0.2350	4.372	1.027
1.35	0.9757	0.2190	4.455	0.2245	4.566	1.025
1.36	0.9779	0.2092	4.673	0.2140	4.779	1.023
1.37	0.9799	0.1994	4.913	0.2035	5.014	1.021
1.38	0.9819	0.1896	5.177	0.1931	5.273	1.018
1.39	0.9837	0.1798	5.471	0.1828	5.561	1.017
	sin	cos	tan	cot	sec	csc

Table 1
(continued)

↓→	sin	cos	tan	cot	sec	csc
1.40	0.9854	0.1700	5.798	0.1725	5.883	1.015
1.41	0.9871	0.1601	6.165	0.1622	6.246	1.013
1.42	0.9887	0.1502	6.581	0.1519	6.657	1.011
1.43	0.9901	0.1403	7.055	0.1417	7.126	1.010
1.44	0.9915	0.1304	7.602	0.1315	7.667	1.009
1.45	0.9927	0.1205	8.238	0.1214	8.299	1.007
1.46	0.9939	0.1106	8.989	0.1113	9.044	1.006
1.47	0.9949	0.1006	9.887	0.1011	9.938	1.005
1.48	0.9959	0.0907	10.983	0.0910	11.029	1.004
1.49	0.9967	0.0807	12.350	0.0810	12.390	1.003
1.50	0.9975	0.0707	14.101	0.0709	14.137	1.003
1.51	0.9982	0.0608	16.428	0.0609	16.458	1.002
1.52	0.9987	0.0508	19.670	0.0508	19.695	1.001
1.53	0.9992	0.0408	24.498	0.0408	24.519	1.001
1.54	0.9995	0.0308	32.461	0.0308	32.476	1.000
1.55	0.9998	0.0208	48.078	0.0208	48.089	1.000
1.56	0.9999	0.0108	92.620	0.0108	92.626	1.000
1.57	1.0000	0.0008	1,255.8	0.0008	1,255.8	1.000
1.58	1.0000	−0.0092	−108.65	−0.0092	−108.65	1.000
1.59	0.9998	−0.0192	−52.067	−0.0192	−52.08	1.000
1.60	0.9996	−0.0292	−34.233	−0.0292	−34.25	1.000
	sin	cos	tan	cot	sec	csc

Table 2 Trigonometric Functions of Decimal Degrees*

θ deg	deg min	sin *θ*	csc *θ*	tan *θ*	cot *θ*	sec *θ*	cos *θ*		
0.0	0 0	0.0000	No value	0.0000	No value	1.0000	1.0000	90 0	90.0
0.1	0 6	0.0017	572.96	0.0017	572.96	1.0000	1.0000	89 54	89.9
0.2	0 12	0.0035	286.48	0.0035	286.48	1.0000	1.0000	89 48	89.8
0.3	0 18	0.0052	190.99	0.0052	190.98	1.0000	1.0000	89 42	89.7
0.4	0 24	0.0070	143.24	0.0070	143.24	1.0000	1.0000	89 36	89.6
0.5	0 30	0.0087	114.59	0.0087	114.59	1.0000	1.0000	89 30	89.5
0.6	0 36	0.0105	95.495	0.0105	95.490	1.0001	0.9999	89 24	89.4
0.7	0 42	0.0122	81.853	0.0122	81.847	1.0001	0.9999	89 18	89.3
0.8	0 48	0.0140	71.622	0.0140	71.615	1.0001	0.9999	89 12	89.2
0.9	0 54	0.0157	63.665	0.0157	63.657	1.0001	0.9999	89 6	89.1
		cos *θ*	sec *θ*	cot *θ*	tan *θ*	csc *θ*	sin *θ*	deg min	*θ* deg

Table 2 *(continued)*

θ deg	deg min	sin θ	csc θ	tan θ	cot θ	sec θ	cos θ		
1.0	1 0	0.0175	57.299	0.0175	57.290	1.0002	0.9998	89 0	89.0
1.1	1 6	0.0192	52.090	0.0192	52.081	1.0002	0.9998	88 54	88.9
1.2	1 12	0.0209	47.750	0.0209	47.740	1.0002	0.9998	88 48	88.8
1.3	1 18	0.0227	44.077	0.0227	44.066	1.0003	0.9997	88 42	88.7
1.4	1 24	0.0244	40.930	0.0244	40.917	1.0003	0.9997	88 36	88.6
1.5	1 30	0.0262	38.202	0.0262	38.188	1.0003	0.9997	88 30	88.5
1.6	1 36	0.0279	35.815	0.0279	35.801	1.0004	0.9996	88 24	88.4
1.7	1 42	0.0297	33.708	0.0297	33.694	1.0004	0.9996	88 18	88.3
1.8	1 48	0.0314	31.836	0.0314	31.821	1.0005	0.9995	88 12	88.2
1.9	1 54	0.0332	30.161	0.0332	30.145	1.0005	0.9995	88 6	88.1
2.0	2 0	0.0349	28.654	0.0349	28.636	1.0006	0.9994	88 0	88.0
2.1	2 6	0.0366	27.290	0.0367	27.271	1.0007	0.9993	87 54	87.9
2.2	2 12	0.0384	26.050	0.0384	26.031	1.0007	0.9993	87 48	87.8
2.3	2 18	0.0401	24.918	0.0402	24.898	1.0008	0.9992	87 42	87.7
2.4	2 24	0.0419	23.880	0.0419	23.859	1.0009	0.9991	87 36	87.6
2.5	2 30	0.0436	22.926	0.0437	22.904	1.0010	0.9990	87 30	87.5
2.6	2 36	0.0454	22.044	0.0454	22.022	1.0010	0.9990	87 24	87.4
2.7	2 42	0.0471	21.229	0.0472	21.205	1.0011	0.9989	87 18	87.3
2.8	2 48	0.0488	20.471	0.0489	20.446	1.0012	0.9988	87 12	87.2
2.9	2 54	0.0506	19.766	0.0507	19.740	1.0013	0.9987	87 6	87.1
3.0	3 0	0.0523	19.107	0.0524	19.081	1.0014	0.9986	87 0	87.0
3.1	3 6	0.0541	18.492	0.0542	18.464	1.0015	0.9985	86 54	86.9
3.2	3 12	0.0558	17.914	0.0559	17.886	1.0016	0.9984	86 48	86.8
3.3	3 18	0.0576	17.372	0.0577	17.343	1.0017	0.9983	86 42	86.7
3.4	3 24	0.0593	16.862	0.0594	16.832	1.0018	0.9982	86 36	86.6
3.5	3 30	0.0610	16.380	0.0612	16.350	1.0019	0.9981	86 30	86.5
3.6	3 36	0.0628	15.926	0.0629	15.895	1.0020	0.9980	86 24	86.4
3.7	3 42	0.0645	15.496	0.0647	15.464	1.0021	0.9979	86 18	86.3
3.8	3 48	0.0663	15.089	0.0664	15.056	1.0022	0.9978	86 12	86.2
3.9	3 54	0.0680	14.703	0.0682	14.669	1.0023	0.9977	86 6	86.1
4.0	4 0	0.0698	14.336	0.0699	14.301	1.0024	0.9976	86 0	86.0
4.1	4 6	0.0715	13.987	0.0717	13.951	1.0026	0.9974	85 54	85.9
4.2	4 12	0.0732	13.654	0.0734	13.617	1.0027	0.9973	85 48	85.8
4.3	4 18	0.0750	13.337	0.0752	13.300	1.0028	0.9972	85 42	87.7
4.4	4 24	0.0767	13.035	0.0769	12.996	1.0030	0.9971	85 36	85.6
4.5	4 30	0.0785	12.746	0.0787	12.706	1.0031	0.9969	85 30	85.5
4.6	4 36	0.0802	12.469	0.0805	12.429	1.0032	0.9968	85 24	85.4
4.7	4 42	0.0819	12.204	0.0822	12.163	1.0034	0.9966	85 18	85.3
4.8	4 48	0.0837	11.951	0.0840	11.909	1.0035	0.9965	85 12	85.2
4.9	4 54	0.0854	11.707	0.0857	11.665	1.0037	0.9963	85 6	85.1
		cos θ	sec θ	cot θ	tan θ	csc θ	sin θ	deg min	θ deg

Table 2 *(continued)*

θ deg	deg min	sin θ	csc θ	tan θ	cot θ	sec θ	cos θ		
5.0	5 0	0.0872	11.474	0.0875	11.430	1.0038	0.9962	85 0	85.0
5.1	5 6	0.0889	11.249	0.0892	11.205	1.0040	0.9960	84 54	84.9
5.2	5 12	0.0906	11.034	0.0910	10.988	1.0041	0.9959	84 48	84.8
5.3	5 18	0.0924	10.826	0.0928	10.780	1.0043	0.9957	84 42	84.7
5.4	5 24	0.0941	10.626	0.0945	10.579	1.0045	0.9956	84 36	84.6
5.5	5 30	0.0958	10.433	0.0963	10.385	1.0046	0.9954	84 30	84.5
5.6	5 36	0.0976	10.248	0.0981	10.199	1.0048	0.9952	84 24	84.4
5.7	5 42	0.0993	10.069	0.0998	10.019	1.0050	0.9951	84 18	84.3
5.8	5 48	0.1011	9.8955	0.1016	9.8448	1.0051	0.9949	84 12	84.2
5.9	5 54	0.1028	9.7283	0.1033	9.6768	1.0053	0.9947	84 6	84.1
6.0	6 0	0.1045	9.5668	0.1051	9.5144	1.0055	0.9945	84 0	84.0
6.1	6 6	0.1063	9.4105	0.1069	9.3573	1.0057	0.9943	83 54	83.9
6.2	6 12	0.1080	9.2593	0.1086	9.2052	1.0059	0.9942	83 48	83.8
6.3	6 18	0.1097	9.1129	0.1104	9.0579	1.0061	0.9940	83 42	83.7
6.4	6 24	0.1115	8.9711	0.1122	8.9152	1.0063	0.9938	83 36	83.6
6.5	6 30	0.1132	8.8337	0.1139	8.7769	1.0065	0.9936	83 30	83.5
6.6	6 36	0.1149	8.7004	0.1157	8.6428	1.0067	0.9934	83 24	83.4
6.7	6 42	0.1167	8.5711	0.1175	8.5126	1.0069	0.9932	83 18	83.3
6.8	6 48	0.1184	8.4457	0.1192	8.3863	1.0071	0.9930	83 12	83.2
6.9	6 54	0.1201	8.3238	0.1210	8.2636	1.0073	0.9928	83 6	83.1
7.0	7 0	0.1219	8.2055	0.1228	8.1444	1.0075	0.9925	83 0	83.0
7.1	7 6	0.1236	8.0905	0.1246	8.0285	1.0077	0.9923	82 54	82.9
7.2	7 12	0.1253	7.9787	0.1263	7.9158	1.0079	0.9921	82 48	82.8
7.3	7 18	0.1271	7.8700	0.1281	7.8062	1.0082	0.9919	82 42	82.7
7.4	7 24	0.1288	7.7642	0.1299	7.6996	1.0084	0.9917	82 36	82.6
7.5	7 30	0.1305	7.6613	0.1317	7.5958	1.0086	0.9914	82 30	82.5
7.6	7 36	0.1323	7.5611	0.1334	7.4947	1.0089	0.9912	82 24	82.4
7.7	7 42	0.1340	7.4635	0.1352	7.3962	1.0091	0.9910	82 18	82.3
7.8	7 48	0.1357	7.3684	0.1370	7.3002	1.0093	0.9907	82 12	82.2
7.9	7 54	0.1374	7.2757	0.1388	7.2066	1.0096	0.9905	82 6	82.1
8.0	8 0	0.1392	7.1853	0.1405	7.1154	1.0098	0.9903	82 0	82.0
8.1	8 6	0.1409	7.0972	0.1423	7.0264	1.0101	0.9900	81 54	81.9
8.2	8 12	0.1426	7.0112	0.1441	6.9395	1.0103	0.9898	81 48	81.8
8.3	8 18	0.1444	6.9273	0.1459	6.8548	1.0106	0.9895	81 42	81.7
8.4	8 24	0.1461	6.8454	0.1477	6.7720	1.0108	0.9893	81 36	81.6
8.5	8 30	0.1478	6.7655	0.1495	6.6912	1.0111	0.9890	81 30	81.5
8.6	8 36	0.1495	6.6874	0.1512	6.6122	1.0114	0.9888	81 24	81.4
8.7	8 42	0.1513	6.6111	0.1530	6.5350	1.0116	0.9885	81 18	81.3
8.8	8 48	0.1530	6.5366	0.1548	6.4596	1.0119	0.9882	81 12	81.2
8.9	8 54	0.1547	6.4637	0.1566	6.3859	1.0122	0.9880	81 6	81.1
		cos θ	sec θ	cot θ	tan θ	csc θ	sin θ	deg min	θ deg

Table 2 *(continued)*

θ deg	deg min	sin θ	csc θ	tan θ	cot θ	sec θ	cos θ		
9.0	9 0	0.1564	6.3925	0.1584	6.3138	1.0125	0.9877	81 0	81.0
9.1	9 6	0.1582	6.3228	0.1602	6.2432	1.0127	0.9874	80 54	80.9
9.2	9 12	0.1599	6.2547	0.1620	6.1742	1.0130	0.9871	80 48	80.8
9.3	9 18	0.1616	6.1880	0.1638	6.1066	1.0133	0.9869	80 42	80.7
9.4	9 24	0.1633	6.1227	0.1655	6.0405	1.0136	0.9866	80 36	80.6
9.5	9 30	0.1650	6.0589	0.1673	5.9758	1.0139	0.9863	80 30	80.5
9.6	9 36	0.1668	5.9963	0.1691	5.9124	1.0142	0.9860	80 24	80.4
9.7	9 42	0.1685	5.9351	0.1709	5.8502	1.0145	0.9857	80 18	80.3
9.8	9 48	0.1702	5.8751	0.1727	5.7894	1.0148	0.9854	80 12	80.2
9.9	9 54	0.1719	5.8164	0.1745	5.7297	1.0151	0.9851	80 6	80.1
10.0	10 0	0.1736	5.7588	0.1763	5.6713	1.0154	0.9848	80 0	80.0
10.1	10 6	0.1754	5.7023	0.1781	5.6140	1.0157	0.9845	79 54	79.9
10.2	10 12	0.1771	5.6470	0.1799	5.5578	1.0161	0.9842	79 48	79.8
10.3	10 18	0.1788	5.5928	0.1817	5.5027	1.0164	0.9839	79 42	79.7
10.4	10 24	0.1805	5.5396	0.1835	5.4486	1.0167	0.9836	79 36	79.6
10.5	10 30	0.1822	5.4874	0.1853	5.3955	1.0170	0.9833	79 30	79.5
10.6	10 36	0.1840	5.4362	0.1871	5.3435	1.0174	0.9829	79 24	79.4
10.7	10 42	0.1857	5.3860	0.1890	5.2924	1.0177	0.9826	79 18	79.3
10.8	10 48	0.1874	5.3367	0.1908	5.2422	1.0180	0.9823	79 12	79.2
10.9	10 54	0.1891	5.2883	0.1926	5.1929	1.0184	0.9820	79 6	79.1
11.0	11 0	0.1908	5.2408	0.1944	5.1446	1.0187	0.9816	79 0	79.0
11.1	11 6	0.1925	5.1942	0.1962	5.0970	1.0191	0.9813	78 54	78.9
11.2	11 12	0.1942	5.1484	0.1980	5.0504	1.0194	0.9810	78 48	78.8
11.3	11 18	0.1959	5.1034	0.1998	5.0045	1.0198	0.9806	78 42	78.7
11.4	11 24	0.1977	5.0593	0.2016	4.9595	1.0201	0.9803	78 36	78.6
11.5	11 30	0.1994	5.0159	0.2035	4.9152	1.0205	0.9799	78 30	78.5
11.6	11 36	0.2011	4.9732	0.2053	4.8716	1.0209	0.9796	78 24	78.4
11.7	11 42	0.2028	4.9313	0.2071	4.8288	1.0212	0.9792	78 18	78.3
11.8	11 48	0.2045	4.8901	0.2089	4.7867	1.0216	0.9789	78 12	78.2
11.9	11 54	0.2062	4.8496	0.2107	4.7453	1.0220	0.9785	78 6	78.1
12.0	12 0	0.2079	4.8097	0.2126	4.7046	1.0223	0.9781	78 0	78.0
12.1	12 6	0.2096	4.7706	0.2144	4.6646	1.0227	0.9778	77 54	77.9
12.2	12 12	0.2113	4.7321	0.2162	4.6252	1.0231	0.9774	77 48	77.8
12.3	12 18	0.2130	4.6942	0.2180	4.5864	1.0235	0.9770	77 42	77.7
12.4	12 24	0.2147	4.6569	0.2199	4.5483	1.0239	0.9767	77 36	77.6
12.5	12 30	0.2164	4.6202	0.2217	4.5107	1.0243	0.9763	77 30	77.5
12.6	12 36	0.2181	4.5841	0.2235	4.4737	1.0247	0.9759	77 24	77.4
12.7	12 42	0.2198	4.5486	0.2254	4.4374	1.0251	0.9755	77 18	77.3
12.8	12 48	0.2215	4.5137	0.2272	4.4015	1.0255	0.9751	77 12	77.2
12.9	12 54	0.2232	4.4793	0.2290	4.3662	1.0259	0.9748	77 6	77.1
		cos θ	sec θ	cot θ	tan θ	csc θ	sin θ	deg min	θ deg

Table 2 (*continued*)

θ deg	deg min	sin θ	csc θ	tan θ	cot θ	sec θ	cos θ		
13.0	13 0	0.2250	4.4454	0.2309	4.3315	1.0263	0.9744	77 0	77.0
13.1	13 6	0.2267	4.4121	0.2327	4.2972	1.0267	0.9740	76 54	76.9
13.2	13 12	0.2284	4.3792	0.2345	4.2635	1.0271	0.9736	76 48	76.8
13.3	13 18	0.2300	4.3469	0.2364	4.2303	1.0276	0.9732	76 42	76.7
13.4	13 24	0.2317	4.3150	0.2382	4.1976	1.0280	0.9728	76 36	76.6
13.5	13 30	0.2334	4.2837	0.2401	4.1653	1.0284	0.9724	76 30	76.5
13.6	13 36	0.2351	4.2528	0.2419	4.1335	1.0288	0.9720	76 24	76.4
13.7	13 42	0.2368	4.2223	0.2438	4.1022	1.0293	0.9715	76 18	76.3
13.8	13 48	0.2385	4.1923	0.2456	4.0713	1.0297	0.9711	76 12	76.2
13.9	13 54	0.2402	4.1627	0.2475	4.0408	1.0302	0.9707	76 6	76.1
14.0	14 0	0.2419	4.1336	0.2493	4.0108	1.0306	0.9703	76 0	76.0
14.1	14 6	0.2436	4.1048	0.2512	3.9812	1.0311	0.9699	75 54	75.9
14.2	14 12	0.2453	4.0765	0.2530	3.9520	1.0315	0.9694	75 48	75.8
14.3	14 18	0.2470	4.0486	0.2549	3.9232	1.0320	0.9690	75 42	75.7
14.4	14 24	0.2487	4.0211	0.2568	3.8947	1.0324	0.9686	75 36	75.6
14.5	14 30	0.2504	3.9939	0.2586	3.8667	1.0329	0.9681	75 30	75.5
14.6	14 36	0.2521	3.9672	0.2605	3.8391	1.0334	0.9677	75 24	75.4
14.7	14 42	0.2538	3.9408	0.2623	3.8118	1.0338	0.9673	75 18	75.3
14.8	14 48	0.2554	3.9147	0.2642	3.7849	1.0343	0.9668	75 12	75.2
14.9	14 54	0.2571	3.8890	0.2661	3.7583	1.0348	0.9664	75 6	75.1
15.0	15 0	0.2588	3.8637	0.2679	3.7321	1.0353	0.9659	75 0	75.0
15.1	15 6	0.2605	3.8387	0.2698	3.7062	1.0358	0.9655	74 54	74.9
15.2	15 12	0.2622	3.8140	0.2717	3.6806	1.0363	0.9650	74 48	74.8
15.3	15 18	0.2639	3.7897	0.2736	3.6554	1.0367	0.9646	74 42	74.7
15.4	15 24	0.2656	3.7657	0.2754	3.6305	1.0372	0.9641	74 36	74.6
15.5	15 30	0.2672	3.7420	0.2773	3.6059	1.0377	0.9636	74 30	74.5
15.6	15 36	0.2689	3.7186	0.2792	3.5816	1.0382	0.9632	74 24	74.4
15.7	15 42	0.2706	3.6955	0.2811	3.5576	1.0388	0.9627	74 18	74.3
15.8	15 48	0.2723	3.6727	0.2830	3.5339	1.0393	0.9622	74 12	74.2
15.9	15 54	0.2740	3.6502	0.2849	3.5105	1.0398	0.9617	74 6	74.1
16.0	16 0	0.2756	3.6280	0.2867	3.4874	1.0403	0.9613	74 0	74.0
16.1	16 6	0.2773	3.6060	0.2886	3.4646	1.0408	0.9608	73 54	73.9
16.2	16 12	0.2790	3.5843	0.2905	3.4420	1.0413	0.9603	73 48	73.8
16.3	16 18	0.2807	3.5629	0.2924	3.4197	1.0419	0.9598	73 42	73.7
16.4	16 24	0.2823	3.5418	0.2943	3.3977	1.0424	0.9593	73 36	73.6
16.5	16 30	0.2840	3.5209	0.2962	3.3759	1.0429	0.9588	73 30	73.5
16.6	16 36	0.2857	3.5003	0.2981	3.3544	1.0435	0.9583	73 24	73.4
16.7	16 42	0.2874	3.4800	0.3000	3.3332	1.0440	0.9578	73 18	73.3
16.8	16 48	0.2890	3.4598	0.3019	3.3122	1.0446	0.9573	73 12	73.2
16.9	16 54	0.2907	3.4399	0.3038	3.2914	1.0451	0.9568	73 6	73.1
		cos θ	sec θ	cot θ	tan θ	csc θ	sin θ	deg min	θ deg

Table 2 *(continued)*

θ deg	deg min	sin θ	csc θ	tan θ	cot θ	sec θ	cos θ		
17.0	17 0	0.2924	3.4203	0.3057	3.2709	1.0457	0.9563	73 0	73.0
17.1	17 6	0.2940	3.4009	0.3076	3.2506	1.0463	0.9558	72 54	72.9
17.2	17 12	0.2957	3.3817	0.3096	3.2305	1.0468	0.9553	72 48	72.8
17.3	17 18	0.2974	3.3628	0.3115	3.2106	1.0474	0.9548	72 42	72.7
17.4	17 24	0.2990	3.3440	0.3134	3.1910	1.0480	0.9542	72 36	72.6
17.5	17 30	0.3007	3.3255	0.3153	3.1716	1.0485	0.9537	72 30	72.5
17.6	17 36	0.3024	3.3072	0.3172	3.1524	1.0491	0.9532	72 24	72.4
17.7	17 42	0.3040	3.2891	0.3191	3.1334	1.0497	0.9527	72 18	72.3
17.8	17 48	0.3057	3.2712	0.3211	3.1146	1.0503	0.9521	72 12	72.2
17.9	17 54	0.3074	3.2536	0.3230	3.0961	1.0509	0.9516	72 6	72.1
18.0	18 0	0.3090	3.2361	0.3249	3.0777	1.0515	0.9511	72 0	72.0
18.1	18 6	0.3107	3.2188	0.3268	3.0595	1.0521	0.9505	71 54	71.9
18.2	18 12	0.3123	3.2017	0.3288	3.0415	1.0527	0.9500	71 48	71.8
18.3	18 18	0.3140	3.1848	0.3307	3.0237	1.0533	0.9494	71 42	71.7
18.4	18 24	0.3156	3.1681	0.3327	3.0061	1.0539	0.9489	71 36	71.6
18.5	18 30	0.3173	3.1515	0.3346	2.9887	1.0545	0.9483	71 30	71.5
18.6	18 36	0.3190	3.1352	0.3365	2.9714	1.0551	0.9478	71 24	71.4
18.7	18 42	0.3206	3.1190	0.3385	2.9544	1.0557	0.9472	71 18	71.3
18.8	18 48	0.3223	3.1030	0.3404	2.9375	1.0564	0.9466	71 12	71.2
18.9	18 54	0.3239	3.0872	0.3424	2.9208	1.0570	0.9461	71 6	71.1
19.0	19 0	0.3256	2.0716	0.3443	2.9042	1.0576	0.9455	71 0	71.0
19.1	19 6	0.3272	3.0561	0.3463	2.8878	1.0583	0.9449	70 54	70.9
19.2	19 12	0.3289	3.0407	0.3482	2.8716	1.0589	0.9444	70 48	70.8
19.3	19 18	0.3305	3.0256	0.3502	2.8556	1.0595	0.9438	70 42	70.7
19.4	19 24	0.3322	3.0106	0.3522	2.8397	1.0602	0.9432	70 36	70.6
19.5	19 30	0.3338	2.9957	0.3541	2.8239	1.0608	0.9426	70 30	70.5
19.6	19 36	0.3355	2.9811	0.3561	2.8083	1.0615	0.9421	70 24	70.4
19.7	19 42	0.3371	2.9665	0.3581	2.7929	1.0622	0.9415	70 18	70.3
19.8	19 48	0.3387	2.9521	0.3600	2.7776	1.0628	0.9409	70 12	70.2
19.9	19 54	0.3404	2.9379	0.3620	2.7625	1.0635	0.9403	70 6	70.1
20.0	20 0	0.3420	2.9238	0.3640	2.7475	1.0642	0.9397	70 0	70.0
20.1	20 6	0.3437	2.9099	0.3659	2.7326	1.0649	0.9391	69 54	69.9
20.2	20 12	0.3453	2.8960	0.3679	2.7179	1.0655	0.9385	69 48	69.8
20.3	20 18	0.3469	2.8824	0.3699	2.7034	1.0662	0.9379	69 42	69.7
20.4	20 24	0.3486	2.8688	0.3719	2.6889	1.0669	0.9373	69 36	69.6
20.5	20 30	0.3502	2.8555	0.3739	2.6746	1.0676	0.9367	69 30	69.5
20.6	20 36	0.3518	2.8422	0.3759	2.6605	1.0683	0.9361	69 24	69.4
20.7	20 42	0.3535	2.8291	0.3779	2.6464	1.0690	0.9354	69 18	69.3
20.8	20 48	0.3551	2.8161	0.3799	2.6325	1.0697	0.9348	69 12	69.2
20.9	20 54	0.3567	2.8032	0.3819	2.6187	1.0704	0.9342	69 6	69.1
		cos θ	sec θ	cot θ	tan θ	csc θ	sin θ	deg min	θ deg

Table 2 (*continued*)

θ deg	deg min	sin θ	csc θ	tan θ	cot θ	sec θ	cos θ		
21.0	21 0	0.3584	2.7904	0.3839	2.6051	1.0711	0.9336	69 0	69.0
21.1	21 6	0.3600	2.7778	0.3859	2.5916	1.0719	0.9330	68 54	68.9
21.2	21 12	0.3616	2.7653	0.3879	2.5782	1.0726	0.9323	68 48	68.8
21.3	21 18	0.3633	2.7529	0.3899	2.5649	1.0733	0.9317	68 42	68.7
21.4	21 24	0.3649	2.7407	0.3919	2.5517	1.0740	0.9311	68 36	68.6
21.5	21 30	0.3665	2.7285	0.3939	2.5386	1.0748	0.9304	68 30	68.5
21.6	21 36	0.3681	2.7165	0.3959	2.5257	1.0755	0.9298	68 24	68.4
21.7	21 42	0.3697	2.7046	0.3979	2.5129	1.0763	0.9291	68 18	68.3
21.8	21 48	0.3714	2.6927	0.4000	2.5002	1.0770	0.9285	68 12	68.2
21.9	21 54	0.3730	2.6811	0.4020	2.4876	1.0778	0.9278	68 6	68.1
22.0	22 0	0.3746	2.6695	0.4040	2.4751	1.0785	0.9272	68 0	68.0
22.1	22 6	0.3762	2.6580	0.4061	2.4627	1.0793	0.9265	67 54	67.9
22.2	22 12	0.3778	2.6466	0.4081	2.4504	1.0801	0.9259	67 48	67.8
22.3	22 18	0.3795	2.6354	0.4101	2.4383	1.0808	0.9252	67 42	67.7
22.4	22 24	0.3811	2.6242	0.4122	2.4262	1.0816	0.9245	67 36	67.6
22.5	22 30	0.3827	2.6131	0.4142	2.4142	1.0824	0.9239	67 30	67.5
22.6	22 36	0.3843	2.6022	0.4163	2.4023	1.0832	0.9232	67 24	67.4
22.7	22 42	0.3859	2.5913	0.4183	2.3906	1.0840	0.9225	67 18	67.3
22.8	22 48	0.3875	2.5805	0.4204	2.3789	1.0848	0.9219	67 12	67.2
22.9	22 54	0.3891	2.5699	0.4224	2.3673	1.0856	0.9212	67 6	67.1
23.0	23 0	0.3907	2.5593	0.4245	2.3559	1.0864	0.9205	67 0	67.0
23.1	23 6	0.3923	2.5488	0.4265	2.3445	1.0872	0.9198	66 54	66.9
23.2	23 12	0.3939	2.5384	0.4286	2.3332	1.0880	0.9191	66 48	66.8
23.3	23 18	0.3955	2.5282	0.4307	2.3220	1.0888	0.9184	66 42	66.7
23.4	23 24	0.3971	2.5180	0.4327	2.3109	1.0896	0.9178	66 36	66.6
23.5	23 30	0.3987	2.5078	0.4348	2.2998	1.0904	0.9171	66 30	66.5
23.6	23 36	0.4003	2.4978	0.4369	2.2889	1.0913	0.9164	66 24	66.4
23.7	23 42	0.4019	2.4879	0.4390	2.2781	1.0921	0.9157	66 18	66.3
23.8	23 48	0.4035	2.4780	0.4411	2.2673	1.0929	0.9150	66 12	66.2
23.9	23 54	0.4051	2.4683	0.4431	2.2566	1.0938	0.9143	66 6	66.1
24.0	24 0	0.4067	2.4586	0.4452	2.2460	1.0946	0.9135	66 0	66.0
24.1	24 6	0.4083	2.4490	0.4473	2.2355	1.0955	0.9128	65 54	65.9
24.2	24 12	0.4099	2.4395	0.4494	2.2251	1.0963	0.9121	65 48	65.8
24.3	24 18	0.4115	2.4301	0.4515	2.2148	1.0972	0.9114	65 42	65.7
24.4	24 24	0.4131	2.4207	0.4536	2.2045	1.0981	0.9107	65 36	65.6
24.5	24 30	0.4147	2.4114	0.4557	2.1943	1.0989	0.9100	65 30	65.5
24.6	24 36	0.4163	2.4022	0.4578	2.1842	1.0998	0.9092	65 24	65.4
24.7	24 42	0.4179	2.3931	0.4599	2.1742	1.1007	0.9085	65 18	65.3
24.8	24 48	0.4195	2.3841	0.4621	2.1642	1.1016	0.9078	65 12	65.2
24.9	24 54	0.4210	2.3751	0.4642	2.1543	1.1025	0.9070	65 6	65.1
		cos θ	sec θ	cot θ	tan θ	csc σ	sin θ	seg min	θ deg

Table 2 *(continued)*

θ deg	deg min	sin θ	csc θ	tan θ	cot θ	sec θ	cos θ		
25.0	25 0	0.4226	2.3662	0.4663	2.1445	1.1034	0.9063	65 0	65.0
25.1	25 6	0.4242	2.3574	0.4684	2.1348	1.1043	0.9056	64 54	64.9
25.2	25 12	0.4258	2.3486	0.4706	2.1251	1.1052	0.9048	64 48	64.8
25.3	25 18	0.4274	2.3400	0.4727	2.1155	1.1061	0.9041	64 42	64.7
25.4	25 24	0.4289	2.3314	0.4748	2.1060	1.1070	0.9033	64 36	64.6
25.5	25 30	0.4305	2.3228	0.4770	2.0965	1.1079	0.9026	64 30	64.5
25.6	25 36	0.4321	2.3144	0.4791	2.0872	1.1089	0.9018	64 24	64.4
25.7	25 42	0.4337	2.3060	0.4813	2.0778	1.1098	0.9011	64 18	64.3
25.8	25 48	0.4352	2.2976	0.4834	2.0686	1.1107	0.9003	64 12	64.2
25.9	25 54	0.4368	2.2894	0.4856	2.0594	1.1117	0.8996	64 6	64.1
26.0	26 0	0.4384	2.2812	0.4877	2.0503	1.1126	0.8988	64 0	64.0
26.1	26 6	0.4399	2.2730	0.4899	2.0413	1.1136	0.8980	63 54	63.9
26.2	26 12	0.4415	2.2650	0.4921	2.0323	1.1145	0.8973	63 48	63.8
26.3	26 18	0.4431	2.2570	0.4942	2.0233	1.1155	0.8965	63 42	63.7
26.4	26 24	0.4446	2.2490	0.4964	2.0145	1.1164	0.8957	63 36	63.6
26.5	26 30	0.4462	2.2412	0.4986	2.0057	1.1174	0.8949	63 30	63.5
26.6	26 36	0.4478	2.2333	0.5008	1.9970	1.1184	0.8942	63 24	63.4
26.7	26 42	0.4493	2.2256	0.5029	1.9883	1.1194	0.8934	63 18	63.3
26.8	26 48	0.4509	2.2179	0.5051	1.9797	1.1203	0.8926	63 12	63.2
26.9	26 54	0.4524	2.2103	0.5073	1.9711	1.1213	0.8918	63 6	63.1
27.0	27 0	0.4540	2.2027	0.5095	1.9626	1.1223	0.8910	63 0	63.0
27.1	27 6	0.4555	2.1952	0.5117	1.9542	1.1233	0.8902	62 54	62.9
27.2	27 12	0.4571	2.1877	0.5139	1.9458	1.1243	0.8894	62 48	62.8
27.3	27 18	0.4586	2.1803	0.5161	1.9375	1.1253	0.8886	62 42	62.7
27.4	27 24	0.4602	2.1730	0.5184	1.9292	1.1264	0.8878	62 36	62.6
27.5	27 30	0.4617	2.1657	0.5206	1.9210	1.1274	0.8870	62 30	62.5
27.6	27 36	0.4633	2.1584	0.5228	1.9128	1.1284	0.8862	62 24	62.4
27.7	27 42	0.4648	2.1513	0.5250	1.9047	1.1294	0.8854	62 18	62.3
27.8	27 48	0.4664	2.1441	0.5272	1.8967	1.1305	0.8846	62 12	62.2
27.9	27 54	0.4679	2.1371	0.5295	1.8887	1.1315	0.8838	62 6	62.1
28.0	28 0	0.4695	2.1301	0.5317	1.8807	1.1326	0.8829	62 0	62.0
28.1	28 6	0.4710	2.1231	0.5339	1.8728	1.1336	0.8821	61 54	61.9
28.2	28 12	0.4726	2.1162	0.5362	1.8650	1.1347	0.8813	61 48	61.8
28.3	28 18	0.4741	2.1093	0.5384	1.8572	1.1357	0.8805	61 42	61.7
28.4	28 24	0.4756	2.1025	0.5407	1.8495	1.1368	0.8796	61 36	61.6
28.5	28 30	0.4772	2.0957	0.5430	1.8418	1.1379	0.8788	61 30	61.5
28.6	28 36	0.4787	2.0890	0.5452	1.8341	1.1390	0.8780	61 24	61.4
28.7	28 42	0.4802	2.0824	0.5475	1.8265	1.1401	0.8771	61 18	61.3
28.8	28 48	0.4818	2.0758	0.5498	1.8190	1.1412	0.8763	61 12	61.2
28.9	28 54	0.4833	2.0692	0.5520	1.8115	1.1423	0.8755	61 6	61.1
		cos σ	sec θ	cot θ	tan θ	csc θ	sin θ	deg min	θ deg

Table 2 *(continued)*

θ deg	deg min	sin θ	csc θ	tan θ	cot θ	sec θ	cos θ		
29.0	29 0	0.4848	2.0627	0.5543	1.8040	1.1434	0.8746	61 0	61.0
29.1	29 6	0.4863	2.0562	0.5566	1.7966	1.1445	0.8738	60 54	60.9
29.2	29 12	0.4879	2.0598	0.5589	1.7893	1.1456	0.8729	60 48	60.8
29.3	29 18	0.4894	2.0434	0.5612	1.7820	1.1467	0.8721	60 42	60.7
29.4	29 24	0.4909	2.0371	0.5635	1.7747	1.1478	0.8712	60 36	60.6
29.5	29 30	0.4924	2.0308	0.5658	1.7675	1.1490	0.8704	60 30	60.5
29.6	29 36	0.4939	2.0245	0.5681	1.7603	1.1501	0.8695	60 24	60.4
29.7	29 42	0.4955	2.0183	0.5704	1.7532	1.1512	0.8686	60 18	60.3
29.8	29 48	0.4970	2.0122	0.5727	1.7461	1.1524	0.8678	60 12	60.2
29.9	29 54	0.4985	2.0061	0.5750	1.7391	1.1535	0.8669	60 6	60.1
30.0	30 0	0.5000	2.0000	0.5774	1.7321	1.1547	0.8660	60 0	60.0
30.1	30 6	0.5015	1.9940	0.5797	1.7251	1.1559	0.8652	59 54	59.9
30.2	30 12	0.5030	1.9880	0.5820	1.7182	1.1570	0.8643	59 48	59.8
30.3	30 18	0.5045	1.9821	0.5844	1.7113	1.1582	0.8634	59 42	59.7
30.4	30 24	0.5060	1.9762	0.5867	1.7045	1.1594	0.8625	59 36	59.6
30.5	30 30	0.5075	1.9703	0.5890	1.6977	1.1606	0.8616	59 30	59.5
30.6	30 36	0.5090	1.9645	0.5914	1.6909	1.1618	0.8607	59 24	59.4
30.7	30 42	0.5105	1.9587	0.5938	1.6842	1.1630	0.8599	59 18	59.3
30.8	30 48	0.5120	1.9530	0.5961	1.6775	1.1642	0.8590	59 12	59.2
30.9	30 54	0.5135	1.9473	0.5985	1.6709	1.1654	0.8581	59 6	59.1
31.0	31 0	0.5150	1.9416	0.6009	1.6643	1.1666	0.8572	59 0	59.0
31.1	31 6	0.5165	1.9360	0.6032	1.6577	1.1679	0.8563	58 54	58.9
31.2	31 12	0.5180	1.9304	0.6056	1.6512	1.1691	0.8554	58 48	58.8
31.3	31 18	0.5195	1.9249	0.6080	1.6447	1.1703	0.8545	58 42	58.7
31.4	31 24	0.5210	1.9194	0.6104	1.6383	1.1716	0.8536	58 36	58.6
31.5	31 30	0.5225	1.9139	0.6128	1.6319	1.1728	0.8526	58 30	58.5
31.6	31 36	0.5240	1.9084	0.6152	1.6255	1.1741	0.8517	58 24	58.4
31.7	31 42	0.5255	1.9031	0.6176	1.6191	1.1753	0.8508	58 18	58.3
31.8	31 48	0.5270	1.8977	0.6200	1.6128	1.1766	0.8499	58 12	58.2
31.9	31 54	0.5284	1.8924	0.6224	1.6066	1.1779	0.8490	58 6	58.1
32.0	32 0	0.5299	1.8871	0.6249	1.6003	1.1792	0.8480	58 0	58.0
32.1	32 6	0.5314	1.8818	0.6273	1.5941	1.1805	0.8471	57 54	57.9
32.2	32 12	0.5329	1.8766	0.6297	1.5880	1.1818	0.8462	57 48	57.8
32.3	32 18	0.5344	1.8714	0.6322	1.5818	1.1831	0.8453	57 42	57.7
32.4	32 24	0.5358	1.8663	0.6346	1.5757	1.1844	0.8443	57 36	57.6
32.5	32 30	0.5373	1.8612	0.6371	1.5697	1.1857	0.8434	57 30	57.5
32.6	32 36	0.5388	1.8561	0.6395	1.5637	1.1870	0.8425	57 24	57.4
32.7	32 42	0.5402	1.8510	0.6420	1.5577	1.1883	0.8415	57 18	57.3
32.8	32 48	0.5417	1.8460	0.6445	1.5517	1.1897	0.8406	57 12	57.2
32.9	32 54	0.5432	1.8410	0.6469	1.5458	1.1910	0.8396	57 6	57.1
		cos θ	sec θ	cot θ	tan θ	csc σ	sin θ	seg min	θ deg

Table 2 (*continued*)

θ deg	deg min	sin θ	csc θ	tan θ	cot θ	sec θ	cos θ		
33.0	33 0	0.5446	1.8361	0.6494	1.5399	1.1924	0.8387	57 0	57.0
33.1	33 6	0.5461	1.8312	0.6519	1.5340	1.1937	0.8377	56 54	56.9
33.2	33 12	0.5476	1.8263	0.6544	1.5282	1.1951	0.8368	56 48	56.8
33.3	33 18	0.5490	1.8214	0.6569	1.5224	1.1964	0.8358	56 42	56.7
33.4	33 24	0.5505	1.8166	0.6594	1.5166	1.1978	0.8348	56 36	56.6
33.5	33 30	0.5519	1.8118	0.6619	1.5108	1.1992	0.8339	56 30	56.5
33.6	33 36	0.5534	1.8070	0.6644	1.5051	1.2006	0.8329	56 24	56.4
33.7	33 42	0.5548	1.8023	0.6669	1.4994	1.2020	0.8320	56 18	56.3
33.8	33 48	0.5563	1.7976	0.6694	1.4938	1.2034	0.8310	56 12	56.2
33.9	33 54	0.5577	1.7929	0.6720	1.4882	1.2048	0.8300	56 6	56.1
34.0	34 0	0.5592	1.7883	0.6745	1.4826	1.2062	0.8290	56 0	56.0
34.1	34 6	0.5606	1.7837	0.6771	1.4770	1.2076	0.8281	55 54	55.9
34.2	34 12	0.5621	1.7791	0.6796	1.4715	1.2091	0.8271	55 48	55.8
34.3	34 18	0.5635	1.7745	0.6822	1.4659	1.2105	0.8261	55 42	55.7
34.4	34 24	0.5650	1.7700	0.6847	1.4605	1.2120	0.8251	55 36	55.6
34.5	34 30	0.5664	1.7655	0.6873	1.4550	1.2134	0.8241	55 30	55.5
34.6	34 36	0.5678	1.7610	0.6899	1.4496	1.2149	0.8231	55 24	55.4
34.7	34 42	0.5693	1.7566	0.6924	1.4442	1.2163	0.8221	55 18	55.3
34.8	34 48	0.5707	1.7522	0.6950	1.4388	1.2178	0.8211	55 12	55.2
34.9	34 54	0.5721	1.7478	0.6976	1.4335	1.2193	0.8202	55 6	55.1
35.0	35 0	0.5736	1.7434	0.7002	1.4281	1.2208	0.8192	55 0	55.0
35.1	35 6	0.5750	1.7391	0.7028	1.4229	1.2223	0.8181	54 54	54.9
35.2	35 12	0.5764	1.7348	0.7054	1.4176	1.2238	0.8171	54 48	54.8
35.3	35 18	0.5779	1.7305	0.7080	1.4124	1.2253	0.8161	54 42	54.7
35.4	35 24	0.5793	1.7263	0.7107	1.4071	1.2268	0.8151	54 36	54.6
35.5	35 30	0.5807	1.7221	0.7133	1.4019	1.2283	0.8141	54 30	54.5
35.6	35 36	0.5821	1.7179	0.7159	1.3968	1.2299	0.8131	54 24	54.4
35.7	35 42	0.5835	1.7137	0.7186	1.3916	1.2314	0.8121	54 18	54.3
35.8	35 48	0.5850	1.7095	0.7212	1.3865	1.2329	0.8111	54 12	54.2
35.9	35 54	0.5864	1.7054	0.7239	1.3814	1.2345	0.8100	54 6	54.1
36.0	36 0	0.5878	1.7013	0.7265	1.3764	1.2361	0.8090	54 0	54.0
36.1	36 6	0.5892	1.6972	0.7292	1.3713	1.2376	0.8080	53 54	53.9
36.2	36 12	0.5906	1.6932	0.7319	1.3663	1.2392	0.8070	53 48	53.8
36.3	36 18	0.5920	1.6892	0.7346	1.3613	1.2408	0.8059	53 42	53.7
36.4	36 24	0.5934	1.6852	0.7373	1.3564	1.2424	0.8049	53 36	53.6
36.5	36 30	0.5948	1.6812	0.7400	1.3514	1.2440	0.8039	53 30	53.5
36.6	36 36	0.5962	1.6772	0.7427	1.3465	1.2456	0.8028	53 24	53.4
36.7	36 42	0.5976	1.6733	0.7454	1.3416	1.2472	0.8018	53 18	53.3
36.8	36 48	0.5990	1.6694	0.7481	1.3367	1.2489	0.8007	53 12	53.2
36.9	36 54	0.6004	1.6655	0.7508	1.3319	1.2505	0.7997	53 6	53.1
		cos θ	sec θ	cot θ	tan θ	csc θ	sin θ	deg min	θ deg

Table 2 **(*continued*)**

θ deg	deg min	sin θ	csc θ	tan θ	cot θ	sec θ	cos θ		
37.0	37 0	0.6018	1.6616	0.7536	1.3270	1.2521	0.7986	53 0	53.0
37.1	37 6	0.6032	1.6578	0.7563	1.3222	1.2538	0.7976	52 54	52.9
37.2	37 12	0.6046	1.6540	0.7590	1.3175	1.2554	0.7965	52 48	52.8
37.3	37 18	0.6060	1.6502	0.7618	1.3127	1.2571	0.7955	52 42	52.7
37.4	37 24	0.6074	1.6464	0.7646	1.3079	1.2588	0.7944	52 36	52.6
37.5	37 30	0.6088	1.6427	0.7673	1.3032	1.2605	0.7934	52 30	52.5
37.6	37 36	0.6101	1.6390	0.7701	1.2985	1.2622	0.7923	52 24	52.4
37.7	37 42	0.6115	1.6353	0.7729	1.2938	1.2639	0.7912	52 18	52.3
37.8	37 48	0.6129	1.6316	0.7757	1.2892	1.2656	0.7902	52 12	52.2
37.9	37 54	0.6143	1.6279	0.7785	1.2846	1.2673	0.7891	52 6	52.1
38.0	38 0	0.6157	1.6243	0.7813	1.2799	1.2690	0.7880	52 0	52.0
38.1	38 6	0.6170	1.6207	0.7841	1.2753	1.2708	0.7869	51 54	51.9
38.2	38 12	0.6184	1.6171	0.7869	1.2708	1.2725	0.7859	51 48	51.8
38.3	38 18	0.6198	1.6135	0.7898	1.2662	1.2742	0.7848	51 42	51.7
38.4	38 24	0.6211	1.6099	0.7926	1.2617	1.2760	0.7837	51 36	51.6
38.5	38 30	0.6225	1.6064	0.7954	1.2572	1.2778	0.7826	51 30	51.5
38.6	38 36	0.6239	1.6029	0.7983	1.2527	1.2796	0.7815	51 24	51.4
38.7	38 42	0.6252	1.5994	0.8012	1.2482	1.2813	0.7804	51 18	51.3
38.8	38 48	0.6266	1.5959	0.8040	1.2437	1.2831	0.7793	51 12	51.2
38.9	38 54	0.6280	1.5925	0.8069	1.2393	1.2849	0.7782	51 6	51.1
39.0	39 0	0.6293	1.5890	0.8098	1.2349	1.2868	0.7771	51 0	51.0
39.1	39 6	0.6307	1.5856	0.8127	1.2305	1.2886	0.7760	50 54	50.9
39.2	39 12	0.6320	1.5822	0.8156	1.2261	1.2904	0.7749	50 48	50.8
39.3	39 18	0.6334	1.5788	0.8185	1.2218	1.2923	0.7738	50 42	50.7
39.4	39 24	0.6347	1.5755	0.8214	1.2174	1.2941	0.7727	50 36	50.6
39.5	39 30	0.6361	1.5721	0.8243	1.2131	1.2960	0.7716	50 30	50.5
39.6	39 36	0.6374	1.5688	0.8273	1.2088	1.2978	0.7705	50 24	50.4
39.7	39 42	0.6388	1.5655	0.8302	1.2045	1.2997	0.7694	50 18	50.3
39.8	39 48	0.6401	1.5622	0.8332	1.2002	1.3016	0.7683	50 12	50.2
39.9	39 54	0.6414	1.5590	0.8361	1.1960	1.3035	0.7672	50 6	50.1
40.0	40 0	0.6428	1.5557	0.8391	1.1918	1.3054	0.7660	50 0	50.0
40.1	40 6	0.6441	1.5525	0.8421	1.1875	1.3073	0.7649	49 54	49.9
40.2	40 12	0.6455	1.5493	0.8451	1.1833	1.3092	0.7638	49 48	49.8
40.3	40 18	0.6468	1.5461	0.8481	1.1792	1.3112	0.7627	49 42	49.7
40.4	40 24	0.6481	1.5429	0.8511	1.1750	1.3131	0.7615	49 36	49.6
40.5	40 30	0.6494	1.5398	0.8541	1.1708	1.3151	0.7604	49 30	49.5
40.6	40 36	0.6508	1.5366	0.8571	1.1667	1.3171	0.7593	49 24	49.4
40.7	40 42	0.6521	1.5335	0.8601	1.1626	1.3190	0.7581	49 18	49.3
40.8	40 48	0.6534	1.5304	0.8632	1.1585	1.3210	0.7570	49 12	49.2
40.9	40 54	0.6547	1.5273	0.8662	1.1544	1.3230	0.7559	49 6	49.1
		cos θ	sec θ	cot θ	tan θ	csc θ	sin θ	deg min	θ deg

Table 2 *(continued)*

θ deg	deg min	sin θ	csc θ	tan θ	cot θ	sec θ	cos θ		
41.0	41 0	0.6561	1.5243	0.8693	1.1504	1.3250	0.7547	49 0	49.0
41.1	41 6	0.6574	1.5212	0.8724	1.1463	1.3270	0.7536	48 54	48.9
41.2	41 12	0.6587	1.5182	0.8754	1.1423	1.3291	0.7524	48 48	48.8
41.3	41 18	0.6600	1.5151	0.8785	1.1383	1.3311	0.7513	48 42	48.7
41.4	41 24	0.6613	1.5121	0.8816	1.1343	1.3331	0.7501	48 36	48.6
41.5	41 30	0.6626	1.5092	0.8847	1.1303	1.3352	0.7490	48 30	48.5
41.6	41 36	0.6639	1.5062	0.8878	1.1263	1.3373	0.7478	48 24	48.4
41.7	41 42	0.6652	1.5032	0.8910	1.1224	1.3393	0.7466	48 18	48.3
41.8	41 48	0.6665	1.5003	0.8941	1.1184	1.3414	0.7455	48 12	48.2
41.9	41 54	0.6678	1.4974	0.8972	1.1145	1.3435	0.7443	48 6	48.1
42.0	42 0	0.6691	1.4945	0.9004	1.1106	1.3456	0.7431	48 0	48.0
42.1	42 6	0.6704	1.4916	0.9036	1.1067	1.3478	0.7420	47 54	47.9
42.2	42 12	0.6717	1.4887	0.9067	1.1028	1.3499	0.7408	47 48	47.8
42.3	42 18	0.6730	1.4859	0.9099	1.0990	1.3520	0.7396	47 42	47.7
42.4	42 24	0.6743	1.4830	0.9131	1.0951	1.3542	0.7385	47 36	47.6
42.5	42 30	0.6756	1.4802	0.9163	1.0913	1.3563	0.7373	47 30	47.5
42.6	42 36	0.6769	1.4774	0.9195	1.0875	1.3585	0.7361	47 24	47.4
42.7	42 42	0.6782	1.4746	0.9228	1.0837	1.3607	0.7349	47 18	47.3
42.8	42 48	0.6794	1.4718	0.9260	1.0799	1.3629	0.7337	47 12	47.2
42.9	42 54	0.6807	1.4690	0.9293	1.0761	1.3651	0.7325	47 6	47.1
43.0	43 0	0.6820	14663	0.9325	1.0724	1.3673	0.7314	47 0	47.0
43.1	43 6	0.6833	1.4635	0.9358	1.0686	1.3696	0.7302	46 54	46.9
43.2	43 12	0.6845	1.4608	0.9391	1.0649	1.3718	0.7290	46 48	46.8
43.3	43 18	0.6858	1.4581	0.9424	1.0612	1.3741	0.7278	46 42	46.7
43.4	43 24	0.6871	1.4554	0.9457	1.0575	1.3762	0.7266	46 36	46.6
43.5	43 30	0.6884	1.4527	0.9490	1.0538	1.3786	0.7254	46 30	46.5
43.6	43 36	0.6896	1.4501	0.9523	1.0501	1.3809	0.7242	46 24	46.4
43.7	43 42	0.6909	1.4474	0.9556	1.0464	1.3832	0.7230	46 18	46.3
43.8	43 48	0.6921	1.4448	0.9590	1.0428	1.3855	0.7218	46 12	46.2
43.9	43 54	0.6934	1.4422	0.9623	1.0392	1.3878	0.7206	46 6	46.1
44.0	44 0	0.6947	1.4396	0.9657	1.0355	1.3902	0.7193	46 0	46.0
44.1	44 6	0.6959	1.4370	0.9691	1.0319	1.3925	0.7181	45 54	45.9
44.2	44 12	0.6972	1.4344	0.9725	1.0283	1.3949	0.7169	45 48	45.8
44.3	44 18	0.6984	1.4318	0.9759	1.0247	1.3972	0.7157	45 42	45.7
44.4	44 24	0.6997	1.4293	0.9793	1.0212	1.3996	0.7145	45 36	45.6
44.5	44 30	0.7009	1.4267	0.9827	1.0176	1.4020	0.7133	45 30	45.5
44.6	44 36	0.7022	1.4242	0.9861	1.0141	1.4044	0.7120	45 24	45.4
44.7	44 42	0.7034	1.4217	0.9896	1.0105	1.4069	0.7108	45 18	45.3
44.8	44 48	0.7046	1.4192	0.9930	1.0070	1.4093	0.7096	45 12	45.2
44.9	44 54	0.7059	1.4167	0.9965	1.0035	1.4118	0.7083	45 6	45.1
45.0	45 0	0.7071	1.4142	1.0000	1.0000	1.4142	0.7071	45 0	45.0
		cos θ	sec θ	cot θ	tan θ	csc θ	sin θ	deg min	θ deg

APPENDIX C

Answers to Selected Odd-Numbered Exercises

CHAPTER 1

EXERCISES 1.2

1.

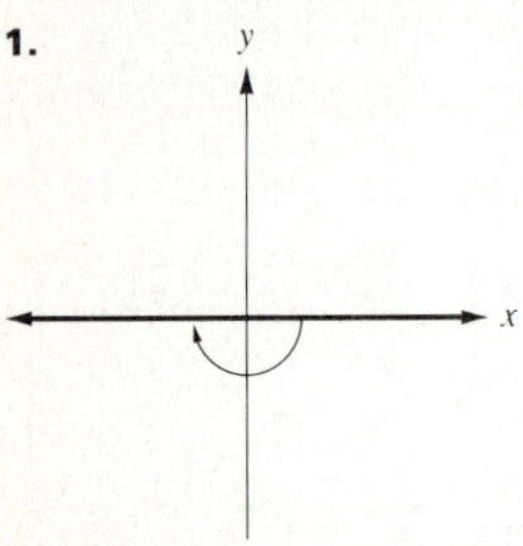

3.

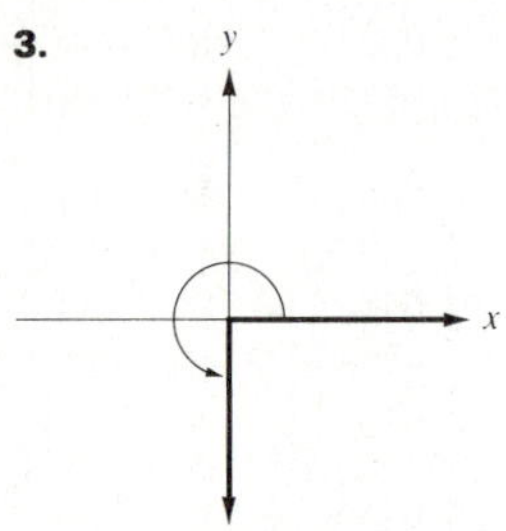

9.

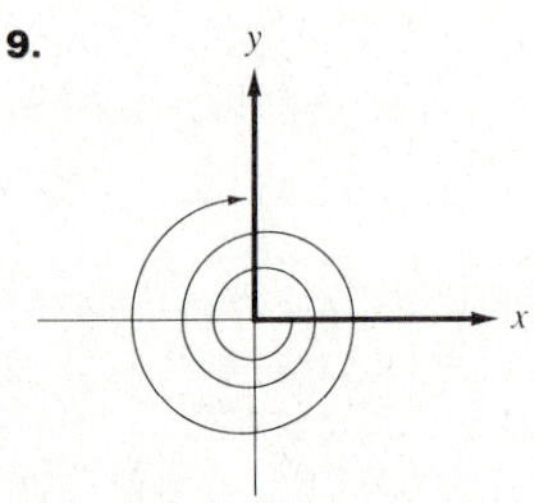

5.

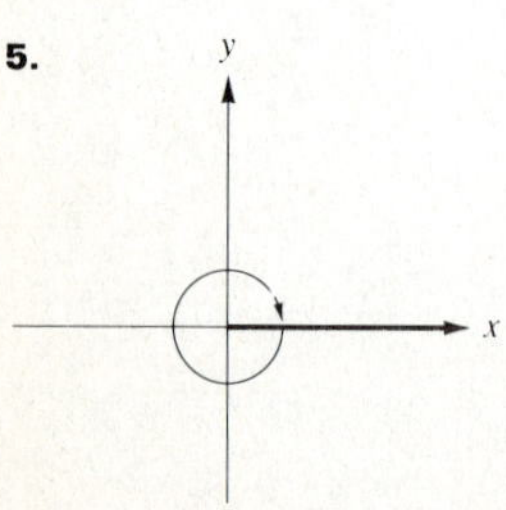

7.

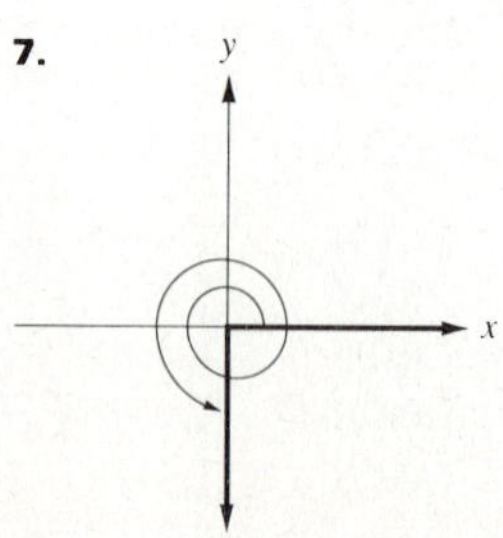

11. Standard position; negative angle **13.** Not in standard position, vertex not at origin; positive angle **15.** Standard position; positive angle **17.** Not in standard position, initial side not along positive x axis; positive angle **19.** Standard position; positive angle

21. Quadrant II **23.** Quadrantal **25.** Quadrant III

27. $\frac{1}{2}$ rev clockwise **29.** $\frac{3}{4}$ rev clockwise

41. Hour hand—$\frac{1}{4}$ rev; minute hand—3 rev; second hand—180 rev; **43.** Hour hand—$\frac{1}{24}$ rev; minute hand—$\frac{1}{2}$ rev; second hand—30 rev; **45.** Hour hand—$\frac{33}{24}$ rev; minute hand—$16\frac{1}{2}$ rev; second hand—990 rev; $\frac{1}{4}$ rev $\frac{5}{8}$ rev

51. 6:00 A.M. **53.** 3:00 P.M. **55.** 1:00 P.M.

57. $\frac{45}{365}$ rev **59.** 9 rev; $\frac{1}{2}$ rev **61.** March 15, 1989
63. **(a)** $-A$ in quadrant II
$2A$ in quadrants I or II
$A/2$ in quadrant II
(b) $-A$ in quadrant I
$2A$ in quadrants III or IV
$A/2$ in quadrant II

EXERCISES 1.3-1

1. 240° **3.** −225° **5.** 150° **7.** −120° **9.** 60° **11.** 45° **13.** 0° **15.** 15°25′ **17.** 5°42′50″ **19.** 8.41° **21.** 58°52′ **23.** 130°27′29″ **25.** 54°59′50″ **27.** 102.1° **29.** 179°42′30″ **31.** 179.75° **33.** −110.17° **35.** 345.84° **37.** 180.00° **39.** 7°13′48″ **41.** −51°21′36″ **43.** 37°56′2″ **45.** 485° and −235° **47.** 180° and −540° **49.** 570.8° and −149.2° **53.** 0.227 rev = 81.9° **55.** 0.948 rev = 341° **57.** 1.11 grad **59.** 17 grad **61.** 225.0° **63.** 312.5 grad **65.** 400 grad; 6400 mil
67. **(a)** $-A$ in quadrant I, $2A$ in quadrants III or IV, $A/2$ in quadrant II
(b) $-A$ in II, $2A$ in I, $A/2$ in II

EXERCISES 1.3-2

1. $\theta = 43.0°$ **3.** $r = 1.4$ yd **9.** **(a)** 1 rad **(b)** 0.85 rad **(c)** 1.50 rad **11.** 39 ft **13.** $\theta = 1.11$ rad **15.** 18.4 in. **17.** **(a)** 2π rad **(b)** π rad **(c)** $\frac{\pi}{2}$ rad

EXERCISES 1.3-3

1. $\frac{5\pi}{18}$ rad **3.** $\frac{5\pi}{4}$ rad **5.** $\frac{-\pi}{4}$ rad **7.** 0.19 rad **9.** 3.49 rad **11.** −0.31 rad **13.** 0.32 rad **15.** 12° **17.** 110° **19.** 900° **21.** −140° **23.** 115° **25.** 199.8° **27.** 36.48° **29.** 565.49° **31.** Quadrant I **33.** Quadrant III **35.** Quadrant III **37.** Quadrant I **39.** Quadrant I **41.** Quadrant III **43.** 31.68° **45.** 92.08° **47.** 0.921° **49.** 3.14 **51.** 3.57 **53.** 10.34 **55.** 5.14 **57.** 90°; $\frac{\pi}{2}$ rad **59.** 390°; $\frac{13}{6}\pi$ rad **61.** 15°; $\frac{\pi}{12}$ rad **63.** 75°; $\frac{5\pi}{12}$ rad **65.** 2 P.M. **67.** 1:30 P.M.
69. **(a)** $-\theta$ in quadrant IV, 2θ in quadrant I, $\theta/2$ in quadrant I
(b) $-\theta$ in quadrant III, 2θ in quadrant III, $\theta/2$ in quadrant I
(c) $-\theta$ in quadrant IV, 2θ in quadrant I or II, $\theta/2$ in quadrant III
(d) $-\theta$ in quadrant I or II, 2θ in quadrants I, II, III, or IV, $\theta/2$ in quadrant IV
71. 126° **73.** 5 red and 11 orange **75.** $r = 56$ cm, $S = 29$ cm **77.** $r = 955$ ft **79.** **(a)** 1020 mil **(b)** 0.00098175 rad **81.** 764 mil

EXERCISES 1.4

1. 18 m **3.** 122 cm **5.** 75.2 cm **7.** 0.98 ft **9.** 243 steps **11.** 3.00 in. **13.** $\pi - 2$ **15.** 1.12×10^7 mi **17.** 29 in. **19.** 0.112 m **21.** 20.6 m **23.** 3.5 mi **25.** 78 rev, 0.13 mi **27.** 125 mi **29.** 400 mi **31.** 34.0° **33.** 7° **35.** 27°46′ east, 10°8′ south **37.** 3960 mi **39.** **(a)** 69 mi, no **(b)** Answer depends on lattitude **41.** 3250 ft **43.** 827 mi, 1770 mi **45.** 590 km **47.** 18.8 mi (Moon), 36.9 mi (Mars) **49.** 32° **51.** 47 times for B, 12.7 times for C, no difference, counterclockwise **53.** 1.4 rev **55.** The other wheel must be 20 in. in diameter. **57.** **(a)** Increases by 20% **(b)** Increases by 9.5% **(c)** Increases by 6.3% **59.** **(a)** 215 cm **(b)** 3.45 ft **(c)** 2.61 m **61.** 2400 in^2 **65.** 2500 ft^2 **67.** 2870 ft **69.** 2.00 rad, 2.25 m^2 **71.** 10.3 oz **73.** It is multiplied by 4. **75.** It is doubled. **77.** **(a)** 20%, **(b)** 5.6% **79.** 280 cm^2 **81.** 180 cm^2, 540 cm^2 **83.** 17,300 cm^3

EXERCISES 1.5

1. 30,000%, 524 rad/s **3.** **(a)** 10,400 rev, $3.73 \times 10^{6\circ}$ **(b)** 72.5 min **5.** 0.071 rad/day **7.** 5.0×10^{-3} s, 0.29 s **9.** 4.0 min, 4.0 s, 0.067 s, 3.8 hr **11.** 77.9° **13.** ≈ 6.7°/yr **15.** In 128 days **17.** 12:16 A.M., 12:49 A.M., 1:22 A.M., 1:55 A.M., 2:27 A.M., 3:00 A.M., 3:33 A.M., etc. repeating every 32.7 min **19.** 0.12 s **21.** 130 ft/s **23.** 6.66×10^4 mph **25.** 1.13×10^{-3} rad/s, 17,200 mph **27.** 940 mi/s **29.** 101 in., 22.5 rad, 16.7 rad **31.** 21 in./s, 9.6 in./s (approx.) **33.** **(a)** 47.6 rad/s **(b)** 1.88 rad/s, 4.30 rad/s **(c)** 281 ft **35.** 23.6 ft/min **37.** 5.65 m/min, rising **39.** 17 in./s **41.** 5.9 in. **43.** 167 rpm **45.** 4.3 in. **47.** 7.0 ft **49.** 46 mph **51.** 60 mph **53.** 1.87 rev/s **55.** 2.2 ft, 3.6 mph

EXERCISES 1.6

1. 0.055 rad, 3.1° **3.** 65 ft **5.** 1.6 mi **7.** 12.6 ft **9.** 26 ft **13.** about 1.9° **15.** 1600 km^2 **17.** about 10° **19.** 2.0 million mi **21.** 103 mi **23.** 0.625 light-year × 0.876 light-year, 3.68×10^{12} mi × 5.15×10^{12} mi **25.** 1.1°, 1.8°, 0.72″ **27.** 91.2 ft **29.** 530 ft **31.** 1.35° **33.** **(a)** 28 ft **(b)** 84 ft **37.** 9.9″ **39.** 25 cm **41.** 21,000 ft **43.** 0.050 cm

EXERCISES 1.7

1. 2860 mi, 28% **3.** 3060 mi, 23% **5.** 3950 mi **7.** 9.3×10^7 mi **9.** 1.42×10^8 mi (Mars), 4.85×10^8 mi (Jupiter) **11.** (3.25×10^{-5}) seconds **13.** 19″ (Sun); (3.65×10^{-4}) seconds (Alpha Centauri) **15.** **(a)** 2.7°, about 5.2 times the size of our moon, small-angle approximation is valid. **(b)** Approx. 43°. Small-angle approximation is of no use here. **17.** 30 light-years **19.** 0.14″ **21.** **(a)** 1.73×10^{14} mi/yr, 55.0 mi/s **(b)** 0.12″/yr **(c)** 3.2×10^8 mi **23.** Yes **25.** 0.28 light-years, no. **27.** 149 light-years **29.** 13′04″ **31.** 8.1×10^6 light-years

EXERCISES 1.8

1. $\pi \approx 3.16$ **3.** $\pi = 4$ **7.** **(a)** $\sqrt{\frac{1}{2} + \frac{1}{2}\sqrt{\frac{1}{2} + \frac{1}{2}\sqrt{\frac{1}{2} + \frac{1}{2}\sqrt{\frac{1}{2}}}}}$ **(b)** 2.83, 3.06, 3.12, 3.137 **9.** **(a)** 3.3397, 3.0418, 3.0916 **(b)** 3.252, 3.194, 3.167 **11.** 3.20 **15.** $\frac{22}{7} = 3.1429$ **17.** **(a)** 3.141592653589793238462643383279 **(f)** 3.14159265...

REVIEW EXERCISES

1. (a) $450°, \frac{5\pi}{2}$ **(b)** $-720°, -4\pi$ **(c)** $-855°, -\frac{19\pi}{4}$ **(d)** $765°, \frac{17\pi}{4}$ **3. (a)** $\frac{\pi}{3}$ **(b)** $\frac{5\pi}{7}$ **(c)** 74°39′46″ **(d)** 10.1° **(e)** 0.37 rad **(f)** 0.18π **(g)** 0.30 rev **5. (a)** 226°, 4300° **(b)** 0.66 cm **(c)** 0.16 rad/s, 3.0 rad/s **7. (a)** 22, 100° **(b)** 884°/s **(c)** 0.16 s **9.** 865,000 mi **11.** 53 m^2

SIGNIFICANT FIGURES

1. (a) 2 **(b)** 3 **(c)** 3 **(d)** 6 **(e)** 1 **(f)** 6 **(g)** 3 **(h)** 3 **(i)** 3 **(j)** 1 **2. (a)** 10 significant figures, 9 decimal places **(b)** (i) 3.1 (ii) 3.142 (iii) 3.141593 **3. (a)** (i) 2.2361 (ii) 2.2360680 (iii) 2.2 **(b)** 9 **4.** 821,000 **5. (a)** 799 **(b)** 30.0 **(c)** 20,650 **(d)** 1.5 **(e)** 3000 **(f)** 1.6 **6.** 34,000 ft^2 **7.** 5400 cm^2 (the first zero is significant) **8.** 7.2 m^3 **10.** $5.635 \le x \le 5.644$

CONVERSION OF UNITS

1. 2000 mm **3.** 610 in. **5.** 30. cm **7.** 1.6 km **9.** 4800 ft/min **11.** 960 in./s **13.** 10^6 m^2 **15.** 46,700 in^3 **17.** 765,000 cm^3 **19.** 2280 cm–mm, 0.0228 m^2 **21.** 8600″ **23.** 0.35 rev

SCIENTIFIC NOTATION

1. (a) 7.5×10^5 **(b)** 2.51×10^4 **(c)** 4.7×10^9 yr **(d)** 8.1×10^2 **(e)** 5.43×10^{-4} **(f)** 9.8×10^{-1} **(g)** 2.5×10^{-10} **(h)** Correct as is **(i)** 4.7×10^8 **(j)** 8.2×10^2 **(k)** 6.09×10^8 **(l)** 1.2×10^{-8} **(m)** 3.8×10^{-6} **(n)** 6.9×10^8
2. (a) 9200 **(b)** 0.072 **(c)** 602,000,000,000,000,000,000,000 **(d)** 160,000 **(e)** 0.000000142 **(f)** 0.0000000001 **(g)** 31,500,000 **(h)** 5,320,000 **(i)** 0.0049
3. (a) 8.88×10^{13} **(b)** 4.0×10^{-8} **(c)** 1.60×10^4 **(d)** 8.66×10^{-14} **(e)** 3.6×10^{11} **(f)** 7.7×10^2 **(g)** 5.1×10^{-2} **(h)** 5.2×10^{-8} **(i)** 5.05×10^9 **(j)** 1.6×10^{-2} **(k)** 1.8×10^{11} **(l)** 1.2×10^{-12} **(m)** 4.1×10^8 **(n)** 2.2×10^{15}

CHAPTER 2

EXERCISES 2.1-1

1. 4.8 ft **3.** $\frac{25}{6}$ **5.** 130 ft **7.** 540 ft **9. (a)** 7.5 ft **(b)** 1.5 ft shorter, 1.5 ft longer **(c)** 40 ft from lamp **11.** 310 ft^2 **13.** 2.0 ft wide, 4.7 ft high **15.** 26 ft

EXERCISES 2.1-2

1. No **3.** Yes **5.** Not a triangle **7.** Yes **9.** Yes **11.** No **13.** 4.0 **15.** 6.2 **17.** Not possible **19.** $\sqrt{2}$ **21.** 3 **23.** $\sqrt{\pi^2 - 9} = 0.93$ **25.** 21.2 cm **27.** 800 in^2 **29.** 15.4 ft, 18.6 ft, 286 sq ft **33. (a)** $R\sqrt{2}$ **(b)** $R(\sqrt{2} - 1)$ **35.** 2.8 ft **37.** $12\sqrt{3}$ ft **39.** 800 ft^2 **41.** 26,000 m^2 **43.** 145 ft **45.** 26.8 ft **47. (b)** $L\sqrt{3}/2, L/2$ **(c)** 30°, 60° **49.** 8300 ft **51.** 1500 ft^2, 17 ft

EXERCISES 2.2-1

1. opp = 2, adj = 5, hyp = $\sqrt{29}$, $\frac{\text{adj}}{\text{hyp}} = \frac{5}{\sqrt{29}}$

3. opp = $\sqrt{17}$, adj = 8, hyp = 9, $\frac{\text{adj}}{\text{hyp}} = \frac{8}{9}$

5. opp = 7, adj = 7, hyp = $7\sqrt{2}$, $\frac{\text{adj}}{\text{hyp}} = \frac{1}{\sqrt{2}}$

7. $\sin A = 0.60, \cos A = 0.80, \tan A = 0.75, \cot A = 1.33, \sec A = 1.25, \csc A = 1.67$
9. $\sin A = 0.47, \cos A = 0.88, \tan A = 0.53, \cot A = 1.88, \sec A = 1.13, \csc A = 2.13$
11. $\sin A = 0.44, \cos A = 0.90, \tan A = 0.49, \cot A = 2.05, \sec A = 1.11, \csc A = 2.28$
13. $\sin A = 0.60, \cos A = 0.80, \tan A = 0.75, \cot A = 1.33, \sec A = 1.25, \csc A = 1.67$
15. $\sin A = 0.77, \cos A = 0.64, \tan A = 1.20, \cot A = 0.83, \sec A = 1.56, \csc A = 1.30$
17. $\sin A = 0.71, \cos A = 0.71, \tan A = 1.00, \cot A = 1.00, \sec A = 1.41, \csc A = 1.41$

19. $\sin A = \frac{2pq}{q^2 + p^2}, \cos A = \frac{q^2 - p^2}{q^2 + p^2}, \tan A = \frac{2pq}{q^2 - p^2}$
$\cot A = \frac{q^2 - p^2}{2pq}, \sec A = \frac{q^2 + p^2}{q^2 - p^2}, \csc A = \frac{q^2 + p^2}{2pq}$

21. $\sin A = \frac{2pq}{q^2 + p^2}, \cos A = \frac{q^2 - p^2}{q^2 + p^2}, \tan A = \frac{2pq}{q^2 - p^2}$
$\cot A = \frac{q^2 - p^2}{2pq}, \sec A = \frac{q^2 + p^2}{q^2 - p^2}, \csc A = \frac{q^2 + p^2}{2pq}$

23. $\sin \theta = \frac{\sqrt{91}}{10}, \cos \theta = \frac{3}{10}, \tan \theta = \frac{\sqrt{91}}{3}$,
$\cot \theta = \frac{3}{\sqrt{91}}, \sec \theta = \frac{10}{3}, \csc \theta = \frac{10}{\sqrt{91}}$

25. $\sin \theta = \frac{11}{13}, \cos \theta = \frac{4\sqrt{3}}{13}, \tan \theta = \frac{11}{4\sqrt{3}}$
$\cot \theta = \frac{4\sqrt{3}}{11}, \sec \theta = \frac{13}{4\sqrt{3}}, \csc \theta = \frac{13}{11}$

27. $\sin\theta = \frac{12}{\sqrt{153}}$, $\cos\theta = \frac{3}{\sqrt{153}}$, $\tan\theta = 4$,

$\cot\theta = \frac{1}{4}$, $\sec\theta = \frac{\sqrt{153}}{3}$, $\csc\theta = \frac{\sqrt{153}}{12}$

29. $\sin A = \frac{2b}{c}$, $\cos A = \frac{b}{c}$, $\tan A = 2$,

$\cot A = \frac{1}{2}$, $\sec A = \frac{c}{b}$, $\csc A = \frac{c}{2b}$

31. $a = 12$ **33.** $a = 9$ **35.** $c = 26$ **37.** $c = 228$

39. $c = \frac{7\sqrt{3}}{3}$; $b = \frac{7\sqrt{3}}{6}$ **41.** $c = 2.5$; $a = 1.5$ **43.** 9 ft^2

45. 160 mm^2 **47.** 0.697 **51.** $\sin\theta = \cos\alpha$, $\csc\theta = \sec\alpha$, $\tan\theta = \cot\alpha$

EXERCISES 2.2-2

1. $-1 + 1/\sqrt{2}$ **3.** 1 **5.** $2\sqrt{3} - 2$ **7.** 1 **9.** 1 **11.** $\frac{3}{2}$
13. True **15.** True **17.** True **19.** True **21.** True
23. True **25.** True **27.** $\theta = 45°$ **29.** $\theta = 30°$
31. $\theta = 60°$ **33.** 18 mi^2 **35.** 26.6 ft^2
37. side $= 2\sqrt{1200}$ cm $= 40\sqrt{3}$ cm $= 69$ cm
area $= 20\sqrt{1200}$ cm^2 $= 400\sqrt{3}$ cm^2 $= 690$ cm^2
39. $a = 3.0$ ft, $b = 5.2$ ft **41.** $b = 16.0$ cm, $c = 22.6$ cm
43. $a = 26$ mm, $c = 52$ mm **45.** $a = 5.1$ ft, $c = 5.4$ ft
47. $b = 128$ cm, $c = 197$ cm **49.** 23 ft 18 in **51.** 1000 ft^2
53. 42.9 ft **55.** 3.9 ft **57.** 5 cm **59.** 57.73 cm **61.** 12 in.
63. Area $= 600\sqrt{3}$ ft^2

EXERCISES 2.2-3

1. $\sin A = \frac{3}{5}$ **3.** $\sec\theta = \frac{13}{5}$ **5.** $\cos\alpha = 0.625$ **7.** sec 40°
9. $\tan(2\pi/9)$ **11.** 1 **13.** 1 **15.** -5 **17.** k
19. $\sin\theta = 0.7661$, $\cos\theta = 0.6428$, $\tan\theta = 1.1918$,
$\cot\theta = 0.8391$, $\sec\theta = 1.5557$, $\csc\theta = 1.3053$
21. $\sin\theta = 0.6428$, $\cos\theta = 0.7660$, $\tan\theta = 0.8392$,
$\cot\theta = 1.1917$, $\sec\theta = 1.3054$, $\csc\theta = 1.5557$
23. $\sin\theta = \frac{1}{4}$, $\cos\theta = \sqrt{15}/4$, $\tan\theta = 1/\sqrt{15}$,
$\cot\theta = \sqrt{15}$, $\sec\theta = 4/\sqrt{15}$, $\csc\theta = 4$
25. $\sin\theta = 3/\sqrt{34}$, $\cos\theta = 5/\sqrt{34}$, $\tan\theta = \frac{3}{5}$,
$\cot\theta = \frac{5}{3}$, $\sec\theta = \sqrt{34}/5$, $\csc\theta = \sqrt{34}/3$
27. $\sin\theta = 2/\sqrt{13}$, $\cos\theta = 3/\sqrt{13}$, $\tan\theta = \frac{2}{3}$,
$\cot\theta = \frac{3}{2}$, $\sec\theta = \sqrt{13}/3$, $\csc\theta = \sqrt{13}/2$
29. $\sin\theta = 0.99$, $\cos\theta = 0.14$, $\tan\theta = 6.90$,
$\cot\theta = 0.14$, $\sec\theta = 7.14$, $\csc\theta = 1.01$
31. $\sin\theta = \sqrt{1 - a^2}$, $\tan\theta = (\sqrt{1 - a^2})/a$,
$\cot\theta = a/\sqrt{1 - a^2}$, $\sec\theta = 1/a$, $\csc\theta = 1/\sqrt{1 - a^2}$

EXERCISES 2.3-1

1. 0.3746 **3.** 0.1405 **5.** 0.0150 **7.** 0.9803 **9.** 0.8972
11. 0.1980 **13.** 0.9825 **15.** 0.8160 **17.** 0.1665 **19.** 0.1977
21. 0.5387 **23.** 0.3148 **25.** 0.0117 **27.** 0.3190 **29.** 0.0002
31. 0.130 **33.** 0.021 **35.** 0.771 **37.** 0.383 **39.** 0.643
41. 0.509 **43.** 0.83910 **45.** 5.24084 **47.** 0.51239
49. 2.84171 **51.** 1.00015 **53.** 1.42110 **55.** 2.29525
57. 0.71511 **59.** 0.90040 **61.** 2.52571 **63.** 0.313 **65.** 0.968
67. 0.466 **69.** -0.623 **71.** 4.096 **73.** 0.143 **75.** 0.409
77. 0.598 **79.** 0.372 **81.** 1.514 **83.** 0.0111 **85.** 1.0000
87. 0.0557 **89.** No, $0.8415 \neq 0.0175$ **91.** No, $0.1017 \neq 0.1086$
93. No, $0.8473 \neq 1.9160$ **95.** No, $0.8992 \neq 0.7302$
97. No, $0.8660 \neq 0.0183$ **99.** False **101.** False **103.** False
105. False **107.** False

EXERCISES 2.3-2

1. $a = 2.9$ ft, $b = 4.0$ ft **3.** 8.2 ft **5.** 65.8 mm, 189 mm
7. 27.7 mi, 21.5 mi **9.** 7.7 ft, 12 ft **11.** 7.8 ft **13.** 19 m^2
15. 38 ft **17.** 38 in. $= b$, 57 in. $= c$, $120° = \theta$ **19.** 3.0 ft
21. 680 ft **23.** No. A is 84 ft tall, B is 570 ft tall
25. Height = 25 m + your height. Distance = 34 m
27. 1000 ft **29.** 135 m **31.** 119 ft **33.** 186 ft **35.** **(a)** 1.1 ft **(b)** 7.2 ft **37.** 1.4321 Å **39.** 3.95 Å, 7.00 Å **41.** 166°
43. 5.0 in. **47.** 22 **49.** **(a)** 51 ft **(b)** 8.1 ft **51.** 2000 ft
53. Where the bisectors of the angle meet, 8.66 ft from each corner **55.** 89 ft **57.** 17.3 ft

EXERCISES 2.4

1. 60° **3.** 45° **5.** 45° **7.** 45° **9.** 60° **11.** 30° **13.** 60°
15. 60° **17.** 14.3° **19.** 64.8° **21.** 6.3° **23.** 67.0°
25. 46.2° **27.** 89.4° **29.** 19.9° **31.** 67.7° **33.** 15.6°
35. 89.9° **37.** 1.400 **39.** 0.695 **41.** 0.277 **43.** 0.178
45. 0.176 **47.** 0.191 **49.** 78.55 **51.** 56.31° **53.** 46.95°
55. 28.96° **57.** 72.34° **59.** 111.40° **61.** 55.21° **63.** 0.97°
65. 124.82° **67.** 16.25° **69.** 84.26° **71.** 17.66° **73.** 71.44°
75. 198.00° **77.** 40.97° **79.** 59.13° **81.** 1.176 **83.** 0.007
85. 1.231 **87.** 0.324 **89.** 0.225 **91.** -0.315 **93.** $1/\sqrt{2}$
95. 1.532 **97.** 1 **99.** $\frac{24}{25}$ **101.** 1 **103.** $\sqrt{3}$ **105.** $\frac{13}{5}$
107. $\frac{\pi}{6}$ rad **109.** 0.71 rad **111.** 3.91 rad **113.** 14.10
115. 2.00 **117.** 0.55 rad **119.** 0.59 rad **121.** 0.92 rad
123. 1.22 rad **125.** 0.44 rad **127.** No **129.** No
131. No **145.** 63 ft^2 **147.** 1370 mm^2 **149.** 14°
153. 54° **155.** **(a)** 19.9° **(b)** 44.0 ft **157.** Yes. Angle of elevation of mountain is 25.5°, whereas the ridges is 20°.
163. 11.8° **165.** 48°,4.9′ **167.** 35.3° **169.** 134.1°
171. 12.7° **173.** 70°

EXERCISES 2.5

1. 4.1°, 39 ft **3.** 33.8 ft
5. **(a)** The tangent of the angle of the slope above the horizontal.
(b) 15% **(c)** 15.2% **(d)** 3.4° **(e)** 317 ft

7. 27° **9.** 86 ft **11.** 39° for 50 mm lens, 15° for 135 mm lens. The object will occupy a larger fraction of the frame with the 135 mm lens, and therefore it will occupy a larger fraction of the picture area. **13.** 520 mm, telephoto lens.
15. (a) 1.9×10^6 ft^2 or ≈ 2 million square feet **(b)** 5600 ft $\simeq$ 1.1 mile **17.** 64.0 ft, 4850 ft^2 **19. (a)** 4.7° **(b)** 21 ft, 24 ft **21.** 21.9 ft, 31.6° **23.** 12,300 km **25.** 34.4 ft **27.** 1.63 m, 70.5° **29.** 719.2 ft **31. (a)** 1.5×10^5 ft^2 **(b)** 7.7×10^4 ft^2 **33. (a)** 120° **(b)** 2.478 Å **35.** 96 ft **37.** 219 yd **39.** Supports: 3.9 ft, 5.7 ft; height: 5.5 ft **41.** No, no **43. (a)** 22 ft **(b)** 23° **45. (a)** 2890 mi **(b)** 4520 mi **47.** Approx. 21 mi **49.** 15.8 in., 31.8° **51.** 53 m^2, 231 m^2, 231 m^2 **53.** 45 cm **55.** 44.8 mm **57.** 12.5 cm **59.** 19° **61.** 8.1°, no **63.** 62, 8 mi high; it is over a point 95.2 mi from station A and 54.8 mi from Station B. **65.** yes **67.** 60° **69.** 27.7 ft^2 **71.** 20° **73.** 26° **75.** 26° **77. (a)** 8.66 m, 5.00 m **(b)** 45.3 cm, 36.3 cm **(c)** 25 ft **(d)** 17.3 mm, 43.9° **(e)** 31.6 in, 45.3° **(f)** 38.9° with x axis, 51.1° with y axis, 2.88 m **79.** $\alpha = 28.8°$, $\beta = 18.9°$, $\theta = 11.1°$ **81.** $(-1.80$ in., 6.51 in.$)$ **83.** $(32.1$ m, -38.3 m$)$ **85. (a)** 3.25 cm **(b)** 142° **87. (a)** 159.8 mm **(b)** 306.7° **89.** Yes **91.** Maximum value is 1 for 0°.

EXERCISES 2.6

1. sin 42° **3.** cos 31°44′ **5.** tan 44°32′ **7.** $\sec(90 - \theta)°$ **9.** sec 66° **11.** $\tan(90 - x)°$ **13.** csc 17.65° **15.** sec 89.987° **17.** sec 69° **19.** $\sin\left(\dfrac{\pi - 2}{2}\right)$ **21.** $\cos\frac{\pi}{3}$ **23.** $\sec\frac{\pi}{10}$ **25.** sin(1.364) **27.** tan 1°1″ **29.** csc 50° **31.** tan 51°36′ **33.** $\cos(90° - y°)$ **35.** $\theta = 45°$ **37.** $\theta = 10°$ **39.** $\theta = 18°$ **41.** $\theta = 12°$ **43.** $\theta = 5 \text{ rad} = \left(\dfrac{900}{\pi}\right)°$ **45.** $\theta = 7.5°$ **47.** Not possible **49.** $\theta = \left(\dfrac{90}{n+1}\right)°$ **51.** $x = \left[\dfrac{90 - 4y}{3}\right]°$ **53.** $x = \left[\dfrac{89 - 2y}{3}\right]°$ **55.** True **57.** True **59.** A and B are complimentary

EXERCISES 2.7

1. 920 ft **3.** 0.10 rad, 0.20 mi **5. (a)** 2.7° **(b)** Decreases by 0.2° **7.** 0.071 in. **9.** 1.4 min **11.** $\frac{20}{13}$ **13.** 3.7° **15.** For 8°; Small-angle approx: $h = 1.68$ m. For sin 8°; $h = 1.67$ m. Percent error = 0.6%. For 15°; Small-angle approx: $h = 3.14$ m. For sin 15°; $h = 3.11$ m. Percent error = 1%. **17. (a)** 7.4 km × 14 km **(b)** 0.053° × 0.10° **19. (a)** 42° from Phobor, 17° from Deimos. Not safe to use small-angle approximation. Approx. 80 times and 34 times our moon's average diameter. **(b)** 3.8 min **21.** 0.37 in. **23.** 3.65 mm **25. (a)** (i) 0.84″ (ii) 0.025″ **(b)** No (it subtends only 0.0071″) **27.** 270 km

REVIEW EXERCISES

1.

	sin	cos	tan	cot	sec	csc
(a)	0.4802	0.8771	0.5475	1.8265	1.1401	2.0824
(b)	0.9857	0.1685	5.8502	0.1709	5.9351	1.0145
(c)	0.8342	0.5514	1.5129	0.6610	1.8136	1.1987
(d)	0.7257	0.6880	1.0547	0.9482	1.4534	1.3780
(e)	0.3827	0.9239	0.4142	2.4142	1.0824	2.6131
(f)	0.9561	0.2932	3.2609	0.3067	3.4108	1.0460
(g)	0.8164	0.5775	1.4136	0.7074	1.7315	1.2249
(h)	0.5128	0.8585	0.5973	1.6742	1.1648	1.9501

3.

(a)	$\frac{8}{17}$	—	$\frac{8}{15}$	$\frac{15}{8}$	$\frac{17}{15}$	$\frac{17}{8}$
(b)	$\frac{5}{9}$	$\sqrt{56}/9$	$5/\sqrt{56}$	$\sqrt{56}/5$	$9/\sqrt{56}$	—
(c)	$3/\sqrt{58}$	$7/\sqrt{58}$	—	$\frac{7}{3}$	$\sqrt{58}/7$	$\sqrt{58}/3$
(d)	0.946	—	2.92	0.342	3.09	1.06

5. 5.34 in. **7. (a)** x **(b)** $\sqrt{1 - x^2}$ **(c)** $x/\sqrt{1 + x^2}$ **9. (a)** 0.59 **(b)** 0.21 **(c)** 0.38 **(d)** 0.38 **(e)** 0.72, 1.23 **(f)** 0.09, 0.78 **11. (a)** 42.80° **(b)** 14.15° **(c)** 45.47° **(d)** 55.52° **(e)** 57.22° **(f)** 11.42° **13.** 7.59°, 18.4° **15.** 143 ft^2 **17.** 17.2° **19.** 18° **21.** 66.8 ft

CHAPTER 3

EXERCISES 3.1-1

1. $\sin\theta = \frac{4}{5}$, $\cos\theta = \frac{-3}{5}$, $\tan\theta = \frac{-4}{3}$, $\cot\theta = \frac{-3}{4}$, $\sec\theta = \frac{-5}{3}$, $\csc\theta = \frac{5}{4}$
3. $\sin\theta = \frac{-4}{5}$, $\cos\theta = \frac{-3}{5}$, $\tan\theta = \frac{4}{3}$, $\cot\theta = \frac{3}{4}$, $\sec\theta = \frac{-5}{3}$, $\csc\theta = \frac{-5}{4}$
5. $\sin\theta = -1/\sqrt{2}$, $\cos\theta = -1/\sqrt{2}$, $\tan\theta = 1$, $\cot\theta = 1$, $\sec\theta = -\sqrt{2}$, $\csc\theta = -\sqrt{2}$
7. $\sin\theta = 2/\sqrt{5}$, $\cos\theta = -1/\sqrt{5}$, $\tan\theta = -2$, $\cot\theta = \frac{-1}{2}$, $\sec\theta = -\sqrt{5}$, $\csc\theta = \sqrt{5}/2$
9. $\sin\theta = -1/\sqrt{2}$, $\cos\theta = 1/\sqrt{2}$, $\tan\theta = -1$, $\cot\theta = -1$, $\sec\theta = \sqrt{2}$, $\csc\theta = -\sqrt{2}$
11. $\sin\theta = -1$, $\cos\theta = 0$, $\tan\theta$ (undefined), $\cot\theta = 0$, $\sec\theta$ (undefined), $\csc\theta = -1$
13. $\sin\theta = -0.92$, $\cos\theta = 0.40$, $\tan\theta = -2.3$, $\cot\theta = 0.43$, $\sec\theta = 2.5$, $\csc\theta = -1.1$
15. $\sin\theta = -0.67$, $\cos\theta = -0.74$, $\tan\theta = 0.91$, $\cot\theta = 1.1$, $\sec\theta = -1.4$, $\csc\theta = -1.5$
17. $\sin\theta = 0$, $\cos\theta = -1$, $\tan\theta = 0$, $\cot\theta$ (undefined), $\sec\theta = -1$, $\csc\theta$ (undefined)
19. $\sin 225° = -\sqrt{2}/2$, $\cos 225° = -\sqrt{2}/2$, $\tan 225° = 1$
21. $\cos 150° = -\sqrt{3}/2$, $\sec 150° = -2/\sqrt{3}$, $\cot 150° = -\sqrt{3}$
23. $\sin 210° = -1/2$, $\cos 210° = -\sqrt{3}/2$, $\tan 210° = 1/\sqrt{3}$
25. $\sin 4\pi/3 = -\sqrt{3}/2$, $\sec 4\pi/3 = -2$, $\tan 4\pi/3 = \sqrt{3}$
27. $\sin 390° = \frac{1}{2}$, $\cos 390° = \sqrt{3}/2$ **29.** $\sin 570° = \frac{-1}{2}$, $\cot 570° = \sqrt{3}$ **31.** $\sin\frac{\pi}{2} = 1$, $\cos\frac{\pi}{2} = 0$, $\tan\frac{\pi}{2}$ undefined) **33.** $\sin 270° = -1$, $\cot 270° = 0$ **35.** $\sin 3\pi = 0$, $\cos 3\pi = 1$

37. $\sec 11\pi = -1$, $\csc 11\pi$ (undefined) **39.** 1 **41.** Undefined
43. -1 **45.** 1 **47.** 0 **49.** $r = 0$

EXERCISES 3.1-2

1. Quadrant I **3.** Quadrant III **5.** Quadrant III
7. Quadrants III and IV **9.** Quadrants I and IV
11. **(a)** Quadrant I **(b)** None

13. $\sin\theta = \frac{3}{4}$, $\cos\theta = \dfrac{\sqrt{7}}{4}$, $\tan\theta = \dfrac{3}{\sqrt{7}}$,

$\cot\theta = \dfrac{\sqrt{7}}{3}$, $\sec\theta = \dfrac{4}{\sqrt{7}}$, $\csc\theta = \frac{4}{3}$

15. $\sin\theta = \frac{3}{5}$, $\cos\theta = \frac{4}{5}$, $\tan\theta = \frac{3}{4}$,
$\cot\theta = \frac{4}{3}$, $\sec\theta = \frac{5}{4}$, $\csc\theta = \frac{5}{3}$

17. $\sin\theta = \frac{5}{13}$, $\cos\theta = \frac{-12}{13}$, $\tan\theta = \frac{5}{-12}$,
$\cot\theta = \frac{-12}{5}$, $\sec\theta = \frac{13}{-12}$, $\csc\theta = \frac{13}{5}$

19. $\sin\theta = \frac{-4}{5}$, $\cos\theta = \frac{-3}{5}$, $\tan\theta = \frac{4}{3}$,
$\cot\theta = \frac{3}{4}$, $\sec\theta = \frac{5}{-3}$, $\csc\theta = \frac{5}{-4}$

21. $\sin\theta = \frac{4}{5}$, $\cos\theta = \frac{3}{5}$, $\tan\theta = \frac{4}{3}$,
$\cot\theta = \frac{3}{4}$, $\sec\theta = \frac{5}{3}$, $\csc\theta = \frac{5}{4}$

23. $\sin\theta = \dfrac{1}{\sqrt{10}}$, $\cos\theta = \dfrac{-3}{\sqrt{10}}$, $\tan\theta = \frac{1}{-3}$,

$\cot\theta = -3$, $\sec\theta = \dfrac{\sqrt{10}}{-3}$, $\csc\theta = \sqrt{10}$

25. Impossible

27. $\sin\theta = -0.657$, $\cos\theta = -0.754$, $\tan\theta = 0.871$,
$\cot\theta = 1.15$, $\sec\theta = -1.33$, $\csc\theta = -1.53$

29. $\sin\theta = 0.909$, $\cos\theta = -0.413$, $\tan\theta = -2.20$,
$\cot\theta = -0.454$, $\sec\theta = -2.42$, $\csc\theta = 1.10$

31. In quadrant III: $\sin\theta = \frac{-1}{3}$, $\cos\theta = \dfrac{-2\sqrt{2}}{3}$, $\tan\theta = \dfrac{\sqrt{2}}{4}$,

$\cot\theta = \dfrac{4}{\sqrt{2}}$, $\sec\theta = \dfrac{-3}{2\sqrt{2}}$, $\csc\theta = \dfrac{3}{-1}$

In quadrant IV: $\sin\theta = \frac{-1}{3}$, $\cos\theta = \dfrac{2\sqrt{2}}{3}$, $\tan\theta = \dfrac{-\sqrt{2}}{4}$,

$\cot\theta = \dfrac{4}{-\sqrt{2}}$, $\sec\theta = \dfrac{3}{2\sqrt{2}}$, $\csc\theta = \frac{3}{-1}$

33. In quadrant II: $\sin\theta = \dfrac{1}{\sqrt{2}}$, $\cos\theta = \dfrac{-1}{\sqrt{2}}$, $\tan\theta = -1$,

$\cot\theta = -1$, $\sec\theta = -\sqrt{2}$, $\csc\theta = \sqrt{2}$

In quadrant IV: $\sin\theta = \dfrac{-1}{\sqrt{2}}$, $\cos\theta = \dfrac{1}{\sqrt{2}}$, $\tan\theta = -1$,

$\cot\theta = -1$, $\sec\theta = \sqrt{2}$, $\csc\theta = -\sqrt{2}$

35. In quadrant II: $\sin\theta = \dfrac{\sqrt{5}}{3}$, $\cos\theta = \frac{-2}{3}$, $\tan\theta = \dfrac{\sqrt{5}}{-2}$,

$\cot\theta = \dfrac{-2}{\sqrt{5}}$, $\sec\theta = \frac{3}{-2}$, $\csc\theta = \dfrac{3}{\sqrt{5}}$

In quadrant III: $\sin\theta = \dfrac{-\sqrt{5}}{3}$, $\cos\theta = \frac{-2}{3}$, $\tan\theta = \dfrac{\sqrt{5}}{2}$,

$\cot\theta = \dfrac{2}{\sqrt{5}}$, $\sec\theta = \frac{3}{-2}$, $\csc\theta = \dfrac{3}{-\sqrt{5}}$

37. In quadrant I: $\sin\theta = \dfrac{1}{\sqrt{26}}$, $\cos\theta = \dfrac{5}{\sqrt{26}}$, $\tan\theta = \frac{1}{5}$,

$\cot\theta = 5$, $\sec\theta = \dfrac{\sqrt{26}}{5}$, $\csc\theta = \sqrt{26}$

In quadrant III: $\sin\theta = \dfrac{-1}{\sqrt{26}}$, $\cos\theta = \dfrac{-5}{\sqrt{26}}$, $\tan\theta = \frac{1}{5}$,

$\cot\theta = 5$, $\sec\theta = \dfrac{\sqrt{26}}{-5}$, $\csc\theta = -\sqrt{26}$

39. In quadrant I: $\sin\theta = 0.521$, $\cos\theta = 0.855$, $\tan\theta = 0.609$,
$\cot\theta = 1.64$, $\sec\theta = 1.17$, $\csc\theta = 1.92$
In quadrant III: $\sin\theta = -0.52$, $\cos\theta = -0.85$,
$\tan\theta = 0.609$, $\cot\theta = 1.64$, $\sec\theta = -1.17$,
$\csc\theta = -1.92$

41. In quadrant I: $\sin\theta = 0.94$, $\cos\theta = 0.35$, $\tan\theta = 2.7$,
$\cot\theta = 0.37$, $\sec\theta = 2.9$, $\csc\theta = 1.1$
In quadrant IV: $\sin\theta = -0.94$, $\cos\theta = 0.35$, $\tan\theta = -2.7$,
$\cot\theta = -0.37$, $\sec\theta = 2.9$, $\csc\theta = -1.1$

43. In quadrant I: $\sin\theta = \dfrac{1}{\sqrt{1+p^2}}$, $\cos\theta = \dfrac{p}{\sqrt{1+p^2}}$,

$\tan\theta = \dfrac{1}{p}$, $\cot\theta = p$, $\sec\theta = \dfrac{\sqrt{1+p^2}}{p}$,

$\csc\theta = \sqrt{1+p^2}$

In quadrant III: $\sin\theta = \dfrac{-1}{\sqrt{1+p^2}}$, $\cos\theta = \dfrac{-p}{\sqrt{1+p^2}}$,

$\tan\theta = \dfrac{1}{p}$, $\cot\theta = p$, $\sec\theta = \dfrac{\sqrt{1+p^2}}{-p}$,

$\csc\theta = -\sqrt{1+p^2}$

45. $\frac{5}{4}$ **47.** 0.58 **49.** 0.72 **51.** **(a)** Yes **(b)** No

EXERCISES 3.1-3

1. $\sqrt{2}/2$ **3.** $-\sqrt{3}/2$ **5.** 2 **7.** $-\sqrt{3}/2$ **9.** $-1/\sqrt{3}$
11. -1 **13.** 0 **15.** -1 **17.** Undefined **19.** True
21. True **23.** Positive **25.** $(\sin\theta - \cos\theta)/(\sin\theta + \cos\theta)$

EXERCISES 3.2

1. 78° **3.** 23° **5.** 23° **7.** 80° **9.** 44° **11.** 82° **13.** 63°

15. 80° **17.** 69°50′ **19.** 14°30′ **21.** 74°17′44″ **23.** 41°40′46″ **25.** 88.8° **27.** 1.6° **29.** 85.8° **31.** 1.4 **33.** 0.53 **35.** 0.83 **37.** 1.15 **39.** 1.3 **41.** 0 **43.** 0.97 **45.** 0.942 **47.** 1.45 **49.** 0.39 **51.** 0.94 **53.** 0.90 **55.** 1.257 **57.** $\pi°$ **59.** 0.63 **61.** 1.14 **63.** $-2/\sqrt{3}, -\sqrt{3}$ **65.** $-1/2, 1/\sqrt{3}$ **67.** $-\sqrt{3}/2, -\sqrt{3}/3$ **69.** $-1/\sqrt{2}, -1/\sqrt{2}$ **71.** $-\sqrt{3}/2, -2$ **73.** $-1, 1/\sqrt{2}$ **75.** $\sqrt{2}/2, \sqrt{2}/2$ **77.** $1, -1/\sqrt{2}$ **79.** $2, -2\sqrt{3}/3$ **81.** $-1, -\sqrt{2}$ **83.** −0.643 **85.** −2.924 **87.** −0.602 **89.** −5.759 **91.** −1.064 **93.** −0.268 **95.** 0.685 **97.** −0.281 **99.** −3.981 **101.** −0.451 **103.** −1.347 **105.** 1.165 **107.** −3.087 **109.** −0.758 **111.** −0.342 **113.** −1.266 **115.** −0.959 **117.** −0.431 **119.** −0.751 **121.** −20.332 **123.** 0.999 **125.** −0.975 **127.** 1.082 **129.** −1.236 **131.** 0.562 **133.** 0.642 **135.** 0.346 **137.** −0.992 **139.** −1.04 **141.** −0.093 **143.** 0.244 **145.** −0.735 **147.** 1.08 **149.** 0.624 **151.** 1.23 **153.** −0.631 **155.** 0.993 **157.** 0.223 **159.** −4.02 **161.** −2.21 **163.** 0.483 **165.** −7.09 **167.** −0.841 **169.** 0.095 **171.** −0.642 **173.** 3.14 **175.** $\sqrt{2}/4$ **177.** $\frac{4}{3}$ **179.** −3 **181.** −6 **183.** $\sqrt{3}/4$ **185.** 2 **187** True **189** True **191.** $x = 0.74, y = 0.67$ **193.** $x = -0.921, y = -0.391$ **195.** $x = 0.968, y = -0.250$ **199.** $x = -0.43, y = 1.3$ **201.** $x = -2.09, y = -1.62$ **203.** $x = -6.40, y = 0$ **205.** $x = 0.00, y = 0.82$

EXERCISES 3.3

1. 60°, 120° **3.** 30°, 150° **5.** 60°, 240° **7.** 0°, 180° **9.** 180° **11.** 45°, 315° **13.** $\frac{4\pi}{3}, \frac{5\pi}{3}$ **15.** $\frac{\pi}{3}, \frac{5\pi}{3}$ **17.** 0 **19.** $\frac{5\pi}{6}, \frac{11\pi}{6}$ **21.** $\frac{3\pi}{2}$ **23.** $\frac{\pi}{2}, \frac{3\pi}{2}$ **25.** $0, \pi$ **27.** 7.00°, 187.00° **29.** 252.31°, 287.69° **31.** 1.30°, 178.70° **33.** 79.40°, 280.60° **35.** 27.20°, 207.20° **37.** 100.40°, 259.60° **39.** 16.20°, 163.80° **41.** 27.90°, 332.10° **43.** 90.84°, 359.16° **45.** 3.81, 5.61 **47.** 1.70, 4.84 **49.** 4.51, 4.91 **51.** 0.92, 5.36 **53.** 0.33, 2.82 **55.** 1.71, 4.57 **57.** 5.95, 1.47 **59.** 2.39, 3.89 **61.** 0.50, 1.07 **63.** 35.0°, 395.0°, 755.0°; 325.0°, 685.0°, 1045.0° **65.** 123.0°, 483.0°, 843.0°; 303.0°, 663.0°, 1023.0° **67.** 3.5, 9.8, 16.1; 5.9, 12.2, 18.5 **69.** 1.1, 7.3, 13.6; 4.2, 10.5, 16.8 **71.** 4.3, 10.6, 16.9; 2.0, 8.3, 14.3 **73.** 25.00°, −335.00°; 335.00°, −25.00° **75.** 248.00°, −112.00°; 292.00°, −68.00° **77.** 1.75, −4.54; 4.89, −1.40 **79.** 0.12, −6.16; 3.02, −3.26 **81.** 4.43, −1.86; 5.00, −1.28 **83.** 0.23, 6.51, 12.79, −6.06, −12.34, −18.62; 3.37, 9.65, 15.93, −2.91, −9.20, −15.48 **85.** 2.06, 8.34, 14.63, −4.22, −10.51, −16.79; 4.22, 10.51, 16.79, −2.06, −8.34, −14.63 **87.** 20.50°, 380.50°, 740.50°, −339.50°, −699.50°, −1059.50°; 159.50°, 519.50°, 879.50°, −200.50°, −560.50°, −920.50° **89.** 15°, 75°, 195°, 255° **91.** 67.5°, 157.5°, 247.5°, 337.5° **93.** 52.5°, 82.5°, 142.5°, 172.5°, 232.5°, 262.5°, 322.5° **95.** 0.82, 4.93, 7.09, 11.21 **97.** 1.44, 3.01, 4.58, 6.15, 7.72, 9.29, 10.86, 12.44 **99.** 4.97, 5.85 **101.** 1.89, 5.03, 1.11, 4.25 **103.** $x = \frac{(2n+1)}{b} \cdot \frac{\pi}{2}, n = 0, 1, 2, \ldots$ **105.** $x = \frac{n\pi}{3}$ and $t = \frac{1.57 \pm 2n\pi}{8}$ **107.** 307°, $y = -0.80$ **109.** 104°, $y = 0.97$ **111.** 139°, $y = 0.66$ **113.** Not possible **115.** $r = 2.7, \theta = 239°$ **117.** $r = 1.3, \theta = 303°$

EXERCISES 3.4-1

1. tan 45° **3.** cos 10° **5.** csc 4 **7.** 6 tan 400° **9.** $\cos\theta = \frac{-3}{4}$ **11.** $\csc\theta = \frac{7}{2}$ **13.** $\cot\theta = 0.5438$ **15.** $\tan(-\theta) = -0.118$ **17.** $\tan\theta = \frac{3}{4}, \cot\theta = \frac{4}{3}, \sec\theta = \frac{5}{4}, \csc\theta = \frac{5}{3}$ **19.** $\tan\theta = 2, \cot\theta = \frac{1}{2}, \sec\theta = \sqrt{5}, \csc\theta = \sqrt{5}/2$ **21.** $\sin\theta = \sqrt{40}/7, \cos\theta = -3/7, \tan\theta = -\sqrt{40}/3, \cot\theta = -3/\sqrt{40}$ **23.** $\tan\theta = 0.770, \cot\theta = 1.30, \sec\theta = -1.26, \csc\theta = -1.64$

EXERCISES 3.4-2

1. $\sin\theta = \frac{15}{17}, \cos\theta = \frac{-8}{17}, \tan\theta = \frac{-15}{8}, \cot\theta = \frac{-8}{15}, \sec\theta = \frac{-17}{8}, \csc\theta = \frac{17}{15}$

3. $\sin\theta = \frac{-4}{5}, \cos\theta = \frac{-3}{5}, \tan\theta = \frac{4}{3}, \cot\theta = \frac{3}{4}, \sec\theta = \frac{-5}{3}, \csc\theta = \frac{-5}{4}$ **5.** $\sin\theta = -0.425, \cos\theta = -0.905, \tan\theta = 0.470, \cot\theta = 2.13, \sec\theta = -1.11, \csc\theta = -2.35$ **7.** $\sin\theta = 1/\sqrt{17}, \cos\theta = -4/\sqrt{17}, \tan\theta = -\frac{1}{4}, \cot\theta = -4, \sec\theta = -\sqrt{17}/4, \csc\theta = \sqrt{17}$ **9.** $\sin\theta = -0.24, \cos\theta = 0.97, \tan\theta = -0.25, \cot\theta = -4.0, \sec\theta = 1.0, \csc\theta = -4.2$ **11.** $\sin\theta = 0.20, \cos\theta = -0.98, \tan\theta = -0.20, \cot\theta = -4.9, \sec\theta = -1.0, \csc\theta = 5.0$ **13.** $\sin\theta = a/b, \cos\theta = (-\sqrt{b^2 - a^2})/b, \tan\theta = -a/(\sqrt{b^2 - a^2}), \cot\theta = (-\sqrt{b^2 - a^2})/a, \sec\theta = -b/(\sqrt{b^2 - a^2}), \csc\theta = b/a$

15. In quadrant I: $\sin\theta = \frac{1}{3}, \cos\theta = 2\sqrt{2}/3, \tan\theta = 1/2\sqrt{2}, \cot\theta = 2\sqrt{2}, \sec\theta = 3/2\sqrt{2}, \cos\theta = 3$
In quadrant II: $\sin\theta = \frac{1}{3}, \cos\theta = -2\sqrt{2}/3, \tan\theta = -1/2\sqrt{2}, \cot\theta = -2\sqrt{2}, \sec\theta = -3/2\sqrt{2}, \csc\theta = 3$

17. In quadrant I: $\sin\theta = 5/\sqrt{29}, \cos\theta = 2/\sqrt{29}, \tan\theta = \frac{5}{2}, \cot\theta = \frac{2}{5}, \sec\theta = \sqrt{29}/2, \csc\theta = \sqrt{29}/5$
In quadrant III: $\sin\theta = -5/\sqrt{29}, \cos\theta = -2/\sqrt{29}, \tan\theta = \frac{5}{2}, \cot\theta = \frac{2}{5}, \sec\theta = -\sqrt{29}/2, \csc\theta = -\sqrt{29}/5$

19. In quadrant II: $\sin\theta = \sqrt{39}/8, \cos\theta = \frac{-5}{8}, \tan\theta = -\sqrt{39}/5, \cot\theta = -5/\sqrt{39}, \sec\theta = \frac{-8}{5}, \csc\theta = 8/\sqrt{39}$
In quadrant III: $\sin\theta = -\sqrt{39}/8, \cos\theta = \frac{-5}{8}, \tan\theta = \sqrt{39}/5, \cot\theta = 5/\sqrt{39}, \sec\theta = \frac{-8}{5}, \csc\theta = -8/\sqrt{39}$

21. In quadrant I: $\sin\theta = 0.467, \cos\theta = 0.884, \tan\theta = 0.529, \cot\theta = 1.89, \sec\theta = 1.13, \csc\theta = 2.14$

In quadrant IV: $\sin\theta = -0.467$, $\cos\theta = 0.884$, $\tan\theta = -0.529$, $\cot\theta = -1.89$, $\sec\theta = 1.13$, $\csc\theta = -2.14$

23. In quadrant I: $\sin\theta = 0.47$, $\cos\theta = 0.88$, $\tan\theta = 0.53$, $\cot\theta = 1.9$, $\sec\theta = 1.1$, $\csc\theta = 2.1$
In quadrant III: $\sin\theta = -0.47$, $\cos\theta = -0.88$, $\tan\theta = 0.53$, $\cot\theta = 1.9$, $\sec\theta = -1.1$, $\csc\theta = -2.1$

25. $(1 - \sin^2 x)/\sin x$ **27.** $\sin x$ **29.** $\sin x/(\pm\sqrt{1 - \sin^2 x})$
31. $(1 + \sin^2 x)/(1 - \sin^2 x)$ **33.** $1/\sin x$ **35.** 1
37. $1 - \cos B$ **39.** $2\cos^2 y$ **41.** $\csc x$ **43.** $\cos x$ **45.** $\sec x$
47. $\sec^2 x$ **49.** 3 **51.** -3 **53.** -0.709 or -0.442
55. Identity **57.** Identity, $\cos\theta \neq 0$ **59.** Identity
61. Not an identity **63.** Identity ($\cos\theta \neq 0$) **65.** $\dfrac{-1}{\sin\theta\cos\theta}$
67. Yes **69.** $\alpha = 60°, 120°, 240°, 300°$ **71.** $x = 135°, 315°$

EXERCISES 3.4-3

1. Period $= 2\pi$ **3.** Period $= \pi$ **5.** Period $= 2\pi$
7. Not periodic **9.** $\sqrt{3}/2$ **11.** $1/\sqrt{3}$ **13.** $2/\sqrt{2}$
15. Undefined **17.** $-\sqrt{3}/2$ **19.** Period $= 180°$ or π rad
21. Period $= 200°$ or 3.49 rad **23.** Period $= 360°$ or 2π rad
25. Period $= 120°$ or 2.09 rad **27.** Period $= 180°$ or π rad
29. Period $= 360°$ or 2π rad **31.** Period $= \pi$

EXERCISES 3.4-4

1. Maximum $= 63.8$, minimum $= -63.8$
3. Maximum $= 3$, minimum $= -3$ **5.** Maximum $= 2.7$, minimum $= -2.7$ **7.** Maximum $= 1$, minimum $= 0$
9. Maximum $= 1$, minimum $= -1$ **11.** Maximum $= 1$, minimum $= -1$

REVIEW EXERCISES

1. (a) $\sin\theta = \frac{-12}{13}$, $\cos\theta = \frac{5}{13}$, $\tan\theta = \frac{-12}{5}$, $\cot\theta = \frac{5}{-12}$, $\sec\theta = \frac{13}{5}$, $\csc\theta = \frac{13}{-12}$
(b) $\sin\theta = \frac{15}{17}$, $\cos\theta = \frac{-8}{17}$, $\tan\theta = \frac{15}{-8}$, $\cot\theta = \frac{-8}{15}$, $\sec\theta = \frac{17}{-8}$, $\csc\theta = \frac{17}{15}$
(c) $\sin\theta = 0$, $\cos\theta = -1$, $\tan\theta = 0$, $\cot\theta$ (undefined), $\sec\theta = -1$, $\csc\theta$(undefined)
(d) $\sin\theta = \dfrac{-\sqrt{2}}{\sqrt{5}}$, $\cos\theta = \dfrac{-\sqrt{3}}{\sqrt{5}}$, $\tan\theta = \dfrac{\sqrt{2}}{\sqrt{3}}$, $\cot\theta = \dfrac{\sqrt{3}}{\sqrt{2}}$, $\sec\theta = \dfrac{\sqrt{5}}{-\sqrt{3}}$, $\csc\theta = \dfrac{\sqrt{5}}{-\sqrt{2}}$
(e) $\sin\theta = \dfrac{1}{\sqrt{37}}$, $\cos\theta = \dfrac{6}{\sqrt{37}}$, $\tan\theta = \frac{1}{6}$, $\cot\theta = 6$, $\sec\theta = \dfrac{\sqrt{37}}{6}$, $\csc\theta = \sqrt{37}$

3. (a) -0.7228 **(b)** 0.3166 **(c)** 2.0732 **(d)** -1.0114
(e) 0.4758 **(f)** -1.5243 **(g)** -0.8746 **(h)** -0.2679
(i) -2.5355 **(j)** 0.7975 **(k)** -1.0668 **(l)** -0.5878
(m) -0.8383 **(n)** -0.4926 **(o)** 0.9526 **(p)** 0.3455
(q) 2.0710 **(r)** 0.9898 **(s)** 1.0699 **5. (a)** 1 **(b)** 1
7. (a) $x = \frac{1}{2}\cos^{-1} y$ **(b)** $x = \frac{1}{3}\cot^{-1} y$
(c) $x = \frac{1}{4}\sin^{-1}\left(\frac{y}{3}\right)$ **(d)** $x = 2\tan^{-1}(y/5)$
(e) $x = (1/m)\cos^{-1}(y/n)$ **(f)** $x = (1/n)\sec^{-1}(y/p)$
9. (a) $\sin\theta = \frac{-5}{13}$ **(b)** $\cos\theta = \frac{12}{13}$ **(c)** $\cot\theta = \frac{-12}{5}$
11. (a) $\perp$ **(b)** $\csc^2\theta$ **(c)** $1 + \sin\theta$ **(d)** $\tan^2\theta$ **(e)** $\cot\theta$
(f) $2\sec\theta$ **13. (a)** $-\sqrt{2}/2$ **(b)** $-1/2$ **(c)** $1/\sqrt{3}$
(d) $1/\sqrt{3}$ **(e)** Undefined **(f)** Undefined **(g)** $\sqrt{2}/2$
(h) $\sqrt{3}/2$ **(i)** $1/\sqrt{3}$ **(j)** -2 **(k)** -1

CHAPTER 4

EXERCISES 4.1

1. 2.8 **3.** 33.9° **5.** 8.0 m **7.** $A = 29.6°$, $B = 104.1$, $C = 46.3°$ **9.** 16.1 **11.** 77° **13.** 8.7 cm, 14 cm **15.** 7.30 ft
17. 21.9 cm **19.** 15.0 m^2 **21.** 164 m **23.** 30° **25.** 117°
31. 17.0 ft **33.** 42 in. **35.** 2000 mi **37.** 24°
39. (a) 2.21 Å, **(b)** 2.22 Å, **(c)** 3.13 Å **41.** 2.36 Å
43. (b) 35.18°, 109.6° **45.** 11 min **47.** 79 cm, 59 cm
49. 79.9 cm **51.** 818 ft^2 **53.** 7.23 ft from the point of tangency, measured *along* the path **55.** 152 ft

EXERCISES 4.2

1. $C = 60°$, $a = 19.7$, $b = 28.8$ **3.** $A = 115°$, $a = 558$ ft, $b = 279$ ft **5.** $C = 123°12'$, $b = 2051$ mm, $c = 2363$ mm
7. $B = \frac{7\pi}{88}$, $a = 0.326$ in., $c = 0.349$ in. **9.** $A = 86.9°$, $C = 45.6°$, $b = 137$ m **11.** $A = 22.33°$, $b = 128$ ft, $c = 200$ ft
13. $B = 64°27'20''$, $a = 59.5$ m, $c = 96.4$ m **15.** $C = 44.2°$. The lengths of the sides cannot be determined. **17.** Not possible
19. $B = 0.674$ rad, $C = 1.68$ rad, $c = 119$ m **21.** Not possible
23. 450 in^2 **25.** 121 m^2 **27.** 56.0 ft^2 **29.** Not possible
31. 79 m^2 **33.** 10 cm^2
35. $\overline{AB} = 4.31$ in., $\overline{BC} = 3.13$ in., $\overline{CD} = 5.43$ in., $\overline{DE} = 4.42$ in., $\overline{EA} = 5.05$ in., $C = 125°06'$, $D = 93°24'$
37. 14 ft, 10 ft **39.** 10.0 ft **41.** 3.37 m **43.** 1.45 m **45.** No
47. $a = \dfrac{x\sin\varphi}{\sin(\varphi - \theta)}$, $b = \dfrac{x\sin\theta}{\sin(\varphi - \theta)}$, $c = \dfrac{x\sin\theta\sin\varphi}{\sin(\varphi - \theta)}$, $d = \dfrac{x\sin\theta\cos\varphi}{\sin(\varphi - \theta)}$ **49.** Vertical: 7.99 ft, 5.39 ft; horizontal: 8.08 ft **51.** 47.3 ft from pitcher's plate; 101 ft from third base **53.** 999 bu **55.** 16.1°, 33.0°, 40.9° **57.** 2.261 Å,

2.057 Å **59.** 30.2°, 42.5°, 107.3° **61.** 19.9 ft **63.** **(a)** 57.2° **(b)** 148×10^6 km **65.** 45.3 cm **67.** 132°

EXERCISES 4.3

1. 2 **3.** 0 **5.** 1 **7.** 0 **9.** 1 **11.** $B = 38.7°$, $C = 111°$, $c = 149$ mi and $B = 141°$, $C = 8.7°$, $c = 24.2$ mi **13.** Not possible **15.** $B = 0.987$ rad, $C = 0.804$ rad, $c = 9.89$ ft **17.** Not possible **19.** $B = 14.8°$, $C = 35.2°$, $c = 113$ units **21.** $B = 90.0°$, $C = 31.4°$, $c = 494$ cm **23.** 92.1 km^2 **25.** Not possible **27.** Not possible **29.** 420 in^2 **31.** No **33.** No **35.** 810 cm^2 or 2800 cm^2

EXERCISES 4.4

1.

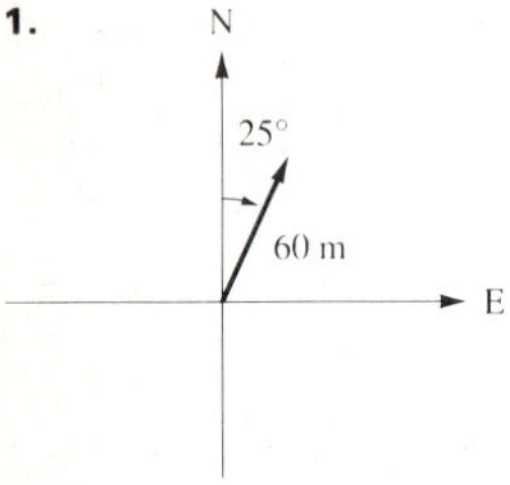

3.

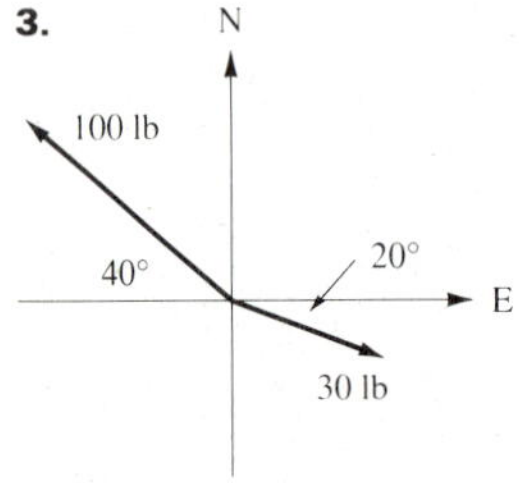

13.

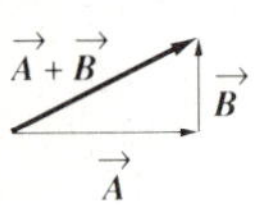

15.

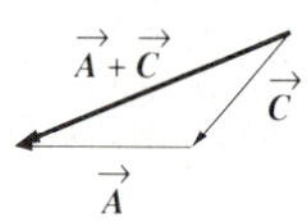

17.

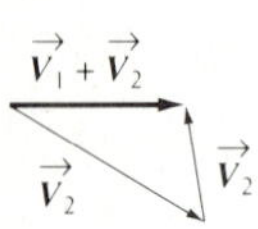

21. 108 m/s, 1.51 rad **23.** 38 mph, 0.61 rad **25.** 120 km/s, 32° N of E **27.** 120 lb; 28.2° E of N **29.** 25 m, 68° S of E **31.** 54 lb, 47° W of N **33.** 170 lb, 0.32 rad W of N **35.** 66.6 mph, 0.122 rad S of E **37.** No

39. **(a)** $-A$ has the same magnitude as A but is in the opposite direction.
(b) 50 lb S
(c) 66 mph at 20° S of E

41. **(a)** 97 ft/s, 66° N of E
(b) 136 lb, 0.266 rad N of W
(c) 440 mi, 0.45 rad S of W
(d) 150 mph, 27° N of W

43. **(a)** $Ax = 7.5$ m, $Ay = 13$ m
(b) $Ax = 22.9$ m/s, $Ay = 9.93$ m/s
(c) $Ax = 61$ mph, $Ay = -22$ mph
(d) $Ax = -29.7$ lb, $Ay = -21.6$ lb
(e) $Ax = -26.05$ cm/hr, $Ay = 96.55$ cm/hr
(f) $Ax = -126$ m/s, $Ay = -72.5$ m/s

45. **(a)** 5.0 ft, $\theta = 37°$
(b) 7.8 m/s, $\theta = 140°$
(c) 160 lb, $\theta = 321°$
(d) 19.6 lb, $\theta = 215°$
(e) 46 km/hr, $\theta = 180°$
(f) 18.2 cm, $\theta = 270°$

47. 112 mi at 28.0° S of W **49.** 28 lb, 1.7° below the horizontal **51.** 36 mph at 20° S of W **53.** 120 lb at 4.6° below the horizontal **55.** 6.3 oz **57.** 71.7 lb, 102 lb **59.** 51 lb, 29° below the horizontal to the right **61.** 32.5 lb, 46.3° above the horizontal to the right

REVIEW EXERCISES

1. **(a)** $B = 44.3°$, $C = 86.6°$, $a = 7.57$ ft
(b) $A = 46.4°$, $B = 104°$, $C = 29.4°$
(c) $A = 63.2°$, $a = 158$ cm, $c = 170$ cm
(d) $B = 63.3°$, $C = 51.9°$, $b = 51.4$ m
(e) $C = \frac{19\pi}{30}$, $a = 11.5$ ft, $b = 13.5$ ft
(f) Not possible

3. **(a)** 12 in^2 **(b)** 62.5 mm^2 **(c)** 123 m^2, 570 m^2 **(d)** Not possible **5.** 42.8 ft **9.** **(a)** 1.84 Å **(b)** 2.31 Å **11.** 110 lb, 40 lb

CHAPTER 5

EXERCISES 5.1

1. Period = 1, maximum value = 1, minimum value = 0
3. Period = 8, maximum value = 4, minimum value = -4
5. No, no, breathing is subject to irregularities; 11 to 12 *s*
9. 30 days (1 month) **11.** Period = 1 year **13.** Period = 12, maximum value = 10

15.

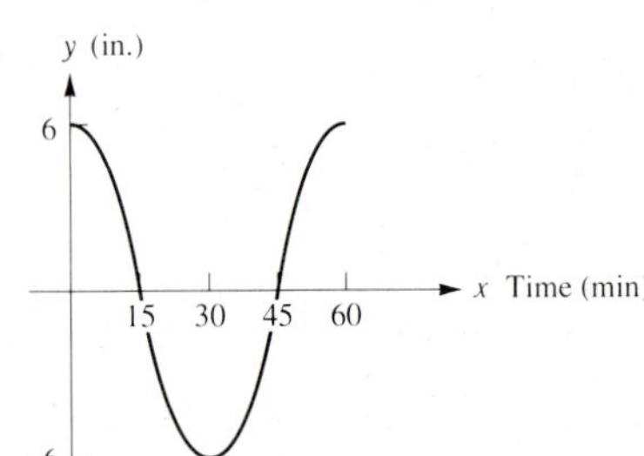

17. Period = perimeter = $2\pi r$; no (height = D, base = πD)

19. **(a)** Frequency = $\frac{1}{60}$ cycles/s, period = 1 min
(b) Frequency = $\frac{1}{86400}$ cycles/s, period = 1 day
(c) Frequency = 3.17×10^{-8} cycles/s, period = $365\frac{1}{4}$ days

21. $1.11\ldots\frac{\text{cycles}}{\text{s}}$ **23.** $58\,\frac{\text{cycles}}{\text{min}}$, 0.017 min

EXERCISES 5.2-1

1. They are identical. **3. (a)** $x = \ldots, -3\pi, -2\pi, -\pi, 0, \pi, 2\pi, 3\pi, \ldots$ **(b)** $y = 0$ **5. (a)** $x = \ldots, \frac{-3\pi}{2}, \frac{\pi}{2}, \frac{5\pi}{2}, \ldots$ **(b)** $y = 1$

7.

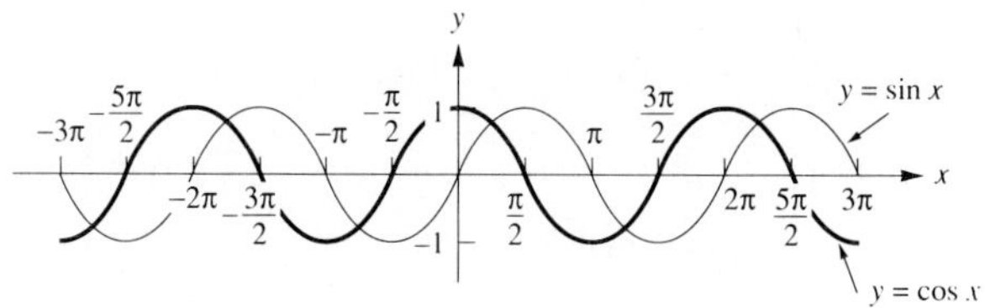

9.

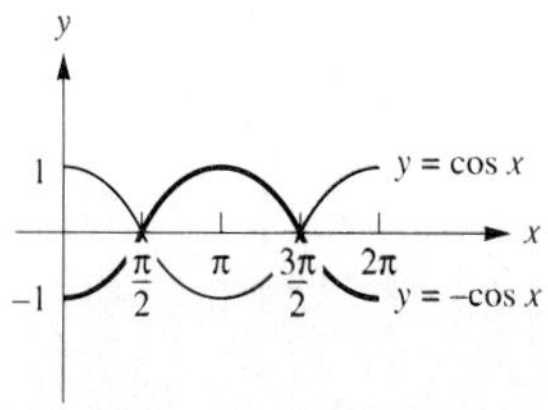

The minus sign has no effect on the period.

11. (a)

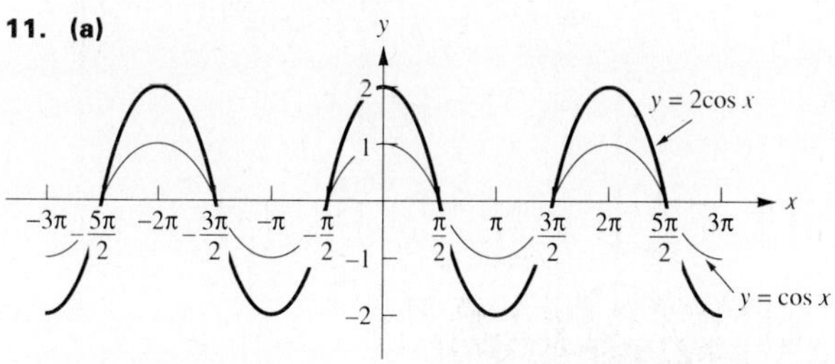

$y = \cos x \begin{cases} \text{Maximum} = 1 \\ \text{Period} = 2\pi \end{cases}$ $\quad y = 2\cos x \begin{cases} \text{Maximum} = 2 \\ \text{Period} = 2\pi \end{cases}$

(b)

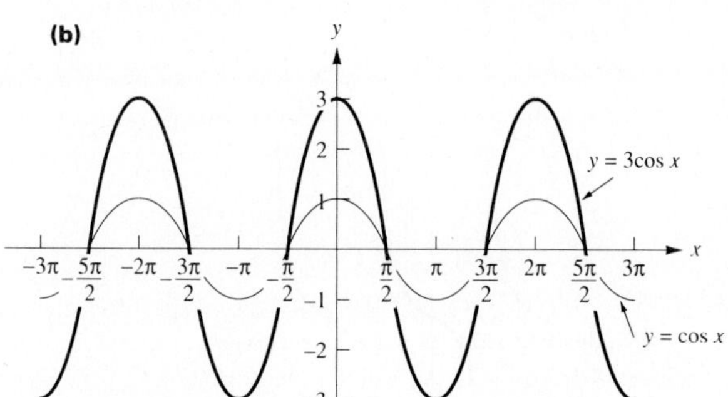

$y = \cos x \begin{cases} \text{Maximum} = 1 \\ \text{Period} = 2\pi \end{cases}$ $\quad y = 3\cos x \begin{cases} \text{Maximum} = 3 \\ \text{Period} = 2\pi \end{cases}$

(c)

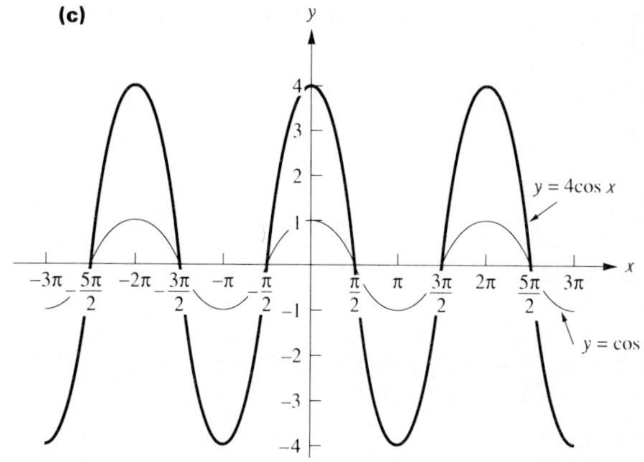

$y = \cos x \begin{cases} \text{Maximum} = 1 \\ \text{Period} = 2\pi \end{cases}$ $\quad y = 4\cos x \begin{cases} \text{Maximum} = 4 \\ \text{Period} = 2\pi \end{cases}$

13. (a) $y = \sin x$ and $y = \sin 2x$

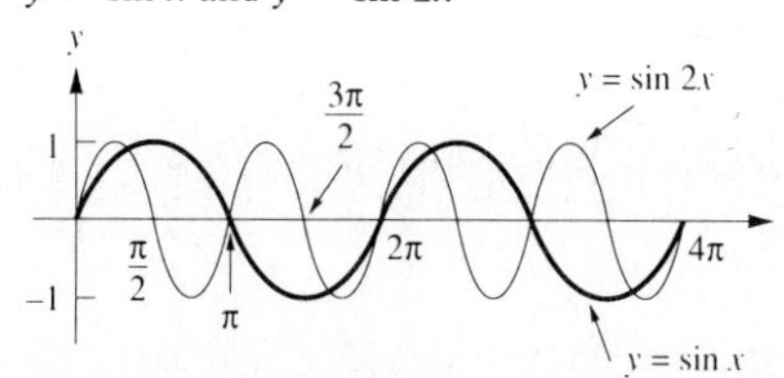

(b) $y = \sin x$ and $y = \sin 3x$

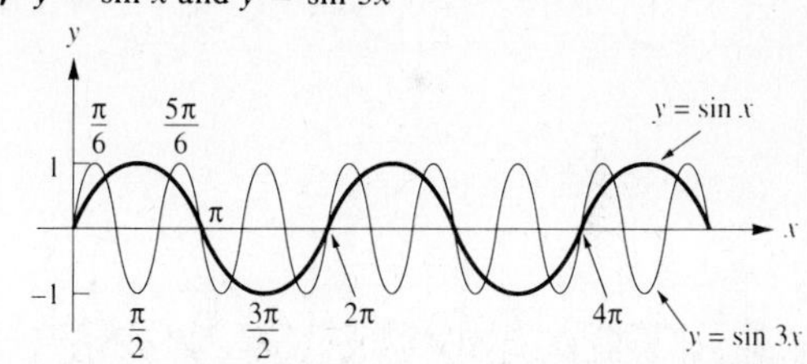

(c) $y = \sin x$ and $y = \sin \frac{x}{2}$

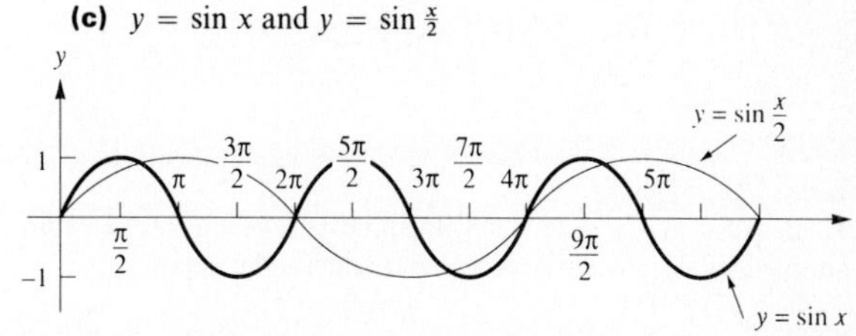

For all of the graphs, the maximum distance from the x axis is 1. The periods are

$$y = \sin x - 2\pi$$
$$y = \sin 2x - \pi$$
$$y = \sin 3x - \frac{2\pi}{3}$$
$$y = \sin \tfrac{x}{2} - 4\pi$$

The period of a graph of $\sin(bx)$, where b is the coefficient of x, is $\frac{2\pi}{b}$.

15. **(a)** Maximum $= 1$, period $= 2\pi$

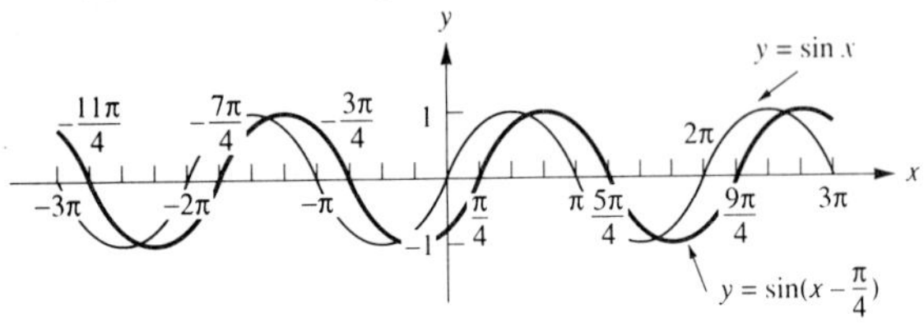

(b) Maximum $= 1$, period $= 2\pi$

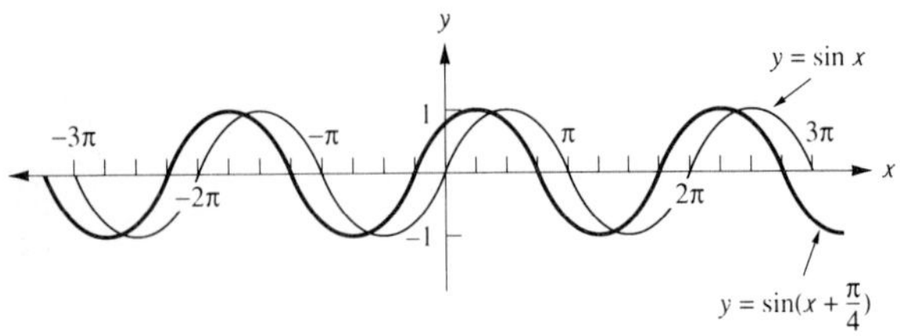

(c) Maximum $= 1$, period $= 2\pi$

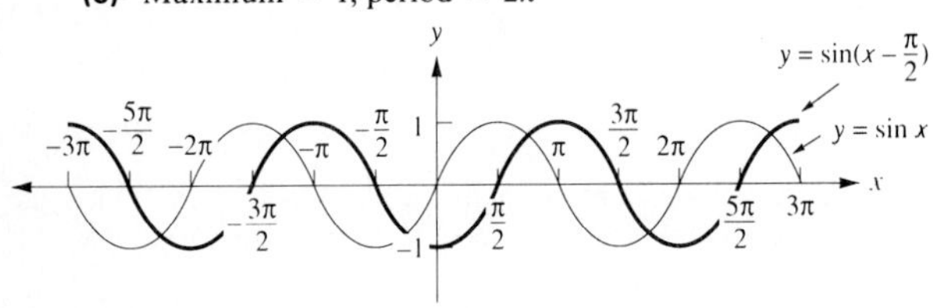

(d) Maximum $= 1$, period $= 2\pi$

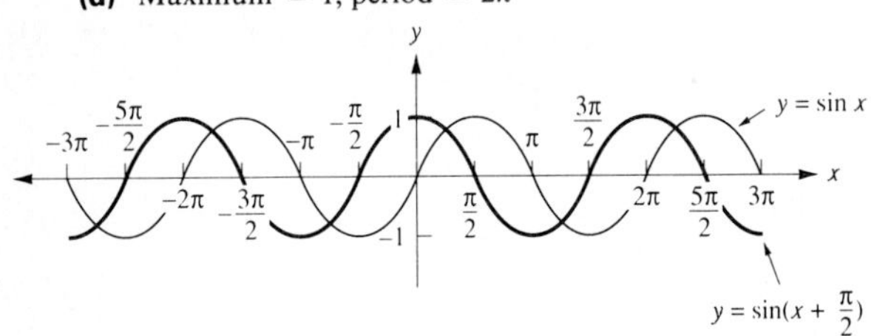

17. Because $\cos^2 x = (\cos x)^2$, it can never give a negative value, and thus will repeat itself each time it reaches the x axis.

19.

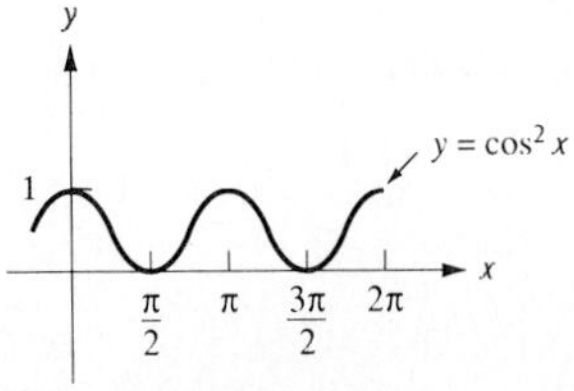

Period $= \pi$

21.

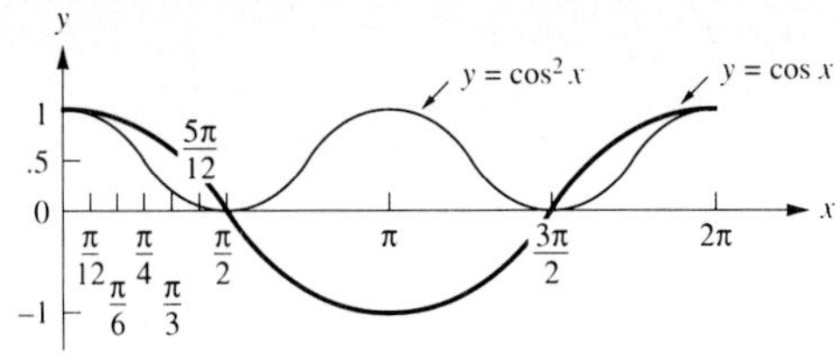

23.

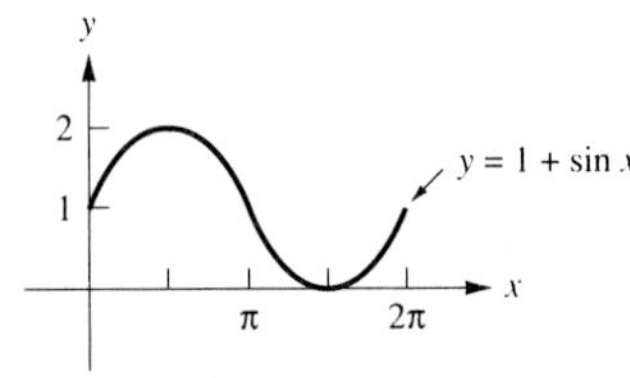

Yes, Period $= 2\pi$.

25. **(a)** $y = -\sin x$ **(d)** $y = -2 \sin x$
(b) $y = 2 \cos x$ **(e)** $y = \cos x$
(c) $y = 4 \sin x$ **(f)** $y = 4 \cos x$

EXERCISES 5.2-2

1. The graph of $y = \tan(x + \pi)$ is the graph of $y = \tan x$ shifted to the left π units. **3.** $x = \cdots \frac{-3\pi}{2}, \frac{-\pi}{2}, \frac{\pi}{2}, \frac{3\pi}{2} \cdots$

5.

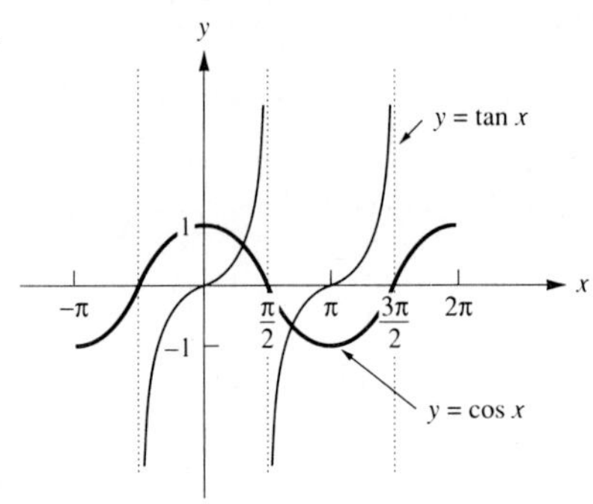

7.

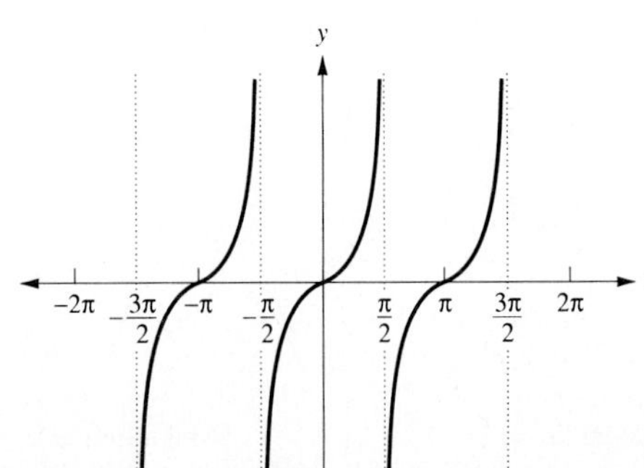

9.

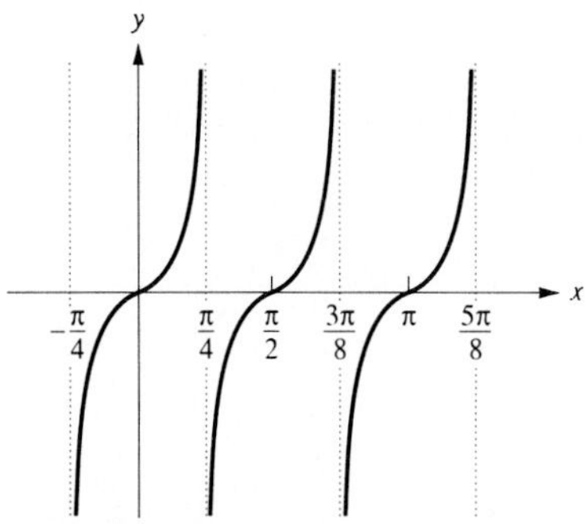

The period of $y = \tan 2x$ is $\frac{\pi}{2}$. This is $\frac{1}{2}$ the period of $y = \tan x$.

11. **(a)** Period: $\frac{\pi}{6}$ for $y = \tan 6x$

(b) Period: 2π for $y = \tan \frac{x}{2}$

(c) Period: 4 for $y = \tan \frac{\pi x}{4}$

EXERCISES 5.2-3

1. (a)

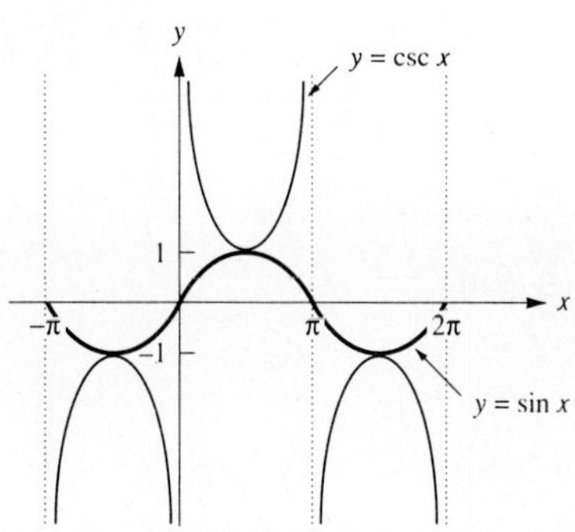

(b)

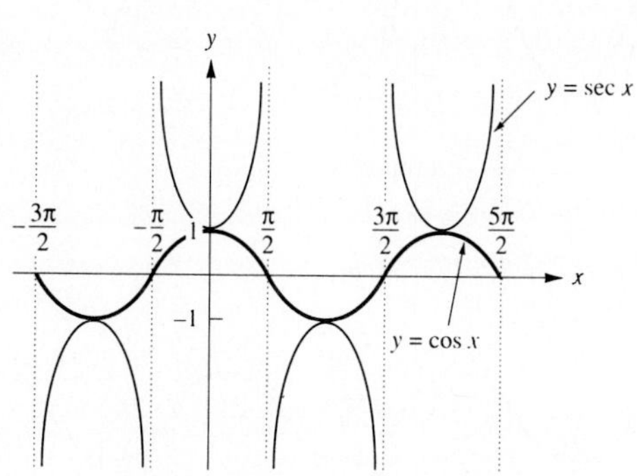

(c)

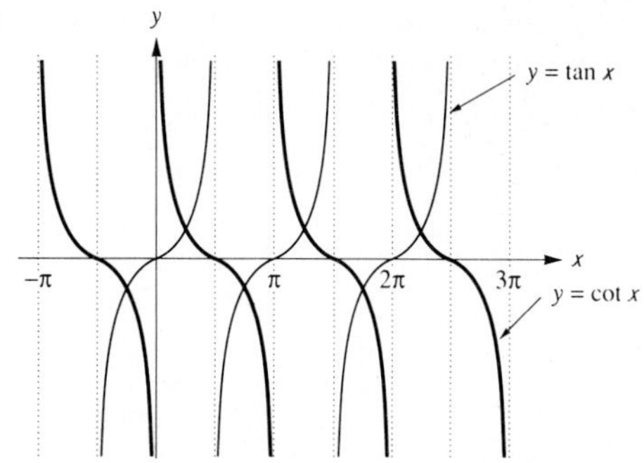

(d)

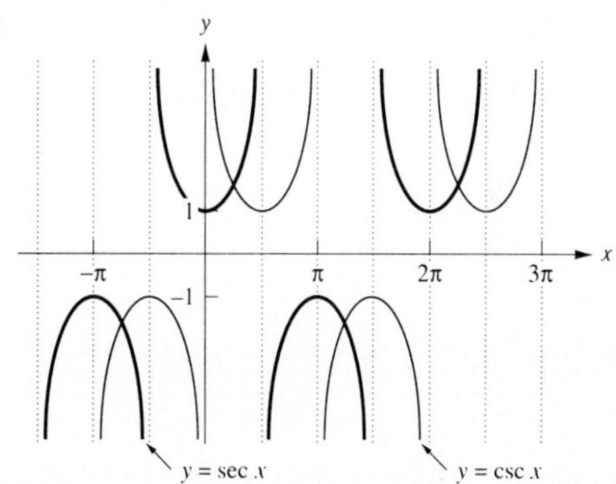

(e)

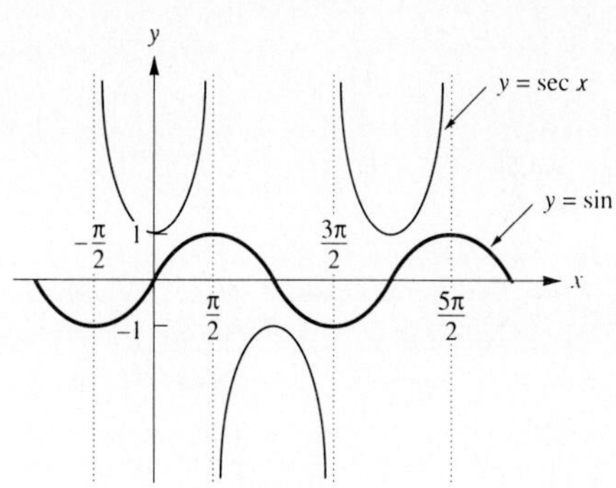

(f)

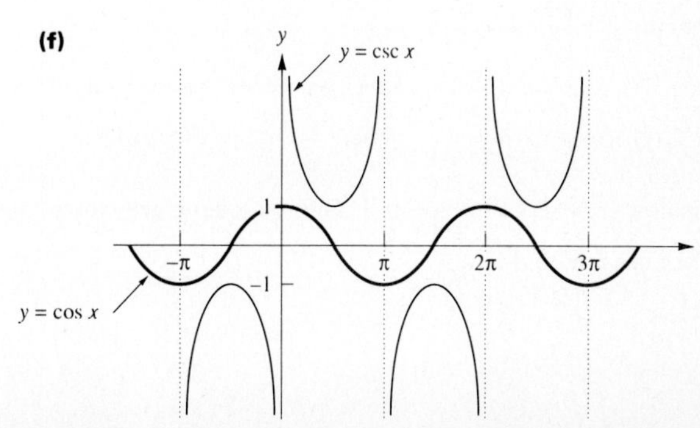

3. cot x approaches $-\infty$ from the left of $0, \pm\pi, \pm 2\pi, \ldots$, cot x approaches $+\infty$ from the right of $0, \pm\pi, \pm 2\pi, \ldots$

5.

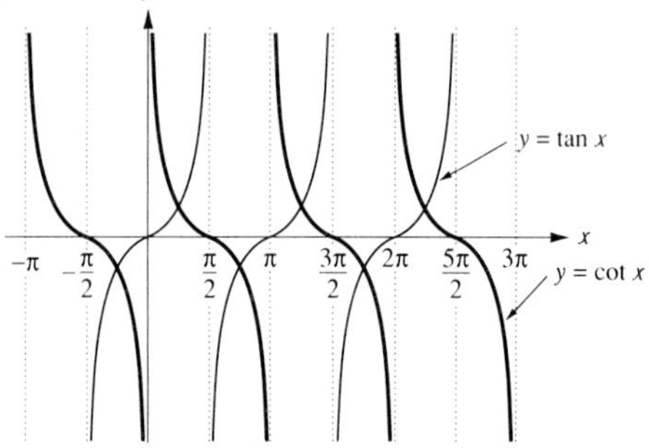

7.

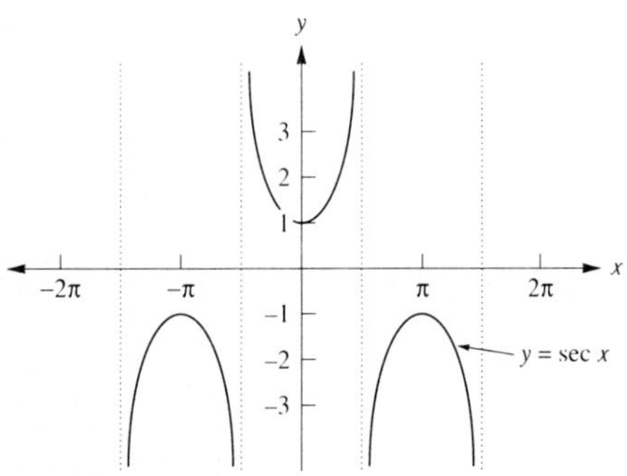

9.

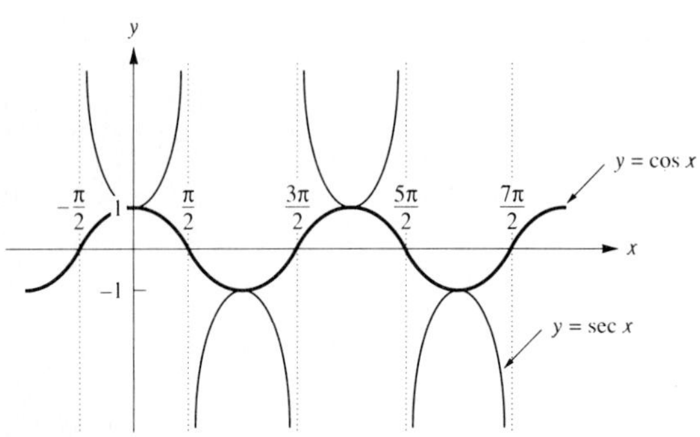

11. The graph of $y = \csc x$ never gets closer than one unit to the x axis because $\csc x = \dfrac{1}{\sin x}$, where $-1 \le \sin x \le 1$ which implies $\dfrac{1}{\sin x} \le -1$ or $\dfrac{1}{\sin x} \ge 1$. Thus csc x never gets closer than one unit to x axis.

13.

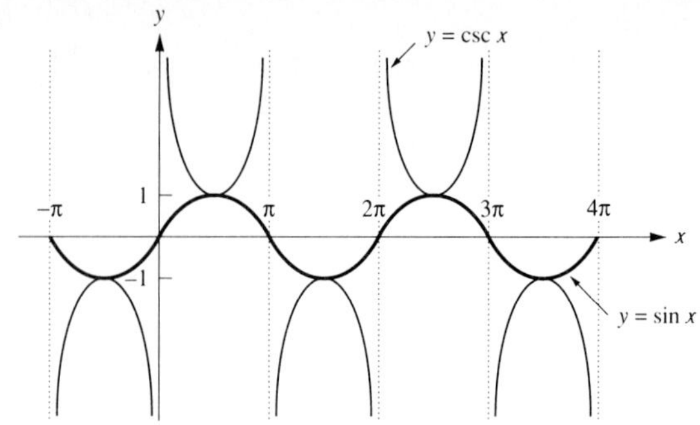

EXERCISES 5.2-4

1.

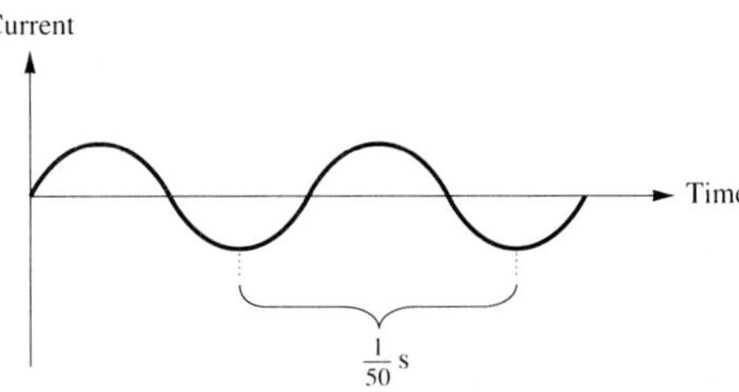

Period $= \frac{1}{50}$ s

3. Yes. Both have the same shape.

EXERCISES 5.3-1

1.

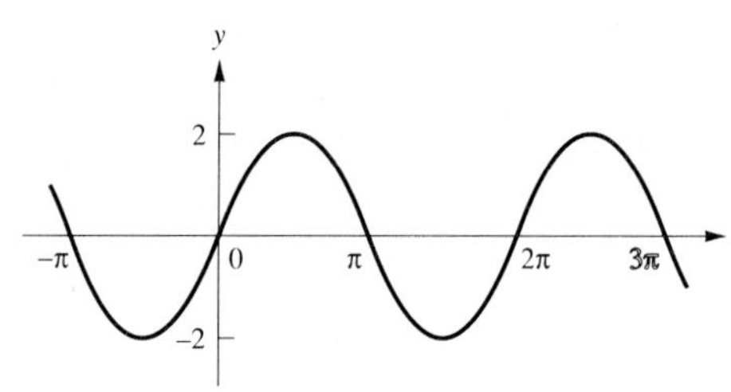

amplitude = 2

3.

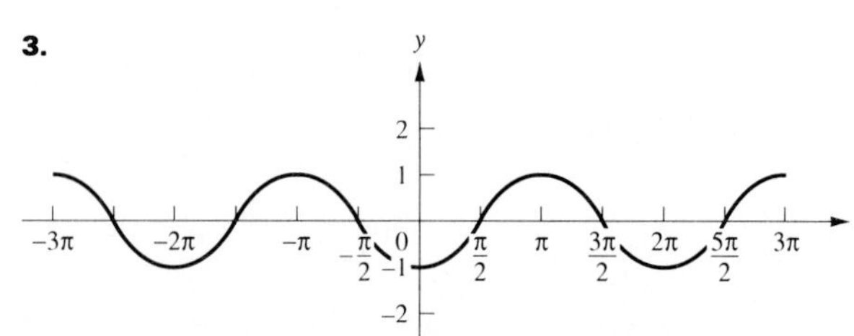

amplitude = 1

5.

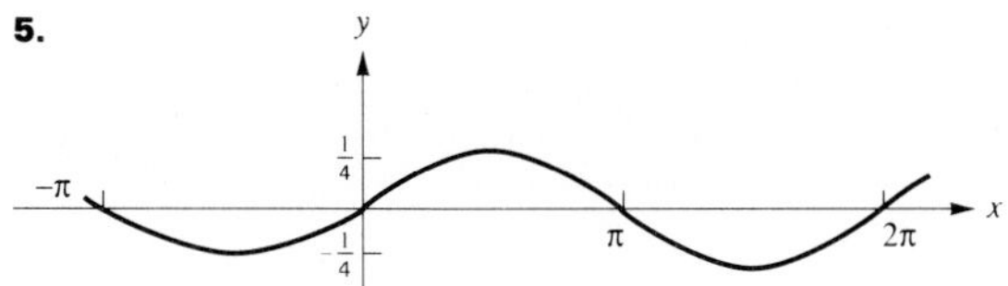

amplitude $= \frac{1}{4}$

7.

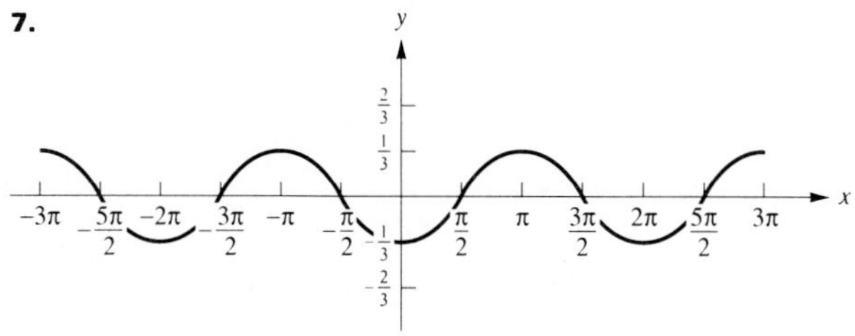

amplitude $= \frac{1}{3}$

9.

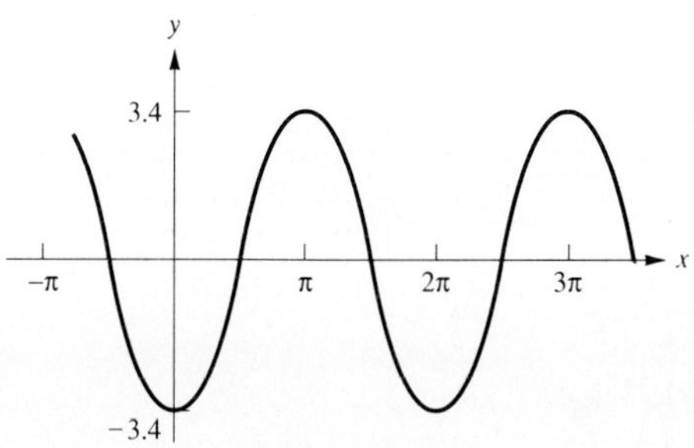

amplitude $= 3.4$

11.

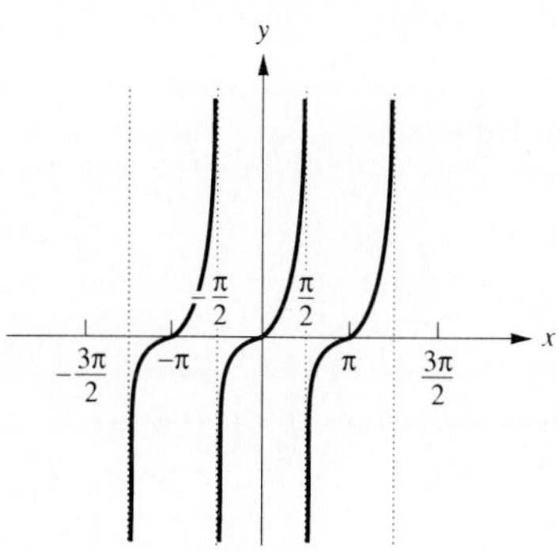

13.

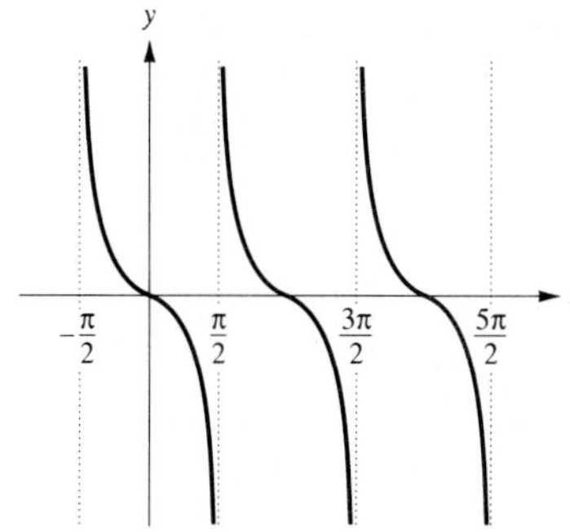

15.

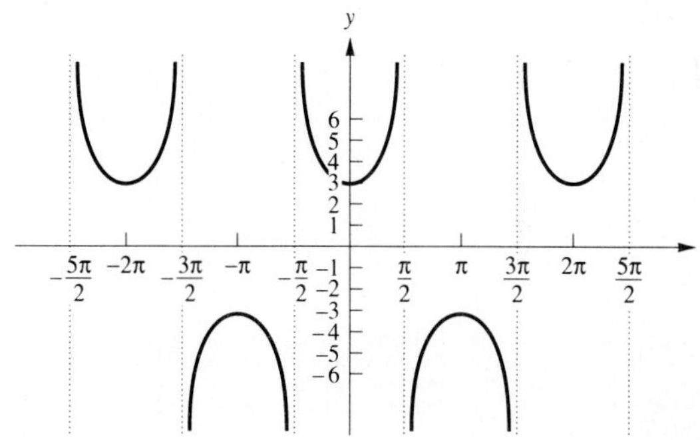

17.

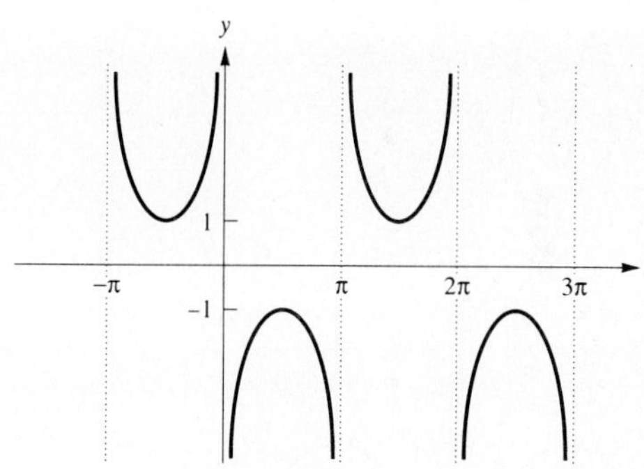

19.

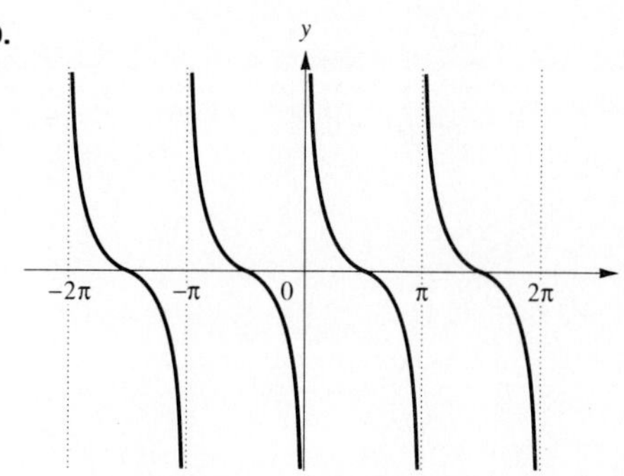

21. $y = 3 \sin x$ **23.** $y = -\cos x$ **25.** $y = 0.75 \cos x$

EXERCISES 5.3-2

1.

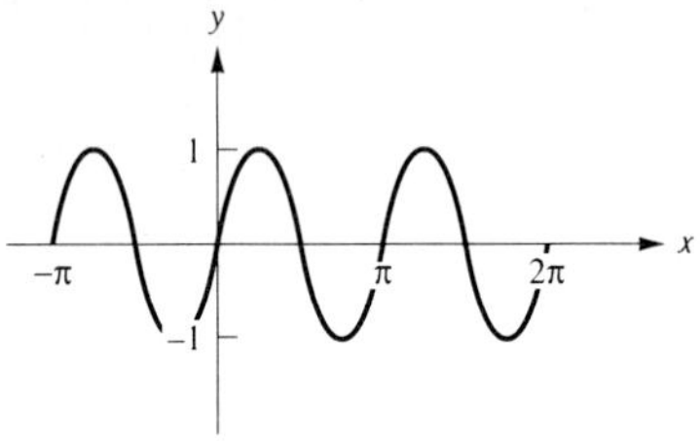

period = π

3.

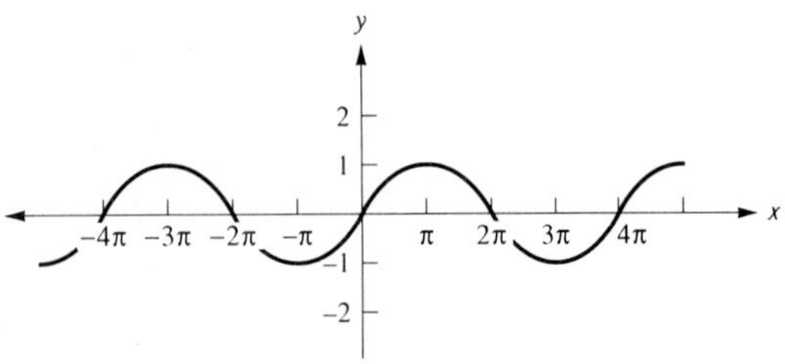

period = 4π

5.

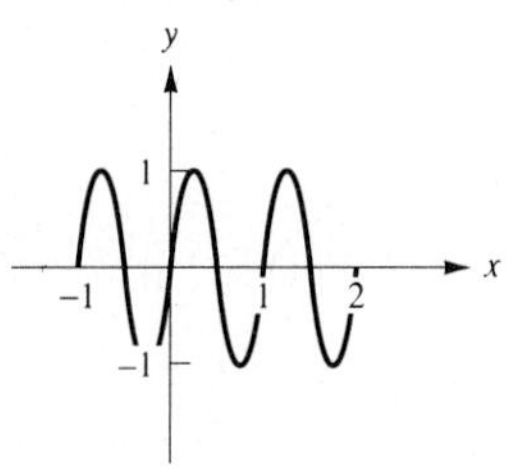

period = 1

7.

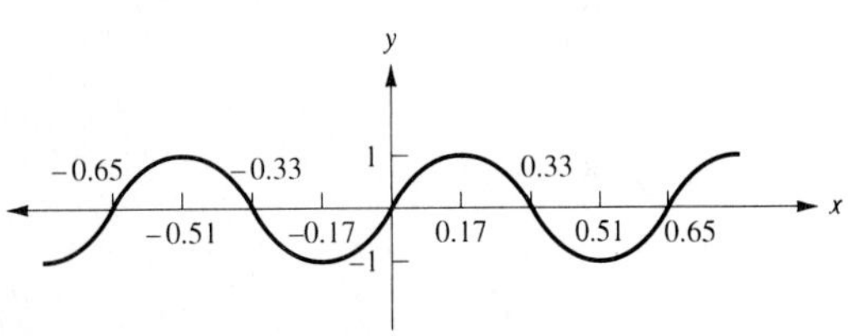

period = 0.65

9.

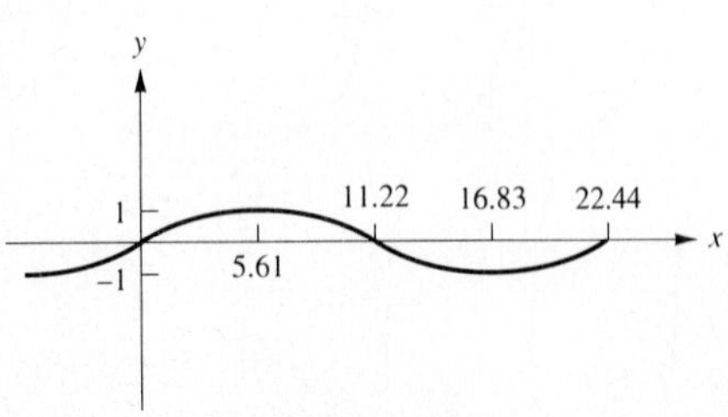

period = 22.44

11.

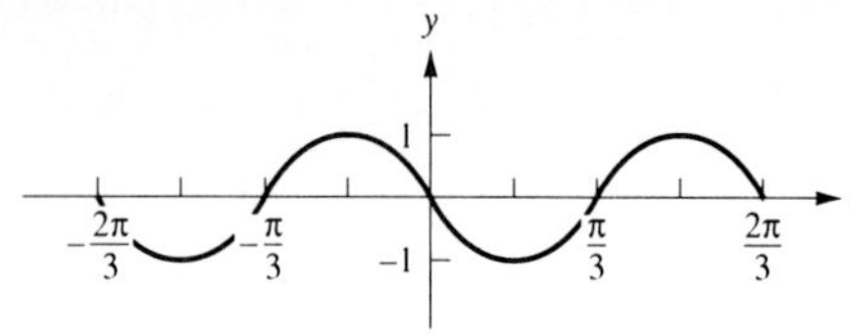

period = $\frac{2\pi}{3}$

13.

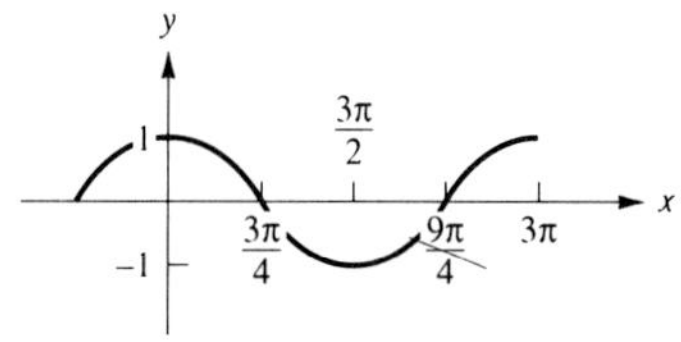

period = 3π

15.

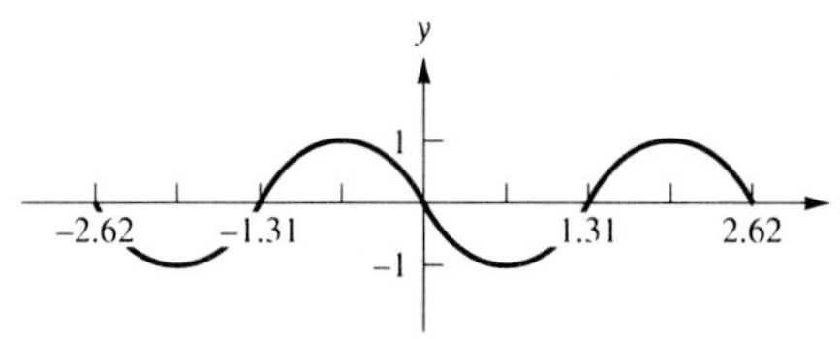

period = 2.62

17.

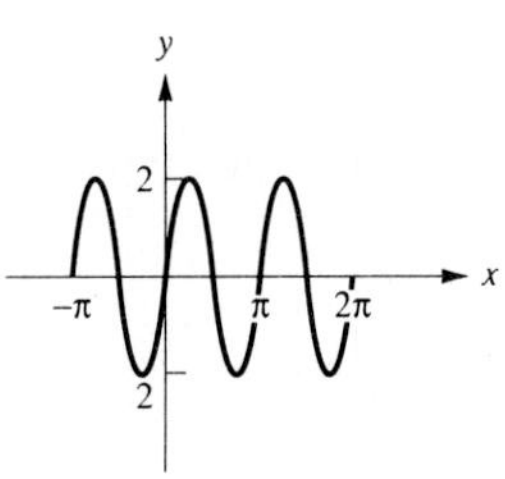

amplitude = 2, period = π

19.

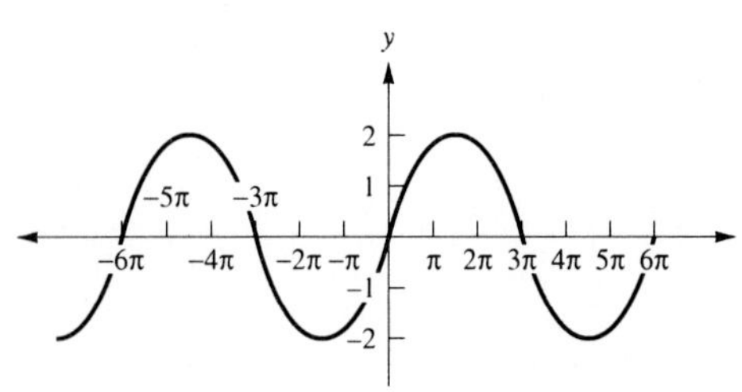

amplitude = 2, period = 6π

21.

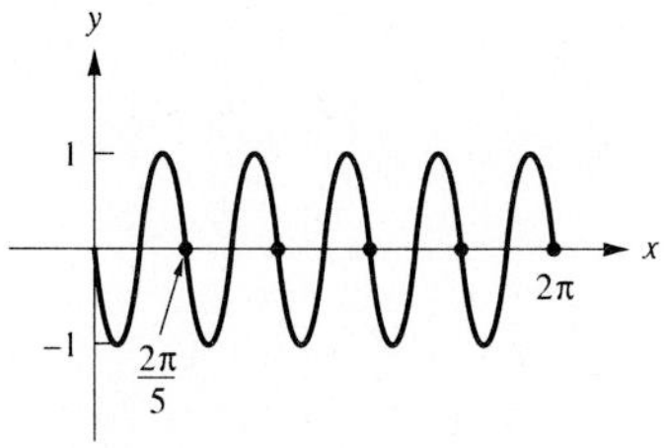

amplitude = 1, period = $\frac{2\pi}{5}$

23.

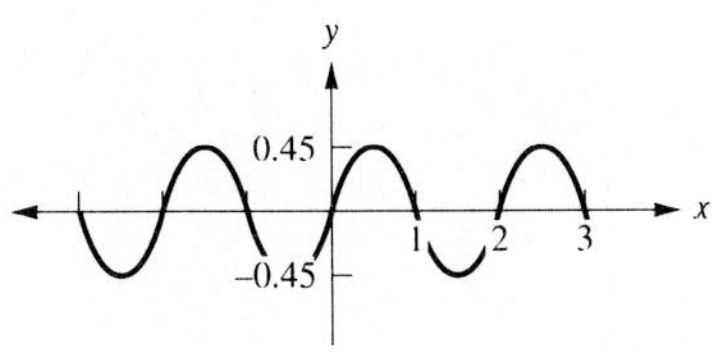

amplitude = 0.45, period = 2

25.

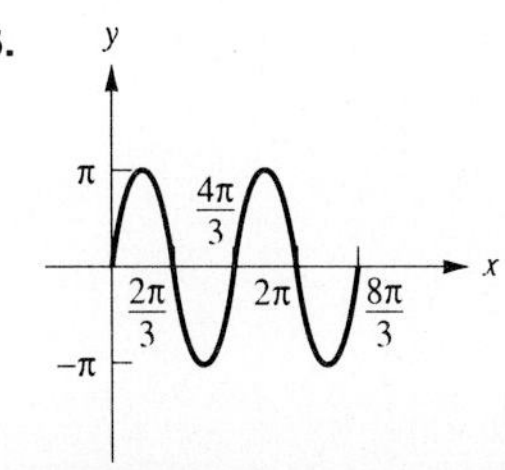

amplitude = π, period = $\frac{4\pi}{3}$

27.

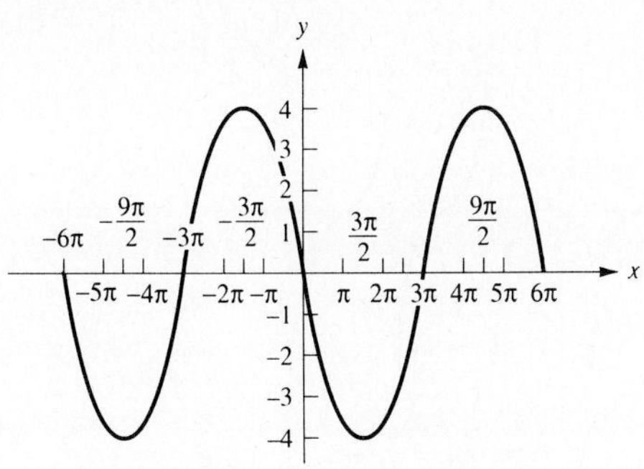

amplitude = 4, period = 6π

29.

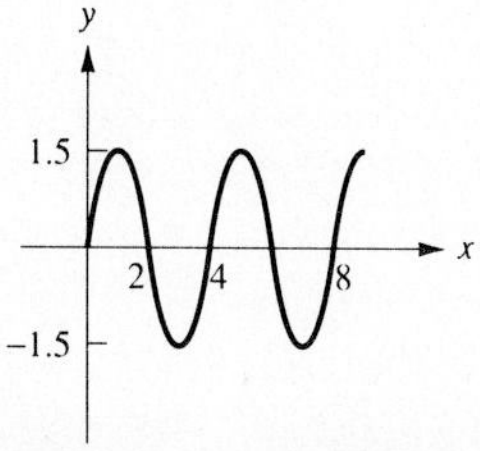

amplitude = 1.5, period = 4

31.

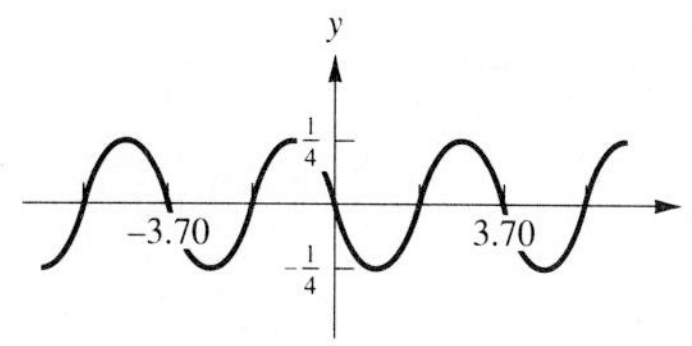

amplitude = $\frac{1}{4}$, period = 3.70

33. $|B| = \frac{1}{2}$; $y = \sin\left(\frac{-x}{2}\right)$

35. $|B| = \frac{\pi}{12}$; $y = \cos\left(\frac{\pi x}{12}\right)$ or $y = \cos\left(\frac{-\pi x}{12}\right)$

37. $|B| = \frac{\pi}{23}$; $y = \sin\left(\frac{\pi x}{2.3}\right)$

39. $|B| = 3\pi$; $y = \sin(-3\pi x)$ or $y = -\sin(3\pi x)$

41. Amplitude = 3.7, $|B| = 4$, $y = 3.7\cos(4x)$ or $y = 3.7\cos(-4x)$

43. Amplitude = $\frac{3}{4}$, $|B| = \frac{4\pi}{3}$, $y = \frac{3}{4}\sin\left(\frac{-4\pi x}{3}\right)$

45. Amplitude = 5.25, $|B| = \frac{3}{4}$, $y = 5.25\cos\left(\frac{3x}{4}\right)$ or $y = 5.25\cos\left(\frac{-3x}{4}\right)$

47. Amplitude = 9.1, $|B| = \frac{\pi}{5}$, $y = 9.1\cos\left(\frac{\pi x}{5}\right)$ or $y = 9.1\cos\left(\frac{-\pi x}{5}\right)$

49. Amplitude = 3.7, $|B| = 4$, $y = 3.7\cos 4x$ or $y = 3.7\cos(-4x)$

51. Amplitude = $\frac{3}{4}$, $|B| = \frac{4\pi}{3}$, $y = \frac{3}{4}\sin\left(\frac{-4\pi x}{3}\right)$

53. Amplitude = 5.25, $|B| = \frac{3}{4}$, $y = 5.25\cos\left(\frac{3x}{4}\right)$ or $y = 5.25\cos\left(\frac{-3x}{4}\right)$

55. Amplitude = 9.1, $|B| = \frac{\pi}{5}$; $y = 9.1\cos\left(\frac{\pi x}{5}\right)$ or $y = 9.1\cos\left(\frac{-\pi x}{5}\right)$

57. Amplitude = 3.7, $|B| = 4$; $y = 3.7\cos(4x)$ or $y = 3.7\cos(-4x)$

59. Amplitude = $\frac{3}{4}$, $|B| = \frac{4\pi}{3}$, $y = -\frac{3}{4}\sin\left(\frac{4\pi x}{3}\right)$ or $y = \frac{3}{4}\sin\left(\frac{-4\pi x}{3}\right)$

61. Amplitude $= 5.25$, $|B| = \dfrac{3}{4}$, $y = 5.25\cos\left(\dfrac{3x}{4}\right)$ or $y = 5.25\cos\left(\dfrac{-3x}{4}\right)$

63. Amplitude $= 9.1$, $|B| = \dfrac{\pi}{5}$, $y = 9.1\cos\left(\dfrac{\pi x}{5}\right)$ or $y = 9.1\cos\left(\dfrac{-\pi x}{5}\right)$

65.

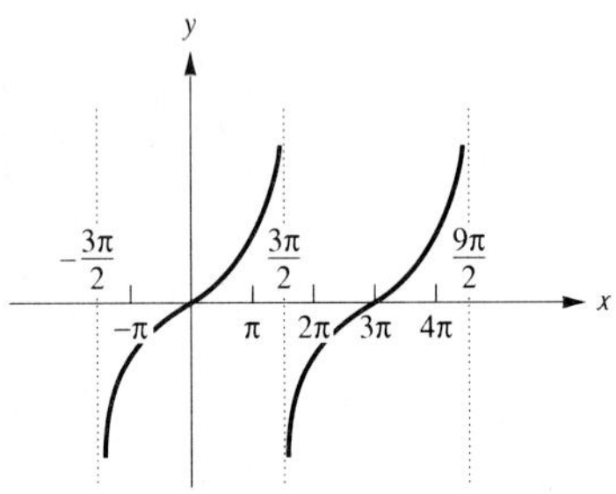

period $= 3\pi$

67.

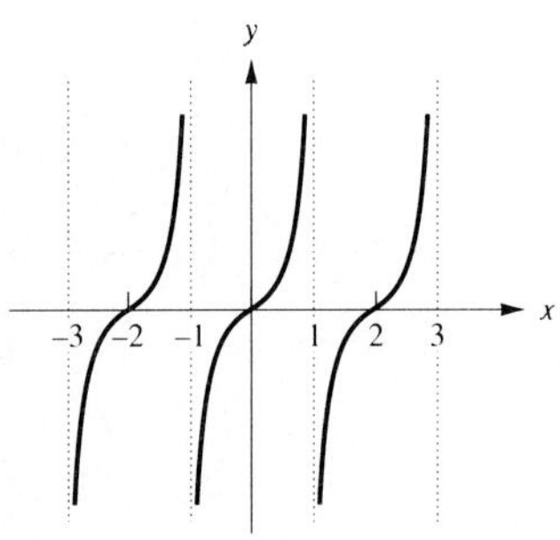

period $= 2$

69.

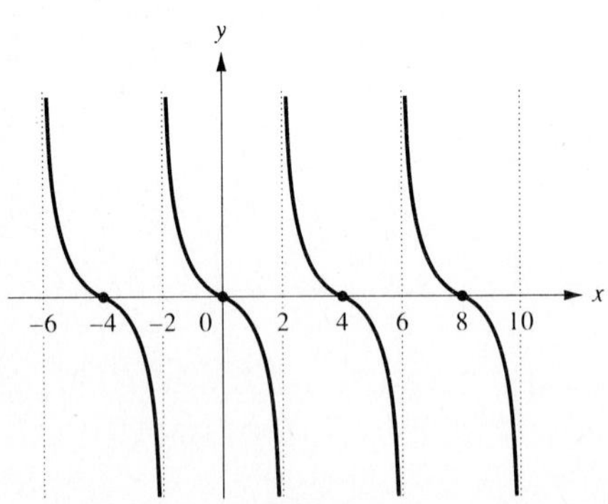

period $= 4$

71.

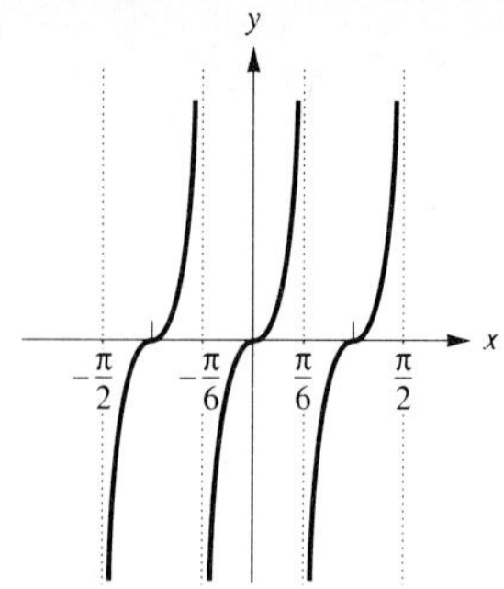

period $= \pi/3$

73.

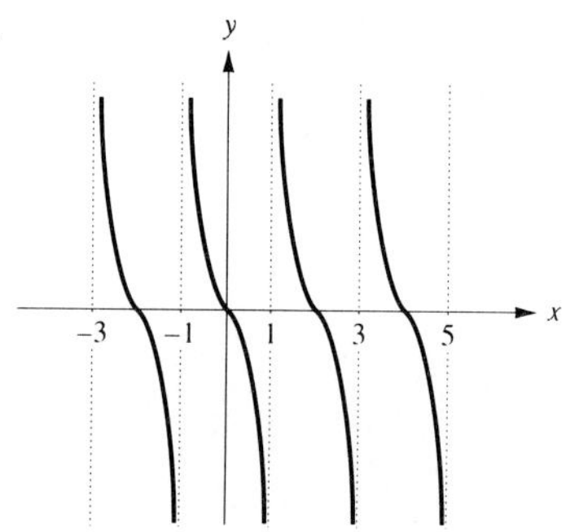

period $= 2$

75.

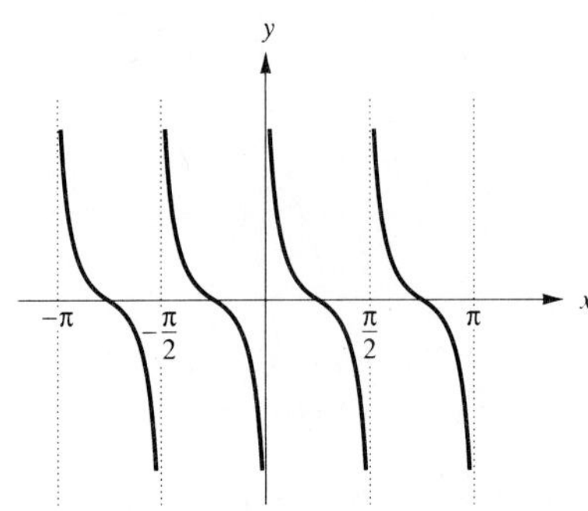

period $= \dfrac{\pi}{2}$

77.

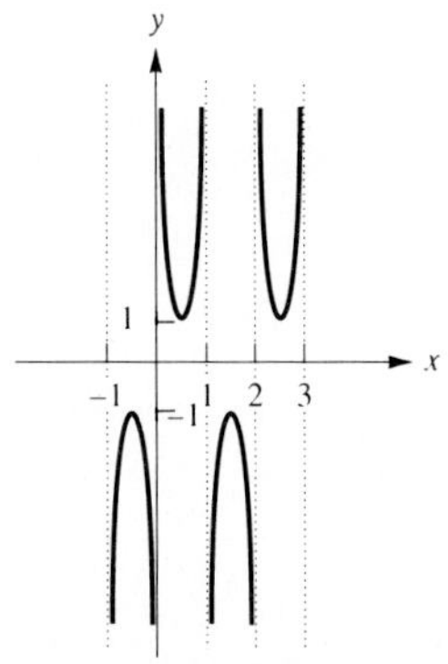

period = 1

79.

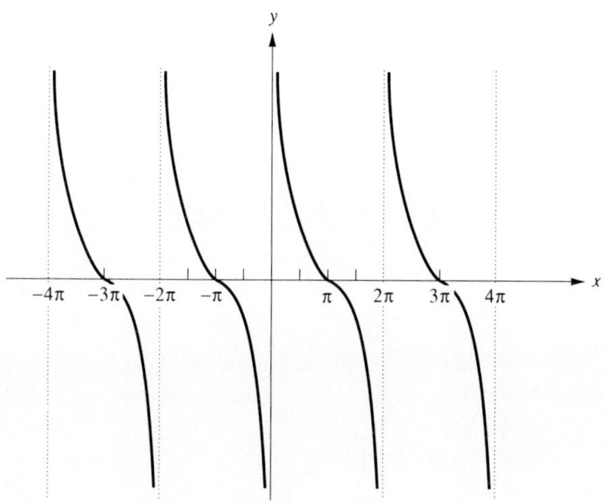

period = 2π

81.

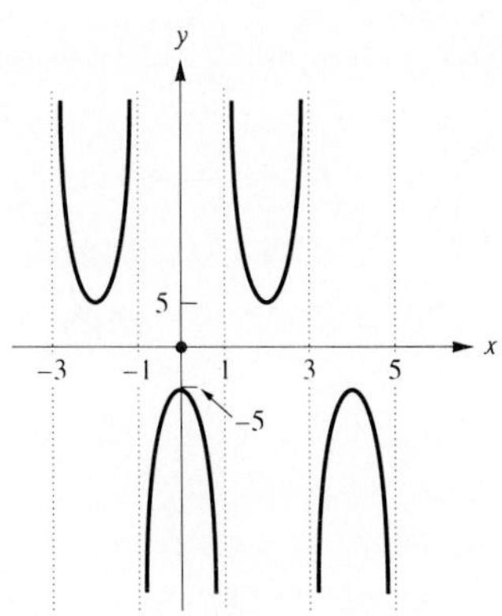

period = 2

83.

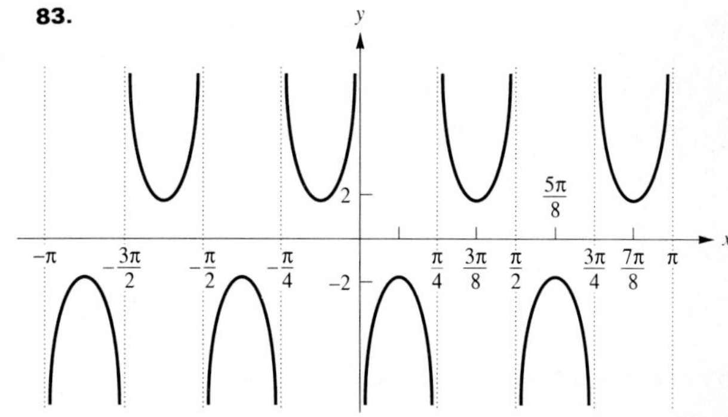

period $= \dfrac{\pi}{4}$

85.

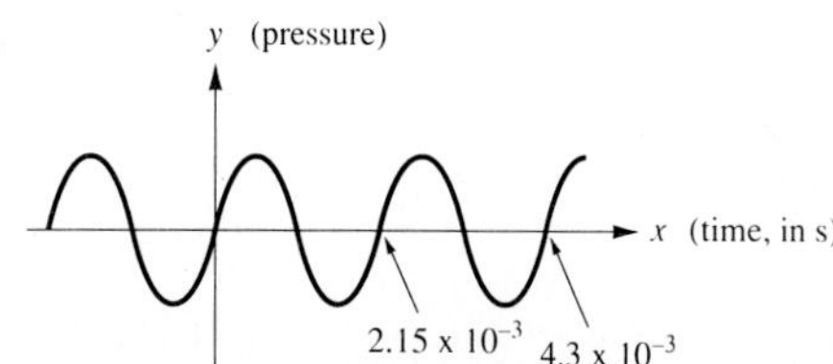

frequency = 466.2 Hz, period = 2.15×10^{-3} s

87. **(a)** Frequency: lowest note = 55 Hz, highest note = 4192 Hz
Period: lowest note = 1.8×10^{-2} s, highest note = 2.4×10^{-4} s

(b)

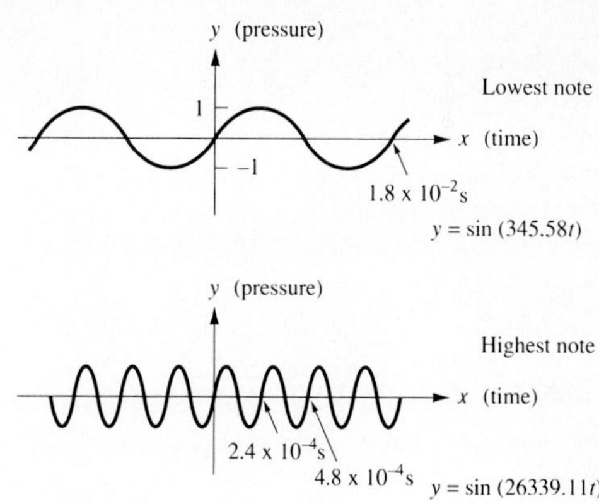

(c) Lowest note: $B = 346\ \text{s}^{-1}$
Highest note: $B = 26000\ \text{s}^{-1}$

89. Frequency $= 2 \times 10^6$ Hz, period $= 5 \times 10^{-7}$ s

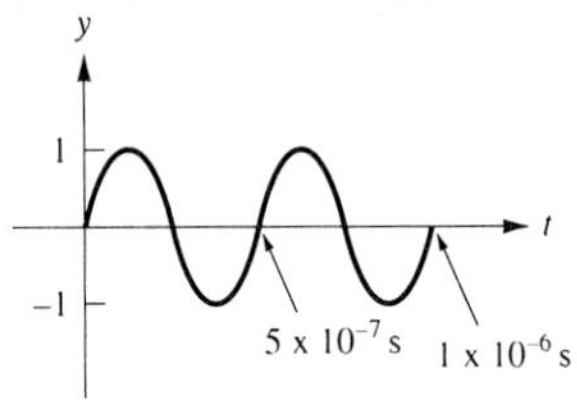

$|B| = 1.26 \times 10^7$ s

91. **(a)** Frequency $= 0.179$ Hz, period $= 5.60$ s

(b)

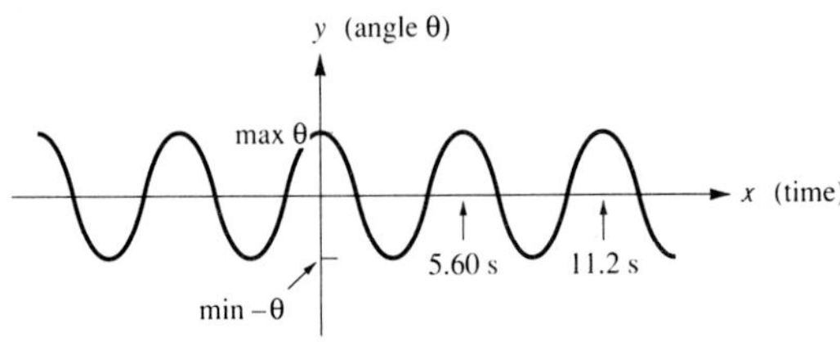

$\pi/6$ to the right

(c) $B = 1.12\ \text{s}^{-1}$

EXERCISES 5.3-3

1.

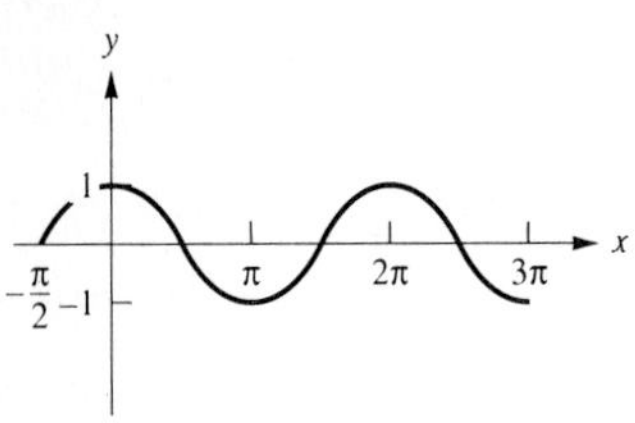

$\pi/2$ to the left

3.

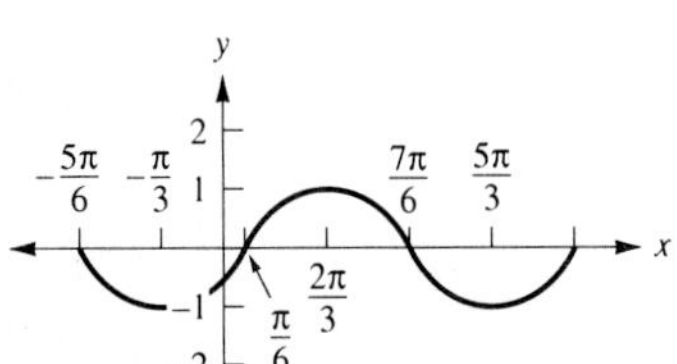

$\pi/6$ to the right

5.

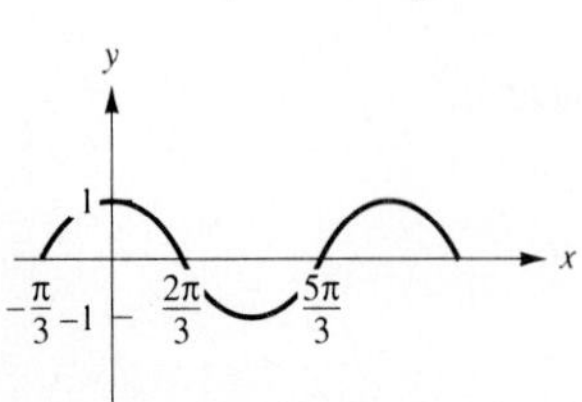

$\pi/3$ to the left

7.

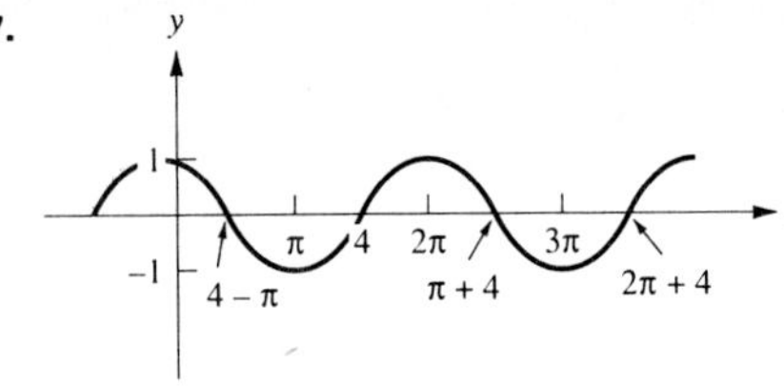

4 to the right

9.

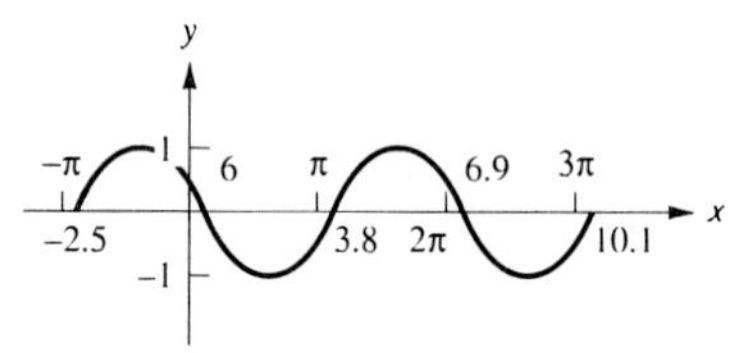

2.5 to the left

11.

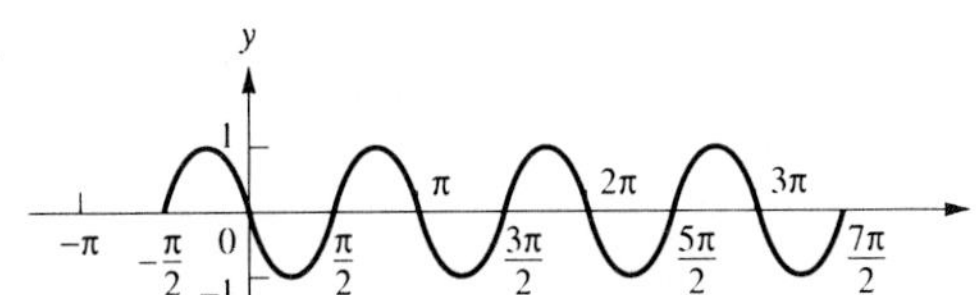

amplitude $= 1$, period $= \pi$, phase shift $= \pi/2$ to the right

13.

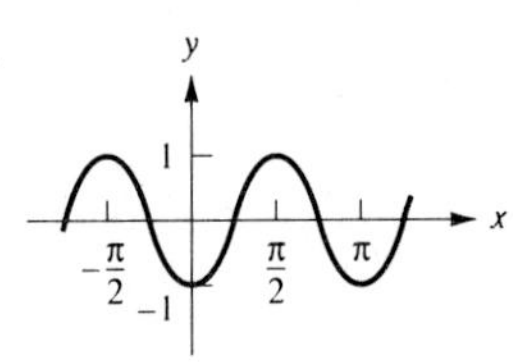

amplitude $= 1$, period $= \pi$, phase shift $= \pi/2$ to the left

15.

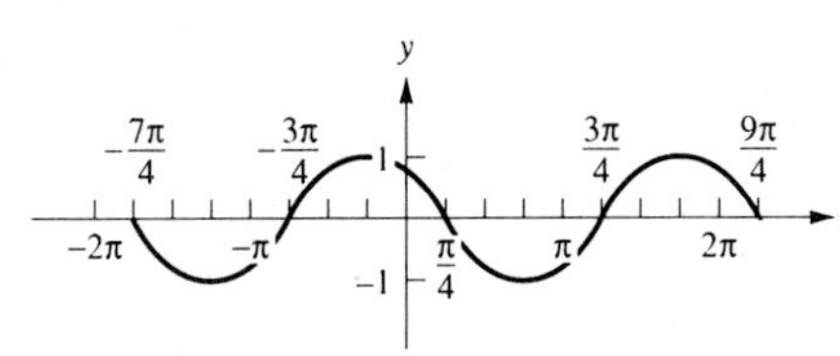

amplitude $= 1$, period $= 2\pi$, phase shift $= \pi/4$ to the right relative to $-\sin x$.

17.

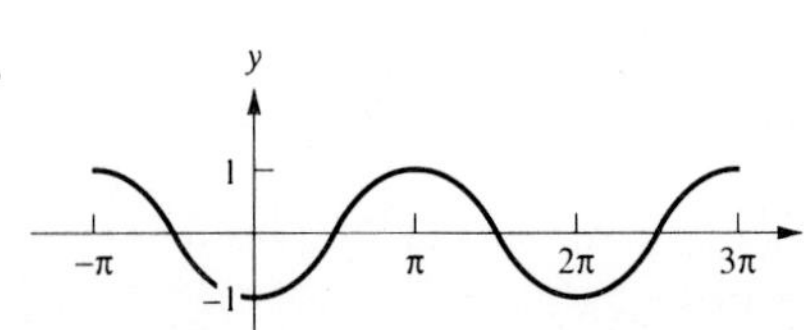

amplitude $= 1$, period $= 2\pi$, phase shift $= \pi/2$ to the left relative to $\sin(-x)$

19.

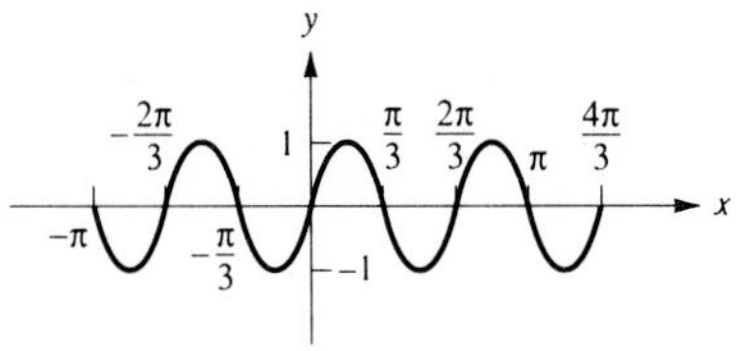

amplitude $= 1$, period $= 2\pi/3$, phase shift $= \dfrac{\pi}{3}$ to the right

relative to $\sin(-3x)$

21.

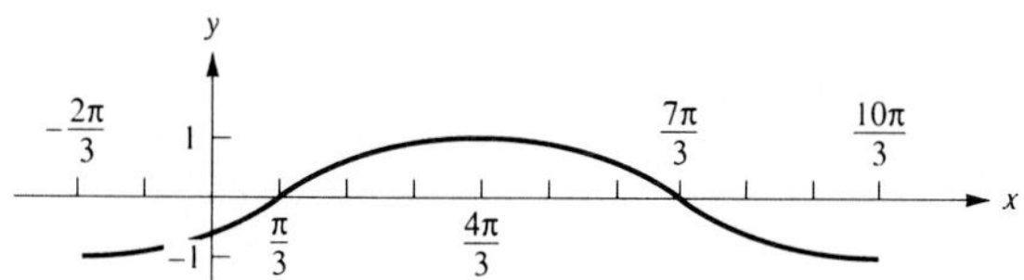

amplitude $= 1$, period $= 4\pi$, phase shift $= \dfrac{4\pi}{3}$ to the right

relative to $\cos\left(\dfrac{x}{2}\right)$

23.

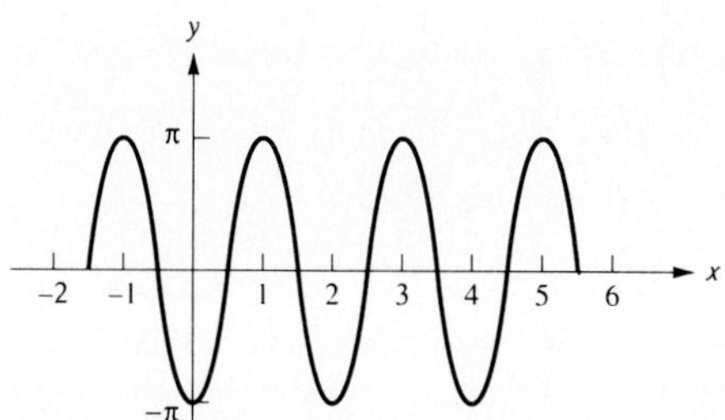

amplitude $= \pi$, period $= 2$, phase shift $= 1$ to the right
relative to $\pi\cos(-\pi x)$

25.

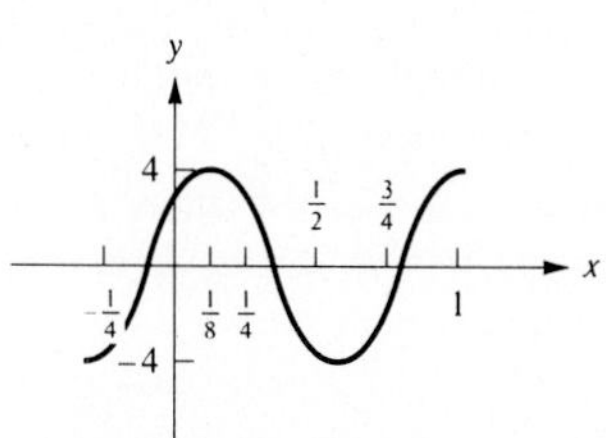

amplitude $= 4$, period $= 1$, phase shift $= 1/8$ to the right
relative to $4\cos(2\pi x)$

27.

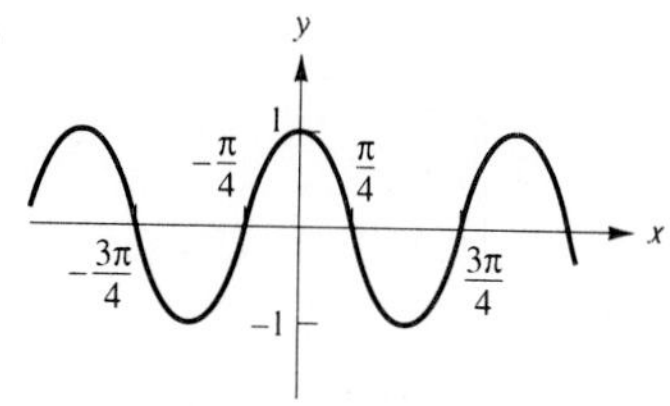

amplitude $= 1$, period $= \pi$, phase shift $= \dfrac{\pi}{4}$ to the right

relative to $\sin(-2x)$

29.

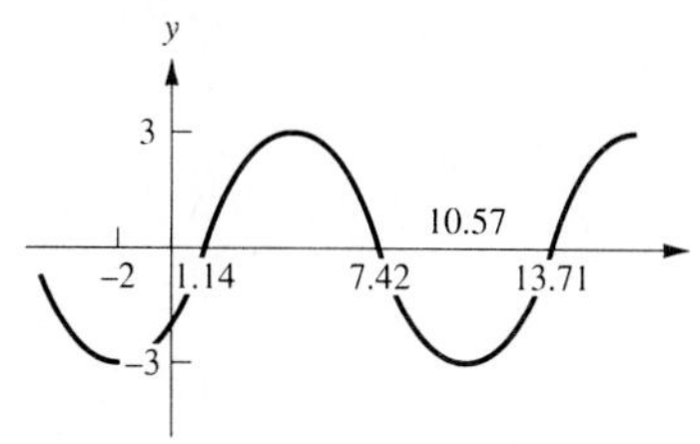

amplitude $= 3$, period $= 4\pi$, phase shift $= 2$ to the left

relative to $-3\cos\left(-\dfrac{x}{2}\right)$

31.

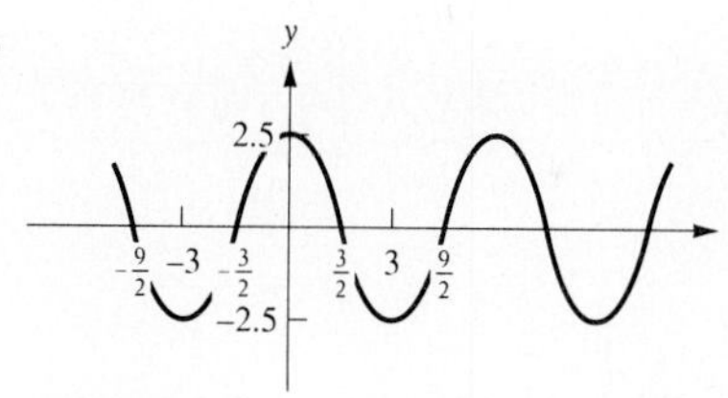

amplitude $= 2.5$, period $= 6$, phase shift $= \dfrac{3}{2}$ to the left

relative to $-2.5\sin\left(\dfrac{-\pi x}{3}\right)$

33.

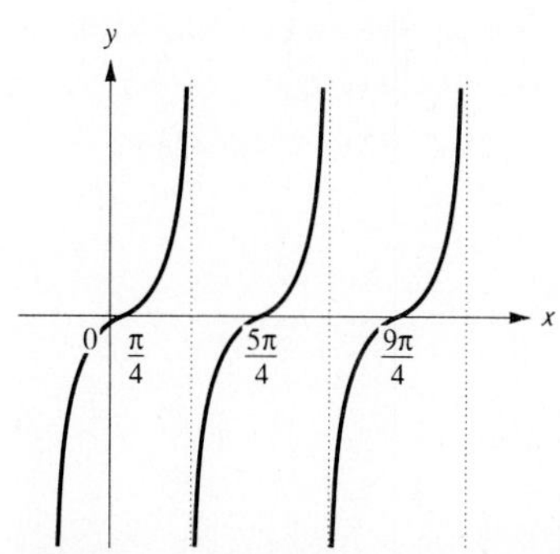

period $= \pi$, phase shift $= \pi/4$ to the right relative to $\tan x$

35.

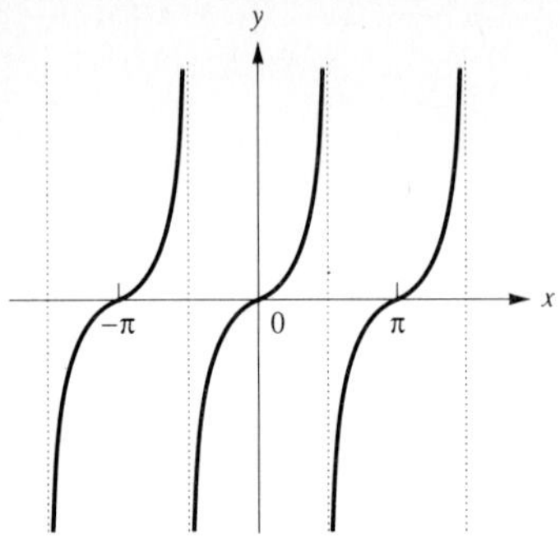

period $= \pi$, phase shift $= \pi$ to the left relative to $\tan x$

37.

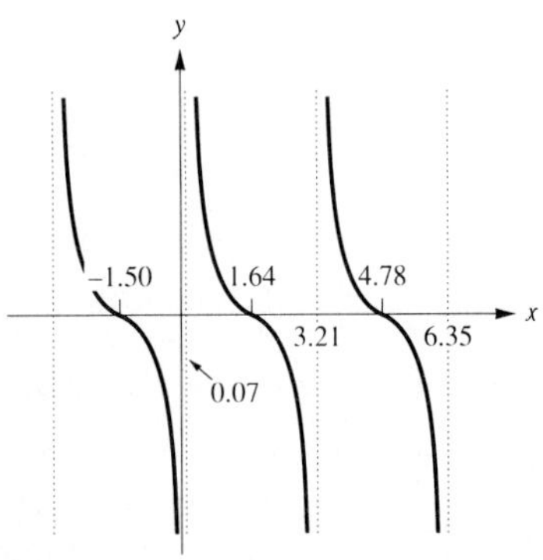

period $= \pi$, phase shift $= 1.50$ to the left relative to $\tan(-x)$

39.

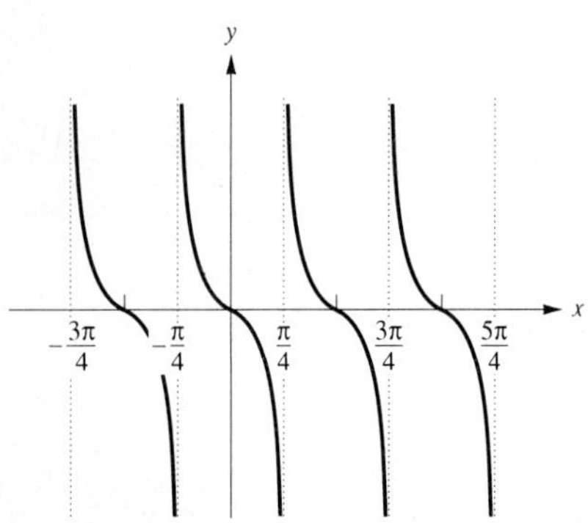

period $= \pi/2$, phase shift $= \pi/2$ to the left relative to $-\tan(2x)$

41.

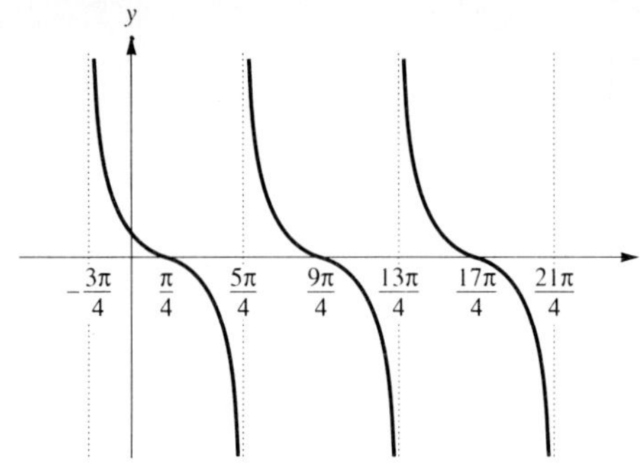

period $= 2\pi$, phase shift $= \pi/4$ to the right relative to $-4\tan\left(\frac{x}{2}\right)$

43.

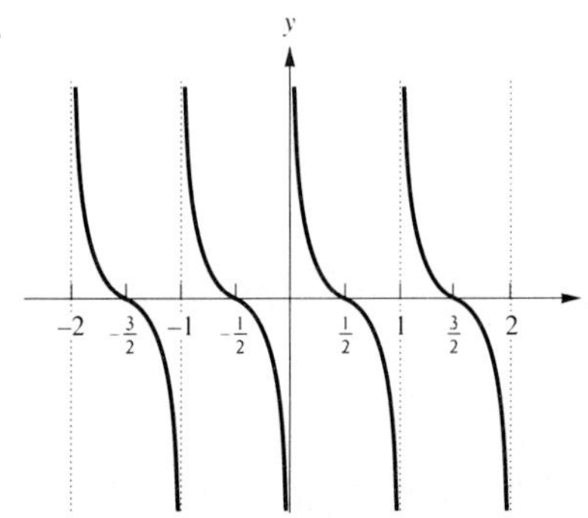

period $= 1$, phase shift $= 1/2$ to the left relative to $\tan(-\pi x)$

45.

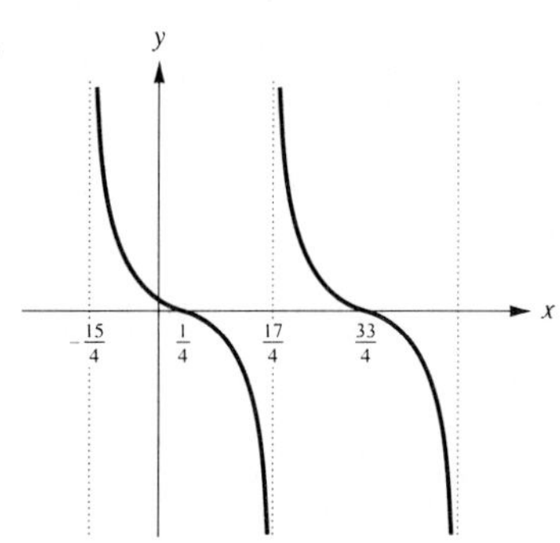

period $= 8$, phase shift $= 1/4$ to the left relative to $\tan\left(\frac{-\pi x}{8}\right)$

47.

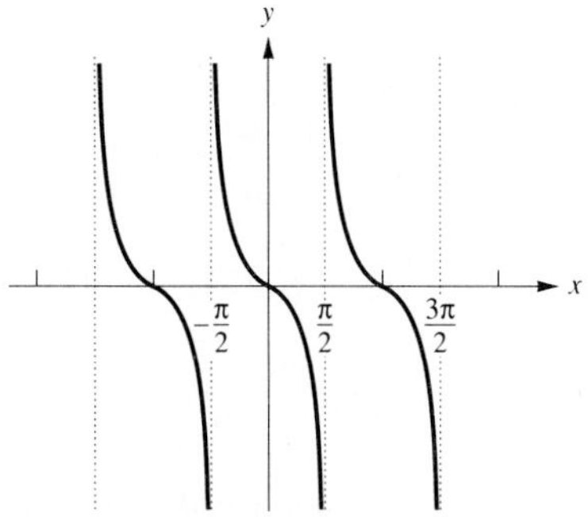

period $= \pi$, phase shift $= \pi/2$ to the right relative to $\cot x$

49.

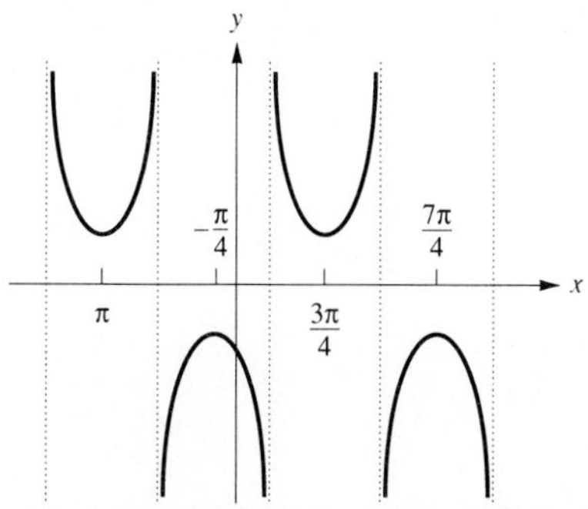

period $= 2\pi$, phase shift $= \pi/4$ to the right relative to $\csc x$

51.

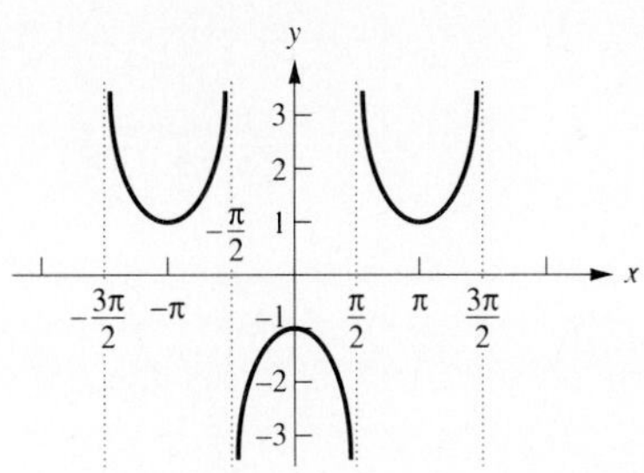

period $= 2\pi$, phase shift $= \pi$ to the right relative to $\sec(-x)$

53.

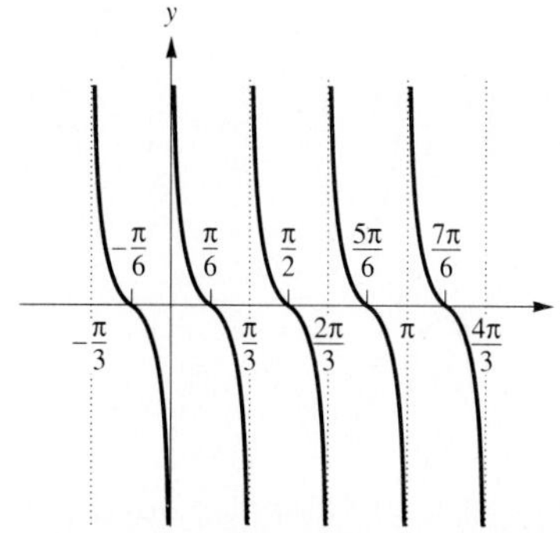

period $= \pi/3$, phase shift $= \pi/3$ to the right relative to $4\cot(3x)$

55.

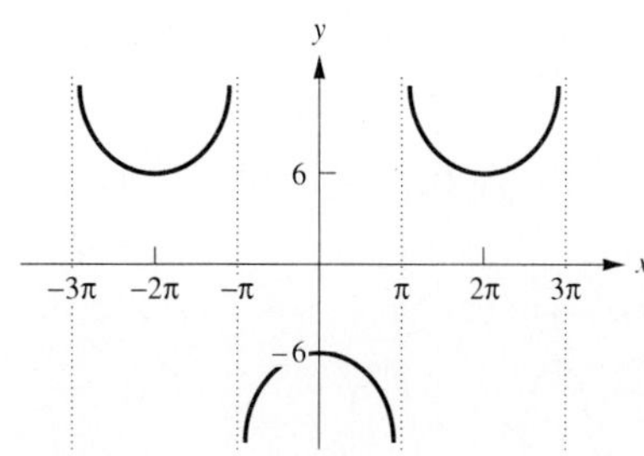

period $= 4\pi$, phase shift $= \pi$ to the right relative to $6\csc\left(\dfrac{x}{2}\right)$

57.

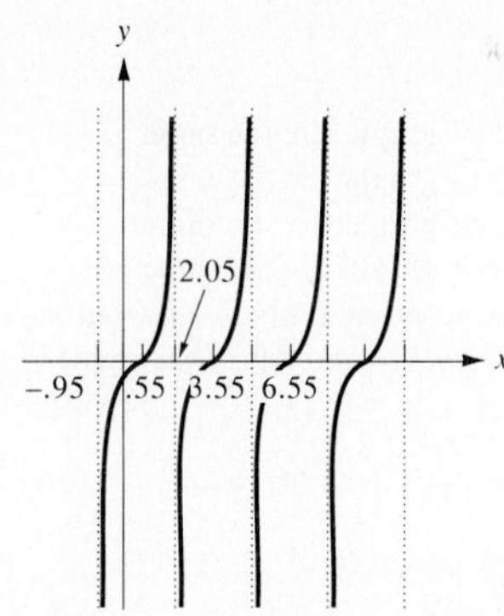

period $= 3$, phase shift $= 3/\pi = 0.95$ to the left relative to $\cot\left(\dfrac{-\pi x}{3}\right)$

59.

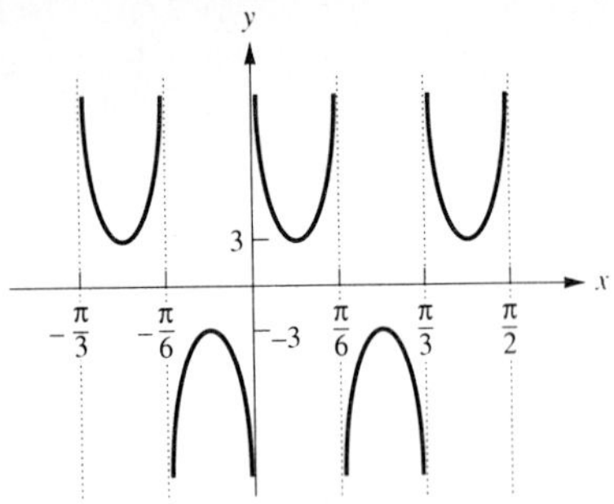

period $= \pi/3$, phase shift $= \pi/3$ to the right relative to $-3\csc(-6x)$

61.

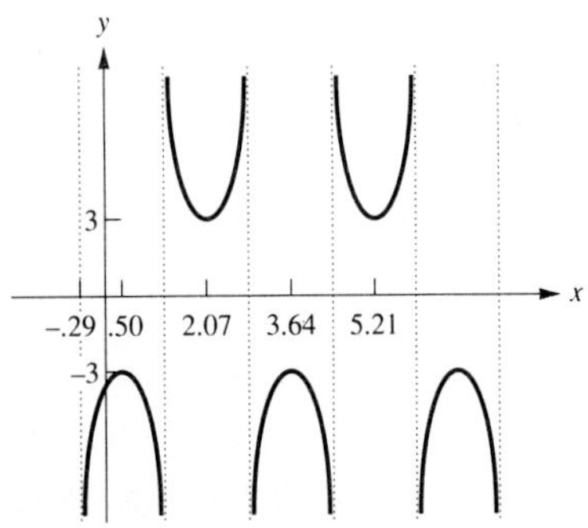

period $= \pi$, phase shift $= 1/2$ to the right relative to $-3\sec(-2x)$

63. True. Identity, because their graphs are the same.
65. True. Identity, because their graphs are the same.
67. True. Identity, because their graphs are the same.
69. True. Identity, because their graphs are the same.
71. $A = 5.0$, $B = 1.0$, $C = 0$ (no phase shift), $y = 5.0 \sin x$
73. $A = 1.0$, $B = 1.0$, $C = -\pi/2$ (phase shift to the right), $y = \sin(x - \pi/2)$ **75.** $A = 1.0$, $B = 1.0$, $C = -1.5$ (phase shift to the right), $y = \sin(x - 1.5)$ **77.** $A = 1.0$, $B = \dfrac{\pi}{5}$, $C = 0$ (no phase shift), $y = \sin\left(\dfrac{\pi x}{5}\right)$ **79.** $A = -7.2$, $B = \pi$, $C = 0$ (no phase shift), $y = -7.2\sin(\pi x)$ **81.** $A = 2.4$, $B = \dfrac{1}{2}$, $C = \dfrac{-\pi}{2}$ (phase shift to right), $y = 2.4\sin\left(\dfrac{x}{2} - \dfrac{\pi}{2}\right)$
83. $A = -3.0$, $B = 1$, $C = 0$ (no phase shift), $y = -3\cos x$
85. $A = 5.8$, $B = 1.0$, $C = -\pi/4$ (phase shift to right), $y = 5.8\cos(x - \pi/4)$ **87.** $A = -2.0$, $B = \dfrac{\pi}{4}$, $C = 0$ (no phase shift), $y = -2.0\cos\left(\dfrac{\pi x}{4}\right)$

89. (a) Phase difference $= 1.21\pi$

(b)

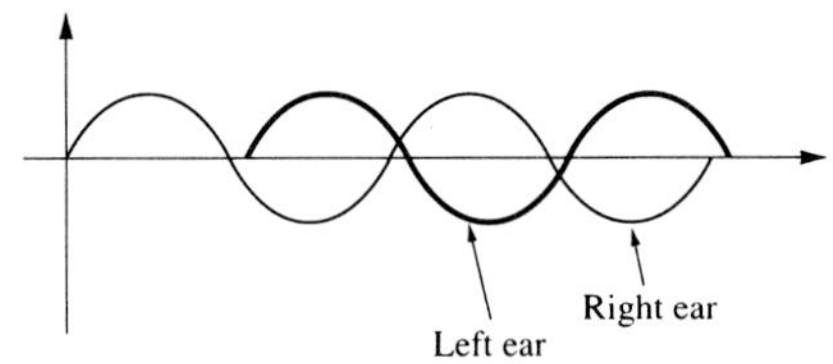

91. (a) 0.2 rad **(b)** 51/3 s, $\dfrac{3}{16}$ Hz **(c)** $\theta = 0.2\cos\left(\dfrac{3\pi}{8}x\right)$

93. $y = (2r\tan\theta)\cos^2\left(\dfrac{x}{2r}\right)$ or $y = r\tan\theta\left[1 + \cos\left(\dfrac{x}{r}\right)\right]$

EXERCISES 5.4

1.

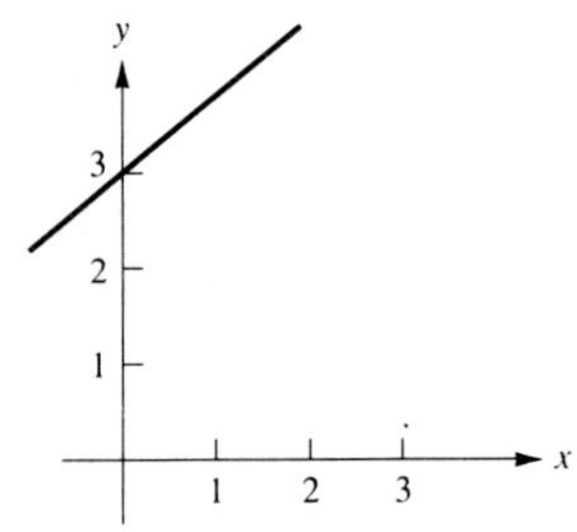

3.

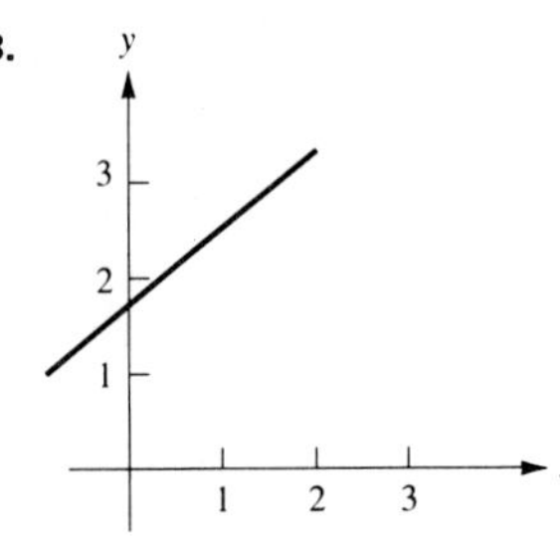

5.

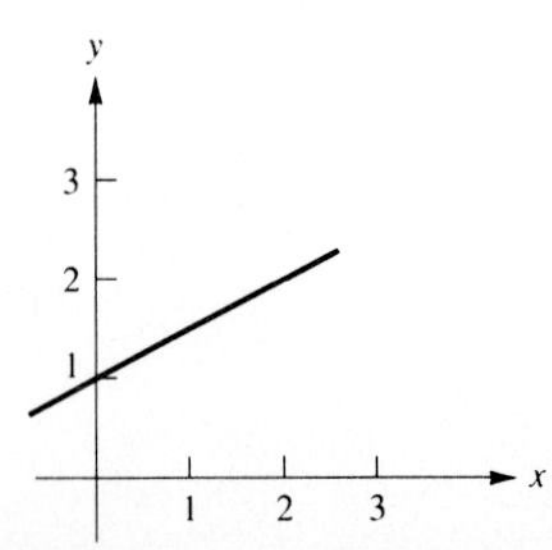

7.

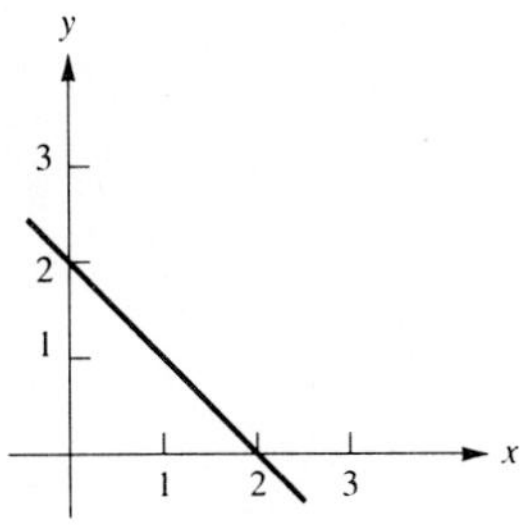

9.

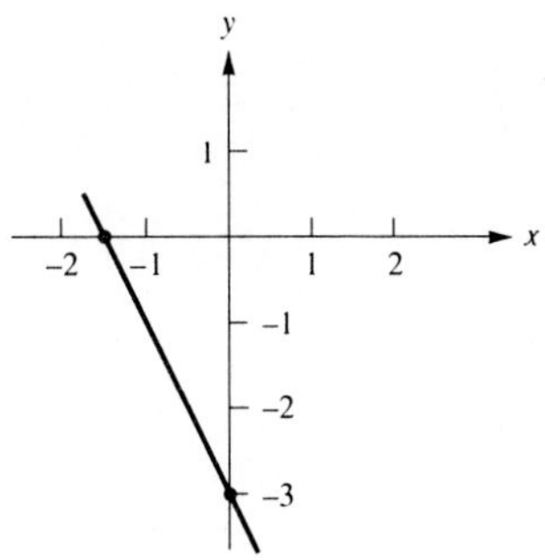

11.

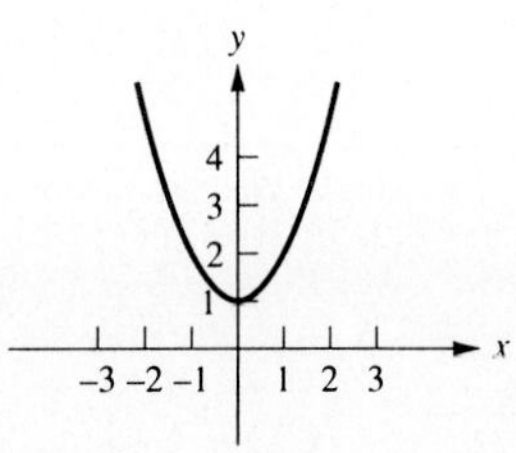

13.

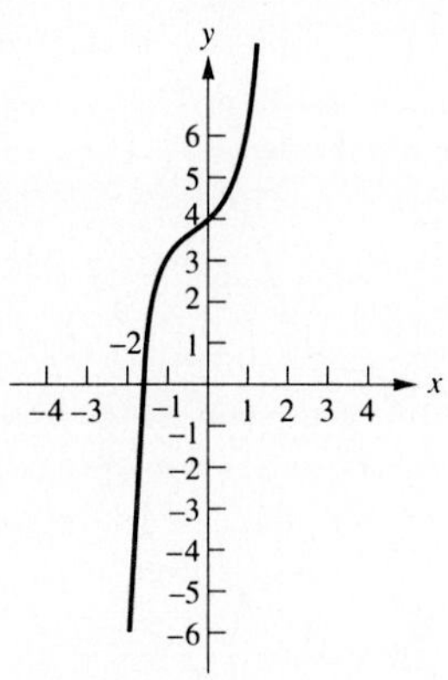

15.

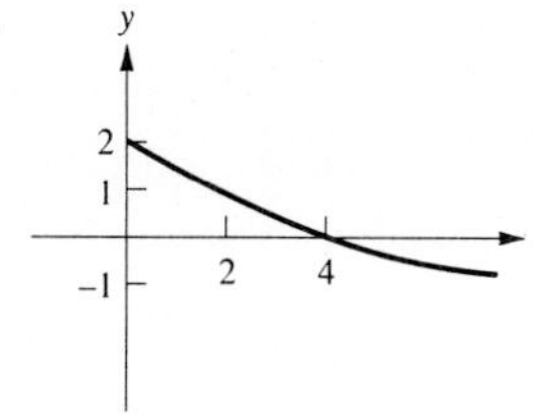

17.

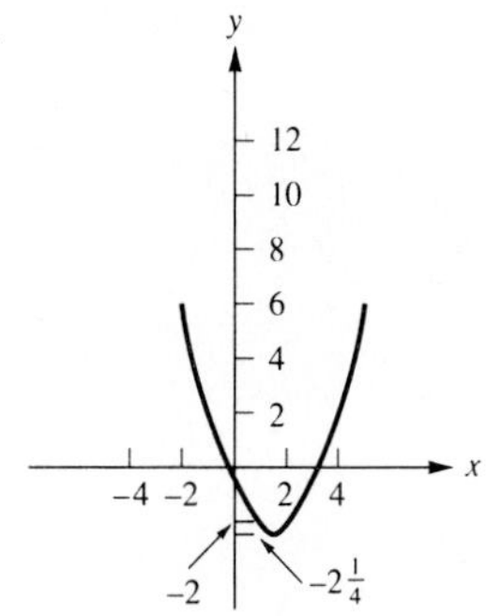

19.

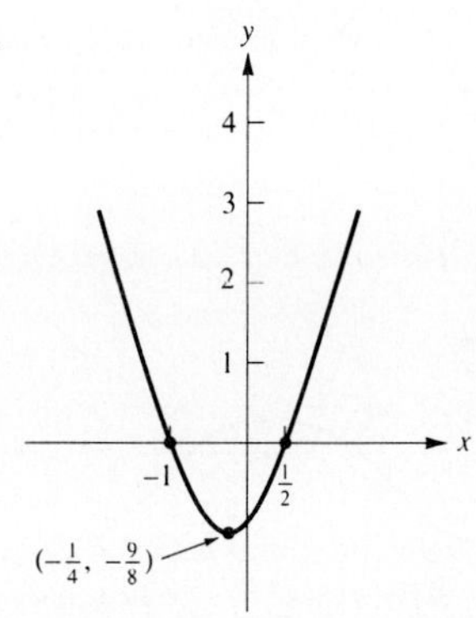

21.

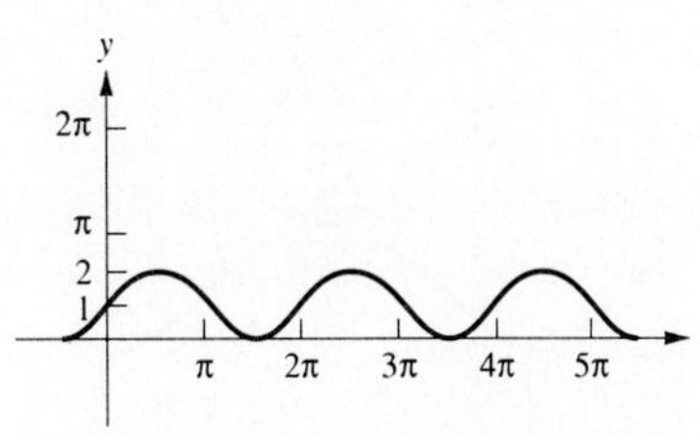

23.

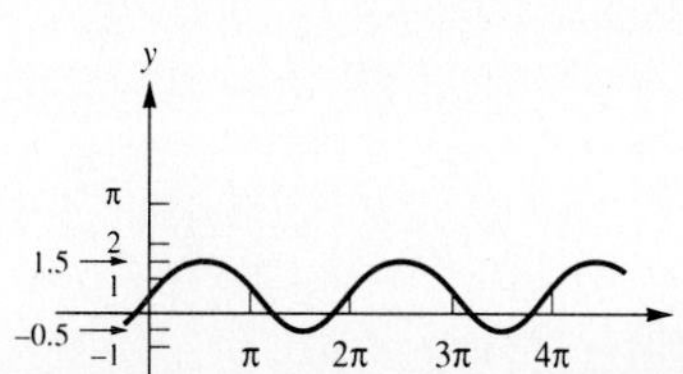

25.

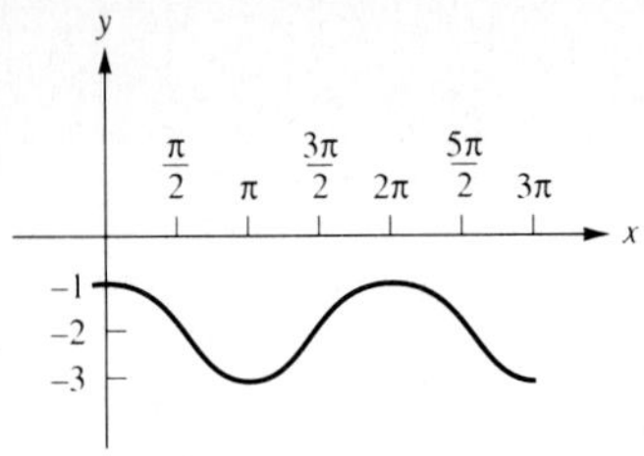

27.

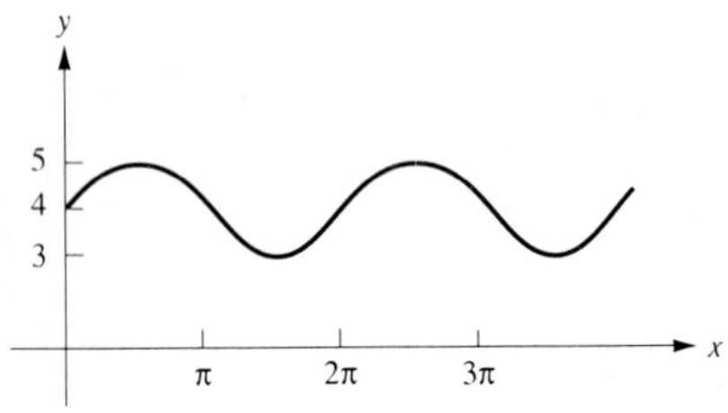

29.

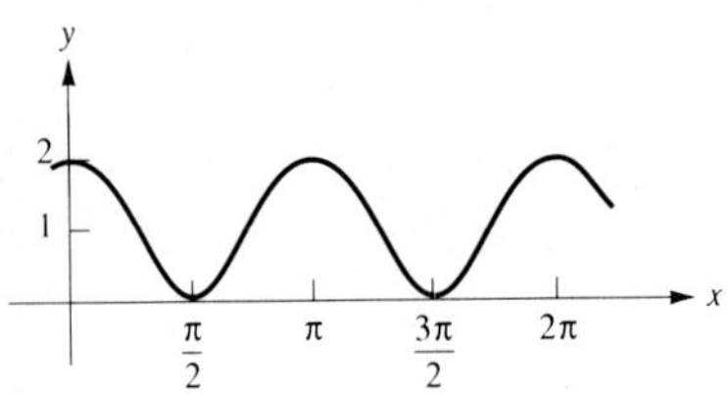

31.

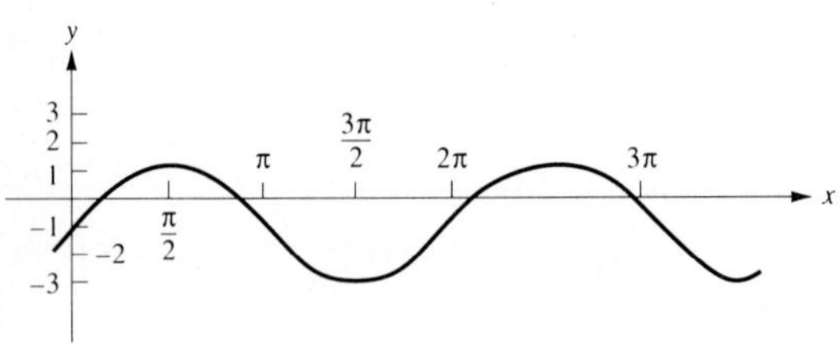

33.

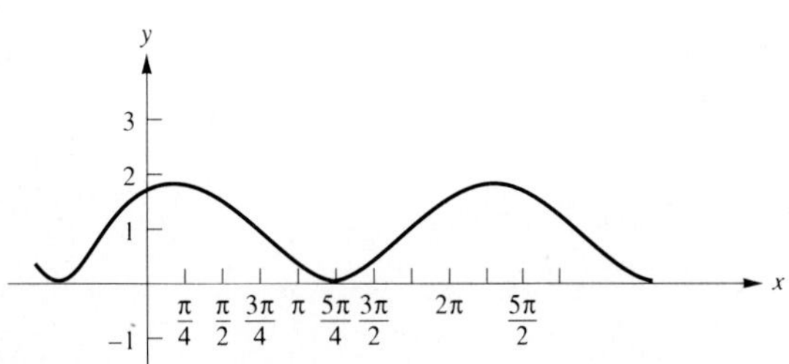

35.

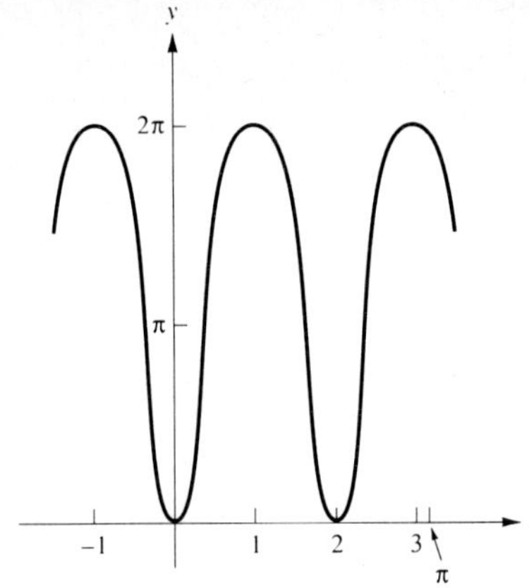

37.

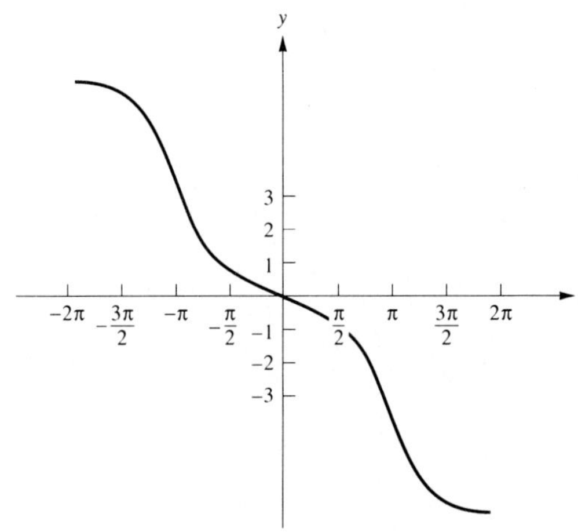

39.

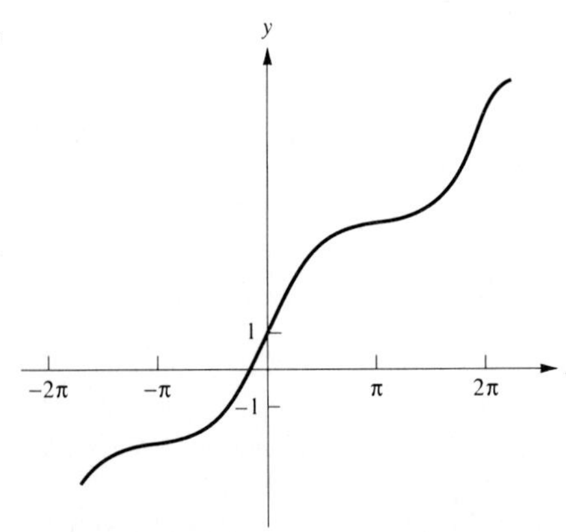

41.

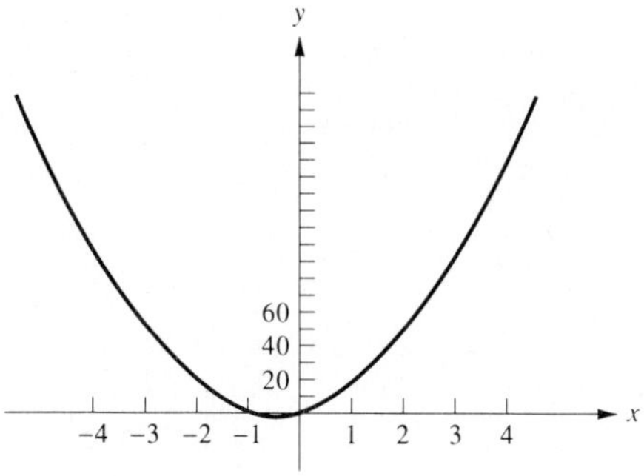

43.

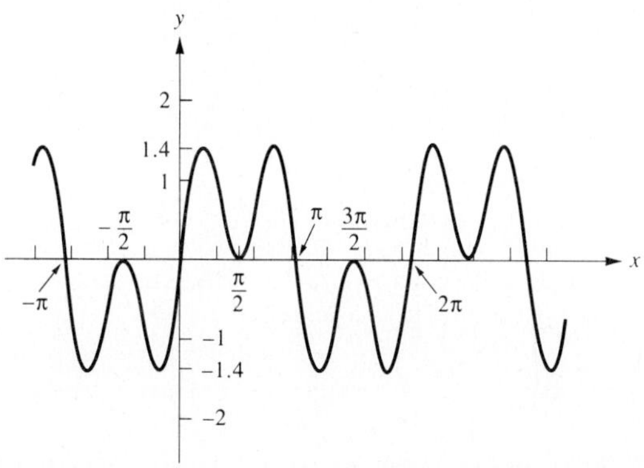

45.

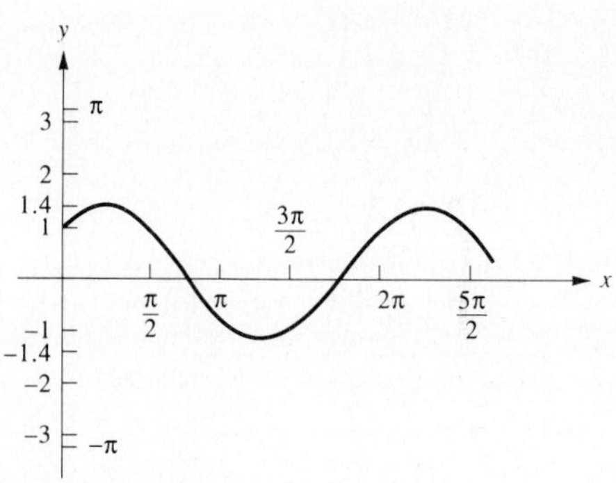

47.

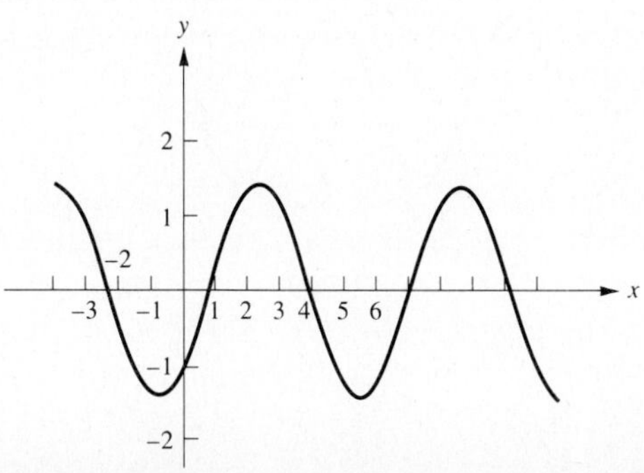

49.

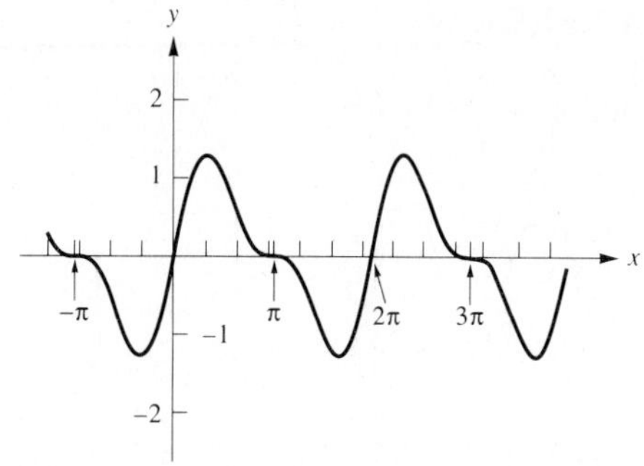

51.

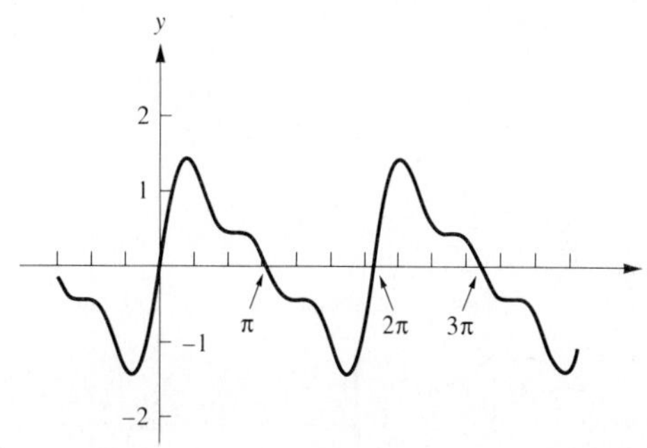

53.

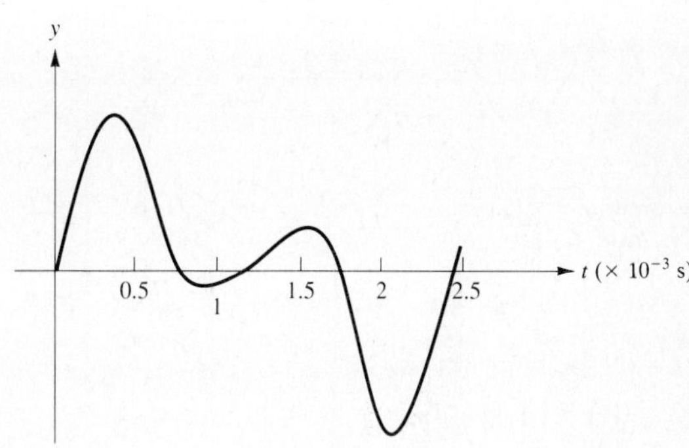

55. Yes, it is possible. The graph is the same as for Exercise 53, in which the frequencies differ by a factor of 2. **57.** Same figure as Exercise 49. **59.** Period $= 2\pi$

REVIEW EXERCISES

1.

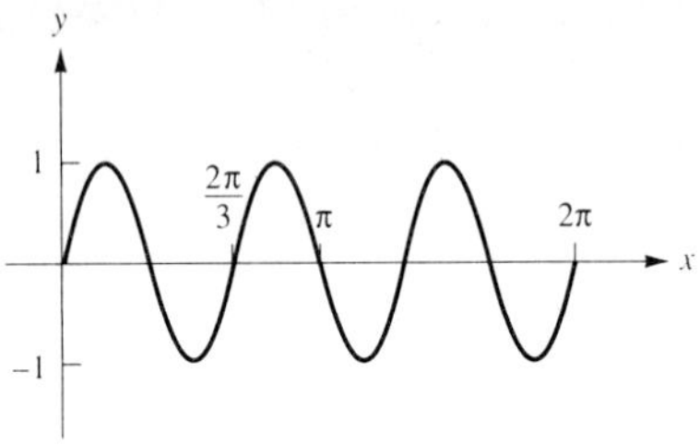

amplitude $= 1$, period $= \frac{2\pi}{3}$, phase shift $= 0$

3.

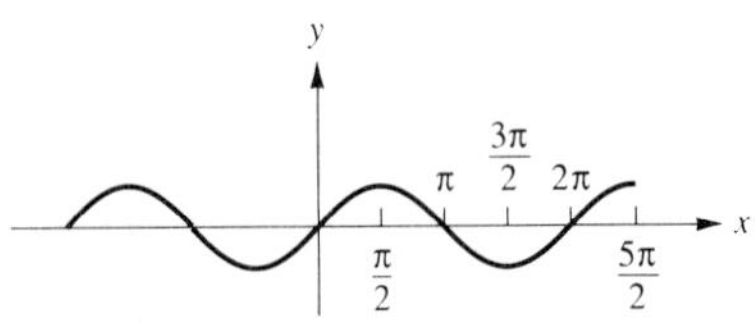

amplitude $= 1$, period $= 2\pi$, phase shift $= \pi/2$ to the right relative to $\cos x$

5.

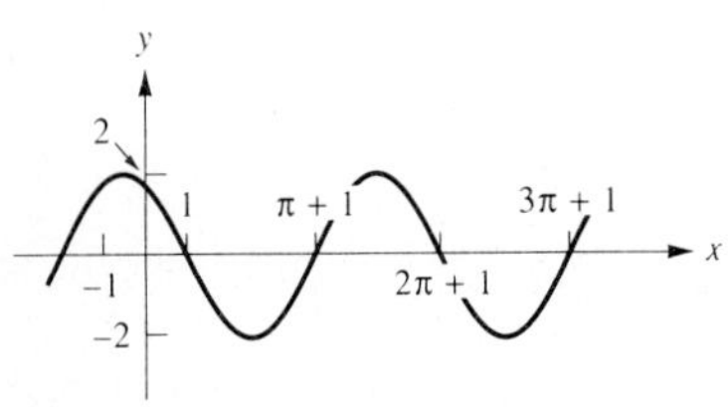

amplitude $= 2$, period $= 2\pi$, phase shift $= 1$ to the right relative to $2\sin(-x)$

7.

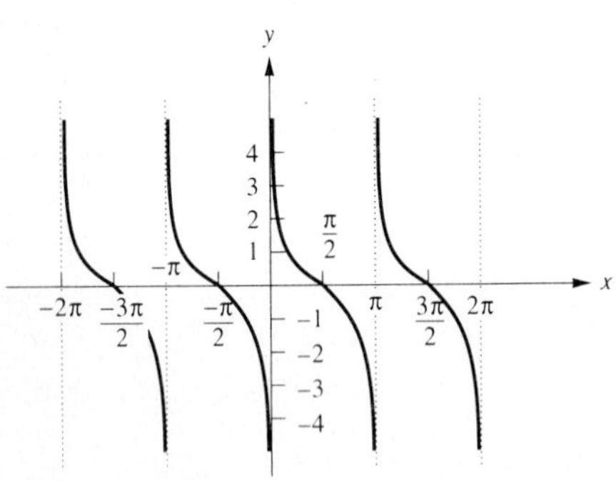

period $= \pi$, phase shift $= \pi/2$ to the right relative to $-\tan x$

9.

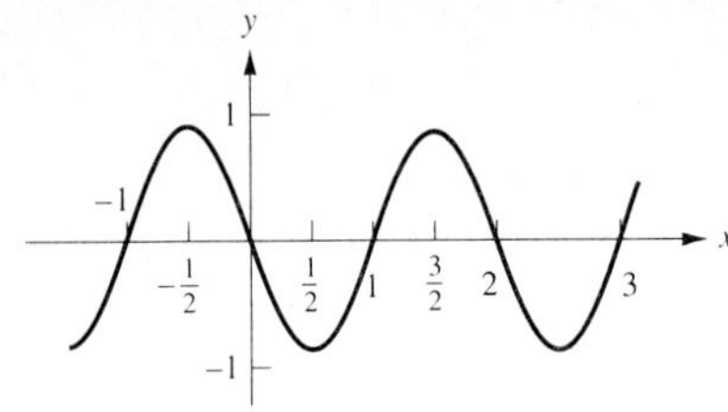

amplitude $= 1$, period $= 2$, phase shift $= 1/2$ to the right relative to $-\cos(-\pi x)$

11.

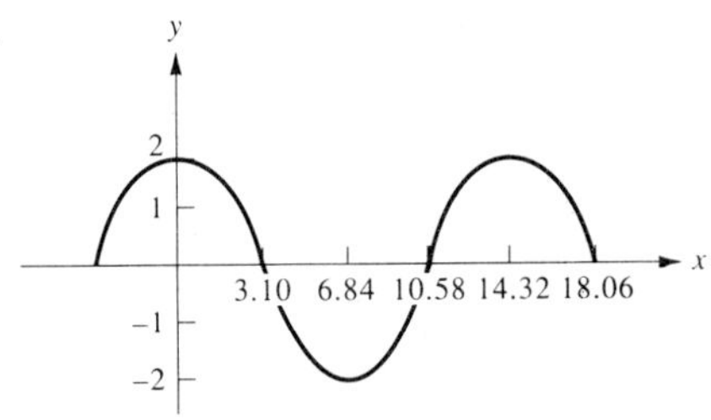

amplitude $= 2$, period $= 4.76\pi = 14.96$, phase shift $= 3.10$ to the right relative to $-2\sin(0.42x)$

13.

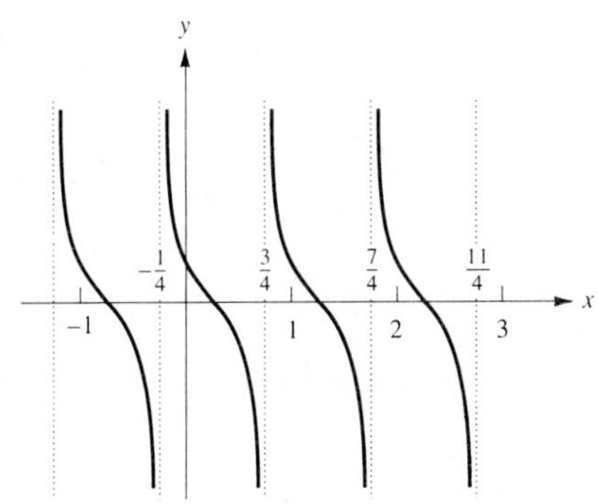

period $= 1$, phase shift $= 1/4$ to the left relative to $1.5\cot(\pi x)$

15. (a)

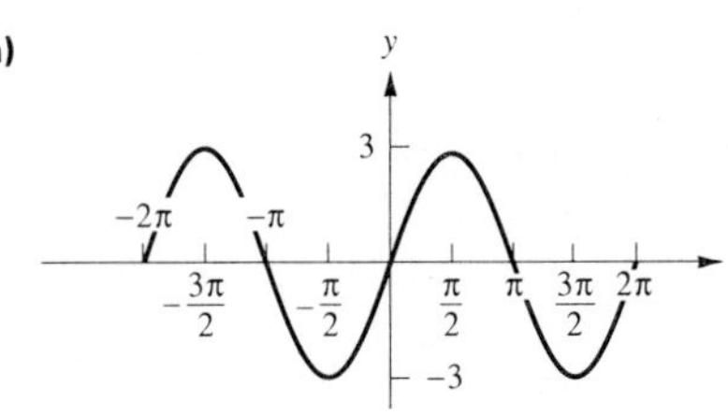

$y = 3\sin x$

(b)

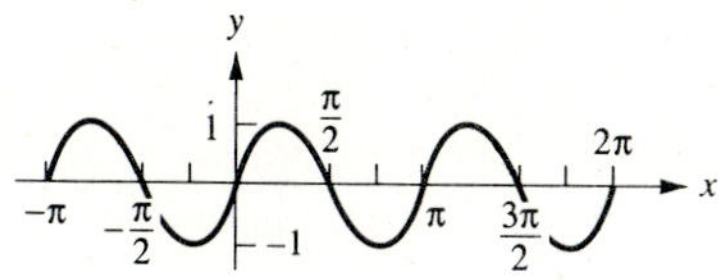

$y = \sin 2x$

(c)

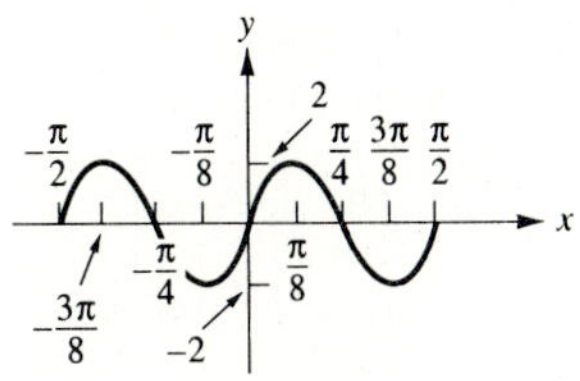

$y = 2 \sin 4x$

(d)

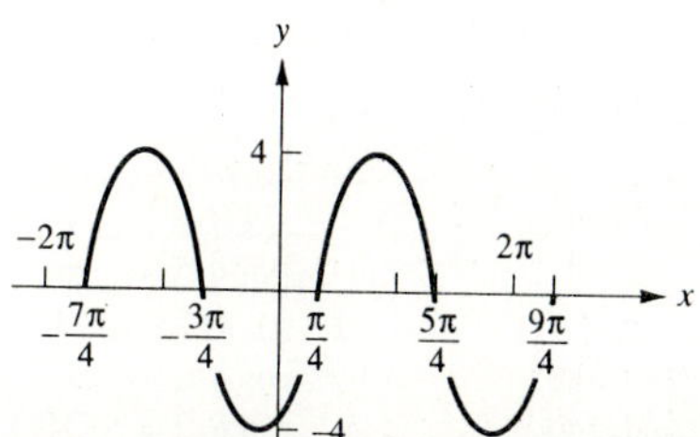

$y = 4 \sin(x - \pi/4)$

(e)

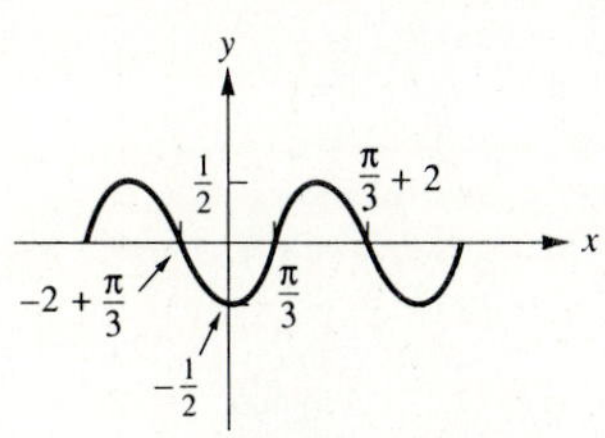

$y = \frac{1}{2} \sin\left(\frac{\pi x}{2} - \frac{\pi^2}{6}\right)$

(f)

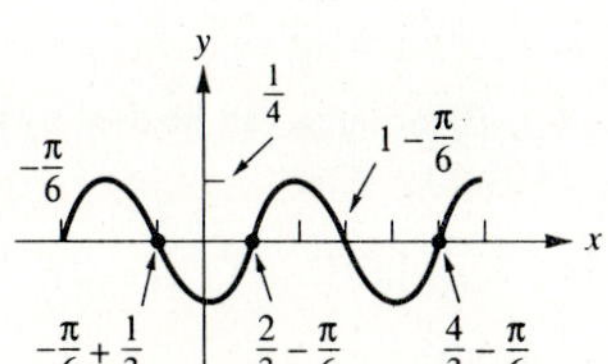

$y = \frac{1}{4} \sin\left(3\pi x + \frac{\pi^2}{2}\right)$

(g)

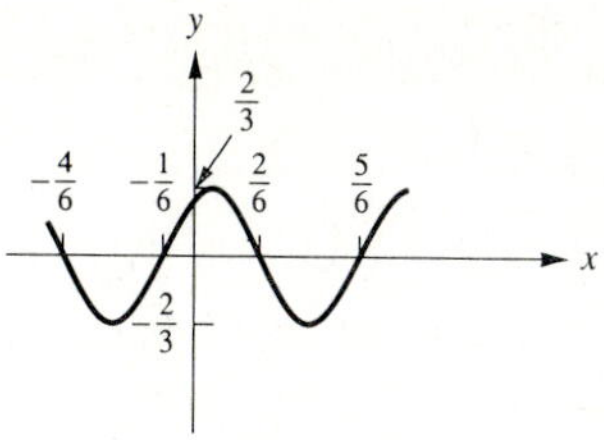

$y = \frac{2}{3} \sin\left(2\pi x + \frac{\pi}{3}\right)$

17. (a) Periodic, period = 6 **(b)** Not periodic **(c)** Not periodic **(d)** Periodic, period = 6 **(e)** Not periodic **19. (a)** Amplitude = 7, phase shift = $\pi/2$ to the left, $y = 7 \sin(x + \pi/2)$ **(b)** Amplitude = 1.8, phase shift = 0, $y = -1.8 \sin x$ **(c)** Amplitude = 3, phase shift = $\pi/4$ to the left, $y = 3 \sin(x + \pi/4)$

CHAPTER 6

EXERCISES 6.1

1. $\frac{3}{4}$ **3. (a)** 0.83 **(b)** 0.83 **5.** 0.396 **7.** 0.00695 **9. (a)** 0.72 **(b)** 0.56 **(c)** 0.72 **11.** 0.262 **13.** 1.4 **15.** -1 **17.** $16 \tan^2 y - 24 \tan y + 9$ **19.** $2 \cot \phi + 10 \cot^3 \phi$ **21.** $(\sin \theta + \cos \theta)(1 + 2 \sin \theta \cos \theta)$ **23.** $\sin x(7 \sin^2 x - 4)$ **25.** $\csc x - 1$ **27.** $\tan x(\tan x - 1)$ **29.** $1 + \sec x$ **31.** -3 **33.** $1, -5$ **35.** $\frac{1}{4}, -\frac{1}{4}$ **37.** None **39.** $(-3 + \sqrt{5})/2$ **41.** $\frac{1}{3}, \frac{\sqrt{2}}{2}, -\frac{\sqrt{2}}{2}$ **43.** $\pi/2$ **45.** $\pi/3, 2\pi/3, 4\pi/3, 5\pi/3$ **47.** $\frac{\pi}{6}, \frac{5\pi}{6}$, 3.99 rad, 5.42 rad **49.** 1.37 rad, 4.91 rad **51.** 0.917 rad, 2.22 rad

EXERCISES 6.2

1. $\pm\sqrt{1 - \sin^2 x} - \sin x$ **3.** $1/[(\sin x)(\pm\sqrt{1 - \sin^2 x})]$ **5.** $\pm\sin x\sqrt{1 - \sin^2 x}$ **7.** $(1 - \sin x)/(\pm\sqrt{1 - \sin^2 x})$ **9.** $1 + \sin x$ **11.** $1 - \sin x$ **13.** $\sin^2 x$ **15.** $2/\sin^2 x$ **17.** $1 - \cos x$ **19.** $1/(\sin x \cos x)$ **21.** $\sin x/\cos x$ **23.** $\cos^6 x/\sin^6 x$ **25.** $\cos^2 x - \sin^2 x$

REVIEW EXERCISES

1. $-(1 + \cos x)$ **3.** $2 \cos^2 x$ **5.** 0 **7.** $4 \tan x$

CHAPTER 7

EXERCISES 7.1-1

1. $(\sqrt{2}+\sqrt{6})/4$ **3.** $(\sqrt{6}-\sqrt{2})/4$ **5.** $-(\sqrt{6}+\sqrt{2})/4$
7. $\sqrt{2}/2$ **9.** $(\sqrt{2}-\sqrt{6})/4$ **11. (a)** 33/65 **(b)** 63/65
13. (a) 0.394 **(b)** 0.741 **15. (a)** -0.910 **(b)** 0.779

EXERCISES 7.1-2

1. $(\sqrt{2}-\sqrt{6})/4$ **3.** $-(\sqrt{6}+\sqrt{2})/4$ **5.** $(\sqrt{2}-\sqrt{6})/4$
7. $-\sqrt{2}/2$ **9.** $(\sqrt{6}+\sqrt{2})/4$ **11. (a)** $\frac{56}{65}$ **(b)** $-\frac{16}{65}$
13. (a) 0.887 **(b)** 0.678 **15. (a)** -0.958 **(b)** -0.392

EXERCISES 7.1-3

1. $(1-\sqrt{3})/(1+\sqrt{3})$ **3.** $(1+\sqrt{3})/(1-\sqrt{3})$
5. $(\sqrt{3}-1)/(\sqrt{3}+1)$ **7.** -1 **9.** $(1+\sqrt{3})/(1-\sqrt{3})$
11. (a) $\frac{56}{33}$ **(b)** $\frac{16}{63}$ **13. (a)** -1.49 **(b)** 0.894
15. (a) -2.38 **(b)** 0.207

EXERCISES 7.1-4

1. $\sin(A+B)=\frac{-33}{65}$, $\sin(A-B)=\frac{-63}{65}$, $\cos(A+B)=\frac{-56}{65}$, $\cos(A-B)=\frac{16}{65}$

3. $\sin(A+B)=\dfrac{\sqrt{2}}{2}$, $\sin(A-B)=\dfrac{\sqrt{2}}{10}$, $\cos(A+B)=\sqrt{2}/2$, $\cos(A-B)=7\sqrt{2}/10$

5. $\sin(A+B)=(6+4\sqrt{21})/25$, $\sin(A-B)=(6-4\sqrt{21})/25$, $\cos(A+B)=(3\sqrt{21}-8)/25$, $\cos(A-B)=(3\sqrt{21}+8)/25$

7. $\sin x=(4\sqrt{24}-3)/25$, $\cos x=(3\sqrt{24}+4)/25$, $\cos x=(6-\sqrt{6})/4$

9. $-\sin x$ **11.** $-\cot x$ **13.** $\sin x$ **17.** $\sin 20°$ **19.** $\cos 19°$
21. $\tan\frac{\pi}{12}$ **23.** $\tan 5\theta$ **25.** $\cos x$
27. $\sin(x-45°)$ or $-\cos(x+45°)$
29. $\sin(30°-x)$ or $\cos(60°+x)$
31. $14.8\sin(x+60°)$ or $14.8\cos(30°-x)$
33. $\sin(x+21°)$ or $\cos(x-69°)$ **35.** $\cos x$ **37.** $\tan(x+45°)$
39. $\tan(x+30°)$ **41.** $\tan(x+60°)$ **43.** $\tan(x+50°)$
45. $6.40\tan(x+27°)$ **47.** 60°, 300° **49.** 0°, 180°
51. 12°, 78°, 192°, 258° **53.** 15°, 75°, 195°, 255° **55.** 150°, 210°
57. 40°, 140° **59.** 13.9°, 346.1° **61.** 0°, 90°
79. $\cot(A+B)=(\cot A\cot B-1)/(\cot A+\cot B)$ **91.** 3
95. $y=3.61\cos(x-0.588)$ **107.** 27.9°, 6.4/12 **109.** 75.7°

EXERCISES 7.2-1

11. (b) No, 90° is not an allowed angle for $\tan\theta$.
13. $2\sin 2x\cos 2x$ **15.** $\cos^2 3x-\sin^2 3x$
17. $(2\tan 2x)/(1-\tan^2 2x)$ **19.** $\cos^2(17.5°)-\sin^2(17.5°)$
21. $[2\tan(x/6)]/[1-\tan^2(x/6)]$
23. $2\sin(45°+x)\cos(45°+x)$ **25.** $4\sqrt{2}/9$ **27.** $3\sqrt{7}/8$
29. $-3/4$ **31.** $4\sqrt{2}/9$ **33.** $-3/4$ **35.** $2a/(1+a^2)$
37. $(2a\sqrt{1-a^2})/(1-2a^2)$ **39.** $(2\sqrt{a^2-1})/(2-a^2)$
41. $2a^2-1$ **43.** $2a\sqrt{1-a^2}$

45. $\sin 2\theta=-3\sqrt{7}/8$, $\cos 2\theta=2/16$, $\tan 2\theta=-3\sqrt{7}$
47. $\sin 2\theta=8/17$, $\cos 2\theta=-15/17$, $\tan 2\theta=-8/15$
49. $\sin 2\theta=2\sqrt{8}/9$, $\cos 2\theta=-7/9$, $\tan 2\theta=-2\sqrt{8}/7$
51. $\sin 2\theta=0$, $\cos 2\theta=1$, $\tan 2\theta=0$

53. $\cos 70°$ **55.** $\sin 64°$ **57.** $\cos x$ **59.** $\frac{1}{2}\sin(10x)$
61. $\sin^2 6x$ **63.** $\tan 2x$ **65.** $\csc 6x$
67. $4\sin\theta\cos\theta(\cos^2\theta-\sin^2\theta)$ **69.** $\sin\theta(4\cos^2\theta-1)$
71. $[4\tan\theta(1-\tan^2\theta)]/[\tan^4\theta-6\tan^2\theta+1]$
73. $(\cos^2\theta-\sin^2\theta)(1-16\sin^2\theta\cos^2\theta)$ **75.** 135°, 315°
77. 22.5°, 112.5°, 202.5°, 292.5° **79.** 30°, 150°, 210°, 330°
81. 17.7°, 72.4°, 107.7°, 162.4°, 197.7°, 252.4°, 287.7°, 342.4°
83. 45°, 225° **85.** 24.6°, 65.4° **87.** 30°, $18\sqrt{3}$ cm^2
91. 93.5 ft, 28.1°, 56.2°

EXERCISES 7.2-2

13. (a) $\sin\theta/2=0.17$, $\cos\theta/2=0.99$, $\tan\theta/2=0.17$
(b) $\sin\theta/2=0.283$, $\cos\theta/2=0.959$, $\tan\theta/2=0.295$
(c) $\sin\theta/2=0.475$, $\cos\theta/2=0.880$, $\tan\theta/2=0.540$
(d) $\sin\theta/2=0.233$, $\cos\theta/2=0.972$, $\tan\theta/2=0.240$
(e) $\sin\theta/2=0.282$, $\cos\theta/2=0.959$, $\tan\theta/2=0.294$
(f) $\sin\theta/2=0.668$, $\cos\theta/2=0.744$, $\tan\theta/2=0.898$

15. $\sin\theta/2=0.447$, $\cos\theta/2=0.894$

17. (a) $\sin\theta/2=-\frac{3}{5}$, $\cos\theta/2=-\frac{4}{5}$, $\tan\theta/2=+\frac{3}{4}$ or $\sin\theta/2=\frac{3}{5}$, $\cos\theta/2=\frac{4}{5}$, $\tan\theta/2=\frac{3}{4}$
(b) $\sin\theta/2=\frac{3}{5}$, $\cos\theta/2=\frac{4}{5}$, $\tan\theta/2=\frac{3}{4}$ or $\sin\theta/2=-\frac{3}{5}$, $\cos\theta/2=-\frac{4}{5}$, $\tan\theta/2=\frac{3}{4}$

19. $\sqrt{\dfrac{3-\sqrt{8}}{6}}=0.169$ **21.** $\dfrac{7\sqrt{2}}{10}=0.990$

23. $\dfrac{1}{2}\sqrt{\dfrac{4-\sqrt{15}}{2}}=0.126$ **33.** 82.8°, 12.4 ft, 8.27 ft

43. $\pm\sqrt{\dfrac{1}{2}-(\pm)\sqrt{\dfrac{1+\cos\theta}{8}}}$

45. The half-angle and $(A\pm B)$ formulas can be used by realizing the following:
(a) $7.5°=\frac{1}{2}[\frac{1}{2}(30°)]$
(b) and **(c)** $3.75°=\frac{1}{2}(7.5°)$ and $18.75°=15°+3.75°$
(d) $73.125°=\frac{1}{2}(150°-3.75°)$
(e) $89.0625°=90°-\frac{1}{2}[\frac{1}{2}(3.75°)]$
(f) $4.6875°=\frac{1}{2}[\frac{1}{2}(18.75°)]$
(g) $78.7500°=45°+\frac{1}{2}[\frac{1}{2}(135°)]$

EXERCISES 7.3

1. $2 \sin 60° \cos 10°$ **3.** $2 \cos 110° \cos 30°$ **5.** $2 \sin 2x \cos x$
7. $2 \sin x \cos y$ **9.** $-2 \sin\left(7kt - \frac{7\pi}{12}\right)\sin\left(-3kt + \frac{7\pi}{12}\right)$
11. $2 \sin 45° \cos 30° = \sqrt{6}/2$ **13.** $2 \cos \pi/2 \cos 2\pi/3 = 0$
15. $-2 \sin 60° \sin 20° = -\sqrt{3} \sin 20°$ **17.** $\tan 10°$
19. $-\tan\left(\frac{x-y}{2}\right)$ **21.** $-\cot\left(\frac{x-y}{2}\right)$
23. $2 \sin[\pi(f_1 + f_2)t]\cos[\pi(f_1 - f_2)t]$ **33.** 45°, 90°, 135°, 225°, 270°, 315° **35.** 30°, 90°, 150°, 270°

REVIEW EXERCISES

1. **(a)** $\sin 255° = -(\sqrt{6} + \sqrt{2})/4$
$\cos 255° = (\sqrt{2} - \sqrt{6})/4$
$\tan 255° = \dfrac{\sqrt{3}+1}{\sqrt{3}-1}$
(b) $\sin 285° = -(\sqrt{6} + \sqrt{2})/4$
$\cos 285° = (\sqrt{6} - \sqrt{2})/4$
$\tan 285° = \dfrac{\sqrt{3}+1}{1-\sqrt{3}}$
(c) $\sin(-195°) = (\sqrt{6} - \sqrt{2})/4$
$\cos(-195°) = -(\sqrt{6} + \sqrt{2})/4$
$\tan(-195°) = \dfrac{1-\sqrt{3}}{1+\sqrt{3}}$
3. **(a)** $\sin(A + B) = \dfrac{2\sqrt{7} + 3\sqrt{5}}{12}$
$\cos(A + B) = (\sqrt{35} - 6)/12$
(b) $\sin(A + B) = 7\sqrt{65}/65$
$\cos(A + B) = 4\sqrt{65}/65$
(c) $\sin(A + B) = -0.954$
$\cos(A + B) = 0.300$
5. $\frac{1}{2}$ **7.** $\sqrt{3}$ **9.** $\sqrt{3}/3$ **11.** 0 **13.** $\frac{1}{2}$ **15.** 0°, 180°
17. 0°, 120°, 240° **19.** 45°, 90°, 135°, 225°, 270°, 315°
21. 45°, 135°, 225°, 315° **27.** 255/1313 or 735/1313 **29.** 24/25
31. $-336/625$ **33.** 4/5 **35.** **(a)** 4/21 **(b)** 225/4
41. $\sin\frac{x}{2} = \frac{\sqrt{10}}{10}$, $\cos\frac{x}{2} = \frac{3\sqrt{10}}{10}$, $\tan\frac{x}{2} = \frac{1}{3}$
43. $\sin\frac{x}{2} = \frac{\sqrt{2}}{10}$, $\cos\frac{x}{2} = \frac{7\sqrt{2}}{10}$, $\tan\frac{x}{2} = \frac{1}{7}$
45. *If x is in quadrant II*: $\sin\frac{x}{2} = \frac{3\sqrt{10}}{10}$, $\cos\frac{x}{2} = \frac{\sqrt{10}}{10}$, $\tan\frac{x}{2} = 3$
If x is in quadrant IV: $\sin\frac{x}{2} = \frac{\sqrt{10}}{10}$, $\cos\frac{x}{2} = \frac{-3\sqrt{10}}{10}$, $\tan\frac{x}{2} = -\frac{1}{3}$
47. **(a)** $0, \frac{\pi}{3}, \pi, \frac{5\pi}{3}$ **(b)** $0, \frac{2\pi}{3}, \frac{4\pi}{3}$ **49.** $0, \pi, \frac{\pi}{6}, \frac{5\pi}{6}, \frac{7\pi}{6}, \frac{11\pi}{6}$
51. $0, \frac{\pi}{2}, \pi, \frac{3\pi}{2}, \frac{\pi}{6}, \frac{5\pi}{6}, \frac{7\pi}{6}, \frac{11\pi}{6}$ **53.** $0, \pi, \frac{\pi}{3}, \frac{2\pi}{3}, \frac{4\pi}{3}, \frac{5\pi}{3}$

CHAPTER 8

EXERCISES 8.1

1. 0 **3.** $\pi/2, 3\pi/2$ **5.** 1.01, 4.15 **7.** 4.22, 5.20
9. $\pi/4, 3\pi/4, 5\pi/4, 7\pi/4$ **11.** $\pi/4, 3\pi/4, 5\pi/4, 7\pi/4$
13. 0.88, 2.26, 4.02, 5.40 **15.** 0.17, 1.40, 3.31, 4.54
17. $\pi/2, 3\pi/4, 3\pi/2, 7\pi/4$ **19.** $\pi/2, 2\pi/3, 4\pi/3$
21. $5\pi/6, 7\pi/6, \pi/3, 5\pi/3$ **23.** $\pi/6, \pi/2, 5\pi/6$
25. $\pi/3, 2\pi/3, 4\pi/3, 5\pi/3$ **27.** $0, 2\pi/3, 4\pi/3$
29. Identity true for all x. **31.** $\pi/4, 3\pi/4, 5\pi/4, 7\pi/4$
33. $7\pi/6, 11\pi/6$ **35.** 0 **37.** $\pi/6, 5\pi/6, 7\pi/6, 11\pi/6$
39. 0.96, 2.18, 4.10, 5.32, $\pi/3, 2\pi/3, 4\pi/3, 5\pi/3$
41. 0.986, 2.16, 3.78, 5.64 **43.** No solutions
45. $\pi/4, 3\pi/4, 5\pi/4, 7\pi/4$ **47.** $\pi/4, 5\pi/4$ **49.** $0, 2\pi/3, 4\pi/3$
51. $0, \pi/2, \pi, 3\pi/2$ **53.** $\pi/8, 5\pi/8, 9\pi/8, 13\pi/8$
55. $\pi/3, 2\pi/3, 7\pi/6, 5\pi/3, \pi/2, 3\pi/2$ **57.** $0, 3\pi/4, \pi, 5\pi/4$
59. $\pi/2, 3\pi/2, 7\pi/6, 11\pi/6$ **61.** $0, \pi$ **63.** $\pi/3, 2\pi/3, 4\pi/3, 5\pi/3$
65. $\pi/2, 3\pi/2$ **67.** $0, \pi/3, 5\pi/3$ **69.** 0.73, 2.41, $7\pi/6, 11\pi/6$
71. $\pi/12, 5\pi/12, 13\pi/12, 17\pi/12, 3\pi/4, 7\pi/4$
73. 1.07, 2.08, 3.48, 5.94 **75.** 0.52, 3.67, 4.71 **77.** 0.39, $3\pi/2$
79. $\pi/4, 5\pi/4$ **81.** No solutions **83.** $0, \pi/2, \pi, 3\pi/2$
85. $0, \pi, \pi/3, 5\pi/3$ **87.** **(a)** $3\pi/4$ **(b)** $7\pi/4$ **(c)** $\pi/4, 5\pi/4$
89. **(a)** $2\pi/3, 5\pi/3$ **(b)** $\pi/6, 7\pi/6$
(c) $5\pi/12, 11\pi/12, 17\pi/12, 23\pi/12$
91. **(a)** $x \to 2\pi$, **(b)** 0 **(c)** $3\pi/2$ **93.** **(a)** 0.301, 3.59
(b) $5\pi/3$, 1.95 **95.** **(a)** None **(b)** 1.5
97. **(a)** $0.91/n$ (n odd) **(b)** 0
99. **(a)** $\pm\sqrt{4 + (1/n)}$ $(n = \pm 1, \pm 2, \pm 3, \ldots)$ **(b)** ± 2
101. $\pm n\pi/2$ $(n = 0, 1, 2, 3, \ldots)$ **103.** $\pm 3n\pi$ $(n = 0, 1, 2, 3, \ldots)$
105. $3/2 \pm 2n$ $(n = 0, 1, 2, 3, \ldots)$
107. $\pi/16 \pm n(\pi/4)$ $(n = 0, 1, 2, 3, \ldots)$
109. $\pm n$ $(n = 0, 1, 2, 3, \ldots)$
111. $(2n + 1)(\pi/6)$ $(n = 0, \pm 1, \pm 2, \pm 3, \ldots)$
113. $k = (n\pi)/L$ $(n = 0, \pm 1, \pm 2, \pm 3, \ldots)$ **115.** $-2 \le b \le 0$
117. $-3 \le b \le -2$ **119.** $-7 \le b \le -4$ **121.** $b = 0$
123. $-\infty < b \le -1/2$ **125.** $b \ge \sqrt{10}$ and $b \le -\sqrt{10}$
127. $\cos^{-1}\left(\dfrac{-a \pm \sqrt{a^2 + 8a + 8}}{4}\right)$
129. 26° **131.** 19.5° **133.** 31.3°, 58.7° **135.** 26.5°
137. **(a)** 0.027 r.t/day at poles
0.038 r.t/day at equator **(b)** $\pm 54.3°$

EXERCISES 8.2

1. $\left.\begin{matrix}0.73 \pm 2n\pi \\ 2.41 \pm 2n\pi\end{matrix}\right\} n = 0, 1, 2, 3, \ldots$ **3.** 0 **5.** -0.51
7. $0, \pm 4.5, \pm 7.72, \ldots$ **9.** $1.1, 3.1, 5.1, \ldots; -1.8, -4.9, \ldots$
11. 0.865 **13.** No solutions **15.** $-0.95, 0, 0.95$
17. $\frac{\pi}{4} \pm n\pi \; (n = 0, 1, 2, \ldots)$ **19.** No solutions
21. $\left.\begin{matrix}0.62 \pm 2n\pi \\ -1.13 \pm 2n\pi\end{matrix}\right\} n = 0, 1, 2, 3, \ldots$ **23.** $-n\pi \; (n = 1, 2, 3, \ldots)$
25. $0, -1.29, \frac{-3\pi}{2}, \frac{-7\pi}{2}, \frac{-11\pi}{2}, \ldots$
27. $\ldots, -1.6, 0.39, \pi, 2\pi, 3\pi, \ldots$ **31.** Infinite **33.** Finite
35. Finite **37.** Finite **39.** Infinite **41.** None **43.** 147°
45. 54.2° **47.** 59.6°

EXERCISES 8.3

1. $2\pi/3$ **3.** $\left.\begin{matrix}4\pi/3 \pm 2n\pi \\ 5\pi/3 \pm 2n\pi\end{matrix}\right\} n = 0, 1, 2, 3, \ldots$
5. $\left.\begin{matrix}\pi/4 \pm n\pi \\ 5\pi/4 \pm n\pi\end{matrix}\right\} n = 0, 1, 2, 3, \ldots$ **7.** $\pi/6$ **9.** $-\pi/6$ **11.** $-\pi/2$
13. 0 **15.** Not possible **17.** $3\pi/10$ **19.** Not possible
21. $-43°$ **23.** $\sqrt{u}$ **25.** $\sqrt{2}/2$ **27.** $-3/8$ **29.** Not possible
31. $-2/\sqrt{5}$ **33.** $\pi/2$ **35.** -9.68 **37.** $9\pi/20$ **39.** $-4\pi/3$
41. $7/25$ **43.** $\pm\frac{1}{2}\sqrt{(4-\sqrt{7})/2}$ **45.** $\sqrt{7}/3$
47. $(1 + 2\sqrt{6})/6$ **49.** $4\sqrt{5}/9$ **51.** $(3 + 4\sqrt{35})/(4 - 3\sqrt{35})$
53. 0.360 **55.** Not possible **57.** 0.955 **59.** 0.405
61. 0.969 **63.** 0.44 **65.** -0.639 **67.** $\pi/2$ **69.** -0.0939
71. 0.882 **73.** 0.399 **75.** 0.892 **77.** $\sqrt{1 - x^2}$
79. $1/x \; (x \neq 0)$ **81.** $2x^2 - 1$ **83.** $\frac{2x}{1 - x^2}$
85. $+\sqrt{(1 + \sqrt{1 - x^2})/2}$ **87.** $\frac{x^2}{2} + \sqrt{(1 - x^2)(4 - x^2)}$
89. $(3x + 2\sqrt{x^2 - 9})/(6 - x\sqrt{x^2 - 9})$ **91.** $x(3 - 4x^2)$
93. x **103.** $\left[-c + \text{Cos}^{-1}\left(\frac{d}{a}\right)\right]\Big/ b$ **105.** No
109. **(a)** $\pi/7$ **(b)** 0.818 **(c)** 1.4 **(d)** $\pi/3$ **(e)** $\pi/4$
111. No **113.** No **115.** Yes **117.** Yes **119.** Yes
121. No **123.** Yes **125.** No **127.** No **129.** Yes
131. **(b)** $-\frac{\pi}{2} < y < \frac{\pi}{2}$ **133.** **(b)** $0 \le y < \pi, y \neq \pi/2$
135. **(a)** Yes. For example, if $x = 1$, one possible result is $\sin^{-1} 1 = -3\pi/2$ **(b)** No
137. Yes. For example, if $x = 1$, one possible result is $\tan^{-1} 1 = -3\pi/4$ **(b)** No

REVIEW EXERCISES

1. $\pi/2, 3\pi/2$ **3.** $0.90, \pi, 5.38$ **5.** $\pi, \pi/3, 5\pi/3$ **7.** $2.82, 5.96, \pi/4, 5\pi/4$ **9.** $\pi/4, 3\pi/4$ **11.** $0.93, 4.07$ **13.** $\pi/4, 5\pi/4$
15. $0.25, 2.89, \pi/6, 5\pi/6$ **17.** $\pi/2, 3\pi/2, 0.464, 3.604$ **19.** $\pi/4, 3\pi/4, 5\pi/4, 7\pi/4$ **21.** $2\pi/3$ **23.** $\pi/2, 5.6365$ **25.** $0, \pi/2, 3\pi/2, 2\pi/3, 4\pi/3$ **27.** $0, \pi, \pi/4, 5\pi/4$ **29.** $0, 3\pi/4, \pi, 7\pi/4$ **31.** $0, \pi, \pi/2, 3\pi/2, 2\pi/3, 4\pi/3$ **33.** $\pi/6, 11\pi/6$ **35.** $0.93, 2.21, 4.07, 5.35$
37. $0, \pi, 3.85, 5.57$ **39.** $+1$ **41.** $2\pi/3, 4\pi/3$ **43.** $0, \pi/4, \pi/2, \pi, 5\pi/4, 3\pi/2$ **45.** $0, \pi, \pi/8, 5\pi/8, 9\pi/8, 13\pi/8$ **47.** $1.99, 5.86$
49. **(a)** Minimum at $x = 6.18$, maximum at $x = 2.18$
(b) Minimum at $x = \pi/6, 7\pi/6$; maximum at $x = 2\pi/3, 5\pi/3$
(c) Minimum at $x = 5.72$, maximum at $x = 0$ **(d)** Minimum at $x = 1.36$, maximum at $x = 4.36$ **51.** **(a)** $\pi/2$ **(b)** $\pm\pi/2, \pm 3\pi/2$ **53.** **(a)** -0.215 **(b)** $3.36, 6.07, -0.215, -2.93$
55. $7\pi/12$ **57.** 0 **59.** $\pi/2$ **61.** 2π **63.** $\pi/2$ **65.** $\sqrt{3}$
67. 0 **69.** $+1$ **71.** 2.22 **73.** -1 **81.** 19.5° **83.** **(a)** 129°
(b) 786 yd **85.** $0.50 \pm n\pi \; (n = 0, 1, 2, 3, \ldots)$ **87.** 0.63

CHAPTER 9

EXERCISES 9.1

1. **(a)** $r = 0, \theta =$ anything **(b)** (3.5, 0°) **(c)** (6, 45°)
(d) (6, 90°) **(e)** (4, 90°) **(f)** (3, 90°) **(g)** (5, 120°)
(h) (7, 150°) **(i)** (4, 180°) **(j)** (6, 210°) **(k)** (5, 240°)
(l) (6, 240°) **(m)** (6, 270°) **(n)** (4, 300°) **(o)** (4, 315°)
(p) (4, 330°)
3. $-23.$

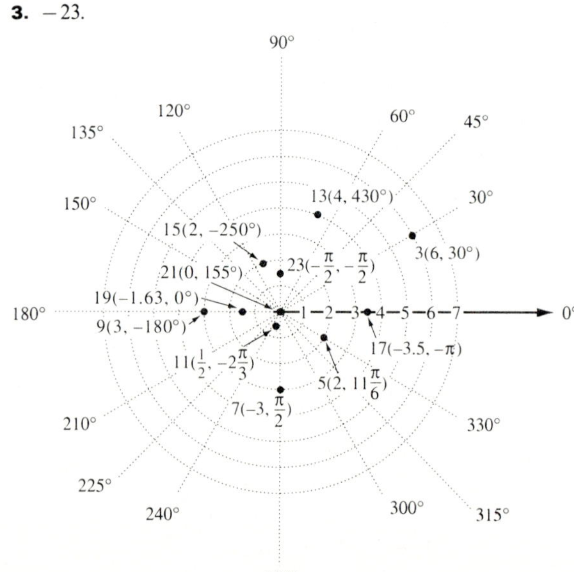

25. $(3, 2\pi)$ **27.** $(-2, 5\pi/4)$ **29.** $(1, 7\pi/6), (1, 19\pi/6)$
31. $(5, \theta)$ where $\theta = 50° \pm n\,360° \; (n = 1, 2, 3, 4, \ldots)$

33. If $r > 0$: $(8, \pi \pm 2n\pi)$ $(n = 1, 2, 3, 4, \ldots)$
If $r < 0$: $(-8, \pm 2n\pi)$ $(n = 0, 1, 2, 3, 4, \ldots)$
35. **(a)** $(9.5, 3\pi/2)$ **(b)** $(9.5, \pi)$ **(c)** $(9.5, 2\pi/3)$ **(d)** $(9.5, 7\pi/4)$

EXERCISES 9.1-2

1. $(2, 2)$ **3.** $(-3\sqrt{2}, -3\sqrt{2})$ **5.** $(-3\sqrt{2}/2, +3\sqrt{2}/2)$
7. $(-3.0, -4.0)$ **9.** $(4, 3)$ or $(-4, -3)$ **11.** $(-0.3210, -1.327)$
13. $(1, 0)$ **15.** **(a)** $(0, 0)$ **(b)** $(7, 0)$ **(c)** $(0, 2)$ **(d)** $(3, 3\sqrt{3})$
(e) $(-7\sqrt{2}/2, 7\sqrt{2}/2)$ **(f)** $(-4, 0)$ **(g)** $(-3\sqrt{3}/2, -3/2)$
(h) $(0, -5)$ **(i)** $(3\sqrt{3}, -3)$ **17.** $(\sqrt{2}, 3\pi/4)$ **19.** $(\sqrt{8}, 7\pi/4)$
21. $(3, \pi/2)$ **23.** $(4, 3\pi/2)$ **25.** $(5, 2.50 \text{ rad})$
27. $(5, 4.07 \text{ rad})$ **29.** $(\pi\sqrt{2}, \pi/4)$
31. **(a)** $(4, 0)$ **(b)** $(\sqrt{20}, 1.11 \text{ rad})$ **(c)** $(0, \text{anything})$
(d) $(\sqrt{5}, 2.68 \text{ rad})$ **(e)** $(6, \pi)$ **(f)** $(4.3, 3.8 \text{ rad})$ **(g)** $(3.5, 3\pi/2)$
(h) $(\sqrt{52}, 5.70 \text{ rad})$ **33.** (10.8 blocks, 56.3°)
35. (2960 km, 126°) **37.** 1300 m from the hive at an angle of 61° relative to the line to the sun **39.** (A) $(4, 0)$ (B) $(\sqrt{8}, \pi/4)$ (C) $(0, \text{anything})$ (D) $(\sqrt{8}, 7\pi/4)$

41. **(a)**

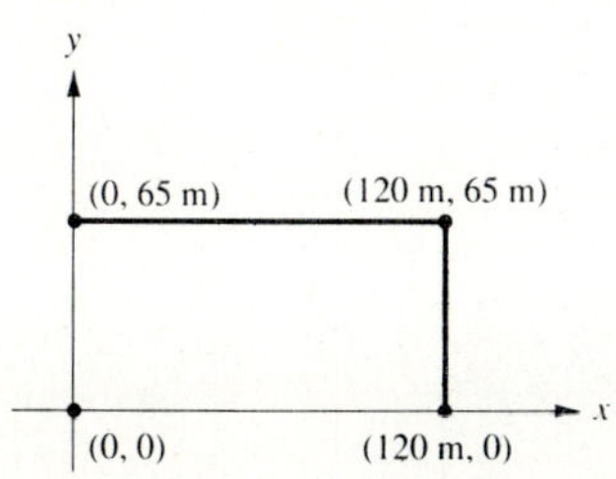

(b)

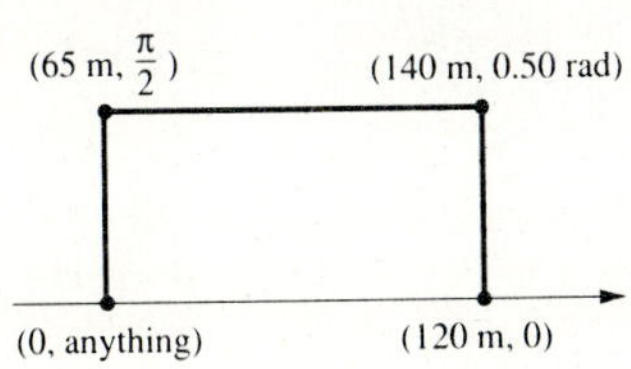

43. **(a)**

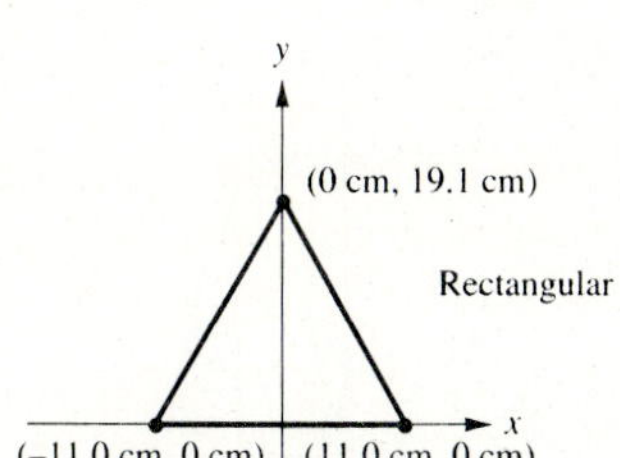

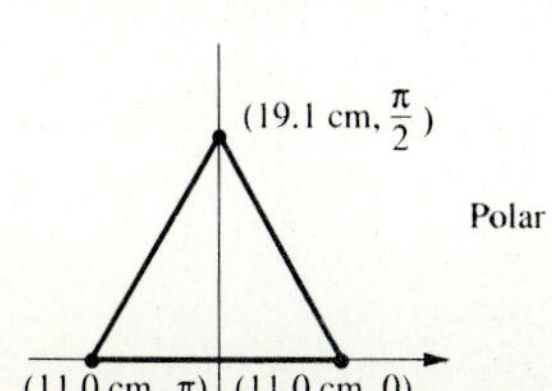

(b)

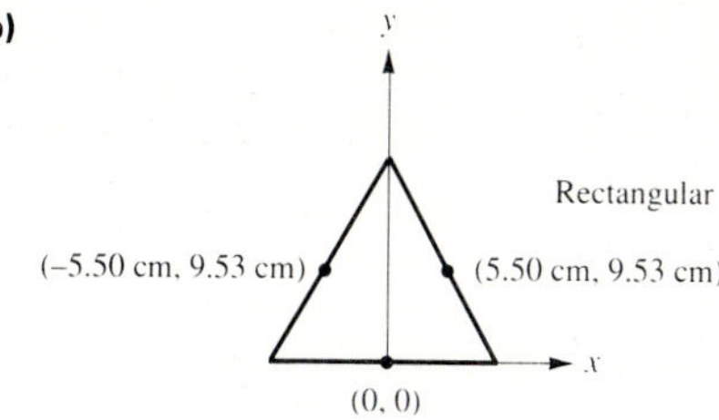

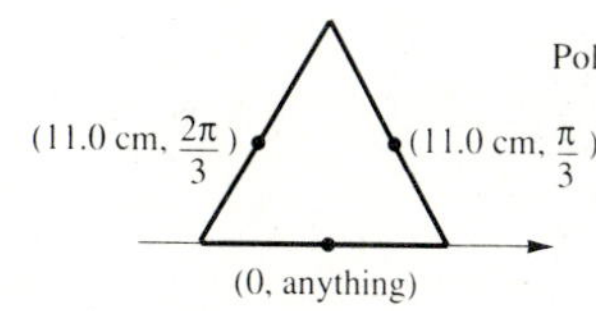

45. First base: (90.0 ft, 0°)
Second base: (127 ft, 45°)
Pitcher's Plate: (60.5 ft, 45°)

47.

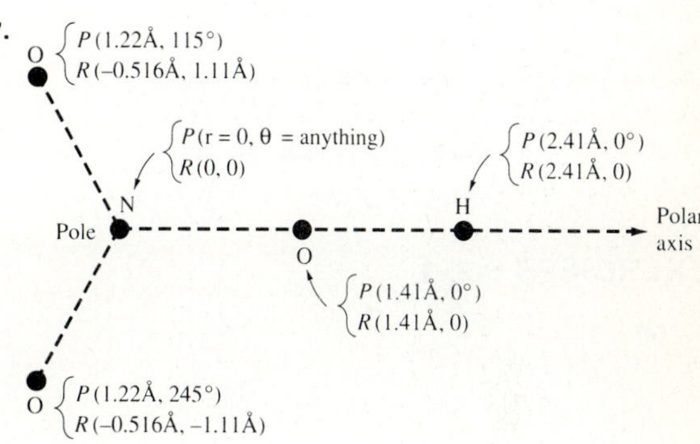

49. **(a)**

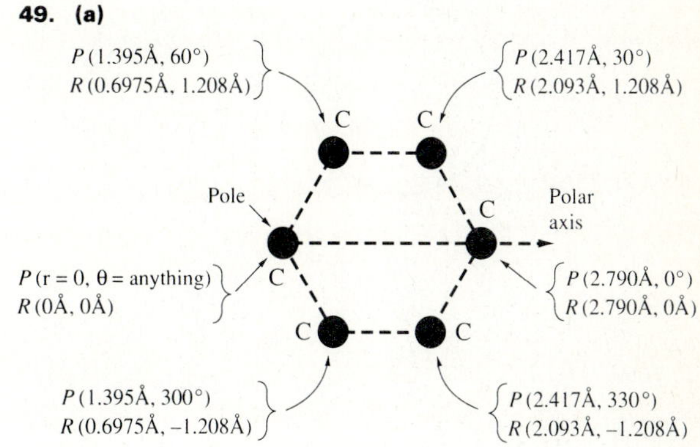

(b)

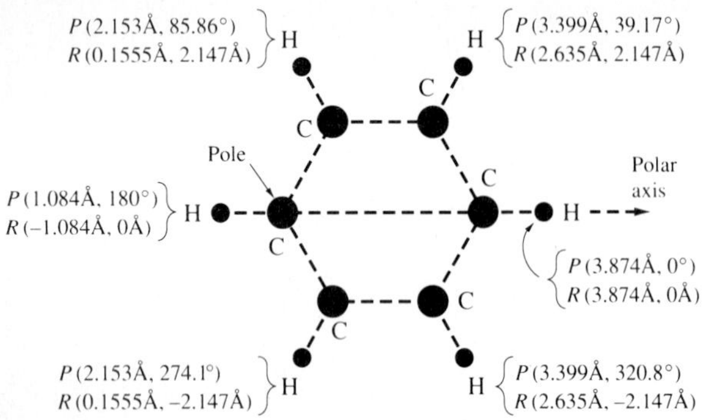

53. **(a)** $(x, -y)$ **(b)** $(-x, -y)$ **(c)** $(-x, y)$

EXERCISES 9.1-3

1. $r \sin \theta = 4$ **3.** $2r(\cos \theta - 2 \sin \theta) = 1$
5. $r \cos^2 \theta = 3 \sin \theta$ **7.** $r^2 = 144$
9. $r^2 - 2r(\cos \theta + 2 \sin \theta) + 1 = 0$
11. $r + (\sin \theta - \cos \theta) = 0$ **13.** $r^2(1 + a \cos^2 \theta)^2 = a^2$
15. $x = -2$ **17.** $x^2 + y^2 = -5x$ **19.** $x = 3$
21. $\sqrt{x^2 + y^2} = 5$ **23.** $y = x$ **25.** $5 = 2xy$
27. $y = \pm x/\sqrt{3}$ **29.** $x^3 + y^2 x = y^2$ **31.** $x^4 + y^4 = 2xy$

EXERCISES 9.2-1

1. **(a)**

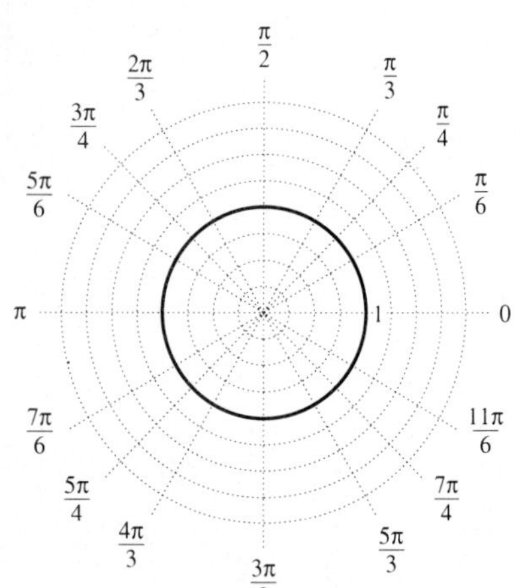

(b) The same as (a).

3.

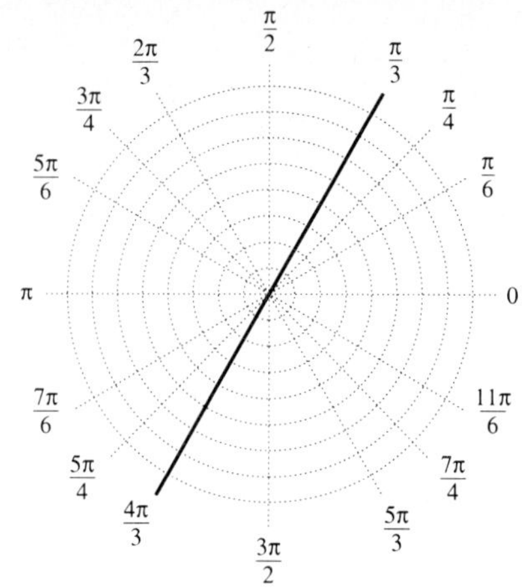

5.

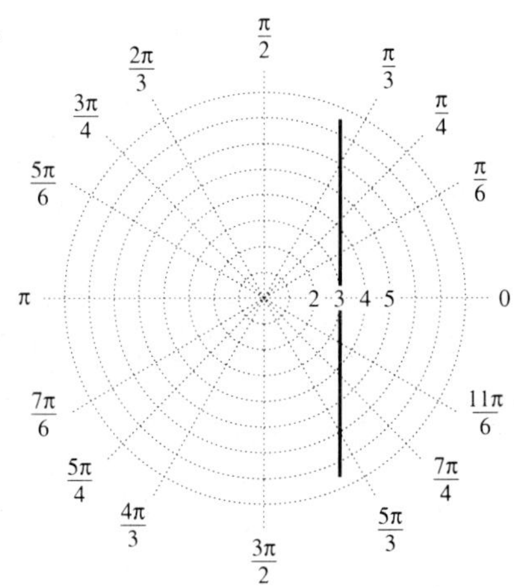

7.

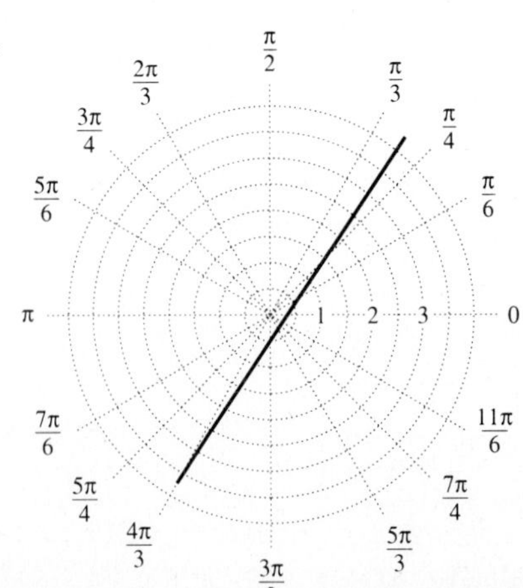

9.

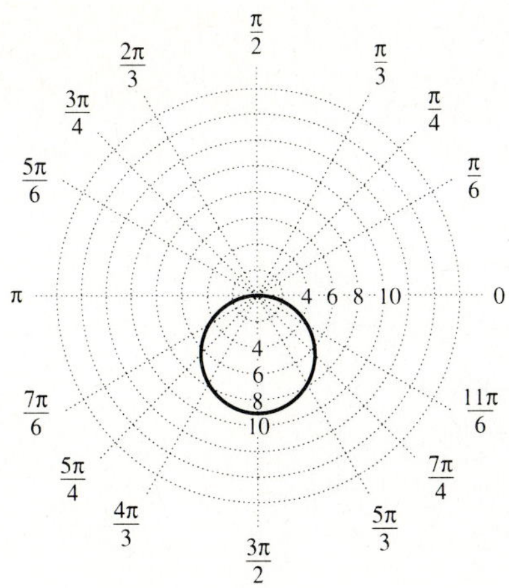

15.

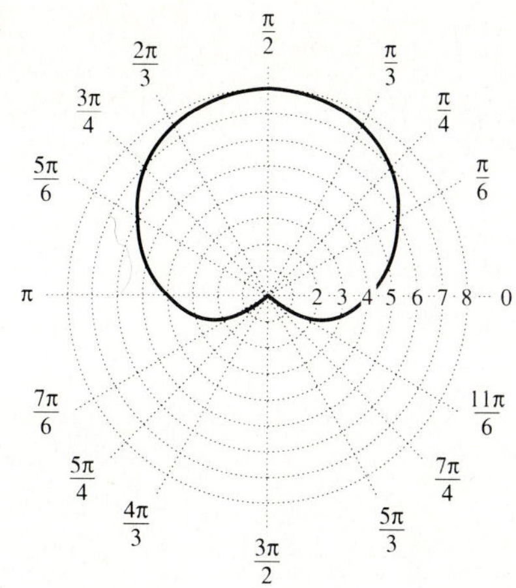

11.

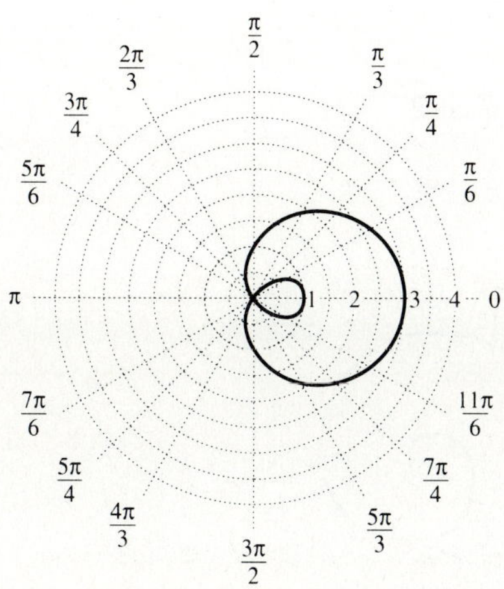

17.

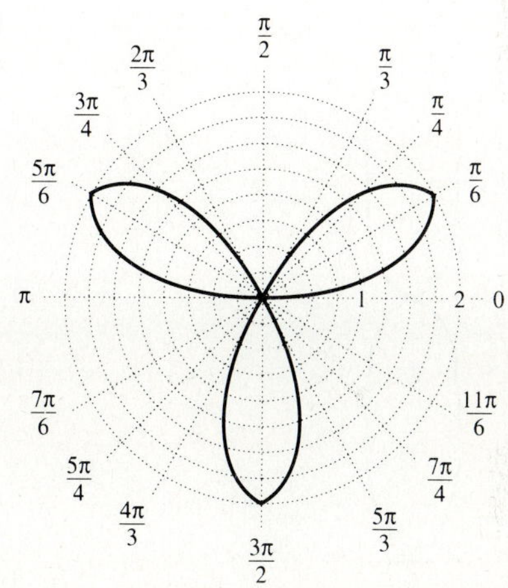

13.

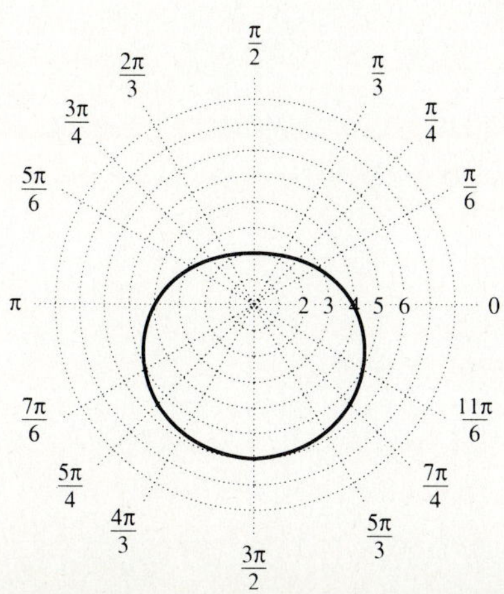

19.

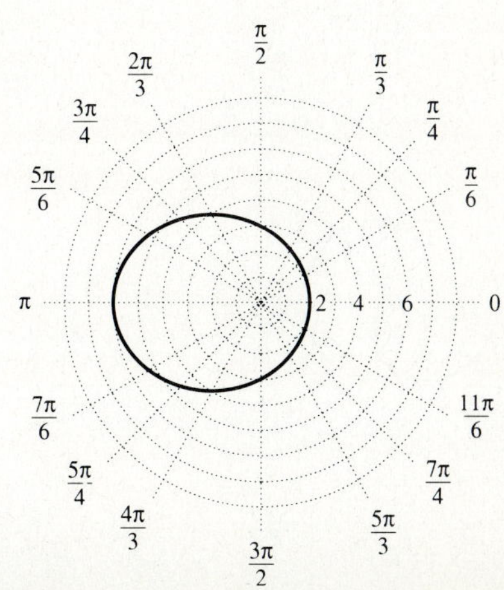

21.

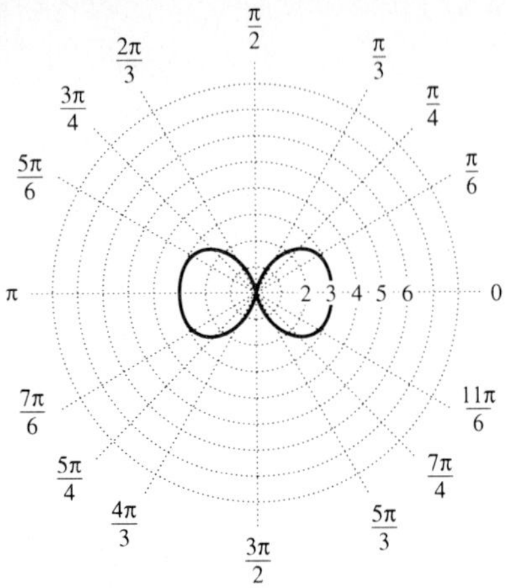

23.

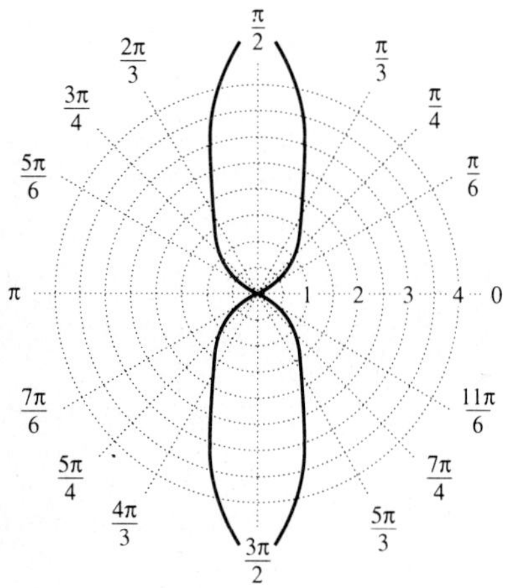

25.

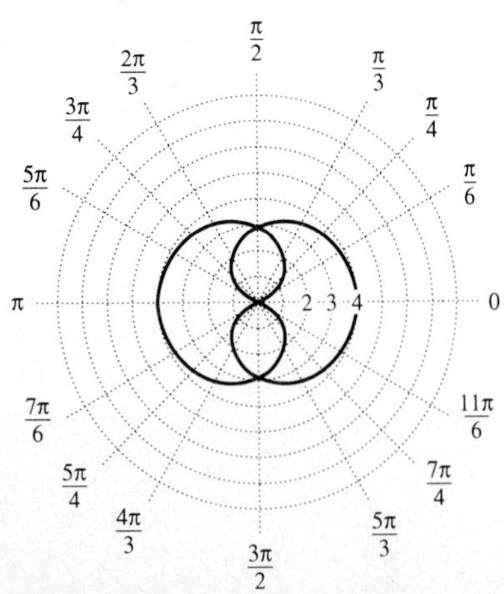

27.

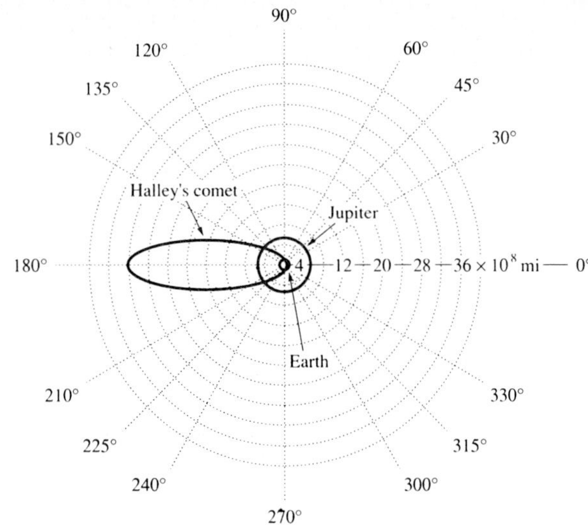

EXERCISES 9.2-2

1.

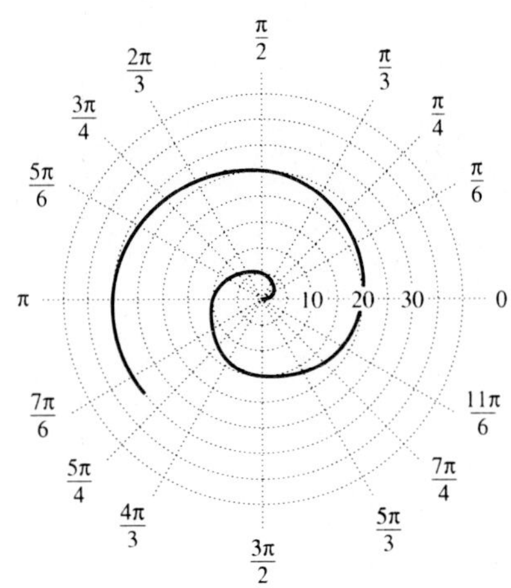

3.

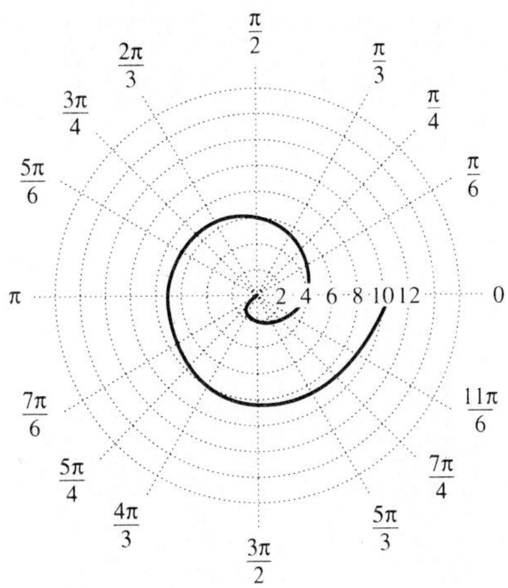

5.

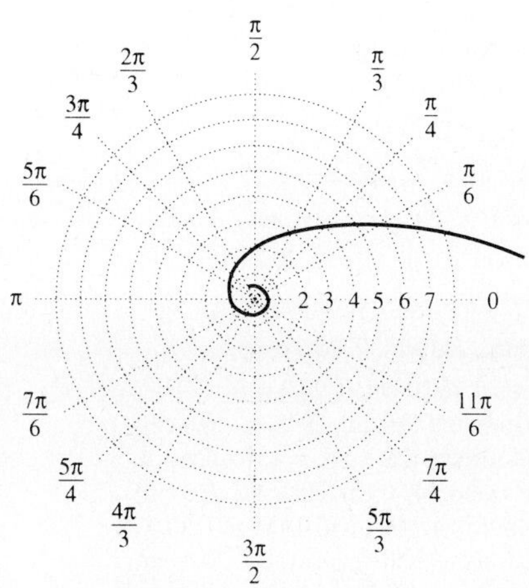

7. (a) and **(b)**

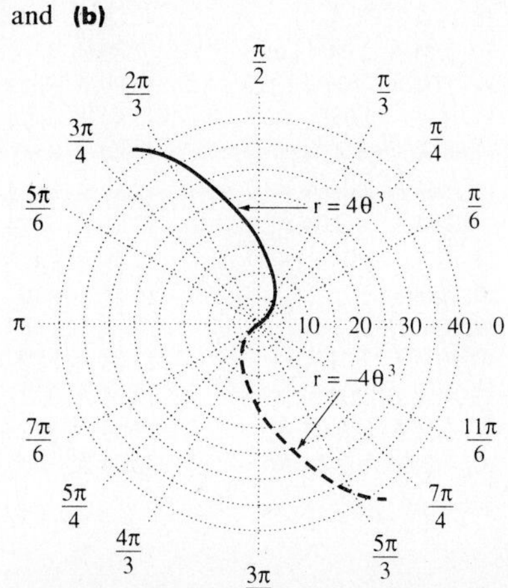

9.

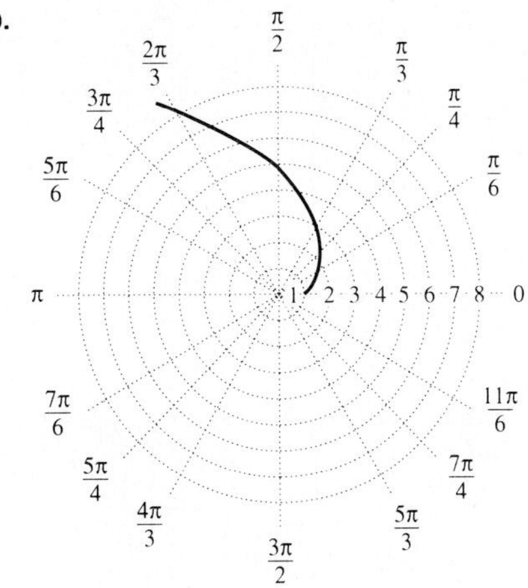

11.

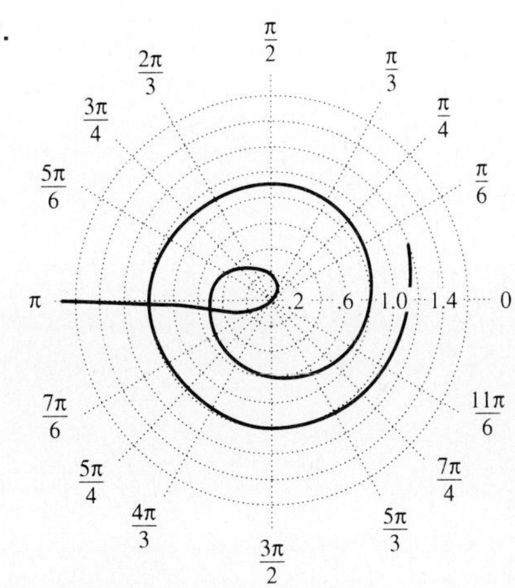

EXERCISES 9.2-3

1.

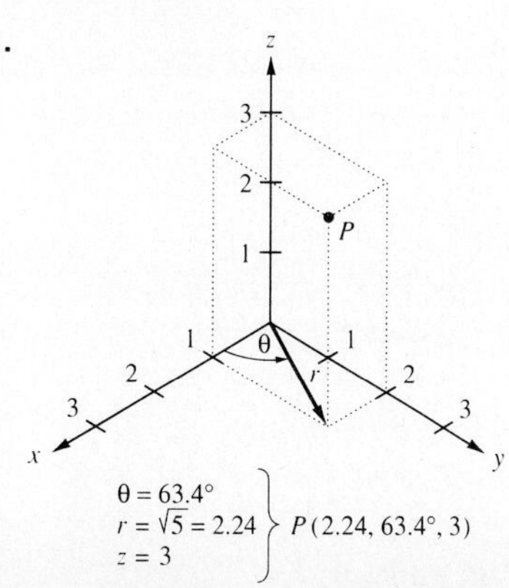

3.

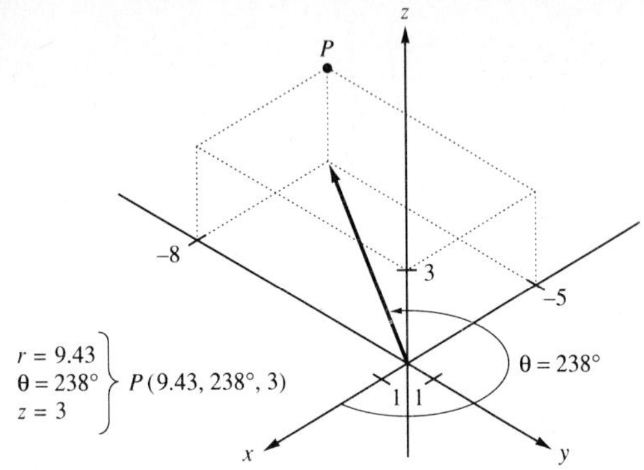

5.

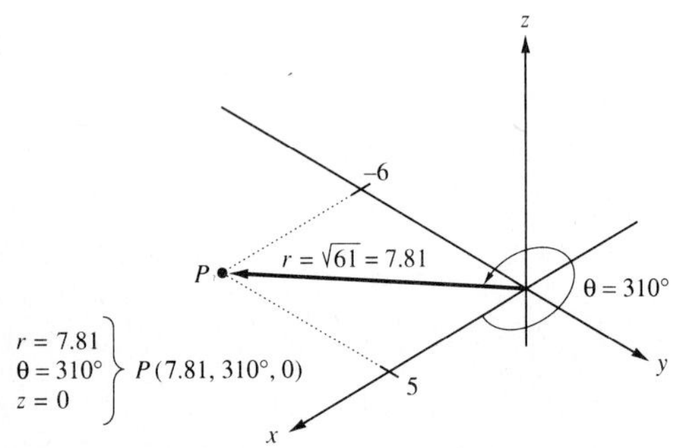

7.

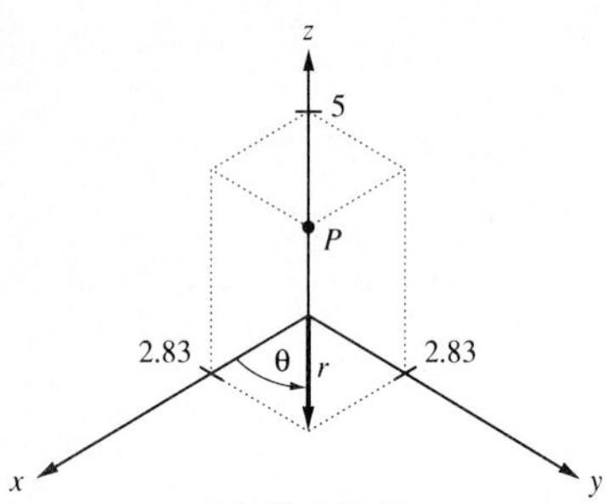

P (2.83, 2.83, 5)

9.

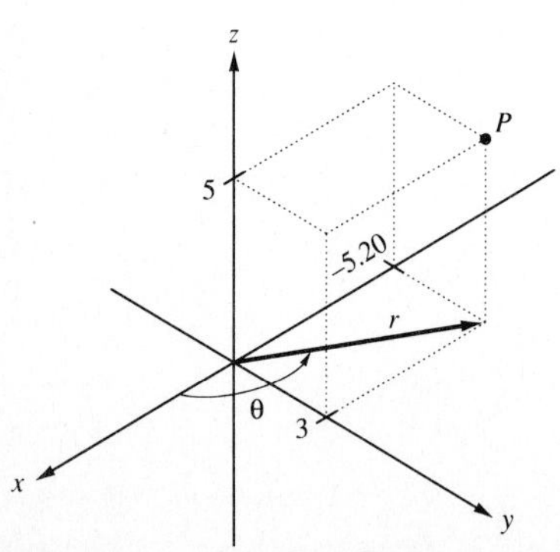

P (–5.20, 3.00, 5)

11.

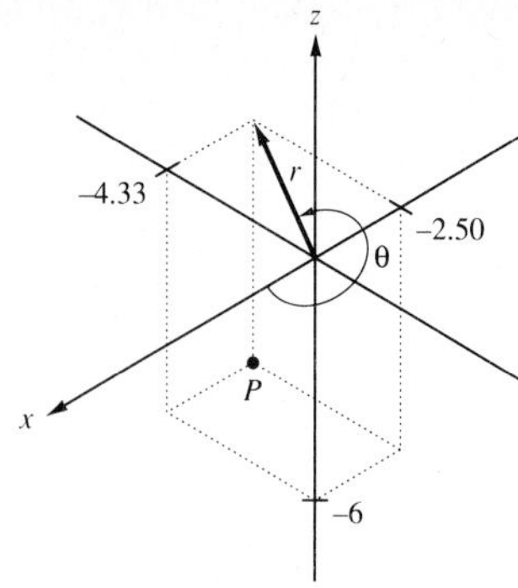

P (–2.50, –4.33, –6)

13. (6.86 km, 235°, 1.5 km) **15.** (70.71 m, 45.00°, 75.0 m)

17. **(a)** Cartesian: (20.0 cm, 20.0 cm, 20.0 cm)
Cylindrical: (28.3 cm, $\pi/4$, 20.0 cm

(b) It touches the x-y plane at:

Cartesian: (20.0 cm, 20.0 cm, 0)
Cyindrical: (28.3 cm, $\pi/4$, 0)

It touches the $y - z$ plane at:

Cartesian: (0, 20.0 cm, 20.0 cm)
Cylindrical: (20.0 cm, $\pi/2$, 20.0 cm)

It touches the $x - z$ plane at:

Cartesian: (20.0 cm, 0, 20.0 cm)
Cylindrical: (20.0 cm, 0, 20.0 cm)

(c) (i) Cartesian: (20.0 cm, 20.0 cm, 40.0 cm)
Cylindrical: (28.3 cm, $\pi/4$, 40.0 cm)
(ii) Cartesian: (40.0 cm, 20.0 cm, 20.0 cm)
Cylindrical: (44.7 cm, 0.464 rad, 20.0 cm)
(iii) Cartesian: (20.0 cm, 40.0 cm, 20.0 cm)
Cylindrical: (44.7 cm, 1.11 rad, 20.0 cm)

19.

	Rectangular	Cylindrical
Corner atoms	(2.04 Å, 2.04 Å, 0)	(2.89 Å, $\pi/4$, 0)
	(−2.04 Å, 2.04 Å, 0)	(2.89 Å, $3\pi/4$, 0)
	(−2.04 Å, −2.04 Å, 0)	(2.89 Å, $5\pi/4$, 0)
	(2.04 Å, −2.04 Å, 0)	(2.89 Å, $7\pi/4$, 0)
	(2.04 Å, 2.04 Å, 4.07 Å)	(2.89 Å, $\pi/4$, 4.07 Å)
	(−2.04 Å, 2.04 Å, 4.07 Å)	(2.89 Å, $3\pi/4$, 4.07 Å)
	(−2.04 Å, −2.04 Å, 4.07 Å)	(2.89 Å, $5\pi/4$, 4.07 Å)
	(2.04 Å, −2.04 Å, 4.07 Å)	(2.89 Å, $7\pi/4$, 4.07 Å)
Face centered atoms	(0, 0, 0)	($r = z = 0$)
	(0, 0, 4.07 Å)	($r = 0$, $z =$ 4.07 Å)
	(0, 2.04 Å, 2.04 Å)	(2.04 Å, $\pi/2$, 2.04 Å)
	(2.04 Å, 0, 2.04 Å)	(2.04 Å, 0, 2.04 Å)
	(−2.04 Å, 0, 2.04 Å)	(2.04 Å, π, 2.04 Å)
	(0, −2.04 Å, 2.04 Å)	(2.04 Å, $3\pi/2$, 2.04 Å)

21.

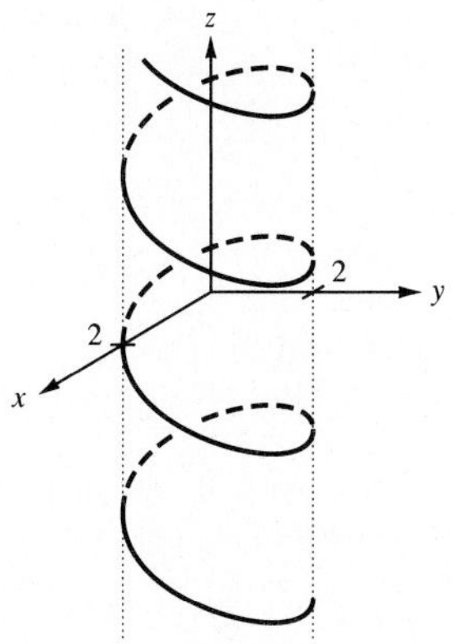

Helix

23.

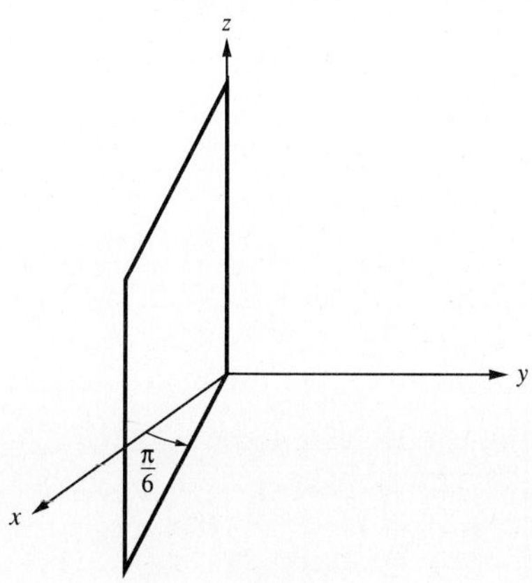

Plane

25.

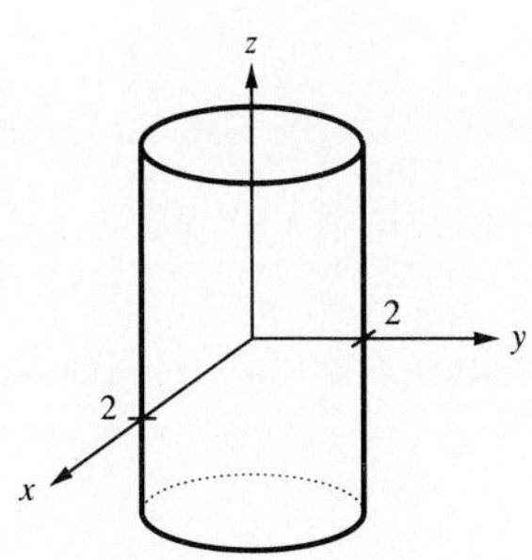

Right circular cylinder

27.

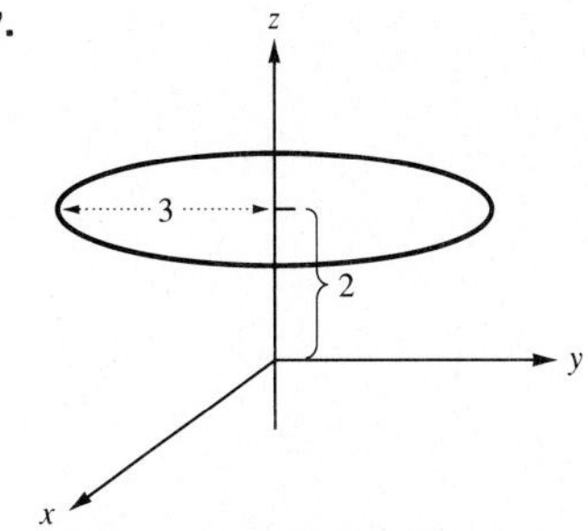

Circle of radius 3 parallel to x-y plane two units above it, centered on z axis

EXERCISES 9.3-1

1. $4i$ **3.** $8\sqrt{7}i$ **5.** $1 + 14\sqrt{3}i$ **7.** $4 - \dfrac{\sqrt{5}}{3}i$ **9.** $0.2i$

11. -1 **13.** $-14\sqrt{15}$ **15.** 1 **17.** -60 **19.** xyi
23. Real $= 2$, imaginary $= -3$, conjugate $= 2 + 3i$
25. Real $= 0$, imaginary $= -2$, conjugate $= 2i$
27. Real $= -7$, imaginary $= 6$, conjugate $= -6i - 7$
29. Real $= 0$, imaginary $= 16$, conjugate $= -16i$
31. Real $= 0$, imaginary $= 1$, conjugate $= -i$
33. Real $= -4$, imaginary $= 3$, conjugate $= -4 - 3i$
35. Real $= -12$, imaginary $= 2$, conjugate $= -12 - 2i$
37. Real $= 4$, imaginary $= 5$, conjugate $= 4 - 5i$ **39.** $4 - 7i$
41. $8 + i$ **43.** -8.90 **45.** $5 + 6i$ **47.** -1
53. Pure real: 26, 30, 36, Pure imaginary: 25, 29, 31, 32
55. $1/4 \pm (\sqrt{7}/4)i$ **57.** $3/4 \pm (\sqrt{15}/12)i$ **59.** $0.584 \pm 0.346i$

EXERCISES 9.3-2

1. $6 - 7i$ **3.** $-1 - 4i$ **5.** $-\dfrac{1}{2} - \dfrac{5}{6}i$ **7.** $\dfrac{1}{6} + \dfrac{5}{6}i$

9. $7 + 26i$ **11.** $\dfrac{13}{36}$ **13.** $-2i$ **15.** -4 **17.** $\dfrac{6}{5} + \dfrac{12}{5}i$

19. $-\dfrac{12}{13} + \dfrac{8}{13}i$ **21.** $\dfrac{2}{3} - \dfrac{5\sqrt{2}}{3}i$ **23.** 1 **25.** $-\dfrac{35}{26} - \dfrac{19}{26}i$

27. $\dfrac{c^2 - d}{c^2 + d} + \dfrac{2c\sqrt{d}}{c^2 + d}i$ **29.** $x = 14, y = 8$ **31.** $x = 3, y = -2$
33. $x = 2, y = -1$ **35.** $x = 0, y = 0$ **37.** $-10 + 5i$
39. $-16 - 30i$ **45. (a)** Pure real **(b)** Pure imaginary
(c) Neither **(d)** Pure real **(e)** Neither **(f)** Neither
(g) Pure imaginary **(h)** Pure real
47. $a + c = 0$, $b + d \neq 0$ **49.** $a - c = 0$, $b - d \neq 0$ **51.** $ac - bd = 0$, $ad + bc \neq 0$
53. $-3 - 2i$ **55.** i **57.** $9.1(-2.4i - 3.7)$
61. $2a^2 - 2b^2 = c^2$

EXERCISES 9.4-1

1.–9.

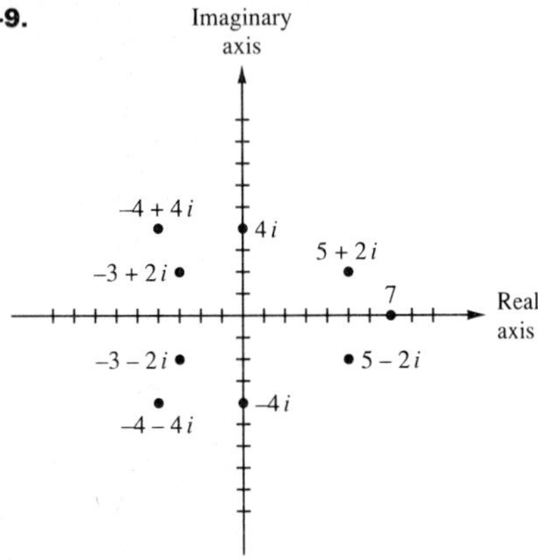

11. $A = 2i, B = 3 + 4i, C = -3 + i, D = -i, E = 3 - 2i, F = -4 - 4i, G = 1 - 4i, H = 2, I = -5, J = -3 + 3i$

EXERCISES 9.4-2

1. $r = 4, \theta = 25°$ **3.** $r = 2.83, \theta = 315°$ **5. (a)** $r = 1, \theta = 0°$ **(b)** $r = 1, \theta = 180°$ **7.** $r = 3.61, \theta = 146.31°$ **9.** $r = 11.4, \theta = 307°$ **11.** $r = 5, \theta = 233.13°$ **13.** $r = 6.13, \theta = 45°$ **15.** $r = 4, \theta = 130°$ **17.** $\sqrt{2}$ cis $\pi/4, r = \sqrt{2}, \theta = \pi/4$ **19.** $\sqrt{2}$ cis $7\pi/4, r = \sqrt{2}, \theta = 7\pi/4$ **21.** 4 cis $\pi/2, r = 4, \theta = \pi/2$ **23.** 8 cis $\pi, r = 8, \theta = \pi$ **25.** $3\sqrt{2}$ cis $5\pi/4, r = 3\sqrt{2}, \theta = 5\pi/4$ **27.** 2 cis $7\pi/4, r = 2, \theta = 7\pi/4$ **29.** 1.40 cis 21.80°, $r = 1.40, \theta = 21.80°$ **31.** $1 + i$ **33.** $-5.20 + 3i$ **35.** $-1.73 - i$ **37.** -5.00 **39.** $0.50 - 0.50i$ **41.** $2.60 + 1.50i$ **43.** $10.61 + 10.61i$ **45.** $21.21 + 15i$ **47.** $5.20 + 3i$

EXERCISES 9.4-3

1. 18 cis 90° **3.** 4 cis $\frac{3\pi}{2}$ **5.** 10 cis $\pi/3$

7. $15[(\cos 25° \cos 15° - \sin 25° \sin 20°) + i(\sin 25° \cos 15° + \cos 25° \sin 20°)]$

9. 104 cis 75° **11.** 2 cis 30° **13.** $\frac{5}{7}$ cis(−35°)

15. 0.83 cis$\left(\frac{-3\pi}{20}\right)$ **17.** 2.6 cis(−5.26°) **19.** 7.58 cis 340°

21. $\sqrt{3}$ cis 289° **23.** $\sqrt{26}$ cis 371.31°, $5.00 + 1.02i$

25. 40 cis 143.01°, $-32 + 24i$ **27.** $\sqrt{\frac{17}{5}}$ cis 102.53°, $-0.40 + 1.80i$ **29.** 11.40 cis 37.88, $9 + 7i$

33. (a) $r^2(\cos 2\theta + i \sin 2\theta)$ **(b)** $r^3(\cos 3\theta + i \sin 3\theta)$, $r^n(\cos n\theta + i \sin n\theta)$ **35.** $r^2 = c^2/(2 \cos 2\theta)$ **39. (a)** 124° **(b)** 292° **(c)** 131° **(d)** 665° (or 305°) **45.** 279.51 cis 174°

EXERCISES 9.5-1

1. $4(\cos 20° + i \sin 20°)$, $3.76 + 1.37i$
3. $4096(\cos 180° + i \sin 180°)$, -4096
5. $2.07(\cos 200° + i \sin 200°)$, $-1.95 - 0.71i$
7. $(1/27)(\cos 2\pi + i \sin 2\pi)$, $1/27$ **9.** $(16/81)(\cos 7\pi + i \sin 7\pi)$, $-16/81$ **11.** $2\sqrt{2}[\cos(-135°) + i \sin(-135°)]$, $-2 - 2i$
13. $8(\cos 270° + i \sin 270°)$, $-8i$ **15.** $\cos 270° + i \sin 270°$, $-i$
17. $\cos 210° + i \sin 210°$, $-0.87 - 0.50i$
19. $(\cos 720° + i \sin 720°)$, 1.00
21. $(1/4)(\cos 180° + i \sin 180°)$, $-1/4$
23. $1/(13\sqrt{13})[\cos(-168.93°) + i \sin(-168.93°)]$, -0.02
25. $4[\cos(-270°) + i \sin(-270°)]$, $2i$

EXERCISES 9.5-2

1. $z_0 = 4$ cis $60° = 2 + 2\sqrt{3}i$ $z_1 = 4$ cis $240° = -2 - 2\sqrt{3}i$

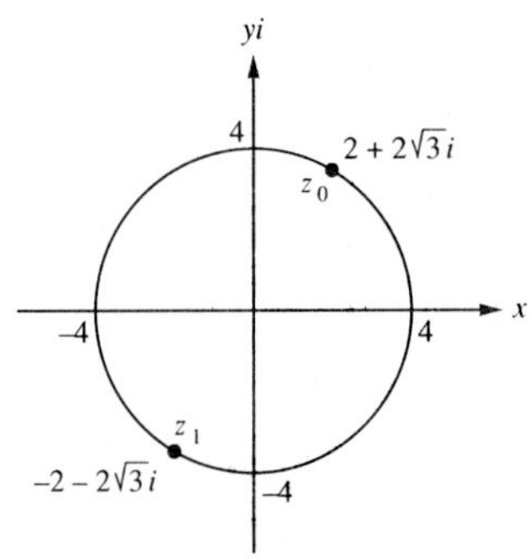

3.
$z_0 =$ cis $59° = 0.5150 + 0.8572i$
$z_1 =$ cis $119° = -0.4848 + 0.8746i$
$z_2 =$ cis $179° = -0.9998 + 0.0175i$
$z_3 =$ cis $239° = -0.5150 - 0.8572i$
$z_4 =$ cis $299° = 0.4848 - 0.8746i$
$z_5 =$ cis $359° = 0.9998 - 0.0175i$

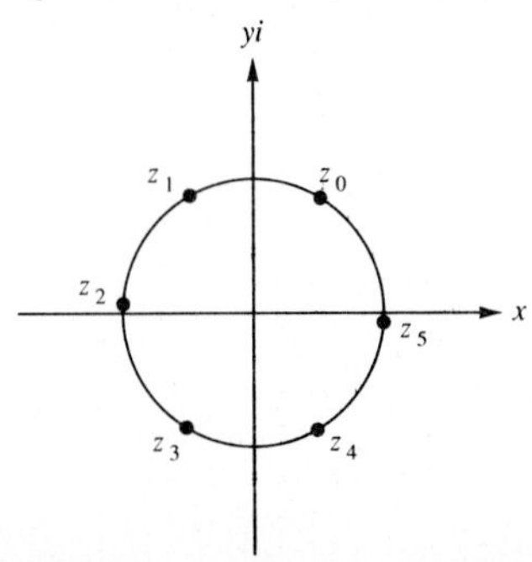

5. $z_0 = 4 \text{ cis } \dfrac{2\pi}{7} = 2.4940 + 3.1273i$

$z_1 = 4 \text{ cis } \dfrac{20\pi}{21} = -3.9553 + 0.5962i$

$z_2 = 4 \text{ cis } \dfrac{34\pi}{21} = 1.4614 - 3.7235i$

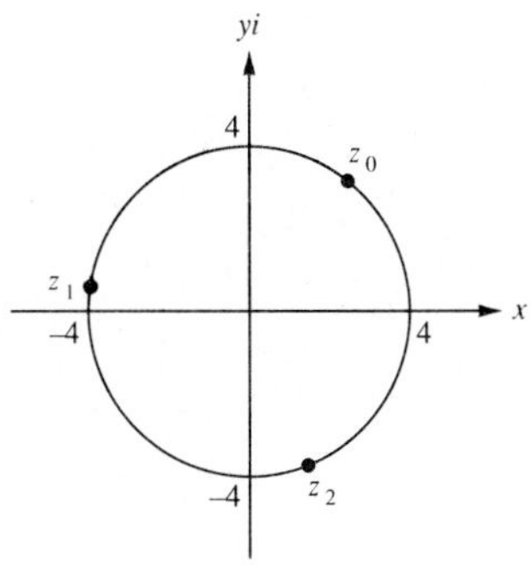

7. $z_0 = \text{cis } 0 = 1 + 0i$

$z_1 = \text{cis } \dfrac{\pi}{3} = \dfrac{1}{2} + \dfrac{\sqrt{3}}{2}i$

$z_2 = \text{cis } \dfrac{2\pi}{3} = \dfrac{-1}{2} + \dfrac{\sqrt{3}}{2}i$

$z_3 = \text{cis } \pi = -1 + 0i$

$z_4 = \text{cis } \dfrac{4\pi}{3} = -\dfrac{1}{2} - \dfrac{\sqrt{3}}{2}i$

$z_5 = \text{cis } \dfrac{5\pi}{3} = \dfrac{1}{2} - \dfrac{\sqrt{3}}{2}i$

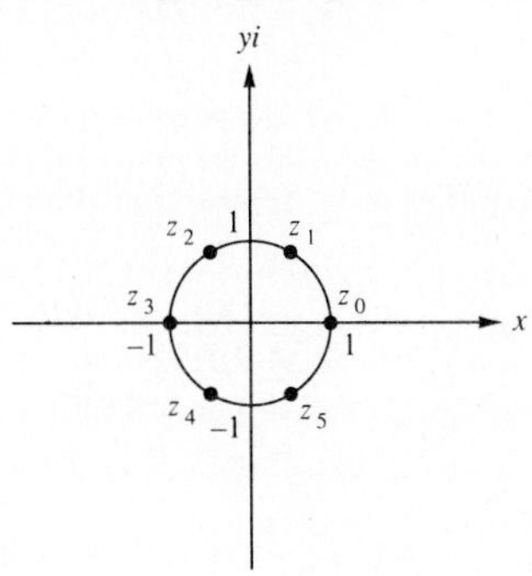

9. $z_0 = \text{cis } \dfrac{\pi}{4} = 0.7071 + 0.7071i$

$z_1 = \text{cis } \dfrac{5\pi}{4} = -0.7071 - 0.7071i$

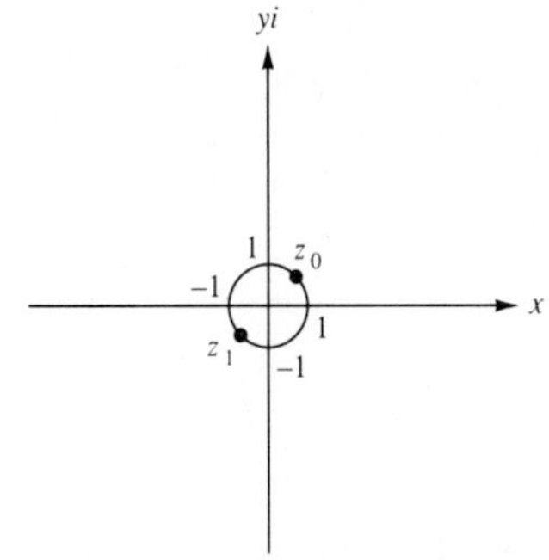

11. $z_0 = \text{cis } 135° = -\dfrac{\sqrt{2}}{2} + \dfrac{\sqrt{2}}{2}i$

$z_1 = \text{cis } 315° = \dfrac{\sqrt{2}}{2} - \dfrac{\sqrt{2}}{2}i$

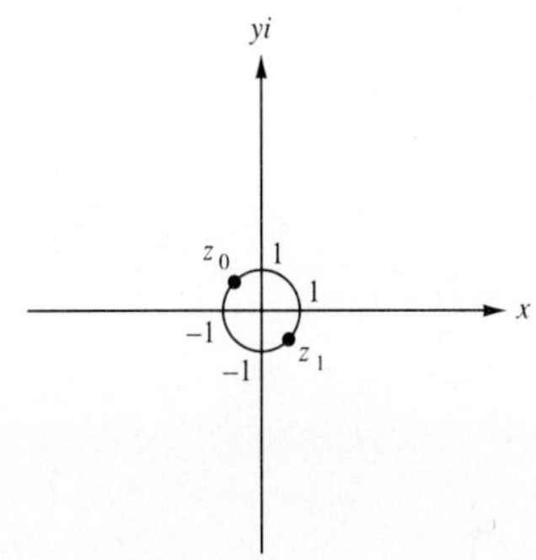

13. $z_0 = \text{cis } \dfrac{\pi}{3} = 0.5 + 0.8660i$

$z_1 = \text{cis } \pi = -1 + 0i$

$z_2 = \text{cis } \dfrac{5\pi}{3} = 0.5 - 0.8660i$

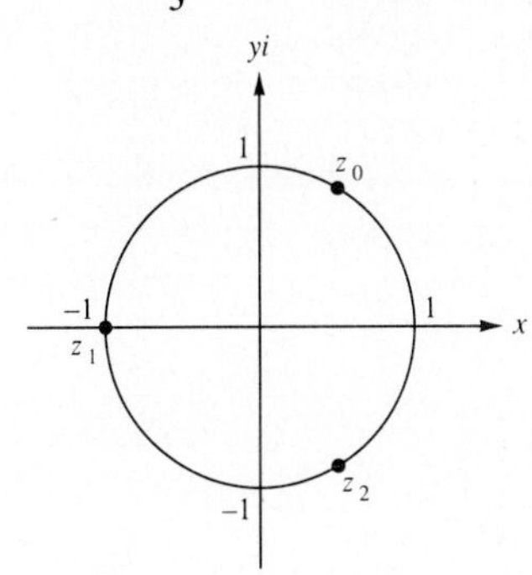

15. $z_0 = \sqrt[4]{2} \text{ cis } 67.5^\circ = 0.4551 + 1.0987i$
$z_1 = \sqrt[4]{2} \text{ cis } 247.5^\circ = -0.4551 - 1.0987i$

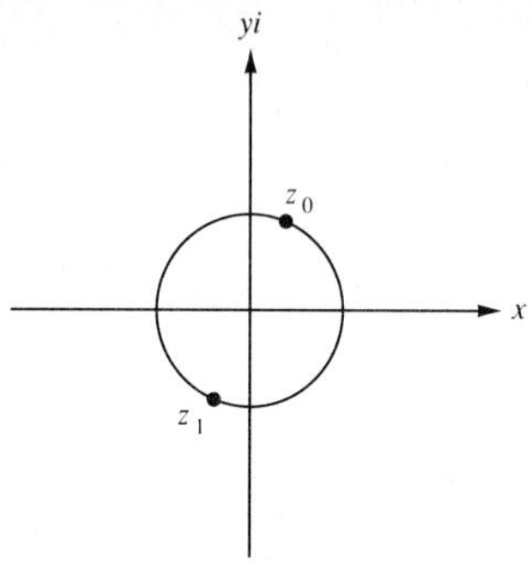

17. $z_0 = \text{cis}\left(\frac{5.236}{4}\right) = 0.2588 + 0.9659i$
$z_1 = \text{cis}(2.8798) = -0.9659 + 0.2588i$
$z_2 = \text{cis}(4.4506) = -0.2588 - 0.9659i$
$z_3 = \text{cis}(6.0214) = 0.9659 - 0.2588i$

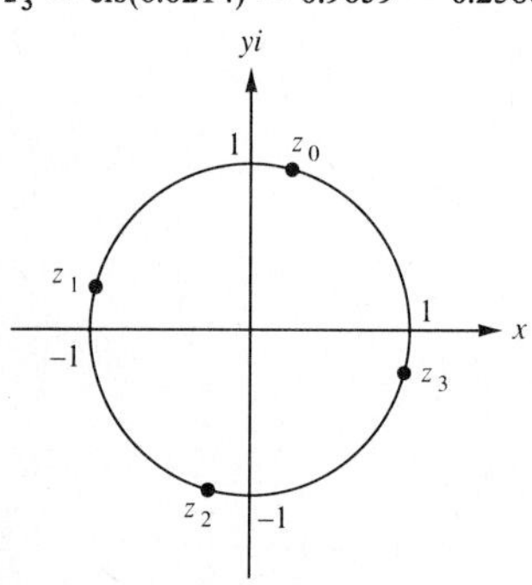

19. $z_0 = \text{cis } 20^\circ = 0.9397 + 0.3420i$
$z_1 = \text{cis } 80^\circ = 0.1736 + 0.9848i$
$z_2 = \text{cis } 140^\circ = -0.7660 + 0.6428i$
$z_3 = \text{cis } 200^\circ = -0.9397 - 0.3420i$
$z_4 = \text{cis } 260^\circ = -0.1736 - 0.9848i$
$z_5 = \text{cis } 320^\circ = 0.7660 - 0.6428i$

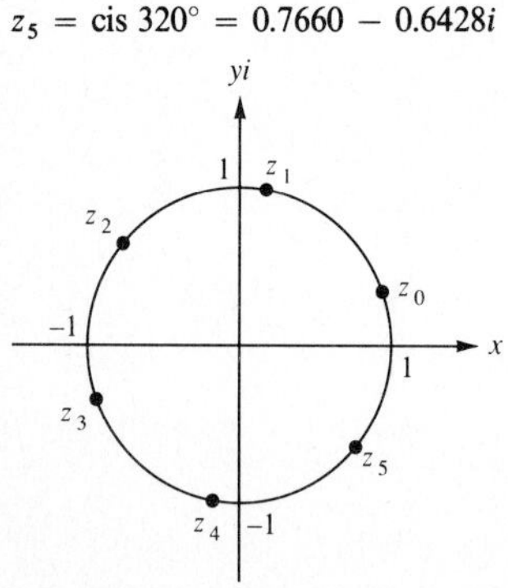

21. $z_{0,0} = 1.975 \text{ cis}(166.1^\circ) = -1.92 + 0.474i$
$z_{0,1} = 1.975 \text{ cis}(346.1^\circ) = 1.92 - 0.474i$
$z_{1,0} = 2.214 \text{ cis}(10.89^\circ) = 2.17 + 0.418i$
$z_{1,1} = 2.214 \text{ cis}(190.9^\circ) = -2.17 - 0.418i$

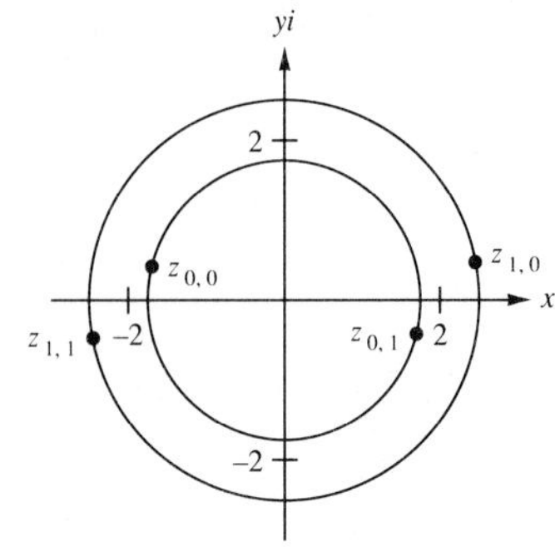

23. $z_0 = \frac{\sqrt{3}}{2} + \frac{1}{2}i$

$z_1 = 0 + i = i$

$z_2 = \frac{-\sqrt{3}}{2} + \frac{1}{2}i$

$z_3 = \frac{-\sqrt{3}}{2} - \frac{1}{2}i$

$z_4 = 0 - i = -i$

$z_5 = \frac{\sqrt{3}}{2} - \frac{1}{2}i$

25. $z_0 = 0.3827 + 0.9239i$

$z_1 = -0.9239 + 0.3827i$

$z_2 = -0.3827 - 0.9239i$

$z_3 = 0.9239 - 0.3827i$

27. $z_0 = \sqrt{3} + i$
$z_1 = 2i$
$z_2 = -\sqrt{3} + i$
$z_3 = -\sqrt{3} - i$
$z_4 = -2i$
$z_5 = \sqrt{3} - i$

29. $z_0 = 0.7071 + 1.2247i$
$z_1 = -1.2247 + 0.7071i$
$z_2 = -0.7071 - 1.2247i$
$z_3 = 1.2247 - 0.7071i$

31. $z_0 = \frac{1}{2} + \frac{\sqrt{3}}{2}i$
$z_1 = -1$
$z_2 = \frac{1}{2} - \frac{\sqrt{3}}{2}i$

35. $z_0 = -4 + 6.9282i$
$z_1 = 6.9282 + 4i$
$z_2 = 4 - 6.9282i$
$z_3 = -6.9282 - 4i$

37. $z_0 = 1.8661 \text{ cis } (3.2908)$
$z_1 = 1.8661 \text{ cis}(7.0686)$
$z_2 = 1.8661 \text{ cis}(10.8385)$
$z_3 = 1.8661 \text{ cis}(14.6084)$
$z_4 = 1.8661 \text{ cis}(15.0796)$

39. $z_0 = -0.1916 + 0.2296i$
$z_1 = -0.2296 - 0.1916i$
$z_2 = 0.1916 - 0.2296i$
$z_3 = 0.2296 + 0.1916i$

REVIEW EXERCISES

1. **(a)** (4.47, 1.11 rad) **(b)** $(2.5, \pi/2)$ **(c)** $r = 0$ **(d)** (5.83, 2.60 rad) **(e)** (3.61, 4.12 rad) **(f)** (5.83, 5.74 rad)
3. (5.4, 110°) **5.** (16, 180°) **7.** (8.7, 68°) **9.** $(\pi\sqrt{2}, 315°)$ or (4.44, 3.15°) **11.** $(4\sqrt{2}, 4\sqrt{2})$ **13.** $(-3\sqrt{2}, -3\sqrt{2})$
15. (2.4, 2.4) **17.** **(a)** (4, 0) **(b)** (0, 0)

19.

Rectangular	Polar
(0, 0)	$(r = 0)$
(−15 ft, 0)	(15 ft, 180°)
(−3.0 ft, 5.2 ft)	(6.0 ft, 120°)
(−18 ft, 5.2 ft)	(18.7 ft, 164°)

21.

	Rectangular	Polar
C	(0, 0)	$(r = 0)$
O	(−1.427 Å, 0)	(1.427 Å, 180°)
H′	(−1.737 Å, 0.904 Å)	(1.96 Å, 153°)

23. $x^2 + y^2 = 49$

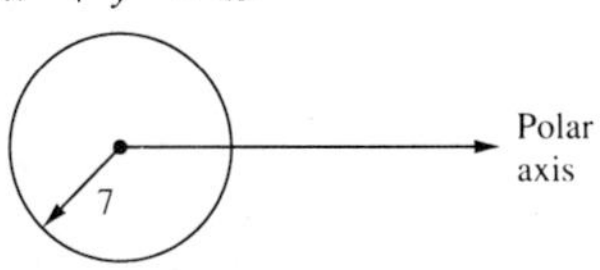

25. $y = -x\sqrt{3}/3$

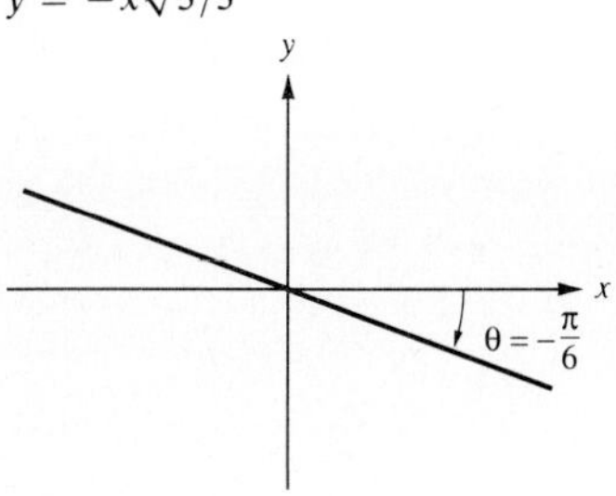

27. $y^2 = 2 + x^2$
29. $\tan\theta = 1$ or $\theta = \pi/4$

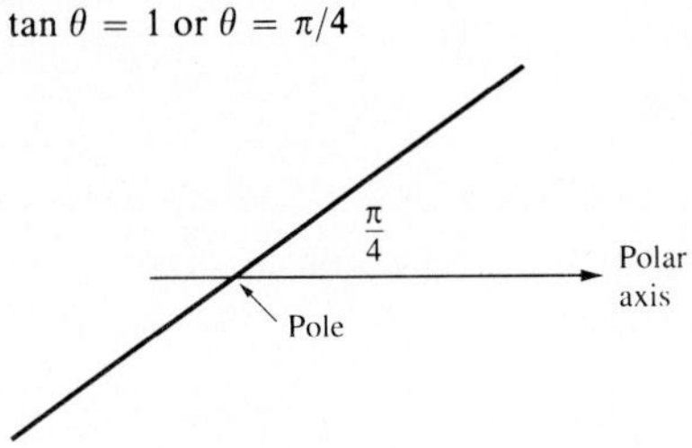

31. $r\sin^2\theta = 4\cos\theta$

33.

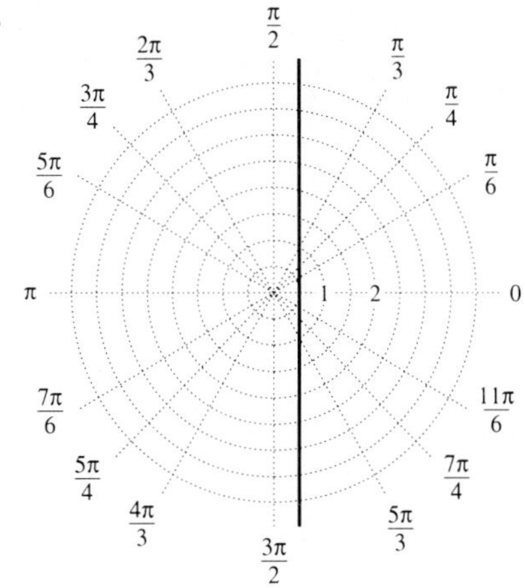

35.

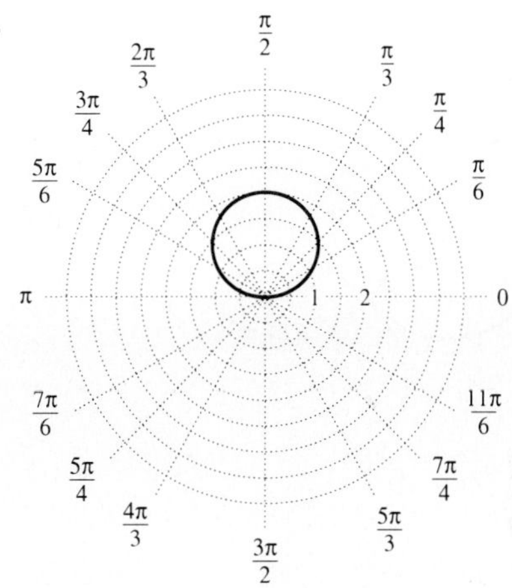

37.

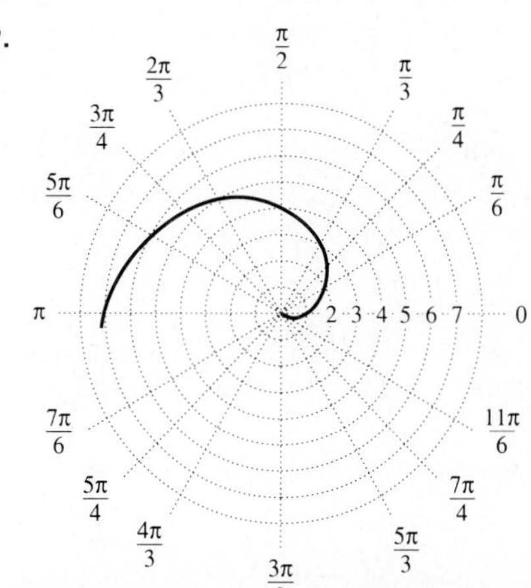

39.

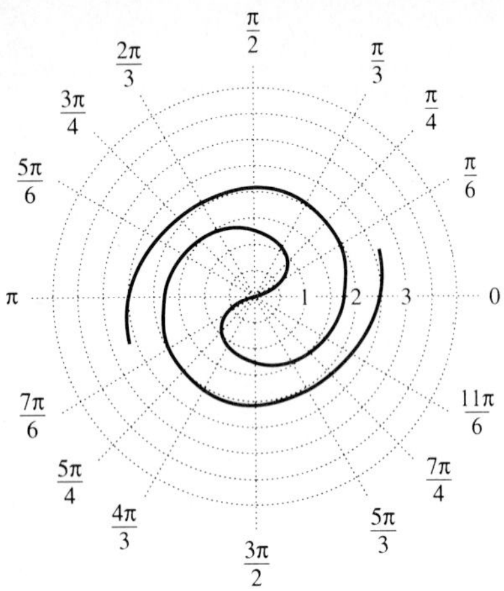

43.

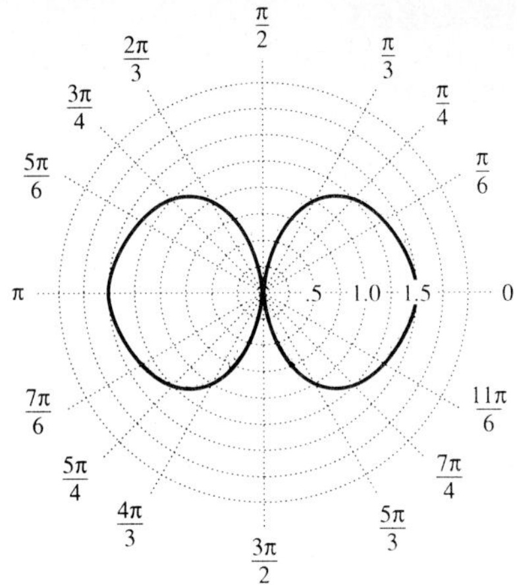

41. (a)

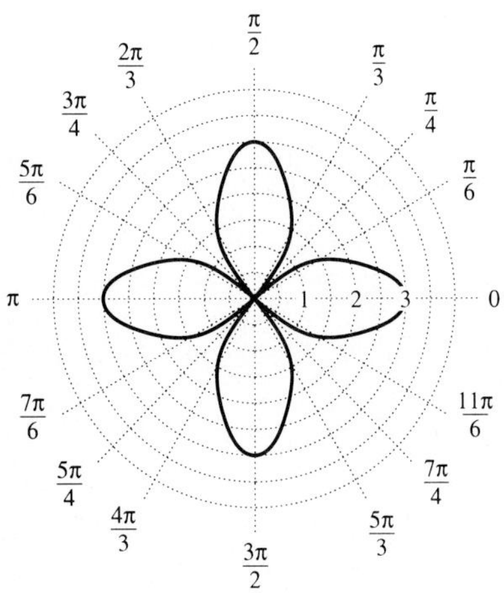

(b)

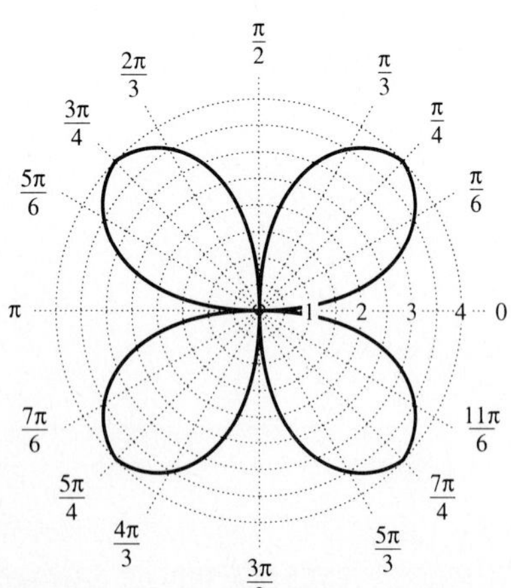

45. (a)

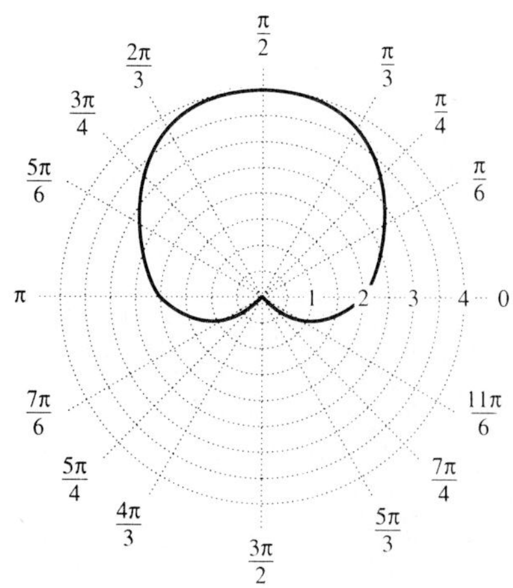

(b)

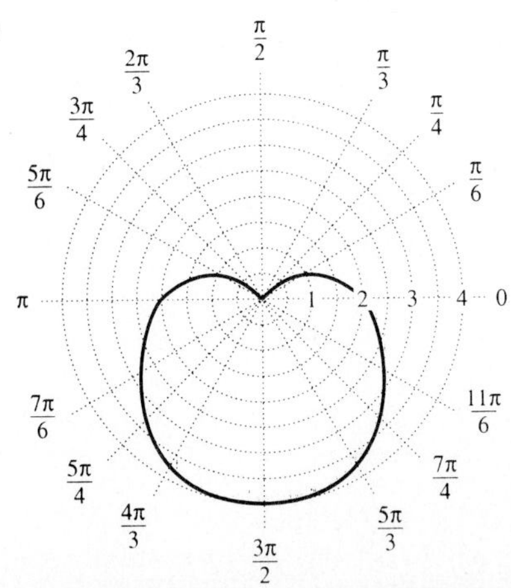

47. **(a)** $(1, \pi/2, 4)$ **(b)** $(2, 3\pi/2, -3)$ **(c)** $(\sqrt{10}, 5.03 \text{ rad}, 5)$ **(d)** $(\sqrt{61}, 2.45 \text{ rad}, 2)$ **(e)** $(10, 3.79 \text{ rad}, 1)$

49.

	Rectangular	Cylindrical
Atoms in the corners	(4.07 Å, 0, 0)	(4.07 Å, 0, 0)
	(4.07 Å, 4.07 Å, 0)	(5.76 Å, $\pi/4$, 0)
	(0, 4.07 Å, 0)	(4.07 Å, $\pi/2$, 0)
	(0, 0, 0)	($r = z = 0$)
	(4.07 Å, 0, 4.07 Å)	(4.07 Å, 0, 4.07 Å)
	(4.07 Å, 4.07 Å, 4.07 Å)	(5.76 Å, $\pi/4$, 4.07 Å)
	(0, 4.07 Å, 4.07 Å)	(4.07 Å, $\pi/2$, 4.07 Å)
	(0, 0, 4.07 Å)	($r = 0, z = 4.07$ Å)
Atoms centered in the faces	(2.04 Å, 2.04 Å, 0)	(2.89 Å, $\pi/4$, 0)
	(2.04 Å, 2.04 Å, 4.07 Å)	(2.89 Å, $\pi/4$, 4.07 Å)
	(4.07, 2.04 Å, 2.04 Å)	(4.55 Å, 0.464 rad, 2.04 Å)
	(2.04 Å, 4.07 Å, 2.04 Å)	(4.55 Å, 1.11 rad, 2.04 Å)
	(0, 2.04 Å, 2.04 Å)	(2.04 Å, $\pi/2$, 2.04 Å)
	(2.04 Å, 0, 2.04 Å)	(2.04 Å, 0, 2.04 Å)

51. **(a)** $z = 5 \text{ cis } \pi/2$, modulus $= 5 =$ magnitude, $\arg \mathbf{z} = \pi/2$, $\bar{z} = -5i$ **(b)** $z = 5 \text{ cis } 3\pi/2$, modulus $=$ magnitude $= 5$, $\arg \mathbf{z} = 3\pi/2$, $\bar{z} = 5i$
53. **(a)** $z = 2 \text{ cis } 7\pi/4$, modulus $=$ magnitude $= 2$, $\arg \mathbf{z} = 7\pi/4$, $\bar{z} = \sqrt{2} + \sqrt{2}i$ **(b)** $z = 2 \text{ cis } \pi/4$, modulus $=$ magnitude $= 2$, $\arg \mathbf{z} = \pi/4$, $\bar{z} = \sqrt{2} - \sqrt{2}i$
55. $z = 7$, $\arg \mathbf{z} = 0°$, modulus $= |z| = 7$
57. $z = 3 + i3\sqrt{3}$, $\arg \mathbf{z} = \dfrac{\pi}{3}$, modulus $= |z| = 6$
59. $z = -1.5\sqrt{2} - 1.5\sqrt{2}i$, $\arg \mathbf{z} = \dfrac{5\pi}{4}$, modulus $= |z| = 3$ **61.** $-31 - 78i$, $\arg \mathbf{z} = 248°$, modulus $= |z| = 84$ **63.** **(a)** $-\sqrt{6}$ **(b)** $\sqrt{6}$ **65.** -18
67. $1 - 4\sqrt{3}i$ **69.** $-24i$ **71.** **(a)** i **(b)** $+1$ **(c)** 0 **(d)** $-i$ **73.** $-54 - 33i$ **75.** $27 - 11i$ **77.** $\dfrac{2}{13} + \dfrac{3}{13}i$
79. $-\dfrac{5}{2} - \dfrac{3}{2}i$ **81.** $-\dfrac{14}{17} + \dfrac{5}{17}i$ **85.** $3 \text{ cis } 105°$
87. $4 \text{ cis } 150°$ **89.** $729 \text{ cis } 2\pi$ (or $729 \text{ cis } 0$)
91. $1024\sqrt{2} \text{ cis } 135°$ **93.** $\dfrac{1}{32} \text{ cis}(-70°)$ or $\dfrac{1}{32} \text{ cis } 290°$
95. $2^{5/4} \text{ cis}\left(\dfrac{\pi}{8}\right), 2^{5/4} \text{ cis}\left(\dfrac{9\pi}{8}\right)$ **97.** $2 \text{ cis } 60°, 2 \text{ cis } 150°, 2 \text{ cis } 240°, 2 \text{ cis } 330°$ **99.** $4 \text{ cis } \pi/9, 4 \text{ cis } 7\pi/9, 4 \text{ cis } 13\pi/9$
101. $\text{cis}\left(\dfrac{\pi}{4}\right), \text{cis}\left(\dfrac{3\pi}{4}\right), \text{cis}\left(\dfrac{5\pi}{4}\right), \text{cis}\left(\dfrac{7\pi}{4}\right)$

Index

ANGLES, ARC LENGTHS AND SECTORS

$$360° = 1 \text{ rev} = 2\pi \text{ rad}$$

If θ is in radians,

$$S = R\theta \qquad A = \tfrac{1}{2}R^2\theta$$

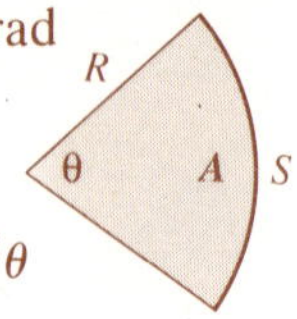

TRIGONOMETRIC FUNCTIONS FOR A RIGHT TRIANGLE

$$\sin\theta = \frac{a}{c} \qquad \tan\theta = \frac{a}{b} \qquad \sec\theta = \frac{c}{b}$$

$$\cos\theta = \frac{b}{c} \qquad \cot\theta = \frac{b}{a} \qquad \csc\theta = \frac{c}{a}$$

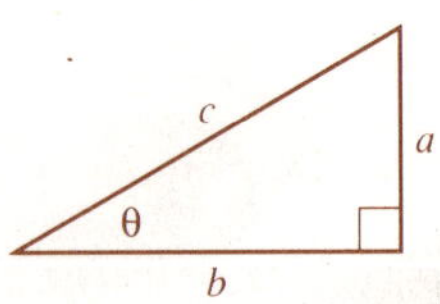

SPECIAL ANGLES

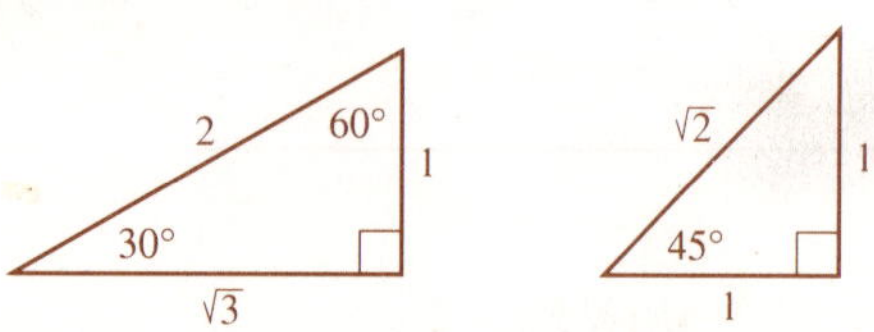

SMALL-ANGLE APPROXIMATION

If θ is much less than 1 rad,

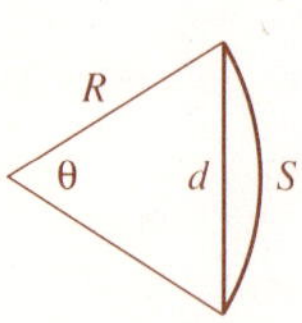

$$\tan\theta \approx \sin\theta \approx \theta \approx \frac{d}{R}$$

OBLIQUE TRIANGLES

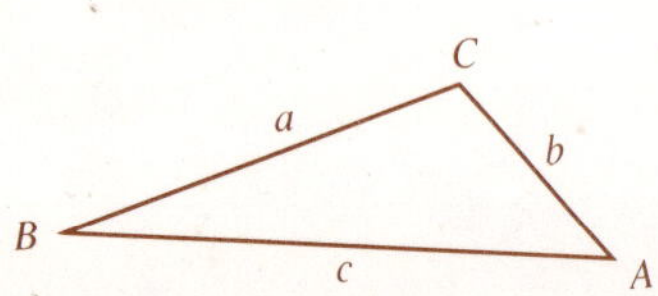

Law of Sines

$$\frac{\sin A}{a} = \frac{\sin B}{b} = \frac{\sin C}{c}$$

Law of Cosines

$$a^2 = b^2 + c^2 - 2bc\cos A$$

$$\text{Area of Triangle} = \tfrac{1}{2}ac\sin B = \tfrac{1}{2}bc\sin A = \tfrac{1}{2}ab\sin C$$

MAJOR IDENTITIES

Reciprocal Identities

$$\csc x = \frac{1}{\sin x} \qquad \sec x = \frac{1}{\cos x} \qquad \cot x = \frac{1}{\tan x}$$

Quotient Identities

$$\tan x = \frac{\sin x}{\cos x} \qquad \cot x = \frac{\cos x}{\sin x}$$

Pythagorean Identities

$$\sin^2 x + \cos^2 x = 1 \qquad \tan^2 x + 1 = \sec^2 x$$

$$\cot^2 x + 1 = \csc^2 x$$

Negative-Angle Identities

$$\sin(-\theta) = -\sin\theta \qquad \cos(-\theta) = \cos\theta$$

$$\tan(-\theta) = -\tan\theta$$

Sum and Difference Identities

$$\sin(A + B) = \sin A\cos B + \cos A\sin B$$

$$\sin(A - B) = \sin A\cos B - \cos A\sin B$$

$$\cos(A + B) = \cos A\cos B - \sin A\sin B$$

$$\cos(A - B) = \cos A\cos B + \sin A\sin B$$

$$\tan(A + B) = \frac{\tan A + \tan B}{1 - \tan A\tan B}$$

$$\tan(A - B) = \frac{\tan A - \tan B}{1 + \tan A\tan B}$$